GENE THERAPY
OF CANCER

SECOND EDITION

GENE THERAPY OF CANCER

Translational Approaches from Preclinical Studies to Clinical Implementation

Edited by

EDMUND C. LATTIME, PhD

Departments of Surgery and Molecular Genetics & Microbiology
The Cancer Institute of New Jersey
UMDNJ–Robert Wood Johnson Medical School
Piscataway, New Jersey

STANTON L. GERSON, MD

Division of Hematology/Oncology
Department of Medicine and Ireland Cancer Center at Case Western Reserve University
and University Hospitals of Cleveland
Cleveland, Ohio

ACADEMIC PRESS

An Elsevier Science Imprint

San Diego   San Francisco   New York   Boston   London   Sydney   Tokyo

Academic Press
An Elsevier Science Imprint
525 B Street, Suite 1900, San Diego, California 92101-4495, USA
http://www.academicpress.com

Academic Press
32 Jamestown Road, London NW1 7BY, UK
http://www.academicpress.com

Library of Congress Catalog Card Number: 2001099437

International Standard Book Number: 0-12-437551-0

PRINTED IN THE UNITED STATES OF AMERICA
02 03 04 05 06 07 EB 9 8 7 6 5 4 3 2 1

To Holly and Deb, who have given us
the enthusiastic support needed to pursue this second
edition, and to our children, Sarah, Ruth, James, and David who
have known of gene therapy from their earliest days. We hope they
experience with us the fulfillment of the latent promises of this field in all its forms.

Contents

6. The Advent of Lentiviral Vectors: Prospects for Cancer Therapy

MICHEL SADELAIN AND ISABELLE RIVIÈRE

PART

II

IMMUNE TARGETED GENE THERAPY

7. Immunologic Targets for the Gene Therapy of Cancer

SUZANNE OSTRAND-ROSENBERG, VIRGINIA K. CLEMENTS,
SAMUDRA DISSANAYAKE, MILEKA GILBERT,
BETH A. PULASKI, AND LING QI

PART

IIa

VACCINE STRATEGIES

8. Development of Epitope-Specific Immunotherapies for Human Malignancies and Premalignant Lesions Expressing Mutated *ras* Genes

SCOTT I. ABRAMS

PART

IIb

DENDRITIC CELL-BASED GENE THERAPY

9. Introduction to Dendritic Cells

PATRICK BLANCO, A. KAROLINA PALUCKA,
AND JACQUES BANCHEREAU

PART

IIc

CYTOKINES AND CO-FACTORS

PART

IId

GENETICALLY MODIFIED EFFECTOR CELLS FOR IMMUNE-BASED IMMUNOTHERAPY

Contributors

Numbers in parentheses indicate the pages on which the authors' contribution begin.

Rafat Abonour (355) Department of Medicine, Indiana University School of Medicine, Indianapolis, Indiana 46202

Scott I. Abrams (145) Laboratory of Tumor Immunology and Biology, Center for Cancer Research, National Cancer Institute, National Institutes of Health, Bethesda, Maryland 20892

Laura K. Aguilar (513) Harvard Gene Therapy Initiative, Harvard Medical School, Boston, Massachusetts 02115

Estuardo Aguilar-Cordova (513) Department of Radiology, Baylor College of Medicine, Houston, Texas 77030 and Harvard Gene Therapy Initiative, Harvard Medical School, Boston, Massachusetts 02115

Steven M. Albelda (493) Thoracic Oncology Research Laboratory, Pulmonary, Allergy, and Critical Care Division, University of Pennsylvania Medical Center, Philadelphia, Pennsylvania 19104

Mark R. Albertini (225) The University of Wisconsin, Comprehensive Cancer Center, Madison, Wisconsin 53792

Gustavo Ayala (513) Department of Pathology, Baylor College of Medicine, Houston, Texas 77030

Jacques Banchereau (167) Baylor Institute for Immunology Research, Dallas, Texas 75204

Christopher Baum (3) Medizinische Hochschule, Abt. Haematologie, 30625 Hannover, Germany

Christian M. Becker (421) Department of Surgery, Children's Hospital, Harvard Medical School, Boston, Massachusetts 02115

Carmela Beger (95) Department of Medicine, University of San Diego, La Jolla, California 92093

Joseph R. Bertino (365) Department of Medicine, Division of Hematologic Oncology and Lymphoma, and Programs of Molecular Pharmacology and Therapeutics, Memorial-Sloan Kettering Cancer Center, New York, New York 10021

Patrick Blanco (167) Baylor Institute for Immunology Research, Dallas, Texas 75204

Tulin Budak-Alpdogan (365) Department of Medicine, Programs of Molecular Pharmacology and Therapeutics, Memorial-Sloan Kettering Cancer Center, New York, New York 10021

E. Brian Butler (513) Department of Radiology, Baylor College of Medicine, Houston, Texas 77030

Lisa H. Butterfield (179) Division of Surgical Oncology, UCLA Medical Center, University of California, Los Angeles, California 90095

Alfred E. Chang (241) Department of Surgery, Division of Surgical Oncology, Comprehensive Cancer Center, University of Michigan, Ann Arbor, Michigan 48109

Saswati Chatterjee (53) Division of Virology, City of Hope National Medical Center, Duarte, California 91010

K. V. Chin (393) Departments of Medicine and Pharmacology, The Cancer Institute of New Jersey, UMDNJ-Robert Wood Johnson Medical School, Piscataway, New Jersey 08901

Virginia K. Clements (127) Department of Biological Sciences, University of Maryland, Baltimore, Maryland 21250

Mark J. Cooper (31) Copernicus Therapeutics, Inc., Cleveland, Ohio 44106

Kenneth Cornetta (355) Department of Medicine, Indiana University School of Medicine, Indianapolis, Indiana 46202

James M. Croop (355) Department of Pediatrics, Indiana University School of Medicine, Indianapolis, Indiana 46202

Samudra Dissanayake (127) Department of Biological Sciences, University of Maryland, Baltimore, Maryland 21250

Stephen L. Eck (505) HUP-Department of Neurosurgery, The University of Pennsylvania Medical Center, Philadelphia, Pennsylvania 19004

James S. Economou (179) Division of Surgical Oncology, and Department of Immunology, Microbiology, and Molecular Genetics, UCLA Medical Center, University of California, Los Angeles, California 90095

Laurence C. Eisenlohr (207) Department of Microbiology and Immunology, Thomas Jefferson University, Philadelphia, Pennsylvania 19107

Wafik S. El-Deiry (273, 279, 299) Laboratory of Molecular Oncology and Cell Cycle Regulation, Howard Hughes Medical Institute, Departments of Medicine and Genetics, Cancer Center and The Institute for Human Gene Therapy, University of Pennsylvania School of Medicine, Philadelphia, Pennsylvania 19104

Filip A. Farnebo (421) Departments of Surgery and Genetics, Children's Hospital, Harvard Medical School, Boston, Massachusetts 02115

Andrew L. Feldman (405) Surgery Branch, National Cancer Institute, Bethesda, Maryland 20892

Judah Folkman (421) Department of Surgery, Children's Hospital, Harvard Medical School, Boston, Massachusetts 02115

Stanton L. Gerson (341) Division of Hematology/Oncology, Department of Medicine and Ireland Cancer Center at Case Western Reserve University and University Hospitals of Cleveland, Cleveland, Ohio 44106

Mileka Gilbert (127) Department of Biological Sciences, University of Maryland, Baltimore, Maryland 21250

Leonard G. Gomella (207) Department of Urology, Thomas Jefferson University, Philadelphia, Pennsylvania 19107

David H. Gorski (435) The Cancer Institute of New Jersey, UMDNJ-Robert Wood Johnson Medical School, New Brunswick, New Jersey 08901

William N. Hait (393) Departments of Medicine and Pharmacology, The Cancer Institute of New Jersey, UMDNJ-Robert Wood Johnson Medical School, Piscataway, New Jersey 08901

Mien-Chie Hung (465) Departments of Molecular and Cellular Oncology and Surgical Oncology, M. D. Anderson Cancer Center, The University of Texas, Houston, Texas 77030

Kevin D. Judy (505) HUP-Department of Neurosurgery, The University of Pennsylvania Medical Center, Philadelphia, Pennsylvania 19004

Dov Kadmon (513) Department of Urology, Baylor College of Medicine, Houston, Texas 77030

Thomas Kearney (393) Department of Surgery, The Cancer Institute of New Jersey, UMDNJ-Robert Wood Johnson Medical School, Piscataway, New Jersey 08901

Edsel U. Kim (257) Department of Otolaryngology, University of Michigan, Ann Arbor, Michigan 48109

David M. King (225) The University of Wisconsin, Comprehensive Cancer Center, Madison, Wisconsin 53792

David Kirn (449) Imperial Cancer Research Fund, Program for Viral and Genetic Therapy of Cancer, Imperial College School of Medicine, Hammersmith Hospital, London W11 OHS, United Kingdom

Omer N. Koç (341) Division of Hematology/Oncology, Department of Medicine and Ireland Cancer Center at Case Western Reserve University and University Hospitals of Cleveland, Cleveland, Ohio 44106

Martin Krüger (95) Department of Medicine, University of San Diego, La Jolla, California 92093

Calvin J. Kuo (421) Departments of Surgery and Genetics, Children's Hospital, Harvard Medical School, Boston, Massachusetts 02115

C. Lampert (81) Department of Hematology and Medical Oncology, St. Peter's University Hospital, New Brunswick, New Jersey 08901

Edmund C. Lattime (207, 393) Departments of Surgery and Molecular Genetics & Microbiology, The Cancer Institute of New Jersey, UMDNJ-Robert Wood Johnson Medical School, Piscataway, New Jersey 08901

Irina V. Lebedeva (315) Department of Medicine and Pharmacology, Columbia University, College of Physicians and Surgeons, New York, New York 10032

Steven K. Libutti (405) Surgery Branch, National Cancer Institute, Bethesda, Maryland 20892

H. Kim Lyerly (199) Department of Surgery, Duke University Medical Center, Durham, North Carolina 27710

Michael J. Mastrangelo (207) Division of Medical Oncology, Department of Medicine, Thomas Jefferson University, Philadelphia, Pennsylvania 19107

Helena J. Mauceri (435) Department of Radiation and Cellular Oncology, University of Chicago Hospitals, Chicago, Illinois 60637

A. M. McCall (81) Fox Chase Cancer Center, Philadelphia, Pennsylvania 91010

Kevin T. McDonagh (241) Department of Internal Medicine, Division of Hematology/Oncology, Comprehensive Cancer Center, University of Michigan, Ann Arbor, Michigan 48109

R. Scott McIvor (383) Gene Therapy Program, Institute of Human Genetics, Department of Genetics, Cell Biology and Development, University of Minnesota, Minneapolis, Minnesota 55455

Raymond D. Meng (273, 279, 299) Laboratory of Molecular Oncology and Cell Cycle Regulation, Howard Hughes Medical Institute, Departments of Medicine and Genetics, Cancer Center and The Institute for Human Gene Therapy, University of Pennsylvania School of Medicine, Philadelphia, Pennsylvania 19104

Brian Miles (513) Department of Urology, Baylor College of Medicine, Houston, Texas 77030

Frederick L. Moolten (481) Edith Nourse Rogers Memorial Veterans Hospital, Bedford, Massachusetts 01730 and Boston University School of Medicine, Boston, Massachusetts 02118

Michael A. Morse (199) Department of Medicine, Division of Medical Oncology and Transplantation, Duke University Medical Center, Durham, North Carolina 27710

Paula J. Mroz (481) Edith Nourse Rogers Memorial Veterans Hospital, Bedford, Massachusetts 01730

James J. Mulé (257) Department of Surgery, University of Michigan, Ann Arbor, Michigan 48109

Smita K. Nair (199) Department of Surgery, Duke University Medical Center, Durham, North Carolina 27710

Owen A. O'Connor (365) Department of Medicine, Division of Hematologic Oncology and Lymphoma, and Developmental Chemotherapy Services, Memorial-Sloan Kettering Cancer Center, New York, New York 10021

Wolfram Ostertag (3) Heinrich-Pette-Institut für Experimentelle Virologie und Immunologie an der Universität Hamburg, 20251 Hamburg, Germany

Suzanne Ostrand-Rosenberg (127) Department of Biological Sciences, University of Maryland, Baltimore, Maryland 21250

A. Karolina Palucka (167) Baylor Institute for Immunology Research, Dallas, Texas 75204

Beth A. Pulaski (127) Department of Biological Sciences, University of Maryland, Baltimore, Maryland 21250

Ling Qi (127) Department of Biological Sciences, University of Maryland, Baltimore, Maryland 21250

Alexander L. Rakhmilevich (225) The University of Wisconsin, Comprehensive Cancer Center, Madison, Wisconsin 53792

Jane S. Reese (341) Division of Hematology/Oncology, Department of Medicine and Ireland Cancer Center at Case Western Reserve University and University Hospitals of Cleveland, Cleveland, Ohio 44106

Michael Reiss (393) Department of Medicine, The Cancer Institute of New Jersey, UMDNJ-Robert Wood Johnson Medical School, Piscataway, New Jersey 08901

Antoni Ribas (179) Division of Surgical Oncology, and Division of Hematology/Oncology, UCLA Medical Center, University of California, Los Angeles, California 90095

Isabelle Rivière (109) Laboratory of Gene Transfer and Gene Expression, Department of Medicine and Immunology Program, Memorial Sloan-Kettering Cancer Center, New York, New York 10021

Justin C. Roth (341) Division of Hematology/Oncology, Department of Medicine and Ireland Cancer Center at Case Western Reserve University and University Hospitals of Cleveland, Cleveland, Ohio 44106

Michel Sadelain (109) Laboratory of Gene Transfer and Gene Expression, Department of Medicine and Immunology Program, Memorial Sloan-Kettering Cancer Center, New York, New York 10021

Ruping Shao (465) Department of Molecular and Cellular Oncology, M. D. Anderson Cancer Center, The University of Texas, Houston, Texas 77030

C. A. Stein (315) Department of Medicine and Pharmacology, Columbia University, College of Physicians and Surgeons, New York, New York 10032

Daniel H. Sterman (493) Thoracic Oncology Research Laboratory, Pulmonary, Allergy, and Critical Care Division, University of Pennsylvania Medical Center, Philadelphia, Pennsylvania 19104

Carol Stocking (3) Heinrich-Pette-Institut für Experimentelle Virologie und Immunologie an der Universität Hamburg, 20251 Hamburg, Germany

Bin S. Teh (513) Department of Radiology, Baylor College of Medicine, Houston, Texas 77030

Timothy C. Thompson (513) Department of Urology, Baylor College of Medicine, Houston, Texas 77030

Deborah Toppmeyer (393) Department of Medicine, The Cancer Institute of New Jersey, UMDNJ-Robert Wood Johnson Medical School, Piscataway, New Jersey 08901

Catherine M. Verfaillie (331) Stem Cell Institute, Cancer Center, and Division of Hematology, Oncology and Transplantation, Department of Medicine, University of Minnesota, Minneapolis, Minnesota 55455

Maria T. Vlachaki (513) Department of Radiology and Veterans Affairs Medical Center, Baylor College of Medicine, Houston, Texas 77030

Dorothee von Laer (3) Chemotherapeutisches Forschungsinstitut, Georg-Speyer-Haus, 60596 Frankfurt, Germany

Ralph R. Weichselbaum (435) Department of Radiation and Cellular Oncology, University of Chicago Hospitals, Chicago, Illinois 60637

L. M. Weiner (81) Fox Chase Cancer Center, Philadelphia, Pennsylvania 91010

Thomas Wheeler (513) Department of Pathology, Baylor College of Medicine, Houston, Texas 77030

Lee G. Wilke (257) Department of Surgery, University of Michigan, Ann Arbor, Michigan 48109

K. K. Wong, Jr. (53) Division of Hematology and Bone Marrow Transplantation, and Division of Virology, City of Hope National Medical Center, Duarte, California 91010

Flossie Wong-Staal (95) Department of Medicine, University of San Diego, La Jolla, California 92093

Duen-Hwa Yan (465) Departments of Molecular and Cellular Oncology and Surgical Oncology, M. D. Anderson Cancer Center, The University of Texas, Houston, Texas 77030

Steven P. Zielske (341) Division of Hematology/Oncology, Department of Medicine and Ireland Cancer Center at Case Western Reserve University and University Hospitals of Cleveland, Cleveland, Ohio 44106

Robert CH Zhao (331) Stem Cell Institute, Cancer Center, and Division of Hematology, Oncology and Transplantation, Department of Medicine, University of Minnesota, Minneapolis, Minnesota 55455

Preface

The second edition of *Gene Therapy of Cancer* comes at a pivotal transition point in the development of this exciting technology. Much has occurred in the past 4 years to catapult preclinical and basic scientific concepts into therapeutic trials. In addition, while the outcome of the initial phase of clinical trials using gene therapy to target cancers has not yielded the amazing results initially hoped for, as with every new therapeutic venture in medicine, initial results provide the fodder for critical experiments, new targets, and new questions that propel the field forward.

We have reorganized the presentations in the second edition to reflect the continued new emerging strategies that will ultimately lead to the success of this therapeutic approach and have added introductory chapters to a number of the sections with the goal of setting the contributions in their proper basic scientific context.

Immune therapeutics takes on added emphasis given some of the recent breakthroughs in vaccine development and targeted delivery. Oncolytic virus therapeutics have also emerged in a very promising light with initial positive results observed in head and neck cancer leading to a number of preclinical advances. Therapies directed towards oncogenes, be it by expression of normal oncogenes, use of ribozymes and antisense therapeutics, and the use of E1A continue to be promising in preclinical and early clinical models. Hematopoietic stem cells are being used in gene therapy both in the antisense setting, for instance use of BCR/ABL antisense to block CML stem cell proliferation, and in genetically modified stem cells as immunotherapies. In addition, bone marrow protection by introduction of a drug resistant gene into hematopoietic stem cells is entering its next phase of clinical trials and appears to have accomplished its goal of achieving stem cell protection in both preclinical and clinical settings.

Gene delivery remains an important aspect of gene therapy of cancer and a number of chapters focus on gene delivery systems using both viral and nonviral approaches.

The reader will find this to be a comprehensive assessment of the current state of gene therapy of cancer offering state of the art research, a review of basic mechanisms and approaches, and a compilation of current clinical trial efforts.

We have not encumbered this text with a number of the current controversies in gene therapy but would note the importance of these issues in the field. Conflict of interest remains an active area of discussion at every major cancer center in the country and has been the recent focus of the American Society of Gene Therapy. The careful monitoring of patients undergoing clinical trials in gene therapy will remain a priority for all clinical trialists in this field. Linking preclinical models to clinical endpoints is an important aspect of this focus and will enable intermediate assessments to define whether the gene therapy effect has been achieved prior to relying on clinical cancer response and will help drive both Phase I and Phase II clinical trial design.

The next 5 years will be explosive in the next generation of preclinical and clinical developments of gene therapy of cancer. We hope this second edition will provide an important reference for investigators and observers alike in this exciting field.

Edmund C. Lattime, PhD
Stanton L. Gerson, MD

VECTORS FOR GENE THERAPY OF CANCER

1

Retroviral Vector Design for Cancer Gene Therapy

CHRISTOPHER BAUM

Medizinische Hochschule
Abt. Haematologie
30625 Hannover, Germany

WOLFRAM OSTERTAG

Heinrich-Pette-Institut für
Experimentelle Virologie und Immunologie
an der Universität Hamburg,
20251 Hamburg, Germany

CAROL STOCKING

Heinrich-Pette-Institut für
Experimentelle Virologie und Immunologie
an der Universität Hamburg,
20251 Hamburg, Germany

DOROTHEE VON LAER

Chemotherapeutisches Forschungsinstitut
Georg-Speyer-Haus
60596 Frankfurt, Germany

I. INTRODUCTION

In the past years, oncology was the center of gene therapy research [1]. However, despite generous support by, for example, the National Institutes of Health and related institutions in Europe, there is still a wide gap between the hopes raised and the results achieved. Most of the failures of gene therapy trials can be attributed to a discordant combination of overinterpreted clinical concepts and immature technol-ogy, including poor vector design [2]. Nevertheless, many former skeptics were turned to true believers, not only due to the enormous public and economical interest [3]. Thus, strong international competition in the field was generated, with an increasing number of researchers following valuable long-term concepts, including improvement of basic vector technology.

An ideal vector should (1) allow efficient and selective transduction of the target cell of interest, (2) be maintained, (3) be expressed at levels necessary for achieving therapeutic effects, and, last but not least, (4) be safe in terms of avoiding unexpected side effects in the host. Viruses are a perfect tool for gene transfer as they have evolved to deliver their genome efficiently to target cells with subsequent high-level gene expression. Vector systems for therapeutic gene transfer have been developed from different virus groups, each system having specific advantages and drawbacks. Retroviruses have several unique features that render them highly suitable for vector development. Retroviral vectors are, therefore, the prevalent system for gene transfer in human cells. Retroviruses integrate and express their genome in a stable manner, thus allowing long-term manipulation with transferred genes. This is a prerequisite for many gene therapy applications, including some approaches in cancer gene therapy. Integration usually does not alter host cell functions and is well tolerated. In the retroviral genome, *cis*-acting elements, responsible for reverse transcription, integration, and packaging, can be well separated from coding sequences. Such a genome structure facilitates the design of safe vectors and packaging cell lines.

However, with the transition to applications in human gene therapy, severe limitations of conventional retroviral vector systems have become apparent. These include low and variable particle titers, lack of appropriate vector targeting to specific cell types and genomic loci, failure to transduce quiescent cells, and relatively inefficient, position-dependent transcription. Fortunately, substantial progress in vector development has been made, based on deeper understanding of the biology of retroviruses and target cells. Here, we review some of the work relevant to cancer gene therapy.

We start with a short overview of potential applications of retroviral vectors in oncology. Then, we describe aspects of retrovirus biology relevant to gene therapy, to create a basis for discussing principles of and specific recent advances in retroviral vector design.

II. APPLICATIONS FOR RETROVIRAL VECTORS IN ONCOLOGY

In oncology, several different strategies involving somatic gene transfer are currently considered (Table 1). We can distinguish between diagnostic and therapeutic approaches; in either case, both healthy tissues or tumor cells may be targeted. Each strategy has special implications for vector design.

Gene marking uses stable retroviral transduction of heterologous genetic sequences to analyze the biological (stem cell function, antiviral effects) or pathogenic (tumor cell contamination, graft-versus-host reaction) capacity of blood cell transplants [4–6]. Here, efficient transduction of long-lived hematopoietic cells is required. Moreover, long-term transgene expression is necessary for follow-up analyses involving phenotyping and preparative sorting of transduced cells, based on cell surface or cytoplasmic markers encoded by the vector [7–9].

Besides this entirely diagnostic approach, several therapeutic strategies target healthy tissues. These strategies are also relevant to gene therapy of some inborn genetic disorders or acquired viral infections, due to the use of marker genes that allow selection of transduced cells *in vivo*. Positive selection is established in the context of drug resistance gene transfer, negative selection in adoptive immunotherapy.

Drug resistance gene transfer in nontumor tissues such as bone marrow is aimed at augmenting the therapeutic index of anticancer chemotherapy [10,11]. Protection at the level of hematopoietic progenitor cells reduces short-term toxicity, and protection at the stem cell level might even prevent

TABLE 1   Somatic Gene Transfer in Oncology and Implications for Vector Design

Approach	Aim	Target cells	Vector requirements	Vector system
Gene marking	Diagnostic	Healthy hematopoietic or lymphocytic cells, tumor cells (both *ex vivo*)	Transduction of long-lived stem cells, stable gene expression	Retroviral vectors
Drug resistance gene transfer	Therapeutic (paradigm for positive selection of transduced cell *in vivo*)	Healthy hematopoietic cells (*ex vivo*)	Transduction of repopulating cells, stable and high gene expression	Retroviral vectors
Adoptive immunotherapy	Therapeutic (paradigm for negative selection of transduced cell *in vivo*)	Donor lymphocytes (*ex vivo*)	Transduction of lymphocytes, stable and high gene expression	Retroviral vectors
Mini-organs	Therapeutic	Healthy autologous or xenogenic cells (*ex vivo*)	Stable or inducible gene expression	Retroviral vectors
Suicide gene transfer	Therapeutic (but not systemic)	Tumor cells (usually *in vivo*)	Applicability *in vivo*, targeting to tumor cells; strong, but not necessarily stable gene expression	Retroviral vectors; alternatively herpes virus vectors or adenoviral vectors
Oncogene antagonism	Therapeutic (but not systemic)	Tumor cells (usually *in vivo*)	Applicability *in vivo*, targeting to tumor cells; strong, but not necessarily stable gene expression	Retroviral vectors; alternatively herpes virus vectors or adenoviral vectors
Tumor vaccination	Therapeutic	Tumor cells, antigen-presenting cells (*ex vivo* or *in vivo*)	Applicability *in vivo*; moderate, but not necessarily stable gene expression	Retroviral vectors; alternatively herpes virus vectors, adenoviral vectors, or physicochemically

long-term toxicity and the mutagenicity of chemotherapy. The benefit for the patient will depend on the numbers of protected cells obtainable. These are expected to increase with each cycle of chemotherapy, because cells acquire a selective advantage upon expression of the drug resistance gene. Thus, this approach sets a paradigm for forced expansion of transduced cells *in vivo*. Similar to gene marking, this approach requires a number of technological improvements: First, helper functions of the vector systems and transduction conditions have to mediate efficient uptake (see Section V.A) and nuclear translocation (see Section V.B) in primitive hematopoietic cells (reviewed in Baum *et al.* [12]). Second, the vector needs to be equipped with *cis*-regulatory elements mediating dominant gene expression levels and thus strong penetrance of the phenotype (see Section V.B) [13,14]. Third, coexpression of a second gene (see Section V.B.7) is important in this approach, because coordinated transfer of two complementary drug resistance genes greatly widens its flexibility [15–18]. Finally, malignant cells must be excluded from productive transduction by cell purging or vector targeting (see Sections V.A.1.b and V.B.1).

A paradigm for negative selection of transduced cells in vivo is established in *adoptive immunotherapy*. Here, *ex vivo* selected populations of allogenic donor lymphocytes are used to elicit an antiviral or antileukemic effect [19]. Gene transfer in lymphocytes serves for both positive and negative selection. After transduction, positive selection of gene-modified lymphocytes is performed *ex vivo* using cell surface markers. After reinfusion, concomitant expression of a negative selection marker (a suicide gene) is instrumental for treating eventually occurring graft-versus-host disease. Here, the key issue is to design vectors with reliable and persisting coexpression of two genes (see Sections V.B.2, V.B.4, and V.B.7). An extension of this approach is the transfer and expression of "designer" T-cell receptors in autologous or allogeneic T cells, in order to generate effector cells with a new, predefined target cell specificity [20].

Positive or negative selection and monitoring of transduced cells *in vivo* is crucial for the development of artificial *mini-organs* (derived from genetically manipulated cells [21]). In oncology, these are of interest for systemic delivery of tumor-antagonistic factors such as immunotoxins or inhibitors of angiogenesis. Further applications extend to genetic or acquired disorders that can be treated by delivery with enzymes, hormones, or ligands. Equipping mini-organs with regulatable promoters might allow the adjustment of supply according to individual clinical requirements (see Section V.B.5) [22].

Other therapeutic concepts rely on direct genetic manipulation of tumor cells. Some of these suffer from poor predictability, mostly due to the tremendous variability of tumor evolution among and within individual patients. Moreover, not all of these strategies acknowledge the systemic character of tumor diseases. Nevertheless, in selected pa-

tients, these strategies might offer interesting perspectives. Usually, vectors have to be applied *in vivo* to become effective, and sometimes even replication-competent vectors will be needed. Here, nonretroviral systems may offer important alternatives, given that the problem of instability of persistence or expression is of minor importance.

Transfer of prodrug-converting enzyme genes (*suicide gene transfer*) is performed to render tumor cells susceptible to cytotoxic compounds requiring activation by a heterologous enzyme [23,24]. Alternatively, toxin genes may be used [25,26]. Another approach to control tumor cells by gene transfer is *oncogene antagonism* [27,28]. Here, one attempts to counteract tumor-promoting mutations of cellular genes. This is achieved by transducing tumor cells with wild-type copies of tumor suppressor genes or dominant negative proteins, antisense nucleotides, or ribozymes directed against oncogenes and their products. In compact tumor masses, some of these strategies may profit from the so-called bystander effect. This refers to cytostatic effects observed in nontransduced cells that result from delivery of proteins or activated cytotoxic drugs through direct intercellular exchange. However, this exchange might also dilute the effects in transduced cells [29]. With either approach, immune responses may be triggered that are expected to promote antitumor efficiency. Transfer of suicide or toxin genes, in contrast to oncogene antagonism, should exclude healthy tissues. Besides direct targeting of retroviral vectors to tumor cells, cellular vehicles may be used to deliver tumor-antagonistic gene products into tumor masses; these may be either cytotoxic T cells [26] or normal progenitor cells with homing capacity [30], tumor cells themselves [31], and possibly also endothelial cells or their precursors [32].

Application of retroviral vectors *in vivo* requires production of complement-resistant particles at high titers. Selectivity can be achieved already at the level of transduction, taking advantage of preferential infection of dividing cells by vectors based on murine retroviruses [33]. Specific targeting using engineered envelope proteins may, however, be superior (see Section V.A.1.b). At the level of transcriptional regulation, selectivity can be accomplished by insertion of promoters preferentially activated in tumor cells or tumor vasculature (see Section V.B.3). Depending on the tumor type, herpesviruses or adenoviruses (some of the latter specifically replicating in p53-negative cells) may represent alternative vectors [34]. Key aspects of targeting using specific receptors or promoters also apply to these nonretroviral systems.

Finally, *tumor vaccination* is performed to evoke a systemic immune response to tumor-specific antigens [35,36]. This is accomplished by transfer and expression of genes that increase antigen presentation or improve effector cell functions. Target cells for transduction are tumor cells, antigen-presenting cells, or tumor-infiltrating T cells. Thus, tumor vaccination strategies are highly variable with respect to the target cell population and to the type and numbers of activating genes to be transferred. Vectors may be applied either *ex vivo* or *in vivo* (after injection in tumor masses), and sustained gene expression in target cells is not necessarily required. For this approach, alternative vector systems (e.g., adenoviral or herpes vectors or physicochemical methods such as biolistics) may also be valuable.

In all these different strategies, important variables to account for in vector design are the route of gene transfer, the target cell population, the efficiency and specificity of transduction, and the level, duration, and specificity of transgene expression (Table 1). Therefore, appropriate components of the vector system have to be identified for each application ("tailored vectors"). Fortunately, substantial progress has been made toward all aspects of vector design relevant to cancer gene therapy. These vector improvements are based on detailed insights into the biology of retrovirus–host interactions.

III. BIOLOGY OF RETROVIRUSES

A. Classification

Sequence data and genome structure are the basis for the classification of retroviruses (Table 2) [37]. Each group contains several virus strains that differ in biological properties, such as receptor utilization and pathogenicity. Until recently, most retroviral vector systems discussed for human gene therapy were based on murine leukemia viruses (MLVs). MLVs belong to the mammalian C-type retroviruses and are further classified according to the species distribution of their receptors. Ecotropic MLVs replicate only in rodent cells and xenotropic MLVs only in nonmurine cells, while polytropic and amphotropic MLVs can infect murine and nonmurine cells. The 10A1 strain has an overlapping but distinct host range, due to the use of the receptor for the gibbon ape leukemia virus (GALV) in addition to the amphotropic

TABLE 2   Retrovirus Genera

Genus	Example viruses
Avian leukosis sarcoma	Rous sarcoma virus (RSV)
Mammalian C type	Murine leukemia virus (MLV), several strains: such as Moloney-, Harvey-, Abelson-, 407A-MLV.
	Feline leukemia virus (FeLV)
	Gibbon ape leukemia virus (GALV)
D-type viruses	Mason-Pfizer monkey virus (MPMV)
B-type viruses	Mouse mammary tumor virus (MMTV)
HTLV-BLV group	Human T cell leukemia virus (HTLV)-1 and 2
Lentivirus	Human immunodeficiency virus (HIV)-1 and -2
Spumavirus	Human foamy virus (HFV)

receptor [38]. Except for ecotropic viruses, gene transfer into human cells is possible with all groups of viruses mentioned. For historical reasons, most retroviral vectors applied thus far in human gene therapy have utilized the amphotropic receptor, although this is not the most efficient envelope for many targets.

B. Retroviral Genes and Their Products

Retroviruses within a group share a very similar proviral structure. In the first three groups (see Table 2), including mammalian C-type retroviruses, the genome codes only for the virion structural proteins Gag, Pol, and Env (Fig. 1) [37]. The *gag* gene products constitute the viral matrix and package the two retroviral RNA genomes into a viral nucleocapsid. Encoded by the *pol* gene, the virion also includes several enzymes necessary for virus replication. These are the reverse transcriptase, the integrase, and the viral protease, which cleaves the Gag and Pol precursors into the individual proteins. Receptor utilization is determined mainly by the glycosylated *env* gene product SU, which is anchored in the viral envelope by the transmembrane protein TM. The viruses of the HTLV–BLV group, the lentiviruses, and the spumaviruses are more complex and also encode specific nonvirion proteins with different regulatory functions [39,40]. Examples are the viral transcriptional activators of gene expression, such as *tax* in HTLV, *tat* in HIV, and *bel*-1 in foamy viruses.

C. Retroviral *cis* Elements

The *cis*-acting elements that regulate viral gene expression, reverse transcription, and integration of the provirus into the cellular DNA are organized very similarly in all retroviruses [37]. The provirus is flanked by the long terminal repeats (LTRs), carrying the terminal *att* sites, which are recognized by the integration machinery. The LTR is further divided into the three sections U3, R, and U5 (Fig. 1). The U3 region carries the viral enhancer and promoter elements. In the 3' LTR, initiation of transcription is suppressed, possibly due to interference with the 5' LTR. The polyadenylation signal resides in the R or U3 region. A recent report suggests that the retroviral splice donor may play an important role in suppressing the utilization of the polyadenylation signal of the 5' LTR [41]. Transcription of viral genomic RNA thus begins at the R region in the 5' LTR and ends with R in the 3' LTR. The RNA genome is thus flanked by identical redundant regions (R), which play an important role during reverse transcription (see Section II.D). The U5 region contains sequences necessary for reverse transcription and terminates with the *att* site.

The untranslated leader comprises R and U5 regions of the 5' LTR and sequences upstream of *gag*, including 18 nucleotides that form the primer binding site (PBS). The PBS is perfectly complementary to the 3' terminus of the tRNA

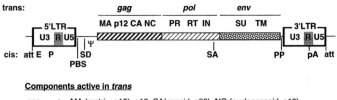

Components active in *trans*

gag ⟶ MA (matrix, p15), p12, CA(capsid, p30), NC (nucleocapsid, p10)

pol ⟶ protease (PR), reverse transcriptase/RNaseH (RT), integrase (IN)

env ⟶ SU (surface protein, gp70), TM(transmembrane protein, p15SE)

Components active in *cis*

att ⟶ integration signal

E P ⟶ enhancer / promoter

PBS ⟶ tRNA-primer binding site

SD, SA ⟶ splice donor, splice acceptor

Ψ ⟶ dimerization and packaging signal

PP ⟶ polypurine tract

pA ⟶ polyadenylation signal

FIGURE 1 Scheme of the proviral form of a replication-competent simple C-type retrovirus. Sequences coding for *trans*-acting proteins are indicated above the drawing, *cis*-acting sequences below.

primer that initiates reverse transcription of the RNA genome into the minus strand of proviral DNA. Leader sequences downstream of the PBS contain the splice donor site for generating the subgenomic RNAs, as well as the packaging and dimerization signal, which directs incorporation of two viral RNA genomes into virions. Optimal packaging was originally reported to require additional sequences, such as the first 400 nucleotides of *gag* in MLV-based vectors (so-called *gag*+ vectors) [42], although more recent data suggest that viral coding sequences are dispensable for high-titer packaging of MLV vectors [43,44]. In human immunodeficiency virus (HIV), the minimal sequences sufficient for packaging have not yet been defined [45,46]; in addition to part of the leader, a region in the 5′ end of the *gag* gene and sequences within *env* encompassing the Rev responsive element (RRE) appear to improve packaging [47].

The untranslated region between *env* and the 3′ LTR contains the polypurine (PP) tract, a run of at least nine A and G residues. Synthesis of the plus strand of proviral DNA is initiated here.

D. Retroviral Life Cycle

The retroviral life cycle is illustrated in Fig. 2. Initially, retroviruses bind through the Env protein SU to a specific viral receptor on the cell surface. All known retroviral receptors are membrane proteins, and several have been cloned [48]. The receptors for amphotropic MLV (Pit-2) and for GALV (Pit-1) are phosphate transporters found on most human cells. Interaction of viral SU with the receptor exposes a fusion peptide in the TM and triggers fusion of the viral and cellular membranes with subsequent release of the nucleocapsid into the cytoplasm [49]. Here, the viral RNA genome is reverse transcribed into the proviral DNA by the viral reverse

transcriptase (RT) [37]. The nucleocapsid protein NC is also required for this process. Reverse transcription is initiated at the PBS. RT synthesizes the negative strand complementary to the U5 and R regions of the 5′ LTR, while the RNAse H activity of RT degrades the genomic RNA. The nascent DNA strand is transferred to the 3′ end of the RNA genome, where, starting with the U5 region, the negative strand is completed with concomitant degrading of the RNA genome. The PP tract escapes digestion and serves as a primer for

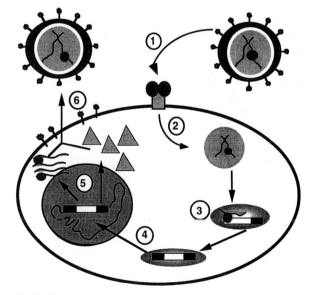

FIGURE 2 Life cycle of a replication-competent retrovirus: (1) virion binding, (2) virion penetration and uncoating, (3) reverse transcription of RNA genome into proviral DNA, (4) nuclear transport of preintegration complex and integration of provirus, (5) transcription of genomic and subgenomic mRNA and translation of viral gene products, and (6) nucleocapsid assembly, budding, and maturation of virion.

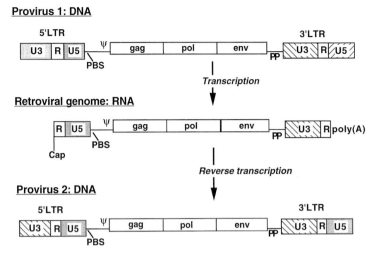

FIGURE 3   Structure of provirus and viral genome of MLV. Provirus 1 is transcribed into the retroviral genome flanked by the R sequences. Two RNA genomes are packaged into a virion and released from the cell. After infection of the target cell, the genomic RNA is reverse transcribed into provirus 2. This again is flanked by two LTRs that contain the U5 region of the 5′ LTR and the U3 region of the 3′ LTR of provirus 1. Abbreviations are explained in Fig. 1.

plus-strand DNA synthesis, which proceeds through U3 and R. The plus strand is then transferred to the 5′ end and transcription is completed. In brief, reverse transcription of the two RNA genomes generates a single provirus with two complete LTRs by duplicating the U3 and U5 regions of the RNA genome; importantly, 3′ LTR and 5′ LTR serve as templates for U3 and U5, respectively (Fig. 3). The infidelity of reverse transcription and recombinations occurring as a consequence of switching between the two RNA templates during reverse transcription lead to a high degree of variability of retroviruses. This represents a potential drawback for vector design, because unpredicatable errors may be introduced during vector infection [50]. However, other viral or nonviral methods of DNA transfer may be associated with even higher rates of recombination. In many instances, this results from the transfer of multiple copies of homologous DNA, which is excluded in retroviral systems.

After reverse transcription, the nucleocapsid proteins remain tightly associated with the proviral DNA. This complex carries factors necessary for the integration of the viral DNA into the genome of the host cell. In MLV, this complex cannot pass through the nuclear pores, and nuclear transport requires mitosis with breakdown of the nuclear membrane [51,52]. Nuclear transport, however, is not the only factor limiting transduction of quiescent cells by vectors based on simple retroviruses [53]. In lentiviruses, such as HIV, mitosis is not required, and the preintegration complex is targeted to the intact nucleus with the help of the matrix, integrase, and possibly the Vpr protein [54,55]. Interestingly, a DNA flap generated during reverse transcription has been identified as a *cis*-acting component required for nulcear import of HIV-1

[56]. Despite the ability to transduce quiescent cells, integration of lenitviruses occurs more efficiently in metabolically activated cells and best in cycling cells [55].

Integration of the provirus is random with regard to position in the genome, with some preference for open chromatin [57]. Local structural features of host DNA rather than specific sequences influence the susceptibility to integration. Rarely, integration can produce alteration of the phenotype of an infected cell by activation or disruption of cellular genes [58]. Infection with actively replicating virus is accompanied by repetitive integration events in different cells and thus increases the possibility of proto-oncogene activation and the development of neoplasias [59]. In therapeutic retroviral gene transfer, the probability of such insertional mutagenesis has been minimized by the use of replication-incompetent retroviruses as vectors that integrate at low copy numbers (usually one or two per cell). For a single integration event, the risk of inducing tumor-promoting mutations is estimated to be in the range of 10^{-6} or lower [60]. Immune responses to altered cellular genes further reduce but do not exclude the likelihood of inducing tumors by retroviral vector integration.

The integrated provirus is transcribed by the cellular RNA polymerase II. In simple retroviruses such as MLV, this process is solely dependent on the cellular transcription machinery. Between viral strains, binding sites for cellular transcription factors in the U3 region differ. Because the expression of many transcription factors is developmentally regulated, cell tropism and pathogenicity of retroviruses are influenced by the composition of *cis* elements in the LTR [61,62]. Complex retroviruses have a number of transcriptional and posttranscriptional transactivators which, in cooperation with cellular factors, influence viral transcription levels, nuclear export of RNAs, and splicing patterns [40].

Viral transcripts are modified by cellular capping enzymes at the 5′ end, and a poly(A) tail is added to the 3′ end at the R–U5 border following specific polyadenylation signals. Retroviral transcripts enter one of three pathways: (1) The RNA is translated into the Gag or Gag–Pol precursor proteins. (2) The RNA is spliced into subgenomic RNA. (3) The full-length viral RNA is packaged as a viral RNA genome into the virion and released from the cell. All subgenomic mRNAs are spliced from the same splice donor generally located in the leader. In simple retroviruses only the *env* transcript is spliced. Complex retroviruses have several splice acceptors in the 3′ half of the genome, where several different smaller spliced transcripts are generated which code for regulatory proteins [37,40].

Initiation of translation of the *gag–pol* or the *env* message may also occur in a cap-independent manner, facilitated by an RNA structure that resembles an internal ribosomal entry signal [63]. The Gag precursor protein is always translated from the full-length viral RNA. Translation is continued past the stop codon to generate a Gag–Pro or Gag–Pro–Pol precursor protein at a low frequency (5–10%). The viruses of the avian sarcoma–leukosis virus (ASLV) complex are an exception, as *gag* has no stop codon and is always translated as a Gag–Pro polyprotein [64]. After virus assembly, the viral protease (PR) is activated by autocatalytic release from the precursor. PR then cleaves the Gag precursors into the matrix (MA), capsid (CA), and nucleocapsid (NC) proteins and several smaller peptides. The Pol precursor is cleaved by PR to yield the reverse transcriptase, which has an RNA- and DNA-directed polymerase and a ribonuclease H activity, and the smaller carboxyl terminal viral integrase. The Env protein is cleaved, most likely during transport to the cell surface, by a cellular protease into the viral surface protein (SU) and a transmembrane protein (TM).

Viral genomic RNA is assembled into the nucleocapsid through specific interaction of the NC portion of the Gag precursor and *cis*-active viral packaging sequences [45]. B- and D-type retroviruses preassemble in the cytoplasm, while C-type viruses, such as MLV, assemble at the cytoplasmic membrane. The MA domain interacts with the inner surface of the cytoplasmic membrane and mediates budding of the virus [65]. The virus acquires the viral envelope by budding through membrane areas that contain Env proteins. However, viral glycoproteins are not necessary for virion formation, and in the absence of Env noninfectious, bald, enveloped particles are released. Proteolytic cleavage of viral polyproteins begins during the budding process and is completed in the released particles, thereby generating the mature infectious virions [66].

IV. PRINCIPLES OF RETROVIRAL VECTOR SYSTEMS

First, we will discuss general rules for designing packaging cell line and vectors. In Section V, we approach specific aspects related to vector entry, integration, and expression.

A. Packaging Cells

Retroviral vector systems are designed to mimic the infectious properties of retroviruses (stable transduction of target cells without inducing rearrangements and relatively stable gene expression) in replication-incompetent vector particles. The latter are produced from packaging cell lines. In these cells, viral coding regions are physically and functionally separated from the vector genome. The strict separation of *cis*-active and *trans*-active components serves purposes of safety, efficiency, and increased flexibility (Fig. 4).

A modern, safety-modified packaging cell contains at least two expression constructs for viral genes, one encoding for *gag–pol* and one for *env* genes [67]. To prevent mobilization in retroviral particles, these mRNAs lack the packaging signal. In the case of MLV, this is easy to achieve, as retroviral

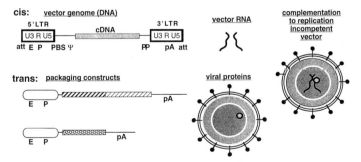

Replication-incompetent vector: separation of *cis* and *trans*

FIGURE 4 Separation of *cis*-acting sequences and *trans*-acting retroviral coding regions to generate safe packaging cell lines for release of replication-incompetent vectors. Abbreviations are explained in Fig. 1.

coding regions are not sufficient for packaging. In HIV, the extended and less well-defined packaging signal might represent a drawback to the development of helper cells (and vectors).

High expression of viral proteins in packaging cells is best accomplished by allowing direct selection for the promoter driving the retroviral genes [68]. This can be managed by linkage with a dominant selectable marker, with coexpression obtained by reinitiation of transcription from the 3′ untranslated region, internal ribosomal entry sites for reinitiation of translation, or alternative splice signals. Alternatively, inducible promoters can be used, representing the method of choice when particle components have cytopathic effects (as in the case of some proteins used for pseudotyping; see Section 5.A.1.a). Encoded from such "packaging constructs," the packaging cell provides all *trans*-acting viral elements required for particle assembly, release, maturation, and transduction of the target cell.

The vector RNA is also generated inside the packaging cell and contains those *cis*-active elements required for a single transduction: cap site and poly(A) signal for the genomic message; packaging signal for incorporation into particles; PBS, PP, and R–U5 sequences for reverse transcription; and *att* sites for integration. Moreover, it contains the transgene cassette(s) of interest including enhancer/promoter sequences for initiation of transcription (Fig. 4).

Packaging cells not expressing RNAs with suitable packaging signals should not release infectious particles. However, depending on the cellular background, there is the potential risk of packaging endogenously expressed retroviral or retrovirus-like sequences. In the case of MLV-based systems, this risk is highest in a rodent background. Here, packaging and transfer of VL30-sequences are observed as frequent events, especially when vector titers are low [69,70], implying that this risk can be reduced when high-titer producer clones are selected for vector production. This is important because packaging of and recombination with endogenous retroviral elements are the most important events leading to the generation of replication-competent retroviruses from safety-modified packaging cells [71,72].

In immunocompromised primates and permissive mouse strains, replication-competent amphotropic MLV can induce lymphoma or leukemia [58,59]. Also, amphotropic retrovirus-induced spongiform encephalomyelopathy has been observed after inoculation in newborn mice [73] (for a thoughtful discussion on safety aspects of non-human, xenogenic viruses, see also Isacson and Brakefield [74]). More recently, many packaging cell lines have been developed or are under construction in a non-rodent background, such as human or canine, not known to express retroviral sequences packaged in MLV particles.

In the human host, most xenogenic retroviruses are complement sensitive, precluding administration *in vivo*.

Complement sensitivity is defined at two levels: first, by specific Env sequences, and, second, by protein modifications characteristic to the species background of the producer cell [75,76]. Complete complement resistance is achieved by producing vector particles with alternative envelopes (e.g., derived from feline leukemia virus) in human packaging cells. However, repetitive administration is likely to be hindered by immunogenicity of retroviral proteins. Moreover, fundamental restrictions to successful transduction *in vivo* are present at the physicochemical level, such as particle concentration and motility (reviewed by Palsson and Andreadis [77]). *Ex vivo*, these can be overcome by suitable transduction protocols.

To produce replication-incompetent vectors, vector genomes are introduced into packaging cells either by transfection or by retroviral transduction. Packaging systems have also been developed to release high-vector titers after transient transfection, or semi permanently from episomally replicating plasmids [78,79]. For clinical applications, stably transfected clonal packaging cell lines still represent the ultimate choice, as these allow vigorous preclinical testing of safety (especially the absence of replication-competent retroviruses [80]) and efficiency, as well as large-scale production of vector stocks from defined cell banks under conditions of good manufacturing practice (GMP). Titers released from retroviral packaging cell lines usually do not exceed 10^6 to 10^7 per milliliter of cell-free supernatant. Currently, this is the minimum required for *ex vivo* approaches, such as transduction of hematopoietic cells. When all components are improved, titers might be as high as 10^8. Concentration of the fragile retroviral particles is alleviated when the membrane is stabilized with non-retroviral components (see Section V.A.1.a).

B. Basic Vector Architecture

The flexibility of the retroviral genome offers a great degree of freedom for insertion of transgene cassettes. Still, retroviral vectors mimicking the basic architecture of their replication-competent ancestors (LTR–leader–gene(s)–LTR) have found the most widespread use (Fig. 5A and B) [81]. These vectors are usually very stable and also mediate reasonable transgene expression in the cellular system of interest, provided that appropriate enhancers are employed (see Section V.B). The gene is expressed either from within the *gag* region [42,82], exactly replacing *gag* [44], or from a subgenomic, spliced RNA which can lead to higher translation efficacy (MFG vector [43,83], GRS vector [44]).

Efficiency as well as safety of retroviral vectors might be further improved when the leader is functionally inactivated or physically deleted in the target cell (Fig. 5C to F). Increased efficiency results from removing the long untranslated leader from the transcript, which might contain repressory elements for transcription or translation. This also excludes the

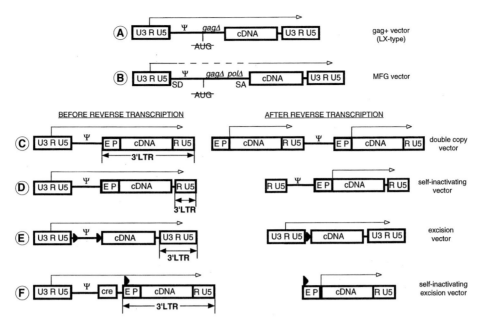

FIGURE 5 Flexibility in basic vector architecture. (A, B) LTR-controlled vectors including the packaging region (Ψ) in the genomic transcript in transduced cells. *gag* Δ and *pol* Δ, residual fragments of viral genes *gag* and *pol*, respectively; destroyed start codon of *gag*. (C–F) Different forms vectors excluding Ψ from transcripts in transduced cells. Plasmid constructions are represented on the left (before reverse transcription), and the status of the proviral form after reverse transcription is shown on the right. In (E) and (F), the status in transduced cells after site-specific recombination is shown. The black filled triangles represent loxP sites recognized by the site-specific recombinase, Cre.

packaging region from vector transcripts in the target cell, thus precluding transmission of the vector in the hypothetical case of accidental superinfection with replication-competent retroviruses.

There are several options for excluding the packaging signal. First, the transgene plus its enhancer-promoter can be placed in the U3 or R region of the LTR, resulting in a "double-copy" vector after completion of reverse transcription [84,85]. Duplication of the transgene cassette is expected to result in higher expression levels. Some sequences, however, are not compatible with this strategy, resulting in a high incidence of recombination. Second, self-inactivating or suicide vectors can be generated by deleting the enhancer–promoter or the promoter only in the U3 region of the LTR and placing the transgene of interest under control of an internal promoter, either in sense or in antisense orientation to the LTRs [86,87]. Third, in LTR-controlled vectors, sequences between PBS and the start codon of the transgene can be flanked by, for example, loxP sites, allowing conditional deletion upon expression of the bacteriophage recombinase cre [88]. Finally, reversion of double-copy vectors to a monocopy vector is possible with a self-contained loxP/cre vector [89]. Given that stability, titer, and expression characteristics of these more sophisticated constructions are better defined, they represent valuable alternatives to conventional, LTR-controlled vectors.

V. ADVANCES IN RETROVIRAL VECTOR TAILORING

Besides the more general aspects of packaging cell line design and vector construction discussed above, specific advances can be noted with respect to distinct stages of the retroviral life cycle. Of special importance are those related to vector entry, integration (both defined by *trans*-active vector components), and expression (defined by *cis*-active elements) relevant to cancer gene therapy. This work should lead to vectors specifically tailored for clinical applications.

A. Components Active in *trans*

1. The Retroviral Envelope

For many reasons, the amphotropic Env, hitherto used in most gene therapy trials involving retroviruses, is not a perfect choice for mediating vector entry. The cognate receptor, Pit-2, is too widely expressed to allow specific cell targeting, with the ironical exception of primitive hematopoietic cells, where expression is too poor to allow efficient transduction [90–93]. Moreover, the amphotropic Env is involved in an unexpected pathogenicity of replication-competent retroviruses: induction of spongiform encephalomyelopathy [73]. Alternative Env proteins like that of the 10A1 strain are also associated with these potential drawbacks [94]. Moreover, the low stability of retroviral particles with conventional retroviral envelopes complicates vector concentration for *in vivo* delivery. Importantly, both vector stability and targeting can be improved by altering the retroviral envelope. The two major approaches are discussed below.

a. Pseudotyped Retroviral Vectors

Coinfection with two viruses generates hybrid virions, which contain the genome and core proteins of one virus and mixed envelope glycoproteins of both viruses. The host range of these "pseudotypes" is determined by both envelope proteins [48,95]. Pseudotyping can be used to alter the host range of retroviral vectors. Pseudotyped MLV-derived vectors thus can transduce cells that are normally resistant to MLV due to lack of functional amphotropic receptor (reviewed by Friedmann and Yee [96]).

The mechanisms that determine whether a foreign viral envelope protein can be incorporated into the viral envelope are not well understood. The cytoplasmic anchor of the transmembrane (TM) Env protein was shown to guide MLV glycoproteins to the envelope of budding virus particles [97]. Thus, homologous Env proteins are efficiently incorporated into the MLV envelope, while heterologous proteins must be expressed at high densities in the cell membrane to allow pseudotype formation [98].

Pseudotyped retroviral vectors are generated by coexpression of vector RNA with retroviral Gag and Pol and the unrelated glycoprotein. Packaging systems for several pseudotypes of MLV have been developed. Pseudotypes that incorporate the glycoprotein of vesicular stomatitis virus (VSV-G) have an extremely broad host range [96,99]. The VSV-G protein enters the cell by interacting with an ubiquitous phospholipid component of cell membranes (Fig. 6) [100]. Mammalian, fish, and insect cells can be transduced

[96,99–103]. CD34$^+$ hematopoietic progenitors were shown to be up to 10-fold more susceptible to a VSV-G than to an amphotropic pseudotype [103]. We found that VSV-G pseudotypes can infect hematopoietic stem cell lines and fibroblasts equally well, while transduction of stem cells with amphotropic vectors was at least 100-fold less efficient [93]. This observation reveals that the receptor deficiency of primitive hematopoietic cells to retroviral transduction [91,92] can be completely overcome by vector pseudotyping [93].

An additional advantage is that VSV-G confers great stability on the retroviral particle, and pseudotypes can be concentrated to high titers by ultracentrifugation. This is of interest also for *in vivo* applications. However, the host range of VSV-G is too broad to allow specific cell targeting, and the high immunogenicity of VSV-G is expected to preclude repetitive administrations *in vivo*. Another major drawback has been that the VSV-G protein is toxic for the cell. Pseudotypes are thus only produced for a limited period from already dying packaging cells [101]. Recently, stable packaging cell lines have been generated by placing the VSV-G gene behind an inducible promoter. In these lines the VSV-G gene is repressed but can be induced for vector production. However, vector production here is also accompanied by cell death [102].

Alternatively, pseudotypes of MLV vectors that incorporate Env proteins of the feline RD114 retrovirus, the gibbon ape leukemia virus (GALV) or the 10A1 mouse retrovirus transduce hematopoietic progenitors and lymphocytes more efficiently than amphotropic pseudotypes [104–107]. Other retroviral Env proteins have also been utilized. Examples are the glycoproteins of mink cell focus-forming (MCF) MLV-strains, HTLV-I, HIV-1, and human foamy virus (HFV) [38,108–111]. The tropism of these pseudotypes generally has not been properly evaluated to show advantage over amphotropic vectors. HFV pseudotypes are especially

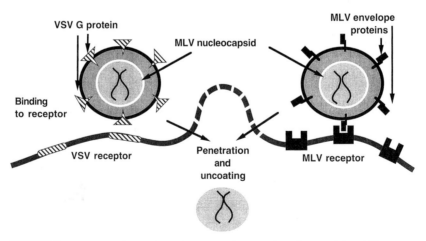

FIGURE 6 VSV-G protein pseudotyped retrovirus. Pseudotyping affects particle stability, receptor targeting, and mode of entry. After uncoating, different envelope pseudotypes follow the same pathway.

promising, as it is assumed that all mammalian cell types are infectable. However, we have observed that the hematopoietic progenitor cell lines FDC-Pmix and FDC-P1 are not only partially resistant to amphotropic MLV but also to HFV infection [93,112].

Recently, we described a novel retroviral pseudotype with the glycoproteins GP-1/-2 of the lymphocytic choriomeningitis virus (LCMV). LCMV GP is not cell toxic and is efficiently incorporated into MLV as well as into lentiviral vectors. This pseudotype has a broad host range and can be concentrated to high titers [113].

Further perspectives in pseudotype development are opened by generating chimeric envelopes. For instance, efficient pseudotyping of MLV with HFV surface proteins is only possible for chimeric envelope proteins containing an unprocessed cytoplasmic tail of MLV TM fused to a truncated HFV envelope protein [108]. Using similar chimeric envelope proteins it may be possible to generate a large panel of different pseudotypes with unrelated viral or nonviral membrane proteins which normally are not incorporated into retroviral envelopes efficiently.

b. Ligand-Directed Targeting

Ideally, vectors should be designed to selectively transduce specific target cells of interest present within mixed cell populations *ex vivo* or even intact organs *in vivo*. In oncogene antagonism and suicide gene transfer (see Section II), specific or at least preferential targeting to tumor cells has to be achieved *in vivo*. Here, binding of virus to non target cells will lead to considerable loss of the effective virus titer and also increases unwanted side effects. In contrast, *drug resistance gene transfer* to hematopoietic cells (see Section II) is performed *ex vivo*, with strict exclusion of malignant cells. Targeting retroviral transduction can be achieved at two levels: first, by colocalization of cells and viruses on a specific matrix (Fig. 7A), and, second, by equipping retroviral particles with cell-specific ligands (Fig. 7B to F).

Colocalization of cells and viruses on a biochemical matrix can only be used *ex vivo*, alleviating vector–cell interactions at the physicochemical level. Colocalization can lead to higher transduction efficiency in cells with poor receptor representation, paradigmatically shown in fibronectin-assisted retroviral transduction of hematopoietic progenitor cells or

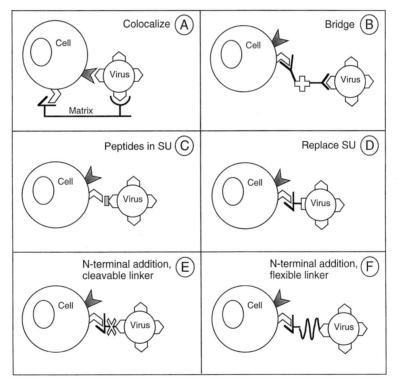

FIGURE 7   Targeting transduction via cellular receptors. The strategies discussed in Section V.A.1.b are schematically represented. Entry either occurs via the differentiation-specific cellular receptor (open symbol on cell surface) or still requires the natural retrovirus receptor (shaded symbol). (A) colocalization of virus and cell via matrix proteins; (B) molecular bridge between virus and cell (here, cross-linked antibodies); (C) a peptide in the binding domain of SU alters its tropism; (D) the binding domain of SU is replaced with a targeting ligand; (E) an N-terminal addition is linked to SU via a protease-cleavable linker or, as shown in (F), via a flexible linker.

lymphocytes [9,114,115]. However, it remains to be seen whether this approach can be elaborated for target-specific virus uptake—for example, by displaying ligands on the matrix which are selectively recognized by the target cell population of interest (Fig. 7A).

The alternative approach is to retarget retroviral entry via specific cell surface molecules by manipulating the viral envelope. To this end, several strategies have been followed (reviewed by Cosset and Russell [116,117]). While specific binding is relatively easy to achieve, virus uptake with engineered envelopes often is much less efficient. One initial approach was to direct specific virus binding by creating a molecular bridge between the virion and the cell surface (Fig. 7B). Here, virus particles are coated with specific antibodies for the surface subunit of Env (SU), and the cells are incubated with an antibody specific for a membrane protein such as the epidermal growth factor receptor or the insulin receptor. Both antibodies are then linked by secondary antibodies or by biotin/streptavidin [118]. In other studies, ecotropic or avian retroviral envelope proteins were modified. Wild-type Env proteins of these viruses do not allow infection of human cells. Therefore, incorporation of specific binding epitopes can selectively retarget the virus to human cells with the complementary membrane protein of choice. Three general strategies have been followed:

1. Small peptides that specifically bind to cellular receptors are introduced into binding domains of the SU protein without affecting natural receptor recognition [119,120]. Human breast cancer cell lines, overexpressing human epidermal growth factor receptors (HER-2 and HER-4) could be specifically targeted by insertion of the heregulin peptide, a ligand for HER-2 and HER-4, into ecotropic SU of MLV [121]. However, titers and efficiency of transduction were too low for *in vivo* applications (Figure 7 C).

2. The complete binding domains of SU are substituted by alternative ligands for cellular receptors (Fig. 7D). Erythroid progenitor cells have been targeted by substituting SU binding domains with erythropoietin [119]. Chimeric SU proteins that contain single-chain antibodies (scAs) attached to the truncated retroviral Env proteins are a versatile system with the potential of targeting cells via specific epitopes of many different membrane proteins. An example is the scA B6.2, which binds to an antigen on breast and colon cancer cells [122]. This strategy works well with the Env of the avian spleen necrosis virus, whereas transduction with MLV-derived chimeric Env proteins is inefficient and associated with low titers [123]. Differences in the flexibility of Env proteins and in the pathways involved in virus internalization might be responsible for this discrepancy. In wild-type Env, virion binding causes a conformational change in SU, thereby exposing fusion domains of the Env transmembrane unit (TM), finally leading to viral penetration. In the chimeric

envelope proteins described so far, conformation is generally altered and fusion processes are not triggered efficiently [124]; therefore, most chimeric Env proteins support efficient binding of virions, but postbinding events are impeded or even completely blocked. This indicates that additional alterations in TM might be required to improve uptake. An alternative would be to incorporate foreign viral envelope glycoproteins, such as the hemagglutinin glycoproteins from fowl plague virus which display a targeting ligand, into the retroviral envelope [125]. The postbinding functions of such nonretroviral envelope glycoproteins may be less sensitive to modification of the protein. Virion infectivity can also be increased by incorporating additional wild-type Env proteins that mediate fusion ("fusion helpers") [126]. Such receptor cooperation is also of central importance for the third targeting strategy.

3. With the aim of improving virus penetration, specific binding domains have been added to the N-terminus of the complete amphotropic Env proteins [127–130] (Fig. 7E and F). Amphotropic Env mediates infection of many human cell types. N-terminal additions, however, can block binding to the amphotropic receptor. Instead, these viruses bind to another membrane receptor of choice, but penetration still might require the amphotropic receptor. Two types of linkers between the retroviral Env and the N-terminal targeting domain allow such a two-step entry mechanism. In one approach, the ligand is fused to the amphotropic Env via a protease-cleavable linker [127] (Fig. 7E). Particles displaying these proteins bind to but do not infect cells. After protease cleavage, the N-terminal extension is released and the amphotropic binding domain exposed. Bound virus can then enter the cell efficiently by the amphotropic receptor. An unsolved problem *in vivo* is the systemic application of protease. An alternative would be to characterize cellular membrane proteases with their specific target sequences. In another approach, cooperation between two receptors is mediated by a proline-rich linker between amphotropic Env and the additional binding domain (Fig. 7F). Here, it is assumed that binding of the added specific ligand to its receptor triggers a conformational change that exposes amphotropic binding domains, which then mediate efficient entry via the amphotropic receptor [131]. A problem with both approaches may be that the amphotropic receptor is not expressed at sufficient levels on all cell types, as evident from studies with early hematopoietic cells [93,107].

While most investigators concentrate on positive targeting, negative targeting of selected cells can also be desirable in cancer gene therapy. An example is drug resistance gene transfer (see Section II), where the transduction of malignant cell is potentially hazardous because clones resistant to chemotherapy might be generated. Here, the specific blockade to retrovirus entry found with many engineered Env proteins can potentially be exploited to increase the safety of vectors.

2. Nuclear Transport and Integration

a. Vectors Derived from Complex Retroviruses

Nuclear transport of the preintegration complex is restricted in C-type retroviruses such as MLV which require mitosis and breakdown of the nuclear membrane for integration into the host cell genome. Unlike MLV vectors, lentiviral vectors can transduce nondividing, yet postmitotic, cells such as neurons and terminally differentiated macrophages [132–135]. Malignant cells, especially in larger tumors where blood supply becomes limiting, are also often quiescent. Similarly, hematopoietic stem cells rarely cycle but can be transduced with lentiviral vectors [136].

Several lentiviral packaging systems have been developed that generally use the G protein of vesicular stomatitis virus (VSV-G) and not the retroviral Env as an envelope glycoprotein [136–138]. A major problem is that several gene products used in lentiviral packaging systems, such as the protease, VSV-G, and vpr, have proven to be toxic [139]. Therefore, vector titers in stable packaging cell lines have been low. Recently, however, an inducible packaging system that produces titers as high as 10^6 per milliliter has been described [140,141]. A problem for vector safety is that packaging sequences are dispersed throughout the HIV genome and are not clearly separated from coding regions [46,47]. Therefore, vectors and packaging constructs share common sequences with the potential to generate replication-competent viruses with pathogenic potential by homologous recombinations. In the latest generation of lentiviral vectors, this risk has been largely eliminated by reducing lentiviral genes to *gag*, *pol*, and *rev*, which are expressed from two separate plasmids [142]. However, there still remains a concern that individuals treated with HIV-derived vectors may exhibit serum conversion to HIV-1 [143]. Another alternative in the future may be vectors derived from animal lentiviruses such as simian immunodeficiency virus, feline immunodeficiency virus, and equine infectious anemia virus. Such vectors would have the ability to transduce quiescent cells but not the pathogenic potential of HIV [40,144–147].

Vectors derived from foamy viruses may have several advantages over lentiviral vectors. Foamy viruses are now generally considered to be apathogenic in humans, although this issue has been controversial in the past [148]. Foamy viruses have an increased packaging capacity (12 kb compared to 9–10 kb in MLV). They infect many mammalian cell types; therefore, the host range is generally considered to be broad [149,150]. However, infection of hematopoietic stem cell was found to be inefficient for cell-free vectors that carry the foamy virus envelope glycoprotein, and

foamy virus capsids require the cognate envelope protein for particle export [112,151]. This indicates that the viral host range may be more restricted than is generally believed. It has been postulated that HFV vectors transduce stationary cells more efficiently than MLV; however, this issue is still controversial [152,153]. Helper virus free vectors have been developed, but safety concerns remain, as the packaging sequences in the viral genome have not been defined clearly and a clear separation of viral *trans* and *cis* elements may not be possible [154,155]. Taken together, knowledge of the biology of foamy viruses and other complex retroviruses is still limited. Extensive studies will be necessary before the value of these vectors for human gene therapy can be assessed.

b. Targeting Integrase

Targeting integrase to selected genomic loci is desirable to completely avoid insertional mutagenesis, to select integration sites supporting long-term expression, and to reduce clonal variability of gene expression. In the context of cancer gene therapy, these considerations are of relevance for gene marking, drug resistance gene transfer, and adoptive immunotherapy (see Section II).

Among the factors influencing site selection are overall DNA confirmation (open chromatin is a better target than heterochromatin), DNA sequence (in terms of local chemical or structural features rather than concrete motifs), DNA bending, and associated nuclear proteins (transcription factors, topoisomerases, replication proteins, matrix proteins) [156–158]. Integrases from various retroviruses differ with respect to target site selection, depending on the central core domain [159]. Systems for targeting retroviral integration to specific sequences are based on fusion of the IN protein with DNA-binding domains of well-characterized transcription factors, resulting in preferred but not specific integration to cognate sites [160–162]. A more efficient alternative might be to exploit the specificity of some yeast retrotransposons (Ty1, Ty3) for genes transcribed by RNA-polymerase III which exist in multiple copies and for which integration of a transgene is not expected to be hazardous [163,164]. Also, some human LINE elements and related retrotransposable sequences from other species encode endonucleases that prefer DNA with certain structural features [165]. It seems attractive to exploit such endonucleases, which are functionally distinct from integrases, for vector packaging systems.

A completely different approach is to block the viral integration process by elimination of *att* sites from the vector. Then, selection can be made for integration via homologous recombination; this process, however, is limited by the cloning capacity of retroviral vectors (9–10 kb) and by the extremely low frequency of gene targeting in somatic cells [166].

B. *cis*-Active Elements

With the exception of gene marking, which theoretically can be performed without introducing active transcription units, all other applications for somatic gene transfer in oncology require a certain strength and duration of vector transcription (Table 1). Choosing the appropriate *cis*-acting elements guarantees full penetrance of the phenotype of interest and thus influences both safety and efficiency of the gene transfer.

As opposed to physicochemical transfection methods, retroviruses are characterized by only moderate integration site dependence of gene expression. This implies that integration occurs at permissive loci or that retroviruses transfer genetic elements that can actively induce conformational or functional changes in their environment [167]. Such elements may reside in the enhancer region and involve yet poorly understood mechanisms, including secondary DNA structures [168]. Residual modulatory influences by the integration site usually lead to about 50-fold variation of gene expression levels among independent clones. However, depending on specific vector sequences and the genetic environment, complete extinction (silencing) of retroviral gene expression can also occur (reviewed by Lund *et al.* [169]). This is most evident in embryonal stem cells [170] but also has been observed in hematopoietic stem cells [171] and more mature tissues such as fibroblasts *in vivo* [172] and several somatic cell lines *in vitro* [169]. Therefore, modifying *cis*-acting elements in retroviral vectors can affect all aspects mentioned: differentiation-dependent gene expression levels, integration-site-dependent modifications, and incidence as well as kinetics of silencing. Thus, it is crucial to equip vectors with enhancer sequences fitting to the host's transcriptional setting.

In simple retroviruses such as MLV and derived vectors, two major targets for transcriptional control have been identified: the dominant enhancer–promoter is located in the U3 region of the LTR, but sequences of the nontranslated leader (especially PBS) also contribute [173,174] (Fig. 8). Retroviral enhancers display recognition sites for a variety of transcription factors intimately involved in differentiation processes of their natural target cell population. Precise consensus sequences, their numbers, and relative orientation are crucial for enhancer strength and specificity [175]. Most retroviral enhancers are poorly expressed in more primitive, uncommitted cells such as embryonic and hematopoietic stem cells [174]. This is mainly because these cells are not fully equipped with transcriptional activators or even express active repressors recognizing the retroviral enhancer. In permissive environments, such as in more mature hematopoietic cells, retroviral *cis*-elements generally act quite autonomously and in a dominant manner, resulting in efficient transcription levels. Here, up to 0.1% of cellular transcripts can be generated from single-copy integrations. But even in more mature cells,

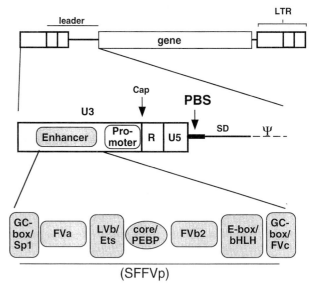

FIGURE 8 Dominant *cis*-acting elements of a murine leukemia virus reside in the U3 region of the LTR (specified in more detail for the strain SFFVp) and in the primer binding site (PBS) of the untranslated leader. SD, splice donor; Ψ, packaging signal. For abbreviations of the enhancer boxes (gray) shown, refer to the text and Baum *et al.* [168].

differences in crucial enhancer elements greatly influence tissue tropism.

1. Early Hematopoietic Cells

These represent a mixed cell population of primitive and uncommitted cells, with a latent, yet enormous potential for proliferation and step-wise differentiation following predefined genetic programs (reviewed by Morrison *et al.* [176] and Weissman [177]). This is the target cell population for drug resistance gene transfer (see Section II), where high levels of transgene expression are crucial for protection from chemotherapeutic side effects [13,14]. Many vectors utilize control elements of the Moloney MLV (MoMLV) or the related Harvey murine sarcoma virus. These elements are strongly recruited in activated T cells but are only moderately active in more mature myeloid and erythroid precursor cells and are repressed to low levels in stem cells. Further repression of MoMLV-based vectors results from inhibitory elements targeting the PBS, both in embryonic stem cells and in early hematopoietic cells [13]. Based on systematic studies of transcription control of murine retroviruses in embryonic and early hematopoietic cells, we developed a series of vectors better adapted to the needs of these cells. The complex genealogy of these vectors is illustrated in Fig. 9.

cis-Active elements of myeloproliferative sarcoma virus (MPSV), differing from MoMLV by mutations in putative repressor sites and in one binding site for the transcription factor Sp1, perform better in hematopoietic progenitor and in

**Evolution of vectors permissive for
potent gene expression in early hematopoietic and embryonic stem cells**

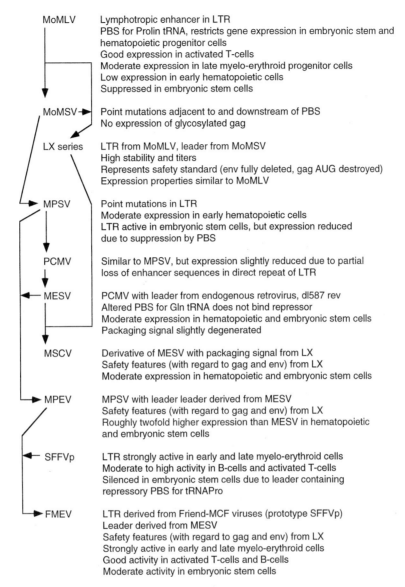

MoMLV — Lymphotropic enhancer in LTR
PBS for Prolin tRNA, restricts gene expression in embryonic stem and hematopoietic progenitor cells
Good expression in activated T-cells
Moderate expression in late myelo-erythroid progenitor cells
Low expression in early hematopoietic cells
Suppressed in embryonic stem cells

MoMSV — Point mutations adjacent to and downstream of PBS
No expression of glycosylated gag

LX series — LTR from MoMLV, leader from MoMSV
High stability and titers
Represents safety standard (env fully deleted, gag AUG destroyed)
Expression properties similar to MoMLV

MPSV — Point mutations in LTR
Moderate expression in early hematopoietic cells
LTR active in embryonic stem cells, but expression reduced due to suppression by PBS

PCMV — Similar to MPSV, but expression slightly reduced due to partial loss of enhancer sequences in direct repeat of LTR

MESV — PCMV with leader from endogenous retrovirus, dl587 rev
Altered PBS for Gln tRNA does not bind repressor
Moderate expression in hematopoietic and embryonic stem cells
Packaging signal slightly degenerated

MSCV — Derivative of MESV with packaging signal from LX
Safety features (with regard to gag and env) from LX
Moderate expression in hematopoietic and embryonic stem cells

MPEV — MPSV with leader leader derived from MESV
Safety features (with regard to gag and env) from LX
Roughly twofold higher expression than MESV in hematopoietic and embryonic stem cells

SFFVp — LTR strongly active in early and late myelo-erythroid cells
Moderate to high activity in B-cells and activated T-cells
Silenced in embryonic stem cells due to leader containing repressory PBS for tRNAPro

FMEV — LTR derived from Friend-MCF viruses (prototype SFFVp)
Leader derived from MESV
Safety features (with regard to gag and env) from LX
Strongly active in early and late myelo-erythroid cells
Good activity in activated T-cells and B-cells
Moderate activity in embryonic stem cells

FIGURE 9   Genealogy of retroviral vectors developed for strong constitutive gene expression in early hematopoietic cells and embryonic stem cells.

embryonic stem cells. PCC4-cell passaged MPSV (PCMV) is an MPSV-variant that has lost one copy of the direct repeat of the enhancer. It arose by forced passage in embryonic carcinoma cells and contains the first retroviral enhancer known to be active in primitive embryonic stem cells [178]. When it was combined with the leader of an endogenous retrovirus displaying an alternative PBS sequence, a vector resulted that allowed LTR-driven gene expression in undifferentiated embryonic stem cells. This chimeric virus is known as murine embryonic stem cell virus (MESV) [173]. The MESV-backbone has been modified to include features of the MoMLV-based LX vectors [82] in the 5′ untranslated

region (UTR) (packaging signal and untranslated *gag* sequences) and in the 3′ UTR (complete deletion of *env*). These modifications were incorporated to increase packaging efficiency and vector safety but did not improve gene expression as compared to MESV (MSCV, murine stem cell vector [179]). Vectors based on MESV (including MSCV) have found widespread use in experimental hematology, being associated with moderate, yet reliable transgene expression in myelo-erythroid progenitor cells and lymphocytes [7].

In the MPSV–MESV hybrid vector (MPEV), the enhancer of MESV was replaced with the corresponding sequences of MPSV, roughly doubling gene expression levels due to the presence of the second copy of the direct repeat. A similar vector has been developed by Kohn and colleagues and was named MD [180]. This group has shown that removal of an enhancer sequence located 5′ to the direct repeat, containing a putative repressor site, may raise the probability for long-term expression in transplanted hematopoietic cells [171,182].

Enhancers of Friend–MCF viruses such as SFFVp (spleen focus-forming virus) were found to allow further increased gene expression levels in myeloerythroid cells [13]. An SFFVp-based vector can mediate sustained multilineage gene expression through serial transplantations in mice [183]. When the Friend–MCF-related U3 regions are combined with the nonrestrictive leader of MESV, novel vectors result which we named FMEV (Friend–MCF–MESV hybrid). These currently represent a reasonable choice for strong transgene expression in hematopoietic cells (Fig. 10A) [13,14].

The importance of improving enhancer strength became evident from comparative vector studies in the context of drug resistance gene transfer. Only MPEV, and, even better, FMEV mediated high-dose drug resistance. Background-free selection of primary hematopoietic cells was thus possible when the human multidrug resistance 1 (MDR1) gene was expressed [13,14] (Fig. 10B). Moreover, intact proliferation and differentiation of transduced hematopoietic progenitor cells were observed in the presence of myeloablative doses of chemotherapeutic agents, indicating complete detoxification [14] (Fig. 10C). FMEV also allows dominant selection with MDR1 when a second gene is coexpressed. This is remarkable because coexpression of a second gene leads to reduced MDR1 expression when compared with the monocistronic counterpart [16]. Strong gene expression from FMEV vectors can also be instrumental for studies employing cell surface markers [184] or cytoplasmic proteins such as green fluorescent protein [185].

In order to further increase the transcriptional strength and specificity of FMEV, we are performing a molecular analysis of Friend–MCF-type enhancers. At least three crucial motifs contributing to strong and relatively lineage-independent activity in hematopoietic cells were identified: recognition sites for the ubiquitous transactivator, Sp1; ETS family members; and AML1/PEBP [168] (Fig. 8). As expected, these are all important transcriptional regulators in hematopoietic cells [186]. Additional activation may result from E-Box binding basic helix–loop–helix factors [187] and Myb [188]. Similar recognition sites are represented in a number of endogenous promoters controlling differentiation-dependent cellular genes. Such cellular motifs can be successfully incorporated in retroviral vectors [189]. Variations in enhancer assembly (e.g., by developing hybrid enhancers composed of distinct modules of retroviral or endogenous enhancers) are expected to result in even higher gene expression levels. Other alterations may lead to more specific and lineage-restricted activity within the hematopoietic system. Thus, it seems possible to develop novel enhancers that are strongly recognized

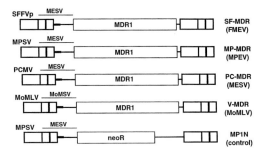

(A) Retroviral vectors

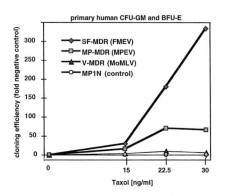

(B) Selective advantage

(C) Colony morphology

FIGURE 10 Vector design determines phenotype, here shown for myeloprotection by drug resistance gene transfer. (A) Different types of retroviral vectors evaluated in the context of transfer of the multidrug resistance gene (MDR1). (B) Relative selective advantage conferred to primary human hematopoietic colony forming units (CFUs) kept under selection with the chemotherapeutic agent Taxol, recognized by the MDR1-encoded efflux pump, P-glycoprotein. Data are calculated from Eckart *et al.* [14] and expressed as cloning efficiency fold negative control (i.e., cells transduced with MP1N). PC-MDR (MESV-type) is only slightly better than V-MDR and therefore not shown. (C) Average colony morphology at selection with 15 ng Taxol/mL reveals importance of complete detoxification. This can only be achieved with vector backbones of strong transcriptional activity.

in hematopoietic progenitors but have low activity in tumor cells (e.g., are of epithelial origin). With such hematopoiesis-specific enhancers, transduction of non hematopoietic tumor cells would have no significant consequences in terms of inducing drug resistance.

2. T Lymphocytes

Although they represent a mature blood cell population, T cells can be very long lived and have the capacity for limited clonal activation and expansion. In the switch between resting and the activated status, chromosomal organization and transcription factor equipment is reordered. Thus, stably integrating retroviral vectors are a perfect tool for genetic manipulation of T cells, but vector expression may vary depending on the cellular activation status [190,191]. All MLV-based vectors described in Fig. 10A and related constructs mediate sufficient expression in activated T lymphocytes for application in adoptive immunotherapy [190,192]. In T cells, however, MPEV is clearly stronger than FMEV. The enhancer of SL3-3, a highly lymphotropic MLV, is an interesting alternative [193]. As discussed for early hematopoietic cells, insights into the molecular mechanisms defining T lymphotropism of retroviral or endogenous enhancers is expected to create the basis for developing artificial transgene enhancers with increased T-cell specificity. Interestingly, as with some endogenous T-lymphocytic promoters, reversible downregulation of retroviral gene expression was observed in resting T cells. This might be prevented by inclusion of scaffold attachment regions in the vector (see Section V.B.4) [191].

3. Tumor Cells

Mechanisms of tumor-specific transcriptional controls are of interest for targeting of tumor cells in suicide gene transfer and oncogene antagonism, as outlined above (see Section II). Generally, the specificity of heterologous promoters in retroviral vectors is increased when more promiscuous retroviral enhancer sequences are deleted. Transcriptional targeting of tumors can be achieved using control elements of genes that are "tumor specific" or over-expressed in tumors. When targeting metastases, control elements of genes specific to the parental tissue of the tumor might also be sufficient. Also, hypoxia-responsive promoters have been proposed for tumor targeting [194]. A more indirect approach is the targeting of endothelial cells involved in tumor angiogenesis using "endotheliotropic" control regions. Thus, an ever-increasing number of candidate promoters is being proposed (reviewed by Sikora [23] and Miller and Whelan [195]). However, for most of them, evidence for tumor specificity *in vivo* is yet to be confirmed.

4. Silencing

Silencing not only reduces the efficiency but can also compromise the safety of gene transfer strategies. This is of special importance for negative selection of transduced cells (as required in suicide gene transfer, adoptive immunotherapy, or mini-organs; see Section II). Here, cells having silenced the vector will escape exogenous control. Silencing results from dominant negative influences of the integration site or may be directly triggered by the integrated vector. Silencing involves functional reorganizations within the chromosome. As a result, vector sequences can be methylated in CpG islands, which may play a role for fixation of downregulation [196,197]. The speed and incidence of silencing depend on the cellular background, the genomic integration site, and (not well defined) on specific vector sequences, including transgene cDNAs (reviewed by Lund *et al.* [169]). This opens perspectives for active prevention of silencing by vector improvements. Studies with housekeeping promoters indicate that Sp1 binding sites can counteract silencing to some extent [198]. The retroviral enhancers of MoMLV, MPSV, PCMV, and Friend–MCF viruses differ with respect to number, affinity, and positioning of Sp1 binding sites [168,199]. The relevance to long-term expression remains to be shown. Furthermore, MESV-derived leader sequences or vectors containing other, even artificial, primer binding sites avoiding transcriptional repression in embryonic and hematopoietic cells (see Section V.B.1) might support long-term expression. However, silencing of MESV-leader-based vectors is also observed upon differentiation of embryonic stem cells permissive to vector expression in the undifferentiated state [170]. Insertion of scaffold attachment regions in retroviral vectors, as described by Bode and colleagues [200], may shield retroviral control regions from negative influences of the integration site and thus support transcriptional autonomy of a chromosomally integrated transgene [191], as demonstrated earlier for stably transfected plasmids [201]. Consequently, downregulation of retroviral enhancers in resting T cells [191] and irreversible silencing in transplanted hematopoietic cells can be prevented to some extent [202]. Similarly, insulator elements derived from the chicken HS4 element may reduce position dependence of retrovirally integrated transgenes [203], and even vectors lacking such elements can exhibit consistent long-term expression in hematopoietic cells [181,182]. However, most studies published so far were conducted under conditions that allowed more than one transgene integration in single repopulating cells. To clarify the probability of silencing from a single integrated transgene, further systematic analyses in appropriate primary cell systems and results from comparative clinical studies are still awaited. Importantly, results achieved with a given reporter cDNA may not necessarily be predictive for vectors containing different inserts, as coding sequences may also exhibit *cis* elements that influence the probability of gene silencing (see Section V.B.8).

5. Regulatable Promoters

Regulatable promoters are of interest for generating artificial mini-organs and also for drug resistance gene transfer (see Section II). Progress in regulatable promoter systems has been revewied by others [195]. Best documented in retroviral vectors is the tetracycline-regulated system, available both for conditional repression and induction of transgene expression [204,205]. Moreover, a number of alternative artificial systems for conditional promoter induction or repression have already been described [195]. The applicability of synthetic inducer/promoter systems has been demonstrated *in vivo* using retroviral vectors expressing erythropoietin from a tetracycline-regulated cassette [21,22]. Further advances in regulated vectors are expected to address potential limitations of the systems: side effects of the drugs administered for regulation, immunogenicity of the synthetic transactivators or repressors employed, toxic squelching effects eventually occurring from overexpressed synthetic transcription factors, clonal variabilities in inducibility related to the retroviral integration site, differentiation dependence of regulation, and maintenance of regulation over time.

6. RNA Elements

Nuclear and cytoplasmic processing of newly transcribed RNA is dependent on *cis*-regulatory RNA elements, which determine the rate of splicing, polyadenylation, nuclear export, RNA stability, and initiation of translation. Most of these processes are functionally coupled. Considering that cytoplasmic accumulation and translation of many cellular RNAs are rate limiting and can be dependent on the presence of appropriate introns, export signals, and/or polyadenylation tails [206,207], there is a growing interest in sequences that improve posttranscriptional processing of a given RNA.

At least three categories of RNA modules may enhance expression from retroviral gene transfer vectors on a posttranscriptional level: splice signals that create an intron in the 5′ untranslated region [43,44,83]; constitutive RNA transport elements, originally discovered in D-type retroviruses [208]; and last, but not least, the posttranscriptional regulatory element of woodchuck hepatitis virus [209]. Importantly, enhancement of expression depends not only on the specific element, but also on the gene and promoter of interest, implying context-dependent activity of RNA elements [210].

Proper combinations of RNA elements can enhance expression of a given cDNA by more than one order of magnitude, and expression of some coding sequences may even be absolutely dependent on the presence of either a constitutive transport element (CTE) or an intron [210]. Thus, comparative analyses are recommended to improve the performance of a given vector by inclusion of RNA elements. Moreover, the efficiency and safety of retroviral gene vectors may be increased by redesigning 5′ untranslated regions to avoid aberrant start codons located 5′ of the cDNA of interest [44].

7. Coexpression Strategies

Vectors expressing more than one transgene greatly widen the perspective of most cancer gene therapy approaches. Depending on the specific application, coexpression is used to combine two selectable marker genes, a selectable marker gene with a nonselectable gene, or two nonselectable genes (Table 3). There are several options for simultaneously expressing different biological functions from a single vector (Fig. 11). In general, type and positioning of transgenes, as well as cellular background and specific experimental conditions (especially the stringency of selection applied) greatly influence the efficacy of the coexpression strategy. Therefore, as with the inclusion of other *cis* elements, systematic comparative studies appear desirable for each coexpression vector developed for a specific clinical use [16,211].

a. Internal Promoters

To express genes from retroviral vectors, promoters can be placed not only in the LTR but also, in either orientation, in the sequences between the leader and the 3′ LTR (Fig. 11A). These internal promoters can be used in vectors where the U3 promoter has been deleted or in addition to an LTR-controlled transcription unit. However, when two promoters are located close to each other, there is the potential of promoter interference, leading to shutdown of one promoter to the advantage

TABLE 3   Reasons for Expressing Two or More Genes from a Single Vector

Combination	Approach (see Section II)	Example
Selectable marker gene only.	Drug resistance gene transfer	Complementary drug resistance genes (to widen spectrum of resistance).
		Drug resistance gene(s) plus suicide gene (to remove transduced cells in case of pathogenicity).
	Adoptive immunotherapy	Surface marker plus suicide gene (to select transduced cells before reinfusion).
	Suicide gene transfer	Two suicide genes (improves efficacy).
Selectable marker gene plus nonselectable gene.	Mini-organs	Suicide gene plus therapeutic gene of interest (to remove transduced cells in case of pathogenicity).
	Oncogene antagonism	Suicide gene plus anti-oncogene (improves efficacy).
Nonselectable genes only.	Oncogene antagonism	Complementary anti-oncogenes (improves efficacy).
	Tumor vaccination	Cooperating immunostimulatory genes (improves efficacy).

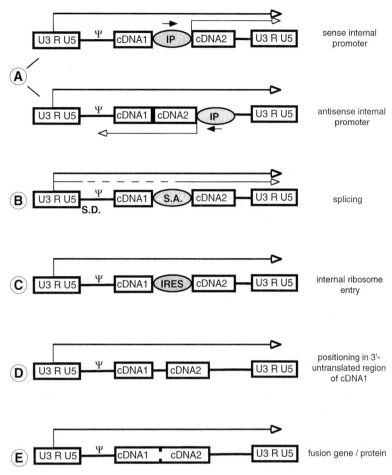

FIGURE 11   Strategies for coexpression of two genes from a retroviral vector (see Section V.B.7). Open arrows indicate mRNAs, and the bold arrow represents the genomic message of the vector. Ψ, packaging signal; IP, internal promoter (orientation indicated by the filled arrow); S.D. and S.A., splice donor and splice acceptor, respectively; IRES, internal ribosomal entry site.

of its neighbor [212–214]. The stronger promoter (or the promoter selected for) either tends to exploit enhancer sequences of the neighboring promoter or inhibits formation of the Pol-II initiation complex at the internal promoter. Here, separation by transcriptional termination signals would be a possible solution [215]; however, this is inappropriate in retroviral vectors, as it would lead to premature termination of genomic messages in packaging cells. Placing the internal promoter in antisense orientation to the LTR might reduce interference at the transcriptional level, but doing so necessarily generates antisense RNA, which is expected to disturb translation of the cotransferred gene. Therefore, vectors containing internal promoters might generate unwanted effects, especially under conditions of dominant selection for only one promoter, as used in adoptive immunotherapy.

b. Alternative Splicing

For reasons not entirely understood, retroviral splice donor and splice acceptor sequences are only partly recognized in host cells. This leads to a defined ratio of genomic and subgenomic messages and can be exploited for constructing splicing vectors that sometimes, but not always, yield good results [16,211,216]. Generating a spliced, subgenomic message can be associated with improved nuclear export, increased half-life of cytoplasmic RNA, or improved translation efficacy. Importantly, type and positioning of the transgenes will affect the efficacy of alternative splicing. Finally, cDNAs inserted in splice vectors must be free of cryptic splice signals (Fig. 11B).

c. Internal Ribosome Entry

The internal ribosomal entry site (IRES) was originally described in picorna viruses. The IRES is a complex domain of the RNA (the size of a few hundred base pairs), generating a specific structure allowing cap-independent initiation of translation. When introduced in front of the start codon of the transgene, bi- or even oligocistronic vectors can be generated [213,217–219]. Compared to internal promoters and alternative splicing, IRES control has the advantage of exploiting a single mRNA for translation of two (or more) proteins (Fig. 11C). However, not every cDNA is fully compatible with translation via an IRES, and sometimes alternative and mutually exclusive rather than simultaneous initiation of translation might predominate. Early reports state that IRES-dependent initiation of translation occurs as efficiently as that from capped RNAs, with capped RNAs referring to transgenes expressed from within the *gag* region of vector mRNAs [213,217–219]. However, it was demonstrated that MLVs also use an IRES-related mechanism for translating *gag–pol* and *env* messages [63], and there is accumulating evidence that initiation of translation from within the *gag* region is suboptimal [44,83]. Accordingly, we and others found that IRES-dependent gene expression may be significantly reduced [211,220]. Also, virus titers may be suboptimal in the presence of a nonretroviral IRES, and even expression of the gene located 5′ to the IRES may be compromised [16,211]. Moreover, it remains to be elucidated as to what extent IRES-dependent translation may show reduced fidelity with respect to the choice of initiation codon and whether it is subject to differentiation-dependent control.

d. Positioning in Untranslated Vector Regions

Sometimes it is sufficient to express RNA without translation, as in approaches utilizing antisense RNA or ribozymes for oncogene antagonism (see Section II). These therapeutic RNAs can be located in untranslated vector regions, preferably in the 3′ untranslated region of another gene coexpressed from the vector (Fig. 11D). The same strategy cannot be recommended for open reading frames: Spontaneous reinitiation of translation from the 3′ untranslated region of a gene occurs at greatly reduced efficiency [68].

e. Fusion Proteins and Protein Cleavage

Multifunctional fusion proteins are a good choice for coexpression provided that the domains of interest are active in similar subcellular localizations (Fig. 11E). Some cytosolic proteins might also function when expressed as the cytoplasmic tail of a membrane-anchored fusion protein [221]. It needs to be determined whether the efficacy of the fusion protein is comparable to those of the individual components. A potential risk of this approach is that the fusion site might give rise to an immunogenic peptide. An interesting extension of this approach is the inclusion of a cleavable linker between the protein domains of interest. The 2A proteinase of the foot and mouth disease virus (FMDV), a short peptide that has the interesting property of inducing cotranslational protein separation when inserted in the frame between two protein domains, can be successfully introduced in retroviral vectors [222]. Further interesting features of the FMDV 2A proteinase are that it does not disturb virus titers and it allows coexpression of two proteins that have different subcellular localizations [223]. Thus, vectors may be generated that express two or more proteins at coordinated levels.

8. *cis* Elements in cDNAs

Even cDNAs can contain *cis*-acting elements, active either at the transcriptional [224] or posttranscriptional level [225–227]. This aspect of vector design is often neglected but can have profound influence on overall vector performance. A cDNA may harbor silencer elements or contain enhancers influencing levels as well as tissue-tropism of vector expression [224]. Some cDNAs may be unstable when expressed from retroviral vectors [225]. The retroviral life cycle implies that aberrant signals for splicing, termination, and polyadenylation; primer binding; or cryptic PP tracts will reduce vector titers or give rise to rearranged vector copies, with unpredictable immunological or toxicological consequences. Examples relevant to cancer gene therapy are the drug resistance genes MDR1 [16,226] and thymidine kinase of herpes simplex virus [227]. A stable, selectable marker gene coexpressed with the unstable sequence can serve as a tool to tag hot spots of recombination, providing the basis for cDNA improvement [16]. Thus, a lot of fine-tuning work may be required to develop stable and, hence, safe vectors suitable for actual clinical use. Evolution has done that work for retroviral genes. Vector designers usually follow empirical approaches, not always the most elegant and effective way to success.

VI. OUTLOOK

Retroviral vector systems have dominated cancer gene therapy research in the past years, and they will certainly continue to play an important role. However, in future clinical trials it will be of outstanding importance to use specifically tailored and highly effective vectors. Only then can the perspectives of gene therapy concepts be evaluated. Based on a deeper understanding of the biology of retroviruses and their target cells, improved vector systems have already been created and now await clinical testing to assess efficacy and safety. Key developments include the advent of complex retrovirus-based systems for transduction of nondividing cells, pseudotyping and envelope engineering to widen or specify the host range at the level of transduction, and higher diversity in enhancer choice based on deeper insights into the transcriptional control of retroviral transgenes. Especially, further progress in the field of transductional and transcriptional targeting will have substantial impact on the therapeutic quality of cancer gene therapy approaches. So far, vector design has been dominated and also limited by deductive analyses of virus–host interactions. Future vector design should also follow a more evolutionary approach, taking advantage of the inherent genetic variability of viruses; therefore, we need to establish intelligent systems for selecting and screening improved mutants. For widely applicable oncologic strategies, tailoring can be performed as an international, multicenter effort. Unfortunately, for more specialized applications with small patient numbers this will be unaffordable. Here, concentration in specific centers of expertise might represent a solution. Importantly, many aspects of vector tailoring worked out using simple retroviral vectors will also be applicable to lentiviral vectors and non-retroviral systems (based on adenovirus, adeno-associated viruses, herpes viruses, or physicochemical methods), which are emerging as important alternatives for some approaches in cancer gene therapy and will substantially widen the perspectives of the field.

References

1. Mulligan, R. C. (1993). The basic science of gene therapy. *Science* **260**, 926–932.

2. Friedmann, T. (1996). Human gene therapy—an immature genie, but certainly out of the bottle. *Nat. Med.* **2**, 144–145.

3. Dickman, S. (1997). Richard Mulligan: from skeptic to true believer. *Curr. Biol.* **7**, R601–R602.

4. Brenner, M. K. (1994). Genetic marking and manipulation of hematopoietic progenitor cells using retroviral vectors. *Immuno-methods* **5**, 204–210.

5. Dunbar, C. E. (1996). Gene transfer to hematopoietic stem cells: implications for gene therapy of human disease. *Annu. Rev. Med.* **47**, 11–20.

6. Rooney, C. M., Smith, C. A., Ng, C. Y., Loftin, S., Li, C., Krance, R. A., Brenner, M. K., and Heslop, H. E. (1995). Use of gene-modified virus-specific T lymphocytes to control Epstein–Barr-virus-related lympho-proliferation. *Lancet* **345**, 9–13.

7. Pawliuk, R., Eaves, C. J., and Humphries, K. R. (1997). Sustained high-level reconstitution of the hematopoietic system by preselected hematopoietic cells expressing a transduced cell-surface antigen. *Hum. Gene Ther.* **8**, 1595–1604.

8. Phillips, K., Gentry, T., McCowage, G., Gilboa, E., and Smith, C. (1996). Cell-surface markers for assessing gene transfer into human hematopoietic cells. *Nat. Med.* **2**, 1154–1156.

9. Fehse, B., Uhde, A., Fehse, N. Eckert, H. G., Clausen, J., Rüger, R., Koch, S., Ostertag, W., Zander, A. R., and Stockschläder, M. (1997). Selective immunoaffinity-based enrichment of CD34+ cells transduced with retroviral vectors containing an intracytoplasmatically truncated version of the human low-affinity nerve growth factor receptor (ΔLNGFR) gene. *Hum. Gene Ther.* **8**, 1815–1827.

10. Gottesman, M. M., Germann, U. A., Aksentijevich, I., Sugimoto, Y., Cardarelli, C. O., and Pastan, I. (1994). Gene transfer of drug resistance genes. Implications for cancer therapy. *Ann. N.Y. Acad. Sci.* **716**, 126–138.

11. Baum, C., Fairbairn, L., Hildinger, M., Lashford, L. S., Hegewisch-Becker, S., and Rafferty, J. R. (1999). New perspectives for cancer chemotherapy by genetic protection of hematopoietic cells. Expert reviews in molecular medicine. *http://www-ermm.cbcu.cam.ac.uk*.

12. Baum, C. (1997). Gene transfer and transgene expression in hematopoietic cells. In: *Concepts in Gene Therapy* (M. Strauss, and J. A. Barranger, eds.), pp. 233–265. DeGruyter, Berlin.

13. Baum, C., Hegewisch-Becker, S., Eckert, H.-G., Stocking, C., and Ostertag, W. (1995). Novel retroviral vectors for efficient expression of the multidrug resistance (*mdr*-1) gene in early hematopoietic cells. *J. Virol.* **69**, 7541–7547.

14. Eckert, H.-G., Stockschläder, M., Just, U., Hegewisch-Becker, S., Grez, M., Uhde, A., Zander, A., Ostertag, W., and Baum, C. (1996). High-dose multidrug resistance in primary human hematopoietic progenitor cells transduced with optimized retroviral vectors. *Blood* **88**, 3407–3415.

15. Galipeau, J., Benaim, E., Spencer, H. T., Blakley, R., and Sorrentino, B. P. (1997). A bicistronic retroviral vector for protecting hematopoietic cells against antifolates and P-glycoprotein effluxed drugs. *Hum. Gene Ther.* **8**, 1773–1783.

16. Hildinger, M., Fehse, B., Hegewisch-Becker, S., John, J., Rafferty, J. R., Ostertag, W., and Baum, C. (1998). Dominant selection of hematopoietic progenitor cells with retroviral MDR1 co-expression vectors. *Hum. Gene Ther.* **9**, 33–42.

17. Sauerbrey, A., McPherson, J. P., Zhao, S. C., Banerjee, D., and Bertino, J. R. (1999). Expression of a novel double-mutant dihydrofolate reductase-cytidine deaminase fusion gene confers resistance to both methotrexate and cytosine arabinoside. *Hum. Gene Ther.* **10**, 2495–2504.

18. Jelinek, J., Rafferty, J. A., Cmejla, R., Hildinger, M., Chinnasamy, D., Lashford, L. S., Ostertag, W., Margison, G. P., Dexter, T. M., Fairbairn, L. J., and Baum, C. (1999). A novel dual function retrovirus expressing multidrug resistance 1 and O^6-alkylguanine-DNA-alkyltransferase for engineering resistance of haemopoietic progenitor cells to multiple chemotherapeutic agents. *Gene Ther.* **6**, 1489–1493.

19. Bonini, C. *et al.* (1997). HSV-TK gene transfer into donor lymphocytes for control of allogeneic graft-versus-leukemia. *Science* **176**, 1719–1724.

20. Clay, T. M., Custer, M. C., Spiess, P. J., and Nishimura, M. I. (1999). Potential use of T cell receptor genes to modify hematopoietic stem cells for the gene therapy of cancer. *Pathol. Oncol. Res.* **5**, 3–15.

21. Bohl, D., and Heard, J.-M. (1997). In vivo secretion of therapeutic proteins from neo-organs. In: *Concepts in Gene Therapy* (M. Strauss and J. A. Barranger, eds.), pp. 297–314. De Gruyter, Berlin.

22. Bohl, D., and Heard, J. M. (1997). Modulation of erythropoietin delivery from engineered muscles in mice. *Hum. Gene Ther.* **8**, 195–204.

23. Greco, O., and Dachs, G.U. (2001). Gene directed enzyme/prodrug therapy of cancer: historical appraisal and future prospectives. *J. Cell. Physiol.* **187**, 22–36.

24. Harris, J. D., Gutierrez, A. A., Hurst, H. C., Sikora, K., and Lemoine, N. R. (1994). Gene therapy for cancer using tumour-specific prodrug activation. *Gene Ther.* **1**, 170–175.

25. Martin, V., Cortes, M. L., de Felipe, P., Farsetti, A., Calcaterra, N. B., and Izquierdo, M. (2000). Cancer gene therapy by thyroid hormone-mediated expression of toxin genes. *Cancer Res.* **60**, 3218–3224.

26. Vallera, D. A., Jin, N., Baldrica, J. M., Panoskaltsis-Mortari, A., Chen, S. Y., and Blazar, B. R. (2000). Retroviral immunotoxin gene therapy of acute myelogenous leukemia in mice using cytotoxic T cells transduced with an interleukin 4/diphtheria toxin gene. *Cancer Res.* **60**, 976–984.

27. Roth, J. A., Nquyen, D., Lawrence, D. D., Kemp, B. L., Carrasco, C. H. *et al.* (1996). Retrovirus-mediated wild-type p53 gene transfer to tumors of patients with lung cancer. *Nat. Med.* **2**, 985–991.

28. Yang, Z. Y., Perkins, N. D., Ohno, T., Nabel, E. G., and Nabel, G. J. (1995). The p21 cyclin-dependent kinase inhibitor suppresses tumorigenicity in vivo. *Nat. Med.* **1**, 1052–1056.

29. Wygoda, M. R., Wilson, M. R., Davis, M. A., Trosko, J. E., Rehemtulla, A., and Lawrence, T. S. (1997). Protection of herpes simplex virus thymidine kinase-transduced cells from ganciclovir-mediated cytotoxicity by bystander cells: the Good Samaritan effect. *Cancer Res.* **57**, 1699–1703.

30. Benedetti, S., Pirola, B., Pollo, B., Magrassi, L., Bruzzone, M. G., Rigamonti, D., Galli, R., Selleri, S., Di Meco, F., De Fraja, C., Vescovi, A., Cattaneo, E., and Finocchiaro, G. (2000). Gene therapy of experimental brain tumors using neural progenitor cells. *Nat. Med.* **6**, 447–450.

31. Tamura, M., Ikenaka, K., Tamura, K., Yoshimatsu, T., Miyao, Y., Kishima, H., Mabuchi, E., and Shimizu, K. (1998). Transduction of glioma cells using a high-titer retroviral vector system and their subsequent migration in brain tumors. *Gene Ther.* **5**, 1698–1704.

32. Gomez-Navarro, J., Contreras, J. L., Arafat, W., Jiang, X. L., Krisky, D., Oligino, T., Marconi, P., Hubbard, B., Glorioso, J. C., Curiel, D. T., and

Thomas, J. M. (2000). Genetically modified CD34+ cells as cellular vehicles for gene delivery into areas of angiogenesis in a rhesus model. *Gene Ther.* **7**, 43–52.

33. Hurford, R. J., Dranoff, G., Mulligan, R. C., and Tepper, R. I. (1995). Gene therapy of metastatic cancer by in vivo retroviral gene targeting. *Nat. Genet.* **10**, 430–435.

34. Bischoff, J. R., Kirn, D. H., Willimas, A., Heise, C., Horn, S., Muna, M., Ng, L., Sampson-Johannes, A., Fattaey, A., and McCormick, F. (1996). An adenovirus mutant that replicates selectively in p53-deficient human tumor cells. *Science* **274**, 373–376.

35. Blankenstein, T., Cayeux, S., and Qin, Z. (1996). Genetic approaches to cancer immunotherapy. *Rev. Physiol. Biochem. Pharmacol.* **129**, 1–49.

36. Simons, J. W., and Mikhak, B. (1998). Ex-vivo gene therapy using cytokine-transduced tumor vaccines: molecular and clinical pharmacology. *Semin. Oncol.* **25**, 661–676.

37. Coffin, J. M. (1996). Retroviridae: the viruses and their replication. In: *Field's Virology* (B. N. Fields, D. M. Knipe, and P. M. Howley, eds.), pp. 763–844. Lippincott-Raven, Philadelphia.

38. Miller, A. D., and Chen, F. (1996). Retrovirus packaging cells based on 10A1 murine leukemia virus for production of vectors that use multiple receptors for cell entry. *J. Virol.* **70**, 5564–5571.

39. Mergia, A., Shaw, K. E., Lowe, E., Barry, P. A., and Luciw, P. A. (1990). Simian foamy virus type 1 is a retrovirus which encodes a transcriptional transactivator. *J. Virol.* **64**, 3598–3604.

40. Luciw, P. A. (1996). Human immunodeficiency viruses and their replication. In: *Field's Virology* (B. N. Fields, D. M. Knipe, and P. M. Howley, eds.), pp. 1881–1975. Lippincott-Raven, Philadelphia.

41. Ashe, M. P., Furger, A., and Proudfoot, N. J. (2000). Stem-loop 1 of the U1 snRNP plays a critical role in the suppression of HIV-1 polyadenylation. *RNA* **6**, 170–177.

42. Bender, M. A., Palmer, T. D., Gelinas, R. E., and Miller, A. D. (1987). Evidence that the packaging signal of Moloney murine leukemia virus extends into the gag region. *J. Virol.* **61**, 1639–1646.

43. Kim, S. H., Yu, S. S., Park, J. S., Robbins, P. D., An, C. S., and Kim, S. (1998). Construction of retroviral vectors with improved safety, gene expression, and versatility. *J. Virol.* **72**, 994–1004.

44. Hildinger, M., Abel, K. L., Ostertag, W., and Baum, C. (1999) Design of 5′ untranslated sequences in retroviral vectors developed for medical use. *J. Virol.* **73**, 4083–4089.

45. Berkowitz, R., Fisher, J., and Goff, S. P. (1996). RNA packaging. *Curr. Top. Microbiol. Immunol.* **214**, 177–218.

46. Berkowitz, R., Hammarskjold, M.-L., Helga-Maria, C., Rekosh, D., and Goff, S. (1995). 5′ regions of HIV-1 RNAs are not sufficient for encapsidation: implications for the HIV-1 packaging signal. *Virology* **212**, 718–723.

47. Richardson, J. H., Child, L. A., and Lever, A. M. (1993). Packaging of human immunodeficiency virus type 1 RNA requires *cis*-acting sequences outside the 5′ leader region. *J. Virol.* **67**, 3997–4005.

48. Weiss, R. A. (1993). Pseudotyped viruses and envelope composition. In: *The Retroviridae* (J. A. Levy, ed.), pp. 5–8. Plenum Press, New York.

49. Hunter, E., and Swanstrom, R. (1990). Retrovirus envelope glycoproteins. *Curr. Top. Microbiol. Immunol.* **157**, 187–253.

50. Temin, H. M. (1993). Retrovirus variation and reverse transcription: abnormal strand transfers result in retrovirus genetic variation. *Proc. Natl. Acad. Sci. USA* **90**, 6900–6903.

51. Roe, T., Reynolds, T., Yu, G., and Brown, P. O. (1993). Integration of murine leukemia virus DNA depends on mitosis. *EMBO J.* **12**, 2099–2108.

52. Miller, D. G., Adam, M. A., and Miller, A. D. (1990). Gene transfer by retrovirus vectors occurs only in cells that are actively replicating at the time of infection. *Mol. Cell. Biol.* **10**, 4239–4242.

53. Lieber, A., Kay, M. A., and Li, Z. Y. (2000). Nuclear import of Moloney murine leukemia virus DNA mediated by adenovirus preterminal protein is not sufficient for efficient retroviral transduction in nondividing cells. *J. Virol.* **74**, 721–734.

54. Bukrinsky, M. I., Haggerty, S., Dempsey, P., Sharova, N., Adzhubei, A., Spitz, L., Lewis, P., Goldfarb, D., Emerman, M., and Stevenson, M. (1993). A nuclear localization signal within HIV-1 matrix protein that governs infection of non-dividing cells. *Nature* **365**, 666–669.

55. Fouchier, R. A., and Malim, M. H. (1999). Nuclear import of human immunodeficiency virus type-1 preintegration complexes. *Adv. Virus Res.* **52**, 275–299.

56. Zennou, V., Petit, C., Guetard, D., Nerhbass, U., Montagnier, L., and Charneau, P. (2000) HIV-1 genome nuclear import is mediated by a central DNA flap. *Cell* **101**, 173–185.

57. Rohdewohld, H., Weiher, H., Reik, W., Jaenisch, R., and Breindl, M. (1987). Retrovirus integration and chromatin structure: Moloney murine leukemia proviral integration sites map near DNase I-hypersensitive sites. *J. Virol.* **61**, 336–343.

58. Jonkers, J., and Berns, A. (1996). Retroviral insertional mutagenesis as a strategy to identify cancer genes. *Biochim. Biophys. Acta* **1287**, 29–57.

59. Donahue, R. E., Kessler, S. W., Bodine, D., McDonagh, K., Dunbar, C., Goodman, S., Agricola, B., Byrne, E., Raffeld, M., Moen, R. *et al.* (1992). Helper virus induced T cell lymphoma in nonhuman primates after retroviral mediated gene transfer. *J. Exp. Med.* **176**, 1125–1135.

60. Stocking, C., Bergholz, U., Friel, J., Klingler, K., Wagener, T., Starke, C., Kitamura, T., Miyajima, A., and Ostertag, W. (1993). Distinct classes of factor-independent mutants can be isolated after retroviral mutagenesis of a human myeloid stem cell line. *Growth Factors* **8**, 197–209.

61. Tsichlis, P. N., and Lazo, P. A. (1991). Virus-host interactions and the pathogenesis of murine and human oncogenic retroviruses. In: *Retroviral Insertion and Oncogene Activation* (H. G. Kung and P. K. Vogt, eds.), pp. 95–173. Springer-Verlag, Berlin.

62. Ostertag, W., Stocking, C., Johnson, G. R., Kluge, N., Kollek, R., Franz, T., and Hess, N. (1987). Transforming genes and target cells of murine spleen focus-forming viruses. *Adv. Cancer Res.* **48**, 193–355.

63. Deffaud, C., and Darlix, J. L. (2000). Characterization of an internal ribosomal entry segment in the 5′ leader of murine leukemia virus env RNA. *J. Virol.* **74**, 846–850.

64. Luciw, P. A., and Leung, N. J. (1992). Mechanisms of retrovirus replication. In: *The Retroviridae 1* (J. A. Levy, ed.), pp. 159–298. Plenum Press, New York.

65. Kräusslich, H.-G., and Welker, R. (1996). Intracellular transport of capsid components. In: *Morphogenesis and Maturation of Retroviruses* (H. G. Kräusslich, ed.), pp. 25–64. Springer-Verlag, Berlin.

66. Einfeld, D. (1996). Maturation and assembly of retroviral glycoproteins. In: *Morphogenesis and Maturation of Retroviruses* (H. G. Kräusslich, ed.), pp. 133–176. Springer-Verlag, Berlin.

67. Miller, A. D. (1990). Retrovirus packaging cells. *Hum. Gene Ther.* **1**, 5–14.

68. Cosset, F.-L., Takeuchi, Y., Battini, J.-L., Weiss, R. A., and Collins, M. K. L. (1995). High-titer packaging cells producing recombinant retrovirus resistant to human serum. *J. Virol.* **69**, 7430–7436.

69. Wagener, T., Stocking, C., and Ostertag, W. (1995). unpublished data.

70. Hatzoglou, M., Hodgson, C. P., Mularo, F., and Hanson, R. W. (1990). Efficient packaging of a specific VL30 retroelement by psi 2 cells which produce MoMLV recombinant retroviruses. *Hum. Gene Ther.* **1**, 385–397.

71. Chong, H., and Vile, R. G. (1996). Replication-competent retrovirus produced by a 'split-function' third generation amphotropic packaging cell line. *Gene Ther.* **3**, 624–629.

72. Vanin, E. F., Kaloss, M., Broscius, C., and Nienhuis, A. W. (1994). Characterization of replication-competent retroviruses from nonhuman primates with virus-induced T-cell lymphomas and observations regarding the mechanism of oncogenesis. *J. Virol.* **68**, 4241–4250.

73. Münk, C., Lohler, J., Prassolov, V., Just, U., Stockschlader, M., and Stocking, C. (1997). Amphotropic murine leukemia viruses induce

spongiform encephalomyelopathy. *Proc. Natl. Acad. Sci. USA* **94**, 5837–5842.

74. Isacson, O., and Brakefield, X. O. (1997). Benefits and risks of hosting animal cells in the human brain. *Nat. Med.* **3**, 964–969.

75. Takeuchi, Y., Cosset, F. L., Lachmann, P. J., Okada, H., Weiss, R. A., and Collins, M. K. (1994). Type C retrovirus inactivation by human complement is determined by both the viral genome and the producer cell. *J. Virol.* **68**, 8001–8007.

76. Takeuchi, Y., Porter, C. D., Strahan, K. M., Preece, A. F., Gustafsson, K., Cosset, F. L., Weiss, R. A., and Collins, M. K. (1996). Sensitization of cells and retroviruses to human serum by (alpha 1–3) galactosyl-transferase. *Nature* **379**, 85–88.

77. Palsson, B., and Andreadis, S. (1997). The physico-chemical factors that govern retrovirus-mediated gene transfer. *Exp. Hematol.* **25**, 94–102.

78. Kinsella, T. M., and Nolan, G. P. (1996). Epsiomal vectors rapidly and stably produce high-titer recombinant retrovirus. *Hum. Gene Ther.* **7**, 1405–1413.

79. Grignani, F., Kinsella, T., Mencarelli, A., Valtieri, M., Riganelli, D., Grignani, F., Lanfrancone, L., Peschle, C., Nolan, G. P., and Pelicci, P. G. (1998). High-efficiency gene transfer and selection of human hematopoietic progenitor cells with a hybrid EBV/retroviral vector expressing the green fluorescence protein. *Cancer Res.* **58**, 14–19.

80. Wilson, C. A., Ng, T. H., and Miller, A. E. (1997). Evaluation of recommendations for replication-competent retrovirus testing associated with use of retroviral vectors. *Hum. Gene Ther.* **8**, 869–874.

81. Correll, P. H., Colilla, S., and Karlsson, S. (1994). Retroviral vector design for long-term expression in murine hematopoietic cells in vivo. *Blood* **84**, 1812–1822.

82. Miller, A. D., and Rosman, G. J. (1989). Improved retroviral vectors for gene transfer and expression. *Biotechniques* **7**, 980–982.

83. Krall, W. J., Skelton, D. C., Yu, X.-J., Riviere, I., Lehn, P., Mulligan, R. C., and Kohn, D. B. (1996). Increased levels of spliced RNA account for augmented expression from the MFG retroviral vector in hematopoietic cells. *Gene Ther.* **3**, 37–48.

84. Hantzopolos, P. A., Sullenger, B. A., Ungers, G., and Gilboa, E. (1989). Improved gene expression upon transfer of the adenosine deaminase minigene outside the transcriptional unit of a retroviral vector. *Proc. Natl. Acad. Sci. U.S.A.* **86**, 3519–3523.

85. Adam, M. A., Osborne, W. R., and Miller, A. D. (1995). R-region cDNA inserts in retroviral vectors are compatible with virus replication and high-level protein synthesis from the insert. *Hum. Gene Ther.* **6**, 1169–1176.

86. Yu, S. F., von Ruden, T., Kantoff, P. W., Garber, C., Seiberg, M., Ruther, U., Anderson, W. F., Wagner, E. F., and Gilboa, E. (1986). Self-inactivating retroviral vectors designed for transfer of whole genes into mammalian cells. *Proc. Natl. Acad. Sci. USA* **83**, 3194–3198.

87. Olson, P., Nelson, S., and Dornburg, R. (1994). Improved self-inactivating retroviral vectors derived from spleen necrosis virus. *J. Virol.* **68**, 7060–7066.

88. Bergemann, J., Kuhlcke, K., Fehse, B., Ratz, I., Ostertag, W., and Lother, H. (1995). Excision of specific DNA-sequences from integrated retroviral vectors via site-specific recombination. *Nucleic Acids Res.* **23**, 4451–4456.

89. Russ, A. P., Friedel, C., Grez, M., and von Melchner, H. (1996). Self-deleting retrovirus vectors for gene therapy. *J. Virol.* **70**, 4927–4932.

90. Beck-Engeser, G., Stocking, C., Just, U., Albritton, L., Dexter, M., Spooncer, E., and Ostertag, W. (1991). Retroviral vectors related to the myeloproliferative sarcoma virus allow efficient expression in hematopoietic stem and precursor cell lines, but retroviral infection is reduced in more primitive cells. *Hum. Gene Ther.* **2**, 61–70.

91. Crooks, G. M., and Kohn, D. B. (1993). Growth factors increase amphotropic retrovirus binding to human CD34+ bone marrow progenitor cells. *Blood* **82**, 3290–3297.

92. Orlic, D., Girard, L. J., Jordan, C. T., Anderson, S. M., Cline, A. P., and Bodine, D. M. (1996). The level of mRNA encoding the amphotropic retrovirus receptor in mouse and human hematopoietic stem cells is low and correlates with the efficiency of retrovirus transduction. *Proc. Natl. Acad. Sci. USA* **93**, 11097–11102.

93. von Laer, D., Thomsen, S., Vogt, B., Donath, M., Kruppa, J., Rein, A., Ostertag, W., and Stocking, C. (1998). Entry of amphotropic and 10A1 pseudotyped murine retroviruses is restricted in hematopoietic stem cell lines. *J. Virol.* **72**, 1424–1430.

94. Münk, C., Thomsen, S., Stocking, C., and Lohler, J. (1998). Murine leukemia virus recombinants that use phosphate transporters for cell entry induce similar spongiform encephalomyelopathies in newborn mice. *Virology* **252**, 318–323.

95. Rubin, H. (1965). Genetic control and cellular susceptibility to pseudotypes of Rous sarcoma virus. *Virology* **26**, 270–282.

96. Friedmann, T., and Yee, J.-K. (1995). Pseudotyped retroviral vectors for studies of human gene therapy. *Nat. Med.* **1**, 275–277.

97. Januszeski, M. M., Cannon, P. M., Chen, D., Rozenberg, Y., and Anderson, W. F. (1997). Functional analysis of the cytoplasmic tail of Moloney murine leukemia virus envelope protein. *J. Virol.* **71**, 3613–3619.

98. Suomalainen, M., and Garoff, H. (1994). Incorporation of homologous and heterologous proteins into the envelope of Moloney murine leukemia virus. *J. Virol.* **68**, 4879–4889.

99. Emi, N., Friedmann, T., and Yee, J.-K. (1991). Pseudotype formation of murine leukemia virus with the G protein of vesicular stomatitis virus. *J. Virol.* **65**, 1202–1207.

100. Conti, C., Mastromarino, P., and Orsi, P. (1991). Role of membrane phospholipids and glycolipids in cell-to-cell fusion of VSV. *Comp. Immun. Microbiol. Infect. Dis.* **14**, 303–313.

101. Burns, J. C., Friedmann, T., Driever, W., Burrascano, M., and Yee, J.-K. (1993). Vesicular stomatitis virus G glycoprotein pseudotyped retroviral vectors: concentration to very high titer and efficient gene transfer into mammalian and nonmammalian cells. *Proc. Natl. Acad. Sci. USA* **90**, 8033–8037.

102. Yang, Y. P., Vanin, E. F., Whitt, M. A., Fornerod, M., Zwart, R., Schneiderman, R. D., Grosveld, G., and Nienhuis, A. W. (1995). Inducible, high-level production of infectious murine leukemia retroviral vector particles pseudotyped with vesicular stomatitis virus G envelope protein. *Hum. Gene Ther.* **6**, 1203–1213.

103. Akkina, R. K., Walton, R. M., Chen, M. L., Li, Q.-X., Planelles, V., and Chen, I. S. Y. (1996). High-efficiency gene transfer into CD34+ cells with a human immunodeficiency virus type 1-based retroviral vector pseudotyped with vesicular stomatitis virus envelope glycoprotein G. *J. Virol.* **70**, 2581–2585.

104. von Kalle, C., Kiem, H.-P., Goehle, S., Darovsky, B., Heimfeld, S., Torok-Storb, B., Storb, R., and Schuening, F. G. (1994). Increased gene transfer into human hematopoietic progenitor cells by extended in vitro exposure to a pseudotyped retroviral vector. *Blood* **84**, 2890–2897.

105. Bunnell, B. A., Mesler Muul, L., Donahue, R. E., Blaese, R. M., and Morgan, R. A. (1995). High-efficiency retroviral-mediated gene transfer into human and nonhuman primate peripheral blood lymphocytes. *Proc. Natl. Acad. Sci. USA* **92**, 7739–7743.

106. Uckert, W., Becker, C., Gladow, M., Klein, D., Kammertoens, T., Pedersen, L., and Blankenstein, T. (2000). Efficient gene transfer into primary human CD8+ T lymphocytes by MuLV-10A1 retrovirus pseudotype. *Hum. Gene Ther.* **11**, 1005–1014.

107. Barrette, S., Douglas, J., Orlic, D., Anderson, S. M., Seidel, N. E., Miller, A. D., and Bodine, D. M. (2000). Superior transduction of mouse hematopoietic stem cells with 10A1 and VSV-G pseudotyped retrovirus vectors. *Molec. Ther.* **1**, 330–338.

108. Lindemann, D., Bock, M., Schweizer, M., and Rethwilm, A. (1997). Efficient pseudotyping of murine leukemia virus particles with chimeric human foamy virus envelope proteins. *J. Virol.* **71**, 4815–4820.

109. Wilson, C., Reitz, M. S., Okayama, H., and Eiden, M. V. (1989). Formation of infectious hybrid virions with gibbon ape leukemia virus and human T-cell leukemia virus retroviral envelope glycoproteins and the Gag and Pol proteins of Moloney murine leukemia virus. *J. Virol.* **63,** 2374–2378.

110. Mammano, F., Salvatori, F., Indraccolo, S., De Rossi, A., Chieco-Bianchi, L., and Göttlinger, H. G. (1997). Truncation of the human immunodeficiency virus type 1 envelope glycoprotein allows efficient pseudotyping of Moloney murine leukemia virus particles and gene transfer into CD4+ cells. *J. Virol.* **71,** 3341–3345.

111. Loiler, S. A., DiFronzo, N. L., and Holland, C. A. (1997). Gene transfer to human cells using retrovirus vectors produced by a new polytropic packaging cell line. *J. Virol.* **71,** 4825–4828.

112. von Laer, D., Lindemann, D., Roscher, S., Herwig, U., Friel, J., and Herchenröder, O. (2001). Low level expression of functional foamy virus receptor on hematopoietic progenitor cells, *Virology* **288,** 139–144.

113. Miletic, H., Bruns, M., Tsiakas, K., Vogt, B., Rezai, R., Baum, C., Kuhlke, K., Cosset, F. L., Ostertag, W., Lother, H., and von Laer, D. (1999). Retroviral vectors pseudotyped with lymphocytic choriomeningitis virus. *J. Virol.* **73,** 6114–6116.

114. Hanenberg, H., Xiao, L. X., Dilloo, D., Hashino, K., Kato, I., and Williams, D. A. (1996). Colocalization of retrovirus and target cells on specific fibronactin fragments increases genetic transduction of mammalian cells. *Nat. Med.* **2,** 876–882.

115. Moritz, T., Patel, V. P., and Williams, D. A. (1994). Bone marrow extracellular matrix molecules improve gene transfer into human hematopoietic cells via retroviral vectors. *J. Clin. Invest.* **93,** 1451–1457.

116. Cosset, F. L., and Russell, S. J. (1996). Targeting retrovirus entry. *Gene Ther.* **3,** 946–956.

117. Russell, S. J., and Cosset, F. L. (1999). Modifying the host range properties of retroviral vectors. *J. Gene Med.* **1,** 300–311.

118. Etienne-Julan, M., Roux, P., Carillo, S., Jeanteur, P., and Piechaczyk, M. (1992). The efficiency of cell targeting by recombinant retroviruses depends on the nature of the receptor and the composition of the artificial cell-virus linker. *J. Gen. Virol.* **73,** 3251–3255.

119. Kasahara, N., Dozy, A. M., and Kan, Y. W. (1994). Tissue-specific targeting of retroviral vectors through ligand-receptor interactions. *Science* **266,** 1373–1376.

120. Valsesia, W. S., Drynda, A., Deleage, G., Aumailley, M., Heard, J. M., Danos, O., Verdier, G., and Cosset, F. L. (1994). Modifications in the binding domain of avian retrovirus envelope protein to redirect the host range of retroviral vectors. *J. Virol.* **68,** 4609–4619.

121. Xiaoliang, H., Kasahara, N., and Wai Kan, Y. (1995). Ligand-directed retroviral targeting of human breast cancer cells. *Proc. Natl. Acad. Sci. USA* **92,** 9747–9751.

122. Chu, T. H., and Dornburg, R. (1995). Retroviral vector particles displaying the antigen-binding site of an antibody enable cell-type-specific gene transfer. *J. Virol.* **69,** 2659–2663.

123. Jiang, A., and Dornburg, R. (1999). In vivo cell type-specific gene delivery with retroviral vectors that display single chain antibodies. *Gene Ther.* **6,** 1982–1987.

124. Zhao, Y., Zhu, L., Lee, S., Li, L., Chang, E., Soong, N. W., Douer, D., and Anderson, W. F. (1999). Identification of the block in targeted retroviral-mediated gene transfer. *Proc. Natl. Acad. Sci. USA* **96,** 4005–4010.

125. Hatziioannou, T., Valsesia-Wittmann, S., Russell, S. J., and Cosset, F. L. (1998). Incorporation of fowl plague virus hemagglutinin into murine leukemia virus particles and analysis of the infectivity of the pseudotyped retroviruses. *J. Virol.* **72,** 5313–5317.

126. Chu, T. H., and Dornburg, R. (1997). Toward highly efficient cell-type-specific gene transfer with retroviral vectors displaying single-chain antibodies. *J. Virol.* **71,** 720–725.

127. Nilson, B. H., Morling, F. J., Cosset, F. L., and Russell, S. J. (1996). Targeting of retroviral vectors through protease-substrate interactions. *Gene Ther.* **3,** 280–286.

128. Somia, N. V., Zoppe, M., and Verma, I. M. (1995). Generation of targeted retroviral vectors by using single-chain variable fragment: an approach to in vivo gene delivery. *Proc. Natl. Acad. Sci. USA* **92,** 7570–7574.

129. Marin, M., Noel, D., Valsesia, W. S., Brockly, F., Etienne, J. M., Russell, S., Cosset, F. L., and Piechaczyk, M. (1996). Targeted infection of human cells via major histocompatibility complex class I molecules by Moloney murine leukemia virus-derived viruses displaying single-chain antibody fragment-envelope fusion proteins. *J. Virol.* **70,** 2957–2962.

130. Russell, S. J., Hawkins, R. E., and Winter, G. (1993). Retroviral vectors displaying functional antibody fragments. *Nucleic Acids Res.* **21,** 1081–1085.

131. Valsesia-Wittmann, S., Morling, F. J., Hatziioannou, T., Russell, S. J., and Cosset, F. L. (1997). Receptor co-operation in retrovirus entry: recruitment of an auxiliary entry mechanism after retargeted binding. *EMBO J.* **16,** 1214–1223.

132. Naldini, L., Blömer, U., Gallay, P., Ory, D., Mulligan, R., Gage, F. H., Verma, I. M., and Trono, D. (1996). In vivo gene delivery and stable transduction of nondividing cells by a lentiviral vector. *Science* **272,** 263–267.

133. Naldini, L., Blömer, U., Gage, F. H., Trono, D., and Verma, I. M. (1996). Efficient transfer, integration, and sustained long-term expression of the transgene in adult rat brains injected with a lentiviral vector. *Proc. Natl. Acad. Sci. USA* **93,** 1382–1388.

134. Reiser, J., Harmison, G., Kluepfel-Stahl, S., Brady, R. O., Karlsson, S., and Schubert, M. (1996). Transduction of nondividing cells using pseudotyped defective high-titer HIV type particles. *Proc. Natl. Acad. Sci. USA* **93,** 15266–15271.

135. Blömer, U., Naldini, L., Kafri, T., Trono, D., Verma, I. M., and Gage, F. H. (1997). Highly efficient and sustained gene transfer in adult neurons with a lentivirus vector. *J. Virol.* **71,** 6641–6649.

136. Miyoshi, H., Smith, K. A., Mosier, D. E., Verma, I. M., and Torbett, B. E. (1999). Transduction of human CD34+ cells that mediate long-term engraftment of NOD/SCID mice by HIV vectors. *Science* **283,** 682–686.

137. Carroll, R., Lin, J.-T., Dacquel, E. J., Mosca, J. D., Burke, D. S., and St. Louis, D. C. (1994). A human immunodeficiency virus type 1 (HIV-1)-based retroviral vector system utilizing stable HIV-1 packaging cell lines. *J. Virol.* **68,** 6047–6051.

138. Richardson, J. H., Kaye, J. F., Child, L. A., and Lever, A. M. L. (1995). Helper virus-free transfer of human immunodeficiency virus type 1 vectors. *J. Gen. Virol.* **76,** 691–696.

139. Konvalinka, J., Litterst, M. A., Welker, R., Kottler, H., Rippmann, F., Heuser, A. M., and Krausslich, H. G. (1995). An active-site mutation in the human immunodeficiency virus type 1 proteinase (PR) causes reduced PR activity and loss of PR-mediated cytotoxicity without apparent effect on virus maturation and infectivity. *J. Virol.* **69,** 7180–7186.

140. Corbeau, P., Kraus, G., and Wong-Staal, F. (1996). Efficient gene transfer by a human immunodeficiency virus type 1 (HIV-1)-derived vector using a stable HIV packaging cell line. *Proc. Natl. Acad. Sci. USA* **93,** 14070–14075.

141. Kafri, T., van Praaag, H., Ouyang, L., Gage, F. H., and Verma, I. M. (1999). A packaging cell line for lentivirus vectors. *J. Virol.* **73,** 576–584.

142. Dull, T., Zufferey, R., Kelly, M., Mandel, R. J., Nguyen, M., Trono, D., and Naldini, L. (1998). A third-generation lentivirus vector with a conditional packaging system. *J. Virol.* **72,** 8463–8471.

143. Romano, G., Michell, P., Pacilio, C., and Giordano, A. (2000). Latest developments in gene transfer technology: achievements, perspectives,

and controversies over therapeutic applications. *Stem Cells* **18,** 19–39.

144. Poeschla, E. M., Wong-Staal, F., and Looney, D. J. (1998). Efficient transduction of nondividing human cells by feline immunodeficiency virus lentiviral vectors. *Nat. Med.* **4,** 354–357.

145. Olsen, J. C. (1998). Gene transfer vectors derived from equine infectious anemia virus. *Gene Ther.* **5,** 1481–1487.

146. Schnell, T., Foley, P., Wirth, M., Munch, J., and Uberla, K. (2000). Development of a self-inactivating, minimal lentivirus vector based on simian immunodeficiency virus. *Hum. Gene Ther.* **11,** 439–447.

147. Metharom, P., Takyar, S., Xia, H. H., Ellem, K. A., Macmillan, J., Shepherd, R. W., Wilcox, G. E., and Wei, M. Q. (2000). Novel bovine lentiviral vectors based on Jembrana disease virus. *J. Gene Med.* **2,** 176–185.

148. Schweizer, M., Turek, R., Hahn, H., Schliephake, A., Netzer, K.-O., Eder, G., Reinhardt, M., Rethwilm, A., and Neumann-Haefelin, D. (1995). Markers of foamy virus infections in monkeys, apes and accidentally infected humans: appropriate testing fails to confirm suspected foamy prevalence in humans. *AIDS Res. Hum. Retroviruses* **11,** 161–170.

149. Hooks, J. J., and Gibbs, C. J. J. (1995). The foamy viruses. *Bacteriolo. Revi.* **39,** 169–185.

150. Mikovits, J. A., Hoffman, P. M., Rethwilm, A., and Ruscetti, F. W. (1996). In vitro infection of primary and retrovirus-infected human leukocytes by human foamy virus. *J. Virol.* **70,** 2774–2780.

151. Pietschmann, T., Heinkelein, M., Heldmann, M., Zentgraf, H., Rethwilm, A., and Lindemann, D. (1999). Foamy virus capsids require the cognate envelope protein for particle export. *J. Virol.* **73,** 2613–2621.

152. Bieniasz, P. D., Weiss, R. A., and McClure, M. O. (1995). Cell cycle dependence of foamy retrovirus infection. *J. Virol.* **69,** 7295–7299.

153. Russell, D. W., and Miller, A. D. (1996). Foamy virus vectors. *J. Virol.* **70,** 217–222.

154. Trobridge, G. D., and Russell, D. W. (1998). Helper-free foamy virus vectors. *Hum. Gene Ther.* **9,** 2517–2525.

155. Heinkelein, M., Thurow, J., Dressler, M., Imrich, H., Neumann-Haefelin, D., McClure, M. O., and Rethwilm, A. (2000). Complex effects of deletions in the 5′ untranslated region of primate foamy virus on viral gene expression and RNA packaging. *J. Virol.* **74,** 3141–3148.

156. Sandmeyer, S. B., Hansen, L. J., and Chalker, D. L. (1990). Integration specificity of retrotransposons and retroviruses. *Annu. Rev. Genet.* **24,** 491–518.

157. Withers-Ward, E. S., Kitamura, Y., Barnes, J. P., and Coffin, J. M. (1994). Distribution of targets for avian retrovirus DNA integration in vivo. *Genes Dev.* **8,** 1473–1487.

158. Muller, H. P., and Varmus, H. E. (1994). DNA bending creates favored sites for retroviral integration: an explanation for preferred insertion sites in nucleosomes. *EMBO J.* **13,** 4704–4714.

159. Shibagaki, Y., and Chow, S. A. (1997). Central core domain of retroviral integrase is responsible for target site selection. *J. Biol. Chem.* **272,** 8361–8369.

160. Katz, R. A., Merkel, G., and Skalka, A. M. (1996). Targeting of retro viral integrase by fusion to a heterologous DNA binding domain: in vitro activities and incorporation of a fusion protein into viral particles. *Virology* **217,** 178–190.

161. Goulavic, H., and Chow, S. A. (1996). Directed integration of viral DNA mediated by fusion proteins consisting of human immunodeficiency virus type 1 integrase and *Escherichia coli* LexA protein. *J. Virol.* **70,** 37–46.

162. Bushman, F. D., and Miller, M. D. (1997). Tethering human immunodeficiency virus type 1 preintegration complexes to target DNA promotes integration at nearby sites. *J. Virol.* **71,** 458–464.

163. Dildine, S. L., and Sandmeyer, S. B. (1997). Integration of the yeast retrovirus-like element Ty3 upstream of a human tRNA gene expressed in yeast. *Gene* **194,** 227–233.

164. Devine, S. E., and Boeke, J. D. (1996). Integration of the yeast retrotransposon Ty1 is targeted to regions upstream of genes transcribed by RNA polymerase III. *Genes Dev.* **10,** 620–633.

165. Feng, Q., Moran, J., Kazazian, H., and Boeke, J. D. (1996). Human L1 retrotransposon encodes a conserved endonuclease required for retrotransposition. *Cell* **87,** 905–916.

166. Ellis, J., and Bernstein, A. (1989). Gene targeting with retroviral vectors: recombination by gene conversion into regions of nonhomology. *Mol. Cell. Biol.* **9,** 1621–1627.

167. Pazin, M. J., Sheridan, P. L., Cannon, K., Cao, Z., Keck, J. G., Kadonga, J. T., and Jones, K. A. (1996). NF-kappa B-mediated chromatin reconfiguration and transcriptional activation of the HIV-1 enhancer in vitro. *Genes Dev.* **10,** 37–49.

168. Baum, C., Itoh, K., Meyer, J., Laker, C., Ito, Y., and Ostertag, W. (1997). The potent enhancer activity of the polycythemic strain of spleen focus-forming virus in hematopoietic cells is governed by a binding site for Sp1 in the upstream control region and by a unique enhancer core motif, creating an exclusive target for PEBP/CBF. *J. Virol.* **71,** 6323–6331.

169. Lund, A. H., Duch, M., and Pedersen, F. S. (1996). Transcriptional silencing of retroviral vectors. *J. Biomed. Sci.* **3,** 365–378.

170. Laker, C., Meyer, J., Schopen, A., Friel, J., Heberlein, C., Ostertag, W., and Stocking, C. (1998). Host *cis*-mediated extinction of a retrovirus permissive for expression in embryonal stem cells during differentiation. *J. Virol.* **72,** 339–348.

171. Challita, P.-M., and Kohn, D. B. (1994). Lack of expression from a retroviral vector after transduction of murine hematopoietic stem cells is associated with methylation in vivo. *Proc. Natl. Acad. Sci. USA* **91,** 2567–2571.

172. Scharfmann, R., Axelrod, J. H., and Verma, I. (1991). Long-term in vivo expression of retrovirus-mediated gene transfer in mouse fibroblast implants. *Proc. Natl. Acad. Sci. USA* **88,** 4626–2630.

173. Grez, M., Akgün, E., Hilberg, F., and Ostertag, W. (1990). Embryonic stem cell virus, a recombinant murine retrovirus with expression in embryonic stem cells. *Proc. Natl. Acad. Sci. USA* **87,** 9202–9206.

174. Stocking, C., Grez, M., and Ostertag, W. (1993). Regulation of retrovirus infection and expression in embryonic and hematopoietic stem cells. In: *Virus Strategies: Molecular Biology and Pathogenesis* (W. Doerfler and P. Böhm, eds), pp. 433–455. VCH Verlagsgesellschaft, Weinheim.

175. Speck, N. A., Renjifo, B. V., Golemis, E., Fredrickson, T. N., Hartley, J. W., and Hopkins, N. (1990). Mutations of the core or adjacent LVb elements of the Moloney leukemia virus enhancer alters disease specificity. *Genes Dev.* **4,** 223–242.

176. Morrison, S. J., Uchida, N., and Weissman, I. L. (1995). The biology of hematopoietic stem cells. *Annu. Rev. Cell Dev. Biol.* **11,** 35–71.

177. Weissman, I. L. (2000). Stem cells: units of development, units of regeneration, and units in evolution. *Cell* **100,** 157–168.

178. Hilberg, F., Stocking, C., Ostertag, W., and Grez, M. (1987). Functional analysis of a retroviral host range mutant: altered long terminal repeat sequences allow expression in embryonal carcinoma cells. *Proc. Natl. Acad. Sci. USA* **84,** 5232–5236.

179. Hawley, R. G., Lieu, F. H., Fong, A. Z., and Hawley, T. S. (1994). Versatile retroviral vectors for potential use in gene therapy. *Gene Ther.* **1,** 136–138.

180. Challita, P. M., Skelton, D., el-Khoueiry, A., Yu, X. J., Weinberg, K., and Kohn, D. B. (1995). Multiple modifications in *cis* elements of the long terminal repeat of retroviral vectors lead to increased expression and decreased DNA methylation in embryonic carcinoma cells. *J. Virol.* **69,** 748–755.

181. Robbins, P. B., Skelton, D. C., Yu, X. J., Halene, S., Leonard, E. H., and Kohn, D. B. (1998). Consistent, persistent expression from modified

retroviral vectors in murine hematopoietic stem cells. *Proc. Natl. Acad. Sci. USA* **95,** 10182–10187.

182. Halene, S., Wang, L., Cooper, R. M., Bockstoce, D. C., Robbins, P. B., and Kohn, D. B. (1999). Improved expression in hematopoietic and lymphoid cells in mice after transplantation of bone marrow transduced with a modified retroviral vector. *Blood* **94,** 3349–3357.

183. Tumas, D. B., Spangrude, G. J., Brooks, D. M., Williams, C. D., and Chesebro, B. (1996). High-frequency cell-surface expression of a foreign protein in murine hematopoietic stem cells using a new retroviral vector. *Blood* **87,** 509–517.

184. Hildinger, M., Eckert, H. G., Schilz, A. J., John, J., Ostertag, W., and Baum, C. (1998). FMEV vectors: both retroviral long terminal repeat and leader are important for high expression in transduced hematopoietic cells. *Gene Ther.* **5,** 1575–1579.

185. van Hennik, P. B., Verstegen, M. M., Bierhuizen, M. F., Limon, A., Wognum, A. W., Cancelas, J. A., Barquinero, J., Ploemacher, R. E., and Wagemaker, G. (1998). Highly efficient transduction of the green fluorescent protein gene in human umbilical cord blood stem cells capable of cobblestone formation in long-term cultures and multilineage engraftment of immunodeficient mice. *Blood* **92,** 4013–4022.

186. Shivdasani, R. A., and Orkin, S. (1996). The transcriptional control of hematopoiesis. *Blood* **87,** 4025–4039.

187. Nielsen, A. L., Pallisgaard, N., Pedersen, F. S., and Jorgensen, P. (1994). Basic helix–loop–helix proteins in murine type C retrovirus transcriptional regulation. *J. Virol.* **68,** 5638–5647.

188. Zaiman, A. L., and Lenz, J. (1996). Transcriptional activation of a retrovirus enhancer by CBF (AML1) requires a second factor: evidence for cooperativity with c-Myb. *J. Virol.* **70,** 5618–5629.

189. Malik, P., Krall, W. J., Yu, X. J., Zhou, C., and Kohn, D. B. (1995). Retroviral-mediated gene expression in human myelomonocytic cells: a comparison of hematopoietic cell promoters to viral promoters. *Blood* **86,** 2993–3005.

190. Plavec, I., Voyovich, A., Moss, K., Webster, D., Hanley, M. B., Escaich, S., Ho, K. E., Boehnlein, E., and DiGiusto, D. L. (1996). Sustained retroviral gene marking and expression in lymphoid and myeloid cells derived from transduced hematopoietic progenitor cells. *Gene Ther.* **3,** 717–724.

191. Agarwal, M., Austin, T. W., Morel, F., Chen, J., Bohnlein, E., and Plavec, I. (1998). Scaffold attachment region-mediated enhancement of retroviral vector expression in primary T cells. *J. Virol.* **72,** 3720–3728.

192. Onodera, M., Nelson, D. M., Yachie, A., Jagadeesh, G. J., Bunnell, B. A., Morgan, R. A., and Blaese, R. M. (1998). Development of improved adenosine deaminase retroviral vectors. *J. Virol.* **72,** 1769–1774.

193. Couture, L. A., Mullen, C. A., and Morgan, R. A. (1994). Retroviral vectors containing chimeric promoter/enhancer elements exhibit cell-type-specific gene expression. *Hum. Gene Ther.* **5,** 667–677.

194. Dachs, G. U., Patterson, A. V., Firth, J. D., Ratcliffe, P. J., Townsend, K. M., Stratford, I. J., and Harris, A. L. (1997). Targeting gene expression to hypoxic tumor cells. *Nat. Med.* **3,** 515–520.

195. Miller, N., and Whelan, J. (1997). Progress in transcriptionally targeted and regulatable vectors for genetic therapy. *Hum. Gene Ther.* **8,** 803–815.

196. Gautsch, J. W., and Wilson, M. C. (1983). Delayed de novo methylation in teratocarcinoma suggests additional tissue-specific mechanisms for controlling gene expression. *Nature* **301,** 32–37.

197. Bird, A. P. (1986). CpG-rich islands and the function of DNA methylation. *Nature* **321,** 209–213.

198. Macleod, D., Charlton, J., Mullins, J., and Bird, A. P. (1994). Sp1 sites in the mouse aprt gene promoter are required to prevent methylation of the CpG island. *Genes Dev.* **8,** 2282–2292.

199. Grez, M., Zörnig, M., Nowock, J., and Ziegler, M. (1991). A single point mutation activates the Moloney murine leukemia virus long terminal repeat in embryonal stem cells. *J. Virol.* **65,** 4691–4698.

200. Schubeler, D., Mielke, C., Maass, K., and Bode, J. (1996). Scaffold/matrix-attached regions act upon transcription in a context-dependent manner. *Biochemistry* **35,** 11160–11169.

201. Phi-Van, L., von Kries, J. P., Ostertag, W., and Strätling, W. H. (1990). The chicken lysozyme matrix attachment region increases transcription from a heterologous promoter in heterologous cells and dampens position effects on the expression of transfected cells. *Mol. Cell. Biol.* **10,** 2302–2307.

202. Dang, Q., Auten, J., and Plavec, I. (2000). Human beta interferon scaffold attachment region inhibits de novo methylation and confers long-term, copy-number-dependent expression to a retroviral vector. *J. Virol.* **74,** 2671–2678.

203. Rivella, S., Callegari, J. A., May, C., Tan, C. W., and Sadelain, M. (2000). The cHS4 insulator increases the probability of retroviral expression at random chromosomal integration sites. *J. Virol.* **74,** 4679–4687.

204. Gossen, M., Bonin, A. L., and Bujard, H. (1993). Control of gene activity in higher eukaryotic cells by prokaryotic regulatory elements. *Trends Biochem. Sci.* **18,** 471–475.

205. Gossen, M., Freundllieb, S., Bender, G., Muller, G., Hillen, W., and Bujard, H. (1995). Transcriptional activation by tetracyclines in mammalian cells. *Science* **268,** 1766–1769.

206. Huang, Y., Wimler, K. M., and Carmichael, G. G. (1999). Intronless mRNA transport elements may affect multiple steps of pre-mRNA processing. *Embo J.* **18,** 1642–1652.

207. Luo, M. J., and Reed, R. (1999). Splicing is required for rapid and efficient mRNA export in metazoans. *Proc. Natl. Acad. Sci. USA* **96,** 14937–14942.

208. Pollard, V. W., and Malim, M. H. (1998). The HIV-1 Rev protein. *Annu. Rev. Microbiol.* **52,** 491–532.

209. Zufferey, R., Donello, J. E., Trono, D., and Hope, T. J. (1999). Woodchuck hepatitis virus post-transcriptional regulatory element enhances expression of transgenes delivered by retroviral vectors. *J. Virol.* **73,** 2886–2892.

210. Schambach, A., Wodrich, H., Hildinger, M., Bohne, J., Kräusslich, H.-G., and Baum, C. (2000). Context-dependence of different modules for post-transcriptional enhancement of gene expression from retroviral vectors. *Mol. Ther.* **2,** 435–445.

211. Hildinger, M., Schilz, A., Eckert, H. G., Bohn, W., Fehse, B., Zander, A., Ostertag, W., and Baum, C. (1999). Bicistronic retroviral vectors for combining myeloprotection with cell-surface marking. *Gene Ther.* **6,** 1222–1230.

212. Emerman, M., and Temin, H. M. (1984). Genes with promoters in retrovirus vectors can be independently suppressed by an epigenetic mechanism. *Cell* **39,** 449–467.

213. Ghattas, I. R., Sanes, J. R., and Majors, J. E. (1991). The encephalomyocarditis virus internal ribosome entry site allows efficient coexpression of two genes from a recombinant provirus in cultures cells and in embryos. *Mol. Cell. Biol.* **11,** 5848–5849.

214. Eggermont, J., and Proudfoot, N. J. (1993). Poly(A) signals and transcriptional pause sites combine to prevent interference between RNA polymerase II promoters. *EMBO J.* **12,** 2539–2548.

215. Proudfoot, N. J. (1986). Transcriptional interference and termination between duplicated alpha-globin gene constructs suggest a novel mechanism for gene regulation. *Nature* **322,** 562–565.

216. Ahlers, N., Hunt, N., Just, U., Laker, C., Ostertag, W., and Nowock, J. (1994). Selectable retrovirus vectors encoding Friend virus gp55 or erythropoietin induce polycythemia with different phenotypic expression and disease progression. *J. Virol.* **68,** 7235–7243.

217. Adam, M. A., Ramesh, N., Miller, A. D., and Osborne, W. R. (1991). Internal initiation of translation in retroviral vectors carrying picornavirus 5′ nontranslated regions. *J. Virol.* **65,** 4985–4990.

218. Boris-Lawrie, K. A., and Temin, H. M. (1993). Recent advances in retrovirus vector technology. *Curr. Opin. Genet. Dev.* **3**, 102–109.

219. Morgan, R. A., Couture, L., Elroy-Stein, O., Ragheb, J., Moss, B., and Anderson, W. F. (1992). Retroviral vectors containing putative internal ribosome entry sites: development of a polycistronic gene transfer system and applications to human gene therapy. *Nucleic Acids Res.* **20**, 1293–1299.

220. Mizuguchi, H., Xu, Z.,Ishii-Watabe, A., Uchida, E., and Hayakawa, T. (2000). IRES-dependent second gene expression is significantly lower than cap-dependent first gene expression in a bicistronic vector. *Molecu. Ther.* **1**, 376–382.

221. Germann, U. A., Chin, K.-V., Pastan, I., and Gottesman, M. M. (1990). Retroviral transfer of a chimeric multidrug resistance-adenosine deaminase gene. *FASEB J.* **4**, 1501–1506.

222. de Felipe, P., Martin, V., Cortes, M. L., Ryan, M., and Izquierdo, M. (1999). Use of the 2A sequence from foot-and-mouth disease virus in the generation of retroviral vectors for gene therapy. *Gene Ther.* **6**, 198–208.

223. Klump, H., Schiedlmeier, B., Vogt, B., Ryan, M., Ostertag, W., and Baum, C. (2000). Retroviral vector-mediated expression of HOXB4 in hematopoietic cells using a novel coexpression strategy. *Gene Ther.* **8**, 811–817.

224. Artelt, P., Grannemann, R., Stocking, C., Friel, J., Bartsch, J., and Hauser, H. (1991). The prokaryotic neomycin-resistance-encoding gene acts as a transcriptional silencer in eukaryotic cells. *Gene* **99**, 249–254.

225. Bunting, K. D., Webb, M., Giorgianni, G., Galipeau, J., Blakley, R. L., Townsend, A., and Sorrentino, B. P. (1997). Coding region-specific destabilization of mRNA transcripts attenuates expression from retroviral vectors containing class 1 aldehyde dehydrogenase cDNAs. *Hum. Gene Ther.* **8**, 1531–1543.

226. Sorrentino, B. P., McDonagh, K. T., Woods, D., and Orlic, D. (1995). Expression of retroviral vectors containing the human multidrug resistance 1 cDNA in hematopoietic cells of transplanted mice. *Blood* **86**, 491–501.

227. Garin, M. I., Garrett, E., Tiberghien, P., Apperley, J. F., Chalmers, D., Melo, J. V., and Ferrand, C. (2001). Molecular mechanism for ganciclovir resistance in human T lymphocytes transduced with retroviral vectors carrying the herpes simplex virus thymidine kinase gene. *Blood.* **97**, 122–129.

2

Noninfectious Gene Transfer and Expression Systems for Cancer Gene Therapy

MARK J. COOPER

Copernicus Therapeutics, Inc.
Cleveland, Ohio 44106

I. INTRODUCTION

Gene therapy provides a significant opportunity to devise novel strategies for the control or cure of cancer. Current approaches to cancer gene therapy typically employ viral-based vectors to express suitable target genes in human cancer cells either *ex vivo* or *in vivo* [1–4]. Therapeutic gene targets currently being evaluated include susceptibility genes, such as herpes simplex thymidine kinase followed by ganciclovir treatment [5–15]; genes that target the immune system to eliminate cancer cells, such as cytokines [16–35], costimulatory molecules [36], foreign histocompatibility genes

[37,38]; anti-sense constructs to insulin-like growth factor I [39,40], and polynucleotide vaccines [41–45]; replacement of wild-type tumor suppressor genes, such as p53 [44–49]; and anti-sense blockade of oncogenes, such as K-*ras* [50–52]. In order to move gene therapy into the mainstream of cancer therapeutics, however, it will ultimately be necessary to devise strategies to administer a gene therapy reagent to a patient in the familiar context of a pharmaceutical and to perform gene transfer *in vivo*. Currently utilized viral-based gene therapy vectors, including retroviral, adenoviral, and adeno-associated viral vectors, fail to realize this potential due to limitations in their expression characteristics, lack of specificity in targeting tumor cells for gene transfer, immunogenicity and other acute and chronic toxicities, and safety concerns regarding induction of secondary malignancies and recombination to form replication-competent virus. These limitations have refocused efforts to develop noninfectious, gene transfer technologies for *in vivo* gene delivery of plasmid-based expression vectors. These vectors exist as extrachromosomal elements in populations of transiently transfected tumor cells. As discussed later, incorporation of transcription control sequences, including tissue-specific enhancers and inducible promoters, and elements permitting controlled vector replication in tumor cells has the potential to yield cancer gene therapy vectors that are both safe and effective for direct *in vivo* gene transfer.

II. ADVANTAGES AND DISADVANTAGES OF INFECTIOUS, VIRAL-BASED VECTORS FOR HUMAN GENE THERAPY

A number of viruses that infect humans, including retrovirus, herpes virus, adenovirus, and adeno-associated virus, have been modified to generate efficient expression vectors. These vectors either integrate into genomic DNA or persist as extrachromosomal elements and have distinct expression characteristics, as summarized in Table 1. The primary advantage of these vectors is the ability to infect a high percentage of target cells *in vitro* and, in some cases, *in vivo* [7,53,54]. Whereas retroviral vectors yield one or several integrated proviral copies per cell, other vectors can introduce higher copy numbers of transcriptional cassettes, thereby enhancing transient levels of gene expression. Some viral-based vectors, such as those derived from recombinant adenovirus, may replicate in transduced cells at a low level, and this feature has usually been interpreted as an undesired feature raising safety concerns regarding unregulated, systemic gene transfer [4,55–57]. More recently, E1B-attenuated adenoviral vectors have been developed that replicate in tumor cells, resulting in tumor cell lysis and virus propagation within the tumor [58,59]. Although replication of this adenovirus construct was initially thought to be restricted to p53-negative tumor cells [58], other studies demonstrate that virus replication is independent of p53 status [60–62]. Intratumoral injections of these vectors have produced localized tumor regression in patients with recurrent head and neck cancers [63,64].

Although viral-based vectors may be particularly useful for gene transfer *ex vivo*, this approach requires costly manipulations of tumor biopsies to yield either transient [65] or stably selected and characterized transfectants [19]. Additionally, the latter approach may prove to be a particularly poor choice for gene targets that stimulate the immune system to eliminate tumor cells, as representation of tumor heterogeneity is likely lost prior to gene transfer.

While high-level infectivity of viral-based vectors remains an attractive feature, multiple safety concerns and technical features limit their applications, including: (1) safety concerns regarding integration of vector DNA into host cell genomic DNA, which may induce secondary malignancies by activation of proto-oncogenes or inactivation of tumor suppressor genes [66]; (2) potential for recombination events to produce an infectious virus able to replicate *in vivo* (recombination could occur either *in vitro* during vector preparation, or possibly *in vivo*, particularly when using vectors derived from pathogenic human viruses, such as adenovirus) [2,3,55–57,67,68]; (3) presentation of viral antigens on the surface of infected human cells, resulting in T-cell recognition and destruction of transduced cells [69]; (4) lack of specificity of cell types recognized by endogenous viral coat proteins, resulting in unintended transduction of nontargeted cell types *in vivo*; (5) heterogeneity of expression of viral coat protein receptors by tumor cell targets, thereby limiting the tumor cell population that can be transduced (viral receptor-negative cells may be selected for during treatment); (6) the fact that retroviral vectors will not express target genes in nonreplicating tumor cells [70]; (7) technical limitations regarding strategies to produce higher levels of gene expression in an infected cell; (8) difficulties in reproducibly producing, concentrating, delivering, and storing high titer viral vectors for clinical use; (9) complement-mediated mechanisms of inactivation may limit use of some viral-based vectors *in vivo* [71]; (10) the potential for some virally encoded proteins to yield undesired toxic effects in addition to immune recognition, leading to altered cell functions or transformation [2,4]; and (11) immunogenicity of viral-based vectors, resulting in incrementally decreased effectiveness during repeated treatments *in vivo* [2,4,72–76]. These safety concerns and limitations in the ability of infectious, viral-based vectors to yield maintained, high-level gene expression in transiently transfected tumor cells have led to the development of alternative, noninfectious gene expression and gene transfer technologies, as reviewed later.

TABLE 1   Infectious, Viral-Based Vectors for Cancer Gene Therapy

Vector	Integration or extrachromosomal distribution	Expression limited to cells undergoing replication at time of infection	Ref.
Retrovirus	I	Yes	70
Adenovirus	E	No	55–57, 293
Adeno-associated virus	I[a]	Yes[b]	294–296
Herpes simplex virus	E	No	297
Vaccinia virus	E	No	298
Autonomous parvovirus (LuIII)	E	Yes	299

Note: Abbreviations: I, integration; E, extrachromosomal.
[a] Integration in replicating cells, transient extrachromosomal persistence in stationary phase cells.
[b] 90% of expression limited to cells traversing S phase.

III. RATIONALE FOR CONSIDERING NONINFECTIOUS, PLASMID-BASED EXPRESSION SYSTEMS

Initial assumptions regarding requirements for effective cancer gene therapy have changed since demonstration of a significant "innocent bystander" effect using gene targets that confer antibiotic susceptibility, such as herpes simplex virus thymidine kinase followed by ganciclovir treatment [8], or genes that activate the immune system to recognize and kill tumor cells [16–43]. It may therefore not be necessary to transfect 50–100% of tumor cells in order to produce a cure. These findings provide an important rationale to consider nonviral-based vectors for gene therapy applications, particularly constructs that yield high levels of gene expression per transfected cell. Moreover, new technical advances in receptor-mediated gene delivery of plasmid-based vectors now yield transient transfection efficiencies *in vivo* that approximate those observed using viral-based vectors [77–80].

IV. GENE TRANSFER TECHNOLOGIES FOR PLASMID-BASED VECTORS: PRECLINICAL MODELS AND CLINICAL CANCER GENE THERAPY TRIALS

Several gene transfer methods yield efficient transient transfection efficiencies following either *in vitro* or *in vivo* applications, as listed in Table 2. Although some of these methods are limited by the target cell type transfected or by the specificity of gene transfer, receptor-mediated gene transfer technologies have the potential to yield efficient and specific gene delivery to targeted tumor cells *in vivo* and therefore may have widespread utility.

A. Direct Injection of DNA

Perhaps the simplest formulation for *in vivo* gene transfer of plasmid vectors into cells is by direct administration of su-

TABLE 2 Gene Transfer Technologies for Plasmid-Based Vectors

Gene transfer method	Gene transfer limited to specific tissues	Ability to target tumor cells
Direct injection of naked DNA	Yes	No
Particle bombardment	Yes	No
Calcium phosphate	No	No
Liposome/DNA complexes	No	Yes
Ligand/DNA conjugates	No	Yes

percoiled DNA into tissues. Early studies demonstrated that DNA can be directly introduced into cells *in vivo* by simply injecting target organs with viral DNA. For example, when polyoma virus [81,82] or ground squirrel hepatitis virus [83] DNA were directly injected into mice or ground squirrels, respectively, the animals developed systemic infection, and active virus particles were recovered. In these studies, however, very inefficient initial levels of *in vivo* gene transfer of purified virion DNA could be detected due to amplification of the gene transfer mechanism via systemic virus infection. In related studies, gene expression was observed in the liver and spleen of newborn rats 2 days following intraperitoneal injection of calcium-phosphate-precipitated plasmid DNA encoding the chloramphenicol acetyltransferase reporter gene [84]. More recently, direct injection of naked plasmid DNA was shown to yield significant levels of gene expression in rat skeletal and cardiac muscle, but not in kidney, lung, liver, or brain [85,86]. For example, direct injection of 25 μg of p-CMVint-lux plasmid DNA encoding the luciferase marker gene driven by the CMV immediate-early promoter into the rectus femoris muscle of mice yielded peak gene expression at day 14, and expression was detectable for up to 120 days [87]. The mechanism by which plasmid DNA is taken up by muscle cells is unclear but does not seem to be related to direct cell injury to the sarcolemmal membrane [88]. In more recent studies, significant gene expression has also been observed following direct injection of naked plasmid DNA into rat or cat liver [89] and rabbit thyroid follicular cells [90], expanding the tissue types that can be transfected using this method.

Gene expression in transfected muscle cells is sufficient to produce antiviral immunity. For example, mice having their quadriceps muscles injected with a plasmid encoding influenza A nucleoprotein developed humoral and cytotoxic T-cell responses to this antigen and were protected from subsequent challenge with influenza A virus [91]. In a similar fashion, direct intramuscular gene transfer of plasmid DNA encoding HIV envelop protein (gp160) in mice confers humoral and cell-mediated immunity against recombinant envelop protein, and sera from these animals neutralizes HIV infectivity *in vitro* [92]. Direct injection of plasmid DNA also results in efficient gene delivery to subcutaneous tissues, including keratinocytes, fibroblasts, and dendritic cells [93]. This later approach may be superior to direct muscle injection for the development of cytotoxic T-cell immunity, perhaps because of antigen presentation by macrophages and dendritic cells in the subcutaneous tissues [93]. Intradermal gene transfer efficiencies can also be enhanced by delivering electric pulses to subcutaneous tissues after a local injection [94,95] or by use of liquid or powder sprays or particle bombardment methods of DNA transfer [96,97].

Intramuscular or intradermal gene transfer of plasmid vectors encoding tumor-associated antigens may yield effective cancer vaccines. This approach requires prior knowledge of potential tumor-associated antigens in a given patient's tumor that presumably have not yet been adequately presented to the host immune system. A variety of tumor-associated antigens have been identified that have the potential to stimulate a cytotoxic T-cell response and are therefore candidate antigens for tumor vaccines. In human melanoma, such tumor-associated antigens include p97, MAGE-1, MAGE-2, MAGE-3, Melan-A, MART-1, gp100, and tyrosinase [98–107]. Cytotoxic T-cell responses also have been demonstrated against mucin products of the MUC-1 gene in patients with pancreatic and breast carcinomas, and antigenicity appears to be related to underglycosylated forms of the protein found in tumor cells [108]. In ideal circumstances, tumor-associated antigens would only be expressed by the tumor and not normal tissues, and a cancer vaccine would generate a tumor-specific immune response. Because peptide fragments of cellular proteins are displayed on the cell surface in conjunction with major histocompatibility antigens by TAP transporter proteins [109], the immune system is able to survey for the presence of gene mutations that generate novel peptide antigens. Tumor-specific cytotoxic T-cell immunity has been demonstrated against peptide fragments from oncogenes or tumor suppressor genes that are mutated during the generation of the malignancy. These vaccines include peptides encoding point mutations in *ras* genes [110–113] and p53 [114] and the unique breakpoint in the *bcr–abl* fusion gene [115].

One example of a successful cancer vaccine model is development of antitumor immunity to tumor cells expressing human carcinoembryonic antigen (CEA). CEA is expressed at high levels in several types of human adenocarcinomas, including colon, breast, gastric, pancreatic, and non-small-cell lung carcinomas [41,42,116]. CEA is also expressed at high levels in human fetal gut and at low levels in normal colonic mucosal cells, but it is not expressed in murine tissues [41]. Therefore, mice immunized with CEA protein would be expected to develop antitumor immunity to syngeneic tumor cells expressing human CEA. This result has been demonstrated by using a recombinant vaccinia virus vector encoding human CEA cDNA to immunize mice [41]. These studies used a murine colon carcinoma cell line, MC38, that had been transduced with a retroviral vector encoding CEA cDNA, generating the modified MC38–CEA-2 cell line. Vaccinia-vector-immunized syngeneic C57BL/6 mice developed humoral and cell-mediated immunity to CEA, and MC38–CEA-2 cells injected in immunized animals were rejected [41].

To extend these studies, Curiel and colleagues have demonstrated that C57BL/6 mice can develop antitumor immunity to MC38–CEA-2 cells by directly injecting plasmid DNA encoding CEA cDNA into striated muscle [43]. In these studies, the tongues of C57BL/6 mice were injected weekly with 100 μg of plasmid DNA encoding CEA. After four doses, these animals produced anti-CEA antibodies and developed cell-mediated immunity to MC38–CEA-2 cells.

Importantly, these immunized mice rejected MC38–CEA-2 cells that were subcutaneously inoculated in these animals 1 week following the last immunization. These results demonstrate the ability to generate an effective cancer vaccine by expressing a tumor-associated antigen following direct *in vivo* gene transfer of plasmid DNA.

Further issues that need to be addressed by the use of polynucleotide vaccines include choice of specific tumor-associated antigens likely to produce antitumor immunity in cancer patients of a given tumor type and the clinical setting in which this approach is likely to be effective. For example, administration of a tumor vaccine in an adjuvant setting following initial surgical removal of the primary mass may improve conditions for success by selecting a patient population that has not yet received immune-suppressive cytotoxic chemotherapy and whose tumor burden is small. In addition, analysis of tumor tissue for expression of relevant tumor antigens may be quite important, as multiple tumor-associated antigens may need to be targeted to address clonal evolution of heterogeneous populations of tumor cells.

Alternatively, antitumor immunity can be achieved by addressing mechanisms of tumor cell immune tolerance [117]. Such approaches are independent of identification of tumor-specific antigens. For examples, many tumors lack expression of the transporters associated with antigen processing (TAP) [118,119], resulting in insufficient presentation of tumor-associated antigens with nascently produced class I MHC molecules. Direct TAP1 gene transfer to such tumors in an animal model results in prolonged survival of tumor-bearing mice [120]. Intratumor gene transfer of other components required for generation of cytotoxic T-cells, including class I MHC molecules [121,122] and β_2-microglobulin [123], may be effective in some tumors. Other approaches include gene transfer of the B7.1 costimulatory molecule and anti-sense or ribozyme strategies to decrease local production of immunosuppressive cytokines and receptors [36,124–132].

B. Particle-Mediated Gene Delivery

An alternative approach for delivery of plasmid constructs into human cells *in vivo* is to coat metallic particles with a DNA vector and then introduce the particles directly into tissues using a "gene gun" to accelerate the particles to a high velocity [133]. Subcutaneous tissues can be directly transfected *in vivo* because the particles can penetrate to this depth. Visceral tissues have also been transfected *in vivo* in animals, although this approach requires an operative procedure to bring the tissue of interest in close approximation to the gene gun instrument. Nevertheless, particle-mediated gene transfer of a plasmid vector encoding influenza virus hemagglutinin subtype 1 has been demonstrated to immunize mice against challenge with a lethal inoculum of influenza virus [96]. This approach has significant potential for development of cancer vaccines, because efficient gene transfer of polynucleotide vaccines into subcutaneous tissues may be particularly effective in presenting antigens to the immune system [93]. Cancer preclinical models using particle-mediated gene transfer into subcutaneous tumor explants have also demonstrated improved survival of tumor-bearing mice using a variety of cytokine targets, including IL-2, IL-6, and IFN-γ [134]. In addition to ballistic gene transfer using metallic particles, devices have been developed for needleless intradermal DNA delivery using powders and liquid sprays [96,97].

C. Gene Transfer of DNA Precipitated with Calcium Phosphate

Plasmid DNA precipitated with calcium phosphate can efficiently transfect cells in tissue culture, as reported by Graham and Van der Eb in 1973 [135]. More then a decade ago, this technique was also used for *in vivo* gene transfer of viral and plasmid DNA into liver and spleen by either direct inoculation into the tissue bed or intraperitoneal instillation [81–84]. Despite these initial promising results and the ease of preparing these DNA precipitates, this method has largely been supplanted by alternative approaches that are thought to yield superior *in vivo* transfection efficiencies. Nevertheless, this method has recently been employed in preclinical cancer gene therapy studies evaluating introduction of herpes simplex virus thymidine kinase (HSV-TK) into melanoma explants [136]. In these studies, plasmid DNA encoding HSV-TK was precipitated with calcium phosphate and directly injected into established B16 melanoma tumor explants in syngeneic C57/BL mice. After administration of intraperitoneal ganciclovir, treated animals achieved a partial tumor regression.

D. Liposome-Mediated Gene Delivery

Polycationic lipids can be mixed with plasmid DNA to form liposome structures that are thought to fuse with the target cell membrane and thereby mediate gene delivery [137]. Several lipid preparations have been formulated for this application, including mixtures of dioleoyl phosphatidylethanolamine (DOPE) with DOTMA (lipofectin), DOSPA (lipofectamine), DDAB (lipofectace), DOGS (transfectam), DOTAP, DMRIE, and DC cholesterol (reviewed in Felgner *et al.* [138]). This approach can yield very high transfection efficiencies *in vitro* and can also be used for direct *in vivo* gene transfer. Plasmid DNA has been delivered to tumor explants in syngeneic mice by injecting the tumor nodule with liposome/DNA complexes, achieving a transient transfection efficiency of approximately 1–10% [37]. A particular advantage of this approach is the ease of preparing DNA/liposome complexes, the stability of the individual components, and the versatility to transfect a variety of tumor types. The liposome/DNA complex can be directly injected into a palpable tumor nodule [38]. Alternatively, visceral

tumor masses can be directly instilled with liposome/DNA complexes by employing radiologic procedures, such as CAT scans, to identify the location of the tumor and assist in percutaneous tumor injection [139,140]. Alternative approaches include use of bronchoscopy, cystoscopy, endoscopy, or laparoscopy to directly inject liposome/DNA complexes into visualized tumor masses. Liposome/DNA complexes have minimal systemic toxicities [141,142] and can be administered repeatedly to the same patient with expectations of equivalent efficiencies of gene transfer.

Liposome/DNA complexes administered intravenously also can deliver plasmid vectors into multiple tissue types. In 1983, Nicolau *et al.* injected rats intravenously with a plasmid vector encoding rat preproinsulin I complexed with liposomes composed of phosphatidylcholine, phosphatidylserine, and cholesterol [143]. In these studies, radioactive labeled liposomes were shown to be taken up specifically by liver and spleen, and 6 hours after injection treated animals experienced a fall in serum glucose and an increase in serum, liver, and splenic insulin levels relative to control animals. More recently, Zhu and Debs demonstrated gene expression in diverse tissue types, including liver, spleen, kidney, lung, heart, lymph nodes, and bone marrow, following intravenous administration of chloramphenicol acetyltransferase reporter plasmids complexed with liposomes composed of DOTMA and DOPE lipids [144]. Gene expression was detected for up to 9 weeks following gene transfer. This widespread gene delivery raises the possibility of using intravenous administration of liposome/DNA complexes to introduce therapeutic genes in multiple foci of metastatic disease. For example, a study employing a p53 mutant human breast cancer xenogeneic model suggests that intravenous administration of liposome/DNA complexes encoding wild-type p53 may reduce the size of primary tumor explants and decrease the development of metastatic disease to lungs [145]. Although the liposome formulations described above do not specifically target tumor cells, ongoing studies suggest that it may be possible to increase the specificity of liposome-mediated gene transfer by conjugating ligands for cell surface receptors to lipid moieties. In recent studies, receptor-mediated gene transfer *in vitro* has been demonstrated for liposome preparations targeting the folate, erbB-2, transferrin, and mannose receptors [146–153]. Additionally, so-called "stealth" liposomes have been developed to avoid rapid clearance by reticuloendothelial cells following an intravenous injection [148,154–158]. In preclinical models, tumor-bearing animals treated with chemotherapeutic agents encapsulated in stealth liposomes had improved survival compared to control groups treated with free drug alone [157]. Stealth liposomes typically include polyethylene glycol to increase serum half-life, and these preparations pool in tissues, such as tumors, that have increased vascular permeability, resulting in passive targeting of the complexes [159–163]. As much as 3–6% of the dose of DNA was reported to localize in the tumor nodule, although

DNA transfer into tumor cells was inefficient [164,165]. Because liposomes incorporating polyethylene glycol tend to be non-fusogenic, pH-sensitive and kinetically unstable linkages are being developed to reversibly release polyethylene glycol from the lipsome, thereby enhancing their gene transfer properties [166,167]. An active area of current research is to develop targeted and stealthy liposome preparations suitable for intravenous delivery of plasmid constructs for cancer gene therapy.

Several preclinical models have demonstrated antitumor responses when mixtures of liposomes and plasmid vectors have been directly transferred into established tumor explants in syngeneic mice. For example, plasmids encoding the murine class I H-2K^s gene have been complexed with cationic liposomes and injected into established CT26 colon carcinoma (H-2K^d) and MCA 106 fibrosarcoma (H-2K^b) cells. As reported by Plautz and Nabel, a cytotoxic T-cell response to H-2K^s antigen was induced, and animals preimmunized to H-2K^s antigen demonstrated significant antitumor activity, with some animals achieving long-term survival [37]. In addition, this antitumor activity was cell line specific, because animals bearing MCA 106 tumors previously cured following injection with H-2K^s plasmid rejected secondary tumor challenges with parental MCA 106 cells but not syngeneic B16BL/6 melanoma cells. These findings suggest that expression of foreign class I histocompatibility antigens by these tumor cells resulted in recognition of heretofore unrecognized tumor-associated antigens by cytotoxic T cells. This hypothesis would account for the observed efficient tumor elimination and prolonged survival despite the fact that only a modest percentage of the tumor cells were transiently transfected following direct tumor inoculation by DNA/liposome complexes.

In a pilot study at the University of Michigan, Nabel and colleagues have extended their preclinical model to a clinical cancer gene therapy protocol by evaluating liposome-mediated gene transfer of plasmids encoding HLA-B7 in patients with metastatic melanoma. In these studies, liposome/DNA complexes were directly injected into subcutaneous, nodal, and visceral masses, and one out of the first five patients evaluated demonstrated a significant response [38]. These encouraging findings have led to several active trials evaluating the expression of a bicistronic plasmid encoding HLA-B7 and β_2-microglobulin in patients with metastatic colon cancer, renal cancer, and melanoma [139,140,168]. In these trials, plasmid DNA complexed with liposomes composed of DIMRIE and DOPE lipids are directly injected into tumor masses. In initial reports, HLA-B7 gene expression has been shown in tumor biopsies after gene transfer [139,140], and antitumor immunity has been observed in local tumor-infiltrating lymphocytes [169]. These clinical trials are currently in progress, and additional data regarding generation of T-cell immunity to HLA-B7 target cells, tumor responses, survival, and toxicities of the treatment are pending.

Due to a lack of cytosine methylase activity in bacteria, typical preparations of plasmid DNA have unmethylated cytosine nucleotides (CpG islands). Such unmethylated CpG islands possess potent adjuvant and immunomodulatory effects and can produce locally elevated levels of cytokines [170–177]. Such unmethlyated CpG islands play a significant role in the effectiveness of DNA vaccines administered as an intramuscular or intradermal injection by stimulating a potent TH1 response. When administered systemically, plasmid DNA alone does not produce systemic levels of cytokines, but complexes of DNA and cationic liposomes can produce elevated levels of TNF-α, IL-12, and IFN-γ [178]. Moreover, an antitumor effect was observed when mice bearing pulmonary metastases received intravenous injections of liposome/plasmid DNA complexes lacking a therapeutic gene; these effects were comparable to liposome/DNA complexes encoding IL-12 or p53 [178]. Similar results were observed in subcutaneous and intraperitoneal syngeneic tumor models [178,179]. This potent antitumor effect was found in immunocompetent mice but not SCID, athymic, or SCID/Beige mice [178,179]. Moreover, these antitumor effects of liposome/DNA complexes lacking a therapeutic gene were inhibited by prior methylation of CpG motifs in the plasmid using SSI methylase [178]. These interesting results may explain, in part, the ability of intravenously administered cationic liposome/DNA complexes to generate greater antitumor effects than predicted based on their more modest gene transfer efficiency.

E. Ligand/DNA Conjugates

Negatively charged plasmid DNA molecules and polycations, such as poly(L-lysine), can form complex structures consisting of either unimolecular or multimolecular complexes (with respect to the DNA) [77,78]. To enable efficient and cell-specific gene transfer, the poly(L-lysine) polymer can be modified by covalently attaching ligands that can subsequently bind to specific cellular receptors [180]. If the DNA/poly(L-lysine) complex contains a suitable ligand, then the DNA/poly(L-lysine) complex can be internalized in the cell when the receptor undergoes endocytosis. Most of these early DNA/poly(L-lysine) formulations were multimolecular complexes, approximately 100–200 nm in diameter [77], which may have limited their ability to enter cells via receptor-mediated endocytosis. Additionally, efficient expression of the internalized plasmid requires several additional steps, including exit from the endosome prior to destruction of the DNA by fusion of the endosome with lysosomes and transfer of the plasmid DNA to the nucleus [77,181].

Initial formulations of poly(L-lysine)/DNA complexes for *in vivo* gene transfer targeted the liver asialoglycoprotein receptor for gene delivery using asialoorosomucoid covalently linked to poly(L-lysine) [180]. Gene expression was transient, although preferential gene transfer to the liver was observed. In later studies, gene expression was improved by performing a partial hepatectomy in association with receptor-mediated gene transfer [182]. Further improvements in gene expression were achieved by using endosomolytic agents, such as defective adenovirus particles or peptides derived from the N-terminal region of influenza virus hemagglutinin HA-2 protein, to enable transferrin-conjugated poly(L-lysine)/DNA complexes to exit the endosome and enter the cytoplasm for eventual transfer to the nucleus [183–185]. This modification has been shown to achieve transient gene expression in lung following direct instillation of ligand/DNA complexes into the airway of rats [186]. Gene transfer *in vitro* has been demonstrated in primary intestinal mucosal cells and the transformed Caco$_2$ colon adenocarcinoma cell line [187], suggesting an approach for gene delivery into tumor cells. However, most of these studies employed rapidly dividing cells in which the nuclear membrane barrier is broken down during mitosis, thereby permitting plasmid to enter the nucleus. Because the nuclear membrane severely restricts transfer of large DNA complexes into the nucleus [188–196], the relevance of these findings for *in vivo* gene transfer in humans is questioned.

Recent studies have focused on formulations of condensed, unimolecular DNA/poly(L-lysine) complexes that efficiently enter the cell via receptor-mediated endocytosis [77,78]. Such complexes consist of a single molecule of plasmid DNA and are spheroids approximately 15–20 nm in diameter; these preparations achieve efficient and specific gene transfer following intravenous gene delivery. For example, condensed, unimolecular galactosylated DNA/poly(L-lysine) complexes encoding human factor IX cDNA efficiently target the hepatic asialoglycoprotein receptor, and transfected rats have detectable human factor IX in their serum for up to 140 days [78]. This result was achieved without the need for partial hepatectomy. Condensed DNA/poly(L-lysine) complexes have also been prepared by coupling the FAB fragment of an antibody recognizing the polymeric immunoglobulin receptor [79]. These complexes have a diameter of approximately 25 nm and yield efficient gene transfer into target rat lung epithelial cells following intravenous administration. Approximately 18% of tracheal epithelial cells were transfected as monitored by expression of the beta-galactosidase marker gene following a single intravenous injection of 300 μg of plasmid DNA formulated in these condensed complexes [79]. Expression was specific for tissues expressing the polymeric immunoglobulin receptor. In other studies, the mannose receptor on macrophages has been targeted for *in vivo* gene delivery by formulating condensed mannosylated DNA/poly (L-lysine) complexes [80]. In these studies, efficient and specific gene transfer was shown to correlate with the formulation of unimolecular, condensed DNA/poly(L-lysine) complexes. Recently, the serpin enzyme complex receptor (SECR) also has been targeted for gene transfer using con-

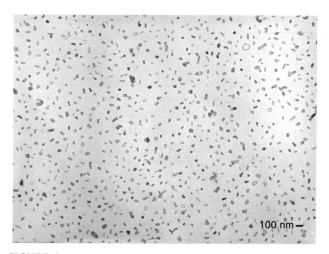

FIGURE 1 Electron micrograph of condensed DNA complexes in normal saline. A 6.7-kbp plasmid was compacted using a polyethylene glycol-substituted polymer consisting of cysteine followed by 30 lysines. Spheroidal particles were observed having an average size of 20 ± 2.5 nm (as assayed by dynamic light scattering analysis). The bar represents 100 nm.

densed DNA/poly(L-lysine) particles [197–199]. Gene transfer *in vitro* correlated closely with the level of cell surface SECR expression, and gene expression *in vivo* following an intravenous injection correlated with SECR-expressing tissues. Together, these studies suggest that coupling poly (L-lysine) to ligands that recognize cellular receptors preferentially expressed by tumor cells may provide an efficient and specific approach for *in vivo* gene transfer of plasmid vectors into cancer cells.

Recently, unimolecularly compacted plasmid DNA complexes have been optimized for stability in physiologic saline and serum at 37°C [200]. These complexes consist of a single molecule of DNA and sufficient polylysine carrier molecules to prepare essentially charge-neutral particles. Based on electron microscopy (Fig. 1) and dynamic light scattering, these complexes have the minimum possible size as predicted by the partial specific volume of DNA and polycation [201]. These preparations of compacted DNA readily transfect nondividing, postmitotic cells [181] and yield very high levels of transgene expression when directly instilled into the lung [202]. Modifications of these complexes to include ligands for receptors that are highly and preferentially expressed by tumor cells may result in an effective and nontoxic gene transfer platform for systemic cancer therapy.

V. PLASMID EXPRESSION VECTORS

Unlike viral-based infectious vectors, plasmid vectors must be introduced into cells by specific gene transfer technologies, as reviewed earlier. Once introduced into a cell, however, plasmids have specific advantages compared to viral vectors, including: (1) no potential to be infectious;

(2) levels of gene expression per cell equivalent to other viral vectors that persist as extrachromosomal elements (see Table 1); (3) lack of immunogenicity (allowing for multiple treatments) [141]; (4) lack of toxicity following intravenous injection [142]; (5) low probability of integration during transient periods of expression, thereby reducing potential for insertional mutagenesis; (6) easy coupling to liposome or receptor-mediated gene delivery systems; and (7) long-term stability, requiring no special preparation or storage requirements. Modifications in vector design, including tissue-specific promoters, inducible promoters, and elements enabling the plasmid to replicate extrachromosomally in tumor cells, further enhance the safety of plasmid vectors and significantly augment the level of expression observed in transiently transfected tumor cells.

A. Tissue-Specific Promoters

The cytomegalovirus (CMV) immediate-early promoter is often utilized in gene therapy studies due to its high level of activity in diverse tissue types [203,204]. Although it is desirable to express target genes at high levels in tumor cells, transcriptionally active promoters, such as CMV, will also direct high-level expression in unintentionally transfected normal cells following *in vivo* gene transfer. To approach current limitations in the ability to specifically target a tumor cell for gene transfer, tissue-specific promoters can be employed that limit expression of the therapeutic gene to tumor cells and normal cells of a specific lineage. Many tissue-specific promoters have been developed [205,206], and a short list includes the insulin promoter (β islet cells of the pancreas) [207], elastase promoter (acinar cells of the pancreas) [208], whey acidic protein promoter (breast) [209], tyrosinase promoter (melanocytes) [210], tyrosine hydroxylase promoter (sympathetic nervous system) [211], neurofilament protein promoter (brain neurons) [212], glial fibrillary acidic protein promoter (brain astrocytes) [213], Ren-2 promoter (kidney) [214], collagen promoter (connective tissues) [215], α-actin promoter (muscle) [216], von Willebrand factor promoter (endothelial cells) [217], α-fetoprotein promoter (hepatoma) [218], albumin promoter (liver) [218], surfactant promoter (lung) [219], CEA promoter (gastrointestinal tract, tumors of colon, breast, lung) [220], uroplakin II promoter (bladder) [221], T-cell receptor promoter (T lymphocytes) [222], immunoglobulin heavy-chain promoter (B lymphocytes) [223], prostatic-specific antigen promoter (prostate) [224], and protamine promoter (testes) [225].

Tissue-specific promoters have been utilized in gene therapy studies to evaluate tumor-specific killing mediated by expression of the herpes simplex thymidine kinase gene followed by exposure to ganciclovir. For example, use of the albumin and α-fetoprotein promoter in retroviral constructs encoding HSV-TK specifically killed hepatoma cell lines but had marginal activity in other tumor cells derived from breast,

colon, or skin [218]. In other studies, Vile and Hart recently reported use of plasmid DNA encoding HSV-TK transcriptionally regulated by the murine tyrosinase promoter to treat B16 melanoma tumors growing as subcutaneous explants in syngeneic mice [136]. Established tumors, approximately 4 mm in diameter, were directly injected with 20 μg of calcium-phosphate-precipitated plasmid DNA, and 2 days later mice were administered daily injections of intraperitoneal ganciclovir for 5 days. A statistically significant reduction in tumor size was observed compared to animals not receiving ganciclovir. No local toxicity was observed in the tissues adjacent to the tumor explant, as expected based on the tissue specificity of the tyrosinase promoter. In similar studies, the CEA promoter also has been utilized to control transcription of HSV-TK [220]. CEA-expressing lung cancer cell lines were highly sensitive to ganciclovir *in vitro* and *in vivo* following gene transfer of these constructs, whereas non-CEA-expressing lung cell lines were resistant to ganciclovir following gene transfer.

Another opportunity to specifically target tumor cells for gene expression is to utilize promoter elements that become activated in chemotherapy-resistant tumor cells. Based on the observation that the metallothionein promoter becomes activated in cisplatin-resistant ovarian carcinoma cells, plasmid DNA encoding the HSV-TK gene transcriptionally controlled by the metallothionein promoter has been introduced into cisplatin-sensitive and -resistant ovarian carcinoma cell lines followed by treatment with ganciclovir [226]. No cytotoxicity was apparent in cisplatin-sensitive, parental 1A9 ovarian carcinoma cells, whereas a cisplatin-resistant subclone was efficiently killed by this treatment. These results suggest a specific approach for gene therapy of cisplatin-resistant ovarian carcinoma cells and underscore the potential of using tumor-specific promoter elements.

B. Inducible Promoters

In addition to using tissue-specific promoters to minimize target gene expression in unintentionally transfected cells, the timing and duration of gene expression also can be modulated by employing inducible promoters that can be externally controlled. Several inducible systems have been developed, and a few appear to be appropriate for use in clinical gene therapy trials due to lack of apparent toxicity and demonstrated effectiveness *in vivo*. For example, a tetracycline-controlled expression system has been developed by Gossen and Bujard [227]. A novel hybrid transcriptional transactivation protein was constructed by ligating the ligand and DNA binding domains of the bacterial tetracycline repressor gene to the C-terminal region of the herpes virus VP16 transcriptional regulator protein containing its transactivation domain. In conjunction with reporter genes containing a heptad repeat of the consensus binding domain of the tetracycline repressor upstream of a minimal core element of the cytomegalovirus immediate-early promoter, tetracycline-controlled expression has been demonstrated *in vitro* and *in vivo* in transgenic mice [227–229]. The hybrid transcriptional transactivator binds to the tet operon in the absence of tetracycline, whereas tetracycline efficiently dissociates the transcription factor from its binding site. Hence, efficient reporter gene expression was observed in the absence of tetracycline, whereas transcription is virtually eliminated in the presence of 0.1–1 μg/mL of tetracycline, a concentration readily attainable in humans. This system has also been used to transiently express target genes following direct *in vivo* gene transfer of these plasmid constructs in rat myocardium [230]. More recently, a tetracycline-on system has been developed utilizing specific point mutations in the tetracycline repressor component of the hybrid transcriptional transactivator [231]. In other studies, tetracycline-controlled transcriptional repressors have been constructed by linking the KRAB transcriptional repressor downstream from the DNA binding domain of the tetracycline repressor [232].

O'Malley and colleagues have also described a novel, regulated transcriptional activator that consists of a truncated ligand binding domain of the human progesterone receptor (which binds tightly to the synthetic progesterone antagonist RU486 but binds very poorly to progesterone), the DNA binding domain of the yeast transcriptional activator GAL4, and a C-terminal fragment of the herpes simplex VP16 transcriptional regulator protein [233]. In conjunction with a target gene containing four copies of the consensus GAL4 binding site, gene expression was activated only in the presence of RU486, and regulation was achieved both *in vitro* and *in vivo* [233,234]. A similar gene switch has been developed by Delort and Capecchi that utilizes different domains of the progesterone receptor and GAL4 binding protein [235]. Wang *et al.* also have developed an inducible repressor system by substituting the KRAB transcriptional repressor domain for the VP16 transactivation domain [236]. The specificity of these inducible systems is dependent upon the presence of the GAL4 consensus sequence upstream of the target gene of interest. Because GAL4-activated genes are not currently known to be present in the human genome, induction of gene expression *in vivo* is predicted to solely activate the therapeutic target gene. In addition, the presence of endogenous progesterone receptors in tumor cells would not be expected to interfere with this expression system.

Other inducible transcriptional activation systems have been developed to control gene expression. These include the *Drosphila ecdysone* receptor gene switch and the rapamycin-controlled transactivation system [237,238]. The latter system utilizes two transcription factor fusion proteins that share a high-affinity binding site for rapamycin. The first element consists of the rapamycin binding protein, FKBP12, fused to the ZFHD1 DNA binding protein. The second element consists of a rapamycin binding protein, FRB, fused to the carboxy terminal portion of the NF-κB transcriptional activator protein. In the presence of rapamycin, these two fusion proteins bind to one another and reconstitute an active transcription factor for target reporter genes placed upstream of a minimal CMV promoter region carrying 12 binding sites for ZFHD1. Highly efficient and specific rapamycin-controlled gene expression has been demonstrated both *in vitro* and in *in vivo* preclinical models [238].

Several issues need to be addressed when considering any of these systems for cancer gene therapy. Although predicted to specifically inactivate or activate the transcription of target genes downstream from their respective consensus binding sequences in the presence of drug, further experimental testing is required to confirm that endogenous cellular genes, such as tumor suppressor genes and proto-oncogenes, are not unexpectedly regulated by these hybrid transcriptional repressors and transactivators, respectively. In addition, these hybrid transcriptional control proteins may very well generate antigenic peptide sequences derived from the bacterial tetracycline repressor, the yeast Gal4 protein, and the herpes simplex virus VP16 protein. An immune response may therefore be generated against tumor and normal cells following *in vivo* gene transfer. Although the toxicity of this immune response may be minimal, it may conceivably limit the duration of target gene expression in tumor cells following repetitive treatments.

Another example of an inducible promoter system utilizes transcriptional control elements that become active following radiation-induced injury. As developed by Weichselbaum and colleagues, the radiation-responsive consensus sequence from the early growth response (EGR-1) gene promoter was ligated upstream from a gene known to significantly enhance radiation injury, TNF-α [239]. This plasmid construct was electroporated into a hematopoietic cell line, HL525, known to be deficient in radiation-induced expression of TNF-α. These gene-modified HL525 cells were injected into established radiation-resistant human squamous carcinoma xenografts in nude mice. Following radiation exposure to the tumor explant, the squamous carcinomas regressed and most of the animals were apparently cured. In contrast, control animals bearing squamous tumor explants that received radiation therapy alone, radiation plus HL525 cells transfected with the neomycin resistance gene, or TNF-α transfected HL525 cells without radiation all developed progressive tumor growth. In other studies, a quartad repeat of specific transcriptional elements within the EGR-1 gene has been combined with the CMV immediate-early core promoter to produce a radiation-responsive chimeric promoter [240]. These studies demonstrate the ability to induce gene expression *in vivo* by focused application of radiation and gene therapy in specific areas known to be involved by tumor.

Several groups have developed inducible transcriptional regulators that are activated by endogenous metabolic conditions or physical stimuli that can be directly applied to the tumor. Because tumor masses often have hypoxic regions [241,242], several groups have developed tumor-activated transcriptional regulators that are stimulated by low oxygen content [243,244]. A number of genes are upregulated by hypoxia, including erythropoietin, vascular endothelial growth factor, and some glycolytic enzymes, and specific hypoxia response elements (HREs) have been identified in 5′ or 3′ flanking regions [245]. Hypoxia-induced transcriptional regulators have been developed using five copies of HREs in conjunction with minimal promoters from E1b or CMV [244,245]. Using the CMV chimeric system, hypoxic conditions induced over a 500-fold increase in gene expression, achieving a level of gene expression comparable to the complete CMV promoter [245]. Using an alternative approach, transcriptional regulators have been developed based on activation of endogenous heat shock genes by hyperthermia [246]. Employing a heat shock gene (*hsp70*) promoter, transgene expression was induced over 10,000-fold in tissue culture cells by a temperature elevation of 39°–43°C [247]. Moreover, growth of syngeneic melanoma tumor explants was delayed following a local injection of adenoviral vectors encoding IL-12 regulated by hsp70 only in limbs treated with hyperthermia.

Finally, cytoplasmic expression systems have been developed that utilize bacteriophage T7 RNA polymerase to control transcription of transgenes regulated by the T7 promoter. As reported by Gao and Huang, co-delivery of purified T7 RNA polymerase and a plasmid containing a T7 promoter upstream of the CAT reporter gene resulted in rapid, high-level transgene expression that lasted about 30 hours [248]. Of note, this cytoplasmic expression system does not require plasmid DNA to enter the nucleus to be transcribed. Further modification of this T7 expression system, as reported by Gao *et al.* [249] and Chen *et al.* [250], utilized a bicistronic plasmid containing a first T7 promoter upstream of the T7 RNA polymerase gene (T7/T7 autogene) and a second T7 promoter upstream of various reporter genes. In this fashion, T7 RNA polymerase protein initiates transgene transcription, and newly synthesized T7 mRNA replenishes and maintains levels of T7 RNA polymerase. In these studies, expression of transfected reporter genes was maintained for up to 5–6 days *in vitro*. Furthermore, direct injection of a T7/T7 autogene vector system into mouse liver, muscle, brain, and connective tissue generated up to 200-fold higher levels of luciferase reporter activity than nuclear gene expression vectors [251]. In subsequent studies, T7 autogene bicistronic vectors encoding HSV-TK have been introduced into established human 143B osteosarcoma xenografts in nude mice, and tumor regression was observed in animals receiving intraperitoneal doses of ganciclovir [252]. More recently, the requirement for co-delivery of T7 RNA polymerase protein was bypassed by designing a plasmid vector incorporating the CMV and T7 promoters upstream of the T7 RNA polymerase gene [253]. In this system, initial CMV-based transcription of T7 RNA polymerase results in further enhancement of T7 polymerase expression by autoregulating the T7 promoter, and high levels of reporter gene expression were maintained for at least 7 days post gene transfer.

C. Replicating Plasmid Vectors: Episomes

Expression of genes encoded by plasmids is generally transient, unless specific modifications are made to enable the plasmid to efficiently integrate into genomic DNA or to replicate in human cells. In dividing tumor cells, plasmid-mediated gene expression falls to very low levels by several days after gene transfer. This decline in gene expression is mediated by several factors, including a logarithmic decline in the percentage of transfected cells during replication of the target population (as the plasmid does not replicate in human cells) [254], potential loss of the transgene by nuclease destruction or by partitioning to non-nuclear compartments, and promoter inactivation by cytokines, chromatin remodeling, or methylation [255–262].

One approach to maintain plasmid copy numbers in transfected tumor cells is to incorporate sequences from human DNA that enable the plasmid to replicate extrachromosomally. Although sequence-specific human DNA origins have been difficult to clone, Calos and colleagues have identified DNA fragments that replicate semiconservatively during S phase of the cell cycle when incorporated into plasmid vectors [263,264]. These vectors replicate once per cell cycle, and the plasmid copy number per cell is therefore dependent upon the initial transfection conditions. In these studies, the size of the DNA fragment is an important factor in conferring replication competence, with random human DNA fragments over 10–15 kb in length having significant activity [265]. Similar sizes of randomly chosen yeast DNA also are replication competent in human 293 cells [266], and large fragments of bacterial DNA have detectable although minor activity [263]. Plasmids containing these DNA fragments will replicate for several months in human cells if the vector additionally includes a portion of the Epstein–Barr virus (EBV) DNA origin (including a tandem array of repeated sequences) and if the transfected cells express the EBV early gene product, EBNA-1. EBNA-1 binds to these tandem repeat sequences and retains plasmid DNA in the nucleus of dividing cells, thereby conferring stable maintenance of the episomal plasmid [267]. In short-term assays, however, these DNA fragments alone enable plasmids to replicate transiently in human cells over several generations, although the copy number of these vectors is low [263]. The expression characteristics of plasmids containing such autonomously replicating human sequences and the potential role of these vectors for cancer gene therapy are currently undefined.

In other studies, human artificial chromosomes have been assembled *in vivo* by transfecting cells with specific fragments of telomeric and centromeric DNA [268–273]. These separate DNA fragments recombine within the cell to form mini-chromosomes approximately 6–10 Mb in size, and stable vertical transfer of these extrachromosomal elements has been demonstrated over multiple generations *in vitro*. These properties make them well suited for introduction into human stem cells, including *ex vivo* gene transfer into hematopoietic progenitor cells. However, the ability to isolate large quantities of homogeneous, unrearranged artificial chromosomes, transfect them into human cells, and then achieve transfer into the nucleus remains to be demonstrated.

Another approach to increase both the peak level and duration of gene expression mediated by plasmid vectors is to include sequences from DNA viruses that enable the plasmid to replicate in human cells. Two elements are required: (1) a viral DNA origin of replication and (2) a viral early gene product. The viral DNA origin alone is not functional in human cells. During the life cycle of DNA viruses, including Epstein–Barr virus and BK virus, an early gene product is synthesized that directly binds to the viral DNA origin [274,275]. This protein/DNA complex is recognized by the infected human cell as a functional DNA origin, and the virus is able to replicate its DNA. In a similar fashion, plasmids encoding a viral DNA origin and its corresponding early gene product can replicate in human cells. Replicating episomal plasmid vectors have two predicted advantages compared to standard plasmid vectors for cancer gene therapy applications: (1) high-level gene expression due to vector amplification and (2) maintenance of gene expression in transiently transfected cells due to efficient vertical transfer of the episome during tumor cell division. These principles are summarized in Table 3 and illustrated in Fig. 2.

Plasmid vectors that replicate in human cells have been constructed from several viruses, including Epstein–Barr virus, BK virus, human papillomavirus, and SV40 [274–277]. For example, EBV episomes replicate in lymphoid cells, achieving a steady-state copy number of approximately 10–50 copies [274]. These plasmids can be stably maintained in cells for many months, and EBV-based vectors containing over 200-kbp inserts have been characterized [278–281]. Constructs derived from BK virus replicate in a wide range of cell types [275,283–284], and stable bladder cell transfectants have been characterized that have approximately 150 copies per cell [284]. In these studies, gene expression was proportional to the episomal plasmid copy number. Additionally, gene expression was maintained in a population of unselected, transiently transfected cells for at least one week

following gene transfer, whereas nonreplicating plasmid-based gene expression fell exponentially at a rate predicted by the doubling time of these cells. In studies by Thierry *et al.*, mice receiving intravenous injections of liposome complexes of BK virus episomes generated transgene expression in multiple tissues up to 3 months post-injection; in contrast, a short duration of expression was observed in animals dosed with nonreplicating plasmids [285]. In summary, the predicted advantages of high-level, maintained gene expression of replicating episomal vectors compared to standard plasmids have been observed using BK virus episomes.

A key distinction among the multiple types of episomal plasmids derived from DNA viruses is their ability to replicate once or multiple times per cell cycle. Some episomal plasmids, including those derived from EBV and BK virus [274,284], replicate once per cell cycle. In this circumstance, the plasmid copy number per cell will never be higher than the level achieved on the day of gene transfer. In circumstances where *in vivo* gene transfer is desired, this replication feature significantly limits transgene expression, as only one or several plasmids per transfected cell nucleus likely can be attained. In contrast, episomes derived from SV40 virus replicate multiple times per cell cycle [286]. In this scenario, transgene expression is not limited by the initial gene transfer efficiency, and episomal replication can generate high-level plasmid copy numbers in dividing cells, thereby optimizing transgene expression.

Despite the clear advantages of replicating plasmid vectors, a significant obstacle to their development is the transformation properties associated with suitable viral early genes that possess replication transactivator function. For example, the Epstein–Barr virus replication transactivator, EBNA-1, has transformation properties in transgenic mice [287]. In addition, papovavirus early gene products, including the large T antigens from BK virus and SV40 virus, have transformation properties thought to be primarily mediated by binding to host tumor suppressor gene products, including p53, RB, and RB-related proteins, such as p107 and p130 [288–291].

To develop replicating episomal vectors for human gene therapy, our laboratory has recently developed a safety-modified, SV40, large T antigen (107/402-T) that lacks detectable binding to human tumor suppressor gene products yet preserves replication competence (Fig. 3A) [286]. This large T antigen mutant has specific point mutations in codons 107 and 402 and lacks detectable binding to p53, RB, and p107 proteins (Fig. 3B and C, Table 4). Episomal vectors incorporating the 107/402-T replicon amplify in a wide range of human and simian cell lines but not in dog or rodent cells (Table 5). In addition to gene transfer *in vitro*, we have observed that 107/402-T-based episomal vectors replicate in human tumor cells following direct *in vivo* gene transfer into human tumor xenografts in nude mice. An example of the replication activity of 107/402-T episomes in human hepatoma (Hep G2) and bladder (HT-1376) cell lines is shown in Fig. 3D and E. Based on transient gene transfer

TABLE 3   Features of Standard Plasmid and Replication-Competent Episomal Vectors

Expression vector	Peak level of gene expression	Sustained expression in dividing tumor cells
Standard plasmid	Low	No
Replication-competent episome	High	Yes

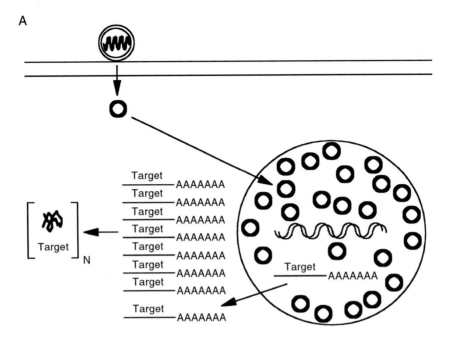

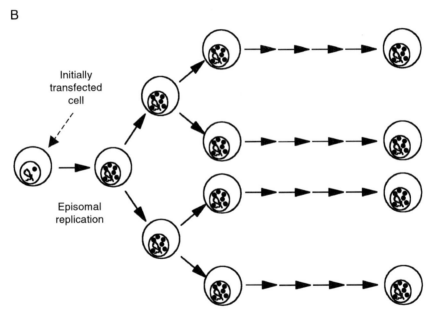

FIGURE 2   (A) Replicating episomal plasmids yield high levels of target gene expression due to vector amplification. Depicted are multiple copies of an episomal plasmid in the nucleus of a transfected cell which have accumulated because of vector replication. The increased copy number of the expression vector produces high levels of target gene mRNA and consequently high levels of target gene protein. (B) High-level gene expression is maintained in transiently transfected tumor cells due to efficient vertical transfer of the episome (•) as these cells divide.

efficiencies and content of genomic DNA per cell, 107/402-T episomes achieve peak copy numbers of approximately 1400 in HT-1376 cells, and 25,000 in Hep G2 cells (Table 5). As a consequence of vector amplification, we observe significantly enhanced levels (>100-fold) of reporter gene ex-

pression when comparing this replicating episomal expression system to analogous, nonreplicating expression vectors (Fig. 4). When transferred into log-phase tumor cells, high levels of reporter gene expression was maintained for at least 1–2 weeks due to efficient vertical transfer of the

TABLE 4   Binding of Wild-Type and Mutant SV40 Large T Antigens to RB, p107, and p53 Tumor Suppressor Gene Products

Tumor suppressor gene	T^a	107-T	402-T	107/402-T
RB	100	0.03	67	0.07
p107	100	0	79	0
p53	100	36.2	0	0

^{a} Shown is the percentage of binding of T antigen mutants compared to wild-type T antigen.

Source: From Cooper, M. J. *et al.* (1997). *Proc. Natl. Acad. Sci. USA* **94**, 6450–6455. With permission.

replicating episomal expression vectors. In ongoing studies, we have developed an externally controlled replicon switch that employs a novel fusion gene consisting of 107/402-T and a portion of the human progesterone receptor [292].

Vector amplification occurs only in the presence of RU486, an FDA-approved synthetic progesterone antagonist. Administration of RU486 to the cancer patient on the day of gene transfer may be sufficient to permit a short burst of vector amplification, thereby boosting levels of transgene expression.

VI. FUTURE DIRECTIONS

Clinical cancer gene therapies have only recently been initiated, and results are currently very preliminary. At present, the optimal delivery system, expression vector, and target genes for a given tumor type are entirely unknown. Success of this modality will ultimately depend upon the ability to express the therapeutic gene of interest at high levels, and being able to target the tumor cell for gene delivery will minimize toxicities. Incorporation of tissue-specific and inducible

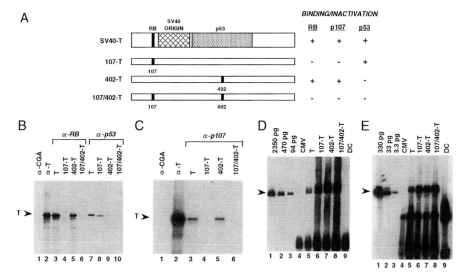

FIGURE 3   107/402-T lacks binding to human tumor suppressor genes and is replication competent. (A) Point mutations in replication-competent, safety-modified, SV40, large T antigen mutants. Highlighted are domains of T antigen that bind to RB, p53, and the SV40 DNA origin. The codon 107 mutation substitutes lysine for glutamic acid, and the codon 402 mutation substitutes glutamic acid for aspartic acid [286]. (B and C) Co-immunoprecipitation analysis of binding of wild-type and mutant T antigens to human tumor suppressor gene products. 2×10^5 dpm of *in vitro* translated T antigens were mixed with CV-1 extracts overproducing human RB protein and anti-RB monoclonal antibody G3-245 (B, lanes 3–6), p53 and anti-p53 monoclonal antibody 1801 (B, lanes 7–10), and p107 and anti-p107 monoclonal antibody SD9 (C, lanes 3–6). As controls, wild-type T antigen is immunoprecipitated with either anti-chromogranin A monoclonal antibody LKH210 (lane 1) or anti-T-antigen monoclonal antibody 416 (lane 2). (D) 107/402-T is replication competent. Hep G2 hepatoma cells (D) were transfected with wild-type and mutant T antigen expression vectors, and total cellular DNA was harvested 2 days post transfection. DNA samples were sequentially digested with ApaI to linearize vector DNA, and then DpnI to distinguish amplified DNA from the input DNA used to transfect these cells. Because human cells lack adenine methylase activity, newly replicated DNA is resistant to digestion by DpnI. Hence, presence of unit-length, linearized plasmid DNA, as indicated by the arrow, demonstrates newly replicated episome. Hybridization probe: pRC/CMV.107/402-T. (E) To evaluate amplification of a cotransfected plasmid in concert with T antigen episomes, HT-1376 bladder carcinoma cells were transfected with T antigen expression vectors and a reporter replication plasmid containing the SV40 DNA origin, pSV2CAT. DNA harvested from cells 4 days post gene transfer was sequentially digested with BamHI to linearize pSV2CAT and then with DpnI. Hybridization probe: BamHI-HindIII CAT fragment. CMV, pRC/CMV transfectants (no T antigen); DC, DpnI digestion control consisting of 5 μg of genomic DNA and 2 ng of either pRC/CMV.107/402-T (D, lane 9) or pSV2CAT (E, lane 9). (From Cooper, M. J. *et al.* (1997). *Proc. Natl. Acad. Sci. USA* **94**, 6450–6455. With permission.)

TABLE 5   Replication Activity of 107/402-T Based Episomes in Human
and Animal Cell Lines

Species	Cell line	Type	Copy number/cell[a]
Human	HT-1376	Bladder	1400
	5637	Bladder	100,000
	MCF-7	Breast	8600
	T98G	Brain	25,000
	SW480	Colon	78
	Hs68	Fibroblast	82
	Hep G2	Hepatoma	25,000
	NCI-H69	Lung	9000
	NCI-H82	Lung	1200
	NCI-H146	Lung	2200
	RAJI	Lymphoma	7000
Simian	CV-1	Kidney	11,000
Dog	MDCK-2	Kidney	<1
	D17	Osteosarcoma	<1
Hamster	BHK	Kidney	<1
	V79	Lung	35
Rat	PC12	Pheochromocytoma	<1
Mouse	F9	Embryonal carcinoma	<1
	3T3	Fibroblast	<1

[a]Peak copy number achieved between days 2 and 6.
Source: From Cooper, M. J. *et al.* (1997). *Proc. Natl. Acad. Sci. USA* **94**, 6450–6455. With permission.

promoters in the vector design will likely permit appropriate control of vector expression and assist in limiting tumor cells for target gene expression. Coupling receptor-mediated gene transfer technologies with safety-modified, replicating episomal vectors may yield high-level gene expression in targeted tumor cells.

Gene therapy may have its most significant impact on patient survival when administered in an adjuvant setting, when tumor burden is at a minimal level. This may be particularly important for gene therapy approaches that attempt to stimulate the immune system to eliminate tumor cells. Appropriately, initial clinical trials are administering gene therapy reagents to patients with either advanced, metastatic cancer or with tumors having a poor prognosis based on local tumor growth, as in glioblastoma multiforme. Current trials will require careful analysis to develop second-generation studies targeting high-risk groups having a lower tumor burden, such as patients with stage II breast cancer having greater than 10 positive axillary lymph nodes or patients with colon, bladder, or lung cancer having local positive lymph node involvement.

The optimal gene targets for a given type of malignancy are unknown. At present, cancer gene therapies are focused on introducing genes in tumor cells to modulate the immune

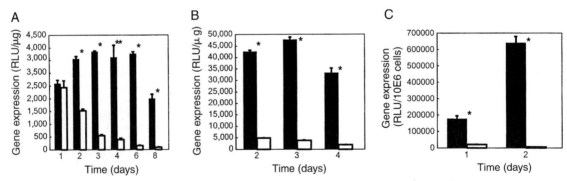

FIGURE 4   Episome-based gene expression in HT-1376 (A), Hep G2 (B), and RAJI (C). Cells were cotransfected with pRSV*lacZII* and either pRC/CMV.107/402-T (solid bars) or pRC/CMV (open bars). Shown are representative results from at least two separate experiments. Significance was determined using an unpaired, one-tailed Student's *t* test; *, *p* < 0.0001; **, *p* = 0.0001. (From Cooper, M. J. *et al.* (1997). *Proc. Natl. Acad. Sci. USA* **94**, 6450–6455. With permission.)

system to achieve antitumor immunity, induce susceptibility to exogenously administered prodrugs, block oncogene expression, express wild-type tumor suppressor gene products, or express chemotherapy resistance genes in normal blood progenitor cells, thereby improving the safety and tolerability of cytotoxic agents. It is anticipated that the success of any of these approaches will be critically dependent upon the type of gene transfer and expression technologies employed. Nevertheless, the potential exists for selection of patient-specific target genes based upon a molecular genetic characterization of gene mutations and an evaluation of the determinants of immunogenicity and tumor tolerance. The promise of cancer gene therapy will be achieved when such tumor-specific analysis is incorporated in the design of clinical trials and highly efficient and specific gene transfer and expression systems are developed.

Acknowledgments

This work was supported in part by National Institutes of Health grants RO1CA72737, R43CA88629, R43IP00005, and Copernicus Therapeutics, Inc.

References

1. Friedmann, T. (1989). Progress toward human gene therapy. *Science* **244**, 1275–1281.
2. Miller, A. D. (1992). Human gene therapy comes of age. *Nature* **357**, 455–460.
3. Anderson, W. F. (1992). Human gene therapy. *Science* **256**, 808–813.
4. Mulligan, R. C. (1993). The basic science of gene therapy. *Science* **260**, 926–932.
5. Moolten, F. L. (1986). Tumor chemosensitivity conferred by inserted herpes thymidine kinase genes: paradigm for a prospective cancer control strategy. *Cancer Res.* **46**, 5276–5281.
6. Borrelli, E., Heyman, R., His, M. *et al.* (1998). Targeting of an inducible toxic phenotype in animal cells. *Proc. Natl. Acad. Sci. USA* **85**, 7572–7576.
7. Culver, K. W., Ram, Z., Wallbridge, S. *et al.* (1992). *In vivo* gene transfer with retroviral vector-producer cells for treatment of experimental brain tumors. *Science* **256**, 1550–1552.
8. Freeman, S. M., Abboud, C. N., Whartenby, K. A. *et al.* (1993). The "bystander effect": tumor regression when a fraction of the tumor mass is genetically modified. *Cancer Res.* **53**, 5274–5283.
9. Barba, D., Hardin, J., Sadelain, M. *et al.* (1994). Development of antitumor immunity following thymidine kinase-mediated killing of experimental brain tumors. *Proc. Natl. Acad. Sci. USA* **91**, 4348–4352.
10. Smythe, W. R., Hwang, H. C., Amin, K. M. *et al.* (1994). Use of recombinant adenovirus to transfer the herpes simplex virus thymidine kinase (HSVtk) gene to thoracic neoplasms: an effective *in vitro* drug sensitization system. *Cancer Res.* **54**, 2055–2059.
11. Chen, S. H., Shine, H. D., Goodman, J. C. *et al.* (1994). Gene therapy for brain tumors: regression of experimental gliomas by adenovirus-mediated gene transfer *in vivo. Proc. Natl. Acad. Sci. USA* **91**, 3054–3057.
12. Mullen, C. A., Kilstrup, M., and Blaese, R. M. (1992). Transfer of the bacterial gene for cytosine deaminase to mammalian cells confers lethal sensitivity to 5′ fluorocytosine: a negative selection system. *Proc. Natl. Acad. Sci. USA* **89**, 33–37.
13. Mullen, C. A., Coale, M. M., Lowe, R. *et al.* (1994). Tumors expressing the cytosine deaminase suicide gene can be eliminated *in vivo* with 5-fluorocytosine and induce protective immunity to wild type tumor. *Cancer Res.* **54**, 1503–1506.
14. Huber, R. E., Austin, E. A., Richards, C. A. *et al.* (1994). Metabolism of 5-fluorocytosine to 5-fluorouracil in human colorectal tumor cells transduced with the cytosine deaminase gene: significant antitumor effects when only a small percentage of tumor cells express cytosine deaminase. *Proc. Natl. Acad. Sci. USA* **91**, 8302–8306.
15. Mroz, P. J., and Moolten, F. L. (1993). Retrovirally transduced *Escherichia coli gpt* genes combine selectability with chemosensitivity capable of mediating tumor eradication. *Hum. Gene Ther.* **4**, 589–595, 1993.
16. Tepper, R. I., Pattengale, P. K., and Leder, P. (1989). Murine interleukin-4 displays potent anti-tumor activity *in vivo. Cell* **57**, 503–512.
17. Watanabe, Y., Kuribayashi, K., Miyatake, S. *et al.* (1989). Exogenous expression of mouse interferon-gamma cDNA in mouse neuroblastoma C1300 cells results in reduced tumorigenicity by augmented anti-tumor immunity. *Proc. Natl. Acad. Sci. USA* **86**, 9456–9460.
18. Fearon, E. R., Pardoll, D. M., Itaya, T. *et al.* (1990). Interleukin-2 production by tumor cells bypasses T helper function in the generation of an antitumor response. *Cell* **60**, 397–403.
19. Gansbacher, B., Zier, K., Daniels, B. *et al.* (1990). Interleukin-2 gene transfer into tumor cells abrogates tumorigenicity and induces protective immunity. *J. Exp. Med.* **172**, 1217–1224.
20. Gansbacher, B., Bannerji, R., Daniels, B. *et al.* (1990). Retroviral vector-mediated gamma-interferon gene transfer into tumor cells generates potent and long lasting antitumor immunity. *Cancer Res.* **50**, 7820–7825.
21. Colombo, M. P., Ferrari, G., Stoppacciaro, A. *et al.* (1991). Granulocyte colony-stimulating factor gene transfer suppresses tumorigenicity of a murine adenocarcinoma *in vivo. J. Exp. Med.* **173**, 889–897.
22. Golumbek, P. T., Lazenby, A. J., Levitsky, H. I. *et al.* (1991). Treatment of established renal cancer by tumor cells engineered to secrete interleukin-4. *Science* **254**, 713–716.
23. Esumi, N., Hunt, B., Itaya, T. *et al.* (1991). Reduced tumorigenicity of murine tumor cells secreting gamma-interferon is due to nonspecific host responses and is unrelated to class I major histocompatibility complex expression. *Cancer Res.* **51**, 1185–1189.
24. Hock, H., Dorsch, M., Diamantstein, T. *et al.* (1991). Interleukin 7 induces CD4⁺ T-cell-dependent tumor rejection. *J. Exp. Med.* **174**, 1291–1298.
25. Ley, V., Langlade-Demoyen, P., Kourilsky, P. *et al.* (1991). Interleukin 2-dependent activation of tumor-specific cytotoxic T lymphocytes *in vivo. Eur. J. Immunol.* **21**, 851–854.
26. Asher, A. L., Mulé, J. J., Kasid, A. *et al.* (1991). Murine tumor cells transduced with the gene for tumor necrosis factor-α. *J. Immunol.* **146**, 3227–3234.
27. Blankenstein, T., Qin, Z., Überla, K. *et al.* (1991). Tumor suppression after tumor cell-targeted tumor necrosis factor α gene transfer. *J. Exp. Med.* **173**, 1047–1052.
28. Pardoll, D. (1992). Immunotherapy with cytokine gene-transduced tumor cells: the next wave in gene therapy for cancer. *Curr. Opin. Oncol.* **4**, 1124–1129.
29. Porgador, A., Tzehoval, E., Katz, A. *et al.* (1992). Interleukin-6 gene transfection into Lewis lung carcinoma tumor cells suppresses the malignant phenotype and confers immunotherapeutic competence against parental metastatic cells. *Cancer Res.* **52**, 3679–3686.
30. Aoki, T., Tashiro, K., Miyatake, S. I. *et al.* (1992). Expression of murine interleukin-7 in a murine glioma cell line results in reduced tumorigenicity *in vivo. Proc. Natl. Acad. Sci. USA* **89**, 3850–3854.
31. Restifo, N. P., Spiess, P. J., Karp, S. E. *et al.* (1992). A nonimmunogenic sarcoma transduced with the cDNA for interferon-gamma elicits CD8⁺ T-cells against the wild-type tumor: correlation with antigen presentation capability. *J. Exp. Med.* **175**, 1423–1431.

32. Tepper, R. I., Coffman, R. L., and Leder, P. (1992). An eosinophil-dependent mechanism for the antitumor effect of interleukin-4. *Science* **257,** 548–551.

33. Dranoff, G., Jaffee, E., Lazenby, A. *et al.* (1993). Vaccination with irradiated tumor cells engineered to secrete murine granulocyte-macrophage colony-stimulating factor stimulates potent, specific, and long-lasting anti-tumor immunity. *Proc. Natl. Acad. Sci. USA* **90,** 3539–3543.

34. Porgador, A., Bannerji, R., Watanabe, Y. *et al.* (1993). Antimetastatic vaccination of tumor-bearing mice with two types of IFN-gamma gene-inserted tumor cells. *J. Immunol.* **150,** 1458–1470, 1993.

35. Rosenthal, F. M., Cronin, K., Bannerji, R. *et al.* (1994). Augmentation of antitumor immunity by tumor cells transduced with a retroviral vector carrying the interleukin-IL2 and interferon-gamma cDNAs. *Blood* **83,** 1289–1298.

36. Townsend, S. E., and Allison, J. P. (1993). Tumor rejection after direct costimulation of CD8$^+$ T cells by B7-transfected melanoma cells. *Science* **259,** 368–370.

37. Plautz, G. E., Yang, Z. Y., Wu, B. Y. *et al.* (1993). Immunotherapy of malignancy by *in vivo* gene transfer into tumors. *Proc. Natl. Acad. Sci. USA* **90,** 4645–4649.

38. Nabel, G. J., Nabel, E. G., Yang, Z. Y. *et al.* (1993). Direct gene transfer with DNA-liposome complexes in melanoma: expression, biologic activity, and lack of toxicity in human. *Proc. Natl. Acad. Sci. USA* **90,** 11307–11311.

39. Trojan, J., Blossey, B. K., Johnson, T. R. *et al.* (1992). Loss of tumorigenicity of rat glioblastoma directed by episome-based antisense cDNA transcription of insulin-like growth factor I. *Proc. Natl. Acad. Sci. USA* **89,** 4874–4878.

40. Trojan, J., Johnson, T. R., Rudin, S. D. *et al.* (1993). Treatment and prevention of rat glioblastoma by immunogenic C6 cells expressing antisense insulin-like growth factor I RNA. *Science* **259,** 94–98.

41. Kantor, J., Irvine, K., Abrams, S. *et al.* (1992). Antitumor activity and immune responses induced by a recombinant carcinoembryonic antigen-vaccinia virus vaccine. *J. Natl. Cancer Inst.* **84,** 1084–1091.

42. Kantor, J., Irvine, K., Abrams, S. *et al.* (1992). Immunogenicity and safety of a recombinant vaccinia virus vaccine expressing the carcinoembryonic antigen gene in a nonhuman primate. *Cancer Res.* **52,** 6917–6925.

43. Conry, R. M., LoBuglio, A. F., Loechel, F. *et al.* (1995). A carcinoembryonic antigen polynucleotide vaccine for human clinical use. *Cancer Gene Ther.* **2,** 33–38.

44. Huang, H. J., Yee, J. K., Shew, J. Y. *et al.* (1988). Suppression of the neoplastic phenotype by replacement of the RB gene in human cancer cells. *Science* **242,** 1563–1566.

45. Chen, P. L., Chen, Y., Bookstein, R. *et al.* (1990). Genetic mechanisms of tumor suppression by the human p53 gene. *Science* **250,** 1576–1580.

46. Baker, S. J., Markowitz, S., Fearon, E. R. *et al.* (1990). Suppression of human colorectal carcinoma cell growth by wild-type p53. *Science* **249,** 912–915.

47. Cai, D. W., Mukhopadhyay, T., Liu, Y. *et al.* (1993). Stable expression of the wild-type p53 gene in human lung cancer cells after retrovirus-mediated gene transfer. *Hum. Gene Ther.* **4,** 617–624.

48. Fujiwara, T., Grimm, E. A., Mukhopadhyay, T. *et al.* (1993). A retroviral wild-type p53 expression vector penetrates human lung cancer spheroids and inhibits growth by inducing apoptosis. *Cancer Res.* **53,** 4129–4133.

49. Wills, K. N., Maneval, D. C., Menzel, P. *et al.* (1994). Development and characterization of recombinant adenovirus encoding human p53 for gene therapy of cancer. *Hum. Gene Ther.* **5,** 1079–1088.

50. Mukhopadhyay, T., Tainsky, M., Cavender, A. C. *et al.* (1991). Specific inhibition of K-*ras* expression and tumorigenicity of lung cancer cells by antisense RNA. *Cancer Res.* **51,** 1744–1748.

51. Zhang, Y., Mukhopadhyay, T., Donehower, L. A. *et al.* (1993). Retroviral vector-mediated transduction of K-*ras* antisense RNA into human lung cancer cells inhibits expression of the malignant phenotype. *Hum. Gene Ther.* **4,** 451–460.

52. Gray, G. D., Hernandez, O. M., Hebel, D. *et al.* (1993). Antisense DNA inhibition of tumor growth induced by c-Ha-ras oncogene in nude mice. *Cancer Res.* **53,** 577–580.

53. Cardoso, J. E., Branchereau, S., Jeyaraj, P. R. *et al.* (1993). *In situ* retrovirus-mediated gene transfer into dog liver. *Hum. Gene Ther.* **4,** 411–418.

54. Li, Q., Kay, M. A., Finegold, M. *et al.* (1993). Assessment of recombinant adenoviral vectors for hepatic gene therapy. *Hum. Gene Ther.* **4,** 403–409.

55. Stratford-Perricaudet, L. D., Makeh, I., Perricaudet, M. *et al.* (1992). Widespread long-term gene transfer to mouse skeletal muscles and heart. *J. Clin. Invest.* **90,** 626–630.

56. LaSalle, G. L., Robert, J. J., Berrard, S. *et al.* (1993). An adenovirus vector for gene transfer into neurons and glia in the brain. *Science* **259,** 988–990.

57. Mitani, K., Graham, F. L., and Caskey, T. (1994). Transduction of human bone marrow by adenoviral vector. *Hum. Gene Ther.* **5,** 941–948.

58. Biscoff, J., Kirn, D., Williams, A. *et al.* (1996). An adenovirus mutant that replicates selectively in p53-deficient human tumor cells. *Science* **274,** 373–376.

59. Heise, C., Williams, A., Olesch, J. *et al.* (1999). Efficacy of a replication-competent adenovirus (ONYX-015) following intratumoral injection: intratumoral spread and distribution effects. *Cancer Gene Ther.* **6,** 499–504.

60. Heise, C., Sampson-Johannes, A., Williams, A. *et al.* (1997). ONYX-015, an E1B gene-attenuated adenovirus, causes tumor-specific cytolysis and antitumoral efficacy that can be augmented by standard chemotherapeutic agents. *Nat. Med.* **3,** 639–645.

61. Rothmann, T., Hengstermann, A., Whitaker, N. *et al.* (1998). Replication of ONYX-015, a potential anticancer adenovirus, is independent of p53 status in tumor cells. *J. Virol.* **72,** 9470–9478.

62. Hay, J., Shapiro, N., Sauthoff, H. *et al.* (1999). Targeting the replication of adenoviral gene therapy vectors to lung cancer cells: the importance of the adenoviral E1b-55kD gene. *Hum. Gene Ther.* **10,** 579–590.

63. Ganly, I., Kirn, D., Eckhardt, S. *et al.* (2000). A phase I study of ONYX-015, an E1B attenuated adenovirus, administered intratumorally to patients with recurrent head and neck cancer. *Clin. Cancer Res.* **6,** 798–806.

64. Khuri, F., Nemunaitis, J., Ganly, I. *et al.* (2000). A controlled trial of intratumoral ONYX-015, a selectively replicating adenovirus, in combination with cisplatin and 5-fluorouracil in patients with recurrent head and neck cancer. *Nat. Med.* **6,** 879–885.

65. Jaffee, E. M., Dranoff, G., Cohen, L. K. *et al.* (1993). High efficiency gene transfer into primary human tumor explants without cell selection. *Cancer Res.* **53,** 2221–2226.

66. Gunter, K. C., Khan, A. S., and Noguchi, P. D. (1993). The safety of retroviral vectors. *Hum. Gene Ther.* **4,** 643–645.

67. Cornetta, K. C., Morgan, R. A., and Anderson, W. F. (1991). Safety issues related to retroviral-mediated gene transfer in humans. *Hum. Gene Ther.* **2,** 5–20.

68. Donahue, R. E., Kessler, S. W., Bodine, D. *et al.* (1992). Helper virus induced T cell lymphoma in nonhuman primates after retroviral mediated gene transfer. *J. Exp. Med.* **176,** 1125–1135.

69. Yang, R., Nunes, F. A., Berencsi, K. *et al.* (1994). Cellular immunity to viral antigens limits E1-deleted adenoviruses for gene therapy. *Proc. Natl. Acad. Sci. USA* **91,** 4407–4411.

70. Miller, D. G., Adam, M. A., and Miller, A. D. (1990). Gene transfer by retrovirus vectors occurs only in cells that are actively replicating at the time of infection. *Mol. Cell. Biol.* **10,** 4239–4242.

71. Cornetta, K. C., Moen, R. C., Culver, K. *et al.* (1990). Amphotropic murine leukemia retrovirus is not an acute pathogen for primates. *Hum. Gene Ther.* **1,** 14–30.

72. Kay, M. A., Holterman, A. X., Meuse, L. *et al.* (1995). Long-term hepatic adenovirus-mediated gene expression in mice following CTLA4Ig administration. *Nat. Genet.* **11,** 191–197.

73. Kay, M. A., Meuse, L., Gown, A. M. *et al.* (1997). Transient immunomodulation with anti-CD40 ligand antibody and CTLA4Ig enhances persistence and secondary adenovirus-mediated gene transfer into mouse liver. *Proc. Natl. Acad. Sci. USA* **94,** 4686–4691.

74. Halbert, C. L., Standaert, T. A., Wilson, C. B. *et al.* (1998). Successful readministration of adeno-associated virus vectors to the mouse lung requires transient immunosuppression during the initial exposure. *J. Virol.* **72,** 9795–9805.

75. Manning, W. C., Zhou, S., Bland, M. P. *et al.* (1998). Transient immunosuppression allows transgene expression following readministration of adeno-associated viral vectors. *Hum. Gene Ther.* **9,** 477–485.

76. Wilson, C. B., Embree, L. J., Schowalter, D. *et al.* (1998). Transient inhibition of CD28 and CD40 ligand interactions prolongs adenovirus-mediated transgene expression in the lung and facilitates expression after secondary vector administration. *J. Virol.* **72,** 7542–7550.

77. Perales, J. C., Ferkol, T., Molas, M. *et al.* (1994). An evaluation of receptor-mediated approaches for the introduction of genes in somatic cells. *Eur. J. Biochem.* **226,** 255–266.

78. Perales, J. C., Ferkol, T., Beegen, H. *et al.* (1994). Gene transfer *in vivo*: sustained expression and regulation of genes introduced into the liver by receptor-targeted uptake. *Proc. Natl. Acad. Sci. USA* **91,** 4084–4090.

79. Ferkol, T., Perales, J. C., Eckman, E. *et al.* (1995). Gene transfer into the airway epithelium of animals by targeting the polymeric immunoglobulin receptor. *J. Clin. Invest.* **95,** 493–502.

80. Ferkol, T., Perales, J. C., Mularo, F. *et al.* (1996). Receptor-mediated gene transfer into macrophages. *Proc. Natl. Acad. Sci. USA* **93,** 101–105.

81. Israel, M. A., Chan, H. W., Hourihan, S. L. *et al.* (1979). Biological activity of polyoma viral DNA in mice and hamsters. *J. Virol.* **29,** 990–996.

82. Dubensky, T. W., Campbell, B. A., and Villarreal, L. P. (1984). Direct transfection of viral and plasmid DNA into the liver or spleen of mice. *Proc. Natl. Acad. Sci. USA* **81,** 7529–7533.

83. Seeger, C., Ganem, D., and Varmus, H. E. (1984). The cloned genome of ground squirrel hepatitis virus is infectious in the animal. *Proc. Natl. Acad. Sci. USA* **81,** 5849–5852.

84. Benvenisty, N., and Reshef, L. (1986). Direct introduction of genes into rats and expression of the genes. *Proc. Natl. Acad. Sci. USA* **83,** 9551–9555.

85. Wolff, J. A., Malone, R. W., Williams, P. *et al.* (1990). Direct gene transfer into mouse muscle *in vivo*. *Science* **247,** 1465–1468.

86. Acsadi, G., Jiao, S., Jani, A. *et al.* (1991). Direct gene transfer and expression into rat heart *in vivo*. *New Biol.* **3,** 71–81.

87. Manthorpe, M., Cornefert-Jensen, F., Hartikka, J. *et al.* (1993). Gene therapy by intramuscular injection of plasmid DNA: studies on firefly luciferase gene expression in mice. *Hum. Gene Ther.* **4,** 419–431.

88. Acsadi, G., Dickson, G., Lover, D. R. *et al.* (1991). Human dystrophin expression in mdx mice after intramuscular injection of DNA constructs. *Nature* **352,** 815–818.

89. Hickman, M. A., Malone, R. W., and Lehmann-Bruinsma, K. (1994). Gene expression following direct injection of DNA into liver. *Hum. Gene Ther.* **5,** 1477–1483.

90. Sikes, M. L., O'Malley, B. W., and Finegold, M. J. (1994). *In vivo* gene transfer into rabbit thyroid follicular cells by direct DNA injection. *Hum. Gene Ther.* **5,** 837–844.

91. Ulmer, J. B., Donnelly, J. J., Parker, S. E. *et al.* (1993). Heterologous protection against influenza by injection of DNA encoding a viral protein. *Science* **259,** 1745–1733.

92. Wang, B., Ugen, K. E., Srikantan, V. *et al.* (1993). Gene inoculation generates immune responses against human immunodeficiency virus type 1. *Proc. Natl. Acad. Sci. USA* **90,** 4156–4160.

93. Raz, E., Carson, D. A., Parker, S. E. *et al.* (1994). Intradermal gene immunization: the possible role of DNA uptake in the induction of cellular immunity to viruses. *Proc. Natl. Acad. Sci. USA* **91,** 9519–9523.

94. Zhang, L., Li, L., Hoffmann, G. A. *et al.* (1996). Depth targeted efficient gene delivery and expression in the skin by pulsed electric fileds: an approach to gene therapy of skin aging and other diseases. *Biochem. Biophys. Res. Com.* **220,** 633–636.

95. Wells, J. M., Li, L. H., Sen, A. *et al.* (2000). Electroporation-enhanced gene delivery in mammary tumors. *Gene Ther.* **7,** 541–547.

96. Chen, D., Endres, R., Erickson, C. *et al.* (2000). Epidermal immunization by a needle-free powder delivery technology: immunogenicity of influenza vaccine and protection in mice. *Nat. Med.* **6,** 1187–1190.

97. Fynan, E. F., Webster, R., Fuller, D. H. *et al.* (1993). DNA vaccines: protective immunizations by parenteral, mucosal, and gene-gun inoculations. *Proc. Natl. Acad. Sci. USA* **90,** 11478–11782.

98. Estin, C. D., Stevenson, U. S., Plowman, G. D. *et al.* (1988). Recombinant vaccinia virus vaccine against the human melanoma antigen p97 for use in immunotherapy. *Proc. Natl. Acad. Sci. USA* **85,** 1052–1056.

99. van der Bruggen, P., Traversati, C., Chomez, P. *et al.* (1991). A gene encoding an antigen recognized by cytolytic T lymphocytes on a human melanoma. *Science* **254,** 1643–1647.

100. Chen, Y. T., Stockert, E., Chen, Y. *et al.* (1994). Identification of the MAGE-1 gene product by monoclonal and polyclonal antibodies. *Proc. Natl. Acad. Sci. USA* **91,** 1004–1008.

101. Coulie, P. G., Brichard, V., Van Pel, A. *et al.* (1994). A new gene coding for a differentiation antigen recognized by autologous cytolytic T lymphocytes on HLA-A2 melanomas. *J. Exp. Med.* **180,** 35–42.

102. Kawakami, Y., Eliyahu, S., Sakaguchi, K. *et al.* (1994). Identification of the immunodominant peptides of the MART-1 human melanoma antigen recognized by the majority of HLA-A2-restricted tumor infiltrating lymphocytes. *J. Exp. Med.* **180,** 347–352.

103. Cox, A. L., Skipper, J., Chen, Y. *et al.* (1994). Identification of a peptide recognized by five melanoma-specific human cytotoxic T cell lines. *Science* **264,** 716–719.

104. Celis, E., Tsai, V., Crimi, C. *et al.* (1994). Induction of anti-tumor cytotoxic T lymphocytes in normal humans using primary cultures and synthetic peptide epitopes. *Proc. Natl. Acad. Sci. USA* **91,** 2105–2109.

105. Brichard, V., Van Pel, A., Wolfel, T. *et al.* (1993). The tyrosinase gene codes for an antigen recognized by autologous cytolytic T lymphocytes on HLA-A2 melanomas. *J. Exp. Med.* **178,** 489–495.

106. Bakker, A. B. H., Schreursm M. W. J., de Boer, A. J. *et al.* (1994). Melanocyte lineage-specific antigen gp100 is recognized by melanoma-derived tumor-infiltrating lymphocytes. *J. Exp. Med.* **179,** 1005–1009.

107. Boon, T., Cerottini, J. C., Van den Eynde, B. *et al.* (1994). Tumor antigens recognized by T lymphocytes. *Annu. Rev. Immunol.* **12,** 337–365.

108. Jerome, K. R., Domenech, N., and Finn, O. J. (1993). Tumor-specific cytotoxic T cell clones from patients with breast and pancreatic adenocarcinoma recognize EBV-immortalized B cells transfected with polymorphic epithelial mucin complementary DNA. *J. Immunol.* **151,** 1654–1662.

109. Hill, A., and Ploegh, H. (1995). Getting the inside out: the transporter associated with antigen processing (TAP) and the presentation of viral antigen. *Proc. Natl. Acad. Sci. USA* **92,** 341–343.

110. Peace, D. J., Chen, W., Nelson, H. *et al.* (1991). T-cell recognition of transforming proteins encoded by mutated *ras* proto-oncogenes. *J. Immunol.* **146,** 2059–2065.

111. Jung, S., and Schluesener, H. J. (1991). Human T lymphocytes recognize a peptide of single point-mutated, oncogenic Ras proteins. *J. Exp. Med.* **173,** 273–276.

112. Skipper, J., and Stauss, H. J. (1993). Identification of two cytotoxic T lymphocyte-recognized epitopes in the Ras protein. *J. Exp. Med.* **177,** 1493–1498.

113. Gedde-Dahl, T., Fossum, B., Eriksen, J. A. *et al.* (1993). T cell clones specific for p21 Ras-derived peptides: characterization of their fine specificity and HLA restriction. *Eur. J. Immunol.* **23**, 754–760.

114. Houbiers, J. G. A., Nijman, H. W., van der Burg, S. H. *et al.* (1993). *In-vitro* induction of human cytotoxic T lymphocyte responses against peptides of mutant and wild-type p53. *Eur. J. Immunol.* **23**, 2072–2077.

115. Chen, W., Peace, D. J., Rovira, D. K. *et al.* (1992). T-cell immunity to the joining region of p210^BCR-ABL protein. *Proc. Natl. Acad. Sci. USA* **89**, 1468–1472.

116. Muraro, R., Wunderlich, D., Thor, A. *et al.* (1985). Definition by monoclonal antibodies of a repertoire of epitopes on carcinoembryonic antigen differentially expressed in human colon carcinomas versus normal adult tissues. *Cancer Res.* **45**, 5769–5780.

117. Levitsky, H. (2000). Augmentation of host immune responses to cancer: overcoming the barrier of tumor antigen-specific T-cell tolerance. *Cancer J.* **6**, S281–S29.

118. Gabathuler, R., Reid, G., Kolaitis, G. *et al.* (1994). Comparison of cell lines deficient in antigen presentation reveals a functional role for TAP-1 alone in antigen processing. *J. Exp. Med.* **180**, 1415–1425.

119. Maeurer, M. J., Gollin, S. M., Martin, D. *et al.* (1996). Tumour escape from immune recognition: lethal recurrent melanoma in a patient associated with downregulation of the peptide transporter protein TAP-1 and loss of expression of the immunodominant MART-1/Melan-A antigen. *J. Clin. Invest.* **98**, 1633–1641.

120. Alimonti, J., Zhang, Q., Gabathuler, R. *et al.* (2000). TAP expression provides a general method for improving the recognition of malignant cells in vivo. *Nat. Biotech.* **18**, 515–520.

121. Tanaka, K., Isselbacher, K. J., Khoury, G. *et al.* (1985). Reversal of oncogenesis by the expression of a major histocompatibility complex class I gene. *Science* **228**, 26–30.

122. Garrido, F., Ruiz-Cabello, F., Cabrera, T. *et al.* (1997). Implications for immunosurveillance of altered HLA class I phenotypes in human tumours. *Immunol. Today* **18**, 89–95.

123. Jefferies, W. A., Kolaitis, G., and Gabathuler, R. (1993). IFN-induced recognition of the antigen-processing variant CMT.64 by cytolytic T cells can be replaced by sequential addition of β^2-microglobulin and antigenic peptides. *J. Immunol.* **151**, 2974–2985.

124. Baskar, S., Ostrand-Rosenberg, S., Nabavi, N. *et al.* (1993). Constitutive expression of B7 restores immunogenicity of tumor cells expressing truncated major histocompatibility complex class II molecules. *Proc. Natl. Acad. Sci. USA* **90**, 5687–5690.

125. Krummel, M. F., and Allison, J. P. (1995). CD28 and CTLA-4 have opposing effects on the response of T cells to stimulation. *J. Exp. Med.* **182**, 459–465.

126. Wu, T. C., Huang, A. Y., Jaffee, E. M. *et al.* (1995). A reassessment of the role of B7-1 expression in tumor rejection. *J. Exp. Med.* **182**, 1415–1421.

127. Leach, D. R., Krummel, M. F., and Allison, J. P. (1996). Enhancement of antitumor immunity by CTLA-4 blockade. *Science* **271**, 1734–1736.

128. Kwon, E. D., Hurwitz, A. A., Foster, B. A. *et al.* (1997). Manipulation of T cell costimulatory and inhibitory signals for immunotherapy of prostate cancer. *Proc. Natl. Acad. Sci. USA* **94**, 8099–8103.

129. Hurwitz, A. A., Yu, T. F., Leach, D. R. *et al.* (1998). CTLA-4 blockade synergizes with tumor-derived granulocyte-macrophage colony-stimulating factor for treatment of an experimental mammary carcinoma. *Proc. Natl. Acad. Sci. USA* **95**, 10067–10071.

130. Teng, Y. T., Gorczynski, R. M., and Hozumi, N. (1998). The function of TGF-beta-mediated bystander suppression associated with physiological self-tolerance in vivo. *Cell Immunol.* **190**, 51–60.

131. Chen, J. J., Sun, Y., and Nabel, G. J. (1998). Regulation of the proinflammatory effects of Fas ligand (CD95L). *Science* **282**, 1714–1717.

132. Conrad, C. T., Ernst, N. R., Dummer, W. *et al.* (1999). Differential expression of transforming growth factor beta 1 and interleukin 10 in progressing and regressing areas of primary melanoma. *J. Exp. Clin. Cancer Res.* **18**, 225–232.

133. Yang, N. S., Burkholder, J., Roberts, B. *et al.* (1990). *In vivo* and *in vitro* gene transfer to mammalian somatic cells by particle bombardment. *Proc. Natl. Acad. Sci. USA* **87**, 9568–9572.

134. Sun, W. H., Burkholder, J. K., Sun, J. *et al.* (1995). *In vivo* cytokine gene transfer by gene gun reduces tumor growth in mice. *Proc. Natl. Acad. Sci. USA* **92**, 2889–2893.

135. Graham, F. L., and Van der Eb, and A. J. (1973). A new technique for the assay of infectivity of human adenovirus-5 DNA. *Virol.* **52**, 456–467.

136. Vile, R. G., and Hart, I. R. (1993). Use of tissue-specific expression of the herpes simplex virus thymidine kinase gene to inhibit growth of established murine melanomas following direct intratumoral injection of DNA. *Cancer Res.* **53**, 3860–3864.

137. Felgner, P. L. and Ringold, G. M. (1989). Cationic liposome-mediated transfection. *Nature* **337**, 387–388.

138. Felgner, P. L., Zaugg, R. H., and Norman, J. A. (1995). Synthetic recombinant DNA delivery for cancer therapeutics. *Cancer Gene Ther.* **2**, 61–65.

139. Stopeck, A. T., Hersh, E. M., Akporiaye, E. T. *et al.* (1997). Phase I study of direct gene transfer of an allogeneic histocompatibility antigen, HLA-B7, in patients with metastatic melanoma. *J. Clin. Oncol.* **15**, 341–349.

140. Rubin, J., Galanis, E., Pitot, H. C. *et al.* (1997). Phase I study of immunotherapy of hepatic metastases of colorectal carcinoma by direct gene transfer of an allogeneic histocompatibility antigen, HLA-B7. *Gene Ther.* **4**, 419, 425.

141. Nabel, E. G., Gordon, D., Yang, Z. Y. *et al.* (1992). Gene transfer *in vivo* with DNA-liposome complexes: lack of autoimmunity and gonadal localization. *Hum. Gene Ther.* **3**, 649–656.

142. Stewart, M. J., Plautz, G. E., delBuono, L. *et al.* (1992). Gene transfer *in vivo* with DNA–liposome complexes: safety and acute toxicity in mice. *Hum. Gene Ther.* **3**, 267–275.

143. Nicolau, C., Le Pape, A., Soriano, P. *et al.* (1983). *In vivo* expression of rat insulin after intravenous administration of the liposome-entrapped gene for rat insulin I. *Proc. Natl. Acad. Sci. USA* **80**, 1068–1072.

144. Zhu, N., Liggitt, D., Liu, Y. *et al.* (1993). Systemic gene expression after intravenous DNA delivery into adult mice. *Science* **261**, 209–211.

145. Lesoon-Wood, L. A., Kim, W. H., Kleinman, H. K. *et al.* (1995). Systemic gene therapy with p53 reduces growth and metastases of a malignant human breast cancer in nude mice. *Hum. Gene Ther.* **6**, 395–405.

146. Wang, S., Lee, R. J., and Cauchon, G. (1995). Delivery of antisense oligodeoxyribonucleotides against the human epidermal growth factor receptor into cultured KB cells with liposomes conjugated to folate via polyethylene glycol. *Proc. Natl. Acad. Sci. USA* **92**, 3318–3322.

147. Lee, R. J., and Huang, L. (1996). Folate-targeted, anionic liposome-entrapped polylysine-condensed DNA for tumor cell-specific gene transfer. *J. Biol. Chem.* **271**, 8481–8487.

148. Goren, D., Horowitz, A. T., and Zalipsky, S. *et al.* (1996). Targeting of stealth liposomes to erbB-2 (Her/2) receptor: in vitro and in vivo studies. *Br. J. Cancer* **74**, 1749–1756.

149. Xu, L., Pirollo, K., Rait, A. *et al.* (1999). Systemic p53 gene therapy in combination with radiation results in human tumor regression. *Tumor Targ.* **4**, 92–104.

150. Xu, L., Pirollo, K., Tang, W. *et al.* (1999). Transferrin-liposome-mediated systemic p53 gene therapy in combination with radiation results in regression of human head and neck cancer xenografts. *Hum. Gene Ther.* **10**, 2941–2952.

151. Kawakami, S., Sato, A. Nishikawa, M. *et al.* (2000). Mannose receptor-mediated gene transfer into macrophages using novel mannosylated cationic liposomes. *Gene Ther.* **7**, 292–299.

152. Gabizon, A., Horowitz, A., Goren, D. *et al.* (1999). Targeting folate receptor with folate linked to extremities of poly(ethylene glycol)-grafted liposomes: in vitro studies. *Biconjug. Chem.* **10**, 289–298.

153. Goren, D., Horowitz, A., Zalipsky, S. *et al.* (1996). Targeting of stealth liposomes to erbB-2 (Her/2) receptor: in vitro and in vivo studies. *Br. J. Cancer* **74**, 1749–1756.

154. Allen, T. M., and Hanson, C. (1991). Pharmacokinetics of stealth versus conventional liposomes: effect of dose. *Biochim. Biophys. Acta* **1068**, 133–141.

155. Mayhew, E. G., Lasic, D., Babbar, S. *et al.* (1992). Pharmacokinetics and antitumor activity of epirubicin encapsulated in long-circulating liposomes incorporating a polyethylene glycol-derivatized phospholipid. *Int. J. Cancer* **51**, 302–309.

156. Wu, N. Z., Da, D., Rudoll, T. L. *et al.* (1993). Increased microvascular permeability contributes to preferential accumulation of stealth liposomes in tumor tissue. *Cancer Res.* **53**, 3765–3770.

157. Yuan, F., Leunig, M., Huang, S. K. *et al.* (1994). Microvascular permeability and interstitial penetration of sterically stabilized (stealth) liposomes in a human tumor xenograft. *Cancer Res.* **54**, 3352–3356.

158. Wheeler, J., Palmer, L., Ossanlou, M. *et al.* (1999). Stabilized plasmid-lipid particles: construction and characterization. *Gene Ther.* **6**, 271–281.

159. Gabizon, A., and Papahadjopoulos, D. (1988). Liposome formulations with prolonged circulation time in blood and enhanced uptake by tumors. *Proc. Natl. Acad. Sci. USA* **85**, 6949–6953.

160. Papahadjopoulos, D., Allen, T., Gabizon, A. *et al.* (1991). Sterically stabilized liposomes: improvements in pharmacokinetics and antitumor therapeutic efficacy. *Proc. Natl. Acad. Sci. USA* **88**, 11460–11464.

161. Gabizon, A., and Papahadjopoulos, D. (1992). The role of surface charge and hydrophilic groups on liposome clearance in vivo. *Biochim. Biophys. Acta* **1103**, 94–100.

162. Longman, S., Tardi, P., Parr, M. *et al.* (1995). Accumulation of protein-coated liposomes in an extravascular site: influence of increasing carrier circulation lifetimes. *J. Pharmacol. Exp. Ther.* **275**, 1177–1184.

163. Zhang, Y. P., Sekirov, L., Saravolac, E. G. *et al.* (1999). Stabilized plasmid-lipid particles for regional gene therapy: formulation and transfection properties. *Gene Ther.* **6**, 1438–1447.

164. Monck, M., Mori, A., Lee, D. *et al.* (2000). Stabilized plasmid-lipid particles: pharmacokinetics and plasmid delivery to distal tumors following intravenous injection. *J. Drug Targeting* **7**, 439–452.

165. Tam, P., Monck, M. Lee, D. *et al.* (2000). Stabilized plasmid-lipid particles for systemic gene therapy. *Gene Ther.* **7**, 1867–1874.

166. Webb, M., Saxon, D., Wong, F. *et al.* (1998). Comparison of different hydrophobic anchors conjugated to poly(ethylene glycol): effects on the pharmacokinetics of liposomal vincristine. *Biochim. Biophys. Acta* **1372**, 272–282.

167. Saravolac, E., Ludkovski, O., Skirrow, R. *et al.* (2000). Encapsulation of plasmid DNA in stabilized plasmid-lipid particles composed of different cationic lipid concentrations for optimal transfection activity. *J. Drug Target.* **7**, 423–437.

168. Clinical protocols. *Cancer Gene Ther.* **2**, 67–74, 1995.

169. Nabel, G. J., Gordon, D., Bishop, D. K. *et al.* (1996). Immune response in human melanoma after transfer of an allogeneic class I major histocompatibility complex gene with DNA–liposome complexes. *Proc. Natl. Acad. Sci. USA* **93**, 15388–15393.

170. Krieg, A. M., Yi, A. K., Matson, S. *et al.* (1995). CpG motifs in bacterial DNA trigger direct B-cell activation. *Nature* **374**, 546–549.

171. Klinman, D., Yi, A. K., Beaucage, S. L. *et al.* (1996). CpG motifs expressed by bacterial DNA rapidly induce lymphocytes to secrete IL-6, IL-12, and IFN. *Proc. Natl. Acad. Sci. USA* **93**, 2879-2783.

172. Davis, J. L., Weeranta, R., Waldschmidt, T. J. *et al.* (1998). CpG DNA is a potent enhancer of specific immunity in mice immunized with recombinant hepatitis B surface antigen. *J. Immunol.* **150**, 870–876.

173. Jones, T. R., Obaldia, N., Gramzinski, R. A. *et al.* (1999). Synthetic oligodeoxynucleotides containing CpG motifs enhance immunogenicity of a peptide malaria vaccine in *Aotus* monkeys. *Vaccine* **17**, 3065–3071.

174. Davis, H. L., Suparto, I., Weeratna, R. *et al.* (2000). CpG DNA overcomes hyporesponsiveness to hepatitis B vaccine in orangutans. *Vaccine* **18**, 1920–1924.

175. Kreig, A. M., Yi, A. Y., and Hartmann, G. (1999). Mechanisms and therapeutic applications of immune stimulatory CpG DNA. *Pharmacol. Therap.* **84**, 113–120.

176. Krieg, A. M. (1999). Mechanisms and applications of immune stimulatory CpG oligodeoxynucleotides. *Biochim. Biophys. Acta* **1489**, 107–116.

177. Krieg, A. M. (1999). Direct immunologic activities of CpG DNA and implications for gene therapy. *J. Gene Med.* **1**, 56–63.

178. Whitmore, M., Li, S., and Huang, L. (1999). LPD lipopolyplex initiates a potent cytokine response and inhibits tumor growth. *Gene Ther.* **6**, 1867–1875.

179. Lanuti, M., Rudginsky, S., Force, S. *et al.* (2000). Cationic lipid:bacterial DNA complexes elicit adaptive cellular immunity in murine intraperitoneal tumor models. *Cancer Res.* **60**, 2955–2963.

180. Wu, G. Y., and Wu, C. H. (1998). Receptor-mediated gene delivery and expression *in vivo. J. Biol. Chem.* **263**, 14621–14624.

181. Li, D., Pasumarthy, M. K., Kowalczyk, T. H. *et al.* (1999). Highly compacted PLAS*min*™ DNA complexes transfect postmitotic cells. *Cancer Gene Ther.* **6**, S12.

182. Wu, G. Y., Wilson, J. M., Shalaby, F. *et al.* (1991). Receptor-mediated gene delivery *in vivo*: partial correction of genetic analbuminemia in Nagase rats. *J. Biol. Chem.* **266**, 14338–14342.

183. Cotton, M., Wagner, W., Zatloukal, K. *et al.* (1992). High-efficiency receptor-mediated delivery of small and large (48) kilobase gene constructs using the endosome-disruption activity of defective or chemically inactivated adenovirus particles. *Proc. Natl. Acad. Sci. USA* **89**, 6094–6098.

184. Wagner, E., Zatloukal, K., Cotton, M. *et al.* (1992). Coupling of adenovirus to transferrin-polylysine/DNA complexes greatly enhances receptor-mediated gene delivery and expression of transfected genes. *Proc. Natl. Acad. Sci. USA* **89**, 6099–6103.

185. Wagner, E., Plank, C., Zatloukal, K. *et al.* (1992). Influenza virus hemagglutinin HA-2 N-terminal fusogenic peptides augment gene transfer by transferrin-polylysine-DNA complexes: toward a synthetic virus-like gene-transfer vehicle. *Proc. Natl. Acad. Sci. USA* **89**, 7934–7938.

186. Gao, L., Wagner, E., Cotton, M. *et al.* (1993). Direct *in vivo* gene transfer to airway epithelium employing adenovirus-polylysine-DNA complexes. *Hum. Gene Ther.* **4**, 17–24.

187. Batra, R. K., Berschneider, H., and Curiel, D. T. (1994). Molecular conjugate vectors mediate efficient gene transfer into gastrointestinal epithelial cells. *Cancer Gene Ther.* **1**, 185–192.

188. Takeshita, S., Gai, D., Leclerc, G. *et al.* (1994). Increased gene expression after liposome-mediated arterial gene transfer associated with intimal smooth muscle cell proliferation. *J. Clin. Invest.* **93**, 652–661.

189. Zabner, J., Fasbender, A. J., Moninger, T. *et al.* (1995). Cellular and molecular barriers to gene transfer by a cationic lipid. *J. Biol. Chem.* **270**, 18997–19007.

190. Wilke, M., Fortunati, E., van den Broek, M. *et al.* (1996). Efficacy of a peptide-based gene delivery system depends on mitotic activity. *Gene Ther.* **3**, 1133–1142.

191. Fasbender, A., Zabner, J., Zeiher, B. G. *et al.* (1997). A low rate of cell proliferation and reduced DNA uptake limit cationic lipid-mediated gene transfer to primary cultures of ciliated human airway epithelia. *Gene Ther.* **4**, 1173–1180.

192. Sebestyen, M. G., Ludtke, J. J., Bassik, M. C. *et al.* (1998). DNA vector chemistry: the covalent attachment of signal peptides to plasmid DNA. *Nat. Biotechnol.* **16**, 80–85.

193. Jiang, C., O'Connor, S. P., Fang, S. L. *et al.* (1998). Efficiency of cationic lipid-mediated transfection of polarized and differentiated airway epithelial cells in vitro and in vivo. *Hum. Gene Ther.* **9**, 531–542.

194. Tseng, W. C., Haselton, F. R., and Giorgio, T. D. (1999). Mitosis enhances transgene expression of plasmid delivered by cationic liposomes. *Biochim. Biophy. Acta* **1445**, 53–64.

195. Mortimer, J., Tam, P., MacLachlan, I. *et al*. (1999). Cationic lipid-mediated transfection of cells in culture requires mitotic activity. *Gene Ther.* **6**, 403–411.

196. Mirzayans, R., Aubin, R., and Paterson, M. (1992). Differential expression and stability of foreign genes introduced into human fibroblasts by nuclear versus cytoplasmic microinjection. *Mutat. Res.* **281**, 115–122.

197. Ziady, A. G., Perales, J. C., Ferkol, T. *et al*. (1997). Gene transfer into hepatoma cell lines via the serpin enzyme complex receptor. *Am. J. Physiol.* **273**, G545–G552.

198. Ziady, A. G., Ferkol, T., Gerken, T. *et al*. (1998). Ligand substitution of receptor targeted DNA complexes affects gene transfer into hepatoma cells. *Gene Ther.* **5**, 1685–1697.

199. Ziady, A. G., Ferkol, T. Dawson, D. V. *et al*. (1999). Chain length of the polylysine in receptor-targeted gene transfer complexes affects duration of reporter gene expression both in vitro and in vivo. *J. Biol. Chem.* **274**, 4908–4916.

200. Kowalczyk, T. H., Pasumarthy, M. K., Gedeon, C. *et al*. (2000). Light scattering by compacted DNA predicts its serum stability. *Mol. Ther.* **1**, S120.

201. Perales, J. C., Grossmann, G. A., Molas, M. *et al*. (1997). Biochemical and functional characterization of DNA complexes capable of targeting genes to hepatocytes via the asialoglycoprotein receptor. *J. Biol. Chem.* **272**, 7398–7407.

202. Gedeon, C., Ziady, A., Miller, T. J. *et al*. (2000). High level expression of compacted DNA complexes following intra-tracheal administration. *Mol. Ther.* **1**, S78.

203. Furth, P. A., Hennighausen, L., Baker, C. *et al*. (1991). The variability in activity of the universally expressed human cytomegalovirus immediate early gene 1 enhancer/promoter in transgenic mice. *Nucl. Acids Res.* **19**, 6205–6208.

204. Cheng, L., Ziegelhoffer, P. R., and Yang, N. S. (1993). *In vivo* promoter activity and transgene expression in mammalian somatic tissues evaluated by using particle bombardment. *Proc. Natl. Acad. Sci. USA* **90**, 4455–4459.

205. Jaenisch, R. (1988). Transgenic animals. *Science* **240**, 1468–1474.

206. Hanahan, D. (1989). Transgenic mice as probes into complex systems. *Science* **246**, 1265–1275.

207. Hanahan, D. (1985). Heritable formation of pancreatic β-cell tumours in transgenic mice expressing recombinant insulin/simian virus 40 oncogenes. *Nature* **315**, 115–122.

208. Ornitz, D. M., Hammer, R. E., Messing, A. *et al*. (1987). Pancreatic neoplasia induced by SV40 T-antigen expression in acinar cells of transgenic mice. *Science* **238**, 188–193.

209. Schoenenberger, C. A., Andres, A. C., Groner, B. *et al*. (1988). Targeted c-*myc* gene expression in mammary glands of transgenic mice induces mammary tumours with constitutive milk protein gene transcription. *EMBO. J.* **7**, 169–175.

210. Vile, R. G., and Hart, I. R. (1993). *In vitro* and *in vivo* targeting of gene expression to melanoma cells. *Cancer Res.* **53**, 962–967.

211. Sasaoka, T., Kobayashi, K., Nagatsu, I. *et al*. (1992). Analysis of the human tyrosine hydroxylase promoter-chloramphenicol acetyltransferase chimeric gene expression in transgenic mice. *Mol. Brain Res.* **16**, 274–286.

212. Julien, J. P., Tretjakoff, I., Beaudet, L. *et al*. (1987). Expression and assembly of a human neurofilament protein in transgenic mice provide a novel neuronal marking system. *Genes Dev.* **1**, 1085–1095.

213. Brenner, M., Kisselberth, W. C., Su, Y. *et al*. (1994). GFAP promoter directs astrocyte-specific expression in transgenic mice. *J. Neurosci.* **14**, 1030–1037.

214. Tronik, D., Dreyfus, M., Babinet, C. *et al*. (1987). Regulated expression of the Ren-2 gene in transgenic mice derived from parental strains carrying only the Ren-1 gene. *EMBO. J.* **6**, 983–987.

215. Stacey, A., Bateman, J., Choi, T. *et al*. (1988). Perinatal lethal osteogenesis imperfecta in transgenic mice bearing an engineered mutant pro-α1(I) collagen gene. *Nature* **332**, 131–136.

216. Shani, M. (1986). Tissue-specific and developmentally regulated expression of a chimeric actin-globin gene in transgenic mice. *Mol. Cell. Biol.* **6**, 2624–2631.

217. Jahroudi, N., and Lynch. D. C. (1994). Endothelial-cell-specific regulation of von Willebrand factor gene expression. *Mol. Cell, Biol.* **14**, 999–1008.

218. Huber, B. E., Richard, C. A., and Krenitsky, T. A. (1991). Retroviral-mediated gene therapy for the treatment of hepatocellular carcinoma: an innovative approach for cancer therapy. *Proc. Natl. Acad. Sci. USA* **88**, 8039–8043.

219. Glasser, S. W., Korfhagen, T. R., Bruno, M. D. *et al*. (1990). Structure and expression of the pulmonary surfactant protein SP-C gene in the mouse. *J. Biol. Chem.* **265**, 21986–21991.

220. Osaki, T., Tanio, Y., Tachibana, I. *et al*. (1994). Gene therapy for carcinoembryonic antigen-producing human lung cancer cells by cell type-specific expression of herpes simplex virus thymidine kinase gene. *Cancer Res.* **54**, 5258–5261.

221. Lin, J. H., Zhao, H., and Sun, T. T. (1995). A tissue-specific promoter that can drive a foreign gene to express in the suprabasal urothelial cells of transgenic mice. *Proc. Natl. Acad. Sci. USA* **92**, 679–683.

222. Krimpenfort, P., de Jong, R., Uematsu, Y. *et al*. (1988). Transcription of T cell receptor β-chain genes is controlled by a downstream regulatory element. *EMBO. J.* **7**, 745–750.

223. Alexander, W. S., Schrader, J. W., and Adams, J. M. (1987). Expression of the c-*myc* oncogene under control of an immunoglobulin enhancer in Eμ-myc transgenic mice. *Mol. Cell. Biol.* **7**, 1436–1444.

224. Murtha, P., Tindall, D. J., and Young, C. Y. F. (1993). Androgen induction of a human prostate-specific kallikrein, hKLK2: characterization of an androgen response element in the 5′ promoter region of the gene. *Biochemistry* **32**, 6459–6464.

225. Peschon, J. J., Behringer, R. R., Brinster, R. L. *et al*. (1987). Spermatid-specific expression of protamine 1 in transgenic mice. *Proc. Natl. Acad. Sci. USA* **84**, 5316–5319.

226. Rixe, O., Calvez, V., Mouawad, R. *et al*. (1995). *trans*-Activation of the metallothionein promoter in cisplatin (CP) resistant cell lines: potential application for a specific gene therapy. *Proc. Am. Assoc. Cancer Res.* **36**, 220.

227. Gossen, M., and Bujard, H. (1992). Tight control of gene expression in mammalian cells by tetracycline-responsive promoters. *Proc. Natl. Acad. Sci. USA* **89**, 5547–5551.

228. Furth, P. A., St Onge, L., Boger, H. *et al*. (1994). Temporal control of gene expression in transgenic mice by a tetracycline-responsive promoter. *Proc. Natl. Acad. Sci. USA* **91**, 9302–9306.

229. Passman, R. S., and Fishman, G. L. (1994). Regulated expression of foreign genes *in vivo* after germline transfer. *J. Clin. Invest.* **94**, 2421–2425.

230. Fishman, G. I., Kaplan, M. L., and Buttrick, P. M. (1994). Tetracycline-regulated cardiac gene expression *in vivo*. *J. Clin. Invest.* **93**, 1864–1868.

231. Gossen, M., Freundlieb, S., Bender, G. *et al*. (1995). Transcriptional activation by tetracyclines in mammalian cells. *Science* **268**, 1766–1769.

232. Deuschle, U., Meyer, W. K., and Thiesen, H. J. (1995). Tetracycline-reversible silencing of eukaryotic promoters. *Mol. Cell. Biol.* **15**, 1907–1914.

233. Wang, Y., O'Malley, B. W., Tsai, S. Y. *et al*. (1994). A regulatory system for use in gene transfer. *Proc. Natl. Acad. Sci. USA* **91**, 8180–8184.

234. Wang, Y., DeMayo, F. J., Tsai, S. Y. *et al*. (1997). Ligand-inducible and liver-specific target gene expression in transgenic mice. *Nature Biotechnol.* **15**, 239–343.

235. Delort, J. P., and Capecchi, M. R. (1996). TAXI/UAS: a molecular switch to control expression of genes in vivo. *Hum. Gene Ther.* **7**, 809–820.

236. Wang, Y., Xu, J., Pierson, T. *et al.* (1997). Positive and negative regulation of gene expression in eukaryotic cells with an inducible transcriptional regulator. *Gene Ther.* **4**, 432–441.

237. No, D., Yao, T. P., and Evans, R. M. (1996). Ecdysone-inducible gene expression in mammalian cells and transgenic mice. *Proc. Natl. Acad. Sci. USA* **93**, 3346–3351.

238. Rivera, V. M., Clackson, T., Natesan, S. *et al.* (1996). A humanized system for pharmacologic control of gene expression. *Nature Med.* **2**, 1028–1032.

239. Weichselbaum, R. R., Hallahan, D. E., Beckett, M. A. *et al.* (1994). Gene therapy targeted by radiation preferentially radiosensitizes tumor cells. *Cancer Res.* **54**, 4266–4269.

240. Marples, B., Scott, S., Hendry, J. *et al.* (2000). Development of synthetic promoters for radiation-mediated gene therapy. *Gene Ther.* **7**, 511–517.

241. Brown, J. M., and Giaccia, A. J. (1998). The unique physiology of solid tumors: opportunities (and problems) for cancer therapy. *Cancer Res.* **58**, 1408–1416.

242. Brizel, D. M., Sibley, G. S., Prosnitz, L. R. *et al.* (1997). Tumor hypoxia adversely affects the prognosis of carcinoma of the head and neck. *Int. J. Radiat. Oncol. Biol. Phys.* **38**, 285–289.

243. Dachs, G. U., Patterson, A. V., Firth, J. D. *et al.* (1997). Targeting gene expression to hypoxic tumor cells. *Nat. Med.* **3**, 515–520.

244. Shibata, T., Akiyama, N., Noda, M. *et al.* (1998). Enhancement of gene expression under hypoxic conditions using fragments of the human vascular endothelial growth factor and the erythropoietin genes. *Int. J. Radiat. Oncol. Biol. Phys.* **42**, 913–916.

245. Shibata, T., Giaccia, A. J., and Brown, J. M. (2000). Development of a hypoxia-responsive vector for tumor-specific gene therapy. *Gene Ther.* **7**, 493–498.

246. Dreano, M., Brochot, J., Myers, A. *et al.* (1986). High-level, heat-regulated synthesis of proteins in eukaryotic cells. *Gene* **49**, 1–8.

247. Huang, Q., Hu, J. K., Lohr, F. *et al.* (2000). Heat-induced gene expression as a novel targeted cancer gene therapy strategy. *Cancer Res.* **60**, 3435–3439.

248. Gao, X., and Huang, L. (1993). Cytoplasmic expression of a reporter gene by co-delivery of T7 RNA polymerase and T7 promoter sequence with cationic liposomes. *Nucleic Acids Res.* **21**, 2867–2872.

249. Gao, X., Jaffurs, D., Robbins, P. D. *et al.* (1994). A sustained, cytoplasmic transgene expression system delivered by cationic liposomes. *Biochem. Biophys. Res. Commun.* **200**, 1201–1206.

250. Chen, X., Li, Y., Xiong, K. *et al.* (1994). A self-initiating eukaryotic transient gene expression system based on contransfection of bacteriophage T7 RNA polymerase and DNA vectors containing a T7 autogene. *Nucleic Acids Res.* **22**, 2114–2120.

251. Chen, X., Li, Y., Xiong, K. *et al.* (1995). A novel nonviral cytoplasmic gene expression system and its implications in cancer gene therapy. *Cancer Gene Ther.* **2**, 281–289.

252. Chen, X., Li, Y., Xiong, K. *et al.* (1998). Cancer gene therapy by direct tumor injection of a nonviral T7 vector encoding a thymidine kinase gene. *Hum. Gene Ther.* **9**, 729–736.

253. Brisson, M., He, Y., Li, S. *et al.* (1999). A novel T7 RNA polymerase autogene for efficient cytoplasmic expression of target genes. *Gene Ther.* **6**, 263–270.

254. Biamonti, G., Della Valle, G., Talarico, D. *et al.* (1985). Fate of exogenous recombinant plasmids introduced into mouse and human cells. *Nucl. Acids Res.* **13**, 5545–5561.

255. Harms, J. S, and Splitter, G. A. (1995). Interferon-gamma inhibits transgene expression driven by SV40 or CMV promoters but augments expression driven by the mammalian MHC I promoter. *Hum. Gene Ther.* **6**, 1291–1297.

256. Qin, L., Ding, Y., Pahud, D. R. *et al.* (1997). Promoter attenuation in gene therapy: interferon-gamma and tumor necrosis factor-alpha inhibit transgene expression. *Hum. Gene Ther.* **8**, 2019–2029.

257. Felsenfeld, G., Boyes, J., Chung, J. *et al.* (1996). Chromatin structure and gene expression. *Proc. Natl. Acad. Sci. USA* **93**, 9384–9388.

258. Pikaart, M. J., Recillas-Targa, F., and Felsenfeld, G. (1998). Loss of transcriptional activity of a transgene is accompanied by DNA methylation and histone deacetylation and is prevented by insulators. *Genes Dev.* **12**, 2852–2862.

259. Bell, A. C., and Felsenfeld, G. (1999). Stopped at the border: bouindaries and insulators. *Curr. Opin. Genet. Dev.* **9**, 191–198.

260. Boyes, J., and Bird, A. (1991). DNA methylation inhibits transcription indirrectly via a methyl-CpG binding protein. *Cell* **64**, 1123–1134.

261. Boyes, J., and Bird, A. (1992). Repression of genes by DNA methylation depends on CpG density and promoter strength: evidence for involvement of a methyl-CpG binding protein. *EMBO J.* **11**, 327–333.

262. Muiznieks, I., and Doerfler, W. (1994). The impact of $5'$-CG-$3'$ methylation on the activity of different eukaryotic promoters: a comparative study. *FEBS Lett.* **344**, 251–254.

263. Krysan, P. J., Haase, S. B., and Calos, M. P. (1989). Isolation of human sequences that replicate autonomously in human cells. *Mol. Cell Biol.* **9**, 1026–1033.

264. Haase, S. B., and Calos, M. P. (1991). Replication control of autonomously replicating human sequences. *Nucl. Acids Res.* **19**, 5053–5058.

265. Heinzel, S. S., Krysan, P. J., Tran, C. T. *et al.* (1991). Autonomous DNA replication in human cells is affected by the size and the source of the DNA. *Mol. Cell. Biol.* **11**, 2263–2272.

266. Tran, C. T., Caddle, M. S., and Calos, M. P. (1993). The replication behavior of *Saccharomyces cerevisiae* DNA in human cells. *Chromosoma* **102**, 129–136.

267. Middleton, T., and Sugden, B. (1994). Retention of plasmid DNA in mammalian cells is enhanced by binding of the Epstein–Barr virus replication protein EBNA1. *J. Virol.* **68**, 4067–4071.

268. Willard, H. (2000). Artificial chromosomes coming to life. *Science* **290**, 1308–1309.

269. Harrington, J. J., Van Bokkelen, G., Mays, R. W. *et al.* (1997). Formation of de novo centromeres and construction of first-generation human artifical microchromosomes. *Nat. Genet.* **15**, 345–355.

270. Ikeno, M., Grimes, B., Okazaki, T. *et al.* (1998). Construction of YAC-based mammalian artificial chromosomes. *Nat. Biotechnol.* **16**, 431–439.

271. Henning, K. A., Novotny, E. A., Compton, S. T. *et al.* (1999). Human artificial chromosomes generated by modification of a yeast artificial chromosome containing both human alpha satellite and single-copy DNA sequences. *Proc. Natl. Acad. Sci. USA* **96**, 592–597.

272. Ebersole, T. A., Ross, A., Clark, E. *et al.* (2000). Mammalian artificial chromosome formation from circular alphoid input DNA does not require telomere repeats. *Hum. Mol. Genet.* **9**, 1623–1631.

273. Csonka, E., Cserpan, I., Fodor, K. *et al.* (2000). Novel generation of human satellite DNA-based artificial chromosomes in mammalian cells. *J. Cell. Sci.* **113**, 3207–3216.

274. Yates, J. L., Warren, N., and Sugden, B. (1985). Stable replication of plasmids derived from Epstein–Barr virus in various mammalian cells. *Nature* **313**, 812–815.

275. Milanesi, G., Barbanti-Brodano, G., Negrini, M. *et al.* (1984). BK virus–plasmid expression vector that persists episomally in human cells and shuttles into *Escherichia coli*. *Mol. Cell Biol.* **4**, 1551–1560.

276. Tsui, L. C., Breitman, M. L., Siminovitch, L. *et al.* (1982). Persistence of freely replicating SV40 recombinant molecules carrying a selectable marker in permissive simian cells. *Cell* **30**, 499–508.

277. Sverdrup, F., Sheahan, L., and Khan, S. (1999). Development of human papillomavirus plasmids capable of episomal replication in human cell lines. *Gene Ther.* **6**, 1317–1321.

278. Vos, J. (1988). Mammalian artificial chromosomes as tools for gene therapy. *Curr. Opin. Genet. Dev.* **8**, 351–359.

279. Kelleher, Z., Fu, H., Livanos, E. *et al.* (1998). Epstein–Barr-based episomal chromosomes shuttle 100 kb of self-replicating circular human DNA in mouse cells. *Nat. Biotechnol.* **16**, 762–768.

280. Westphal, E., Sierakowska, H., Livanos, E. *et al.* (1998). A system for shuttling 200-kb BAC/PAC clones into human cells: stable extrachromosomal persistence and long-term ectopic gene activation. *Hum. Gene Ther.* **9**, 1863–1873.

281. Wade-Martins, R., Frampton, J., and James, M. (1999). Long-term stability of large insert genomic DNA episomal shuttle vectors in human cells. *Nucleic Acids Res.* **27**, 1674–1682.

282. Grossi, M. P., Caputo, A., Rimessi, P. *et al.* (1988). New BK virus episomal vector for complementary DNA expression in human cells. *Arch. Virol.* **102**, 275–283.

283. Grossi, M. P., Caputo, A., Paolinim L. *et al.* (1988). Factors affecting amplification of BK virus episomal vectors in human cells. *Arch. Virol.* **99**, 249–259.

284. Cooper, M. J., and Miron, S. M. (1993). Efficient episomal expression vector for human transitional carcinoma cells. *Hum. Gene Ther.* **4**, 557–566.

285. Thierry, A., Lunardi-Iskandar, Y., Bryant, J. *et al.* (1995). Systemic gene therapy: biodistribution and long-term expression of a transgene in mice. *Proc. Natl. Acad. Sci. USA* **92**, 9742–9746.

286. Cooper, M. J., Lippa, M., Payne, J. M. *et al.* (1997). Safety-modified episomal vectors for human gene therapy. *Proc. Natl. Acad. Sci. USA* **94**, 6450–6455.

287. Wilson, J. B., Bell, J. L., and Levine, A. J. (1996). Expression of Epstein–Barr virus nuclear antigen-1 induces B cell neoplasia in transgenic mice. *EMBO J.* **15**, 3117–3126.

288. Linzer, D. H., and Levine, A. J. (1979). Characterization of a 54K dalton cellular SV40 tumor antigen present in SV40-transformed cells and uninfected embryonal carcinoma cells. *Cell* **17**, 43–52.

289. DeCaprio, J. A., Ludlow, J. W., Figge, J. *et al.* (1988). SV40 large tumor antigen forms a specific complex with the product of the retinoblastoma susceptibility gene. *Cell* **54**, 275–283.

290. Ewen, M. E., Xing, Y., Lawrence, J. B. *et al.* (1991). Molecular cloning, chromosomal mapping, and expression of the cDNA for p107, a retinoblastoma gene product-related protein. *Cell* **66**, 1155–1164.

291. Claudio, P. P., Howard, C. M., Baldi, A. *et al.* (1994). p130/pRb2 has growth suppressive properties similar to yet distinctive from those of retinoblastoma family members pRb and p107. *Cancer Res.* **54**, 5556–5560.

292. Miller, T. J., and Cooper, M. J. (2000). Externally controlled, safety-modified replicon (REPLI*some*™ Switch) for viral and non-viral vectors. *Mol. Ther.* **1**, S241.

293. Quentin, B., Perricaudet, L. D., Tajbakhsh, S. *et al.* (1992). Adenovirus as an expression vector in muscle cells *in vivo*. *Proc. Natl. Acad. Sci. USA* **89**, 2581–2584.

294. Russell, D. W., Miller, A. D., and Alexander, I. E. (1994). Adeno-associated virus vectors preferentially transduce cells in S-phase. *Proc. Natl. Acad. Sci. USA* **91**, 8915–8919.

295. Duan, D., Sharma, P., Yang, J. *et al.* (1998). Circular intermediates of recombinant adeno-associated virus have defined structural characteristics responsible for long-term episomal persistence in muscle tissue. *J. Virol.* **72**, 8568–8577.

296. Malik, A. K., Monahan, P. E., Allen, D. L. *et al.* (2000). Kinetics of recombinant adeno-associated virus-mediated gene transfer. *J. Virol.* **74**, 3555–3565.

297. Fink, D. J., Sternberg, L. R., Weber, P. C. *et al.* (1992). *In vivo* expression of β-galactosidase in hippocampal neurons by HSV-mediated gene transfer. *Hum. Gene Ther.* **3**, 11–19.

298. Moss, B., and Flexner, C. (1987). Vaccinia virus expression vectors. *Annu. Rev. Immunol.* **5**, 305–324.

299. Maxwell, I. H., Maxwell, F., Rhode, S. L. *et al.* (1993). Recombinant LuIII autonomous parvovirus as a transient transducing vector for human cells. *Hum. Gene Ther.* **4**, 411–450.

CHAPTER

3

Parvovirus Vectors for the Gene Therapy of Cancer

K. K. WONG, JR.

*Division of Hematology and Bone
Marrow Transplantation, and
Division of Virology
City of Hope National Medical Center
Duarte, California 91010*

SASWATI CHATTERJEE

*Division of Virology
City of Hope National Medical Center
Duarte, California 91010*

I. INTRODUCTION

Intensive investigation has demonstrated that a wide variety of clinically important diseases including viral infections and cancer arise as a direct result of aberrant gene expression and thus may be considered "genetic" disorders [1]. Furthermore, the rapid development of recombinant DNA technology, in conjunction with expanding knowledge of these underlying pathogenic mechanisms, has led to strategies designed to actually correct disease at the molecular level. Gene transfer approaches have been developed to correct primary inherited diseases, such as cystic fibrosis or adenosine deam-inase deficiency, as well as acquired disorders, such as cancer [2–4]. Although a variety of techniques are available for the introduction of genes into cells, few approach the efficiency necessary for actual gene therapy trials. To address this issue, viral vectors have been developed to exploit the natural ability of viruses to efficiently infect and transfer their genetic material into cells. Within the past few years, a variety of DNA and RNA virus vectors have been developed for gene transfer. Perhaps the most commonly used viral vector system is based upon murine oncoretroviruses, particularly murine leukemia retroviruses (MLVs). MLV-based vectors have a wide host range and efficiently integrate into cellular DNA. However, they require proliferating target cells for efficient transduction which potentially limits their effectiveness [5]. Thus, important nonproliferating populations such as primary neural and hematopoietic stem cells may be refractory to MLV gene transfer. Furthermore, replication-competent retroviruses (RCRs), which may be generated during the vector packaging process and thereby contaminate vector stocks, have been causally linked to the development of T-cell lymphomas in a primate model of hematopoietic progenitor transplantation with genetically modified cells [6,7]. Vectors based upon human adenovirus efficiently transduce a wide variety of cells, including nonproliferating targets, but do not integrate at high efficiency and are highly immunogenic [8–10]. Recently, vectors based upon the lentivirus family of retroviruses, which are related to human immunodeficiency virus (HIV), have been shown to transduce a wide variety of cells [11,12]. These vectors may preferentially transduce targets in the G1b portion of the cell cycle, and safety issues are currently being resolved [13,14]. Thus, safe and efficient gene transfer vectors that can circumvent these problems are continually being sought.

In this chapter, we will briefly review the basic biology of parvoviruses, specifically as it relates to the development of gene transfer vectors. We will focus upon potential advantages and disadvantages of parvovirus-based vectors and the latest developments in vector development, in addition to applications for the treatment of cancer actively being pursued. We refer the reader to several excellent recently published reviews describing the field of cancer gene therapy for more detailed discussions of the rationale and strategies currently undergoing evaluation (see elsewhere in this volume) [15–19].

II. BIOLOGY OF PARVOVIRIDAE AND VECTOR DEVELOPMENT

The family Parvoviridae is subdivided into three distinct groups: the densoviruses, which are exclusive to arthropods; the autonomous parvoviruses, which are lytic and replicate in proliferating target cells; and the dependo- or adeno-associated viruses (AAV), which require coinfection with another DNA virus, or "helper" virus, for productive infection [20]. The latter two groups infect a broad spectrum of vertebrates, including humans. Parvoviruses are small, nonenveloped icosahedral viruses, approximately 18–26 nm in size and containing a 5-kb, single-stranded DNA genome flanked

by palindromic inverted terminal repeats (ITRs). The ITRs serve as *cis*-active elements necessary for viral DNA replication and encapsidation and, in the case of AAV, have weak promoter activity [21].

A. Adeno-Associated Virus

1. Biology

Historically, adeno-associated virus was originally identified as a contaminant of human adenovirus stocks being propagated for development of vaccines [22,23]. AAV are currently subdivided into six serotypes; serotypes 2, 3, and 5 were isolated from humans, while serotypes 1 and 4 were isolated from nonhuman primates. An additional serotype, designated AAV6, was isolated as a contaminant from an adenovirus 5 stock and was believed to represent a variant of AAV1 [24]. AAV2 has been the most extensively studied to date, and the following discussion pertains to AAV2 unless otherwise noted.

Genetic analysis of the AAV2 genome has defined two major open reading frames (ORFs) (Fig. 1). The left ORF encodes functions necessary for AAV *ori*-dependent replication (*rep*), site-specific integration into the primate genome, and regulatory control of AAV promoters. It is composed of two promoters at map positions 5 (p5) and 19 (p19) which encode four colinear proteins arising from spliced and unspliced transcripts. The p5 promoter encodes proteins of 68 and 78

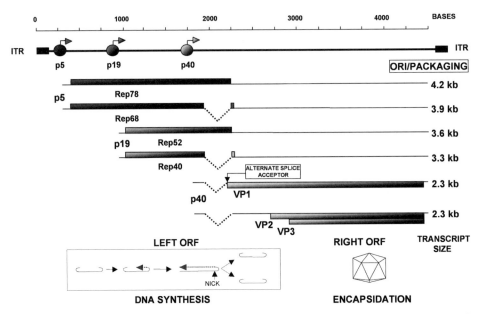

FIGURE 1 Transcriptional map of AAV2. Shown are the two major open reading frames of AAV2, the left consisting of the p5 and p19 promoters and responsible for AAV genomic replication and integration (Rep), and the right consisting of the p40 promoter, which controls expression of the capsid genes (Cap). The inverted terminal repeats (ITRs), corresponding transcripts with splice sites, and proteins with sizes and molecular weights, respectively, are also depicted.

kDa, collectively termed Rep68/78 or simply "Rep," which are the primary mediators of the left ORF function. The right ORF is under control of a promoter at map position 40 (p40) and encodes three colinear structural proteins, VP1, VP2, and VP3, which arise from alternative translational start sites and are necessary for virion encapsidation (*cap*). The AAV genome is flanked by palindromic ITRs which, along with the non-ITR "D" sequence, are essential for viral DNA replication (*ori*), virion encapsidation, and host chromosomal integration [25–28].

Adeno-associated virus exhibits a wide host (from avians to primates) and tissue range. It is likely that cellular infection involves a specific receptor, as in the case of the autonomous parvovirus B19 (Fig. 2, step 1) [29]. Recently, heparin sulfate proteoglycan (HSP) has been implicated as the primary receptor for AAV2, although potentially contradictory data have also been reported [30,31]. Furthermore, additional coreceptors, including basic fibroblast growth factor receptor (bFGFr) and $\alpha V\beta 5$ integrins have been identified, but, again, the findings are not entirely supported by other laboratories [32–34]. After receptor binding, AAV appears to enter the cell via clathrin-coated pits (step 2) in a fashion partially dependent upon dynamin, a clathrin-associated GTPase known to be important for internalization of receptors and their ligands, and then traverses the cytoplasm to the nucleus (step 3), where virus uncoating occurs (step 4) [35,36]. Transport to the nucleus may be facilitated by specific nuclear localization signals on capsid proteins [37,38]. Within

the cell, the single-stranded AAV genome is converted to a double-stranded form using cellular enzymes and the hairpin ITR as a self-annealing primer (step 5). In the absence of helper virus coinfection, transcription and translation of the rep gene occur (step 6), which mediate AAV genomic integration into the host chromosome (step 7). For wild-type AAV, integration is site specific, inserting in human chromosome 19q13.3-qter into a region termed AAVS1 [39–41]. Alternatively, AAV could persist extrachromosomally as an episome. Latent AAV infections have been maintained in tissue culture for more than 100 serial passages in the absence of selective pressure, attesting to the stability of viral integration [42]. Productive AAV infection requires defined "helper" functions provided by another DNA virus, classically adeno- or herpes viruses [25,26]. For adenovirus, these functions include E1a, a transcriptional activator; E1b and E4, which promote mRNA transport from the nucleus to cytoplasm; and E2A and VA RNAs, which stabilize transcripts and enhance translation [43]. The AAV-encoded Rep68/78 protein, which possesses DNA binding, helicase, and endonuclease activity, binds to a specific region of the ITR termed the Rep binding site (rbs), and site-specifically nicks the parental strand at nucleotide number 124 at a sequence known as the terminal resolution site (trs) (step 8) [25,26,43]. Additional DNA replicative intermediates are formed which may also serve as templates for transcription and expression of the cap genes (step 9). Mutational studies have demonstrated that encapsidation of single-stranded DNA viral genomes is specifically

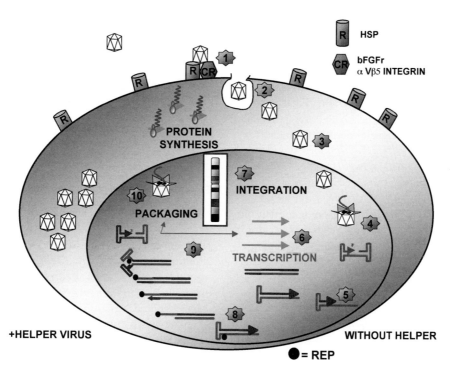

FIGURE 2   Schematic depiction of AAV2 biology. See text for a detailed discussion. HSP = heparin sulfate proteoglycan, bFGFr = basic fibroblast growth factor receptor.

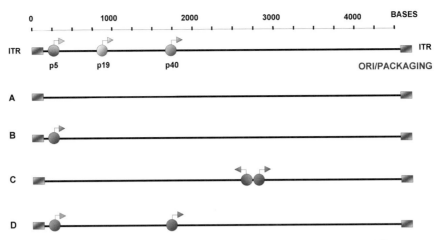

FIGURE 3 Comparison of wild-type AAV-2 sequence with recombinant constructs. All vectors are flanked by AAV2 ITRs, which are necessary for DNA replication, encapsidation, and host-cell integration. The major rAAV motifs include: (A) a vector in which the transgene is expressed from the AAV ITR alone; (B) a single promoter, generally heterologous (non-AAV) which can include internal ribosome entry site (IRES) sequences; and two promoters in either (C) antilinear or (D) colinear fashion. Again, promoters are usually heterologous, although several earlier vectors have included AAV promoters as well.

linked to products of the p19 promoter [44]. Completion of the replicative cycle results in relatively equal encapsidation of plus (similar to coding sequence or mRNA) or minus single-stranded DNA species (step 10).

Biologically, the tissue tropisms of different AAV serotypes are not identical, implying that receptors or coreceptors for the different serotypes may vary. For example, AAV4 and AAV5 do not appear to use HSP as a receptor, and vectors based upon these viruses transduce respiratory and specific neural subpopulations with greater affinity than AAV2-based vectors [45,46]. Similarly, AAV1-based vectors appear to transduce muscle, and AAV3 vectors have been reported to transduce hematopoietic progenitors more efficiently than AAV2-based vectors [47,48]. Finally, host immune responses to one serotype should not preclude subsequent transduction with a vector derived from another serotype [49].

Importantly, since its original discovery, AAV has not been identified as a cause of disease in animals or humans [25,26]. In fact, infection with wild-type AAV may exert an onco-protective effect [50,51]. AAV infection inhibits growth of human melanoma and cervical carcinoma cells and transformation by the activated H-*ras* oncogene and bovine and human papillomaviruses *in vitro* [52–54]. These anti-oncogenic properties have been mapped to the p5 product (Rep68/78), in conjunction with the inverted terminal repeats [50]. Furthermore, wild-type AAV2 infection of tumor cells has been reported to sensitize them to the cytotoxic effects of certain chemotherapeutic agents [55].

Recombinant AAV-based vectors have been developed using several strategies [43,56]. In one early strategy, the transgene of interest was placed under control of an endogenous AAV promoter, typically p40. Another strategy was to remove all endogenous AAV transcriptional units and replace them with strong, heterologous viral or cellular promoters (Fig. 3). This is currently the most common form, as removal of AAV sequences minimizes the opportunity to generate wild-type AAV by homologous recombination during the encapsidation process. For particularly large transgenes, such as the cystic fibrosis transmembrane conductance regulator (CFTR), the ITRs themselves, which have weak intrinsic promoter activity, have been used to drive gene expression [21]. Finally, vectors have been developed that incorporate features of each of the above strategies (e.g., mixing AAV with heterologous, non-AAV promoters, etc.). In most instances, the p5 ORF has been removed to provide additional space for transgene insertion and because the p5 product (Rep68/78) has been demonstrated to inhibit either heterologous promoters or transformation efficiency [57].

2. Vector Packaging, Purification and Titration

The wild-type AAV genome is 4.7 nucleotides in length. Vectors greater than 115–120% of the wild-type size are generally not packaged efficiently [43,58,59]; however, it should be emphasized that many cDNAs are sufficiently small to meet this size constraint. Recently, a strategy to circumvent these size limitations has been developed using a novel "split vector" approach that increases the encapsidation capacity to about 10 kb (Fig. 4). Recombinant AAV (rAAV) vectors have been shown to form intracellular concatamers after transduction [60]. Several groups have exploited this property by physically dividing the vector into two separate, or split, vectors. One vector (the 5′ vector) contains the promoter

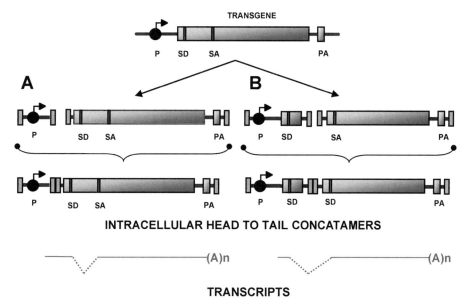

FIGURE 4 Split or dual vector strategy for expression of large transgenes using rAAV vectors. In this strategy, a large expression cassette is split into two rAAV vectors. The 5' vector may contain (A) the promoter/enhancer or (B) the promoter/enhancer with part of the transgene including a splice donor site (SD). The corresponding 3' vector contains the polyadenylation signal (PA) and the transgene (A) or part of the transgene with splice acceptor site (SA) (B). Transduction of cells with both vectors of a vector set generates intracellular concatamers, some of which are in the correct head-to-tail orientation, permitting expression of the correct transcript. This strategy increases the encapsidation capacity of rAAV vectors to about 10 Kb.

and transcriptional enhancer elements with or without the 5' region of the transgene of interest with a splice donor (SD) site. The other vector (the 3' vector) contains either the entire transgene or the 3' portion of the transgene with a splice acceptor (SA) and the polyadenylation (polyA) signal. For split transgenes, splice acceptor or donor sites would be engineered to suitably reconstruct the appropriate transgene mRNA intracellularly. Transduction of cells with both vectors could result in the formation of "head-to-tail" concatameric forms with formation of the appropriate expression cassettes. Studies have demonstrated that this approach can work both *in vitro* and *in vivo* in murine models with surprising efficiencies, with up to 60–80% of that corresponding to transduction with a single vector. These studies have been performed using both "reporter" genes such as β-galactosidase and luciferase and clinically relevant transgenes such as erythropoietin, while additional studies involving clotting factor FVIII, dystrophin, and CFTR are underway [61 65].

A full description of rAAV encapsidation, purification, and titer determination methodologies is beyond the scope of this review; several more detailed discussions are available [66–68]. Briefly, AAV vectors are generally doubly defective; disruption or removal of AAV-encoded genes necessary for replication and encapsidation by insertion of recombinant transgenes necessitates provision of AAV functions *in trans* in addition to "helper" virus functions for productive replication (Fig. 5). AAV vectors are currently encapsidated using a

variety of methods: (1) cotransfection of vectors with an AAV "helper" plasmid encoding AAV *rep* and/or *cap* functions (but lacking ITRs so they cannot be packaged) into helper virus infected cells; (2) transfection of vectors into helper-virus-infected cells engineered to express AAV *rep* and *cap*; (3) transfection of cells with the vector, a separate plasmid encoding AAV *rep* and *cap*, and another plasmid encoding helper virus "helper" genes (e.g., adenovirus E1, E2, E4, and VA RNAs); or (4) various combinations of the above [43,56]. The approach in which only helper virus genes are provided has the added advantage of not using an actual helper virus, per se, essentially eliminating the risk of helper virus contamination of rAAV stocks and significantly simplifying safety and regulatory issues [69,70]. Robust expression of *cap* and appropriately regulated expression of *rep* appear important for efficient encapsidation [71,72]. Encapsidation efficiency may be dependent upon not only the size of the vector but also the encoded transgene [43]. The quality and purity of DNA, cells, and helper virus; transfection efficiency; and presence of wild-type AAV contamination of helper virus stocks can all affect rAAV encapsidation.

Sequence homology between the vector and plasmid encoding AAV *rep* or *cap* functions is minimized or eliminated to prevent generation of wild-type AAV by homologous recombination. However, depending upon the method of encapsidation, replication-competent "wild-type-like" AAV can be generated during the encapsidation process and can

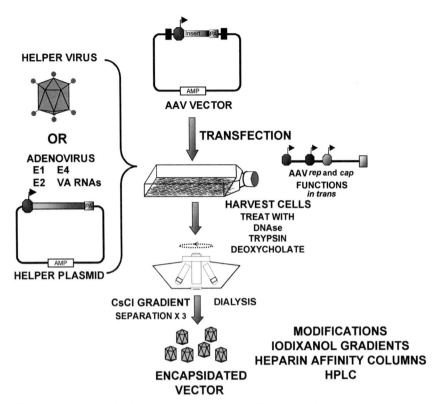

FIGURE 5   rAAV packaging scheme. Currently used rAAV vectors are doubly defective and require AAV *rep* and *cap* provided *in trans*, in addition to classical "helper" virus functions for productive replication. Helper virus functions can be provided by either infection with the helper virus or by transfection with plasmids expressing the appropriate genes. The rAAV vector is transfected into cells with helper virus and AAV *rep/cap*, harvested, and processed through an isopycnic cesium chloride density gradient. Newer methods include Iodixanol gradients, heparin sulfate column purification, and HPLC.

contaminate vector stocks [73]. Although there has been a report that wild-type AAV coinfection may result in improved transduction with rAAV vectors, most laboratories have sought to minimize wild-type AAV contamination for a number of reasons [74]. Generally, wild-type AAV replicates preferentially and at the expense of rAAV and helper viruses, thereby adversely affecting the titers of rAAV and helper virus stocks. Contamination of rAAV stocks with wild-type AAV could affect the frequency and site(s) of vector integration. Finally, in an *in vivo* setting, coinfection with rAAV and wild-type AAV would increase the likelihood of vector rescue and spread after superinfection with helper virus.

Traditionally, rAAV vectors are purified by several rounds of isopycnic CsCl gradient centrifugation. Stocks are then dialyzed to remove the CsCl and heated to inactivate potentially contaminating helper virus. This approach remains cumbersome and labor intensive and can be highly variable. Thus, other strategies including more efficient transfection protocols [75,76], better vector purification systems [77], use of different helper viruses such as herpes simplex virus (HSV) [78,79], and packaging cell lines somewhat analogous to those for retroviral vectors are under development in several laboratories (Wong and Chatterjee, unpubl. data)

[80–82]. Wild-type AAV is known to inhibit adenovirus replication and production, and this inhibition is probably mediated by Rep68/78. While at the National Institutes of Health, we noted that coinfection of cells with AAV and herpes simplex (HSV-1), even at AAV-to-HSV ratios as high as 1000:1, did not significantly affect the number of HSV-1 plaques, although plaque size was somewhat smaller for coinfection with larger amounts of AAV (Wong and Chatterjee, unpubl. data). Reasoning that rAAV vector production was dependent upon both Rep68/78 expression and robust replication of the helper virus, we have used HSV as a helper for almost 10 years [83]. More recently, other purification strategies that have been developed include iodixanol gradients, heparin sulfate column purification, and high-pressure liquid chromatography (HPLC) [67,68,84–88]. These newer strategies have streamlined vector production, resulting in >95% pure vector stocks with fewer contaminating nontransducing defectives (particle-to-infectivity ratios of less than 100). Generally, recombinant AAV vector titers of 10^9 to 10^{12} particles per milliliter are obtained, with minimal wild-type contamination.

Characterization of rAAV vector titers has not been standardized, and different laboratories employ a variety of methods. Titers of rAAV stocks have been determined using techniques that measure (1) DNA content relative to a known standard (DNA assay), (2) measurement of expression of a "reporter" transgene such as β-galactosidase or placental alkaline phosphatase following transduction of cells (functional assay), or (3) using an "infectious" titer that measures the ability of rAAV vectors to infect and replicate in the presence of AAV *rep* and *cap* functions and helper virus (infectious center assay). As with the encapsidation of most DNA viruses, defective, nontransducing particles can be generated during the rAAV encapsidation process. The measurement of these particles or contaminating plasmid DNA using a quantitative DNA analysis results in an overestimation of the vector titer. In contrast, because gene expression is dependent upon formation of a double-stranded DNA template by synthesis of the second strand of the vector, a process that may be impaired in some cell types, measurement of "reporter" gene expression may underestimate the actual vector titer. Currently, the infectious center assay, which measures replication- and transduction-competent vector, is considered the most accurate and reliable assay, while other techniques, including real-time polymerase chain reaction (PCR), are being evaluated [86]. Recombinant stocks are also routinely checked for the presence of infectious helper virus using standard plaque assays and for contaminating wild-type AAV using hybridization or complementation or marker rescue assays of an rAAV vector encoding a "reporter" gene.

What are the potential advantages of AAV vectors (Table 1)? AAV vectors retain the stability, wide host range, and lack of pathogenicity of the parental virus. Pretreatment of cells with inhibitors of proliferation, such as aphidicolin, fluorodeoxyuridine, methotrexate, or nocodazole, does not affect vector transduction, implying that, unlike MLV-based vectors, mitosis is not essential for rAAV transduction [89,90]. These findings are supported by demonstration of efficient AAV vector transduction of nonproliferating

TABLE 1   Advantages and disadvantages of recombinant AAV vectors

Advantages
Safe
Wide host range
Transduce nonproliferating cells
Stable genomic integration
Capable of long-term gene expression *in vivo*
Low immunogenicity
Potential for site-specific integration into human cellular DNA
Stable virions
Disadvantages
Comparatively small encapsidation capacity (5 kb)
Difficult to produce in large quantities

respiratory epithelial cells [91] and cellular populations that are normally nonproliferating, including postmitotic neurons [92–94], retinal [95–96], cochlear [97], muscle (smooth [98–99], cardiac [100], skeletal [101–104]), noncytokine-stimulated hematopoietic progenitors [105–107], human monocyte macrophages [108,109], and hepatocytes [110] (reviewed in McKeon and Samulski [111]). Furthermore, a preliminary study by Wu *et al.* utilizing Alu–PCR supported rAAV integration into nonproliferating neuronal cells [112]. Thus, important cellular populations of either slowly proliferating or quiescent cells may now be amenable to genetic manipulation.

Recently, rAAV vectors have been shown to efficiently transduce murine skeletal muscle, with evidence of transgene expression for 18–24 months [102–104]. Of note is that transgenes were often under control of the cytomegalovirus (CMV) immediate-early (IE) promoter, a highly active viral promoter frequently silenced in murine *in vivo* transduction models. In addition, immune responses to rAAV in these studies were low when compared to adenoviral vectors [103,104]. Finally, several potentially clinically relevant genes, including erythropoietin [102], and Factor IX [113], could be expressed at therapeutic levels following muscle transduction with rAAV vectors. These studies were initially piloted in mice but have subsequently been confirmed in a canine model and have resulted in a promising clinical trial for Factor IX deficiency (hemophilia B) [114]. Recent developments in tightly regulated, chimeric inducible promoters, particularly the tetracycline- and rapamycin-regulatable systems, have increased the overall utility and flexibility of this approach, allowing expression of transgenes to be essentially turned on and off at will [115–118].

Short, defined transcripts can readily be expressed from AAV vectors, a highly desirable feature when expressing either antisense RNA or ribozymes whose function could be significantly compromised by the addition of adventitious sequences [66]. All virus-encoded genes have been removed from most currently available rAAV vectors, making them less immunogenic than vectors, such as adenovirus, that express multiple immunogenic viral proteins after transduction [104]. Additionally, AAV vectors frequently integrate as multi-copy tandem repeats [104,119,120], potentially enhancing transgene expression.

Removal of AAV-encoded genes in the process of vector development can alter the biology of recombinant vectors relative to the wild-type virus. For example, the p5 products (Rep68/78) mediate site-specific integration of wild-type AAV into human chromosomal DNA [25,26,121]. Rep68/78 also mediates site-specific integration of rAAV vectors and DNA containing the AAV ITRs [122,123]. Although the exact mechanism of Rep68/78-mediated, site-specific integration is as yet undefined, Rep has been shown to bind to both AAV ITRs and AAVS1 at analogous sequences known as Rep binding sites. However, the *rep* gene product also downregulates expression from AAV and a variety of heterologous promoters and has been removed from currently used rAAV vectors. This was done both to provide additional space for transgene insertion and to minimize risks for generation of wild-type AAV during the encapsidation process. Using Southern blot analyses, we demonstrated that AAV vectors lacking the p5 ORF do indeed integrate into both cell lines and primary human hematopoietic cells [106–107]. Similarly, Nakai *et al.* [124] demonstrated *rep*-deleted rAAV integration into primary murine hepatocytes, and Omori *et al.* [125] into leukemic cell lines. However, evidence is accumulating that p5-deleted, wild-type-AAV-free rAAV vectors do not integrate site specifically into human AAVS1 and may have reduced integration efficiencies (Chatterjee, Fisher-Adams, and Wong, unpubl. data) [122,126–129]. Indeed, nonintegrated episomal forms have been described following AAV vector transduction [91,129]. Significant efforts are now underway to reinstate this desirable property to rAAV vectors by provision of AAV Rep *in trans*. Lamartina *et al.* delivered Rep68 to cells complexed with a polycationic liposome (lipofectamine) and were able to demonstrate site-specific integration [130]. Palombo *et al.* have constructed baculovirus and adenovirus/AAV hybrid vectors that encode AAV *rep* and have demonstrated vector integration into AAVS1 in 293 and hepatoma cells, respectively [131,132]. Finally, Rinaudo *et al.* [133] fused a C-terminally deleted form of Rep68 to the truncated form of the hormone binding domain of the human progesterone receptor and inserted the construct into cells. Rep expression was dependent upon treatment with RU486, a progesterone analog, and promoted site-specific rAAV integration [133].

Recombinant AAV vectors encapsidate single-stranded DNA, while a double-stranded form of the vector genome is necessary for transcription and transgene expression after transduction. Conversion to double-stranded forms could result from one of two pathways. The more commonly accepted pathway is intracellular second-strand synthesis using a self-priming mechanism involving the ITR. A block in second-strand DNA synthesis occurring after vector entry has been demonstrated by some investigators and could be complemented by the adenovirus E4 ORF6 gene product [134,135]. Augmentation of second-strand synthesis might also contribute to enhanced transgene expression or transduction following treatment of rAAV-transduced cells with "DNA-damaging" or chemotherapeutic agents [135,136], findings that mirror reports from the late 1980s documenting limited wild-type AAV replication in cells exposed to various "genotoxic" stresses [137–140]. Whether these findings are limited to specific vectors (the requirement for second-strand synthesis was achieved by using vectors encoding β-galactosidase) or cell types or are universally applicable has yet to be determined. Transduction has been reported using highly purified, helper-virus-free rAAV stocks, implying that the target cell may also influence the efficiency of transduction [96].

Because AAV encapsidates equal quantities of both plus and minus strands, another potential pathway for generating double-stranded intracellular rAAV genomes is via intracellular hybridization of complementary plus and minus strands. This mechanism requires that sufficient quantities of both plus and minus strands enter the same cell, hybridize within the nucleus, and form double-stranded templates subsequently used for transcription. This mechanism has been postulated to occur following rAAV transduction of liver cells [141].

Finally, because the majority of currently used rAAV vectors encode only inserted transgenes and no AAV genes, immune responses to rAAV transduced cells have generally been comparatively limited. In animal models, systemic rAAV administration can generate an immune response that has been primarily humoral and may preclude successful administration of another vector of the same serotype as the initial challenge [142]. Humoral responses can also be directed against the transgene, as well, when expressed following intramuscular administration [143]. Poor transduction of antigen-presenting cells, particularly dendritic cells (DCs), has been postulated as a potential mechanism to explain the comparatively low immunogenicity of rAAV vectors [144]. These results have recently been modified. Zhang *et al.* reported that immature murine dendritic cells could be transduced with an rAAV vector encoding β-galactosidase and could elicit cytotoxic responses [145]. We and others have shown that primary human DCs are transducible with rAAV [146,147]. Subsequent studies have shown that the route of vector administration affects the type of host immune response generated against the inserted transgene. Brockstedt *et al.* administered an rAAV vector encoding ovalbumin to C57BL/6 mice intraperitoneally (IP), intravenously (IV), subcutaneously (SC), or intramuscularly (IM) and monitored immune responses to the transgene and vector. Cytotoxic response to ovalbumin were detected following IP, IV, and SC vector administration, while humoral responses were noted after IM administration [148]. Xiao *et al.* also noted that host immune responses were affected by the route of rAAV administration. Vector administered IV by tail vein resulted in CD4-dependent anticapsid B-cell responses that were long lived. In contrast, vector administered into the portal circulation generated B-cell responses that were only partially T-cell dependent and were short lived [149].

B. Autonomous Parvoviruses

Unlike the AAV, autonomous parvoviruses productively replicate in target cells utilizing factors provided by the host cell. The genomic organization of autonomous parvoviruses is similar to AAV with two major ORFs flanked by nonidentical ITRs (Fig. 6). The left ORF is comprised of a promoter at map position 4 (p4) and encodes nonstructural proteins (NSs) necessary for DNA replication and transactivation of the promoter for the right ORF. The right ORF is under control

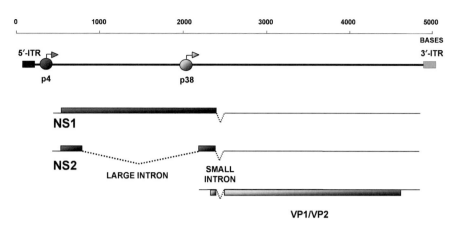

FIGURE 6 Minute virus of mice (MVM) transcriptional map. The transcriptional map of MVM, an autonomous parvovirus, is similar to AAV but lacks a second promoter and has an additional major intron in the left open reading frame.

of a promoter at map position 38 (p38) and encodes capsid proteins. Replication is limited to cells that enter S phase following virus entry. Consequently, autonomous parvoviruses are not known to develop latent infection; genomic integration has not been detected [150]. Like AAV, two autonomous parvoviruses, the prototypic minute virus of mice (MVMp) and the rodent parvovirus H-1, have wide host ranges which include human cells. H-1 is known to infect humans without apparent clinical sequelae [151]. Among the Parvoviridae infecting humans, only the autonomous parvovirus B19 is known to be pathogenic.

The antineoplastic activity of parvoviruses was originally described in 1968 by Toolan [152] and subsequently resulted in a clinical trial of H-1 in patients with osteosarcoma [153]. Extensive studies have subsequently demonstrated that MVMp and H-1 exert a preferential killing of transformed cells and, thus, may be uniquely suited for the gene therapy of cancer [153,154]. MVMp is oncotropic, binding to a cell surface receptor known to be sensitive to neuraminidase and trypsin [155], and is cytolytic to transformed but not normal cells [154]. This cytotoxic effect has been mapped to both the amino and carboxyl terminal regions of the major nonstructural protein, NS-1, which is homologous to the AAV *rep78* gene product and may be modulated by NS-2 [156,157]. The oncolytic effect of the NSs may be due to induction of apoptosis in tumor cells [158]. *In vivo* experiments demonstrated MVMp-mediated suppression of Ehrlich ascites tumor growth in mice when virus was injected at a site distant from the tumor inoculum [159]. Similar studies have demonstrated selective inhibition of tumor growth following infection with H-1 both *in vitro* and *in vivo* [160]. Experiments with mutants of the p53 tumor suppressor gene suggest that a functional wild-type p53 product contributes to cellular resistance to H-1-mediated cytotoxicity [161]. Interestingly, these results are somewhat analogous to those of recently engineered, conditionally replicative, oncolytic adenoviruses. These viruses replicate poorly in cells with a wild-type p53 but are able

to replicate and lyse tumor cells with mutant p53 [162,163]. Thus, the autonomous parvoviruses MVMp and H-1 are well suited as vectors for *in situ* tumor injection as they both are oncotropic and specifically lytic to neoplastic cells.

Recombinant vectors based upon MVMp and LuIII, another autonomous parvovirus that efficiently infects human cells, have been described [164,165] (reviewed in Corsini *et al.* [166]). In these constructs, transgenes have been placed under control of viral p4 or p38 promoters or, more recently, under control of tissue-specific or inducible promoters [167]. Vectors are encapsidated by providing NS and/or capsid functions *in trans* as described above for AAV vectors [168,169].

III. APPLICATIONS OF RECOMBINANT PARVOVIRUS VECTORS TO CANCER GENE THERAPY

Certainly, the recognized strengths of rAAV vectors lie in their ability to safely and stably transduce nonproliferating cellular targets, including neural, muscle, and stem cells. Several strategies to express potential antitumor transgenes via intramuscular gene delivery, including development of antitumor vaccines, expression of inhibitors of insulin-like growth factors, or angiogenesis, have been proposed; however, studies utilizing this approach with parvovirus vectors have yet to be reported [170]. In contrast, a variety of strategies employing combinations of suicide and immunostimulatory cytokine genes expressed from parvovirus vectors have been developed for potential treatment of neural tumors and will be discussed under the heading of each major strategic approach.

A. Gene Transfer to Hematopoietic Cells

Genetic modification of hematopoietic cells is being activity pursed for a number of reasons. Genetic "marking" of hematopoietic progenitors has been used to study the biology of hematopoiesis *in vivo* [171]. A variety of common malignancies arise from hematopoietic cells, including leukemias and lymphomas, or may metastasize to the marrow. Gene transfer strategies have been developed to identify potential contaminating tumor cells within autologous stem/progenitor grafts and for "purging" of contaminating malignant cells [172–175]. Cells of the immune system originate from the marrow and could be harnessed for generation of antitumor responses. Furthermore, the hematopoietic system is a major site of antitumor-treatment-related toxicity, and gene transfer strategies to ameliorate chemo- and radiotherapy-associated myelosuppression have been developed. These issues will be discussed in greater detail later.

Although initial studies were encouraging, it became apparent that the overall marking efficiency of hematopoietic progenitors using standard oncoretrovirus vectors in large animal models and in human clinical trials was relatively low [172–175]. This may result from the strict requirement of oncoretrovirus vectors for actively proliferating cellular targets for efficient gene transfer and proviral integration [5], combined with the nonproliferating state of true hematopoietic stem cells [176,177]. As a result, attempts to enhance retroviral vector transduction of hematopoietic progenitors have incorporated co-culture with retroviral packaging cell lines, use of fibronectin [178], or stimulation of target cells with a variety of cytokines to promote proliferation [179]. However, cytokine stimulation may shift stem cells along differentiation pathways, resulting in transduction of lineage-committed cells with limited self-renewal capacity [180].

Previous studies from our laboratory demonstrated that cellular proliferation was not a requisite for efficient AAV transduction [89]. We subsequently demonstrated that purified rAAV vectors were capable of transducing marrow- or cord-blood-derived, $CD34^+$-enriched cells isolated from over 40 individuals. Furthermore, efficient gene transfer occurred with or without prior cytokine stimulation of target cells and resulted in stable vector integration as determined by both Southern blot and FISH (fluorescent *in situ* hybridization) analyses [106,107]. These studies were extended by examination of rAAV-mediated gene transfer to primitive human myeloid progenitor cells derived from marrow and cord blood in long-term cultures and long-term culture-initiating cell (LTC-IC) assays *without* selection for transduced cells. Single-colony analyses demonstrated that AAV vectors transduced $CD34^+$ and the more primitive $CD34^+/CD38^-$ subpopulations. Gene transfer was observed in LTC-ICs derived from 5-, 8-, and 12-week cultures. Although transduction was observed in every donor analyzed, a wide range of gene transfer (5–100%) was noted. AAV transduction of LTC-ICs was stable, with week-8 and -12 LTC-ICs showing comparable or better transduction relative to week-5 LTC-ICs. FISH analyses showed that 9–28% of $CD34^+$ and $CD34^+/CD38^-$ cells contained stable vector integrants as evidenced by chromosome-associated signals in metaphase

spreads. Comparisons of interphase and metaphase FISH suggested that a fraction of cells also contained episomal vector at early time points after transduction. Despite the apparent loss of the episomal forms with continued culture, the number of metaphases containing integrated vector genomes remained stable long term. Furthermore, transgene (placental alkaline phosphatase) transcription and expression were observed in $CD34^+$ and $CD34^+/CD38^-$ LTC-ICs [107].

As true hematopoietic stem cells are believed to be quiescent or nonproliferating [176,177], we examined the ability of rAAV vectors to transduce nonproliferating hematopoietic progenitors. We noted that >96% of $CD34^{+++}/CD38^-$ cells resided in G0 as determined by staining with Hoechst 33342 (for DNA) and pyronin Y (for RNA). These cells were transduced overnight with an rAAV vector (particle MOI [multiplicity of infection] of 1500), and gene transfer efficiencies were determined by amplification of vector sequences from individual hematopoietic colonies derived from (LTC-ICs). In the absence of selective pressure, 10–83% of colonies arising from extended week-8 and -12 LTC-ICs from the $CD34^{+++}/CD38^-$ were found to contain vector-specific signals, and the number of cells containing integrated vector remained stable over the ensuing 10 weeks of study. Previous studies have shown that progenitors in extended 12-week LTC-ICs were not readily transduced with standard oncoretroviruses [181].

We next tested the potentials of rAAV vectors for the gene transfer to primitive human hematopoietic stem/progenitor cells capable of long-term *in vivo* multi-lineage reconstitution of immune-deficient NOD/SCID mice. It should be noted that the "SCID repopulating" fraction of human progenitors has been refractory to transduction with standard oncoretrovirus vectors [182]. Fresh cord blood (CB)-derived $CD34^+/CD38^-$ cells were sorted by flow cytometry, transduced with a CsCl-purified rAAV vector at a particle MOI of 400, and infused into sublethally irradiated NOD/SCID mice. Comparable levels of engraftment were observed in mice transplanted with either transduced or untransduced cells, indicating that rAAV vectors did not compromise engraftment. Human cells were identified in these xenografts by the expression of CD45, a common lymphocyte antigen. rAAV sequences were demonstrable in the $CD45^+$ fraction sorted from the marrow of each mouse transplanted with transduced cells. Furthermore, rAAV vector sequences were also demonstrable in the myeloid ($CD14^+$) and B ($CD19^+$) cell subpopulations sorted from human cells recovered from the marrow of NOD/SCID xenografts, indicating that transduced $CD34^+/CD38^-$ CB cells were capable of contributing to both lymphoid and myeloid lineages. Significantly the $CD34^+$ human progenitor cell fractions recovered from the marrow of NOD/SCID mice at 8, 12 and 24 weeks post transplantation also contained rAAV-transduced cells, suggesting that rAAV-transduced $CD34^+/CD38^-$ cells were capable of either surviving as, or giving rise to, hematopoietic progenitors for extended periods *in vivo*. Importantly, analysis of rAAV sequences in CD34$^+$ human clonogenic progenitor cells derived from the marrow at 12 weeks post-transplantation revealed that 65–75% of myeloid colonies were derived from transduced CD34$^+$/CD38$^-$ cells, suggesting that rAAV-transduced progenitors retained their clonogenic and hematopoietic properties *in vivo*.

Similar findings have been obtained from other laboratories (reviewed in Chatterjee and Wong [183]). Miller *et al.* [184] reported transgene expression in 20–40% of hematopoietic colonies following transduction with a recombinant AAV vector encoding human γ-globin. Zhou and colleagues [105, 185] described transduction efficiencies of 33–75% and 50–80% to human umbilical cord blood and murine marrow hematopoietic progenitors, respectively, by utilizing an AAV vector encoding neomycin phosphotransferase (Neo^R, conferring cellular resistance to G418). Again, efficient transduction of human umbilical cord progenitors occurred in the absence of cytokine prestimulation [105]. Walsh [186] described the use of an AAV vector encoding the Fanconi anemia complementation group C (FACC) gene to phenotypically correct the Fanconi anemia defect *in vitro* within peripheral blood CD34$^+$ cells isolated from a patient with this disorder. In this study, transduction efficiency was up to 60%, and transduced cells could be engrafted into a SCID-hu murine model [187], Luhovy *et al.* [188] demonstrated that rAAV vectors mediated transgene expression in 25% of human hematopoietic cells in long-term culture, a system which may assay more primitive progenitors; these are findings that we have noted as well [107]. Schimmenti *et al.* studied rAAV transduction of CD34$^+$ cells in a rhesus stem cell transplantation model. rAAV was detected in bone marrow, in granulocytes, and in purified populations of cells, supporting multi-lineage repopulation of stem cells, although the overall frequency of vector marked cells was low (about 1:10^5) [189].

It should be emphasized, however, that not all laboratories have reported such encouraging results [190]. Malik *et al.* [129] constructed an rAAV vector encoding a truncated version of the nerve growth factor receptor (tNGFr), and studied transduction of K562 cells. Transduced cells expressed tNGFr on their surface and could be sorted and purified by flow cytometry using an NGFr-specific antibody. At an MOI of 130, almost all cells expressed tNGFr, but the proportion declined to 62% after 2 months in culture and was associated with concomitant loss of vector genomes. Transduction of primary CD34 cells was less efficient (8–15%). Gardner *et al.* [191] transduced CD34$^+$ progenitors with an rAAV vector encoding LacZ and analyzed the ability of the CD34$^+$ cells to differentiate to T cells in an *in vitro* T-lymphopoiesis system. Approximately 71–79% of transduced cells from either adult marrow or umbilical cord blood expressed LacZ, and there was no evidence of toxicity. However, LacZ expression decayed over 35 days, and vector sequences could not be detected in CD3$^+$ cells by PCR. van Os *et al.* [192] noted

that 60% of primitive cobble-stone-forming cells in a murine model were transduced with an rAAV vector, but transgene expression could not be detected and fewer than 0.1% of marrow cells contained vector sequences by PCR at 6 months after transduction. Similarly, Nathwani *et al.* [193] reported that approximately 42% of CD34$^+$ and 22% of CD34$^+$/CD38$^-$ umbilical cord blood cells were transduced with an rAAV vector, but gene expression, as measured by drug (trimetrexate) resistance, was transient. Finally, Hanazono *et al.* [194] compared rAAV and oncoretrovirus transduction of primate peripheral blood T lymphocytes using vectors that did not encode transgenes so as to minimize the elimination of vector-containing cells by a host immune response. rAAV vectors demonstrated better short-term marking than retroviral vectors, somewhat mirroring the results of Gardner *et al.* [191]. Of note is that cells were cultured under conditions that maximized cell cycling (in the presence of IL-2, 500 IU/mL). This was necessary for MLV-based retroviral transduction but could have been adverse to rAAV transduction (see following discussion).

These disparities may reflect multiple factors, including differing vector purity (crude lysates vs. purified vectors), constructs, vector production and purification strategies, source, method of purification, and culture conditions (amount and type of cytokines) of the hematopoietic progenitors. For example, we and others have noted individual variability in transduction efficiencies, with cells from some individuals being poorly transduced (Chatterjee and Wong, unpubl. data) [195]. However, what is clear from an analysis of multiple reports, is that rAAV vector entry of CD34 cells is not a major impediment. Thus, although there have been discussions about poor heparin sulfate proteoglycan density on CD34$^+$ cells, most laboratories have reported that rAAV is capable of entry into these cells. Furthermore, reporter gene expression (*lacZ*, placental alkaline phosphatase, neomycin phosphotransferase, trimethrexate resistance, etc.) has also been documented independently, implying that second-strand synthesis is not necessarily impaired in this population. The common theme is efficient initial transduction, followed by a decay of transgene expression paralleled by vector genome loss.

The situations where rAAV has excelled for gene transfer have been characterized by transduction of either non- or slowly proliferating target cells. Thus, transduction of neural or adult muscle cells, populations with a low proliferative capacity, has resulted in long-term vector persistence and transgene expression. The biology of rAAV differs from MLV-based retroviruses in that the vector genome may either integrate or persist as a nonintegrated episome. Factors that promote integration versus episomal persistence have not yet been defined, although Balague *et al.* have reported that Rep enhances the frequency of rAAV integration [128]. Theoretically, rAAV might require specific conditions after cellular entry that are necessary for genomic integration. If target cells proliferate rapidly before rAAV has an opportunity to integrate, it probably is lost as an episomal form, particularly in the absence of selective pressure. Of note is that most hematopoietic cell transduction studies were based upon protocols developed for oncoretroviruses in which the target cells were induced to proliferate rapidly in order to enhance transduction. We reasoned that slowly proliferating targets might be better for rAAV transduction and subsequently transduced in the absence of cytokines and maintained cells in the lowest concentration of cytokines necessary to maintain viability. We have also noted that stable transduction is better with CD34 cells isolated from the marrow than from mobilized peripheral blood stem cells, which may be reflective of the proliferating status of the cells (Chatterjee and Wong, unpubl. data).

To determine the frequency of potentially confounding wild-type AAV in hematopoietic tissues, Anderson et al. [196] analyzed 106 human bone marrow samples using PCR but were unable to detect wild-type AAV in CD34+ progenitors; this suggests that vector rescue in transduced cells might be less likely in vivo. Thus, for a variety of reasons, including high transduction efficiencies even in the absence of cytokine stimulation, AAV vectors appear potentially promising for gene transfer to hematopoietic progenitors and are currently undergoing intensive evaluation for in vivo efficacy.

B. Tumor Purging

Several investigators have proposed that AAV2-based vectors might preferentially transduce tumor cells contaminating HSC samples. Veldwijk et al. [197,198] used an rAAV2-based vector encoding green fluorescent protein (GFP) to analyze relative transduction efficiencies of tumor cell lines compared to primary CD34+ cells. Sixty-five percent of the cervical carcinoma Hela and 80% of the sarcoma cell line HS-1 expressed GFP after transduction, while 3–12% of other tumor lines were GFP positive. In contrast, 1.4% of primary peripheral blood CD34+ cells were GFP positive [197]. Similar results were found with breast cancer cell lines, in which 36% of T47D and 41% of MCF-7 cells were GFP positive after transduction [198]. Prior studies by Hoerer et al. [199] demonstrated that some freshly isolated primary tumors, specifically melanoma and ovarian carcinoma, could be readily transduced with rAAV vectors encoding either luciferase or LacZ. These results must be interpreted with the caveat that most of the studies were done with cell lines, and a large number of freshly isolated tumors have not yet been examined. Furthermore, the number of contaminating tumor cells is often reduced by several logs following CD34 selection. Therefore, the ratio of normal CD34+ cells to tumor cells will be quite high, so that even though a small percentage of CD34+ cells might be transduced their overall numbers might be sufficiently high to potentially compete with the tumor cells for transduction.

In a somewhat similar fashion, Itou et al. [200] examined gene transfer to seven leukemia cell lines using an rAAV vector encoding neomycin resistance (NeoR) and assessed transduction efficiencies using colony formation and limiting dilution assays. Transduction was most efficient in K562 cells. Vector integration was confirmed by both FISH and Southern blot analyses but occurred randomly [200]. Omori et al. [125] analyzed rAAV transduction of the acute myeloid leukemia cell line AML5 and the non-Hodgkin's lymphoma cell line OCI-LY18. FISH analyses demonstrated vector integration, which was stable for more than 12 months after transduction [125].

C. Chemoprotection of Hematopoietic Cells

High-dose chemotherapy coupled with stem cell rescue has shown promise in the treatment of a variety of advanced tumors [201]. However, complications directly attributable to myelosuppression, including infection or hemorrhage, still pose potentially life-threatening risks for this form of therapy. Conferring resistance of hematopoietic progenitors to the toxic effects of dose-intensified chemotherapy is one approach to this problem. The human multidrug resistance (MDR1) gene encodes a 170-kDa glycoprotein, P-glycoprotein, which actively exports a variety of toxic substances from the cell, including drugs commonly used in this setting such as anthracyclines, epipodophyllotoxins, and paclitaxel [202]. Although MDR1 expression can be detected among primitive human hematopoietic progenitors [203], expression wanes as cells mature, and P-glycoprotein is seldom found among mature myeloid cells. Thus, granulocytes, myeloid cells that form the first line of defense against bacterial infection, are exquisitely sensitive to the effects of chemotherapy. Augmented MDR1 expression among hematopoietic progenitors and their progeny may therefore abrogate the myelosuppressive effects of dose-intensified chemotherapy and permit more intensive and effective regimens.

The feasibility of such an approach was established through the development of murine models. Mice transgenic for the human MDR1 gene were protected from the myelosuppressive effects of chemotherapeutic agents effluxed by the P-glycoprotein transporter. Myeloprotection correlated with specific, increased P-glycoprotein surface expression on hematopoietic cells [204] and could be transplanted to syngeneic mice [205]. Using a more clinically relevant approach, transplantation of murine marrow mononuclear cells transduced with a retroviral vector encoding human MDR1 conferred myeloprotection against the leukopenic effects of taxol and permitted selection of transduced cells in vivo [206]. The myeloprotective effect of MDR1 could be serially transplanted to syngeneic animals [207], suggesting the genetic modification of true hematopoietic stem cells. Ward et al. [208] transduced primary CD34+ human marrow progenitors prestimulated with interleukin-3 (IL-3), IL-6, and stem cell factor with a retroviral vector encoding *MDR1 in vitro*. Vector sequences could be identified by PCR analysis in 18–70% of erythroid (BFU-E) and 30–60% of granulocyte–monocyte (CFU-GM) colonies grown in methylcellulose [208]. These studies have presaged actual clinical trials using *MDR1* encoding oncoretrovirus vectors [209,210]. Although, a myeloproliferative disorder has been described following transplantation of murine progenitors transduced with a retroviral vector encoding *MDR1* [211], this problem has not been observed in human clinical trials.

Encouraged by our previous studies with AAV-mediated gene transfer to hematopoietic progenitors, we constructed an AAV-based vector encoding the minimal ORF of the *MDR1* cDNA under control of the RSV (Rous sarcoma virus LTR) promoter. This vector, CWRSP-MDR, was approximately 200 bases larger than wild-type AAV yet demonstrated titers comparable to smaller AAV vectors. Flow cytometric analyses of CWRSP-MDR-transduced CD34$^+$-enriched normal human marrow demonstrated increased specific P-glycoprotein expression as determined by both antigenic (MRK-16 or UIC2 anti-*MDR1* monoclonal antibodies) and functional (rhodamine 123 dye exclusion) assays (Shaughnessy, Chatterjee, and Wong, unpubl. data). Transduction efficiencies of CD34$^+$ marrow-derived cells from six different donors ranged from 20–100%, values consistent with transduction of this cell population with other AAV-based vectors. Southern hybridization analyses of high-molecular-weight cellular DNA isolated from CD34$^+$ cells, digested with a restriction enzyme that cleaves once within the vector and hybridized with a vector-specific probe demonstrated bands larger than the unit-length vector and were consistent with vector integration [106]. Importantly, no cellular toxicity was detected following vCWRSP-MDR transduction. However, vector encapsidation, perhaps due to the size of the insert, has been inefficient, necessitating development of alternative constructs. A similar rAAV vector employing a CMV promoter has been constructed that confers colchicine and taxol resistance to KB cells and is currently undergoing evaluation (Juty, de Claro, Shaughnessy, Chatterjee, and Wong, unpubl. data). Additional studies are underway to determine the utility of this vector employing the NOD/SCID *in vivo* murine model described earlier.

Baudard *et al.* reported upon the development of a liposome-encapsidated AAV bicistronic plasmid encoding the MDR1 gene and glucocerebrosidase (GC). Administration of this construct to mice resulted in delivery of *MDR1* and GC cDNAs in all the organs tested [212,213].

D. Delivery of Transdominant Molecules

The discovery of oncogenes and tumor suppressor and mutator genes has opened new areas of research in oncology aimed at discovering agents that could selectively inhibit the biological effects of oncogene products or restore the function of tumor suppressors and DNA repair genes. Thus, a new field of research has arisen focusing upon the development of transdominant molecules designed to selectively inhibit gene expression. Transdominants can be easily classified into two main categories: (1) RNA-based (antisense, sense decoys, and catalytic ribozymes), and (2) protein-based (transdominant proteins and single-chain antibodies). We will focus primarily on antitumor transdominants that have been expressed from parvovirus vectors.

1. Anti-sense Transcripts

During the last decade, synthetic antisense oligonucleotides designed to bind selectively to a complementary RNA sequence have been administered to interrupt targeted gene expression both in tissue culture and in animal models *in vivo*. The mechanisms of action of these antisense molecules appear to differ in different systems and include accelerated degradation of sense/antisense hybrids and interference with trafficking, processing, or intra- or intermolecular interactions [214–216].

Numerous studies demonstrate the anti-oncogenic effects of anti-sense molecules targeting specific oncogenes as assessed by changes in morphology, growth rate, ability to form colonies in soft agar, and *in vivo* tumorigenicity. Molecular targets have included cellular oncogenes such as *bcl2* [217], *p210$^{bcr-abl}$* [218], *erbB-2* [219], *fos* [220], and *myc* [221]; regulators of cell proliferation such as cyclin D1 [222], and virus-encoded transforming genes (e.g., E6/E7 ORFs from human papillomavirus, HPV [223]).

The overall utility of anti-sense RNA as therapeutic tools depends not only upon the optimal design and intracellular expression of this RNA but also upon the availability of safe and highly efficient delivery systems. Previous studies have demonstrated that AAV vectors can function as efficient vehicles for anti-sense RNA delivery. AAV vectors can be constructed that express short, distinct transcripts, a property that is useful for RNA-mediated inhibition of gene expression [66]. As a gene transfer approach to abrogate herpes simplex virus infection, we developed an AAV-based vector that encoded both neomycin phosphotransferase and an anti-sense transcript complementary to a portion of the 5' noncoding leader and first coding "AUG" of the HSV-1 ICP4 transcript [224]. Vector-transduced clonal murine cells expressing this transcript were protected from cytopathogenicity, restricted HSV-1 production by 1000–10,000-fold (99–99.9%), and demonstrated prolonged survival (75–80% viability by trypan blue dye exclusion vs. 0% by day 4 postinfection) in comparison to control cells following viral challenge (MOI 0.1). Similarly, an AAV-based vector encoding an anti-sense transcript complementary to both the HIV-1 *TAR* region and polyadenylation signal, common to all HIV strains, and conferring G418 resistance was constructed [83]. Transduced, clonally derived, G418-resistant 293 cells expressing this anti-sense transcript specifically restricted chloramphenicol acetyltransferase (CAT) expression directed from the HIV-1

LTR, as well as HIV replication following transfection with HIV$_{IIIB}$, an infectious molecular clone. Importantly, transduced, nonclonal but G418-resistant human CD4 T-cell lines (H9 and A3.01) expressing the anti-sense RNA demonstrated prolonged protection with a 1000-fold reduction of titratable virus production after challenge with HIV$_{IIIB}$ (MOI 1). Finally, Ponnazhagan and colleagues utilized an AAV vector encoding an antisense transcript complementary to the human α-globin gene to inhibit gene expression by up to 91% [225].

2. Ribozymes

Ribozymes are a class of RNA discovered in the early 1980s that catalytically cleave RNA molecules in a sequence-specific fashion [226]. A variety of ribozyme motifs such as hammerhead, hairpin, axhead, group I intron, and RNAse P are available for use as *trans*-acting catalysts, among which the hammerhead and hairpin ribozymes are the most commonly used. The catalytic ribozyme core is targeted to specific regions of the target transcript by flanking antisense sequences. Ribozymes (Rz) have been designed to efficiently cleave their targets *in trans* and to inhibit gene expression from either cellular oncogenes or virus-encoded genes requisite for productive infection. Ribozymes may be more effective than anti-sense molecules as inhibitors of gene expression; not only do they cleave the target transcript, but, because they act catalytically, each ribozyme can also theoretically destroy multiple targets [226,227].

Several studies have demonstrated the efficacy of ribozyme-mediated inhibition of oncogene expression, with subsequent reversal of the transformed phenotype. Targets have included cellular oncogenes such as activated H-*ras* both *in vitro* [228] and *in vivo* [229], $p210^{bcr-abl}$ [230], c-*fos* [231], vascular endothelial growth factor (VEGF) receptors [232], and virus-encoded oncogenes (discussed later). In addition, hammerhead ribozymes targeting c-*fos* have reversed cisplatin resistance [233] and *MDR1*-mediated resistance to chemotherapeutic agents [234] (reviewed in Irie *et al.* [235]).

Human papillomavirus is an attractive target for rAAV vectors for several reasons. HPV is a common infection, is a recognized cofactor in the development of several clinically important epithelial tumors (most notably carcinoma of the cervix), and no effective antiviral therapy is currently available [236]. Epidemiologic studies have linked coinfection with HPV and wild-type AAV with a lower risk for the development of cervical carcinoma. Walz *et al.* [237] demonstrated that transfected HPV-16 DNA could rescue infectious AAV2, implying that HPV could act as a "helper" virus for productive AAV replication, a conclusion supported by the study of Ogston *et al.* [238]. Both wild-type AAV and an episomal form of AAV inserted within an Epstein–Barr virus plasmid inhibited growth of HPV transformed cells [239,240]. HPV inhibition has been correlated with expression of Rep78, which has been shown to bind to and functionally interfere

with expression from the HPV-16 p97 promoter, as well as part of the origin of replication [241]. Su *et al.* demonstrated that Rep78 interfered with the binding of TATA-binding protein to the p97 promoter, disrupting transcription [242].

Several studies have demonstrated that either anti-sense transcripts or hammerhead or hairpin ribozymes targeting the E6/E7 transforming regions of HPV can bind to their targets intracellularly and disrupt HPV-mediated transformation in tissue culture [243,244]. Kunke *et al.* [245] constructed rAAV vectors encoding anti-sense transcripts or ribozymes targeting the HPV E6/E7 transforming genes or the monocyte chemoattractant protein-1 (MCP-1), which had been shown to inhibit E6/E7 gene expression. HPV-transformed, tumorigenic HeLa or SiHa cells were transduced and implanted in nude mice. Transduction with the MCP-1 encoding vector resulted in the greatest degree of tumor inhibition. Similar results were obtained when *in situ* SiHa tumors were injected with the vector, supporting the potential use of this system for the treatment of HPV-related tumors [245]. Finally, using a different strategy, Sun *et al.* [246] demonstrated that transduction of HPV-transformed cervical carcinoma cells with an rAAV vector encoding folate receptor cDNA resulted in lower cell proliferation and smaller tumor volume in a murine xenograft model.

E. Complementation of Tumor Suppressor Genes

The p53 tumor suppressor gene product acts a regulator of cell proliferation and is the most commonly mutated gene in the development of cancer [247]. Complementation of mutated p53 with a copy of the wild type is being actively explored as a potential gene therapeutic approach to a number of tumors, particularly carcinoma of the lung [248]. Qazilbash *et al.* [249] inserted the wild-type p53 tumor suppressor gene into an rAAV vector and used it to transduce p53-deficient H-358 cells. Analyses of rAAV/p53-transduced cells confirmed wild-type p53 expression, cellular arrest in the G1-S checkpoint of the cell cycle, and augmented apoptosis. Furthermore, direct injection of rAAV/p53 into H-358 tumors implanted subcutaneously into immunodeficient mice resulted in tumor regression in three of five animals but none of the controls, demonstrating that this may be a viable strategy for accessible tumors with known p53 mutations [249].

F. Modulation of Antitumor Immunity/Tumor Vaccines

The development and progression of malignancy is often considered a failure of the host immune system to recognize or to effectively respond to cancer antigens. A rationale for an immunological approach to cancer treatment can be found in reports of spontaneous regressions of advanced tumors, including melanoma and renal cell carcinoma [250], both immunogenic tumors. Furthermore, lymphocytic infiltrates within tumors, implying an antitumor cellular immune response, often signify a better prognosis [251]. Antigenic determinants must be solely or selectively expressed on the tumor as compared to normal cells for an antitumor immune response to be mounted. These antigens must then be presented in the context of the major histocompatibility complex (MHC) in conjunction with costimulatory signals, such as B7.1 or B7.2, so as to be recognized by T cells and eliminated [252,253].

A variety of gene transfer strategies have been developed to potentiate antitumor immune responses. Rosenberg and colleagues [254] have pioneered the use of genetically modified tumor-infiltrating lymphocytes (TILs) for specific delivery to malignant cells. In this strategy, autologous TILs are harvested from the tumor, genetically modified to express a cytokine gene (e.g., tumor necrosis factor, TNF), expanded *ex vivo*, and readministered to the patient. Theoretically, modified TILs home to tumor-bearing areas and serve as sources for localized cytokine production for either immunomodulation or tumor toxicity, thus circumventing significant toxicities seen after systemic cytokine administration [254]. Chimeric T-cell receptors (TCR) composed of the invariant zeta chain of the TCR coupled to a single-chain antibody against a specific tumor-related antigen have been developed. In this strategy, engagement of the single chain antibody with its target results in a proliferative/cytotoxic signal mediated via the TCR zeta chain. Generated cytotoxic responses are not MHC restricted [255,256]. Education of professional antigen-presenting cells to induce a specific antitumor response is also being investigated (see later discussion) [257]. Finally, tumor cells themselves may be genetically modified to express cytokine genes, such as IL-2, IL-4, interferon-gamma (IFN-γ), TNF-α, granulocyte–macrophage colony-stimulating factor (GM-CSF) [258,259], costimulatory signals (B7-CD28) [260], or alloreactive antigens [261], to augment their immunogenicity. As with the case of TILs bearing cytokine genes, introduction of cytokine genes directly into tumor cells results in augmented localized cytokine expression in the area of disease, potentially enhancing tumor immunogenicity while limiting systemic toxicities.

1. Introduction of Cytokine Genes

Interleukin-2 is a pleiotropic cytokine with a central role in the immune response, inducing the proliferation of antigen-stimulated T cells and natural killer (NK) and lymphokine-activated killer (LAK) cells [262]. We and others have developed rAAV vectors that encode IL-2 for immune stimulation. In murine models, expression of IL-2 within otherwise non-immunogenic malignant cells has resulted in induction of cytotoxic antitumor responses, accompanied by protection against subsequent tumor challenge. This has been demon-

strated using CT26, a murine colon cancer cell line [263]; MBT-2, a murine bladder cancer cell line [264]; CMS-5, a sarcoma line [265]; and P815 mastocytoma cells [266]. This approach has been useful in the prevention of tumor spread in murine breast and lung cancer metastasis models [267–269].

Philip and colleagues [270] demonstrated that AAV vector transduction is not required for sustained transgene expression. They described expression of rhu-IL-2 for more than 30 days in certain cells following transfection with an AAV plasmid–liposome complex. Transfection efficiencies ranged from 10–50% and included a wide variety of target cells, such as primary tumors (breast, ovarian, and lung) and T lymphocytes [270]. Although specific mechanisms are not yet known, prolonged transgene expression may result from increased plasmid stability resulting from the GC-rich AAV ITRs; however, stable integrants using this system have not yet been described. This approach has been used to introduce cytokine genes into primary human breast and ovarian cancer cells *in vitro* and both IL-2 or IFN-γ into tumor cells in an *in vivo* murine tumor model to enhance their immunogenicity [271]. In one study, vaccination with 4T1 breast cancer cells transfected with an AAV plasmid construct encoding IL-2 specifically reduced the extent of pulmonary metastases compared to unmodified or control transfected tumor cells [268]. A study using a similar murine model demonstrated that lung tumor cells transfected with an AAV plasmid-encoding INF-γ proliferated at a slower rate following inoculation. Furthermore, 57% of animals vaccinated with these cells did not develop pulmonary metastases compared to controls, supporting the potential of this approach for the development of tumor vaccines [269].

Okada *et al.* [272] reported upon the use of a bicistronic rAAV vector encoding both the herpes simplex thymidine kinase and IL-2 genes (AAV-tk-IRES-IL2). Transduction of U-251SP cells, a human glioma cell line, with AAV-tk-IRES-IL2 rendered them sensitive to ganciclovir and resulted in expression of IL-2 in a dose-dependent fashion. Stereotactic delivery of 6×10^{10} AAV-tk-IRES-IL2 particles into day-7 tumors in nude mice followed by administration of GCV for 6 days resulted in a 35-fold reduction in the mean volume of tumors compared with controls [272]. Similarly, Mizuno *et al.* [273] employed a human glioma, nude mice xenograft model and noted complete tumor regression and prolonged survival following intratumoral injection of rAAV-tk, followed by treatment with ganciclovir. Rosenfeld *et al.* [274] employed a medulloblastoma nude rat model of leptomeningeal disease and demonstrated that intraventricular injection of an rAAV vector encoding LacZ resulted in transduction of implanted tumor cells but not in underlying normal brain parenchyma, supporting a potential role for the treatment of leptomeningeal tumors. Hybrid vectors have also been studied for the potential gene therapy of central nervous system (CNS) tumors. Johnston *et al.* [275] constructed a hybrid vector consisting of the HSV-1 amplicon which included the origin of replication (oriS) and the DNA cleavage/packaging signal (*pac*), into which a *LacZ* reporter gene was inserted under CMV IE promoter control flanked by AAV ITRs. *LacZ* expression in HSV/AAV-transduced U87 glioma cells persisted for a longer period when compared to a control amplicon vector that lacked the AAV ITRs. Interestingly, transduction with an HSV/AAV hybrid vector that also encoded AAV Rep resulted in the longest expression [275]. Extension of these studies to a rat *in vivo* model showed that the HSV/AAV hybrid vector transduced primary neurons within the striatum, with efficiencies equivalent to or higher than HSV amplicon vectors and without recognized toxicity [276].

Zhang *et al.* [277] constructed an rAAV vector encoding a synthetic human type I interferon gene in addition to neomycin phosphotransferase. A variety of human tumor cell lines (293, HeLa, K562, Eskol) were transduced, and G418 was selected. All cell lines expressed interferon *in vitro*. In contrast to untransduced controls, transduced tumor cells failed to induce tumor formation in nude mice. Furthermore, transduction of established Eskol cells in nude mice resulted in tumor regression [277]. Clary *et al.* [269] constructed an AAV plasmid encoding the murine IFN-γ gene (pMP6A-mIFN-gamma) and used liposome-mediated delivery to transfer it to the murine lung cancer cell line D122. Interferon-γ and augmented major histocompatibility class I expression were documented in transformed cells, which were irradiated and used to vaccinate animals challenged with D122 cells. Animals vaccinated with IFN-γ-expressing cells demonstrated a delay in primary (footpad) tumor growth and displayed a 57% reduction in pulmonary metastases versus a 0, 7, and 15% reduction in untreated controls, animals vaccinated with cells containing an empty AAV vector, or a retroviral vector encoding IFN-γ, respectively [269].

Both the human IL-2 and murine IL-4 genes have been inserted into the oncotropic strain of MVM under control of the endogenous p38 promoter [278]. The tumorcidal NS genes were retained in these vectors. Intratumor IL-4 expression induces a potent antitumor eosinophilic response in mice [279]. Consistent with the biology of MVM, approximately 200 times more IL-2 was expressed in two transformed compared to untransformed parental fibroblasts following transduction. These vectors have the advantages of being oncotropic, directly tumorcidal due to expression of NS genes, and capable of enhancing antitumor immune responses by virtue of IL-2 or IL-4 expression.

Haag *et al.* [280] constructed recombinant H1 vectors that encoded either human IL-2 (rH1/IL-2) or monocyte chemotactic protein (MCP-1) (rH1/MCP-1). HeLa cells transduced with these vectors produced 4–10 μg of cytokine per 10^6 cells over 5 days. Importantly, tumor formation in immunodeficient mice from rH1/IL-2-transduced HeLa cells was reduced 90% when compared to control animals, with histopathologic evidence of immune cell tumor infiltration [280].

Anderson *et al.* [281] constructed an rAAV vector encoding a single-chain version of IL-12 (Flexi-12), a cytokine known to induce antitumor immunity. The expressed monomeric form of IL-12 was shown to be functional *in vitro*, inducing proliferation and INF-γ secretion from phytohemagglutinin-treated blasts. Furthermore, acute myeloid leukemic blasts were efficiently transduced with the vector and expressed Flexi-12 [281]. Paul *et al.* [282] constructed an rAAV vector encoding murine IL-12 under RSV promoter control using the encephalomyocarditis internal ribosomal entry site (IRES) to express the p35 and p40 IL-12 heterodimers. The construct also contained a cassette that conferred resistance to G418. G418-selected clones were shown to express antigenic, as determined by ELISA assays, and functional IL-12 as determined by production of INF-γ and lymphocyte and cytotoxicity assays [282].

2. Costimulatory Molecules

Tumor cells may evade immune recognition because they lack accessory or costimulatory signals (B7.1 or B7.2) for efficient antigen presentation [283]. Chiorini *et al.* [284] constructed an rAAV vector encoding the B7-2 costimulatory molecule. Transduction of the nonadherent human lymphoid cell line LP-1 resulted in increased B7-2 expression from 6.8–78.0%. These studies were extended by Wendtner *et al.* [285], who utilized rAAV vectors encoding B7-1 or B7-2 and used them to transduce LP-1 and RPMI 8226, two multiple myeloma cell lines. Incubation of allogeneic T cells with rAAV/B7-1 or rAAV/B7-2 transduced LP-1 or RPMI 8226 cells resulted in T-cell proliferative responses, IL-2 and IFN-γ cytokine secretion, and a T-cell cytolytic response [285]. The ability to express costimulatory cells on tumor cells may permit the development of specific tumor vaccines.

3. Tumor Antigen-Specific Vaccines

Cancer is an acquired genetic disease. Tumor cells often express mutated genes (*p53*, *ras*, *bcr–abl*), occasionally totally unique genes as occurs after viral oncogenesis (HPV E6 or E7), or aberrantly or overexpressed cellular genes (Her2–*neu*, carcinoembryonic antigen [CEA], mucin) which could be recognized by the host immune system and used to eliminate them. Recombinant AAV vectors could be useful for the generation of immune responses because of their low intrinsic immunogenicity (see previous discussion), and they have been employed to generate immune responses against specific antigens, including HSV glycoproteins [286].

Dendritic cells, currently the most potent antigen-presenting cells known, originate from the bone marrow and occupy a central position in the generation of immune responses [287,288]. A variety of strategies to exploit DCs to generate antitumor immune responses are being actively investigated, including peptide pulsing with tumor antigens;

cell fusion with tumor cells; transfection with DNA, RNA, or proteins; and transduction with viral vectors encoding tumor antigens [289–291].

We and others have shown that primary human DCs are transducible with rAAV vectors encoding reporter genes [146,147]. Chronic myelogenous leukemia (CML) is characterized by a (9,22) translocation, which results in the fusion of the *bcr* region with the cellular *abl* gene, a tyrosine kinase [292]. This fusion results in the expression of a unique transcript, and the protein product, the *bcr–abl* oncogene specific to the cancer cells, is requisite for leukomogenesis and represents a potential neoantigen that could be recognized by the immune system. We have constructed an rAAV vector encoding the unique fusion domain of the *p210^bcr–abl* oncogene. We first demonstrated that the transgene was expressed in tissue culture cells at the mRNA and protein level. We then encapsidated the vector and used it to transduce human-peripheral-blood-derived DCs. Autologous T cells coincubated with these DCs proliferated in an antigen-specific fashion, and T-cell clones were identified that were cytolytic to *bcr–abl* pulsed autologous B cells in an antigen-specific, MHC-restricted fashion. We are in the process of extending these studies to primary CML cells and hope that these studies will lead to a strategy to lessen the risk of relapse following hematopoietic stem cell transplantation for CML [146].

Another approach is to develop strategies to generate or enhance host immunity against oncogenic viruses or their transforming gene products. As discussed previously, HPV is recognized as an important cofactor in the development of cervical and other epithelial cancers. Liu *et al.* [293] developed an rAAV vector encoding the HPV-16 E7 CTL epitope protein fused with heat shock protein as an immunologic adjuvant. Intramuscular administration of this DNA vaccine in mice induced a CD4- and CD8-dependent CTL response and protected animals *in vivo* from tumor challenge with an HPV-16-transformed cell line [293]. We are currently also examining the potential for an rAAV vector encoding the HPV-16 E6 or E7 transforming genes to induce anti-HPV-specific immune responses in a fashion similar to that described for the *bcr–abl* oncogene as described above.

Miyamura *et al.* [294] engineered expression of hen egg white lysozyme (HEL) to the surface of empty capsids of the B19 parvovirus produced in a baculovirus expression system. Surface HEL was detectable from purified recombinant capsids by ELISA, immunoprecipitation, and immune electron microscopy and enzymatically. Furthermore, rabbits inoculated with the recombinant capsids generated an anti-HEL immune response, implying that B19 empty capsids may be useful as platforms to induce immunity against specific proteins, potentially including tumor antigens [294]. Sedlik *et al.* subsequently demonstrated that porcine parvovirus particles could be used in a similar fashion to present epitopes of poliovirus [295].

4. Suicide Genes

Suicide genes convert a prodrug to a toxic metabolite [296]. For example, the herpes simplex virus thymidine kinase (HSV-TK) gene sensitizes cells to the cytotoxic effects of specific drugs, including ganciclovir and acyclovir. Culver and colleagues exploited the transduction requirement of retroviral vectors for proliferating target cells to specifically introduce HSV-TK into rapidly proliferating brain tumors (gliomas) in rats [297]. Gliomas regressed completely in ganciclovir-treated animals. Interestingly, untransduced tumor cells in close proximity to transduced cells were also killed following ganciclovir treatment ("bystander effect") which may result from a diffusible metabolite of ganciclovir which passes from cell to cell via gap junctions [298]. As a prelude to the development of parvovirus vectors encoding suicide genes, Koering and colleagues inserted the HSV-tk gene within MVMp under control of the p38 promoter [299]. Acyclovir treatment of TK cells stably transfected with this construct followed by infection with MVMp reduced cell survival 3.5- to 5-fold compared to controls. Su *et al.* [301] constructed an rAAV vector encoding the herpes simplex thymidine kinase gene (rAAV/TK) under control of the liver-specific albumin promoter coupled with the human alpha fetoprotein (AFP) enhancer. This construct also contained the Neo^R gene. Only liver cells expressing AFP and albumin were sensitive to the effects of ganciclovir (GCV) [300]. Tumors generated by subcutaneous implantation of rAAV/TK-transduced hepatocellular carcinoma cells in nude mice shrank dramatically following treatment with ganciclovir. Tumor regression was also noted after challenge with mixtures of rAAV/TK-transduced and -untransduced cells followed by GCV treatment, or direct vector injection into established tumors followed by GCV treatment [301]. In an attempt to enhance this strategy, Su *et al.* [302] also inserted a transgene encoding IL-2 into rAAV/TK and compared protection of immunodeficient nude and immunocompetent C57L/J mice following challenge with the Hepa 1–6 hepatocellular carcinoma cell line. rAAV/TK/IL-2-transduced Hepa 1–6 cells were more susceptible to GCV treatment than rAAV/TK-transduced cells and displayed tumor shrinkage even without GCV treatment. In fact, the best results were seen with challenge with rAAV/TK/IL-2-transduced cells without GCV treatment, so perhaps treatment with GCV resulted in early elimination of IL-2 expressing cells [302].

5. Interruption of Tumor Vascular Supply

Tumors secrete a number of "angiogenesis" factors, including vascular endothelial growth factor (VEGF), transforming growth factor (TNF)-β1, pleiotropin, fibroblast growth factor (FGF), placental growth factor, and platelet-derived endothelial cell growth factor, which function to induce formation of supporting vasculature [303]. Provision of a suitable tumor blood supply is necessary for sustained tumor growth beyond several millimeters in size. Thus, interruption of this sometimes tenuous vascular supply could provide a therapeutic target to enhance tumor killing. AAV vectors have recently been shown to efficiently transduce endothelial and vascular smooth muscle cells from rodents and nonhuman primates both *in vitro* and *in vivo*, as well as from humans *in vitro* [98,99]. This property could be exploited for expression of tumor angiogenesis inhibitors or thymidine kinase within tumor vessels [304,305]. Expression plasmids encoding angiostatin, endostatin, and TIMP-2 have recently been constructed and will be studied in the context of rAAV vectors [306]. Nguyen *et al.* [307] constructed rAAV vectors coding for angiostatin, endostatin, and an antisense transcript against VEGF. Transduction with the vector encoding antisense to VGEF resulted in diminished VEGF production from tumor cells. Conditioned media from transduced cells resulted in diminished capillary proliferation when compared to controls, implying that this approach could have value in the inhibition of tumor-related angiogenesis [307].

6. Vector Targeting

The ability to target vectors to specific tissues or to either broaden or restrict vector host range is an area of active research for many gene delivery systems. To facilitate transduction of human hematopoietic cells, Yang and colleagues described the construction of a chimeric AAV capsid protein fused with a single-chain antibody against the human CD34 molecule and demonstrated enhanced transduction of KG-1, a CD34-bearing cell line [308]. Bartlett *et al.* [309] used a bispecific F(ab')$_2$ antibody that targeted a surface receptor expressed on human megakaryocytes and AAV. Targeted rAAV vectors were able to transduce otherwise nonpermissive megakaryocytic cell lines at levels equivalent to permissive cells [309]. Girod *et al.* [310] specifically engineered a 14-amino-acid peptide containing an RGD integrin binding motif into six regions of the AAV2 capsid that were predicted to be external capsid epitopes by analogy to the known three-dimensional structure of the canine parvovirus virion. One construct, I-587, was able to mediate specific infection of otherwise nonpermissive B16F10 cells by pseudotyped wild-type AAV or a recombinant rAAV vector encoding LacZ [310]. Finally, Wu *et al.* [311] constructed 93 mutants at 59 different positions in the AAV capsid gene using site-directed mutagenesis. Mutants were classified according to effects upon the infectious titer which varied from no effect or partial or temperature-sensitive effects to noninfectious. Mutants were identified that did not affect infectivity, potentially presaging capsid modifications that could affect cellular targeting [311].

B19 is an autonomous parvovirus that specifically infects cells of the erythroid, or red blood cell, lineage. Ponnazhagan

et al. [312] pseudotyped an rAAV vector within a B19 capsid and demonstrated efficient transduction of erythroid cell lines and primary erythroid marrow progenitors. Transduction could be blocked by anti-B19, but not anti-AAV antisera [312]. Kurpad *et al.* extended these studies and demonstrated that the B19 p6 and globin promoters were active primarily in erythroid cells, supporting the potential for erythroid-specific transduction and gene expression using this approach [313].

7. Oncotropic Vectors

The natural tropism for MVM and H1 could be exploited to develop oncotropic vectors. To this end, Dupont *et al.* [314] described the construction of an MVM-based vector (MVM/p38cat) retaining the oncolytic NS proteins and encoding the CAT reporter gene under p38 promoter control. Encapsidated MMV/p38cat was used to transduce both primary nonmalignant and transformed cells. Vector DNA replication and CAT expression were limited to transformed cells (including human fibroblasts, epithelial cells, T lymphocytes, and macrophages), although a direct correlation between DNA replication and transgene expression was not always apparent [314]. These investigators subsequently extended these studies using an MVM-based vector encoding GFP or HSV-tk. Transduction with MVM/GFP again demonstrated that tumor cells were readily transduced while their nonmalignant counterparts were not. Furthermore, tumor cells were also readily transduced with MVM/HSV-tk and were susceptible to cytolysis following treatment with ganciclovir [315].

Van Pachterbeke *et al.* [316] examined the sensitivity of cells derived from several breast cancer specimens to lysis following infection with H-1. H-1-mediated oncolysis was positively correlated with the presence of the estrogen receptor and with the proliferative status of the cells; the more rapidly proliferating cells were also more sensitive to H-1 lysis [316]. Faisst *et al.* [317] used a murine tumor xenograft model to examine the ability of H-1 to lyse tumors *in vivo*. In their model, immunodeficient SCID mice were inoculated subcutaneously with rapidly proliferating human HeLa cervical carcinoma cells and then challenged with different doses of H-1. A dose-dependent tumor regression was noted [317].

8. Safety Issues

To date, rAAV vectors have been used clinically in phase I trials for cystic fibrosis and for hemophilia B (factor IX deficiency). Pilot studies of rAAV transduction of pulmonary cells in primate models of cystic fibrosis failed to generate evidence of toxicity, and phase I human studies were initiated [318,319]. Wagner *et al.* [320] reported on a prospective, randomized, unblinded, dose-escalation study of individuals receiving an rAAV vector encoding the cystic fibrosis transmembrane regulator (CFTR) administered to maxillary sinuses. The highest level of gene transfer, as determined by

PCR, was 0.1–1 AAV/CFTR vector copy per cell in specimens obtained 2 weeks after transduction. Few or no inflammatory responses or immune responses were noted, and there was objective evidence of a dose-related transduction effect [320]. Kay *et al.* [114] reported upon the initial findings from of a phase I, dose-escalation trial of an rAAV vector encoding human factor IX (rAAV/FIX) delivered intramuscularly in three adults with severe hemophilia B. Vector sequences were demonstrable in muscle after vector administration by PCR and Southern analyses, and FIX expression was documented by immunohistochemistry. No evidence of local toxicity, formation of inhibitory FIX antibodies, or germline vector transmission were noted. Although initial vector doses of 2×10^{11} vector genomes per kg were calculated to be subtherapeutic by extrapolation from dose escalation studies in animal models, FIX could be detected in the serum, and there was a trend towards a lower utilization of prophylactic FIX concentrates [114].

As mentioned previously, a human clinical trial consisting of the systemic administration of the H-1 autonomous parvovirus in patients with osteogenic sarcoma was initiated by Toolan and colleagues in the 1960s. No significant toxicity was noted, although these studies predated the development of molecular techniques that might have addressed questions of virus spread [152,153].

IV. PERSPECTIVES, PROBLEMS, AND FUTURE CONSIDERATIONS

Given the recognized limitations of current modalities, novel approaches are needed for more effective and nontoxic treatments for cancer. As outlined in the above discussion, parvovirus vectors offer a variety of advantages for the development of gene therapy strategies for cancer treatment, and several approaches are in development for potential clinical applications. Vectors can be developed that transiently express transgenes (MVM, H-1) or stably integrate into cellular DNA (AAV), that have an extremely wide host range (AAV), or that have specific oncotropic and tumorcidal activity (MVM, H-1). Newer aspects of parvovirus vector development and applications for their use are continually being described. Will it be possible to provide AAV Rep in such a fashion that rAAV vectors can integrate site-specifically into human chromosomal DNA? Will it be possible to use rAAV vectors to directly correct genomic mutations, including those that cause cancer [321]? As rAAV vectors efficiently transduce cells of the intestinal tract after oral administration, could they be used to introduce genes that protect against mucositis or be useful as an oral vaccine for the generation of humoral responses against tumor antigens [322,323]? Can rAAV capids be modified so that vectors can be targeted to specific cell types? Will different AAV serotypes be more useful for gene transfer to different tissues? Can long-term transgene expression following intramuscular administration of rAAV vectors be used to develop better antitumor strategies? Can the oncotropic and oncolytic properties of H1 or MVM be harnessed in a clinically relevant manner? These are just a few of the potential questions that remain to be addressed.

Also, vector safety issues, including potential risks of horizontal or vertical vector transmission, vector shedding, vector distribution within the host, and vector rescue, are just beginning to be addressed within human trials. Preliminary studies indicate that rAAV vector administration via intramuscular or respiratory routes is safe, with minimal risk of germline transmission [114,320,321]. Additional *in vivo* animal studies and actual human clinical trials will help to resolve many of these issues. We and others are actively studying AAV-mediated gene transfer in several animal models, with the anticipation of advancing to human clinical trials. Finally, efficient, simple, reproducible encapsidation strategies must be developed, particularly for the production of large quantities of wild-type, virus-free, clinical-grade vector. This has become a major focus of several laboratories, including our own.

It is perhaps of note that almost exactly 35 years ago, Helen Toolan at Sloan–Kettering administered H1 parvovirus intravenously to two patients with incurable, disseminated osteosarcoma in an attempt to eradicate their disease [152,153]. Although only a modest therapeutic effect was noted, this important study laid the groundwork for the study and development of parvovirus vectors in our continuing struggle against cancer.

Acknowledgments

We thank the members of our laboratory who have contributed countless hours of hard work and innumerable helpful discussions. We also thank Stephen Forman, John Zaia, Christine Wright, and the City of Hope Bone Marrow Transplantation team, whose support made this work possible. This work was supported in part by grants AI40001, CA75186, CA71947, PO1 HL60898-01A1, CA59308, CA33572, and AI25959 from the National Institutes of Health.

References

1. Bishop, J. M. (1995). Cancer: the rise of the genetic paradigm. *Genes Develop.* **9**, 1309–1315.
2. Sokol, D. L., and Gewirtz, A. M. (1996). Gene therapy: basic concepts and recent advances. *Crit. Rev. Eukaryot. Gene Expr.* **6**, 29–57.
3. Verma, I. M., and Somia, N. (1997) Gene therapy—promises, problems and prospects. *Nature* **389**, 239–242.
4. Anderson, W. F. (1998). Human gene therapy. *Nature* **392**(suppl.), 25–30.
5. Miller, D. G., Adam, M. A., and Miller, A. D. (1990). Gene transfer by retrovirus vectors occurs only in cells that are actively replicating at the time of infection. *Mol. Cell. Biol.* **10**, 4239–4242.

6. Donahue, R. E., Kessler, S. W., Bodine, D. *et al.* (1992). Helper virus induced T cell lymphoma in nonhuman primates after retroviral mediated gene transfer. *J. Exp. Med.* **176,** 1125–1135.

7. Purcell, D. F., Broscius, C. M., Vanin, E. F. *et al.* (1996). An array of murine leukemia virus-related elements is transmitted and expressed in a primate recipient of retroviral gene transfer. *J. Virol.* **70,** 887–897.

8. Yang, Y., Ertl, H. C., and Wilson, J. M. (1994). MHC class I-restricted cytotoxic T lymphocytes to viral antigens destroy hepatocytes in mice infected with E1-deleted recombinant adenoviruses. *Immunity* **1,** 433–442.

9. Worgall, S., Wolff, G., Falck-Pedersen, E. *et al.* (1997). Innate immune mechanisms dominate elimination of adenoviral vectors following in vivo administration. *Hum. Gene Ther.* **8,** 37–44.

10. Hitt, M. M., Addison, C. L., and Graham, F. L. (1997). Human adenovirus vectors for gene transfer into mammalian cells. *Adv. Pharmacol.* **40,** 137–206.

11. Miyoshi, H., Smith, K. A., Mosier, D. E. *et al.* (1999). Transduction of human CD34+ cells that mediate long-term engraftment of NOD/SCID mice by HIV vectors. *Science* **283,** 682–686.

12. Buchschacher, G. L., Jr., and Wong-Staal, F. (2000). Development of lentiviral vectors for gene therapy for human diseases. *Blood* **95,** 2499–2504.

13. Korin, Y. D., and Zack, J. A. (1999). Progression to the G1b phase of the cell cycle is required for completion of human immunodeficiency virus type 1 reverse transcription in T cells. *J. Virol.* **72,** 3161–3168.

14. Sutton, R. E., Reitsma, M. J., Uchida, N. *et al.* (1999). Transduction of human progenitor hematopoietic stem cells by human immunodeficiency virus type 1-based vectors is cell cycle dependent. *J. Virol.* **73,** 3649–3660.

15. Dranoff, G., and Mulligan, R. C. (1995). Gene transfer as cancer therapy. *Adv. immunol.* **58,** 417–454.

16. Zhang, W-W., Fujiwara, T., Grimm, E. A. *et al.* (1995) Advances in cancer gene therapy. *Adv. Pharm.* **32,** 289–341.

17. Gomez-Navarro, J., Curiel, D. T., and Douglas, J. T. (1999). Gene therapy for cancer. *Eur. J. Cancer* **35,** 2039–2057.

18. Vile, R. G., Russell, S. J., and Lemoine, N. R. (2000). Cancer gene therapy: hard lessons and new courses. *Gene Ther.* **7,** 2–8.

19. (2000). Cancer gene therapy. *Adv. Exp. Med. Biol.* **465.**

20. Siegl, G., Bates, R. C., Berns, K. I. *et al.* (1985). Characterization and taxonomy of parvoviridae. *Intervirology* **23,** 61–73.

21. Flotte, T. R., Afione, S. A., Solow, R. *et al.* (1993). Expression of the cystic fibrosis transmembrane conductance regulator from a novel adeno-associated virus promoter. *J. Biol. Chem.* **268,** 3781–3790.

22. Atchinson, R. W., Casto, B. C., and Hammond, W. M. (1965). Adenovirus-associated defective virus particles. *Science* **149,** 754–756.

23. Hoggan, M. D., Blacklow, N. R., and Rowe, W. P. (1966). Studies of small DNA viruses found in various adenovirus preparations: physical, biological, and immunological characteristics. *Proc. Natl. Acad. Sci. USA* **55,** 1457–1471.

24. Rutledge, E. A., Halbert, C. L., and Russell, D. W. (1998). Infectious clones and vectors derived from adeno-associated virus (AAV) serotypes other than AAV type 2. *J. Virol.* **72,** 309–319.

25. Berns, K. I., and Giraud, C. (1996). Biology of adeno-associated virus. *Curr. Top. Microbiol. Immunol.* **218,** 1–23.

26. Berns, K. I. (1996). The Parvoviridae: the viruses and their replication, in *Fields Virology* (B. N. Fields, D. M. Knipe, and P. M. Howley, eds.), 3rd ed., pp. 2173–2197, Lippincott-Raven, Philadelphia.

27. Wang, X. S., Srivastava, A., Ponnazhagan, S. *et al.* (1997). Adeno-associated virus type 2 DNA replication in vivo: mutation analyses of the D sequence in viral inverted terminal repeats. *J. Virol.* **71,** 3077–82.

28. Xiao, X., Samulski, R. J., Li, J. *et al.* (1997). A novel 165-base-pair terminal repeat sequence is the sole *cis* requirement for the adeno-associated virus life cycle. *J. Virol.* **71,** 941–948.

29. Brown, K. E., Anderson, S. M., and Young, N. S. (1993). Erythrocyte P antigen: cellular receptor for B19 parvovirus. *Science* **262,** 114–117.

30. Summerford, C., and Samulski, R. J. (1998). Membrane-associated heparan sulfate proteoglycan is a receptor for adeno-associated virus type 2 virions. *J. Virol.* **72,** 1438–1445.

31. Qiu, J., Handa, A, Kirby, M. *et al.* (2000). The interaction of heparin sulfate and adeno-associated virus 2. *Virology* **269,** 137–147.

32. Qing, K., Mah, C., Hansen, J. *et al.* (1999). Human fibroblast growth factor receptor 1 is a co-receptor for infection by adeno-associated virus 2. *Nat. Med.* **5,** 71–7.

33. Summerford, C., Bartlett, J. S., and Samulski, R. J. (1999). AlphaV-beta5 integrin, a co-receptor for adeno-associated virus type 2 infection. *Nat. Med.* **5,** 78–82, 1999.

34. Qiu, J., and Brown, K. E. (1999). Integrin alphaVbeta5 is not involved in adeno-associated virus type 2 (AAV2) infection. *Virology* **264,** 436–440.

35. Duan, D., Li, Q., Kao, A. W. *et al.* (1999). Dynamin is required for recombinant adeno-associated virus type 2 infection. *J. Virol.* **73,** 10371–10376.

36. Bartlett, J. S., Wilcher, R., and Samulski, R. J. (2000). Infectious entry pathway of adeno-associated virus and adeno-associated virus vectors. *J. Virol.* **74,** 2777–2785.

37. Ruffing, M., Zentgraf, H., and Kleinschmidt, J. A. (1992). Assembly of viruslike particles by recombinant structural proteins of adeno-associated virus type 2 in insect cells. *J. Virol.* **66,** 6922–6930.

38. Qiu, J., and Brown, K. E. (1999). A 110-kDa nuclear shuttle protein, nucleolin, specifically binds to adeno-associated virus type 2 (AAV-2) capsid. *Virology* **257,** 373–82.

39. Kotin, R. M., Siniscalco, M., Samulski, R. J. *et al.* (1990). Site-specific integration by adeno-associated virus. *Proc. Natl. Acad. Sci. USA* **87,** 2211–2215.

40. Kotin, R. M., Linden, R. M., and Berns, K. I. (1992). Characterization of a preferred site on human chromosome 19q for integration of adeno-associated virus DNA by non-homologous recombination. *EMBO J.* **11,** 5071–5078.

41. Samulski, R. J., Zhu, X., Xiao, X. *et al.* (1991). Targeted integration of adeno-associated virus (AAV) into human chromosome 19. *EMBO J.* **10,** 3941–3950.

42. Berns, K. I., Pinkerton, T. C., Thomas, G. F. *et al.* (1975). Detection of adeno-associated virus (AAV)-specific nucleotide sequences in DNA isolated from latently infected Detroit 6 cells. *Virology* **68,** 556–560.

43. Muzyczka, N. (1992). Use of AAV as a general transduction vector for mammalian cells. *Curr. Top. Micro. Immunol.* **158,** 97–129.

44. Chejanovsky, N., and Carter, B. J. (1989). Mutagenesis of an AUG codon in the adeno-associated virus rep gene: effects on viral DNA replication. *Virology* **173,** 120–128.

45. Zabner, J., Seiler, M., Walters, R. *et al.* (2000). Adeno-associated virus type 5 (AAV5) but not AAV2 binds to the apical surfaces of airway epithelia and facilitates gene transfer. *J. Virol.* **74,** 3852–3858.

46. Davidson, B. L., Stein, C. S., Heth, J. A. *et al.* (2000). Recombinant adeno-associated virus type 2, 4, and 5 vectors: transduction of variant cell types and regions in the mammalian central nervous system. *Proc. Natl. Acad. Sci. USA* **97,** 3428–3432.

47. Xiao, W., Chirmule, N., Berta, S. C. *et al.* (1999) Gene therapy vectors based on adeno-associated virus type 1. *J. Virol.* **73,** 3994–4003.

48. Handa, A., Muramatsu, Si., Qiu, J. *et al.* (2000). Adeno-associated virus (AAV)-3-based vectors transduce haematopoietic cells not susceptible to transduction with AAV-2-based vectors. *J. Gen. Virol.* **81,** 2077–2084.

49. Halbert, C. L., Rutledge, E. A., Allen, J. M. *et al.* (2000). Repeat transduction in the mouse lung by using adeno-associated virus vectors with different serotypes. *J. Virol.* **74,** 1524–1532.

50. Schlehofer, J. R. (1994). The tumor suppressive properties of adeno-associated viruses. *Mutat. Res.* **305,** 303–313.

51. Mayor, H. D. (1993). Defective parvoviruses may be good for your health! *Prog. Med. Virol.* **40**, 193–205.

52. Bantel-Schaal, U. (1990). Adeno-associated parvoviruses inhibit growth of cells derived from malignant human tumors. *Int. J. Cancer* **45**, 190–194.

53. Hermonat, P. L. (1994). Down-regulation of the human c-*fos* and c-*myc* proto-oncogene promoters by adeno-associated virus Rep78. *Cancer Lett.* **81**, 129–136.

54. Hermonat, P. L. (1994). Adeno-associated virus inhibits human papillomavirus type 16, a viral interaction implicated in cervical cancer. *Cancer Res.* **54**, 2278–2281.

55. Hillgenberg, M., Schlehofer, J. R., von Knebel Doeberitz, M. *et al.* (1999). Enhanced sensitivity of small cell lung cancer cell lines to cisplatin and etoposide after infection with adeno-associated virus type 2. *Eur. J. Cancer* **35**, 106–110.

56. Kotin, R. M. (1994). Prospects for the use of adeno-associated virus as a vector for human gene therapy. *Hum. Gene Ther.* **5**, 793–801.

57. Labow, M. A., Graf, L. H., and Berns, K. I. (1987). Adeno-associated virus gene expression inhibits cellular transformation by heterologous genes. *Mol. Cell Biol.* **7**, 1320–1325.

58. Dong, J. Y., Frizzell, R. A., and Fan, P. D. (1996). Quantitative analysis of the packaging capacity of recombinant adeno-associated virus. *Hum. Gene Ther.* **7**, 2101–2112.

59. Hermonat, P. L., Han, L., Bishop, B. M. *et al.* (1997). The packaging capacity of adeno-associated virus (AAV) and the potential for wild-type-plus AAV gene therapy vectors. *FEBS Lett.* **407**, 78–84.

60. Yang, J., Zhou, W., Zhang, Y. *et al.* (1999). Concatamerization of adeno-associated virus circular genomes occurs through intermolecular recombination. *J. Virol.* **73**, 9468–9477.

61. Duan, D., Yue, Y., Yan, Z. *et al.* (2000). A new dual-vector approach to enhance recombinant adeno-associated virus-mediated gene expression through intermolecular cis activation. *Nat. Med.* **6**, 595–598.

62. Sun, L., Li, J., and Xiao, X. (2000). Overcoming adeno-associated virus vector size limitation through viral DNA heterodimerization. *Nat. Med.* **6**, 599–602.

63. Nakai, H., Storm, T. A., and Kay, M. A. (2000). Increasing the size of rAAV-mediated expression cassettes in vivo by intermolecular joining of two complementary vectors. *Nat. Biotech.* **18**, 527–532.

64. Yan, Z., Zhang, Y., Duan, D. *et al.* (2000). *trans*-Splicing vectors expand the utility of adeno-associated virus for gene therapy. *Proc. Natl. Acad. Sci. USA* **97**, 6716–6721.

65. Samulski, R. J. (2000). Expanding the AAV package. *Nat. Biotech.* **18**, 497–498.

66. Chatterjee, S., and Wong, K. K., Jr. (1993). Adeno-associated viral vectors for the delivery of antisense RNA. *Methods: A Companion to Meth. Enzymol.* **5**, 51.

67. Chatterjee, S. and Wong, K. K. Jr. (1999). *rAAV vectors in Intracellular Ribozyme Applications: Principles and Protocols* (J. J. Rossi and L. Couture, eds.), Horizon Press.

68. Hauswirth, W. W., Lewin, A. S., Zolotukhin, S. *et al.* (2000). Production and purification of recombinant adeno-associated virus. *Meth. Enzymol.* **316**, 743–761.

69. Xiao, X., Li, J., and Samulski, R. J. (1998). Production of high-titer recombinant adeno-associated virus vectors in the absence of helper adenovirus. *J. Virol.* **72**, 2224–2232.

70. Collaco, R. F., Cao, X., and Trempe, J. P. (1999). A helper virus-free packaging system for recombinant adeno-associated virus vectors. *Gene* **238**, 397–405.

71. Fan, P. D., and Dong, J. Y. (1997). Replication of *rep–cap* genes is essential for the high-efficiency production of recombinant AAV. *Hum. Gene Ther.* **8**, 87–98.

72. Li, J., Xiao, X., and Samulski, R. J. (1997). Role for highly regulated *rep* gene expression in adeno-associated virus vector production. *J. Virol.* **71**, 5236–5243.

73. Wang, X. S., Khuntirat, B., Qing, K. *et al.* (1998). Characterization of wild-type adeno-associated virus type 2-like particles generated during recombinant viral vector production and strategies for their elimination. *J. Virol.* **72**, 5472–5480.

74. Koeberl, D. D., Alexander, I. E., Halbert, C. L. *et al.* (1997). Persistent expression of human clotting factor IX from mouse liver after intravenous injection of adeno-associated virus vectors. *Proc. Natl. Acad. Sci. USA* **94**, 1426–1431.

75. Mamounas, M., Leavitt, M., Yu, M. *et al.* (1995). Increased titer of recombinant AAV vectors by gene transfer with adenovirus coupled to DNA-polylysine complexes. *Gene Ther.* **2**, 429–432.

76. Maxwell, F., Maxwell, I. H., and Harrison, G. S. (1997). Improved production of recombinant AAV by transient transfection of NB324K cells using electroporation. *J. Virol. Meth.* **63**, 129–36.

77. Tamayose, K., Shimada, T., and Hirai, Y. (1996). A new strategy for large-scale preparation of high-titer recombinant adeno-associated virus vectors by using packaging cell lines and sulfonated cellulose column chromatography. *Hum. Gene Ther.* **7**, 507–513.

78. Conway, J. E., Rhys, C. M., Zolotukhin, I. *et al.* (1999). High-titer recombinant adeno-associated virus production utilizing a recombinant herpes simplex virus type I vector expressing AAV-2 Rep and Cap. *Gene Ther.* **6**, 986–993.

79. Zhang, X., De Alwis, M., Hart, S. L. *et al.* (1999). High-titer recombinant adeno-associated virus production from replicating amplicons and vectors deleted for glycoprotein H. *Hum. Gene Ther.* **10**, 2527–37.

80. Flotte, T. R., Barraza-Ortiz, X., Solow, R. *et al.* (1995). An improved system for packaging recombinant adeno-associated virus vector capable of in vivo transduction. *Gene Ther.* **2**, 29–37.

81. Clark, K. R., Johnson, P. R., Fraley, D. M. *et al.* (1995). Cell lines for the production of recombinant adeno-associated virus. *Hum. Gene Ther.* **6**, 1329–1341.

82. Trempe, J. P. (1996). Packaging systems for adeno-associated virus vectors. *Curr. Top. Microbiol. Immunol.* **218**, 35–50.

83. Chatterjee, S., Johnson, P. R., Wong, K. K., Jr. (1992). Dual target inhibition of HIV-1 in vitro by means of an adeno-associated virus antisense vector. *Science* **258**, 1485–1488.

84. Zolotukhin, S., Byrne, B. J., Mason, E. *et al.* (1999). Recombinant adeno-associated virus purification using novel methods improves infectious titer and yield. *Gene Ther.* **6**, 973–985.

85. Hermens, W. T., ter Brake, O., Dijkhuizen, P. A. *et al.* (1999). Purification of recombinant adeno-associated virus by iodixanol gradient ultracentrifugation allows rapid and reproducible preparation of vector stocks for gene transfer in the nervous system. *Hum. Gene Ther.* **10**, 1885–1891.

86. Clark, K. R., Liu, X., McGrath, J. P. *et al.* (1999). Highly purified recombinant adeno-associated virus vectors are biologically active and free of detectable helper and wild-type viruses. *Hum. Gene Ther.* **10**, 1031–1039.

87. Grimm, D., and Kleinschmidt, J. A. (1999). Progress in adeno-associated virus type 2 vector production: promises and prospects for clinical use. *Hum. Gene Ther.* **10**, 2445–2450.

88. Drittanti, L., Rivet, C., Manceau, P. *et al.* (2000). High throughput production, screening and analysis of adeno-associated viral vectors. *Gene Ther.* **7**, 924–929.

89. Podsakoff, G., Wong, K. K., Jr., and Chatterjee, S. (1994). Stable and efficient gene transfer into non-dividing cells by adeno-associated virus (AAV)-based vectors. *J. Virol.* **68**, 5656–5666.

90. Alexander, I. E., Russell, D. W., and Miller, A. D. (1994). DNA-damaging agents greatly increase the transduction of nondividing cells by adeno-associated virus vectors. *J. Virol.* **68**, 8282–8287.

91. Flotte, T. R., Afione, S. A., and Zeitlin, P. L. (1994). Adeno-associated virus vector gene expression occurs in nondividing cells in the absence of vector DNA integration. *Am. J. Respir. Cell. Mol. Biol.* **11**, 517–521.

92. Kaplitt, M. G., Leone, P., Samulski, R. J. *et al.* (1994). Long-term gene expression and phenotypic correction using adeno-associated virus vectors in the mammalian brain. *Nat. Gen.* **8**, 148–153.

93. Du, B., Terwilliger, E. F., Boldt-Houle, D. M. *et al.* (1996). Efficient transduction of human neurons with an adeno-associated virus vector. *Gene Ther.* **3**, 254–261.

94. Peel, A. L., Reier, P. J., Muzyczka, N. *et al.* (1997). Efficient transduction of green fluorescent protein in spinal cord neurons using adeno-associated virus vectors containing cell type-specific promoters. *Gene Ther.* **4**, 16–24.

95. Ali, R. R., Bhattacharya, S. S., Hunt, D. M. *et al.* (1996). Gene transfer into the mouse retina mediated by an adeno-associated viral vector. *Hum. Mol. Genet.* **5**, 591–594.

96. Flannery, J. G., Hauswirth, W. W., Muzyczka, N. *et al.* (1997). Efficient photoreceptor-targeted gene expression in vivo by recombinant adeno-associated virus. *Proc. Natl. Acad. Sci. USA* **94**, 6916–6921.

97. Lalwani, A. K., Mhatre, A. N., Muzyczka, N. *et al.* (1996). Development of in vivo gene therapy for hearing disorders: introduction of adeno-associated virus into the cochlea of the guinea pig. *Gene Ther.* **3**, 588–592.

98. Lynch, C. M., Geary, R. L., Dean, R. H. *et al.* (1997). Adeno-associated virus vectors for vascular gene delivery. *Circ. Res.* **80**, 497–505.

99. Arnold, T. E., Bahou, W. F., and Gnatenko, D. (1997). In vivo gene transfer into rat arterial walls with novel adeno-associated virus vectors. *J. Vasc. Surg.* **25**, 347–355.

100. Kaplitt, M. G., Diethrich, E. B., Strumpf, R. K. *et al.* (1996). Long-term gene transfer in porcine myocardium after coronary infusion of an adeno-associated virus vector. *Ann. Thorac. Surg.* **62**, 1669–1676.

101. Bartlett, R. J., Ricordi, C., Sharma, K. *et al.* (1996). Long-term expression of a fluorescent reporter gene via direct injection of plasmid vector into mouse skeletal muscle: comparison of human creatine kinase and CMV promoter expression levels in vivo. *Cell Transplant.* **5**, 411–419.

102. Kessler, P. D., Byrne, B. J., Kurtzman, G. J. *et al.* (1996). Gene delivery to skeletal muscle results in sustained expression and systemic delivery of a therapeutic protein. *Proc. Natl. Acad. Sci. USA* **93**, 14082–14087.

103. Xiao, X., Samulski, R. J., and Li, J. (1996). Efficient long-term gene transfer into muscle tissue of immunocompetent mice by adeno-associated virus vector. *J. Virol.* **70**, 8098–8108.

104. Fisher, K. J., Wilson, J. M., Raper, S. E. *et al.* (1997). Recombinant adeno-associated virus for muscle directed gene therapy. *Nat. Med.* **3**, 306–312.

105. Zhou, S. Z., Cooper, S., Kang, L. Y. *et al.* (1994). Adeno-associated virus 2-mediated high efficiency gene transfer into immature and mature subsets of hematopoietic progenitors cells in human umbilical cord blood. *J. Exp. Med.* **179**, 1867–1875.

106. Fisher-Adams, G., Wong, K. K. Jr., Forman, S. *et al.* (1996). Integration of adeno-associated virus vector genomes in human CD34 cells following transduction. *Blood* **88**, 492–504.

107. Chatterjee, S., Li, W., Wong, C. A. *et al.* (1999). Transduction of primitive human marrow and cord blood-derived hematopoietic progenitor cells with adeno-associated virus vectors. *Blood* **93**, 1882–1894.

108. Chatterjee, S., Podsakoff, G., and Wong, K. K., Jr. (1994) Gene transfer into terminally differentiated primary human peripheral blood-derived mononuclear cells by adeno-associated virus. *Blood* **84**, 360a.

109. Inouye, R. T., Terwilliger, E. F., Pomerantz, R. J. *et al.* (1997). Potent inhibition of human immunodeficiency virus type 1 in primary T cells and alveolar macrophages by a combination anti-Rev strategy delivered in an adeno-associated virus vector. *J. Virol.* **71**, 4071–4078.

110. Miao, C. H., Nakai, H., Thompson, A. R. *et al.* (2000). Nonrandom transduction of recombinant adeno-associated virus vectors in mouse hepatocytes in vivo: cell cycling does not influence hepatocyte transduction. *J. Virol.* **74**, 3793–3803.

111. McKeon, C., and Samulski, R. J. (1996). NIDDK workshop on AAV vectors: gene transfer into quiescent cells. *Hum. Gene Ther.* **7**, 1615–1619.

112. Wu, P., Phillips, M. I., Bui, J. *et al.* (1998). Adeno-associated virus vector-mediated transgene integration into neurons and other nondividing cell targets. *J. Virol.* **72**, 5919–5926.

113. Herzog, R. W., High, K. A., Fisher, K. J. *et al.* (1997). Stable gene transfer and expression of human blood coagulation factor IX after intramuscular injection of recombinant adeno-associated virus. *Proc. Natl. Acad. Sci. USA* **94**, 5804–5809.

114. Kay, M. A., Manno, C. S., Ragni, M. V. *et al.* (2000). Evidence for gene transfer and expression of factor IX in haemophilia B patients treated with an AAV vector. *Nat. Genet.* **24**, 257–261.

115. Rendahl, K. G., Leff, S. E., Otten, G. R. *et al.* (1998). Regulation of gene expression in vivo following transduction by two separate rAAV vectors. *Nat. Biotech.* **16**, 757–761.

116. Rivera, V. M., Ye, X., Courage, N. L. *et al.* (1999). Long-term regulated expression of growth hormone in mice after intramuscular gene transfer. *Proc. Natl. Acad. Sci. USA* **96**, 8657–8662.

117. Ye, X., Rivera, V. M., Zoltick, P. *et al.* (1999). Regulated delivery of therapeutic proteins after in vivo somatic cell gene transfer. *Science* **283**, 88–91.

118. Agha-Mohammadi, S., and Lotze, M. T. (2000). Regulatable systems: applications in gene therapy and replicating viruses. *J. Clin. Invest.* **105**, 1177–1183.

119. McLaughlin, S. K., Collis, P., Hermonat, P. L. *et al.* (1988). Adeno-associated virus general transduction vectors: analysis of proviral structures. *J. Virol.* **62**, 1963–1973.

120. Hargrove, P. W., Nienhuis, A. W., Kurtzman, G. J. *et al.* (1997). High-level globin gene expression mediated by a recombinant adeno-associated virus genome that contains the 3′ gamma globin gene regulatory element and integrates as tandem copies in erythroid cells. *Blood* **89**, 2167–2175.

121. Urcelay, E., Ward, P., Wiener, S. M. *et al.* (1995). Asymmetric replication in vitro from a human sequence element is dependent on adeno-associated virus rep protein. *J. Virol.* **69**, 2038–2046.

122. Shelling, A. N., and Smith, M. G. (1994). Targeted integration of transfected and infected adeno-associated virus vectors containing the neomycin resistance gene. *Gene Ther.* **1**, 165–169.

123. Surosky, R. T., Urabe, M., Godwin, S. G. *et al.* (1997). Adeno-associated virus Rep proteins target DNA sequences to a unique locus in the human genome. *J. Virol.* **71**, 7951–7959.

124. Nakai, H., Iwaki, Y., Kay, M. A. *et al.* (1999). Isolation of recombinant adeno-associated virus vector-cellular DNA junctions from mouse liver. *J. Virol.* **73**, 5438–5447.

125. Omori, F., Messner, H. A., Ye, C. *et al.* (1999). Nontargeted stable integration of recombinant adeno-associated virus into human leukemia and lymphoma cell lines as evaluated by fluorescence in situ hybridization. *Hum. Gene Ther.* **10**, 537–543.

126. Kearns, W. G., Cutting, G. R., Flotte, T. R. *et al.* (1996). Recombinant adeno-associated virus (AAV-CFTR) vectors do not integrate in a site-specific fashion in an immortalized epithelial cell line. *Gene Ther.* **3**, 748–755.

127. Ponnazhagan, S., Erikson, D., Kearns, W. G. *et al.* (1997). Lack of site-specific integration of the recombinant adeno-associated virus 2 genomes in human cells. *Hum. Gene Ther.* **8**, 275–284.

128. Balague, C., Zhang, W. W., and Kalla, M. (1997). Adeno-associated virus Rep78 protein and terminal repeats enhance integration of DNA sequences into the cellular genome. *J. Virol.* **71**, 3299–3306.

129. Malik, P., Kohn, D. B., Kurtzman, G. J. *et al.* (1997). Recombinant adeno-associated virus mediates a high level of gene transfer but less efficient integration in the K562 human hematopoietic cell line. *J. Virol.* **71**, 1776–1783.

130. Lamartina, S., Roscilli, G., Rinaudo, D. et al. (1998). Lipofection of purified adeno-associated virus Rep68 protein: toward a chromosome-targeting nonviral particle. J. Virol. **72,** 7653–7658.

131. Palombo, F., Monciotti, A., Recchia, A. et al. (1998). Site-specific integration in mammalian cells mediated by a new hybrid baculovirus-adeno-associated virus vector. J. Virol. **72,** 5025–5034.

132. Recchia, A., Parks, R. J., Lamartina, S. et al. (1999). Site-specific integration mediated by a hybrid adenovirus/adeno-associated virus vector. Proc. Natl. Acad. Sci. USA **96,** 2615–2620.

133. Rinaudo, D., Lamartina, S., Roscilli, G. et al. (2000). Conditional site-specific integration into human 19 by using a ligand-dependent chimeric adeno-associated virus/Rep protein. J. Virol. **74,** 281–294.

134. Fisher, K. J., Gao, G. P., Weitzman, M. D. et al. (1996). Transduction with recombinant adeno-associated virus for gene therapy is limited by leading-strand synthesis. J. Virol. **70,** 520–532.

135. Ferrari, F. K., Samulski, R. J., Shenk, T., et al. (1996). Second-strand synthesis is a rate-limiting step for efficient transduction by recombinant adeno-associated virus vectors. J. Virol. **70,** 3227–34.

136. Russell, D. W., Miller, A. D., and Alexander I. E. (1995). DNA synthesis and topoisomerase inhibitors increase transduction by adeno-associated virus vectors. Proc. Natl. Acad. Sci. USA **92,** 5719–5723.

137. Yakobson, B., Koch, T., and Winocour, E. (1987). Replication of adeno-associated virus in synchronized cells without the addition of a helper virus. J. Virol. **61,** 972–981.

138. Yakobson, B., Hrynko, T. A., Peak, M. J. et al. (1989). Replication of adeno-associated virus in cells irradiated with UV light at 254 nm. J. Virol. **63,** 1023–1030.

139. Yalkinoglu, A. Ö., Heilbronn, R., Bürkle, A. et al. (1988). DNA amplification of adeno-associated virus as a response to cellular genotoxic stress. Cancer Res. **48,** 3123–3129.

140. Yalkinoglu A. Ö., Zentgraf, H., and Hubscher, U. (1991). Origin of adeno-associated virus DNA replication is a target of carcinogen-inducible DNA replication. J. Virol. **65,** 3175–3184.

141. Nakai, H., Storm, T. A., and Kay, M. A. (2000). Recruitment of single-stranded recombinant adeno-associated virus vector genomes and intermolecular recombination are responsible for stable transduction of liver in vivo. J. Virol. **74,** 9451–9463.

142. Hernandez, Y. J., Wang, J., Kearns, W. G. et al. (1999). Latent adeno-associated virus infection elicits humoral but not cell-mediated immune responses in a nonhuman primate model. J. Virol. **73,** 8549–8558.

143. Fields, P. A., Kowalczyk, D. W., Arruda, V. R. et al. (2000). Role of vector in activation of T cell subsets in immune responses against the secreted transgene product factor IX. Mol. Ther. **1,** 225–235.

144. Jooss, K., Yang, Y., Fisher, K. J. et al. (1998). Transduction of dendritic cells by DNA viral vectors directs the immune response to transgene products in muscle fibers. J. Virol. **72,** 4212–4223.

145. Zhang, Y., Chirmule, N., Gao, G. P. et al. (2000). CD40 ligand-dependent activation of cytotoxic t-lymphocytes by adeno-associated virus vectors in vivo: role of immature dendritic cells. J. Virol. **74,** 8003–8010.

146. Sun, J. Y., Krouse, R. S., Forman, S. J. et al. (2001). Immunogenicity of a p210$^{bcr-abl}$ fusion domain candidate DNA vaccine targeted to dendritic cells by an rAAV vector in vitro, (submitted).

147. Liu, Y., Santin, A. D., Mane, M. et al. (2000). Transduction and utility of the granulocyte-macrophage colony-stimulating factor gene into monocytes and dendritic cells by adeno-associated virus. J. Interferon Cytokine Res. **20,** 21–30.

148. Brockstedt, D. G., Podsakoff, G. M., Fong, L. et al. (1999). Induction of immunity to antigens expressed by recombinant adeno-associated virus depends on the route of administration. Clin. Immunol. **92,** 67–75.

149. Xiao, W., Chirmule, N., Schnell, M. A. et al. (2000). Route of administration determines induction of T-cell-independent humoral responses to adeno-associated virus vectors. Mol. Ther. **1,** 323–329.

150. Cotmore, S. F., and Tattersall, P. (1987). The autonomously replicating parvoviruses of vertebrates. Adv. Virus Res. **33,** 91–174.

151. Toolan, H. W., Saunders, E. L., Southam, C. M. et al. (1965). H-1 virus viremia in the human. Proc. Soc. Exp. Biol. Med. **119,** 711–715.

152. Toolan, H. W., and Ledinko, N. (1968). Inhibition by H-1 virus on the incidence of tumors produced by adenovirus 12 in hamsters. Virology **35,** 475–478.

153. Van Pachterbeke, C., Tuynder, M., Cosyn, J. P. et al. (1993). Parvovirus H-1 inhibits growth of short-term tumor-derived but not normal mammary tissue cultures. Int. J. Cancer **55,** 672–677.

154. Rommelaere, J., and Cornelis, J. J. (1991). Anti-neoplastic activity of parvoviruses. J. Virol. Meth. **33,** 233–251.

155. Linser, P., Bruning, H., and Armentrout, R. W. (1977). Specific binding sites for a parvovirus minute virus of mice on cultured mouse cells. J. Virol. **24,** 211–221.

156. Legendre, D., and Rommelaere, J. (1992). Terminal regions of the NS-1 protein of the parvovirus minute virus of mice are involved in cytotoxicity and promoter trans inhibition. J. Virol. **66,** 5705–5713.

157. Legrand, C., Rommelaere, J., and Caillet-Fauquet, P. (1993). MVM(p) NS-2 protein expression is required with NS-1 for maximal cytotoxicity in human transformed cells. Virology **195,** 149–155.

158. Rayet, B., Lopez-Guerrero, J. A., Rommelaere, J. et al. (1998). Induction of programmed cell death by parvovirus H-1 in U937 cells: connection with the tumor necrosis factor alpha signalling pathway. J. Virol. **72,** 8893–8903.

159. Guetta, E., Graziani, Y., and Tal, J. (1986). Suppression of Ehrlich ascites tumors in mice by minute virus of mice. J. Natl. Cancer Inst. **76,** 1177–1780.

160. Dupressoir, T., Vanacker, J. M., Cornelis, J. et al. (1989). Inhibition by parvovirus H-1 of the formation of tumors in nude mice and colonies in vitro by transformed human mammary epithelial cells. Cancer Res. **49,** 3203–3208.

161. Telerman, A., Tuynder, M., Dupressoir, T. et al. (1993). A model for tumor suppression using H-1 parvovirus. Proc. Natl. Acad. Sci. USA **90,** 8702–8706.

162. Bischoff, J. R., Kirn, D. H., Williams, A. et al. (1996). An adenovirus mutant that replicates selectively in p53-deficient human tumor cells. Science **274,** 373–376.

163. Heise, C., and Kirn, D. H. (2000). Replication-selective adenoviruses as oncolytic agents. J. Clin. Invest. **105,** 847–851.

164. Russell, S. J., Brandenburger, A., Flemming, C. L. et al. (1992). Transformation-dependent expression of interleukin genes delivered by a recombinant parvovirus. J. Virol. **66,** 2821–2828.

165. Maxwell, I. H., Maxwell, F., Rhode, S. L., III et al. (1993). Recombinant LuIII autonomous parvovirus as a transient transducing vector for human cells. Hum. Gene Ther. **4,** 441–450.

166. Corsini, J., Carlson, J. O., Maxwell, I. H. et al. (1996). Autonomous parvovirus and densovirus gene vectors. Adv. Virus Res. **47,** 303–351.

167. Maxwell, I. H., Maxwell, F., Long, C. J. et al. (1996). Autonomous parvovirus transduction of a gene under control of tissue-specific or inducible promoters. Gene Ther. **3,** 28–36.

168. Brandenburger, A., and Russell, S. (1996). A novel packaging system for the generation of helper-free oncolytic MVM vector stocks. Gene Ther. **3,** 927–931.

169. Avalosse, B., Burny, A., Mine, N. et al. (1996). Method for concentrating and purifying recombinant autonomous parvovirus vectors designed for tumour-cell-targeted gene therapy. J. Virol. Meth. **62,** 179–83.

170. Coe, S., Harron, M., Winslet, M. et al. (2000). The use of skeletal muscle to express genes for the treatment of cancer. Adv. Exp. Med. Biol. **465,** 95–111.

171. Lemischka, I. R. (1992). What we have learned from retroviral marking of hematopoietic stem cells. Curr. Top. Micro. Immunol. **177,** 59–71.

172. Brenner, M. K., Rill, D. R., Moen, R. C. *et al.* (1993). Gene-marking to trace origin of relapse after autologous bone-marrow transplantation. *Lancet* **341**, 85–86.

173. Brenner, M. K., Rill, D. R., Holladay, M. S. *et al.* (1993). Gene marking to determine whether autologous marrow infusion restores long-term haemopoiesis in cancer patients. *Lancet* **342**, 1134–1137.

174. Deisseroth, A. B., Zu, Z., Claxton, D. *et al.* (1994). Genetic marking shows that Ph+ cells present in autologous transplants of chronic myelogenous leukemia (CML) contribute to relapse after autologous bone marrow in CML. *Blood* **83**, 3068–3076.

175. Dunbar, C. E., Cottler-Fox, M., O'Shaughnessy, J. A. *et al.* (1995). Retrovirally marked CD34-enriched peripheral blood and bone marrow cells contribute to long-term engraftment after autologous transplantation. *Blood* **85**, 3048–3057.

176. Spangrude, G. J., and Johnson, G. R. (1990). Resting and activated subsets of mouse multipotent hematopoietic stem cells. *Proc. Natl. Acad. Sci. USA* **87**, 7433–7437.

177. Ogawa, M. (1993). Differentiation and proliferation of hematopoietic stem cells. *Blood* **81**, 2844–2853.

178. Moritz, T., Dutt, P., Xiao, X. *et al.* (1996). Fibronectin improves transduction of reconstituting hematopoietic stem cells by retroviral vectors: evidence of direct viral binding to chymotryptic carboxy-terminal fragments. *Blood* **88**, 855–862.

179. Luskey, B. D., Rosenblatt, M., Zsebo, K. *et al.* (1992). Stem cell factor, interleukin-3, and interleukin-6 promote retroviral-mediated gene transfer into murine hematopoietic stem cells. *Blood* **80**, 396–402.

180. Williams, D. A. (1993). Ex vivo expansion of hematopoietic stem and progenitor cells—robbing Peter to pay Paul? *Blood* **81**, 3169–3172.

181. Hao, Q. L., Thiemann, F. T., Petersen, D. *et al.* (1996). Extended long-term culture reveals a highly quiescent and primitive human hematopoietic progenitor population. *Blood* **88**, 3306–3313.

182. Larochelle, A., Vormoor, J., Hanenberg, H. *et al.* (1996). Identification of primitive human hematopoietic cells capable of repopulating NOD/SCID mouse bone marrow: implications for gene therapy. *Nat. Med.* **2**, 1329–1337.

183. Chatterjee, S., and Wong, K. K., Jr. (1996). Adeno-associated virus vectors for gene therapy of the hematopoietic system. *Curr. Top. Microbiol. Immunol.* **218**, 61–73.

184. Miller, J. L., Donahue, R. E., Sellers, S. E. *et al.* (1994). Recombinant adeno-associated virus (rAAV)-mediated expression of a human gamma-globin gene in human progenitor-derived erythroid cells. *Proc. Natl. Acad. Sci. USA* **91**, 10183–10187.

185. Zhou, S. Z., Broxmeyer, H. E., Cooper, S. *et al.* (1993). Adeno-associated virus 2-mediated gene transfer in murine hematopoietic progenitor cells. *Exp Hematol.* **21**, 928–933.

186. Walsh, C. E., Nienhuis, A. W., Samulski, R. J. *et al.* (1994). Phenotypic correction of Fanconi anemia in human hematopoietic cells with a recombinant adeno-associated virus vector. *J. Clin. Invest.* **94**, 1440–1448.

187. Walsh, C. E., Liu, I. M., Wang, S. *et al.* (1994). In vivo gene transfer with a novel adeno-associated virus vector to human hematopoietic cells engrafted in SCID-hu mice. *Blood* **84**, 256a.

188. Luhovy, M., Prchal, J. T., Townes, T. M. *et al.* (1996). Stable transduction of recombinant adeno-associated virus into hematopoietic stem cells from normal and sickle cell patients. *Biol. Blood Marrow Transplant.* **2**, 24–30.

189. Schimmenti, S., Boesen, J., Claassen, E. A. *et al.* (1998). Long-term genetic modification of rhesus monkey hematopoietic cells following transplantation of adenoassociated virus vector-transduced CD34+ cells. *Hum. Gene Ther.* **9**, 2727–2734.

190. Russell, D. W., and Kay, M. A. (1999). Adeno-associated virus vectors and hematology. *Blood* **94**, 864–874.

191. Gardner, J. P., Zhu, H., Colosi, P. C. *et al.* (1997). Robust, but transient expression of adeno-associated virus-transduced genes during human T lymphopoiesis. *Blood* **90**, 4854–4864.

192. van Os, R., Avraham, H., Banu, N. *et al.* (1999). Recombinant adeno-associated virus-based vectors provide short-term rather than long-term transduction of primitive hematopoietic stem cells. *Stem Cells* **17**, 117–120.

193. Nathwani, A. C., Hanawa, H., Vandergriff, J. *et al.* (2000). Efficient gene transfer into human cord blood CD34+ cells and the CD34+CD38–subset using highly purified recombinant adeno-associated viral vector preparations that are free of helper virus and wild-type AAV. *Gene Ther.* **7**, 183–195.

194. Hanazono, Y., Brown, K. E., Handa, A. *et al.* (1999). In vivo marking of rhesus monkey lymphocytes by adeno-associated viral vectors: direct comparison with retroviral vectors. *Blood* **94**, 2263–2270.

195. Ponnazhagan, S., Mukherjee, P., Wang, X. S. *et al.* (1997). Adeno-associated virus type 2-mediated transduction in primary human bone marrow-derived CD34+ hematopoietic progenitor cells: donor variation and correlation of transgene expression with cellular differentiation. *J. Virol.* **71**, 8262–8267.

196. Anderson, R. J., Prentice, H. G., Corbett, T. J. *et al.* (1997). Detection of adeno-associated virus type 2 in sorted human bone marrow progenitor cells. *Exp. Hematol.* **25**, 256–262.

197. Veldwijk, M. R., Schiedlmeier, B., Kleinschmidt, J. A. *et al.* (1999). Superior gene transfer into solid tumour cells than into human mobilised peripheral blood progenitor cells using helpervirus-free adeno-associated viral vector stocks. *Eur. J. Cancer* **35**, 1136–1142.

198. Veldwijk, M. R., Fruehauf, S., Schiedlmeier, B. *et al.* (2000). Differential expression of a recombinant adeno-associated virus 2 vector in human CD34+ cells and breast cancer cells. *Cancer Gene Ther.* **7**, 597–604.

199. Hoerer, M., Bogedain, C., Scheer U. *et al.* (1997). The use of recombinant adeno-associated viral vectors for the transduction of epithelial tumor cells. *Int. J. Immunopharmacol.* **19**, 473–479.

200. Itou, T., Miyamura, K., Abe, A. *et al.* (1998). Recombinant adeno-associated virus-mediated gene transfer into human leukemia cell lines. *Int. J. Hematol.* **67**, 27–35.

201. Antman, K. H., Elias, A., and Fine, H. A. (1994). Dose-intensive therapy with autologous bone marrow transplantation in solid tumors, in *Bone Marrow Transplantation* (S. J. Forman, K. G. Blume, and E. D. Thomas, eds.), pp. 767–788. Blackwell Scientific, Cambridge, MA.

202. Gottesman, M. M., and Pastan, I. (1993). Biochemistry of multidrug resistance mediated by the multidrug transporter. *Annu. Rev. Biochem.* **62**, 385–427.

203. Chaudhary, P. M., and Roninson, I. B. (1991). Expression and activity of P-glycoprotein, a multidrug efflux pump, in human hematopoietic stem cells. *Cell* **66**, 85–94.

204. Mickisch, G. H., Merlino, G. T., Galski, H. *et al.* (1991). Transgenic mice that express the human multidrug-resistance gene in bone marrow enable a rapid identification of agents that reverse drug resistance. *Proc. Natl. Acad. Sci. USA* **88**, 547–551.

205. Mickisch, G. H., Aksentijevich, I., Schoenlein, P. V. *et al.* (1992). Transplantation of bone marrow cells from transgenic mice expressing the human MDR1 gene results in long-term protection against the myelosuppressive effect of chemotherapy in mice. *Blood* **79**, 1087–1093.

206. Sorrentino, B. P., Brandt, S. J., Bodine, D. *et al.* (1992). Selection of drug-resistant bone marrow cells in vivo after retroviral transfer of human MDR1. *Science* **257**, 99–103.

207. Hanania, E. G., and Deisseroth, A. B. (1994). Serial transplantation shows that early hematopoietic precursor cells are transduced by MDR-1 retroviral vector in a mouse gene therapy model. *Cancer Gene Ther.* **1**, 21–25.

208. Ward, M., Richardson, C., Pioli, P. *et al.* (1994). Transfer and expression of the human multiple drug resistance gene in human CD34+ cells. *Blood* **84,** 1408–1414.

209. Hesdorffer, C., Ayello, J., Ward, M. *et al.* (1998). Phase I trial of retroviral-mediated transfer of the human MDR1 gene as marrow chemoprotection in patients undergoing high-dose chemotherapy and autologous stem-cell transplantation. *J. Clin. Oncol.* **16,** 165–172.

210. Abonour, R., Williams, D. A., Einhorn, L. *et al.* (2000). Efficient retrovirus-mediated transfer of the multidrug resistance 1 gene into autologous human long-term repopulating hematopoietic stem cells. *Nat. Med.* **6,** 652–658.

211. Bunting, K. D., Galipeau, J., Topham, D. *et al.* (1998). Transduction of murine bone marrow cells with an MDR1 vector enables ex vivo stem cell expansion, but these expanded grafts cause a myeloproliferative syndrome in transplanted mice. *Blood* **92,** 2269–2279.

212. Baudard, M., Gottesman, M. M., Kearns, W. G. *et al.* (1996). Expression of the human multidrug resistance and glucocerebrosidase cDNAs from adeno-associated vectors: efficient promoter activity of AAV sequences and in vivo delivery via liposomes. *Hum. Gene Ther.* **7,** 1309–1322.

213. Baudard, M. (1998). Construction of MDR1 adeno-associated virus vectors for gene therapy. *Meth. Enzymol.* **292,** 538–545.

214. Neckers, L., Whitesell, L., Rosolen, A. *et al.* (1992). Antisense inhibition of oncogene expression. *Crit. Rev. Oncogol.* **3,** 175–231.

215. Hélène, C. (1994). Control of oncogene expression by antisense nucleic acids. *Eur. J. Cancer* **30A,** 1721–1726.

216. Zhang, W. W. (1996). Antisense oncogene and tumor suppressor gene therapy of cancer. *J. Mol. Med.* **74,** 191–204.

217. Reed, J. C., Cuddy, M., Haldar, S. *et al.* (1990). BCL2-mediated tumorigenicity of a human T-lymphoid cell line: synergy with MYC and inhibition by BCL2 antisense. *Proc. Natl. Acad. Sci. USA* **87,** 3660–3664.

218. Skorski, T., Nieborowska-Skorska, M., Nicolaides, N. C. *et al.* (1994). Suppression of Philadelphia leukemia cell growth in mice by *bcr–abl* antisense oligodeoxynucleotide. *Proc. Natl. Acad. Sci. USA* **91,** 4504–4508.

219. Colomer, R., Lupu, R., Bacus, S. S. *et al.* (1994). erb B-2 antisense oligonucleotides inhibit the proliferation of breast carcinoma cells with erbB-2 oncogene amplification. *Br. J. Cancer* **70,** 819–825.

220. Mercola, D., Rundell, A., Westwick, J. *et al.* (1987). Antisense RNA to the c-*fos* gene, restoration of density dependent growth arrest in a transformed cell line. *Biochem. Biophys. Res. Commun.* **147,** 288–294.

221. Yokoyama, K., and Imamoto, F. (1987). Transcriptional control of endogenous *myc* protooncogene by antisense RNA. *Proc. Natl. Acad. Sci. USA* **84,** 7363–7367.

222. Sauter, E. R., Herlyn, M., Liu, S. C. *et al.* (2000). Prolonged response to antisense cyclin D1 in a human squamous cancer xenograft model. *Clin. Cancer Res.* **6,** 654–660.

223. Steele, C., Cowsert, L. M., Shillitoe, E. J. (1993). Effects of human papillomavirus type 18-specific antisense oligonucleotides on the transformed phenotype of human carcinoma cell lines. *Cancer Res.* **53,** 2330–2337.

224. Wong, K. K., Jr., Rose, J. A., and Chatterjee, S. (1991). Restriction of HSV-1 production in cell lines transduced with an antisense viral vector targeting the HSV-1 ICP4 gene, in *VACCINE '91* (F. Brown, R. Chanock, H. Ginsberg, and R. Lerner, eds.), pp. 183–189. Cold Spring Harbor Press, New York.

225. Ponnazhagan, S., Nallari, M. L., and Srivastava, A. (1994). Suppression of human alpha-globin gene expression mediated by the recombinant adeno-associated virus 2-based antisense vectors. *J. Exp. Med.* **179,** 733–738.

226. Castanotto, D., Rossi, J. J., and Sarver, N. (1994). Antisense catalytic RNAs as therapeutic agents. *Adv. Pharm.* **25,** 289–317.

227. Rossi, J. J. (1999). Ribozymes, genomics and therapeutics. *Chem. Biol.* **6,** R33–R37.

228. Kashani, S. M., Funato, T., Tone, T. *et al.* (1992). Reversal of the malignant phenotype by an anti-*ras* ribozyme. *Antisense Res. Dev.* **2,** 3–15.

229. Tone, T., Kashani-Sabet, M., Funato, T. *et al.* (1993). Suppression of EJ cells tumorigenicity. *In Vivo* **7,** 471–476.

230. Snyder, D. S., Wu, Y., Wang, J. L. *et al.* (1993). Ribozyme-mediated inhibition of *bcr–abl* gene expression in Philadelphia chromosome-positive cell line. *Blood* **82,** 600–605.

231. Scanlon, K. J., Jiao, L., Funato, T. *et al.* (1991). Ribozyme-mediated cleavage of c-*fos* mRNA reduces gene expression of DNA synthesis enzymes and metallothionein. *Proc. Natl. Acad. Sci. USA* **88,** 10591–10595.

232. Pavco, P. A., Bouhana, K. S., Gallegos, A. M. *et al.* (2000). Antitumor and antimetastatic activity of ribozymes targeting the messenger RNA of vascular endothelial growth factor receptors. *Clin. Cancer Res.* **6,** 2094–2103.

233. Funato, T., Yoshida, E., Jiao, L. *et al.* (1992). The utility of an anti-*fos* ribozyme in reversing cisplatin resistance in human carcinomas. *Adv. Enzyme Regul.* **32,** 195–209.

234. Holm, P. S., Scanlon, K. J., and Dietal, M. (1994). Reversion of multidrug resistance in the P-glycoprotein-positive human pancreatic cell line (EPP85-181RDB) by introduction of a hammerhead ribozyme. *Br. J. Cancer* **70,** 239–243.

235. Irie, A., Kijima, H., Ohkawa, T. *et al.* (1997). Anti-oncogene ribozymes for cancer gene therapy. *Adv. Pharmacol.* **40,** 207–257.

236. Lowy, D. R., Kirnbauer, R., and Schiller, J. T. (1994). Genital human papillomavirus infection. *Proc. Natl. Acad. Sci. USA* **91,** 2436–2440.

237. Walz, C., Deprez, A., Dupressoir, T. *et al.* (1997). Interaction of human papillomavirus type 16 and adeno-associated virus type 2 co-infecting human cervical epithelium. *J. Gen. Virol.* **78,** 1441–1452.

238. Ogston, P., Raj, K., and Beard, P. (2000). Productive replication of adeno-associated virus can occur in human papillomavirus type 16 (HPV-16) episome-containing keratinocytes and is augmented by the HPV-16 E2 protein. *J. Virol.* **74,** 3494–3504.

239. Su, P. F., and Wu, F. Y. (1996). Differential suppression of the tumorigenicity of HeLa and SiHa cells by adeno-associated virus. *Br. J. Cancer* **73,** 1533–1537.

240. Wu, F. Y., Wu, C. Y., Lin, C. H. *et al.* (1999). Suppression of tumorigenicity in cervical carcinoma HeLa cells by an episomal form of adeno-associated virus. *Int. J. Oncol.* **15,** 101–106.

241. Zhan, D., Santin, A. D., Liu, Y. *et al.* (1999). Binding of the human papillomavirus type 16 p97 promoter by the adeno-associated virus Rep78 major regulatory protein correlates with inhibition. *J. Biol. Chem.* **274,** 31619–31624.

242. Su, P. F., Chiang, S. Y., Wu, C. W. *et al.* (2000). Adeno-associated virus major rep78 protein disrupts binding of TATA-binding protein to the p97 promoter of human papillomavirus type 16. *J. Virol.* **74,** 2459–2465.

243. Lu, D., Chatterjee, S., Brar, D. *et al.* (1994). High efficiency in vitro cleavage of transcripts arising from the major transforming genes of human papillomavirus type 16 mediated by ribozymes transcribed from an adeno-associated virus-based vector. *Cancer Gene Ther.* **1,** 267–277.

244. Alvarez-Salas, L. M., Cullinan, A. E., Siwkowski, A. *et al.* (1998). Inhibition of HPV-16 E6/E7 immortalization of normal keratinocytes by hairpin ribozymes. *Proc. Natl. Acad. Sci. USA* **95,** 1189–1194.

245. Kunke, D., Grimm, D., Denger, S. *et al.* (2000). Preclinical study on gene therapy of cervical carcinoma using adeno-associated virus vectors. *Cancer Gene Ther.* **7,** 766–777.

246. Sun, X. L., Antony, A. C., Srivastava, A. *et al.* (1995). Transduction of folate receptor cDNA into cervical carcinoma cells using recombinant adeno-associated virions delays cell proliferation in vitro and in vivo. *J. Clin. Invest.* **96,** 1535–1547.

247. Somasundaram, K. (2000). Tumor suppressor p53: regulation and function. *Front. Biosci.* **1**, D424–D437.
248. Swisher, S. G., Roth, J. A., Nemunaitis, J. *et al.* (1999). Adenovirus-mediated p53 gene transfer in advanced non-small-cell lung cancer. *J. Natl. Cancer Inst.* **91**, 763–771.
249. Qazilbash, M. H., Xiao, X., Seth, P. *et al.* (1997). Cancer gene therapy using a novel adeno-associated virus vector expressing human wild-type p53. *Gene Ther.* **4**, 675–682.
250. Challis, G. B., and Stam, H. J. (1990). The spontaneous regression of cancer: a review of cases 1900–1987. *Acta Oncol.* **29**, 545–550.
251. Medeiros, L. J., Picker, L. J., Gelb, A. B. *et al.* (1989). Number of 'host' helper T cells and proliferating cells predict survival in diffuse small-cell lymphomas. *J. Clin. Oncol.* **7**, 1009–1017.
252. Zinkernagel, R. M., and Doherty, P. C. (1979). MHC-restricted cytotoxic T cells: studies on the biological role of polymorphic major transplantation antigen determining T cell restriction specificity, function and responsiveness. *Adv. Immunol.* **27**, 51–177.
253. June, C., Bluestone, J., Nadler, L. M. *et al.* (1994). The B7 and CD28 receptor families. *Immunol. Today* **15**, 321–331.
254. Hwu, P., and Rosenberg, S. A. (1994). The use of gene-modified tumor-infiltrating lymphocytes for cancer therapy. *Ann. N.Y. Acad. Sci.* **716**, 188–197.
255. Abken, H., Hombach, A., Heuser, C. *et al.* (1997). Chimeric T-cell receptors: highly specific tools to target cytotoxic T-lymphocytes to tumour cells. *Cancer Treat. Rev.* **23**, 97–112.
256. Paillard, F. (1999). Immunotherapy with T cells bearing chimeric antitumor receptors. *Hum. Gene Ther.* **10**, 151–153.
257. Flamand, V., Sornasse, T., Thielemans, K. *et al.* (1994). Murine dendritic cells pulsed in vitro with tumor antigen induce tumor resistance in vivo. *Eur. J. Immunol.* **24**, 605–610.
258. Pardoll, D. M. (1998). Cancer vaccines. *Nat Med* **4**(5 Suppl), 525–31.
259. Tepper, R. I., and Mule, J. (1994). Experimental and clinical studies of cytokine gene-modified tumor cells. *Hum. Gene Ther.* **5**, 153–164.
260. Townsend, S., and Allison, J. (1993). Tumor rejection after direct costimulation of CD8+ T cells by B7 transfected melanoma cells. *Science* **259**, 368–372.
261. Nabel, G. J., Nabel, E. G., Yang, Z-Y. *et al.* (1993). Direct gene transfer with DNA-liposome complexes in melanoma: expression, biologic activity, and lack of toxicity in humans. *Proc. Natl. Acad. Sci.* **90**, 11307–11311.
262. Bruton, J. K., and Koeller, J. M. (1994). Recombinant interleukin-2. *Pharmacotherapy* **14**, 635–656.
263. Fearon, E. R., Pardoll, D. M., Itaya, T. *et al.* (1990). Interleukin-2 production by tumor cells bypasses T helper function in the generation of an antitumor response. *Cell* **60**, 397–403.
264. Connor, J., Bannerji, R., Saito, S. *et al.* (1993). Regression of bladder tumors in mice treated with interleukin-2 gene-modified tumor cells. *J. Exp. Med.* **177**, 1127–1134.
265. Gansbacher, B., Zier, K., Daniels, B. *et al.* (1990). Interleukin-2 gene transfer into tumor cells abrogates tumorigenicity and induces protective immunity. *J. Exp. Med.* **172**, 1217–1224.
266. Haddada, H., Ragot, T., Cordier, L. *et al.* (1993). Adenoviral interleukin-2 gene transfer into P815 tumor cells abrogates tumorigenicity and induces antitumoral immunity in mice. *Hum. Gene Ther.* **4**, 703–711.
267. Coveney, E., Clary, B., DiMaio, J. M. *et al.* (1994). Inhibition of breast cancer metastasis by cytokine gene-modified tumor vaccination in tumor-bearing mice. *Surgical Forum* **XV**, 540–542.
268. Coveney, E., Clary, B., Iacobucci, M. *et al.* (1996). Active immunotherapy with transiently transfected cytokine-secreting tumor cells inhibits breast cancer metastases in tumor-bearing animals. *Surgery* **120**, 265–272.
269. Clary, B. M., Coveney, E. C., Blazer, D. G. III *et al.* (1997). Active immunization with tumor cells transduced by a novel AAV plasmid-based gene delivery system. *J. Immunother.* **20**, 26–37.
270. Philip, R., Brunette, E., Kilinski, L. *et al.* (1994). Efficient and sustained gene expression in primary T lymphocytes and primary and cultured tumor cells mediated by adeno-associated virus plasmid DNA complexed to cationic liposomes. *Mol. Cell. Biol.* **14**, 2411–2418.
271. Philip, R., Clary, B., Brunette, E. *et al.* (1996). Gene modification of primary tumor cells for active immunotherapy of human breast and ovarian cancer. *Clin. Cancer Res.* **2**, 59–68.
272. Okada, H., Yoshida, J., Kurtzman, G. *et al.* (1996). Gene therapy against an experimental glioma using adeno-associated virus vectors. *Gene Ther.* **3**, 957–964.
273. Mizuno, M., Yoshida, J., Colosi, P. *et al.* (1998). Adeno-associated virus vector containing the herpes simplex virus thymidine kinase gene causes complete regression of intracerebrally implanted human gliomas in mice, in conjunction with ganciclovir administration. *Jpn. J. Cancer Res.* **89**, 76–80.
274. Rosenfeld, M. R., Bergman, I., Schramm, L. *et al.* (1997). Adeno-associated viral vector gene transfer into leptomeningeal xenografts. *J. Neurooncol.* **34**, 139–144.
275. Johnston, K. M., Jacoby, D., Pechan, P. A. *et al.* (1997). HSV/AAV hybrid amplicon vectors extend transgene expression in human glioma cells. *Hum. Gene Ther.* **8**, 359–370.
276. Costantini, L. C., Jacoby, D. R., Wang, S. *et al.* (1999). Gene transfer to the nigrostriatal system by hybrid herpes simplex virus/adeno-associated virus amplicon vectors. *Hum. Gene Ther.* **10**, 2481–2494.
277. Zhang, J. F., Taylor, M. W., Blatt, L. M. *et al.* (1996). Gene therapy with an adeno-associated virus carrying an interferon gene results in tumor growth suppression and regression. *Cancer Gene Ther.* **3**, 31–38.
278. Russell, S. J., Brandenburger, A., Flemming, C. L. *et al.* (1992). Transformation-dependent expression of interleukin genes delivered by a recombinant parvovirus. *J. Virol.* **66**, 2821–2828.
279. Tepper, R. I., Pattengale, P. K., and Leder, P. (1989). Murine interleukin-4 displays potent anti-tumor activity in vivo. *Cell* **57**, 503–512.
280. Haag, A., Menten, P., Van Damme, J. *et al.* (2000). Highly efficient transduction and expression of cytokine genes in human tumor cells by means of autonomous parvovirus vectors: generation of antitumor responses in recipient mice. *J. Hum. Gene. Ther.* **11**, 597–609.
281. Anderson, R., Macdonald, I., Corbett, T. *et al.* (1997). Construction and biological characterization of an interleukin-12 fusion protein (Flexi-12): delivery to acute myeloid leukemic blasts using adeno-associated virus. *Hum. Gene Ther.* **8**, 1125–1135.
282. Paul, D., Qazilbash, M. H., Song, K. *et al.* (2000). Construction of a recombinant adeno-associated virus (rAAV) vector expressing murine interleukin-12 (IL-12). *Cancer Gene Ther.* **7**, 308–315.
283. Chen, L., Ashe, S., Brady, W. A. *et al.* (1992). Costimulation of antitumor immunity by the B7 counterreceptor for the T lymphocyte molecules CD28 and CTLA-4. *Cell* **71**, 1093–1102.
284. Chiorini, J. A., Kotin, R. M., Hallek, M. *et al.* (1995). High-efficiency transfer of the T cell co-stimulatory molecule B7-2 to lymphoid cells using high-titer recombinant adeno-associated virus vectors. *Hum. Gene Ther.* **6**, 1531–1541.
285. Wendtner, C. M., Nolte, A., Mangold, E. *et al.* (1997). Gene transfer of the costimulatory molecules B7-1 and B7-2 into human multiple myeloma cells by recombinant adeno-associated virus enhances the cytolytic T cell response. *Gene Ther.* **4**, 726–735.
286. Manning, W. C., Paliard, X., Zhou, S. *et al.* (1997). Genetic immunization with adeno-associated virus vectors expressing herpes simplex virus type 2 glycoproteins B and D. *J. Virol.* **71**, 7960–7962.
287. Hart, D. N. (1997). Dendritic cells: unique leukocyte populations which control the primary immune response. *Blood* **90**, 3245–3287.
288. Banchereau, J., and Steinman, R. M. (1998). Dendritic cells and the control of immunity. *Nature* **392**, 245–252.

289. Morse, M. A., and Lyerly, H. K. (2000). Dendritic cell-based immunization for cancer therapy. *Adv. Exp. Med. Biol.* **465,** 335–346.

290. Fong, L., and Engleman, E. G. (2000). Dendritic cells in cancer immunotherapy. *Annu. Rev. Immunol.* **18,** 245–273.

291. Kirk, C. J., and Mule, J. J. (2000). Gene-modified dendritic cells for use in tumor vaccines. *Hum. Gene Ther.* **11,** 797–806.

292. Faderl, S., Talpaz, M., Estrov, Z. *et al.* (1999). The biology of chronic myeloid leukemia. *N. Engl. J. Med.* **341,** 164–172.

293. Liu, D. W., Tsao, Y. P., Kung, J. T. *et al.* (2000). Recombinant adeno-associated virus expressing human papillomavirus type 16 E7 peptide DNA fused with heat shock protein DNA as a potential vaccine for cervical cancer. *J. Virol.* **74,** 2888–2894.

294. Miyamura, K., Kajigaya, S., Momoeda, M. *et al.* (1994). Parvovirus particles as platforms for protein presentation. *Proc. Natl. Acad. Sci. USA* **91,** 8507–8511.

295. Sedlik, C., Casal, I., Leclerc, C. *et al.* (1995). Immunogenicity of poliovirus B and T cell epitopes presented by hybrid porcine parvovirus particles. *J. Gen. Virol.* **76,** 2361–2368.

296. Lal, S., Lauer, U. M., Niethammer, D. *et al.* (2000). Suicide genes: past, present and future perspectives. *Immunol. Today* **21,** 48–54.

297. Culver, K. W., Ram, Z., Wallbridge, S. *et al.* (1992). In vivo gene transfer with retroviral vector-producer cells for treatment of experimental brain tumors. *Science* **256,** 1550–1552.

298. Mesnil, M., and Yamasaki, H. (2000). Bystander effect in herpes simplex virus-thymidine kinase/ganciclovir cancer gene therapy: role of *gap*-junctional intercellular communication. *Cancer Res.* **60,** 3989–3999.

299. Koering, C. E., Dupressoir, T., Plaza, S. *et al.* (1994). Induced expression of the conditionally cytotoxic herpes simplex virus thymidine kinase gene by means of a parvoviral regulatory circuit. *Hum. Gene Ther.* **5,** 457–463.

300. Su, H., Kan, Y. W., Xu, S. M. *et al.* (1996). Selective killing of AFP-positive hepatocellular carcinoma cells by adeno-associated virus transfer of the herpes simplex virus thymidine kinase gene. *Hum. Gene Ther.* **7,** 463–470.

301. Su, H., Lu, R., Chang, J. C. *et al.* (1997). Tissue-specific expression of herpes simplex virus thymidine kinase gene delivered by adeno-associated virus inhibits the growth of human hepatocellular carcinoma in athymic mice. *Proc. Natl. Acad. Sci. USA* **94,** 13891–13896.

302. Su, H., Lu, R., Ding, R. *et al.* (2000). Adeno-associated viral-mediated gene transfer to hepatoma: thymidine kinase/interleukin 2 is more effective in tumor killing in non-ganciclovir (GCV)-treated than in GCV-treated animals. *Mole. Ther.* **1,** 509–515.

303. Pluda, J. M. (1997). Tumor-associated angiogenesis, mechanisms, clinical implications, and therapeutic strategies. *Semin. Oncol.* **24,** 203–218.

304. Saleh, M., Wilks, A. F., and Stacker, S. A. (1996). Inhibition of growth of C6 glioma cells in vivo by expression of antisense vascular endothelial growth factor sequence. *Cancer Res.* **56,** 393–401.

305. Ozaki, K., Terada, M., Sugimura, T. *et al.* (1996). Use of von Willebrand factor promoter to transduce suicidal gene to human endothelial cells, HUVEC. *Hum. Gene Ther.* **7,** 1483–1490.

306. Indraccolo, S., Minuzzo, S., Gola, E. *et al.* (1999). Generation of expression plasmids for angiostatin, endostatin and TIMP-2 for cancer gene therapy. *Int. J. Biol. Markers* **14,** 251–256.

307. Nguyen, J. T., Wu, P., Clouse, M. E. *et al.* (1998). Adeno-associated virus-mediated delivery of antiangiogenic factors as an antitumor strategy. *Cancer Res.* **58,** 5673–5677.

308. Yang, Q., Mamounas, M., Yu, G. *et al.* (1998). Development of novel cell surface CD34-targeted recombinant adenoassociated virus vectors for gene therapy. *Hum. Gene Ther.* **9,** 1929–1937.

309. Bartlett, J. S., Kleinschmidt, J., Boucher, R. C. *et al.* (1999). Targeted adeno-associated virus vector transduction of nonpermissive cells mediated by a bispecific F(ab'gamma)2 antibody. *Nat. Biotech.* **17,** 181–186.

310. Girod, A., Ried, M., Wobus, C. *et al.* (1999). Genetic capsid modifications allow efficient re-targeting of adeno-associated virus type 2. *Nat. Med.* **5,** 1052–1056.

311. Wu, P., Xiao, W., Conlon, T. *et al.* (2000). Mutational analysis of the adeno-associated virus type 2 (AAV2) capsid gene and construction of AAV2 vectors with altered tropism. *J. Virol.* **74,** 8635–8647.

312. Ponnazhagan, S., Weigel, K. A., Raikwar, S. P. *et al.* (1998). Recombinant human parvovirus B19 vectors: erythroid cell-specific delivery and expression of transduced genes. *J. Virol.* **72,** 5224–5230.

313. Kurpad, C., Mukherjee, P., Wang, X. S. *et al.* (1999). Adeno-associated virus 2-mediated transduction and erythroid lineage-restricted expression from parvovirus B19p6 promoter in primary human hematopoietic progenitor cells. *J. Hematother. Stem Cell Res.* **8,** 585–592.

314. Dupont, F., Tenenbaum, L., Guo, L. P. *et al.* (1994). Use of an autonomous parvovirus vector for selective transfer of a foreign gene into transformed human cells of different tissue origins and its expression therein. *J. Virol.* **68,** 1397–1406.

315. Dupont, F., Avalosse, B., Karim, A. *et al.* (2000). Tumor-selective gene transduction and cell killing with an oncotropic autonomous parvovirus-based vector. *Gene Ther.* **7,** 790–796.

316. Van Pachterbeke, C., Tuynder, M., Brandenburger, A. *et al.* (1997). Varying sensitivity of human mammary carcinoma cells to the toxic effect of parvovirus H-1. *Eur. J. Cancer* **33,** 1648–1653.

317. Faisst, S., Guittard, D., Benner, A. *et al.* (1998). Dose-dependent regression of HeLa cell-derived tumours in SCID mice after parvovirus H-1 infection. *Int. J. Cancer* **75,** 584–589.

318. Conrad, C. K., Flotte, T. R., Guggino, W. B. *et al.* (1996). Safety of single-dose administration of an adeno-associated virus (AAV)-CFTR vector in the primate lung. *Gene Ther.* **3,** 658–668.

319. Flotte, T., Wetzel, R., Walden, S. *et al.* (1996). A phase I study of an adeno-associated virus-CFTR gene vector in adult CF patients with mild lung disease. *Hum. Gene Ther.* **7,** 1145–1159.

320. Wagner, J. A., Messner, A. H., Moran, M. L. *et al.* (1999). Safety and biological efficacy of an adeno-associated virus vector-cystic fibrosis transmembrane regulator (AAV-CFTR) in the cystic fibrosis maxillary sinus. *Laryngoscope* **109**(2, part 1), 266–274.

321. Inoue, N., Hirata, R. K., and Russell, D. W. (1999). High-fidelity correction of mutations at multiple chromosomal positions by adeno-associated virus vectors. *J. Virol.* **73,** 7376–7380.

322. During, M. J., Xu, R., Young, D. *et al.* (1998). Peroral gene therapy of lactose intolerance using an adeno-associated virus vector. *Nat. Med.* **4,** 1131–1135.

323. During, M. J., Symes, C. W., Lawlor, P. A. *et al.* (2000). An oral vaccine against NMDAR1 with efficacy in experimental stroke and epilepsy. *Science* **287,** 1453–1460.

CHAPTER

4

Antibody-Targeted Gene Therapy

C. LAMPERT

*Department of Hematology and
Medical Oncology
St. Peter's University Hospital
New Brunswick, New Jersey 08901*

A. M. McCALL

*Fox Chase Cancer Center
Philadelphia, Pennsylvania 91010*

L. M. WEINER

*Fox Chase Cancer Center
Philadelphia, Pennsylvania 91010*

I. INTRODUCTION

In its purest form, gene therapy seeks to reverse the somatic gene mutations that may predate and lead to malignant change within cells. Though efficient gene transfer is an essential first step in any gene therapy strategy, it remains one of the biggest barriers to success. There are several issues that make this a daunting task. The first relates to the difficulty in delivering sufficient copies of a gene to *all* tumor sites. Unlike diseases such as hemophilia, where it would be necessary to replace only one gene in a portion of cells in one organ, the biology of neoplastic disease would dictate that all disease sites and potentially multiple genes must be addressed. The second barrier to an effective gene strategy is the difficulty in ensuring that the gene is actually introduced into the target cell. Currently available gene delivery

systems are capable of introducing genes into cells *in vivo* but are nonselective in their targeting. With the incorporation of antibodies or antibody fragments into the different delivery systems, cell-specific targeting has been achieved *in vivo*. While specific DNA transfer to tumor cells has been achieved with, for example, antibody DNA/poly-L-lysine complexes, antibody-mediated gene transfer remains extremely inefficient and new constructs are needed if antibody-mediated gene therapy is to be a viable option.

II. BACKGROUND: MONOCLONAL ANTIBODIES AND CANCER THERAPY

A. Monoclonal Antibodies and Antibody Fragments

The primary drive behind the interest in monoclonal antibodies relates to the specific targeting abilities of these molecules. This unique target specificity has created multiple opportunities for diagnostic and therapeutic applications. By specifically binding to tumor-associated antigens, antibodies can initiate or perpetuate immune responses against tumors. Antibodies are also capable of delivering toxins, chemotherapeutic agents, isotopes, and genes directly to tumor cells while selectively avoiding normal tissues.

Each molecule consists of two light and two heavy chains, which are divided into variable and constant regions. Variable regions (V_H and V_L) contain three complementarity-determining regions (CDRs) flanked on either side by framework regions. It is the six CDRs (three from the light chain and three from the heavy chain) that make up the binding site of the antibody. This binding site is considered to be the idiotype of the antibody. The heavy-chain-constant regions contain a minimum of three domains (CH1, CH2, and CH3), whereas the light-chain-constant regions consist of only one domain (CK or Cλ). Domains CH2 and CH3 form the Fc region of the antibody which determines its effector functions. For example, complement is activated by the Fc regions and leukocytes recognize other epitopes of this region through their Fc receptors to mediate phagocytosis and antibody-dependent cellular cytotoxicity (ADCC).

Early workers were hampered by the relative impurity of antibodies, and early antibodies were of limited use in the treatment of cancer. With the advent of hybridoma technology by Kohler and Milstein [1], large quantities of antibodies could be generated in monoclonal form. New techniques have enabled workers to molecularly define tumor-associated antigens that can be used as targets for monoclonal antibodies. Monoclonal antibodies directed against tumor antigens have been very successful in the treatment of leukemias and lymphomas [2–4]. Antibodies directed against solid tumors have, however, met with less success [5]. Several factors have limited the efficacy of antibody therapy in solid tumors. There is limited distribution and retention of antibodies within tumors, and this phenomenon might in part be related to the abnormal vascular skeleton found within tumors [6]. Vascular tortuosity, high interstitial tumor pressures, and an abnormal vascular interstitium all play a role in limiting the distribution of antibodies. Furthermore, renal and hepatic disposition and uptake must be considered in the design of antibody therapies. For example, smaller antibody structures infiltrate tumors more rapidly and more evenly than IgG and have exhibited promise as diagnostic agents. However, the potential of these molecules in clinical therapeutics has been limited by their rapid and primarily renal clearance [7]. The heterogeneous expression of tumor antigens by malignant and normal cells is another potentially important barrier to the effective use of antibodies [8]. Biological response modifiers such as recombinant interferons have been shown to selectively upregulate tumor antigens and offer one means to improve antibody retention in tumors [9]. Shed antigen can inhibit the binding of antibodies to neoplastic cells [10]. This is especially problematic for antibodies that mediate their effect at the cell surface. This is also true for antigens that are internalized. Finally, should an antibody reach and bind to its target antigen, there is little *in vivo* evidence that it is able to induce an antibody-mediated cytotoxic immune response engaging macrophages, natural killer (NK) cells, and cytotoxic T cells [11]. The majority of monoclonal antibodies tested in humans have been derived from mice. These antibodies induce human antimouse antibodies (HAMAs), the appearance of which usually limits continued therapy [12]. One means of circumventing this immune response is to humanize the antibody. In this process,

the amino acid sequences essential for antigen binding are retained in a recombinant immunoglobulin containing a human antibody framework. Many clinically active antibodies, such as trastuzomab (Herceptin™), are produced in this fashion.

The antibody armamentarium is no longer limited to conventional immunoglobulin molecules. For example, bispecific antibodies contain two antibodies with specificity for distinct targets [13]. The bispecific molecule 2B1 binds to the extracellular domain of the HER2/*neu* oncogene product on tumor cells and to the Fc[γ] receptor on natural killer cells and macrophages [14]. This bispecific antibody promotes the *in vitro* lysis of HER2/*neu*-expressing cells by NK cells.

Smaller molecules such as the Fab or F(ab)₂ molecules do not contain Fc binding domains [15,16]. The Fab and F(ab)₂ molecules are made by digesting whole antibodies with the enzymes papain and pepsin. Single-chain Fv (scFv) molecules contain only the heavy and light chains of the binding site joined by a short linker [17]. Smaller antibodies have been shown to infiltrate a tumor more rapidly and more evenly [18] but are cleared from the circulation with greater rapidity. More recently, bispecific scFv molecules have been produced and may improve tumor targeting and penetration while minimizing toxicity imposed by the Fc domains of antibody. The absence of the Fc domain in these fragments leads to a loss of Fc-dependent effector functions.

B. Unconjugated Antibodies

Antibodies are able to achieve cell kill either through binding alone or through Fc-mediated effector functions. p185^{HER2} is a transmembrane glycoprotein member of the epidermal growth factor (EGF) receptor family. It is overexpressed in several cancers and imparts a poor prognosis in carcinoma of the breast. Recently, humanized antibodies to p185^{HER2} provided the first compelling evidence of a clinical benefit with the use of unconjugated monoclonal antibodies in solid tumors [19]. Besides having single-agent activity, antibodies to p185^{HER2} enhance the activity of chemotherapy agents such as paclitaxel. Experimental *in vivo* data suggest that tumors overexpressing HER2/*neu* are resistant to paclitaxel-induced apoptosis. Randomized clinical trials showing a poor response to paclitaxel in HER2/*neu*-positive patients show improved response rates and disease-free survival for patients treated with paclitaxel in combination with antibodies to p185^{HER2}. The relevant mechanisms of interaction remain unclear but might include inhibition of angiogenesis [20]; downmodulation of receptor–ligand complexes [21]; complement-dependent cytotoxicity; antibody-dependent, cell-mediated cytotoxicity [22]; and interference with the growth signaling properties of the HER2 system. Unconjugated antibodies directed against the lymphocyte antigen CD20 have likewise shown significant activity in patients with previously treated low-grade, lymphomas and are currently indicated in the treatment of relapsed or refractory, CD20+, B-cell, low-grade, or follicular non-Hodgkin's lymphoma. In a large multicenter trial involving 166 patients, the overall response rate was 48%, with 6% complete and 42% partial responses. Median time to progression for responders was 13.2 months, and median duration of response was 11.6 months [23]. A 40% response rate has been observed on retreatment with rituximab.

The precise mechanisms underlying the activity of these antibodies have not been fully elucidated, but it is likely that they mediate complement-dependent cell lysis and antibody-dependent cellular cytotoxicity. Antibodies directed against p185HER2 and the lymphocyte antigen CD20 are now regarded as standard care for patients with progressive disease. It is hoped that moving these two agents into the setting of early stage disease will yield a meaningful impact on survival, and studies in such settings are currently underway. Treatment with monoclonal antibodies directed against the GD2 ganglioside antigen in patients with metastatic neuroblastoma resulted in responses to therapy in several patients [24]. The same is true for monoclonal antibodies directed against the GD3 ganglioside in patients with melanoma [25]. Treatment with unconjugated murine monoclonal antibody 17-1A resulted in improved outcomes in patients with Dukes'C colonic neoplasms. In a randomized phase III clinical trial, treatment led to a 30% reduction in death ($p = 0.04$) and 27% reduction in recurrence ($p = 0.03$) of the cancer when compared with the control group [26].

Bispecific monoclonal antibodies contain two antibodies with specificity for distinct targets [27] and are thus able to directly conjugate the immunological effector cells and the tumor cells. Bispecific antibodies can be generated by chemical conjugation, gene fusion, and fusion of the two hybridomas to create quadromas that secrete bispecific monoclonal antibodies. The bispecific molecule 2B1 binds to the extracellular domain of the HER2/neu oncogene product on tumor cells and to the Fcγ receptor on natural killer cells and macrophages [28]. This union induces potent cytokine release and in vitro promotes the lysis of HER2/neu-expressing cells by NK cells. Additionally, a bispecific monoclonal antibody targeting HER2/neu and human CD64 expressed by monocytes and activated neutrophils has been shown to be immunologically active and induces tumor inflammation [29].

Antibodies can also work through the idiotype networks and exert their influence by acting as tumor vaccines [30]. In this setting, an antibody (Ab1) with specificity for a particular tumor antigen is injected into mice in order to generate Ab2-secreting B cells that recognize the idiotype or binding site of Ab1. When patients are immunized with Ab2, they produce anti-anti-idiotype antibodies (Ab3), which recognize the original tumor antigen. Herlyn and colleagues treated 30 patients with advanced colorectal carcinoma with serial injections of polyclonal goat antibodies induced by

immunizing the animals with murine 17-1A monoclonal antibody [31]. Six patients developed brief clinical responses, and all 30 developed antibodies directed against the immunizing goat antibody. Mittleman and colleagues treated 15 patients with metastatic melanoma using a murine anti-idiotype monoclonal antibody directed against an antibody recognizing a high-molecular-weight human melanoma-associated antigen [32]. In this trial, Ab3 was identified in seven patients, three of which showed a reduction in the size of metastases in the skin or lungs. Similar findings have been observed by Foon and colleagues who were able to counteract immune tolerance to carcinoembryonic antigen (CEA) by vaccinating patients with a monoclonal anti-idiotype antibody that is the internal image of CEA (3H1). While no patients manifested a clinical response, hyperimmune sera from 17 of 23 patients demonstrated an anti-anti-idiotypic Ab3 response, and 13 of these responses were demonstrated to be true anti-CEA responses (Ab1') [33].

C. Conjugated Antibodies

Antibodies are also capable of delivering toxins, chemotherapeutic agents, isotopes, and genes directly to tumor cells while selectively avoiding normal tissues. Toxins that have been used in the treatment of cancer include the plant toxin ricin, Pseudomonas exotoxin, and Diphtheria toxin. The toxin ricin is derived from the seeds of the castor bean. It is composed of a B chain responsible for cell binding and internalization and an A chain that causes cell death by catalytically inactivating ribosomal elongation factor 2. By utilizing a monoclonal antibody conjugated to the A chain, binding and internalization will preferentially occur in antigen-positive target cells. A similar principle has been employed with other plant toxins such as gelonin and saporin. Like ricin, pseudomonas exotoxin catalyzes the ADP ribosylation and inactivation of elongation factor 2, thereby inhibiting protein synthesis and causing cell death. The toxin requires internalization before it can cause cell death. LMB-1 is an antibody construct (B3) conjugated to PE38 and directed against the Lewisy antigen. PE38 is engineered by deleting the cell-binding domain from Pseudomonas exotoxin. In a recent phase I trial, 38 patients with metastatic solid tumors expressing the Lewisy antigen received LMB-1. Objective antitumor activity was observed in five patients, and 18 had stable disease. A complete remission was observed in one patient with metastatic breast cancer, and a greater than 75% tumor reduction was observed in a patient with colon cancer. Of the 38 patients, 33 developed neutralizing antibodies to LMB-1 and could not be retreated [34]. "Capillary leak syndrome" is the major toxicity associated with the use of the toxin moieties. The syndrome is manifest by decreased serum albumin and fluid accumulation within the interstitial tissues and can result in pulmonary edema [35]. LMB-2 is an anti-CD25 recombinant immunotoxin that contains an antibody Fv fragment fused to truncated *Pseudomonas* exotoxin. Thirty-five patients with CD25(+) refractory hematologic malignancies were treated with 59 cycles of LMB-2 Only six of 35 patients developed significant neutralizing antibodies after the first cycle. One patient with hairy cell leukemia achieved a sustained complete remission. Seven partial responses were observed. Toxicities were transient and included fevers and transaminitis [36].

Chemotherapy agents can be conjugated to antibodies to confer specific tumor targeting. Sjogren et al conjugated the monoclonal antibody BR96 to doxorubicin using an acid-labile hydrazone bond to doxorubicin. The resulting conjugate, termed BR96-DOX, bound the tumor associated Lewis(y) antigen which is expressed on the surface of many human carcinoma cells. When used against the human colon carcinoma line, RCA, BR96-DOX resulted in complete resolution of subcutaneous deposits athymic mice and rats. BR96-DOX also resulted in resolution of both subcutaneous and intrahepatic deposits of the human cancer line, BN7005, in immunocompetent Brown Norway rats. In contrast, unconjugated doxorubicin, administered at maximum tolerated dose, and matching doses of nonbinding IgG-DOX conjugate were not active against RCA or BN7005 carcinomas [37]. Similar results were obtained by Apelgren *et al.* utilizing the KS1/4 monoclonal antibody, conjugated to derivatives of the vinca alkaloid desacetylvinblastine hydrazide [38]. KS1/4 targets human ovarian carcinoma cells lines. In general, chemo-immunoconjugates require that the chemo-immuno-antigen complex be internalised in order to have activity. Recently, calicheamicin-conjugated antibodies have been shown to have promising clinical activity in acute leukemias [39].

While the bystander effect eliminates the need for standard monoclonal antibodies to target each and every tumor cell, adequate cell kill still requires that the majority of cells in a tumor be targeted. Radioimmunoconjugates have the potential to kill large numbers of surrounding cells as a consequence of the long path lengths of the radionuclides. The longer track lengths also aid in addressing the limitations imposed by antigenic heterogeneity. The most promising results have been obtained in lymphoma, where both Press [40] and Kaminski [41] have shown the clinical activity of radioimmunotherapy in poor-prognosis patients. The radiolabeled antibodies Lym-1 and OKB7 have resulted in durable responses in patients with lymphoma. More recently, radiolabeled antibodies targeting CD20 have demonstrated significant clinical activity in patients with chemotherapy-pretreated lymphomas [42]. CEA has been targeted in several trials utilizing radionuclide-conjugated monoclonal antibodies. Dose-limiting toxicity was largely hematologic and felt to be related to the effect of unbound circulating radiolabeled antibodies on marrow reserves [43]. Results have been less promising in solid tumors. Radiolabeled antibodies administered in an intracavitary fashion were evaluated with two phase I trials that targeted human milk fat globulin (breast and pancreas) and OC 125 (ovarian). Higher energy radionuclides, such as ^{90}Yttrium, have longer track lengths that allow for the delivery of lethal radiation over multiple cell diameters.

Antibodies have recently found new roles in the gene therapy of cancer. Elaborate gene delivery systems have utilized antibodies in the cell-specific transfer of genes, and intracellular antibodies or intrabodies have been used to block the expression of proteins related to the oncogenic process. These results are discussed in the remainder of this chapter.

III. RECENT ADVANCES: MONOCLONAL-ANTIBODY-MEDIATED TARGETING AND CANCER GENE THERAPY

A. The Role of Antibodies in Nonviral Gene Delivery

1. DNA/poly-L-lysine Complexes

DNA/poly-L-lysine complexes deliver DNA into cells by taking advantage of receptor-mediated endocytosis. One particular receptor that has been useful for the delivery of DNA/poly-L-lysine complexes is the asialoglycoprotein receptor found exclusively on the surface of hepatocytes [44]. Wu and Wu [45] covalently coupled galactose-terminal (asialo-)glycoprotein, asialoorosomucoid (AsOR), to poly-L-lysine. The conjugate was then complexed in a 2:1 molar ratio to the plasmid pSV2CAT, which contained the gene for the bacterial enzyme chloramphenicol acetyltransferase (CAT). When the DNA carrier system was labeled with ^{32}P and injected intravenously into rats, it was determined that the liver selectively took up 85% of the DNA–polylysine–AsOR. Using this gene delivery system Wu and colleagues [46] attempted to correct a genetic inborn error of metabolism in the Nagase analbuminemic rat. As a result of a splicing defect in serum albumin mRNA, this strain has almost undetectable serum albumin levels. A plasmid containing the gene for human serum albumin was complexed with the AsOR–poly-L-lysine conjugate and used to target hepatocytes in this model. Two weeks postinjection, human serum albumin was detected at levels of 34 μg/mL, for up to 4 weeks. Clearly, the transient nature of this response is a major hurdle facing the field. De Marco *et al.* [47] used magnetic resonance imaging (MRI) to demonstrate transfection by using DNA encoding for humanized green fluorescent protein (GFP) conjugated to aminated (poly-L-lysine-conjugated) dextran chains anchored together with a central superparamagnetic core to show that DNA constructs induced signal intensity changes that colocalized with ^{33}P-labeled plasmid distribution at autoradiography. After injection of the constructs into the corpus callosum of rats, weak GFP expression of neuronal and glial cells could be detected at immunohisto-logic examination.

The mannose receptor has been an effective target for receptor-mediated gene transfer. This receptor recognizes glycoproteins with mannose, glucose, fucose, and N-acetyl glucosamine residue in exposed, nonreducing positions and is expressed on the surfaces of tissue macrophages [48,49]. A molecular conjugate, consisting of mannosylated polylysine complexed with an expression plasmid containing the *Photinus pyralis* luciferase receptor gene, was used to transfect macrophages in the liver and spleen of adult mice. Luciferase activity was detected for up to 16 days posttransfection [50].

In order to improve the efficacy of uptake of the DNA/ligand-poly-L-lysine, Perales and colleagues [51] devised a method for producing unimolecular complexes (i.e., complexes that contain a single molecule of DNA). By making the DNA/ligand-poly-L-lysine complexes smaller, it was felt that there would be a greater uptake by endocytic pathways. A plasmid containing the human factor IX gene was condensed with galactosylated poly-L-lysine and titrated with sodium chloride to form complexes of defined size (10–12 nm in diameter) and shape. The molecular conjugate was injected into adult rats and was found to specifically target the liver. Treatment with these unimolecular complexes led to prolonged expression of human factor IX, which could be detected up to 140 days after administration.

While adequate receptor mediated targeting and subsequent endocytosis are essential first steps in gene transfer, the endocytosed gene must safely make its way into the nucleus. A major stumbling block to adequate gene transfer relates to lysosomal destruction of the endocytosed DNA. DNA particles that undergo receptor-mediated endocytosis are typically trapped in intracellular vesicles only to undergo lysosomal digestion. Cristiano and colleagues highlighted this problem when they reported that only 0.1% of hepatocytes were transfected by a *Escherichia coli* β-galactosidase gene condensed with AsOR–poly-L-lysine [52]. In an attempt to overcome this problem, workers have exploited the ability of adenoviridae to disrupt endosomes, thus allowing the DNA to enter the cytoplasm. Wagner and colleagues [53] incubated the replication-defective adenovirus with the AsOR–β-galactosidase gene/poly-L-lysine complex with the result that 100% of hepatocytes were found to have been transfected. This resulted in a 1000-fold enhancement of β-galactosidase activity. The adenovirus was typically covalently or noncovalently coupled to the ligand–DNA/poly-L-lysine complex. Because the adenovirus expresses its own ligands for binding of cellular receptors, the result of such a fusion is the loss of cell-specific targeting by the ternary complex. Michael and colleagues were able to block the binding of serotype 5 adenovirus using a monoclonal antibody specific for the fiber protein and hence restore specific targeting [54].

Nguyen and colleagues were able to deliver p53, a tumor suppressor gene, to lung cancer cells. Transfection of the p53-negative human lung cancer cell line H1299 with the adenovirus/DNA-poly-L-lysine complex resulted in high levels of p53 protein and induction of apoptosis. Subcutaneous tumor sites were injected with the complex, resulting in significant inhibition of tumorigenicity as measured by the number and size of tumors that developed 21 days after treatment. Three and six injections of the complex carrying the p53 gene into H1299 subcutaneous tumor nodules led to significant dose-related tumor growth suppression 18 days after the first injection compared with control-treated tumors [55].

By determining the viral components responsible for disruption of the endosomes, it is possible to avoid using the entire virus in the ternary complex. One such structure is the 20N-terminal residue of the influenza virus hemaglutinin subunit HA-2, which was synthesized by Wagner and colleagues [56] and chemically coupled to poly-L-lysine. At a peptide/poly-L-lysine molar ratio of 8:1, the peptide transferrin–DNA/poly-L-lysine complex produced up to a 1000-fold increase in luciferase gene expression in the murine hepatocyte cell line BNL CL.2 compared to the transferrin–DNA/poly-L-lysine alone. A dimeric derivative of the N-terminal peptide was found by Plank *et al.* [57] to lead to a 5000-fold increase in luciferase gene expression in BNL CL.2 cells. Zauner and colleagues [58] observed a similar endosome-disruptive capacity with the use of peptides derived from human rhinovirus serotype 2 (HRV2). A peptide comprising 24 amino acids and derived from the N-terminus of VP1 of the HRV2 was ionically bound to poly-L-lysine and conjugated to a transferrin–DNA/poly-L-lysine complex. When 150 μg of peptide was used as the ternary complex, there was a 500-fold increase in the activity of luciferase within the transfected NIH 3T3 cells.

Hogset and colleagues have used photochemical transfection to achieve a 20-fold increase in transfection efficiency with DNA/poly-L-lysine. This technology essentially uses photosensitizing compounds that localize to the membranes of endosomes and lysosomes. On illumination, these membrane structures are destroyed and release endocytosed DNA into the cell cytosol. This has resulted in a 50% transfection level in a melanoma cell line.

Concerns regarding possible cytotoxicity related to the use of poly-L-lysine prompted Choi and colleagues [59] to construct polymers of poly-L-lysine grafted to lactose-poly(ethylene glycol). The resulting polymer (Lac–PEG–PLL), when complexed with plasmid DNA, was shown to efficiently deliver DNA to a hepatoma cell line *in vitro* which may be related to the specific targeting ability of Lac-PEG. By increasing the lactose PEG content, transfection efficiency was improved. Furthermore, no significant cytotoxicity was observed due to Lac–PEG–PLL or its complex with DNA. Other polymer blocks currently being investigated include poly(ethylene glycol) (pEG), dextran, and poly[N-(2-hydroxypropyl)methacrylamide] (pHPMA). These complexes tend to show greater aqueous solubility than simple DNA/poly-L-lysine complexes.

2. Antibodies and DNA/poly-L-lysine Complexes

Ligand-directed DNA/poly-L-lysine complexes provide only a limited amount of flexibility with respect to specific tumor targeting. The transferrin receptor, for example, can be found on almost all cells and is therefore of little use in targeting cancer cells. Unfortunately, few cell specific ligands are currently known to allow for improved targeting. Coupling tumor-cell-specific antibodies to DNA complexes might allow for improved targeting of the DNA complexes to cancer cells. Coll et al. [60] utilized the chimeric mouse–human antibody chCE7, which has specificity for CEA. The antibody was covalently linked to poly-L-lysine, and the resulting conjugate was then condensed with a plasma-encoding IFN-γ. With the aid of chloroquine, the complex was then transfected into neuroblastoma cells. Chloroquine, acts to lyse endosomes to allow for plasmid release. Successful transfection was indicated by HLA–ABC expression on the neuroblastoma cells induced by the IFN-γ, thus enabling activation of autologous cytotoxic T lymphocytes in vitro.

Antibodies have also been used in the receptor-mediated transfer of genes into normal cells. The epithelial cells of rat airways have been targeted using the polymeric immunoglobulin receptor (pIgR) [61]. Fab fragments of polyclonal antibodies raised against the rat secretory component (SC) of the extracellular portion of pIgR were covalently linked to poly-L-lysine. The anti-SC Fab–poly-L-lysine was condensed with plasmid pGL2 containing the luciferase gene and injected into the caudal vena cava of a rat. Luciferase enzyme activity in protein extracts from the liver and lung produced maximum values of approximately 14,000 and 350,000 integrated light units (ILU) per milligram of protein extract, respectively. As the lung and liver express pIgR, this demonstrated tissue-specific delivery of the luciferase gene by the anti-SC Fab–poly-L-lysine/DNA complex. Using an expression plasmid encoding the β-galactosidase gene, it was determined that the genes were transferred to the surface epithelium of the airways and the submucosal glands. A follow-up study by Ferkol et al. [62] looked at the effect of multiple injections of anti-SC Fab–poly-L-lysine/DNA complex within the lungs of mice. Animals that received one injection of ternary complex exhibited luciferase activity of approximately 17,000 ILU/mg, whereas those that received three injections of ternary complex produced luciferase activity of about 3800 ILU/mg. It was found that in the mice that had received three injections, a humoral immune response was mounted against the rabbit anti-SC Fab which resulted in decreased luciferase gene transfer. Thus, the immune response

may reduce the efficiency of gene transfer when antibody–DNA/poly-L-lysine complexes are injected more than once.

Human T lymphocytes have also been transfected by receptor-mediated gene transfer using anti-CD3 antibodies [63]. Biotinylated adenovirus d1312 and streptavidin–poly-L-lysine were used to enhance the endosomal release of the anti-CD3 monoclonal antibody–DNA/poly-L-lysine complex. Following prestimulation with IL-2 and phytohemagglutinin, primary peripheral blood lymphocytes (>95% CD2$^+$ cells) were treated with the ternary complex, which resulted in 5% of the cells expressing the β-galactosidase gene. It was found that up to 50% of Jurkat E6 cells (human acute T-cell leukemia line) were transfected when the synthetic influenza peptide IFN5 was used as an endosome-disruptive agent. Thus, like the neuroblastoma and airway epithelial cells, human T lymphocytes can be selectively targeted by antibodies delivering transgenes.

Ohtake and colleagues [64] studied suicide genes such as herpes simplex virus thymidine kinase (HSV-TK). The product of the HSV-TK gene is responsible for the phosphorylation of ganciclovir, which in turn results in the accumulation of toxic ganciclovir phosphates. In animal models, the injection of retroviral vectors expressing HSV-TK followed by the infusion of ganciclovir resulted in the death of affected cells and of neighboring cells, as well (bystander effect). Several mechanisms account for the bystander effect, including the movement of toxic products across gap junctions, induction of a local immune response, and phagocytosis of apoptotic vesicles of dead tumor cells by live cells. Ohtake et al. [64] made use of the Fab fragment of monoclonal antibody B4G7 directed against human epidermal growth factor receptor. The antibody was conjugated with poly-L-lysine, and the resulting conjugate was further complexed with reporter genes or therapeutic genes. The reporter gene utilized was the β-galactosidase gene, and the therapeutic genes utilized were the "suicide genes" HSV-TK or Escherichia coli cytosine deaminase (ECCD). The in vitro transfer of this complex was mediated via the EGF receptors in two melanoma cell lines. The frequency of cells expressing β-galactosidase reporter gene, was approximately 1%. The induction of suicide effects after Fab immunogene transfer of the HSV-TK gene or Escherichia coli cytosine deaminase gene was significant, and in the presence of ganciclovir or 5-fluorocytosine the growth of melanoma cells was inhibited for more than 7 days. Similarly, when melanoma cells treated in vitro with the Fab immunogene carrying HSV-TK or ECCD were transplanted into the back of nude mouse, subsequent systemic administration of ganciclovir or 5-fluorocytosine effectively suppressed the growth of tumors.

Foster and colleagues [65] proved that known tumor-associated antigens could serve as immunological targets for gene transfer strategies. To this end, a luciferase expression vector (pRSVLuc) was noncovalently linked to a humanized HER2 antibody (rhuMAbHER2) that was further covalently

TABLE 1 Tumor-Associated Antigens

Antigen	Site
Differentiation related	
GM2, GD2	Melanoma/sarcoma
TF, sTn	Epithelial cancers
MUC-1	Breast and pancreas cancers
MAGE 1-3	Melanoma
Tyrosinase	Melanoma
Mart-1/Melan-A	Melanoma
gp100, gp75	Melanoma
TRP-1, TRP-2	Melanoma
Oncogene/suppresser gene products	
HER2/*neu*	Breast cancers
p53	Most cancers
K-*ras*	
BCR/Abl	
Viral transformation antigens	
Epsteint–barr virus	Hodgkin's disease/nasopharyngeal cancer
Human papillomavirus	Cervical carcinoma

modified with poly-L-lysine bridges. NIH3T3 (HER2 nonexpressing) and NIH3T3.HER2 (HER2 expressing) cell lines were exposed to the pRSVLuc–poly-L-lysine/rhuMAbHER2 complex. Twenty-four hours after exposing NIH3T3 cells to chloroquine and the pRSVLuc–poly-L-lysine/rhuMAbHER2 complex, luciferase expression was found to be 180-fold higher than that obtained from a conjugate made with an isotype-matched antibody against an irrelevant target. Exposing the HER2-expressing adenocarcinoma cell lines BT474 and SKBR3 to the HER2-targeted complexes also resulted in successful gene transfer and expression. These studies suggest that HER2 may be an appropriate target for selective gene transfer and that PL–rhuMAbHER2–DNA complexes may be useful vehicles for directing gene transfer to cells that express HER2. Furthermore, it is likely that other known tumor-associated antigens might also serve as immunological targets for tissue-specific gene transfer strategies. Table 1 lists some antigens that have the potential to act as immunological targets.

3. Antibodies and DNA/GAL4 Complexes

Fominaya and Wels [66] adopted a novel approach in the design of a nonviral gene delivery system. Instead of employing the endosome-disruptive function as a separate component of the ternary complex, Fominaya and Wels incorporated the *Pseudomonas* exotoxin A translocation domain in a chimeric multidomain protein. Targeted cell specificity of the chimeric protein was conferred by the FRP5 anti-HER2/*neu* scFv. The second unique feature of the nonviral gene delivery system was the inclusion of a DNA binding domain derived from the yeast GAL4 protein in the chimeric protein. A plasmid containing the luciferase receptor gene

was designed with the GAL4-specific recognition sequence $5'$–$CGGN_3(T/A)N_5CCG$–$3'$ [67]. In order to use the complex for the transfection of HER2/*neu* positive cells, the excess negative charge of the DNA was neutralized with poly-L-lysine. Transfection of HER2/*neu*-positive COS-1 cells with 240 ng of chimeric protein and 4 μg of pSV2G4LUC luciferase reporter plasmid DNA resulted in luciferase activity of nearly 5×10^6 relative light units/mg protein. This indicated the potential of a multidomain protein for the cell-specific delivery of genes.

4. Immunoliposome/DNA Complexes

An alternative method for targeted gene delivery makes use of liposomal technology. Liposomes are self-assembling colloidal particles in which a lipid bilayer encapsulates a fraction of the surrounding aqueous medium. By using liposomes that have been complexed with plasmids, selected genes can be delivered to targeted cells [68]. Besides being nontoxic, liposomes are stable *in vitro* and most importantly lack immunogenicity [69,70]. Cationic liposomes are able to directly fuse to cell surface membranes and therefore lack specificity in targeting. Antibodies provide a means of specifically directing liposome/DNA complexes to target neoplastic cells. Such liposome–antibody conjugates are known as immunoliposomes. Immunoliposomes have demonstrated specific targeting to HER2/*neu*-overexpressing breast cancer cells using the humanized mouse anti-HER2/*neu* antibody 4D5 (rhuMabHER2) [71]. Nonspecific uptake of anti-HER2/*neu* immunoliposomes cells by negative control breast cancer cells was less than 0.2% of the uptake of HER2/*neu*-overexpressing breast cancer cells. One disadvantage to the use of immunoliposomes relates to their relatively rapid clearance by the cells of the reticuloendothelial system. As this is thought to be mediated by Fc or C3b receptors interacting with the constant regions of the antibody molecules [72], the use of antibody fragments may solve part of this problem. Sterically stabilized liposomes, which are synthesized by conjugation of PEG to the liposome surface, have lower reticuloendothelial uptake [73]. The anti-HER2/*neu* immunoliposomes used by Kirpotin *et al.* [71] were sterically stabilized and had rhuMAbHER2 covalently linked to the termini of the PEG molecule. A further disadvantage is that to some degree the liposome–antibody complex may attain a degree of immunogenicity.

De Kruif and colleagues [74] developed another way of coupling antibodies or antibody fragments to liposomes. By fusing bacterial lipoprotein nucleotide sequences to the DNA encoding the scFv, the scFv can be fatty acylated and hence incorporated into liposomes. De Kruif and colleagues lipid-tagged human anti-CD22 scFv and fused them to liposomes. The resulting anti-CD22 immunoliposomes bound specifically to and were internalized by CD22$^+$ cell lines and CD22$^+$ peripheral blood lymphocytes.

B. The Role of Antibodies in Retroviral Gene Delivery

To date, the majority of gene therapy strategies have been based on retroviral vector technology [75]. Retroviruses are RNA viruses that allow for stable integration of DNA into the target cell genome. After entering a cell, the RNA is reverse-transcribed to DNA by reverse transcriptase. The DNA is then transported to the nucleus, where it is integrated into the host cell genome. Because viral protein-coding regions are absent, the virus is incapable of replicating. Theoretical concerns about the use of retroviruses are related to the potential for mutagenesis [76]. Unlike other viruses, such as adeno-associated virus, there is little control over where the transcribed DNA is integrated and it is possible that the random insertion of a gene into the host genome might inactivate a tumor suppressor gene. Reverse transcriptase itself is frequently responsible for mutations in transcribed genes. Retroviral vectors are highly efficient at attaching to and internalizing into cells and therefore represent a potentially excellent means of gene transfer. One of the major drawbacks of retroviral vectors in gene therapy relates to their lack of target specificity. Improved targeting would not only allow for specific targeting of cancer cells but would also allow systemic administration of the vector. To date, the greatest success has been with strategies involving local delivery of vector, but this does not address the systemic nature of malignant disease. As with DNA/poly-L-lysine complexes, antibodies may represent a means of targeting retroviral vectors for delivery of specific genes to cells of interest.

Retroviral vectors carrying the transgene of interest are constructed in packaging cell lines. The packaging cell lines provide all the proteins required to assemble a retrovirus, including the products of the *gag, pol,* and *env* genes [77]. A vector containing a retroviral-packaging signal bounded by long terminal repeats enables the transgene to be incorporated into a retrovirus. The product of the *env* gene dictates the binding specificity or tropism of the retrovirus. Because the infection of a packaging cell line by the produced virus is blocked due to the downregulation of the retrovirus receptor by the envelope glycoprotein, the vector genome should be introduced by a virus with a host tropism different from the one of the packaging cell line. An ecotropic envelope glycoprotein restricts infection of the retrovirus to rodent cells, whereas an amphotropic envelope glycoprotein permits infection of most mammalian cells. The ecotropic envelope glycoprotein of the Moloney murine leukemia virus (MoMLV) can be made as a precursor (Pr80env). This molecule is proteolytically cleaved to yield the mature surface (SU) (gp70) and transmembrane (TM) (p15E) proteins [78]. On the virion surface, three SU proteins and three TM proteins associate to form a homotrimeric complex [79]. The tropism of the ecotropic envelope glycoprotein lies within the SU protein.

One of the first uses of antibodies in the cell-specific targeting of retroviral vectors involved a bispecific antibody complex [80]. Anti-gp70 monoclonal antibodies were biotinylated and allowed to bind to the surface of ecotropic murine retroviruses, and human HeLa cells were coated with biotinylated B.9.12.1 anti-human major histocompatibility complex (MHC) class I antibodies. Using the ψ2 packaging cell line, a plasmid containing a neomycin resistance gene was inserted into the retroviruses. Streptavidin was then used to bridge the ectropic murine retroviruses to the MHC class I molecules on the HeLa cells, and 50–80 G418-resistant clones were isolated from 2×10^5 cells. When the anti-class-I antibodies were substituted for anti-class-II antibodies in the bispecific complexes, 100–350 G418 clones were isolated. Although this approach of redirecting retroviruses was successful, the efficiency of infection was low and the system is impractical for *in vivo* applications. Lorimer and colleagues [81] utilized a mutant EGF receptor, EGFRvIII, expressed in several malignancies but not found on normal tissues. An expression plasmid was constructed by inserting a scFv antibody, specific for EGFRvIII, at a novel position within a disulfide-bonded surface loop. This was inserted in proximity to the native receptor-binding site of the MoMMV ecotropic envelope glycoprotein. This fusion protein was expressed and incorporated into retroviral particles as efficiently as normal envelope glycoprotein. Retroviral vectors made with the fusion protein were able to bind peptide antigen and EGFRvIII expressed on the surface of human glioblastoma cells. The retroviral vectors had normal levels of infectivity on mouse cells, suggesting that the envelope glycoprotein tolerated an insertion at this site, but did not show significant infectivity to human cells expressing EGFRvIII. However, despite being able to redirect retrovirus binding to the tumour-specific target while preserving the normal function of the ecotropic envelope glycoprotein, sufficient infectivity could not be achieved.

Genetic manipulation of the ecotropic envelope glycoprotein was pursued as a means of changing the tropism of retroviral vectors while maintaining a high level of infection; however, altering ecotropic envelope glycoprotein without the loss of infectivity has proven to be difficult. No viral titer was obtained when Schnierle and colleagues [82] attempted to infect SKBR-3 and MDA-MB-453 cells with MoMLV retroviruses expressing chimeric envelope proteins. In this study, the cloning of the FRP5 anti-HER2/*neu* scFv between amino acids 6 and 7 of the MoMLV SU protein completely ablated the infectivity of the SU protein (Table 2). FACS analysis of the CB25 packaging cells revealed that the chimeric envelope proteins had not been inserted into the cell membranes and were therefore unavailable for incorporation into the retroviral particles.

Marin and colleagues [83] encountered similar difficulties, producing very low retroviral titers from human TE671 cells when the B.9.12.1 anti-human MHC class I scFv gene was inserted at position 6 of the MoMLV SU gene in the

TABLE 2   Antibody-Mediated Retroviral Infection of Nonmurine Cell Lines

Antibody fragment	Envelope fusion protein partner	Wild-type envelope protein	Cell line	Retroviral titer (CFu/mL)	Ref.
FRP5 anti-HER2 scFV	Residues 6+ of Moloney murine leukemia virus (MoMLV) SU protein		SKBR-3, MDA-MB-453 (both human)	0	[82]
OKT3 anti-CD3 scFv; A10, B3, C215, and F1 scFv	Residues 1+ or 7+ of MoMLV SU protein		Jurkat T cells, Colo 205 cells (both human)	0	[84]
B.9.12.1 anti-human MHC class I scFv	Residues 6+ of MoMLV SU protein		TE671 (human)	3–112	[83]
Anti-DNP scFv	292–398 of spleen necrosis virus (SNV) SU protein		CHO	30	[85]
	SNV TM proteins		CHO	20	[85]
B6.2 scFv (specific for antigen expression on human colon carcinoma cells)	Two thirds SNV SU protein		HeLa	10	[86]
	One third SNV SU protein		HeLa	5	[86]
	SNV TM protein		HeLa	20	[86]
	Two thirds SNV SU protein	+	HeLa	9×10^2	[86]
	One third SNV SU protein	+	HeLa	1.3×10^3	[86]
	SNV TM protein	+	HeLa	9.4×10^2	[86]
	SNV TM protein	+	HeLa	7×10^2	[87]
	SNV TM protein	(Lower levels)	HeLa	7×10^3	[87]
Anti-low-density-lipoprotein receptor scFv	Residues 6+ MoMLV SU protein	+	HeLa	1×10^4	[89]

plasmid pMB34. The plasmid pMB34, which carries an antibiotic phleomycin selection marker, was transfected into the *env*-deficient packaging cell line TelCeb6 along with an nl-sLACZ reporter-gene-carrying retroviral vector. Retroviruses derived from the TelCeb6 cells produced viral titers of 3–112 CFU/mL on human TE671 cells (Table 2). Interestingly, the ecotropic retroviruses were still able to infect murine NIH3T3 cells, having produced viral titers of 2×10^3 CFU/mL. This suggests that, although the inserted scFv functions poorly in mediating the internalization of viral particles, the SU part of the chimeric envelope protein is still capable of recognizing its murine receptor and initiating infection.

In an attempt to improve the efficiency of infection, Ager and colleagues [84] fused the scFv to different positions of the SU envelope glycoprotein. In addition, linker sequences of different lengths were inserted between the scFv and the SU envelope glycoprotein. Despite these alterations, the ecotropic retroviruses expressing the chimeric envelope glycoprotein variants failed to infect human cells (Table 2). Nevertheless, Ager and colleagues did demonstrate that varying the position of the scFv and utilizing different length

linker sequences can have an effect on the functionality of the SU envelope glycoprotein as reflected by the infection of mouse cells. Five different scFv proteins were tested in the chimeric envelope glycoprotein construct. The first scFv (OKT3) was specific for T-cell surface marker CD3; the remaining scFv proteins (A10, B3, C215, and F1) were specific for distinct antigens expressed on colonic cancer cells. Four different constructs of the OKT3 scFv–SU fusion protein were made. The scFv proteins were fused to residues 1 and 7 of SU, employing either a noncleavable three-residue linker or the above linker with an additional factor Xa cleavage linker tetrapeptide. Eight different constucts of the anticolonic cancer cell scFv–SU fusion proteins were made. Retroviral particles expressing scFv–SU chimeric constructs were harvested from TELCeB.6 packaging cells and were assayed for infectivity on murine NIH3T3 fibroblasts. The investigators found a direct correlation between both the insertion portion and the linker on the infectivity of the vectors. The greatest degree of infectivity was observed when the scFv was inserted at position +1 employing the heptapeptide linker. Thus, the insertion portion in the SU envelope glycoprotein and the choice of linker can have an impact on the ability of a virus to infect cells.

Chu and colleagues [85] have adopted similar approaches to optimizing the cell-specific targeting of retroviral vectors, focusing on the avian spleen necrosis virus (SNV), which produces a 70-kDa SU and 20-kDa TM from a 90-kDa precursor (PR90env) [86]. Two chimeric constructs were made using an anti-DNP scFv. In the first construct, the anti-DNP scFv was fused to residues 292–398 of SU; in the second construct, the anti-DNP scFv was fused directly to TM. scFv–SU$_{292-398}$ and scFv–TM chimeric envelope glycoproteins were expressed on the surface of retroviral particles harvested from the D17 dog osteosarcoma cell line. Upon the infection of DNP-conjugated CHO cells, the scFv–SU$_{292-398}$ and scFv–TM chimeric envelope glycoproteins produced retroviral titers of 30 and 20 CFU/mL, respectively (Table 2). Because the retroviral titers are low, the remaining parts of the retroviral envelope glycoprotein have little activity.

Using the spleen necrosis virus, Chu and Dornburg [87] were able to improve the efficiency of infection by coexpressing wild-type envelope glycoprotein with chimeric envelope glycoprotein in the homotrimeric complex on the viral surface. For cell-specific targeting, an scFv that recognized an antigen on human colon carcinoma cells was used. Constructs were designed with either one third or two thirds of the SU envelope glycoprotein removed. The removed envelope glycoprotein was replaced with the B6.2 scFv. A third construct was made in which the scFv was directly fused to the TM envelope glycoprotein. The construct with one third SU removed was cloned into the vector pTC26, the construct with two thirds SU removed was cloned into the vector pTC24, and the construct with complete SU removal was cloned into the vector pTC25. These were then transfected into the packaging cell lines DSgp13 and DSH-cxl. Retroviruses produced by DSH-cxl cells expressed both wild-type and chimeric envelope glycoprotein, whereas retroviruses produced by DSgp13 cells expressed only chimeric envelope glycoprotein. The harvested viral particles were used to infect DLD-1 (human colon carcinoma cell line), HeLa, and HOS (human osteosarcoma cell line) cells. DSgp13-derived retroviruses expressing the three different constructs exhibited similar viral titers on HeLa cells (pTC26, 10; pTC24, 5; and pTC25, 20 CFU/mL) (Table 2). Thus, in terms of viral infectivity, the SU envelope glycoprotein had little impact. Compared to the DSgp13-derived retroviruses, the DSH-cxl-derived retroviruses expressing the three different constructs produced higher viral titers on HeLa cells (pTC26, 900; pTC24, 1300; and pTC25, 940 CFU/mL) (Table 2). The retroviral titers of the DLD-1 and HOS cells reflected the findings of the HeLa cells. From these increases in retroviral titer, it would appear that unmodified wild-type envelope glycoproteins have an important role to play in viral infectivity. Wild-type envelope glycoproteins may mediate their effects by allowing the viral

membrane to efficiently fuse with the membrane of the target cell. Any changes in the envelope glycoprotein may nullify this effect.

Further investigations by Chu and Dornburg [87] reconfirmed that only a fully functional wild-type envelope glycoprotein can assist the chimeric envelope glycoprotein in efficient virus penetration. Moreover, the ratio of wild-type envelope glycoprotein to chimeric envelope glycoprotein in the viral membrane appeared to determine the efficiency of infection. These investigators were able to attain different levels of wild-type envelope glycoprotein by using the SNV envelope glycoprotein-expressing constructs that either contained (pRD134) or lacked (pIM29) the adenovirus tripartite leader sequence. The adenovirus tripartite leader sequence enhances envelope expression in D17 cells about 10-fold [88]. The above plasmids were cotransfected with the plasmid pTC25 into DSH cells. When used to infect HeLa cells, retrovirus-expressing pRD134-derived envelope glycoprotein produced viral titers that were 10-fold lower than those achieved with retroviruses-expressing pIM29-derived envelope glycoprotein (700 and 7000 CFU/mL, respectively) (Table 2). This demonstrated that lower levels of wild-type envelope glycoprotein allowed higher levels of infection (Table 2). To test the effects of nonfunctional wild-type envelope glycoprotein on infection, two mutant constructs were designed. The first construct (pTC12) contained two point mutations (arg$_{398}$ → trp, ala$_{399}$ → pro) at the SU/TM cleavage site; the second construct (pTC76) carried the point mutation, asp$_{192}$ → arg in the middle of SU. The latter change reduced the efficiency of wild-type SNV infection by about 4000-fold. These mutations led to negligible or dramatically reduced retroviral titers on HeLa cells. Retroviruses that expressed the pTC12-derived envelope glycoproteins produced viral titers of <1 CFU/mL, whereas retroviruses that expressed the pTC76-derived envelope glycoproteins exhibited viral titers of 80 CFU/mL. Interestingly, when the retroviral stocks were concentrated 25-fold, a 200-fold rise in viral titers was observed. This observation suggests that the efficiency of infection increases at higher viral concentrations.

Wild-type envelope glycoproteins have also been coexpressed with chimeric envelope glycoproteins in MoMLV [89]. An anti-low-density-lipoprotein receptor scFv was inserted between residues 3 and 4 of the ecotropic SU envelope protein. $\psi 2$ packaging cells were contransfected with the plasmid pC7Env, which contains the gene for the scFv–envelope fusion protein, and a hygromycin B phosphotransferase expression vector. Hygromycin-B resistant clones were transfected with the gene for β-galactosidase. Retroviruses were produced with chimeric and wild-type envelope glycoproteins on their surface and neomycin and β-galactosidase genes packaged inside. These viruses were harvested and used to infect HeLa cells, which resulted in a viral titer of 1000 CFU/mL (Table 2). Thus, like Chu and Dornburg,

Somia and colleagues [89] found that wild-type envelope glycoproteins help to stabilize the chimeric envelope glycoproteins and increase the efficiency of infection.

C. The Role of Antibodies in Non-Retroviral Viral Gene Delivery

Use of a matrix metalloprotease (MMP) by Martin and colleagues [90] facilitated improved cell targeting and infectivity. A scFv fragment directed against the surface glycoprotein high-molecular-weight melanoma-associated antigen (HMW-MAA) was fused to the amphotropic murine leukemia virus envelope. A proline-rich hinge and MMP cleavage site linked the two proteins. Because inclusion of the proline-rich hinge prevented viral binding to the amphotropic viral receptor, the modified viruses bound only to HMW-MAA-expressing cells. Following attachment to HMW-MAA, MMP cleavage of the envelope at the melanoma cell surface removed the scFv and proline-rich hinge, allowing for infection. In a cell mixture, 40% of HMW-MAA-positive cells but less than 0.01% of HMW-MAA-negative cells were found to be infected. By complexing targeted retroviruses with 2,3-dioleoyloxy-N-[2(spermine-carboxamido)ethyl]N,N-dimethyl-1-propanaminium trifluoroacetate-dioleoyl phosphatidylethanolamine, liposome infectivity was greatly enhanced and cell specificity was unaffected.

Antibodies also play a role in the genetic modification of dendritic cells by adenoviruses. Current vector systems have only a limited efficacy in gene delivery to these cells. The use of bispecific antibodies has resulted in a dramatic increase in the efficacy of gene transfer to monocyte-derived dendritic cells. Tillman and colleagues [90], using antibodies directed against CD40, not only were able to dramatically increase the efficacy of gene transfer but were also able to show induction and maturation of the dendritic cell. This was demonstrated phenotypically by the increased expression of CD83 and MHC, as well as by the production of IL-12. This was thought to be a direct result of CD40 targeting and not a function of the adenoviral particle.

The promiscuous tropism of adenoviral vectors has limited their usefulness as efficient gene delivery systems. Curiel and colleagues [92] have been able to alter viral tropism via immunologic retargeting. This has been achieved with the aid of conjugates composed of an antifiber knob Fab and a targeting moiety consisting of a ligand or antireceptor antibody. This has facilitated gene delivery via receptors for folate, fibroblast growth factor (FGF), and EGF. This process does not require the presence of the native adenoviral receptor (CAR). Furthermore, modifying or replacing fiber adenoviral tropism has successfully achieved adenoviral-mediated, cell-specific gene delivery. For a further discussion on the role of adenoviridae as vectors, please see the chapter by Curiel and colleagues.

IV. FUTURE DIRECTIONS

Significant advances are still required if gene therapy is to become a viable treatment modality. While targeting and efficiency of infection continue to improve, significant obstacles remain to be overcome before gene therapy can be considered a viable option for patients with cancer. This is most evident in the areas of endosome-mediated transfer and retroviral gene delivery. Antibodies are well suited to this task through their innate abilities to target and bind specific antigens.

Of the different mechanisms of gene transfer, retroviral gene delivery holds the most potential in the treatment of cancer. However, as previously discussed, attempts to redirect the retroviral envelope towards specific cellular receptors have yielded poor infectivity. Reasonable infection of targeted cells has only been achieved in the presence of wild-type retroviral glycoproteins. As it has been difficult to modify the retroviral envelope without the loss of activity, a viable option would be to use an adaptor molecule containing both the retroviral receptor and a domain capable of specifically binding the targeted cell population. Ohno and colleagues [93] have utilized such an approach. A fusion protein containing the extracellular domain 3 of the modified H14 protein and transforming growth factor (TGF)-α was used to cross-link an ecotropic AKR virus with the epidermal growth factor receptor (EGFR) expressed on the target cell surface. In principle, the human modified H13 would be equivalent to the murine ecotropic retrovirus receptor. As the adaptor molecule contains the retroviral receptor, there is no need to modify the retroviral envelope glycoprotein. Thus, the high efficiency of the viral transduction could be retained. However, in this study, the ability of the adaptor molecule to promote the infection of EGFR-positive human A431 cells by ecotropic AKR virus was not investigated.

One of the factors that has been overlooked during the development of cell-specific retroviral targeting is the choice of cell receptors. From a number of reports, it is apparent that cell receptors determine the number of retroviruses that are able to enter a cell. Some cell receptors are more efficient than others in allowing retroviral entry into a cell (Table 3). For example, MHC class I molecules have permitted viral infection within target cells, albeit it at low levels (Table 2) [80,83]. In contrast, retroviral infection did not occur when the transferrin receptor on Hep G2 cells was targeted by antitransferrin receptor antibodies bridged to anti-gp70 antibodies [94]. Choosing the correct cell receptor can optimize the efficiency of viral infection. Infection may further be hampered by a retrovirus becoming trapped within an endosome. This can potentially happen following internalization of certain receptors. The trapped retrovirus is rapidly routed to the lysosome, where it undergoes degradation. This phenomenon was observed by Cosset [95] with ecotropic MLV expressing EGF–SU fusion proteins. As would be expected, treatment

TABLE 3 Ligand-Mediated Retroviral Infection of Human Cell Lines

Receptor	Targeting ligand	Cell line	Titer
High-density-lipoprotein receptor	ApoA1–proteinA	Hep G2	0^a
Galactose receptor	Biotinylated asialofetuin	Hep G2	0^a
Insulin receptor	Biotinylated insulin	HeLa	8^a
Epidermal growth factor receptor	Biotinylated EGF	A431	$8–23^a$
Erythropoeitin receptor	EPO	HEL	50^b
HER3, HER4	Heregulin	SKBR-3	$<1^c$
		MDA-MB-453	$<1^c$
EGF-R	53 amino acids of EGF	A431	225^d
		TE671	46^d

^{a}G418-resistant cells [94].
^{b}Foci/mL [97].
^{c}CFU/mL [82].
^{d}CFU/mL [95].

with chloroquine reduced retroviral degradation in lysosomes and significant increased infection in human A431 and TF671 cells. It has been suggested by Weiss and Taylor [96] that in certain cell types the SU envelope glycoprotein of ecotropic MLV requires cleavage by pH-dependent cathepsins for endosomal escape. If these pH-dependent cathepsins were absent, then the ecotropic MLV would become trapped and undergo degradation. This problem may be overcome by incorporating a translocation domain into the antibody–SU fusion proteins. The *Pseudomonas* exotoxin translocation domain is ideally suited to this purpose.

Improvements in the design of antibody–DNA/poly-L-lysine complexes would lead to improvements in gene transfer efficiency. One such improvement could include the utilization of nuclear localization signals, which could direct a higher degree of transcription of the transgene. Anti-HIV-Tat scFv proteins containing carboxy terminal SV40 nuclear localization signals have been shown to direct heterologous proteins into the nucleus. Similarly, SV40 nuclear localization signals could be fused into the C termini of antibodies or antibody fragments directed against cell receptors.

Many of the barriers that therapeutic antibodies have encountered in penetrating tumors will also have to be faced by gene delivery systems. These include limited vascular penetration, heterogeneous tumor antigen expression, and elevated interstitial tumor pressures. For many gene delivery systems, the obstacles may be more difficult to overcome due to the far greater size of the DNA complexes and virus particles

when compared with antibodies. These issues will have to be addressed if high-efficiency gene transfer systems are to become part of the therapeutic armamentarium in cancer.

References

1. Kohler, G., and Milstein C. (1975). Continuous cultures of fused cells secreting antibody of predefined specificity. *Nature* **256,** 495–497.
2. Ghetie, M. A., Ghetie, V., and Vitetta, E. S. (1997). Immunotoxins for the treatment of B-cell lymphoma. *Mol. Med.* **3,** 420–427.
3. Caron, P. C., and Schienberg, D. A. (1993). Anti-CD33 monoclonal antibody M195 for the therapy of myeloid leukemia. *Leuk. Lymphoma* **11**(suppl. 2), 1–6.
4. Kaminski, M. S., Zasadny, K. R., Francis, I. R. *et al.* (1993). Radioimmunotherapy of B-cell lymphoma with [131I]anti-B1 (anti-CD20) antibody. *N. Engl. J. Med.* **329,** 459–465.
5. Bodey, B., Siegel, S. E., and Kaiser, H. E. (1996). Human cancer detection and immunotherapy with conjugated and non-conjugated monoclonal antibodies. *Anticancer Res.* **16,** 661–674.
6. Jain, R. K., and Baxter, L. T. (1988). Mechanisms of heterogenous distribution of monoclonal antibodies and other macromolecules in tumors: significance of elevated interstitial pressure. *Cancer Res.* **48,** 7022–7032.
7. Milenic, D. E., Yokota, T., Filpula, D. R. *et al.* (1991). Construction, binding properties, metabolism, and tumor targeting of a single chain Fv derived from the pancarcinoma monoclonal antibody CC49. *Cancer Res.* **51,** 6363–6371.
8. Ingvar, C., Jakobsson, B., and Brodin, T. (1990). Tumor antigen heterogeneitywithin melanoma metastases—an evaluation by immunohistochemistry. *Anticancer Res.* **10,** 219–223.
9. Guadagni, F., Schlom, J., and Greiner, J. W. (1991). In vitro and in vivo regulation of tumor antigen expression by human recombinant interferons. *Int. J. Rad. Appl. Instrum. B* **18,** 409–412.
10. Miller, R. A., Oseroff, A. R., Stratte, P. T. *et al.* (1983). Monoclonal antibody therapeutic trials in seven patients with T-cell lymphoma. *Blood* **62,** 988–995.
11. Holland, G., and Zlotnik, A: Interleukin-10 and cancer. *Cancer Invest.* **11,** 751–758.
12. Khazaeli, M. B., Conry, R. M., and LoBuglio, A. F. (1994). Human immune response to monoclonal antibodies. *J. Immunother.* **15,** 42–52.
13. Fanger, M. W., Morganelli, P. M., and Guyre, P. M. (1992). Bispecific antibodies. *Crit. Rev. Immunol.* **12,** 101–124.
14. Carter, P., Presta, L., Gorman, C. M. *et al.* (1992). Humanization of an anti-p185HER2 antibody for human cancer therapy. *Proc. Natl. Acad. Sci.* **89,** 4285–4289.
15. Porter, R. R. (1958). Separation of fractions of rabbit gamma-globulin containing the antibody and antigenic combining sites. *Nature* **182,** 670–671.
16. Nisonoff, A., Wissler, F. C., and Lipman, L. N. Properties of the major component of a peptic digest of rabbit antibody. *Science* **132,** 1770–1771.
17. Bird, R. E., Hardman, K. D., Jacobson, J. W. *et al.* (1988). Single-chain antigen-binding proteins. *Science* **242,** 423–426.
18. Yokota, T., Milenic, D. E., Whitlow, M. *et al.* (1992). Rapid tumor penetrance of a single chain Fv and comparison with other immunoglobulin forms. *Cancer Res.* **52,** 3402–3408.
19. Baselga, J., Tripathy, D., Mendelsohn, J. *et al.* (1996). Phase II study of weekly intravenous recombinant humanized anti-p185HER2 monoclonal antibody in patients with HER2/*neu*-overexpressing metastatic breast cancer. *J. Clin. Oncol.* **14,** 737–744.
20. Petit, A. M., Rak, J., Hung, M. C. *et al.* (1997). Neutralizing antibodies against epidermal growth factor and ErbB-2/*neu* receptor tyrosine kinases down-regulate vascular endothelial growth factor production

by tumor cells in vitro and in vivo: angiogenic implications for signal transduction therapy of solid tumors. *Am. J. Pathol.* **151,** 1523–1530.

21. Sarup, J. C., Johnson, R. M., King, K. L. *et al.* (1991). Characterization of an anti-p185HER2 monoclonal antibody that stimulates receptor function and inhibits tumor cell growth. *Growth Regul.* **1,** 72–82.

22. Lewis, G. D., Figari, I., Fendly, B. *et al.* (1993). Differential responses of human tumor cell lines to anti-p185HER2 monoclonal antibodies. *Cancer Immunol. Immunother.* **37,** 255–263.

23. McLaughlin, P., Grillo-Lopez, A. J., Link, B. K. *et al.* (1998). Rituximab chimeric anti-CD20 monoclonal antibody therapy for relapsed indolent lymphoma: half of patients respond to a four dose treatment program. *J. Clin. Oncol.* **16,** 2825–2833.

24. Handgretinger, R., Baader, P., Dopfer *et al.* (1992). A phase I study of neuroblastoma with the anti-ganglioside GD2 antibody 14.G2a. *Cancer Immunol. Immunother.* **35,** 199–204.

25. Vadhan-Raj, S., Cordon-Cardo, C., Carswell, E. *et al.* (1988). Phase I trial of a mouse monoclonal antibody against GD3 ganglioside in patients with melanoma: induction of an inflammatory response at tumor sites. *J. Clin. Oncol.* **6,** 1636–1648.

26. Reithmuller, G., Schneider-Gadicke, E., Schlimock, G. *et al.* (1994). Randomised trial of monoclonal antibody for adjuvant therapy of resected Duke's C colorectal carcinoma. *Lancet* **343,** 1172–1174.

27. Fanger, M. W., Morganelli, P. M., and Guyre, P. M. (1992). Bispecific antibodies. *Crit. Rev. Immunol.* **12,** 101–124.

28. Carter, P., Presta, L., Gorman, C. M. *et al.* (1992). Humanization of an anti-p185HER2 antibody for human cancer therapy. *Proc. Natl. Acad. Sci.* **89,** 4285–4289.

29. Valone, F. H., Kaufman, P. A., Guyre, P. M. *et al.* (1995). Phase Ia/Ib trial of bispecific antibody MDX-210 in patients with advanced breast or ovarian cancer that overexpresses the proto-oncogene HER2/*neu*. *J. Clin. Oncol.* **13,** 2281–2292.

30. Jerne, N. K. (1974). Towards a network theory of the immune system. *Ann. Immunol.* **125,** 373–389.

31. Herlyn, D., Wettendorff, M., Schmoll, E. *et al.* (1987). Anti-idiotype immunization of cancer patients: modulation of the immune response. *Proc. Natl. Acad. Sci.* **84,** 8055–8059.

32. Mittleman, A., Chen, Z. J., Kageshita, T. *et al.* (1990). Active specific immunotherapy in patients with melanoma. A clinical trial with mouse antiidiotypic monoclonal antibodies elicited with syngeneic anti-high-molecular-weight melanoma-associated antigen monoclonal antibodies. *J. Clin. Invest.* **86,** 2136–2144.

33. Foon, K. A., Chakraborty, M., John, W. J. *et al.* (1995). Immune response to the carcinoembryonic antigen in patients treated with anti-idiotype antibody vaccine. *J. Clin. Invest.* **96,** 334–342.

34. Pai, L. H., Wittes, R., Setser, A. *et al.* (1996). Treatment of advanced solid tumors with immunotoxin LMB-1: an antibody linked to *Pseudomonas* exotoxin. *Nat. Med.* **2,** 350–353.

35. Soler-Rodriguez, A. M., Ghetie, M. A., Oppenheimer-Marks, N. *et al.* (1993). Ricin-A chain and ricin-A chain immunotoxins rapidly damage human endothelial cells: implications for vascular leak syndrome. *Exp. Cell. Res.* **206,** 227–234.

36. Kreitman, R. J., Wilson, W. H., White, J. D. *et al.* (2000). Phase I trial of recombinant immunotoxin anti-Tac(Fv)-PE38 (LMB-2) in patients with hematologic malignancies. *J. Clin. Oncol.* **18,** 1622–1636.

37. Sjogren, H. O., Isaksson, M., Willner, D. *et al.* (1997). Antitumor activity of carcinoma-reactive BR96-doxorubicin conjugate against human carcinomas in athymic mice and rats and syngeneic rat carcinomas in immunocompetent rats. *Cancer Res.* **57,** 4530–4536.

38. Apelgren, L. D., Zimmerman, D. L., Briggs, S. L. *et al.* (1990). Antitumor activity of the monoclonal antibody-Vinca alkaloid immunoconjugate LY203725 (KS1/4-4-desacetylvinblastine-3-carboxhydrazide) in a nude mouse model of human ovarian cancer. *Cancer Res.* **50,** 3540–3544.

39. Sievers, E. L., Bernstein, I. D., Spielberger, R. T. *et al.* (1997). Dose escalation phase I study of recombinant engineered human anti-

CD33 antibody-calicheamicin drug conjugate (CMA-676) in patients with relapsed or refractory acute myeloid leukemia (AML) (meeting abstract). *Proc. Annu. Meet. Am. Soc. Clin. Oncol.* **16,** A8.

40. Press, O. W., Eary, J. F., Appelbaum, F. R. *et al.* (1993). Radiolabeled-antibody therapy of B-cell lymphoma with autologous bone marrow support. *N. Engl. J. Med.* **329,** 1219–1224.

41. Kaminski, M. S., Zasadny, K. R., and Francis, I. R. (1993). Radioimmunotherapy of B cell lymphoma with [^{131}I]anti-B1 (anti-CD20) antibody. *N. Engl. J. Med.* **329,** 459–465.

42. Wahl, R. L., Zasadny, K. R., Macfarlane, D. *et al.* (1998). Iodine-131 anti-B1 antibody for B-cell lymphoma: an update on the Michigan phase I experience. *J. Nucl. Med.* **39**(8, suppl.), 21S–27S.

43. Behr, T. M., Sharkey, R. M., Juweid, M. E. *et al.* (1997). Phase I/II clinical radioimmunotherapy with iodine-131-labeled anti-carcinoembryonic antigen murine monoclonal antibody IgG. *J. Nucl. Med.* **38,** 858–870.

44. Ashwell, G., and Morell, A. G. (1974). The role of surface carbohydrates in the hepatic recognition and transport of circulating glycoproteins. *Adv. Enzymol. Relat. Areas Mol. Biol.* **41,** 99–128.

45. Wu, G. Y., and Wu, C. H. (1988). Receptor mediated gene delivery and expression *in vivo*. *J. Biol. Chem.* **263,** 14621–14624.

46. Wu, G. Y., Wilson, J. M., Shalaby, F. *et al.* (1991). Receptor mediated gene delivery and expression *in vivo*. Partial correction of genetic analbuminemia in Nagase rats. *J. Biol. Chem.* **266,** 14388–14342.

47. De, Marco, G., Bogdanov, A., Marecos, E. *et al.* (1998). MR imaging of gene delivery to the central nervous system with an artificial vector. *Radiology* **208,** 65–71.

48. Achord, D., Brot, F., and Sly, W. (1977). Inhibition of the rat clearance system for agalacto-orosomucoid by yeast mannans and by mannose. *Biochem. Biophys. Res. Commun.* **77,** 409–415.

49. Wileman, T., Boshans, R., and Stahl, P. (1985). Uptake and transport of mannosylated ligands by alveolar macrophages. Studies on ATP-dependent receptor-ligand dissociation. *J. Biol. Chem.* **260,** 7387–7393.

50. Ferkols, T., Perales, J. C., and Mukaro, F. (1996). Receptor-mediated gene transfer into macrophages. *Proc. Natl. Acad. Sci.* **93,** 101–105.

51. Perales, J. C., Ferkol, T., Beegen, H. *et al.* (1994). Gene transfer *in vivo*: sustained expression and regulation of genes introduction into the liver by receptor-targeted uptake. *Proc. Natl. Acad. Sci. USA* **91,** 4086–4090.

52. Cristiano, R. J., Smith, L. C., and Woo, SLC. (1993). Hepatic gene therapy: adenovirus enhancement of receptor-mediated gene delivery and expression in primary hepatocytes. *Proc. Natl. Acad. Sci. USA* **90,** 2122–2126.

53. Wagner, E., Zatloukal, K., Cotton, M. *et al.* (1992). Coupling of adenovirus to transferring-polylysine/DNA complexes greatly enhances receptor-mediated gene delivery and expression of transfected genes. *Proc. Natl. Acad. Sci. USA* **89,** 6099–6103.

54. Michael, S. I., Huang, C., Romer, M. U. *et al.* (1993). Binding-incompetent adenovirus facilitates molecular conjugate-mediated gene transfer by the receptor-mediated endocytosis pathway. *J. Biol. Chem.* **268,** 6866–6869.

55. Nguyen, D. M., Wiehle, S. A., Koch, P. E. *et al.* (1997). Delivery of the p53 tumor suppressor gene into lung cancer cells by an adenovirus/DNA complex. *Cancer Gene Ther.* **4,** 191–198.

56. Wagner, E., Plank, C., Zatloukal, K. *et al.* (1992). Influenza virus hemagglutinin HA-2 N-terminal fusogenic peptides augment gene transfer by transferring-polylysine/DNA complexes: toward a synthetic virus-like gene transfer vehicle. *Proc. Natl. Acad. Sci. USA* **89,** 7934–7938.

57. Plank, C., Oberhauser, B., Mechtler, K. *et al.* (1994). The influence of endosome-disruptive peptides on gene transfer using synthetic virus-like gene transfer systems. *J. Biol. Chem.* **269,** 12918–12924.

58. Zauner, W., Blaas, D., Kuechler, E. *et al.* (1995). Rhinovirus-mediated endosomal release of transfection complexes. *J. Virol.* **69,** 1085–1092.

59. Choi, Y. H., Liu, F., Park, J. S. *et al.* (1998). Lactose-poly(ethylene glycol)-grafted poly-L-lysine as hepatoma cell-targeted gene carrier. *Bioconjug. Chem.* **9,** 708–718.

60. Coll, J. L., Wagner, E., Combaret, V. *et al.* (1997). *In vitro* targeting and specific transfection of human neuroblastoma cells by chCE7 antibody-mediated gene transfer. *Gene Ther.* **4,** 156–161.

61. Ferkol, T., Perales, J. C., Eckman, E., *et al.* (1995). Gene transfer into the airway epithelium of animals by targeting the polymeric immunoglobulin receptor. *J. Clin. Invest.* **95,** 493–502.

62. Ferkol, T., Pellicena-Palle, A., Eckman, E. *et al.* (1996). Immunologic responses to gene transfer into mice via the polymeric immunoglobulin receptor. *Gene Ther.* **3,** 669–678.

63. Buschle, M., Cotton, M., Kirlappos, H. *et al.* (1995). Receptor-mediated gene transfer into human T lymphocytes via binding of DNA/CD3 antibody particles to the CD3 T cell receptor complex. *Hum. Gene Ther.* **6,** 753–761.

64. Ohtake, Y., Chen, J., Gamou, S. *et al.* (1999). Ex vivo delivery of suicide genes into melanoma cells using epidermal growth factor receptor-specific Fab immunogene. *Jpn. J. Cancer Res.* **90,** 460–468.

65. Foster, B. J., and Kern, J. A. (1997). HER2-targeted gene transfer. *Hum. Gene Ther.* **8,** 719–727.

66. Fominaya, J., and Wels, W. (1996). Target cell-specific DNA transfer mediated by a chimeric multidomain protein. Novel non-viral gene delivery system. *J. Biol. Chem.* **271,** 10560–10568.

67. Kodadek, T. (1993). How does the GAL4 transcription factor recognise the appropriate DNA binding sites *in vivo? Cell. Mol. Biol. Res.* **39,** 355–360.

68. Yosida, J., and Mizuno, M. (1994). Simple method to prepare cationic multilamellar liposomes for efficient transfection of human interferon-β gene to human glioma cells. *J. Neurooncol.* **19,** 269–274.

69. Nabel, E. G., Gordon, D., Yang, Z. Y. *et al.* (1992). Gene transfer in vivo with DNA–liposome complexes: lack of autoimmunity and gonadal localization. *Hum. Gene Ther.* **3,** 649–656.

70. Stewart, M. J., Plautz, G. E., del Buono, L. *et al.* (1992). Gene transfer *in vivo* with DNA-liposome complexes: safety of autoimmunity and gonadal localization. *Hum. Gene Ther.* **3,** 267–275.

71. Kirpotin, D., Park, J. W., Hong, K. *et al.* (1997). Sterically stabilized anti-HER2 immunoliposomes: design and targeting to human breast cancer cells *in vitro. Biochemistry* **36,** 66–75.

72. Debs, R. J., Heath, T. D., and Papahadjopoulos, D. (1987). Targeting of anti-Thy 1.1 monoclonal antibody conjugated liposomes in Thy 1.1 mice after intravenous administration. *Biochem. Biophys. Acta* **901,** 183–190.

73. Woodle, M. C., and Lasic, D. D. (1992). Sterically stabilized liposomes. *Biochem. Biophys. Acta* **1113,** 171–199.

74. De, Kruif, J., Storm, G., Van Bloois, L. *et al.* (1996). Biosynthetically lipid-modified human scFv fragments from phage display libraries as targeting molecules for immunoliposomes. *FEBS Lett.* **399,** 232–236.

75. Cirielli, C., Capogrossi, M. C., and Passaniti, A. (1997). Anti-tumor gene therapy. *J. Neurooncol.* **31,** 217–223.

76. Boris-Lawrie, K., and Temin, H. M. (1994). The retroviral vector: replication cycle and safety considerations for retrovirus-mediated gene therapy. *Ann. N.Y. Acad. Sci.* **716,** 59.

77. Vile, R. G., and Russel, S. J. (1990). Retroviruses as vectors. *Br. Med. Bull.* **51,** 12–30.

78. Hunter, E., and Swanstrom, E. Retrovirus envelope glycoproteins. *Curr. Top. Microbiol. Immunol.* **157,** 187–253.

79. Faas, D., Harrison, S. C., Kim, P. S. *et al.* (1996). Retrovirus envelope domain at 1.7A resolution. *Nat. Struct. Biol.* **3,** 465–469.

80. Roux, P., Jeanteur, P., and Piechaczyk, M. (1989). A versatile and potentially general approach to the targeting of specific types by retroviruses: application to the infection of human cells by means of major histocompatibility complex class I and class II antigens by mouse ectropic murine leukemia virus-derived viruses. *Proc. Natl. Acad. Sci. USA* **86,** 9079–9083.

81. Lorimer, I. A., and Lavictoire, S. J. (2000). Targeting retrovirus to cancer cells expressing a mutant EGF receptor by insertion of a single chain antibody variable domain in the envelope glycoprotein receptor binding lobe. *J. Immunol. Meth.* **237,** 147–157.

82. Schnierle, B. S., Moritz, D., Jeschke, M. *et al.* (1996). Expression of chimeric envelope proteins in helper cell lines and intergration into Moloney murine leukemia virus particles. *Gene Ther.* **3,** 334–342.

83. Marin, M., Noel, D., Valseia-Wittman, S. *et al.* (1996). Targeted infection of human cells via major histocompatibility complex class I molecules by Moloney murine leukemia virus-derived viruses displaying single-chain antibody fragment-envelope fusion proteins. *J. Virol.* **70,** 2957–2962.

84. Ager, S., Nilson, B. H. K., Morling, F. J. *et al.* (1996). Retroviral display of antibody fragments: interdomain spacing strongly influences vector infectivity. *Hum. Gene Ther.* **7,** 2157–2164.

85. Chu, T. T., Martinez, I., Sheay, W. C. *et al.* (1994). Cell targeting with retroviral vector particles containing antibody-envelope fusion proteins. *Gene Ther.* **1,** 292–299.

86. Chu, T. T., and Dornburg, R. (1995). Retroviral vector particles displaying the antigen-binding site of an antibody enable cell-type specific gene transfer. *J. Virol.* **69,** 2659–2663.

87. Chu, T. T., and Dornburg, R. (1997). Toward highly efficient cell-type-specific gene transfer with retroviral vectors displaying single-chain antibodies. *J. Virol.* **71,** 720–725.

88. Martinez, I., and Dornburg, R. (1995). Improved retroviral packaging lines derived from spleen necrosis virus. *Virology* **208,** 234–241.

89. Somia, N. V., Zoppe, M., and Verma, I. M. (1995). Generation of targeted retroviral vectors by using single-chain variable fragment: an approach to *in vivo* gene delivery. *Proc. Natl. Acad. Sci. USA* **92,** 7570–7574.

90. Martin, F., Neil, S., Kupsch, J. *et al.* (1999). Retrovirus targeting by tropism restriction to melanoma cells. *J. Virol.* **73,** 6923–6929.

91. Tillman, B. W., de, Gruijl, T. D., Luykx-de Bakker, S. A. *et al.* (1999). Maturation of dendritic cells accompanies high-efficiency gene transfer by a CD40-targeted adenoviral vector. *J. Immunol.* **162,** 6378–6383.

92. Curiel, D. T. (1999). Strategies to adapt adenoviral vectors for targeted delivery. *Ann. N.Y. Acad. Sci.* **886,** 158–171.

93. Ohno, K., Brown, G. D., and Meruelo, D. (1995). Cell targeting for gene therapy: use of fusion protein containing the modified human receptor for ecotropic murine leukemia virus. *Biochem. Mol. Med.* **56,** 172–175.

94. Etienne-Julan, M., Roux, P., Carillo, S. *et al.* (1992). The efficacy of cell targeting by recombinant retroviruses depends on the nature of the receptor and the composition of the artificial cell-virus linker. *J. Gen. Virol.* **73,** 3251–3255.

95. Cosset, F., Morling, F. J., Takeuchi, Y. *et al.* (1995). Retroviral retargeting by envelopes expressing an N-terminal binding domain. *J. Virol.* **69,** 6314–6322.

96. Weiss, R. A., and Taylor, C. S. (1995). Retrovirus receptors. *Cell* **82,** 531–533.

97. Kasahara, N., Dozy, A. M., and Kan, Y. W. (1995). Tissue-specific targeting of retroviral vectors through ligand-receptor interactions. *Science* **266,** 1373–1376.

Ribozymes in Cancer Gene Therapy

CARMELA BEGER

Department of Medicine
University of San Diego
La Jolla, California, 92093

MARTIN KRÜGER

Department of Medicine
University of San Diego
La Jolla, California, 92093

FLOSSIE WONG-STAAL

Department of Medicine
University of San Diego
La Jolla, California, 92093

I. INTRODUCTION

Ribozymes are small RNA molecules with endoribonuclease activity that hybridize to complementary sequences of a particular target mRNA transcript through Watson–Crick base pairing. Under appropriate conditions, ribozymes exhibit catalytic sequence-specific cleavage of the target. The cleaved target mRNA is destabilized and subject to intracellular degradation; consequently, the expression of this specific gene and the synthesis of the encoded protein are prevented. Since their discovery in the 1980s [1], the conserved sequences, secondary structure, and biochemistry of different groups of ribozymes have been well characterized, including group I ribozymes, hammerhead ribozymes, hairpin ribozymes, ribonuclease P (RNase P), and hepatitis delta virus ribozymes (for recent reviews, see references [2–4]). Their simple structures and ability to cleave RNA molecules

in trans in a site-specific manner have important therapeutic implications for a variety of diseases where well-defined key RNA molecules are involved in causing or maintaining a disease state. Theoretically, any RNA involved in a disease state is a potential target for ribozyme cleavage. Practically, the ribozyme approach is limited by certain requirements for the specific recognition sequence of the catalytic center of the ribozyme. The binding arms that hybridize to the sequence flanking the cleavage site within the target RNA determine specificity of the recognized target sequence. The ability to cleave RNA and thereby selectively inhibit the expression of a gene of interest can be used as a tool for manipulation of RNAs *in vitro* and for the inactivation of gene expression and function *in vivo*. The spectrum of potential targets for ribozyme-mediated gene modulation currently ranges from genes involved in malignant diseases to those causing infectious diseases. Most studies have demonstrated cleavage activity of particular ribozymes *in vitro*. An increasing number of studies are providing additional data about ribozyme activity in preclinical cellular or animal models. These studies have significantly improved the knowledge about target-specific optimization, delivery, stability, and intracellular localization of ribozymes as a requirement for successful clinical application. Most investigators have utilized the hammerhead and hairpin ribozymes, because their small sizes (35–50 nucleotides) are easily manipulated or synthesized chemically. This chapter will focus on these two classes of ribozymes, explain their unique features and specific characteristics, review recent advances using these ribozymes in the field of cancer gene therapy, and briefly outline current limitations and recent developments in the application of ribozymes.

II. RIBOZYME STRUCTURES AND FUNCTIONS

Hammerhead ribozymes are a group of self-cleaving RNAs characterized by a two-dimensional structural motif known as the "hammerhead," which enables site-specific cleavage activity. Haseloff and Gerlach originally described the hammerhead structure with three base-paired stems (helices 1, 2, and 3) flanking the catalytic center and two highly conserved single-stranded regions with defined sequence (see Fig. 1A) [1]. To facilitate hybridization between ribozyme and substrate, helices 1 and 2 of the hammerhead ribozyme sequence have to align to the substrate sequence (target) via complementary Watson–Crick base pairing. Mutagenesis studies defined important nucleotides and functional groups for efficient catalysis. The target site can be any NUH sequence within the substrate (N: any nucleotide; H: adenine [A], cytosine [C], or uridine [U] but not a guanine [G]). Length and composition of the ribozyme binding arms control a cascade of reactions during hammerhead-mediated cleavage. The length of the flanking sequences (helices 1 and 3) determines not only site specificity but also the secondary structure of the RNA molecule and kinetic profile of the ribozyme-substrate interaction [5]. The optimal length and composition of the binding arms can differ between *in vitro* and inside the cell and should therefore be determined individually for each ribozyme/target combination. Usually a length between 9 and 12 nucleotides for the flanking sequences is considered as a good experimental starting point for optimization. In addition, while choosing a suitable target site for ribozyme cleavage, functional importance and secondary structure of the potential target region should also be considered. Although it is not the focus of this review, it should be mentioned that parts of the hammerhead RNA sequence have been successfully substituted by DNA, thereby increasing the stability of the ribozyme. Hendry *et al.* have shown that ribozymes with DNA in helices 1 and 3 showed threefold enhanced cleavage activity compared with all-RNA ribozymes [6]. The site-specific mechanism of a hammerhead-mediated RNA cleavage has been studied in detail [7] and enables us to successfully discriminate substrate RNAs with a single-base mutation [8].

The hairpin ribozyme was originally isolated from the negative strand of the satellite RNA of tobacco ringspot virus [9]. It folds into a two-dimensional "hairpin" structure, consisting of a small catalytic region of four helical domains and five loops. Based on Watson–Crick base-pairing, two helices (helix 1 and 2) form between the substrate and ribozyme (see Fig. 1B). Helix 2 has to be four nucleotides long, whereas the length of helix 1 can be 4 or more nucleotides, with a functional ribozyme typically having between 6 and 10 nucleotides. The combination of these two helices facilitates binding of the substrate and determines specificity of binding for *trans*-acting hairpin ribozymes. Between the two helices (helix 1 and 2) in the substrate (Fig. 1B, loop 5) is the cleavage site, consisting of four nucleotides: N*GUC (cleavage occurs at *), where GUC on the 3′ side of the cleavage site is the only sequence required in the substrate for maximal cleavage and N is any nucleotide [10]. Further information about structure and properties, including a practical approach for the design of hairpin ribozymes, can be found in a recent article published by Yu and Burke [11]. Most researchers use disabled hairpin ribozymes, as a control to assess antisense rather than cleavage-specific effect of the ribozymes. Compared with the hammerhead ribozyme, the requirements for the hairpin target sequence are more restricted, but still, within a given substrate RNA of interest, numerous potential target sites should be found.

Both groups of ribozymes are usually engineered to cleave the RNA of interest *in trans*. Following substrate cleavage, the two products are released and the ribozyme is free to bind and cleave another substrate. *In vitro*, the ribozymes can be shown to be truly catalytic in that more than one substrate molecule is processed per ribozyme molecule [9]. This "recycling" of ribozymes is considered to be one of the major advances over antisense technology.

Substantial progress and better understanding of ribozyme design and delivery have already led to the first clinical trials of ribozyme safety in humans [12,13]. During recent years, the site-specific cleavage mechanism of ribozymes became of particular relevance to identify and reduce activity of gene products associated with stimulation of cell growth. The tremendous increase in information about pathways that determine growth signal transduction, regulation of the cell cycle, mechanisms of action of oncogenes and tumor suppressors, and mechanisms of programmed cell death now allows the design of sequence-specific ribozymes. Ribozymes can be designed to target oncogenes and to transport proteins or growth factors to specifically inhibit tumor cell proliferation, drug resistance, or angiogenesis. Genes known to be sufficient for malignant transformation (such as *ras, raf,* and *bcr–abl*) and genes for which expression is important but not sufficient for malignant transformation (e.g., *fos*) are both potential targets for ribozymes. Ribozymes that cleave or ligate a particular RNA target sequence can be expressed in tumor cells to prevent or promote expression and translation of RNA molecules comprising the target sequence, thereby leading to a better understanding of the role and importance of the targeted RNA of interest. In comparison to conventional drug therapy, the ribozyme approach offers considerable advantages. Once genes involved in tumorgenicity or tumor pathology are identified, cloned, sequenced, and ideally functionally characterized, this information immediately allows the design of sequence-specific ribozymes aimed to target mRNA molecules, thereby modulating the activity of these genes or their protein products involved in the malignant state of the tumor cell. The following sections summarize the current experimental approaches and data regarding ribozymes targeted against a variety of mRNAs involved in

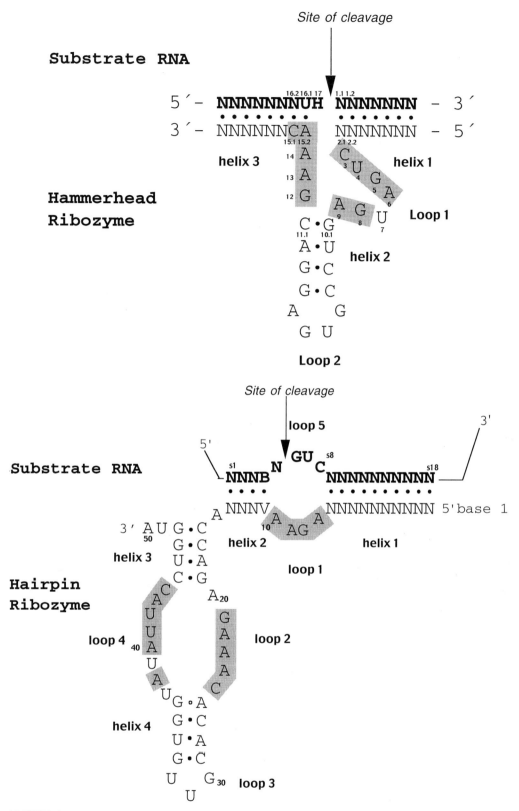

FIGURE 1   Sequence of the hammerhead (A) and the hairpin (B) ribozyme with the corresponding substrate RNA (target). Dots indicate Watson–Crick base pairs. Conserved sequences required for optimal ribozyme activity are shaded. The target site of the hammerhead ribozyme can be any NUH sequence (N: any nucleotide; H: adenine [A], cytosine [C] or uridine [U] but not a guanine [G]), with cleavage occurring on the 3′ side of H_{17} as shown by the arrow. Numbering of the hammerhead ribozyme is according to Hertel *et al.* [173]. The hairpin ribozyme requires a GUC triplet on the 3′ side of the cleavage site (marked by an arrow) for optimal activity [10]. The B nucleotide is C, G, or U (not A), and the V nucleotide is G, C, or A (not U). Nucleotides of the hairpin ribozyme are numbered from 1–50 [174]. Substrate nucleotides are numbered consecutively.

TABLE 1   Malignant Disorders as Potential Targets for Ribozyme Gene Therapy

Target gene	Gene product	Ribozyme-induced change of function	Ref.
bcr–abl	Tyrosine kinase	Inhibition of cell proliferation and colony formation	[17–27]
*PML/RAR*α	Transcriptional regulator	Inhibition of cell proliferation; induction of apoptosis; increase in sensitivity against ATRA	[28–30]
AML1/MTG8	Transcription factor	Inhibition of cell proliferation; induction of apoptosis	[33–35]
N-*ras*, H-*ras*, K-*ras*	Signal transduction pathway	Inhibition of cell proliferation and colony formation; change in morphology, enhanced melanin synthesis; decrease of *in vivo* tumorigenicity	[41–56]
EGFR	Receptor tyrosine kinase	Inhibition of cell proliferation and colony formation; decrease of *in vivo* tumorigenicity	[58,59]
c-*erbB-2* (HER2/*neu*)	Receptor tyrosine kinase	Inhibition of cell proliferation; decrease of *in vivo* tumorigenicity	[62–66]
c-*erbB-4*	Receptor tyrosine kinase	Inhibition of mitogenesis and colony formation; decrease of *in vivo* tumorigenicity	[67,68]
Estrogen receptor	Transcriptional regulator	Inhibition of cell cycle progression	[69]
Androgen receptor	Transcriptional regulator	Inhibition of androgen receptor transcriptional activity	[70]
c-*fms*	Growth factor receptor	Inhibition of cell proliferation	[71]
RET	Receptor tyrosine kinase	Inhibition of colony formation	[72]
mdr-1	Drug-efflux pump	Reduction in resistance to chemotherapeutic drugs	[74–84]
c-*fos*	Transcriptional regulator	Change in morphology; reduction in resistance to chemotherapeutic drugs	[86]
CD44	Cell adhesion molecule	n.d.	[94]
VLA-6	Adhesion receptor	Decrease of *in vitro* invasion and *in vivo* metastatic ability	[95]
MMP-9	Matrix metalloproteinase	Decrease of *in vivo* metastatic ability	[96,97]
CAPL (S100A4)	Calcium-binding protein	Decrease of *in vitro* invasion; reduction in expression of MMP-2, MT1-MMP, and TIMP-1; decrease of *in vivo* metastatic ability	[98,99]
Pleiotrophin	Growth factor	Inhibition of colony formation; decrease of *in vivo* tumor growth, tumor angiogenesis, and metastatic ability	[63,102,103]
VEGF-R1/VEGF-R2	Growth factor receptors	Decrease of *in vivo* tumor growth; decrease of *in vivo* metastatic ability (VEGF-R2 only)	[106]
VEGF	Growth factor	n.d.	[107]
bFGF-BP	bFGF-binding protein	Reduction of release of biologically active bFGF; decrease of *in vivo* tumor growth and tumor angiogenesis	[108]
Telomerase	Ribonucleoprotein	Suppression of telomerase activity; inhibition of cell proliferation; change in morphology; induction of apoptosis	[112–117]
bcl-2	Anti-apoptotic protein	Induction of apoptosis	[119–121]
PKC-α	Anti-apoptotic protein	Induction of apoptosis	[123,124]

malignant cell transformation and proliferation, multidrug resistance, tumor angiogenesis, and metastasis, as well as malignant complications of viral infections (see Tables 1 and 2).

III. CANCER DISEASE MODELS FOR RIBOZYME APPLICATION

A. Chromosomal Translocations

One possible application of ribozymes for cancer therapy is to target aberrant genes (e.g., fusion transcripts resulting from chromosomal translocations). The best known example is the Philadelphia translocation t(9;22), which can be detected in more than 95% of patients with chronic myelogenous leukemia [14]. The chimeric *bcr–abl* hybrid gene encodes a fusion protein with a deregulated tyrosine kinase activity, which leads to cell transformation in hematopoietic cells and causes a leukemia-like phenotype in mice [15,16]. Various ribozymes have been constructed to cleave within the region of the *bcr–abl* fusion point. As shown in extracellular cleavage assays, these ribozymes can efficiently cleave chimeric RNAs [17–26]. However, in addition to specific cleavage of the chimeric transcripts, these ribozymes often also nonspecifically cleaved wild-type *bcr* or *abl* sequences. Although this unwanted cleavage usually occurred at reduced efficiency, it could limit the application of ribozymes against chimeric fusion genes, in which normal expression of the

TABLE 2 Viral Disorders with Malignant Complications as Potential Targets for Ribozyme Gene Therapy

Virus	Malignant complication	Target gene for ribozymes	Ribozyme-induced change of function	Ref.
Human papilloma virus (HPV)	Cervical cancer, oral cancer	E6, E7	Inhibition of cell proliferation and colony formation	[125–127]
Epstein–Barr virus (EBV)	Burkitt's lymphoma, nasopharyngeal carcinoma, lymphoproliferative disorders in immuno-suppressed patients (AIDS, transplant recipients)	EBNA-1	Inhibition of cell proliferation	[128]
Hepatitis B virus (HBV)	Hepatocellular carcinoma	Pregenomic RNA	Inhibition of viral gene expression	[131–134]
Hepatitis C virus (HCV)	Hepatocellular carcinoma	5′ untranslated/core region	Inhibition of viral gene expression	[135–139]

wild-type genes are important to the cell. To achieve higher target selectivity, different strategies of designing anti-*bcr–abl* ribozymes have been developed by utilizing an anchor sequence in helix 3 of the ribozyme [19], introducing a single base mismatch in helix 3 [24], and varying possible target sites in the 5′ (*bcr*) and 3′ (*abl*) portions of the fusion point, as well as testing various lengths of helices 1 and 3 within the ribozymes [21], creating helix-1- or helix-3-forming antisense arms such that binding and cleavage occur on opposite sides of the *bcr–abl* fusion point [23], or constructing a triple-unit ribozyme consisting of three hammerhead ribozymes [17]. Intracellular applications of anti-*bcr–abl* ribozymes demonstrate their efficiency within t(9;22)-positive cells [17,20,22,25–27], as measured by reduction of *bcr–abl* mRNA or protein expression, growth inhibition, decreased ability for colony formation, and increase in apoptosis.

Another translocation, the t(15;17), is found in acute promyelocytic leukemia and is characterized by the expression of the fusion transcript *PML/RARα*. *PML/RARα* expression is linked to leukemogenesis and clinical sensitivity to all-*trans* retinoic acid (ATRA). Hammerhead ribozymes have been designed against the fusion RNA [28–30]. Intracellular application of an anti-*PML/RARα* ribozyme resulted in higher sensitivity against ATRA, growth inhibition, and induction of apoptosis but did not lead to cellular differentiation [29,30].

Furthermore, ribozymes have been developed to target the fusion gene resulting from the translocation (8;21). This translocation is detected in 7–17% of patients with acute non-lymphocytic leukemia and involves the *AML1* and the *MTG8* genes [31]. The fusion protein forms the chimeric transcriptional factor AML1/MTG8, which is part of the core binding factor (CBF) oncoprotein [32]. Several groups have constructed hammerhead ribozymes against the chimeric RNA [33–35]. By using *in vitro* cleavage assays they have demonstrated cleavage activity for these ribozymes. In addition, introduction of the anti-*AML1/MTG8* ribozymes in t(8;21)-

positive cells demonstrated inhibition of cell growth and induction of apoptosis.

B. Malignant Cell Proliferation

The signal transduction pathways regulate cell growth and differentiation. Several genes in the signal transduction cascades are known to be altered in cancer cells and have been targeted by ribozymes. The *ras* gene family illustrates some of these studies. The three *ras* genes (N-*ras*, H-*ras*, K-*ras*) code for proteins that are members of the supergene family of GTP/GDP-binding proteins [36]. Mutations in the *ras* oncogene have frequently been found in a variety of tumors [37,38–40]. These point mutations cause structural changes in the GTP binding site leading to activation of the Ras protein. Recent studies have investigated the potential of using ribozymes to inhibit such deregulated signal transduction. Hammerhead ribozymes targeting mutated *ras* oncogenes showed selective cleavage of mutated *ras* RNA *in vitro* [41–45] as well as inhibition of *in vivo* function of the oncogene [41–52]. Intracellular application of ribozymes that specifically target mutated *ras* oncogene in H-*ras*-transformed NIH3T3 resulted in abrogation of the transformed phenotype and suppression of *in vivo* tumorigenicity or tumor regression [41,42,46,51,53,54]. Similar results were obtained when mutant H-*ras* was targeted by ribozymes in a bladder cancer model [43,47,50,52,55] or in a human melanoma cell line (FEM) that contains a heterozygous H-*ras* gene [48,49]. Several groups used pancreatic cancer cells either homozygous or heterozygous for this mutation or containing an unrelated K-*ras* mutation and applied hammerhead ribozymes specifically directed against mutated K-*ras* oncogene [45,56]. A ribozyme-mediated effect was observed that was dependent on the characteristics of the target gene. These studies confirmed the oncogenic potential of activated *ras* genes and demonstrated the capability of anti-*ras* ribozymes to inhibit gene expression, reverse the transformed phenotype, and suppress *in vivo* tumorigenicity. However, additional studies will have to define the activity of anti-*ras* ribozymes in primary tumor cells and the optimal ribozyme delivery system before anti-*ras* ribozymes can be evaluated as a potential clinical application in the context of current chemotherapeutic protocols.

Epidermal growth factor receptor (EGFR) and EGF-like receptor tyrosine kinases are important genes involved in malignant cell proliferation. Amplification and rearrangement of the EGFR gene are frequently associated with malignant gliomas [57]. Ribozymes directed against aberrant EGFR transcripts effectively inhibited the tumorigenic ability of NIH3T3 cells transformed by overexpression of aberrant EGFR [58]. Similarly, glioblastoma cells overexpressing aberrant EGFR were inhibited in their proliferative characteristics and anchorage-independent growth following introduction of anti-EGFR ribozymes [59]. The family of EGF-like receptor tyrosine kinases (c-erbB) consists of several cellular oncoproteins involved in human cancers [60]. c-*erbB-2* (HER2/*neu*) is a well-characterized member of this family, which is often found to be overexpressed in breast cancers [61]. Anti-c-*erbB-2* ribozymes were shown to inhibit the function of the oncoprotein [62–66]. Anti-c-*erbB-2* ribozymes decrease target gene expression, cell proliferation, and tumor formation in breast or ovarian cancer cells [62–66]. Further experiments using these ribozymes suggest that the oncogenic activity of c-*erbB-2* in human tumors is in part related to its ability to increase the growth response to stroma-derived EGF-like growth factors [66]. Ribozymes targeting c-*erbB-4* decreased c-*erbB-4*-mediated mitogenesis and inhibition of colony formation as well as tumor formation in cells overexpressing c-*erbB-4* [67,68]. The results of these studies along with data derived from genes involved in malignant cell proliferation such as estrogen receptors [69], androgen receptor [70], or the oncogenes c-*fms* [71] and *RET* [72] facilitate the development of new therapeutic approaches based on ribozyme technology.

C. Multidrug Resistance

The development of multidrug resistance (MDR) significantly limits the effectiveness of cytostatic agents and the success of cancer therapy. Chemotherapy-resistant cancer cells may result from alterations at any step in the cell-killing pathway, which includes drug transport, drug metabolism, drug target, cellular repair mechanisms, and toxicity-induced apoptosis. One well-documented mechanism is the overexpression of P-glycoprotein (P-GP), usually resulting from amplification of the multidrug-resistance gene (*mdr-1*). P-GP is localized in the cellular membrane and functions as an ATP-dependent drug-efflux pump, resulting in decreased intracellular levels of structurally unrelated cytotoxic agents [73]. Several groups have designed ribozymes against *mdr-1* and tested their activities [74–84]. When expressed in cells

overexpressing *mdr-1* in a variety of tumor or leukemia cells, ribozymes targeting *mdr-1* transcripts caused suppression of *mdr-1* mRNA expression associated with a significant reduction in drug resistance. A recent study utilized the carcinoembryonic-antigen (CEA) promoter for ribozyme expression and demonstrated a ribozyme-mediated suppression of *mdr-1* expression and increased drug-efflux activity selectively in cells overexpressing CEA [84].

The nuclear oncogene c-*fos* has been suggested to regulate downstream enzymes associated with DNA synthesis and repair. c-*fos* overexpression is strongly involved in cancer cell resistance to cisplatin chemotherapy [85]. Anti-*fos* ribozymes lead to decreased expression of c-*fos* mRNA, accompanied by reduced mRNA levels of thymidylate (dTMP) synthase, DNA polymerase β, topoisomerase I, and metallothionein IIA. Expression of ribozymes against c-*fos* in cisplatin-resistant cells was successful in restoring their cisplatin-sensitivity [86–88]. Furthermore, ribozyme activity resulted in decreased expression of c-*fos* as well as *mdr-1*, c-*jun*, and mutant *p53* mRNAs. Phenotypically, the transfomed cells showed altered morphology and restored sensitivity to chemotherapeutic agents comprising the MDR phenotype [86]. Interestingly, anti-*fos* ribozymes induced a more rapid reversal of the MDR phenotype in comparison to an anti-*mdr-1*-ribozyme, implicating the central role of c-*fos* in drug resistance. Fos is thought to mediate its various effects through transcriptional activation after interaction with the Jun protein to form the AP-1 complex [89,90]. The upstream promoter of the *mdr-1* gene contains an AP-1 binding site [91], which is required for full promoter activity in Chinese hamster ovary cells [92] and is suggested to be active in cell lines that overexpress *mdr-1* RNA without gene amplification [73,93]. Therefore, c-*fos* may play an essential role in regulating *mdr-1* gene expression.

D. Tumor Angiogenesis and Metastasis

Invasion and metastasis still represent the greatest obstacles to successful cancer treatment. Metastasis development involves sequential events, such as detachment of tumor cells from the primary site, invasion of the basement membrane and the underlying extracellular matrix, neovascularization, and tumor cell proliferation. In theory, the process of angiogenesis and metastasis can be stopped by targeting molecules at different levels within this cascade. Herein, ribozymes may offer a great potential to target single or multiple steps within this cascade. Ribozyme-mediated inhibition of metastasis has been developed for the following genes involved in tumor cell invasion and metastasis: CD44 [94], VLA-6 integrin [95], MMP-9 [96,97], and *CAPL*/S100A4 [98,99]. In all cases, ribozyme expression reduced intracellular expression of the target gene. Furthermore, ribozymes targeting S100A4 resulted in decreased expression of the metalloproteinases MMP-2, MT1–MMP, and TIMP-1 [99].

In addition, in most applications ribozyme activity has been linked to suppression of the *in vitro* invasive ability of target cells or the metastatic potential after inoculation in nude mice [95,96,98,99].

Besides factors that directly influence the metastatic potential of tumor cells, clinical and experimental evidence suggests that spreading of malignant cells from a localized tumor is directly related to the number of microvessels in the primary tumor [100]. Tumor angiogenesis is thought to be mediated by tumor-cell-derived growth factors. The secreted polypeptide pleiotrophin (PTN) is one of such tumor-derived growth factors [101]. The role of pleiotrophin in tumor growth was recently assessed utilizing anti-PTN ribozymes [63,102,103]. These ribozymes inhibited PTN-induced colony formation in melanoma cells and decreased their ability for tumor formation and growth as well as their metastatic spread in nude mice. Another gene correlated to angiogenesis and cell growth in a variety of cancers is vascular endothelial growth factor (VEGF) and its receptor [104,105]. Ribozymes targeting VEGF-receptor-1 (VEGF-R1) or VEGF-receptor-2 (VEGF-R2) resulted in decreased tumor growth [106]. However, only anti-VEGF-R1 ribozymes, but not anti-VEGF-R2 ribozymes, had the potential to reduce the metastatic ability of tumor cells in nude mice. Ke and colleagues developed anti-VEGF ribozymes and observed a reduction of VEGF RNA and protein expression [107]. Furthermore, ribozymes targeting a secreted fibroblast growth factor (FGF)-binding protein (FGF-BP) were applied to reduce the release of biologically active basic FGF (bFGF) [108]. Ribozyme-mediated suppression of FGF-BP resulted in reduced cell growth and tumor angiogenesis of xenograft tumors in mice, suggesting that human tumors are capable of utilizing FGF-BP as an angiogenic switch molecule.

These studies underline the potential of ribozymes to specifically reduce the expression of genes involved in angiogenesis or metastasis development which might ultimately translate into a therapeutic benefit for a variety of tumors.

E. Telomerase

Telomeres are structures at chromosome ends consisting of highly conserved $(TTAGGG)_n$ repeats that are physiologically lost as a function of cell division. Progressive telomere shortening is linked to the limited proliferative capacity of normal somatic cells [109]. Telomeres are exclusively replicated (elongated) by telomerase, a ribonucleoprotein complex with telomere-specific reverse transcriptase activity [110]. Indefinite proliferation and malignant progression have been found to be associated with high telomerase activity. While telomerase has been found in a majority of malignancies (85–100%), no activity has been detected in most nonmalignant tissue [111]. Various groups have developed ribozymes directed against the RNA component of human telomerase [112–117]. These ribozymes showed a specific *in*

vitro cleavage activity [112,113,116], as well as an inhibitory effect on telomerase activity in cellular extracts [112,116]. Intracellular application of anti-telomerase ribozymes in cancer cells has reduced telomerase activity and cell proliferation, in part accompanied by morphological changes and induction of apoptosis [113–117].

F. Apoptosis

Apoptosis is a cellular mechanism regulated by a complex network of multiple cellular gene families and gene pathways. One major gene-inhibiting apoptosis is the oncogene *bcl-2,* which has been found to be expressed in various tumor tissues [118]. Overexpression of *bcl-2* results in an increased resistance of prostate cancer cells against a variety of therapeutic agents and hormonal ablation. The intracellular application of anti-*bcl-2* ribozymes was followed by reduced *bcl-2* expression and a significant increase in the apoptotic ability of these cells [119,120]. Similar effects were observed with an anti-*bcl-2* ribozyme introduced in oral cancer cells [121].

Another gene linked to apoptosis is protein kinase C-alpha (PKC-α) [122]. Anti-PKC-α ribozymes reduced cell growth and induced cell death in glioma cell lines [123,124]. Interestingly, ribozyme-mediated downregulation of PKC-α expression was accompanied by a significant reduction in gene expression of BCL-xL, a protein that inhibits apoptosis and is overexpressed in malignant cells [124].

G. Malignant Complications of Viral Infections

A variety of infectious agents are associated with the development of human malignancies. Frequently, continued foreign gene expression following viral infection alters cellular pathways and thereby induces or promotes a malignant cellular phenotype. By their selective inhibition of viral gene expression, ribozymes have contributed to our understanding of virus–host interactions, a requirement for the development of new therapeutics aimed at selectively targeting cancer cells while sparing normal cells.

Human papilloma virus (HPV) RNA has been found in many cervical and oral cancers and is likely to be an important cofactor in the development of these malignancies. HPV RNA from a HeLa cervical cancer cell line was effectively cleaved *in vitro* by hammerhead ribozymes targeted against the E6 and E7 genes of HPV type 18 (nucleotides 123, 309, and 671 of the viral transcript) [125], genes which are capable of efficiently immortalizing a broad spectrum of cell types. The transfected HeLa cells also exhibited reduced growth rates, increased serum dependency, and reduced colony formation in soft agar. An anti-E6 ribozyme reduced the growth rate and significantly affected the immortalizing ability of E6/E7 genes in normal human keratinocytes [126]. However, an inactive version of the ribozyme also was able to significantly prevent immortalization, probably through passive hybridization and antisense effect. Recently, a transacting ribozyme was found to reduce the E7 RNA transcript by 90% in CV-1 cells after transient transfection and ribozyme expression mediated via the Rous sarcoma virus–long terminal repeat (RSV–LTR) promoter [127].

Epstein–Barr virus (EBV) infection is associated with the development of several human malignancies, such as Burkitt's lymphoma, AIDS-associated lymphomas, nasopharyngeal carcinoma, and lymphoproliferative disorders in transplant recipients. Under these conditions, which all share a situation of immunosuppression of the host, the proliferative potential of EBV-infected cells and lifetime exposure through viral persistence are major predisposing factors for the development of EBV-associated neoplasms. Recent studies suggest that expression of the EBV nuclear antigen-1 (EBNA-1) is the major determinant of EBV-associated B-cell neoplasia [128]. EBNA-1 protein is required for the replication and maintenance of the EBV genome. EBV-transformed lymphocytes upregulated αV integrin expression and, in contrast to non-transformed B cells, were susceptible to adenovirus infection. Ribozymes targeted against functional EBNA-1 delivered via adenoviral gene transfer downregulated EBNA-1 RNA and protein expression, thereby inhibiting EBV-induced cell proliferation in these cells [128]. EBV-specific ribozymes might also be useful to further elucidate the mechanism of EBV-induced B-cell activation, transformation, and viral replication by allowing one to target specific steps in these developments.

Hepatocellular carcinoma is one of the most common malignancies and causes an estimated one million deaths per year worldwide [129]. Chronic infections with hepatitis B (HBV) and hepatitis C virus (HCV) are identified major risk factors for the development of primary hepatocellular carcinoma [130]. Chronic active hepatitis is induced by both infections (HBV, 5–10%; HCV, 50–75%), and in a significant percentage of patients long-term liver inflammation is followed by cirrhosis, leading to the development of liver tumors. It is anticipated that effective treatment regimens against these viral diseases will successfully decrease the incidence of primary liver cancer associated with chronic viral infection worldwide. Both viruses undergo replication through an RNA intermediate and thereby offer the potential for ribozymes to interfere within their replication cycle. The pregenomic RNA transcribed from HBV genomic DNA has been targeted *in vitro* by three different hammerhead ribozymes derived from a single DNA template [131]. In addition to confirming the feasibility of an *in vitro* cleavage of HBV RNA intermediates, Welch *et al.* demonstrated the intracellular effectiveness of HBV-directed hairpin ribozymes [132]. Recently, hammerhead ribozymes have been developed to target the X protein (HBx) of HBV, a well-established transcriptional activator protein. HBx protein plays an important role for viral replication and establishment of HBV infection and is implicated

in chronic viral hepatitis and hepatocellular carcinoma. Following cotransfection, hammerhead-mediated cleavage not only reduced HBx RNA but also dramatically reduced the transactivation activity of HBx protein [133] and decreased HBsAg and HBeAg secretion from cells transfected with the ribozymes and an HBV replication-competent plasmid [134]. Several recent reports also provide evidence for an antireplicative effect of ribozymes on hepatitis C virus, a hepatotropic virus that replicates entirely through an RNA life cycle. The successful use of hammerhead ribozymes [135–138] and hairpin ribozymes [139] targeting highly conserved regions of the 5' end of the 9.4-kb HCV genome underlines the great potential of ribozymes to decrease or eventually eliminate HCV RNA within infected cells. Despite recent promising developments towards a tissue-culture infection system for HCV [140], it is premature to draw conclusions about the feasibility of a clinical application of these ribozymes in humans.

IV. CHALLENGES AND FUTURE DIRECTIONS

The relatively recent discovery of ribozymes and investigation of their therapeutic use in humans are reflected by only a handful of clinical gene therapy protocols planned or approved worldwide. However, given the flexibility in design of ribozymes and their unique potential, an increasing number of clinical protocols will most likely be presented over the next several years. As ribozymes are being tested in their first clinical trials [12,13,141,142] to determine their safety and efficacy *in vivo*, researchers are continuing to optimize their intracellular stability, and to target specificity and catalytic activities.

The primary structural characteristics of ribozymes can be changed for optimization of ribozyme cleavage activity on a given substrate. In the substrate cleavage region of a hammerhead or hairpin ribozyme, only minor base changes are tolerated to preserve enzyme activity. In contrast, the composition and length of the bases complementary to the ribozyme binding arms can be altered to improve cleavage activity. *In vitro* evolution and selection of ribozymes starting from random-sequence nucleotides have led to improved catalytic capabilities of particular ribozymes in comparison to wild-type sequences [143–145] and significantly improved target-specific cleavage activity [146–149]. However, even these results might not ultimately correlate with *in vivo* efficacy of a ribozyme on a particular substrate cleavage site. Furthermore, recent developments have improved ribozyme activity and specificity by developing artificial DNA enzymes or "deoxyribozymes" [150], hammerhead-based dimeric maxizymes [151,152], and a hairpin-derived twin ribozyme [153].

Even with an optimally designed ribozyme against a suitable target sequence, gene delivery and expression are still two of the most difficult steps towards successful application of ribozymes to human gene therapy. Generally, there are two types of delivery strategies: exogenous delivery of synthetic or *in vitro* transcribed ribozymes or endogenous expression using a plasmid or viral vector containing a transcriptional unit. To achieve optimal expression of ribozymes from a vector system, an expression cassette is necessary to maximize ribozyme transcription. Self-cleaving ribozymes flanking either site of the therapeutic ribozymes have been used to enhance ribozyme activity [154,155]. In addition, multiple ribozymes under a single promoter have been constructed to increase intracellular ribozyme expression [156]. The ribozyme coding sequence is inserted into the untranslated regions of genes transcribed by RNA polymerase II (pol II) (e.g., SV early promoter), retroviral long terminal repeats (LTRs), or pol III promoters such as tRNA [157]. Recent developments have been achieved mainly using endogenous delivery systems, in which the ribozyme expression cassette can be regulated using inducible promoters [158,159]. The choice of vector and promoter constructs or transcriptional units for endogenous delivery is also dependent on cell-type-specific differences. Ideally, a delivery vehicle should selectively target the organ or tissue of interest by exploiting the tissue tropism of different vectors or tissue-specific promoters [84]. Colocalization of ribozyme and target RNA in the same cellular compartment is an important requirement for efficient ribozyme-mediated RNA cleavage [160]. The transduction by retroviral, adenoviral, or adeno-associated virus constructs might include signals that localize ribozymes to certain cellular compartments (e.g., utilizing antigen-binding sites found on most snRNAs). Such localization signals can enhance the intracellular effectiveness of hammerhead ribozymes [161].

In a variety of studies, ribozymes have been developed against overexpressed or aberrant transcripts in order to decrease or eliminate the expression of the corresponding gene product. Besides this distinct therapeutic application of ribozymes, several investigators have applied ribozymes to validate and characterize the function of specific genes and gene pathways ("target validation") either with a therapeutic potential (e.g., ribozymes targeting cellular growth factors [162–167], or without any therapeutic purposes, such as by targeting genes for which a ribozyme-mediated inhibition is associated with a transformed phenotype of cells (e.g., the retinoblastoma gene [155], p16INK4a [168], or Bard1 [169]).

Recently, a novel ribozyme-based technology was developed that allows the identification of genes that are directly or indirectly involved in the development of a particular phenotype. Here, a specific cellular function or phenotype is changed by the application of an unknown ribozyme. For this purpose, a library of hairpin ribozymes with randomized substrate-binding sequences was developed to potentially cleave any RNA substrate containing a GUC. Expression of the ribozyme library in cells and subsequent selection for a given phenotype would enable the selection of single ribozymes responsible for this particular phenotype. Subsequently, the target-recognition sequences of ribozymes that reproducibly confer phenotypic changes are used to identify the corresponding target gene(s). Utilizing this powerful "inverse genomics" approach, several genes were identified that are involved variably in internal ribosome entry site- (IRES)-mediated translation of hepatitis C virus [170], anchorage-independent cellular growth [171], or fibroblast transformation [172].

As demonstrated by the increasing number of publications on ribozymes in cancer gene therapy, significant progress has been made in the development of ribozymes as a potential platform technology for human gene therapy. Ribozymes should become valuable tools for targeting cellular transcripts for destruction, alteration, or repair of genetic information. In the field of cancer gene therapy, ribozymes might be particularly useful to protect cells, either by *ex vivo* or *in vivo* gene therapy, with transduced, protected, expanded autologous cells or hematopoietic stem cell precursors. However, more experience is must be gained from relevant preclinical models of diseases and innovative strategies are needed to improve ribozyme delivery, expression, colocalization, target specificity, and catalysis properties of ribozymes. Relevant animal models also need to be developed that will allow assessment of the different parameters. Many of these issues are common to the field of gene therapy in general. As the hurdles of gene delivery, expression, and targeting are removed, ribozyme gene therapy can become part of the daily routine in clinics in the prophylaxis and treatment of malignant disorders.

References

1. Haseloff, J., and Gerlach, W. L. (1988). Simple RNA enzymes with new and highly specific endoribonuclease activities. *Nature* **334**, 585–591.
2. Amarzguioui, M., and Prydz, H. (1998). Hammerhead ribozyme design and application. *Cell. Mol. Life Sci.* **54**, 1175–1202.
3. Shippy, R., Lockner, R., Farnsworth, M. *et al.* (1999). The hairpin ribozyme. Discovery, mechanism, and development for gene therapy. *Mol. Biotechnol.* **12**, 117–129.
4. Muotri, A. R., da Veiga Pereira, L., dos Reis Vasques, L. *et al.* (1999). Ribozymes and the anti-gene therapy: how a catalytic RNA can be used to inhibit gene function. *Gene* **237**, 303–310.
5. Fedor, M. J., and Uhlenbeck, O. C. (1990). Substrate sequence effects on "hammerhead" RNA catalytic efficiency. *Proc. Natl. Acad. Sci. USA* **87**, 1668–1672.
6. Hendry, P., McCall, M. J., Santiago, F. S. *et al.* (1992). A ribozyme with DNA in the hybridising arms displays enhanced cleavage ability. *Nucleic Acids Res.* **20**, 5737–5741.
7. Ruffner, D. E., Stormo, G. D., and Uhlenbeck, O. C. (1990). Sequence requirements of the hammerhead RNA self-cleavage reaction. *Biochemistry* **29**, 10695–10702.
8. Hertel, K. J., Herschlag, D., and Uhlenbeck, O. C. (1996). Specificity of hammerhead ribozyme cleavage. *EMBO J.* **15**, 3751–3757.
9. Hampel, A., Nesbitt, S., Tritz, R. *et al.* (1993). The hairpin ribozyme. *Methods: A Companion to Meth. Enzymol.* **5**, 37–42.
10. Anderson, P., Monforte, J., Tritz, R. *et al.* (1994). Mutagenesis of the hairpin ribozyme. *Nucleic Acids Res.* **22**, 1096–1100.

11. Yu, Q., and Burke, J. (1997). Design of hairpin ribozymes for *in vitro* and cellular applications, in *Methods in Molecular Biology*, (B. Turner, ed.), Vol. 74, pp. 161–169. Humana Press, Totowa, NJ.

12. Amado, R. G., Mitsuyasu, R. T., Symonds, G. *et al.* (1999). A phase I trial of autologous CD34+ hematopoietic progenitor cells transduced with an anti-HIV ribozyme. *Hum. Gene Ther.* **10**, 2255–2270.

13. Wong-Staal, F., Poeschla, E. M., and Looney, D. J. (1998). A controlled, phase 1 clinical trial to evaluate the safety and effects in HIV-1 infected humans of autologous lymphocytes transduced with a ribozyme that cleaves HIV-1 RNA. *Hum. Gene Ther.* **9**, 2407–2425.

14. Kurzrock, R., Gutterman, J. U., and Talpaz, M. (1998). The molecular genetics of Philadelphia chromosome-positive leukemias. *N. Engl. J. Med.* **319**, 990–998.

15. Lugo, T. G., Pendergast, A. M., Muller, A. J. *et al.* (1990). Tyrosine kinase activity and transformation potency of *bcr–abl* oncogene products. *Science* **247**, 1079–1082.

16. Daley, G. Q., Van Etten, R. A., and Baltimore, D. (1990). Induction of chronic myelogenous leukemia in mice by the *P210bcr/abl* gene of the Philadelphia chromosome. *Science* **247**, 824–830.

17. Leopold, L. H., Shore, S. K., Newkirk, T. A. *et al.* (1995). Multi-unit ribozyme-mediated cleavage of *bcr–abl* mRNA in myeloid leukemias. *Blood* **85**, 2162–2170.

18. Wright, L., Wilson, S. B., Milliken, S. *et al.* (1993). Ribozyme-mediated cleavage of the *bcr/abl* transcript expressed in chronic myeloid leukemia. *Exp. Hematol.* **21**, 1714–1718.

19. Pachuk, C. J., Yoon, K., Moelling, K. *et al.* (1994). Selective cleavage of *bcr–abl* chimeric RNAs by a ribozyme targeted to non-contiguous sequences. *Nucleic Acids Res.* **22**, 301–307.

20. Snyder, D. S., Wu, Y., Wang, J. L. *et al.* (1993). Ribozyme-mediated inhibition of *bcr–abl* gene expression in a Philadelphia chromosome-positive cell line. *Blood* **82**, 600–605.

21. James, H., Mills, K., and Gibson, I. (1996). Investigating and improving the specificity of ribozymes directed against the *bcr–abl* translocation. *Leukemia* **10**, 1054–1064.

22. Shore, S. K., Nabissa, P. M., and Reddy, E. P. (1993). Ribozyme-mediated cleavage of the *bcr–abl* oncogene transcript: in vitro cleavage of RNA and in vivo loss of P210 protein-kinase activity. *Oncogene* **8**, 3183–3188.

23. Kronenwett, R., Haas, R., and Sczakiel, G. (1996). Kinetic selectivity of complementary nucleic acids: *bcr–abl*-directed antisense RNA and ribozymes. *J. Mol. Biol.* **259**, 632–644.

24. Kearney, P., Wright, L. A., Milliken, S. *et al.* (1995). Improved specificity of ribozyme-mediated cleavage of *bcr–abl* mRNA. **23**, 986–989.

25. Lange, W., Cantin, E. M., Finke, J. *et al.* (1993). In vitro and in vivo effects of synthetic ribozymes targeted against BCR/ABL mRNA. *Leukemia* **7**, 1786–1794.

26. Snyder, D. S., Wu, Y., McMahon, R. *et al.* (1997). Ribozyme-mediated inhibition of a Philadelphia chromosome-positive acute lymphoblastic leukemia cell line expressing the p190 *bcr–abl* oncogene. *Biol. Blood Marrow Transplant.* **3**, 179–186.

27. Wright, L. A., Milliken, S., Biggs, J. C. *et al.* (1998). Ex vivo effects associated with the expression of a *bcr–abl*-specific ribozyme in a CML cell line. *Antisense Nucleic Acid Drug Develop.* **8**, 15–23.

28. Pace, U., Bockman, J. M., MacKay, B. J. *et al.* (1994). A ribozyme which discriminates in vitro between PML/RAR alpha, the t(15;17)-associated fusion RNA of acute promyelocytic leukemia, and PML and RAR alpha, the transcripts from the nonrearranged alleles. *Cancer Res.* **54**, 6365–6369.

29. Nason-Burchenal, K., Takle, G., Pace, U. *et al.* (1998). Targeting the PML/RAR alpha translocation product triggers apoptosis in promyelocytic leukemia cells. *Oncogene* **17**, 1759–1768.

30. Nason-Burchenal, K., Allopenna, J., Begue, A. *et al.* (1998). Targeting of PML/RARalpha is lethal to retinoic acid-resistant promyelocytic leukemia cells. *Blood* **92**, 1758–1767.

31. Koeffler, H. P. (1987). Syndromes of acute nonlymphocytic leukemia. *Ann. Intern. Med.* **107**, 748–758.

32. Friedman, A. D. (1999). Leukemogenesis by CBF oncoproteins. *Leukemia* **13**, 1932–1942.

33. Matsushita, H., Kobayashi, H., Mori, S. *et al.* (1995). Ribozymes cleave the AML1/MTG8 fusion transcript and inhibit proliferation of leukemic cells with t(8;21). *Biochem. Biophys. Res. Commun.* **215**, 431–437.

34. Kozu, T., Sueoka, E., Okabe, S. *et al.* (1996). Designing of chimeric DNA/RNA hammerhead ribozymes to be targeted against AML1/MTG8 mRNA. *J. Cancer Res. Clin. Oncol.* **122**, 254–256.

35. Matsushita, H., Kizaki, M., Kobayashi, H. *et al.* (1999). Induction of apoptosis in myeloid leukaemic cells by ribozymes targeted against AML1/MTG8. *Br. J. Cancer* **79**, 1325–1331.

36. Kiefer, P. E., Bepler, G., Kubasch, M. *et al.* (1987). Amplification and expression of protooncogenes in human small cell lung cancer cell lines. *Cancer Res.* **47**, 6236–6242.

37. Almoguera, C., Shibata, D., Forrester, K. *et al.* (1988). Most human carcinomas of the exocrine pancreas contain mutant c-K-*ras* genes. *Cell* **53**, 549–554.

38. Forrester, K., Almoguera, C., Han, K. *et al.* (1987). Detection of high incidence of K-*ras* oncogenes during human colon tumorigenesis. *Nature* **327**, 298–303.

39. Rodenhuis, S., Slebos, R. J., Boot, A. J. *et al.* (1988). Incidence and possible clinical significance of K-*ras* oncogene activation in adenocarcinoma of the human lung. *Cancer Res.* **48**, 5738–5741.

40. Ball, N. J., Yohn, J. J., Morelli, J. G. *et al.* (1994). Ras mutations in human melanoma: a marker of malignant progression. *J. Invest. Dermatol.* **102**, 285–290.

41. Li, M., Lonial, H., Citarella, R. *et al.* (1996). Tumor inhibitory activity of anti-*ras* ribozymes delivered by retroviral gene transfer. *Cancer Gene Ther.* **3**, 221–229.

42. Koizumi, M., Kamiya, H., and Ohtsuka, E. (1993). Inhibition of c-Ha-*ras* gene expression by hammerhead ribozymes containing a stable C(UUCG)G hairpin loop. *Biol. Pharm. Bull.* **16**, 879–883.

43. Kashani-Sabet, M., Funato, T., Tone, T. *et al.* (1992). Reversal of the malignant phenotype by an anti-*ras* ribozyme. *Antisense Res. Dev.* **2**, 3–15.

44. Scherr, M., Grez, M., Ganser, A. *et al.* (1997). Specific hammerhead ribozyme-mediated cleavage of mutant N-*ras* mRNA in vitro and ex vivo. Oligoribonucleotides as therapeutic agents. *J. Biol. Chem.* **272**, 14304–14313.

45. Tsuchida, T., Kijima, H., Oshika, Y. *et al.* (1998). Hammerhead ribozyme specifically inhibits mutant K-*ras* mRNA of human pancreatic cancer cells. *Biochem. Biophys. Res. Commun.* **253**, 368–373.

46. Koizumi, M., Kamiya, H., and Ohtsuka, E. (1992). Ribozymes designed to inhibit transformation of NIH3T3 cells by the activated c-Ha-*ras* gene. *Gene* **117**, 179–184.

47. Feng, M., Cabrera, G., Deshane, J. *et al.* (1995). Neoplastic reversion accomplished by high-efficiency adenoviral-mediated delivery of an anti-*ras* ribozyme. *Cancer Res.* **55**, 2024–2028.

48. Ohta, Y., Kijima, H., Kashani-Sabet, M. *et al.* (1996). Suppression of the malignant phenotype of melanoma cells by anti-oncogene ribozymes. *J. Invest. Dermatol.* **106**, 275–280.

49. Ohta, Y., Kijima, H., Ohkawa, T. *et al.* (1996). Tissue-specific expression of an anti-*ras* ribozyme inhibits proliferation of human malignant melanoma cells. *Nucleic Acids Res.* **24**, 938–942.

50. Eastham, J. A., and Ahlering, T. E. (1996). Use of an anti-*ras* ribozyme to alter the malignant phenotype of a human bladder cancer cell line. *J. Urol.* **156**, 1186–1188.

51. Chang, M. Y., Won, S. J., and Liu, H. S. (1997). A ribozyme specifically suppresses transformation and tumorigenicity of Ha-*ras*-oncogene-transformed NIH/3T3 cell lines. *J. Cancer Res. Clin. Oncol.* **123**, 91–99.

52. Irie, A., Anderegg, B., Kashani-Sabet, M. *et al.* (1999). Therapeutic efficacy of an adenovirus-mediated anti-H-*ras* ribozyme in experimental bladder cancer. *Antisense Nucleic Acid Drug Dev.* **9,** 341–349.

53. Kashani-Sabet, M., Funato, T., Florenes, V. A. *et al.* (1994). Suppression of the neoplastic phenotype in vivo by an anti-*ras* ribozyme. *Cancer Res.* **54,** 900–902.

54. Funato, T., Shitara, T., Tone, T. *et al.* (1994). Suppression of H-*ras*-mediated transformation in NIH3T3 cells by a *ras* ribozyme. *Biochem. Pharmacol.* **48,** 1471–1475.

55. Tone, T., Kashani, S. M., Funato, T. *et al.* (1993). Suppression of EJ cells tumorigenicity. *In Vivo* **7,** 471–476.

56. Kijima, H., and Scanlon, K. J. (2000). Ribozyme as an approach for growth suppression of human pancreatic cancer. *Mol. Biotechnol.* **14,** 59–72.

57. Ciesielski, M. J., and Fenstermaker, R. A. (2000). Oncogenic epidermal growth factor receptor mutants with tandem duplication: gene structure and effects on receptor function. *Oncogene* **19,** 810–820.

58. Yamazaki, H., Kijima, H., Ohnishi, Y. *et al.* (1998). Inhibition of tumor growth by ribozyme-mediated suppression of aberrant epidermal growth factor receptor gene expression. *J. NaH. Cancer Inst.* **90,** 581–587.

59. Halatsch, M. E., Schmidt, U., Botefur, I. C. *et al.* (2000). Marked inhibition of glioblastoma target cell tumorigenicity in vitro by retrovirus-mediated transfer of a hairpin ribozyme against deletion-mutant epidermal growth factor receptor messenger RNA. *J. Neurosurg.* **92,** 297–305.

60. Pinkas-Kramarski, R., Alroy, I., and Yarden, Y. (1997). ErbB receptors and EGF-like ligands: cell lineage determination and oncogenesis through combinatorial signaling. *J. Mammary Gland Biol. Neoplasia* **2,** 97–107.

61. Ross, J. S., and Fletcher, J. (1999). A. HER-2/*neu* (c-*erb-B2*) gene and protein in breast cancer. *Am. J. Clin. Pathol.* **112,** S53–S67.

62. Juhl, H., Downing, S., Hssieh, S. *et al.* (1997). Ribozyme targeting reduces *erb-B2* expression of SKOV3 carcinoma cells and inhibits tumor growth in vivo. *Langenbecks Arch. Chir. Suppl. Kongressbd.* **114,** 41–45.

63. Czubayko, F., Downing, S. G., Hsieh, S. S. *et al.* (1997). Adenovirus-mediated transduction of ribozymes abrogates HER-2/*neu* and pleiotrophin expression and inhibits tumor cell proliferation. *Gene Ther.* **4,** 943–949.

64. Wiechen, K., Zimmer, C., and Dietel, M. (1998). Selection of a high activity c-*erbB-2* ribozyme using a fusion gene of c-*erbB-2* and the enhanced green fluorescent protein. *Cancer Gene Ther.* **5,** 45–51.

65. Suzuki, T., Anderegg, B., Ohkawa, T. *et al.* (2000). Adenovirus-mediated ribozyme targeting of HER-2/*neu* inhibits in vivo growth of breast cancer cells. *Gene Ther.* **7,** 241–248.

66. Hsieh, S. S., Malerczyk, C., Aigner, A. *et al.* (2000). *erbB*-2 expression is rate-limiting for epidermal growth factor-mediated stimulation of ovarian cancer cell proliferation. *Int. J. Cancer* **86,** 644–651.

67. Tang, C. K., Goldstein, D. J., Payne, J. *et al.* (1998). *erbB*-4 ribozymes abolish neuregulin-induced mitogenesis. *Cancer Res.* **58,** 3415–3422.

68. Tang, C. K., Concepcion, X. Z., Milan, M. *et al.* (1999). Ribozyme-mediated down-regulation of *erbB*-4 in estrogen receptor-positive breast cancer cells inhibits proliferation both in vitro and in vivo. *Cancer Res.* **59,** 5315–5322.

69. Lavrovsky, Y., Tyagi, R. K., Chen, S. *et al.* (1999). Ribozyme-mediated cleavage of the estrogen receptor messenger RNA and inhibition of receptor function in target cells. *Mol. Endocrinol.* **13,** 925–934.

70. Chen, S., Song, C. S., Lavrovsky, Y. *et al.* (1998). Catalytic cleavage of the androgen receptor messenger RNA and functional inhibition of androgen receptor activity by a hammerhead ribozyme. *Mol. Endocrinol.* **12,** 1558–1566.

71. Yokoyama, Y., Morishita, S., Takahashi, Y. *et al.* (1997). Modulation of c-*fms* proto-oncogene in an ovarian carcinoma cell line by a hammerhead ribozyme. *Br. J. Cancer* **76,** 977–982.

72. Parthasarathy, R., Cote, G. J., and Gagel, R. F. (1999). Hammerhead ribozyme-mediated inactivation of mutant RET in medullary thyroid carcinoma. *Cancer Res.* **59,** 3911–3914.

73. Gottesman, M. M. (1993). How cancer cells evade chemotherapy: sixteenth Richard and Hinda Rosenthal Foundation Award lecture. *Cancer Res.* **53,** 747–754.

74. Palfner, K., Kneba, M., Hiddemann, W. *et al.* (1995). Improvement of hammerhead ribozymes cleaving *mdr*-1 mRNA. *Biol. Chem. Hoppe Seyler* **376,** 289–295.

75. Holm, P. S., Scanlon, K. J., and Dietel, M. (1994). Reversion of multidrug resistance in the P-glycoprotein-positive human pancreatic cell line (EPP85–181RDB) by introduction of a hammerhead ribozyme. *Br. J. Cancer* **70,** 239–243.

76. Scanlon, K. J., Ishida, H., and Kashani, S. M. (1994). Ribozyme-mediated reversal of the multidrug-resistant phenotype. *Proc. Natl. Acad. Sci. USA* **91,** 11123–11127.

77. Kobayashi, H., Dorai, T., Holland, J. F. *et al.* (1994). Reversal of drug sensitivity in multidrug-resistant tumor cells by an MDR1 (PGY1) ribozyme. *Cancer Res.* **54,** 1271–1275.

78. Bertram, J., Palfner, K., Killian, M. *et al.* (1995). Reversal of multiple drug resistance in vitro by phosphorothioate oligonucleotides and ribozymes. *Anticancer Drugs* **6,** 124–134.

79. Daly, C., Coyle, S., McBride, S. *et al.* (1996). *mdr1* ribozyme mediated reversal of the multi-drug resistant phenotype in human lung cell lines. *Cytotechnology* **19,** 199–205.

80. Matsushita, H., Kizaki, M., Kobayashi, H. *et al.* (1998). Restoration of retinoid sensitivity by MDR1 ribozymes in retinoic acid-resistant myeloid leukemic cells. *Blood* **91,** 2452–2458.

81. Masuda, Y., Kobayashi, H., Holland, J. F. *et al.* (1998). Reversal of multidrug resistance by a liposome-MDR1 ribozyme complex. *Cancer Chemother. Pharmacol.* **42,** 9–16.

82. Wang, F. S., Kobayashi, H., Liang, K. W. *et al.* (1999). Retrovirus-mediated transfer of anti-MDR1 ribozymes fully restores chemosensitivity of P-glycoprotein-expressing human lymphoma cells. *Hum. Gene Ther.* **10,** 1185–1195.

83. Kobayashi, H., Takemura, Y., Wang, F. S. *et al.* (1999). Retrovirus-mediated transfer of anti-MDR1 hammerhead ribozymes into multidrug-resistant human leukemia cells: screening for effective target sites. *Int. J. Cancer* **81,** 944–950.

84. Gao, Z., Fields, J. Z., and Boman, B. M. (1999). Tumor-specific expression of anti-*mdr1* ribozyme selectively restores chemosensitivity in multidrug-resistant colon-adenocarcinoma cells. *Int. J. Cancer* **82,** 346–352.

85. Kashani-Sabet, M., Lu, Y., Leong, L. *et al.* (1990). Differential oncogene amplification in tumor cells from a patient treated with cisplatin and 5-fluorouracil. *Eur. J. Cancer* **26,** 383–390.

86. Scanlon, K. J., Jiao, L., Funato, T. *et al.* (1991). Ribozyme-mediated cleavage of c-*fos* mRNA reduces gene expression of DNA synthesis enzymes and metallothionein. *Proc. Natl. Acad. Sci. USA* **88,** 10591–10595.

87. Funato, T., Yoshida, E., Jiao, L. *et al.* (1992). The utility of an anti-*fos* ribozyme in reversing cisplatin resistance in human carcinomas. *Adv. Enzyme Regul.* **32,** 195–209.

88. Funato, T., Ishii, T., Kanbe, M. *et al.* (1997). Reversal of cisplatin resistance in vivo by an anti-*fos* ribozyme. *In Vivo* **11,** 217–220.

89. Ransone, L. J., and Verma, I. M. (1990). Nuclear proto-oncogenes *fos* and *jun*. *Annu. Rev. Cell Biol.* **6,** 539–557.

90. Rauscher, F. J. D., Sambucetti, L. C., Curran, T. *et al.* (1988). Common DNA binding site for Fos protein complexes and transcription factor AP-1. *Cell* **52,** 471–480.

91. Ueda, K., Pastan, I., and Gottesman, M. M. (1987). Isolation and sequence of the promoter region of the human multidrug-resistance (P-glycoprotein) gene. *J. Biol. Chem.* **262,** 17432–17436.

92. Teeter, L. D., Eckersberg, T., Tsai, Y. *et al.* (1991). Analysis of the Chinese hamster P-glycoprotein/multidrug resistance gene *pgp1* reveals that the AP-1 site is essential for full promoter activity. *Cell Growth Differ.* **2,** 429–437.

93. Shen, D. W., Fojo, A., Chin, J. E. *et al.* (1986). Human multidrug-resistant cell lines: increased *mdr1* expression can precede gene amplification. *Science* **232,** 643–645.

94. Ge, L., Resnick, N. M., Ernst, L. K. *et al.* (1995). Gene therapeutic approaches to primary and metastatic brain tumors. II. Ribozyme-mediated suppression of CD44 expression. *J. Neurooncol.* **26,** 251–257.

95. Yamamoto, H., Irie, A., Fukushima, Y. *et al.* (1996). Abrogation of lung metastasis of human fibrosarcoma cells by ribozyme-mediated suppression of integrin alpha6 subunit expression. *Int. J. Cancer* **65,** 519–524.

96. Hua, J., Muschel, and R. J. (1996). Inhibition of matrix metalloproteinase 9 expression by a ribozyme blocks metastasis in a rat sarcoma model system. *Cancer Res.* **56,** 5279–5284.

97. Sehgal, G., Hua, J., Bernhard, E. J. *et al.* (1998). Requirement for matrix metalloproteinase-9 (gelatinase B) expression in metastasis by murine prostate carcinoma. *Am. J. Pathol.* **152,** 591–596.

98. Maelandsmo, G. M., Hovig, E., Skrede, M. *et al.* (1996). Reversal of the in vivo metastatic phenotype of human tumor cells by an anti-*CAPL* (*mts1*) ribozyme. *Cancer Res.* **56,** 5490–5498.

99. Bjornland, K., Winberg, J. O., Odegaard, O. T. *et al.* (1999). S100A4 involvement in metastasis: deregulation of matrix metalloproteinases and tissue inhibitors of matrix metalloproteinases in osteosarcoma cells transfected with an anti-S100A4 ribozyme. *Cancer Res.* **59,** 4702–4708.

100. Weidner, N. (1995). Intratumor microvessel density as a prognostic factor in cancer. *Am. J. Pathol.* **147,** 9–19.

101. Li, Y. S., Milner, P. G., Chauhan, A. K. *et al.* (1990). Cloning and expression of a developmentally regulated protein that induces mitogenic and neurite outgrowth activity. *Science* **250,** 1690–1694.

102. Czubayko, F., Riegel, A. T., and Wellstein, A. (1994). Ribozyme targeting elucidates a direct role of pleiotrophin in tumor growth. *J. Biol. Chem.* **269,** 21358–21363.

103. Czubayko, F., Schulte, A. M., Berchem, G. J. *et al.* (1996). Melanoma angiogenesis and metastasis modulated by ribozyme targeting of the secreted growth factor pleiotrophin. *Proc. Natl. Acad. Sci. USA* **93,** 14753–14758.

104. McMahon, G. (2000). VEGF receptor signaling in tumor angiogenesis. *Oncologist* **5,** 3–10.

105. Nguyen, J. T. (2000). Vascular endothelial growth factor as a target for cancer gene therapy. *Adv. Exp. Med. Biol.* **465,** 447–456.

106. Pavco, P. A., Bouhana, K. S., Gallegos, A. M. *et al.* (2000). Antitumor and antimetastatic activity of ribozymes targeting the messenger RNA of vascular endothelial growth factor receptors. *Clin. Cancer Res.* **6,** 2094–2103.

107. Ke, L. D., Fueyo, J., Chen, X. *et al.* (1998). A novel approach to glioma gene therapy: down-regulation of the vascular endothelial growth factor in glioma cells using ribozymes. *Int. J. Oncol.* **12,** 1391–1396.

108. Czubayko, F., Liaudet-Coopman, E. D., Aigner, A. *et al.* (1997). A secreted FGF-binding protein can serve as the angiogenic switch in human cancer. *Nat. Med.* **3,** 1137–1140.

109. Chiu, C. P., and Harley, C. B. Replicative senescence and cell immortality: the role of telomeres and telomerase. *Proc. Soc. Exp. Biol. Med.* **214,** 99–106.

110. Feng, J., Funk, W. D., Wang, S. S. *et al.* (1995). The RNA component of human telomerase. *Science* **269,** 1236–1241.

111. Kim, N. W., Piatyszek, M. A., Prowse, K. R. *et al.* (1994). Specific association of human telomerase activity with immortal cells and cancer. *Science* **266,** 2011–2015.

112. Kanazawa, Y., Ohkawa, K., Ueda, K. *et al.* (1996). Hammerhead ribozyme-mediated inhibition of telomerase activity in extracts of human hepatocellular carcinoma cells. *Biochem. Biophys. Res. Commun.* **225,** 570–576.

113. Yokoyama, Y., Takahashi, Y., Shinohara, A. *et al.* (1998). Attenuation of telomerase activity by a hammerhead ribozyme targeting the template region of telomerase RNA in endometrial carcinoma cells. *Cancer Res.* **58,** 5406–5410.

114. Qu, Y., Liu, S., Zhang, C. *et al.* (1999). Study on the inhibition of nude mice transplantation tumor growth by telomerase ribozyme [transl.]. *Chung Hua I. Hsueh I. Chuan Hsueh Tsa Chih* **16,** 368–370.

115. Qu, Y., OuYang, X., Liu, S. *et al.* (1999). Study on telomerase inhibition by ribozyme targeted to telomerase RNA component [transl.]. *Chung Hua I. Hsueh I. Chuan Hsueh Tsa Chih* **16,** 133–137.

116. Folini, M., Colella, G., Villa, R. *et al.* (2000). Inhibition of telomerase activity by a hammerhead ribozyme targeting the RNA component of telomerase in human melanoma cells. *J. Invest. Dermatol.* **114,** 259–267.

117. Yokoyama, Y., Takahashi, Y., Shinohara, A. *et al.* (2000). The 5′-end of hTERT mRNA is a good target for hammerhead ribozyme to suppress telomerase activity. *Biochem. Biophys. Res. Commun.* **273,** 316–321.

118. Lu, Q. L., Abel, P., Foster, C. S. *et al.* (1996). *bcl-2*: role in epithelial differentiation and oncogenesis. *Hum. Pathol.* **27,** 102–110.

119. Dorai, T., Olsson, C. A., Katz, A. E. *et al.* (1997). Development of a hammerhead ribozyme against *bcl-2*. 1. Preliminary evaluation of a potential gene therapeutic agent for hormone-refractory human prostate cancer. *Prostate* **32,** 246–258.

120. Dorai, T., Perlman, H., Walsh, K. *et al.* (1999). A recombinant defective adenoviral agent expressing anti-*bcl-2* ribozyme promotes apoptosis of *bcl-2*-expressing human prostate cancer cells. *Int. J. Cancer* **82,** 846–852.

121. Gibson, S. A., Pellenz, C., Hutchison, R. E. *et al.* (2000). Induction of apoptosis in oral cancer cells by an anti-*bcl-2* ribozyme delivered by an adenovirus vector. *Clin. Cancer Res.* **6,** 213–222.

122. Li, W., Zhang, J., Flechner, L. *et al.* (1999). Protein kinase C-alpha overexpression stimulates Akt activity and suppresses apoptosis induced by interleukin 3 withdrawal. *Oncogene* **18,** 6564–6572.

123. Sioud, M., and Sorensen, D. R. (1998). A nuclease-resistant protein kinase C alpha ribozyme blocks glioma cell growth. *Nat. Biotechnol.* **16,** 556–561.

124. Leirdal, M., and Sioud, M. (1999). Ribozyme inhibition of the protein kinase C alpha triggers apoptosis in glioma cells. *Br. J. Cancer* **80,** 1558–1564.

125. Chen, Z., Kamath, P., Zhang, S. *et al.* (1995). Effectiveness of three ribozymes for cleavage of an RNA transcript from human papillomavirus type 18. *Cancer Gene Ther.* **2,** 263–271.

126. Alvarez-Salas, L. M., Cullinan, A. E., Siwkowski, A. *et al.* (1998). Inhibition of HPV-16 E6/E7 immortalization of normal keratinocytes by hairpin ribozymes. *Proc. Natl. Acad. Sci. USA* **95,** 1189–1194.

127. Huang, Y. Z., Kong, Y. Y., Wang, Y. *et al.* (1996). Identification of activity of anti-Hpv16 E7 ribozyme expressed in Cv-1 cells. *Chinese Sci. Bull.* **41,** 1291–1296.

128. Huang, S., Stupack, D., Mathias, P. *et al.* (1997). Growth arrest of Epstein–Barr virus immortalized B lymphocytes by adenovirus-delivered ribozymes. *Proc. Natl. Acad. Sci. USA* **94,** 8156–8161.

129. Di Bisceglie, A. M., Rustgi, V. K., Hoofnagle, J. H. *et al.* (1988). NIH conference. Hepatocellular carcinoma. *Ann. Intern. Med.* **108,** 390–401.

130. Johnson, P. J. (1996). The epidemiology of hepatocellular carcinoma. *Eur. J. Gastroenterol. Hepatol.* **8,** 845–849.

131. von Weizsacker, F., Blum, H. E., and Wands, J. R. (1992). Cleavage of hepatitis B virus RNA by three ribozymes transcribed from a single DNA template. *Biochem. Biophys. Res. Commun.* **189,** 743–748.

132. Welch, P. J., Tritz, R., Yei, S. *et al.* (1997). Intracellular application of hairpin ribozyme genes against hepatitis B virus. *Gene Ther.* **4,** 736–743.

133. Kim, Y. K., Junn, E., Park, I. *et al.* (1999). Repression of hepatitis B virus X gene expression by hammerhead ribozymes. *Biochem. Biophys. Res. Commun.* **257,** 759–765.

134. Weinberg, M., Passman, M., Kew, M. *et al.* (2000). Hammerhead ribozyme-mediated inhibition of hepatitis B virus X gene expression in cultured cells. *J. Hepatol.* **33,** 142–151.

135. Sakamoto, N., and Wu, C. H., and Wu, G. Y. (1996). Intracellular cleavage of hepatitis C virus RNA and inhibition of viral protein translation by hammerhead ribozymes. *J. Clin. Invest.* **98,** 2720–2728.

136. Ohkawa, K., Yuki, N., Kanazawa, Y. *et al.* (1997). Cleavage of viral RNA and inhibition of viral translation by hepatitis C virus RNA-specific hammerhead ribozyme in vitro. *J. Hepatol.* **27,** 78–84.

137. Lieber, A., He, C. Y., Polyak, S. J. *et al.* (1996). Elimination of hepatitis C virus RNA in infected human hepatocytes by adenovirus-mediated expression of ribozymes. *J. Virol.* **70,** 8782–8791.

138. Macejak, D. G., Jensen, K. L., Jamison, S. F. *et al.* (2000). Inhibition of hepatitis C virus (HCV)-RNA-dependent translation and replication of a chimeric HCV poliovirus using synthetic stabilized ribozymes. *Hepatology* **31,** 769–776.

139. Welch, P. J., Tritz, R., Yei, S. *et al.* (1996). A potential therapeutic application of hairpin ribozymes: in vitro and in vivo studies of gene therapy for hepatitis C virus infection. *Gene Ther.* **3,** 994–1001.

140. Lohmann, V., Korner, F., Koch, J. *et al.* (1999). Replication of subgenomic hepatitis C virus RNAs in a hepatoma cell line [see comments]. *Science* **285,** 110–113.

141. Leavitt, M. C., Yu, M., Wong-Staal, F. *et al.* (1996). Ex vivo transduction and expansion of CD4+ lymphocytes from HIV+ donors: prelude to a ribozyme gene therapy trial. *Gene Ther.* **3,** 599–606.

142. Rowe, P. M. (1996). Ribozymes enter clinical trials for HIV-1 treatment [news]. *Lancet* **348,** 1302.

143. Berzal-Herranz, A., Joseph, S., and Burke, J. M. (1992). In vitro selection of active hairpin ribozymes by sequential RNA-catalyzed cleavage and ligation reactions. *Genes Dev.* **6,** 129–134.

144. Barroso-del Jesus, A., Tabler, M., and Berzal-Herranz, A. (1999). Comparative kinetic analysis of structural variants of the hairpin ribozyme reveals further potential to optimize its catalytic performance. *Antisense Nucleic Acid Drug Dev.* **9,** 433–440.

145. Juneau, K., and Cech, T. R. (1999). In vitro selection of RNAs with increased tertiary structure stability. *RNA* **5,** 1119–1129.

146. Joseph, S., Berzal-Herranz, A., and Chowrira, B. M. *et al.* (1993). Substrate selection rules for the hairpin ribozyme determined by in vitro selection, mutation, and analysis of mismatched substrates. *Genes Dev.* **7,** 130–138.

147. Ho, S. P., Britton, D. H., Stone, B. A. *et al.* (1996). Potent antisense oligonucleotides to the human multidrug resistance-1 mRNA are rationally selected by mapping RNA-accessible sites with oligonucleotide libraries. *Nucleic Acids Res.* **24,** 1901–1907.

148. Vaish, N. K., Heaton, P. A., and Eckstein, F. (1997). Isolation of hammerhead ribozymes with altered core sequences by in vitro selection. *Biochemistry* **36,** 6495–6501.

149. Yu, Q., Pecchia, D. B., Kingsley, S. L. *et al.* (1998). Cleavage of highly structured viral RNA molecules by combinatorial libraries of hairpin ribozymes. The most effective ribozymes are not predicted by substrate selection rules. *J. Biol. Chem.* **273,** 23524–23533.

150. Li, Y., and Breaker, R. R. (1999). Deoxyribozymes: new players in the ancient game of biocatalysis. *Curr. Opin. Struct. Biol.* **9,** 315–323.

151. Kuwabara, T., Warashina, M., Tanabe, T. *et al.* (1998). A novel allosterically *trans*-activated ribozyme, the maxizyme, with exceptional specificity in vitro and in vivo. *Mole. Cell* **2,** 617–627.

152. Kuwabara, T., Warashina, M., Orita, M. *et al.* (1998). Formation of a catalytically active dimer by tRNA(Val)-driven short ribozymes. *Nat. Biotechnol.* **16,** 961–965.

153. Schmidt, C., Welz, R., and Muller, S. (2000). RNA double cleavage by a haripin-derived twin ribozyme. *Nucleic Acids Res.* **28,** 886–894.

154. Ruiz, J., Wu, C. H., Ito, Y. *et al.* (1997). Design and preparation of a multimeric self-cleaving hammerhead ribozyme. *Biotechniques* **22,** 338–345.

155. Benedict, C. M., Pan, W. H., Loy, S. E. *et al.* (1998). Triple ribozyme-mediated down-regulation of the retinoblastoma gene. *Carcinogenesis* **19,** 1223–1230.

156. Ohkawa, J., Yuyama, N., Takebe, Y. *et al.* (1993). Importance of independence in ribozyme reactions: kinetic behavior of trimmed and of simply connected multiple ribozymes with potential activity against human immunodeficiency virus. *Proc. Natl. Acad. Sci. USA* **90,** 11302–11306.

157. Yu, M., Ojwang, J., Yamada, O. *et al.* (1993). A hairpin ribozyme inhibits expression of diverse strains of human immunodeficiency virus type 1. *Proc. Natl. Acad. Sci. USA* **90,** 6340–6344.

158. Fraisier, C., Irvine, A., Wrighton, C. *et al.* (1998). High level inhibition of HIV replication with combination RNA decoys expressed from an HIV-Tat inducible vector. *Gene Ther.* **5,** 1665–1676.

159. Bramlage, B., Alefelder, S., Marschall, P. *et al.* (1999). Inhibition of luciferase expression by synthetic hammerhead ribozymes and their cellular uptake. *Nucleic Acids Res.* **27,** 3159–3167.

160. Sullenger, B. A. (1995). Colocalizing ribozymes with substrate RNAs to increase their efficacy as gene inhibitors. *Appl. Biochem. Biotechnol.* **54,** 57–61.

161. Lee, N. S., Bertrand, E., and Rossi, J. (1999). mRNA localization signals can enhance the intracellular effectiveness of hammerhead ribozymes. *RNA* **5,** 1200–1209.

162. Kintner, R. L., and Hosick, H. L. (1998). Reduction of Cripto-1 expression by a hammerhead-shaped RNA molecule results from inhibition of translation rather than mRNA cleavage. *Biochem. Biophys. Res. Commun.* **245,** 774–779.

163. Iversen, P. O., and Sioud, M. (1998). Modulation of granulocyte-macrophage colony-stimulating factor gene expression by a tumor necrosis factor specific ribozyme in juvenile myelomonocytic leukemic cells. *Blood* **92,** 4263–4268.

164. Sato, K., Nishi, T., Takeshima, H. *et al.* (1999). Expression of p120 nucleolar proliferating antigen in human gliomas and growth suppression of glioma cells by p120 ribozyme vector. *Int. J. Oncol.* **14,** 417–424.

165. Xu, Z. D., Oey, L., Mohan, S. *et al.* (1999). Hammerhead ribozyme-mediated cleavage of the human insulin-like growth factor-II ribonucleic acid in vitro and in prostate cancer cells. *Endocrinology* **140,** 2134–2144.

166. Shelburne, C. P., and Huff, T. F. (1999). Inhibition of *kit* expression in P815 mouse mastocytoma cells by a hammerhead ribozyme. *Clin. Immunol.* **93,** 46–58.

167. Abounader, R., Ranganathan, S., Lal, B. *et al.* (1999). Reversion of human glioblastoma malignancy by U1 small nuclear RNA/ribozyme targeting of scatter factor/hepatocyte growth factor and c-*met*. *J. Natl. Cancer Inst.* **91,** 1548–1556.

168. Nylandsted, J., Rohde, M., Bartek, J. *et al.* (1998). Expression of a p16INK4a-specific ribozyme downmodulates p16INK4a abundance and accelerates cell proliferation. *FEBS Lett.* **436,** 41–45.

169. Irminger-Finger, I., Soriano, J. V., Vaudan, G. *et al.* (1998). In vitro repression of Brca1-associated RING domain gene, Bard1, induces phenotypic changes in mammary epithelial cells. *J. Cell Biol.* **143,** 1329–1339.

170. Kruger, M., Beger, C., Li, Q. X. *et al.* (2000). Identification of eIF2Bgamma and eIF2gamma as cofactors of hepatitis C virus internal ribosome entry site-mediated translation using a functional genomics approach. *Proc. Natl. Acad. Sci. USA* **97,** 8566–8571.

171. Welch, P. J., Marcusson, E. G., Li, Q. X. *et al.* (2000). Identification and validation of a gene involved in anchorage-independent cell growth control using a library of randomized hairpin ribozymes. *Genomics* **66,** 274–283.

172. Li, Q. X., Robbins, J. M., Welch, P. J. *et al.* (2000). A novel functional genomics approach identifies mTERT as a suppressor of fibroblast transformation. *Nucleic Acids Res.* **28,** 2605–2612.

173. Hertel, K. J., Pardi, A., Uhlenbeck, O. C. *et al.* (1992). Numbering system for the hammerhead. *Nucleic Acids Res.* **20,** 3252.

174. Earnshaw, D. J., and Gait, M. J. (1997). Progress toward the structure and therapeutic use of the hairpin ribozyme. *Antisense Nucleic Acid Drug Dev.* **7,** 403–411.

C H A P T E R

6

The Advent of Lentiviral Vectors: Prospects for Cancer Therapy

MICHEL SADELAIN

Laboratory of Gene Transfer and Gene Expression
Department of Medicine and Immunology Program
Memorial Sloan–Kettering Cancer Center
New York, New York 10021

ISABELLE RIVIÈRE

Laboratory of Gene Transfer and Gene Expression
Department of Medicine and Immunology Program
Memorial Sloan–Kettering Cancer Center
New York, New York 10021

I. INTRODUCTION

Recombinant lentiviruses are the latest category of viral vectors to enter the gene transfer arena. Following in the footsteps of the oncoretroviruses, they represent the second instance of Retroviridae being solicited as gene transfer tools. Like vectors derived from oncoretroviruses and adeno-associated viruses, lentiviral vectors integrate in host-cell chromosomes. They are therefore relevant to gene transfer in cells that undergo clonal expansion and for applications that require long-term transgene expression. The successful elimination of viral open reading frames from the recombinant vectors offers the prospect of nonimmunogenic approaches to gene delivery (notwithstanding the possible immunogenicity of the transgene product itself and that of the virion-

associated antigens that enter cells at the time of infection). Stable integration and reduced immunogenicity are properties shared with adeno-associated and oncoretroviral vectors. The attractiveness of lentiviral vectors stems from their capacity to transduce non dividing cells, either *ex vivo* or *in vivo,* as well as their stable and relatively large genomic capacity. By accommodating large inserts of genetic material, including genes and genomic regulatory elements, they hold the promise to help improve regulation of transgene expression.

The potential advantages of lentiviral vectors are reviewed and discussed in this chapter. They are often compared to oncoretroviral vectors, which are reviewed in Chapter 1. The use of replication-defective mammalian oncoretroviruses goes back some 20 years [1]. Within a decade, packaging systems used to generate recombinant viral particles with human tropism and improved safety features became available [2–4]. The use of replication-defective lentiviruses as tools for gene transfer is much more recent [5,6]. Owing to the guidance provided by 15 years of experience with oncoretroviral vectors, progress in developing lentiviral vectors has been much facilitated. Nonetheless, the greater complexity of lentiviral genomes and associated biosafety concerns hindered some of the earlier developments. It has only been 5 years since a relatively simple and safe methodology was established for the generation of replication-defective vectors derived from human immunodeficiency virus-1 (HIV-1) [7]. Not surprisingly, HIV-1-derived vectors are the most investigated to date. HIV-1 is the best known of all lentiviruses in terms of gene function, gene regulation, pathogenesis, and treatment. Other lentiviruses, such as feline immunodeficiency virus (FIV), simian immunodeficiency virus (SIV), and equine infectious anemia virus (EIAV), also warrant investigation as replication-defective vectors. However, the overall knowledge of these non-human pathogens is, at this time, less advanced. This chapter will focus on vectors derived from HIV-1, taken to represent a paradigm for all lentiviral vectors. Three general questions are addressed: What are the potential advantages of lentiviral vectors? What is the current status of technologies for generating lentiviral vectors? What applications could lentiviral vectors find in cancer therapy?

II. STRUCTURE AND FUNCTION OF LENTIVIRUSES

A. Classification

Lentiviruses are one of the seven genera amongst the Retroviridae (Table 1). They include exogenous viruses of humans, primates, domestic cats, and a variety of livestock (sheep, cattle, horses). No isolates from rodents have been described, nor have any closely related endogenous viruses [8]. Infection by lentiviruses can result in a variety of diseases, including immunodeficiencies, neurological degeneration, and arthritis.

Similarly to oncoretroviruses, lentiviruses are enveloped particles that bud from the plasma membrane. The cores of mature virions have a distinctive, truncated, cone-like shape. In addition to the Gag, Pol, and Env proteins present in all Retroviridae, lentiviruses express at least two additional regulatory proteins (Tat, Rev). The primate lentiviruses produce other accessory proteins, such as Nef, Vpr, Vpu, and Vif. The different groups of retroviruses (Table 1) are recognized based on serology and genome organization and correspond to particular host ranges of the viruses. The genomic structure of the most studied lentivirus, HIV-1, is represented in Fig. 1 next to that of one of the most studied oncoretroviruses, the Moloney murine leukemia virus (MoMLV).

TABLE 1   Classification and Size of Retroviruses

Genus	Example	Genome (nucleotides)
Avian sarcoma and leukosis viral group	Rous sarcoma virus (RSV)	9,300
Mammalian B-type viral group	Mouse mammary tumor virus (MMTV)	8,600
Murine leukemia-related viral group	Moloney murine leukemia virus (MoMLV)	8,300
Human T-cell leukemia, bovine leukemia viral group	Human T-cell leukemia virus-1 (HTLV-I)	8,500
D-type viral group	Mason–Pfizer monkey virus (M–PMV)	7,500
Lentiviruses	Human immune deficiency virus-1 (HIV-1)	9,200
Spumaviruses	Human foamy virus (HFV)	11,200

Note: See references 8, 12, and 15 for additional information.

B. Retroviral Life Cycle and Gene Function

Infectious particles bind to their target cell via a receptor, the expression of which determines the host range of the infectious particle. Receptor selection is determined by the sequence of the *env*-encoded glycoprotein. Ecotropic particles bind to an amino-acid tranporter expressed in mouse and rat cells, and amphotropic particles to a phosphate transporter present on numerous mammalian cells, including rodent and primate cells [9]. HIV-1 binds to CD4 and chemokine receptors [10]. After fusion of the viral and cellular membranes, the core particle enters the cytoplasm, where the single-stranded RNA retroviral genome is reverse-transcribed into a double-stranded DNA copy bearing direct long terminal repeats (LTRs) at each end. After transport to the nucleus, the viral DNA integrates into the host-cell genome. The roles of different proteins encoded by retroviral genes are briefly summarized below.

1. Gag Proteins

The matrix (MA) protein is a structural protein derived from the aminoterminal domain of the Gag polyprotein. MA is myristylated and associated with the inside of the virus envelope [11]. The capsid (CA) is the main structural protein in the virion core and it is derived from the central region of Gag. The nucleocapsid (NC), derived from the carboxy-terminal domain of Gag, is a basic protein that forms a ribonucleoprotein complex with the genomic RNA inside the core [12].

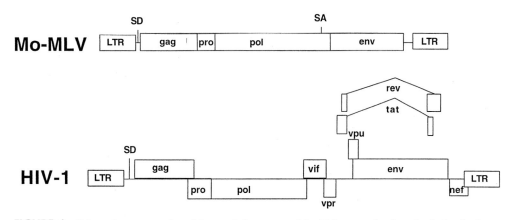

FIGURE 1   Schematic representation of the proviral structure of the Moloney strain of murine leukemia viruses (MoMLV) and of human immunodeficiency virus-1 (HIV-1). The essential function of each retroviral gene is summarized in the text. LTR: long terminal repeat; SD: splice donor; SA: splice acceptor. The HIV-1 genome comprises several additional SD and SA sites [12].

2. Pro and Pol Enzymes

The protease (PR) cleaves the Gag–Pro–Pol polyprotein to produce viral proteins in their mature forms [12]. The reverse transcriptase (RT) is a DNA polymerase that can copy RNA or DNA templates. It has RNAse H activity and uses a specific cellular tRNA primer to initiate minus-strand DNA synthesis. RNAse-H-resistant polypurine tracts (PPTs) are used to initiate plus-strand DNA synthesis [13]. The integrase (IN) inserts the linear double-stranded DNA copy of the retroviral genome into host chromosomal DNA, yielding the provirus.

3. Env Proteins

The surface (SU) subunit mediates viral adsorption by binding to a specific cell surface receptor (e.g., CD4 and chemokine receptors acting as coreceptors in the case of HIV-1) [10]. The carboxy terminus of SU is defined by a cellular furin protease cleavage site that separates SU from the transmembrane (TM) subunit. The associated TM subunit, encoded in the carboxy-terminal region of the *env* open reading frame, mediates virus entry by triggering virus–host cell membrane fusion [14].

4. Accessory Proteins

Lentiviruses use a variety of accessory proteins that regulate replication and infectivity [12,15–17]. Tat is a small *trans*-activator protein that activates transcription by binding to the element within the LTR. Rev binds to the Rev-responsive element (RRE), thereby facilitating the transport of unspliced or partially spliced mRNAs to the cytoplasm [18,19]. Nef is a myristylated intracellular protein that can reduce the level of cell surface CD4 expression and even-

tually promote cell division [20]. Viral protein r, or Vpr, is present in the virions. Vpr causes infected cells to arrest in G_2 and may also promote nuclear import of the preintegration complex. Viral protein u, or Vpu, appears to play a role in viral assembly and release. Virion infectivity factor, or Vif, also aids in the production of infectious virions [16,17].

III. FEATURES THAT DISTINGUISH LENTIVIRAL FROM ONCORETROVIRAL VECTORS

The potential usefulness and efficacy of any viral vectors are determined by the biological characteristics of the parental virus. Relevant features distinguishing lentiviral and oncoretroviral vectors are listed in Table 2 and summarized below.

TABLE 2   Relevant Features Distinguishing Lentiviral from Oncoretroviral Vectors

Feature	MLV/MLV-derived vector	HIV-1/HIV-1-derived vector
Cell-cycle requirement for productive infection:	S phase	Transition to G_1 phase
Preintegration complex stability:	~4-hour half-life	Greater than for MLV
Estimated maximum packaging size:	~7000 nucleotides	~9000 nucleotides or may be more
Reverse transcriptase fidelity *in vitro*:	~1/29,000 error rate	~1/5000 error rate

A. Cell-Cycle Requirements for Successful Transduction

Retroviral vectors derived from murine leukemia virus require that the target cell divide to allow proviral integration [21,22]. In contrast, lentiviruses such as HIV-1 are able to successfully infect nondividing cells [23]. Indeed, the HIV preintegration complex, which includes the integrase, the Vpr gene product, the matrix protein, and other elements [13], is able to translocate to the nucleus in the absence of nuclear membrane breakdown [23].

Unlike oncoretroviral vectors, HIV-based vectors can transduce nondividing, terminally differentiated cells such as neurons [24], retinal photoreceptors [25], and dendritic cells [26,27]. However, the potential advantages are not as sweeping as they may appear at first glance. Indeed, lentiviruses cannot efficiently transduce G_0 cells due to a block of reverse transcription [28]. Completion of reverse transcription requires progression through at least the G_{1b} stage of the cell cycle [29]. The RT enzyme requires sufficient cytoplasmic deoxynucleotide (dNTP) pools for efficient processive action to generate the viral DNA. Manipulation of the intracellular dNTP concentration can have dramatic effects on virus production. Thus, HIV-1 infection has been shown to be arrested at the stage of minus-strand synthesis in quiescent lymphocytes [30], possibly resulting from a low concentration of cytoplasmic dNTPs [31]. Exogenous addition of dNTPs prior to infection of these resting cells can improve reverse transcription of HIV-1 [32,33].

B. Vector Stability

The retroviral genome is made of RNA. Accordingly, the fate of recombinant retroviruses is highly dependent on the stability of their full-length RNA transcript. Inserted sequences that promote RNA splicing prior to encapsidation are one major cause of vector rearrangements. These rearrangements are relatively infrequent when cDNAs only are inserted into the vector. They are much more common when genes and other genomic elements are incorporated [34,35]. The genomic stability afforded by lentiviral vectors may be enhanced relative to that of oncoretroviral vectors [36]. Unlike oncoretroviral vectors, which rely on balanced splicing to generate different transcripts, lentiviral vectors regulate splicing of their own genomic transcripts. This property is important for the regulation of protein synthesis throughout the lentiviral life cycle and for the packaging of full-length, unspliced genomic transcripts [15]. The binding of the Rev protein to the RRE is essential in this regard [18,19]. The recent breakthrough in regulating ß-globin gene expression in hematopoietic cells is made possible in part by the stable packaging of large regulatory genomic elements in lentiviral vectors [37].

Another aspect of lentiviral vector stability occurs at the level of the preintegration complex. The intracellular half-life of this complex is short for MuLV, on the order of 4 hours [38], but longer for HIV-1 [39]. This level of stability may be extremely valuable, possibly enhancing gene transfer efficiency in certain cell types such as hematopoietic stem cells [40].

C. Packaging Sizes

The size of different retroviral genomes varies bewteen 7500 and 11200 nucleotides (Table 1). This is likely to translate into a greater packaging capacity for lentiviral vectors compared to MoMLV-based vectors. The latter allow for insertion of up to ~7000 nucleotides. The maximum packaging capacity of HIV-1-derived vectors has not yet been determined, but in principle it should be larger by at least another 1000 nucleotides. The human ß-globin gene (2.8 kb), along with 5 kb of additional genomic sequence, has been successfully transmitted without vector rearrangement (unpubl. observ.).

D. Reverse Transcriptase Mutation Rates

Retroviral vectors are reverse-transcribed by RT in the cytoplasm of the infected cells. The fidelity of this enzyme, which cannot correct errors by exonucleolytic proofreading like other DNA polymerases, is directly linked to the mutation rate and evolution of this class of viruses. The introduction of mutations in vector sequences represents a hindrance to safe and efficient gene delivery. The fidelity of RNA-dependent and DNA-dependent DNA polymerase activity of RT can be measured in various *in vitro* assays. MoMLV RT has an error rate of about one in 28–30,000 polymerized nucleotides [41–43]. HIV-1 RT has an error rate of one in 2–7000 [42–46], mostly G-to-A transition mutations [47]. This could result in multiple copying errors per genome per round of replication. Thus, according to these *in vitro* measurements, HIV-1 RT is substantially more error-prone than its MLV counterpart by a factor of 5–10. The fidelity of HIV-1 RT needs to be further assessed in transduced cells.

E. Vector Silencing

Oncoretroviruses and oncoretroviral vectors are highly susceptible to transcriptional inactivation and position effects imparted by chromosomal sequences at their integration site [48,49]. While silencing of viral and transposed elements may protect the genome from insertional mutagenesis [50], it also acts to repress the expression of retrovirally transduced genes [48,49,51–53]. Retrovirus expression is also greatly influenced by endogenous enhancers and heterochromatic regions near the integration site [54]. These effects hamper the correct regulation and sustained expression of retrovirally transduced genes and thus represent a major obstacle for the therapeutic use of recombinant retroviruses [48,49]. Retroviral methylation is commonly associated with transcriptional silencing [50,55–57]. Methylation of retroviral vectors occurs in a number of tissues and is likewise associated with decreased transgene expression *in vivo* [51–53]. The role of methylation as a causative or secondary event associated with vector silencing is a matter still under investigation. Studies in transgenic *Drosophila* clearly indicate that transgene silencing can be induced by retroviral sequences in the absence of methylation [58].

There is still little information on lentiviral vector silencing. While it has been suggested that the HIV LTR is as prone as the MoMLV LTR to silencing in some settings [58], other reports allow for more optimism. In embryonal carcinoma cells and stem cells, in which oncoretroviral elements are commonly silenced [49,55–57,59], lentiviral vectors sustain expression of a reporter gene under the control of an internal phosphoglycerate kinase promoter [60]. *In vivo*, in bone marrow chimeras engrafted with a lentiviral vector bearing an erythroid-specific gene and erythroid-specific transcriptional control elements, transgene expression was sustained over at least 24 weeks [36]. These encouraging reports need to be bolstered by additional studies. It remains to be established as to what extent lentiviral sequences are susceptible to silencing in primate cells *in vivo*.

F. Other Properties

Other potentially important characteristics distinguish oncoretroviral from lentiviral vectors. For example, integration sites may differ in significant ways because integration occurs in different phases of the cell cycle. Such a difference may in turn affect expression patterns and susceptibility to vector silencing [61]. Another difference may be in the risk of recombination, which is in part determined by the presence of homologous endogenous or exogenous sequences. Different types of lentiviral vectors will likely differ in this respect.

IV. MANUFACTURE OF LENTIVIRAL VECTORS

A. Principles of Vector Design

Gene transfer systems based on replication-defective viral vectors typically comprise two components: a vector that bears the sequences to be transferred and a helper virus or subelements derived therefrom that provides *in trans* the transcriptional and packaging functions required for propagation of the recombinant genome. This functional complementation takes place within a cell called the packaging cell. The guiding principle of retroviral vector design is to retain within

the vector itself the minimal *cis*-acting sequences sufficient for genomic transcription, packaging, reverse transcription, and integration. Thus, delineation of dispensable and indispensable genes and sequences within the parental virus is essential. In HIV-1-derived vectors, the four accessory genes, *vpr, vpu, vif,* and *nef,* have been removed without precluding gene transfer, at least in certain cell types. In others, however, absence of accessory functions can diminish infectivity of wild-type HIV-1 and replication-defective recombinants. This is particularly relevant in quiescent T lymphocytes (see below). Studies on HIV-1 infection of macrophages show that the Vpr protein plays a crucial role in the nuclear import of the preintegration complex [62–64]. Dependence on accessory proteins is likewise observed with HIV-1-derived lentiviral vectors in macrophages [63] and in hepatocytes [64,65]. However, Vpr turns out to be dispensable for lentiviral vector-mediated gene delivery into neurons and growth-arrested cell lines [63,66].

In addition to extending the packaging capacity within the vector, the purpose of removing any nonessential viral sequence from the transfer vector is to avoid recombination between the vector and other viral sequences that could generate a replication-competent genome. The probability of recombination critically depends on the packaging cell system and its possible interactions with the vector to be packaged. Recombination events are minimized when the Gag, Pol, and Env polypeptides, coexpressed in the packaging cell, are transcribed from at least two separate plasmids lacking the packaging signal and displaying as little homology as possible with the vector.

Packaging systems and vectors improve in incremental steps. The development of oncoretroviral and lentiviral vectors has followed the same three prototypic stages, referred to as generations (Table 3). In the first generation, a minimally deleted genome, usually lacking *env*, encodes several retroviral genes and a transgene; *env* is encoded by a separate plasmid. A single recombination event reintroducing *env* into the vector would result in a replication-competent retrovirus (RCR). Such systems are unacceptable in the clinical setting. In the second generation, the vector does not encode any retroviral gene and *gag–pol* and *env* are encoded by separate plasmids. At least two productive rearrangements would be required to reintroduce *gag–pol* and *env* in the transfer vector and thus generate RCR. Second-generation oncoretroviral packaging systems have been quite extensively used in humans. No replication competent retrovirus has been reported so far in clinical trials [67]. In the third generation, a series of additional modifications are introduced in the vector, the packaging constructs, and eventually the packaging cell to minimize the probability of recombination. The multiplication of plasmids without sequence overlap or homology aims to eliminate the risk of DNA recombination between plasmids.

TABLE 3   Generational Evolution of Oncoretroviral and Lentiviral Vector Systems

Generation	Essential characteristics	Examples of packaging cells	
		MLV-based	HIV-1-based
First	Ψ-deleted helper genome High risk of RCR generation	—[a]	—[a]
Second	Split helper genome Low risk of RCR generation	Stable, amphotropic[b] Stable, VSV-G–pseudotyped[c]	Transient, VSV-G–pseudotyped; multiply attenuated[e]
Third	Additionally split helper genome Maximal deletion of viral genes and viral sequences Minimal homology between plasmids	Stable, amphotropic[d]	Transient, VSV-G–pseudotyped;[f] Stable, VSV-G–pseudotyped[g]

[a]Not clinically relevant.
[b]See Miller and Buttimore [2], Danos and Mulligan [3], and Markowitz *et al.* [4].
[c]See Ory *et al.* [75].
[d]See Sheridan *et al.* [175].
[e]See Naldini *et al.* [7] and Zufferey *et al.* [63].
[f]See Dull *et al.* [68] and Wu *et al.* [69].
[g]See Klages *et al.* [74].

B. Current Packaging Systems for HIV-1-Based Vectors

Recombinant lentiviral vectors are generated following the same principles as oncoretroviral vectors. Structural genes are provided in *trans* by either transient transfection or by stable expression in a packaging cell line. Multiple nonoverlapping plasmids are used to generate recombinant virions with a minimal risk of regenerating replication-competent retrovirus. The common features found in third-generation packaging systems are (1) elimination of nonessential viral genes, (2) multiplication of the number of plasmids encoding different virion constituents, and (3) excision of noncoding viral sequences from all plasmids to further avoid any overlap (for example, the 5′ LTR is replaced by alternative enhancer/promoters and the 3′ LTR by alternative polyadenylation signals).

Transient systems require triple or even quadruple cotransfections. Recombinant virions are harvested 2 to 3 days later. The human embryonic kidney 293 or 293T cell lines are frequently used as packaging cells because they are highly transfectable. Zufferey *et al.* [63] described a multiply attenuated system from which Vpr, Vpu, Nef, and Vif are excluded. Gag–Pol, Tat, and Rev are, however, encoded by a single plasmid (Fig. 2a). In the system described by Dull *et al.* [68], three plasmids are used to express Gag–Pol, Rev, and VSV-G (Fig. 2b). Wu et al. split the lentiviral genome following different lines [69]. As packaging constructs, they used one plasmid that encodes Gag–Pro, as well as Vif, Tat, and Rev, and another that encodes Vpr, RT, and IN (Fig. 2c). The protease encoded by *pol* is therefore physically dissociated from RT and IN, thus making it very unlikely that a functional Gag–Pol could reform. This system, less attenuated

than the former, provides some advantages in terms of monitoring recombination events and the risk of generating RCR.

While transient systems can be optimized and scaled-up, stable packaging cell lines offer several advantages. The plasmids are stably integrated, diminishing the risk of DNA recombination, and the stable producer bearing a given vector is amenable to repeated biosafety testing and optimization of culture conditions for the production of large vector stocks. The generation of stable packaging cells requires the identification of stable clones that express high enough levels of each viral constituent. As some may be toxic for the packaging cell [70–72], a strategy must be devised to modulate the production of the toxic product. This is commonly achieved via tetracycline-mediated repression of transcription [73]. Thus, Klages *et al.* [74] generated the LV[G] packaging cell by expressing Gag–Pol and Rev in 293G cells which stably express the tTA activator and doxycycline-repressible vesicular stomatitis virus G glycoprotein (VSV-G) [75]. Rev was placed under the control of a tetracycline-inducible promoter, allowing modulation of Gag–Pol expression.

C. Infectious Spectrum of Recombinant Virions

The first step towards retroviral infection is the binding of the infectious particle to a cell surface receptor. A number of different retroviral envelopes have been identified and their receptors cloned [9]. The HIV-1 envelope, however, restricts virus entry to a limited number of cell types [10]. The natural envelopes can be substituted with heterologous retroviral envelopes or alternative fusogenic molecules, a process referred to as pseudotyping. Recombinant lentiviruses are

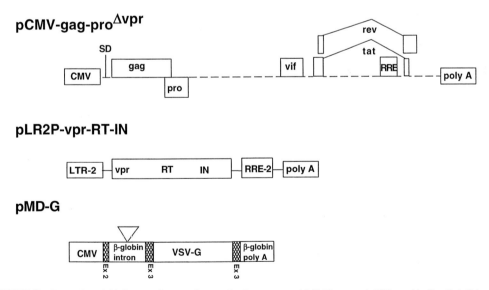

FIGURE 2 Examples of third-generation transient packaging systems. (a) Zufferey *et al.* [63] provide Gag-Pol, Tat, and Rev from a single plasmid. (b) Dull *et al.* [68] separate Gag–Pol and Tat from Rev. (c) Wu *et al.* [69] separate Gag–Pro from RT-IN; unlike the two former techniques, this packaging system retains two accessory proteins in addition to Tat and Rev, Vpr and Vif. All these systems make use of pMD-G to express the VSV-G glycoprotein. pMD plasmids incorporate the B-X fragment derived from Mß6L [35] and TNS9 (see Fig. 3d) [36]. VSV-G and Gag–Pol–Tat are cloned into the EcoRI site in exon 2 of the human ß-globin gene (see Fig. 3d).

a. pHR' (Naldini et al, Science 1996)

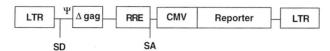

b. pHR' SIN-18 (Zufferey et al, J. Virol. 1998)

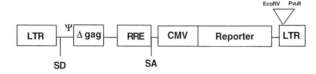

c. pRRLSin (Follenzi et al, Nature Gen. 2000)

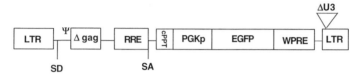

d. TNS9 (May et al, Nature 2000)

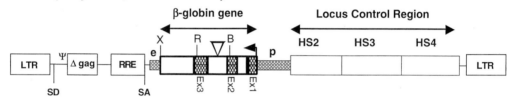

FIGURE 3   Examples of improvements in lentiviral vectors. (a) The original pHR' vector [7]; (b) introduction of a deletion in the U3 region of the 3' LTR [82]; (c) addition of the central polypyrimidine tract [84] and the WPRE [86]; (d) inclusion of the entire human ß-globin gene, including introns, polyadenylation signal, extended promoter (p), 3' enhancer (e), and HS2, HS3, and HS4 elements of its locus control region (which encompasses additional enhancers and elements with chromatin opening activity [36]). X, R, and B are restriction enzyme sites used in the pMD plasmids (see Fig. 2).

often pseudotyped with the (VSV-G), conferring a broad infectious spectrum [72]. Other envelopes may prove to be more effective for particular cell types. For example, the amphotropic envelope of the murine leukemia virus (MLV strain 4070A), the envelope of the gibbon ape leukemia virus (GaLV strain SEATO), and the envelope of feline endogenous retrovirus (RD-114) may be more effective in cells of hematopoietic origin. The *Ebola* virus envelope (Zaire strain) may be useful for certain epithelia [76]. There are still few reports on the pseudotyping of lentiviral vectors [76–79]. Such studies are warranted and will certainly be undertaken to selectively optimize gene transfer in different tissues.

D. Improvements of Transfer Vectors

The transfer vector itself is the object of countless variations that aim to improve transgene expression and safety features. A number of published studies have used the pHR' vector [7] or derivatives thereof. Derived from HIV-1, pHR'

retains both HIV-1 LTRs, the primer binding site, the packaging region and adjacent INS sequences overlapping *gag* sequences [80], and a portion of the *env* sequence spanning the RRE. An enhancer/promoter and cDNA are typically cloned downstream of the RRE, generating an intronless transcript that makes use of the 3' LTR polyadenylation signal (Fig. 3a). The human cytomegalovirus (CMV) immediate/early (IE) enhancer/promoter is commonly used, despite its unreliable activity following integration. Deletion of the enhancer/promoter sequence in the 3' LTR has long been used in oncoretroviral vectors to decrease transcriptional interference and improve safety [81]. Such a deletion decreases competition between an internal enhancer/promoter and the flanking 5' and 3' LTRs. With respect to safety, LTR inactivation reduces the potential risk of transcribing endogenous genes lying downstream of the integrated LTR. Insertional mutagenesis caused by gene disruption, however, remains unaffected by this vector modification. Deletions in the 3' HIV LTR have been reported that apparently do not cause decreases in vector titers (Fig. 3b) [82,83]. Incorporation of the central PPT element in the vector has been shown to ameliorate infectivity of recombinant virions. This element is thought to enhance nuclear translocation of wild-type HIV [13] and of recombinant vectors (Fig. 3c) [84].

A number of enhancer/promoters and cDNAs will eventually be tested in lentiviral vectors. However, general means to increase transgene expression are emerging. For example, incorporation of the woodchuck hepatitis virus post-transcriptional regulatory element (WPRE) has been shown to increase transgene expression obtained with a number of enhancer/promoters (Fig. 3c). This element is thought to work at the post-transcriptional level, by increasing cytoplasmic export of mRNA transcripts [85] and has been successfully utilized in oncoretroviral vectors [86,87].

The ultimate goal in a number of gene transfer applications is to achieve tissue-specific, regulated, position-independent, and sustained transgene expression. It is unlikely that this goal will be obtained by simply incorporating enhancers and promoters into a vector. Transgene expression is indeed highly sensitive to the site of integration (accounting for the position effect) and the nature of the target cell [49]. To better control the fate of randomly integrated vectors, we believe it will be necessary to incorporate determinants of chromatin structure, such as locus control regions, insulators, and matrix attachment regions [49]. This view is supported by the recent findings of May et al. who showed for the first time that regulated and sustained transgene expression could be achieved in the progeny of virally transduced stem cells [36].

There are important lessons to learn from vectors encoding the human ß-globin gene, the expression of which has to be erythroid specific, differentiation stage specific, and elevated to be therapeutically relevant. The incorporation into a viral vector of the entire human ß-globin gene along with its promoter and two proximal enhancers does not suffice to obtain therapeutic gene expression [88–94]. These vectors permit expression of the ß-globin polypeptide in bone marrow chimeras that is tissue specific but low and variable, varying between 0 and 2% of endogenous ß-globin expression. Studies in transgenic mice demonstrated that inclusion of distal genetic elements, referred to as the locus control region (LCR), were needed to achieve high-level expression [95,96] and raised the hope that a juxtaposed LCR could allow expression of randomly integrated transgenes in a chromosomal-position-independent fashion. Most viral vectors cannot accommodate genomic sequences as large as the 20-kb ß–globin LCR. A number of studies carried out by different investigators defined smaller segments of the LCR, or core sites, that appeared to retain part of the LCR function in transgenic mice (see Sadelain [97] for review). In the context of retroviral vectors, the HS2, HS3, and HS4 core sites of the human ß-globin LCR acted as powerful erythroid enhancers but failed to provide position-independent expression [34,35]. In long-term bone marrow chimeras transplanted with ß-globin

vectors harboring the HS2, HS3, and HS4 core sites, three out of 12 long-term chimeras expressed some level of human ß-globin in one study [98] and none out of seven in another [49].

Using a lentiviral vector, May et al. recently succeeded in stably transmitting a vector harboring larger segments of the LCR [36]. Following integration in mouse hematopoietic stem cells, human ß-globin expression was erythroid specific and elevated enough to ameliorate red cell morphology and anemia in ß-thalassemic mice. In primary and secondary chimeras, human ß-globin expression remained undiminished after 6 and 3 months, respectively. These results illustrate the need to appropriately combine the human ß-globin gene, its promoter and enhancers, and locus control region elements to achieve therapeutic benefits. Future studies will determine whether the ß-globin model is a paradigm for achieving regulated transgene expression in all genetically modified stem cells.

V. POSSIBLE APPLICATIONS OF LENTIVIRAL VECTORS IN CANCER THERAPY

A. Hematopoietic Stem Cells

Gene transfer efficiency in hematopoietic stem cells (HSCs) remains rate-limiting, despite continuing improvements [40,99]. Because oncoretroviral vectors require target cell division for efficient gene transfer, the induction of cell division in HSC has been a major research objective [40,99]. A multitude of combinations of cytokines and stromal cells have been evaluated for their ability to activate quiescent HSCs and induce cell division. A number of newer cytokine cocktails and optimized culture conditions appear to yield significant improvements [100–108] relative to the more classic combination of IL-3, IL-6, and stem cell factor.

However, there are major caveats in having to induce cell division in stem cells. While manipulation of HSC cycling status may favor vector integration, it may also alter fundamental cellular functions and ultimately jeopardize the graft. One complication is the risk of inducing differentiation and/or restricting the differentiation potential of stem cells. Cytokine combinations may differ in this respect. For instance, the beneficial effect of IL-3 on expansion is controversial [109] and requires further investigation [101,103,110]. Another risk is hematopoietic exhaustion. The dangers of unrestricted HSC cell division are illustrated in mice deficient in p21/WAF1, a cyclin-dependent kinase inhibitor acting as G_1 checkpoint regulator. When exposed to hematopoietic stress, these mice die prematurely and present with HSC exhaustion [111]. Another risk is that of antagonizing homing and limiting engraftment [112], thus transduced stem cells may not home suitably because they are cycling. It is also established that homing and mobilization of HSCs are in part determined by chemokines. The chemokine stromal cell-derived factor-1 (SDF-1) and its receptor, CXCR-4, have been implicated in the homing and mobilization of human CD34$^+$ cells [113,114]. Preservation of the self-renewing, differentiation, and homing potentials of manipulated HSCs is therefore essential and must be accounted for during the manufacture of genetically modified cells [40].

The advent of lentiviral vectors raised the hope that cell division would not be a prerequisite for vector integration. Indeed, recombinant lentiviral vectors derived from HIV-1 have been shown to transduce CD34$^+$CD38$^-$ hematopoietic cells [115–122], including CD34$^+$ cells capable of engrafting and differentiating into multiple hematopoietic cell lineages in NOD-*scid/scid* mice. Transduction efficiency is clearly greater than that obtained with MLV–based oncoretroviral vectors under conditions that do not promote cell division [115]. However, while lentiviral vectors compare favorably to oncoretroviral vectors in overall G$_0$/G$_1$ CD34$^+$ populations [118], it is clearly established that cells in G$_1$ are more readily transduced than cells in G$_0$ and that integration may be confined in the latter group to cells that are exiting G$_0$ and entering G$_1$ [120].

Thus, the purported advantage of lentiviral vectors may seem greatly reduced if G$_0$ cells remain out of reach. However, the requirement for transition to G$_{1b}$ rather than passage through S phase for the formation of an active preintegration complex and integration may well be crucial. Importantly, this lesser activation requirement may diminish the duration of infection and the need for cytokine induction. Consequently, if transplanted cells harbor the vector genome in their cytoplasm, and have the *time* to engraft and undergo activation *in vivo* under physiological conditions before degradation of the vector, efficient gene transfer may take place in a cell that was minimally manipulated *ex vivo* [40]. Preliminary results in rhesus macaques suggest that lentiviral vectors may allow efficient gene transfer in mobilized CD34$^+$ cells without addition of cytokines [123] and perhaps improve engraftment of transduced cells by avoiding active cell cycling at the time of reinfusion [124]. Ultimately, it will be important to compare results achieved with lentiviral and oncoretroviral vectors used under their respective optimal conditions, an analysis that is yet to be performed.

B. T Lymphocytes

The adoptive transfer of T cells in cancer patients is an interesting therapeutic approach for certain conditions (e.g., Epstein–Barr-virus-associated lymphoproliferative disease and chronic myelogenous leukemia) [125,126]. Genetic alterations of these T cells may be useful to increase their safety [127,128] and efficacy [129]. Oncoretroviral-mediated gene transfer requires prior T-cell activation to induce proliferation. This is sometimes advantageous, allowing, for example, preferential transduction of antigen-specific T cells within bulk populations [130]. In other instances, transduction of resting T lymphocytes may be preferable, in order to transduce to naive T cells or avoid detrimental consequences of *in vitro* T cell activation.

There are conflicting reports on the ability of HIV-1 to stably integrate in quiescent primary T cells. Some indicate that reverse-transcription is partial and followed by rapid degradation of the nonintegrated genetic material [28,131]. One study reports that progression through the cell cycle from G$_0$ to G$_1$ is necessary for efficient and complete reverse transcription of the HIV-1 genome [29], while another suggests that this is not absolutely required [132]. The accessory proteins play an important role in HIV infectivity [16,17,20,71,133–136]. This is reflected in recent reports on the use of replication-defective HIV-1-derived vectors to transduce human T cells. One study suggests that resting T cells can be stably infected using a nonattenuated vector, but not if Vif, Vpr, Vpu, and Nef are removed [137]. Another study, however, suggests that elimination of these auxiliary proteins reduces but does not abrogate transduction efficiency [138]. These observations are consistent with observations by Unutmaz *et al.* who found that prolonged pretreatment with cytokines is required to enable transduction by an attenuated lentiviral vector [139]. Taken together, these observations indicate that the molecular mechanisms underlying HIV-1 infectivity need to be better understood to enable rational vector design. The current data suggest that some degree of T-cell activation is required to permit lentiviral transduction, albeit at a lower level than is needed to prompt cell division. The relevance of lentiviral vectors for gene transfer in nondividing primary T cells will be determined to a large extent by these biological requirements and their consequences for T cell function.

C. Dendritic Cells

Dendritic cells (DCs) are specialized antigen-presenting cells with a remarkable ability to stimulate naive T lymphocytes and generate memory T lymphocytes. Immature DCs effectively take up antigen and upon maturation upregulate expression of a number of molecules critical for effective antigen presentation [140]. DCs are thus currently viewed as ideal adjuvants for immunizing against pathogens or tumor antigens [141,142]. Genetic approaches to antigen loading, based on the transduction of DCs with cDNA- or mRNA-encoding tumor antigens, offer several advantages over pulsing with peptide or cellular extracts. A set of selected antigens can be expressed at high levels in the DCs; antigens can be processed into human leukocyte antigen (HLA)-restricted peptides, irrespective of the HLA type of each individual and without prior knowledge of the peptide best suited for that particular HLA; antigen presentation should be more sustained if the antigen is expressed endogenously rather than pulsed.

Oncoretroviral vectors based on MLVs have been used successfully to genetically modify CD34$^+$ DC progenitors from various sources (e.g., bone marrow, cord blood, and cytokine-elicited peripheral blood) [143–148]. This approach is thus not applicable to DCs generated from peripheral blood mononuclear cells (PBMCs), a common and practical source of DCs, as they do not show any substantial proliferation [149–151]. Replication-defective lentiviruses have been shown to stably transduce nondividing monocyte-derived DCs [26,27,152,153]. Other vector systems have been used to transduce blood-derived DCs, including those based on adenovirus [154], herpesvirus [155], or poxvirus [156–158]. All of these vectors encode viral proteins that modify the antigenicity and/or function of the transduced DC. In contrast, recombinant lentiviruses permit efficient and stable gene delivery using a vector that does not itself encode any immunogenic proteins of viral origin. The advent of a stable gene delivery system based on a multiply attenuated lentivirus that does not encode any viral protein and allows efficient and sustained antigen-presentation by DCs derived from blood monocytes may ultimately be useful for the improvement of immunotherapies.

D. Tumor Cells

Gene transfer in tumor cells is the basis for a number of therapeutic strategies that aim to directly eliminate tumor cells or augment their immunogenicity. It is uncertain whether lentiviral vectors will play an important role in this setting because transient gene expression is sufficient in most applications. Lentiviral vectors may nonetheless be called upon because of their ability to transduce nondividing cells. Indeed, the vast majority of cells within tumor masses are typically not cycling. Furthermore, certain tumor cells cannot be propagated *ex vivo,* although they can be maintained in culture for some time.

There are at this time few reports on lentiviral-mediated gene transfer in primary tumor cells. Leukemia cells have been successfully transduced with multiply attenuated HIV-1-derived vectors encoding costimulatory ligands [159–161]. Several studies have demonstrated efficient gene transfer in growth-arrested tumor cell lines [63,162]. There are still no reports of *in vivo* gene transfer into tumor masses. On the other hand, *in vivo* gene transfer has been reported in several normal tissues. It remains to be established whether sensitivity of recombinant lentiviral vectors to human serum [163] will hinder *in vivo* gene delivery strategies if they are eventually pursued. Interestingly, pseudotyped lentiviral vectors harboring oncogenes have been shown to target different tissues *in vivo,* offering the prospect of creating valuable animal tumor models [79].

E. Other Cell Types

Recombinant vectors derived from HIV-1 have successfully been used for *in vivo* or *in vitro* transduction of term-

inally differentiated cells, including neurons, hepatocytes, retinocytes, endothelial cells, pancreatic islet ß cells, and macrophages [24,25,63,65,66,164–168]. These studies fall beyond the scope of this chapter. Nonetheless, these results are encouraging and may be helpful in developing novel therapeutic approaches in cancer patients.

VI. CONCLUSIONS

The use of recombinant lentiviruses for gene transfer purposes is recent. The excitement surrounding their development is due to three key properties: (1) the ability to integrate in nondividing cells, *in vitro* and *in vivo* [7]; (2) the availability of multiply attenuated and minimally immunogenic vectors [63]; and (3) the stable transmission of large genomic fragments, which will allow to improve transgene expression [36]. Recent advances in the development of third-generation packaging systems derived from HIV-1 will make possible broader investigation of lentiviral-mediated gene transfer. The availability of such vectors is one of several important steps toward their approval for human use. Indeed, oncoretroviral vectors generated in a similar fashion have a very good safety track record so far [67]. However, these new and improved vectors have not yet been evaluated in relevant experimental settings. There is an imperative need for developing sensitive and reliable assays for the detection of RCR, as well as recombinant intermediates that are only one step away from a productive rearrangement that could create an RCR.

A number of further developments are anticipated, one of which is the availability of advanced versions of other lentiviral vectors of simian, feline, or equine origin [85,169–171]. These may turn out to have advantageous features in terms of their safety [172,173], depending on what their relative risks of recombination turn out to be. Another anticipated development is the improved targeting of lentiviral vectors, using virions pseudotyped with either alternative envelopes or genetically engineered envelope substitutes [174]. These would ensure safer and more efficient gene transfer in specific cell types, possibly including *in vivo* applications. Lentiviral vectors are also of great interest for the development of complex, regulated vectors. Much of the future of gene therapy depends on the availability of "smarter" vectors that provide controlled transgene expression. The availability of stable vectors with extended packaging capacity will allow greater flexibility in vector design, making it possible to better regulate transgene expression, attenuate position effects, and decrease the likelihood of transcriptional inactivation and vector silencing.

The next edition is likely to have vastly expanded sections on non-primate lentiviral vectors, biosafety testing, and the use of lentiviral vectors in disease models. On the other hand, it is not yet determined when lentiviral vectors will be approved for human use; nonetheless, this category of vectors is poised to open the way for important advances in cancer therapy, especially in therapeutic strategies that depend on efficient gene transfer in stem cells, T lymphocytes, and tumor cells.

Acknowledgments

The authors thank Ms. A. Castillo for assistance with preparation of the manuscript and Ms. C. Sadelain for helpful review. This work is supported by grants CA-59350, HL-57612, HL-66952, and the DeWitt Wallace Fund at MSKCC.

References

1. Mann, R., Mulligan, R. C., and Baltimore, D. (1983). Construction of a retrovirus packaging mutant and its use to produce helper-free defective retrovirus. *Cell* **33**, 153–159.
2. Miller, A. D., and Buttimore, C. (1986). Redesign of retrovirus packaging cell lines to avoid recombination leading to helper virus formation. *Mol. Cell. Biol.* **6**, 2895–2902.
3. Danos, O., and Mulligan, R. C. (1988). Safe and efficient generation of recombinant retroviruses with amphotropic and ecotropic host range. *Proc. Natl. Acad. Sci. USA* **85**, 6460–6464.
4. Markowitz, D., Goff, S. P., and Bank, A. (1988). A safe packaging cell line for gene transfer: separating viral genes on two different plasmids. *J. Virol.* **62**, 1120–1124.
5. Carroll, R., Lin, J. T., Dacquel, E. J., Mosca, J. D., Burke, D. S., and St. Louis, D. C. (1994). A HIV-1-based retroviral vector system utilizing stable HIV-1 packaging cell lines. *J. Virol.* **68**, 6047–6051.
6. Yu, H., Rabson, A. B., Kaul, M., Ron, Y., and Dougherty, J. P. (1996). Inducible HIV-1 packaging cell lines. *J. Virol.* **70**, 4530–4537.
7. Naldini, L., Blomer, U., Gallay, P., *et al.* (1996). In vivo gene delivery and stable transduction of nondividing cells by a lentiviral vector. *Science* **272**, 263–267.
8. Coffin, J. M., Hughes, S. H., and Varmus, H. E., eds. (1997). *Retroviruses,* pp. 802–804. Cold Spring Harbor Laboratory Press. Cold Spring Harbor, NY.
9. Miller, A. D. *et al.* (1996). Cell-surface receptors for retroviruses and implications for gene transfer. *Proc. Natl. Acad. Sci. USA* **93**, 11407–11413.
10. Atchison, R. E., Gosling, J., Monteclaro, F. S. *et al.* (1996). Multiple extracellular elements of CCR5 and HIV-1 entry: dissociation from response to chemokines. *Science* **274**, 1924–1926.
11. Tritel, M., and Resh, M. D. (2000). Kinetic analysis of HIV-1 assembly reveals presence of sequential intermediates. *J. Virol.* **74**, 5845–5855.
12. Coffin, J. M., Hughes, S. H., and Varmus, H. E., eds. (1997). *Retroviruses,* pp. 42–48. Cold Spring Harbor Laboratory Press. Cold Spring Harbor, NY.
13. Zennou, V., Petit, C., Guetard, D., Nehrbass, U., Montagnier, L., Charneau, P. (2000). HIV-1 genome nuclear import is mediated by a central DNA flap. *Cell* **101**, 173–185.
14. Doms, R. W., and Trono, D. (2000). The plasma membrane as a combat zone in the HIV battlefield. *Genes Dev.* **14**, 2677–2688.
15. Coffin, J. M., Hughes, S. H., Varmus, H. E., eds. (1997). *Retroviruses,* pp. 33–42. Cold Spring Harbor Laboratory Press, Cold Spring Harbor, NY.
16. Emerman, M., and Malim, M. H. (1998). HIV-1 regulatory/accessory genes: keys to unraveling viral and host cell biology. *Science* **280**, 1880–1884.
17. Cullen, B. R. (1998). HIV-1 auxiliary proteins: making connections in dying cell. *Cell* **93**, 685–692.
18. Malim, M. H., Hauber, J., Le, S. Y., Maizel, J. V., and Cullen, B. R. (1989). The HIV-1 Rev *trans*-activator acts through a structured target sequence to activate nuclear export of unspliced viral mRNA. *Nature* **338**, 254–257.
19. Felber, B. K., Hadzopoulou-Cladaras, M., Cladaras, C., Copeland, T., and Pavlakis, G. N. (1989). Rev protein of human immunodeficiency virus type 1 affects the stability and transport of the viral mRNA. *Proc. Natl. Acad. Sci. USA* **86**, 1495–1499.
20. Baur, A. S., Sawai, E. T., Dazin, P. *et al.* (1994). HIV-1 Nef leads to inhibition or activation of T cells depending on its intracellular localization. *Immunity* **1**, 373–384.
21. Roe, T., Reynolds, T. C., Yu, G., and Brown, P. O. (1993). Integration of murine leukemia virus DNA depends on mitosis. *EMBO J.* **12**(5), 2099–2108.
22. Hajihosseini, M., Iavachev, L., and Price, J. (1993). Evidence that retroviruses integrate into post-replication host DNA. *EMBO J.* **12**(13), 4969–4974.
23. Lewis, P. F., and Emerman, M. (1994). Passage through mitosis is required for oncoretroviruses but not for the human immunodeficiency virus. *J. Virol.* **68**, 510–516.
24. Naldini, L., Blomer, U., Gage, F. H., Trono, D., and Verma, I. M. (1996). Efficient transfer, integration, and sustained long-term expression of the transgene in adult rat brains injected with a lentiviral vector. *Proc. Natl. Acad. Sci. USA* **93**(21), 11382–11388.
25. Miyoshi, H., Takahashi, M., Gage, F. H., and Verma, I. M. (1997). Stable and efficient gene transfer into the retina using an HIV-based lentiviral vector. *Proc. Natl. Acad. Sci. USA* **94**(19), 10319–10323.
26. Chimmasamy, N., Chimmasamy, D., Toso, J. F., Lapointe, R., Candotti, F., Morgan, R. A., and Hwu, P. (2000). Efficient gene transfer to human peripheral blood moncyte-derived dendritic cells using human immunodeficiency virus type 1-based lentiviral vectors. *Hum. Gene Ther.* **11**, 1901–1909.
27. Dyall, J., Latouche, J. B., Schnell, S., and Sadelain, M. (2001). Lentivirus-transduced human monocyte-derived dendritic cells efficiently stimulate antigen-specific cytotoxic T lymphocytes. *Blood* **97**, 114–121.
28. Zack, J. A. *et al.* (1990). HIV-1 entry into quiescent primary lymphocytes: molecular analysis reveals a labile, latent viral structure. *Cell* **61**, 213.
29. Korin, Y. D., and Zack, J. A. (1998). Progression to the G_{1b} phase of the cell cycle is required for completion of HIV1 reverse transcription. *J. Virol.* **72**, 3161–3168.
30. Zack, J. A., Haislip, A. M., Krogstad, P., and Chen, I. S. (1992). Incompletely reverse-transcribed human immunodeficiency virus type 1 genomes in quiescent cells can function as intermediates in the retroviral life cycle. *J. Virol.* **66**(3), 1717–1725.
31. Gao, W. Y., Cara, A., Gallo, R. C., and Lori, F. (1993). Low levels of deoxynucleotides in peripheral blood lymphocytes: a strategy to inhibit human immunodeficiency virus type 1 replication. *Proc. Natl. Acad. Sci. USA* **90**(19), 8925–8928.
32. Meyerhans, A., Vartanian, J. P., Hultgren, C. *et al.* (1994). Restriction and enhancement of human immunodeficiency virus type 1 replication by modulation of intracellular deoxynucleoside triphosphate pools. *J. Virol.* **68**(1), 535–540.
33. Zhang, H., Duan, L. X., Dornadula, G., Pomerantz, R. J. (1995). Increasing transduction efficiency of recombinant murine retrovirus vectors by initiation of endogenous reverse transcription: potential utility for genetic therapies. *J. Virol.* **69**(6), 3929–3932.
34. Lebouch, P. *et al.* (1994). Mutagenesis of retroviral vectors transducing human beta-globin gene and beta-globin locus control region derivatives results in stable transmission of an active transcriptional structure. *EMBO J.* **13**, 3065–3076.

35. Sadelain, M., Wang, C. H., Antoniou, M., Grosveld, F., and Mulligan, R. C. (1995). Generation of a high-titer retroviral vector capable of expressing high levels of the human beta-globin gene. *Proc. Natl. Acad. Sci. USA* **92**(15), 6728–6732.

36. May, C., Rivella, S., Callegari, J. *et al.* (2000). Therapeutic hemoglobin synthesis in ß-thalassemic mice expressing lentivirus-encoded human ß-globin. *Nature* **406**, 82–86.

37. Bodine, D. (2000). Globin gene therapy: one (seemingly) small vector change, one giant leap in optimism. *Mol. Ther.* **2**, 101–102.

38. Andreadis, S. T., Brott, D., Fuller, A. O. *et al.* (1997). Mo-MLV-derived retroviral vectors decay intracellularly with a half-life in the range of 5.5 to 7.5 hours. *J. Virol.* **71**, 7541–7548.

39. Chang, L. J., Urlacher, V., Iwakuma, T., Cui, Y., and Zucali, J. (1999). Efficacy and safety analyses of a recombinant human immunodeficiency virus type 1 derived vector system. *Gene Ther.* **6**, 715–728.

40. Sadelain, M., Frassoni, F., and Rivière, I. (2000). Issues in the manufacture and transplantation of genetically modified hematopoietic stem cells. *Curr. Opin. Hematol.* **7**, 364–377.

41. Roberts, J. D., Preston, B. D., Johnston, L. A., Soni, A., Loeb, L. A., and Kunkel, T. A. (1989). Fidelity of two retroviral reverse transcriptases during DNA-dependent DNA synthesis in vitro. *Mol. Cell. Biol.* **9**, 469–476.

42. Bakhanashvili, M., and Hizi, A. (1992). Fidelity of the reverse transcriptase of human immunodeficiency virus type 2. *FEBS Lett.* **306**, 151–156.

43. Ji, J. P., and Loeb, L. A. (1992). Fidelity of HIV-1 reverse transcriptase copying RNA in vitro. *Biochemistry* **31**, 954–958.

44. Preston, B. D., Poiesz, B. J., and Loeb, L. A. (1988). Fidelity of HIV-1 reverse transcriptase. *Science* **242**, 1168–1171.

45. Roberts, J. D., Bebenek, K., and Kunkel, T. A. (1988). The accuracy of reverse transcriptase from HIV-1. *Science* **242**, 1171–1173.

46. Ji, J., and Loeb, L. A. (1994). Fidelity of HIV-1 reverse transcriptase copying a hypervariable region of the HIV-1 *env* gene. *Virology* **199**, 323–330.

47. Mansky, L. M., and Temin, H. M. (1995). Lower in vivo mutation rate of human immunodeficiency virus type 1 than that predicted from the fidelity of purified reverse transcriptase. *J. Virol.* **69**, 5087–5094.

48. Verma, I. M., and Somia, N. (1997). Gene therapy—promises, problems and prospects. *Nature* **389**, 239–242.

49. Rivella, S., and Sadelain, M. (1998). Genetic treatment of severe hemoglobinopathies: the combat against transgene variegation and transgene silencing. *Semin. Hematol.* **35**, 112–125.

50. Bestor, T. H., and Tycko, B. (1996). Creation of genomic methylation patterns. *Nat. Genet.* **12**, 363–367.

51. Palmer, T. D., Rosman, G. J., Osborne, W. R., and Miller, A. D. (1991). Genetically modified skin fibroblasts persist long after transplantation but gradually inactivate introduced genes. *Proc. Natl. Acad. Sci. USA* **88**, 1330–1334.

52. Challita, P. M., and Kohn, D. B. (1994). Lack of expression from a retroviral vector after transduction of murine hematopoietic stem cells is associated with methylation in vivo. *Proc. Natl. Acad. Sci. USA* **91**, 2567–2571.

53. Rivella, S., Callegari, J., May, C. *et al.* (2000). The cHS4 insulator increases the probability of retroviral expression at random chromosomal integration sites. *J. Virol.* **74**, 4679–4687.

54. Jaenisch, R., Jahner, D., Nobis, P., Simon, I., Loher, J., Harbers, K., and Grotkopp, D. (1981). Chromosomal position and activation of retroviral genomes inserted into the germ line of mice. *Cell* **24**, 519–529.

55. Jaenisch, R., Harbers, K., Jahner, D., Stewart, C., and Stuhlmann, H. (1982). DNA methylation, retroviruses, and embryogenesis. *J. Cell. Biochem.* **20**, 331–336.

56. Jahner, D., and Jaenisch, R. (1985). Retrovirus-induced de novo methylation of flanking host sequences correlates with gene inactivity. *Nature* **315**, 594–597.

57. Laker, C., Meyer, J., Schopen, A., Friel, J., Heberlein, C., Ostertag, W., and Stocking, C. (1998). Host *cis*-mediated extinction of a retrovirus permissive for expression in embryonal stem cells during differentiation. *J. Virol.* **72**, 339–348.

58. Pannell, D., Osborne, C. S., Yao, S. *et al.* (2000). Retrovirus vector silencing is de novo methylase-independent and marked by a repressive histone code. *EMBO J.* **19**, 5884–5894.

59. Cherry, S. R., Biniskiewicz, D., van Parijs, L., Baltimore, D., and Jaenisch, R. (2000). Retroviral expression in embryonic stem cells and hematopoietic stem cells. *Mol. Cell. Biol.* **20**, 7419–7426.

60. Hamaguchi, I., Woods, N. B., Panagopoulos, I., Andersson, E., Mikkola, H., Fahlman, C., Zufferey, R., Carlsson, L., Trono, D., and Karlsson, S. (2000). Lentivirus vector gene expression during ES cell-derived hematopoietic development in vitro. *J. Virol.* **74**, 10778–10784.

61. Mikkola, H., Woods, N. B., Sjogren, M., Helgadottir, H., Hamaguchi, I., Jacobsen, S. E., Trono, D., and Karlsson, S. (2000). Lenitivirus gene transfer in murine hematopoietic progenitor cells is compromised by a delay in proviral integration and results in transduction mosaicism and heterogeneous gene expression in progeny cells. *J. Virol.* **74**, 11911–11918.

62. Heinzinger, N. K., Bukinsky, M. I., Haggerty, S. A. *et al.* (1994). The Vpr protein of human immunodeficiency virus type 1 influences nuclear localization of viral nucleic acids in nondividing host cells. *Proc. Natl. Acad. Sci. USA* **91**, 7311–7315.

63. Zufferey, R., Nagy, D., Mandel, R. J., Naldini, L., and Trono, D. (1997). Multiply attenuated lentiviral vector achieves efficient gene delivery in vivo. *Nat. Biotechnol.* **15**, 871–875.

64. Popov, S., Rexach, M., Ratner, L., Blobel, G., and Bukrinsky, M. (1998). Viral protein, R regulates docking of the HIV-1 preintegration complex to the nuclear pore complex. *J. Biol. Chem.* **273**, 13347–13352.

65. Kafri, T., Blomer, U., Peterson, D. A., Gage, F. H., and Verma, I. M. (1997). Sustained expression of genes delivered directly into liver and muscle by lentiviral vectors. *Nat. Genet.* **17**, 314–317.

66. Park, F., Ohashi, K., Chiu, W., Naldini, L., and Kay, M. A. (2000). Efficient lentiviral transduction of liver requires cell cycling in vivo. *Nat. Genet.* **24**, 49–52.

67. Anderson, W. F. (1998). Human gene therapy. *Nature* **392**, 25–30.

68. Dull, T., Zufferey, R., Kelly, M. *et al.* (1998). A third-generation lentivirus vector with a conditional packaging system. *J. Virol.* **72**, 8463–8471.

69. Wu, X., Wakefield, J. K., Liu, H. *et al.* (2000). Development of a novel trans-lentiviral vector that affords predictable safety. *Mol. Ther.* **2**, 47–55.

70. Kaplan, A. H., and Swanstrom, R. (1991). The HIV-1 gag precursor is processed via two pathways: implications for cytotoxicity. *Biomed. Biochem. Acta* **50**, 647–653.

71. Emerman, M. (1996). HIV-1, Vpr and the cell cycle. *Curr. Biol.* **6**, 1096–1103.

72. Burns, J. C., Friedmann, T., Driever, W., Burrascano, M., and Yee, J. K. (1993). Vesicular stomatitis virus G glycoprotein pseudotyped retroviral vectors: concentration to very high titer and efficient gene transfer into mammalian and nonmammalian cells. *Proc. Natl. Acad. Sci. USA* **90**, 8033–8037.

73. Baron, U., and Bujard, H. (2000). Tet repressor-based system for regulated gene expression in eukaryotic cells: principles and advances. *Meth. Enzymol.* **327**, 401–421.

74. Klages, N., Zufferey, R., and Trono, D. (2000). A stable system for the high-titer production of multiply attenuated lentiviral vectors. *Mol. Ther.* **2**, 170–176.

75. Ory, D. S., Neugeboren, B. A., and Mulligan, R. C. (1996). A stable human-derived packaging cell line for production of high titer retrovirus/vesicular stomatitis virus G pseudotypes. *Proc. Natl. Acad. Sci. USA* **93**, 11400–11406.

76. Kobinger, G. P., Weiner, D. J., Yu, Q. C., Wilson, J. M. (2001). Filovirus-pseudotyped lentiviral vector can efficiently and stably transduce airway epithelia in vivo. *Nature Biotech* **19**, 225–230.

77. Stitz, J., Buchholz, C. J., Engelstadter, M., Uckert, W., Bloemer, U., Schmitt, I., and Cichutek, K. (2000). Lentiviral vectors pseudotyped with envelope glycoproteins derived from gibbon ape leukemia virus and murine leukemia virus 10A1. *Virology* **273**, 16–20.

78. Salmon, P., Negre, D., Trono, D., and Cosset, F. L. (2000). A chimeric GALV-derived envelope glycoprotein harboring the cytoplasmic tail of MLV envelope efficiently pseudotypes HIV-1 vectors. *J. Gene Med.* **2**(5), S10.

79. Chinnasamy, N., Lewis, B. C., Morgan, R. A., and Varmus, H. E. (2000). Development of an avian leukosis virus envelope-pseudotyped HIV vector. *Mol. Ther.* **1**(5), S870.

80. Mikaelian, I., Kreig, M., Gait, M. J., and Karn, J. (1996). Interactions of INS (CRS) elements and the splicing machinery regulate the production of Rev-responsive mRNAs. *J. Mol. Biol.* **257**, 246–264.

81. Riviere, I., and Sadelain, M. (1997). Methods for the construction of retroviral vectors and the generation of high titer producers, in *Gene Therapy Protocols* (R. D. Robbins, ed.), pp. 59–78. Humana Press, Totowa, NJ.

82. Miyoshi, H., Blömer, U., Takahashi, M., Gage, F. H., and Verma, I. M. (1998). Development of a self-inactivating lentivirus vector. *J. Virol.* **10**, 8150–8157.

83. Zufferey, R., Dull, T., Mandel, R. J., Bukovsky, A., Quiroz, D., Naldini, L., and Trono, D. (1998). Self-inactivating lentivirus vector for safe and efficient in vivo gene delivery. *J. Virol.* **12**, 9873–9880.

84. Follenzi, A., Ailles, L. E., Bakovic, S., Geuna, M., and Naldini, L. (2000). Gene transfer by lentiviral vectors is limited by nuclear translocation and rescued by HIV-1 pol sequences. *Nat. Genet.* **25**, 217–222.

85. Tang, T., Kuhen, K. L., and Wong-Staal, F. (1999). Lentivirus replication and regulation. *Annu. Rev. Genet.* **33**, 133–170.

86. Zufferey, R., Donello, J. E., Trono, D., and Hope, T. J. (1999) Woodchuck hepatitis virus posttranscriptional regulatory element enhances expression of transgenes delivered by retroviral vectors. *J. Virol.* **73**, 2886–2892.

87. Schambach, A., Wodrich, H., Hildinger, M. *et al.* (2000). Context dependence of different modules for post-transcriptional enhancement of gene expression from retroviral vectors. *Mol. Ther.* **2**, 435–445.

88. Cone, R. D., Weber-Benarous, A., Baorto, D., and Mulligan, R. C. (1987). Regulated expression of a complete human beta-globin gene encoded by a transmissible retrovirus vector. *Mol. Cell. Biol.* **7**(2), 887–897.

89. Karlsson, S., Papayannopoulou, T., Schweiger, S. G., Stamatoyannopoulos, G., and Nienhuis, A. W. (1987). Retroviral-mediated transfer of genomic globin genes leads to regulated production of RNA and protein. *Proc. Natl. Acad. Sci. USA* **84**, 2411–2415.

90. Dzierzak, E. A., Papayannopoulou, T., and Mulligan, R. C. (1988). Lineage-specific expression of a human ß-globin gene in murine bone marrow transplant recipients reconstituted with retrovirus-transduced stem cells. *Nature* **331**, 35–41.

91. Bender, M. A., Miller, A. D., and Gelinas, R. E. (1988). Expression of the human beta-globin gene after retroviral transfer into murine erythroleukemia cells and human BFU-E cells. *Mol. Cell. Biol.* **8**(4), 1725–1735.

92. Karlsson, S., Bodine, D. M., Perry, L., Papayannopoulou, T., and Nienhuis, A. W. (1988). Expression of the human beta-globin gene following retroviral-mediated transfer into multipotential hematopoietic progenitors of mice. *Proc. Natl. Acad. Sci. USA* **85**(16), 6062–6066.

93. Bodine, D. M., Karlsson, S., and Nienhuis, A. W. (1989). Combination of interleukins 3 and 6 preserves stem cell function in culture and

enhances retrovirus-mediated gene transfer into hematopoietic stem cells. *Proc. Natl. Acad. Sci. USA* **86**(22), 8897–8901.

94. Bender, M. A., Gelinas, R. E., and Miller, A. D. (1989). A majority of mice show long-term expression of a human beta-globin gene after retrovirus transfer into hematopoietic stem cells. *Mol. Cell. Biol.* **9**(4), 1426–1434.

95. Grosveld, F., van Assendelft, G. B., Greaves, D. R., and Kollias, G. (1987). Position-independent, high-level expression of the human beta-globin gene in transgenic mice. *Cell* **51**(6), 975–985.

96. van Assendelft, G. B., Hanscombe, O., Grosveld, F., and Greaves, D. R. (1989). The beta-globin dominant control region activates homologous and heterologous promoters in a tissue-specific manner. *Cell* **56**(6), 969–977.

97. Sadelain, M. (1997). Genetic treatment of the haemoglobinopathies: recombinations and new combinations. *Br. J. Haematol.* **98**, 247–253.

98. Raftopoulos, H., Ward, M., Leboulch, P., and Bank, A. (1997). Long-term transfer and expression of the human beta-globin gene in a mouse transplant model. *Blood* **90**(9), 3414–3422.

99. Halene, S., and Kohn, D. B. (2000). Gene therapy using hematopoietic stem cells: Sisyphus approaches the crest. *Hum. Gene Ther.* **11**, 1259–1267.

100. Hanenberg, H. *et al.* (1997). Optimization of fibronectin-assisted retroviral gene transfer into human CD34+ hematopoietic cells. *Hum. Gene Ther.* **8**, 2193–2206.

101. Hennemann, B. *et al.* (1999). Optimization of retroviral-mediated gene transfer to human NOD/SCID mouse repopulating cord blood cells through a systematic analysis of protocol variables. *Exp. Hematol.* **27**, 817–825.

102. Goerner, M. *et al.* (1999). The use of granulocyte colony-stimulating factor during retroviral transduction on fibronectin fragment CH-296 enhances gene transfer into hematopoietic repopulating cells in dogs. *Blood* **94**(7), 2287–2292.

103. Murray, L. *et al.* (1999). Optimization of retroviral gene transduction of mobilized primitive hematopoietic progenitors by using thrombopoietin, Flt3, and Kit ligands and RetroNectin culture. *Hum. Gene. Ther.* **10**(11), 1743–1752.

104. Dao, M. A. *et al.* (1998). Engraftment and retroviral marking of CD34+ and CD34+CD38–human hematopoietic progenitors assessed in immune-deficient mice. *Blood* **91**(4), 1243–1255.

105. Dao, M. A. *et al.* (1997). FLT3 ligand preserves the ability of human CD34+ progenitors to sustain long-term hematopoiesis in immune-deficient mice after ex vivo retroviral-mediated transduction. *Blood* **89**(2), 446–456.

106. Malech, H. L. *et al.* (2000). Multiple cycles of ex vivo gene therapy for X-linked chronic granulomatous disease (CGD) sustain production of oxidase-normal peripheral blood neutrophils. *Mol. Ther.* **1**(5), S146.

107. Malech, H. L. (2000). Use of serum-free medium with fibronectin fragment enhanced transduction in a system of gas permeable plastic containers to achieve high levels of retrovirus transduction at clinical scale. *Stem Cells* **18**(2), 155–156.

108. Abonour, R. *et al.* (2000). Efficient retrovirus-mediated transfer of the multidrug resistance 1 gene into autologous human long-term repopulating hematopoietic stem cells. *Nat. Med.* **6**, 652–658.

109. Matsunaga, T. *et al.* (1998). Negative regulation by interleukin-3 (IL-3) of mouse early B-cell progenitors and stem cells in culture: transduction of the negative signals by β and βIL-3 proteins of IL-3 receptor and absence of negative regulation by granulocyte–macrophage colony-stimulating factor. *Blood* **92**(3), 901–907.

110. Huhn, R. D. *et al.* (1999). Retroviral marking and transplantation of rhesus hematopoietic cells by nonmyeloablative conditioning. *Hum. Gene Ther.* **10**(11), 1783–1790.

111. Cheng, T., Rodrigues, N., Shen, H., Yang, Y., Dombkowski, D., Sykes, M., and Scadden, D. T. Hematopoietic stem cell quiescence maintained by p21cip1/waf1. *Science* **287**, 1804–1808.

112. Habibian, II. K., Peters, S. O., Hsieh, C. C. *et al.* (1998). The fluctuating phenotype of the lymphohematopoietic stem cell with cell cycle transit. *J. Exp. Med.* **188**, 393.

113. Lataillade, J. J., Clay, D., Dupuy, C., Rigal, S., Jasmin, C., Bourin, P., and Le Bousse-Kerdiles, M. C. (2000). Chemokine SDF-1 enhances circulating CD34(+) cell proliferation in synergy with cytokines: possible role in progenitor survival. *Blood* **95**(3), 756–768.

114. Peled, A., Kollet, O., Ponomaryov, T., Petit, I., Franitza, S., Grabovsky, V., Slav, M. M., Nagler, A., Lider, O., Alon, R., Zipori, D., and Lapidot, T. The chemokine SDF-1 activates the integrins LFA-1, VLA-4, and VLA-5 on immature human CD34(+) cells: role in transendothelial/stromal migration and engraftment of NOD/SCID mice. *Blood* **95**(11), 3289–3296.

115. Miyoshi, H., Smith, K. A., Mosier, D. E., Verma, I. M., and Torbett, B. E. (1999). Transduction of human CD34+ cells that mediate long-term engraftment of NOD/SCID mice by HIV vectors. *Science* **283**, 682–686.

116. Sutton, R. E., Wu, H. T. M., Rigg, R., Bohnlein, E., and Brown, P. O. (1998). HIV type 1 vectors efficiently transduce human hematopoietic stem cells. *J. Virol.* **72**, 5781–5788.

117. Case, S. S., Price, M. A., Jordan, C. T., Yu, X. J., Wang, L., Bauer, G., Haas, D. L., Xu, D., Stripecke, R., Naldini, L., Kohn, D. B., and Crooks, G. M. (1999). Stable transduction of quiescent CD34(+)CD38(–) human hematopoietic cells by HIV-1-based lentiviral vectors. *Proc. Natl. Acad. Sci. USA* **96**, 2988–2993.

118. Uchida, N., Sutton, R. E., Friera, A. M. *et al.* (1998). HIV, but not MLV, vectors mediate high efficiency gene transfer into freshly isolated G0/G1 human hematopoietic stem cells. *Proc. Natl. Acad. Sci. USA* **95**, 11939–11944.

119. Evans, J. T., Kelly, P. F., O'Neill, and E., Garcia, J. V, (1999). Human cord blood CD34+CD38–cell transduction via lentivirus-based gene transfer vectors. *Hum. Gene Ther.* **10**, 1479–1489.

120. Sutton, R. E., Reitsman, M. J., Uchida, N., and Brown, P. O. (1999). Transduction of human progenitor hematopoietic stem cells by human immunodeficiency virus type 1-based vectors is cell cycle dependent. *J. Virol.* **73**, 3649–660.

121. Guenechea, G., Gan, O. I., Inamitsu, T. *et al.* (2000). Transduction of human CD34+CD38–bone marrow and cord blood-derived SCID-repopulating cells with third-generation lentiviral vectors. *Mol. Ther.* **1**, 566–573.

122. Sirven, A., Pflumio, F., Zennou, V. *et al.* (2000). The GIV-1 central DNA flap is a critical determinant for lentiviral vector nuclear import and gene transduction of human hematopoietic stem cells. *Blood* **96**, 4103–4110.

123. An, D. S., Kung, S. K. P., Wersto, R. P., Agricola, B. A., Metzger, M. E., Mao, S. H., Chen, I. S. Y., and Donahue, R. E. (2000). *Mol. Ther.* **1**(5), S44.

124. Takatoku, M., Sellers, S., Agricola, B. A., Metzger, M. E., Donahue, R. E., and Dunbar, C. E. (2000). Avoidance of active cell cycle improves hematopoietic stem cell retroviral gene marking in non-human primates. *Mol. Ther.* **1**(5), S32.

125. O'Reilly, R. J. *et al.* (1998). Adoptive immunotherapy for Epstein–Barr virus-associated lymphoproliferative disorders complicating marrow allografts. *Springer Semin. Immunopathol.* **20**, 455–491.

126. Brenner, M. K., Heslop, H. E., and Rooney, C. M. (1998). Gene and cell transfer for specific immunotherapy. *Vox Sang.* **2**, 87–90.

127. Bonini, C. *et al.* (1997). HSV-tk gene transfer into donor lymphocytes for control of allogeneic graft-versus-leukemia. *Science* **276**, 1719–1724.

128. Sadelain, M., and Luzzatto, L. (1997). Immunogene therapy comes into its own. *Gene Ther.* **4**, 1129–1131.

129. Sadelain, M. (2000). A friendly merger in the war on cancer. *Mol. Ther.* **1**, 115–116.

130. Koehne, G., Gallardo, H. F., Sadelain, M., and O'Reilly, R. J. (2000). Rapid selection of antigen-specific T lymphocytes by retroviral transduction. *Blood* **96**, 109–117.

131. Stevenson, M., Stanwick, T. L., Dempsey, M. P. *et al.* (1990). HIV-1 replication is controlled at the level of T cell activation and proviral integration. *EMBO J.* **9**, 1551–1560.

132. Spina, C. A., Guatelli, J. C., Richman, D. D. *et al.* (1995). Establishment of a stable, inducible form of human HIV-1 DNA in quiescent CD4 lymphocytes in vitro. *J. Virol.* **69**, 2977–2988.

133. Trono, D. (1995). HIV accessory proteins: leading roles for the supporting cast. *Cell* **82**, 189–192.

134. Swingler, S., Mann, A., Jacque, J. *et al.* (1999). HIV-1 Nef mediates lymphocyte chemotaxis and activation by infected macrophages. *Nat. Med.* **5**, 997–1003.

135. Von Schwedler, U., Song, J., Aiken, C. *et al.* (1993). Vif is crucial for human immunodeficiency virus type 1 proviral DNA synthesis in infected cells. *J. Virol.* **67**, 4945–4955.

136. Simon, J. H., and Malim, M. H. (1996). The human immunodeficiency virus type 1 Vif protein modulates the postpenetration stability of viral nucleoprotein complexes. *J. Virol.* **70**, 5297–5305.

137. Chinnasamy, D., Chinnasamy, N., Enriquez, M. J., Otsu, M., Morgan, R. A., and Candotti, F. (2000). Lentiviral-mediated gene transfer into human lymphocytes: role of HIV-1 accessory proteins. *Blood* **96**, 1309–1316.

138. Costello, E., Munoz, M., Buetti, E., Meylan, P. R., Diggelmann, H., and Thali, M. (2000). Gene transfer into stimulated and unstimulated T lymphocytes by HIV-1-derived lentiviral vectors. *Gene Ther.* **7**, 596–604.

139. Unutmaz, D., KewalRamani, V. N., Marmon, S., and Littman, D. R. (1999). Cytokine signals are sufficient for HIV-1 infection of resting human T lymphocytes. *J. Exp. Med.* **189**, 1735–1746.

140. Banchereau, J., and Steinman, R. M. (1998). Dendritic cells and the control of immunity. *Nature* **392**, 245–252.

141. Bell, D., Young, J. W., and Banchereau, J. (1999). Dendritic cells. *Adv. Immunol.* **72**, 255–324.

142. Timmerman, J. M., and Levy, R. (1999). Dendritic cell vaccines for cancer immunotherapy. *Annu. Rev. Med.* **50**, 507–529.

143. Reeves, M. E., Royal, R. E., Lam, J. S., Rosenberg, S. A., and Hwu, P. (1996). Retroviral transduction of human dendritic cells with a tumor-associated antigen gene. *Cancer Res.* **56**, 5672–5677.

144. Szabolcs, P., Gallardo, H. F., Ciocon, D. H., Sadelain, M., and Young, J. W. (1997). Retrovirally transduced human dendritic cells express a normal phenotype and potent T-cell stimulatory capacity. *Blood* **90**, 2160–2167.

145. Bello-Fernandez, C., Matyash, M., Strobl, H. *et al.* (1997). Efficient retrovirus-mediated gene transfer of dendritic cells generated from CD34+ cord blood cells under serum-free conditions. *Hum. Gene Ther.* **8**, 1651–1658.

146. Henderson, R. A., Konitsky, W. M., Barratt-Boyes, S. M., Soares, M., Robbins, P. D., and Finn, O. J. (1998). Retroviral expression of MUC-1 human tumor antigen with intact repeat structure and capacity to elicit immunity in vivo. *J. Immunother.* **21**, 247–256.

147. Chischportich, C., Bagnis, C., Galindo, R., and Mannoni, P. Expression of the nlsLacz gene in dendritic cells derived from retrovirally transduced peripheral blood CD34+ cells. *Haematologica* **84**, 195–203.

148. Movassagh, M., Baillou, C., Cosset, F. L., Klatzmann, D., Guigon, M., and Lemoine, F. M. (1999). High level of retrovirus-mediated gene transfer into dendritic cells derived from cord blood and mobilized peripheral blood CD34+ cells. *Hum. Gene Ther.* **10**, 175–187.

149. O'Doherty, U., Steinman, R. M., Peng, M. *et al.* (1993). Dendritic cells freshly isolated from human blood express CD4 and mature into typical immunostimulatory dendritic cells after culture in monocyte-conditioned medium. *J. Exp. Med.* **178**, 1067–1076.

150. Granelli-Piperno, A., Delgado, E., Finkel, V., Paxton, W., and Steinman, R. M. (1998). Immature dendritic cells selectively replicate macrophage-tropic (M-tropic) human immunodeficiency virus type 1, while mature cells efficiently transmit both M- and T-tropic virus to T cells. *J. Virol.* **72,** 2733–2737.

151. Schreurs, M. W., Eggert, A. A., de Boer, A. J., Figdor, C. G., Adema, G. J. (1999). Generation and functional characterization of mouse monocyte-derived dendritic cells. *Eur. J. Immunol.* **29,** 2835–2841.

152. Gruber, A., Kan-Mitchell, J., Kuhen, K. L. *et al.* (2000). Dendritic cells transduced by multiply deleted HIV-1 vectors exhibit normal phenotypes and functions and elicit an HIV-specific CTL response in vitro. *Blood* **96,** 1327–1333.

153. Schroers, R., Sinha, I., Segall, H. *et al.* (2000). Transduction of human PBMC-derived dendritic cells and macrophages by an HIV-1-based lentiviral vector system. *Mol. Ther.* **1,** 171–179.

154. Dietz, A. B., and Vuk-Pavlovic, S. (1998). High efficiency adenovirus-mediated gene transfer to human dendritic cells. *Blood* **91,** 392–398.

155. Coffin, R. S., Thomas, S. K., Thomas, N. S. *et al.* (1998). Pure populations of transduced primary human cells can be produced using GFP expressing herpes virus vectors and flow cytometry. *Gene Ther.* **5,** 718–722.

156. Subklewe, M., Chahroudi, A., Schmaljohn, A., Kurilla, M. G., Bhardwaj, N., and Steinman, R. M. (1999). Induction of Epstein–Barr virus-specific cytotoxic T-lymphocyte responses using dendritic cells pulsed with EBNA-3A peptides or UV-inactivated, recombinant EBNA-3A vaccinia virus. *Blood* **94,** 1372–1381.

157. Chaux, P., Luiten, R., Demotte, N. *et al.* (1999). Identification of five MAGE-A1 epitopes recognized by cytolytic T lymphocytes obtained by in vitro stimulation with dendritic cells transduced with MAGE-A1. *J. Immunol.* **163,** 2928–2936.

158. Brown, M., Davies, D. H., Skinner, M. A. *et al.* (1999). Antigen gene transfer to cultured human dendritic cells using recombinant avipoxvirus vectors. *Cancer Gene Ther.* **6,** 238–245.

159. Mascarenhas, L., Stripecke, R., Case S. S., Xu, D., Weinberg, K. I., and Kohn, D. B. (1998). Gene delivery to human B-precursor acute lymphoblastic leukemia cells. *Blood* **92,** 3527–3545.

160. Stipecke, R., Cardoso, A. A., Pepper, K. A., Skelton, D. C., Yu, X. J., Mascarenhas, L., Weinber, K. I., Nadler, L. M., and Kohn, D. B. (2000). Lentiviral vectors for efficient delivery of CD80 and granulocyte–macrophage colony-stimulating factor in human acute lymphoblastic leukemia and acute myeloid leukemia cells to induce antileukemic immune responses. *Blood* **96,** 1317–1326.

161. Koehne, G., Gallardo, H. F., Latouche, J.-B., *et al.* (2000). Controlled graft-versus-leukemia reaction after adoptive transfer of hsvrtk modi-fied allogeneic donor T lymphocytes sensitized with B7.1-transduced ALL cells. *Blood* **96,** 550a.

162. Gerolami, R., Uch, R., Jordier, F., *et al.* (2000). Gene transfer to hepatocellular carcinoma: transduction efficacy and transgene expression kinetics by using retroviral and leutiviral vectors. *Cancer Gene Therapy* **7,** 1286–1292.

163. DePolo, N. J., Reed, J. D., Sheridan, P. L., Townsend, K., Sauter, S. L., Jolly, D. J., and Dubensky, Jr., T. W. (2000). VSV-G pseudotyped lentiviral vector particles produced in human cells are inactivated by human serum. *Mol. Ther.* **2,** 218–222.

164. Blomer, U., Naldini, L., Kafri, T. *et al.* (1997). Highly efficient and sustained gene transfer in adult neurons with a lentiviral vector. *J. Virol.* **71,** 6641–6649.

165. Naldini, L. (2000). In vivo gene delivery by lentiviral vectors. *Thromb. Haemost.* **82,** 552–554.

166. Leibowitz, G., Beattie, G. M., Kafri, T., Cirulli, V., Lopez, A. D., Hayek, A., and Levine, F. (1999). Gene transfer to human pancreatic endocrine cells using viral vectors. *Diabetes* **48,** 745–753.

167. Giannoukakis, N., Mi, Z., Gambotto, A. *et al.* (1999). Infection of intact human islets by a lentiviral vector. *Gene Ther.* **6,** 1545–1551.

168. Kordower, J. H., Emborg, M. E., Bloch, J. *et al.* (2000). Neurodegeneration prevented by lentiviral vector delivery of GDNF in primate models of Parkinson's disease. *Science* **290,** 767–773.

169. Poeschla, E. M., Wong-Staal, F., and Looney, D. J. (1998). Efficient transduction of nondividing human cells by feline immunodeficiency virus lentiviral vectors. *Nat. Med.* **4,** 354–357.

170. White, S., Renda, M., Nam, N. Y. *et al.* (1999). Lentivirus vectors using human and simian immunodeficiency virus elements. *J. Virol.* **73,** 2832–2840.

171. Curran, M. A., Kaiser, S. M., Achacoso, P. L., and Nolan, G. P. (2000). Efficient transduction of nondividing cells by optimized FIV vectors. *Mol. Ther.* **1,** 31–38.

172. Ismail, S. I., Kingsman, S. M., Kingsman, A. J., and Uden, M. (2000). Split-intron retroviral vectors: enhanced expression with improved safety. *J. Virol.* **74,** 2365–2371.

173. Wagner, R., Graf, M., Bieler, K., Wolf, H., Grunwald, T., Foley, P., and Uberla, K. (2000). Rev-independent expression of synthetic *gag–pol* genes of human immunodeficiency virus type 1 and simian immunodeficiency virus: implications for the safety of lentiviral vectors. *Hum. Gene Ther.* **20,** 2403–2413.

174. Russell, S. J., and Cosset, F. L. (1999). Modifying the host range properties of retroviral vectors. *J. Gene Med.* **1,** 300–311.

175. Sheridan, P. L., Bodner, M., Lynn, A., *et al.* (2000). Generation of retroviral packaging and producer cell lines for large-scale vector production and clinical application: improved safety and high titer. *Mol. Therapy* **2,** 262–275.

PART II

IMMUNE TARGETED GENE THERAPY

CHAPTER

7

Immunologic Targets for the Gene Therapy of Cancer

SUZANNE OSTRAND-ROSENBERG
Department of Biological Sciences
University of Maryland
Baltimore, Maryland 21250

VIRGINIA K. CLEMENTS
Department of Biological Sciences
University of Maryland
Baltimore, Maryland 21250

SAMUDRA DISSANAYAKE
Department of Biological Sciences
University of Maryland
Baltimore, Maryland 21250

MILEKA GILBERT
Department of Biological Sciences
University of Maryland
Baltimore, Maryland 21250

BETH A. PULASKI
Department of Biological Sciences
University of Maryland
Baltimore, Maryland 21250

LING QI
Department of Biological Sciences
University of Maryland
Baltimore, Maryland 21250

I. INTRODUCTION

For at least 100 years immunologists have proposed activating the immune system to specifically target and eradicate autologous tumor cells. Until the advent of molecular gene transfer techniques and increased knowledge of the basic pathways of lymphocyte activation, however, effective methods for harnessing the immune system as a therapeutic agent were unsuccessful, despite numerous efforts and enthusiasm for the approaches. During the last 10–15 years, however, the field of immunotherapy of cancer has been inundated with novel strategies for the treatment of malignancies. As a result, there is considerable optimism in both the basic science and clinical arenas for pursing immunotherapy strategies. The renewed enthusiasm for immunotherapy is largely due to the recent and cumulative expansion of knowledge in basic immunology and molecular biology and the rapidly improving understanding of regulation of the immune response. Acquisition of this basic information has allowed tumor immunologists to design approaches to control immune responses against tumors, leading, at least in some cases, to impressive antitumor responses in experimental systems. Tumor immunologists are now at the cusp of applying what has been learned and successfully applied in animal models to the treatment of human cancers.

Many of the current immunotherapy/gene therapy strategies are presented in detail in subsequent chapters of this book. To fully appreciate and understand these strategies, it is necessary to understand how the immune system responds to antigen and to appreciate the characteristics of tumor cells that make them potential targets for an immune response. The goal of this chapter, therefore, is to provide a concise overview of the induction of immunity and to clarify how an immune response can be specifically enhanced and directed towards tumor cells.

II. CELLULAR (T-LYMPHOCYTE-MEDIATED) VERSUS HUMORAL (ANTIBODY-MEDIATED) IMMUNE RESPONSES TO TUMOR CELLS

The immune system mediates two basic types of responses: Antibody-mediated (B-cell) responses and cell-mediated (T-cell) responses. Antibodies usually bind to conformational (three-dimensional) epitopes of intact molecules and, therefore, are very effective agents against extracellular and soluble pathogens. They also bind to denatured protein or small peptides, although their binding affinity for these structures is usually lower than for conformational epitopes. Because antibodies cannot penetrate a plasma membrane, antibodies are effective antitumor reagents only if conformational tumor antigens (TAs) and/or antigenic fragments are expressed at the tumor cell surface.

Monoclonal antibodies (mAbs) and recombinant antibodies have been used in cancer immunotherapy in a variety of settings. They have been tagged with markers such as radioisotopes or fluorescent probes and used as imaging reagents to localize tumors. They have also been coupled to toxic agents, such as radioisotopes or toxins, in an attempt to specifically deliver a "lethal hit" to targeted tumor cells. Most of the antibody approaches involve passive transfer of *in vitro* generated reagents, which can mediate immediate antitumor responses but which do not result in long-term immunological memory. Two recombinant, humanized mAbs—Herceptin (trastuzamab, specific for HER2/*neu* on breast and other tumors) and Rituximab (specific for CD20 on non-Hodgkin's lymphoma)—were recently approved for clinical use [1]. Additional antibodies have shown treatment potential in patients with established tumor and are undergoing continual development to decrease their inherent immunogenicity and to increase their targeting specificity [1–5].

The utility of antibodies as therapeutic agents is obviously dependent on their specific binding to tumor antigens and to their binding affinity. Until recently, the identification of new tumor-specific or tumor-reactive antibodies depended on conventional immunization techniques and the availability of tumor cells for screening the resulting monoclonal antibodies. The recently developed SEREX approach, however, may greatly increase the number and repertoire of serologically defined tumor antigens. SEREX, an acronym for serological analysis of tumor antigens by recombinant expression cloning, uses cancer patients' sera to screen autologous tumor cell cDNA expression libraries [6,7]. SEREX has already been used to identify several human tumor antigens that were previously identified via T-cell reactivity, as well as to identify previously unknown antigens from a variety of human cancers [6–10]. Because SEREX uses patients' sera as the detection system, the tumor antigens characterized are immunogenic in tumor-bearing individuals and, therefore, may be useful targets for antibody-mediated immunotherapy.

T lymphocytes, which mediate cell-mediated immunity, are effective agents against intracellular pathogens and malignant cells because they recognize antigens synthesized within target cells. Many therapeutic strategies involving T-cell activation are also aimed at inducing active antitumor immunity in the tumor-bearing patient, such that immunological memory will be induced. Induction of immunological memory to tumor antigens has several advantages in a therapy setting. For example, many metastatic lesions are refractory to conventional treatments. Induction of protective immunity against metastatic cells has the obvious benefit of controlling malignant cell growth and providing long-term protection against distant metastases that are otherwise not treatable. Likewise, active antitumor immunity would provide long-term protection against recurrence of primary tumor. Many studies, therefore, have explored the harnessing and enhancing of T-lymphocyte-mediated immunity to malignant cells, and most of these approaches involve gene therapy. The remainder of this chapter will focus on T-cell responses to tumor cells.

III. RESPONSE OF CD4+ AND CD8+ T LYMPHOCYTES TO TUMOR ANTIGENS PRESENTED IN THE CONTEXT OF MOLECULES ENCODED BY THE MAJOR HISTOCOMPATIBILITY COMPLEX

T lymphocytes recognize target cells via their T-cell receptor for antigen (TcR). The TcR is a membrane-anchored heterodimeric molecule with specificity for antigen and is associated with additional protein molecules known as the CD3 complex. The TcR/CD3 complex of T cells interacts with antigen bound to a major histocompatibility complex (MHC)-encoded molecule on the antigen-presenting cell (APC) or target cell. CD8+ cytotoxic T lymphocytes (CTLs) typically recognize antigen bound to MHC class I molecules, while CD4+ T-helper lymphocytes (T_h cells) recognize antigen bound to MHC class II molecules. Although specificity of antigen recognition is accomplished by the TcR, additional accessory molecules expressed by the T cell, such as CD4, CD8, ICAMs (intracellular adhesion molecules) and LFAs (lymphocyte function associated antigens) are also involved in T-cell activation. By binding to their cognate receptors on the APC, these accessory molecules stabilize the binding of the T cell to its target cell, and in some cases also deliver specific activation signals to the T lymphocyte [11,12].

A. CD8+ CTLs Recognize Endogenously Synthesized Antigens

The antigen bound by the MHC and TcR is a processed peptide fragment derived from a larger intact protein. MHC-class-I-bound peptides are typically derived from proteins synthesized within the target cell and hence are called "endogenously" synthesized proteins. These molecules are processed by the proteosome of the cell into peptides that are transported into the endoplasmic reticulum (ER) where they are bound to newly synthesized MHC class I molecules. The resulting MHC class I/peptide complex then traffics through the default secretory pathway to the target cell membrane, where it is anchored via its hydrophobic region. Figure 1 shows a schematic diagram of the endogenous trafficking pathway (also called the "cytosolic pathway").

Because of this pathway, most endogenously synthesized self proteins can be presented by MHC class I molecules and hence be available as potential targets for CD8+ CTLs. CD8+ T cells, therefore, are particularly good candidates for recognizing tumor cells, because many tumor antigens are endogenously synthesized molecules that are not expressed at the cell surface in intact form but are presented as peptides bound to endogenously synthesized MHC class I molecules [13,14].

B. CD4+ T Lymphocytes Recognize Exogenously Synthesized Antigens

Similar to CD8+ T lymphocytes, CD4+ T cells recognize processed peptide fragments of antigen; however, in the case of CD4+ T cells, processed peptide is bound by MHC class II instead of MHC class I molecules. In addition, the peptides recognized by CD4+ T cells are from a different source than those bound by CD8+ T cells, and they associate with MHC class II molecules via a different pathway than peptides presented by MHC class I molecules. MHC class II molecules are usually expressed by a subset of cells called "professional" antigen-presenting cells, such as B lymphocytes, dendritic cells (DCs), and macrophages. Only these cells are able to bind peptide antigen to their MHC class II molecules and "present" it to CD4+ T lymphocytes, which are subsequently activated (see Section III.C). In order to present class-II-restricted antigen, professional APCs phagocytose or endocytose soluble antigen. CD4+ T cells, therefore, are typically activated by antigen synthesized outside of the APC, or exogenous antigen. Following internalization, antigen is enzymatically degraded or "processed" within the endosomal compartment of the APC, and the resulting peptides then bind to class II molecules. The antigenic peptide/MHC class II complex is then transferred and inserted into the plasma membrane. Unlike class I molecules, class II molecules traffic to an endosomal compartment because they associate in the endoplasmic reticulum with a chaperon molecule called invariant (Ii) chain. Ii chain has two functions: (1) When tightly bound to class II in the ER, Ii prevents binding of endogenously synthesized peptides to class II molecules; and (2) Ii contains a trafficking motif that guides the class II/Ii complex to the endosomal compartment. Once in the endosomal compartment, the Ii chain is progressively degraded and dissociates from the class II molecule. Ii chain dissociation exposes the peptide-binding region of class II molecules. Processed, exogenously synthesized peptides, which have entered the endosome by phagocytosis, are then able to bind to the peptide-binding site of class II molecules. Peptide binding is facilitated by the accessory molecule, HLA-DM [13–15]. (See Fig. 1 for a schematic diagram of the exogenous antigen presentation pathway.)

The MHC class I and class II antigen processing and presentation pathways, therefore, yield very different repertoires of antigenic peptides presented to CD8+ versus CD4+ T cells.

C. Activation of Tumor-Specific CD4+ and CD8+ T Lymphocytes Requires Two Signals and Is Usually Mediated by Professional Antigen-Presenting Cells

Extensive studies have demonstrated that activation of T lymphocytes requires two signals, as shown in Fig. 2A. The first signal is the antigen-specific signal received by the

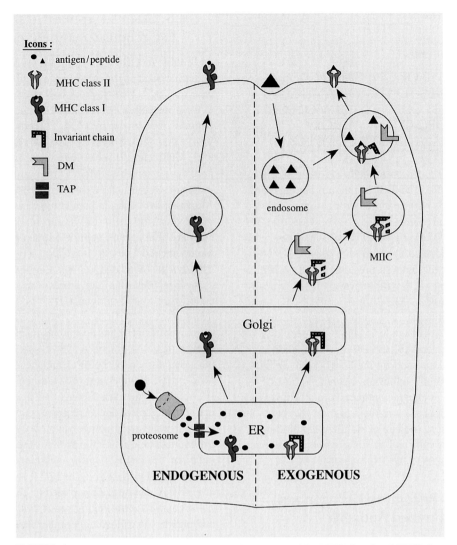

FIGURE 1 Trafficking pathways of antigenic peptides presented by MHC class I and MHC class II molecules. Molecules synthesized within the APC (endogenously synthesized antigen) traffic via the endogenous or cytosolic pathway. They are digested in the proteosome into peptides and transported by TAP into the ER where they bind to newly synthesized MHC class I molecules, which themselves are bound to another protein called β2-microglobulin. The MHC class I/peptide complex then travels by the secretory pathway to the plasma membrane, where the hydrophobic region of the class I molecule anchors the complex in the membrane, and the peptide/external domain of the class I molecule is displayed. Molecules synthesized outside of the APC (exogenously synthesized antigen) are internalized via endocytosis and digested within endosomal compartments. Newly synthesized MHC class II molecules bind the class II-associated Ii chain in the ER, and the class II/Ii complexes are directed to endosomal compartments by trafficking signals contained in the Ii chain. Within the increasingly acidic environment of endosomes, the Ii chain is progressively degraded and dissociates from the class II molecule, revealing the peptide-binding cleft of the class II molecule. With the help of the accessory molecule HLA-DM, peptides are then loaded onto class II. The class II/peptide complex is then shuttled to the cell surface, where it is anchored by its hydrophobic region into the plasma membrane, and the antigenic peptide/external domain of the class II molecule is available for binding to a CD4[+] T lymphocyte.

responding CD4[+] or CD8[+] T cell when its TcR binds to the MHC/peptide complex on the APC. The second or costimulatory signal is delivered to the T lymphocyte when a costimulatory molecule, such as CD80 (B7.1) or CD86 (B7.2) on the APC binds to its cognate receptor, CD28, on the responding T cell. Because only professional APCs express costimulatory molecules, only these APCs are capable of activating T cells. Recent studies with DCs indicate these cells are particularly effective at priming and activating tumor-specific T cells [16,17]. CD80 and CD86 are constitutively expressed

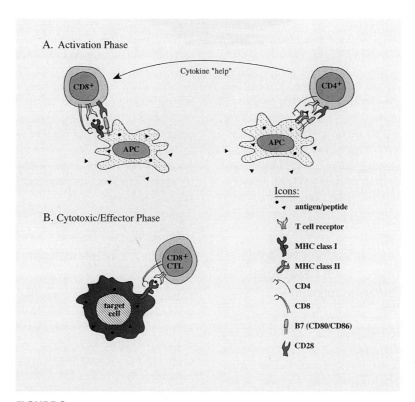

FIGURE 2   Role of MHC class I, class II, and costimulatory molecules in the activation and effector function of CD8$^+$ and CD4$^+$ T lymphocytes. (A) CD8$^+$ and CD4$^+$ T lymphocytes require two signals for activation. The first signal is an antigen-specific signal delivered when the peptide/MHC class I or peptide/MHC class II complex binds to the TcR of CD8$^+$ or CD4$^+$ T cells, respectively. The second signal is delivered when a costimulatory molecule on the APC binds to its counter-receptor, CD28, on the responding CD8$^+$ or CD4$^+$ T lymphocyte. (B) Activated CD8$^+$ CTLs only need to recognize and bind to the appropriate peptide/MHC class I complex on the target cell to initiate cytolysis. (*Note*: Other accessory molecules, such as ICAM and LFA, which stabilize T cell/APC or T cell/target cell binding, are not shown.)

on dendritic cells and are inducible on B lymphocytes. On B cells, CD86 levels reach their maximum expression between 24 and 96 hours after B-cell activation, while CD80 levels peak between 48 and 72 hours after B-cell activation. CD80 and CD86, as well as the adhesion molecules ICAM and LFA-3, are induced on APC following binding of APC-expressed CD40 with its ligand CD154 expressed on T cells [18–21]. Expression of T-cell costimulatory molecules, therefore, is fluid and depends on the activation state of the APC and interaction of the APC with the responding T cell [22,23].

If the antigen-specific signal is delivered in the absence of the costimulatory signal, then the responding T lymphocytes are inactivated (anergized) rather than activated. Although different APCs can deliver the antigen-specific and costimulatory signals, T cells are most efficiently activated if both signals are delivered by the same APC [24,25]. As a result, somatic cells that present MHC class I molecules with self peptides and do not express costimulatory ligands may anergize T cells because they are unable to deliver the requisite costimulatory signal. Indeed, it has been hypothesized that the

CD8$^+$ T cells of tumor-bearing individuals are "anergized" or "tolerized" to their tumors because MHC class I$^+$ tumor cells present tumor antigens, but do not deliver a costimulatory signal [26,27]. Tolerization of tumor-specific CD4$^+$ T lymphocytes is less likely to be a problem than tolerization of CD8$^+$ T cells because professional APCs that activate CD4$^+$ T cells usually coexpress MHC class II molecules and costimulatory molecules.

Although costimulatory signals are required to activate naive T lymphocytes, they are not needed to reactivate memory T cells, nor are they needed as target molecules for T-cell-mediated cytolysis [28], as shown in Fig. 2B. Cell-based vaccines consisting of autologous tumor cells transfected/transduced with CD80, CD86, and/or CD40 genes, therefore, have been generated. Such vaccines activate T cells, which subsequently destroy unmodified, wild-type tumor cells *in situ* [26,27,29]. Although these vaccines were initially designed to directly activate naive tumor-specific T cells, some studies suggest that host-derived professional APCs, rather than the tumor cells themselves, are the relevant APCs *in vivo* [30,31].

In addition to interacting with CD28, B7.1 and B7.2 have a second receptor on T cells: cytotoxic T lymphocyte antigen 4 (CTLA4). In contrast to B7/CD28 interactions, which are stimulatory, B7/CTLA4 interactions block activation and promote T-cell death or apoptosis. The success of T-cell activation, therefore, hinges on signaling via CD28 instead of via CTLA4. One reported antitumor strategy capitalized on the differential function of CD28 versus CTLA4 and demonstrated that mice given anti-CTLA4 mAbs develop a potent antitumor immunity. Presumably, the anti-CTLA4 antibodies sterically inhibited signaling through CTLA4, thus favoring signalling and subsequent T-cell activation through CD28 [32].

D. Effective Antitumor Immunity Probably Requires Activation of Both CD4$^+$ and CD8$^+$ T Lymphocytes

Although CD8$^+$ T cells can be activated in the absence of CD4$^+$ T helper lymphocytes, optimal CD8$^+$ T-cell activation and long-term immunological memory probably require coactivation of CD4$^+$ T lymphocytes. The added benefit of CD4$^+$ T-cell help is particularly apparent in tumor immunity, where numerous studies have demonstrated that improved CD4$^+$ T cell help facilitates tumor rejection [33–37].

Differentiated CD4$^+$ T lymphocytes have been categorized into either T helper type 1 or type 2 (T_h1 or T_h2, respectively). T_h1 cells predominantly secrete the cytokines IL-2 and interferon-gamma (IFN-γ, whereas T_h2 cells secrete interleukin-4 (IL-4), IL-10, and IL-13. T_h1 cells are generally considered to provide cytokine help to CD8$^+$ T cells, whereas T_h2 cells provide cytokine help to B lymphocytes. Because of this functional dichotomy, tumor immunologists have hypothesized that CD4$^+$ T_h1 lymphocytes facilitate CD8$^+$ T-cell-mediated tumor immunity [38]. Induction of optimal antitumor immunity, therefore, should involve activation of both CD4$^+$ T_h1 and CD8$^+$ tumor-specific T cells.

IV. RESPONSE OF TUMOR-BEARING INDIVIDUALS TO TUMOR ANTIGENS

The active immunization therapies currently under development are dependent on the tumor-bearing patient's capacity to generate an immune response against the tumor. The immunocompetence of the host, therefore, is very important in assessing if an individual is a reasonable candidate for immunotherapy. Both anecdotal and experimental data indicate that tumor-bearing individuals, especially ones with advanced disease, have significantly reduced immunocompetence. The extent and onset of the immunosuppression varies depending on the type of tumor and tumor burden [39,40]. For example, in some cases, patients or animals with experimental tumors are fully immunocompetent for all antigens except for antigens present on the autologous tumor [41,42], suggesting that the tumor-bearing individual is immunologically tolerant to tumor while retaining the ability to respond to nontumor-associated antigens. In other cases, tumor-bearers are generally immunosuppressed, as measured by lack of alloreactivity and the inability to generate CTL responses [43]. In still other situations, the host may be fully immunocompetent, but tumor cells have evolved to evade the immune response.

A. Tumor Bearers May Be Specifically Tolerant to their Tumors

As discussed in Section V, many tumor antigens are molecules expressed by nonmalignant tissue and are also expressed and/or overexpressed by tumor cells. As a result, the immune system of the tumor-bearing host has been exposed to the TAs before onset of the malignancy and may be tolerant to the TAs. Several studies suggest that tolerance to tumor antigens is typical [44,45]. In contrast, other studies suggest that lack of response in tumor-bearing individuals is not due to active tolerance induction, but rather to immunological ignorance of tumor antigens because the host's immune system has not been exposed to TAs in a manner that results in effective immunization [46,47]. If the host is tolerant to TAs, however, induction of tumor-specific immunity clearly requires overcoming tolerance, and several immunization strategies aimed at this goal have been tested and shown to be efficacious in experimental animal systems [48].

B. Defects in Intracellular Signaling Pathways May Inhibit Tumor-Specific T Cells from Responding to Tumor Cells

Although some individuals with large tumor burdens are significantly immunosuppressed, there is no consensus about the mechanism of suppression. Earlier studies indicated that some advanced tumor bearers contain suppressor T cells that downregulate effector T cells [49,50]. In more recent experiments, advanced tumor bearers were shown to lack certain components in the T-cell signaling pathway, such as the T-cell receptor ζ chain, p56^{lck}, and p59^{fyn}, and in some cases displayed abnormal intracellular signaling. A wide variety of signaling molecules and T-cell response parameters have been examined, including levels of CD3ζ, p56^{lck}, p59^{fyn}, and ZAP-70 proteins; NF-κB expression and/or translocation; tyrosine phosphorylation patterns; protein tyrosine kinase activity; and T-cell proliferative capacity. In many cases, reductions or defects were found in some tumor-bearing individuals [51–58]. The implication of these studies is that the immunodeficiency in tumor bearers is due to aberrant or insufficient T-cell signaling during T-cell activation [51–53,59]. The physiological relevance and universality of these signaling abnormalities are unclear, however, because other studies using different tumors in both mice and patients do not show consistent defects in these signal-transduction molecules despite decreased protein tyrosine phosphorylation or reduced T-cell activation and CTL activity [43,56,60].

The discrepancies between these studies have been attributed to differences between tumors, as well as technical differences in the way the experiments were performed [61]. Regardless of the reasons for the discrepancies between the individual studies, it seems likely that at least some tumor-bearing individuals have reduced levels of proteins involved in signal transduction in T lymphocytes. Whether these reductions contribute to tumor growth by preventing T-cell activation and the development of effective antitumor immunity remains unclear.

C. Malignant Cells May Mutate and Evade the Host's Immune Response

As discussed in Section III.A, CD8$^+$ T lymphocytes recognize peptides bound to MHC class I molecules. Lack of tumor cell MHC I expression, therefore, results in malignant cells that are "invisible" to CD8$^+$ T lymphocytes, and numerous human tumors are deficient for MHC I molecules. All types of human cancers display MHC I losses, and deficiencies can be for all human leukocyte antigen (HLA) class I genes, one haplotype, a specific locus, a particular allele, or any combination of these. Typically, 10–50% of tumors display some type of HLA class I deficiency [62], so these mutations are likely to have a significant impact on immune responses to malignancies.

MHC I deficiencies are usually the result of one of three types of mutations: (1) mutations in MHC class I heavy-chain genes causing deletion of class I heavy chain or expression of aberrant class I molecules; (2) mutations in the β2-microglobulin (β2m) gene causing deletion of β2m protein and subsequent dysfunction of the class I heavy chain; or (3) mutations in proteins required for antigen processing and presentation (e.g., functional deletion of the transporter for antigen processing [TAP] proteins required for importing peptides from the cytosol into the ER; see Fig. 1) [62,63]. In addition, class I and class II levels can be downregulated by viral proteins following viral infection [64].

Natural killer (NK) cells are also cytotoxic for tumor cells and, in contrast to CD8$^+$ T lymphocytes, are most effective against tumors expressing reduced levels of MHC class I molecules. NK cells have an inhibitory receptor (KIR) and are inactivated when the KIR receives an inhibitory signal. HLA-A, -B, and -C molecules deliver such an inhibitory signal.

Tumor cells expressing HLA-A, -B, or -C class I molecules, therefore, block NK-mediated cytolysis. In contrast, MHC class I$^-$ tumors which do not deliver an inhibitory signal, should be targets for NK cells [65]. Some tumors, however, that do not express HLA-A, -B, or -C molecules do express the nonclassical HLA-G class I molecule, and recent studies suggest that HLA-G also provides an NK inhibitory signal [66]. Therefore, although tumor-bearing hosts may be fully immunocompetent, genetic alterations in tumor cells may enable them to circumvent their immune response.

V. TUMOR-ASSOCIATED PEPTIDES AS CANDIDATE TARGETS FOR TUMOR-SPECIFIC T LYMPHOCYTES

If tumor cells are to be targets for tumor-specific T lymphocytes, then the tumor cells must express TAs that are recognized by T cells. Recent studies have identified a variety of such tumor antigens that activate CD4$^+$ and/or CD8$^+$ T cells. The antigens fall into several categories, including: (1) normal (i.e., unmutated) self antigens expressed by tumor cells and minimally expressed by some normal cells; (2) tissue-specific differentiation antigens expressed by both tumor cells and normal cells that may be overexpressed by malignant cells; (3) mutated self proteins unique to tumor cells; (4) oncogenic proteins inappropriately expressed by tumor cells; and (5) virally encoded antigens. Although the antigens in each of these categories are potential targets for activating tumor-specific T cells, the therapeutic efficacy of these epitopes remains to be demonstrated.

A. Normal (Unmutated) Self Antigens Expressed by Tumor Cells and Minimally Expressed by Some Normal Cells

Typically, this class of tumor antigen is a nonmutated self molecule expressed on tumor cells but also on a very limited range of normal, nonmalignant cells. The first antigens were identified using CTLs from melanoma patients and cells transfected with genomic melanoma DNA [67]. Because CD8$^+$ CTLs were used, only MHC-class-I-restricted TAs were identified. Subsequent studies using cytokine-secreting CD4$^+$ T cells identified MHC-class-II-restricted TAs. These and other studies (reviewed in Van den Eynde and Van der Bruggen [68], Van den Eynde and Boon [69], Robbins and Kawakami [70], and Wang and Rosenberg [71]) identified self antigens that are shared among many melanomas, as well as being expressed on other types of tumors. Subsequent studies have identified additional MHC-class-I and MHC-class-II-restricted normal antigens from other tumor types. For many of these antigens, multiple antigenic epitopes have been characterized. Table 1 lists many of these antigens and

TABLE 1 MHC Class I and Class II-Restricted Self Antigens Shared Among Tumors and
Expressed at Minimal Levels on Some Normal Cells[a]

Antigen	Primary tumor[b]	Other tumors[b,c]	Normal cells	MHC restriction[d]	Ref.
MAGE-1	Melanoma	A, B, F, G, H, L, O, Q	Testis,[e] trophoblast	A1 Cw16	[127,128]
MAGE-3	Melanoma	A, B, C, H, M, Q	Testis,[e] trophoblast	A1 A2 B44 DR11 DR13	[129–131]
BAGE	Melanoma	C	Testis,[e] trophoblast	Cw16	[132]
GAGE 1,2	Melanoma	A, B, D, L, Q	Testis,[e] trophoblast	Cw6	[133]
RAGE-1	Renal carcinoma	D, R, S	Retina[e]	B7	[134]
MUC-1[f]	Breast	J, L, P	Lactating breast[g]		[135–137]
NY-ESO-1/CAG3	Esophogus	R, E, I, C, U	Testis,[e] ovary	A2 A31	[10,138,139]

[a] Table based on data from Van den Eynde and Van der Bruggen [68], Van den Eynde and Boon [69], Wang and Rosenberg [71], and Restifo [140].

[b] ≥25% of tumors tested express antigen.

[c] A, non-small-cell lung carcinoma (NSCLC); B, head and neck cancers; C, bladder; D, sarcoma; E, mammary carcinoma; F, gastric carcinoma; G, hepatocellular carcinoma; H, esophogeal carcinoma; I, prostatic carcinoma; J, colon carcinoma; K, renal cell carcinoma; L, ovarian carcinoma; M, neuroblastoma; N, leukemias and lymphomas; O, T-cell leukemias; P, pancreas; Q, testicular seminoma; R, melanoma; S, mesothelioma; T, small cell lung cancer; U, lung carcinoma.

[d] Tumor antigen peptides restricted to these MHC class I (HLA-A, -B, and/or -C) or class II (HLA-DR) alleles have been identifed.

[e] These tissues do not express MHC class I molecules, so antigen is not presented.

[f] Underglycosylated mucin.

[g] Lactating breast is not accessible to T cells, so antigen may not be seen by the immune system.

shows their distribution on normal tissue and tumor cells, as well as the HLA class I and class II alleles (HLA restriction) for which epitopes have been identified.

B. Tissue-Specific Differentiation Antigens Expressed by Both Tumor Cells and Normal Cells That May Be Overexpressed by Malignant Cells

This group of antigens is shared between tumor cells and their normal cellular counterpart and were identified because CD8+ CTL or CD4+ T cells reacted with both the tumor cells and normal cells. Representative antigens of this class are listed in Table 2 and include antigens such as carcinoembryonic antigen (CEA), which is shared between carcinomas of the gut and normal colon, and immunoglobulin idiotypes, which are shared between malignant B-cell tumors and normal B lymphocytes. Several melanocyte differentiation antigens (e.g., tyrosinase, MART1) are also in this category.

Because these molecules are common and expressed at high levels by normal tissue, any immunotherapy strategies targeting these antigens must take into account the possibility of reactivity to normal tissues. Tumor rejection accompanied by autoimmunity against normal tissue is undesirable. How-

ever, it has been reported that prolonged survival and tumor regression in melanoma patients treated with immunotherapy targeting a common antigen is accompanied by transient vitiligo, an autoimmune response to normal melanocytes [72,73]. The potential danger of concomitant autoimmunity, therefore, is unclear and may depend on the distribution and function of the targeted normal tissue.

C. Mutated Self Proteins Unique to Tumor Cells

This class of tumor antigens includes mutated self proteins, some of which are unique to individual tumors and some of which have been found on several tumors. Some of these molecules, such as mutated β-catenin and CASP-8, are likely to be involved in the transformation process either as stimulators of cell proliferation or as inhibitors of apoptosis [74–77].

The Ras oncogene and the p53 tumor suppressor gene are also frequently mutated in many cancers, and several mutations of these genes are shared among tumors. For some of these mutations, CTLs can be generated that are cytotoxic for tumor cells expressing the corresponding wild-type antigen [78–80]; however, in other cases, the CTLs are not cytotoxic for tumor cells expressing wild-type antigens [81,82].

TABLE 2 Tumor Antigens That Are Differentiation Antigens[a]

Antigen	Primary tumor[b]	Other tumors[b,c]	Normal cells	MHC restriction[d]	Ref.
Tyrosinase	Melanoma	—	Melanocytes	A1, A2, A24, B44, DR4, DR15	[141–148]
CEA	Colon, GI malignancies	E, T	Colon	—	[149]
Ig idiotype	B-cell lymphoma	—	B Lymphocytes	—	[150,151]
gp100	Melanoma	—	Melanocytes, retina[e]	A2, A3	[146,152–154]
melan A/Mart1	Melanoma	—	Melanocytes, retina[e]	A2, B45	[155–158]
gp75/TRP-1	Melanoma	—	Melanocytes	A31	[159]
TRP-2	Melanoma	—	Melanocytes	—	[160–163]
PSA	Prostate	—	Prostate	A2, A31, A33, A68	[164, 165]

[a]Table based on data of Van den Eynde and Van der Bruggen [68], Wang and Rosenberg [71], and Restifo [140].
[b–e]See footnotes for Table 1.

Although some of the p53 mutations are widely expressed on human tumors, many mutated tumor antigens are restricted in their expression to individual or a small subset of tumors (see Chapter 8). Immunotherapeutic strategies targeting these antigens, therefore, would have to be customized to individual patients. The utility of these antigens as generalized immunotherapeutic targets, therefore, is unclear and may be limited.

D. Oncogenic Proteins Inappropriately Expressed and Overexpressed by Tumor Cells

This class of tumor antigens includes nonmutated oncogenic molecules whose expression is at least partially responsible for the malignant phenotype of the tumor. These proteins are frequently growth factors or growth-factor receptors and can be expressed at low levels by normal cells, sometimes only during certain stages of differentiation but are constitutively expressed (usually at high levels) by particular malignant cells. Classical examples of these molecules include HER2/*neu*, a growth factor receptor overexpressed in approximately 30% of breast and ovarian cancers as well as numerous other adenocarcinomas [83], and epidermal growth factor receptor (EGFR), a cell surface molecule also expressed by breast carcinoma cells [84]. Spontaneous anti-HER2/*neu* T-cell responses have been detected in some patients with HER2/*neu*-expressing cancers [83], and animal studies have demonstrated that immunity to HER2/*neu* can be induced by immunization with selected peptides derived from the HER2/*neu* protein [85]. Antibody responses to HER2/*neu* have also been noted in patients [83], and animal studies using adoptive transfer of anti-HER2/*neu* monoclonal antibodies into HER2/*neu* transgenic mice significantly reduced the onset of breast cancer [86].

As target antigens for immunotherapy, nonmutated oncogene products have a major advantage: Immunotherapy frequently selects for tumor antigen loss variants that are resistant to the host's immune response. If an overexpressed oncogenic protein is the target antigen, its loss would produce a less malignant or nonmalignant tumor cell [87]. Immunity to these self proteins, therefore, is inducible, and because of their relationship to the malignant phenotype such molecules may be desirable target antigens for immunotherapy.

E. Viral Antigens

Viruses are clearly associated with some human tumors, and antigens expressed by the viruses may serve as target molecules for immunotherapy. The most prominent association is between three of the human papilloma virus strains (HPV-16, -18, and -45) which are found in more than 90% of human uterine and cervical tumors. Because the E6 and E7 proteins are essential to maintain the transformed state, these molecules are favored as targets for immunotherapy [88]. Vaccines are currently under development for prophylactic administration prior to HPV infection, as well as for therapeutic treatment after HPV infection and viral integration [88–90].

VI. IMMUNOTHERAPEUTIC STRATEGIES FOR THE TREATMENT OF CANCER

New immunotherapy strategies using gene therapy are rapidly being developed based on an improved understanding of the immune response and the availability of sophisticated gene manipulation techniques. Immunotherapy is unlikely to replace existing treatments that are curative (e.g., successful surgical resections), although it may significantly contribute to the treatment of cancers that are unresponsive to conventional treatments. Disseminated, metastatic lesions, which are largely unresponsive to existing treatments, may be particularly amenable to immunotherapy because of the systemic scope of the immune response. As novel strategies are designed and implemented, however, various questions must be considered.

A. Questions To Consider When Designing Immunotherapy Strategies for the Treatment of Cancer

1. *What type of immunity is desired?* Immunotherapy can be directed at activating antibody responses and/or cell-mediated responses. Because the various arms of the immune response require selective activation of different lymphocyte subsets, any immunotherapy must be targeted to the appropriate cell population. If antibody-mediated immunity is desired, then activation of T_h2 helper T lymphocytes should be targeted; if cell-mediated immunity is desired, then activation of T_h1 helper T cells should be targeted [38,91].

2. *Is the patient immunosuppressed or tolerant to autologous tumor?* Some tumor-bearing patients are systemically immunosuppressed because of their tumors, while other patients may be immunocompromised because of chemotherapy and/or radiation therapy treatments. If tumor burden is the reason for immunosuppression, then immunotherapy may only be warranted once the tumor burden is reduced by conventional treatments. Alternatively, immunocompetent patients may be actively tolerized to their tumors. Any decrease in immunocompetence or increase in tolerance will obviously complicate immunotherapy, and initiation of immunotherapy treatments may need to be phased with respect to other therapies (i.e., radiation and chemotherapy) that potentially compromise the induction of immunity.

3. *Is the goal of the immunotherapy short-term tumor rejection or prolonged antitumor immunity?* The goal of the immunotherapy may dictate the type of immunity to be induced. If the role of immunotherapy is to eliminate residual primary or metastatic tumor that is not immediately accessible to conventional treatments, then a potent short-term response should be targeted. Alternatively, if the immunotherapy goal is to protect against outgrowth of micrometastases or recurrence of primary tumor, then long-term immunological memory should be induced.

4. *Is the tumor antigen known?* If a particular tumor antigen or peptide has been previously implicated in tumor rejection, then the antigen or peptide could be the immunizing agent. However, relatively few tumor antigens have been identified, and none of these has definitively been shown to be an efficacious target for immunotherapy. The use of multiple tumor antigens/peptides, therefore, might improve the chances of successful immunotherapy. Immunization with multiple epitopes may also overcome the *in vivo* selection of antigen loss variants, as it is less likely that single mutations could result in the loss of all targeted tumor antigens.

5. *Will optimal immunotherapy require "customized" therapy for each patient or will a generic, tumor-type of*

TABLE 3 Tumor Antigens That Are Oncopeptides Overexpressed on Tumors[a]

Antigen	Initial tumor[b]	Additional tumors[b,c]	Ref.
HER2/*neu*	Mammary carcinoma, ovary	L	[166,167]
p53 (wild-type)	Squamous cell carcinoma	Many	[168]

[a] Table based on data from Van den Eynde and Van der Bruggen [68] and Restifo [140].
[b,c] See footnotes for Table 1.

immunotherapy be used? If shared tumor antigens like those described in Tables 1, 2, and 3 are effective immunotherapy targets, then generic antitumor vaccines may be feasible. Patients could be typed for their MHC alleles, and the same allele-specific tumor peptides could be used for patients sharing the same type of tumor and HLA alleles. However, if immunotherapy is most effective when the target antigens are mutated self proteins (antigens in Table 4) or other tumor antigens unique to individuals, then therapy may need to be customized for each patient.

6. *How will the therapeutic genes be delivered in the immunotherapy setting?* A major problem for gene therapy is the delivery method for the therapeutic genes. At present, effective gene delivery methods for long-term, stable gene expression are very limited. Likewise, methods for delivering genes to selected locales in the body are limited. Although they are a problem for gene therapy of genetic diseases, these limitations may not be problematic for immunotherapy of cancer. If the immunotherapy goal is activation of the host's

TABLE 4 Tumor Antigens That Are Mutated Self-Proteins[a]

Antigen	Initial tumor[b]	Additional tumors[b,c]	Ref.
CDK4	Melanoma		[126]
β-catenin	Melanoma		[52,127]
CASP-8	Squamous cell carcinoma		[53]
p53	Many	Many	[57]
ras	Colon, lung, pancreas	~30% all malignancies	[55,56]
MUM-1	Melanoma		[128]
bcr/abl[d]	Chronic myelogenous leukemia		[129]

[a] Table based on data from Van den Eynde and Van der Bruggen [47] and Restifo [107].
[b,c] See footnotes for Table 1.
[d] Chimeric protein from a chromosomal translocation.

immune response, then transient rather than long-term, sustained gene expression may be sufficient.

B. A Variety of Novel Immunotherapeutic Strategies for the Treatment of Cancer Are Currently Being Studied and Developed

This section briefly reviews some of the novel immunotherapeutic strategies currently in development. Many of these approaches plus additional strategies are discussed in detail in the following chapters of this book.

1. *Immunization with tumor cell-based vaccines*. Both autologous (syngeneic) and allogeneic tumor cells have been transfected/transduced with a variety of different genes and used as cell-based vaccines. Because the transfected/transduced genes are not the target molecules for the induced immune response, vaccination with the genetically modified tumor cells should induce immunity against the wild-type, unmodified tumor. Genes encoding the following molecules have been used: (a) costimulatory molecules, such as CD80, CD86, and CD40, whose expression provides the "second signal" for T-cell activation [21,92–94]; (b) cytokines, such as IL-2, IL-3, IL-4, IL-6, IL-7, IL-10, IL-12, granulocyte–macrophage colony-stimulating factor (GM-CSF), tumor necrosis factor (TNF), and IFN-γ, whose expression should facilitate differentiation/activation of effector cells requiring specific cytokines for maturation [92,93]; (c) allogeneic MHC class I molecules, such as HLA-B7, whose expression induces immunity to the alloantigen that is cross-reactive on the wild-type tumor cells [95]; and (d) syngeneic MHC class II molecules, whose expression enables the tumor cell to directly present antigen to CD4$^+$ T lymphocytes and stimulate T_h lymphocyte activation [21,94]. Although this approach is based on the hypothesis that the genetically modified tumor cell functions as the APC, mechanistic studies show that in some situations host cells are the APCs [30,96–98], while in other situations the genetically modified cells are APC [31,97,99].

2. *Immunization with tumor peptides or with peptide-pulsed professional antigen presenting cells*. As listed in Tables 1–4, a variety of MHC-class-I-restricted tumor antigens/peptides that are putative regression antigens have been identified. Studies in both animal systems and patients are in progress to determine if these molecules are tumor-regression antigens. The overall strategy is to immunize with the antigen/peptide and determine if tumor immunity is induced and/or tumor regression occurs [73,100]. Two immunization strategies are being tested: (a) immunization with antigen/peptide, either alone or in adjuvant [101]; and (b) immunization with peptide-pulsed professional APCs such as DCs [17,102–108]. Many of these studies have shown tumor regression in animal systems; however, in one study peptide vaccination led to enhanced tumor

growth by induction of tumor-specific immunological tolerance [109].

3. *Immunization with DNA or RNA encoding tumor antigen/peptide*. As an alternative to immunization with protein, DNA encoding tumor antigen/peptide has been used [17,110,111]. The DNA can be introduced either via direct injection into the target tissue [112,113] (see Chapter 8) or following immunization with DCs that are stably transfected with tumor-antigen-encoding genes [114]. Several types of viruses, including poxvirus [115,116], adenovirus [117,118], and retrovirus [119], have been used to insert tumor-antigen genes into DCs. In another strategy, DCs have been pulsed with tumor cell RNA and used as immunogens (see Chapter 11). This approach has the advantage that it is not necessary to characterize relevant tumor antigen(s) and that only a small number of autologous tumor cells is necessary to provide the required RNA [120].

4. *Enhancement of antitumor immunity by antibody treatment*. Improved understanding of T-cell activation has led to two novel antibody treatment approaches for enhancing antitumor immunity in the tumor-bearing host. One approach involves blocking of the inhibitory receptor, CTLA4, on T cells. T lymphocytes express two counter-receptors for CD80 and CD86: CD28 and CTLA4. Interaction of CD80 or CD86 with CTLA4 leads to T-cell anergy or apoptosis, while interaction with CD28 results in T-cell activation. Administration of anti-CTLA4 mAbs to tumor-bearing mice induces a potent antitumor immunity, presumably by blocking the inhibitory pathway and favoring the T-cell activation pathway [32]. In a second approach, mAbs have been used to directly activate tumor-reactive T cells. 4-1BB is the receptor on T lymphocytes for the costimulatory molecule 4-1BB ligand (4-1BBL). Tumor-bearing mice given anti-4-1BB mAbs reject their tumors, presumably by providing the second signal necessary for T-cell activation [121] and a long-term survival signal for CD8$^+$ T lymphocytes [122].

5. *Adoptive immunotherapy with tumor-specific T lymphocytes*. Adoptive transfer of *in vitro* activated tumor-specific T lymphocytes into tumor-bearing individuals has been shown to mediate tumor rejection in animal systems [33]. In one such approach, patients have been immunized with autologous tumor, draining lymph nodes (LNs) removed, and LN cells expanded *in vitro* prior to treatment (see Chapter 14). With the characterization of human tumor antigens, it has become feasible to antigen-stimulate autologous human T cells *in vitro* and reinfuse them into the tumor-bearing patient [100,123]. Because various studies have demonstrated that maximal antitumor responses occur when both CD4$^+$ and CD8$^+$ T cells are generated, optimal immunity will probably involve adoptive transfer of both lymphocyte populations [124].

6. *Immunization with heat shock proteins*. Many cells produce heat shock proteins (HSPs) in response to environmental stress. These proteins can function as chaperons for peptides,

and it has been demonstrated that HSPs are involved in loading immunogenic peptides onto MHC class I molecules. Recent studies suggest that HSPs from tumor cells bind TA, and that immunization of tumor-bearing individuals with HSP/peptide complexes enhances the antitumor immune response [122,125,126]

VII. CONCLUSIONS

Recent advances in understanding T-cell activation and in defining tumor antigens suggest a variety of novel approaches for using immunotherapy in the treatment of cancer. Technical advances in molecular biology and gene therapy have made these approaches feasible. Many of these novel immunotherapeutic strategies have shown efficacy in animal studies; however, determining their ultimate clinical efficacy awaits their testing in clinical situations. Patients participating in phase I and phase II clinical trials typically have advanced disease and high tumor burden, and it is unlikely that immunotherapy will be effective in this setting. A more accurate assessment of the efficacy of immunotherapy, therefore, may only be made when patients with more moderate tumor burden or less advanced disease are treated.

Acknowledgments

Original studies from the authors' laboratory were supported by grants from the National Institutes of Health (NIH) (R01CA52527, R01CA84232) and U.S. Army Medical Research and Material Command (DAMD 17-94-J-4323 and DAMD 17-01-0312). B. Pulaski is supported by a postdoctoral fellowship from the U.S. Army Breast Cancer Program (DAMD 17-97-1-7152).

References

1. Weiner, L. (1999). Monoclonal antibody therapy of cancer. *Semin. Oncol.* **26,** 43–51.
2. Pai, L., and Pastan, I. (1997). Immunotoxin therapy, in *Cancer: Principles and Practice of Oncology,* vol. 2 (V. DeVita, S. Hellman, and S. Rosenberg, eds.), pp. 3045–3057. Lippincott-Raven, Philadelphia, PA.
3. Scott, A., and Welt, S. (1997). Antibody-based immunological therapies. *Curr. Opin. Immunol.* **9,** 717–722.
4. Scott, A., and Cebon, J. (1997). Clinical promise of tumor immunology. *Lancet* **349,** 19–22.
5. Van de Winkel, J., Bast, B., and de Gast, G. (1997). Immunotherapeutic potential of bispecific antibodies. *Immunol. Today* **18,** 562–564.
6. Sahin, U., Tureci, O., Schmitt, H., Cochlovius, B., Johannes, T., Schmits, R., Stenner, F., Luo, G., Schobert, I., and Pfreundschuh, M. (1995). Human neoplasms elicit multiple immune responses in the autologous host. *Proc. Natl. Acad. Sci. USA* **92,** 11810–11813.
7. Sahin, U., Tureci, O., and Pfreundschuh, M. (1997). Serological identification of human tumor antigens. *Curr. Opin. Immunol.* **9,** 709–716.
8. Brass, N., Heckel, D., Sahin, U., Pfreundschuh, M., Sybrecht, G., and Meese, E. (1997). Translation initiation factor eIF-4gamma is encoded by an amplified gene and induces an immune response in squamous cell lung carcinoma. *Hum. Mol. Genet.* **6,** 33–39.
9. Tureci, O., Sahin, U., Schobert, I., Koslowski, M., Schmitt, H., Schild, H., Stenner, F., Seitz, G., and Rammensee, H. (1996). The SSX2 gene, which is involved in the t(X,18) translocation of synovial sarcomas, codes for the human tumor antigen HOM-Mel-40. *Cancer Res.* **56,** 4766–4772.
10. Chen, Y., Scanlan, M., Sahin, U., Tureci, O., Gure, A., Tsang, S., Williamson, B., Stockert, E., Pfreundschuh, M., and Old, L. (1997). A testicular antigen aberrantly expressed in human cancers detected by autologous antibody screening. *Proc. Natl. Acad. Sci. USA* **94,** 1914–1918.
11. Bentley, G., and Mariuzza, R. (1996). The structure of the T cell antigen receptor. *Annu. Rev. Immunol.* **14,** 591–618.
12. Cantrell, D. (1996). T cell antigen receptor signal transduction pathways. *Annu. Rev. Immunol.* **14,** 259–274.
13. York, I., and Rock, K. (1996). Antigen processing and presentation by the class I major histocompatibility complex. *Annu. Rev. Immunol.* **14,** 369–396.
14. Lanzavecchia, A. (1996). Mechanisms of antigen uptake for presentation. *Curr. Opin. Immunol.* **8,** 348–354.
15. Watts, C. (1997). Capture and processing of exogenous antigens for presentation on MHC molecules. *Annu. Rev. Immunol.* **15,** 821–850.
16. Bancherau, J., Briere, F., Caux, C., Davoust, J., Lebecque, S., Liu, Y., Pulendran, B., and Palucka, K. (2000). Immunobiology of dendritic cells. *Annu. Rev. Immunol.* **18,** 767–812.
17. Fong, L., and Engleman, E. (2000). Dendritic cells in cancer immunotherapy. *Annu. Rev. Immunol.* **18,** 245–273.
18. Mackey, M., Barth, R., and Noelle, R. (1998). The role of CD40/CD154 interactions in the priming, differentiation, and effector function of helper and cytotoxic T cells. *J. Leukoc. Biol.* **63,** 418–428.
19. Mackey, M., Gunn, J., Maliszewski, C., Kikutani, H., Noelle, R., and Barth, R. (1998). Cutting edge: dendritic cells require maturation via CD40 to generate protective antitumor immunity. *J. Immunol.* **161,** 2094–2098.
20. Mackey, M., Gunn, J., Ting, P., Kikutani, H., Dranoff, G., Noelle, R., and Barth, Jr., R. (1997). Protective immunity induced by tumor vaccines requires interaction between CD40 and its ligand, CD154. *Canc. Res.* **57,** 2569–2574.
21. Costello, R., Gastaut, J., and Oliver, D. (1999). What is the real role of CD40 in cancer immuotherapy? *Immunol. Today* **20,** 488–493.
22. Chambers, C., and Allison, J. (1997). Co-stimulation in T cell responses. *Curr. Opin. Immunol.* **9,** 396–404.
23. Lenschow, D., Walunas, T., and Bluestone, J. (1996). CD28/B7 system of T cell costimulation. *Annu. Rev. Immunol.* **14,** 233–258.
24. Liu, Y., and Janeway, C. (1992). Cells that present both specific ligand and costimulatory activity are the most efficient inducers of clonal expansion of normal CD4 T cells. *Proc. Natl. Acad. Sci. USA* **89,** 3845–3849.
25. Baskar, S., Glimcher, L., Nabavi, N., Jones, R. T., and Ostrand-Rosenberg, S. (1995). Major histocompatibility complex class II$^+$B7-1$^+$ tumor cells are potent vaccines for stimulating tumor rejection in tumor-bearing mice. *J. Exp. Med.* **181,** 619–629.
26. Chen, L., Ashe, S., Brady, W. A., Hellstrom, I., Hellstrom, K. E., Ledbetter, J. A., McGowan, P., and Linsley, P. S. (1992). Costimulation of antitumor immunity by the B7 counterreceptor for the T lymphocyte molecules CD28 and CTLA-4. *Cell* **71,** 1093–1102.
27. Townsend, S. E., and Allison, J. P. (1993). Tumor rejection after direct costimulation of CD8+ T cells by B7-transfected melanoma cells. *Science* **259,** 368–370.
28. Harding, F. A., and Allison, J. P. (1993). CD28-B7 interactions allow the induction of CD8+ cytotoxic T lymphocytes in the absence of exogenous help. *J. Exp. Med.* **177,** 1791–1796.

29. Baskar, S., Ostrand-Rosenberg, S., Nabavi, N., Nadler, L. M., Freeman, G. J., and Glimcher, L. H. (1993). Constitutive expression of B7 restores immunogenicity of tumor cells expressing truncated major histocompatibility complex class II molecules. *Proc. Natl. Acad. Sci. USA* **90**, 5687–5690.

30. Huang, A., Bruce, A., Pardoll, D., and Levitsky, H. (1996). Does B7-1 expression confer antigen-presenting cell capacity to tumors in vivo? *J. Exp. Med.* **183**, 769–776.

31. Armstrong, T., Pulaski, B., and Ostrand-Rosenberg, S. (1998). Tumor antigen presentation: Changing the rules. *Canc. Immunol. Immunother.* **46**, 70–74.

32. Leach, D., Krummel, M., and Allison, J. (1996). Enhancement of antitumor immunity by CTLA-4 blockade. *Science* **271**, 1734–1736.

33. Greenberg, P. (1991). Adoptive T cell therapy of tumors: mechanisms operative in the recognition and elimination of tumor cells. *Adv. Immunol.* **49**, 281–355.

34. Ostrand-Rosenberg, S., Thakur, A., and Clements, V. (1990). Rejection of mouse sarcoma cells after transfection of MHC class II genes. *J. Immunol.* **144**, 4068–4071.

35. Hung, K., Hayashi, R., Lafond-Walker, A., Lowenstein, C., Pardoll, D., and Levitsky, H. (1998). The central role of CD4+ T cells in the antitumor immune response. *J. Exp. Med.* **188**, 2357–2368.

36. Toes, R., Ossendorp, F., Offringa, R., and Melief, C. (1999). CD4 T cells and their role in antitumor immune responses. *J. Exp. Med.* **189**, 753–756.

37. Pardoll, D., and Topalian, S. (1998). The role of CD4+ T cell responses in antitumor immunity. *Curr. Opin. Immunol.* **10**, 588–594.

38. Shurin, M., Lu, L., Kalinski, P., Stewart-Akers, A., and Lotze, M. (1999). Th1/Th2 balance in cancer, transplantation and pregnancy. *Springer Semin. Immunopathol.* **21**, 339–359.

39. Velders, M., Schreiber, H., and Kast, W. (1998). Active immunization against cancer cells: impediments and advances. *Semin. Oncol.* **25**, 697–706.

40. Kiessling, R., Wasserman, K., Horiguchi, S., Kono, K., Sjoberg, J., Pisa, P., and Petersson, M. (1999). Tumor-induced immune dysfunction. *Cancer Immunol. Immunother.* **48**, 353–362.

41. Perdrizet, G., Ross, S., Stauss, H., Singh, S., Koeppen, H., and Schreiber, H. (1990). Animals bearing malignant grafts reject normal grafts that express through gene transfer the same antigen. *J. Exp. Med.* **171**, 1205–1220.

42. Wick, M., Dubey, P., Koeppen, H., Siegel, C., Fields, P., Chen, L., Bluestone, J., and Schreiber, H. (1997). Antigenic cancer cells grow progressively in immune hosts without evidence for T cell exhaustion or systemic anergy. *J. Exp. Med.* **186**, 229–238.

43. Levey, D., and Srivastava, P. (1995). T cells from late tumor-bearing mice express normal levels of p56^{lck}, p59^{fyn}, ZAP-70, and CD3-zeta despite suppressed cytolytic activity. *J. Exp. Med.* **182**, 1029–1036.

44. Sotomayor, E., Borrello, I., and Levitsky, H. (1996). Tolerance and cancer: a critical issue in tumor immunology. *Crit. Rev. Oncol.* **7**, 433–456.

45. Staveley-O'Carroll, K., Sotomayor, E., Montgomery, J., Borrello, I., Hwang, L., Fein, S., Pardoll, D., and Levitsky, H. (1998). Induction of antigen-specific T cell anergy: An early event in the course of tumor progression. *Proc. Natl. Acad. Sci. USA* **95**, 1178–1183.

46. Melero, I., Bach, N., and Chen, L. (1997). Costimulation, tolerance and ignorance of cytolytic T lymphocytes in immune responses to tumor antigens. *Life Sci.* **60**, 2035–2041.

47. Ochsenbein, A., Klenerman, P., Karrer, U., Ludewig, B., Pericin, M., Hengartner, H., and Zinkernagel, R. (1999). Immune surveillance against a solid tumor fails because of immunological ignorance. *Proc. Natl. Acad. Sci. USA* **96**, 2233–2238.

48. Sotomayor, E., Borrello, I., Tubb, E., Rattis, F., Bien, H., Lu, Z., Fein, S., Schoenberger, S., and Levitsky, H. (1999). Conversion of tumor-specific CD4+ T cell tolerance to T cell priming through in vivo ligation of CD40. *Nat. Med.* **5**, 780–787.

49. Bursuker, I., and North, R. (1984). Generation and decay of the immune response to a progressive fibrosarcoma. II. Failure to demonstrate postexcision immunity after the onset of T cell-mediated suppression of immunity. *J. Exp. Med.* **159**, 1312–1321.

50. North, R., and Bursuker, I. (1984). Generation of decay of the immune response to a progressive fibrosarcoma. I. Ly-1$^+$2$^-$ suppressor T cells down-regulate the generation of Ly-1$^-$2$^+$ effector cells. *J. Exp. Med.* **159**, 1295–1231.

51. Mizoguchi, H., O'Shea, J., Longo, D., Loeffler, C., McVicar, D., and Ochoa, A. (1992). Alterations in signal transduction molecules in T lymphocytes from tumor-bearing mice. *Science* **258**, 1795–1798.

52. Nakagomi, H., Peterson, M., Magnusson, I., Jublin, C., Matsuda, M., Mellstedt, H., Taupin, J., Vivier, E., Anderson, P., and Kiessling, R. (1993). Decreased expression of the signal transducing zeta chains in tumor-infiltrating T cells and NK cells of patients with colorectal carcinoma. *Cancer Res.* **53**, 5613–5616.

53. Finke, J., Zea, A., Stanley, J., Longo, D., Mizoguchi, H., Tubbs, R., Wiltrout, R., O'Shea, J., Kudoh, S., Klein, E., and Ochoa, A. (1993). Loss of T cell receptor zeta chain and p56^{lck} in T cells infiltrating human renal cell carcinoma. *Cancer Res.* **53**, 5613–5616.

54. Zier, K., Gansbacher, B., and Salvadori, S. (1996). Preventing abnormalities in signal transduction of T cells in cancer: the promise of cytokine gene therapy. *Immunol. Today* **39**, 39–45.

55. Zea, A., Curti, B., and Longo, D. (1995). *Clin. Canc. Res.* **1**, 1327–1335.

56. Tartour, E., Latour, S., and Mathiot, C. (1995). *Int. J. Cancer* **63**, 205–212.

57. Aoe, T., Okamoto, Y., and Saito, T. (1995). Activated macrophages induce structural abnormalities of the T cell receptor-CD3 complex. *J. Exp. Med.* **181**, 1881–1886.

58. Matusuda, N., Petersson, M., and Lenkei, R. (1995). *Int. J. Cancer* **61**, 765–772.

59. Salvadori, S., Gansbacher, B., Pizzimenti, A., and Zier, K. (1994). Abnormal signal transduction by T cells of mice with parenteral tumors is not seen in mice bearing IL-2-secreting tumors. *J. Immunol.* **153**, 5176–5182.

60. Wang, Q., Stanley, J., Kudoh, S., Myles, J., Kolenko, V., Yi, T., Tubbs, R., Bukowski, R., and Finke, J. (1995). T cells infiltrating non-Hodgkin's B cell lymphomas show altered tyrosine phosphorylation pattern even though T cell receptor/CD3-associated kinases are present. *J. Immunol.* **155**, 1382–1392.

61. Levey, D., and Srivastava, P. (1996). Alterations in T cells of cancer bearers: whence specificity? *Immunol. Today* **17**, 365–368.

62. Algarra, I., Cabrera, T., and Garrido, F. (2000). The HLA crossroad in tumor immunology. *Human Immunol.* **61**, 65–73.

63. Yewdell, J., Norbury, C., and Bennink, J. (1999). Mechanisms of exogenous antigen presentation by MHC class I molecules in vitro and in vivo: implications for generating CD8+ T cell responses to infectious agents, tumors, transplants, and vaccines. *Adv. Immunol.* **73**, 1–77.

64. Brodsky, F., Lem, L., Solache, A., and Bennett, E. (1999). Human pathogen subversion of antigen presentation. *Immunol. Rev.* **168**, 199–215.

65. Garrido, F., Ruiz-Cabello, F., Cabrera, T., Perez-Villar, J., Lopez-Botct, M., Duggan-Keen, M., and Stern, P. (1997). Implications for immunosurveillance of altered class I phenotypes in human tumors. *Immunol. Today* **18**, 89–93.

66. Cabestre, F., Lefebvre, S., Moreau, P., Rouas-Friess, N., Dausset, J., Carosella, E., and Paul, P. (1999). HLA-G expression: immune privilege for tumor cells? *Semin. Cancer Biol.* **9**, 27–36.

67. Van der Bruggen, P., Traversari, C., Chomez, P., Lurquin, C., De Plaen, E., Van den Eynde, B., Knuth, A., and Boon, T. (1991). A gene encoding an antigen recognized by cytolytic T lymphocytes on a human melanoma. *Science* **254**, 1643–1647.

68. Van den Eynde, B., and Van der Bruggen, P. (1997). T cell-defined tumor antigens. *Curr. Opin. Immunol.* **9,** 684–693.

69. Van den Eynde, B., and Boon, T. (1997). Tumor antigens recognized by T lymphocytes. *Int. J. Clin. Lab. Res.* **27,** 81–86.

70. Robbins, P., and Kawakami, Y. (1996). Human tumor antigens recognized by T cells. *Curr. Opin. Immunol.* **8,** 628–636.

71. Wang, R., and Rosenberg, S. (1999). Human tumor antigens for cancer vaccine development. *Immunol. Rev.* **170,** 85–100.

72. Rosenberg, S., and White, D. (1996). Vitiligo in patients with melanoma: normal tissue antigens can be targets for cancer immunotherapy. *J. Immunotherapy* **19,** 81–84.

73. Rosenberg, S. (1999). A new era for cancer immunotherapy based on the genes that encode cancer antigens. *Immunity* **10,** 281–287.

74. Peifer, M. (1997). Beta-catenin as oncogene: the smoking gun. *Science* **275,** 1752–1753.

75. Rubinfeld, B., Robbins, P., El-Gamil, M., Albert, I., Porfiri, E., and Polakis, P. (1997). Stabilization of beta-catenin by genetic defects in melanoma cell lines. *Science* **275,** 1790–1792.

76. Mandruzzato, S., Brasseur, F., Andry, G., Boon, T., and Van der Bruggen, P. (1997). A CASP-8 mutation recognized by cytolytic T lymphocytes on a human head and neck carcinoma. *J. Exp. Med.* **186,** 785–793.

77. Boon, T., and Old, L. (1997). Tumor antigens. *Curr. Opin. Immunol.* **9,** 681–683.

78. Skipper, J., and Stauss, H. (1993). Identification of two cytotoxic T lymphoycte-recognized epitopes in the Ras protein. *J. Exp. Med.* **177,** 1493–1498.

79. Peace, D., Smith, J., Chen, W., You, S., Cosand, W., Blake, J., and Cheever, M. (1994). Lysis of *ras* oncogene-transformed cells by specific cytotoxic T lymphocytes elicited by primary in vitro immunization with mutated *ras* peptide. *J. Exp. Med.* **179,** 473–479.

80. Noguchi, Y., Chen, Y., and Old, L. (1994). A mouse mutant p53 product recognized by CD4$^+$ and CD8$^+$ T cells. *Proc. Natl. Acad. Sci. USA* **91,** 3171–3175.

81. Houbiers, J., Nijman, H., Van Der Burg, S., Drijfhout, J., Kenemans, P., Van De Velde, C., Brand, A., Momberg, F., Kast, W., and Melief, C. (1993). In vitro induction of human cytotoxic T lymphocyte responses against peptides of mutant and wild-type p53. *Eur. J. Immunol.* **23,** 2072–2077.

82. Elas, A., Nijman, H., Van Der Minne, C., Mourer, J., Kast, M., Melief, C., and Schrier, P. (1995). Induction and characterization of cytotoxic T lymphoyctes recognizing a mutated p21 RAS peptide presented by HLA-A2010. *Int. J. Cancer,* 389–396.

83. Cheever, M., Disis, M., Bernhard, H., Gralow, J., Hand, S., Huseby, E., Qin, H., Takahashi, M., and Chen, W. (1995). Immunity to oncogenic proteins. *Immunol. Rev.* **145,** 33–59.

84. Baselga, J., and Mendelsohn, J. (1994). The epidermal growth factor receptor as a target for therapy in breast carcinoma. *Breast Canc. Res. Treat.* **29,** 127–138.

85. Disis, M., Gralow, J., Bernhard, H., Hand, S., Rubin, W., and Cheever, M. (1996). Peptide-based, but now whole protein, vaccines elicit immunity to HER-2/*neu,* an oncogenic self-protein. *J. Immunol.* **156,** 3151–3158.

86. Katsumata, M., Okudaira, T., Samanta, A., Clark, D., Drebin, J., Jolicoeur, P., and Greene, M. (1995). Prevention of breast tumour development in vivo by downregulation of the p185neu receptor. *Nature Med.* **1,** 644–648.

87. Disis, M., and Cheever, M. (1996). Oncogenic proteins as tumor antigens. *Curr. Opin. Immunol.* **8,** 637–642.

88. Tindle, R. (1997). Human papillomavirus vaccines for cervical cancer. *Curr. Opin. Immunol.* **8,** 643–650.

89. Ressing, M., de Jong, J., Brandt, R., Drijfhout, J., Benckhuijsen, W., Schreuder, G., Offringa, R., Kast, W., and Melief, C. (1999). Differential binding of viral peptides to HLA-A2 alleles. Implications for human papillomavirus type 16 E7 peptide-based vaccination against cervical carcinoma. *Eur. J. Immunol.* **29,** 1292–1303.

90. van Driel, W., Ressing, M., Kenter, G., Brandt, R., Krul, E., van Rossum, A., Schurring, E., Offringa, R., Bauknecht, T., Tamm-Hermelink, A., van Dam, P., Fleuren, G., Kast, W., Melief, C., and Tribmos, J. (1999). Vaccination with HPV16 peptides of patients with advanced cervical carcinoma: clinical evaluation of a phase I–II trial. *Eur. J. Cancer* 946–952.

91. Constant, S., and Bottomly, K. (1997). Induction of the TH1 and TH2 CD4$^+$ T cell responses: alternative approaches. *Annu. Rev. Immunol.* **15,** 297–322.

92. Blankenstein, T., Cayeux, S., and Qin, Z. (1996). Genetic approaches to cancer immunotherapy. *Rev. Phys. Biochem. Pharm.* **129,** 1–49.

93. Musiani, P., Modesti, A., Giovarelli, M., Cavallo, F., Colombo, M., Lollini, P., and Forni, G. (1997). Cytokines, tumour-cell death and immunogenicity: a question of choice. *Immunol. Today* **18,** 32–26.

94. Ostrand-Rosenberg, S., Pulaski, B., Clements, V., Qi, L., Pipeling, M., and Hanyok, L. (1999). Cell-based vaccines for the stimulation of immunity to metastatic cancers. *Immunol. Rev.* **170,** 101–114.

95. Nabel, G., Gordon, D., Bishop, D., Nickoloff, B., Yang, Z., Aruga, A., Cameron, M., Nabel, E., and Chang, A. (1996). Immune response in human melanoma after transfer of an allogeneic class I major histocompatibility complex gene with DNA-liposome complexes. *Proc. Natl. Acad. Sci. USA* **93,** 15388–15393.

96. Huang, A., Golumbek, P., Ahmadzadeh, M., Jaffee, E., Pardoll, D., and Levitsky, H. (1994). Role of bone marrow-derived cells in presenting MHC class I-restricted tumor antigens. *Science* **264,** 961–965.

97. Cayeux, S., Richter, G., Noffz, G., Dorken, B., and Blankenstein, T. (1997). Influence of gene-modified (IL-7, IL-4, and B7) tumor cell vaccines on tumor antigen presentation. *J. Immunol.* **158,** 2834–2841.

98. Pulaski, B., Yeh, K., Shastri, N., Maltby, K., Penney, D., Lord, E., and Frelinger, J. (1996). IL-3 enhances CTL development and class I MHC presentation of exogenous antigen by tumor-infiltrating macrophages. *Proc. Natl. Acad. Sci. USA* **93,** 3669–3674.

99. Armstrong, T., Clements, V., and Ostrand-Rosenberg, S. (1998). MHC class II-transfected tumor cells directly present antigen to tumor-specific CD4$^+$ T lymphocytes. *J. Immunol.* **160,** 661–666.

100. Rosenberg, S. (1997). Cancer vaccines based on the identification of genes encoding cancer regression antigens. *Immunol. Today* **18,** 175–182.

101. Melief, C., Offringa, R., Toes, R., and Kast, M. (1996). Peptide-based cancer vaccines. *Curr. Opin. Immunol.* **8,** 651–657.

102. Mukherji, B., Chakraborty, N., Yamasaki, S., Okino, T., Yamase, H., Sporn, J., Kurtzman, S., Ergin, M., Ozols, J., Meehan, J., and Mauri, F. (1995). Induction of antigen-specific cytolytic T cells in situ in human melanoma by immunization with synthetic peptide-pulsed autologous antigen presenting cells. *Proc. Natl. Acad. Sci. USA* **92,** 8078–8082.

103. Zitvogel, L., Mayordomo, J., Tjandrawan, T., DeLeo, A., Clarke, M., Lotze, M., and Storkus, W. (1996). Therapy of murine tumors with tumor-peptide-pulsed dendritic cells: dependence on T cells, B7 costimulation, and T helper cell 1-associated cytokines. *J. Exp. Med.* **183,** 87–97.

104. Schreurs, M., Eggert, A., Punt, C., Figdor, C., and Adema, G. (2000). Dendritic cell-based vaccines: from mouse models to clinical cancer immunotherapy. *Crit. Rev. Oncog.* **11,** 1–17.

105. Kirk, C., and Mule, J. (2000). Gene-modified dendritic cells for use in tumor vaccines. *Hum. Gene Ther.* **11,** 797–806.

106. Tarte, K., and Klein, B. (1999). Dendritic cell-based vaccine: a promising approach for cancer immunotherapy. *Leukemia* **13,** 653–663.

107. Timmerman, J., and Levy, R. (1999). Dendritic cell vaccines for cancer immunotherapy. *Annu. Rev. Med.* **50,** 507–529.

108. Gilboa, E. (1999). How tumors escape immune destruction and what we can do about it. *Cancer Immunol. Immunother.* **48,** 382–385.

109. Toes, R., Offringa, R., Blom, R., Melief, C., and Kast, M. (1996). Peptide vaccination can lead to enhanced tumor growth through specific T-cell tolerance induction. *Proc. Natl. Acad. Sci. USA* **93,** 7855–7860.

110. Gurunathan, S., Klinman, D., and Seder, R. (2000). DNA vaccines: immunology, application, and optimization. *Annu. Rev. Immunol.* **18,** 927–974.

111. Stevenson, F. (1999). DNA vaccines against cancer: from genes to therapy. *Ann. Oncol.* **10,** 1413–1418.

112. Irvine, K., Rao, R., Rosenberg, S., and Restifo, N. (1996). Cytokine enhancement of DNA immunization leads to effective treatment of established pulmonary metastases. *J. Immunol.* **156,** 238–245.

113. Doe, B., Selby, M., Barnett, S., Baenziger, J., and Walker, C. (1996). Induction of cytotoxic T lymphocytes by intramuscular immunization with plasmid DNA is facilitated by bone marrow-derived cells. *Proc. Natl. Acad. Sci. USA* **93,** 8578–8583.

114. Alijagic, S., Moller, P., Artuc, M., Jurgovsky, K., Czarnetzki, B., and Schadendorf, D. (1995). Dendritic cells generated from peripheral blood transfected with human tyrosinase induce specific T cell activation. *Eur. J. Immunol.* **25,** 3100–3107.

115. Restifo, N. (1996). The new vaccines: building viruses that elicit antitumor immunity. *Curr. Opin. Immunol.* **8,** 658–663.

116. Bronte, V., Charroll, M., Goletz, T., Wang, M., Overwijk, W., Marincola, F., Rosenberg, S., Moss, B., and Restifo, N. (1997). Antigen expression by dendritic cells correlates with the therapeutic effectiveness of a model recombinant poxvirus tumor vaccine. *Proc. Natl. Acad. Sci. USA* **94,** 3183–3188.

117. Arthur, J., Butterfield, L., Kiertscher, S., Roth, M., Bui, L., Lau, R., Dubinett, S., Glaspy, J., and Economou, J. (1997). A comparison of gene transfer methods in human dendritic cells. *Cancer Gene Ther.* **4,** 17–25.

118. Ribas, A., Butterfield, L., McBride, W., Jilani, S., Bui, L., Vollmer, C., Lau, R., Dissette, V., Hu, B., Chen, A., Glaspy, J., and Economou, J. (1997). Genetic immunization for the melanoma antigen MART-1/Melan-A using recombinant adenovirus-transduced murine dendritic cells. *Cancer Res.* **57,** 2865–2869.

119. Reeves, M., Royalo, R., Lam, J., Rosenberg, S., and Hwu, P. (1996). Retroviral transduction of human dendritic cells with a tumor-associated antigen gene. *Cancer Res.* **56,** 5672–5677.

120. Boczkowski, D., Nair, S., Snyder, D., and Gilboa, E. (1996). Dendritic cells pulsed with RNA are potent antigen-presenting cells in vitro and in vivo. *J. Exp. Med.* **184,** 465–472.

121. Melero, I., Shuford, W., Newby, S., Aruffo, A., Ledbetter, J., Hellstrom, K., Mittler, R., and Chen, L. (1997). Monoclonal antibodies against the 4-1BB T-cell activation molecule eradicate established tumors. *Nat. Med.* **3,** 682–685.

122. Takahashi, C., Mittler, R., and Vella, A. (1999). 4-1BB is a bona fide CD8 T cell survival signal. *J. Immunol.* **162,** 5037–5040.

123. Li, Q., and Chang, A. (1999). Adoptive T-cell immunotherapy of cancer. *Cytokines Cell. Mol. Ther.* **5,** 105–117.

124. Yee, C., Gilbert, M., Riddell, S., Brichard, V., Fefer, A., Thompsom, J., Boon, T., and Greenberg, P. (1997). *J. Immunol.* **157,** 4079–4086.

125. Przepiorka, D., and Srivastava, P. (1998). Heat shock protein–peptide complexes as immunotherapy for human cancer. *Mol. Med. Today* **4,** 478–484.

126. Srivastava, P., Menoret, A., Basu, S., Binder, R., and McQuade, K. (1998). Heat shock proteins come of age: primitive functions acquire new roles in an adaptive world. *Immunity* **8,** 657–665.

127. Traversari, C., Van der Bruggen, P., Luescher, I., Lurquin, C., Chomez, P., Van Pel, A., De Plaen, E., Amar-Costecec, A., and Boon, T. (1992). A nonapeptide encoded by human gene MAGE-1 is recognized on HLA-A1 by cytolytic T lymphocytes directed against tumor antigen MZ2-E. *J. Exp. Med.* **176,** 1453–1457.

128. Van der Bruggen, P., Szikora, J., Boel, P., Wildmann, C., Somville, M., Sensi, M., and Boon, T. (1994). Autologous cytolytic T lymphocytes recognize a MAGE-1 nonapeptide on melanomas expressing HLA-Cw*1601. *Eur. J. Immunol.* **24,** 2134–2140.

129. Gaugler, B., Van den Eynde, B., Van der Bruggen, P., Romero, P., Gaforio, J., De Plaen, E., Lethe, B., Brasseur, F., and Boon, T. (1994). Human gene MAGE-3 codes for an antigen recognized on a melanoma by autologous cytolytic T lymphocytes. *J. Exp. Med.* **179,** 921–930.

130. Herman, J., Van der Bruggen, P., Luescher, I., Mandruzzato, S., Romero, P., Thonnard, J., Fleischhauer, K., Boon, T., and Coulie, P. (1996). A peptide encoded by human gene MAGE-3 and presented by HLA-B44 induces cytolytic T lymphocytes that recognize tumor cells expressing MAGE-3. *Immunogenetics* **43,** 377–383.

131. Van der Bruggen, P., BAstin, J., Gajewski, T., Coulie, P., Boel, P., De Smet, C., Traversari, C., Townsend, A., and Boon, T. (1994). A peptide encoded by human gene MAGE-3 and presented by HLA-A2 induces cytolytic T lymphocytes that recognize tumor cells expressing MAGE-3. *Eur. J. Immunol.* **24,** 3038–3043.

132. Boel, P., Wildmann, C., Sensi, M., Brasseur, R., Renauld, J., Coulie, P., Boon, T., and Van der Bruggen, P. (1995). BAGE, a new gene encoding an antigen recognized on human melanomas by cytolytic T lymphocytes. *Immunity* **2,** 167–175.

133. Van den Eynde, B., Peeters, O., De Backer, O., Gaugler, B., Lucas, S., and Boon, T. (1995). A new family of genes coding for an antigen recognized by autologous cytolytic T lymphocytes on a human melanoma. *J. Exp. Med.* **182,** 689–698.

134. Gaugler, B., Brouwenstijn, N., Vantomme, V., Szikora, J.-P., Van der Spek, C., Patard, J., Boon, T., Schrier, P., and Van den Eynde, B. (1996). A new gene coding for an antigen recognized by autologous cytolytic T lymphocytes on a human renal carcinoma. *Immunogenetics* **44,** 323–330.

135. Jerome, K., Wakefield, D., and Watkins, S. (1988). Tumor-specific cytotoxic T cell clones from patients with breast and pancreatic adenocarcinoma recognize EBV-immortalized B cells transfected with polymorphic epithelial mucin cDNA. *J. Immunol.* **151,** 1654–1662.

136. Barratt-Boyes, S. (1996). Making the most of mucin: a novel target for tumor immunotherapy. *Cancer Immunol. Immunother.* **43,** 142–151.

137. Finn, O., Jerome, K., Henderson, A., Pecher, G., Domenech, N., Magarian-Blander, J., and Barratt-Boyes, S. (1995). MUC1 epithelial tumor mucin-based immunity and cancer vaccines. *Immunol. Rev.* **87,** 982–990.

138. Jager, E., Chen, Y., Drijfhout, J., Karbach, J., Ringhoffer, M., Jager, D., Arand, M., Wada, H., Noguchi, Y., Stockert, E., Old, L., and Knuth, A. (1998). Simultaneous humoral and cellular immune response against cancer-testis antigen NY-ESO-1: definition of human histocompatibility leukocyte antigen (HLA)-A2-binding peptide epitopes. *J. Exp. Med.* **187,** 265–270.

139. Wang, R., Johnston, S., Zeng, G., Schwartzentruber, D., and Rosenberg, S. (1998). A breast and melanoma-shared tumor antigen: T cell responses to antigenic peptides translated from different open reading frames. *J. Immunol.* **161,** 3596–3606.

140. Restifo, N. (1997). Cancer vaccines, in *Cancer: Principles and Practice of Oncology,* Vol. 2 (V. DeVita, S. Hellman, and S. Rosenberg, eds.), pp. 3023–3043. Lippincott-Raven, Philadelphia, PA.

141. Topalian, S., Gonzales, M., Parkhurst, M., Li, Y., Southwood, S., Sette, A., Rosenberg, S., and Robbins, P. (1996). Melanoma-specific CD4+ cells recognize nonmutated HLA-DR restricted tyrosinase epitopes. *J. Exp. Med.* **183,** 1965–1971.

142. Brichard, V., Van Pel, A., Wolfel, T., Wolfel, C., De Plaen, E., Lethe, B., Coulie, P., and Boon, T. (1993). The tyrosinase gene codes for an antigen recognized by autologous cytolytic T lymphocytes on HLA-A2 melanomas. *J. Exp. Med.* **178,** 489–495.

143. Wolfel, T., Van Pel, A., Brichard, V., Schneider, J., Seliger, B., Meyer zum Buschenfelde, K., and Boon, T. (1994). Two tyrosinase nonapeptides recognized on HLA-A2 melanomas by autologous cytolytic T lymphocytes. *Eur. J. Immunol.* **24,** 759–764.

144. Kang, X., Kawakami, Y., El-Gamil, M., Wang, R., Sakaguchi, K., Yannelli, J., Appella, E., Rosenberg, S., and Robbins, P. (1995). Identification of a tyrosinase epitope recognized by HLA-A24-restricted, tumor-infiltrating lymphocytes. *J. Immunol.* **155,** 1343–1348.

145. Brichard, V., Herman, J., Van Pel, A., Wildmann, C., Gaugler, B., Wolfel, T., Boon, T., and Lethe, B. (1996). A tyrosinase nonapeptide presented by HLA-B44 is recognized on a human melanoma by autologous cytolytic T lymphocytes. *Eur. J. Immunol.* **26,** 224–230.

146. Kawakami, Y., Robbins, P., Wang, X., Tupesis, J., Parkhurst, M., Kang, X., Sakaguchi, K., Appella, E., and Rosenberg, S. (1998). Identification of new melanoma epitopes on melanosomal proteins recognized by tumor infiltrating T lymphocytes restricted by HLA-A1, -A2, and -A3 alleles. *J. Immunol.* **161,** 6985–6992.

147. Kobayashi, H., Kokubo, T., Takahashi, M., Sato, K., Miyokawa, N., Kimura, S., Kinouchi, R., and Katagiri, M. (1998). Tyrosinase epitope recognized by an HLA-DR-restricted T-cell line from a Vogt–Koyanagi–Harada disease patient. *Immunogenetics* **47,** 398–403.

148. Kobayashi, H., Kokubo, T., Sato, K., Kimura, S., Asano, K., Takahashi, H., Iizuka, H., Miyokawa, N., and Katagiri, M. (1998). CD4+ T cells from peripheral blood of a melanoma patient recognize peptides derived from nonmutated tyrosinase. *Cancer Res.* **58,** 296–301.

149. Gold, P., and Freeman, S. (1965). Specific carcinomembryonic antigens of the human digestive system. *J. Exp. Med.* **122,** 467.

150. Hsu, F., Benike, C., Fagnoni, F., Liles, T., Czerwinski, D., Taidi, B., Engelman, E., and Levy, R. (1996). Vaccination of patients with B-cell lymphoma using autologous antigen-pulsed dendritic cells. *Nature Med.* **2,** 52–58.

151. Hsu, F., Casper, C., Czerwinski, D., Kwak, L., Liles, T., Syrengelas, A., Taidi-Laskowski, B., and Levy, R. (1997). Tumor-specific idiotype vaccines in the treatment of patients with B cell lymphoma—long-term results of a clinical trial. *Blood* **89,** 3129–3135.

152. Bakker, A., Schreurs, M., Tafazzul, G., De Boer, A., Kawakami, Y., Adema, G., and Figdor, C. (1995). Identification of a novel peptide derived from the malanocyte-specific gp100 antigen as the dominant epitope recognized by an HLA-A2.1-restricted anti-melanoma CTL line. *Int. J. Cancer* **62,** 97–102.

153. Kawakami, Y., Eliyahu, S., Jennings, C., Sakaguchi, K., Kang, X., Southwood, S., Robbins, P., Sette, A., Appella, E., and Rosenberg, S. (1995). Recognition of multiple epitopes in the human melanoma antigen gp100 by tumor-infiltrating T lymphocytes associated with in vivo tumor regression. *J. Immunol.* **154,** 3961–3968.

154. Robbins, P., El-Gamil, M., Li, Y., Fitzgerald, E., Kawakami, Y., and Rosenberg, S. (1997). The intronic region of an incompletely spliced gp100 gene transcript encodes an epitope recognized by melanoma-reactive tumor-infiltrating lymphocytes. *J. Immunol.* **159,** 303–308.

155. Kawakami, Y., Eliyahu, S., Sakaguchi, K., Robbins, P., Rivoltini, L., Yannelli, J., Appella, E., and Rosenberg, S. (1994). Identification of the immunodominant peptides of the MART-1 human melanoma antigen recognized by the majority of HLA-A2-restricted tumor infiltrating lymphocytes. *J. Exp. Med.* **180,** 347–352.

156. Castelli, C., Storkus, W., Maeurer, M., Martin, D., Huang, E., Pramanik, B., Nagabhushan, T., Parmiani, G., and Lotze, M. (1995). Mass spectrometric identification of a naturally processed melanoma peptide recognized by CD8[+] cytotoxic T lymphocytes. *J. Exp. Med.* **181,** 363–368.

157. Coulie, P., Brichard, V., Van Pel, A., Wolfel, T., Schneider, J., Traversari, C., Mattei, S., De Plaen, E., Lurquin, C., Szikora, J., Renauld, J., and Boon, T. (1994). A new gene coding for a differentiation antigen recognized by autologous cytolytic T lymphocytes on HLA-A2 melanomas. *J. Exp. Med.* **180,** 35–42.

158. Schneider, J., Brichard, V., Boon, T., Meyer zum Buschenfelde, K., and Wolfel, T. (1998). Overlapping peptides of melanocyte differentiation antigen Melan-A/MART-1 recognized by autologous cytolytic T lymphocytes in association with HLA-B45.1 and HLA-A2.1. *Int. J. Cancer* **75,** 451–458.

159. Wang, R., Parkhurst, M., Kawakami, Y., Robbins, P., and Rosenberg, S. (1996). Utilization of an alternative open reading frame of a normal gene in generating a novel human cancer antigen. *J. Exp. Med.* **183,** 1131–1140.

160. Wang, R., Appella, E., Kawakami, Y., Kang, X., and Rosenberg, S. (1996). Identification of TRP-2 as a human tumor antigen recognized by cytotoxic T lymphocytes. *J. Exp. Med.* **184,** 2207–2216.

161. Wang, R., Johnston, S., Southwood, S., Sette, A., and Rosenberg, S. (1998). Recognition of an antigenic peptide derived from TRP-2 by cytotoxic T lymphocytes in the context of HLA-A31 and A33. *J. Immunol.* **160,** 890–897.

162. Parkhurst, M., Fitzgerald, E., Southwood, S., Sette, A., Rosenberg, S., and Kawakami, Y. (1998). Identification of a shared HLA-A*0201-restricted T cell epitope from the melanoma antigen tyrosinase-realted protein 2 (TRP2). *Cancer Res.* **58,** 4895–4901.

163. Lupetti, R., Pisarra, P., Verrecchia, A., Farina, C., Nicolini, G., Anichini, A., Bordignon, C., Sensi, M., Parmiani, G., and Traversari, C. (1998). Translation of a retained intron in tyrosinase-related protein (TRP) 2 mRNA generates a new cytotoxic T lymphocyte (CTL)-defined and shared human melanoma antigen not expressed in normal cells of the melanocytic lineage. *J. Exp. Med.* **188,** 1005–1016.

164. Correale, P., Walmsley, K., Nieroda, C., Zaremba, S., Zhu, M., Schlom, J., and Tsang, K. (1997). In vitro generation of human cytotoxic T lymphocytes specific for peptides derived from prostate-specific antigen. *J. Natl. Cancer Inst.* **89,** 293–300.

165. Alexander, R., Brady, F., Leffell, M., Tsai, V., and Celis, E. (1998). Specific T cell recognition of peptides derived from prostate specific antigen in patients with prostatic cancer. *Urology,* **51,** 150–157.

166. Peoples, G., Goedegebuure, P., Smith, R., Linehan, D., Yoshino, I., and Eberlein, T. (1995). Breast and ovarian cancer-specific cytotoxic T lymphocytes recognize the same HER-2/*neu*-derived peptide. *Proc. Natl. Acad. Sci. USA* **92,** 432–436.

167. Fisk, B., Blevins, T., Wharton, J., and Ionnides, C. (1995). Identification of an immunodominant peptide of HER2/*neu* protooncogene recognized by ovarian tumor-specific cytotoxic T lymphocyte lines. *J. Exp. Med.* **181,** 2109–2117.

168. Ropke, M., Hald, J., Guldberg, P., Zeuthen, J., Norgaard, L. F., Svejgaard, A., Van Der Burg, S., Nijman, M., Melief, C., and Claesson, M. (1996). Spontaneous human squamous cell carcinomas are killed by a human cytotoxic T lymphocyte clone recognizing a wild type p53-derived peptide. *Proc. Natl. Acad. Sci. USA* **93,** 14704–14707.

VACCINE STRATEGIES

Development of Epitope-Specific Immunotherapies for Human Malignancies and Premalignant Lesions Expressing Mutated *ras* Genes

SCOTT I. ABRAMS

Laboratory of Tumor Immunology and Biology
Center for Cancer Research
National Cancer Institute
National Institutes of Health
Bethesda, Maryland 20892

I. INTRODUCTION

Understanding the basic cellular and molecular elements important for T-lymphocyte recognition of the major histocompatibility complex (MHC)/peptide ligand expressed by the antigen (Ag)-bearing target cell is fundamental toward our understanding of the development of cell-mediated immunity against neoplasia. The trimolecular complex formed among the MHC molecule, peptide ligand, and T-cell receptor (TCR) represents the elementary structural unit ultimately responsible for cellular immune recognition and activation against pathogenic challenges [1,2]. (For a further discussion of antitumor immune mechanisms, see Chapter 7.) It is generally thought that cancer cells are potentially antigenic but lack adequate intrinsic immunogenic properties necessary to elicit protective cell-mediated and antitumor immunity in tumor-bearing hosts [3–7]. Thus, an important experimental prerequisite for the effective design of immunotherapies for human neoplasms, particularly for solid tumors, is the identification of tumor-associated antigens (TAAs) or tumor-specific antigens (TSAs) that express potential cancer regression (rejection) epitopes. Once defined and characterized, purified or recombinant forms of TAAs or TSAs can then be employed for use as immunogens for activation of the relevant immune effector mechanisms by two major strategies: (1) active immunotherapy, which involves *in vivo* immunization or vaccination; and (2) passive or adoptive cellular immunotherapy, which involves *ex vivo* expansion of the *in vivo* primed, autologous tumor-specific effector cells, which can then be returned to the host.

Oncogenes and their products have now clearly been implicated in the molecular and biochemical pathways of neoplastic transformation and development [8,9]. The *ras* p21 oncogenes represent one such prominent example. Somatic point mutations in these genes are commonly found in a wide range and high proportion of human malignancies, as well as in preneoplastic lesions (see later discussion; also further reviewed in Bos [10,11], Kiaris and Spandidos [12], and Abrams *et al.* [13]). The overall rationale for the study of the interaction between the host immune system and oncogene products, such as those encoded by mutated *ras* genes, is based on the fact the resulting oncoproteins are distinct from those encoded by the normal protooncogenes. From an immunologic perspective, these altered self-proteins may present unique immunodominant and/or subdominant peptide epitopes surrounding the sites of mutations for immune recognition and activation of the appropriate tumor-specific T-cell precursors. The expression of *ras* oncogenes and their products thus represents potentially attractive tumor-specific target molecules for cancer immunotherapy, not only for clinical situations of overt metastatic disease but also, perhaps, in patients with minimal residual disease or premalignant lesions that are detectable during the early stages of neoplastic formation. The aim of this chapter is to focus on an understanding of the nature of the biologic interactions between the host cellular immune system and products of activated *ras* genes, mainly in human systems, which may have clinical applications for the development of cell-mediated and antitumor immunity in human neoplasia.

II. CELLULAR IMMUNE RESPONSE AND ANTIGEN RECOGNITION

The cellular immune response has been implicated in a diversity of experimental models as the prominent effector mechanism important for the control and/or elimination of intracellular or endogenous pathogens, such as viruses and neoplasia [3–7,14,15]. Key to this biologic paradigm is the cell–cell interaction between the thymus-derived T lymphocyte and its Ag-bearing target cell [1,2,16]. T lymphocytes, through their clonotypic $\alpha\beta$-TCRs, recognize cognate Ags as short linear peptides produced from degraded proteins, bound to self-major histocompatibility complex (MHC) class I or class II molecules that are displayed on the extracellular surfaces of antigen-presenting cells (APCs) or aberrant target cells [1,2,17]. It is this unique and basic TCR–MHC/peptide ligand interaction that underlies the hallmark of cellular immune recognition and its exquisite specificity and, thus, defines a crucial characteristic feature in support of the concepts of immune surveillance, MHC restriction, discrimination of "self versus nonself," and maintenance of peripheral tolerance against normal self-tissues [18,19]. Although TCR engagement of MHC/peptide complexes is essential for immune recognition, initiation and regulation of the immune response additionally requires the contribution of Ag-independent, receptor–ligand interactions [20–23]. These cell-surface interactions involve a complex set of accessory molecules that participate in other aspects of MHC recognition, adhesion (conjugation), and costimulation important for the production of cytokines involved in clonal differentiation and expansion.

III. PATHWAYS OF ANTIGEN PROCESSING, PRESENTATION, AND EPITOPE EXPRESSION

Generally, two independent pathways of Ag processing—exogenous and endogenous—have been characterized [24–26]. The outcome of each is thought to have a profound influence on the nature of the resulting cellular immune response. Native protein is processed in specialized APCs, such as dendritic cells, macrophages, or B cells, by intracellular degradation pathways leading to the cell-surface display of potential antigenic peptides for T-cell recognition. Generally, the exogenous pathway involves the uptake of soluble, nonreplicative Ag from the extracellular milieu by receptor-mediated endocytosis or phagocytosis with intracellular degradation in endocytic/lysosomal compartments. APC populations that infiltrate tumor deposits, for example, may encounter exogenous sources of Ag as byproducts of proliferating cells in circulation (plasma) or at sites of necrosis or apoptosis. The resulting peptide fragments may associate with MHC class II molecules, which are synthesized within the same vesicles. In contrast, the endogenous pathway is active within the cytoplasm of the cell, where biosynthetic Ags, including those of viral or neoplastic origin, are processed. Peptide fragments generated in the cytosol of the cell may associate with MHC class I molecules, which are also recycled and synthesized in the same compartment. The endogenous pathway of Ag processing and presentation is not restricted to professional APCs and is likely a universal characteristic of any MHC-class-I-bearing cell type, including tumor cells. Nevertheless, the peptide/MHC class I or II complexes generated by these various pathways are then transported to the cell surface, where they become accessible for selective immune recognition.

The binding association between a given MHC molecule and its peptide, whether it be a self or foreign peptide for either a class I or II interaction, is a genetically restricted event [1,24,27,28]. Thus, peptides defined as agretopes in a given mouse strain or individual of a specific H-2 or HLA haplotype may be completely inactive in other members of that same species lacking those alleles. Certain residues within the peptide are critical for MHC binding, while others are important for TCR recognition [24,29,30]. The MHC contact residues include dominant and secondary anchors and may vary for different HLA molecules [1,27,28,31,32]. In both murine and human systems, such "consensus anchor motifs" have been defined for a variety of MHC molecules. In contrast to class I/peptide interactions, class II/peptide interactions involve anchor sites that appear to be more degenerate in specificity. Consequently, a given class II ligand may exhibit broader MHC reactivity and bind to multiple class II alleles [1,28]. Additionally, MHC class I and II molecules bind and present peptides of different sizes, which generally are 8–10 and 13–18 amino acid residues in length, respectively. Although peptide binding to MHC molecules is a prerequisite for TCR recognition, not all MHC-reactive peptides are antigenic, which may reflect fundamental mechanisms controlling peripheral tolerance or clonal deletion (i.e., a "hole" in the T-cell repertoire).

IV. T-LYMPHOCYTE SUBSETS

Originally, two major subpopulations of peripheral T lymphocytes were phenotypically defined and classified: $CD4^+$ and $CD8^+$. Subset dichotomy, in part, reflects intrinsic differences in MHC/peptide recognition requirements and the nature of the resulting cellular immune responses. $CD4^+$ lymphocytes have been proposed to play an important and central role in immunoregulation through the production and action of lymphokines [16,33], while $CD8^+$ lymphocytes have been described as cytotoxic T lymphocytes (CTLs) that mediate the destruction of Ag-bearing targets [34,35]. Although both $CD4^+$ and $CD8^+$ T cells independently recognize antigenic determinants expressed by an APC, optimal development and regulation of the cellular immune response typically requires the cellular cooperation between these two subpopulations [16,33,36–38]. It is becoming clearer, however, that the seemingly simple division of T cells into $CD4^+$ and $CD8^+$ T-cell subsets is actually more complex. In fact, multiple functional subtypes of $CD4^+$ and $CD8^+$ T cells have been described in model systems [33,39–46]. These subtypes have been termed type 1 (i.e., $CD4^+$ T_h1; $CD8^+$ Tc1) and type 2 (i.e., $CD4^+$ T_h2; $CD8^+$ Tc2) and predominantly reflect differences in their cytokine secretion patterns following TCR stimulation. Type 1 cells have been shown to selectively secrete interleukin-2 (IL-2), interferon-gamma (IFN-γ), and tumor necrosis factor-beta (TNF-β) (lymphotoxin), whereas type 2 cells have been described to preferentially produce IL-4, IL-5, IL-6, IL-10, and IL-13. Some cytokines, such as IL-3, granulocyte–macrophage colony-stimulating factor

(GM-CSF) and TNF-α, may be secreted by both type 1 and type 2 T cells.

In addition to having a positive effect on the development of a specific immune response, some cytokines have been documented to play a role in the coordination or downregulation of the opposing response. For example, IFN-γ produced by type 1 cells inhibits a type 2 response, whereas IL-4 or IL-10 produced by type 2 cells inhibits a type 1 response. Thus, the functional role of a $CD4^+$ T_h1 or T_h2 cell or of a $CD8^+$ Tc1 or Tc2 cell in the development of cell-mediated or humoral immunity may be largely governed by the cytokine secretion phenotype. Furthermore, the "type 1/type 2" balance may be important, not only for the maintenance of immune homeostasis under normal conditions, but also for the induction of an inappropriate immune response that may be associated with certain pathogenic states [33,39,41,44–47]. Because much of what is known about $CD4^+$ and $CD8^+$ T cell subtypes has been determined in model systems of infectious disease, allergy, autoimmunity, and alloreactivity, their precise roles in cancer remain to be fully understood.

The TCR of $CD8^+$ CTLs recognize antigenic peptide epitopes displayed on the cell surface of the APC/target cell in the context of self-MHC class I molecules. The resulting effector cell response, whether it be of a Tc1 or Tc2 phenotype, is the death of the Ag-bearing target cell [34,35], although the mechanisms by which these CTL subtypes mediate lysis may be different [48]. The TCR of $CD4^+$ T cells recognize antigenic peptide epitopes displayed on the cell surface of the APC/target cell in the context of self-MHC class II molecules. The $CD4^+$ T_h1 subtype has been reported to express lytic activity [49,50], which may be important for the termination of an immune response through direct elimination of MHC class II$^+$ APCs or, perhaps, for lysis of tumor cells that may expose antigenic peptides in the context of MHC class II molecules. Furthermore, $CD4^+$ T cells (T_h1 subtype) have been shown to exert antitumor effects *in vivo* in various models of active or adoptive immunotherapy [4,51–53]. Antitumor reactivity may result from the release of cytokines that modify tumor-cell viability directly (i.e., TNF-α/β) or indirectly (i.e., interleukins, IFN-γ, GM-CSF) by recruitment and further activation of other cytotoxic effector cells, such as $CD8^+$ T cells, macrophages, or natural killer (NK) cells. Moreover, IFN-γ may induce or upregulate target cell surface expression of MHC class II molecules, which might facilitate direct interactions between $CD4^+$ CTLs and Ag-bearing tumor cells.

V. ras ONCOGENES IN NEOPLASTIC DEVELOPMENT

The *ras* p21 protooncogenes encode a family of evolutionarily conserved, 21-kDa intracellular guanosine triphosphate (GTP) binding proteins (189 amino acids in length) that are integral for cellular signal transduction and that

TABLE 1   General Characteristics of *ras* Genes

Mammalian cells express three *ras* protooncogenes: K-*ras*, H-*ras*, N-*ras*
    Highly conserved in evolution and ubiquitous in diverse cell types
    Encode a family of intracellular proteins (21-kDa, 189 amino acids
      in length)
    Serve as guanine nucleotide binding proteins
    Play a central role in cellular signal transduction, growth, and function
    Activity regulated by phosphorylation patterns
Point mutations (at codons 12, 13, 59, and 61) are linked to oncogenesis
    Alteration of phosphorylation/dephosphorylation cycle; proteins re-
      main in a prolonged state of activation, resulting in constitutive
      transduction of "growth" promoting signals to downstream target
      molecules
    Early event in the pathogenesis of neoplastic development
    Found in a broad array and high proportion of different tumor types:
      carcinomas, melanomas, and hematopoietic-derived malignancies

ultimately regulate cellular differentiation, proliferation, and function in a wide diversity of cell types [54,55]. In mammalian cells, the *ras* protooncogenes consist of three highly homologous members, K-*ras*, H-*ras*, and N-*ras*. Each gene has the capacity to become oncogenic as a consequence of a somatic mutation (Table 1) [9–12,56]. Under normal physiologic conditions, the basic function of *ras* p21 is to bind one molecule of GTP or guanosine diphosphate (GDP) per molecule of *ras* p21 protein. With respect to normal signal transduction and cellular function, *ras* p21 acts as a binary switch and is in the active ("on") state when bound to GTP and in the inactive ("off") state when bound to GDP. Mutations at affected codons result in oncogenic proteins containing single amino acid substitutions at those corresponding positions. In a functional sense, such *ras* proteins are aberrant as they become constitutively GTP-bound due to a loss both in intrinsic GTPase activity and responsiveness to GAPs (GTPase-activating proteins), which ordinarily help to facilitate the hydrolysis or dephosphorylation of GTP to GDP. Hence, *ras* oncoproteins in the activated form appear to be in a constitutively "on" state or signal transmitting mode, which has been proposed to represent an early event in the initiation of the neoplastic process. This is most notable, for example, in the pathogenesis and progression of human colorectal and pancreatic adenocarcinoma. Mutations of *ras* have been found in, and associated with, benign adenomas/polyps and pancreatic intraductal lesions (PILs) of colorectal and pancreatic tumorigenesis, respectively, suggesting that the mutation precedes the development of malignancy [57,58].

    ras p21 was originally discovered as an oncogene product (i.e., the activated form of *ras* was identified as a transforming gene of an acute transforming murine sarcoma virus). Subsequently, genomic DNA from human tumors and tumor cell lines were demonstrated to contain activated *ras* genes. Cloning and sequencing of *ras* oncogenes from acute transforming retroviruses, human tumor DNA, and protooncogenes from normal cells revealed that the activated forms of the *ras* genes contained mutations restricted to specific sites, principally at codons 12, 13, 59, or 61. In human tumors, the vast majority of *ras* mutations are found at codon 12. In rodent tumors, such as those induced by chemical carcinogens, the majority of *ras* mutations are found at codon 61. Originally, the most widely used assay system for detection of *ras* mutations was the NIH3T3 transfection assay, which was based on the ability of *ras* genes to transform the murine NIH3T3 cells. Although it was able to detect point mutations, it was not suitable for screening large numbers of tumor samples. The development of more sophisticated and rapid assay systems for the identification of mutated *ras* genes has facilitated the analysis of large numbers of human tumor samples. These methods include: (1) selective hybridization of tumor DNA with synthetic oligodeoxynucleotide probes specific for mutations in codons 12, 13, and 61 of the *ras* genes [59,60]; (2) RNase mismatch cleavage assays [61]; and (3) use of polymerase chain reaction (PCR) to amplify segments of the *ras* genes from DNA derived from tumor [62,63] and, more recently, plasma specimens from patients with colorectal and pancreatic carcinomas [64,65].

    As mentioned earlier, in human cancers, the vast majority of point mutations in the *ras* genes are linked to codon 12. Mutations in codon 12 result in the single amino acid substitution of the wild-type Gly residue at position 12 with an Asp, Val, Cys, Ser, Arg, or Ala residue. Carcinomas, especially adenocarcinomas, and hematologic malignancies of the myelomonocytic lineage are more frequently associated with *ras* mutations than tumors of neuroectodermal origin or differentiated lymphoid malignancies, while carcinomas of the breast and cervix rarely contain *ras* mutations. Furthermore, there does appear to be an association between the expression of a given mutated *ras* gene (K-, H-, or N-*ras*) and certain tumor types. For example, K-*ras* mutations are predominantly expressed in pancreatic, colorectal, and lung adenocarcinomas; H-*ras* mutations are mainly found in thyroid, renal, and bladder carcinomas; and N-*ras* mutations are largely associated with melanomas, hepatocellular carcinoma, lymphomas, and myeloid malignancies. An extensive survey of the literature [13] has revealed that a large percentage of carcinomas of the pancreas, colon/rectum, lung, endometrium, and thyroid contain *ras* position 12 mutations, as well as a proportion of several other histologic tumor types just described. Collectively, for a range of such prevalent tumor types, the nature, pattern, and frequency of codon 12 mutations reflect the substitution of Gly to Asp (~40%) > Val (~28%) > Cys (~16%) > Ala (~6%) > Arg (~5%) > Ser (~5%) [13]. In the U.S. alone, for example, more than 140,000 new cases of cancer are reported each year in patients whose tumors harbor *ras* codon 12 mutations. Based on these estimates, more than 800,000 cancer patients in the United States whose tumors contain codon 12 *ras* mutations are currently available for treatment.

VI. CELLULAR IMMUNE RESPONSES INDUCED BY *ras* ONCOGENE PEPTIDES

A number of investigations have been conducted in both murine and human systems to examine the potential immunogenicity of mutated *ras* peptides for the generation of Ag-specific CD4$^+$ and/or CD8$^+$ T cell reactivities. Murine models have become instrumental in exploring a diversity of biologic and immunologic principles, which can perhaps be conceptually and mechanistically translated to clinical situations. Such principles include: (1) immune *discrimination* of "self" (i.e., oncogene) versus "nonself" (i.e., protooncogene) forms; (2) immune *induction* of both CD4$^+$ and CD8$^+$ T-cell responses; (3) nature of immune *regulatory* and immune *effector* functions; (4) immune *recognition* of exogenously or endogenously processed forms of *ras* oncoproteins by APCs/tumor cells; and (5) immune *modulation* of immunogenic properties to enhance sensitivity and potency of cellular and antitumor immune responses. With these objectives in mind, preclinical findings in animal models [50,66–74] strongly support the underlying hypothesis that *ras* oncogenes encode for T-cell epitopes that surround the sites of mutations and that activate both cellular (CD4$^+$ and/or CD8$^+$) and antitumor responses, without inducing significant immunologic cross- or autoreactivity against self-determinants. Thus, the collective observations from these preclinal studies demonstrate that T-cell responses (using lines/clones) can distinguish recognition of the mutant from wild-type *ras* forms, supporting the contention that products of activated *ras* genes represent tumor-specific, neo-antigenic targets.

A. Human CD4$^+$ T-Cell Responses

In human *in vitro* studies, Jung and Schluesener [75] were the first to describe the immunogenicity of mutated *ras* peptides. Human leukocyte antigen (HLA) class-II-restricted, CD4$^+$ T cell lines were established from normal individuals using a point-mutated *ras* 12-mer peptide spanning positions 5–16 and containing the substitution of Gly to Val at position 12. Subsequently, Gedde-Dahl *et al.* [76–80] characterized in detail functional properties of the peptide-induced CD4$^+$ T-cell response from normal individuals and from a colorectal carcinoma patient with localized disease against a variety of *ras* mutations at codons 12, 13, or 61, including the fine specificity requirements for peptide/HLA class II interactions. In fact, all three major HLA class II alleles, DR, DP and DQ, were shown to serve as restriction elements for *ras* peptide presentation. Tsang *et al.* [81] established several peptide-specific CD4$^+$ T-cell lines *in vitro* from normal donors using point-mutated *ras* 13-mer peptides spanning positions 5–17 and containing the substitution of Gly at position 12 to a Val, Cys, or Asp residue. No T-cell lines were generated

against the wild-type peptide sequence. Cytokine analysis revealed the production of IFN-γ and IL-6, but not IL-4, in response to Ag-specific stimulation. In a Val12 model, Ag-specific cytotoxicity was observed against targets incubated with exogenously supplied mutant, but not wild-type, peptide. Furthermore, specific lysis was observed against autologous Epstein–Barr virus (EBV)-transformed B-cell targets expressing the full-length human K-*ras* oncogene via retrovirus transduction [81].

B. Human CD8$^+$ T-Cell Responses

In contrast to the host of reports on the *in vitro* generation of human anti-*ras* CD4$^+$ T-cell reactivity, less has been described and characterized regarding the *in vitro* induction of human CD8$^+$ CTL responses. Fossum *et al.* [82] generated CD8$^+$ CTL clones from a colon carcinoma patient, harboring a Gly to Asp mutation at codon 13, by peptide stimulation *in vitro* using a 25-mer peptide. The resulting CTL clones lysed a carcinoma cell line harboring the corresponding *ras* mutation in a HLA class I B12(44)-restricted manner following IFN-γ pretreatment of the target cells. In a human melanoma model, Van Elsas *et al.* [83] and Juretic *et al.* [84] defined a HLA-A2-reactive peptide sequence reflecting an N-*ras* oncogene mutation, Gln to Leu at codon 61, which led to the generation of anti-*ras* CD8$^+$ T cells from normal individuals *in vitro*. These HLA-A2-restricted CTLs could lyse target cells pulsed with exogenous peptide, but not transfected melanoma cells expressing the *ras* oncogene. In a subsequent study [85], however, if peptide-specific CTLs were established using a more aggressive strategy involving peripheral blood dendritic cells as APCs, then CTL-mediated lysis against these *ras*-transfected tumor cells could be achieved.

VII. IDENTIFICATION OF MUTANT *ras* CD4$^+$ AND CD8$^+$ T-CELL EPITOPES REFLECTING CODON 12 MUTATIONS

In both murine [50,71,72,86] and human [81,87,88] systems, our laboratory has been exploring the underlying hypothesis that *ras* oncogenes encode for tumor-specific CD4$^+$ and/or CD8$^+$ CTL epitopes reflecting codon 12 mutations (Table 2). The model or system has focused on codon 12, as the vast majority of human cancers that contain *ras* oncogenes harbor point mutations at that "hot spot". The identification of both CD4$^+$ and CD8$^+$ T-cell epitopes in an overlapping or nested configuration surrounding the same *ras* mutation [50,71] may have important implications for the activation of both anti-*ras* immune effector mechanisms against malignancies harboring the same neo-antigenic determinant, resulting in a more coordinated, comprehensive, and potent antitumor response. Thus, the specific goals of these studies include: (1) identify peptide sequences reflecting *ras* codon

TABLE 2   Mutant *ras* Peptide Sequences Reflecting the Substitution of Gly at Position 12 Identified as Human T-Cell Epitopes[a]

ras Peptides	Amino acid sequence														Immune response
	4	5	6	7	8	9	10	11	12	13	14	15	16	17	
ras 4–17(Gly12)[b]	Tyr–Lys–Leu–Val–Val–Val–Gly–Ala–Gly–Gly–Val–Gly–Lys–Ser														Wild-type sequence
ras 5–17(Val12)	Lys–Leu–Val–Val–Val–Gly–Ala–Val–Gly–Val–Gly–Lys–Ser														CD4+
ras 5–17(Cys12)	Lys–Leu–Val–Val–Val–Gly–Ala–Cys–Gly–Val–Gly–Lys–Ser														CD4+
ras 5–17(Asp12)	Lys–Leu–Val–Val–Val–Gly–Ala–Asp–Gly–Val–Gly–Lys–Ser														CD4+
ras 5–14(Val12)	Lys–Leu–Val–Val–Val–Gly–Ala–Val–Gly–Val														CD8+
ras 4–12(Val12)	Tyr–Lys–Leu–Val–Val–Val–Gly–Ala–Val														CD8+
ras 5–14(Asp12)	Lys–Leu–Val–Val–Val–Gly–Ala–Asp–Gly–Val														CD8+
ras 5–14(Cys12)	Lys–Leu–Val–Val–Val–Gly–Ala–Cys–Gly–Val														Not tested[c]

[a]Based on ability to generate mutant *ras*-specific cellular immune responses and T-cell lines (CD4+ and/or CD8+) from either normal individuals by IVS methodologies (see Abrams *et al.* [13], Tsang *et al.* [81], and Bergmann-Leitner *et al.* [87]), or from appropriately immunized carcinoma patients in a phase I clinical trial (see Khleif *et al.* [88] and Abrams *et al.* [97]).

[b]Normal sequence of *ras* p21 from positions 4–17 of the protein, which encompasses the various peptides analyzed in these studies.

[c]Peptide sequence binds to HLA-A2 (by T2 bioassays), although immunogenicity remains to be determined.

12 mutations as putative tumor-specific, neo-determinants for the induction of epitope-specific CD4+ and/or CD8+ T-cell responses *in vitro* and *in vivo* (i.e., in clinical trial settings); (2) analyze and characterize the phenotypic and functional properties of the resulting T-cell responses, with emphasis on their capacity to recognize exogenously or endogenously derived tumor Ags; and (3) modify the *in vitro* stimulation (IVS) conditions and/or *in vivo* immunization parameters (i.e., in clinical trial settings) as prospective considerations to potentially accelerate the generation and enhance the potency and sensitivity of the induced T-cell response. The production and propagation of *ras* oncogene-specific CD4+ and/or CD8+ T-cell populations may then be expanded *ex vivo* and potentially used in adoptive cellular immunotherapy, alone or in concert with further active immunization.

A. Identification of Mutant *ras* Peptides That Bind to HLA-A2

In the vast majority of human studies involving the identification of CD8+ CTL epitopes, whether they be of viral or tumor origin, interest has been focused on defining peptide epitopes presented by the HLA-A2 subtypes [5,6,14,89–92], as these molecules are shared by a substantial percentage (average 43.1%) of the human population [93,94] and the peptide binding motifs are well-characterized [27,31,32,95]. Thus, due to the prevalence of both HLA-A2 expression in the human population and the incidence of codon 12 *ras* mutations in human cancers, codon 12 of the *ras* oncogenes was examined for potential expression of HLA-A2-restricted, CD8+ CTL epitopes. As an experimental system to explore these hypotheses, the mutant *ras* protein sequence reflecting the Gly to Val substitution at position 12 was scanned at that site for putative HLA-A2 consensus anchor motifs (Table 2

and Fig. 1). The preferred dominant anchor motif for HLA-A2 originally described Leu, Ile, or Met at position 2 and Val, Leu, or Ile at the C-terminus of a 9-mer or 10-mer sequence. Tyr at position 1 has been shown to have a favorable impact as a secondary anchor [32]. Two nominal *ras* sequences were identified as potential HLA-A2-reactive peptides: the *ras* 10-mer sequence 5–14, as it contained the dominant anchors Leu and Val at the second (i.e., Leu6) and C-terminus (i.e., Val14) positions of the peptide, respectively; and the *ras* 9-mer sequence 4–12, as it appeared to satisfy the motif for binding interactions at positions 1 (i.e., Tyr4) and 9 (i.e., Val12). However, the N-terminal location of Leu, the first dominant anchor, was found at position 3 (i.e., Leu6).

The capacity of these *ras* peptides to bind to HLA-A2 was first examined by T2 bioassays (Fig. 1) [96]. Peptide binding to HLA-A2 on T2 cells was weak, as compared with both negative and positive control peptides, and required co-incubation with exogenous human β2-microglobulin to facilitate and maximize complex formation. However, in contrast to the wild-type *ras* 5–14(Gly12) peptide, the wild-type *ras* 4–12(Gly12) peptide failed to bind to HLA-A2, even in the presence of exogenous β2-microglobulin. This observation was consistent with the hypothesis that in the 9-mer sequence, the Val12 substitution led to the creation of a C-terminus anchor. This fundamental difference thus provided the rationale at that time to explore the immunogenicity of the *ras* 4–12(Val12) peptide as a human CD8+ CTL epitope [87]. Support for this hypothesis comes from our earlier study in mice [71], which defined the same peptide sequence as a mutant *ras* CD8+ CTL epitope peptide at position 12, reflecting its capacity (but not the homologous wild-type sequence) to specifically bind to the MHC class I H-2K[d] molecule. This did not exclude the possibility, however, that the mutant *ras* 5–14 sequence encoded for CD8+ CTL epitopes (i.e., Val12,

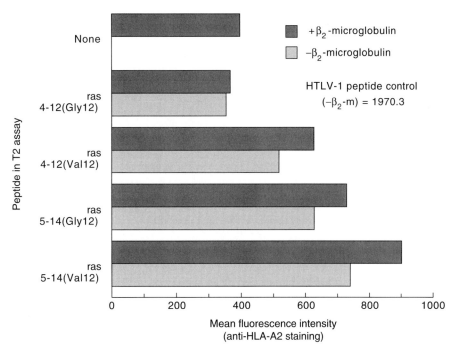

FIGURE 1   Identification of *ras* peptides reflecting the Val12 substitution at codon 12 that bind to HLA-A2. *ras* peptide sequences 4–12 and 5–14 containing the mutated Val amino acid at position 12 were identified as putative HLA-A2-reactive peptides, as they contained predicted MHC anchor residues. Their capacity to bind to HLA-A2, as compared with the counterpart wild-type (Gly12) sequences, was determined by T2 bioassays. T2 cells were incubated at 37°C overnight with the appropriate peptide (50 μg/mL) $\pm$ purified human β2-microglobulin (2 μg/mL). Treatment labeled as "None" was tested with β2-microglobulin only, while the positive "control" peptide (derived from an HTLV-1 sequence) was tested without β2-microglobulin. Cells were then stained with an anti-HLA-A2 mAb and evaluated by flow cytometry for mean fluorescence intensity. Results were expressed as the mean of three separate experiments.

Asp12, or Cys12). In fact, these possibilities were explored in detail in separate studies using lymphocyte populations from vaccinated carcinoma patients [88,97].

B. Generation of a Human CD8$^+$ CTL Line Specific for the *ras* 4-12(Val12) Epitope

To evaluate the potential immunogenicity of the *ras* 4–12 (Val12) sequence as a HLA-A2-restricted, CD8$^+$ T-cell epitope for the production of CTLs (from naive precursors), a model APC system was developed *in vitro* consisting of peptide/β2-microglobulin-pulsed T2 cells as APCs plus IL-2 and IL-12 as cytokines to maximize lymphoid maturation and growth [87]. Based on the observation that T2 cells pulsed with exogenous *ras* 4–12(Val12) peptide plus β2-microglobulin resulted in enhanced and peptide-specific HLA-A2 binding, this form of Ag presentation was adapted for the initiation and activation of potential CD8$^+$ CTL precursors. The rationale of T2 cells as APCs was largely based on the fact that they express "empty" class I heavy-chain molecules and fail to present endogenously derived peptides due to a defect in TAP (transporter associated with antigen processing)-dependent mechanisms [96]; thus, these cells become an efficient vehicle under serum-free conditions for loading and presentation of high densities of exogenously

supplied peptides, which can be potentiated by coincubation with β2-microglobulin. This becomes particularly important with weaker binding peptide immunogens, as their interaction with unoccupied class I molecules may be facilitated in the absence of competing serum-derived exogenous peptides or cellular endogenous pathways of Ag presentation. This model system was then applied for the generation of a mutant *ras* peptide-specific CTL response. Indeed, a CD8$^+$ CTL line was produced *in vitro* from a HLA-A2$^+$ normal donor, which displayed peptide-specific and HLA-A2-restricted cytotoxicity against peptide-pulsed targets (Fig. 2A). No lysis was detectable in the presence of the wild-type *ras* peptide, indicating the absence of "autoimmune" recognition (Fig. 2A legend).

C. Human CD8$^+$ CTL-Mediated Lysis of Tumor Cells Harboring the K-*ras* Oncogene

An important immunologic issue was to determine whether such peptide-induced CD8$^+$ CTLs could also recognize a processed form of the corresponding mutant *ras* protein. To test this hypothesis, the anti-*ras* 4–12(Val12)-derived CD8$^+$ CTL line was examined for its capacity to lyse the

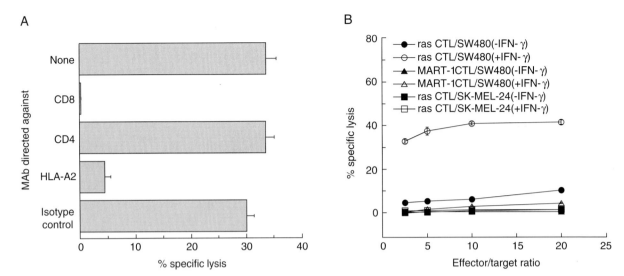

FIGURE 2 Functional characterization of the anti-*ras* 4–12(Val12) CTL line. (**A**) Role of CD8 and HLA-A2 molecules in cytotoxicity by anti-*ras* 4–12(Val12) T-cell line using peptide-coated (1 μg/mL) autologous EBV-B cells as targets (effector/target ratio, 10/1; 6 hr assay), in the absence and presence of the different mAbs. Lytic activity in the absence of peptide or in the presence of the control *ras* 4–12(Gly12) peptide was less than 2%. (**B**) Ability of anti-*ras* 4–12(Val12) CD8$^+$ CTL to recognize and lyse SW480 tumor cells ($\pm$ IFN-γ pretreatment, 250 U/mL for 24 hours) expressing the *ras* Val12 oncogene, in the absence of exogenous peptide. Control combinations included anti-*ras* CTLs with irrelevant targets (i.e., SK-MEL-24 melanoma: HLA-A2$^+$, *ras* Val12$^-$) or irrelevant CTLs (i.e., HLA-A2-restricted, MART-1$_{27-35}$ peptide-specific) with SW480 targets. Percent specific lysis against IFN-γ-pretreated SW480 cells mediated by anti-*ras* Val12 CTLs or anti-MART CTLs (effector/target ratio, 10/1) in the presence of additional saturable amounts of exogenous *ras* 4–12(Val12) peptide or MART-1$_{27-35}$ peptide (5 μg/mL) was 66 $\pm$ 3 and 79 $\pm$ 2, respectively. In a separate experiment, anti-CD8 and anti-HLA-A2 mAbs inhibited CTL-mediated lysis of IFN-γ-pretreated SW480 tumor targets, demonstrating the importance of those molecular interactions in T-cell recognition of endogenously expressed mutant *ras* epitopes. In both panels, results were expressed as the mean $\pm$SEM of triplicate cultures. (Data from Bergmann-Leitner *et al.* [87].)

SW480 colon adenocarcinoma cell line (Fig. 2B), which endogenously expresses both the appropriate MHC restriction element (HLA-A2) and *ras* mutation at codon 12 [98,99]. Furthermore, SW480 cells were tested either untreated or after a short-term, low-dose pretreatment with IFN-γ, which was initially thought, in part, to upregulate the expression of HLA-A2, as well as other events potentially associated with epitope processing and presentation. In the absence of IFN-γ pretreatment, little CTL lysis was detectable; however, following IFN-γ pretreatment, demonstrable cytotoxicity was observed at multiple effector/target ratios, which was maximal by 12 hr of incubation (Fig. 2B). Specificity of lysis was revealed by absence of lysis against a HLA-A2$^+$ melanoma (SK-MEL-24; with or without IFN-γ), which lacked the Val12 mutation, and the inability of an irrelevant HLA-A2-restricted, CD8$^+$ CTL line (i.e., MART-1$_{27-35}$ peptide-specific) to lyse SW480 cells (with or without IFN-γ; Fig. 2B), unless the appropriate exogenous peptide was added (Fig. 2B legend). Finally, anti-*ras* Val12 CTL-mediated lysis against IFN-γ-pretreated SW480 tumor cells could be boosted even further in the presence of exogenously supplied and saturable amounts of *ras* 4–12(Val12) peptide (Fig. 2B legend). These findings demonstrated that tumor cells may express mutant *ras* epitopes, such as *ras* 4–12(Val12), albeit in limiting amounts.

D. *In Vivo* Immunogenicity of *ras* Oncogene Peptides in Carcinoma Patients

Although a host of immunologic principles and concepts can be explored in animal models and *in vitro* human systems, such as those described here, the most compelling information regarding immunogenicity and the potential for development of cell-mediated and antitumor immunity could only be determined from vaccinated patients in clinical trials. Gjertsen *et al.* [100–104] were the first to report on the vaccination of metastatic cancer patients with mutated *ras* peptides. In that clinical study, pancreatic carcinoma patients were vaccinated with autologous peripheral blood mononuclear cells (PBMCs) prepulsed *ex vivo* with mutant *ras* peptides (spanning positions 5–21) reflecting codon 12 mutations. Of five patients receiving multiple vaccinations, two showed evidence of anti-*ras* T-cell responses, as measured by proliferation. In one patient, the T-cell response was specific for the immunizing peptide (Val12). In the other patient, the T-cell response reacted with both the immunizing peptide (Asp12) and, to a lesser extent, the non-mutated peptide (Gly12), based on precursor frequency differences [102]. The anti-*ras* Val12-specific T-cell proliferative response was later shown to be mediated by CD4$^+$ T cells, which were HLA class II restricted [102]. Two peptide-specific CD4$^+$

TABLE 3 Patient Profile and Immune Responses Induced After Vaccination with Mutated *ras* Peptide Immunogens[a]

Dose level (peptide dose)	Patient number	Age	Cancer type	*ras* mutation (Gly12→)[b]	Number of vaccines[c]	Immune response[d]
I (0.1 mg/injection)	1	55	Colon	Asp	3	Not detectable
	2	49	Colon	Asp	3	CD8$^+$ T cell
	3	46	Pancreas	Val	1	Not evaluated
II (0.5 mg/injection)	4	58	NSCLC	Cys	3	CD4$^+$ T cell
	5	48	Colon	Val	3	Not detectable
	6	76	Pancreas	Asp	3	Not detectable
III (1.0 mg/injection)	7	42	Appendix	Val	3	Not detectable
	8	29	Duodenal	Val	3	CD4$^+$ and CD8$^+$ T cell
	9	67	Rectal	Val	3	Not detectable
IV (1.5 mg/injection)	10	69	Colon	Asp	2	Not evaluated
	11	42	Colon	Asp	3	Not evaluated
	12	62	Colon	Val	1	Not evaluated
V (5.0 mg/injection)	13	50	Colon	Asp	3	Not detectable
	14	50	Colon	Asp	2	Not evaluated
	15	51	Colon	Asp	3	Not detectable

[a] Data summarized from Khleif *et al.* [88] and Abrams *et al.* [97].

[b] *ras* codon 12 mutation defined in autochthonous primary tumor specimen by PCR analysis of paraffin-embedded section. Based on identity of mutation, patients were vaccinated subcutaneously with the appropriate mutated *ras* 13-mer peptide (in Detox™ adjuvant) corresponding to that specific mutation. Mutations reflected the substitution of Gly at position 12 to an Asp, Val, or Cys residue.

[c] Number of completed vaccinations given to each patient. Although three vaccinations were scheduled for each patient, some patients (where indicated) received fewer than three because of progressive disease and, subsequently, were dismissed from the clinical trial.

[d] Nature of peptide-induced T-cell response(s) resulting from the vaccination, as determined by phenotypic and functional analyses of lymphocyte cultures obtained after the third vaccination. In those three patients displaying evidence of CD4$^+$ and/or CD8$^+$ T-cell responses, no specific cellular immune responses were observed in prevaccine lymphocytes examined in parallel. "Not detectable" indicates the absence of induction of peptide-specific, cell-mediated immunity, as determined using lymphocyte preparations obtained after the third vaccination. "Not evaluated" indicates that immunologic testing has not been conducted because lymphocyte preparations following the third vaccine were unavailable (for clinical reasons; see footnote c) and that, pending availability, lymphocyte preparations obtained after the first and/or second vaccines from those same patient have not yet been analyzed.

T-cell clones were established *in vitro;* one was determined to be HLA-DR6 restricted, and the other HLA-DQ2 restricted. Further characterization of the HLA-DR6-restricted, anti-*ras* Val12 CD4$^+$ T-cell clone revealed specific cytotoxicity *in vitro* against autologous EBV-B-cell targets pulsed with the mutant, but not the wild-type, *ras* peptide [102]. Anti-*ras* Val12 CD4$^+$ T-cell-mediated lysis was also observed against autologous pancreatic tumor cells following IFN-γ pretreatment of the targets, which enhanced MHC class II (HLA-DR) expression [103]. Furthermore, from this same immunized patient, an anti-*ras* Val12-specific CD8$^+$ CTL clone was produced, which lysed this same autologous pancreatic tumor cell line *in vitro*. The anti-*ras* Val12 CD8$^+$ CTL clone was HLA class I restricted and mapped to the HLA-B35 allele, and it recognized the mutant *ras* peptide sequence 7–15(Val12) nested within the original immunogen [103].

Similarly, a peptide-based phase I immunotherapy trial was initiated at the National Cancer Institute (NCI; Bethesda, MD) in metastatic carcinoma patients harboring K-*ras* codon 12 mutations, encoding the substitution of Gly at position 12 to Asp, Cys, or Val [88,97]. Patients were vaccinated subcutaneously three times, separated at monthly intervals, with

Detox™ adjuvant admixed with a mutated *ras* 13-mer peptide spanning positions 5–17 and corresponding to the specific mutation at codon 12 found in an autochthonous primary tumor specimen, as determined by PCR analysis of the paraffin-embedded section (see Table 2 for peptide sequences and Table 3 for patient profile). The selection of this particular peptide sequence was based, in part, on earlier preclinical *in vitro* human studies [13,81]. The Detox™ adjuvant (kindly supplied by RIBI ImmunoChem Research, Inc.; Hamilton, MT) consists of two active immunostimulants—cell wall skeleton from *Mycobacterium phlei* and monophosphoryl lipid A from *Salmonella minnesota* R595—and is prepared as an oil-in-water emulsion with squalane and Tween 80. We compared patient immune status before the first and after the third vaccination and characterized in detail the phenotypic and functional properties of the T-cell lines that displayed evidence of cell-mediated immunity. It is noteworthy that all patients included in this study at the time of entry generally displayed immune competence, as assessed by *in vivo* delayed-type hypersensitivity (DTH) reactivity to at least one Ag in a skin anergy panel and by *in vitro* responses to T- or B-cell mitogens.

Patient lymphocyte populations were incubated *in vitro* at 7- to 14-day intervals by continuous stimulation with an autologous source of APCs (i.e., PBMCs), mutant *ras* peptide as Ag, and low-dose IL-2 and then were retested for peptide-specific reactivity at or toward the end of an IVS cycle. We reasoned that these culture conditions, in contrast to the model APC system previously described (for Fig. 2), reflected the nominal requirements for Ag-specific expansion of *in vivo* peptide-primed lymphocytes (and their subsequent testing as evidence for "recall" responses to vaccination), as well as for the production of T-cell lines for detailed characterization. Initially, emphasis was placed on the impact of peptide vaccination on the development of the CD4$^+$ T-cell response, as previous studies supported the hypothesis that these or closely related mutant *ras* peptide sequences selectively stimulated CD4$^+$ reactivity [75-77,81,100-102]. However, we also explored the hypothesis for the induction of CD8$^+$ T-cell responses, which may have occurred as a result of immunogen processing *in vivo*. Based on an earlier finding that the *ras* 5–14 peptide sequences (independent of *ras* mutation) displayed binding to HLA-A2 (Fig. 1), we explored their capacity to serve as immunogens *in vitro* to generate Ag-specific CD8$^+$ CTL lines from HLA-A2$^+$ patients, pre- versus post-vaccination. Overall, 3 of 10 evaluable patients have demonstrated the development of Ag-specific, cell-mediated immunity resulting from the vaccination (Table 3).

First, an Ag-specific, MHC class II (HLA-DP)-restricted CD4$^+$ T-cell line was established *in vitro* from post-vaccinated lymphocytes of a nonsmall-cell lung carcinoma (NSCLC) patient whose primary tumor contained a Cys12 mutation when cultured on the immunizing peptide (Table 3). Moreover, CD4$^+$ proliferation was inducible against the corresponding mutant K-*ras* protein (derived from the Calu-1 lung carcinoma cell line), suggesting productive TCR recognition of exogenously processed Ag (Table 4). Second, an Ag-specific, MHC class I (HLA-A2)-restricted CD8$^+$ CTL line was established *in vitro* from post-vaccinated lymphocytes of a colon carcinoma patient whose primary tumor contained an Asp12 mutation (Tables 3 and 5). To that end, a 10-mer peptide, nested within the 13-mer immunizing peptide, was identified (i.e., *ras* 5–14(Asp12)) which was shown to bind to HLA-A2 and display specific functional capacity for expansion of the *in vivo* primed CD8$^+$ CTL precursors (Table 5). Third, both Ag-specific, MHC class II (HLA-DQ)-restricted CD4$^+$ and MHC class I (HLA-A2) restricted CD8$^+$ (Fig. 3A) T-cell lines were generated from a single patient with duodenal carcinoma whose primary tumor contained a Val12 mutation when cultured on the immunizing 13-mer peptide or a nested 10-mer peptide (i.e., *ras* 5–14(Val12)), respectively (Table 3). It is important to emphasize that in all three patients no specific responses were detectable against the normal proto-*ras* sequence (Gly12), and no T-cell lines

TABLE 4 CD4$^+$ T-Cell Response to Mutant K-*ras* Protein

Antigen in assay	Concentration	K-*ras* mutation	^{3}H-thymidine uptake
Cys12 peptide	30	Cys12	112,354 ± 413
	10		109,519 ± 4,942
	3		116,529 ± 4,947
	1		48,732 ± 473
Gly12 peptide	30	None	8,130 ± 314
None (T cell + APC)	—	—	6,836 ± 307
Calu-1 lysate	250	Cys12	64,357 ± 2,824
	75		55,540 ± 1,172
	25		6,884 ± 445
SW480 lysate	250	Val12	6,163 ± 1,073
	75		4,963 ± 1,075
HT-29 lysate	250	None	6,955 ± 384
	75	(Gly12)	7,223 ± 179

Note: Proliferation of anti-*ras* 5–17(Cys12) peptide-specific CD4$^+$ T-cell line (from patient 4; Table 3) in response to stimulation with irradiated, autologous EBV-B cells as APCs plus the indicated concentrations of tumor-derived lysates (i.e., Calu-1, lung carcinoma; SW480 and HT-29, colon carcinomas). Peptide expressed as μg/mL; cellular extract (lysate) expressed as μg protein. Three-day assay, with cultures pulsed with ^{3}H-thymidine during the final 18 hours of the incubation period. Data expressed as the mean cpm ± SEM of triplicate cultures (cpm of irradiated EBV-B cells = 4181 ± 330). (Data from Abrams et al. [97].)

were derived from preimmune lymphocytes of the same patients (see Table 5 for example). Finally, as with CD4$^+$ T-cell function, an important immunologic issue was to determine whether such peptide-induced CD8$^+$ CTLs could also recognize a processed form of the corresponding mutant *ras* protein. To test this hypothesis, the anti-*ras* Val12 CD8$^+$ CTL line was examined for its ability to lyse the SW480 colon adenocarcinoma cell line, either untreated or following pretreatment with IFN-γ (Fig. 3B), as described earlier (in Fig. 2B). In the absence of IFN-γ pretreatment, little CTL lysis was detectable; however, following IFN-γ pretreatment, demonstrable cytotoxicity was observed at multiple effector/target ratios (Fig. 3B), which was consistent with the earlier observations with the anti-*ras* 4–12(Val12) CTL line.

Indeed, evidence has been presented that vaccination with *ras* oncogene peptides in adjuvant induced anti-*ras* cell-mediated immunity at both CD4$^+$ and CD8$^+$ T-cell levels in some patients. Although investigators have described in other studies [75–83,85] the capacity to generate *ras* peptide-specific T-cell responses from unvaccinated individuals, it has not yet been possible to do so using the pre-vaccine preparations of these patients from this clinical trial. Although the exact reasons remain unclear, the advanced nature of disseminated metastatic disease in this patient population and/or differences in IVS culture conditions may be important factors

TABLE 5   Generation of an Anti-*ras* Asp12 Peptide-Specific CD8[+] CTL Line from
Post-Vaccinated Lymphocytes[a]

Culture[b]	IVS cycle[c]	Recovery[d] (fold increase)	% Specific lysis[e] with *ras* peptide		
			5–14(Asp12)	5–14(Gly12)	None
Pre-vaccine	6	2.3	14 ± 0	10 ± 1	13 ± 1
	7	1.3	0	0	0
	8	0.2	NA[f]	NA	NA
Post-vaccine	6	6.5	24 ± 2	8 ± 0	13 ± 1
	7	3.5	26 ± 2	8 ± 1	2 ± 0
	8	7.2	67 ± 2	10 ± 1	8 ± 0
	9	2.5	68 ± 4	21 ± 2	22 ± 1

[a] Data summarized from Khleif *et al.* [88] and Abrams *et al.* [97].

[b] CD8[+] T cells were isolated from patient PBMCs (from patient 2; Table 3) pre-vaccination and after the third vaccination with mutant *ras* 5–17(Asp12) peptide in Detox[TM] adjuvant.

[c] Cultures were initiated and maintained by weekly IVS using autologous PBMCs as APCs (up to cycle 3) or EBV-B cells (thereafter) incubated with mutant *ras* peptide/β2-microglobulin and IL-2. The mutant *ras* peptide represented a nested HLA-A2-binding, 10-mer sequence, *ras* 5–14(Asp12).

[d] Cell growth as depicted by the ratio of cells recovered before and after each IVS cycle.

[e] Cytotoxicity was determined by a ^{51}Cr-release assay using the C1R-A2 cell line as a target, incubated with and without *ras* 10-mer peptides (mutant = 5–14(Asp12), wild-type = 5–14(Gly12); each @ 10 μg/mL). Effector/target ratio shown, 20/1.

[f] NA, no cells available for assay due to lack of growth.

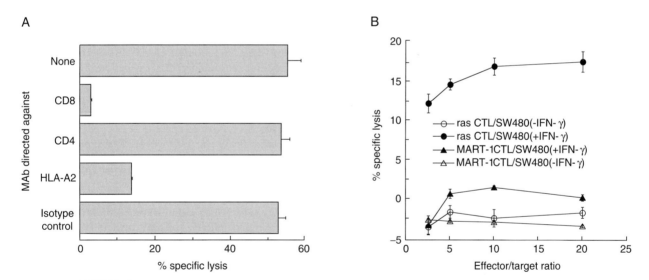

FIGURE 3   Functional characterization of the anti-*ras* 5–14(Val12) CTL line. (**A**) Role of CD8 and HLA-A2 molecules in cytotoxicity by anti-*ras* 5–14(Val12) T-cell line (from patient 8; Table 3), using peptide-coated (3 μg/mL) autologous EBV-B cells as targets (effector/target ratio, 10/1, 6 hr assay) in the absence and presence of the different mAbs. Lytic activity in the absence of peptide or in the presence of the control *ras* 5–14(Gly12) peptide was less than 3%. (**B**) Ability of anti-*ras* 5–14(Val12) CD8[+] CTLs to recognize and lyse SW480 tumor cells (± IFN-γ pretreatment, 250 U/mL for 24 hours) expressing the *ras* Val12 oncogene, in the absence of exogenous peptide (6 hr assay). Control combinations included irrelevant CTLs (i.e., HLA-A2-restricted, MART-1$_{27-35}$ peptide-specific) with SW480 targets or anti-*ras* CTLs with irrelevant targets (i.e., HLA-A2[+], *ras* Val12[−]) (in other experiments, not shown). Percent specific lysis against IFN-γ-pretreated SW480 cells mediated by anti-*ras* Val12 CTLs or anti-MART CTLs (effector/target ratio, 10/1) in the presence of additional saturable amounts of exogenous *ras* 5–14(Val12) peptide or MART-1$_{27-35}$ peptide (5 μg/mL) was 40 ± 2 and 85 ± 4, respectively. In both panels, results were expressed as the mean ± SEM of triplicate cultures. (Data from Khleif *et al.* [88] and Abrams *et al.* [97].)

influencing the efficacy of *in vitro* sensitization of precursor T cells. In fact, multiple IVS cycles of post-vaccinated lymphocytes were required to detect peptide-specific reactivity, typically ≤3 for CD4$^+$ T-cell responses and ≥6 for CD8$^+$ T-cell responses, further suggesting low precursor frequencies of the Ag-reactive lymphocytes (at least in PBMCs); Ag-specific peripheral T-cell anergy, perhaps partially impairing TCR/CD3 structure and/or signaling events [105]; weak *in vivo* immunogenicity of the peptides; suboptimal immunization parameters or IVS conditions; or combinations of these and other possibilities. Thus, efforts to understand the limitations of this system to potentially modify and improve the sensitivity, scope, and potency of the induced immune response represent important considerations of future studies.

Nevertheless, these studies [88,97] led to the identification of human CD8$^+$ CTL epitope peptides reflecting specific mutations (i.e., Val12 and Asp12) in the K-*ras* oncogenes at codon 12 restricted by the HLA-A2 molecules. Similarly, Gjertsen *et al.* [103] defined an anti-*ras* Val12 CD8$^+$ CTL epitope peptide restricted by the HLA-B35 molecule. Taken collectively, the average phenotypic frequency of HLA-A2 alleles in the human population is approximately 43.1%, while that of HLA-B35 is approximately another 16.7% [94]. Furthermore, the finding that a single mutant *ras* peptide immunogen contains both CD4$^+$ and CD8$^+$ T-cell epitopes in an overlapping or nested configuration, as described in our laboratory [71,97] and elsewhere [103], may have important implications for the generation of a more efficient antipathogen immune response. The potential importance of overlapping CD4$^+$ and CD8$^+$ T-cell epitopes in biologic systems is supported further by other studies that have defined their existence in experimental models of influenza [106], human immunodeficiency virus (HIV) [107], and p53 [108].

VIII. ANTI-*ras* IMMUNE SYSTEM INTERACTIONS: IMPLICATIONS FOR TUMOR IMMUNITY AND TUMOR ESCAPE

A. Multiple CD8$^+$ CTL Epitopes Reflecting the Same Neo-Antigenic Determinant

The ability to produce two independent anti-*ras* Val12 CTL lines, one derived on the *ras* 4–12(Val12) peptide and the other derived on the *ras* 5–14(Val12) peptide, supports the hypothesis for multiple CD8$^+$ CTL epitopes reflecting the same point-mutated Val12 neo-determinant. The immunologic basis for these peptides as separate CTL epitopes reflects the central observations that the primary MHC anchor residues appear to be situated at different positions in the peptide sequences and that the TCR contact residues also appear to be distinct. For example, in the *ras* 4–12(Val12) sequence, the point-mutated Val12 residue appears to serve as

the C-terminus anchor, because the wild-type 9-mer peptide (Gly12) fails to bind to HLA-2 (Fig. 1). Therefore, the CTL response reactive with the *ras* 4–12(Val12) epitope is unlikely to involve TCR recognition of the Val12 amino acid residue, which is consistent with what was observed in our murine study [71]. In contrast, in the *ras* 5–14(Val12) sequence, the normal Val14 residue (but not the mutant Val12 residue) appears to be the C-terminus anchor residue, as both the mutant and wild-type (Gly12) *ras* 10-mer peptides bind to HLA-A2. The prediction here is that the CTL response directed against the *ras* 5–14(Val12) epitope does involve TCR recognition of the Val12 amino acid residue, because these CTLs failed to specifically lyse HLA-A2$^+$ target cells pulsed with the wild-type 10-mer peptide (Fig. 3A legend); similar results were observed with the anti-*ras* Asp12 CTL line (Table 5).

Although future investigations are necessary to precisely map both MHC and TCR contact sites in these two peptide sequences and to explore the potential nature of clonal diversity (i.e., TCR gene and chain usage) between these different anti-*ras* Val12 CTL lines, these findings still have important scientific and clinical implications for (1) the identification, characterization, and potential clinical use of immunodominant/subdominant *ras* oncogene epitope peptides; and (2) the generation and expansion of a "polyclonal" anti-*ras* Val12 CTL response expressing a range of TCR specificities and affinities for recognition of multiple processed forms of the endogenously derived Val12 determinant. The observation that cytotoxicity by both anti-*ras* Val12 CTL lines could be enhanced by IFN-γ indicates a potentially important role for that cytokine in immunotherapy which may be provided endogenously by immune system interactions (i.e., type 1 CD4$^+$ or CD8$^+$ cells) and/or exogenously by the application of recombinant IFN-γ or IFN-γ-inducing cytokines, such IL-12 [109,110].

B. Mechanisms of Tumor Escape

Indeed, a variety of mechanisms have been presented to account for the generation of weak T-cell responses and poor tumor immunogenicity, and the capacity of tumor cells to escape immune recognition and attack (reviewed in Melero *et al.* [111] and Walker *et al.* [112]). These mechanisms may occur at multiple levels of the effector/target interaction, including TCR–MHC/peptide recognition, cell–cell adhesion (conjugate formation) or the cytotoxic mechanism. Specifically, it has been proposed that tumor cells may escape immune recognition as a consequence of (1) a paucity of expression of MHC class I or II alleles, β2-microglobulin protein, or adhesion (i.e., ICAM-1 [CD54] or LFA-3 [CD58 in the human system or CD48 in the murine system]) and/or costimulatory (i.e., B7.1 [CD80] or B7.2 [CD86]) molecules; (2) suboptimal expression of the relevant rejection/regression epitope(s) (in the context of the appropriate MHC molecules), which may reflect intracellular defects in

TABLE 6   Upregulation of MHC Class I, ICAM-1, and Fas Molecules on SW480 Colon
Carcinoma Cells Following IFN-γ Treatment

IFN-γ pretreatment	mAb directed against % positive cells (MFI)			
	HLA-A2	ICAM-1	LFA-3	Fas
Untreated	72 ± 8	39 ± 6	96 ± 0.4	13 ± 4
	(68 ± 7)	(42 ± 6)	(114 ± 9)	(30 ± 2)
Treated	**97 ± 1**	**98 ± 1**	96 ± 0.4	**65 ± 4**
	(544 ± 32)	**(289 ± 35)**	(90 ± 9)	(41 ± 3)

Note: SW480 cells were incubated in the absence (untreated) or presence (treated) of recombinant human IFN-γ (250 U/mL for 24 hours). After incubation, untreated and IFN-γ-treated SW480 cells were stained with mAbs directed against the indicated cell surface markers and subsequently evaluated by flow cytometry for both the percentage of specific positive cells and mean fluorescence intensity (MFI), as a relative measurement of Ag density (data shown in parentheses). Results were expressed as the mean ± SEM of 8 (for HLA-A2, ICAM-1), 4 (for LFA-3), and 12 (for Fas) separate experiments. Data in bold type indicate statistically significant responses, based on comparison with the corresponding untreated preparation. (Data from Bergmann-Leitner *et al.* [87].)

endogenous Ag processing and presentation pathways (i.e., TAP-dependent or proteosome activities) or unstable (short-lived) MHC/peptide ligand complexes that rapidly disassemble from the cell surface; (3) production of tumor-derived inhibitory factors, such as IL-10, TGF-β, and prostaglandins (i.e., PGE$_2$), which downregulate type 1 cell-mediated immune reactions; (4) failure to activate Ag-specific T-cell responses due to defects in TCR structure/assembly, signaling, and function; (5) induction of inappropriate T-cell responses, such as type 2, which may be facilitated by IL-10 exposure; (6) resistance to cell-mediated cytotoxicity, due to the inability of effector T cells to mediate Fas-dependent apoptosis; or (7) cell-surface expression of Fas ligand (FasL or CD95L), which has been suggested to eliminate infiltrating Fas$^+$ CD4$^+$ or CD8$^+$ T lymphocytes, although this latter possibility remains controversial [113,114].

Clearly, these and other mechanisms represent potentially significant obstacles and challenges confronting successful cancer immunotherapy. It is important to keep in perspective, however, that because human cancer is indeed complex it is likely that different and distinct combinations of factors, as opposed to any single factor alone, may be operative and unique for different host–tumor interactions *in vivo*. Accordingly, each model system or clinical situation needs to be independently examined to best understand and predict the potential physiologic benefit of immune intervention. In the context of human *ras* oncogene studies, we have determined thus far that tumor cell "escape" from efficient immune recognition and attack may occur, at least in part, at the levels of TCR–MHC/peptide complex formation, ICAM-1-based cell–cell adhesion, and susceptibility to the CTL lytic mechanism(s). These possibilities became evident from experiments involving: (1) IFN-γ and its influence on CTL-mediated lysis; and (2) analysis of MHC–peptide epitope binding interactions.

In regard to the first possibility, the SW480 tumor targets required pretreatment with IFN-γ to elicit measurable anti-*ras* Val12 CTL responses in the absence of exogenous peptide (Figs. 2 and 3). Although the exact mechanisms by which IFN-γ acted in this model remain to be fully elucidated, cytotoxicity did correlate with increased expression of HLA-A2, ICAM-1, and Fas molecules (Table 6), consistent with the following hypotheses: (1) the endogenously produced mutant *ras* epitopes were limiting but could be enhanced by increasing the density of class I/peptide complexes available for more efficient TCR recognition; (2) the avidity of the CTL–target interaction was weak but could be enhanced or strengthened by increasing the density of accessory molecules important for secondary or Ag-independent interactions; or (3) tumor cell sensitivity to lysis via a Fas/FasL-dependent apoptotic pathway was initially defective or weak but could be restored or enhanced by increasing the density of cell-surface Fas. Although future studies are necessary to delineate the relative contributions of each of these possibilities, recent experiments support the hypothesis that lytic efficiency correlates with the expression by tumor cells of Fas and their resultant sensitivity to a Fas-dependent pathway of apoptosis, as monoclonal antibodies (MAbs) directed against Fas on the tumor cell surface or FasL on the CTL surface strongly abrogate cytotoxicity [115,116]. These data suggest that at least one mechanism by which tumor cells may resist cell death involves downregulation of cell-surface Fas molecules and/or presenting deficiencies in the Fas intracellular signal transduction pathway, which may be partially modulated or restored by cytokine interactions, such as with IFN-γ. In the context of a physiologic situation and an *in vivo* source of IFN-γ, recent data indicate that IFN-γ may be supplied endogenously by Ag-activated CD4$^+$ T cells. Supernatants isolated from anti-*ras* Val12-specific CD4$^+$ T-cell cultures (from patient 8 of Table 3), following stimulation with exogenous *ras* 5–17(Val12) peptide, were shown to produce substantial amounts of IFN-γ. Upon subsequent transfer and incubation with fresh SW480 tumor cells, these supernatants were found to induce the same phenotypic and functional effects as those observed in parallel with recombinant IFN-γ (i.e., upregulation of HLA-A2, ICAM-1, and Fas expression and sensitization of these target cells to anti-*ras* Val12 CD8$^+$ CTL-mediated lysis).

In regard to the second possibility, we have determined that both *ras* 4–12(Val12) and *ras* 5–14(Val12) peptides (as well as *ras* 5–14(Asp12)) appeared to be intrinsically weak immunogens or epitopes. The degree of immunogenic potency correlated with weak functional binding to HLA-A2, which may have been due to the suboptimal nature (composition) or position of primary and/or secondary MHC anchor residues. Nevertheless, these results support the hypothesis that, although the *ras* 4–12(Val12) and *ras* 5–14(Val12) peptides may be intrinsically weak immunogens, they remain potentially relevant CD8$^+$ epitopes, as CTLs produced against these peptides recognized and lysed tumor cells expressing the naturally occurring mutation. Thus, the rather weak but specific nature of immunogenic activity may have important implications for (1) understanding potential mechanisms of tumor escape from productive immune system interactions, which may reflect the relative stability or half-life properties of the MHC/peptide ligand complex resulting in a more rapid dissociation of the complex, poorer TCR recognition and generation of a weaker affinity/avidity T-cell response; and (2) targeted modifications in peptide immunogen design, such as amino acid substitutions at MHC anchor sites, which may improve MHC binding affinity and immunogenic potency for the induction and amplification of a potentially higher affinity/avidity, antigenically relevant T-cell response, as described in other systems [30,117–121].

C. Considerations for Development of Mutant *ras* CD8$^+$ CTL Epitope Peptide Variants

Although mutant *ras* peptides were identified as HLA-A2-reactive CD8$^+$ CTL epitopes, they appeared to exhibit rather weak binding to HLA-A2 and required exogenous β2-microglobulin to further improve MHC/peptide interaction. The identification of epitope peptide variants that strengthen or increase the stability or half-life of the MHC/peptide complex (*in vivo* and *in vitro*) may enhance the potency of intrinsically weak immunogenic peptides for the induction and amplification of the relevant T-cell response. This concept was originally described in a murine CD4$^+$ T-cell model using HIV peptides [30,120] and now has been adapted to, for example, human-melanoma-associated peptides that have been modified to enhance the potential generation of anti-gp100-specific CD8$^+$ CTL responses [118]. Because the consensus anchor motif for HLA-A2

is well characterized, putative *ras* peptide variants can be synthesized, reflecting single or multiple amino acid substitutions at either primary and/or secondary MHC anchor positions, while maintaining the integrity of TCR recognition for the naturally expressed epitope.

In comparative studies of the unmodified epitope peptide sequences, these variants can be examined and screened for their ability to functionally bind to HLA-A2. Variants that display enhanced binding to HLA-A2 can then be assessed, in comparison with the unmodified epitope peptide sequences, for their capacity to (1) sensitize targets for lysis using established Ag-specific CTL lines, (2) stimulate proliferation and expansion of established Ag-specific CTL lines, and (3) generate anti-*ras* CTL lines/clones from bulk populations of immune lymphocytes, with emphasis on both the *quantity* and *quality* of the resultant T-cell response. In terms of quantitative effects, if the peptide variant is superior to that of the unmodified sequence, then the ability to generate, propagate, and expand the Ag-specific CTL precursors would be predicted to be accelerated and enhanced. Furthermore, in terms of qualitative effects, if the peptide variant is superior to that of the unmodified sequence as an immunogen *in vitro,* then the resulting CTL lines/clones would be predicted to exhibit enhanced potency and sensitivity for recognition of tumor cells expressing extremely low densities of endogenously produced mutant *ras* epitopes.

Taken collectively, these studies may provide important insights into the hypothesis that targeted modifications at primary and, perhaps, secondary MHC amino acid anchor residues can be introduced into oncogene-derived, tumor-specific epitope peptides to enhance both MHC binding activity and immunogenicity for improved induction, expansion (*quantity*), and potency (*quality*) of the antigenically relevant T-cell response without cross-reaction against self (or proto-*ras* forms). Peptide variants may be used *in vivo* and/or *in vitro* to improve sensitization and expansion of Ag-specific T-cell lines/clones for potential clinical use in adoptive cellular immunotherapy. Moreover, the notion that this concept can be applied directly to oncogene-derived peptides at sites of mutations that create MHC anchor positions, producing surrogate peptide epitopes with enhanced immunogenic potency, may help broaden our understanding of the development and augmentation of not only CD8$^+$ but also CD4$^+$ T-cell-mediated immunity against malignancies expressing neo-epitopes encoded by a variety of mutated cellular protooncogenes and/or tumor suppressor genes.

IX. PARADIGM FOR ANTI-*ras* IMMUNE SYSTEM INTERACTIONS IN CANCER IMMUNOTHERAPY

Based on these preclinical and clinical findings, the following paradigm can be proposed for potential anti-*ras* oncogene immune system interactions operative in metastatic cancer immunotherapy, and, for that matter, for the immunotherapy of many other cancer target Ags. Active immunization or vaccination of the tumor-bearing host with mutated *ras* immunogens reflecting CD4$^+$ and/or CD8$^+$ T-cell epitopes may result in the *in vivo* priming and expansion of the Ag-specific T-cell precursor pools. *In vivo* sensitized CD4$^+$ and/or CD8$^+$ T-cell populations may then also be isolated *ex vivo* from immunized hosts and expanded by IVS methodologies to achieve larger quantities of immune effector cells for adoptive transfer. The combination of both active and passive immunotherapies may thus have a more comprehensive impact on control of the progression of the metastatic burden. Moreover, the activation of both Ag-specific T-cell subpopulations may enhance the diversity, breadth, and repertoire of antitumor immune mechanisms, as described later.

The Ag-specific CD4$^+$ T lymphocyte may be a central player important for the optimal induction and development of cell-mediated and, perhaps, antitumor immune reactions. At the tumor site, the peptide-induced CD4$^+$ T cell may be stimulated by specialized APC populations, such as activated macrophages, dendritic cells, or B lymphocytes, that infiltrate these metastatic lesions. Such APC populations, expressing a spectrum of adhesion and costimulatory molecules, may exogenously process mutated *ras* proteins derived from tumor cells and present them as peptide epitopes in association with self-MHC class II molecules for initiation of the cellular immune response. Lymphokines produced by the Ag-primed CD4$^+$ T-cell response, such as IL-2, may further amplify the clonal expansion of the epitope- or Ag-sensitized CD8$^+$ T-cell population.

During the effector phase of the immune response, the activated CD8$^+$ CTL or CD4$^+$ T$_h$1-type cell may lyse a susceptible target population *directly,* if displaying the relevant MHC/peptide ligand complex via Fas-dependent and/or -independent (i.e., perforin/granzymes/TNF) mechanisms. The production and availability of IFN-γ are noteworthy in this context, as this particular cytokine was found to be crucial for sensitization of Ag-bearing tumor targets for MHC-class-I-restricted, anti-*ras* CTL-mediated lysis [87,116]. Cytotoxicity correlated with the upregulation of HLA class I (HLA-A2), adhesion (i.e., ICAM-1), and Fas molecules on the tumor cell surface. Monoclonal antibodies directed against the Fas/FasL interaction substantially inhibited tumor cell lysis, supporting the following hypotheses: (1) apoptosis via a Fas-dependent pathway is important for human CTL-mediated lysis of certain populations of Ag-bearing carcinoma cells; (2) the low level of cell-surface Fas expression and/or defects in the Fas trigger molecule or Fas intracellular signaling events endogenous to the tumor cell may contribute to the mechanism of tumor escape from CTL attack; and (3) IFN-γ, perhaps in concert with other cytokines such as TNF-α, may be important components for improving the overall efficacy of cancer immunotherapy. IFN-γ-induced augmentation of MHC class I and adhesion molecules may thus serve to facilitate and strengthen both specific (i.e., TCR–

MHC/peptide) and nonspecific (i.e., LFA-1/ICAM-1) aspects of the effector/target interaction, leading to enhanced T-cell activation and triggering of the cytolytic pathways. Such cytolytic activity may then result from both Fas–FasL interactions (i.e., upregulation of FasL on the activated CTL for engagement with Fas on the tumor cell surface) and Fas-independent mechanisms (i.e., rapid exocytosis of cytoplasmic granules containing lytic molecules). In the context of the metastatic tumor microenvironment, IFN-γ and other potentially relevant cytokines may be provided endogenously by immune system interactions, most notably by Ag-activated CD4$^+$ T$_h$1 cells following interaction with MHC class II$^+$ Ag-bearing APCs.

In addition to IFN-γ, CD4$^+$ T$_h$1-derived cytokines such as IL-2 and GM-CSF may further influence the recruitment, activation, and expansion of various cytotoxic effector cells, including Ag-primed CD8$^+$ CTLs, macrophages, and NK cells. Furthermore, CD8$^+$ CTLs or CD4$^+$ T$_h$1-type cells may lyse susceptible tumor cells *indirectly* through the release of cytotoxic-lymphokines, such as TNF-α, TNF-β, or perhaps secreted soluble FasL [122] following Ag-specific immune stimulation. In this context, the initiation of the immune response remains Ag specific, while the effector mechanisms become Ag independent and bystander in nature. These cell-contact-independent lytic mechanisms may represent biologically significant pathways for the elimination of Ag-negative tumor cells and, thus, circumvent, at least in part, tumor antigenic heterogeneity associated with the loss or downregulation of MHC/peptide ligand expression.

X. FUTURE DIRECTIONS

Because of their prevalence in human cancer and functional association with the initiation, development, and support of the malignant phenotype, *ras* oncogenes represent clinically promising and specific targets for the activation of tumor-specific cellular immune mechanisms in settings of both overt disease and preneoplastic conditions. Scientifically, studies in murine models and in human *in vitro* systems strongly support the hypothesis that although *ras* p21 may not be exposed on the APC/target cell surface as integral membrane proteins, *ras* oncoproteins may be endogenously or exogenously processed and presented as immunodominant/subdominant peptide epitopes in the context of the appropriate MHC class I and II molecules for Ag-specific immune recognition by both CD8$^+$ and CD4$^+$ T-cell subsets. Moreover, the finding that peptide-induced CD4$^+$ and CD8$^+$ T cells can recognize APCs/target cells expressing activated *ras* molecules provides the continuing rationale and impetus for the design of immunotherapies directed against human malignancies that harbor products of *ras* oncogenes.

Indeed, in independent clinical trials, evidence has been accumulating that vaccination of some cancer patients with *ras* oncogene peptides reflecting codon 12 mutations—given with adjuvant or pulsed onto autologous PBMCs or dendritic cells—can induce anti-*ras* cell-mediated immunity at CD4$^+$ and/or CD8$^+$ T-cell levels. *In vivo* priming of the CD8$^+$ T-cell response likely resulted from *in vivo* mechanisms of Ag processing (i.e., extracellular or intracellular), with the subsequent generation of MHC-class-I-reactive epitopes. Moreover, because multiple T-cell epitopes and HLA restriction patterns were characterized from these different clinical studies, these findings suggest that mutant *ras* peptide-based "cancer vaccines" may induce cell-mediated immunity and potential antitumor effects across broader segments of the human population. Thus, further insights into potential patient immune responsiveness to these immunogens in these and future clinical trials may be provided by a detailed analysis and understanding of HLA restriction patterns for peptide presentation and by identification of putative consensus anchor motifs for those alleles.

The fundamental basis of effective cancer immunotherapy and the capacity to overcome a potential diversity of tumor escape mechanisms will likely require a synergism between both quantitative and qualitative aspects of the T-lymphocyte-mediated immune response. This becomes particularly important in settings such as those outlined here, where defined immunogens/epitopes may be weakly reactive (with MHC molecules on the APC/tumor cell surface) and/or weak T-cell responses may be induced that lack adequate potency to achieve clinical efficacy. Accordingly, targeted modifications in immunogen design and specific immunization parameters, in concert with the *ex vivo* isolation and expansion of epitope-specific CD4$^+$ and/or CD8$^+$ T lymphocytes for adoptive transfer, may be essential to the production of tumor-specific T-cell clones that express sufficiently high sensitivity for Ag recognition and the potential to mediate desirable antitumor reactions. Such modifications in immunization parameters, for analysis in future clinical trials, may involve: (1) altering the number of immunogen doses and the dose schedule, and (2) analyzing the potency of different adjuvant systems or using peptide-pulsed dendritic cells. Studies in animal models have shown that Ag-specific cellular and antitumor immune responses can be markedly enhanced by employing dendritic cells as carriers for peptide immunogens [123,124]. Furthermore, studies in human *in vitro* systems demonstrate the efficient priming and generation of tumor Ag-specific CTLs using peptide-pulsed dendritic cells, particularly for subdominant epitopes [125]. Additional modifications may include: (1) designing and using epitope peptide "variants" and formulating peptide mixtures (i.e., "cocktails") for optimal activation of CD4$^+$ and/or CD8$^+$ T-cell responses (in the appropriate subsets of patients), and (2) coadministering peptide-based immunogens with exogenous sources of recombinant cytokine proteins, such as IL-2, IL-12, or GM-CSF, to enhance both the development and potency of the cellular immune response and potential clinical benefit.

References

1. Rothbard, J. B., and Gefter, M. L. (1991). Interactions between immunogenic peptides and MHC proteins. *Annu. Rev. Immunol.* **9**, 527–565.
2. Jorgensen, J. L., Reay, P. A., Ehrich, E. W., and Davis, M. M. (1992). Molecular components of T-cell recognition. *Annu. Rev. Immunol.* **10**, 835–873.
3. Boon, T., and Van der Bruggen, P. (1996). Human tumor antigens recognized by T lymphocytes. *J. Exp. Med.* **183**, 725–729.
4. Greenberg, P. D., Finch, R. J., Gavin, M. A., Kalos, M., Lewinsohn, D. A., Lonergan, M., Lord, J. D., Nelson, B. H., Ohlen, C., Sing, A. P., Warren, E. H., Yee, C., and Riddell, S. R. (1998). Genetic modification of T-cell clones for therapy of human viral and malignant diseases. *Cancer J. Sci. Am.* **4**(suppl. 1), S100–S105.
5. Melief, C., Annels, N., Dunbar, R., Forni, G., Kiessling, R., Ricciardi-Castagnoli, P., Rickinson, A., and Van den Eynde, B. (1998). Protective T-cell responses against tumours. *Res. Immunol.* **149**, 877–879.
6. Rosenberg, S. A. (1999). A new era for cancer immunotherapy based on the genes that encode cancer antigens. *Immunity* **10**, 281–287.
7. Toes, R. E., Ossendorp, F., Offringa, R., and Melief, C. J. (1999). CD4 T cells and their role in antitumor immune responses. *J. Exp. Med.* **189**, 753–756.
8. Bishop, J. M. (1991). Molecular themes in oncogenesis. *Cell* **64**, 235–248.
9. Spandidos, D. A., Sourvinos, G., and Koffa, M. (1997). *ras* genes, p53 and HPV as prognostic indicators in human cancer (review). *Oncol. Reports* **4**, 211–218.
10. Bos, J. L. (1988). The *ras* gene family and human carcinogenesis. *Mut. Res.* **195**, 255–271.
11. Bos, J. L. (1989). *ras* oncogenes in human cancer: a review. *Cancer Res.* **49**, 4682–4689.
12. Kiaris, H., and Spandidos, D. A. (1995). Mutations of *ras* genes in human tumors. *Int. J. Oncol.* **7**, 413–421.
13. Abrams, S. I., Hand, P. H., Tsang, K. Y., and Schlom, J. (1996). Mutant *ras* epitopes as targets for cancer vaccines. *Semin. Oncol.* **23**, 118–134.
14. Rosenberg, S. A., Yang, J. C., Schwartzentruber, D. J., Hwu, P., Marincola, F. M., Topalian, S. L., Restifo, N. P., Dudley, M. E., Schwarz, S. L., Spiess, P. J., Wunderlich, J. R., Parkhurst, M. R., Kawakami, Y., Seipp, C. A., Einhorn, J. H., and White, D. E. (1998). Immunologic and therapeutic evaluation of a synthetic peptide vaccine for the treatment of patients with metastatic melanoma. *Nat. Med.* **4**, 321–327.
15. Brodie, S. J., Lewinsohn, D. A., Patterson, B. K., Jiyamapa, D., Krieger, J., Corey, L., Greenberg, P. D., and Riddell, S. R. (1999). *In vivo* migration and function of transferred HIV-1-specific cytotoxic T cells. *Nat. Med.* **5**, 34–41.
16. Swain, S. L., Croft, M., Dubey, C., Hayes, L., Rogers, P., Zhang, X., and Bradley, L. M. (1996). From naive to memory T cells. *Immunol. Rev.* **150**, 143–167.
17. Germain, R. N., and Margulies, D. H. (1993). The biochemistry and cell biology of antigen processing and presentation. *Annu. Rev. Immunol.* **11**, 403–450.
18. Schwartz, R. H. (1996). Models of T cell anergy: is there a common molecular mechanism? *J. Exp. Med.* **184**, 19–29.
19. Zinkernagel, R. M., and Doherty, P. C. (1997). The discovery of MHC restriction. *Immunol. Today* **18**, 14–17.
20. Bierer, B. E., and Burakoff, S. J. (1988). T cell adhesion molecules. *FASEB J.* **2**, 2584–2590.
21. Allison, J. P., and Krummel, M. F. (1995). The Yin and Yang of T cell costimulation. *Science* **270**, 932–933.
22. Wingren, A. G., Parra, E., Varga, M., Kalland, T., Sjogren, H. O., Hedlund, G., and Dohlsten, M. (1995). T cell activation pathways: B7, LFA-3, and ICAM-1 shape unique T cell profiles. *Crit. Rev. Immunol.* **15**, 235–253.

23. Chambers, C. A., Krummel, M. F., Boitel, B., Hurwitz, A., Sullivan, T. J., Fournier, S., Cassell, D., Brunner, M., and Allison, J. P. (1996). The role of CTLA-4 in the regulation and initiation of T-cell responses. *Immunol. Rev.* **153**, 27–46.

24. Fairchild, P. J. (1998). Presentation of antigenic peptides by products of the major histocompatibility complex. *J. Peptide Sci.* **4**, 182–194.

25. Pamer, E., and Cresswell, P. (1998). Mechanisms of MHC class I-restricted antigen processing. *Ann. Rev. Immunol.* **16**, 323–358.

26. Germain, R. N., and Stefanova, I. (1999). The dynamics of T cell receptor signaling: complex orchestration and the key roles of tempo and cooperation. *Annu. Rev. Immunol.* **17**, 467–522.

27. Falk, K., Rotzschke, O., Stevanovic, S., Jung, G., and Rammensee, H.-G. (1991). Allele-specific motifs revealed by sequencing of self-peptides eluted from MHC molecules. *Nature* **351**, 290–296.

28. Rammensee, H.-G., Friede, T., and Stevanovic, S. (1995). MHC ligands and peptide motifs: first listing. *Immunogenetics* **41**, 178–228.

29. Bjorkman, P. J., Saper, M. A., Samraoui, B., Bennett, W. S., Strominger, J. L., and Wiley, D. C. (1987). The foreign antigen binding site and T cell recognition regions of class I histocompatibility antigens. *Nature* **329**, 512–518.

30. Boehncke, W.-H., Takeshita, T., Pendleton, C. D., Sadegh-Nasseri, S., Racioppi, L., Houghten, R. A., Berzofsky, J. A., and Germain, R. N. (1993). The importance of dominant negative effects of amino acid side chain substitution in peptide-MHC molecule interactions and T cell recognition. *J. Immunol.* **150**, 331–341.

31. Parker, K. C., Bednarek, M. A., Hull, L. K., Utz, U., Cunningham, B., Zweerink, H. J., Biddison, W. E., and Coligan, J. E. (1992). Sequence motifs important for peptide binding to the human MHC class I molecule, HLA-A2. *J. Immunol.* **149**, 3580–3587.

32. Ruppert, J., Sidney, J., Celis, E., Kubo, R. T., Grey, H. M., and Sette, A. (1993). Prominent role of secondary anchor residues in peptide binding to HLA-A2.1 molecules. *Cell* **74**, 929–937.

33. Abbas, A. K., Murphy, K. M., and Sher, A. (1996). Functional diversity of helper T lymphocytes. *Nature* **383**, 787–793.

34. Henkart, P. A. (1994). Lymphocyte-mediated cytotoxicity: two pathways and multiple effector molecules. *Immunity* **1**, 343–346.

35. Kagi, D., Vignaux, F., Ledermann, B., Burki, K., Depraetere, V., Nagata, S., Hengartner, H., and Golstein, P. (1994). Fas and perforin pathways as major mechanisms of T cell-mediated cytotoxicity. *Science* **265**, 528–530.

36. Abrams, S. I., Hodge, J. W., McLaughlin, J. P., Steinberg, S. M., Kantor, J. A., and Schlom, J. (1997). Adoptive immunotherapy as an *in vivo* model to explore antitumor mechanisms induced by a recombinant anticancer vaccine. *J. Immunother.* **20**, 48–59.

37. Stuhler, G., and Schlossman, S. F. (1997). Antigen organization regulates cluster formation and induction of cytotoxic T lymphocytes by helper T cell subsets. *Proc. Natl. Acad. Sci. USA* **94**, 622–627.

38. Lu, Z., Yuan, L., Zhou, X., Sotomayor, E., Levitsky, H. I., and Pardoll, D. M. (2000). CD40–independent pathways of T cell help for priming of CD8(+) cytotoxic T lymphocytes. *J. Exp. Med.* **191**, 541–550.

39. Salgame, P., Abrams, J. S., Clayberger, C., Goldstein, H., Convit, J., Modlin, R. L., and Bloom, B. R. (1991). Differing lymphokine profiles of functional subsets of human CD4 and CD8 T cell clones. *Science* **254**, 279–282.

40. Croft, M., Carter, L., Swain, S. L., and Dutton, R. W. (1994). Generation of polarized antigen-specific CD8 effector populations: reciprocal action of interleukin (IL)-4 and IL-12 in promoting type 2 versus type 1 cytokine profiles. *J. Exp. Med.* **180**, 1715–1728.

41. Romagnani, S. (1994). Lymphokine production by human T cells in disease states. *Annu. Rev. Immunol.* **12**, 227–257.

42. Sad, S., Marcotte, R., and Mosmann, T. R. (1995). Cytokine-induced differentiation of precursor mouse CD8$^+$ T cells into cytotoxic CD8$^+$ T cells secreting Th1 or Th2 cytokines. *Immunity* **2**, 271–279.

43. Halverson, D. C., Schwartz, G. N., Carter, C., Gress, R. E., and Fowler, D. H. (1997). *In vitro* generation of allospecific human CD8$^+$ T cells of Tc1 and Tc2 phenotype. *Blood* **90**, 2089–2096.

44. Hu, H.-M., Urba, W. J., and Fox, B. A. (1998). Gene-modified tumor vaccine with therapeutic potential shifts tumor-specific T cell response from a type 2 to a type 1 cytokine profile. *J. Immunol.* **161**, 3033–3041.

45. Fallon, P. G., Smith, P., and Dunne, D. W. (1998). Type 1 and type 2 cytokine-producing CD4$^+$ and CD8$^+$ T cells in *Schistosoma mansoni* infection. *Eur. J. Immunol.* **28**, 1408–1416.

46. Kobayashi, M., Kobayashi, H., Pollard, R. B., and Suzuki, F. (1998). A pathogenic role of Th2 cells and their cytokine products on the pulmonary metastasis of murine B16 melanoma. *J. Immunol.* **160**, 5869–5873.

47. Clerici, M., Shearer, G. M., and Clerici, E. (1998). Cytokine dysregulation in invasive cervical carcinoma and other human neoplasias: time to consider the Th1/Th2 paradigm. *J. Natl. Cancer Inst.* **90**, 261–263.

48. Carter, L. L., and Dutton, R. W. (1995). Relative perforin- and Fas-mediated lysis in T1 and T2 CD8 effector populations. *J. Immunol.* **155**, 1028–1031.

49. Chang, J. C., Zhang, L., Edgerton, T. L., and Kaplan, A. M. (1990). Heterogeneity in direct cytotoxic function of L3T4 T cells. Th1 clones express higher cytotoxic activity to antigen-presenting cells than Th2 clones. *J. Immunol.* **145**, 409–416.

50. Abrams, S. I., Dobrzanski, M. J., Wells, D. T., Stanziale, S. F., Zaremba, S., Masuelli, L., Kantor, J. A., and Schlom, J. (1995). Peptide-specific activation of cytolytic CD4$^+$ T lymphocytes against tumor cells bearing mutated epitopes of K-*ras* p21. *Eur. J. Immunol.* **25**, 2588–2597.

51. Nagarkatti, M., Clary, S. R., and Nagarkatti, P. S. (1990). Characterization of tumor-infiltrating CD4$^+$ T cells as Th1 cells based on lymphokine secretion and functional properties. *J. Immunol.* **144**, 4898–4905.

52. Kahn, M., Sugawara, H., McGowan, P., Okuno, K., Nagoya, S., Hellstrom, H. E., Hellstrom, I., and Greenberg, P. (1991). CD4$^+$ T cell clones specific for the human p97 melanoma-associated antigen can eradicate pulmonary metastases from a murine tumor expressing the p97 antigen. *J. Immunol.* **146**, 3235–3241.

53. Ostrand-Rosenberg, S., Pulaski, B. A., Clements, V. K., Qi, L., Pipeling, M. R., and Hanyok, L. A. (1999). Cell-based vaccines for the stimulation of immunity to metastatic cancers. *Immunol. Rev.* **170**, 101–114.

54. Grand, R. J. A., and Owen, D. (1991). The biochemistry of *ras* p21. *Biochemistry* **279**, 609–631.

55. Satoh, T., and Kaziro, Y. (1992). *ras* in signal transduction. *Semin. Cancer Biol.* **3**, 169–177.

56. Minamoto, T., Mai, M., and Ronai, Z. (2000). K-*ras* mutation: early detection in molecular diagnosis and risk assessment of colorectal, pancreas, and lung cancers—a review. *Cancer Detect. Prev.* **24**, 1–12.

57. Bos, J. L., Fearon, E. R., Hamilton, S. R., Verlaan-de Vries, M., van Boom, J. H., van der Eb, A. J., and Vogelstein, B. (1987). Prevalence of *ras* gene mutations in human colorectal cancers. *Nature* **327**, 293–297.

58. Moskaluk, C. A., Hruban, R. H., and Kern, S. E. (1997). p16 and K-*ras* gene mutations in the intraductal precursors of human pancreatic adenocarcinoma. *Cancer Res.* **57**, 2140–2143.

59. Bos, J. L., Verlaan-de Vries, M., Jansen, A. M., Veeneman, G. H., van Boom, J. H., and van der Eb, A. J. (1984). Three different mutations in codon 61 of the human N-*ras* gene detected by synthetic oligonucleotide hybridization. *Nucleic Acids Res.* **12**, 9155–9163.

60. Verlaan-de Vries, M., Bogaard, M. E., van den Elst, H., van Boom, J. H., van der Eb, A. J., and Bos, J. L. (1986). A dot-blot screening procedure for mutated *ras* oncogenes using synthetic oligodeoxynucleotides. *Gene* **50**, 313–320.

61. Winter, E., Yamamoto, F., Almoguera, C., and Perucho, M. (1985). A method to detect and characterize point mutations in transcribed

162 Scott I. Abrams

genes: amplification and overexpression of the mutant c-Ki-*ras* allele in human tumor cells. *Proc. Natl. Acad. Sci. USA* **82**, 7575–7579.

62. McMahon, G., Davis, E., and Wogan, G. N. (1987). Characterization of c-Ki-*ras* oncogene alleles by direct sequencing of enzymatically amplified DNA from carcinogen-induced tumors. *Proc. Natl. Acad. Sci. USA* **84**, 4974–4978.

63. Collins, S. J. (1988). Direct sequencing of amplified genomic fragments documents N-*ras* point mutations in myeloid leukemia. *Oncogene Res.* **3**, 117–123.

64. Kopreski, M. S., Benko, F. A., Borys, D. J., Khan, A., McGarrity, T. J., and Gocke, C. D. (2000). Somatic mutation screening: identification of individuals harboring K-*ras* mutations with the use of plasma DNA. *J Natl. Cancer Inst.* **92**, 918–923.

65. Sorenson, G. D. (2000). Detection of mutated KRAS2 sequences as tumor markers in plasma/serum of patients with gastrointestinal cancer. *Clin Cancer Res.* **6**, 2129–2137.

66. Peace, D. J., Chen, W., Nelson, H., and Cheever, M. A. (1991). T cell recognition of transforming proteins encoded by mutated *ras* proto-oncogenes. *J. Immunol.* **146**, 2059–2065.

67. Peace, D. J., Smith, J. W., Disis, M. L., Chen, W., and Cheever, M. A. (1993). Induction of T cells specific for the mutated segment of oncogenic p21 *ras* protein by immunization *in vivo* with the oncogenic protein. *J. Immunother.* **14**, 110–114.

68. Fenton, R. G., Taub, D. D., Kwak, L. W., Smith, M. R., and Longo, D. L. (1993). Cytotoxic T-cell response and *in vivo* protection against tumor cells harboring activated *ras* proto-oncogenes. *J. Natl. Cancer Inst.* **85**, 1294–1302.

69. Skipper, J., and Stauss, H. J. (1993). Identification of two cytotoxic T lymphocyte-recognized epitopes in the *ras* protein. *J. Exp. Med.* **177**, 1493–1498.

70. Peace, D. J., Smith, J. W., Chen, W., You, S.-G., Cosand, W. L., Blake, J., and Cheever, M. A. (1994). Lysis of *ras* oncogene-transformed cells by specific cytotoxic T lymphocytes elicited by primary *in vitro* immunization with mutated *ras* peptide. *J. Exp. Med.* **179**, 473–479.

71. Abrams, S. I., Stanziale, S. F., Lunin, S. D., Zaremba, S., and Schlom, J. (1996). Identification of overlapping epitopes in mutant *ras* onco-gene peptides that activate CD4+ and CD8+ T cell responses. *Eur. J. Immunol.* **26**, 435–443.

72. Bristol, J. A., Schlom, J., and Abrams, S. I. (1998). Development of a murine mutant Ras CD8+ CTL peptide epitope variant that possesses enhanced MHC class I binding and immunogenic properties. *J. Immunol.* **160**, 2433–2441.

73. Bristol, J. A., Schlom, J., and Abrams, S. I. (1999). Persistence, immune specificity, and functional ability of murine mutant *ras* epitope-specific CD4+ and CD8+ T lymphocytes following *in vivo* adoptive transfer. *Cell. Immunol.* **194**, 78–89.

74. Luo, Y., Chen, X., Han, R., Chorev, M., Dewolf, W. C., and O'Donnell, M. A. (1999). Mutated *ras* p21 as a target for cancer therapy in mouse transitional cell carcinoma. *J. Urol.* **162**, 1519–1526.

75. Jung, S., and Schluesener, H. J. (1991). Human T lymphocytes recognize a peptide of single point-mutated, oncogenic *ras* proteins. *J. Exp. Med.* **173**, 273–276.

76. Gedde-Dahl, III., T., Eriksen, J. A., Thorsby, E., and Gaudernack, G. (1992). T-cell responses against products of oncogenes: generation and characterization of human T-cell clones specific for p21 *ras*-derived synthetic peptides. *Hum. Immunol.* **33**, 266–274.

77. Fossum, B., Gedde-Dahl, III., T., Hansen, T., Eriksen, J. A., Thorsby, E., and Gaudernack, G. (1993). Overlapping epitopes encompassing a point mutation (12Gly→Arg) in p21 *ras* can be recognized by HLA-DR, -DP and -DQ restricted T cells. *Eur. J. Immunol.* **23**, 2687–2691.

78. Fossum, B., Breivik, J., Meling, G. I., Gedde-Dahl, III, T., Hansen, T., Knutsen, I., Rognum, T. O., Thorsby, E., and Gaudernack, G. (1994). A K-*ras* 13Gly→Asp mutation is recognized by HLA-DQ7 restricted T cells in a patient with colorectal cancer. Modifying effect of DQ7 on established cancers harbouring this mutation? *Int. J. Cancer* **58**, 506–511.

79. Fossum, B., Gedde-Dahl, III, T., Breivik, J., Eriksen, J. A., Spurkland, A., Thorsby, E., and Gaudernack, G. (1994). p21-*ras*-peptide-specific T-cell responses in a patient with colorectal cancer. CD4+ and CD8+ T cells recognize a peptide corresponding to a common mutation (13Gly→Asp). *Int. J. Cancer* **56**, 40–45.

80. Gedde-Dahl, III, T., Nilsen, E., Thorsby, E., and Gaudernack, G. (1994). Growth inhibition of a colonic adenocarcinoma cell line (HT29) by T cells specific for mutant p21 *ras*. *Cancer Immunol. Immunother.* **38**, 127–134.

81. Tsang, K. Y., Nieroda, C. A., DeFilippi, R., Chung, Y. K., Yamaue, H., Greiner, J. W., and Schlom, J. (1994). Induction of human cytotoxic T cell lines directed against point-mutated p21 *ras*-derived synthetic peptides. *Vaccine Res.* **3**, 183–193.

82. Fossum, B., Olsen, A. C., Thorsby, E., and Gaudernack, G. (1995). CD8+ T cells from a patient with colon carcinoma, specific for a mutant p21-*Ras*-derived peptide (Gly13→Asp), are cytotoxic towards a carcinoma cell line harbouring the same mutation. *Cancer Immunol. Immunother.* **40**, 165–172.

83. Van Elsas, A., Nijman, H. W., Van der Minne, C. E., Mourer, J. S., Kast, W. M., Melief, C. J. M., and Schrier, P. I. (1995). Induction and characterization of cytotoxic T-lymphocytes recognizing a mutated p21 *ras* peptide presented by HLA-A*0201. *Int. J. Cancer* **61**, 389–396.

84. Juretic, A., Jurgens-Gobel, J., Schaefer, C., Noppen, C., Willimann, T. E., Kocher, T., Zuber, M., Harder, F., Heberer, M., and Spagnoli, G. C. (1996). Cytotoxic T-lymphocyte responses against mutated p21 *ras* peptides: an analysis of specific T-cell-receptor gene usage. *Int. J. Cancer* **68**, 471–478.

85. Van Elsas, A., Scheibenbogen, C., van der Minne, C., Zerp, S. F., Keilholz, U., and Schrier, P. I. (1997). UV-induced N-*ras* mutations are T-cell targets in human melanoma. *Melanoma Res.* **7**(Suppl. 2), S107–S113.

86. Schott, M. E., Wells, D. T., Schlom, J., and Abrams, S. I. (1996). Comparison of linear and branched peptide forms (MAPs) in the induction of T helper responses to point-mutated *ras* immunogens. *Cell. Immunol.* **174**, 199–209.

87. Bergmann-Leitner, E. S., Kantor, J. A., Shupert, W. L., Schlom, J., and Abrams, S. I. (1998). Identification of a human CD8+ T lymphocyte neo-epitope created by a *ras* codon 12 mutation which is restricted by the HLA-A2 allele. *Cell. Immunol.* **187**, 103–116.

88. Khleif, S. N., Abrams, S. I., Hamilton, J. M., Bergmann-Leitner, E., Chen, A., Bastian, A., Bernstein, S., Chung, Y., Allegra, C. J., and Schlom, J. (1999). A phase I vaccine trial with peptides reflecting *ras* oncogene mutations of solid tumors. *J. Immunother.* **22**, 155–165.

89. Nijman, H. W., Houbiers, J. G. A., Vierboom, M. P. M., van der Burg, S. H., Drijfhout, J. W., D'Amaro, J., Kenemans, P., Melief, C. J. M., and Kast, W. M. (1993). Identification of peptide sequences that potentially trigger HLA-A2.1-restricted cytotoxic T lymphocytes. *Eur. J. Immunol.* **23**, 1215–1219.

90. Sette, A., Vitiello, A., Reherman, B., Fowler, P., Nayersina, R., Kast, W. M., Melief, C. J., Oseroff, C., Yuan, L., Ruppert, J., Sidney, J., del Guercio, M.-F., Southwood, S., Kubo, R. T., Chestnut, R. W., Grey, H. M., and Chisari, F. V. (1994). The relationship between class I binding affinity and immunogenicity of potential cytotoxic T cell epitopes. *J. Immunol.* **153**, 5586–5592.

91. Rosenberg, S. A., Zhai, Y., Yang, J. C., Schwartzentruber, D. J., Hwu, P., Marincola, F. M., Topalian, S. L., Restifo, N. P., Seipp, C. A., Einhorn, J. H., Roberts, B., and White, D. E. (1998). Immunizing patients with metastatic melanoma using recombinant adenoviruses encoding MART-1 or gp100 melanoma antigens. *J. Natl. Cancer Inst.* **90**, 1894–1900.

92. Ressing, M. E., de Jong, J. H., Brandt, R. M., Drijfhout, J. W., Benckhuijsen, W. E., Schreuder, G. M., Offringa, R., Kast, W. M., and Melief, C. J. (1999). Differential binding of viral peptides to HLA-A2 alleles. Implications for human papillomavirus type 16 E7 peptide-based vaccination against cervical carcinoma. *Eur. J. Immunol.* **29**, 1292–1303.

93. Browning, M., and Krausa, P. (1996). Genetic diversity of HLA-A2: evolutionary and functional significance. *Immunol. Today* **4**, 165–169.

94. Sidney, J., Grey, H. M., Kubo, R. T., and Sette, A. (1996). Practical, biochemical and evolutionary implications of the discovery of HLA class I supermotifs. *Immunol. Today* **17**, 261–266.

95. del Guercio, M.-F., Sidney, J., Hermanson, G., Perez, C., Grey, H. M., Kubo, R. T., and Sette, A. (1995). Binding of a peptide to multiple HLA alleles allows definition of an A2–like supertype. *J. Immunol.* **154**, 685–693.

96. Salter, R., and Cresswell, P. (1986). Impaired assembly and transport of HLA-A and -B antigens in a mutant T × B cell hybrid. *EMBO J.* **5**, 943–949.

97. Abrams, S. I., Khleif, S. N., Bergmann-Leitner, E. S., Kantor, J. A., Chung, Y., Hamilton, J. M., and Schlom, J. (1997). Generation of stable CD4$^+$ and CD8$^+$ T cell lines from patients immunized with *ras* oncogene-derived peptides reflecting codon 12 mutations. *Cell. Immunol.* **182**, 137–151.

98. Leibovitz, A., Stinson, J. C., McCombs, 3rd, W. B., McCoy, C. E., Mazur, K. C., and Mabry, N. D. (1976). Classification of human colorectal adenocarcinoma cell lines. *Cancer Res.* **36**, 4562–4569.

99. Capon, D. J., Seeburg, P. H., McGrath, J. P., Hayflick, J. S., Edman, U., Levinson, A. D., and Goeddel, D. V. (1983). Activation of Ki-*ras*2 gene in human colon and lung carcinomas by two different point mutations. *Nature* **304**, 507–513.

100. Gjertsen, M. K., Bakka, A., Breivik, J., Saeterdal, I., Solheim, B. G., Soreide, O., Thorsby, E., and Gaudernack, G. (1995). Vaccination with mutant ras peptides and induction of T-cell responsiveness in pancreatic carcinoma patients carrying the corresponding RAS mutation. *Lancet* **346**, 1399–1400.

101. Gjertsen, M. K., Bakka, A., Breivik, J., Saeterdal, I., Gedde-Dahl, III, T., Stokke, K. T., Solheim, B. G., Egge, T. S., Soreide, O., Thorsby, E., and Gaudernack, G. (1996). *Ex vivo ras* peptide vaccination in patients with advanced pancreatic cancer: results of a phase I/II study. *Int. J. Cancer* **65**, 450–453.

102. Gjertsen, M. K., Saeterdal, I., Thorsby, E., and Gaudernack, G. (1996). Characterisation of immune responses in pancreatic carcinoma patients after mutant p21 *ras* peptide vaccination. *Br. J. Cancer* **74**, 1828–1833.

103. Gjertsen, M. K., Bjorheim, J., Saeterdal, I., Myklebust, J., and Gaudernack, G. (1997). Cytotoxic CD4$^+$ and CD8$^+$ T lymphocytes, generated by mutant p21-*ras* (12Val) peptide vaccination of a patient, recognize 12Val-dependent nested epitopes present within the vaccine peptide and kill autologous tumour cells carrying this mutation. *Int. J. Cancer* **72**, 784–790.

104. Gjertsen, M. K., and Gaudernack, G. (1998). Mutated Ras peptides as vaccines in immunotherapy of cancer. *Vox Sang.* **74**, 489–495.

105. Nakagomi, H., Petersson, M., Magnusson, I., Juhlin, C., Matsuda, M., Mellstedt, H., Taupin, J. L., Vivier, E., Anderson, P., and Kiessling, R. (1993). Decreased expression of the signal-transducing zeta chains in tumor-infiltrating T-cells and NK cells of patients with colorectal carcinoma. *Cancer Res.* **53**, 5610–5612.

106. Carreno, B. M., Turner, R. V., Biddison, W. E., and Coligan, J. E. (1992). Overlapping epitopes that are recognized by CD8$^+$ HLA class I-restricted and CD4$^+$ class II-restricted cytotoxic T lymphocytes are contained within an influenza nucleoprotein peptide. *J. Immunol.* **148**, 894–899.

107. Takeshita, T., Takahashi, H., Kozlowski, S., Ahlers, J. D., Pendleton, C. D., Moore, R. L., Nakagawa, Y., Yokomuro, K., Fox, B. S., Margulies, D. H., and Berzofsky, J. A. (1995). Molecular analysis of the same HIV peptide functionally binding to both a class I and a class II MHC molecule. *J. Immunol.* **154**, 1973–1986.

108. Noguchi, Y., Chen, Y.-T., and Old, L. J. (1994). A mouse mutant p53 product recognized by CD4$^+$ and CD8$^+$ T cells. *Proc. Natl. Acad. Sci. USA* **91**, 3171–3175.

109. Hendrzak, J. A., and Brunda, M. J. (1995). Biology of disease, interleukin-12: biologic activity, therapeutic utility, role in disease. *Lab. Invest.* **72**, 619–637.

110. Tahara, H., and Lotze, M. T. (1995). Antitumor effects of interleukin-12 (IL-12): applications for the immunotherapy and gene therapy of cancer. *Gene Ther.* **2**, 96–106.

111. Melero, I., Bach, N., and Chen, L. (1997). Costimulation, tolerance and ignorance of cytolytic T lymphocytes in immune responses to tumor antigens. *Life Sci.* **60**, 2035–2041.

112. Walker, P. R., Saas, P., and Dietrich, P.-Y. (1997). Role of Fas ligand (CD95L) in immune escape: the tumor cell strikes back. *J. Immunol.* **158**, 4521–4524.

113. Favre-Felix, N., Fromentin, A., Hammann, A., Solary, E., Martin, F., and Bonnotte, B. (2000). Cutting edge: the tumor counterattack hypothesis revisited: colon cancer cells do not induce T cell apoptosis via the Fas (CD95, APO-1) pathway. *J. Immunol.* **164**, 5023–5027.

114. Restifo, N. P. (2000). Not so Fas: re-evaluating the mechanisms of immune privilege and tumor escape. *Nat. Med.* **6**, 493–495.

115. Bergmann-Leitner, E. S., and Abrams, S. I. (2000). Differential role of Fas/Fas ligand interactions in cytolysis of primary and metastatic colon carcinoma cell lines by human antigen-specific CD8$^+$ CTL. *J. Immunol.* **164**, 4941–4954.

116. Bergmann-Leitner, E. S., and Abrams, S. I. (2000). Influence of interferon-γ on modulation of Fas expression by human colon carcinoma cells and their subsequent sensitivity to antigen-specific CD8$^+$ cytotoxic T lymphocyte attack. *Cancer Immunol. Immunother.* **49**, 193–207.

117. Alexander-Miller, M. A., Leggatt, G. R., and Berzofsky, J. A. (1996). Selective expansion of high- or low-avidity cytotoxic T lymphocytes and efficacy for adoptive transfer. *Proc. Natl. Acad. Sci. USA* **93**, 4102–4107.

118. Parkhurst, M. R., Salgaller, M. L., Southwood, S., Robbins, P. F., Sette, A., Rosenberg, S. A., and Kawakami, Y. (1996). Improved induction of melanoma-reactive CTL with peptides from the melanoma antigen gp100 modified at HLA-A*0201–binding residues. *J. Immunol.* **157**, 2539–2548.

119. Tourdot, S., Oukka, M., Manuguerra, J. C., Magafa, V., Vergnon, I., Riche, N., Bruley-Rosset, M., Cordopatis, P., and Kosmatopoulos, K. (1997). Chimeric peptides: a new approach to enhancing the immunogenicity of peptides with low MHC class I affinity. *J. Immunol.* **159**, 2391–2398.

120. Berzofsky, J. A., Ahlers, J. D., Derby, M. A., Pendleton, C. D., Arichi, T., and Belyakov, I. M. (1999). Approaches to improve engineered vaccines for human immunodeficiency virus and other viruses that cause chronic infections. *Immunol. Rev.* **170**, 151–172.

121. Abrams, S. I., and Schlom, J. (2000). Rational antigen modification as a strategy to upregulate or downregulate antigen recognition. *Curr. Opin. Immunol.* **12**, 85–91.

122. Tanaka, M., Suda, T., Takahashi, T., and Nagata, S. (1995). Expression of the functional soluble form of human Fas ligand in activated lymphocytes. *EMBO J.* **14**, 1129–1135.

123. Mayordomo, J. I., Zorina, T., Storkus, W. J., Zitvogel, L., Celluzzi, C., Falo, L. D., Melief, C. J., Ildstad, S. T., Kast, W. M., Deleo, A. B. *et al.* (1995). Bone marrow-derived dendritic cells pulsed with synthetic tumour peptides elicit protective and therapeutic antitumour immunity. *Nat. Med.* **1**, 1297–1302.

124. Schuler, G., and Steinman, R. M. (1997). Dendritic cells as adjuvants for immune-mediated resistance to tumors. *J. Exp. Med.* **186**, 1183–1187.

125. Tsai, V., Southwood, S., Sidney, J., Sakaguchi, K., Kawakami, Y., Appella, E., Sette, A., and Celis, E. (1997). Identification of subdominant CTL epitopes of the GP100 melanoma-associated tumor antigen by primary in vitro immunization with peptide-pulsed dendritic cells. *J. Immunol.* **158**, 1796–1802.

DENDRITIC CELL BASED GENE THERAPY

Introduction to Dendritic Cells

PATRICK BLANCO

Baylor Institute for Immunology Research
Dallas, Texas 75204

A. KAROLINA PALUCKA

Baylor Institute for Immunology Research
Dallas, Texas 75204

JACQUES BANCHEREAU

Baylor Institute for Immunology Research
Dallas, Texas 75204

I. INTRODUCTION

Dendritic cells (DCs) are professional antigen-presenting cells (APCs) which have the unique ability to induce and sustain immune responses, and they are now recognized as an integral part of the lymphohematopoietic system. DC progenitors in the bone marrow give rise to circulating precursors that home to the tissue, where they reside as immature cells with high phagocytic capacity (Fig. 1). Upon tissue damage, immature DCs capture antigen (Ag) and subsequently migrate to the lymphoid organs, where they select rare Ag-specific T cells, thereby initiating immune responses. DCs

present Ag to $CD4^+$ T cells, which in turn regulate other immune effectors, including Ag-specific $CD8^+$ T cells and B cells, as well as non-Ag-specific macrophages, eosinophils, and natural killer (NK) cells. Just as lymphocytes are composed of different subsets with specific effector functions (B cells, NK cells, and T cells), DCs are composed of distinct subsets with specific regulatory functions. The picture is further complicated, as four stages of DC development have been delineated. Thus, DCs appear as a complex system of cells which, under different microenvironmental conditions, can induce such contrasting states as immunity and tolerance. Given this central role in controlling immunity, DCs represent logical targets for many clinical situations, including resistance to tumors. In this chapter, we will discuss the recent progress in DC biology and the potential implications of DCs in the context of immunization against cancer.

II. FEATURES OF DENDRITIC CELLS

A. Discovery and Function

Dendritic cells were first identified in the epidermis in 1868 and were termed Langerhans cells. A century later, Steinman and Cohn [1] identified their presence in other tissues, but their functions remained poorly understood for nearly 20 more years. Knowledge of DC physiology has progressed considerably because of the discovery, in the early 1990s, of culture systems that permitted *in vitro* generation of large numbers of mouse [2] and human [3] DCs. Currently, human DCs are generated *in vitro* from (1) bone marrow progenitors cultured in granulocyte–macrophage colony-stimulating factor (GM-CSF) and tumor necrosis factor (TNF), and

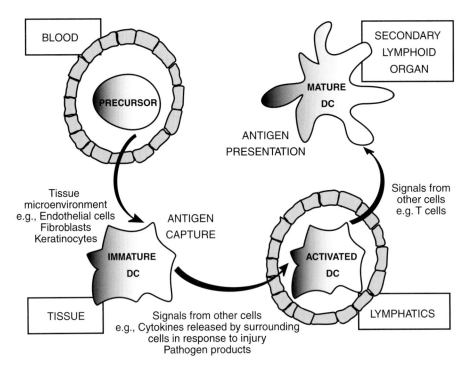

FIGURE 1 Dendritic cells originate from hematopoietic progenitors. Circulating precursors give rise to immature tissue-residing, antigen-capturing DCs, the differentiation of which is subject to microenvironmental regulation. Following antigen capture and activation by either signals from surrounding cells or pathogen products, DCs migrate to lymphoid organs. Mature antigen-presenting DCs display peptide/MHC complexes and costimulatory molecules, allowing selection, expansion, and differentiation of antigen-specific lymphocytes.

(2) blood monocytes cultured with GM-CSF and interleukin (IL)-4 or IL-13 (reviewed in Banchereau *et al.* [4]).

The life span of DCs consists of several differentiation stages, including: (1) bone marrow progenitors; (2) circulating precursor DCs; (3) tissue-residing immature DCs, whose main property is to capture antigen(s); and (4) mature DCs, present within secondary lymphoid organs, whose main property is antigen presentation (Fig. 1) [4–6]. Besides replenishing the pool of tissue-residing immature DC, circulating DC precursors play a critical role in the immediate reaction to pathogens and in the shaping of immune response. Indeed, monocytes, the most abundant precursors of myeloid DCs in blood, have long been recognized as initial effectors of lipopolysaccharide (LPS)-related inflammatory responses. Another circulating precursor population, $CD11c^-IL-3R\alpha^+$ plasmacytoid DCs, has recently been shown to be a major source of type I interferons in response to virus [7,8]. The emerging picture is that of the plasticity of the DC system revealed by: (1) specialization of DC precursors to respond to different pathogens, virus, or bacteria; and (2) the dual function of these cells at two distinct stages of differentiation as exemplified by the ability of precursor DCs to secrete large amounts of proinflammatory and/or antiviral cytokines

and the ability of mature DCs to activate and modulate T-cell responses, as discussed later.

B. Different Maturation Stages

1. Immature Dendritic Cells

All tissues, with the possible exception of brain and testis, contains immature DCs capable of capturing Ags by several pathways, such as: (1) macropinocytosis, (2) receptor-mediated endocytosis via C-type lectins or Fcγ receptors, (3) phagocytosis of particulate live and non-live antigens (reviewed in Banchereau *et al.* [4] and Banchereau and Steinman [6]), and (4) internalization of heat shock proteins [9] (Fig. 2). Captured antigens are targeted to endosomal compartments for MHC class II loading and subsequent presentation to CD4 T cells [10,11]. In most cells, MHC class I molecules associate only with peptides derived from endogenous proteins, but not with peptides from exogenous proteins taken up by the cell. Here again, DCs are unique as they have evolved a property of loading class I molecules with peptides derived from exogenous antigens. This mechanism, called cross-presentation or cross-priming, has now been shown

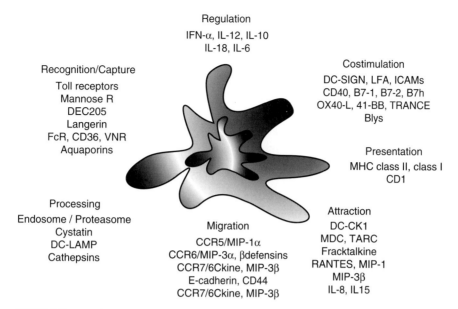

Regulation
IFN-α, IL-12, IL-10
IL-18, IL-6

Recognition/Capture
Toll receptors
Mannose R
DEC205
Langerin
FcR, CD36, VNR
Aquaporins

Costimulation
DC-SIGN, LFA, ICAMs
CD40, B7-1, B7-2, B7h
OX40-L, 41-BB, TRANCE
Blys

Presentation
MHC class II, class I
CD1

Processing
Endosome / Proteasome
Cystatin
DC-LAMP
Cathepsins

Migration
CCR5/MIP-1α
CCR6/MIP-3α, βdefensins
CCR7/6Ckine, MIP-3β
E-cadherin, CD44
CCR7/6Ckine, MIP-3β

Attraction
DC-CK1
MDC, TARC
Fracktalkine
RANTES, MIP-1
MIP-3β
IL-8, IL15

FIGURE 2 Dendritic cells display an array of molecules at different stages of differentiation and/or maturation. Given the enormous progress in genomic and proteomic approaches, we are likely to uncover the molecular pathways regulating DC functions and explaining their crucial role in the induction, regulation, and maintenance of immune responses.

to be relevant for presentation of antigens from immune complexes or dying cells [12–15].

2. Mature Dendritic Cells

Numerous factors induce and/or regulate DC maturation, including: (1) pathogen-related molecules (e.g., LPS [10,16,17], double-stranded RNA [18], or CpG nucleotides [19,20]; (2) the balance between proinflammatory and antiinflammatory signals in the local microenvironment, including TNF, IL-1, IL-6, IL-10, TGF-β, and prostaglandins [4,21]; and (3) T-cell-derived signals [4]. All mediators of DC maturation trigger their migration to secondary lymphoid organs, where the DCs can present peptide MHC complexes to naïve T cells. This migration of maturing DCs also involves a coordinated action of several chemokines. After antigen uptake, inflammatory stimuli turn off the response of immature DCs to MIP-3α (and other chemokines specific for immature DC) through either receptor downregulation or receptor desensitization, depending on autocrine chemokine production [22–24]. Consequently, maturing DCs escape from the local gradient of MIP-3α. Upon maturation, DCs upregulate a single known chemokine receptor, CCR7 [25], and accordingly acquire responsiveness to MIP-3β (ELC, Exodus 3) and 6Ckine (SLC, Exodus 2) [22,26]. Consequently, maturing DCs will leave the inflamed tissues and enter the lymph stream, potentially directed by 6Ckine expressed on lymphatic vessels [27,28]. Mature DCs entering in the draining lymph nodes will be directed into the paracortical area in

response to MIP-3β and/or 6Ckine by cells spread over the T-cell zone [22,29]. The newly arriving DCs might themselves become a source of MIP-3β and 6Ckine [22,23,29], allowing an amplification and/or a persistence of the chemotactic signal. The initial contact between DCs and resting T cells seems to be mediated by a transient, high-affinity interaction between DC-SIGN on DCs and the adhesion molecule ICAM-3 on T cells [30] and is followed by involvement of other adhesion molecules and their corresponding ligands (ICAM-1/LFA-1, LFA-3/CD2). Following TCR engagement, an intimate contact often referred to as the *immunological synapse* evolves, where multiple interactions between costimulatory molecules on DC and their ligands on T cells result in final DC maturation and T-cell activation [31]. The induced antigen-specific CD4 T cells then orchestrate other effectors of the immune system, including CD8 T cells, B cells, and NK cells. However, DCs can directly present antigens to CD8 T cells as well as induce proliferation of naïve B cells and their differentiation into plasma cells [4]. Thus, DCs induce a diverse immune response involving multiple effectors of both cellular and humoral immunity.

III. DENDRITIC CELL SUBSETS

A. Mice

As of today, three distinct pathways of DC development have been identified; myeloid DC, Langerhans cells (LCs), and lymphoid DCs. Evidence for the myeloid origin of DCs comes mainly from *in vitro* studies where myeloid-committed precursors give rise to both granulocytes/monocytes and myeloid DCs under the influence of GM-CSF [2]. DC can also arise from lymphoid-committed precursors (reviewed in Shortman [32]). To date, the most reliable marker distinguishing "myeloid" and "lymphoid" subsets is $CD8\alpha$, which is expressed as a homodimer on the lymphoid DC but is absent from the myeloid subset [33–36]. Although the lymphoid origin of DCs has only been demonstrated for $CD8\alpha^+$ thymic DCs, the similar phenotype of $CD8\alpha^+$ splenic and lymph node DCs [36] suggests a common origin. Nevertheless, a clonal analysis permitting us to conclude that DCs arise from the same precursor cells as lymphocytes has not yet been done. Moreover, recent data suggest that myeloid clonal progenitors can give rise to both $CD8\alpha^+$ or $CD8\alpha^-$ DCs [37].

$CD8\alpha^+$ and $CD8\alpha^-$ DCs differ in phenotype, localization, and function. Both subsets express high levels of CD11c, class II MHC, and the costimulatory molecules CD86 and CD40. $CD8\alpha^+$ DCs are localized in the T-cell-rich areas of the periarteriolar lymphatic sheaths (PALSs) in the spleen and lymph nodes [34,38,39]. In contrast, $CD8\alpha^-$ DCs are in the marginal zone [38,40]. The $CD8\alpha^+$ DCs make higher levels of IL-12 and are more phagocytic than $CD8\alpha^-$ DCs [34,41,42]. While *in vitro*, the $CD8\alpha^+$ DCs prime allogeneic CD4 and CD8 T cells less efficiently than $CD8\alpha^-$ DCs

[43,44]; *in vivo* both subsets appear to prime antigen-specific $CD4^+$ T cells efficiently [41,45].

B. Humans

In humans, revealing the existence of distinct DC subsets came from several directions based on analyses of skin DCs [46], DCs generated *in vitro* by culture of $CD34^+$ hematopoietic progenitors (HPCs) [47], and blood DC precursors [48] (Fig. 3). Human skin contains two subsets with distinct localization: LCs within the epidermis, characterized by the expression of CD1a and Birbeck granules, and interstitial (dermal) DCs (intDCs), lacking Birbeck granules but expressing coagulation factor XIIIa. These two subsets also emerge in cultures of $CD34^+$ HPC driven by GM-CSF and TNF [47], and, most interestingly, these subsets have common as well as unique functions [49]. For example, intDCs, but not LCs, are able to induce the differentiation of naïve B cells into immunoglobulin-secreting plasma cells. Interstitial DCs demonstrate a high efficiency of Ag capture that is about tenfold higher than that of LC. They also express high levels of nonspecific esterases, markers of the lysosomal compartment, while LCs do not. While no unique function has yet been formally attributed to LCs, there are hints that they may be particularly efficient activators of cytotoxic CD8 T cells.

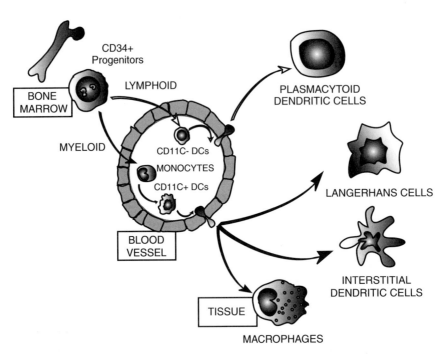

FIGURE 3 Dendritic cell progenitors give rise to myeloid (monocytes and $CD11c^+$ DCs) and lymphoid ($CD11c-$ plasmacytoid DCs) precursors. Upon interaction with inflamed endothelium, monocytes differentiate into $CD11c^+$ blood DCs which give rise to Langerhans cells, interstitial DCs, and macrophages. Differentiation of plasmacytoid DCs from $CD34^+$ progenitors [61] can be blocked by Id2 and Id3 overexpression, suggesting their lymphoid origin [62].

Three subsets of DC precursors circulate in the blood: $CD14^+$ monocytes, lineage negative $CD11c^+$ precursor DCs, and $CD11c^-IL-3R\alpha^+$ plasmacytoid DCs [4,50]. Both monocyte and $CD11c^+$ subsets can give rise to intDCs (under the influence of GM-CSF and IL-4 or TNF), LCs (in the presence of TGF-β), and macrophages (in the presence of M-CSF or GM-CSF) [49,51,52]. Distinct factors regulate the survival and differentiation of $CD11c^-$ DC precursors, originally described as plasmacytoid T cells or plasmacytoid monocytes [7,53–57]. These cells die rapidly after isolation and are critically dependent on IL-3 for survival and CD40-L for maturation. Plasmacytoid DCs (pDCs) can be further distinguished from $CD11c^+$ DCs by differential expression of immunoglobulin-like transcripts (ILTs), with pDCs being $ILT1^-/ILT3^+$ and $CD11c^+$ DCs being $ILT1^+/ILT3^+$. They have unique properties such as the expression of lymphoid antigens [54], the ability to produce large amounts of type I interferon [7], and the ability to polarize a fraction of the T cells towards IL-4 and IL-5 production (type 2 cells), hence the nomenclature DC2 [58].

Two recent sets of findings will accelerate research on human pDCs. First, systemic administration of both Flt3-L and G-CSF increases the number of pDCs in blood [50,59,60], and, second, it may be possible to generate those cells *in vitro*. Blom *et al.* [61] demonstrated the existence of CD34low cells with the phenotype and function of pDCs, as well as a simple culture system in which early $CD34^+$ progenitors yield pDCs [61]. At the same time, Spits *et al.* [62] provided another culture system in which early $CD34^+$ progenitors yielded pDCs and which demonstrated that their differentiation can be blocked by overexpression of Id2 and Id3 proteins [62].

IV. FUNCTIONAL HETEROGENEITY OF DENDRITIC CELL SUBSETS

A. Type 1/Type 2 T Cells

Dendritic Cell subsets may provide T cells with the different cytokine/molecule microenvironments that determine the classes of immune response (e.g., T_h1 versus T_h2). In mice, the splenic $CD8\alpha^+$ DCs prime naïve CD4 T cells to make T_h1 cytokines in a process involving IL-12, while the splenic $CD8\alpha^-$ DCs prime naïve CD4 T cells to make T_h2 cytokines [41,45]. In humans, the picture is less clear. CD40-ligand (CD40-L)-activated monocyte-derived DCs prime T_h1 responses via an IL-12-dependent mechanism, while IL-3^+CD40-L-activated pDCs have been shown to secrete negligible amounts of IL-12 and prime T_h2 responses [58]. However, the polarizing effects of DC subsets may be susceptible to microenvironmental signals, which could instruct a given DC subset to elicit different T_h responses. Indeed, monocyte-derived DCs can induce T cells to make IL-4 rather than IFN-γ when the DCs are used at low numbers [63]; when the

DCs are exposed to factors such as prostaglandin E_2 (PGE_2), corticosteroids, or IL-10; or upon prolonged activation *in vitro* [64–67]. Furthermore, pDCs stimulated by virus secrete IFN-α, which drives T_h1 responses in humans [68] and matures into DCs that can induce T cells to produce IFN-γ and IL-10 [8]. Thus, the types of DC subset and microenvironmental signals are important for T_h polarization.

B. Regulatory T Cells

While T_h1 and T_h2 cells have been considered as the two extremes of T-cell polarization [69], other subsets of CD4 T cells with regulatory function have been identified: TGF-β producing T_h3 cells [70,71] and IL10-producing T_r1 cells [72]. Regulatory T cells specific for tumor antigens have been found in melanoma [73,74]. Furthermore, repetitive *in vitro* stimulation with Ag-loaded APCs (immature DCs in particular [75]) can lead to the emergence of regulatory T cells producing large amounts of IL-4 and IL-10, whose supernatant can block the activation of fresh T cells. IL-10 may also convert DC function to the induction of antigen-specific anergy, thus leading to the state of tolerance against tumor tissue [76,77].

C. Regulation of B Lymphocytes

Beside activating naïve T cells, DCs can directly activate naïve and memory B cells. DCs enhance differentiation of CD40-activated memory B cells toward IgG-secreting cells through secretion of the soluble IL-6Rα gp80, which complexes to IL-6 [78]. Interstitial DCs, but not Langerhans cells, also help in differentiation of activated naive B cells to plasma cells and the secretion of both IgA_1 and IgA_2 subclasses [79].

The germinal center is the microenvironment that allows the generation of B cell memory. There, B cells proliferate and undergo somatic mutation, isotype switching, affinity selection, and differentiation into memory B cells or plasmablasts. The germinal center also contains T cells, follicular DC (FDC) and germinal center DC (GCDC). It is now clear that GCDC are quite different from FDC in phenotype and function [80,81]. GCDC stimulate, in an IL-12-dependent manner, CD40-activated germinal center B cell proliferation and drive their differentiation towards plasma cells. In addition, GCDC induce IL10-independent isotype switching towards IgG_1.

Thus, DC subsets have the capacity to directly regulate B-cell responses. In order to generate a humoral immune response, antigen-specific $CD4^+$ T-helper and antigen-specific B cells must interact. Within paracortical areas of the secondary lymphoid organs, interdigitating DCs select the rare antigen-specific T and B cells. As recently demonstrated *in vivo* in the rat, DCs can also capture and retain unprocessed antigen then transfer it to naive B cells to initiate a specific T_h2-associated antibody response [82]. This could be the role of the GCDC population localized within germinal centers and originally described as "antigen-transporting cells" [83,84] which could display the antigen to both T and B cells. One could consider that a conditioned DC can be a temporal bridge between a CD4$^+$ T helper and a B lymphocyte by analogy to recent models where DCs offer costimulatory signals to CD4 T helper cells and CD8 T cells [85–87]. During the extrafollicular reaction, intDCs could play a role in the induction of an IL2-dependent IgM plasma cell differentiation. Germinal center formation starts with the migration of GC founder cells in the follicles and involves T_h2 CD4$^+$ T cells. CD40 activation upregulates OX40L expression on DC and B cells [26,88,89], and early OX40 ligation promotes T_h2 cytokine secretion [90] and causes CD4 T-cell migration within B-cell follicles [91]. Thus, GCDC may contribute to the germinal center reaction, and the role of OX40/OX40L needs to be analyzed.

D. Effectors of the Innate Immunity

Dendritic cells at different stages of differentiation can regulate effectors of innate immunity such as NK cells and NK T cells. Both direct cell–cell interactions and indirect cytokine mediated interactions have been implicated. Precursors of pDCs may activate NK cells through the release of IFN-α, hereby leading to enhanced antiviral and antitumor activity of NK cells [7,53]. DCs at later stages of differentiation may regulate the activity of NK and NK T cells [92] through the release of IL-12, IL-15, and IL-18 [93,94].

V. DENDRITIC CELLS IN TUMOR IMMUNOLOGY

The immune system evolved to protect us from harmful pathogens. This formidable task relies upon a concerted action of both Ag-nonspecific innate immunity and Ag-specific adaptive immunity (reviewed in Fearon and Locksley [95]). Those two systems are linked by DCs. The innate system includes phagocytic cells, NK cells, complement, and interferons and is characterized by the ability to rapidly recognize pathogens and/or tissue injury and the ability to signal the presence of "danger" to cells of the adaptive immune system. In turn, the adaptive immunity is characterized by the ability to rearrange genes of the immunoglobulin family, permitting creation of a large diversity of Ag-specific clones, and immunological memory. The inflammatory reaction occurring upon pathogen invasion (the "danger" signal [96]) leads to DC activation, migration, and maturation, culminating in the induction of immunity and pathogen elimination. In this context, tumors are "silent." As they do not provide a "danger" signal, there is no reason for DCs to migrate to the lymphoid organs and initiate an immune response. Furthermore, much like normal tissues, tumors themselves do not provide cos-

timulation, and, finally, most of the tumor-associated antigens are derived from self-antigens, expressed, for instance, during tissue differentiation [97,98]. Hence, the immune system is tolerized against such antigens.

Thus, inducing effective antitumor immunity, as yet an elusive goal, must be seen as inducing autoimmunity [99,100]. Indeed, although rare, tumor-related autoimmune responses exist and manifest themselves as paraneoplastic neurologic disorders (PNDs). Discovery of onconeural antigens [101] and the identification of onconeural antibodies led to the proposal that paraneoplastic cerebellar degeneration (PCD), associated with breast and ovarian cancer, is an autoimmune disorder mediated by the humoral arm of the immune system. Furthermore, the presence of cdr-2-specific CD8$^+$ CTLs circulating in the blood of these patients has been demonstrated [102].

A. Animal Models of Tumor Immunotherapy

In animal models, DCs loaded with tumor-associated antigens (TAAs) are able to induce protective/rejection antitumor responses [103–109]. Several systems have been employed to deliver TAAs to DCs, including: (1) defined peptides of known sequences, (2) undefined acid-eluted peptides from autologous tumor, (3) whole tumor lysates, (4) retroviral and adenoviral vectors, (5) tumor-cell-derived RNA, (6) fusion of DCs with tumor cells, and, recently, (7) tumor-peptide-pulsed DC-derived exosomes (subcellular structures containing high levels of MHC molecules and peptides), which have been successfully used to prime specific CTLs *in vivo* and eradicate established murine tumors [110]. Most recent studies show that combination therapy of DCs and biological response modifiers (i.e, IL-2) increases the efficacy of inducing protective/rejection antitumor responses in mice, thus providing the rationale for using similar approaches in humans [111].

B. Human Trials

Human trials reported to date have proven the safety and tolerability of administration of TAA-loaded DCs in cancer patients, as have limited clinical responses. Furthermore, DCs loaded with influenza matrix peptide and keyhole limpet hemocyanin (KLH) have been shown to be safe and immunogenic in healthy volunteers [112,113]. Clinical responses that have been observed in preliminary trials include: (1) a pioneer study based on injection of blood-derived DCs loaded with lymphoma idiotype [114], (2) administration of peptide-pulsed APC generated by culturing monocytes with GM-CSF [115]; (3) vaccination with monocyte-derived DCs loaded with melanoma peptides [116,117], and (4) vaccination of patients with prostate cancer using monocyte-derived DCs pulsed with prostate-specific membrane antigen (PSMA) peptide [118] or blood DCs loaded with recombinant protein consisting of tumor antigen prostate acid phosphatase (PAP) and GM-CSF [119]. In particular, intranodal injection of immature monocyte-derived DCs pulsed with synthetic melanoma peptides or tumor lysate induced delayed-type hypersensitivity (DTH) toward vaccine antigens in 11 patients [116]. The most recent study has demonstrated that vaccination with melanoma-peptide-loaded, mature-monocyte-derived DCs leads to substantial immune responses in blood [117]. Preliminary results of our study demonstrate that vaccination of patients with stage IV melanoma with DCs derived from CD34$^+$ HPC and pulsed with multiple antigens (including KLH protein, flu-matrix peptide, and melanoma peptides) leads to significant primary and recall immune responses in blood [120].

In view of these encouraging results, several parameters should be established, such as: (1) the subset of DCs and the method of generation (*ex vivo* culture or *in vivo* mobilization); (2) the DC dose, route, and frequency of injection; (3) the optimal DC activation/maturation status; (4) the source and preparation of TAA as well as DC loading strategy; (5) the combination of DC vaccination with other therapies (for instance, biological response modifiers such as IL-2 or IFN-α); (6) the evaluation of vaccine efficacy, particularly at these early stages of DC vaccine development; (7) determination of vaccine potency (for instance, the feasibility of using TAA-specific T-cell lines to determine TAA presentation by loaded DCs in large-scale trials); and, finally, (8) the duration of vaccination.

VI. DENDRITIC CELLS AND GENE THERAPY

In recent years, several reports have shown that genetically engineered dendritic cells can be a powerful tool for inducing an antigen-specific immune response (reviewed in Kirk and Mule [121]). In an attempt to generate a persistent endogenous production of antigenic epitopes, the cDNA or mRNA coding for tumor antigens can be delivered to dendritic cells. Indeed, Gilboa *et al.* have demonstrated that DCs transfected with tumor-derived RNA generate tumor-specific cellular responses [103,122,123]. This strategy, although technically challenging, has several benefits: (1) practically unlimited supply of TAA, (2) possibility of retrieving TAA from archival pathology material, (3) expression of several TAAs both known and unknown, (4) possibility for combined expression of patient-specific (autologous tumor) and shared (allogeneic tumor) TAA, and (5) possibility of targeting the antigen to lysosomal compartments.

A strategy more widely tested is the delivery of tumor antigens in viral vectors [121]. The induction of transgene-specific MHC-class-I- and -II-restricted responses by transduced DCs has been shown in both animal models and human *in vitro* systems. The ideal vectors for TAA delivery must (1) be efficient in transducing DCs to allow expression, processing, and presentation of the transgene without blocking or skewing DC functions; (2) not be the target of established immune responses that would eliminate the vector (for instance, preexisting neutralizing antibodies are likely to

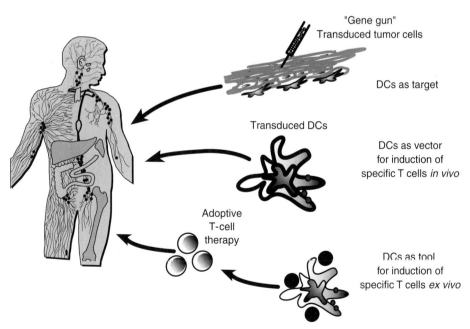

FIGURE 4 Three levels of DC-based immune intervention in cancer involving genetic approaches: (1) "Gene gun" and/or transduced tumor cell vaccines that target DCs randomly, (2) DCs transduced with either tumor RNA or viral vectors expressing tumor antigens as vectors for induction of tumor immunity *in vivo*, and (3) DCs transduced with either tumor RNA or viral vectors expressing tumor antigens as vectors for induction of tumor immunity *ex vivo* and for subsequent adoptive T-cell therapy.

eliminate adenoviral vectors); and, (3) be poorly immunogenic so that transgene-specific responses are prevalent over vector-specific responses.

Finally, gene transfer techniques have been used to convert tumor cells into better APCs [98,124]. A central hypothesis is that tumor cells already expressed signal 1 (the antigen), but lack signal 2 (costimulation) or the production of immune-stimulatory cytokines. For example, the introduction of the costimulatory molecule B7 into tumor cells, provides signal 2 and converts the tumor cells into their own APCs. In a serial testing of these tumor vaccines, Mach and Dranoff [124] noted that the introduction of the gene for GM-CSF was superior to the insertion of other genes into tumor cells. GM-CSF production by the tumor vaccines attracted a greater number of host APCs, with antigen being taken up from the dying tumor vaccines and cross-presented to the host's immune system. Several tumor vaccines produced by the insertion of immune stimulatory genes into autologous tumor cells are currently in clinical trials [124].

VII. CONCLUSIONS

Given the fact that it has taken the immune system millions of years to evolve efficient pathogen countermeasures, the progress in our efforts to use the immune system to combat cancer is encouraging, if slow. Dendritic cells are crucial players in several immunotherapy approaches either as targets or as vectors for induction of tumor-specific immunity, both *in vivo* and *ex vivo* (Fig. 4). Everything that we learn from current studies will permit us in the future to develop an "intelligent missile," a generic cancer vaccine equipped with tumor antigens, chaperons, and DC activation molecules, as well as with specific ligands that would permit targeting of desired DC subsets. This will keep us busy for a while.

References

1. Steinman, R. M., and Cohn, Z. A. (1973). Identification of a novel cell type in peripheral lymphoid organs of mice. I. Morphology, quantitation, tissue distribution. *J. Exp. Med.* **137,** 1142–1162.
2. Inaba, K., Inaba, M., Romani, N., Aya, H., Deguchi, M., Ikehara, S., Muramatsu, S., and Steinman, R. M. (1992). Generation of large numbers of dendritic cells from mouse bone marrow cultures supplemented with granulocyte/macrophage colony-stimulating factor. *J. Exp. Med.* **176,** 1693–1702.
3. Caux, C., Dezutter-Dambuyant, C., Schmitt, D., and Banchereau, J. (1992). GM-CSF and TNF-alpha cooperate in the generation of dendritic Langerhans cells. *Nature* **360,** 258–261.
4. Banchereau, J., Briere, F., Caux, C., Davoust, J., Lebecque, S., Liu, Y., Pulendran, B., and Palucka, K. (2000). Immunobiology of dendritic cells. *Ann. Rev. Immunol.* **18,** 767–812.
5. Steinman, R. M. (1991). The dendritic cell system and its role in immunogenicity. *Ann. Rev. Immunol.* **9,** 271–296.
6. Banchereau, J., and Steinman, R. M. (1998). Dendritic cells and the control of immunity. *Nature* **392,** 245–252.
7. Siegal, F. P., Kadowaki, N., Shodell, M., Fitzgerald-Bocarsly, P. A., Shah, K., Ho, S., Antonenko, S., and Liu, Y. J. (1999). The nature of the principal type 1 interferon-producing cells in human blood. *Science* **284,** 1835–1837.
8. Kadowaki, N., Antonenko, S., Lau, J. Y., and Liu, Y. J. (2000). Natural interferon-alpha/beta-producing cells link innate and adaptive immunity. *J. Exp. Med.* **192,** 219–226.
9. Srivastava, P. K., Menoret, A., Basu, S., Binder, R. J., and McQuade, K. L. (1998). Heat shock proteins come of age: primitive functions acquire new roles in an adaptive world. *Immunity* **8,** 657–665.
10. Cella, M., Engering, A., Pinet, V., Pieters, J., and Lanzavecchia, A. (1997). Inflammatory stimuli induce accumulation of MHC class II complexes on dendritic cells. *Nature* **388,** 782–787.
11. Inaba, K., Turley, S., Yamaide, F., Iyoda, T., Mahnke, K., Inaba, M., Pack, M., Subklewe, M., Sauter, B., Sheff, D., Albert, M., Bhardwaj, N., Mellman, I., and Steinman, R. M. (1998). Efficient presentation of phagocytosed cellular fragments on the major histocompatibility complex class II products of dendritic cells. *J. Exp. Med.* **188,** 2163–2173.
12. Albert, M. L., Sauter, B., and Bhardwaj, N. (1998). Dendritic cells acquire antigen from apoptotic cells and induce class I- restricted CTLs. *Nature* **392,** 86–89.
13. Rodriguez, A., Regnault, A., Kleijmeer, M., Ricciardi-Castagnoli, P., and Amigorena, S. (1999). Selective transport of internalized antigens to the cytosol for MHC class I presentation in dendritic cells. *Nat. Cell Biol.* **1,** 362–368.
14. Nouri-Shirazi, M., Banchereau, J., Bell, D., Burkeholder, S., Kraus, E. T., Davoust, J., and Palucka, K. A. (2000). Dendritic cells capture killed tumor cells and present their antigens to elicit tumor-specific immune responses. *J. Immunol.* **165,** 3797–3803.
15. Berard, F., Blanco, P., Davoust, J., Neidhart-Berard, E. M., Nouri-Shirazi, M., Taquet, N., Rimoldi, D., Cerottini, J. C., Banchereau, J., and Palucka, A. K. (2000). Cross-priming of naive CD8 T cells against melanoma antigens using dendritic cells loaded with killed allogeneic melanoma cells. *J. Exp. Med.* **192,** 1535–1544.
16. Cella, M., Sallusto, F., and Lanzavecchia, A. (1997). Origin, maturation and antigen presenting function of dendritic cells. *Curr. Opin. Immunol.* **9,** 10–16.
17. Rescigno, M., Granucci, F., and Ricciardi-Castagnoli, P. (2000). Molecular events of bacterial-induced maturation of dendritic cells. *J. Clin. Immunol.* **20,** 161–166.
18. Cella, M., Salio, M., Sakakibara, Y., Langen, H., Julkunen, I., and Lanzavecchia, A. (1999). Maturation, activation, and protection of dendritic cells induced by double-stranded RNA. *J. Exp. Med.* **189,** 821–829.
19. Sparwasser, T., Koch, E. S., Vabulas, R. M., Heeg, K., Lipford, G. B., Ellwart, J. W., and Wagner, H. (1998). Bacterial DNA and immunostimulatory CpG oligonucleotides trigger maturation and activation of murine dendritic cells. *Eur. J. Immunol.* **28,** 2045–2054.
20. Hartmann, G., Weiner, G. J., and Krieg, A. M. (1999). CpG DNA: a potent signal for growth, activation, and maturation of human dendritic cells. *Proc. Natl. Acad. Sci. USA* **96,** 9305–9310.
21. Kalinski, P., Hilkens, C. M., Wierenga, E. A., and Kapsenberg, M. L. (1999). T-cell priming by type-1 and type-2 polarized dendritic cells: the concept of a third signal. *Immunol. Today* **20,** 561–567.
22. Dieu, M. C., Vanbervliet, B., Vicari, A., Bridon, J. M., Oldham, E., Ait-Yahia, S., Briere, F., Zlotnik, A., Lebecque, S., and Caux, C. (1998). Selective recruitment of immature and mature dendritic cells by distinct chemokines expressed in different anatomic sites. *J. Exp. Med.* **188,** 373–386.
23. Sallusto, F., Schaerli, P., Loetscher, P., Schaniel, C., Lenig, D., Mackay, C. R., Qin, S., and Lanzavecchia, A. (1998). Rapid and coordinated switch in chemokine receptor expression during dendritic cell maturation. *Eur. J. Immunol.* **28,** 2760–2769.
24. Sozzani, S., Allavena, P., Vecchi, A., and Mantovani, A. (1999). The role of chemokines in the regulation of dendritic cell trafficking. *J. Leukoc. Biol.* **66,** 1–9.

25. Yoshida, R., Imai, T., Hieshima, K., Kusuda, J., Baba, M., Kitaura, M., Nishimura, M., Kakizaki, M., Nomiyama, H., and Yoshie, O. (1997). Molecular cloning of a novel human CC chemokine EBI1-ligand chemokine that is a specific functional ligand for EBI1, CCR7. *J. Biol. Chem.* **272,** 13803–13809.

26. Chan, V. W., Kothakota, S., Rohan, M. C., Panganiban-Lustan, L., Gardner, J. P., Wachowicz, M. S., Winter, J. A., and Williams, L. T. (1999). Secondary lymphoid-tissue chemokine (SLC) is chemotactic for mature dendritic cells. *Blood* **93,** 3610–3616.

27. Gunn, M. D., Tangemann, K., Tam, C., Cyster, J. G., Rosen, S. D., and Williams, L. T. (1998). A chemokine expressed in lymphoid high endothelial venules promotes the adhesion and chemotaxis of naive T lymphocytes. *Proc. Natl. Acad. Sci. USA* **95,** 258–263.

28. Saeki, H., Moore, A. M., Brown, M. J., and Hwang, S. T. (1999). Cutting edge: secondary lymphoid-tissue chemokine (SLC) and CC chemokine receptor 7 (CCR7) participate in the emigration pathway of mature dendritic cells from the skin to regional lymph nodes. *J. Immunol.* **162,** 2472–2475.

29. Ngo, V. N., Tang, H. L., and Cyster, J. G. (1998). Epstein–Barr–virus-induced molecule 1 ligand chemokine is expressed by dendritic cells in lymphoid tissues and strongly attracts naive T cells and activated B cells. *J. Exp. Med.* **188,** 181–191.

30. Geijtenbeek, T. B., Torensma, R., van Vliet, S. J., van Duijnhoven, G. C., Adema, G. J., van Kooyk, Y., and Figdor, C. G. (2000). Identification of DC-SIGN, a novel dendritic cell-specific ICAM-3 receptor that supports primary immune responses. *Cell* **100,** 575–585.

31. Lanzavecchia, A. and Sallusto, F. (2000). From synapses to immunological memory: the role of sustained T cell stimulation. *Curr. Opin. Immunol.* **12,** 92–98.

32. Shortman, K. (2000). Burnet oration: dendritic cells: multiple subtypes, multiple origins, multiple functions. *Immunol. Cell Biol.* **78,** 161–165.

33. Maraskovsky, E., Brasel, K., Teepe, M., Roux, E. R., Lyman, S. D., Shortman, K., and McKenna, H. J. (1996). Dramatic increase in the numbers of functionally mature dendritic cells in Flt3 ligand-treated mice: multiple dendritic cell subpopulations identified. *J. Exp. Med.* **184,** 1953–1962.

34. Pulendran, B., Lingappa, J., Kennedy, M. K., Smith, J., Teepe, M., Rudensky, A., Maliszewski, C. R., and Maraskovsky, E. (1997). Developmental pathways of dendritic cells in vivo: distinct function, phenotype, and localization of dendritic cell subsets in Flt3 ligand-treated mice. *J. Immunol.* **159,** 2222–2231.

35. Wu, L., Li, C. L., and Shortman, K. (1996). Thymic dendritic cell precursors: relationship to the T lymphocyte lineage and phenotype of the dendritic cell progeny. *J. Exp. Med.* **184,** 903–911.

36. Vremec, D., Lieschke, G. J., Dunn, A. R., Robb, L., Metcalf, D., and Shortman, K. (1997). The influence of granulocyte/macrophage colony-stimulating factor on dendritic cell levels in mouse lymphoid organs. *Eur. J. Immunol.* **27,** 40–44.

37. Traver, D., Akashi, K., Manz, M., Merad, M., Miyamoto, T., Engleman, E. G., and Weissman, I. L. (2000). Development of CD8alpha-positive dendritic cells from a common myeloid progenitor. *Science* **290,** 2152–2154.

38. De Smedt, T., Pajak, B., Muraille, E., Lespagnard, L., Heinen, E., De Baetselier, P., Urbain, J., and Moser, M. (1996). Regulation of dendritic cell numbers and maturation by lipopolysaccharide in vivo. *J. Exp. Med.* **184,** 1413–1424.

39. Steinman, R. M., Pack, M., and Inaba, K. (1997). Dendritic cells in the T-cell areas of lymphoid organs. *Immunol. Rev.* **156,** 25–37.

40. Reis e Sousa, C., Hieny, S., Scharton-Kersten, T., Jankovic, D., Charest, H., Germain, R. N., and Sher, A. (1997). in vivo microbial stimulation induces rapid CD40 ligand-independent production of interleukin 12 by dendritic cells and their redistribution to T cell areas. *J. Exp. Med.* **186,** 1819–1829.

41. Maldonado-Lopez, R., De Smedt, T., Michel, P., Godfroid, J., Pajak, B., Heirman, C., Thielemans, K., Leo, O., Urbain, J., and Moser, M. (1999). CD8alpha+ and CD8alpha− subclasses of dendritic cells direct the development of distinct T helper cells in vivo. *J. Exp. Med.* **189,** 587–592.

42. Ohteki, T., Fukao, T., Suzue, K., Maki, C., Ito, M., Nakamura, M., and Koyasu, S. (1999). Interleukin 12-dependent interferon gamma production by CD8alpha+ lymphoid dendritic cells. *J. Exp. Med.* **189,** 1981–1986.

43. Suss, G., and Shortman, K. (1996). A subclass of dendritic cells kills CD4 T cells via Fas/Fas-ligand-induced apoptosis. *J. Exp. Med.* **183,** 1789–1796.

44. Kronin, V., Winkel, K., Suss, G., Kelso, A., Heath, W., Kirberg, J., von Boehmer, H., and Shortman, K. (1996). A subclass of dendritic cells regulates the response of naive CD8 T cells by limiting their IL-2 production. *J. Immunol.* **157,** 3819–3827.

45. Pulendran, B., Smith, J. L., Caspary, G., Brasel, K., Pettit, D., Maraskovsky, E., and Maliszewski, C. R. (1999). Distinct dendritic cell subsets differentially regulate the class of immune response in vivo. *Proc. Natl. Acad. Sci. USA* **96,** 1036–1041.

46. Cerio, R., Griffiths, C. E., Cooper, K. D., Nickoloff, B. J., and Headington, J. T. (1989). Characterization of factor XIIIa positive dermal dendritic cells in normal and inflamed skin. *Br. J. Dermatol.* **121,** 421–431.

47. Caux, C., Vanbervliet, B., Massacrier, C., Dezutter-Dambuyant, C., de Saint-Vis, B., Jacquet, C., Yoneda, K., Imamura, S., Schmitt, D., and Banchereau, J. (1996). CD34+ hematopoietic progenitors from human cord blood differentiate along two independent dendritic cell pathways in response to GM-CSF + TNF-alpha. *J. Exp. Med.* **184,** 695–706.

48. O'Doherty, U., Peng, M., Gezelter, S., Swiggard, W. J., Betjes, M., Bhardwaj, N., and Steinman, R. M. (1994). Human blood contains two subsets of dendritic cells, one immunologically mature and the other immature. *Immunology* **82,** 487–493.

49. Caux, C., Massacrier, C., Vanbervliet, B., Dubois, B., Durand, I., Cella, M., Lanzavecchia, A., and Banchereau, J. (1997). CD34+ hematopoietic progenitors from human cord blood differentiate along two independent dendritic cell pathways in response to granulocyte–macrophage colony-stimulating factor plus tumor necrosis factor alpha: II. Functional analysis. *Blood* **90,** 1458–1470.

50. Pulendran, B., Banchereau, J., Burkholder, S., Kraus, E., Guinet, E., Chalouni, C., Caron, D., Maliszewski, C., Davoust, J., Fay, J., and Palucka, K. (2000). Flt3-ligand and granulocyte colony-stimulating factor mobilize distinct human dendritic cell subsets in vivo. *J. Immunol.* **165,** 566–572.

51. Palucka, K. A., Taquet, N., Sanchez-Chapui, F., and Gluckman, J. C. (1998). Dendritic cells as the terminal stage of monocyte differentiation. *J. Immunol.* **160,** 4587.

52. Ito, T., Inaba, M., Inaba, K., Toki, J., Sogo, S., Iguchi, T., Adachi, Y., Yamaguchi, K., Amakawa, R., Valladeau, J., Saeland, S., Fukuhara, S., and Ikehara, S. (1999). A CD1a+/CD11c+ subset of human blood dendritic cells is a direct precursor of Langerhans cells. *J. Immunol.* **163,** 1409–1419.

53. Cella, M., Jarrossay, D., Facchetti, F., Alebardi, O., Nakajima, H., Lanzavecchia, A., and Colonna, M. (1999). Plasmacytoid monocytes migrate to inflamed lymph nodes and produce large amounts of type I interferon. *Nat. Med.* **5,** 919–923.

54. Grouard, G., Rissoan, M. C., Filgueira, L., Durand, I., Banchereau, J., and Liu, Y. J. (1997). The enigmatic plasmacytoid T cells develop into dendritic cells with interleukin (IL)-3 and CD40-ligand. *J. Exp. Med.* **185,** 1101–1111.

55. Olweus, J., BitMansour, A., Warnke, R., Thompson, P. A., Carballido, J., Picker, L. J., and Lund-Johansen, F. (1997). Dendritic cell ontogeny: a human dendritic cell lineage of myeloid origin. *Proc. Natl. Acad. Sci. USA* **94,** 12551–12556.

56. Palucka, K. and Banchereau, J. (1999). Linking innate and adaptive immunity. *Nat. Med.* **5,** 868–870.

57. Rissoan, M. C., Soumelis, V., Kadowaki, N., Grouard, G., Briere, F., de Waal Malefyt, R., and Liu, Y. J. (1999). Reciprocal control of T helper cell and dendritic cell differentiation. *Science* **283** 1183–1186.

58. Rissoan, M. C., Soumelis, V., Kadowaki, N., Grouard, G., Briere, F., de Waal Malefyt, R., and Liu, Y. J. (1999). Reciprocal control of T helper cell and dendritic cell differentiation. *Science* **283**, 1183–1186.

59. Arpinati, M., Green, C. L., Heimfeld, S., Heuser, J. E., and Anasetti, C. (2000). Granulocyte-colony stimulating factor mobilizes T helper 2-inducing dendritic cells. *Blood* **95**, 2484–2490.

60. Maraskovsky, E., Daro, E., Roux, E., Teepe, M., Maliszewski, C. R., Hoek, J., Caron, D., Lebsack, M. E., and McKenna, H. J. (2000). *In vivo* generation of human dendritic cell subsets by Flt3 ligand. *Blood* **96**, 878–884.

61. Blom, B., Ho, S., Antonenko, S., and Liu, Y. J. (2000). Generation of IFN alpha producing pre-DC2 from human CD34+ hematopoietic stem cells. *J. Exp. Med.* **192**, 1785–1796.

62. Spits, H., Couwenberg, F., Bakker, A. Q., Weijer, K., and Uittenbogaart, C. H. (2000). Id2 and Id3 inhibit development of CD34+ stem cells into pre-DC2 but not into pre-DC1: evidence for a lymphoid origin of pre-DC2. *J. Exp. Med.* **192**, 1775–1784.

63. Tanaka, H., Demeure, C. E., Rubio, M., Delespesse, G., and Sarfati, M. (2000). Human monocyte-derived dendritic cells induce naive T cell differentiation into T helper cell type 2 (T$_h$2) or T$_h$1/T$_h$2 effectors. Role of stimulator/responder ratio. *J. Exp. Med.* **192**, 405–412.

64. Kalinski, P., Schuitemaker, J. H., Hilkens, C. M., Wierenga, E. A., and Kapsenberg, M. L. (1999). Final maturation of dendritic cells is associated with impaired responsiveness to IFN-gamma and to bacterial IL-12 inducers: decreased ability of mature dendritic cells to produce IL-12 during the interaction with Th cells. *J. Immunol.* **162**, 3231–3236.

65. Vieira, P. L., de Jong, E. C., Wierenga, E. A., Kapsenberg, M. L., and Kalinski, P. (2000). Development of T$_h$1-inducing capacity in myeloid dendritic cells requires environmental instruction. *J. Immunol.* **164**, 4507–4512.

66. Langenkamp, A., Messi, M., Lanzavecchia, A., and Sallusto, F. (2000). Kinetics of dendritic cell activation: impact on priming of T$_h$1, T$_h$2 and nonpolarized T cells. *Nat. Immunol.* **1**, 311–316.

67. Lanzavecchia, A., and Sallusto, F. (2000). Dynamics of T lymphocyte responses: intermediates, effectors, and memory cells. *Science* **290**, 92–97.

68. Parronchi, P., Mohapatra, S., Sampognaro, S., Giannarini, L., Wahn, U., Chong, P., Maggi, E., Renz, H., and Romagnani, S. (1996). Effects of interferon-alpha on cytokine profile, T cell receptor repertoire and peptide reactivity of human allergen-specific T cells. *Eur. J. Immunol.* **26** 697–703.

69. Mosmann, T. R. and Coffman, R. L. (1989). T$_h$1 and T$_h$2 cells: different patterns of lymphokine secretion lead to different functional properties. *Ann. Rev. Immunol.* **7**, 145–173.

70. Fukaura, H., Kent, S. C., Pietrusewicz, M. J., Khoury, S. J., Weiner, H. L., and Hafler, D. A. (1996). Induction of circulating myelin basic protein and proteolipid protein-specific transforming growth factor-beta1-secreting T$_h$3 T cells by oral administration of myelin in multiple sclerosis patients. *J. Clin. Invest.* **98**, 70–77.

71. Kitani, A., Chua, K., Nakamura, K., and Strober, W. (2000). Activated self-MHC-reactive T cells have the cytokine phenotype of T$_h$3/T regulatory cell 1 T cells. *J. Immunol.* **165** 691–702.

72. Groux, H., O'Garra, A., Bigler, M., Rouleau, M., Antonenko, S., de Vries, J. E., and Roncarolo, M. G. (1997). A CD4+ T-cell subset inhibits antigen-specific T-cell responses and prevents colitis. *Nature* **389**, 737–742.

73. Mukherji, B., Guha, A., Chakraborty, N. G., Sivanandham, M., Nashed, A. L., Sporn, J. R., and Ergin, M. T. (1989). Clonal analysis of cytotoxic and regulatory T cell responses against human melanoma. *J. Exp. Med.* **169**, 1961–1976.

74. Chakraborty, N. G., Li, L., Sporn, J. R., Kurtzman, S. H., Ergin, M. T., and Mukherji, B. (1999). Emergence of regulatory CD4+ T cell response to repetitive stimulation with antigen-presenting cells *in vitro*:

implications in designing antigen-presenting cell-based tumor vaccines. *J. Immunol.* **162**, 5576–5583.

75. Jonuleit, H., Schmitt, E., Schuler, G., Knop, J., and Enk, A. H. (2000). Induction of interleukin-10-producing, nonproliferating CD4(+) T cells with regulatory properties by repetitive stimulation with allogeneic immature human dendritic cells. *J. Exp. Med.* **192**, 1213–1222.

76. Enk, A. H., Jonuleit, H., Saloga, J., and Knop, J. (1997). Dendritic cells as mediators of tumor-induced tolerance in metastatic melanoma. *Int. J. Cancer* **73**, 309–316.

77. Steinbrink, K., Jonuleit, H., Muller, G., Schuler, G., Knop, J., and Enk, A. H. (1999). Interleukin-10-treated human dendritic cells induce a melanoma-antigen-specific anergy in CD8(+) T cells resulting in a failure to lyse tumor cells. *Blood* **93**, 1634–1642.

78. Dubois, B., Vanbervliet, B., Fayette, J., Massacrier, C., Van Kooten, C., Briere, F., Bancherau, J., and Caux, C. (1997). Dendritic cells enhance growth and differentiation of CD40-activated B lymphocytes. *J. Exp. Med.* **185**, 941–951.

79. Fayette, J., Dubois, B., Vandenabeele, S., Bridon, J. M., Vanbervliet, B., Durand, I., Bancherau, J., Caux, C., and Briere, F. (1997). Human dendritic cells skew isotype switching of CD40-activated naive B cells towards IgA1 and IgA2. *J. Exp. Med.* **185**, 1909–1918.

80. Liu, Y. J., Grouard, G., de Bouteiller, O., and Bancherau, J. (1996). Follicular dendritic cells and germinal centers. *Int. Rev. Cytol.* **166**, 139–179.

81. Chaplin, D. D., and Fu, Y. (1998). Cytokine regulation of secondary lymphoid organ development. *Curr. Opin. Immunol.* **10**, 289–297.

82. Wykes, M., Pombo, A., Jenkins, C., and MacPherson, G. G. (1998). Dendritic cells interact directly with naive B lymphocytes to transfer antigen and initiate class switching in a primary T-dependent response. *J. Immunol.* **161**, 1313–1319.

83. Grouard, G., Durand, I., Filgueira, L., Bancherau, J., and Liu, Y. J. (1996). Dendritic cells capable of stimulating T cells in germinal centres. *Nature* **384**, 364–367.

84. Szakal, A. K., Kosco, M. H., and Tew, J. G. (1989). Microanatomy of lymphoid tissue during humoral immune responses: structure function relationships. *Ann. Rev. Immunol.* **7**, 91–109.

85. Bennett, S. R., Carbone, F. R., Karamalis, F., Flavell, R. A., Miller, J. F., and Heath, W. R. (1998). Help for cytotoxic-T-cell responses is mediated by CD40 signalling. *Nature* **393**, 478–480.

86. Ridge, J. P., Di Rosa, F., and Matzinger, P. (1998). A conditioned dendritic cell can be a temporal bridge between a CD4+ T-helper and a T-killer cell. *Nature* **393**, 474–478.

87. Schoenberger, S. P., Toes, R. E., van der Voort, E. I., Offringa, R., and Melief, C. J. (1998). T-cell help for cytotoxic T lymphocytes is mediated by CD40–CD40L interactions. *Nature* **393**, 480–483.

88. Stuber, E., Neurath, M., Calderhead, D., Fell, H. P., and Strober, W. (1995). Cross-linking of OX40 ligand, a member of the TNF/NGF cytokine family, induces proliferation and differentiation in murine splenic B cells. *Immunity* **2**, 507–521.

89. Ohshima, Y., Tanaka, Y., Tozawa, H., Takahashi, Y., Maliszewski, C., and Delespesse, G. (1997). Expression and function of OX40 ligand on human dendritic cells. *J. Immunol.* **159**, 3838–3848.

90. Flynn, S., Toellner, K. M., Raykundalia, C., Goodall, M., and Lane, P. (1998). CD4 T cell cytokine differentiation: the B cell activation molecule, OX40 ligand, instructs CD4 T cells to express interleukin 4 and upregulates expression of the chemokine receptor, Blr-1. *J. Exp. Med.* **188**, 297–304.

91. Brocker, T. (1999). The role of dendritic cells in T cell selection and survival. *J. Leukoc. Biol.* **66**, 331–335.

92. Fernandez, N. C., Lozier, A., Flament, C., Ricciardi-Castagnoli, P., Bellet, D., Suter, M., Perricaudet, M., Tursz, T., Maraskovsky, E., and Zitvogel, L. (1999). Dendritic cells directly trigger NK cell functions: cross-talk relevant in innate anti-tumor immune responses *in vivo. Nat. Med.* **5**, 405–411.

93. Shah, P. D. (1987). Dendritic cells but not macrophages are targets for immune regulation by natural killer cells. *Cell. Immunol.* **104,** 440–445.

94. Geldhof, A. B., Moser, M., Lespagnard, L., Thielemans, K., and De Baetselier, P. (1998). Interleukin-12-activated natural killer cells recognize B7 costimulatory molecules on tumor cells and autologous dendritic cells. *Blood* **91,** 196–206.

95. Fearon, D. T., and Locksley, R. M. (1996). The instructive role of innate immunity in the acquired immune response. *Science* **272,** 50–53.

96. Matzinger, P. (1998). An innate sense of danger. *Semin. Immunol.* **10,** 399–415.

97. Sogn, J. A. (1998). Tumor immunology: the glass is half full. *Immunity* **9,** 757–763.

98. Allison, J. (2000). Cancer. *Curr. Opinion Immunol.* **12,** 569–570.

99. Bowne, W. B., Srinivasan, R., Wolchok, J. D., Hawkins, W. G., Blachere, N. E., Dyall, R., Lewis, J. J., and Houghton, A. N. (1999). Coupling and uncoupling of tumor immunity and autoimmunity. *J. Exp. Med.* **190,** 1717–1722.

100. Pardoll, D. M. (1999). Inducing autoimmune disease to treat cancer. *Proc. Natl. Acad. Sci. USA* **96,** 5340–5342.

101. Darnell, R. B. (1996). Onconeural antigens and the paraneoplastic neurologic disorders: at the intersection of cancer, immunity, and the brain. *Proc. Natl. Acad. Sci. USA* **93,** 4529–4536.

102. Albert, M. L., Darnell, J. C., Bender, A., Francisco, L. M., Bhardwaj, N., and Darnell, R. B. (1998). Tumor-specific killer cells in paraneoplastic cerebellar degeneration. *Nat. Med.* **4,** 1321–1324.

103. Boczkowski, D., Nair, S. K., Snyder, D., and Gilboa, E. (1996). Dendritic cells pulsed with RNA are potent antigen-presenting cells *in vitro* and *in vivo. J. Exp. Med.* **184,** 465–472.

104. Flamand, V., Sornasse, T., Thielemans, K., Demanet, C., Bakkus, M., Bazin, H., Tielemans, F., Leo, O., Urbain, J., and Moser, M. (1994). Murine dendritic cells pulsed *in vitro* with tumor antigen induce tumor resistance *in vivo. Eur. J. Immunol.* **24,** 605–610.

105. Mayordomo, J. I., Zorina, T., Storkus, W. J., Zitvogel, L., Celluzzi, C., Falo, L. D., Melief, C. J., Ildstad, S. T., Kast, W. M., Deleo, A. B., *et al.* (1995). Bone marrow-derived dendritic cells pulsed with synthetic tumour peptides elicit protective and therapeutic antitumour immunity. *Nat. Med.* **1,** 1297–1302.

106. Porgador, A. and Gilboa, E. (1995). Bone marrow-generated dendritic cells pulsed with a class I-restricted peptide are potent inducers of cytotoxic T lymphocytes. *J. Exp. Med.* **182,** 255–260.

107. Song, W., Kong, H. L., Carpenter, H., Torii, H., Granstein, R., Rafii, S., Moore, M. A., and Crystal, R. G. (1997). Dendritic cells genetically modified with an adenovirus vector encoding the cDNA for a model antigen induce protective and therapeutic antitumor immunity. *J. Exp. Med.* **186,** 1247–1256.

108. Specht, J. M., Wang, G., Do, M. T., Lam, J. S., Royal, R. E., Reeves, M. E., Rosenberg, S. A., and Hwu, P. (1997). Dendritic cells retrovirally transduced with a model antigen gene are therapeutically effective against established pulmonary metastases. *J. Exp. Med.* **186,** 1213–1221.

109. Toes, R. E., Blom, R. J., van der Voort, E., Offringa, R., Melief, C. J., and Kast, W. M. (1996). Protective antitumor immunity induced by immunization with completely allogeneic tumor cells. *Cancer Res.* **56,** 3782–3787.

110. Zitvogel, L., Regnault, A., Lozier, A., Wolfers, J., Flament, C., Tenza, D., Ricciardi-Castagnoli, P., Raposo, G., and Amigorena, S. (1998). Eradication of established murine tumors using a novel cell-free vaccine: dendritic cell-derived exosomes. *Nat. Med.* **4,** 594–600.

111. Shimizu, K., Fields, R. C., Giedlin, M., and Mule, J. J. (1999). Systemic administration of interleukin 2 enhances the therapeutic efficacy of dendritic cell-based tumor vaccines. *Proc. Natl. Acad. Sci. USA* **96,** 2268–2273.

112. Dhodapkar, M. V., Steinman, R. M., Sapp, M., Desai, H., Fossella, C., Krasovsky, J., Donahoe, S. M., Dunbar, P. R., Cerundolo, V., Nixon, D. F., and Bhardwaj, N. (1999). Rapid generation of broad T-cell immunity in humans after a single injection of mature dendritic cells. *J. Clin. Invest.* **104,** 173–180.

113. Dhodapkar, M. V., Krasovsky, J., Steinman, R. M., and Bhardwaj, N. (2000). Mature dendritic cells boost functionally superior CD8(+) T-cell in humans without foreign helper epitopes. *J. Clin. Invest.* **105,** R9–R14.

114. Hsu, F. J., Benike, C., Fagnoni, F., Liles, T. M., Czerwinski, D., Taidi, B., Engleman, E. G., and Levy, R. (1996). Vaccination of patients with B-cell lymphoma using autologous antigen-pulsed dendritic cells. *Nat. Med.* **2,** 52–58.

115. Mukherji, B., Chakraborty, N. G., Yamasaki, S., Okino, T., Yamase, H., Sporn, J. R., Kurtzman, S. K., Ergin, M. T., Ozols, J., Meehan, J., *et al.* (1995). Induction of antigen-specific cytolytic T cells in situ in human melanoma by immunization with synthetic peptide-pulsed autologous antigen presenting cells. *Proc. Natl. Acad. Sci. USA* **92,** 8078–8082.

116. Nestle, F. O., Alijagic, S., Gilliet, M., Sun, Y., Grabbe, S., Dummer, R., Burg, G., and Schadendorf, D. (1998). Vaccination of melanoma patients with peptide- or tumor lysate-pulsed dendritic cells. *Nat. Med.* **4,** 328–332.

117. Thurner, B., Haendle, I., Roder, C., Dieckmann, D., Keikavoussi, P., Jonuleit, H., Bender, A., Maczek, C., Schreiner, D., von den Driesch, P., Brocker, E. B., Steinman, R. M., Enk, A., Kampgen, E., and Schuler, G. (1999). Vaccination with mage-3A1 peptide-pulsed mature, monocyte-derived dendritic cells expands specific cytotoxic T cells and induces regression of some metastases in advanced stage IV melanoma. *J. Exp. Med.* **190,** 1669–1678.

118. Salgaller, M. L., Tjoa, B. A., Lodge, P. A., Ragde, H., Kenny, G., Boynton, A., and Murphy, G. P. (1998). Dendritic cell-based immunotherapy of prostate cancer. *Crit. Rev. Immunol.* **18,** 109–119.

119. Small, E. J., Fratesi, P., Reese, D. M., Strang, G., Laus, R., Peshwa, M. V., and Valone, F. H. (2000). Immunotherapy of hormone-refractory prostate cancer with antigen-loaded dendritic cells. *J. Clin. Oncol.* **18,** 3894–3903.

120. Banchereau, J., Palucka, K., Dhodapkar, M., Burkeholder, S., Taquet, N., Rolland, A., Taquet, S., Coquery, S., Wittkowski, K., Bhardwaj, N., Pineiro, L., Steiman, R., and Fay, J. (2001). Immunological and clinical responses to CD34+ hematopoietic progenitors-derived DC in patients with stage IV melanoma, *Cancer Res.* **61,** 6451–6458.

121. Kirk, C. J., and Mule, J. J. (2000). Gene-modified dendritic cells for use in tumor vaccines. *Hum. Gene. Ther.* **11,** 797–806.

122. Nair, S. K., Hull, S., Coleman, D., Gilboa, E., Lyerly, H. K., and Morse, M. A. (1999). Induction of carcinoembryonic antigen (CEA)-specific cytotoxic T-lymphocyte responses *in vitro* using autologous dendritic cells loaded with CEA peptide or CEA RNA in patients with metastatic malignancies expressing CEA. *Int. J. Cancer* **82,** 121–124.

123. Nair, S. K., Heiser, A., Boczkowski, D., Majumdar, A., Naoe, M., Lebkowski, J. S., Vieweg, J., and Gilboa, E. (2000). Induction of cytotoxic T cell responses and tumor immunity against unrelated tumors using telomerase reverse transcriptase RNA transfected dendritic cells. *Nat. Med.* **6,** 1011–1017.

124. Mach, N., and Dranoff, G. (2000). Cytokine-secreting tumor cell vaccines. *Curr. Opin. Immunol.* **12,** 571–575.

C H A P T E R

10

DNA and Dendritic Cell-Based Genetic Immunization Against Cancer

LISA H. BUTTERFIELD

Division of Surgical Oncology
UCLA Medical Center
University of California
Los Angeles, California 90095

ANTONI RIBAS

Division of Surgical Oncology, and
Division of Hematology/Oncology
UCLA Medical Center
University of California
Los Angeles, California 90095

JAMES S. ECONOMOU

Division of Surgical Oncology, and
Department of Immunology,
Microbiology, and Molecular Genetics
UCLA Medical Center
University of California
Los Angeles, California 90095

I. INTRODUCTION

Genetic immunization is a powerful technique in cancer gene therapy that utilizes DNA, usually in plasmid or virus form, to stimulate a tumor-specific immune response. The target of this immune response is a protein expressed in peptide epitope form on the surface of the tumor cells and derived from an intracellular protein. The tumor protein can be a normal self-protein or a mutated tumor-specific protein. With the identification and cloning of several human tumor rejection antigens and a better understanding of the molecular requirements for induction of T-cell immune responses and the use of dendritic cells, genetic immunization has become a promising new treatment for cancer.

II. BACKGROUND

A. Tumor-Antigen Targets

The identification of genes encoding cancer rejection antigens has allowed the development of new strategies for the immunotherapy of human cancer. In malignant melanoma, these antigens were identified by tumor-infiltrating lymphocytes (TIL) capable of recognizing shared major histocompatibility complex (MHC)-class-I-restricted epitopes present on tumors from different patients with the same HLA type [1,2]. The genes for the common melanocyte/melanoma lineage antigens MART-1/Melan-A (MART-1), gp100, tyrosinase, and tyrosinase-related protein (TRP-1) were cloned by cDNA library transfection and TIL screening [3–7]. The MAGE family of tumor antigens, with at least 14 genes, has been cloned and analyzed [8,9]. Their expression occurs in a wide variety of tumor types [10–12], while in normal tissues, MAGE expression appears to be

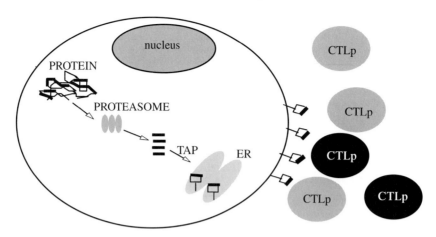

FIGURE 1 MHC class I peptide epitope processing. A cytosolic protein is digested by the proteosome into peptide fragments that are transported into the endoplasmic reticulum (ER) by TAP molecules. There, the peptides interact with newly synthesized class I molecules and are transported to the cell surface. The epitopes on the cell surface are available to interact with T-cell receptors of CTL precursors.

restricted to testes and perhaps early stages of wound healing [13]. Many MAGE epitopes have been identified and used to induce cytotoxic T lymphocytes (CTLs) from normal donors and melanoma patients, in both *in vitro* cultures and clinical trials [14–19]. For common and well-characterized class I alleles (e.g, A1, A2, A3, A24, A31), the immunodominant peptides for many melanoma tumor antigens have been defined by a variety of techniques, including screening of synthetic overlapping peptides and high-performance liquid chromatography (HPLC) fractionation of eluted peptides. Some of these immunogenic tumor-antigen epitopes are currently in clinical testing [19–28].

Many of the described melanoma-associated antigens are normal nonmutated differentiation antigens present in melanomas and normal melanocytes, which are present in the skin and retina. Any immune response generated to these antigens on tumor cells should also induce a response to the same antigens if they are present on normal tissues. Thus, the almost 29% incidence of vitiligo (patchy loss of skin pigmentation) in responding patients undergoing IL-2 immunotherapy is now believed to be the result of a cell-mediated immune response directed towards these normally expressed antigens on cutaneous melanocytes [29–32]. In different immunotherapy strategies tested in melanoma thus far, vitiligo appears to be the only manifestation of autoimmunity, while no vision impairment or other toxicities in responding patients have been observed. Interestingly, there has been one report of the involvement of melanoma-antigen-reactive CTLs in a melanocyte autoimmunity disorder, Vogt–Koyanagi–Harada disease [33].

B. MHC Restriction

In order to study the tumor antigens as they appear on the cell surface and are seen by T cells, analysis of the processed and presented peptides from these antigens has been under-

way for several years. The HLA-A2.1 genotype is the most common allele among Caucasians (approximately 50% are HLA-A2.1), and it is also well represented in other populations [34]. HLA-A2.1 has been crystallized and its binding groove well characterized [35], so the sequences of naturally processed and presented peptides from HLA-A2.1 have been most thoroughly analyzed. Many peptides have been sequenced from immunoprecipitated A2 peptide complexes from JY lymphoblastoid cells (HLA-A2.1 homozygous) [36]. Similarly, A2.1 peptide complexes have been purified and sequenced from lymphoblastoid cells transfected with A2.1 (Fig. 1).

These studies have defined the sequences of HLA-A2.1 binding peptides. Synthetic peptides have been used in binding competition studies [37–39], and various HLA-A2 subtypes have been analyzed [40,41]. This extensive body of work has defined important peptide motifs. A2.1 binds peptides from 8–10 amino acids in length, but primarily 9-mers. Amino acids L, I, and M are considered important anchor residues in peptide position 2, while V, L, and I are anchors in position 9 (or 10, depending on peptide length) [36,39,42,43]. To expand the number of patients that can be treated with peptide-based therapies and to analyze the specificity of T cells induced in treated patients, peptides presented by other HLA types have been identified [7,44–52].

A number of class I epitopes are derived from the protein sequence of tumor antigens that have been shown to bind to class I molecules due to the presence of either one or two of the optimal anchor residues and other important amino acids in their sequence. These epitopes, derived from intracellular proteins that are naturally processed and presented by the intracellular machinery, are digested by the proteasome to free candidate epitopes, transported to the endoplasmic reticulum (ER) antigen-processing via transporters, bound to nascent MHC molecules, and displayed on the cell surface. However, many epitopes that are displayed on the surface of a

TABLE 1 Immunogenic Epitopes for
Melanoma Antigens

Antigen	HLA type	Peptides
MART-1	A2	AAGIGILTV
gp100	A2	YLEPGPVTA
		KTWGQYWQV
		ITDQVPFSV
		LLDGTATLRL
	A3	ALLAVGATK
Tyrosinase	A2	MLLAVLYCL
		YMNGTMSQV
	A1	KCDICTDEY
	A24	AFLPWHRLF
	B44	SEIWRDIDF
MAGE-1	A1	EADPTGHSY
	Cw1601	SAYGEPRKL
MAGE-3	A1	EVDPIGHLY
	A2	FLYGPRALV
	B44	MEVDPIGHLY
TRP1(gp75)	A31	MSLQRQFLR
TRP-2	A2	SVYDFFVWL
	A31	LLGPGRPYR

tumor cell may not be sufficiently immunogenic to stimulate T cells without some change in the immunosuppressive tumor environment. Some examples of epitopes that are naturally processed and presented and have been demonstrated to be able to generate antigen-specific immune responses are provided in Table 1 [53].

C. Melanoma Peptides

The melanoma story has significantly revised our thinking about the human T-cell repertoire and immunological tolerance. It is clear that there are T-cell receptors capable of recognizing these 'self'-peptides and that these T cells have not been deleted from the immune system [54]. For MART-1 and gp100 immunodominant peptides, the HLA-A2 binding affinities are categorized as intermediate due to the absence of optimal residues at either the second or ninth anchor positions [55,56]. No high-affinity peptides from these two tumor antigens appear to serve as targets recognized by TILs. One hypothesis that has been forwarded is that high-affinity peptides, expressed at high levels on the cell surface, may generate tolerance. Lower binding or "subdominant" determinants may be capable of stimulating peptide-responsive T cells not

deleted from the T-cell repertoire [57]. In contrast, tyrosinase contains immunogenic epitopes restricted by HLA-A2 that contain both anchor residues. MAGE-3 also contains an immunogenic two-anchor epitope for A2. It is possible that a strong-binding, stable peptide is also required in an immunogenic epitope and that tolerance to such peptides could be broken if high enough antigen density could be attained in an immunostimulatory environment. In the case of lower affinity peptides, some have been shown to make up in stability, or slow off-kinetics, what they lack in initial binding capacity [58]. An informative example has been the melanoma antigen gp100. One of the HLA-A2.1-restricted peptides, gp100$_{209}$ ITDQVPFSV, has one anchor. In a series of $in vitro$ and $in vivo$ studies, the responses to a two-anchor residue containing version 209(2M) IMDQVPFSV, which retains the proper gp100 antigen specificity, has been shown to stimulate a stronger T-cell response than the native peptide. In another example, by studying the hepatocellular cancer (HCC) tumor antigen alpha-fetoprotein (AFP), we have found examples of both weak affinity/slow off-kinetics and strong binding/intermediate off-kinetics peptide epitopes [59–61].

These results have spawned investigations of 'self'-antigens expressed by other human cancers that might serve as suitable targets for immunotherapy (see Table 2). A number of these proteins, generally overexpressed by human cancer cells, have been identified, including carcinoembryonic antigen (CEA) in colorectal cancer [62], prostate-specific antigen (PSA) in prostate cancer [63], HER2/neu in breast cancer [64], AFP in hepatocellular carcinoma [59–61], mutated Ras in many cancers [65], $bcr–abl$ in chronic myelogenous leukemia (CML) (ref), and p53 (wt and mutant) in many cancers [66,67]. Both peptide-based and DNA-based genetic immunization clinical trials are in progress testing a number of these putative antigens.

1. Carcinoembryonic Antigen

Carcinoembryonic antigen (CEA) is expressed in the majority of colorectal, gastric, and pancreatic cancers and also in some breast cancer and non-small-cell lung cancer (NSCLC) (68). It is also expressed in some normal colon epithelium. Many different strategies have been investigated for therapy against CEA+ tumors, including generation of anti idiotype antibodies, vaccination with vaccinia and avipox virus containing the CEA cDNA, and vaccination with plasmid CEA cDNA and recombinant CEA protein, as well as use of the immunodominant HLA-A2-restricted CEA peptide.

TABLE 2 Types of Tumor Antigens

Normal differentiation antigens	MART-1/ Melan-A	gp100	Tyrosinase	TRP-1	TRP-2
Tumor-specific antigens	MAGE family (1–14)	BAGE	GAGE		
Overexpressed proteins	HER2/neu	CEA	PSA	p53	AFP
Mutated oncogenes	p21 ras	p53	$bcr–abl$		

In a clinical trial where patients received vaccinations with recombinant vaccinia–CEA, an increase in CEA reactive T cells after treatment was found [62]. In this trial, *in vitro* restimulation of T cells from patients could generate CEA peptide (CAP-1)/HLA-A2-specific killing of CEA+/A2+ tumor cells. Since then, other trials have been conducted in which patients have been treated with autologous dendritic cells pulsed with CEA peptide or RNA [193], vaccinia–CEA [69], and avipox–CEA vaccines [70], all with CEA-specific immunological responses. These results indicate that CEA is a very promising target for a variety of potential genetic immunization strategies.

2. Prostate-Specific Antigen

Prostate-specific antigen (PSA) and prostate-specific membrane antigen (PSMA) are overexpressed in many prostate tumors. Use of PSA as a tumor antigen has been reported in a murine model [63] in which vaccination of mice with a human PSA-transfected tumor led to PSA-specific killing by the T cells generated *in vivo*. In humans, normal donor blood could be used *in vitro* to generate PSA peptide-specific killing of peptide-pulsed HLA-A2+/PSA+ tumors, but not peptide pulsed, non-A2 tumors [71]. A series of PSMA peptide-based phase I and II trials have been performed using HLA-A2-restricted peptides, a variable number of dendritic cells, or both in prostate cancer patients who were either HLA-A2+ or non-A2 [72–74]. PSA serum levels, delayed-type hypersensitivity (DTH), and T-cell proliferation were followed. While some PSA level decreases were seen, non-antigen-specific immune status seemed to best correlate with clinical response.

3. HER2/*neu*

HER2/*neu* is over expressed by a subset of both breast and ovarian tumors. *In vitro*, CTLs have been generated from both breast and ovarian cancer patients which cross-react with common HLA-A2-restricted peptides. Some of these peptides have been HLPC fractionated, purified, and partially sequenced [64,75]. This has led to the identification of several potential peptide targets restricted by HLA-A2 for HER2/*neu*. An *in vivo* study in which rats were immunized with either rat Neu peptides or the whole Neu protein [76] demonstrated that tolerance to rat HER2/*neu* could be broken to this self-protein with peptide-based immunization but not whole protein. This result also supports the notion that, for some proteins, tolerance to an immunodominant peptide can mask an immune reaction to a subdominant epitope unless that subdominant epitope is separated from the whole protein before intracellular processing. (This was shown for the model antigen hen egg lysozyme [77].) DNA-based immunization has also been used in a mouse model with good antitumor effects observed [78]. More recently, a clinical trial

has been performed with HER2/*neu* peptides delivered intradermally with granulocyte–macrophage colony-stimulating factor (GM-CSF) as adjuvant to breast and ovarian cancer patients [76]. Most of these patients developed HER2/*neu* antigen-specific T-cell proliferative responses.

4. Alpha Fetoprotein

Alpha-fetoprotein (AFP) is the most abundant serum protein before birth, and the levels decrease to very low but detectable levels after birth. AFP is reactivated by approximately 80% of HCC. Like PSA, levels of serum AFP are an important diagnostic tool for detection of HCC. Extensive epitope mapping has been performed which has identified four immunodominant and ten subdominant epitopes restricted by HLA-A2.1 [61]. We have found that T-cell responses to AFP can be generated in both murine *in vivo* [60] and human *in vitro* [59] systems. This work has shown that AFP peptides are processed and presented by the cellular machinery and that AFP antigen-specific effector T cells can be expanded. A peptide-based clinical trial has been initiated to test the safety, toxicity and ability of A2.1+/AFP+ HCC patients to generate T-cell responses to the peptides *in vivo*.

5. Ras

The mutated *ras* oncogene has been detected in a wide variety of tumors. The fact that three common mutations in the Ras protein create the constitutively active form of p21 Ras makes it a particularly attractive target for genetic immunization. This should preclude any concerns about autoimmunity and tolerance because the immune reaction would be directed toward a tumor-specific mutated epitope. To address the feasibility of using mutated *ras* as a target, CTLs were initially raised in mice against the human Ras-12 mutation [65]. *In vitro*, Ras-61 mutant-specific human CTLs have been raised that are HLA-A2 restricted, but none of the CTLs have exhibited mutant Ras+ tumor cell killing [80]. A problematic finding has been that the precursor frequency of CTLs against mutant *ras* is extremely low in the donors analyzed thus far. A CD4+ T-cell clone was isolated from a gastric cancer patient who exhibited a proliferative response and specific cytokine release in response to mutated *ras* epitopes or the mutant protein [81]. For more details on *ras* strategies, see Chapter 8.

6. p53

The most commonly mutated (and often overexpressed) gene in all cancer appears to be the tumor suppressor gene p53. Unlike *ras*, p53 has a wide range of mutational hot spots, located over 11 exons. This makes peptide epitope-based strategies more difficult to conceive of. To begin the study of using p53 as a tumor target, peptides from the wild-type p53 sequence were screened for binding to class I molecules [82]. Several epitopes were found that were sufficiently immunogenic to generate peptide-specific CTLs *in vitro*. Peptides (also from the wild-type p53 sequence) that bind HLA-A2 were used to generate CTLs from a normal donor which could lyse an allogeneic mutant p53+ tumor line in an HLA-A2-restricted fashion [83], which indicates further that tolerance can be broken to the normal self-protein p53. It is somewhat troubling that the CTLs generated from normal p53 epitopes can kill mutant p53 tumor cells. Because expression of normal p53 is essential to the normal growth control function of most cells in the body, the issue of autoimmunity against normally expressed p53 is raised. Perhaps the short half-life of normal p53 will protect nonmalignant cells from CTL killing. In addition, it has been proposed that wt p53 epitopes would have to be used due to the wide potential mutation spectrum present across a large patient population. Some recent studies [84,85] utilize HLA-A2 transgenic mice to generate high-affinity CTL against human p53 epitopes, trying to circumvent tolerance [85]. More recently, using normal p53 peptide-pulsed dendritic cells, human p53-specific CTL were generated *in vitro* [86] from normal donor peripheral blood mononuclear cells (PBMCs), further indicating that breaking tolerance is possible. For more details on p53, see Chapter 27.

In addition to "shared" tumor antigens, "private" antigens from abberantly expressed or newly mutated proteins have been identified in individual tumors, for example β-catenin [87], MUM-1 [88], and MUM-2 [89]. A strength of utilizing the large number of nonmutated shared antigens is that tumor-specific therapies can be developed with them that apply to a large number of patients. Conversely, a strength of targeting an individual patient's specific mutations in the tumor is that such private antigens are completely unique to the tumor, and an immune response directed only toward them would not have the potential to generate an autoimmune response.

While attempts to target private antigens are underway, use of such antigens requires either a great deal of work to identify them for every patient to be treated or the use of tumor lysates or heat shock protein gp96 mixtures of unidentified proteins and peptides [90,91], which necessarily include normal self-proteins.

III. RECENT ADVANCES: METHODS OF GENETIC IMMUNIZATION

A powerful means of inducing both cellular and humoral immunity is the expression of tumor antigen-encoding DNA sequences in host antigen-presenting cells (APCs). Two methods, DNA immunization and dendritic cell pulsing or transduction, appear to be effective in inducing antitumor immunity (see Fig. 2).

A. Plasmid Immunization

Antigen-encoding DNA plasmids administered intramuscularly (i.m.) or intradermally (i.d.) can induce cellular and humoral immune responses to pathogenic viruses, parasites, bacteria, and tumors [92–94]. The primary advantages of using DNA as the immunogen are purity, ease of production, stability of a DNA vaccine, and the long-term immunity that can be generated. The preferential stimulation of CD8$^+$ T cells and T$_h$1-based responses [95] is another benefit that is often observed with i.m. injection of DNA. In an early study using commercially available plasmids containing the human CEA gene [96], vaccinated mice were able to generate anti-CEA antibodies as well as T cells that proliferated and secreted IL-2 and IL-4 in response to CEA. In a follow-up study [97], improvements made to the plasmid constructs

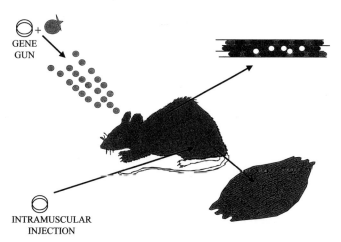

FIGURE 2 DNA immunization methods. Plasmid DNA can be coated onto the surface of gold beads and pulsed into the skin layers by helium gas. Once in the skin cells, the DNA dissociates from the gold. Alternatively, plasmid DNA can be injected into muscle cells, where it is taken up by both muscle cells and local APCs.

included the addition of the B7-1 costimulatory molecule to increase antigen presentation by the transfected muscle cells (which had to be on the same DNA construct as the CEA antigen gene to obtain optimal results). A second improvement was injecting a GM-CSF-containing plasmid 3 days before treatment with the CEA plasmid in order to increase potential APC numbers in the injected muscle environment, which would better present the CEA molecule to the immune system. These changes led to increased T-cell responses and slowed the growth of a CEA+ tumor in the mice. These studies demonstrate not only the feasibility of plasmid immunization, but also the importance of stimulating APCs with either antigen plus B7-1 costimulation in a single transfected cell or drawing professional APCs to the vaccination site to optimize antigen presentation.

The plasmids employed in these studies have been found to contain immune-system-stimulating sequences in the plasmid backbone. The hypomethylated state of CpG motifs in plasmids grown in bacteria can yield a nonspecific inflammatory response to the plasmid itself. This can be important for the overall immune response to the specific antigen encoded by the plasmid by increasing the number and activation state of the cells drawn to the vaccination site. In another sense, the bacterially derived DNA is seen as a "danger" signal by the immune system, initiating innate immunity. The general sequence motif Pu–Pu–CpG–Py–Py has been shown to be an important context for CpGs [98–103]. The immunogenicity of these unmethylated CpGs may be due to their ability to activate NFκB and to stimulate IL-12, TNF-α, and IFN-α/β production by monocytes and macrophages, among other effects. In addition, CpG dinucleotides that occur in inhibitory contexts have also been identified. In order to make plasmid backbones that can be used *in vivo* for human genetic immunization therapies, many avenues to maximize both safety and efficacy have been investigated. These include replacement of the ampicillin resistance gene for bacterial propagation with the kanamycin resistance gene (to protect against a possible penicillin allergic response, a feature recommended by the Food and Drug Administration "Points to Consider" for DNA-based vaccines). To increase expression of the antigen gene, promoters, enhancers, intron sequences for optimal processing, termination, and poly-adenylation signals continue to be thoroughly investigated and optimized [104].

A number of reports present persuasive evidence that coexpression of cytokines or costimulatory molecules or injection of a DNA vaccine into injured/regenerating muscle will enhance T-cell responses to tumor antigen plasmid immunization [105–109]. Injection of DNA into regenerating skeletal muscle following injection of agents such as snake venom toxins, bupivacaine, or edematous muscle (after an injection of a hypertonic 25% sucrose solution) [110] may improve expression greater than 80-fold in some models.

Irvine *et al.* [107] have shown that intraperitoneal treatment with recombinant cytokines IL-2, IL-6, IL-7, and IL-12 enhanced antitumoral responses after particle-mediated gene delivery to the epidermis. Geissler *et al.* [106] have further shown that i.m. coimmunization with IL-2, IL-4, and GM-CSF DNA expression constructs enhances the response elicited by a plasmid encoding a viral epitope. Insertion of certain cytokine genes into the same expression cassette as the antigen-encoding DNA plasmid might attract APCs into the site of DNA delivery. As mentioned, this was shown by Conry *et al.* [97] to be an optimal strategy to enhance the immune response to CEA using an i.m. injection model. In our murine model of AFP DNA immunotherapy, coinjection of plasmids encoding hIL-2, hIL-7, mIL-12, mB7-1, or mB7-2 with AFP plasmid failed to improve protection [60] or peptide-specific responses in A2.1/K^b transgenic mice (Meng *et al.*, in press). Improved protection was also not seen in bupivicaine-injured muscle [60]. Although these APC-differentiating and Th1-skewing strategies are conceptually attractive, they have not been successful in our model.

Several reports have described the improvement in antitumor responses when tumor antigen delivery is coupled to expression of GM-CSF [132,134,136]. The hypothesis is that GM-CSF recruits additional APCs to the site of antigen which results in increased T-cell stimulation. We have recently found that plasmid mGM-CSF delivery is capable of increasing the frequency of AFP-specific T cells compared to plasmid AFP immunization alone (Meng *et al.*, in press) in A2.1/K^b mice by ELISPOT. To date, GM-CSF and IL-12 strategies are the most investigated cytokines.

Optimal processing of antigens can exert an important effect upon the presentation of peptide epitopes. The rate of antigen processing, which can lead to enhanced presentation of epitopes [111,112], can be regulated at the step of ubiquitinylation. Ubiquitinylation is an ATP-dependent process in which the small protein ubiquitin is conjugated to lysine residues in a protein. The conjugation of ubiquitin to cellular proteins targets the protein for degradation by the proteasome and subsequent generation of immunogenic peptide epitopes. The rate of ubiquitinylation is dependent upon both the N-terminal aa (stabilizing or destabilizing, the "N-end rule") and accessibility of the target protein lysine residues. Strategies in which the tumor antigen is conjugated to a ubiquitin moiety are under investigation.

Recent work [114,115] has demonstrated improved immunogenicity of DNA vectors based on the alpha viruses Semliki Forest and Sinbis. Vectors based on alpha viruses encode their own RNA replicase, which is an autoproteolytic poly-protein that generates the four required nonstructural proteins for rapid, high-level RNA replication. These vectors have been shown with a model tumor antigen (β-galactosidase) to yield considerably higher levels of mRNA in transfected cells. This increase in RNA does not necessarily lead to an increase in the amount of encoded protein in the cells [114,115]. The hypothesis is that the increased immunogenicity with these vectors may be related to the inherent immunogenicity of the RNA replication and the dsRNA template molecules, dsRNA is a "danger signal" which induces interferon production by alpha-virus-infected cells.

The mechanism of immune response induction by genetic immunization, particularly MHC-class-I-restricted CTL, is still being defined. Gene expression can be detected for many months in myocytes, but it is unlikely that these transfected cells can serve as the primary APCs. Because the epidermis is a tissue known to contain professional APCs and is easily accessible, this tissue has also been investigated as a route for genetic immunization. Several studies have addressed the mechanism of the immune response generated by two plasmid DNA-based methods of genetic immunization. Many of these reports implicate bone-marrow-derived professional APCs in the induction of protective immunity, whether the DNA immunization was via i.m. injection or gene gun to the epidermis [105,115–118]. While these mechanistic studies indicate that the most important cells are professional APCs, other studies using transfection of myocytes with both antigens and the costimulatory molecules B7-1 and/or B7-2 [119–121] support the notion that direct antigen presentation by nonprofessional APCs does occur and can be made more efficient by giving a "second signal" of costimulation. This second signal is most often B7-1 in experimental systems, but B7-2 and ICAM-1 are also being investigated.

Extensive use has been made of the helium-driven 'gene gun', primarily for epidermal delivery of plasmid DNA coated onto gold beads to an organism, as well as for *in vitro* transfection of cells (for more details on this subject, see Chapter 13). For *in vivo* use, this method is based on the observation that the skin is a rich source of the APCs known as Langerhans cells (LCs). Therefore, direct transfection of LCs using the gene gun could yield APCs that would be more efficient than muscle cells at generating an immune response to the tumor antigen of interest. In one recent study, the gene gun was used to deliver either mutant p53 or HIV gp120 epitopes to murine ear epidermis [122]. This strategy utilized the adenovirus E3 gene leader sequence to target the antigen epitope to the endoplasmic reticulum for potentially more efficient presentation by class I molecules. These investigators were able to observe epitope-specific T-cell responses to the gene-gun-delivered epitopes. Use of surrogate tumor antigens such as β-galactosidase (β-gal) in gene-gun-based experiments has also generated clearly defined antigen-specific responses. Irvine *et al.* [107] showed that immunization with a β-gal plasmid yielded both cytotoxic and antibody responses against the antigen, as well as slowed tumor growth of a β-gal-expressing tumor line. This response was improved with

the addition of recombinant cytokines. Another observation has been a type 2 cytokine bias in some models using gene-gun administration of DNA for vaccination [123].

Another method of directly delivering tumor antigen genes to an organism is systemic delivery by vaccinia or adenovirus (AdV). In a phase I trial at the Surgery Branch of the National Cancer Institute (NCI), AdV type 2 viruses encoding MART-1 or gp100 melanoma antigens with or with out systemic IL-2 were tested [24]. Due to preexisting immunity to AdV from normal environmental exposure, there was initial concern over immediate neutralization of the vector by serum antibodies. One complete response was observed in the AdVMART1 virus-only group and other objective responses were observed, but interpretation is difficult in patients treated with IL-2 due to the antimelanoma effects of IL-2 alone. Five of 23 patients showed some MART-1 specificity in IFN-γ synthesis of PBMC post-treatment, but overall the results were not strong enough for further trials at this time.

The sequential combination of naked DNA immunization followed by boosting with viral vectors expressing the same antigen has resulted in generation of high levels of specific immunity. This strategy, referred to as "prime-boost," seems to have a greater effect than either type of immunization protocol alone. It is believed that this is due, in part, to the ability of DNA vaccines to stimulate T cells of great affinity and generate a long-lived antigen deposit, while the subsequent viral immunization is a more powerful stimulus for T-cell expansion than repeated DNA vaccinations [124,125].

B. Dendritic Cells

Dendritic cells (DCs) are among the most potent APCs described [126,127]. DCs are at least 100× more potent in presenting antigen than the monocyte/macrophage [126,127]. They are bone-marrow derived and characterized by dendritic morphology (elongated, stellate processes), high mobility, and the ability to present antigen to naive T cells and to stimulate primary T-cell responses in an MHC-restricted fashion [128,129]. DCs express high levels of MHC class I and II molecules; costimulatory molecules B7-1, B7-2, CD40 receptor, and CD1a; and adhesion molecules ICAM-1 (CD54), LFA-3 (CD58), and CD11a, b, and c [127]. DCs do not express classical T, B, NK, or monocyte/macrophage lineage markers. DCs are present in low numbers in the circulation and peripheral tissues but can migrate from tissues to lymphoid organs and stimulate antigen-reactive T cells [130,131].

Dendritic Cells acquire antigen in at least three major ways: (1) through constitutive macropinocytosis, in which many soluable antigens can be taken up in the fluid phase; (2) via receptor-mediated endocytosis with the mannose receptor (in which the yeast cell wall constituent zymosan has been shown to be taken up) and FcγII (for antigen–antibody complex uptake); and (3) by utilizing phagocytosis, for example, to take up either bacteria or latex beads [132,133].

Once an antigen is taken up, DCs process intracellular proteins within the proteosome complex into 8–11 amino acid peptides and transport these to the endoplasmic reticulum by transporter molecules TAP-1 and TAP-2 (Fig. 1). There they associate within synthesized MHC class I molecules, and this complex is expressed on the cell surface [134]. The class I pathway has recently been shown to be accessible by the nonclassical endocytic pathway via macropinocytosis in addition to the classical cytoplasmic endogenous antigen processing pathway [135]. Processing for class II expression is also very efficient [136–138]. For more details on DC biology, see Chapter 9.

It is not difficult to prepare large numbers of DCs from murine and human myeloid progenitors *in vitro*. This has made possible the explosion in DC research as well as the potential to treat patients with therapeutic doses of autologous DCs. DCs may be differentiated from loosely adherent peripheral blood mononuclear leukocytes in 7-day cultures in the presence of GM-CSF and IL-4 [139]. CD34 progenitor cells, cultured in stem-cell factor (SCF), GM-CSF, and TNF-α, likewise will yield enriched DC populations in humans [140]. In mice, DCs can be differentiated from bone marrow precursors in a 7-day culture with GM-CSF and IL-4 [141]. Lymphoid-derived "DC2" have also been reported [142] but will not be discussed further here.

Another method of obtaining DCs is harvesting them *in vivo* [143]. In humans, this can be achieved using cell-separation machines that are able to collect circulating DCs (usually less than 0.5% of blood cells). Although yields are still limiting, some of these machines have become efficient enough to generate DCs for clinical testing [144]. This approach may become more feasible if the circulating number of DCs can be increased. In mice and humans, numbers of DCs can be dramatically increased with *in vivo* administration of Flt3 ligand [145,146] or GM-CSF and IL-4 delivered systemically [147].

C. Delivering Tumor Antigen Genes, Proteins, and Peptides to Dendritic Cells

There are four major methods for utilizing DCs in genetic immunization strategies. First, pulsing the DCs with tumor cell extracts containing a complex mixture of known and unknown tumor antigens and proteins or pulsing with a purified, recombinant protein for a known tumor antigen allows the DCs to endocytose, process, and present any number of epitopes in an MHC-restricted fashion, without requiring a specific MHC type for the patient. Second, DCs can be pulsed with mild-acid-eluted tumor peptides containing a mixture of known and unknown tumor epitopes or with a high concentration of a purified, synthetic peptide epitope. This requires either eluting peptides off of the

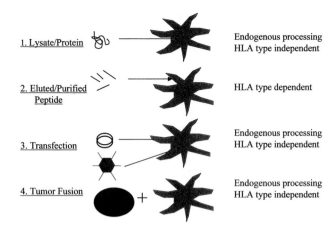

FIGURE 3 Dendritic cell-based strategies. Four methods of utilizing DCs to present antigenic epitopes are shown. (1) Tumor cell lysates or purified proteins are added to DCs, which take up, process, and present the peptide epitopes; (2) mild-acid-eluted peptides or purified, synthetic peptides are pulsed onto the surface of the DCs for direct presentation; (3) plasmids or viruses can be used to transfect DCs with entire tumor antigen genes, leading to protein synthesis, processing, and presentation by the DCs; and (4) DCs can be fused to tumor cells to allow the cellular proteins of the tumor cell to be processed and presented by the DCs. Only the use of eluted or synthetic peptides depends upon a specific HLA type for correct presentation by the DCs.

same patient's tumor or having a human leukocyte antigen (HLA)-matched tumor. Employing the synthetic peptide approach requires knowing the tumor antigen and the immunogenic epitope for the HLA type of each patient. Because a number of shared tumor antigen epitopes have been identified for common HLA types, use of synthetic peptides is a promising strategy with DCs. Third, DCs can be pulsed or transfected with mRNA or cDNA via lipids or viruses or without carriers. The mRNA or cDNA can be either synthetic from a specific tumor antigen or from the patient's tumor, and this leads to multiple class I and class II epitopes being presented (see also Chapter 11). Finally, DCs can also be fused with tumor directly, allowing the DC antigen-presentation machinery to present known and unknown tumor antigens from the tumor. Each of these strategies has been demonstrated to be feasible experimentally and all are currently in clinical testing. Examples of each strategy are diagrammed in Fig. 3.

D. Tumor Lysates

To investigate the use of unfractionated tumor extracts as an immunogen, Nair and colleagues compared DCs and macrophages as APCs after pulsing each with poorly immunogenic murine tumor lysates [148]. They were able to demonstrate the presence of tumor-specific CTLs after as little as one immunization. In addition, they observed increased survival of tumor-challenged mice. This demonstrates how effective a complex, unpurified mix of normal and tumor antigens can be when used in conjunction with DCs. Whole, purified protein antigens used in the context of DC-based immunization include the model tumor antigen β-gal [149,150]. These studies used murine DCs loaded *in vitro* with soluable β-gal protein as a vaccine against stably transfected β-gal-expressing tumors. CTLs were generated specific to β-gal antigen, and tumor immunity resulted.

E. Peptides

Dendritic Cells can be pulsed with peptide antigens, bypassing the need for processing. DCs pulsed with peptide of genuine or surrogate tumor antigens can induce protective immunity, even to established tumors. For example, Mayordomo *et al.* [150,151] investigated three different tumor antigen systems; in one, they pulsed murine DCs with an ovalbumin (OVA) peptide that protected mice against a lethal challenge of OVA-transfected tumor cells. MUT-1 and an E7 human papillomavirus (HPV) viral peptide were also used successfully, and the effects were dependent upon CD8^{+} cells. Using mutant p53 peptide-pulsed DCs as a vaccine, Gabrilovich and coworkers [152] were able to protect mice from a lethal challenge of mutant p53-transfected tumor line. This effect was better than that seen with mice treated with systemic recombinant IL-12 but was weaker than the effect of administering both mutant p53-peptide-pulsed DCs and systemic IL-12. This provides another example of the value of vaccination with a tumor antigen or its epitope at the same time as adjuvant cytokine treatment. These initial studies have been sufficiently substantiated to allow clinical trials to begin.

F. Acid-Eluted Peptides

In the setting in which the tumor antigens or peptides are unknown, several groups have described the antitumoral effect of DCs pulsed with peptides obtained by mild acid treatment of immunogenic tumor cells. Zitvogel and coworkers [153] have shown that peptides eluted from three weakly immunogenic tumor lines (MCA205 sarcoma, CL8.1 melanoma, and TS/A mammary carcinoma) induced transient tumor stasis when pulsed onto cultured syngeneic DCs. Only immunization with eluted peptides derived from the more immunogenic C3 tumor elicited true tumor regressions. Nair *et al.* [148,154] used a similar protocol to elute peptides from the OVA-transfected E.G7–OVA cell line, the EL-4 parental thymoma cell line, and the F10.9 clone of the B6 melanoma. The eluted peptides were pulsed onto RMA-S cells (a TAP peptide-transporter-defective line) or onto APC purified from adherent splenocytes. They observed a significant reduction in lung metastases when immunizing with peptides eluted from the OVA-transfected cell line and the F10.9 clone, but not with peptides derived from the EL-4 thymoma. In our experience, only eluted peptide-pulsed

DCs could effectively immunize mice against immunogenic tumors; no protection was observed when using peptides from the non immunogenic NFSA tumor [155]. Genetically engineering the peptide-pulsed DC to produce IL-2 was not able to elicit a greater protective in the immunogenic tumors, nor was it able to alter the non-immunogenicity of peptides eluted from NFSA tumors.

One of the potential drawbacks of this approach, compared to the use of defined peptide epitopes, is a reduced efficiency due to the low concentration of the effective tumor antigens in the eluted peptide mixture. However, it has been shown in the OVA system that peptides eluted from an OVA-transfected cell line are as effective as purified, synthetic OVA peptide in generating OVA-specific, class-I-restricted CTLs [148,154]. Another potential drawback is the possibility of loading the most powerful APCs known with autologous peptides to which an autoimmune response could be generated. However, we and others [148,153,154] have not observed any adverse side effects in treated mice that might suggest an autoimmune response.

G. Dendritic Cell Transduction and Transfection

We have investigated many potential methods of DC transfection [156]. While we could not detect transgene expression with any physical method of gene transfer utilized (calcium phosphate, electroporation, lipofection), we found that adenovirus was a particularly easy, reliable, and (most importantly) highly efficient vehicle for transgene expression in human and murine DCs. Adenoviral vectors can transduce up to 100% of the DCs in a population by β-galactosidase transduction and staining, and, when transduced with IL-2 or IL-7 AdV vectors, DCs synthesize up to nanogram amounts of cytokine per milliliter of culture medium per 24 hours. Others have also found adenovirus to be an easy and efficient vector for DCs, [157–159] although high multiplicities of infection (mois) are required.

Lipofection has been used to transfect DCs with antigen genes [160], although with this less efficient gene transfer method direct demonstration of gene expression can be difficult to come by. Perhaps by using "nature's adjuvant," the DC, a small amount of antigen transgene expression may be sufficient to stimulate T cells. Indeed, Alijagic and coworkers were able to demonstate antigen-specific T-cell clustering around only tumor antigen-transfected DCs, not chloramphenicol acetyl transferase (CAT)-transfected DC.

Other viruses have also been used to transduce DCs efficiently. Human DCs have been transduced with retroviruses carrying the β-gal reporter gene [161]. With three cycles of viral infection, 35–67% of 7-day-cultured DCs were positive for β-gal, and expression lasted for at least 20 days as shown by polymerase chain reaction (PCR). The genomic PCR demonstrates that the gene was incorporated into the genomic DNA of the DCs. A retrovirus encoding the melanoma tumor antigen MART-1 has been used to transduce CD34+ hematopoietic progenitors [162]. Transduced cells were then differentiated *in vitro* into DCs. In addition, recombinant vaccinia viruses encoding β-gal [163], MART-1 [164], or gp100 [165] have been used to transduce murine and human DCs, which were able to stimulate antigen-specific CTL *in vitro*. Recently, recombinant bacteria have also been used as vehicles for antigen delivery to DCs [166].

We [59,61,167–172] and others [162,165] have demonstrated that both murine and human DCs, genetically engineered to express an entire tumor antigen protein, could induce immunity by processing and presenting relevant immunodominant peptides. Transduction of the OVA gene into murine DCs using replication-defective adenoviral vectors protects against the challenge with an OVA-expressing tumor cell line. This antitumoral protection has been shown to be superior to the one elicited by direct injection of the adenoviral vector alone. While there was initial concern that use of adenovirus vectors might induce an overwhelming adenovirus antigen response that could mask the desired tumor antigen response, work by ourselves and others has shown that this is not an insurmountable problem.

We have developed murine models of genetic immunotherapy using adenoviruses that encode the human melanoma antigen MART-1 (AdVMART1) or the HCC antigen AFP (AdVmAFP). In these models, vaccination of mice with adenovirus-transduced DCs gave antitumor protection superior to systemic injection of the adenovirus alone or intramuscular injection of naked antigen-encoding DNA (Fig. 4).

This MART-1 animal model has provided us with a valuable tool to understand the biology of DC-based genetic immunotherapy. We have made the following observations: (1) induction of immunity requires CD8 and CD4 T cells; (2) antigen-specific, perforin-dependent, CTL- and IFN-γ-producing T cells are generated; (3) CD4 help can be replaced by ligation of CD40 on DCs; (4) CD28-mediated costimulation is required for the generation of antigen-specific T cells; (5) multiple immunizations in certain mouse strains generate progressively lower immunity and a T_h1 to T_h2 shift due to FasL/Fas-dependent, activation-induced cell death; and (6) measures that do *not* appear to improve the level of protective immunity include transduction of DC with IL-2, IL-7, GM-CSF, or IL-12 or blocking IL-4 and IL-10 in recipient mice. In other murine tumor models, IL-12 and IL-7 have improved the effects of tumor-antigen-expressing DCs [173,174].

We have performed a series of preclinical *in vitro* studies to test the utility of our AdVMART1 and AdVhAFP vectors in a human system (Fig. 5). First, transduction with these adenoviruses resulted in tumor-associated antigen (TAA) mRNA transcription in antigen-negative cells, and increas-

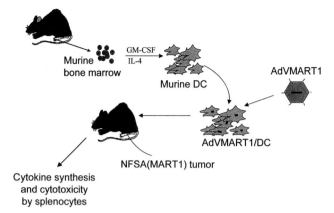

FIGURE 4 Murine model using DCs transduced with AdVMART1 or AdVmAFP. Mice are vaccinated with murine bone-marrow-derived DCs transduced with the MART-1- or AFP-expressing adenovirus. Splenocytes from the vaccinated mice are harvested and restimulated *in vitro* with NFSA cells stably transfected with MART-1, or 'NFSA(MART1)' (in a C3H model), or EL4(MART) or EL4(mAFP) in a BL/6 model. These splenocytes from AdV/DC-vaccinated animals synthesize T_h1 cytokines IL-2, IFN-γ, and TNF-α, in contrast to little or no synthesis from animals vaccinated with nothing, DC only, or DC transduced with an irrelevant adenovirus (RR5), as shown by RT-PCR and ELISPOT. In addition, the splenocytes from AdV/DC-vaccinated animals specifically lyse tumor-antigen-expressing targets, while control animal splenocytes do not.

ing the multiplicity of infection (MOI) correlates with increasing amounts of TAA mRNA. A time-course analysis demonstrates that this mRNA synthesis continues for at least 8 days *in vitro*. Second, TAA protein is made in transduced cells, and the immunodominant HLA-A2.1-restricted MART-1$_{27-35}$ peptide is correctly processed and presented, as are the four immunodominant AFP-derived peptides (as shown by the ability of AdV–TAA-transduced cells to become sensitized to lysis by peptide-specific CTLs). These AdV can be used to efficiently generate CTLs by transduction of both normal donor and cancer-patient-derived DCs and coincubation with autologous CD8+ responder T cells. After as little as one week, anti-MART-1 or anti-AFP-specific killing has been observed against TAA+ tumor cells. An important finding of this method of CTL generation has been that anti-adenovirus antigen responses do not overwhelm the desired anti-MART-1 tumor antigen response [59,61,168].

H. Tumor-Dendritic Cell Fusion

Dendritic Cells can be fused with tumor cells using techniques similar to those used to create hybridomas. Gong *et al.* [175] have reported that murine DCs fused with a murine adenocarcinoma cell line expressing the MUC-1 TAA were able to induce MUC-1-specific CTLs *in vivo* and reject established lung metastases. In this system, tumor cells would supply TAAs that would then be optimally presented to the

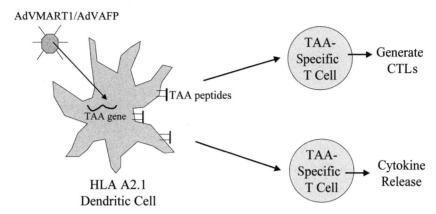

FIGURE 5 AdVMART1 and AdVhAFP transduction of human DCs *in vitro*. The AdV–TAA adenoviruses (with a map of the TAA expression cassette shown above) have been used to transduce HLA-A2.1 DCs. These transduced DCs are used to generate CTL *in vitro* that can lyse TAA-positive, HLA-matched tumor cells, but not TAA-negative, HLA-matched tumor cells. In addition, these TAA-transduced cells can elicit cytokine release from immunodominant TAA peptide-specific CTL lines (MART$_{27-35}$ for MART-1, hAFP$_{137-145}$, hAFP$_{158-166}$, hAFP$_{325-334}$, and hAFP$_{542-550}$ for hAFP).

host immune system by the DC. Benefits of this approach are that TAAs do not need to be previously identified, the antigenic peptides would be physiologically processed and presented by the DCs, and, once the fused population is cloned, the cells used for vaccination would be immortal. Problems include the requirement of this technique for a fusion for each patient, the time necessary to select and grow the fused cells, and the extensive *in vitro* manipulation required to differentiate the fused cells from the contaminating tumor cells. Another approach might be to establish allogeneic DC–tumor-fused cell lines using DCs from common HLA subtypes and tumors that express common shared tumor antigens [176]. Despite these technical issues, tumor–DC fusions have been successfully translated to the clinic with impressive results. In a trial treating renal cell cancer patients, 7 of 17 patients responded to the DC–tumor cell fusion vaccine [177].

Naked DNA intramuscular immunization and DNA/gene gun approaches continue to be of great interest for use with self tumor antigens. If these methods could be further refined (with the best cytokine adjuvant and improved plasmid backbones), they would be a very attractive strategy for genetic immunization because of their safety, ease of use, and inexpensive and stable materials. And, as with all whole-gene/whole-protein methods, they are not restricted to HLA type. However, to date, the impressive immune-stimulating

capacity of DCs, particularly with regard to naive T cells, makes them the most promising vehicle for genetic immunization. *Ex vivo* differentiation and expansion of these cells has the added benefit of overcoming the inhibitory state of the DCs that has been observed in the tumor-bearing host [178,179]. Even in prostate cancer patients who have received prior radiation therapy, DCs can be expanded and differentiated *in vitro* which have the important immunostimulatory functions of those obtained from healthy donors [180]. The field of genetic immunization has already demonstrated impressive tumor immunotherapeutic results with not only model tumor antigens such as β-galactosidase, ovalbumin, and sperm whale myoglobin, but also known human tumor antigens. These methods can generate immunity to these known human tumor-rejection antigens in mice, where there is only partial homology in protein sequence, and in humans, both patients and normal donors, *in vitro* and *in vivo*.

I. Potential Problems

An emerging issue to be addressed with any of these tumor-antigen-driven immunotherapies is the emergence of antigen-loss varients. Subpopulations of antigen-negative tumor cells have been demonstrated to be resistant to tumor antigen epitope CTL induction methods [20,181]. One potential way to overcome this is to immunize with multiple peptide antigens or to use the entire tumor antigen (as a naked DNA injection or as transfected DCs) as an immunogen to allow multiple epitopes from the same antigen to be presented. In some cases, lack of CTL activity is due to defects in the T-cell receptor (TCR) zeta chain signal transduction that can be overcome with adjuvant cytokine therapy such as IL-2. Treatment with IFN-γ can cause sufficient upregulation of class I antigen presentation in target cells to cause CTL killing susceptibility. Again, these situations support the use of a combination treatment including cytokines as part of the antitumor arsenal.

IV. PRECLINICAL DEVELOPMENT AND TRANSLATION TO THE CLINIC

Several issues must be addressed in order to take these new strategies to the clinic. The cytokines needed to culture DCs for clinical trials must be prepared under 'good manufacturing practice' (GMP) conditions. Sources for cytokines prepared this way are limited. DCs must be cultured either serum-free or in autologous serum which may require additional blood draws. This is a particular concern for translation of preclinical murine models to human trials. Most murine DCs are grown in medium containing fetal bovine serum where foreign "helper" antigens can increase antitumor effects. Use of serum-free DC generation conditions has been under investigation, but it can be difficult to assess whether the DCs obtained have the same functional capabilities and maturation status. With regard to synthetic peptides and proteins, these must also be of sufficient purity and safety for patient use, and the criteria for direct infusion are more strict than the criteria for *ex vivo* culture, with extensive rinsing. When considering use of patient tumor cells, there must be sufficient live tumor available to elute peptides, obtain tumor lysate, or fuse with autologous DC.

For *ex vivo* expansion of DCs to be reinfused, the amount of blood needed for monocyte-lineage IL-4/GM-CSF DCs can be daunting despite the DC yields that can be obtained. For example, to infuse a patient with 10^7 DCs in three weekly doses, one must obtain at least 2×10^8 PBMCs for each dose, which can necessitate 100 mL of blood from a healthy donor each of 3 weeks and possibly more from a cancer patient. Leukapheresis is a requirement for most trials.

To use a virus for direct administration or to transduce DCs, even with an E1-deleted, replication-deficient adenovirus, extensive and expensive pharmacology and toxicity testing must be performed. Perhaps, if the results of the early trials with these reagents continue to indicate a total lack of toxicity and side effects, some of the safety concerns will disappear. This could reduce the time and cost of performing these trials and allow faster progress in the genetic immunization field.

V. PROPOSED AND CURRENT CLINICAL TRIALS

These methods of genetic immunization have been extended to human trials. In one phase I trial, HLA-A1 melanoma patients carrying MAGE-1-positive tumors were treated with MAGE-1/A1 peptide-pulsed autologous GM-CSF-treated APCs [21,182]. These patients showed an increased CTL precursor frequency for MAGE-1-reactive T cells, as well as MAGE-1-reactive and autologous melanoma-reactive CTLs at both the vaccination site and distant metastases. In another early trial, Hsu *et al.* [144] vaccinated four patients with low-grade B lymphoma with autologous DCs expressing their own tumor-specific B-lymphoma idiotype protein. All four patients had evidence of an idiotype-specific immune response. Also, measurable clinical responses were observed in each patient, with no adverse events. Nestle *et al.* used melanoma peptides and patient tumor lysate-pulsed DCs with KLH marker/nonspecific helper protein. This study found KLH responses in all patients and tumor antigen peptide DTH responses in 11 of 16. In another early trial designed to test both the safety and efficiency of DC-mediated T-cell stimulation, normal subjects were vaccinated with autologous DCs pulsed with a foreign "priming" protein (KLH), a "boosting" protein (tetanus toxoid), and the influenza M1 peptide. There was clear evidence of vaccine immunogenicity to all three antigens and no toxicity [183]. In another clinical trial, six of 11 HLA-A1 patients with MAGE-3-positive melanoma responded to immunizations with mature autologous DCs. Interestingly, immune responses were detected after the first subcutaneous or intradermal injections but decreased after the subsequent intravenous (i.v.) injections [184]. These encouraging data indicate that antigen-pulsed DC have detectable physiological effects in advanced cancer patients, generating cytotoxic T cells with the desired specificity at the desired site. Although trials are still in an early stage, it is important to know that these strategies are headed in the right direction.

Since these initial reports, other phase I trials have been published. In lymphoma and myeloma, idiotype responses have been detected from protein-pulsed DCs [185,186]. In a trial where the use of tumor lysate-pulsed autologous DCs plus KLH was investigated, some KLH-specific responses were seen *in vitro,* and, importantly, *in vitro* T-cell responses to both tumor cell lysate and normal kidney cell lysates were observed [187]. Many melanoma trials have been performed targeting melanoma tumor antigens. Most of these involve administration of synthetic peptides alone and with systemic IL-2 [23,25,188,189]. Interestingly, *in vitro* immunological monitoring results have been inversely correlated with clinical outcome in many of these studies. Using melanoma antigen peptide-pulsed DCs, we and others have observed transient increases in antigen-specific T cells and some mixed clinical responses [61,184].

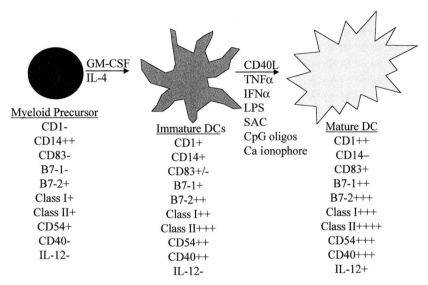

FIGURE 6 Properties of immature and mature myeloid DCs. When DCs are expanded from myeloid precursors, they upregulate many antigen-presentation-related markers and become specialized for antigen capture. When these immature DCs are further differentiated into mature DCs by CD40 ligation, exposure to bacterial products, or other means, they become further specialized for optimal antigen presentation. Some of the markers for immature and mature DCs are listed.

VI. FUTURE DIRECTIONS

We can look forward to a variety of improvements in the near future. In addition to creating better DNA backbones for plasmid-based strategies, there is considerable heterogeneity in studies with cytokines used in different tumor systems. Much work needs to be done either to prove the universality in cytokine use or to show which particular cytokines might be best for particular tumor types.

There is also room for improvement in DC-generation methodology, as well as continuing controversy over the use of "immature" DCs, which are still proficient at antigen uptake, versus "mature" DCs, which have greater T-cell stimulatory activity and increased display of MHC molecules (Fig. 6). One important difference is that immature DCs express chemokine receptors (e.g., CCR6) that cause homing to the periphery, while mature DCs preferentially express chemokine receptors (e.g., CCR7) that lead to lymph-node homing [190,191]. While there are valid arguments on both sides, it should be kept in mind that there will be little control over the cytokine environment of the reinfused DCs encountered in the patient. In an *in vivo* trafficking study in rhesus macaques comparing intradermal (i.d.) immature and mature DC vaccinations, it was found that both immature and mature DCs trafficked properly to draining lymph nodes [192]. Looking at routes of administration in cancer patients using In[111]-labeled DC, i.v. administration was found to lead to DC accumulation in the lungs, then in the liver, spleen, and bone marrow; i.d. or s.c. delivery resulted in partial clearance from the injection site, but only i.d. DCs were found in the lymph nodes, not s.c. DCs [193]. This is in contrast to a mouse study in which i.v.-delivered DCs trafficked to the spleen and s.c. DCs were found in lymph nodes. B16 murine melanoma tumor responses were greater with s.c.-delivered DC than by i.v. [194]. This points to the limits of extrapolation from murine models to human studies.

The area of virus improvement is an active one. "Third generation" adenoviral vectors are being developed that express fewer and fewer viral genes. These will be used to maintain the transduction efficiency and high transgene expression levels of viruses while eliminating the virus-associated antigenicity. Other methods of transgene expression in professional APCs, such as adeno-associated viruses and lentiviruses, are also actively being studied. Finally, an important improvement in AdV-mediated transduction of DCs involves retargeting the virus by engineering new, CAR receptor-independent viral attachment regions. The studies in animal models, the preclinical laboratory data, and the results of pioneer clinical trials suggest that tumor-antigen-based genetic immunization holds promise as a novel immunotherapy approach for cancer.

References

1. Nair, S. K., Hull, S., Coleman, D., Gilboa, E., Lyerly, H. K., and Morse, M. A. (1999). Induction of carcinoembryonic antigen (CEA)-specific cytotoxic T-lymphocyte responses in vitro using autologous dendritic cells loaded with CEA peptide or CEA RNA in patients with metastatic malignancies expressing CEA. *Int. J. Cancer* **82,** 1:21.
2. Brossart, P., Goldrath, A. W., Butz, E. A., Martin, S., and Bevan, M. J. (1997). Virus-mediated delivery of antigenic epitopes into dendritic cells as a means to induce CTL. *J. Immunol.* **158,** 7:3270.

3. Nestle, F. O., Alijagic, S., Gilliet, M., Sun, Y., Grabbe, S., Dummer, R., Burg, G., and Schadendorf, D. (1998). Vaccination of melanoma patients with peptide- or tumor lysate-pulsed dendritic cells [see comments]. *Nat. Med.* **4**, 3:328.

4. Coulie, P. G., Brichard, V., Van Pel, A., Wolfel, T., Schneider, J., Traversari, C., Mattei, S., De Plaen, E., Lurquin, C., Szikora, J. P., Renauld, J. P., and Boon, T. (1994). A new gene coding for a differentiation antigen recognized by autologous cytolytic T lymphocytes on HLA-A2 melanomas [see comments]. *J. Exp. Med.* **180**, 35.

5. Bakker, A. B., Schreurs, M. W., de Boer, A. J., Kawakami, Y., Rosenberg, S. A., Adema, G. J., and Figdor, C. G. (1994). Melanocyte lineage-specific antigen gp100 is recognized by melanoma-derived tumor-infiltrating lymphocytes. *J. Exp. Med.* **179**, 1005.

6. Brichard, V., Van Pel, A., Wolfel, T., Wolfel, C., De Plaen, E., Lethe, B., Coulie, P., and Boon, T. (1993). The tyrosinase gene codes for an antigen recognized by autologous cytolytic T lymphocytes on HLA-A2 melanomas. *J. Exp. Med.* **178**, 489.

7. Wang, R. F., Robbins, P. F., Kawakami, Y., Kang, X. Q., and Rosenberg, S. A. (1995). Identification of a gene encoding a melanoma tumor antigen recognized by HLA-A31-restricted tumor-infiltrating lymphocytes [published *erratum* appears in *J. Exp. Med.* 181(3), 1261, 1995]. *J. Exp. Med.* **181**, 799.

8. Coulie, P. G., Weynants, P., Lehmann, F., Herman, J., Brichard, V., Wolfel, T., Van Pel, A., De Plaen, E., Brasseur, F., and Boon, T. (1993). Genes coding for tumor antigens recognized by human cytolytic T lymphocytes. *J. Immunother.* **14**, 104.

9. Itoh, K., Hayashi, A., Nakao, M., Hoshino, T., Seki, N., and Shichijo, S. (1996). Human tumor rejection antigens MAGE. *J. Biochem. (Tokyo)* **119**, 385.

10. Chen, Y. T., Stockert, E., Chen, Y., Garin-Chesa, P., Rettig, W. J., Van der Bruggen, P., Boon, T., and Old, L. J. (1994). Identification of the MAGE-1 gene product by monoclonal and polyclonal antibodies. *Proc. Natl. Acad. Sci. USA* **91**, 1004.

11. Rimoldi, D., Romero, P., and Carrel, S. (1993). The human melanoma antigen-encoding gene, MAGE-1, is expressed by other tumour cells of neuroectodermal origin such as glioblastomas and neuroblastomas [letter]. *Int. J. Cancer* **54**, 527.

12. Brasseur, F., Marchand, M., Vanwijck, R., Herin, M., Lethe, B., Chomez, P., and Boon, T. (1992). Human gene MAGE-1, which codes for a tumor-rejection antigen, is expressed by some breast tumors [letter]. *Int. J. Cancer* **52**, 839.

13. Becker, J. C., Gillitzer, R., and Brocker, E. B. (1994). A member of the melanoma antigen-encoding gene (MAGE) family is expressed in human skin during wound healing. *Int. J. Cancer* **58**, 346.

14. Van der Bruggen, P., Bastin, J., Gajewski, T., Coulie, P. G., Boel, P., De Smet, C., Traversari, C., Townsend, A., and Boon, T. (1994). A peptide encoded by human gene MAGE-3 and presented by HLA-A2 induces cytolytic T lymphocytes that recognize tumor cells expressing MAGE-3. *Eur. J. Immunol.* **24**, 3038.

15. Van der Bruggen, P., Szikora, J. P., Boel, P., Wildmann, C., Somville, M., Sensi, M., and Boon, T. (1994). Autologous cytolytic T lymphocytes recognize a MAGE-1 nonapeptide on melanomas expressing HLA-Cw*1601. *Eur. J. Immunol.* **24**, 2134.

16. Celis, E., Tsai, V., Crimi, C., DeMars, R., Wentworth, P. A., Chesnut, R. W., Grey, H. M., Sette, A., and Serra, H. M. (1994). Induction of anti-tumor cytotoxic T lymphocytes in normal humans using primary cultures and synthetic peptide epitopes. *Proc. Natl. Acad. Sci. USA* **91**, 2105.

17. Tanaka, F., Fujie, T., Go, H., Baba, K., Mori, M., Takesako, K., and Akiyoshi, T. (1997). Efficient induction of antitumor cytotoxic T lymphocytes from a healthy donor using HLA-A2-restricted MAGE-3 peptide in vitro. *Cancer Immunol. Immunother.* **44**, 21.

18. Traversari, C., Van der Bruggen, P., Luescher, I. F., Lurquin, C., Chomez, P., Van Pel, A., De Plaen, E., Amar-Costesec, A., and Boon, T. (1992). A nonapeptide encoded by human gene MAGE-1 is recognized on HLA-A1 by cytolytic T lymphocytes directed against tumor antigen MZ2-E. *J. Exp. Med.* **176**, 1453.

19. Weber, J. S., Hua, F. L., Spears, L., Marty, V., Kuniyoshi, C., and Celis, E. (1999). A phase I trial of an HLA-A1 restricted MAGE-3 epitope peptide with incomplete Freund's adjuvant in patients with resected high-risk melanoma. *J. Immunother.* **22**, 431.

20. Jaeger, E., Bernhard, H., Romero, P., Ringhoffer, M., Arand, M., Karbach, J., Ilsemann, C., Hagedorn, M., and Knuth, A. (1996). Generation of cytotoxic T-cell responses with synthetic melanoma-associated peptides in vivo: implications for tumor vaccines with melanoma-associated antigens. *Int. J. Cancer* **66**, 162.

21. Hu, X., Chakraborty, N. G., Sporn, J. R., Kurtzman, S. H., Ergin, M. T., and Mukherji, B. (1996). Enhancement of cytolytic T lymphocyte precursor frequency in melanoma patients following immunization with the MAGE-1 peptide loaded antigen presenting cell-based vaccine. *Cancer Res.* **56**, 2479.

22. Rosenberg, S. A., Yang, J. C., Schwartzentruber, D. J., Hwu, P., Marincola, F. M., Topalian, S. L., Restifo, N. P., Sznol, M., Schwarz, S. L., Spiess, P. J., Wunderlich, J. R., Seipp, C. A., Einhorn, J. H., Rogers-Freezer, L., and White, D. E. (1999). Impact of cytokine administration on the generation of antitumor reactivity in patients with metastatic melanoma receiving a peptide vaccine. *J. Immunol.* **163**, 1690.

23. Lee, K. H., Wang, E., Nielsen, M. B., Wunderlich, J., Migueles, S., Connors, M., Steinberg, S. M., Rosenberg, S. A., and Marincola, F. M. (1999). Increased vaccine-specific T cell frequency after peptide-based vaccination correlates with increased susceptibility to in vitro stimulation but does not lead to tumor regression. *J. Immunol.* **163**, 6292.

24. Rosenberg, S. A., Zhai, Y., Yang, J. C., Schwartzentruber, D. J., Hwu, P., Marincola, F. M., Topalian, S. L., Restifo, N. P., Seipp, C. A., Einhorn, J. H., Roberts, B., and White, D. E. (1998). Immunizing patients with metastatic melanoma using recombinant adenoviruses encoding MART-1 or gp100 melanoma antigens. *Natl. Cancer Inst.* **90**, 1894.

25. Pass, H. A., Schwarz, S. L., Wunderlich, J. R., and Rosenberg. S. A. (1998). Immunization of patients with melanoma peptide vaccines: immunologic assessment using the ELISPOT assay [see comments]. *Cancer J. Sci. Am.* **4**, 316.

26. Kawakami, Y., Robbins, P. F., Wang, R. F., Parkhurst, M., Kang, X., and Rosenberg, S. A. (1998). The use of melanosomal proteins in the immunotherapy of melanoma. *J. Immunother.* **21**, 237.

27. Wang, R. F., Johnston, S. L., Southwood, S., Sette, A., and Rosenberg, S. A. (1998). Recognition of an antigenic peptide derived from tyrosinase-related protein-2 by CTL in the context of HLA-A31 and -A33. *J. Immunol.* **160**, 890.

28. Cormier, J. N., Abati, A., Fetsch, P., Hijazi, Y. M., Rosenberg, S. A., Marincola, F. M., and Topalian, S. L. (1998). Comparative analysis of the in vivo expression of tyrosinase, MART-1/Melan-A, and gp100 in metastatic melanoma lesions: implications for immunotherapy. *J. Immunother.* **21**, 27.

29. Rosenberg, S. A. (1992). Gene therapy for cancer [clinical conference]. *JAMA* **268**, 2416.

30. Kawakami, Y., Robbins, P. F., Wang, R. F., and Rosenberg, S. A. (1996). Identification of tumor-regression antigens in melanoma. *Important Adv. Oncol.* 3.

31. Overwijk, W. W., Lee, D. S., Surman, D. R., Irvine, K. R., Touloukian, C. E., Chan, C. C., Carroll, M. W., Moss, B., Rosenberg, S. A., and Restifo, N. P. (1999). Vaccination with a recombinant vaccinia virus encoding a "self" antigen induces autoimmune vitiligo and tumor cell destruction in mice: requirement for CD4(+) T lymphocytes. *Proc. Natl. Acad. Sci. USA* **96**, 2982.

32. Rosenberg, S. A., and White, D. E. (1996). Vitiligo in patients with melanoma: normal tissue antigens can be targets for cancer immunotherapy. *J. Immunother. Emphasis Tumor Immunol.* **19**, 81.

33. Sugita, S., Sagawa, K., Mochizuki, M., Shichijo, S., and Itoh, K. (1996). Melanocyte lysis by cytotoxic T lymphocytes recognizing the MART-1 melanoma antigen in HLA-A2 patients with Vogt–Koyanagi–Harada disease. *Int. Immunol.* **8**, 799.

34. Lee, T. (1990). Distribution of HLA antigens in North American Caucasians, North American Blacks and Orientals, in *The HLA System* (J. Lee, ed.), p. 14, Springer-Verlag, NewYork.

35. Saper, M. A., Bjorkman, P. J., and Wiley, D. C. (1991). Refined structure of the human histocompatibility antigen HLA-A2 at 2.6 A resolution. *J. Mol. Biol.* **219**, 277.

36. Falk, K., Rotzschke, O., Stevanovic, S., Jung, G., and Rammensee, H. G. (1991). Allele-specific motifs revealed by sequencing of self-peptides eluted from MHC molecules. *Nature* **351**, 290.

37. Celis, E., Fikes, J., Wentworth, P., Sidney, J., Southwood, S., Maewal, A., Del Guercio, M. F., Sette, A., and Livingston, B. (1994). Identification of potential CTL epitopes of tumor-associated antigen MAGE-1 for five common HLA-A alleles. *Mol. Immunol.* **31**, 1423.

38. Hunt, D. F., Henderson, R. A., Shabanowitz, J., Sakaguchi, K., Michel, H., Sevilir, N., Cox, A. L., Appella, E., and Engelhard, V. H. (1992). Characterization of peptides bound to the class I MHC molecule HLA-A2.1 by mass spectrometry [see comments]. *Science* **255**, 1261.

39. Ruppert, J., Sidney, J., Celis, E., Kubo, R. T., Grey, H. M., and Sette, A. (1993). Prominent role of secondary anchor residues in peptide binding to HLA-A2.1 molecules. *Cell* **74**, 929.

40. Santamaria, P., Lindstrom, A. L., Boyce-Jacino, M. T., Myster, S. H., Barbosa, J. J., Faras, A. J., and Rich, S. S. (1993). HLA class I sequence-based typing [published *erratum* appears in *Hum. Immunol.* 41(4), 292, 1994]. *Hum. Immunol.* **37**, 39.

41. Fruci, D., Rovero, P., Falasca, G., Chersi, A., Sorrentino, R., Butler, R., Tanigaki, N., and Tosi, R. (1993). Anchor residue motifs of HLA class-I-binding peptides analyzed by the direct binding of synthetic peptides to HLA class I alpha chains. *Hum. Immunol.* **38**, 187.

42. Drijfhout, J. W., Brandt, R. M., Kast, W. M., and Melief, C. J., (1995). Detailed motifs for peptide binding to HLA-A*0201 derived from large random sets of peptides using a cellular binding assay. *Hum. Immunol.* **43**, 1.

43. Kubo, R. T., Sette, A., Grey, H. M., Appella, E., Sakaguchi, K., Zhu, N. Z., Arnott, D., Sherman, N., Shabanowitz, J., Michel, H., Bodnar, W. M., Davis, T. A., and Hunt, D. F. (1994). Definition of specific peptide motifs for four major HLA-A alleles. *J. Immunol.* **152**, 3913.

44. Brichard, V. G., Herman, J., Van Pel, A., Wildmann, C., Gaugler, B., Wolfel, T., Boon, T., and Lethe, B. (1996). A tyrosinase nonapeptide presented by HLA-B44 is recognized on a human melanoma by autologous cytolytic T lymphocytes. *Eur. J. Immunol.* **26**, 224.

45. Farina, C., Van der Bruggen, P., Boel, P., Parmiani, G., and Sensi, M. (1996). Conserved TCR usage by HLA-Cw*1601-restricted T cell clones recognizing melanoma antigens. *Int. Immunol.* **8**, 1463.

46. Fleischhauer, K., Fruci, D., Van Endert, P., Herman, J., Tanzarella, S., Wallny, H. J., Coulie, P., Bordignon, C., and Traversari, C. (1996). Characterization of antigenic peptides presented by HLA-B44 molecules on tumor cells expressing the gene MAGE-3. *Int. J. Cancer* **68**, 622.

47. Herman, J., Van der Bruggen, P., Luescher, I. F., Mandruzzato, S., Romero, P., Thonnard, J., Fleischhauer, K., Boon, T., and Coulie, P. G. (1996). A peptide encoded by the human MAGE3 gene and presented by HLA-B44 induces cytolytic T lymphocytes that recognize tumor cells expressing MAGE3. *Immunogenetics* **43**, 377.

48. Kang, X., Kawakami, Y., el-Gamil, M., Wang, R., Sakaguchi, K., Yannelli, J. R., Appella, E., Rosenberg, S. A., and Robbins, P. F. (1995). Identification of a tyrosinase epitope recognized by HLA-A24-restricted, tumor-infiltrating lymphocytes. *J. Immunol.* **155**, 1343.

49. Mazzocchi, A., Storkus, W. J., Traversari, C., Tarsini, P., Maeurer, M. J., Rivoltini, L., Vegetti, C., Belli, F., Anichini, A., Parmiani, G., and Castelli, C. (1996). Multiple melanoma-associated epitopes recognized by HLA-A3-restricted CTLs and shared by melanomas but not melanocytes. *J. Immunol.* **157**, 3030.

50. Nukaya, I., Yasumoto, M., Iwasaki, T., Ideno, M., Sette, A., Celis, E., Takesako, K., and Kato, I. (1999). Identification of HLA-A24 epitope peptides of carcinoembryonic antigen which induce tumor-reactive cytotoxic T lymphocyte. *Int. J. Cancer* **80**, 92.

51. Robbins, P. F., el-Gamil, M., Li, Y. F., Topalian, S. L., Rivoltini, L., Sakaguchi, K., Appella, E., Kawakami, Y., and Rosenberg, S. A. (1995). Cloning of a new gene encoding an antigen recognized by melanoma-specific HLA-A24-restricted tumor-infiltrating lymphocytes. *J. Immunol.* **154**, 5944.

52. Sidney, J., Grey, H. M., Southwood, S., Celis, E., Wentworth, P. A., del Guercio, M. F., Kubo, R. T., Chesnut, R. W., and Sette, A. (1996). Definition of an HLA-A3-like supermotif demonstrates the overlapping peptide-binding repertoires of common HLA molecules. *Hum. Immunol.* **45**, 79.

53. Maeurer, M. J., Storkus, W. J., Kirkwood, J. M., and Lotze, M. T. (1996). New treatment options for patients with melanoma: review of melanoma-derived T-cell epitope-based peptide vaccines. *Melanoma Res.* **6**, 11.

54. Cole, D. J., Weil, D. P., Shilyansky, J., Custer, M., Kawakami, Y., Rosenberg, S. A., and Nishimura, M. I. (1995). Characterization of the functional specificity of a cloned T-cell receptor heterodimer recognizing the MART-1 melanoma antigen. *Cancer Res.* **55**, 748.

55. Kawakami, Y., Eliyahu, S., Sakaguchi, K., Robbins, P. F., Rivoltini, L., Yannelli, J. R., Appella, E., and Rosenberg, S. A. (1994). Identification of the immunodominant peptides of the MART-1 human melanoma antigen recognized by the majority of HLA-A2-restricted tumor infiltrating lymphocytes. *J. Exp. Med.* **180**, 347.

56. Kawakami, Y., Eliyahu, S., Jennings, C., Sakaguchi, K., Kang, X., Southwood, S., Robbins, P. F., Sette, A., Appella, E., and Rosenberg, S. A. (1995). Recognition of multiple epitopes in the human melanoma antigen gp100 by tumor-infiltrating T lymphocytes associated with in vivo tumor regression. *J. Immunol.* **154**, 3961.

57. van der Burg, S. H., Visseren, M. J., Brandt, R. M., Kast, W. M., and Melief, C. J. (1996). Immunogenicity of peptides bound to MHC class I molecules depends on the MHC-peptide complex stability. *J. Immunol.* **156**, 3308.

58. van der Burg, S. H., Visseren, M. J., Offringa, R., and Melief, C. J. (1997). Do epitopes derived from autoantigens display low affinity for MHC class I? [letter]. *Immunol. Today* **18**, 97.

59. Butterfield, L. H., Koh, A., Meng, W., Vollmer, C. M., Ribas, A., Dissette, V. B., Lee, E., Glaspy, J. A., McBride, W. H., and Economou, J. S. (1999). Generation of human T cell responses to an HLA-A2. 1-restricted peptide epitope from alpha fetoprotein. *Cancer Res.* **59**, 3134.

60. Vollmer, C. M., Eilber, F. C., Butterfield, L. H., Ribas, A., Dissette, V. B., Koh, A., Montejo, L., Andrews, K., McBride, W. H., Glaspy, J. A., and Economou, J. S. (1999). Alpha fetoprotein-specific immunotherapy for hepatocellular carcinoma. *Cancer Res.* **59**, 3064.

61. Butterfield, L. H., Meng, W. S., Koh, A., Vollmer, C. M., Ribas, A., Dissette, V. B., Faull, K., Glaspy, J. A., McBride, W. H., and Economou, J. S. (2000). T cell responses to HLA-A*0201-restricted peptides derived from alpha fetoprotein. *J. Immunol.* **166**, 5300.

62. Tsang, K. Y., Zaremba, S., Nieroda, C. A., Zhu, M. Z., Hamilton, J. M., and Schlom, J. (1995). Generation of human cytotoxic T cells specific for human carcinoembryonic antigen epitopes from patients immunized with recombinant vaccinia-CEA vaccine [see comments]. *J. Natl. Cancer Inst.* **87**, 982.

63. Wei, C., Storozynsky, E., McAdam, A. J., Yeh, K. Y., Tilton, B. R., Willis, R. A., Barth, R. K., Looney, R. J., Lord, E. M., and Frelinger,

J. G. (1996). Expression of human prostate-specific antigen (PSA) in a mouse tumor cell line reduces tumorigenicity and elicits PSA-specific cytotoxic T lymphocytes. *Cancer Immunol. Immunother.* **42**, 362.

64. Peoples, G. E., Goedegebuure, P. S., Smith, R., Linehan, D. C., Yoshino, I. and Eberlein, T. J. (1995). Breast and ovarian cancer-specific cytotoxic T lymphocytes recognize the same HER2/neu-derived peptide. *Proc. Natl. Acad. Sci. USA* **92**, 432.

65. Peace, D. J., Smith, J. W., Chen, W., You, S. G., Cosand, W. L., Blake, J., and Cheever, M. A. (1994). Lysis of *ras* oncogene-transformed cells by specific cytotoxic T lymphocytes elicited by primary in vitro immunization with mutated Ras peptide. *J. Exp. Med.* **179**, 473.

66. Mayordomo, J. I., Loftus, D. J., Sakamoto, H., De Cesare, C. M., Appasamy, P. M., Lotze, M. T., Storkus, W. J., Appella, E., and DeLeo, A. B. (1996). Therapy of murine tumors with p53 wild-type and mutant sequence peptide-based vaccines. *J. Exp. Med.* **183**, 1357.

67. Theobald, M., Biggs, J., Dittmer, D., Levine, A. J., and Sherman, L. A. (1995). Targeting p53 as a general tumor antigen. *Proc. Natl. Acad. Sci. USA* **92**, 11993.

68. Hodge, J. W. (1996). Carcinoembryonic antigen as a target for cancer vaccines. *Cancer Immunol. Immunother.* **43**, 127.

69. Conry, R. M., Khazaeli, M. B., Saleh, M. N., Allen, K. O., Barlow, D. L., Moore, S. E., Craig, D., Arani, R. B., Schlom, J., and LoBuglio, A. F. (1999). Phase I trial of a recombinant vaccinia virus encoding carcinoembryonic antigen in metastatic adenocarcinoma: comparison of intradermal versus subcutaneous administration. *Clin. Cancer Res.* **5**, 2330.

70. Zhu, M. Z., Marshall, J., Cole, D., Schlom, J., and Tsang, K. Y. (2000). Specific cytolytic T-cell responses to human CEA from patients immunized with recombinant avipox-CEA vaccine. *Clin. Cancer Res.* **6**, 24.

71. Xue, B. H., Zhang, Y., Sosman, J. A., and Peace, D. J. (1997). Induction of human cytotoxic T lymphocytes specific for prostate-specific antigen. *Prostate* **30**, 73.

72. Tjoa, B. A., Simmons, S. J., Bowes, V. A., Ragde, H., Rogers, M., Elgamal, A., Kenny, G. M., Cobb, O. E., Ireton, R. C., Troychak, M. J., Salgaller, M. L., Boynton, A. L., and Murphy, G. P. (1998). Evaluation of phase I/II clinical trials in prostate cancer with dendritic cells and PSMA peptides. *Prostate* **36**, 39.

73. Murphy, G., Tjoa, B., Ragde, H., Kenny, G., and Boynton, A. (1996). Phase I clinical trial: T-cell therapy for prostate cancer using autologous dendritic cells pulsed with HLA-A0201-specific peptides from prostate-specific membrane antigen. *Prostate* **29**, 371.

74. Murphy, G. P., Tjoa, B. A., Simmons, S. J., Jarisch, J., Bowes, V. A., Ragde, H., Rogers, M., Elgamal, A., Kenny, G. M., Cobb, O. E., Ireton, R. C., Troychak, M. J., Salgaller, M. L., and Boynton, A. L. (1999). Infusion of dendritic cells pulsed with HLA-A2-specific prostate-specific membrane antigen peptides: a phase II prostate cancer vaccine trial involving patients with hormone-refractory metastatic disease. *Prostate* **38**, 73.

75. Linehan, D. C., Goedegebuure, P. S., Peoples, G. E., Rogers, S. O., and Eberlein, T. J. (1995). Tumor-specific and HLA-A2-restricted cytolysis by tumor-associated lymphocytes in human metastatic breast cancer. *J. Immunol.* **155**, 4486.

76. Disis, M. L., Gralow, J. R., Bernhard, H., Hand, S. L., Rubin, W. D., and Cheever, M. A. (1996). Peptide-based, but not whole protein, vaccines elicit immunity to HER-2/neu, oncogenic self-protein. *J. Immunol.* **156**, 3151.

77. Moudgil, K. D., and Sercarz, E. E. (1994). The T cell repertoire against cryptic self determinants and its involvement in autoimmunity and cancer. *Clin. Immunol. Immunopathol.* **73**, 283.

78. Amici, A., Venanzi, F. M., and Concetti, A. (1998). Genetic immunization against neu/erbB2 transgenic breast cancer. *Cancer Immunol. Immunother.* **47**, 183.

79. Disis, M. L., Grabstein, K. H., Sleath, P. R., and Cheever, M. A. (1999). Generation of immunity to the HER-2/neu oncogenic protein

in patients with breast and ovarian cancer using a peptide-based vaccine. *Clin. Cancer Res.* **5**, 1289.

80. Juretic, A., Jurgens-Gobel, J., Schaefer, C., Noppen, C., Willimann, T. E., Kocher, T., Zuber, M., Harder, F., Heberer, M., and Spagnoli, G. C. (1996). Cytotoxic T-lymphocyte responses against mutated p21 ras peptides: an analysis of specific T-cell-receptor gene usage. *Int. J. Cancer* **68**, 471.

81. Yokomizo, H., Matsushita, S., Fujisao, S., Murakami, S., Fujita, H., Shirouzu, M., Yokoyama, S., Ogawa, M., and Nishimura, Y. (1997). Augmentation of immune response by an analog of the antigenic peptide in a human T-cell clone recognizing mutated Ras-derived peptides. *Hum. Immunol.* **52**, 22.

82. Nijman, H. W., Van der Burg, S. H., Vierboom, M. P., Houbiers, J. G., Kast, W. M., and Melief, C. J. (1994). p53, a potential target for tumor-directed T cells. *Immunol. Lett.* **40**, 171.

83. Ropke, M., Hald, J., Guldberg, P., Zeuthen, J., Norgaard, L., Fugger, L., Svejgaard, A., van der Burg, S., Nijman, H. W., Melief, C. J., and Claesson, M. H. (1996). Spontaneous human squamous cell carcinomas are killed by a human cytotoxic T lymphocyte clone recognizing a wild-type p53-derived peptide. *Proc. Natl. Acad. Sci. USA* **93**, 14704.

84. Yu, Z., Liu, X., McCarty, T. M., Diamond, D. J., and Ellenhorn, J. D. (1997). The use of transgenic mice to generate high affinity p53 specific cytolytic T cells. *J. Surg. Res.* **69**, 337.

85. Theobald, M., Biggs, J., Hernandez, J., Lustgarten, J., Labadie, C., and Sherman, L. A. (1997). Tolerance to p53 by A2.1-restricted cytotoxic T lymphocytes. *J. Exp. Med.* **185**, 833.

86. Barfoed, A. M., Petersen, T. R., Kirkin, A. F., Thor Straten, P., Claesson, M. H., and Zeuthen, J. (2000). Cytotoxic T-lymphocyte clones, established by stimulation with the HLA-A2 binding p5365–73 wild type peptide loaded on dendritic cells in vitro, specifically recognize and lyse HLA-A2 tumour cells overexpressing the p53 protein. *Scand. J. Immunol.* **51**, 128.

87. Robbins, P. F., El-Gamil, M., Li, Y. F., Kawakami, Y., Loftus, D., Appella, E., and Rosenberg, S. A. (1996). A mutated beta-catenin gene encodes a melanoma-specific antigen recognized by tumor infiltrating lymphocytes. *J. Exp. Med.* **183**, 1185.

88. Coulie, P. G., Lehmann, F., Lethe, B., Herman, J., Lurquin, C., Andrawiss, M., and Boon, T. (1995). A mutated intron sequence codes for an antigenic peptide recognized by cytolytic T lymphocytes on a human melanoma. *Proc. Natl. Acad. Sci. USA* **92**, 7976.

89. Chiari, R., Foury, F., De Plaen, E., Baurain, J. F., Thonnard, J., and Coulie, P. G. (1999). Two antigens recognized by autologous cytolytic T lymphocytes on a melanoma result from a single point mutation in an essential housekeeping gene. *Cancer Res.* **59**, 5785.

90. Yedavelli, S. P., Guo, L., Daou, M. E., Srivastava, P. K., Mittelman, A., and Tiwari, R. K. (1999). Preventive and therapeutic effect of tumor derived heat shock protein, gp96, in an experimental prostate cancer model. *Int. J. Mol. Med.* **4**, 243.

91. Ménoret, A., Peng, P., and Srivastava, P. K. (1999). Association of peptides with heat shock protein gp96 occurs in vivo and not after cell lysis. *Biochem. Biophy. Res. Comm.* **262**, 813.

92. Wolff, J. A., Malone, R. W., Williams, P., Chong, W., Acsadi, G., Jani, A., and Felgner, P. L. (1990). Direct gene transfer into mouse muscle in vivo. *Science* **247**, 1465.

93. Parkhurst, M. R., Salgaller, M. L., Southwood, S., Robbins, P. F., Sette, A., Rosenberg, S. A., and Kawakami, Y. (1996). Improved induction of melanoma-reactive CTL with peptides from the melanoma antigen gp100 modified at HLA-A*0201-binding residues. *J. Immunol.* **157**, 2539.

94. Liu, Y., Liggitt, D., Zhong, W., Tu, G., Gaensler, K., and Debs, R. (1995). Cationic liposome-mediated intravenous gene delivery. *J. Biol. Chem.* **270**, 24864.

95. Kumar, V., and Sercarz, E. (1996). Genetic vaccination: the advantages of going naked [comment]. *Nat. Med.* **2**, 857.

96. Conry, R. M., LoBuglio, A. F., Kantor, J., Schlom, J., Loechel, F., Moore, S. E., Sumerel, L. A., Barlow, D. L., Abrams, S., and Curiel, D. T. (1994). Immune response to a carcinoembryonic antigen polynucleotide vaccine. *Cancer Res.* **54,** 1164.

97. Conry, R. M., Widera, G., LoBuglio, A. F., Fuller, J. T., Moore, S. E., Barlow, D. L., Turner, J., Yang, N. S., and Curiel, D. T. (1996). Selected strategies to augment polynucleotide immunization. *Gene Ther.* **3,** 67.

98. Sato, Y., Roman, M., Tighe, H., Lee, D., Corr, M., Nguyen, M. D., Silverman, G. J., Lotz, M., Carson, D. A., and Raz, E. (1996). Immunostimulatory DNA sequences necessary for effective intradermal gene immunization. *Science* **273,** 352.

99. Krieg, A. M. (1996). Lymphocyte activation by CpG dinucleotide motifs in prokaryotic DNA. *Trends Microbiol.* **4,** 73.

100. Wooldridge, J. E., Ballas, Z., Krieg, A. M., and Weiner, G. J. (1997). Immunostimulatory oligodeoxynucleotides containing CpG motifs enhance the efficacy of monoclonal antibody therapy of lymphoma. *Blood* **89,** 2994.

101. Krieg, A. M., Yi, A. K., Schorr, J., and Davis, H. L. (1998). The role of CpG dinucleotides in DNA vaccines. *Trends Microbiol.* **6,** 23.

102. Liu, H. M., Newbrough, S. E., Bhatia, S. K., Dahle, C. E., Krieg, A. M., and Weiner, G. J. (1998). Immunostimulatory CpG oligodeoxynucleotides enhance the immune response to vaccine strategies involving granulocyte-macrophage colony-stimulating factor. *Blood* **92,** 3730.

103. Jakob, T., Walker, P. S., Krieg, A. M., von Stebut, E., Udey, M. C., and Vogel, J. C. (1999). Bacterial DNA and CpG-containing oligodeoxynucleotides activate cutaneous dendritic cells and induce IL-12 production: implications for the augmentation of Th1 responses. *Int. Arch. Allergy Immunol.* **118,** 457.

104. Hartikka, J., Sawdey, M., Cornefert-Jensen, F., Margalith, M., Barnhart, K., Nolasco, M., Vahlsing, H. L., Meek, J., Marquet, M., Hobart, P., Norman, J., and Manthorpe, M. (1996). An improved plasmid DNA expression vector for direct injection into skeletal muscle. *Hum. Gene Ther.* **7,** 1205.

105. Corr, M., Lee, D. J., Carson, D. A., and Tighe, H. (1996.) Gene vaccination with naked plasmid DNA: mechanism of CTL priming. *J. Exp. Med.* **184,** 1555.

106. Geissler, M., Gesien, A., Tokushige, K., and Wands, J. R. (1997). Enhancement of cellular and humoral immune responses to hepatitis C virus core protein using DNA-based vaccines augmented with cytokine-expressing plasmids. *J. Immunol.* **158,** 1231.

107. Irvine, K. R., Rao, J. B., Rosenberg, S. A., and Restifo, N. P. (1996). Cytokine enhancement of DNA immunization leads to effective treatment of established pulmonary metastases. *J. Immunol.* **156,** 238.

108. Iwasaki, A., Stiernholm, B. J., Chan, A. K., Berinstein, N. L., and Barber, B. H. (1997). Enhanced CTL responses mediated by plasmid DNA immunogens encoding costimulatory molecules and cytokines. *J. Immunol.* **158,** 4591.

109. Vitadello, M., Schiaffino, M. V., Picard, A., Scarpa, M., and Schiaffino, S. (1994). Gene transfer in regenerating muscle. *Hum. Gene Ther.* **5,** 11.

110. Davis, H. L., Whalen, R. G., and Demeneix, B. A. (1993). Direct gene transfer into skeletal muscle in vivo: factors affecting efficiency of transfer and stability of expression. *Hum. Gene Ther.* **4,** 151.

111. Wu, Y., and Kipps, T. J. (1997). Deoxyribonucleic acid vaccines encoding antigens with rapid proteasome-dependent degradation are highly efficient inducers of cytolytic T lymphocytes. *J. Immunol.* **159,** 6037.

112. Fu, T. M., Guan, L., Friedman, A., Ulmer, J. B., Liu, M. A., and Donnelly, J. J. (1998). Induction of MHC class I-restricted CTL response by DNA immunization with ubiquitin-influenza virus nucleoprotein fusion antigens. *Vaccine* **16,** 1711.

113. Leitner, W. W., Ying, H., and Restifo, N. P. (1999). DNA and RNA-based vaccines: principles, progress and prospects. *Vaccine* **18,** 765.

114. Leitner, W. W., Ying, H., Driver, D. A., Dubensky, T. W., and Restifo, N. P. (2000). Enhancement of tumor-specific immune response with plasmid DNA replicon vectors. *Cancer Res.* **60,** 51.

115. Doe, B., Selby, M., Barnett, S., Baenziger, J., and Walker, C. M. (1996). Induction of cytotoxic T lymphocytes by intramuscular immunization with plasmid DNA is facilitated by bone marrow-derived cells. *Proc. Natl. Acad. Sci. USA* **93,** 8578.

116. Iwasaki, A., Torres, C. A., Ohashi, P. S., Robinson, H. L., and Barber, B. H. (1997). The dominant role of bone marrow-derived cells in CTL induction following plasmid DNA immunization at different sites. *J. Immunol.* **159,** 11.

117. Schirmbeck, R., Bohm, W., and Reimann, J. (1996). DNA vaccination primes MHC class I-restricted, simian virus 40 large tumor antigen-specific CTL in H-2d mice that reject syngeneic tumors. *J. Immunol.* **157,** 3550.

118. Torres, C. A., Iwasaki, A., Barber, B. H., and Robinson, H. L. (1997). Differential dependence on target site tissue for gene gun and intramuscular DNA immunizations. *J. Immunol.* **158,** 4529.

119. Cayeux, S., Richter, G., Noffz, G., Dorken, B., and Blankenstein, T. (1997). Influence of gene-modified (IL-7, IL-4, and B7) tumor cell vaccines on tumor antigen presentation. *J. Immunol.* **158,** 2834.

120. Schultze, J., Nadler, L. M., and Gribben, J. G. (1996). B7-mediated costimulation and the immune response. *Blood Rev.* **10,** 111.

121. Baskar, S., Clements, V. K., Glimcher, L. H., Nabavi, N., and Ostrand-Rosenberg, S. (1996). Rejection of MHC class II-transfected tumor cells requires induction of tumor-encoded B7-1 and/or B7-2 costimulatory molecules. *J. Immunol.* **156,** 3821.

122. Ciernik, I. F., Berzofsky, J. A., and Carbone, D. P. (1996). Induction of cytotoxic T lymphocytes and antitumor immunity with DNA vaccines expressing single T cell epitopes. *J. Immunol.* **156,** 2369.

123. Feltquate, D. M., Heaney, S., Webster, R. G., and Robinson, H. L. (1997). Different T helper cell types and antibody isotypes generated by saline and gene gun DNA immunization. *J. Immunol.* **158,** 2278.

124. Ramshaw, I. A., and Ramsay, A. J. (2000). The prime-boost strategy: exciting prospects for improved vaccination. *Immunol. Today* **21,** 163.

125. Ramsay, A. J., Kent, S. J., Strugnell, R. A., Suhrbier, A., Thomson, S. A., and Ramshaw, I. A. (1999). Genetic vaccination strategies for enhanced cellular, humoral and mucosal immunity. *Immunol. Rev.* **171,** 27.

126. Steinman, R. M. (1991). The dendritic cell system and its role in immunogenicity. *Annu. Rev. Immunol.* **9,** 271.

127. Steinman, R. M., Pack, M., and Inaba, K. (1997). Dendritic cell development and maturation. *Adv. Exp. Med. Biol.* **417,** 1.

128. Macatonia, S. E., Taylor, P. M., Knight, S. C., and Askonas, B. A. (1989). Primary stimulation by dendritic cells induces antiviral proliferative and cytotoxic T cell responses in vitro. *J. Exp. Med.* **169,** 1255.

129. Inaba, K., Metlay, J. P., Crowley, M. T., Witmer-Pack, M., and Steinman, R. M. (1990). Dendritic cells as antigen presenting cells in vivo. *Int. Rev. Immunol.* **6,** 197.

130. Lotze, M. T. (1997). Getting to the source: dendritic cells as therapeutic reagents for the treatment of patients with cancer [editorial; comment]. *Ann. Surg.* **226,** 1.

131. Morse, M. A., Zhou, L. J., Tedder, T. F., Lyerly, H. K., and Smith, C. (1997). Generation of dendritic cells in vitro from peripheral blood mononuclear cells with granulocyte-macrophage-colony-stimulating factor, interleukin-4, and tumor necrosis factor-alpha for use in cancer immunotherapy [see comments]. *Ann. Surg.* **226,** 6.

132. Lanzavecchia, A. (1996). Mechanisms of antigen uptake for presentation. *Curr. Opin. Immunol.* **8,** 348.

133. Shurin, M. R. (1996). Dendritic cells presenting tumor antigen. *Cancer Immunol. Immunother.* **43,** 158.

134. Heemels, M. T., and Ploegh, H. (1995). Generation, translocation, and presentation of MHC class I-restricted peptides. *Annu. Rev. Biochem.* **64,** 463.

135. Norbury, C. C., Chambers, B. J., Prescott, A. R., Ljunggren, H. G., and Watts, C. (1997). Constitutive macropinocytosis allows TAP-dependent major histocompatibility complex class I presentation of exogenous soluble antigen by bone marrow-derived dendritic cells. *Eur. J. Immunol.* **27,** 280.

136. Inaba, K., Metlay, J. P., Crowley, M. T., and Steinman, R. M. (1990). Dendritic cells pulsed with protein antigens in vitro can prime antigen-specific, MHC-restricted T cells in situ [published *erratum* appears in *J. Exp. Med.* 172(4), 1275, 1990]. *J. Exp. Med.* **172,** 631.

137. Svensson, M., Stockinger, B., and Wick, M. J. (1997). Bone marrow-derived dendritic cells can process bacteria for MHC-I and MHC-II presentation to T cells. *J. Immunol.* **158,** 4229.

138. Liu, L. M., and MacPherson, G. G. (1993). Antigen acquisition by dendritic cells: intestinal dendritic cells acquire antigen administered orally and can prime naive T cells in vivo. *J. Exp. Med.* **177,** 1299.

139. Romani, N., Gruner, S., Brang, D., Kampgen, E., Lenz, A., Trockenbacher, B., Konwalinka, G., Fritsch, P. O., Steinman, R. M., and Schuler, G. (1994). Proliferating dendritic cell progenitors in human blood. *J. Exp. Med.* **180,** 83.

140. Young, J. W., Szabolcs, P., and Moore, M. A. (1995). Identification of dendritic cell colony-forming units among normal human CD34+ bone marrow progenitors that are expanded by c-kit-ligand and yield pure dendritic cell colonies in the presence of granulocyte/macrophage colony-stimulating factor and tumor necrosis factor alpha. *J. Exp. Med.* **182,** 1111.

141. Inaba, K., Inaba, M., Romani, N., Aya, H., Deguchi, M., Ikehara, S., Muramatsu, S., and Steinman, R. M. (1992). Generation of large numbers of dendritic cells from mouse bone marrow cultures supplemented with granulocyte/macrophage colony-stimulating factor. *J. Exp. Med.* **176,** 1693.

142. Rissoan, M. C., Soumelis, V., Kadowaki, N., Grouard, G., Briere, F., de Waal Malefyt, R., and Liu, Y. J. (1999). Reciprocal control of T helper cell and dendritic cell differentiation [see comments]. *Science* **283,** 1183.

143. McLellan, A. D., Starling, G. C., and Hart, D. N. (1995). Isolation of human blood dendritic cells by discontinuous Nycodenz gradient centrifugation. *J. Immunol. Meth.* **184,** 81.

144. Hsu, F. J., Benike, C., Fagnoni, F., Liles, T. M., Czerwinski, D., Taidi, B., Engleman, E. G., and Levy, R. (1996). Vaccination of patients with B-cell lymphoma using autologous antigen-pulsed dendritic cells. *Nat. Med.* **2,** 52.

145. Maraskovsky, E., Brasel, K., Teepe, M., Roux, E. R., Lyman, S. D., Shortman, K., and McKenna, H. J. (1996). Dramatic increase in the numbers of functionally mature dendritic cells in Flt3 ligand-treated mice: multiple dendritic cell subpopulations identified. *J. Exp. Med.* **184,** 1953.

146. Jacobsen, S. E., Okkenhaug, C., Myklebust, J., Veiby, O. P., and Lyman, S. D. (1995). The FLT3 ligand potently and directly stimulates the growth and expansion of primitive murine bone marrow progenitor cells in vitro: synergistic interactions with interleukin (IL)-11, IL-12, and other hematopoietic growth factors. *J. Exp. Med.* **181,** 1357.

147. Roth, M. D., Gitlitz, B. J., Kiertscher, S. M., Park, A. N., Mendenhall, M., Moldawer, N., and Figlin, R. A. (2000). Granulocyte macrophage colony-stimulating factor and interleukin 4 enhance the number and antigen-presenting activity of circulating CD14+ and CD83+ cells in cancer patients. *Cancer Res.* **60,** 1934.

148. Nair, S. K., Boczkowski, D., Snyder, D., and Gilboa, E. (1997). Antigen-presenting cells pulsed with unfractionated tumor-derived peptides are potent tumor vaccines. *Eur. J. Immunol.* **27,** 589.

149. Paglia, P., Chiodoni, C., Rodolfo, M., and Colombo, M. P. (1996). Murine dendritic cells loaded in vitro with soluble protein prime cytotoxic T lymphocytes against tumor antigen in vivo [see comments]. *J. Exp. Med.* **183,** 317.

150. Mayordomo, J. I., Zorina, T., Storkus, W. J., Zitvogel, L., Celluzzi, C., Falo, L. D., Melief, C. J., Ildstad, S. T., Kast, W. M., Deleo, A. B. *et al.* (1995). Bone marrow-derived dendritic cells pulsed with synthetic tumour peptides elicit protective and therapeutic antitumour immunity. *Nat. Med.* **1,** 1297.

151. Celluzzi, C. M., Mayordomo, J. I., Storkus, W. J., Lotze, M. T., and Falo, Jr., L. D. (1996). Peptide-pulsed dendritic cells induce antigen-specific CTL-mediated protective tumor immunity [see comments]. *J. Exp. Med.* **183,** 283.

152. Gabrilovich, D. I., Cunningham, H. T., and Carbone, D. P. (1996). IL-12 and mutant P53 peptide-pulsed dendritic cells for the specific immunotherapy of cancer. *J. Immunother. Emphasis Tumor Immunol.* **19,** 414.

153. Zitvogel, L., Mayordomo, J. I., Tjandrawan, T., DeLeo, A. B., Clarke, M. R., Lotze, M. T., and Storkus, W. J. (1996). Therapy of murine tumors with tumor peptide-pulsed dendritic cells: dependence on T cells, B7 costimulation, and T helper cell 1-associated cytokines [see comments]. *J. Exp. Med.* **183,** 87.

154. Nair, S. K., Snyder, D., Rouse, B. T., and Gilboa, E. (1997). Regression of tumors in mice vaccinated with professional antigen-presenting cells pulsed with tumor extracts. *Int. J. Cancer* **70,** 706.

155. Ribas, A., Bui, L. A., Butterfield, L. H., Vollmer, C. M., Jilani, S., Dissette, V. B., Glaspy, J. A., McBride, W. H., and Economou, J. S. (1999). Antitumor protection using murine dendritic cells pulsed with acid-eluted peptides from in vivo grown tumors of different immunogenicities. *Anticancer Res.* **19,** 1165.

156. Arthur, J. F., Butterfield, L. H., Roth, M. D., Bui, L. A., Kiertscher, S. M., Lau, R., Dubinett, S., Glaspy, J., McBride, W. H., and Economou, J. S. (1997). A comparison of gene transfer methods in human dendritic cells. *Cancer Gene Ther.* **4,** 17.

157. Dietz, A. B., and Vuk-Pavlovic, S. (1998). High efficiency adenovirus-mediated gene transfer to human dendritic cells. *Blood* **91,** 392.

158. Jonuleit, H., Tüting, T., Steitz, J., Brück, J., Giesecke, A., Steinbrink, K., Knop, J., and Enk, A. H. (2000). Efficient transduction of mature CD83+ dendritic cells using recombinant adenovirus suppressed T cell stimulatory capacity. *Gene Ther.* **7,** 249.

159. Zhong, L., Granelli-Piperno, A., Choi, Y., and Steinman, R. M. (1999). Recombinant adenovirus is an efficient and non-perturbing genetic vector for human dendritic cells. *Eur. J. Immunol.* **29,** 964.

160. Alijagic, S., Moller, P., Artuc, M., Jurgovsky, K., Czarnetzki, B. M., and Schadendorf, D. (1995). Dendritic cells generated from peripheral blood transfected with human tyrosinase induce specific T cell activation. *Eur. J. Immunol.* **25,** 3100.

161. Aicher, A., Westermann, J., Cayeux, S., Willimsky, G., Daemen, K., Blankenstein, T., Uckert, W., Dorken, B., and Pezzutto, A. (1997). Successful retroviral mediated transduction of a reporter gene in human dendritic cells: feasibility of therapy with gene-modified antigen presenting cells. *Exp. Hematol.* **25,** 39.

162. Reeves, M. E., Royal, R. E., Lam, J. S., Rosenberg, S. A., and Hwu, P. (1996). Retroviral transduction of human dendritic cells with a tumor-associated antigen gene. *Cancer Res.* **56,** 5672.

163. Bronte, V., Carroll, M. W., Goletz, T. J., Wang, M., Overwijk, W. W., Marincola, F., Rosenberg, S. A., Moss, B., and Restifo, N. P. (1997). Antigen expression by dendritic cells correlates with the therapeutic effectiveness of a model recombinant poxvirus tumor vaccine. *Proc. Natl. Acad. Sci. USA* **94,** 3183.

164. Kim, C. J., Prevette, T., Cormier, J., Overwijk, W., Roden, M., Restifo, N. P., Rosenberg, S. A., and Marincola, F. M. (1997). Dendritic cells infected with poxviruses encoding MART-1/Melan A sensitize T lymphocytes in vitro. *J. Immunother.* **20,** 276.

165. Yang, S., Kittlesen, D., Slingluff, Jr., C. L., Vervaert, C. E., Seigler, H. F., and Darrow, T. L. (2000). Dendritic cells infected with a

vaccinia vector carrying the human gp100 gene simultaneously present multiple specificities and elicit high-affinity T cells reactive to multiple epitopes and restricted by HLA-A2 and -A3. *J. Immunol.* **164,** 4204.

166. Corinti, S., Medaglini, D., Cavani, A., Rescigno, M., Pozzi, G., Ricciardi-Castagnoli, P., and Girolomoni, G. (1999). Human dendritic cells very efficiently present a heterologous antigen expressed on the surface of recombinant gram-positive bacteria to CD4+ T lymphocytes. *J. Immunol.* **163,** 3029.

167. Ribas, A., Butterfield, L. H., McBride, W. H., Jilani, S. M., Bui, L. A., Vollmer, C. M., Lau, R., Dissette, V. B., Hu, B., Chen, A. Y., Glaspy, J. A., and Economou, J. S. (1997). Genetic immunization for the melanoma antigen MART-1/Melan-A using recombinant adenovirus-transduced murine dendritic cells. *Cancer Res.* **57,** 2865.

168. Butterfield, L. H., Jilani, S. M., Chakraborty, N. G., Bui, L. A., Ribas, A., Dissette, V. B., Lau, R., Gamradt, S. C., Glaspy, J. A., McBride, W. H., Mukherji, B., and Economou, J. S. (1998). Generation of melanoma-specific cytotoxic T lymphocytes by dendritic cells transduced with a MART-1 adenovirus. *J. Immunol.* **161,** 5607.

169. Ribas, A., Butterfield, L. H., Dissette, V. B., Ho, B., Chen, A. Y., Andrews, K. J., Eibler, F. C., Glaspy, J. A., Economou, J. S., and McBride, W. H. (1998). Generation of anti-tumor immunity using dendritic cells genetically modified to express tumor specific antigen and cytokines. *Proc. 17th Intl. Cancer Congress,* Rio de Janeiro.

170. Ribas, A., Butterfield, L. H., McBride, W. H., Dissette, V. B., Koh, A., Vollmer, C. M., Hu, B., Chen, A. Y., Glaspy, J. A., and Economou, J. S. (1999). Characterization of antitumor immunization to a defined melanoma antigen using genetically engineered murine dendritic cells. *Cancer Gene Ther.* **6,** 523.

171. Ribas, A., Butterfield, L. H., Hu, B., Dissette, V. B., Chen, A. Y., Koh, A., Amarani, S. N., Glaspy, J. A., McBride, W. H., and Economou, J. S. (1999). Generation of T cell immunity to a murine melanoma using MART-1 engineered dendritic cells. *J. Immunother.* (in press).

172. Ribas, A., Butterfield, L. H., Hu, B., Dissette, V. B., Koh, A., Lee, M., Andrews, K. J., Meng, W., Glaspy, J. A., McBride, W. H., and Economou, J. S. (2000). Immune deviation and Fas-mediated deletion limit antitumor activity after multiple dendritic cell vaccinations in mice. *Cancer Res.* **60,** 2218.

173. Miller, P. W., Sharma, S., Stolina, M., Butterfield, L. H., Luo, J., Lin, Y., Dohadwala, M., Batra, R. K., Wu, L., Economou, J. S., and Dubinett, S. M. (2000). Intratumoral administration of adenoviral interleukin 7 gene-modified dendritic cells augments specific anti-tumor immunity and achieves tumor eradication. *Hum. Gene Ther.* **11,** 53.

174. Zitvogel, L., Couderc, B., Mayordomo, J. I., Robbins, P. D., Lotze, M. T., and Storkus, W. J. (1996). IL-12-engineered dendritic cells serve as effective tumor vaccine adjuvants in vivo. *Ann. N.Y. Acad. Sci.* **795,** 284.

175. Gong, J., Chen, D., Kashiwaba, M., and Kufe, D. (1997). Induction of antitumor activity by immunization with fusions of dendritic and carcinoma cells. *Nat. Med.* **3,** 558.

176. Hart, I., and Colaco, C. (1997). Immunotherapy. Fusion induces tumour rejection [news]. *Nature* **388,** 626.

177. Kugler, A., Stuhler, G., Walden, P., Zöller, G., Zobywalski, A., Brossart, P., Trefzer, U., Ullrich, S., Müller, C. A., Becker, V., Gross, A. J., Hemmerlein, B., Kanz, L., Müller, G. A., and Ringert, R. H. (2000). Regression of human metastatic renal cell carcinoma after vaccination with tumor cell-dendritic cell hybrids [see comments]. *Nat. Med.* **6,** 332.

178. Gabrilovich, D. I., Ciernik, I. F., and Carbone, D. P. (1996). Dendritic cells in antitumor immune responses. I. Defective antigen presentation in tumor-bearing hosts. *Cell Immunol.* **170,** 101.

179. Kiertscher, S. M., Luo, J., Dubinett, S. M., and Roth, M. D. (2000). Tumors promote altered maturation and early apoptosis of monocyte-derived dendritic cells. *J. Immunol.* **164,** 1269.

180. Tjoa, B., Erickson, S., Barren, 3rd, R., Ragde, H., Kenny, G., Boynton, A., and Murphy, G. (1995). In vitro propagated dendritic cells from prostate cancer patients as a component of prostate cancer immunotherapy. *Prostate* **27,** 63.

181. Van Waes, C., Monach, P. A., Urban, J. L., Wortzel, R. D., and Schreiber, H. (1996). Immunodominance deters the response to other tumor antigens thereby favoring escape: prevention by vaccination with tumor variants selected with cloned cytolytic T cells in vitro. *Tissue Antigens* **47,** 399.

182. Mukherji, B., Chakraborty, N. G., Yamasaki, S., Okino, T., Yamase, H., Sporn, J. R., Kurtzman, S. K., Ergin, M. T., Ozols, J., Meehan, J. *et al.* (1995). Induction of antigen-specific cytolytic T cells in situ in human melanoma by immunization with synthetic peptide-pulsed autologous antigen presenting cells. *Proc. Natl. Acad. Sci. USA* **92,** 8078.

183. Dhodapkar, M. V., Steinman, R. M., Sapp, M., Desai, H., Fossella, C., Krasovsky, J., Donahoe, S. M., Dunbar, P. R., Cerundolo, V., Nixon, D. F., and Bhardwaj, N. (1999). Rapid generation of broad T-cell immunity in humans after a single injection of mature dendritic cells. *J. Clin. Invest.* **104,** 173.

184. Thurner, B., Haendle, I., Roder, C., Dieckmann, D., Keikavoussi, P., Jonuleit, H., Bender, A., Maczek, C., Schreiner, D., von den Driesch, P., Brocker, E. B., Steinman, R. M., Enk, A., Kampgen, E., and Schuler, G. (1999). Vaccination with mage-3A1 peptide-pulsed mature, monocyte-derived dendritic cells expands specific cytotoxic T cells and induces regression of some metastases in advanced stage IV melanoma. *J. Exp. Med.* **190,** 1669.

185. Reichardt, V. L., Okada, C. Y., Liso, A., Benike, C. J., Stockerl-Goldstein, K. E., Engleman, E. G., Blume, K. G., and Levy, R. (1999). Idiotype vaccination using dendritic cells after autologous peripheral blood stem cell transplantation for multiple myeloma—a feasibility study. *Blood* **93,** 2411.

186. Cull, G., Durrant, L., Stainer, C., Haynes, A., and Russell, N. (1999). Generation of anti-idiotype immune responses following vaccination with idiotype-protein pulsed dendritic cells in myeloma. *Br. J. Haematol.* **107,** 648.

187. Höltl, L., Rieser, C., Papesh, C., Ramoner, R., Herold, M., Klocker, H., Radmayr, C., Stenzl, A., Bartsch, G., and Thurnher, M. (1999). Cellular and humoral immune responses in patients with metastatic renal cell carcinoma after vaccination with antigen pulsed dendritic cells. *J. Urol.* **161,** 777.

188. Cormier, J. N., Salgaller, M. L., Prevette, T., Barracchini, K. C., Rivoltini, L., Restifo, N. P., Rosenberg, S. A., and Marincola, F. M. (1997). Enhancement of cellular immunity in melanoma patients immunized with a peptide from MART-1/Melan A [see comments]. *Cancer J. Sci. Am.* **3,** 37.

189. Rosenberg, S. A., Yang, J. C., Schwartzentruber, D. J., Hwu, P., Marincola, F. M., Topalian, S. L., Restifo, N. P., Dudley, M. E., Schwarz, S. L., Spiess, P. J., Wunderlich, J. R., Parkhurst, M. R., Kawakami, Y., Seipp, C. A., Einhorn, J. H., and White, D. E. (1998). Immunologic and therapeutic evaluation of a synthetic peptide vaccine for the treatment of patients with metastatic melanoma [see comments]. *Nat. Med.* **4,** 321.

190. Dieu-Nosjean, M. C., Vicari, A., Lebecque, S., and Caux, C. (1999). Regulation of dendritic cell trafficking: a process that involves the participation of selective chemokines. *J. Leukocyte Biol.* **66,** 252.

191. Dieu, M. C., Vanbervliet, B., Vicari, A., Bridon, J. M., Oldham, E., Aït-Yahia, S., Brière, F., Zlotnik, A., Lebecque, S., and Caux, C. (1998). Selective recruitment of immature and mature dendritic cells by distinct chemokines expressed in different anatomic sites. *J. Exp. Med.* **188,** 373.

192. Barratt-Boyes, S. M., Zimmer, M. I., Harshyne, L. A., Meyer, E. M., Watkins, S. C., Capuano, 3rd, S., Murphey-Corb, M., Falo, Jr., L. D., and Donnenberg, A. D. (2000). Maturation and trafficking of monocyte-derived dendritic cells in monkeys: implications for dendritic cell-based vaccines. *J. Immunol.* **164,** 2487.

193. Morse, M. A., Coleman, R. E., Akabani, G., Niehaus, N., Coleman, D., and Lyerly, H. K. (1999). Migration of human dendritic cells after injection in patients with metastatic malignancies. *Cancer Res.* **59,** 56.

194. Eggert, A. A., Schreurs, M. W., Boerman, O. C., Oyen, W. J., de Boer, A. J., Punt, C. J., Figdor, C. G., and Adema, G. J. (1999). Biodistribution and vaccine efficiency of murine dendritic cells are dependent on the route of administration. *Cancer Res.* **59,** 3340.

CHAPTER

11

RNA-Transfected Dendritic Cells as Immunogens

MICHAEL A. MORSE
Department of Medicine
Division of Medical Oncology and Transplantation
Duke University Medical Center
Durham, North Carolina 27710

SMITA K. NAIR
Department of Surgery
Duke University Medical Center
Durham, North Carolina 27710

H. KIM LYERLY
Department of Surgery
Duke University Medical Center
Durham, North Carolina 27710

I. INTRODUCTION

The field of active immunotherapy has blossomed in the last few years due to its appeal as a method for directly targeting tumors using naturally occurring pathways and the possibility of achieving promising results with minimal toxicity. Active immunotherapy has become feasible as some of the requirements for inducing a potent tumor-specific immune response have been elucidated. These requirements include the presence of cytolytic T cells (CTLs) capable of recognizing tumor antigens, presentation of antigens by tumors as peptide fragments bound in the groove of the tumor major histocompatibility complex (MHC) class I molecule [1], and adequate stimulation of CTLs achieved by presenting antigen in conjunction with costimulatory molecules. Although

tumors generally lack costimulatory molecules, dendritic cells (DCs), bone-marrow-derived antigen-presenting cells, express high levels of costimulatory and human leukocyte antigen (HLA) molecules, making them particularly potent inducers of T-cell activity [2]. Previously, the limited availability of DCs slowed progress in this field, but over the last 5 years, many methods for obtaining DCs for immunization strategies have been developed (summarized in Morse and Lyerly [3].) DCs loaded *ex vivo* with tumor antigens induce antigen-specific immune responses *in vitro* (summarized in Table 2 of Esche *et al.* [4]) and *in vivo* in animal models of metastatic tumor (summarized in Table 1 of Tarte and Klein [5]). Preliminary human clinical trials have demonstrated promising immunologic responses with these approaches.

An important issue in the field of DC-based immunotherapy is what antigens should be used and in what form they should be provided to the DCs. Antigens may be defined or undefined and provided in the form of tumor cells, cell extracts, apoptotic bodies, protein, peptides, and genetic material. This review will focus on the use of mRNA-transfected dendritic cells and the advantages of this approach when applied to the immunotherapy of cancer.

II. ADVANTAGES OF LOADING DENDRITIC CELLS WITH GENETIC MATERIAL

The specificity of the immune response induced by DCs lies in the presentation of antigen in the form of peptide fragments within HLA molecules to T cells with receptors capable of recognizing the particular epitope; therefore, the final common pathway for antigen presentation requires peptide to complex with HLA molecules. This can occur within the cell, when peptide complexes with HLA molecules in the endoplasmic reticulum or other endosomal sites, or exogenously, when peptides applied to the milieu surrounding the DCs bind directly to HLA molecules on the cell surface. All the methods for loading DCs utilize these pathways, therefore, the choice of antigenic loading strategy depends more on factors such as availability of antigen, whether defined or undefined antigens are desired, the necessity for class II helper epitopes, HLA restrictions, and the effect of "bystander" antigens.

The use of genetic material allows production of full-length proteins that, when processed, may provide multiple different epitopes, both class I and class II. Antigen loading does not depend on HLA type because the peptide fragments that can fit within the particular HLA molecule will be "chosen" within the cell. In contrast, peptide loading requires the use of amino acid sequences compatible with the individual's HLA type, and each peptide only provides a single epitope. Genetic material is easier to produce than proteins, and it can be used in all cases, even when the actual protein, encoded by the genetic material is not available in sufficient quantity with adequate purity. Genetic material can be readily reengineered to add additional sequences that improve translation and expression of the protein, target the protein towards the desired HLA molecules (e.g., lysosome-associated membrane protein [LAMP] sequences [6]), or increase the immunogenicity of the encoded epitopes. Tumor cells, tumor extracts, and apoptotic bodies contain other material that is not antigenic and could potentially interfere with immune stimulation or cause undesirable immune stimulation against self-antigens. These approaches also require adequate amounts of tumor tissue, which is often in short supply. Genetic material can theoretically be amplified from small quantities of tumor to overcome this problem.

III. VIRAL VERSUS NONVIRAL METHODS OF GENE TRANSFER

Genetic material may be delivered to cells within viral or bacterial vectors or by mechanical methods. Viral methods have features that may be desirable under some circumstances. Retroviral vectors [7] stably integrate into the host-cell genome, resulting in constitutive expression of full-length proteins, and may theoretically cause prolonged antigen presentation in transduced DCs. Adenovirus can be produced in high titer, and the genome accomodates large gene constructs [8]. Fowlpox are unable to replicate in human cells, making them fairly safe vectors [9]. Viral vectors also have a number of undesirable features. Some would argue that stable integration risks inducing carcinogenesis in retrovirally transduced cells; adenovirus can induce potent anti-adenoviral immune responses that may interfere with induction of im-

mune responses to less immunogenic epitopes [10]. Finally, immune responses against viral antigens may cause rejection of the dendritic cells by viral-antigen-specific T cells before they are able to present antigen to tumor-antigen-specific T cells [11].

Nonviral methods of gene transfer, while generally resulting in transient gene expression, eliminate many of these concerns. While it is true that nonviral methods achieve lower gene transduction efficiency of DCs than adenoviral vectors [12], even low levels of expression of antigen are sufficient to stimulate T-cell responses [13]. When *in vitro*-generated DCs were transfected with the tyrosinase gene with lipofectin, low levels of tyrosinase were observed, but tyrosinase-expressing DCs were able to activate tyrosinase-specific T cells [13].

The most common form of nonviral transfer utilizes plasmids containing the gene of interest, a promotor, and usually reporter and selection sequences. DCs have been successfully transfected with plasmids encoding genes of interest and, despite low transfection efficiency, plasmid-transfected DCs are potent immune stimulators. Philip and colleagues [14], using monocyte-derived DCs transiently transfected with MART-1, stimulated naïve CD8$^+$ T cells with cytolytic activity against human HLA-matched tumor cells expressing MART-1. Similarly Tuting and colleagues [15] used DCs transfected with plasmid encoding the human papilloma virus E7 antigen or p53 to induce protective responses against tumors bearing these antigens. Yang *et al.* [16] transfected murine DCs with plasmid DNA containing the gp100 gene and observed that immunization with the DCs protected mice from subsequent challenge with tumors expressing gp100. Thus, it is well established that nonviral transfer of genetic material to DCs is adequate enough to result in stimulation of detectable immune responses.

IV. RNA VERSUS DNA LOADING OF DENDRITIC CELLS

Both DNA and RNA have distinct advantages. DNA is more stable than RNA and can be produced in large quantities. It is also possible to control the expression of genes encoded in DNA by including target tissue-specific promoters or other sequences in the plasmid, that allow selection of transfected cells. Multiple copies of the gene can be placed into one plasmid, permitting greater gene expression [17]. In contrast, RNA can be directly translated into protein in the cytoplasm, thus eliminating the extra step of DNA transcription to RNA in the nucleus. Second, if the total content of tumor genetic material is to be used for DC loading, mRNA allows one to narrow the antigenic pool to those proteins actually expressed by the tumor cell. Third, many different mRNAs can be loaded into a cell, but usually only a few DNA-encoded genes can be expressed in any one cell. Fourth, production of large quantities of DNA requires cloning in bacteria, but a cDNA can be produced from mRNA by polymerase chain reaction (PCR) and then the cDNA can be transcribed into large amounts of RNA *in vitro*.

The apparent disadvantage of RNA due to its short half-life, even in serum-free medium, does not seem to affect levels of gene expression in tumor cell models [18]. Furthermore, even though it has been difficult to demonstrate protein production within RNA-transfected dendritic cells, these cells do induce T-cell responses specific for the antigen (as we will discuss further). It is not known whether this reflects low uptake or translation of the RNA, rapid RNA degradation, or rapid protein degradation. Finally, some authors have suggested that high levels of antigenic presentation lead to induction of low-affinity T cells, whereas low levels of antigen presentation result in higher affinity T cells [18].

V. RNA LOADING OF DENDRITIC CELLS

RNA transfection of cells generally requires a physical method. Previously, liposomes [19–22] and ballistic incorporation using a gene gun [23] were used to incorporate mRNA into cells. Liposomes present a number of challenges, including their toxic effects on target cells, and the best choice for loading RNA into dendritic cells is not known. For tumor cells, 1,2-dioleoyl-3-trimethylammonium-propane (DOTAP) was found to result in the best gene expression [19]. Efficiency may also be increased by incorporating targeting molecules into the liposome. For example, Glenn and colleagues [20] incorporated glycophorin into liposomes in order to target the gene delivery to influenza-bearing cells, thus taking advantage of the binding of glycophorin to influenza hemagglutinin.

More recently, the use of naked RNA has been suggested. This approach may be difficult, in theory, because of the ubiquitous presence of RNAses, but naked RNA vectors have been tested *in vivo*. A single intramuscular injection of a self-replicating RNA immunogen elicited antigen-specific antibody and CD8$^+$ T-cell responses at doses as low as 0.1 μg. Preimmunization with a self-replicating RNA vector protected mice from tumor challenge, and therapeutic immunization prolonged the survival of mice with established tumors [24]. The immune response appeared to involve apoptotic cell death of the transfected cells and apoptotic body uptake by *in situ* DCs. Recently, the direct transfection of DCs by mRNA has been achieved [25]. Nair and colleagues observed that naked mRNA, although not leading to easily detectable protein expression, was still capable of loading DCs and stimulating antigen-specific T-cell responses [26]. The results achieved were similar to those observed when mRNA was mixed with lipid complexes, except that a greater amount of mRNA was required when used in the absence of lipid.

An important issue in using RNA for transfection of DCs is the requirement that the DCs be immature [27]. We observed

that, if DCs are matured first and then exposed to mRNA, they do not stimulate CTL activity as effectively does as exposure to mRNA prior to maturation [27]. This is likely due to the need for DCs to actively take up the mRNA, an ability dramatically downregulated once the cells become mature.

VI. AMPLIFICATION OF RNA USED TO LOAD DENDRITIC CELLS

One advantage of using RNA as a source of antigens is that the minute quantities extracted from small tumor specimens may be amplified into amounts adequate for the transfection procedure. Previously, a limitation to the use of total tumor RNA has been the availability of adequate specimens, but RNA amplification potentially obviates this need. Boczkowski and colleagues have demonstrated that CTL responses may be stimulated by DCs loaded with RNA amplified from tumors [28]. In this study, tumor cells were microdissected from pathologic slides and the RNA was isolated, amplified, and used to transfect DCs. This group has determined a number of important issues related to the amplification, including the fact that primers for the reverse transcription procedure must include 64T so that mRNA with long polyA tails will be produced.

VII. USES OF RNA-LOADED DENDRITIC CELLS

RNA-transfected DCs may be used as therapeutic vaccines, for discovery of new antigens, and in bioassays of immunologic responses to vaccination in clinical trials. Boczkowski *et al.* [25] were the first to describe the potential therapeutic use of RNA-transfected DCs. Murine DCs transfected with RNA encoding the chicken ovalbumin antigen were capable of stimulating potent primary CTL responses *in vitro*, and tumor RNA-transfected DCs were effective in preventing tumor progression in a postsurgical metastasis model [25]. Other murine models have demonstrated the feasibility of loading DCs with total tumor-derived RNA. In a murine model of B16/F10 murine melanoma localized to the central nervous system (CNS), vaccination with bone-marrow-generated DCs, pulsed with B16 total RNA protected animals from tumor challenge. DCs-based vaccines also led to prolongation of survival in mice with established tumors [29]. In murine models, both mRNA of defined tumor antigens and total tumor-derived RNA have been used to modify the DCs. Zhang *et al.* [30] vaccinated mice with DCs transfected with tumor RNA and demonstrated potent CTL activity specific for the tumor cells, protection against subsequent tumor challenge, a reduction in pulmonary metastases, and prolonged survival in the 3LL and B16 tumor models. As few as 40,000 cells were effective for immunization. In this study, the DCs were also modified with an adenoviral vector encoding lymphotactin (Lptn) that increases chemotaxis to T cells, therefore, it is not possible to determine fully the effect of the RNA immunizations alone.

In vitro, human DCs transfected with RNA encoding tumor antigens are capable of stimulating primary CTL responses [26,31]. In our studies, dendritic cells from patients with advanced cancers have been loaded with mRNA encoding carcinoembryonic antigen (CEA) and have been demonstrated to induce CTL activity specific for CEA [32]. We have initiated clinical trials to assess the safety of and immunologic response to immunization with DCs loaded with mRNA encoding tumor antigens. Preliminary experience with these cells demonstrates that they are feasible to prepare and safe to administer intravenously and intradermally [33].

A. Antigen Discovery

As noted previously, either known (defined) or undefined antigens may be used to load DCs. Although the number of described antigens has increased, it is still likely that the true rejection antigens for a particular individuals tumor are undescribed and may even differ between people. By using subtraction hybridization, it may be possible to obtain populations of mRNA encoding antigens restricted to tumor cells. Transfection of DCs with these mRNA species would permit induction of CTL specific for these antigens. Subsequently, clones of CTL against individual antigens may be produced and used to search for these new tumor-specific antigens.

B. Immunologic Monitoring

Previously reported clinical trials of immunotherapy have used relatively crude methods of monitoring immunologic responses. One particularly vexing limitation has been the need for targets expressing the antigens of interest. If tumor cells are available in abundance, they would be the preferred choice, but because they are usually limiting other methods of obtaining target cells expressing the antigens of interest must be pursued. One particularly useful approach would be to load DCs with the total tumor RNA from a small sample of autologous tumor or with defined tumor-antigen RNA. We have observed that these loaded DCs may serve as targets for antigen-specific CTL [32]. Thus, DCs loaded with RNA may serve as a surrogate target for tumor cells.

VIII. FUTURE DIRECTIONS

There are ongoing efforts to improve the efficacy of RNA transfection of dendritic cells. RNA transfection efficiency is low, although apparently functionally effective. It is possible that better transfection efficiency may further improve induction of immune responses. Modifications to the RNA to improve translation efficiency and stability of the message and

to include signals that affect protein processing may increase antigenicity. Also necessary are improvements in the amplification technology to make it more readily available and to encourage amplification of all mRNA species instead of preferential amplification. Finally, clinical trials using modern methods of immune analysis to detect the immune response to these vaccines will help speed along the progress.

References

1. Darrow, T., Slingluff, C., and Seigler, H. (1989). The role of HLA class I antigens in recognition of melanoma cells by tumor-specific cytotoxic T lymphocytes. Evidence for shared tumor antigens. *J. Immunol.* **42,** 3329–3335.

2. Liu, Y., and Janeway, C. A. (1992). Cells that present both specific ligand and costimulatory activity are the most efficient inducers of clonal expansion of normal CD4 T cells. *Proc. Natl. Acad. Sci. USA* **89,** 3845–3849.

3. Morse, M. A., and Lyerly, H. K. (1999). The isolation and culture of human dendritic cells, in *Human Cell Culture, Vol. IV. Primary Hematopoietic Cells* (M. F. Koller, B. O. Palsson, and J. R. W., eds.), pp. 171–192. Kluver Academic Publishers, Dordrecht.

4. Esche, C., Shurin, M. R., and Lotze, M. T. (1999). The use of dendritic cells for cancer vaccination. *Curr. Opin. Mol. Ther.* **1,** 72–81.

5. Tarte, K., and Klein, B. (1999). Dendritic cell-based vaccine: a promising approach for cancer immunotherapy. *Leukemia* **13,** 653–663.

6. Ruff, A. L., Guarnieri, F. G., Staveley-O'Carroll, K., Siliciano, R. F., and August, J. T. (1997). The enhanced immune response to the HIV gp160/LAMP chimeric gene product targeted to the lysosome membrane protein trafficking pathway. *J. Biol. Chem.* **272,** 8671–8678.

7. Palu, G., Parolin, C., Takeuchi, Y., and Pizzato, M. (2000) Progress with retroviral gene vectors. *Rev. Med. Virol.* **10,** 185–202.

8. Zhang, W. W. (1999). Development and application of adenoviral vectors for gene therapy of cancer. *Cancer Gene Ther.* **6,** 113–138.

9. Tartaglia, J., Pincus, S., and Paoletti, E. (1990). Poxvirus-based vectors as vaccine candidates. *Crit. Rev. Immunol.* **10,** 13–30.

10. Pion, S., Christianson, G. J., Fontaine, P., Roopenian, D. C., and Perreault, C. (1999). Shaping the repertoire of cytotoxic T-lymphocyte responses: explanation for the immunodominance effect whereby cytotoxic T lymphocytes specific for immunodominant antigens prevent recognition of nondominant antigens. *Blood* **93,** 952–962.

11. Ingulli, E., Mondino, A., Khoruts, A., and Jenkins, M. K. (1997). In vivo detection of dendritic cell antigen presentation to CD4(+) T cells. *J. Exp. Med.* **185,** 2133–2141.

12. Arthur, J. F., Butterfield, L. H, Roth, M. D., Bui, L. A., Kiertscher, S. M., Lau, R., Dubinett, S., Glaspy, J., McBride, W. H., and Economou, J. S. (1997). A comparison of gene transfer methods in human dendritic cells. *Cancer Gene Ther.* **4,** 17–25.

13. Alijagic, S., Moller, P., Artuc, M., Jurgovsky, K., Czarnetzki, B. M., and Schadendorf. D. (1995). Dendritic cells generated from peripheral blood transfected with human tyrosinase induce specific T cell activation. *Eur. J. Immunol* **25,** 3100–3107.

14. Philip, R., Brunette, E., Ashton, J., Alters, S., Gadea, J., Sorich, M., Yau, J., O'Donoghue, G., Lebkowski, J., Okarma, T., and Philip, M. (1998). Transgene expression in dendritic cells to induce antigen-specific cytotoxic T cells in healthy donors. *Cancer Gene Ther.* **5,** 236–246.

15. Tuting, T., DeLeo, A. B., Lotze, M. T., and Storkus, W. J. (1997). Genetically modified bone marrow-derived dendritic cells expressing tumor-associated viral or "self" antigens induce antitumor immunity in vivo. *Eur. J. Immunol.* **27,** 2702–2707.

16. Yang, S., Vervaert, C. E., Burch, Jr., J., Grichnik, J., Seigler, H. F., and Darrow, T. L. (1999). Murine dendritic cells transfected with human

GP100 elicit both antigen-specific CD8(+) and CD4(+) T-cell responses and are more effective than DNA vaccines at generating anti-tumor immunity. *Int. J. Cancer* **83,** 532–540.

17. Yarovoi, S. V., Mouawad, R., Colbere-Garapin, F., Soubrane, C., Khayat, D., and Rixe, O. (1996). In vitro sensitization of the B16 murine melanoma cells to ganciclovir by different RNA and plasmid DNA constructions encoding HSVtk. *Gene Ther.* **3,** 913–918.

18. Vitiello, A., Sette, A., Yuan, L., Farness, P., Southwood, S., Sidney, J., Chesnut, R. W., Grey, H. M., and Livingston, B. (1997). Comparison of cytotoxic T lymphocyte responses induced by peptide or DNA immunization: implications on immunogenicity and immunodominance. *Eur. J. Immunol.* **27,** 671–678.

19. Lu, D., Benjamin, R., Kim, M., Conry, R. M., and Curiel, D. T. (1994). Optimization of methods to achieve mRNA-mediated transfection of tumor cells in vitro and in vivo employing cationic liposome vectors. *Cancer Gene Ther.* **1,** 245–252.

20. Glenn, J. S., Ellens, H., and White, J. M. (1993). Delivery of liposome-encapsulated RNA to cells expressing influenza virus hemagglutinin. *Methods Enzymol.* **221,** 327–339.

21. Wolff, J. A., Malone, R. W., Williams, P., Chong, W., Acsadi, G., Jani, A., Felgner, and P. L. (1990). Direct gene transfer into mouse muscle in vivo. *Science* **247,** 1465–1468.

22. Hoerr, I., Obst, R., Rammensee, H. G., and Jung, G. (2000). In vivo application of RNA leads to induction of specific cytotoxic T lymphocytes and antibodies. *Eur. J. Immunol.* **30,** 1–7.

23. Qiu, P., Ziegelhoffer, P., Sun, J., and Yang, N. S. (1996). Gene gun delivery of mRNA in situ results in efficient transgene expression and genetic immunization. *Gene Ther.* **3,** 262–268.

24. Ying, H., Zaks, T. Z., Wang, R. F., Irvine, K. R., Kammula, U. S., Marincola, F. M., Leitner, W. W., and Restifo, N. P. (1999). Cancer therapy using a self-replicating RNA vaccine. *Nat. Med.* **5,** 823–827.

25. Boczkowski, D., Nair, S., Snyder, D., and Gilboa, E. (1996). Dendritic cells pulsed with RNA are potent antigen presenting cells in vitro and in vivo. *J. Exp. Med.* **184,** 465–472.

26. Nair, S. K., Boczkowski, D., Morse, M., Cumming, R. I., Lyerly, H. K., and Gilboa, E. (1998). Induction of primary carcinoembryonic antigen (CEA)-specific cytotoxic T lymphocytes in vitro using human dendritic cells transfected with RNA. *Nat. Biotechnol.* **16,** 364–369.

27. Morse, M. A., Lyerly, H. K., Gilboa, E., Thomas, E. K., and Nair, S. K. (1998). Optimization of the sequence of antigen loading and CD40-ligand-induced maturation of dendritic cells. *Cancer Res.* **58,** 2965–2968.

28. Boczkowski, D., Nair, S. K., Nam, J. H., Lyerly, H. K., and Gilboa, E. (2000). Induction of tumor immunity and cytotoxic T lymphocyte responses using dendritic cells transfected with messenger RNA amplified from tumor cells. *Cancer Res.* **60,** 1028–1034.

29. Ashley, D. M., Faiola, B., Nair, S., Hale, L. P., Bigner, D. D., and Gilboa, E. (1997). Bone marrow-generated dendritic cells pulsed with tumor extracts or tumor RNA induce antitumor immunity against central nervous system tumors. *J. Exp. Med.* **186,** 1177–1182.

30. Zhang, W., He, L., Yuan, Z., Xie, Z., Wang, J., Hamada, H., and Cao, X. (1999). Enhanced therapeutic efficacy of tumor RNA-pulsed dendritic cells after genetic modification with lymphotactin. *Hum. Gene Ther.* **10,** 1151–1161.

31. Heiser, A., Dahm, P., Yancey, D., Maurice, M. A., Boczkowski, D., Nair, S. K., Gilboa, E., and Vieweg, J. (2000). Human dendritic cells transfected with RNA encoding prostate-specific antigen stimulate prostate-specific CTL responses in vitro. *J. Immunol.* **164,** 5508–5514.

32. Nair, S. K., Hull, S., Coleman, D., Gilboa, E., Lyerly, H. K., and Morse, M. A. (1999). Induction of carcinoembryonic antigen (CEA)-specific cytotoxic T-lymphocyte responses in vitro using autologous dendritic cells loaded with CEA peptide or CEA RNA in patients with metastatic malignancies expressing CEA. *Int. J. Cancer.* **82,** 121–124.

33. Morse, M. A., Coleman, R. E., Akabani, G., Niehaus, N., Coleman, D., and Lyerly, H. K. (1999). Migration of human dendritic cells after injection in patients with metastatic malignancies. *Cancer Res.* **59,** 56–58.

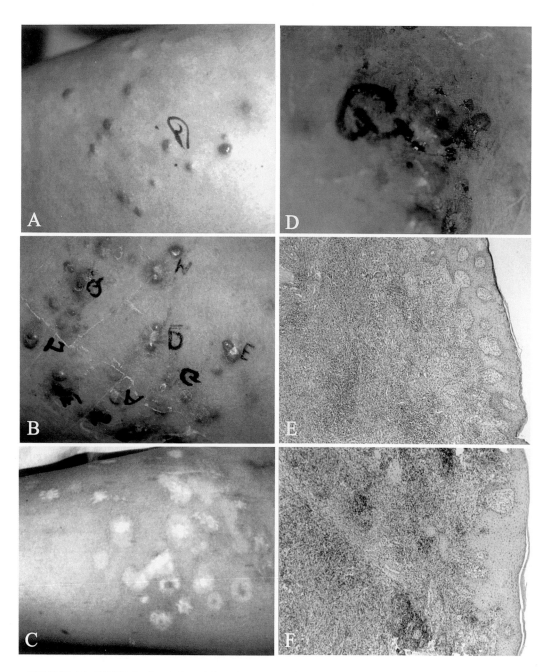

CHAPTER 12, FIGURE 4 Resolution of dermal metastases following intralesional injection of recombinant vac-cinia–granulocyte–macrophage colony-stimulating factor (GM-CSF). Patient 3 is a 32-year-old female with extensive dermal metastases of the left thigh before treatment (A), on day 81 (B), and on day 600, 150 days following cessation of treatment (C). Regression was accompanied by gross (D) and histologic (E) evidence of inflammation, including significant T-cell (CD3+) involvement (F). (From Mastrangelo, M. J. *et al.*, *Cancer Gene Ther.*, 6(5):409–422, 1999. With permission.)

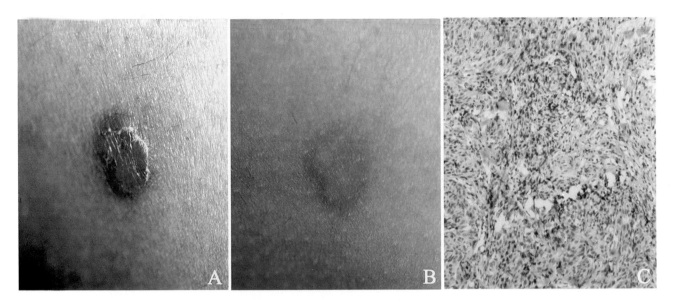

CHAPTER 12, FIGURE 5   Regression of uninjected lesions accompanied by T-cell infiltration. A representative unin-jected distant regressing lesion prior to (A) and following (B) patient treatment demonstrated T-cell (CD8) infiltration (C). (From Mastrangelo, M. J. *et al., Cancer Gene Ther.,* 6(5):409–422, 1999. With permission.)

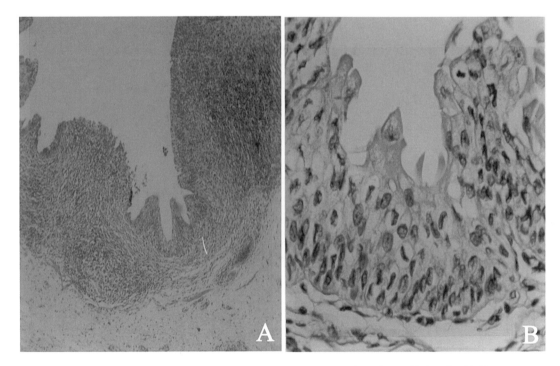

CHAPTER 12, FIGURE 7   Intravesical administration of vaccinia vector in patients with invasive bladder cancer. H&E-stained section of bladder from patient 2, taken at time of cystectomy 24 hours after the third of three intravesical instillations of vaccinia, shows widespread inflammation (A) and infection of the urothelium (B).

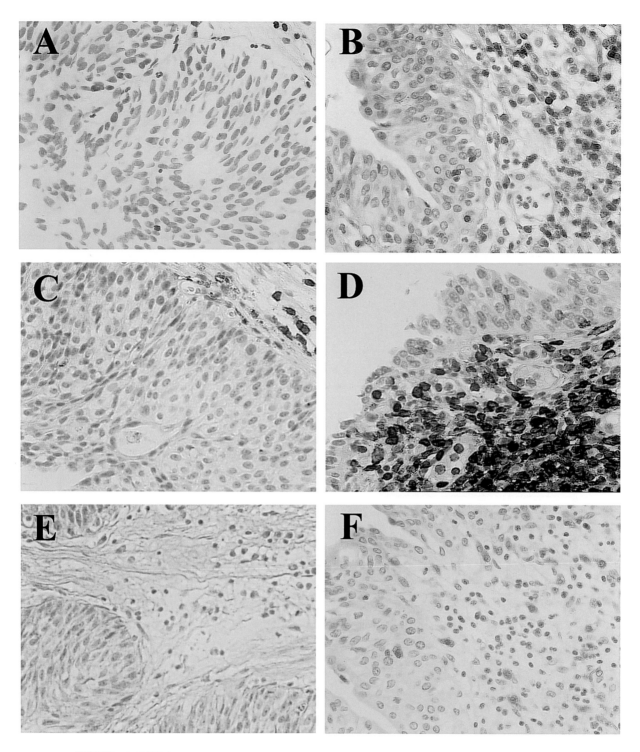

CHAPTER 12, FIGURE 8   Immunohistochemical staining of bladder from patient 2. Pretreatment biopsies (A, C, E) and post-treatment cystectomy sections (B, D, F) were stained for CD3 (A, B), CD45RO (C, D), and Factor XIIIa (dendritic cells) (E, F). (From Mastrangelo, M. J. *et al.*, *J. Clin. Invest.*, 105(8):1031, 2000. With permission.)

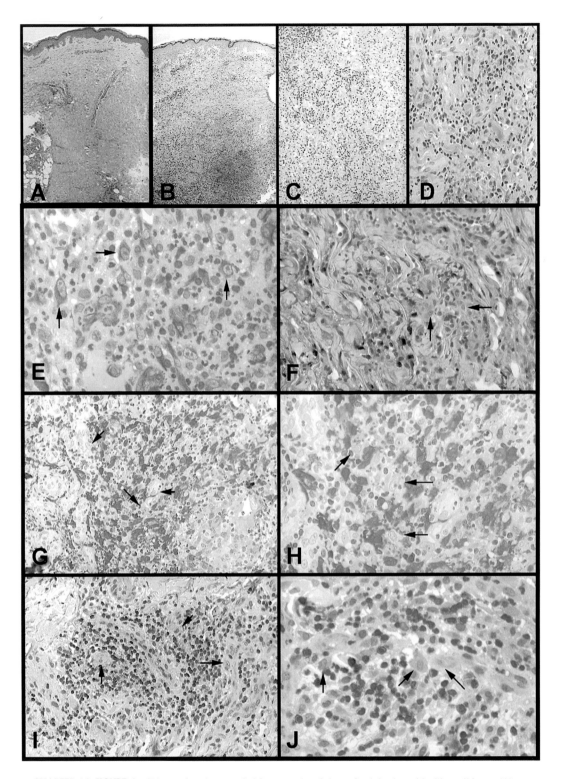

CHAPTER 14, FIGURE 4   Skin vaccine sites sampled from a patient 8 days after injection with either wild-type (A) or granulocyte–macrophage colony-stimulating factor (GM-CSF)-transfected (B to J) autologous melanoma cells. (A) Routine light microscopic appearance of skin injected with mock-transfected melanoma cells with sparse superficial and deep perivasular lymphocyte infiltrate. (B to D) Routine light microscopic appearance demonstrating dense mononuclear cell infiltrate beginning in upper papillary dermis and extending into deep reticular dermis. This inflammatory infiltrate included lymphocyte, neutrophils, and eosinophils. (E) Presence of melanoma cells confirmed by vimentin immunoreactivity. Note that the malignant cells have enlarged nuclei with vesicular chromatic and prominent nucleoli (arrows). (F) Occasional MAC387-positive macrophages are present near melanoma cells (arrows). (G and H) Extensive increase in Factor XIIIa-postive dermal dendrocytes with elongated cytoplasmic processes in mid- and deep dermis closely associated with melanoma tumor cells (arrows). (I and J) Numerous CD45RO-positive "memory" T cells are present throughout the dermis admixed with melanoma tumor cells (arrows).

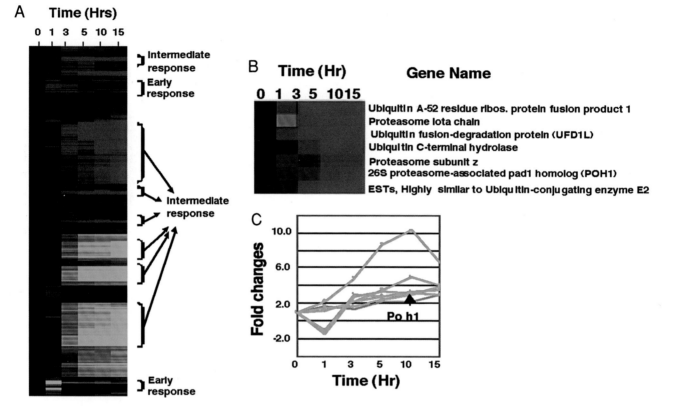

CHAPTER 25, FIGURE 2   Cluster image showing distinct patterns of gene expression. (A) Approximately 500 genes that showed alterations in expression in response to doxorubicin were selected for hierarchical clustering analysis using a software suite developed by Eisen *et al*. The altered genes were clustered into groups on the basis of their similarity in expression patterns and the results are displayed using TreeView® software. Transition of color for each gene from brown to green indicates a gradual decrease in expression with time, and changes in color from brown to red indicate increased expression. (B) The insert shows a TreeView® display of the ubiquitin-proteasome gene cluster with altered expression following exposure to doxorubicin. (C) Plot of the ubiquitin-proteasome genes showing their gradual changes with time. The entire data set analyzed in this experiment can be found at: *http://cinj.umdnj.edu/drug resistance*.

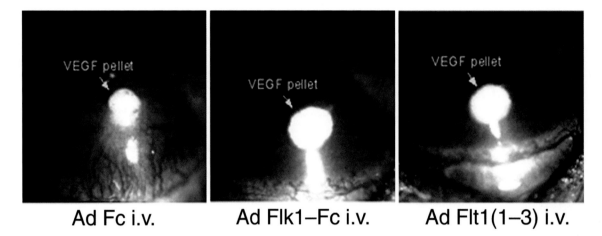

Ad Fc i.v.   Ad Flk1–Fc i.v.   Ad Flt1(1–3) i.v.

CHAPTER 27, FIGURE 3   Systemic inhibition of angiogenesis produced by remote infection with anti-angiogenic adenoviruses. C57B1/6 mice received injections of 10^9 plaque-forming units of Ad Flk1–Fc, Ad Flt1(1–3), or the control virus
Ad Fc by tail vein, a procedure that predominantly produces infection of the liver. Two days later, vascular endothelial
growth factor (VEGF)-containing hydron pellets were implanted in the cornea, and corneal neovascularization was
assessed 5 days post virus administration. Note the robust angiogenic response toward the pellet in Ad Fc mice but not
Ad Flk1–Fc or Ad Flt1(1–3) mice.

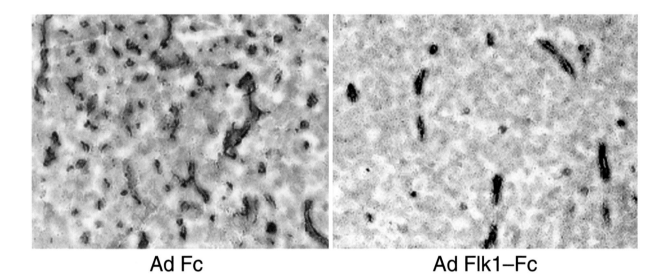

Ad Fc   Ad Flk1–Fc

CHAPTER 27, FIGURE 5   Reduction in microvessel density in tumors treated with Ad Flk1–Fc. Mice bearing Lewis
lung carcinoma tumors of approximately 100 mm^3 received intravenous injections of 10^9 plaque-forming units of the
control virus Ad Fc or the anti-angiogenic adenoviruses Ad Flk1–Fc. Tumors were harvested for immunohistochemistry
with anti-CD31 antibody after 4 days. The density of CD31 immunostaining is reduced in the tumor treated with Ad
Flk1–Fc (right panel) relative to the Ad Fc control virus (left panel).

A

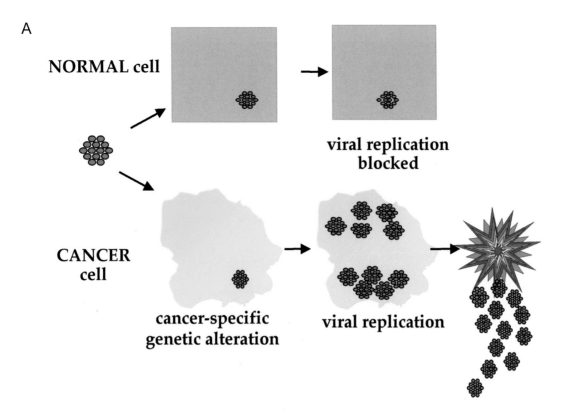

NORMAL cell

viral replication blocked

CANCER cell

cancer-specific genetic alteration

viral replication

B

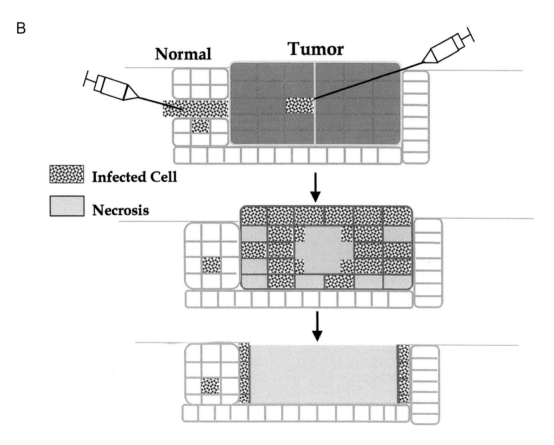

Normal

Tumor

Infected Cell

Necrosis

CHAPTER 29, FIGURE 1   Schematic representation of tumor-selective viral replication and cell killing (panel A) and tumor-selective tissue necrosis (panel B).

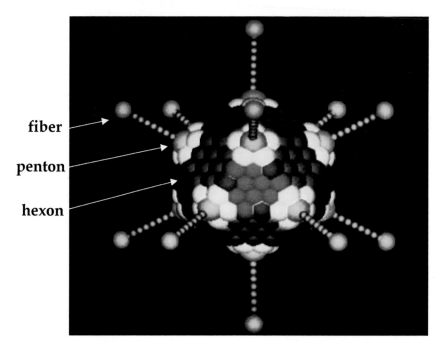

CHAPTER 29, FIGURE 2   Human adenovirus coat structure.

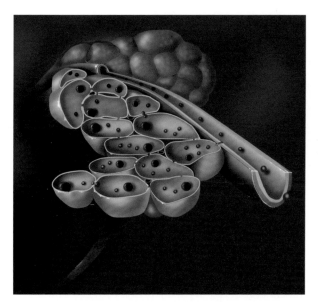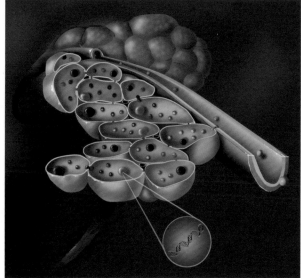

CHAPTER 33, FIGURE 1   Conventional chemotherapy for brain tumors (left panel) requires that a systemically distributed drug reach the tumor in concentrations sufficient to exert a tumoricidal effect. This approach is typically limited by systemic toxicity that prohibits dose escalation to levels sufficient for tumor eradication. Gene-directed enzyme prodrug therapy (right panel) permits the systemic administration of relatively nontoxic drugs (e.g., ganciclovir) which are only converted to their active form in cells that have been transduced to express the enzyme (e.g., HSV-TK) needed to activate them. Moreover, the activated drug can be locally redistributed within the tumor to nontransduced cells, achieving a "bystander effect." This limits systemic exposure to the active form of the drug, which accumulates selectively within the tumor. Prodrugs can be selected for their ability to cross the blood–brain barrier, even though their activated forms may lack this ability.

PART IIc

CYTOKINES AND CO-FACTORS

12

In Situ Immune Modulation Using Recombinant Vaccinia Virus Vectors: Preclinical Studies to Clinical Implementation

EDMUND C. LATTIME
Departments of Medicine and Molecular & Microbiology The Cancer Institute of New Jersey UMDNJ–Robert Wood Johnson Medical School Piscataway, New Jersey 08901

LAURENCE C. EISENLOHR
Department of Microbiology and Immunology Thomas Jefferson University Philadelphia, Pennsylvania 19107

LEONARD G. GOMELLA
Department of Urology Thomas Jefferson University Philadelphia, Pennsylvania 19107

MICHAEL J. MASTRANGELO
Division of Medical Oncology Department of Medicine Thomas Jefferson University Philadelphia, Pennsylvania 19107

I. INTRODUCTION

The concept that the immune response may be manipulated so as to eliminate established neoplasms is an appealing one that has been under study for decades. This concept stemmed from the early suggestion of Lewis Thomas in 1959 that the immune response might be useful in ridding the body of aberrant cells [1] and was later refined into the immune surveillance hypothesis of Burnet in 1970 which in its simplest form hypothesized that the immune system would recognize incipient tumors as foreign and reject them and that only those tumors that evaded this surveillance mechanism would persist and grow [2]. Whereas the broader interpretation of the immune surveillance hypothesis and its role in incipient tumors was disproved, primarily in studies by Stutman [3], who showed that severely immunocompromised mice failed to display an increased incidence of most solid tumors, investigators continue to explore the role of induced antitumor immunity in mediating tumor regression. The immunogenicity of chemically induced tumors is supported by studies of Prehn and Main, who demonstrated in mice that immunization of a syngeneic (inbred) mouse with a given tumor protected against a subsequent challenge with the same, but not other, tumors similarly derived [4]. Clinical support for the existence of tumor antigens is based on the reports that a number of human tumors, especially melanoma and renal cell, spontaneously regress, presumably by the development of an antitumor immune response [5,6]. Based on these and other findings as well as the rapidly evolving understanding of basic immune regulation, studies by numerous investigators in preclinical and clinical settings have continued to focus on the harnessing of the immune response as a therapeutic for malignancy.

To date, investigators primarily have approached the immunotherapy of tumors in three ways. Local therapy with immune-active adjuvants has been shown to be highly effective in the case of localized tumors of the skin and bladder [7–9]. To the extent that such adjuvants lead to the generation of a systemic cell-mediated response via recruitment of immune effector cells or production of cytokines, these studies have led to the use of cytokine-gene-transfected tumor vaccines and our approach of *in situ* tumor transfection with cytokine genes. Second, studies from a number of investigators have focused on the generation of tumor vaccines. In their earliest manifestations, these included the use of either whole tumor cells given unmodified or following modification with viral antigens or haptens (reviewed in Mastrangelo *et al.* [10,11]). These vaccines, often given with adjuvants such as Bacille Calmette-Guerin (BCG) to provide an enhanced immune environment, are still used clinically. Success in inducing regression of clinically evident disease using these first-generation vaccines has been quite limited; however, in the adjuvant setting, where one would expect minimal residual disease, prolongation of disease-free survival has been reported using these first-generation approaches. More recently, these have given rise to studies of tumor extracts and, most recently, defined protein antigens and peptides [12] used as vaccines directly or used to pulse antigen-presenting cells, which are subsequently used as vaccines (see Chapter 7). Such antigen-primed dendritic cells (DCs) have shown significant activity in preclinical models and are currently being studied in early clinical trials (see Chapters 10 and 11 and References [13–15]). Of note, significant antitumor activity has been seen in two such trials using tumor-pulsed DCs [15] and tumor–DC fusions [16], both of which combine optimal antigen presentation by DCs with autologous tumor antigens.

The use of genetic means for traditional immunization strategies has centered on three approaches. First, molecular means have been used to produce protein antigens for immunization [17,18]. Second, viral vectors encoding tumor antigens have been used as immunogens (see Chapter 8). Third, naked DNA or plasmid vectors that give rise to tumor or allogeneic antigen expression have also been used (see Chapters 10, 11, and 13).

II. GENERATION OF CELL-MEDIATED IMMUNE RESPONSES

Given the systemic or disseminated nature of most tumors, the goal of immunotherapeutic approaches must be the generation of a tumor-specific immune response in the host. Prior to discussing the current approaches to immunologically based gene therapy for cancer and our intralesional/intravesical approach, it would be helpful to discuss in some detail the cells and mechanisms involved in the generation of cell-mediated and predominantly T-cell-mediated immunity. We will focus almost exclusively on the cellular immune response with little discussion of the humoral (antibody)-based antitumor responses. While a number of laboratories are focused on the generation of antitumor antibody responses [19], and monoclonal-antibody-based therapies continue to be studied [20] and reviewed [21–24], the preponderance of immunologically based gene therapy approaches are aimed at generating antitumor T-cell responses.

For a comprehensive discussion of immune mechanisms and targets for gene therapy, see Chapter 7. A more restricted discussion follows here and will serve as a more specific background to our studies. While there are certain nuances regarding tumors, it is important to point out that the underlying cells and mechanisms are the same whether one is generating an immune response to virus-infected cells, parasite-infected cells, or tumor cells. Central to the generation of any T-cell-mediated response are (1) tumor antigen presented by antigen-presenting cells (APCs) in the context of major histocompatibility complex (MHC) antigens and costimulatory molecules such as B7 (CD80) [25], (2) responder T cells with the appropriate T-cell receptor or recognition structures, and (3) resultant cytokine production required to regulate the nature of the response and drive the expansion of the resultant effector T cells. To understand the approaches to therapy discussed here, it will be of value to highlight aspects of these three facets of the immune response with particular relevance to the generation of antitumor immunity.

A fundamental underlying requirement for the success of any form of immunotherapy is the expression of either unique or shared tumor antigens, which may include antigens shared with normal tissue but which, by virtue of their overexpression or modified expression, may function as targets. Antigens currently under study include mutated or overexpressed oncogene products (see Chapter 8) [26–29], as well as antigens also present on normal cells such as melanin-pigment-associated antigens in melanoma (see Chapter 7) [30–32] and prostate specific antigen (PSA) expressed on prostate cancer and normal prostate cells [33,34]. In the case of antigens shared with normal tissues, a logical concern is the generation of a concomitant autoimmune response. While this concern may seem inconsequential when one is attempting to eradicate a life-threatening malignancy, autoimmune complications must be considered.

Targetable tumor antigens must be presented in an appropriate fashion, most efficiently by professional antigen-presenting cells such as dendritic cells and macrophages. In this light, it is important to note that responder T cells recognize antigen not in isolation but in the complex of antigenic peptides and self-MHC antigens (MHC class II in the case of CD4$^+$ helper cells, and MHC class I in the case of

CD8$^+$ cytotoxic cells). Thus, expression of a shared tumor antigen among a number of patients may not elicit cross-reacting T-cell responses due to differences in MHC antigen expression. This MHC restriction phenomenon limits the application of immunotherapy strategies that utilize allogeneic tumor vaccines—that is, the use of tumor cells from one individual to immunize another. While allogeneic tumors may share some antigens, and thus one might think that such an approach would be feasible, unless the donor and recipient are MHC matched these shared antigens will not be recognized because they are expressed in the context of dissimilar MHC molecules. These restrictions, then, favor approaches that utilize the patient's own tumor for immunization. For a more comprehensive review of MHC restriction, see Chapter 7 and Klein and Sato [35,36]. In addition to the MHC–antigen complex, optimal T-cell activation requires expression of costimulatory molecules, with the most studied being B7.1 (CD80), which is the ligand for CD28 on the responder T cell. In the absence of B7 expression, the APC fails to stimulate a T-cell response and may even stimulate a state of tolerance or anergy to the expressed antigen (reviewed in Linsley and Ledbetter [37], Lenschow and Bluestone [38], and Gimmi *et al.* [39]).

Fundamental to stimulating an antitumor response is the need for T cells capable of recognizing the tumor antigen. In the case of antigens shared between normal and malignant tissues, one is faced with the necessity of overcoming or "breaking" tolerance to these self-antigens. A discussion of tolerance mechanisms can be found elsewhere (see Chapter 7 and Janeway and Travers [40]). Here, it will suffice to point out that when one isolates lymphocytes from tumor masses, and in a some cases from peripheral blood, and stimulates them *in vitro* under the appropriate conditions, one is able to demonstrate the existence of such T lymphocytes that recognize autologous tumor. Such antigen-reactive T cells have been identified in patients with a variety of tumor types [35,41–47] and shown to recognize tumor-specific antigens such as oncogene-encoded proteins [26–28] as well as antigens also found on normal cells such as the melanin-pigment-associated antigens [30–32] alluded to above. Demonstrating a similar response *in vivo* has been more difficult to do and may be the result of tolerance and/or suppression mechanisms not found *in vitro*. In sum, however, it is clear that tumor-reactive T cells are present in patients based on *in vitro* analyses.

With regard to the requirement for the production of appropriate cytokine based help for the development of an antitumor response, studies have focused on approaches that enhance proimmune cytokine production in the context of tumors either as adjuncts to vaccine strategies or, in the case of our studies, the *in situ* introduction of cytokine genes and/or cytokine regulatory constructs into tumors. In fact, the hypothesis underlying the majority of ongoing gene therapy approaches is that by modulating the immune milieu at the local tumor or tumor vaccination sites using cytokine manipulation one may induce systemic antitumor immunity. Toward this end, multiple strategies are being employed which include: (1) the use of recombinant viral vectors encoding the genes for tumor-specific antigens plus immune active cytokines (see Chapter 8); (2) immunization using vaccines that incorporate cytokine-gene-transfected tumor cells (see Chapters 7 and 14) or fibroblasts [48,49]; and (3) as we have proposed, *in situ* introduction of cytokine genes or cytokine-producing cells at the tumor site *in vivo* using viral vectors. Studies have examined a panel of cytokines that target multiple regulatory mechanisms, including: (1) the upregulation of MHC and perhaps tumor-associated antigens, such as tumor necrosis factor (TNF) or interferon-gamma (IFN-γ); (2) macrophage and APC recruitment and activation with subsequent enhanced antigen processing and presentation, (such as granulocyte–macrophage colony-stimulating factor (GM-CSF) or IFN-γ; (3) cytokines that direct the T-cell helper response toward a T$_h$1 response and thus enhance the generation of delayed-type hypersensitivity (DTH) and cytotoxic T-lymphocyte (CTL) responses, such as IFN-γ and IL-12, to be described further; and (4) cytokines that drive the expansion of activated T cells (IL-2, IL-4).

Regarding the cytokine direction of cell-mediated responses, studies have shown that the profile of cytokines produced in an immune response correlates with and may direct the response to either a cellular response expressing DTH, termed T$_h$1, or a humoral response resulting in antibody production, termed T$_h$2. In studies of experimental and clinical parasite infections (leishmania and leprosy, respectively), the presence of a T$_h$1 response is associated with DTH and a good clinical course, while a T$_h$2 response is associated with an antibody response and a less favorable course [50,51]. Of particular interest therapeutically, T$_h$1 and T$_h$2 responses have been shown to cross-regulate one another, with T$_h$1-associated cytokines such as IFN-γ inhibiting T$_h$2 immunity and T$_h$2-associated cytokines such as IL-10 suppressing the generation of cellular responses. For this reason, therapeutic strategies are being developed to maximize expression of T$_h$1 cytokines at the tumor site which, it is hoped, will drive the antitumor responses toward the cell-mediated arm.

In addition to the cytokine regulatory networks in play during the generation of an immune response, the tumor itself can play a role in directing the nature of the ongoing response. In the first case, expression of tumor antigens in the context of MHC but in the absence of the costimulatory activity conferred by B7 (CD80) expression not only may fail to stimulate a response but, as has been shown in a number of other systems, may also actually confer tolerance or anergy [25,37–39,52,53]. A second characteristic of a number of tumors may drive the response towards T$_h$2 with resultant suppression of cell-mediated immunity (DTH and cytotoxic T-cell generation). Studies from our laboratory and others have reported the production of the cytokine IL-10 by a number of human [54–59] and murine [60–62] tumor types. Initial studies by Hersey *et al.* [56] found IL-10 production by human melanoma cell lines, while we and others subsequently found that biopsies from metastatic melanoma lesions expressed IL-10 mRNA and that tumor cell lines derived from these biopsies produced IL-10 protein [54,55]. Subsequently, we have also found IL-10 mRNA expression in biopsies of human transitional cell carcinoma of the bladder [57] and by bladder tumor cell lines [63]. The production of IL-10 by tumor and/or tumor-associated cells would be expected to drive any immune response towards T_h2 and operationally suppress cell-mediated immunity to the tumor. In fact, we have described just such an IL-10-driven inhibition of DTH to a tumor-associated antigen in our murine studies [64] and have demonstrated that elimination of the IL-10-mediated suppression using IL-10 knockout mice or anti-IL-10 antibodies *in vivo* "unmasks" tumor-specific responses with effective tumor rejection [62,64,65]. At least one mechanism of the IL-10 suppressive response may also include blockade of the upregulation of the expression of the costimulatory antigen, B7 and MHC antigen, thus tying together these two putative tumor-induced suppressive mechanisms [66–69].

In sum, the development of a productive cellular immune response toward tumor relies on the activation of a cascade of cells and cytokines. During the initiation of such a response, production of suppressive factors by tumor and/or immune cells can result in the inhibition of antitumor cellular immunity. Elucidation of these mechanisms has led a number of investigators, including ourselves, to put forth strategies for enhancing the generation of cellular responses directly or via overcoming suppressive influences. A discussion of a number of such strategies follows and is also presented by additional contributors to this volume.

III. CYTOKINE GENE TRANSFER STUDIES IN ANTITUMOR IMMUNITY

In an attempt to overcome a hypothesized lack of immune-stimulated cytokine production and/or further stimulate antitumor responses, a number of laboratories [70–76] have stably transfected murine and more recently human tumor with a variety of cytokine genes for use as vaccines (see Chapter 14) [77–79]. Inoculation of syngeneic mice with experimental tumors transfected with the genes for TNF [71], IL-2 [70,74], IFN-γ [80], IL-4 [81], and GM-CSF [75,76] has resulted in rejection of the injected tumor. In some cases, mice were shown to generate a measurable systemic antitumor response based on rejection of subsequent challenge with the nontransfected tumor [70,71,81–83]. In a limited number of cases, "vaccination" with such cells resulted in the elimination or reduced growth of preexisting tumor [74,75,81]. While these findings have been less than overwhelming in regard to the

effects on existing tumors, they do show that localized cytokine/lymphokine production can enhance the generation of tumor-specific immunity. Most recently, this approach has been translated to clinical trials in both renal carcinoma and prostate cancer [77–79] and melanoma (Chapter 14). While there were limited indications of clinical antitumor response, both studies showed positive immunologic findings [77–79].

In addition to systemic T-cell-dependent antitumor responses resulting from such treatment, local T-cell-independent antitumor activity has also been demonstrated [73]. IL-4-producing tumors, and admixed normal tumor cells, when injected into T-lymphocyte-deficient nude (nu/nu) mice have been shown to regress as a result of a localized inflammatory response characterized by infiltration of eosinophils and macrophages [73]. Thus, while the goal of such genetic intervention is to optimize for the production of tumor-specific T-cell immunity, local antitumor effects may also contribute to a positive clinical outcome.

We should point out two limitations of the *in vitro* cytokine gene transfection approach to clinical translation. First, current cytokine studies have focused on *in vitro* transfection of tumor cell lines or fibroblasts prior to injection into mice or patients. While this is a reasonable approach in preclinical studies, extension of this system to clinical trials will severely limit its availability. The requirement that autologous tumor, based on the need for proper antigen and MHC expression, be available, removed, transfected, cloned, etc. severely limits the number of suitable patients.

In addition to the cytokine-gene transfectant vaccines previously mentioned, three additional approaches have been studied that have reached differing stages in their move towards clinical trials. With the goal of overcoming limitations in antigen presentation due to the lack of B7 (CD80), the ligand for CD28, a number of investigators have shown in preclinical animal models that transfection of tumors with B7 and their subsequent use as vaccines enhance the generation of antitumor responses and subsequent protection against challenge with nontransfected tumor [38,39]. Additional support for the use of this approach in therapy comes from studies by Ostrand-Rosenberg's group [84] which demonstrated in a murine sarcoma model that preexisting nonmodified tumor could be eradicated with vaccination using B7-transfected tumor. More recently, genes encoding combinations of costimulatory molecules have been engineered into a single viral recombinant and shown to be superior to any one given singly [85].

IV. *IN SITU* CYTOKINE GENE TRANSFER TO ENHANCE ANTITUMOR IMMUNITY

As outlined above, murine tumors transfected with a variety of cytokine and accessory molecule genes have been

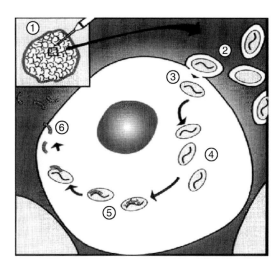

FIGURE 1 Initiating cytokine production around a tumor site. (1) Virus including gene encoding for cytokine is injected into tumor. (2) Virus binds with tumor cell. (3) Virus and gene enter cell. (4) Virus undergoes replication. (5) Cytokines are manufactured. (6) Cytokines are released, initiating a local cytokine-mediated host immune response.

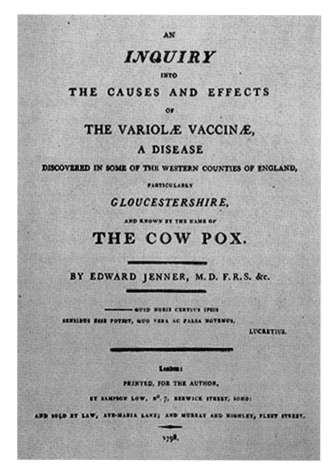

FIGURE 2 Cowpox (vaccinia) as a therapy for smallpox.

shown to induce variable levels of antitumor immunity and in some cases tumor regression when injected into syngeneic mice with preexisting tumors. While these results show promise for such an approach to developing immunotherapeutic modalities in humans, limitations in the ability to harvest, transfect, and reinject a variety of human tumors on a patient-by-patient basis raise questions about its feasibility in a number of human tumors. For this reason, we have developed a strategy of directly inserting the desired cytokine gene into the tumor utilizing vaccinia virus recombinants, as shown in Fig. 1. Injection of the virus intralesionally or intravesically in the case of bladder cancer would result in the infection of the tumor cells and subsequently the production of cytokine mRNA and secretion of biologically active protein. Supported by the preclinical tumor transfection studies also described above, it is our hypothesis that production of proimmune cytokines locally at the tumor site in this way would enhance the generation of systemic tumor-specific immunity and resultant tumor destruction.

A. Vaccinia Virus Vectors

We have chosen vaccinia virus vectors for our studies for a number of reasons. Members of the poxviridae family of viruses, including vaccinia, are unusual in that replication and transcription of the genome occur in the cytosol of infected cells, with virally encoded polymerases driving these processes. Thus, recombination of viral DNA into the genome is not of concern with vaccinia, as it is with other vectors, particularly retroviruses. The infectious cycle is divided into three phases. Early phase genes, typically encoding proteins

with enzymatic function, are expressed prior to replication. The expression of a small number of intermediate genes is dependent upon replication of the genome, and intermediate gene expression in turn drives expression of a large set of late genes, typically encoding structural proteins [86]. Generally speaking, "late" vaccinia promoters drive stronger gene expression than "early" vaccinia promoters. Some genes, such as that encoding the 7.5-kDa protein, have both "early" and "late" promoters and are expressed during both phases of infection, with the late promoter component accounting for approximately 75% of total expression [87].

Vaccinia was best known 20 years ago for its critical role in the eradication of smallpox (caused by the variola poxvirus), first described in 1978 by Jenner [88] (see Fig. 2). This points to one obvious advantage of vaccinia as a vector: its extensive use in humans and characterization in the field. A second key characteristic is its stability. Vaccinia can be carried to remote regions and reconstituted from desiccated material to highly infectious stock. Over time, vaccinia has been recognized as a relatively safe agent, with infrequent serious side effects, although it does induce a vigorous immune response and can be lethal for those who are immunocompromised or have eczema [89]. While variola is highly contagious, transmission being mainly via the respiratory tract, vaccinia is much less so and can be easily confined under standard Biosafety Level 2 practices.

Today, as vaccination of the general population against smallpox has been discontinued, vaccinia is best known as a vector for transient expression of proteins. A clear advantage of vaccinia in this regard is its wide tropism. With variable efficiency vaccinia infects most mammal-derived permanent cell lines and any of the common laboratory animals including mice, rabbits, and monkeys. Its large genome (approximately 200 kb) allows for the stable insertion of very large fragments of DNA (25 kb has been reported [90]) into the genome at a single site. This is well above the range of many other vectors. Such recombinants have been used for a wide range of studies, including those concerned with folding and oligomerization of proteins [91], signal transduction [92], identifying human immunodeficiency virus coreceptor molecules [93,94], elucidating the antiviral effects of various cytokines [95], and eliciting protective immune responses against pathogens and transformed cells [96,97]. Most recently, vaccinia recombinants encoding PSA and carcinoembryonic antigen (CEA) have been used in clinical trials of patients with prostate and colorectal cancer, respectively [34,98,99]. Vaccinia is also capable of rendering some cells very receptive to transfection, and "infection/plasmid transfection" protocols yield high levels of gene expression in a large percentage of cells, bypassing the need for generating a recombinant. In this case, the plasmid need be delivered only to the cytosol of the cell, not the nucleus, provided that the gene of interest can be transcribed in that location. This has been accomplished by preceding the gene with a vaccinia-specific promoter [100] and by utilizing the T7 and T3 bacteriophage promoters and infection by T7- or T3-expressing vaccinia recombinants [101,102]. Even if a true recombinant is needed, the procedure can be helpful in confirming the integrity of a construct prior to generation of the recombinant stock which requires at least several weeks.

Generating a vaccinia recombinant is relatively straightforward. The gene of interest is inserted into a plasmid that minimally contains an origin of replication, an antibiotic resistance gene for cloning purposes, a vaccinia virus promoter to drive expression of the inserted gene, and segments of the vaccinia virus genome, flanking the promoter and inserted gene, to direct site-specific recombination. The $P_{7.5}$ promoter, active during early and late phases of infection, has most often been employed to drive heterologous gene expression. The most popular site of recombination is the viral thymidine kinase (TK) gene, which is disrupted by the recombination event. For a graphical representation of the production of recombinant vaccinia, see Fig. 3. Recombination is achieved by

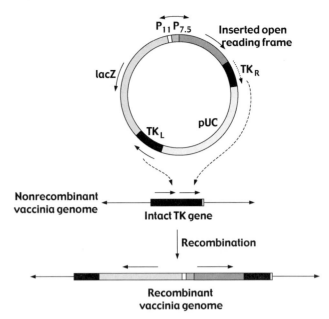

FIGURE 3 Schematic of a typical vaccinia recombination strategy. Shown at top is a recombination plasmid as developed by Chakrabarti *et al.* [131] featuring the fragmented vaccinia thymidine kinase gene (TK_L and TK_R flanking the *lacZ* and inserted open reading frame driven by the P_{11} [late] and $P_{7.5}$ [early/late] promoters). The pUC-derived segment of the plasmid provides the origin of replication and β-lactamase gene to confer ampicillin resistance. The result of recombination into the vaccinia genome at the thymidine kinase (TK) gene is depicted at the bottom.

infection of cells (often the African-green-monkey-derived CV-1 cells), with wild-type vaccinia stock followed shortly thereafter by transfection of the infected cells with the recombination plasmid. The resultant vaccinia stock will contain recombinant virus at a low frequency, on the order of 0.1%, which is then selected and plaque-purified in subsequent passages. Early vaccinia recombination protocols suffered from the consequences of spontaneous inactivation of the TK locus. These TK$^-$ mutants cannot be distinguished from true recombinants without assaying for the recombinant gene or protein. The selection process is now straightforward, as current vectors are usually designed to allow incorporation of a reporter gene, such as *lacZ,* driven by a second vaccinia promoter. Addition of the substrate X-gal during plaque purification causes β-galactosidase-producing plaques (true recombinants) to turn dark blue. Such plaques are then amplified and carefully titered prior to their use. It is critical that production of the gene of interest be confirmed prior to experimentation, as incorporation of a reporter gene into the genome indicates recombination but certainly not integrity of the gene of interest. As mentioned, other loci have been targeted for recombination, and these necessarily require other means of selection, such as neomycin resistance [103]. On occasion, a double recombinant is needed. One means of achieving this is by reinserting a TK gene into the vaccinia genome via recombination, as a reporter, and selecting under conditions that require a viable TK gene [104]. Considerable control of heterologous gene expression is possible with the vaccinia recombinant system. Not only can one select among the three types of promoters, but expression driven by each can also be varied by modification of the promoter sequence ranging from levels that are barely detectable to those that are extraordinarily high [105–107]. Finally, it has been shown recently that, with sufficient care, heterologous DNA, also as large as 25 kb in length, can be directly ligated into the vaccinia genome, obviating the need for recombination and the associated procedures [108].

Wild-type vaccinia infection is quite cytolytic and, as mentioned previously, can be lethal in immunocompromised individuals. Much effort has been expended in generating less virulent poxvirus vectors. These include use of viruses such as modified vaccinia ankara (MVA) or fowlpox virus, both of which replicate in avian cells and will infect but not complete replication in most mammalian cells [109–111]. This inhibition of replication can also be achieved with recombinants based upon wild-type vaccinia by treatment of purified virions with psoralen and ultraviolet light [112]. An alternative approach to reducing virulence has involved the systematic inactivation of nonessential genes recognized as contributing to virulence [111].

Vaccinia and poxviruses in general are clearly vectors of great utility, and continuing advancements will undoubtedly broaden their applicability. There are some situations for which poxviruses will probably never be suitable. Vaccinia expresses on the order of 100 different proteins, at least some of which stimulate a vigorous immune response, as anyone receiving the smallpox vaccine can attest to. Thus, expression of the recombinant gene may be limited in individuals with prior exposure. It must be noted, particularly with respect to the subject of this chapter, that expression is not completely prevented, even following multiple inoculations with the same poxvirus. In fact, our clinical studies using both wild-type vaccinia virus [113] and GM-CSF-encoding recombinant vaccinia [114] demonstrate significant infection/transfection of cells in the presence of high titers of neutralizing antibodies to vaccinia, the result of multiple intralesional injections. This could be due to vaccinia-specific antibodies interfering with the release of newly assembled virions from the cell rather than with the attachment and entry phase [115].

B. Tumor Transfection by Vaccinia Recombinants

The overall hypothesis behind our studies is that by modulating the immune milieu at the local tumor site and thus recruiting antigen-presenting and effector cell populations it will be possible to engender a systemic tumor-specific immune response. The result would be to eliminate both localized and disseminated tumor. We have pursued both preclinical and clinical studies to determine the feasibility of the use of recombinant vaccinia as a vector for *in situ* transfection, with the result that highly supportive data have been generated that enhance our enthusiasm for the approach.

Prior to developing recombinant vaccinia for our gene therapy studies, it was necessary to demonstrate that vaccinia virus recombinants are capable of transfecting murine and human tumor cells. A panel of cell lines including the murine melanoma B16, bladder tumors MBT2 and MB49 [116,117], and human melanoma lines produced from our patients [113,118] and bladder (T24) and prostate carcinoma (LNCAP, PC3) cell lines [119] were examined for their ability to be infected/transfected with vaccinia recombinants. Cell lines were exposed *in vitro* to vaccinia virus recombinants containing the genes for influenza hemagglutinin and nuclear protein antigens termed reporter genes, which allow one to stain for productively infected/transfected cells. Without exception, all cell lines tested were highly susceptible to infection/transfection [117,120], data not shown.

To determine if recombinant vaccinia are able to infect/transfect tumor *in vivo,* vaccinia recombinants containing reporter constructs (HA, NP, or the *lacZ* gene) were injected intralesionally into murine B16 melanoma lesions or instilled via urethral catheters into the bladders of C57BL/6 mice bearing the MB49 tumor [116,117,121; unpubl. results]. Eight hours following administration, tumors were processed and stained for expression of the reporter genes as a measure of tumor transfection. As noted above, given the immunogenicity of vaccinia and the possibility that immunity to the virus would prevent infection/transfection following *in vivo* administration, mice were preimmunized to vaccinia prior to use in these studies which would model the human state where patients have received the smallpox vaccine. Our studies using intralesional injection of B16 melanoma and intravesical administration in the MB49 bladder model demonstrated high levels of infection/transfection of tumor as measured by expression of the influenza HA gene product and/or encoded β-galactosidase [117,120]. Thus, systemic immunity to vaccinia, which would be expected to be present in adult patients and following initial vaccinia treatments, does not prevent *in vivo* tumor infection/transfection.

C. Cytokine Gene Delivery Using Vaccinia Recombinants

To determine if vaccinia recombinants could be used to transfect tumors with resultant cytokine production, we have established a panel of vaccinia recombinants expressing murine IL-4, IL-5, IFN-γ, and GM-CSF [121,122]. At the time at which these were being produced, a report by

Ramshaw *et al.* [123] demonstrated cytokine production by such recombinants and their positive effects on antiviral immunity. In our studies, using methods described by the Moss group (reviewed earlier and in Mackett *et al.* [124]), we inserted the various cytokine genes into a plasmid containing sequences from the thymidine kinase gene behind the early and late $P_{7.5}$ vaccinia promoter (see earlier description of recombinant vaccinia production and Fig. 3). The plasmid was then incorporated into the vaccinia using homologous recombination into the thymidine kinase site, and recombinant virus was isolated, propagated, and tested for activity [121,122,124]. A schematic of the virus-generating protocol is presented in Fig. 3. ELISA and functional cytokine analyses have shown that the vaccinia recombinant-infected tumor cells produce significant levels of cytokine protein [122; data not shown]. Subsequently, studies using vaccinia-specific primers designed in our laboratory that allow elucidation of encoded cytokine mRNA *in vivo* have added to our earlier findings demonstrating prolonged cytokine gene expression following intralesional injection (our unpublished results), and our preliminary studies demonstrate significant retardation of tumor growth in mice bearing the B16 melanoma treated with recombinant vaccinia expressing GM-CSF (data not shown). Parallel studies by three additional laboratories are consistent with our results demonstrating efficient *in vivo* tumor transfection by vaccinia recombinants in preclinical models [125–127].

D. Intralesional Vaccinia Vector in Patients with Melanoma

As a prelude to studying the effects of intralesional recombinant vaccinia in human melanoma, we obtained an Investigational New Drug (IND) approval from the Food and Drug Administration (FDA) to inject the Wyeth strain of vaccinia (the vaccine used in the U.S. for smallpox immunization and our nonrecombinant parent) intralesionally in patients with recurrent superficial melanoma [113,128]. Following the demonstration of systemic immunity to vaccinia via an intradermal administration of vaccine to the patients, increasing doses of vaccinia were injected intratumorally. In a representative patient, 10^6 plaque-forming units (PFU) of Wyeth vaccinia was injected into four sites in a 3-cm superficial melanoma lesion. The lesion was biopsied at 6 hours, and 4 days following administration the biopsies were processed as frozen sections and stained using the monoclonal antibody TW2-3, which is specific for an early viral protein product of the EL3 gene present at sites of viral replication. The biopsy taken 6 hours following intralesional vaccinia contained diffuse numbers of melanoma cells staining intracytoplasmically for the viral antigen [113]. To determine if increasing immunity to vaccinia induced by multiple treatments would block productive infection/transfection, addi-

tional biopsies were similarly analyzed over the course of therapy in one patient who received 19 biweekly injections of as much as 10^7 PFU of virus (total cumulative dose of 13×10^7 PFU). While the duration of expression was diminished with increasing immunity, as measured by antiviral antibody titer, productive infection was seen throughout the treatment course [113,128]. It should be noted that minimal systemic side effects were seen in the trial. These findings demonstrate, as did our murine studies, that vaccinia recombinants are able to infect/transfect tumor *in vivo* following intralesional injection, even in the face of systemic immunity to the virus. It is our conclusion from these studies that systemic immunity to the virus acts to protect patients from toxicity but does not prevent local gene expression. Our demonstration of sustained infection in virus-immune individuals strongly supports our approach, suggesting that infection/transfection using cytokine gene-encoding vaccinia should result in cytokine production for a prolonged period.

E. Intralesional Vaccinia–GM-CSF Recombinant in Patients with Melanoma

Satisfied that the vaccinia vector met our requirements of safety and efficacy, we carried out a phase I trial of intralesional vaccinia–GM-CSF, produced in our laboratory and grown under good manufacturing practice (GMP) conditions, in patients with therapy-refractory recurrent melanoma. All patients were required to have accessible dermal and/or subcutaneous disease, and a number also had visceral disease. Following the demonstration of immune competence (important, given the replicative nature of the vector), patients received twice weekly intralesional injections of the recombinant with dose escalation within each patient. Table 1 summarizes the results seen in the first seven patients which are described in detail in Mastrangelo *et al.* [114]. At the highest doses, patients received 2×10^7 PFU per lesion, with the injection of multiple lesions resulting in doses as high as 8×10^7 per session. For comparison, vaccinia was used as a smallpox immunization at a scarification dose of 2.5×10^5. Figure 4 (see also color insert) demonstrates the complete resolution of dermal metastases seen in patient 3 of the study following treatment accompanied by recruitment of large numbers of CD3$^+$ T cells of both CD4 and CD8 phenotype (not shown) into injected lesions. In addition to the injected lesions, four of the seven patients experienced regression of uninjected lesions following treatment (Fig. 5; see also color insert). It was this latter regression of distant uninjected lesions that we have taken as evidence of the possible induction of tumor-specific immunity. Laboratory studies have demonstrated that patients develop high levels of immunity to both vaccinia and the included β-galactosidase gene product [114]. Also, in the face of

TABLE 1   Intratumoral Injection of the Vaccinia–GM-CSF Recombinant Virus in Patients with Malignant Melanoma

Patient	Age/Sex	Metastases	Prior treatment	Total lesions/sessions	Total dose ($\times 10^7$ pfu)	Anti vaccinia titer	Anti β-galactosidase titer	Response
1	81/F	Dermal, LN	Radiation	12/26	74.73	180	220	Dermal, partial; LN, none
2	68/M	SC, LN, and lung	BCDT, Taxol	12/27	82.0	80	30	None
3	32/F	Dermal, breast and	Radiation, BCDT, DCV+IL-2+ IFN, Taxol	Dermal, 45/56; breast, 2/8	Dermal, 207.3; breast, 20.5	225	100	Dermal, complete; breast, unknown
4	61/F	Dermal and SC; LN and lung	BCDT, IFN	9/11	49	400	200	Dermal and SC, partial; LN and lung, none
5	71/F	Dermal, LN	Limb perfusion GP 100, MARTI, IL-12, BCDT	17/12	47	200	75	Dermal, partial; LN, none
6	67/M	SC, LN, lung	BCDT	13/13	64	>300	80	None
7	75/M	Dermal	None	13/11	47.1	100	50	Complete

Note: pfu, plaque forming units; LN, lymph node; SC, subcutaneous; IL, interleukin; BCDT, BCNU + cicplatin + DTIC + tamoxifen; IFN, interferon; DCV, DTIC + cisplatin + vinblastine.

maximal antibody titers, we continued to be successful at achieving recombinant gene expression, measured both as vaccinia-encoded GM-CSF (V-GM-CSF) (RT-PCR using primers designed to specifically identify viral encoded GMCSF) and viral thymidine kinase gene (V-TK) mRNA expression (Fig. 6).

F. Intravesical Vaccinia in Patients with Bladder Cancer

As the first step of our planned expansion of this strategy to the localized treatment of bladder cancer (our preclinical data demonstrated significant infection/transfection of the orthotopically growing murine bladder tumor MB49 following intravesical administration of recombinant vaccinia [117]), we have completed a phase I study of intravesical vaccinia vector in patients with advanced transitional cell carcinoma. As with our phase I of vector alone in melanoma [113], we used the vaccinia in a dose escalation study, with each patient receiving three intravesical doses over a 2-week period. Given safety concerns, this study focused on patients with invasive transitional cell carcinoma scheduled for cystectomy the day following the third dose. Table 2 summarizes patient characteristics, doses employed, and toxicity. As noted in our prior clinical trials, patients developed high titers of antivaccinia antibody, although maximal titers were measured after cystectomy in patients given the shortened

course of therapy (data not shown). Also as noted earlier, treatment was associated with a significant degree of inflammation (Fig. 7; see also color insert) and recruitment of activated T lymphocytes (CD3$^+$, CD45RO$^+$) (Fig. 8; see also color insert), as well as dendritic cells (Factor XIIIa$^+$) that we feel will enhance prospects for the induction of immunity to tumor.

V. FUTURE DIRECTIONS

We have demonstrated (1) that vaccinia virus can be effectively used to infect/transfect tumor cells *in vivo* in both preclinical melanoma and bladder systems, (2) that the vaccinia virus vector and recombinants can be given safely and with continued infectivity in patients despite preexisting or developing immunity to vaccinia, and (3) that vaccinia recombinants expressing the genes for a panel of cytokines effectively induce infected cells to produce high levels of biologically active cytokines. We have translated our preclinical results into the clinical use of a GM-CSF-encoding vaccinia recombinant given intralesionally to patients with melanoma and have seen encouraging clinical responses (long-term remission in two of seven patients and rejection of uninjected lesions in four of the seven; see Table 1).

Our initial decision to focus on *in situ* cytokine modulation in an attempt to enhance immune recognition was

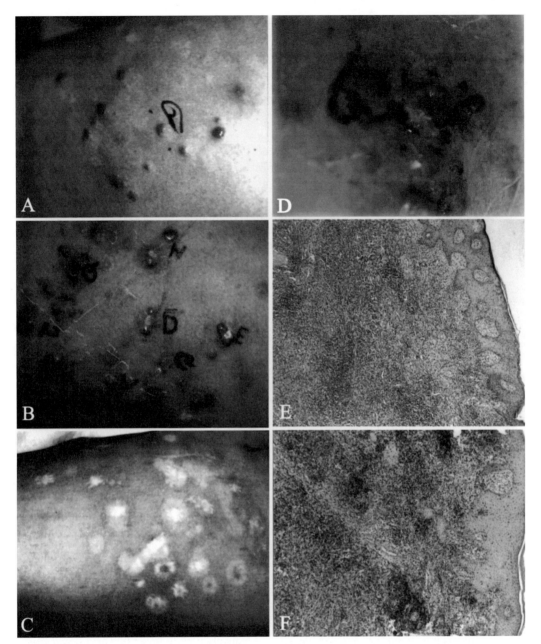

FIGURE 4   Resolution of dermal metastases following intralesional injection of recombinant vaccinia–granulocyte–macrophage colony-stimulating factor (GM-CSF). Patient 3 is a 32-year-old female with extensive dermal metastases of the left thigh before treatment (A), on day 81 (B), and on day 600, 150 days following cessation of treatment (C). Regression was accompanied by gross (D) and histologic (E) evidence of inflammation, including significant T-cell (CD3+) involvement (F). (From Mastrangelo, M. J. *et al., Cancer Gene Ther.,* 6(5):409–422, 1999. With permission.) (See color insert.)

based on our early laboratory studies demonstrating that in both melanoma and bladder tumors the tumor-host environment was rich in IL-10, an immune suppressive cytokine [54,129,130]. Having made this clinical observation, we have focused on demonstrating the feasibility of modulating the tumor-host environment with immune-enhancing cytokines described in this chapter. While doing this, our laboratory studies focused on validating IL-10 and other suppressive molecules as targets for modulation [62]. We are now focused on the production and cloning of a new series of recombinants with the ability not only to instill positive mediators but also to neutralize the negative.

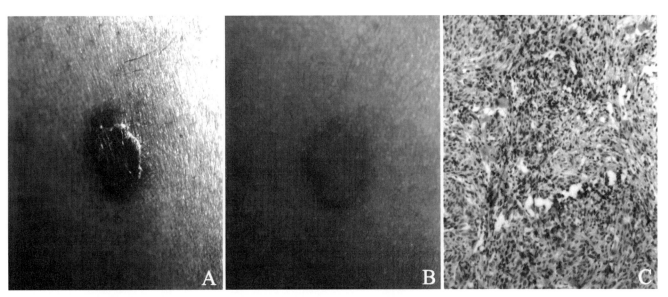

FIGURE 5   Regression of uninjected lesions accompanied by T-cell infiltration. A representative uninjected distant regressing lesion prior to (A) and following (B) patient treatment demonstrated T-cell (CD8) infiltration (C). (From Mastrangelo, M. J. *et al., Cancer Gene Ther.,* 6(5):409–422, 1999. With permission.) (See color insert.)

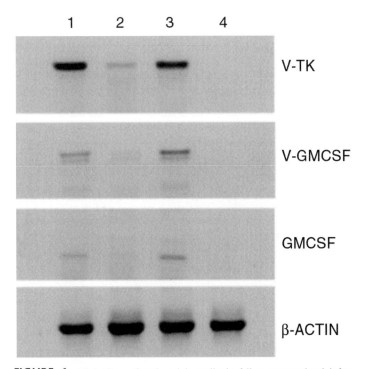

FIGURE 6   High titers of antivaccinia antibody fail to prevent local infection/transfection following injection with vaccinia–GM-CSF recombinant. RT-PCR of mRNA from melanoma biopsies for vaccinia, thymidine kinase (V-TK), vaccinia-encoded human GM-CSF (V-GMCSF), human GM-CSF (GMCSF) and actin control (β-actin). Biopsies from injected lesions (lanes 1–3) and an uninjected lesion (lane 4). Lane 1, biopsy 18 hours following the last of a series of multiple injections; lanes 2 and 3, biopsies 18 hours following a single injection; lane 4, uninjected lesion. All biopsies were taken from patient 3 at week 31. From *J. Clin. Invest.* **105(8);** 1031, 2000. With permission.)

TABLE 2   Intravesical Vaccinia Vector Prior to Cystectomy in Patients
with Muscle-Invasive Bladder Cancer

Patient	Age/Sex	Dose per treatment ($\times 10^6$ pfu)	Toxicity	Bladder inflammation
1	57/F	1, 5, 10	Mild dysuria	Slight
2	36/M	10, 25, 100	Mild dysuria	Significant
3	64/M	25, 100, 100	Mild dysuria	Significant
4	52/M	25, 100, 100	Mild dysuria	Significant

VI. CONCLUSIONS

In summary, we have developed an approach to immuno-logically based gene therapy logically designed from the requirements to generate a productive cellular immune response. We have translated the approach to clinical trials, with positive antitumor activity seen in a number of patients. As outlined in this review, numerous strategies have been hypothesized and tested in both preclinical and clinical settings with this goal as an endpoint. It is our hypothesis that *in situ* tumor transfection with cytokine genes will provide a logical extension of the vaccine strategies that have been previously studied. By incorporating genes selected based on their known contribution to the generation of systemic immune responses, we anticipate the ability to optimize the generation of an antitumor response. In addition to this logical *in vivo* vaccine design, this methodology will allow the generation of a single reagent in a bottle that will be of use in any tumor type provided it is accessible to injection. This will preclude the need to have sufficient autologous tumor for harvest and subsequent vaccine production and will overcome the significant limitation of the *in vitro* transfectants for tumor transfection and selection in the lab. As noted above, the use of the patient's own tumor as a source of antigens in our system optimizes the generation of a T-cell response and has significant advantages over allogeneic vaccine strategies that rely on shared antigens restricted by common MHC antigens.

Acknowledgments

The authors wish to thank Arvin Yang and Faryal Mahmud for assistance in developing the graphics and text of this chapter. Our research is supported by ACS grants IM-742 and EDT-78842; USPHS grants CA-42908, CA-55322, CA-69253, CA-74543; and the Nat Pincus Trust.

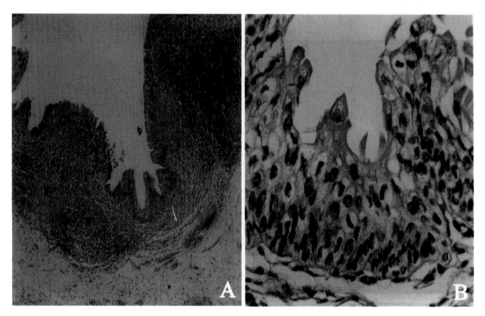

FIGURE 7   Intravesical administration of vaccinia vector in patients with invasive bladder cancer. H&E stained section of bladder from patient 2, taken at time of cystectomy 24 hours after the third of three intravesical instillations of vaccinia, shows widespread inflammation (A) and infection of the urothelium (B). (See color insert.)

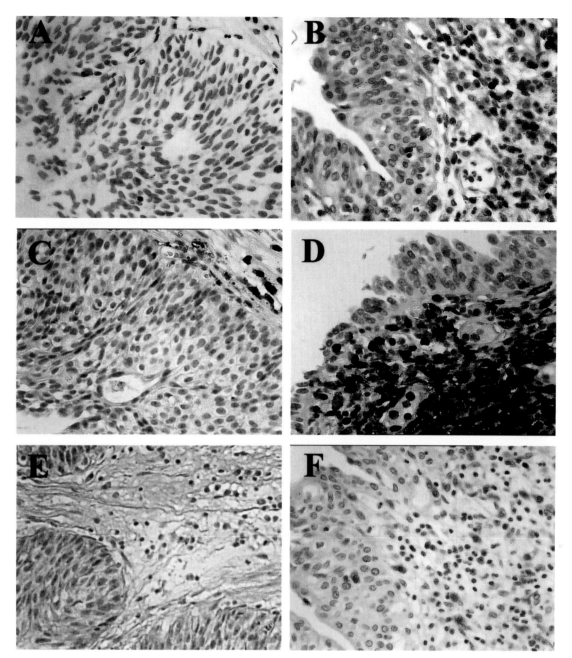

FIGURE 8   Immunohistochemical staining of bladder from patient 2. Pretreatment biopsies (A, C, E) and post-treatment cystectomy sections (B, D, F) were stained for CD3 (A, B), CD45RO (C, D), and Factor XIIIa (dendritic cells) (E, F). (From Mastrangelo, M. J. *et al., J. Clin. Invest.,* 105(8):1031, 2000. With permission.) (See color insert.)

References

1. Thomas, L. (1959). In *Cellular and Humoral Aspects of the Hypersensitive States* (H. S. Lawrence, ed.), pp. 529–532. Hoeber-Harper, New York.
2. Burnet, F. M. (1970). The concept of immunological surveillance. *Prog. Exp. Tumor. Res.* **13,** 1.
3. Stutman, O. (1975). Immunodepression and malignancy. *Adv. Cancer Res.* **22,** 261–422.
4. Prehn, R. T., and Main, J. M. (1957). Immunity to methylcholanthrene-induced sarcomas. *J. Natl. Cancer Inst.* **18,** 769–778.
5. Bodurtha, A. J., Berkelhammer, J., Kim, Y. H., Laucius, J. F., and Mastrangelo, M. J. (1976). A clinical, histologic, and immunologic study of a case of metastatic malignant melanoma undergoing spontaneous remission. *Cancer* **37,** 735–742.
6. *Spontaneous Remission: An Annotated Bibliography,* Institute of Noetic Sciences, Savsalito, CA, pp. 1–710.
7. Bornstein, R. S., Mastrangelo, M. J., Sulit, H., Chee, D., Yarbro, J. W.,

Prehn, L. M., and Prehn, R. T. (1973). Immunotherapy of melanoma with intralesional BCG. *Natl. Cancer Inst. Monogr.* **39**, 213–220.

8. Laucius, J. F., Bodurtha, A. J., Mastrangelo, M. J., and Creech, R. H. (1974). Bacillus Calmette-Guerin in the treatment of neoplastic disease. *J. Reticuloendothelial Soc.* **16**, 347–373.

9. Lamm, D. L., Thor, D. E., Harris, S. C., Reyna, J. A., Stogdill, V. D., and Radwin, H. M. (1980). Bacillus Calmette-Guerin immunotherapy of superficial bladder cancer. *J. Urol.* **124**, 38–42.

10. Mastrangelo, M. J., Maguire, H. C. J., Lattime, E. C., and Berd, D. (1995). Whole cell vaccines, in *Biological Therapy of Cancer,* 2nd ed. (V. T. DaVita, S. Hellman, and S. A. Rosenberg, eds.), pp. 648–658. Lippincott, Philadelphia, PA.

11. Mastrangelo, M. J., Sato, T., Lattime, E. C., Maguire, Jr., H. C., and Berd, D. (1998). Cellular vaccine therapies for cancer, in *Biological and Hormonal Therapies of Cancer* (K. A. Foon and H. B. Muss, eds.), pp. 35–50. Kluwer, Boston.

12. Hu, X., Chakraborty, N. G., Sporn, J. R., Kurtzman, S. H., Ergin, M. T., and Mukherji, B. (1996). Enhancement of cytolytic T lymphocyte precursor frequency in melanoma patients following immunization with MAGE-1 peptide loaded antigen presenting cell-based vaccine. *Cancer Res.* **56**, 2479–2483.

13. Mule, J. J. (2000). Tumor vaccine strategies that employ dendritic cells and tumor lysates: experimental and clinical studies. *Immunol. Invest.* **29**, 127–129.

14. Geiger, J., Hutchinson, R., Hohenkirk, L., McKenna, E., Chang, A., and Mule, J. (2000). Treatment of solid tumours in children with tumour-lysate-pulsed dendritic cells. *Lancet* **356**, 1163–1165.

15. Holtl, L., Rieser, C., Papesh, C., Ramoner, R., Herold, M., Klocker, H., Radmayr, C., Stenzl, A., Bartsch, G., and Thurnher, M. (1999). Cellular and humoral immune responses in patients with metastatic renal cell carcinoma after vaccination with antigen pulsed dendritic cells. *J. Urol.* **161**, 777–782.

16. Kugler, A., Stuhler, G., Walden, P., Zoller, G., Zobywalski, A., Brossart, P., Trefzer, U., Ullrich, S., Muller, C. A., Becker, V., Gross, A. J., Hemmerlein, B., Kanz, L., Muller, G. A., and Ringert, R. H. (2000). Regression of human metastatic renal cell carcinoma after vaccination with tumor cell-dendritic cell hybrids. *Nat. Med.* **6**, 332–336.

17. Ciborowski, P., and Finn, O. J. (1995). Recombinant epithelial cell mucin (MUC-1) expressed in baculovirus resembles antigenically tumor associated mucin, target for immunotherapy. *Biomed. Peptides Proteins Nucleic Acids* **1**, 193–198.

18. Bei, R., Kantor, J., Kashmiri, S. V., Abrams, S., and Schlom, J. (1994). Enhanced immune responses and anti-tumor activity by baculovirus recombinant carcinoembryonic antigen (CEA) in mice primed with the recombinant vaccinia CEA. *J. Immunother. Emphasis Tumor Immunol.* **16**, 275–282.

19. Livingston, P. O., Natoli, E. J., Jones Calves, M., Stockert, E., Oettgen, H. F., and Old, L. J. (1987). Vaccines containing purified GM2 ganglioside elicit GM2 antibodies in melanoma patients. *Proc. Natl. Acad. Sci. USA* **84**, 2911–2915.

20. Maloney, D. G., Liles, T. M., Czerwinski, D. K., Waldichuk, C., Rosenberg, J., Grillo-Lopez, A., and Levy, R. (1994). Phase I clinical trial using escalating single-dose infusion of chimeric anti-CD20 monoclonal antibody (IDEC-C2B8) in patients with recurrent B-cell lymphoma. *Blood* **84**, 2457–2466.

21. Vitetta, E., and Ghetie, V. (1994). Immunotoxins in the therapy of cancer: from bench to clinic. *Pharmacol. Ther.* **63**, 209–234.

22. Pai, L. H., and Pastan, I. (1994). Immunotoxins and recombinant toxins for cancer treatment. *Important Adv. Oncol.* 3–19.

23. Green, M. C., Murray, J. L., and Hortobagyi, G. N. (2000). Monoclonal antibody therapy for solid tumors. *Cancer Treat. Rev.* **26**, 269–86.

24. White, C. A., Weaver, R. L., and Grillo-Lopez, A. J. (2001). Antibody-targeted immunotherapy for treatment of malignancy. *Annu. Rev. Med.* **52**, 125–145.

25. Schwartz, R. H. (1992). Costimulation of T lymphocytes: the role of CD28, CTLA-4, and B7/BB1 in interleukin-2 production and immunotherapy. *Cell* **71**, 1065–1068.

26. Disis, M. L., Smith, J. W., Murphy, A. E., Chen, W., and Cheever, M. A. (1995). In vitro generation of human cytolytic T-cells specific for peptides derived from the HER-2/*neu* protooncogene protein. *Cancer Res.* **54**, 1071–1076.

27. Peace, D. J., Smith, J. W., Chen, W., You, S. G., Cosand, W. L., Blake, J., and Cheever, M. A. (1994). Lysis of *ras* oncogene-transformed cells by specific cytotoxic T lymphocytes elicited by primary in vitro immunization with mutated Ras peptide. *J. Exp. Med.* **179**, 473–479.

28. Peoples, G. E., Goedegebuure, P. S., Smith, R., Linehan, D. C., Yoshino, I., and Eberlein, T. J. (1995). Breast and ovarian cancer-specific cytotoxic T lymphocytes recognize the same HER2/*neu*-derived peptide. *Proc. Natl. Acad. Sci. USA* **92**, 432–436.

29. Van Elsas, A., Nijman, H. W., Van der Minne, C. E., Mourer, J. S., Kast, W. M., Melief, C. J., and Schreir, P. I. (1995). Induction and characterization of cytotoxic T-lymphocytes recognizing a mutated p21ras peptide presented by HLA-A*0201. *Int. J. Cancer* **61**, 389–396.

30. Storkus, W. J., Zeh, H. J., Maeurer, M. J., Salter, R. D., and Lotze, M. T. (1993). Identification of human melanoma peptides recognized by class I restricted tumor infiltrating T lymphocytes, *J. Immunol.* **151**, 3719–3727.

31. Slingluff, C. L., Hunt, D. F., and Engelhard, V. H. (1994). Direct analysis of tumor-associated peptide antigens. *Curr. Opin. Immunol.* **6**, 733–740.

32. Cox, A. L., Skipper, J., Chen, Y., Henderson, R. A., Darrow, T. L., Shabanowitz, J., Engelhard, V. H., Hunt, D. F., and Slingluff, C. L. (1994). Identification of a peptide recognized by five melanoma-specific human cytotoxic T cell lines. *Science* **264**, 716–719.

33. Correale, P., Walmsley, K., Nieroda, C., Zaremba, S., Zhu, M., Schlom, J., and Tsang, K. Y. (1997). In vitro generation of human cytotoxic T lymphocytes specific for peptides derived from prostate-specific antigen. *J. Natl. Cancer Inst.* **89**, 293–300.

34. Eder, J. P., Kantoff, P. W., Roper, K., Xu, G. X., Bubley, G. J., Boyden, J., Gritz, L., Mazzara, G., Oh, W. K., Arlen, P., Tsang, K. Y., Panicali, D., Schlom, J., and Kufe, D. W. (2000). A phase I trial of a recombinant vaccinia virus expressing prostate-specific antigen in advanced prostate cancer. *Clin. Cancer Res.* **6**, 1632–1638.

35. Klein, J., and Sato, A. (2000). The HLA system [second of two parts]. *N. Engl. J. Med.* **343**, 782–786.

36. Klein, J., and Sato, A. (2000). The HLA system [first of two parts]. *N. Engl. J. Med.* **343**, 702–709.

37. Linsley, P. S., and Ledbetter, J. A. (1993). The role of the CD28 receptor during T cell responses to antigen. *Ann. Rev. Immunol.* **11**, 191–212.

38. Lenschow, D. J., and Bluestone, J. A. (1993). T cell co-stimulation and in vivo tolerance. *Curr. Opin. Immunol.* **5**, 747–752.

39. Gimmi, C. D., Freeman, G. J., Gribben, J. G., Gray, G., and Nadler, L. M. (1993). Human T-cell clonal anergy is induced by antigen presentation in the absence of B7 costimulation. *Proc. Natl. Acad. Sci. USA* **90**, 6586–6590.

40. Janeway, C. A., and Travers, P. (1995). *Immunobiology: The Immune System in Health and Disease,* Garland Publishing, New York.

41. Finke, J. H., Rayman, P., Alexander, J., Edinger, M., Tubbs, R. R., Connelly, R., Pontes, E., and Bukowski, R. (1990). Characterization of the cytolytic activity of CD4$^+$ and CD8$^+$ tumor-infiltrating lymphocytes in human renal cell carcinoma. *Cancer Res.* **50**, 2363–2370.

42. Li, W. Y., Lusheng, S., Kanbour, A., Herberman, R. B., and Whiteside, T. L. (1989). Lymphocytes infiltrating human ovarian tumors: synergy between tumor necrosis factor α and interleukin 2 in the generation of CD8$^+$ effectors from tumor-infiltrating lymphocytes. *Cancer Res.* **49**, 5979–5985.

43. Schoof, D. D., Jung, S.-E., and Eberlein, T. J. (1989). Human tumor-infiltrating lymphocyte (TIL) cytotoxicity facilitated by anti-T-cell receptor antibody. *Int. J. Cancer* **44,** 219–224.

44. Balch, C. M., Riley, L. B., Bae, Y. J., Salmeron, M. A., Platsoucas, C. D., Von Eschenbach, A., and Itoh, K. (1990). Patterns of human tumor-infiltrating lymphocytes in 120 human cancers. Arch. Surg. **125,** 200–205.

45. Haas, G. P., Solomon, D., and Rosenberg, S. A. (1990). Tumor-infiltrating lymphocytes from nonrenal urological malignancies. *Cancer Immunol. Immunother.* **30,** 342–350.

46. Disis, M. L., and Cheever, M. A. (1998). HER-2/*neu* oncogenic protein: issues in vaccine development. *Crit. Rev. Immunol.* **18,** 37–45.

47. Disis, M. L., and Cheever, M. A. (1996). Oncogenic proteins as tumor antigens. *Curr. Opin. Immunol.* **8,** 637–42.

48. Zitvogel, L., Tahara, H., Robbins, P. D., Storkus, W. J., Clarke, M. R., Nalesnik, M. A., and Lotze, M. T. (1995). Cancer immunotherapy of established tumors with IL-12: effective delivery by genetically engineered fibroblasts. *J. Immunol.* **155,** 1393–1403.

49. Kurane, S., Arca, M. T., Aruga, A., Krinock, R. A., Krauss, J. C., and Chang, A. E. (1997). Cytokines as an adjuvant to tumor vaccines: efficacy of local methods of delivery. *Ann. Surg. Oncol.* **4,** 579–585.

50. Yamamura, M., Utemura, K., Deans, R. J., Weinberg, K., Rea, T. H., Bloom, B. R., and Modlin, R. L. (1991). Defining protective responses to pathogens: cytokine profiles in leprosy lesions. *Science* **254,** 277–279.

51. Scott, P., Pearce, E., Cheever, A. W., Coffman, R. L., and Sher, A. (1989). Role of cytokines and CD4+ T-cell subsets in the regulation of parasite immunity and disease [review]. *Immunol. Rev.* **112,** 161–182.

52. Townsend, S. E., and Allison, J. P. (1993). Tumor rejection after direct costimulation of CD8+ T cells by B7-transfected melanoma cells. *Science* **259,** 368–370.

53. Chen, L., Ashe, S., Brady, W. A., Hellstrom, I., Hellstrom, K. E., Ledbetter, J. A., McGowan, P., and Linsley, P. S. (1992). Costimulation of anti-tumor immunity by B7 counterreceptor for the T lymphocyte molecules CD28 and CTLA-4. *Cell* **71,** 1093–1102.

54. Lattime, E. C., Mastrangelo, M. J., Bagasra, O., Li, W., and Berd, D. (1995). Expression of cytokine mRNA in human melanoma tissues. *Cancer Immunol. Immunother.* **41,** 151–156.

55. Kruger-Krasagakes, S., Krasagakis, K., Garbe, C., Schmitt, E., Huls, C., Blankenstein, T., and Diamantstein, T. (1994). Expression of interleukin 10 in human melanoma. *Br. J. Cancer* **70,** 1182–1185.

56. Chen, Q., Daniel, V., Maher, D. W., and Hersey, P. (1994). Production of IL-10 by melanoma cells: examination of its role in immunosuppression mediated by melanoma. *Int. J. Cancer* **56,** 755–760.

57. Lattime, E. C., McCue, P. A., Keeley, F. X., Li, W., and Gomella, L. G. (1995). Expression of IL10 mRNA in biopsies of superficial and invasive TCC of the human bladder. *Proc. Am. Assoc. Cancer Res.* **36,** 462.

58. Gastl, G. A., Abrams, J. S., Nanus, D. M., Oosterkamp, R., Silver, J., Liu, F., Chen, M., Albino, A. P., and Bander, N. H. (1993). Interleukin-10 production by human carcinoma cell lines and its relationship to interleukin-6 expression. *Int. J. Cancer* **55,** 96–101.

59. Pisa, P., Halapi, E., Pisa, E. K., Gerdin, E., Hising, C., Bucht, A., Gerdin, B., and Kiessling, R. (1992). Selective expression of interleukin 10, interferon γ, and granulocyte–macrophage colony-stimulating factor in ovarian cancer biopsies. *Proc. Natl. Acad. Sci. USA* **89,** 7708–7712.

60. McAveney, K. M., Gomella, L. G., and Lattime, E. C. (1994). Induction of TH1 and TH2 associated cytokine mRNA in mouse bladder following intravesical growth of the murine bladder tumor MB49 and BCG immunotherapy. *Clin. Immunol. Immunother.* **39,** 401–406.

61. Gorelik, L., Prokhorova, A., and Mokyr, M. B. (1994). Low-dose melphalan-induced shift in the production of a Th2-type cytokine to a Th1-type cytokine in mice bearing a large MOPC-315 tumor. *Clin. Immunol. Immunother.* **39,** 117–126.

62. Halak, B. K., Maguire, Jr., H. C., and Lattime, E. C. (1999). Tumor-induced interleukin-10 inhibits type 1 immune responses directed at a tumor antigen as well as a non-tumor antigen present at the tumor site. *Cancer Res.* **59,** 911–917.

63. Monken, C. E., Gomella, L. G., Li, W., Fink, E., and Lattime, E. C. (1996). IL10 is produced by human transitional cell carcinoma lines immortalized by retroviral transfection with human papilloma virus E6/E7 genes. *Proc. Am. Assoc. Cancer Res.* **37,** 451.

64. Maguire, Jr., H. C., Ketcha, K. A., Halak, B. K., Holmes, K. L., and Lattime, E. C. (1997). Tumor-induced IL10 production in-vivo suppresses the development of delayed type hypersensitivity (DTH) to tumor associated antigens. *Proc. Am. Assoc. Cancer Res.* **38,** 358.

65. Halak, B. K., Maguire, Jr., H. C., and Lattime, E. C. (1998). Tumor-associated IL-10 inhibits immune responses directed at a tumor specific antigen as well as a non-tumor antigen present at the tumor site. *Proc. Am. Assoc. Cancer Res.* **39,** 652.

66. Ding, L., Linsley, P. S., Huang, L. Y., Germain, R. N., and Shevach, E. M. (1993). IL-10 inhibits macrophage costimulatory activity by selectively inhibiting the up-regulation of B7 expression. *J. Immunol.* **151,** 1224–1234.

67. Steinbrink, K., Jonuleit, H., Muller, G., Schuler, G., Knop, J., and Enk, A. H. (1999). Interleukin-10-treated human dendritic cells induce a melanoma-antigen-specific anergy in CD8(+) T cells resulting in a failure to lyse tumor cells. *Blood* **93,** 1634–1642.

68. Qin, Z., Noffz, G., Mohaupt, M., and Blankenstein, T. (1997). Interleukin-10 prevents dendritic cell accumulation and vaccination with granulocyte–macrophage colony-stimulating factor gene-modified tumor cells. *J. Immunol.* **159,** 770–776.

69. De Smedt, T., Van Mechelen, M., De Becker, G., Urbain, J., Leo, O., and Moser, M. Effect of interleukin-10 on dendritic cell maturation and function. *Eur. J. Immunol.* **27,** 1229–1235.

70. Fearon, E. R., Pardoll, D. M., Itaya, T., Golumbek, P., Levitsky, H. I., Simons, J. W., Karasuyama, H., Vogelstein, B., and Frost, P. (1990). Interleukin-2 production by tumor cells bypasses T helper function in the generation of an antitumor response. Cell. **60,** 397–403.

71. Asher, A. L., Mulé, J. J., Kasid, A., Restifo, N. P., Salo, J. C., Reichert, C. M., Jaffe, G., Fendly, B., Kriegler, M., and Rosenberg, S. A. (1991). Murine tumor cells transduced with the gene for tumor necrosis factor-α: evidence for paracrine immune effects of tumor necrosis factor against tumors. *J. Immunol.* **146,** 3227–3234.

72. Watanabe, Y., Kuribayashi, K., Miyatake, J., Nishihara, K., Nakayama, E., Taniyama, T., and Sakata, T. (1989). Exogenous expression of mouse interferon-gamma cDNA in mouse neuroblastoma C1300 cells results in reduced tumorigenicity by augmented anti-tumor immunity. *Proc. Natl. Acad. Sci. USA* **86,** 9456–9460.

73. Tepper, R. I., Pattengale, P. K., and Leder, P. (1989). Murine interleukin-4 displays potent anti-tumor activity in vivo. *Cell* **57,** 503–512.

74. Connor, J., Bannerji, R., Saito, S., Heston, W., Fair, W., and Gilboa, E. (1993). Regression of bladder tumors in mice treated with interleukin 2 gene-modified tumor cells. *J. Exp. Med.* **177,** 1127–1134.

75. Saito, S., Bannerji, R., Gansbacher, B., Rosenthal, F. M., Romanenko, P., Heston, W. D. W., Fair, W. R., and Gilboa, E. (1994). Immunotherapy of bladder cancer with cytokine gene-modified tumor vaccines. *Cancer Res.* **54,** 3516–3520.

76. Dranoff, G., Jaffee, E., Lazenby, A., Golumbek, P., Levitsky, H., Brose, K., Jackson, V., Hamada, H., Pardoll, D. M., and Mulligan, R. C. (1993). Vaccination with irradiated tumor cells engineered to secrete murine granulocyte–macrophage colony-stimulating factor stimulates potent, specific, and long lasting anti-tumor immunity. *Proc. Natl. Acad. Sci. USA* **90,** 3539–3543.

77. Simons, J. W., Mikhak, B., Chang, J. F., DeMarzo, A. M., Carducci, M. A., Lim, M., Weber, C. E., Baccala, A. A., Goemann, M. A., Clift, S. M., Ando, D. G., Levitsky, H. I., Cohen, L. K., Sanda, M. G., Mulligan, R. C., Partin, A. W., Carter, H. B., Piantadosi, S., Marshall, F. F., and

Nelson, W. G. (1999). Induction of immunity to prostate cancer antigens: results of a clinical trial of vaccination with irradiated autologous prostate tumor cells engineered to secrete granulocyte–macrophage colony-stimulating factor using ex vivo gene transfer. *Cancer Res.* **59**, 5160–5168.

78. Simons, J. W., Jaffee, E. M., Weber, C. E., Levitsky, H. I., Nelson, W. G., Carducci, M. A., Lazenby, A. J., Cohen, L. K., Finn, C. C., Clift, S. M., Hauda, K. M., Beck, L. A., Leiferman, K. M., Owens, Jr., A. H., Piantadosi, S., Dranoff, G., Mulligan, R. C., Pardoll, D. M., and Marshall, F. F. (1997). Bioactivity of autologous irradiated renal cell carcinoma vaccines generated by ex vivo granulocyte–macrophage colony-stimulating factor gene transfer. *Cancer Res.* **57**, 1537–1546.

79. Nelson, W. G., Simons, J. W., Mikhak, B., Chang, J. F., DeMarzo, A. M., Carducci, M. A., Kim, M., Weber, C. E., Baccala, A. A., Goeman, M. A., Clift, S. M., Ando, D. G., Levitsky, H. I., Cohen, L. K., Sanda, M. G., Mulligan, R. C., Partin, A. W., Carter, H. B., Piantadosi, S., and Marshall, F. F. (2000). Cancer cells engineered to secrete granulocyte–macrophage colony-stimulating factor using ex vivo gene transfer as vaccines for the treatment of genitourinary malignancies. *Cancer Chemother. Pharmacol.* **46**, S67–S72.

80. Sigal, R. K., Lieberman, M. D., Reynolds, J. V., Williams, N., Ziegler, M. M., and Daly, J. M. (1990). Tumor immunization: improved results after vaccine modified with recombinant interferon gamma. *Arch. Surg.* **125**, 308–312.

81. Golumbek, P. T., Lazenby, A. J., Levitsky, H. I., Jaffee, L. M., Karasuyama, H., Baker, M., and Pardoll, D. M. (1991). Treatment of established renal cancer by tumor cells engineered to secrete interleukin-4. *Science* **254**, 713–716.

82. Lattime, E. C., McCue, P. A., Ross, R. P., Baltish, M. A., and Gomella, L. (1992). T cells bearing τ/δ receptor in human transitional cell carcinoma of the bladder. *Proc. Am. Assoc. Cancer Res.* **33**, 334.

83. Perussia, B., Chan, S. H., D'Andrea, A., Tsuji, K., Santoli, D., Pospisil, M., Young, D., Wolf, S. F., and Trinchieri, G. (1992). Natural killer cell stimulatory factor or interleukin-12 has differential effects on the proliferation of TCR$\alpha\beta$+, TCR$\tau\delta$+ T lymphocytes and NK cells. *J. Immunol.* **149**, 3495–3502.

84. Basker, S., Glimcher, L., Nabavi, N., Jones, R. T., and Ostrand-Rosenberg, S. (1995). Major histocompatibility complex class II+ B7-1+ tumor cells are potent vaccines for stimulating tumor rejection in tumor bearing mice. *J. Exp. Med.* **181**, 619–629.

85. Hodge, J. W., Rad, A. N., Grosenbach, D. W., Sabzevari, H., Yafal, A. G., Gritz, L., and Schlom, J. (2000). Enhanced activation of T cells by dendritic cells engineered to hyperexpress a triad of costimulatory molecules. *J. Natl. Cancer Inst.* **92**, 1228–1239.

86. Baldick, C. J. J., Keck, J. G., and Moss, B. (1992). Mutational analysis of the core, spacer, and initiator regions of vaccinia virus intermediate-class promoters. *J. Virol.* **66**, 4710–4719.

87. Cochran, M. A., Puckett, C., and Moss, B. (1985). In vitro mutagenesis of the promoter region for a vaccinia virus gene: evidence for tandem early and late regulatory signals. *J. Virol.* **54**, 30–37.

88. Fenner, F., Henderson, D. A., and Anita, I. (1988). *Smallpox and Its Eradication,* World Health Organization, Geneva.

89. Williams, N. R., and Cooper, B. M. (1993). Counselling of workers handling vaccinia virus. *Occup. Med.* **43**, 125–127.

90. Smith, G. L., and Moss, B. (1983). Infectious poxvirus vectors have capacity for at least 25,000 base pairs of foreign DNA. *Gene* **25**, 21–28.

91. Earl, P. L., Moss, B., and Doms, R. W. (1991). Folding, interaction with GRP78-BiP, assembly, and transport of human immunodefficiency virus type 1 envelope protein. *J. Virol.* **65**, 2047–2055.

92. Scharenberg, A., Lin, S., Cuenod, B., Yamamura, H., and King, F. (1995). Reconstitution of interactions between tyrosine kinases and the high affinity IgE receptor which are controlled by recepter clustering. *EMBO J.* **14**, 3385–3394.

93. Feng, Y., Broder, C. C., Kennedy, P. E., and Berger, M. (1996). IIIV-1 entry cofactor: functional cDNA cloning of a seven transmembrane G protein-coupled receptor. *Science* **272**, 872–877.

94. Doranz, B. J., Rucker, Yi, Y., Smyth, R. J., Samson, M., Peiper, S. C., Parmentier, M., Collman, R. G., and Domzig, W. (1996). A dual tropic primary HIV-1 isolate that uses fusin and the beta chemokine receptors CKR-5, CKR-3, and CKR-2b as fusion cofactors. *Cell* **85**, 1149–1158.

95. Kapuiah, G., Woodhams, C. E., Blanden, R. V., and Ramshaw, I. A. (1991). Immunobiology of infection with recombinant vaccinia virus encoding murine IL2. *J. Immunol.* **147**, 4327–4332.

96. Moss, B. (1996). Genetically engineered poxviruses for recombinant gene expression, vaccination, and safety. *Proc. Natl. Acad. Sci. USA* **93**, 11341–11348.

97. Restifo, N. P. (1996). The new vaccines: building viruses that elicit antitumor immunity. *Curr. Opin. Immunol.* **8**, 658–663.

98. Marshall, J. L., Hoyer, R. J., Toomey, M. A., Faraguna, K., Chang, P., Richmond, E., Pedicano, J. E., Gehan, E., Peck, R. A., Arlen, P., Tsang, K. Y., and Schlom, J. (2000). Phase I study in advanced cancer patients of a diversified prime-and-boost vaccination protocol using recombinant vaccinia virus and recombinant nonreplicating avipox virus to elicit anti-carcinoembryonic antigen immune responses. *J. Clin. Oncol.* **18**, 3964–3973.

99. Conry, R. M., Khazaeli, M. B., Saleh, M. N., Allen, K. O., Barlow, D. L., Moore, S. E., Craig, D., Arani, R. B., Schlom, J., and LoBuglio, A. F. (1999). Phase I trial of a recombinant vaccinia virus encoding carcinoembryonic antigen in metastatic adenocarcinoma: comparison of intradermal versus subcutaneous administration. *Clin. Cancer Res.* **5**, 2330–2337.

100. Cochran, M. S., Mackett, M., and Moss, B. (1985). Eukaryotic transient expression system dependent on transcription factors and regulatory DNA sequences of vaccinia virus. *Proc. Natl. Acad. Sci. USA* **82**, 19–23.

101. Fuerst, T. R., Niles, E. G., Studier, F. W., and Moss, B. (1986). Eukaryotic transient-expression system based on recombinant vaccinia virus that synthesizes bacteriophage T7 RNA polymerase. *Proc. Natl. Acad. Sci. USA* **83**, 8122–8126.

102. Rodriguez, D., Zhou, Y., Durbin, R. K., Jimenez, V., McAllister, W. T., and Esteban, M. (1990). Regulated expression of nuclear genes by T3 RNA polymerase and *lac* repressor using recombinant vaccinia virus vectors. *J. Virol.* **64**, 4851–4857.

103. Perkus, M. E., Limbach, K., and Paoletti, E. (1989). Cloning and expression of foreign genes in vaccinia virus using a host range selection system. *J. Virol.* **63**, 3829–3836.

104. Coupar, B. E. H., Andrew, M. E., and Boyle, D. B. (1988). A general method for the construction of recombinant vaccinia viruses expressing multiple foreign genes. *Gene* **68**, 1–10.

105. Davison, A. J., and Moss, B. (1989). Structure of vaccinia virus early promoters. *J. Mol. Biol.* **210**, 749–769.

106. Davison, A. J., and Moss, B. (1989). Structure of vaccinia virus late promoters. *J. Mol. Biol.* **210**, 771–784.

107. Davison, A. J., and Moss, B. (1990). New vaccinia virus recombination plasmids incorporating a synthetic late promoter for high level expression of foreign proteins. *Nucleic Acids Res.* **18**, 4285–4286.

108. Merchlinsky, M., and Moss, B. (1992). Introduction of foreign DNA into the vaccinia virus genome by *in vitro* ligation: recombination-independent selectable cloning vectors. *Virology* **190**, 522–526.

109. Sutter, G., and Moss, B. (1992). Nonreplicating vaccinia vector efficiently expresses recombinant genes. *Proc. Natl. Acad. Sci. USA* **89**, 10847–10851.

110. Carroll, M. W., and Moss, B. (1997). Host range and cytopathogenicity of the highly attenuated MVA strain of vacinia virus: propagation and generation of recombinant viruses in a nonhuman mammalian cell line. *Virology* **238**, 198–211.

111. Paoletti, E. (1996). Application of pox virus vectors to vaccination: an update. *Proc. Natl. Acad. Sci. USA* **93**, 11349–11353.
112. Tsung, K., Yim, J. H., Marti, W., Buller, M. L., and Norton, J. A. (1996). Gene expression and cytopathic effect of vaccinia virus inactivated by psoralen and long-wave UV light. *J. Virol.* **70**, 165–171.
113. Mastrangelo, M. J., Maguire, Jr., H. C., McCue, P. A., Lee, S. S., Alexander, A., Nazarian, L. N., Eisenlohr, L. C., Nathan, F. E., Berd, D., and Lattime, E. C. (1995). A pilot study demonstrating the feasibility of using intratumoral vaccinia injections as a vector for gene transfer. *Vaccine Res.* **4**, 55–69.
114. Mastrangelo, M. J., Maguire, Jr., H. C., Eisenlohr, L. C., Laughlin, C. E., Monken, C. E., McCue, P. A., Kovatich, A. J., and Lattime, E. C. (1999). Intratumoral recombinant GM-CSF-encoding virus as gene therapy in patients with cutaneous melanoma. *Cancer Gene Ther.* **6**, 409–422.
115. Vanderplasschen, A., Hillinshead, M., and Smith, G. L. (1997). Antibodies against vaccinia virus do not neutralize extracellular enveloped virus but prevent virus release from infected cells and comet formation. *J. Gen. Virol.* **78**, 2041–2048.
116. Lee, S. S., Eisenlohr, L. C., McCue, P. A., Mastrangelo, M. J., and Lattime, E. C. (1993). Intravesical gene therapy: vaccinia virus recombinants transfect murine bladder tumors and urothelium. *Proc. Am. Assoc. Cancer Res.* **34**, 337.
117. Lee, S. S., Eisenlohr, L. C., McCue, P. A., Mastrangelo, M. J., and Lattime, E. C. (1994). Intravesical gene therapy: in-vivo gene transfer using vaccinia vectors. *Cancer Res.* **54**, 3325–3328.
118. Lattime, E. C., Maguire, Jr., H. C., McCue, P. A., Eisenlohr, L. C., Berd, D., Lee, S. S., and Mastrangelo, M. J. (1994). Infection of human melanoma cells by intratumoral vaccinia, *J. Invest. Dermatol.* **102**, 568.
119. Gomella, L. G., Mastrangelo, M. J., Eisenlohr, L. C., McCue, P. A., Lee, S. S., and Lattime, E. C. (1995). Localized gene therapy for prostate cancer: strategies for intraprostatic cytokine gene transfection using vaccinia virus vectors. *J. Urol.* **153**, 308A.
120. Lattime, E., Eisenlohr, L., Gomella, L., and Mastrangelo, M. (1999). The use of vaccinia virus vectors for immunotherapy via in-situ tumor transfection, in *Gene Therapy of Cancer: Translational Approaches from Preclinical Studies to Clinical Implementation* (E. Lattime and S. Gerson, eds.), pp. 125–137. Academic Press, San Diego, CA.
121. Lee, S. S., Eisenlohr, L. C., McCue, P. A., Mastrangelo, M. J., Fink, E., and Lattime, E. C. (1995). In-vivo gene therapy of murine tumors using recombinant vaccinia virus encoding GM-CSF. *Proc. Am. Assoc. Cancer Res.* **36**, 248.
122. Lee, S. S., Eisenlohr, L. C., McCue, P. A., Mastrangelo, M. J., and Lattime, E. C. (1994). Vaccinia virus vector mediated cytokine gene transfer for in vivo tumor immunotherapy. *Proc. Am. Assoc. Cancer Res.* **35**, 514.
123. Ramshaw, I., Ruby, J., Ramsay, A., Ada, G., and Karupiah, G. (1992). Expression of cytokines by recombinant vaccinia viruses: a model for studying cytokines in virus infections in-vivo, *Immunol. Rev.* **127**, 157–182.
124. Mackett, M., Smith, G. L., and Moss, B. (1984). General method for production and selection of infectious vaccinia virus recombinants expressing foreign genes. *J. Virol.* **49**, 857–864.
125. Elkins, K. L., Ennist, D. L., Winegar, R. K., and Weir, J. P. (1994). In-vivo delivery of interleukin-4 by a recombinant vaccinia prevents tumor development in mice. *Hum. Gene Ther.* **5**, 809–820.
126. Whitman, E. D., Tsung, K., Paxson, J., and Norton, J. (1994). In vitro and in vivo kinetics of recombinant vaccinia virus cancer-gene therapy. *Surgery* **116**, 183–188.
127. Qin, H., and Chatterjee, S. K. (1996). Recombinant vaccinia expressing interleukin-2 for cancer gene therapy. *Cancer Gene Ther.* **3**, 163–167.
128. Lattime, E. C., Maguire, H. C. J., McCue, P. A., Eisenlohr, L. C., Berd, D., Lee, S. S., and Mastrangelo, M. J. (1994). Gene therapy using vaccinia vectors: repeated intratumoral injections result in tumor infection in the presence of anti-vaccinia immunity. *Proc. Am. Soc. Clin. Oncol.* **13**, 397.
129. Lattime, E. C., Mastrangelo, M. J., and Berd, D. (1994). Human metastatic melanoma lesions and cell lines express mRNA for IL-10. *Proc. Am. Assoc. Cancer Res.* **35**, 489.
130. Lattime, E. C., McCue, P. A., Keeley, F. X., Li, W., Baltish, M. A., and Gomella, L. G. Biopsies of superficial and invasive TCC of the bladder express mRNA for the immunosuppressive cytokine IL10. *J. Urol.* **153**, 488a.
131. Chakrabarti, S., Brechling, K., and Moss, B. (1985). Vaccinia virus expression vector: coexpression of β-galactosidase provides visual screening of recombinant virus plaques. *Mol. Cell. Biol.* **5**, 3403–3409.

CHAPTER

13

The Use of Particle-Mediated Gene Transfer for Immunotherapy of Cancer

MARK R. ALBERTINI
The University of Wisconsin
Comprehensive Cancer Center
Madison, Wisconsin 53792

DAVID M. KING
The University of Wisconsin
Comprehensive Cancer Center
Madison, Wisconsin 53792

ALEXANDER L. RAKHMILEVICH
The University of Wisconsin
Comprehensive Cancer Center
Madison, Wisconsin 53792

interest; (3) transfect resting, nondividing cells, irrespective of cell lineage; and (4) simultaneously deliver multiple candidate therapeutic genes into one cell or neighboring cells. The latter feature of PMGT can be beneficial for genetic immunization strategies when a gene for a tumor-associated antigen is employed in combination with cytokine genes. Exciting preclinical results suggest potential clinical strategies to stimulate *in vivo* antitumor T-cell immunity. Several recently completed clinical studies have demonstrated the safety of *ex vivo* and *in vivo* PMGT. Rigorous clinical evaluation of PMGT, with careful immunological monitoring, is needed to determine the potential clinical importance of this technology.

II. BACKGROUND

A. Uses of PMGT

Particle-mediated gene transfer provides a physical means for intracellular delivery of biologically active molecules. With the use of high-voltage electric discharge, PMGT was originally utilized for genetic transformation of various plants [1,2]. Because of its physical nature, this method displays properties distinct from characteristic properties of chemical and biological gene transfer agents [3]. During the past decade, PMGT technology has been utilized for *in vivo* transfection of various mammalian somatic tissues, including skin, liver, pancreas, muscle, spleen, and other organs [4,5]. In addition, PMGT has been utilized for *ex vivo* transfection of brain, mammary, and leukocyte primary cultures or tissue explants [6–9] as well as a wide range of *in vitro* cell lines [4,8–10]. A potential advantage of PMGT is its applicability to cells *in vivo,* which is the primary focus of this review.

I. INTRODUCTION

The particle-mediated gene transfer (PMGT) technology is a relatively new approach for mammalian gene transfer. This method of gene delivery uses a burst of helium to accelerate DNA-coated gold particles into target cells. Resulting transgene expression levels are often significantly higher than those achieved by other direct DNA delivery methods. Moreover, it takes only several seconds to complete a single treatment by PMGT. Significant advantages of PMGT include the ability to (1) physically confer gene expression by nonviral means; (2) direct the particle delivery to the anatomic site of

B. Technical Aspects

To achieve PMGT, microscopic gold particles are coated with plasmid DNA containing the gene(s) of interest and are accelerated by a motive force to sufficient velocities to penetrate the target cells and provide intracellular delivery of the transgene [11]. The amount of DNA required for PMGT is relatively low, and an amount of 5000 copies of cDNA delivered per cell was found to be optimal for commonly used reporter genes [4,5]. The motive force for PMGT was originally generated by high-voltage electric discharge [1,2]. Subsequently, a helium shock wave was produced by a hand-held version of the original device (formerly the *Accell* gene delivery device; currently known as the Dermal PowderJect-XR device, PowderJect Vaccines, Inc., Madison, WI; also commercially available for research purposes from Bio-Rad under the name Helios). DNA/gold particles are loaded in a small Teflon tube that acts as a cartridge; the research hand-held helium -pulse transfection device can hold 12 cartridges in a revolving cylinder (Fig. 1A). Each gene transfer can be performed in less than 5 seconds, thus making PMGT an efficient method for repeated, multiple gene deliveries. The device designed for clinical applications accepts pre-filled, single-use, disposable nozzles that can be manufactured under good manufacturing practice (GMP) procedures. The design of the clinical device utilized in a recent clinical trial at the University of Wisconsin is shown in Fig. 1B.

Parameters for PMGT into skin that require consideration include the composition, size, and shape of the particles utilized for gene transfer. Particles made from dense materials, including gold, tungsten, iridium, and platinum, are all capable of effectively delivering DNA via PMGT. However, gold particles are most commonly chosen for two reasons: Elemental gold is chemically inert with no significant cytotoxic effects [12], and, owing to the common use of gold in the electronics industry, uniformly sized gold particles are commercially available. A wide range of gold particle sizes have been evaluated for PMGT; for mammalian skin tissues, the 2–3-μm gold particle size is frequently utilized [4,5,13]. Gold particles are readily available in different forms (e.g., as round particles, crystals, or even aggregates), for gene transfer into epidermal cells, crystal and spherical gold particles were found to be similarly effective in gene delivery.

The DNA loading rate per particle, particle loading rate per target surface area, and the physical acceleration rate for particle penetration into skin are additional important parameters for PMGT into skin. More than 5000 copies of 5–10-kb plasmid DNA can be effectively coated onto a single 1-3-μm gold particle with a Ca^{2+}/spermidine or poly(ethylene glucol) (PEG) formulation in precipitated form [11]. With a predetermined gold particle loading rate of 0.1 mg/cm^2, approximately 1–2 gold particles (1–3 μm) per cell can be delivered via random distribution into epidermis containing stratified epithelial cells on the order of 15 μm in

diameter [13]. Excessive particle loading rates could cause trauma to transfected tissues. On the other hand, too low a particle load could result in low gene transfer efficiency. For transfecting murine skin tissues, a pressure of 300–500 psi for the helium pulse device has been found to confer high levels of transgene expression [13].

Another important feature of PMGT is the much reduced restriction on the size of the DNA vectors. Plasmid DNA, genomic DNA ($\sim$23 kb), and reporter genes cloned in lambda phage genomic libraries ($\sim$44 kb) can all be effectively delivered into mammalian cells by PMGT [6]. This capability offers new opportunities for transferring multiple genes, large-size genomic DNA sequences, or multiple tandem genes into mammalian somatic tissues. In addition, cotransfection of multiple genes on different plasmids has also been shown to be efficiently achieved by using the PMGT method [14–16]. Furthermore, RNA molecules can be similarly delivered as DNA vectors by PMGT [17].

The major current disadvantages of PMGT *in vivo* are the limited transfection efficiency for certain tissue systems (particularly if permanent gene transfer to the target cells is necessary) and the depth of tissue that can be accessed. Although transient transfection efficiencies from 3% to approximately 50% are possible *in vitro* [11], efficiencies for stable (i.e., integrative) gene transfer *in vivo* are apparently low and have not been clearly established in various transfected somatic tissues. Long-term transgene expression following PMGT has been observed in muscle and dermis, but these tissues seem to be the exception rather than the rule [13]. More general long-term gene expression may be possible through a combination of PMGT and replicating or actively integrating vector systems, but for the present it appears that the technique is most suitable in applications where short- to medium-term transgene expression is sufficient or desirable, such as DNA vaccine applications. The PMGT method at present also cannot deliver genes systemically to cell fractions scattered in large, three-dimensional tissues such as liver or brain, as can certain other gene transfer systems when administered through the circulatory system [18].

C. *In Vivo* Applications

Among the applications of PMGT technology, transfection of skin tissues of live animals has resulted in some of the most interesting findings. High levels of transgene expression were first demonstrated by *in vivo* PMGT of skin epidermal tissues in rodents [5,19]. These results were highly reproducible for various large animals, including turkeys [20], rabbits [21], dogs [22], pigs [23], horses [24], and rhesus monkeys [25]. Safe and effective skin transfection by PMGT has also been recently demonstrated in humans [26]. Efficient delivery and expression of transgenes in skin tissues have been extended to several reporter genes [4,6], candidate tumor-antigen genes [27–30], cytokine genes [31,32], viral

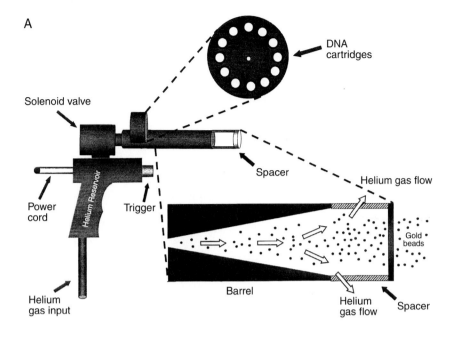

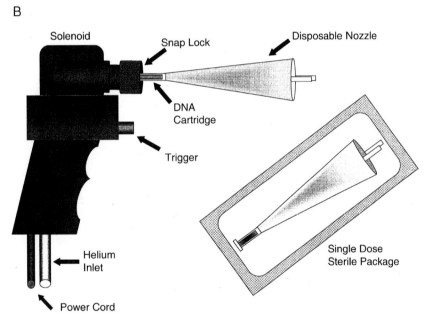

FIGURE 1 Design and operation of the PMGT device. (A) Diagram of the research hand-held helium pulse transfection device. (Artwork by Dr. K. Barton. From Rakhmilevich, A. and Yang, N., in *Methods in Molecular Medicine.* Vol. 35. *Gene Therapy: Methods and Protocols,* W. Walther and U. Stein, eds., Humana Press, Totowa, NJ, 2000. With permission.) (B) Diagram of the clinical device utilized in a recent clinical trial at the University of Wisconsin. (Artwork provided by PowderJect Vaccines, Inc., Madison, WI. With permission.)

antigen genes [25,33–36], and bacterial antigen genes [37,38], demonstrating the wide-ranging applicability of this gene transfer strategy. PMGT into skin tissues has been used for developing genetic immunization approaches [25,27–29,34], gene therapy of subcutaneous tumors [31,32,39], wound healing [40], delivery of RNA as transgenes or immunogens [17], analysis of transcriptional promoters and other regulatory sequences in gene expression vectors [5,41], and the study of various cell types in transgene expression and migration following DNA immunization [42,43].

Histological studies show that cDNA expression vectors can be introduced into and expressed in different cell layers of epidermal or dermal tissues by adjusting the ballistic variables. These include the motive force for particle acceleration, gold particle size, pretreatment and manipulations of epidermal and dermal tissues prior to transfection, and the DNA formulations for coating gold particles. In general, topical skin gene delivery results in high-level transient expression almost exclusively in the epidermal cell layers [4,5,39] which may vary considerably in different animals (e.g., 3–4 cell layers and 10–30 cell layers for mouse and pig skin, respectively). Treatment of dermal tissues via exposure of the underside of a skin flap can result in long-term transgene expression, though at a reduced level [5,31]. Because easily accessible epidermis is competent for eliciting both a humoral and a cell-mediated immune response and is efficient in the synthesis and secretion of transgenic proteins, it is an attractive gene transfer target for vaccination approaches.

D. Vector Considerations

Several vector design strategies to enhance the antigenicity of target cells following PMGT include maximizing transgene expression and the use of multiple vectors or vectors encoding several transgenes. Additional strategies include the enhancement of proteosome-dependent protein degradation, production of secreted cytokine to enhance the immune response, and use of immunostimulatory sequences. The use of a strong promoter is important to optimize transgene expression in mammalian cells. Viral promoters from cytomegalovirus (CMV) and simian virus 40 are commonly used. In addition, incorporation of polyadenylation sequences can be used for stabilization of mRNA transcripts [44].

Plasmid vectors can be used to modify an immune response depending on whether the expressed protein is retained in the cell or is membrane bound or secreted. Incorporation of DNA sequences into the vector can direct intracellular processing to target major histocompatibility complex (MHC) class I or class II pathways [44,45]. N-terminal ubiquitination signals target expressed proteins to proteosomes for rapid cytoplasmic degradation and presentation via the class I pathway. The E3 leader sequence from adenovirus can be used to facilitate transport of antigens directly into the endoplasmic reticulum for class I presentation. Both these methods appear to increase cytotoxic T lymphocyte (CTL) activity following vaccination in vivo [42,46]. If the specific antigen for CTL activation is known, its gene can be incorporated as a "minigene" for antigen expression through the class I pathway. Epitope-specific CTL activation can be seen with both PMGT and intramuscular delivery of plasmid [47].

Tumors are heterogenous and express many different antigens. Thus, potential benefit may be found following vaccination with a plasmid encoding several antigenic peptides. Expression of two or more genes or "minigenes" in a cell

following transfection can be undertaken with dicistronic or multicistronic vectors with internal ribosome entry sites. This multigene strategy can also combine cytokine genes to enhance the activation of CTLs [48]. Cytokine delivery at the site of plasmid vaccination can be used to influence the predominate expression of either a T_{h1} or a T_{h2} response [49,50]. Several in vivo vaccine studies with PMGT describe a predominately T_{h2} response following vaccination [34,51].

Another strategy for gene delivery involves vectors containing minimal amounts of DNA. These vectors incorporate a conditional origin of replication that can only replicate in limited hosts and eliminate antibiotic resistance markers. Reducing the amount of bacterial sequences and severely limiting the ability of the plasmid to replicate provide another vector strategy for nonviral gene therapy [45,52].

Plasmid vectors can also contribute to the immunogenicity of the transgene by utilizing immunostimulatory sequences (ISSs) of DNA in the vector. Cytosine–phosphate–guanosine (CpG) motifs, found in increased numbers in bacterial DNA, can act as potent immunostimulatory molecules [53]. CpG-oligodeoxynucleotide sequences have been shown to activate and mobilize dendritic cells (DCs), inducing the increased production of interleukin-12 (IL-12) that may account for the shift toward Th1-predominant immune responses [54]. The effects of ISSs appear to be less apparent following PMGT than after direct DNA injection (intramuscular or intradermal).

III. RECENT ADVANCES

A. In Vitro Modification of Human Cells

The stimulation of selective recognition and destruction of tumor cells by components of the immune system is a central goal of tumor immunotherapy. Because the T cell is felt to be an important component in the immune response against cancer, several strategies to activate T cells are being considered as candidate vaccine approaches. T-cell stimulation and activation by antigen require interaction of a T cell with a tumor cell or antigen-presenting cell (APC) that can provide appropriate MHC presentation of the antigen as well as necessary costimulation of the T cell [55–58] (Fig. 2). The use of PMGT to directly modify tumor cells or APC with one or more transgenes presents an opportunity to rapidly evaluate promising strategies to activate immune-mediated destruction of human cancer. One of these strategies is to make a tumor cell express costimulatory molecules, and this strategy has been shown to enhance the immunogenicity of tumor cells in murine studies [59].

1. Modification of Melanoma Cells

Human melanoma cells with no detectable baseline surface expression of the B7-1 costimulatory molecule can

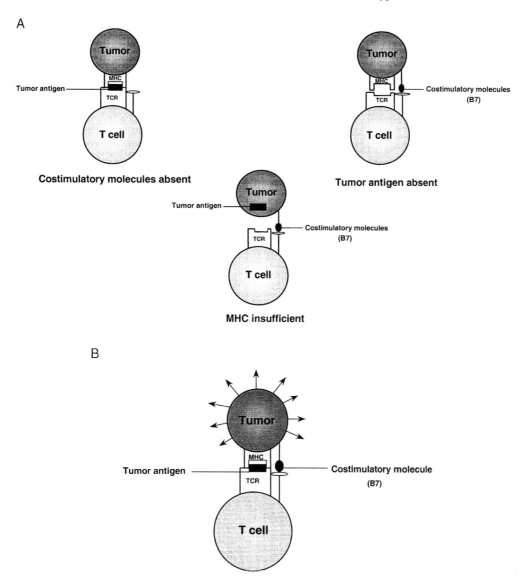

FIGURE 2   The requirements for T-cell activation by tumor cells include T-cell recognition of tumor antigen that is presented by the appropriate MHC molecule and the delivery of an additional costimulatory signal to the T cell. T-cell activation for tumor cell lysis requires each of these components to be present (A), and the lack of any of these components in the stimulating cell could result in a failure of the tumor cell to activate a T-cell response (B). (From Albertini, M. R. and Sondel, P. M., in *Clinical Oncology,* 2nd ed., Abeloff, M. D. *et al.,* eds., Saunders, Philadelphia, PA, 2000, pp. 214–241. With permission.)

have 8–31% of cells become B7-1 positive with no selection procedure after *in vitro* PMGT with human cDNA for B7-1 [16]. To evaluate whether the antigenicity of B7-1-expressing melanoma cells would vary with the level of B7-1 expression, melanoma cell lines with different levels of stable B7-1 expression were obtained [60]. This was accomplished by initially transfecting cells by PMGT with a B7-1-neo cDNA vector and proceeding with subsequent selection in media containing G418. Cells were then sorted on a FACStar Plus flow cytometer by brightness of B7-1 staining (Fig. 3), and cell lines that maintained different levels of stable B7-1 expression (85, 62, 26, and 14%) were obtained.

Functional studies determined the antigenicity of the human melanoma cells with transient (me15-B7, M-21-B7) or stable (me15-B7-neo, M-21-B7-neo) expression of B7-1. Allogeneic normal donor peripheral blood mononuclear cells (PBMCs) secreted greater amounts of granulocyte–macrophage colony-stimulating factor (GM-CSF) when incubated with B7-1-transfected melanoma cells than did PBMCs incubated with unmodified melanoma cells. Similarly, cell-mediated cytotoxicity against unmodified melanoma cells was greater in PBMCs cultured for 5 days with B7-1-transfected cells in comparison with PBMCs cultured with unmodified melanoma cells, and the level of

A

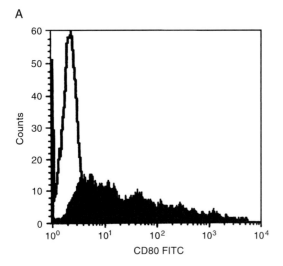

B

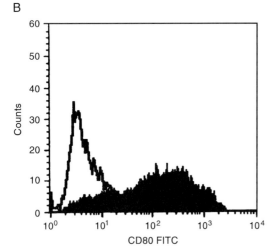

C

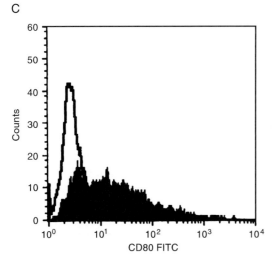

FIGURE 3 Flow cytometric analysis of B7-1 expression by M-21 cells. The B7-1 expression is shown for nontransfected M-21 cells (panel A), as well as for unsorted M-21–B7 stable transfectants (Panel B) and for M-21–B7 (bright) cells sorted to contain M-21 cells with high B7-1 expression (panel C). (From McCarthy, D. *et al.*, *Cancer Immunol. Immunother.*, 49, 85–93, 2000. With permission.

TABLE 1   GP-100 Expression by Transfected B-cell Lines

B-cell Lines[a]	Transgene[b]	HLA-A phenotype[c]	GP-100 expression[d]
721 poly	Neo/B7-gp 100	1,2	10%
.221 CL.5	Neo/B7-gp 100	-,-	60%
.221(A1)CL.8	Hygro/gp 100	1,-	90%
.221(A1)CL.9	Hygro/gp 100	1,-	60%
.221(A1)N-CL.5	Hygro-vector	1,-	0%
.221(A2)CL.10	Hygro/gp 100	2,-	70%
.221(A2)N-CL.29	Hygro-vector	2,-	0%

[a]B-cell line variants of LCL-721 (721, .221, .221(A1), and .221(A2)) were provided by Dr. Robert DeMars for these experiments. The HLA class 1 negative 221 cells were derived from the parental 721 cells by gamma ray irradiation followed by immunoselection with appropriate anlisera or monoclonal antibodies. Transfer of cDNA for HLA-A1 or HLA-A2 into .221 cells was performed for the .221(A1) and .221(A2) cells, respectively.
[b]The indicated transgene was stably transfected into each B-cell line.
[c]The HLA-A phenotype is shown for each of the B-cell lines.
[d]The GP-100 expression of each cell line was determined by flow cytometry.

cytotoxicity was proportional to the level of B7-1 expression on the stimulating cells [60]. Thus, PMGT of cDNA for B7-1 into human melanoma cells increased expression of functional B7-1 and enhanced the antigenicity of the gene-modified cells proportional to the level of B7-1 expression by the modified tumor cells.

As melanoma cells with enhanced expression of both HLA and B7-1 molecules may better stimulate T-cell immunity than unmodified melanoma cells, additional experiments have determined that enhanced expression for each of these molecules could be obtained within the same melanoma cell following PMGT (Fig. 4). The expression of two genes in the same cell after PMGT of separate genetic constructs provides an efficient means to evaluate distinct molecular modifications to stimulate T-cell immunity to human melanoma [16].

2. Modification of B-Cell Lines

As our initial experiments determined conditions to enhance human leukocyte antigen (HLA) expression as well as conditions to enhance or induce B7-1 expression by melanoma cells, subsequent experiments determined the ability to transfect cells with the gene for the melanoma-associated antigen gp100. We proceeded to use PMGT to gene-modify B-cell lines with defined HLA phenotypes[1] to have stable expression of gp100 (Table 1; unpublished results). Because these gp100-expressing B-cell lines are expected to present epitopes of gp100 protein in the context of different HLA molecules, they will be used as target cells in upcoming functional experiments to determine the

[1]The B-cell lines with various HLA phenotypes were obtained from Dr. Robert DeMars.

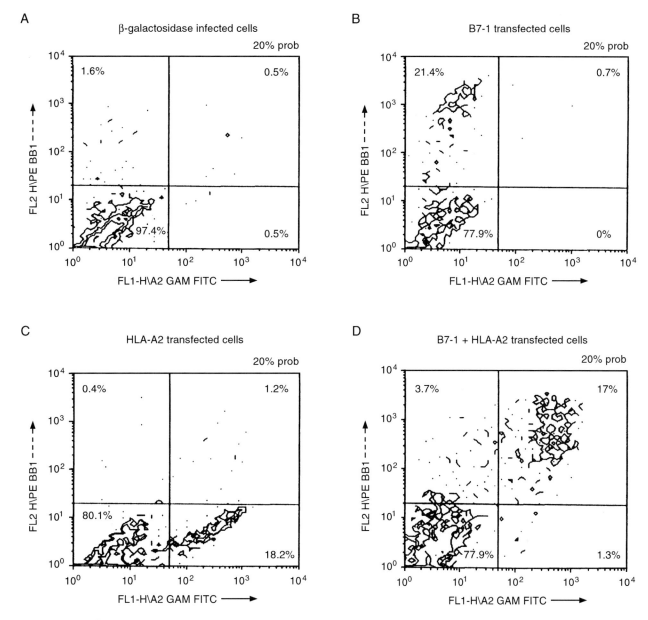

FIGURE 4 Expression of both B7-1 and HLA-A2 following dual delivery of the cDNAs for B7-1 and HLA-A2. Fluorescence intensity of Mel 117 melanoma cells following particle-mediated gene transfer with gold beads coated with (A) β-galactosidase DNA, (B) human B7-1 DNA, (C) human HLA-A2 DNA, or (D) the combination of human B7-1 and human HLA-A2 DNA. Cells were dually stained with anti-BB1 plus goat anti-mouse (GAM) FITC plus anti-HLA-A2 antibody and then analyzed on a FACScan flow cytometer. The percentage of cells in each quadrant is reported. (From Albertini, M. R. *et al., Cancer Gene Ther.,* 3(3), 192–201, 1996. With permission.).

HLA restriction of T cells following stimulation with gp100-expressing cells [61].

B. Antitumor Efficacy in Murine Models

Particle-mediated gene transfer has been used to generate murine tumor vaccines by genetically modifying tumor cells *in vitro* or *ex vivo,* primarily with cytokine genes. In addition, *in vivo* gene delivery using PMGT into skin has been utilized as a means for DNA vaccination with cDNA for tumor-associated molecules and with cDNA for various cytokines [11].

1. DNA Vaccination

This approach involves the delivery into the skin of genes encoding tumor-associated antigens (TAAs) in an attempt to induce antitumor immune responses. Using PMGT, successful vaccinations against tumors have been achieved with the genes encoding model TAAs such as β-galactosidase [27,30] or ovalbumin [30] and natural TAAs such as mutant p53 oncoprotein [28,62], carcinoembryonic antigen [29], or human papillomavirus E7 antigen [63]. In a study by Ross *et al.,* vaccination by PMGT was found to be more effective than vaccination with a peptide even in combination with an adjuvant [30]. In several studies in both mice and monkeys, PMGT required 100–500-fold less DNA than intramuscular or intradermal DNA inoculations to induce comparable or higher levels of cellular or antibody responses [64–69].

In an attempt to elucidate the mechanism of induction of immune response by PMGT into the skin, trafficking of cutaneous DCs carrying DNA/gold beads to draining lymph nodes has been demonstrated [42]. These cutaneous DCs directly transfected with DNA-encoded antigen via PMGT have been shown to play a predominant role in antigen presentation to CD8$^+$ T cells [64]. The results suggest that enhanced antigen presentation by bone-marrow-derived [43] DCs to T lymphocytes is responsible for induction of antitumor immunity in cutaneous gene immunization experiments. PMGT for co-delivery of adjuvant cytokine genes with tumor-antigen genes has been successfully used to further augment antigen presentation and induce enhanced antitumor immunity in mouse models. This was achieved by co-delivering plasmids encoding such cytokines as IL-12 [62,65], IFN-γ [70], and GM-CSF [29,70].

We have recently established a model of gene vaccination against melanoma using PMGT for gp100 gene delivery in the skin of mice [71]. Before evaluating the antitumor efficacy of gp100 DNA vaccination, it was necessary to determine whether B16 melanoma transfection with the human gp100 (hgp100) cDNA expression plasmid results in production of transgenic hgp100 protein. Murine B16 cells were transfected with hgp100 cDNA, selected *in vitro* for positive clones (B16–gp100), and then analyzed for hgp100 expression using flow cytometry. The control wild-type B16 cells had a low-level, positive expression of a molecule cross-reactive with hgp100 (and recognized by HMB-45 mAb against hgp100), reflecting the 80% homology of murine gp100 (mgp100) with hgp100. However, B16-gp100 cells showed a much greater fluorescence intensity with the HMB-45 mAb, documenting the greater expression of gp100 induced by transfection with the hgp100 DNA. Specificity of B16 tumor staining by HMB-45 mAb was confirmed in separate studies, where a nonmelanoma murine cell line, 4T1 adenocarcinoma, did not stain with HMB-45 mAb. In a second approach, we looked at the expression of hgp100 in B16–gp100 cells by using reverse transcription–polymerse chain reaction (RT–PCR) analysis. RT-PCR using the human-specific primer set for the human gp100 gene showed that only the human gp100-transfected cell line, B16–gp100, resulted in an amplification product.

The hgp100 DNA vaccination, either alone or in combination with GM-CSF DNA, also resulted in transgene transcription in the skin tissues 24 hours post transfection.

The gp100 DNA vaccination was performed by PMGT into the skin of mice with hgp100-encoding plasmid DNA, alone or in combination with mGM-CSF-encoding plasmid DNA. This resulted in a total delivery of 2.5 μg of each plasmid. Seven days later, vaccinated or naive mice were challenged intradermally with 5×10^4 B16–gp100 cells, and survival of mice was followed. The gp100 gene vaccination resulted in substantial protection against B16–gp100 tumors, with 40% of mice remaining tumor-free for at least 2 months. Importantly, co-delivery of GM-CSF cDNA with hgp100 cDNA resulted in complete tumor protection in all vaccinated mice (Fig. 5). The ability of GM-CSF gene co-delivery to enhance the effect of gp100 DNA vaccination was consistently reproducible in 12 subsequent experiments, although the degree of protection varied.

We next investigated whether T cells were responsible for the protection induced by gp100–GM-CSF gene vaccination. Our results demonstrated that *in vivo* depletion of T cells with anti-CD4 and anti-CD8 mAbs abrogated the protection induced by the gp100–GM-CSF DNA vaccine, indicating that this protection is T-cell mediated [71]. In addition, vaccinated mice that remained tumor-free for more than a month after tumor challenge were partially or completely immune to a secondary tumor challenge.

The experiments described above demonstrate that gp100 gene vaccination of naive mice, especially in combination

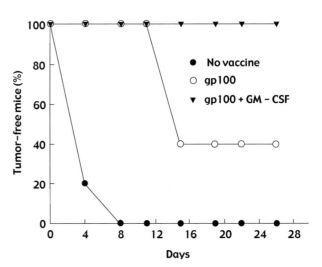

FIGURE 5 GM-CSF cotransfection enhances the effect of gp100 gene vaccination against B16–gp100 melanoma. Skin of C97BL/6 mice was transfected at four sites with hgp100 cDNA (2.5 μg/mouse) or hgp100 cDNA in combination with GM-CSF cDNA (2.5 μg/mouse of each) using PMGT. Expression of transgenic GM-CSF in the skin was confirmed by ELISA (not shown). Seven days following vaccination, mice were injected intradermally with 5×10^4 B16–gp100 tumor cells, and the survival of mice was followed. Each group included five mice.

with GM-CSF, can result in induction of a T-cell-mediated immune response that is capable of protecting mice from a subsequent tumor challenge. To allow the experimental conditions to better simulate the clinical situation, we investigated the therapeutic effect of gp100–GM-CSF gene therapy in mice bearing established tumors. The results demonstrated that the gp100–GM-CSF gene combination induced suppression of tumor growth when compared to control mice ($p < 0.05$ starting from day 17 of tumor growth). The effect of gp100–GM-CSF nucleic acid immunization against established tumors was further confirmed by the extended survival of mice (Fig. 6). The survival was calculated based on the number of days it took for the tumor diameter to reach 15 mm, at which time the mice were sacrificed. Thus, untreated tumor-bearing mice and mice treated with the empty vector–GM-CSF cDNA survived for 26.11 ± 0.96 and 25.5 ± 1.42 days, respectively, whereas mice treated with gp100–GM-CSF cDNA survived for 37.5 ± 2.72 days ($p < 0.005$). This experiment was repeated two times, with the treatment started on day 4 or on day 7 post tumor cell implantation, and similar results were obtained [71].

2. Cytokine Gene Therapy

A second approach using PMGT involves the treatment of already established subcutaneous tumors with cytokine genes. We have reported a powerful T-cell-mediated tumor regression following skin transfection with IL-12 gene in several mouse tumor models [39]. Moreover, the localized IL-12 gene delivery into the skin overlaying the immunogenic tumor has resulted in a systemic effect against visceral metastases [39] and a distant solid tumor [32]. Histological analysis showed that IL-12 transgene expression was readily detectable in the epidermis, although the DNA-coated gold beads did not reach the implanted tumor. These data suggest that a gradual, continuous release of small doses of transgenic IL-12 protein by transfected epidermal cells in the vicinity of a tumor can effectively result in activation of systemic antitumor immunity. This gene therapy approach using PMGT was later shown to be applicable and effective not only against cutaneous tumors, but also against tumors in visceral organs such as liver [72].

In contrast to immunogenic tumors, IL-12 gene therapy of poorly immunogenic tumors via PMGT may elicit different antitumor mechanisms. Thus, IL-12 gene transfer into the skin surrounding and overlaying a poorly immunogenic 4T1 mammary adenocarcinoma has resulted in a significant reduction of spontaneous lung metastases while having no effect on the growth of the primary intradermal tumor [73]. In contrast to the above-described antitumor effect of IL-12 gene therapy with immunogenic tumors, this antimetastatic effect was not mediated by T cells, but involved natural killer

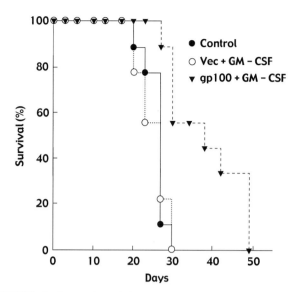

FIGURE 6   Suppression of the growth of established tumors by the gp100–GM-CSF DNA vaccine. C97BL/6 mice were injected intradermally with 5×10^4 B16–gp100 cells. On days 7, 10, 13, and 17, the skin of tumor-bearing mice was transfected at two sites of the abdominal area by PMGT with hgp100 cDNA in combination with GM-CSF cDNA (625 ng of each plasmid per transfection site). Control mice were untreated or received empty vector DNA plus GM-CSF DNA. The survival of mice was determined (mice were sacrificed on the day the tumor diameter reached 15 mm). Each group consisted of five mice.

(NK) cells and IFN-γ. We have also shown that co-delivery of additional gene(s) with IL-12 cDNA, a task that can be easily achieved by PMGT, may enhance the antitumor and antimetastatic activity of IL-12 gene therapy [74].

C. Antitumor Activity of a Canine Tumor Vaccine

Hogge et al. [75] have utilized the dog as a valid translational model for human cancers. A pilot study of seven research dogs demonstrated that a tumor vaccine composed of irradiated canine melanoma cells transfected ex vivo by PMGT with cDNA for human GM-CSF was safe, nontoxic, and well tolerated [75]. Vaccine-site biopsies demonstrated production of human GM-CSF at the vaccine sites, and histological analysis revealed an influx of neutrophils and macrophages. Dogs with spontaneous tumors were evaluated in an additional study. These dogs had tumors surgically excised and processed with mechanical and enzymatic digestion. The tumor cells were irradiated, received PMGT with cDNA for human GM-CSF, and were injected intradermally as a vaccine for the dog. Biopsy samples revealed biologically active human GM-CSF protein, and neutrophil and macrophage infiltration was seen at the vaccine sites. Objective antitumor responses were described in 3 of the 16 dogs in this trial [22].

IV. ISSUES REGARDING EVALUATION IN CLINICAL TRIALS

A. General Considerations

A primary issue for further clinical development of PMGT will be the ability to achieve clinically relevant levels of transgene expression following both *ex vivo* and *in vivo* PMGT. Identification of informative *in vivo* biomarkers of a successful immune response following PMGT will be essential for this technology, as well as other vaccine approaches, to move forward.

B. Skin Penetration Considerations

The ability to transfect a sufficient number of Langerhans cells, as well as other APCs in the skin, will be a key accomplishment for *in vivo* PMGT. The epidermis of human skin is made up of five layers: (1) stratum germinativum (stratum basale), the germinal layer of the epidermis where cell division occurs; (2) stratum spinosum (prickle cell layer), which contains cells in the process of growth and early keratin synthesis; (3) stratum granulosum (granular layer), characterized by the presence of granules within the cells which contribute to the process of keratinization; (4) stratum lucidum, found only in thick skin, and (5) stratum corneum, which consists of dead cell remnants flattened and fused and composed primarily of the fibrous protein keratin. The epidermis ranges in thickness from 0.075–0.15 mm in thin skin and 0.4–0.6 mm in thick skin. Keratinocytes are the predominate cell type within the epidermis. Another cell type found is the Langerhans cell. Langerhans cells, mainly found in the stratum germinativum, are from a monocyte lineage and act as cutaneous DCs [76]. Keratinocytes and deeper myocytes can also act as APCs though they lack costimulatory molecules.

Preliminary observations from PMGT in humans reveal beads present in the epidermis and papillary dermis with a depth of penetration ranging between 0.1 and 0.4 mm [77]. The largest concentration of gold particles was found in the stratum corneum (greater than 90%), the outermost layer of the epidermis [77]. The optimal number of beads needed in the stratum germinativum for an effective immune response is not known. Again, Langerhans cells, which act as professional APCs, are located in the lower layers of the epidermis layer and not in the stratum corneum. In order to increase the penetration of gold particles into Langerhans cells and into viable keratinocytes, a decrease or thinning of the stratum corneum may be needed.

Multiple different established techniques exist for decreasing the stratum corneum [78,79], including mechanical removal utilizing cellophane adhesive tape [80,81] and chemical peels. The primary use of chemical peels is the treatment of photo-aged skin and precancerous lesions and to reduce wrinkles. Agents used in chemical peel include salicylic acid [82,83], glycolic acid [84,85], and retinoic acid [86, 87]. All of these provide mild to moderate removal of the stratum corneum and result in increased growth of the stratum germativum. Additionally, treatment of skin with retinoic acid appears to increase the number of Langerhans cells at the treatment site [88]. These skin pretreatment strategies could be investigated in clinical trials evaluating PMGT.

V. RECENT CLINICAL TRIALS

A. *Ex Vivo* PMGT

The potential advantage of cancer immunotherapy is the ability of the immune system to selectively recognize and attack tumor cells throughout the body. It is important to note, in this regard, that many human cancers express TAAs that can be recognized by the immune system [89]. Immunization using whole tumor cell vaccine, TAA peptides, or TAA genes is a promising therapeutic strategy for cancers [90]. However, vaccination with tumor cells alone, without adjuvants, has rarely achieved a substantial antitumor effect in either experimental models or clinical practice due to insufficient antigen presentation or costimulation [89].

Possibilities for clinical cancer immunotherapy are suggested by the exciting findings that some nonimmunogenic tumors, as defined based on conventional vaccination methods, may be genetically modified to allow them to induce immune responses that can be detected following certain experimental approaches [90,91]. PMGT has been successfully used to transfect tumor cells with costimulatory genes [16] or cytokine genes. Transfection of weakly immunogenic tumor cells with GM-CSF DNA, followed by vaccination of mice with these cells, resulted in the generation of an immune response that was able to control growth of parental tumors in the mouse B16 melanoma model [92]. Whereas viral methods of gene delivery require establishing tumor cell cultures from patients' samples, PMGT can be used with tumor cells freshly harvested from a patient [92]. Based on these findings, the first cancer clinical study using PMGT was conducted utilizing the method of *ex vivo* transfection of tumor cells with GM-CSF DNA [93]. Preliminary results from this trial (led by Dr. David Mahvi at the University of Wisconsin Comprehensive Cancer Center) demonstrate that this technique of gene transfer is safe and feasible. No treatment-related toxicity was seen in patients receiving this treatment. The entire procedure of autologous tumor biopsy and vaccination by PMGT of irradiated tumor cells could be accomplished in less than 6 hours (D. Mahvi, pers. comm.).

Other investigators recently confirmed the efficacy of PMGT for transfecting human tumors using primary renal carcinoma cell lines. These cells were obtained from patient tumor tissues and produced high levels of GM-CSF following transfection with GM-CSF DNA using PMGT [94]. A separate approach of transfecting DC with TAA genes via

PMGT for the induction of immune response has also been recently described [95].

B. *In Vivo* PMGT

Multiple modifications of APCs may be required to activate effective *in vivo* anti-melanoma T-cell immunity. We recently conducted a phase I study of *in vivo* PMGT of cDNAs for gp100 and GM-CSF into uninvolved skin of melanoma patients to evaluate clinical toxicity, transgene expression, and immunological activation with this treatment [77]. Three treatment groups of six patients were originally planned for this trial. Treatment group I (gp100 alone) has been completed, and subsequent treatment groups were planned to receive PMGT with cDNA for GM-CSF administered either 3 days prior to cDNA for gp100 or on the day of gp100 administration. In treatment group I, PMGT was administered with helium at 500 psi using the PowderJect XR-1 device. Our initial treatment group of six patients received gp100 PMGT (0.25 μg DNA and 250 μg gold/treatment) at two (three patients) or four (three patients) separate vaccine sites each 3-week cycle of treatment. Vaccine site reactions consisted of localized erythema that resolved within 1–2 weeks and a brief tingling sensation at the vaccine site. Vaccine site biopsies 2 and 4 days after PMGT revealed beads primarily in the epidermis, but also in the papillary dermis, and with a depth of bead penetration ranging from 0.1–0.4 mm. The greatest bead concentration was in the stratum corneum, and a perivascular lymphoid infiltrate was present in the dermis. Gold bead localization and gp100 transgene expression were detected in large cells of the epidermis, most likely Langerhans cells. One of three patients in dose level 1 and two of the three patients in dose level 2 had stable disease following two 3-week cycles of treatment. Immunological assays to determine the magnitude and specificity of T-cell reactivity and antibody responses are in progress. Our preliminary conclusions are that PMGT of cDNA for gp100 has minimal toxicity and can achieve gp100 transgene expression in normal human skin. Additional clinical evaluation is needed to determine the immune activation that can be achieved with PMGT of gold beads carrying cDNA for gp100, either alone or in combination with gold beads carrying cDNA for GM-CSF.

VI. POTENTIAL NOVEL USES AND FUTURE DIRECTIONS

Various cytokine genes, including IL-2, IL-4, IL-12, IFN-γ, and GM-CSF, have been effective in mediating either T-cell-dependent or inflammatory responses which lead to tumor regression [96,97]. Importantly, the synergistic antimelanoma effects of melanoma peptide vaccines and systemic administration of cytokines such as GM-CSF have been reported [98]. In our murine experiments, we have established a model of gene vaccination against melanoma using PMGT for gene delivery in the skin. The results obtained so far show that coadministration of human gp100 and murine GM-CSF plasmid DNAs into mice led to a much greater vaccination effect than delivery of the gp100 cDNA vaccine alone. Moreover, this DNA combination was found effective in suppression of melanoma growth and extended survival of mice bearing established B16 melanoma that genetically expressed human gp100 [71]. Further plans for investigation are based on the hypothesis that selective activation of immunologic recognition mechanisms against melanoma, followed by activation of the effector mechanisms downstream from the immune recognition event, might provide preferential destruction of the melanoma cells *in vivo*.

The use of PMGT to directly modify APCs with one or more transgenes presents an opportunity to translate successful *in vitro* findings into treatment approaches for patients in the clinic. Exciting preclinical results have demonstrated anti-melanoma efficacy following vaccination with cDNAs for gp100 and GM-CSF by PMGT. Rigorous clinical evaluation of PMGT, with careful immunological monitoring, is required to determine the potential clinical importance of this technology. In addition, the evaluation of combination immunotherapy strategies in preclinical models will provide a foundation for subsequent clinical trials to enhance antitumor T-cell immunity. Primary issues for further clinical development of PMGT will be the ability to achieve clinically relevant levels of transgene expression following both *ex vivo* and *in vivo* PMGT. Identification of informative *in vivo* biomarkers of a successful immune response following PMGT will be essential for this technology, as well as other vaccine approaches, to more forward.

Acknowledgments

The authors thank Sandy Keller for preparation of this manuscript and PowderJect Vaccines, Inc., for providing a dermal PowderJect-XR device for this research. We thank Drs. Paul Sondel, David Mahvi, Jacquelyn Hank, KyungMann Kim, and Ning-Sun Yang for stimulating discussions about PMGT. Our research investigating PMGT was supported by the National Institutes of Health (R29-CA68466; U01-CA61498), the Jay Van Sloan Memorial from the Steven C. Leuthold Family Foundation, gifts to the University of Wisconsin Comprehensive Cancer Center, and the University of Wisconsin General Clinical Research Center (M01 RR03186).

References

1. McCabe, D., Swain, W., Martinell, B., and Christou, P. (1988). Stable transformation of soybean (hylcine max) by particle acceleration. *Bio/Technology* **6**, 923–926.

2. Christou, P., McCabe, D., Martinell, B., and Swain, W. (1990). Soybean genetic engineering-commercial production of transgenic plants. *Trends Biotechnol.* **8,** 145–151.

3. Heiser, W. (1994). Gene transfer into mammalian cells by particle bombardment. *Analyt. Biochem.* **217,** 185–196.

4. Yang, N., Burkholder, J., Martinell, B., and McCabe, D. (1990). In vivo and in vitro gene transfer to mammalian somatic cells by particle bombardment. *Proc. Natl. Acad. Sci. USA* **87,** 9568–9572.

5. Cheng, L., Ziegelhoffer, P., and Yang, N. (1993). In vivo promoter activity and transgene expression in mammalian somatic tissues evaluated by using particle bombardment. *Proc. Natl. Acad. Sci. USA* **90,** 4455–4459.

6. Yang, N.-S., and Ziegelhoffer, P., (1994). The particle bombardment system for mammalian gene transfer, in Yang N-S and Christou P, eds. *Particle Bombardment Technology for Gene Transfer* (N.-S. Yang and Christou, p., eds.), pp. 117–141. Oxford University Press, New York.

7. Jiao, S., Cheng, L., Wolff, J., and Yang, N.-S. (1993). Particle bombardment-mediated gene transfer and expression in rat brain tissues. *J. Immunol.* **159,** 11–14.

8. Thompson, T., Gould, M., Burkholder, J., and Yang, N.-S. (1993). Transient promoter activity in prmary rat mammary epithelial cells evaluated using particle bombardment gene transfer. *In Vitro Cell Dev. Biol.* **29A,** 165–170.

9. Burkholder, J., Decker, J., and Yang, N.-S. (1993). Transgene expression in lymphocyte and macrophage primary cultures after particle bombardment. *J. Immunol. Meth.* **165,** 149–156.

10. Ye, Z., Qiu, P., Burkholder, J., Turner, J., Culp, J., Roberts, T., Shahidi, N., and Yang, N. (1998). Cytokine transgene expression and promoter usage in primary CD34+ cells using particle-mediated gene delivery. *Hum. Gene Ther.* **9,** 2197–2205.

11. Rakhmilevich, A., and Yang, N. (2000). In vivo particle-mediated gene transfer for cancer therapy, in *Methods in Molecular Medicine.* Vol. 35. *Gene Therapy: Methods and Protocols* (W. Walther and U. Stein, eds.), Humana Press, Totowa, NJ.

12. Merchant, B. (1998). Gold, the noble metal and the paradoxes of its toxicology. *Biologicals* **26,** 49–59.

13. Yang, N.-S., Burkholder, J., McCabe, D., Neumann, V., and Fuller, D. (1997). Particle-mediated gene delivery in vivo and in vitro. *Curr. Protocols Hum. Genet.* Suppl. 12, 12.6.1–12.6.14.

14. Chen, R. and Ruggle, S. (1993). In situ cDNA polymerase chain reaction: a novel technique for detecting mRNA expression. *Am. J. Pathol.* **143,** 1527–1534.

15. Williams, R., Johnston, S., Riedy, M., DeVit, M., McElligot, S., and Sanford, J. (1991). Introduction of foreign genes into tissues of living mice by DNA-coated microprojectiles. *Proc. Natl. Acad. Sci. USA* **88,** 2726–2730.

16. Albertini, M. R., Emler, C. A., Schell, K., Tans, K. J., King, D. M., and Sheehy, M. J. (1996). Dual expression of human leukocyte antigen molecules and the B7-1 costimulatory molecule (CD80) on human melanoma cells after particle-mediated gene transfer. *Cancer Gene Ther.* **3**(3), 192–201.

17. Qiu, P., Ziegelhoffer, P., Sun, J., and Yang, N.-S. (1996). Gene gun delivery of mRNA in situ results in efficient transgene expression and immunization. *Gene Ther.* **3,** 262–268.

18. Perales, J., Perkol, T., Beegen, H., Ratnoff, O., and Hanson, R. (1994). Gene transfer in vivo: sustained expression and regulation of genes introduced into the liver by receptor-targeted uptake. *Proc. Natl. Acad. Sci. USA* **91,** 4086–4090.

19. Tang, D., DeVit, M., and Johnston, S. (1992). Genetic immunization is a simple method for eliciting an immune response. *Nature* **356,** 152–154.

20. Vanrompay, D., Cox, E., Vandenbussche, F., Volckaert, G., and Goddeeris, B. (1999). Protection of turkeys against *Chlamydia psittaci* challenge by gene gun-based DNA immunizations. *Vaccine* **17,** 2628–2635.

21. Han, R., Cladel, N., Reed, C., Peng, X., and Christensen, N. (1999). Protection of rabbits from viral challenge by gene gun-based intracutaneous vaccination with a combination of cottontail rabbit papillomavirus E1, E2, E6, and E7 genes. *J. Virol.* **73,** 7039–7043.

22. Hogge, G., Burkholder, F. J., Culp, J., Dubielzig, R., Albertini, M., Keller, E., Yang, N.-S., and MacEwen, E. (1998). Development of human granulocyte macrophage-colony stimulating factor transfected tumor cell vaccines for the treatment of spontaneous canine cancer. *Hum. Gene Ther.* **9,** 1851–1861.

23. Macklin, M., McCabe, D., Mcgregor, M., Neumann, V., Meyer, T., Callan, R., Hinshaw, V., and Swain, W. (1998). Immunization of pigs with a particle-mediated DNA vaccine to influenza A virus protects against challenge with homologous virus. *J. Virol.* **72,** 1491–1496.

24. Lunn, D., Soboll, G., Schram, B., Quass, J., McGregor, M., Drape, R., Macklin, M., McCabe, D., Swain, W., and Olsen, C. (1999). Antibody responses to DNA vaccination of horses using the influenza virus hemagglutinin gene. *Vaccine* **17,** 2245–2258.

25. Fuller, D., Corb, M., Barnett, S., Steimer, K., and Haynes, J. (1997). Enhancement of immunodeficiency virus-specific immune responses in DNA-immunized rhesus macaques. *Vaccine* **15,** 924–926.

26. Tacket, C., Roy, M., Widera, G., Swain, W., Broome, S., and Edelman, R. (1999). Phase I safety and immune response studies of a DNA vaccine encoding hepatitis B surface antigen delivered by a gene delivery device. *Vaccine* **17,** 2826–2829.

27. Irvine, K. R., Rao, J. B., Rosenberg, S. A., and Restifo, N. P. (1996). Cytokine enhancement of DNA immunization leads to effective treatment of established pulmonary metastases. *J. Immunol.* **156**(1), 238–245.

28. Ciernik, F., Berzofsky, J., and Carbone, D. (1996). Induction of cytoxic T lymphocytes and antitumor immunity with DNA vaccines expressing single T cell epitopes. *J. Immunol.* **156,** 2369–2375.

29. Conry, R., Widera, G., LoBuglio, A. *et al.* (1996). Selected strategies to augment polynucleotide immunization. *Gene Ther.* **3,** 67–74.

30. Ross, H., Weber, L., Wang, S. *et al.* (1997). Priming for T-cell-mediated rejection of established tumors by cutaneous DNA immunization. *Clin. Cancer Res.* **3,** 2191–2196.

31. Sun, W., Burkholder, J., Decker, J., Turner, J., Lu, X., Pugh, T., Ershler, W., and Yang, N. (1995). In vivo cytokine gene transfer by particle bombardment reduces tumor growth in mice. *Proc. Natl. Acad. Sci. USA* **92,** 2889–2893.

32. Rakhmilevich, A., Janssen, K., Turner, J., Culp, J., and Yang, N.-S. (1997). Cytokine gene therapy of cancer using gene gun technology: superior antitumor activity of IL-12. *Hum. Gene Ther.* **8,** 1303–1311.

33. Chen, H., Pan, C., Liau, M., Jou, R., Tsai, C., Wu, J., Lin, Y., and Tao, M. (1999). Screening of protective antigens of Japanese encephalitis virus by DNA immunization: a comparative study with conventional viral vaccines. *J. Virol.* **73,** 10137–10145.

34. Feltquate, D., Heaney, S., Webster, R., and Robinson, H. (1997). Different T helper cell types and antibody isotypes generated by saline and gene gun DNA immunization. *J. Immunol.* **163,** 4510–4518.

35. Fomsgaard, A., Nielsen, H., Kirkby, N., Bryder, K., Corbet, S., Nielsen, C., Hinkula, J., and Buus, S. (1999). Induction of cytotoxic T-cell responses by gene gun DNA vaccination with minigenes encoding influenza A virus HA and NP CTL-epitopes. *Vaccine* **18,** 681–691.

36. Hanke, T., Neumann, V., Blanchard, T., Sweeney, P., Hill, A., Smith, G., and McMichael, A. (1999). Effective induction of HIV-specific CTL by multi-epitope using gene gun in combined vaccination regime. *Vaccine* **17,** 589–596.

37. Belperron, A., Feltquate, D., Fox, B., Horii, T., and Bzik, D. (1999). Immune responses induced by gene gun or intramuscular injection of DNA vaccines that express immunogenic regions of the serine repeat antigen from *Plasmodium falciparum. Infect. Immun.* **67,** 5163–5169.

38. Fensterle, J., Grode, L., Hess, J., and Kaufmann, S. (1999). Effective DNA vaccination against listeriosis by prime/boost innoculation with the gene gun. *J. Immunol.* **163,** 4510–4518.

39. Rakhmilevich, A., Turner, J., Ford, M., McCabe, D., Sun, W., Sondel, P., Grota, K., and Yang, N. (1996). Gene gun-mediated skin transfection with interleukin 12 gene results in regression of established primary and metastatic murine tumors. *Proc. Natl. Acad. Sci. USA* **93**, 6291–6296.

40. Andree, C., Swain, W., Page, C., Macklin, M., Slama, J., Hatzis, D., and Eriksson, E. (1994). In vivo transfer and expression of an EGF gene accelerated wound repair. *Proc. Natl. Acad. Sci. USA* **91**, 12188–12192.

41. Rajagopalan, L., Burkholder, J., Turner, J., Culp, J., Yang, N.-S., and Malter, J. (1995). Targeted mutagenesis of GM-CSF cDNA increases transgenic mRNA stability and protein expression in normal cells. *Blood* **86**, 2551–2558.

42. Condon, C., Watkins, S., Celluzzi, C., Thompson, K., and Falo, L. (1996). DNA-bsed immunization by in vivo transfection of dendritic cells. *Natu. Med.* **2**(10), 1122–1128.

43. Iwasaki, A., Torres, C., Ohashi, P., Robinson, H., and Barber, B. (1997). The dominant role of bone marrow-derived cells in CTL induction following plasmid DNA immunization at different sites. *J. Immunol.* **159**, 11–14.

44. Gurunathan, S., Klinman, D., and Seder, R. (2000). DNA vaccines: immunology, application, and optimization. *Annu. Rev. Immunol.* **18**, 927–974.

45. Soubrier, F., Cameron, B., Manse, B., Somarriba, S., Dubertret, C., Jaslin, G., Jung, G., LeCaer, C., Dang, D., Mouvault, J., Scherman, D., Mayaux, J., and Crouzet, J. (1999). pCOR: a new design of plasmid vectors for nonviral gene therapy. *Gene Ther.* **6**, 1482–1488.

46. Wu, Y., and Kipps, T. (1997). Deoxyribonucleic acid vaccines encoding antigens with rapid proteasome-dependent degradation are highly efficient inducer of cytolotic T lymphocytes. *J. Immunol.* **159**, 6037–6043.

47. Iwasaki, A., DelaCruz, C., Young, A., and Barber, B. (1999). Epitope-specific cytoxic T lymphocyte induction by minigene DNA immunization. *Vaccine* **17**, 2081–2088.

48. Thomson, S., Sherritt, M., Medveczky, J., Elliot, S., Moss, D., Fernando, G., Brown, L., and Suhrbier, A. (1998). Delivery of multiple CD8 cytotoxic T cell epitopes by DNA vaccination. *J. Immunol.* **160**, 1717–1723.

49. O'Garra, A. (1998). Cytokines induce the development of functionally heterogeneous T helper cell subsets. *Immunity* **8**, 275–283.

50. Murphy, K., Ouyang, W., Farrar, J., Yang, J., Ranganath, S., Asnagli, H., Afkarian, M., and Murphy, T. (2000). Signaling and transcription in T helper development. *Annu. Rev. Immunol.* **18**, 451–494.

51. Torres, C., Iwasaki, A., Barber, B., and Robinson, H. (1997). Differential dependence on target site tissue for gene gun and intramuscular DNA immunizations. *J. Immunol.* **158**, 4529–4532.

52. Darquet, A.-M., Rangara, R., Kreiss, P., Schwartz, B., Naimi, S., Delaere, P., Crouzet, J., and Scherman, D. (1999). Minicircle: an improved DNA molecule for in vitro and in vivo gene transfer. *Gene Ther.* **6**, 209–218.

53. Sato, Y., Roman, M., Tighe, H., Lee, D., Corr, M., Nguyen, M.-D., Silverman, G., Lotz, M., Carson, D., and Raz, E. (1996). Immunostimulatory DNA sequences necessary for effective intradermal gene immunization. *Science* **273**, 352–354.

54. Jakob, T., Walker, P., Krieg, A., Udey, M., and Vogel, J. (1998). Activation of cutaneous dendritic cells by CpG-containing oligodeoxynucleotides: a role for dendritic cells in the augmentation of Th1 responses by immunostimulatory DNA. *J. Immunol.* **161**, 3042–3049.

55. Jenkins, M. K., and Johnson, J. G. (1993). Molecules involved in T-cell costimulation. *Curr. Opin. Immunol.* **5**(3), 361–367.

56. Bluestone, J. A. (1995). New perspectives of CD28-B7-mediated T cell costimulation. *Immunity* **2**(6), 555–559.

57. Linsley, P. S., and Ledbetter, J. A. (1993). The role of the CD28 receptor during T cell responses to antigen. *Annu. Rev. Immunol.* **11**, 191–212.

58. Lenschow, D. J., Walunas, T. L., and Bluestone, J. A. (1996). CD28/B7 system of T cell costimulation. *Annu. Rev. Immunol.* **14**, 233–258.

59. Townsend, S., and Allison, J. (1993). Tumor rejection after direct costimulation of CD8+ cells by B7-transfected melanoma cells. *Science* **259**, 368–370.

60. McCarthy, D., Glowacki, N., Schell, K., Emler, C., and Albertini, M. (2000). Antigenicity of human melanoma cells transfected to express the B7-1 co-stimulatory molecule (CD80) varies with the level of B7-1 expression. *Cancer Immunol. Immunother.* **49**, 85–93.

61. King, D., Ye, Z.-Q., Roberts, T., Glowacki, N., and Albertini, M. (2000). Cytokine release by thiguanine-resistant T cell clones from melanoma patients. *Proc. Am. Assoc. Cancer. Res.* **41**, 697.

62. Tuting, T., Gambotto, A., Robbins, P., Storkus, J., and Deleo, A. (1999). Co-delivery of T helper 1-biasing cytokine genes enhances the efficacy of gene gun immunization of mice: studies with the model tumor antigen beta-galactosidase and the BALB/c Meth A p53 tumor-specific antigen. *Gene Ther.* **6**, 629–636.

63. Chen, C., Ji, H., Suh, K., Choti, M., Pardoll, D., and Wu, T. (1999). Gene gun-mediated DNA vaccination induces antitumor immunity against human papillomavirus type 16 E7-expressing murine tumor metastases in the liver and lungs. *Gene Ther.* **6**, 1972–1981.

64. Porgador, A., Irvine, K., Iwasaki, A., Barber, B., Restifo, N., and Germain, R. (1998). Predominant role for directly transfected dendritic cells in antigen presentation to CD8+ T cells after gene gun immunization. *J. Exp. Med.* **188**, 1075–1082.

65. Tan, J., Yang, N., Turner, J., Niu, G., Maassab, H., Sun, J., Herlocher, M., Chang, A., and Yu, H. (1999). Interleukin-12 cDNA skin transfection potentiates human Yu papillomavirus E6 DNA vaccine-induced antitumor immune response. *Cancer Gene Ther.* **6**, 331–339.

66. Leitner, W. W., Seguin, M. C., Ballou, W. R., Seitz, J. P., Schulz, A. M., Sheehy, M. J., and Lyon, J. A. (1997). Immune responses induced by intramuscular or gene gun injection of protective deoxyribonucleic acid vaccines that express the circumsporozoite protein from *Plasmodium berghei* malaria parasites. *J. Immunol.* **159**, 6112–6119.

67. Fynan, E. F., Webster, R. G., Fuller, D. H., Haynes, J. R., Santoro, J. C., and Robinson, H. L. (1993). DNA vaccines: protective immunizations by parenteral, mucosal, and gene-gun inoculations. *Proc. Natl. Acad. Sci. USA* **90**, 11478–11482.

68. McCluskie, M. J., Millan, C. L. B., Gramzinski, R. A., Robinson, H. L., Santoro, J. C., Fuller, J. T., Widera, G., Haynes, J. R., Purcell, R. H., and Davis, H. L. (1999). Route and method of delivery of DNA vaccine influence immune responses in mice and non-human primates. *Mol. Med.* **5**, 287–300.

69. Pertmer, T. M., Eisenbraun, M. D., McCabe, D., Prayaga, S. K., Fuller, D. H., and Haynes, J. R. (1995). Gene gun-based nucleic acid immunization: elicitation of humoral and cytotoxic T lymphocyte responses following epidermal delivery of nanogram quantities of DNA. *Vaccine* **13**(15), 1427–1430.

70. Charo, J., Ciupitu, A., LeChevalier De Preville, A., Trivedi, P., Klein, G., Hinkula, J., and Kiessling, R. (1999). A long-term memory obtained by genetic immunization results in full protection from a mammary adenocarcinoma expressing an EBV gene. *J. Immunol.* **163**, 5913–5919.

71. Rakhmilevich, A., Imboden, M., Hao, Z., Macklin, M., Roberts, T., Wright, K., Albertini, M., Yang, N.-S., and Sondel, P. (2001). Effective particle-mediated vaccination against mouse melanoma by co-administration of plasmid DNA encoding gp100 and granulocyte-macrophage colony-stimulating factor. *Clin. Cancer Res.* **7**, 952–961.

72. Weber, S., Shi, F., Heise, C., Warner, T., and Mahvi, D. (1999). Interleukin-12 gene transfer results in CD8-dependent regression of murine CT26 liver tumors. *Ann. Surg. Oncol.* **6**, 186–194.

73. Rakhmilevich, A., Janssen, K., Hao, Z., Sondel, P., and Yang, N.-S. (2000). Interleukin 12 gene therapy of a weakly immunogenic mouse mammary carcinoma results in reduction of spontaneous lung metastases via a T cell-independent mechanism. *Cancer Gene Ther.* **7**, 826–838.

74. Oshikawa, K., Shi, F., Rakhmilevich, A., Sondel, P., Mahvi, D., and Yang, N.-S. (1999). Synergistic inhibition of tumor growth in a murine mammary adenocarcinoma model by combinational gene therapy using interleukin-12, pro-interleukin-18 and IL-1β-converting enzyme cDNA. *Proc. Natl. Acad. Sci. USA* **96**, 13351–13356.

75. Hogge, G., Burkholder, J., Culp, J., Dubielzig, R., Albertini, M., Yang, N.-S., and MacEwen, E. (1999). Preclinical development of hGM-CSF transfected melanoma cell vaccine using established canine cell lines and normal canines. *Cancer Gene Ther.* **6**, 26–36.

76. Wheater, P., Burkitt, H., and Daniels, V. (1987). *Functional Histology,* Churchill-Livingstone, London, pp. 130–141.

77. Albertini, M., King, D., Mahvi, D., Warner, T., Glowacki, N., Roberts, T., Schalch, H., Kim, K., Schiller, J., Rakhmilevich, A., Yang, N., Roy, M., Swain, W., Hank, J., and Sondel, P. (2000). Phase I trial of immunization using particle-mediated transfer of genes for gp100 and GM-CSF into uninvolved skin of patients with melanoma. *Proc. Am. Soc. Clin. Oncol.* **19**, 467a.

78. Moy, L., Peace, S., and Moy, R. (1996). Comparison of the effect of various chemical peeling agents in a mini-pig model. *Dermatol. Surg.* **22**, 429–432.

79. Tse, Y., Ostad, A., Lee, H.-S., Levine, V., Koenig, K., Kamino, H., and Ashinoff, R. (1996). A clinical and histologic evaluation of two medium-depth peels. *Dermatol. Surg.* **22**, 781–786.

80. Tanaka, M., Zhen, Y., and Tagami, H. (1997). Normal recovery of the stratum corneum barrier function following damage induced by tape stripping in patients with atopic dermatitis. *Br. J. Derm.* **136**, 966–967.

81. Marttin, E., Neelissen-Subnel, M., De Haan, F., and Bodde, H. (1996). A critical comparison of methods to quantify stratum corneum removed by tape stripping. *Skin Pharmacol.* **9**, 69–77.

82. Kligman, D., and Kligman, A. (1998). Salicyclic acid peels for the treatment of photoaging. *Dermatol. Surg.* **24**, 325–328.

83. Loden, M., Bostrom, P., and Kneczke, M. (1995). Distribution and keratolytic effect of salicyclic acid and urea in human skin. *Skin Pharmacol.* **8**, 173–178.

84. Piaquadio, D., Dobry, M., Hunt, S., Andree, C., Grove, G., and Hollenbach, K. (1996). Short contact 70% glycolic acid peels as a treatment for photodamaged skin. *Dermatol. Surg.* **22**, 449–452.

85. Newman, N., Newman, A., Moy, L., Babapour, R., Harris, A., and Moy, R. (1996). Clinical improvement of photoaged skin with 50% glycolic acid. *Dermatol. Surg.* **22**, 455–460.

86. Creidi, P., and Humbert, P. (1999). Clinical use of topical retinaldehyde on photoaged skin. *Dermatology* **199** (suppl. 1), 49–52.

87. Sachsenberg-Studer, E. (1999). Tolerance of topical retinaldehyde in humans. *Dermatology* **199** (suppl. 1), 61–63.

88. Meunier, L., Bohjanen, K., Voorhees, J., and Cooper, K. (1994). Retinoic acid upregulates human Langerhans cell antigen presentation and surface expression of HLA-DR and CD11c, a B2 integrin critically involved in T-cell activation. *J. Inv. Dermatol.* **103**, 775–779.

89. Pardoll, D. M. (1998). Cancer vaccines. *Nat. Med.* **4**(5, suppl.), 525–531.

90. Roth, J., and Cristiano, R. (1997). Gene therapy for cancer: what have we done and where are we going? *J. Natl. Cancer Inst.* **89**, 21–39.

91. Colombo, M., Modelsti, A., Parmiani, G., and Forni, G. (1992). Local cytokine availability elicits tumor rejection and systemic immunity through granulocyte-T-lymphocyte cross-talk. *Cancer Res.* **52**, 4853–4857.

92. Mahvi, D. M., Burkholder, J. K., Turner, J., Culp, J., Malter, J. S., Sondel, P. M., and Yang, N.-S. (1996). Particle-mediated gene transfer of granulocyte–macrophage colony-stimulating factor cDNA to tumor cells: implications for a clinically relevant tumor vaccine. *Hum. Gene Ther.* **7**(13), 1535–1543.

93. Mahvi, D. M., Sondel, P. M., Yang, N.-S., Albertini, M. R., Schiller, J. H., Hank, J., Heiner, J., Gan, J., Swain, W., and Logrono, R. (1997). Phase I/IB study of immunization with autologous tumor cells transfected with the GM-CSF gene by particle-mediated transfer in patients with melanoma or sarcoma. *Hum. Gene Ther.* **8**(7), 875–891.

94. Seigne, J., Turner, J., Diaz, J., Hackney, J., Pow-Sang, J., Helal, M., Lockhart, J., and Yu, H. (1999). A feasibility study of gene gun mediated immunotherapy for renal cell carcinoma. *J. Urol.* **162**, 1259–1263.

95. Tuting, T., and Albers, A. (2000). Walther, W., and Stein, U., Eds. Particle-mediated gene transfer into dendritic cells: a novel strategy for the induction of immune responses against tumor antigens, in *Gene Therapy of Cancer: Methods and Protocols,* Vol. 35, pp. 27–47, Totow, N.J.: Humana Press.

96. Tahara, H., Zitvogel, L., Storkus, W., Zeh, H., McKinney, T., Schreiber, R., Bubler, U., Robbins, P., and Lotze, M. (1995). Effective eradication of established murine tumors with IL-12 gene therapy using a polycistronic retroviral vector. *J. Immunol.* **154**, 6466–6474.

97. Cavallo, F., Di Pierro, F., Giovarelli, M., Gulino, A., Vacca, A., Stoppacciaro, A., Forni, M., Modesti, A., and Forni, G. (1993). Protective and curative potential of vaccination with interleukin-2 gene-transfected cells from a spontaneous mouse mammary adenocarcinoma. *Cancer Res.* **53**, 5067–5070.

98. Jager, E., Ringhoffer, M., Dienes, H., Arand, M., Karbach, J., Jager, D., Ilsemann, Hagerdorn, M., Oesch, F., and Knuth, A. (1996). Granulocyte–macrophage-colony-stimulating factor enhances immune responses to melanoma-associated peptides in vivo. *Int. J. Cancer* **67**, 54–62.

GENETICALLY MODIFIED EFFECTOR CELLS FOR IMMUNE-BASED IMMUNOTHERAPY

14

Applications of Gene Transfer in the Adoptive Immunotherapy of Cancer

KEVIN T. McDONAGH

Department of Internal Medicine
Division of Hematology/Oncology
Comprehensive Cancer Center
University of Michigan
Ann Arbor, Michigan 48109

ALFRED E. CHANG

Department of Surgery
Division of Surgical Oncology
Comprehensive Cancer Center
University of Michigan
Ann Arbor, Michigan 48109

I. INTRODUCTION

This chapter focuses on the applications of gene transfer to adoptive immunotherapy of malignancy. This form of cellular therapy refers to the infusion of tumor-reactive immune cells into the tumor-bearing host to mediate, directly or indirectly, regression of established tumor. This review is divided into three areas, each one involving different methods of genetic transfer to generate immune cells into the tumor-bearing host for subsequent adoptive transfer: (1) the use of gene-modified tumor to serve as immunogens to generate effector T cells, (2) genetic manipulation of T cells to enhance antitumor reactivity, and (3) genetic modulation of dendritic cells (DCs).

The feasibility of adoptive immunotherapy for cancer is predicated on two fundamental observations derived from animal models. The first is that tumor cells express antigens that are qualitatively or quantitatively different from normal cells and can elicit an immune response within the syngeneic host. The second is that the immune rejection of established tumors can be mediated by the adoptive transfer of appropriately sensitized lymphoid cells.

In 1943, Gross [1] was the first to recognize that inbred mice could be immunized against a tumor that was developed in a mouse of the same inbred strain, thus documenting the existence of tumor-associated antigens. Over the years, it has become apparent that individual tumors vary greatly in the nature of their "immunogenicity." The immunogenicity of a tumor has a direct influence on the ability to develop cellular antitumor immune responses.

The transfer of immunity to a naive host by the use of cells was first described by Landsteiner and Chase [2] in 1942. They reported that hypersensitivity to simple compounds could be transferred to normal rats by the transfer of peritoneal exudate cells of sensitized donor animals. In 1954, Billingham *et al.* [3] documented the ability to transfer skin allograft immunity to a normal murine host by the use of regional lymph node cells from animals that had rejected primary skin allografts. These investigators developed the term *adoptive immunotherapy* to describe the acquisition of immunity in a normal subject as a result of the transference, not of preformed antibody, but of immunologically competent cells. In 1955, Mitchison [4] was the first to report about the adoptive immunotherapy of tumors in a rodent model. In this study, the adoptive transfer of lymph node cells from mice that rejected tumor allografts conferred accelerated rejection of the same tumor allografts in naive hosts. However, more germane to clinical therapy is the ability to transfer immunity to autologous tumors (i.e., syngeneic tumors in inbred rodent strains) using lymphocytes. Borberg *et al.* [5] were the first to clearly show that the infusion of syngeneic immune cells from hyperimmunized donor animals was capable of mediating the regression of established tumor. During the ensuing years, several other investigators have documented the ability to successfully treat established syngeneic tumors by the adoptive transfer of immune "effector" cells [6,7].

One of the major obstacles in extrapolating the concepts developed in the animal models to the clinical treatment of cancers was the inability to generate adequate numbers of immune cells for therapy. Methods to grow or expand lymphoid cells while retaining their immunologic reactivity were limited; however, in 1976, the discovery of interleukin-2 (IL-2) as a T-cell growth factor made it possible to culture activated T cells in large quantities [8]. In 1981, Cheever *et al.* [9] showed that tumor-reactive T cells could be expanded in IL-2 and still maintain their therapeutic efficacy in adoptive immunotherapy of tumors in mice.

The first successful clinical application of cellular therapy in humans was reported by Rosenberg *et al.* [10] in 1985. These investigators generated large quantities of IL-2-activated peripheral blood lymphocytes (approximately 1–2 $\times 10^{11}$ cells/patient), which were infused along with the concomitant administration of IL-2. These lymphokine-activated killer (LAK) cells were nonspecifically cytolytic to tumor cells by *in vitro* measurements. In these early clinical trials, significant tumor burdens regressed in a subset of patients, and the feasibility of generating large numbers of cells for clinical therapy was realized. Based on subsequent animal studies [11,12], tumor infiltrating lymphocyte (TILs) were found to be an alternative population of cells that were more potent than LAK cells in mediating tumor regression. Rosenberg and co-workers were the first to report significant clinical responses using TILs in the treatment of patients with advanced melanoma [13,14].

Despite the progress in this field, cellular therapy is still in its infancy. The isolation and expansion of potent immune effector cells derived from the tumor-bearing host remains a formidable task. The use of genetic approaches to alter the host-tumor immune response offers potential opportunities to develop more potent cellular reagents (Table 1).

II. USE OF GENE-MODIFIED TUMORS TO GENERATE ANTITUMOR-REACTIVE T CELLS

Genetic modification of tumors to secrete or express immunomodulatory peptides has been found to significantly alter the biology of the tumor cells when these are inoculated into the syngeneic host. A majority of these studies have

TABLE 1   Genetic Approaches to Adoptive Immunotherapy

Generation of effector cells using gene-modified tumors
   TILs isolated from gene-modified tumors
   Lymph node cells draining gene-modified tumors

Gene modification of T cells
   Marking studies of effector cells
   T cells as delivery vehicles for immunoregulatory molecules
   Reengineering T cells with tumor-antigen-specific receptors
      Antigen-specific T-cell receptors
      Chimeric T-cell receptor

Genetic modification of dendritic cells
   Genes encoding tumor-associated antigens
   Genes encoding cytokines/chemokines

involved the use of various cytokine genes introduced into tumor cells. Many of the observed changes are related to an enhanced cellular immune response to tumor-associated antigens expressed on the parental tumor. In selected animal models, regression of established tumors by the inoculation of genetically modified tumor cells that secrete IL-4 or interferon-gamma (IFN-γ) administered as a tumor vaccination has been observed [15–17]. These studies have served as a rationale to initiate vaccination trials in humans for the therapy of cancers. Based on these early animal studies, investigators have used gene-modified tumors to develop cellular reagents for adoptive immunotherapy. The generation of TILs and vaccine-primed lymph node cells using gene-modified tumors as immunogens is described in this section.

A. TILs Derived from Gene-Modified Tumors

Tumor-infiltrating lymphocytes represent lymphoid cells derived from tumors that are disaggregated *ex vivo* and cultured in IL-2 [11]. A significant impediment to generating therapeutic TILs resides in the inherent immunogenicity of the tumor from which the TILs are derived. In animal models using poorly immunogenic tumors, the therapeutic efficacy of the TILs is limited [11]. These observations suggest that TILs represent a heterogeneous population of cells and that a significant portion of these cells are not appropriately sensitized within "poorly" or nonimmunogenic tumors. Because human tumors are postulated to be nonimmunogenic based on their spontaneous origins and ability to escape the host immune system, methods to enhance the isolation of therapeutic TILs from these tumors could have significant clinical applications.

Tumors genetically engineered to produce certain cytokines have been found to contain TILs with enhanced *in vitro* and *in vivo* antitumor reactivity. Using the poorly immunogenic methylcholanthrene (MCA)-induced 101 murine fibrosarcoma, Restifo *et al.* [18] showed that therapeutic TILs could be generated if these tumor cells were genetically engineered to secrete IFN-γ. Transduction of MCA 101 tumor cells to secrete IFN-γ resulted in upregulated expression of major histocompatibility complex (MHC) class I molecules. TILs derived from IFN-γ-secreting tumors mediated regression of established parental tumor metastases compared with TILs derived from wild-type tumor, which were ineffective. In additional studies, they found that TILs from the transduced tumors were capable of presenting viral antigen to sensitized T cells, whereas TILs from the parental tumor could not. These studies showed that tumors can be genetically altered with the IFN-γ gene to become "nonprofessional" antigen-presenting cells and that such tumors appeared to be a more reliable source for therapeutically effective TILs.

In a model of established lung metastases, Marincola *et al.* [19] evaluated the immunologic response of the host to an inoculation of syngeneic tumor modified to secrete tumor necrosis factor alpha (TNF-α). TNF-α can stimulate T-cell proliferation and cytotoxic T lymphocyte (CTL) activity. In these studies, the poorly immunogenic MCA 102 sarcoma was modified to secrete TNF-α. The tumorigenicity of the TNF-α-secreting tumor cells was not different from that of wild-type tumor in normal animals. However, TNF-α-secreting tumor cells inoculated subcutaneously regressed in the presence of established wild-type pulmonary metastases, in contrast to inocula of wild-type tumor cells, which grew progressively. In this setting, TILs that were generated from TNF-α-secreting tumors mediated significant antitumor reactivity in adoptive transfer studies compared with wild-type tumor cells, which could not. These observations are important in documenting that therapeutic TILs can be generated from poorly immunogenic tumors modified to secrete TNF-α; moreover, this was accomplished in hosts bearing significant tumor burden.

IL-7 is another cytokine that has been shown to enhance recovery of TILs from tumors. When the gene for murine IL-7 was retrovirally transferred into an immunogenic murine fibrosarcoma, the tumorigenicity of the tumor was significantly diminished [20]. Moreover, by flow cytometry the IL-7-secreting tumor showed a greater than fivefold increase in infiltrating T cells compared with wild-type tumor. These infiltrating cells were primarily CD8+ cells that mediated enhanced *in vitro* cytotoxicity against wild-type tumor compared with T cells isolated from control tumors transfected with a neomycin-resistance gene. These studies indicated that IL-7 secretion by the gene-modified tumors can promote the recruitment of induction by cytolytic T cells to the site of that same tumor.

A novel variation to alter TIL reactivity has been developed at the University of Michigan. In earlier animal studies, tumors treated by the *in vivo* transfer of an allogeneic MHC class I gene complexed with liposomes resulted in expres-

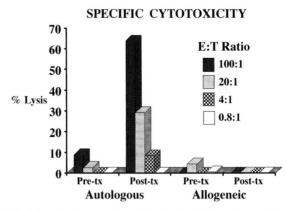

FIGURE 1 *In vitro* cytolytic activity of TILs derived from a melanoma nodule before (pretreatment) and after (posttreatment) intralesional *HLA-B7*/liposome inoculations given three times every 2 weeks. Antitumor reactivity was assessed against autologous and allogeneic melanoma targets at various effector/target (E/T) ratios. Tumor-specific TIL cytolytic reactivity was enhanced after *HLA-B7* inoculation.

sion of the class I molecules by tumor cells [21]. This also resulted in tumor regression and induction of T cells that were reactive not only to transfected tumor cells but also to unmodified cancer cells. We have also been able to achieve gene expression in advanced melanoma patients treated by the *in vivo* inoculation of tumor with DNA/liposome complexes containing a foreign MHC class I gene [22]. Based upon these initial observations, we are currently evaluating the immune reactivity of TILs derived from tumors modified by direct *in vivo* gene transfer utilizing an allogeneic MHC class I gene, *HLA-B7*, in patients with stage IV melanoma. We have confirmed, in *in vitro* assays, an enhanced reactivity of TIL derived from patients inoculated with the foreign class I gene (Fig. 1). We postulate that the allogeneic response induced by the expression of *HLA-B7* results in elaboration of other cytokines that enhance TIL reactivity to tumor-associated antigens. The advantage of this approach is that it does not employ a viral vector and does not require the establishment of a cultured tumor line to accomplish gene transfer.

B. Lymph Node Cells Sensitized with Gene-Modified Tumor Vaccines

Our laboratory has had a long-standing interest in the use of tumor draining lymph nodes (TDLNs) or vaccine-primed lymph nodes (VPLN) as a source of T cells for adoptive immunotherapy in murine models and in human trials [23–31]. We have shown that lymphoid cells derived from TDLNs or VPLNs by themselves do not possess antitumor reactivity but require secondary activation *in vitro* to gain functional antitumor activity. We have called these TDLN or VPLN *pre-effector* cells. Secondary *in vitro* activation may be accomplished by coculture with irradiated tumor and IL-2

[23,24,26] or the sequential activation with anti-CD3 mono-clonal antibody (mAb) and expansion in IL-2 [25,27]. Systemic micrometastases from weakly immunogenic murine tumors such as MCA 205 can be successfully treated using TDLN cells derived from either method of secondary activation. Poorly immunogenic tumors such as the B16–BL6 melanoma, however, fail to sensitize the TDLN cells to become pre-effector cells. To overcome this problem, we have previously shown that the admixture of the B16–BL6 tumor with the potent immunologic adjuvant *Corynebacterium parvum* primes draining lymph nodes to develop pre-effector cells [28]. These cells, upon secondary *in vitro* activation, mediated regression of established metastases in murine models. This reinforces the premise that poorly immunogenic tumors can be genetically modified to be more immunogenic.

Another method that we have explored to sensitize T cells within the draining lymph node has included the transfection of tumor with an allogeneic MHC class I gene [29]. The B16–BL6 melanoma (H-2^b) was transfected *in vivo* with an allogeneic MHC class I (H-2^d) gene using lipofection techniques. Cells from the TDLN were sequentially cultured in an anti-CD3 mAb and IL-2. When adoptively transferred into animals with wild-type pulmonary metastases, these activated TDLN cells showed significant antitumor activity compared to parental TDLN cells, which had minimal therapeutic effect.

We have found that sensitization of TDLN using tumor cells that have been genetically modified to secrete cytokines has been useful in generating T cells reactive to poorly immunogenic tumors. B16–BL6 tumor that has been transfected with the murine IL-4 gene was used to sensitize the regional TDLN. The pre-effector cells sensitized by the IL-4-secreting tumor were effective in mediating regression of preexisting lung micrometastases in a tumor-bearing animal. The *in vivo* antitumor reactivity of these lymphocytes was comparable to that of lymphocytes sensitized by wild-type tumor admixed with *C. parvum*. Another cytokine that we found enhanced the sensitization of the TDLN cells was granulocyte–macrophage colony-stimulating factor (GM-CSF) [30]. GM-CSF is a potent stimulator of macrophages and DCs, which are important antigen-presenting cells involved in the induction of immune responses (See Chapters 7 and 10). Using the B16–BL6 tumor, GM-CSF-secreting tumors were associated with a significant influx of tissue macrophages within the tumor and TDLNs. Activated TDLN cells from mice inoculated with GM-CSF-secreting tumors mediated significant regression of established tumor in adoptive immunotherapy compared with parental TDLN cells, which had no activity. More important, TDLN cells primed with GM-CSF-secreting tumors were more effective in adoptive immunotherapy than were those sensitized by parental tumor admixed with *C. parvum* or tumor cells transduced to secrete other cytokines (e.g., IL-2, IFN-γ, IL-4) (Fig. 2) [31].

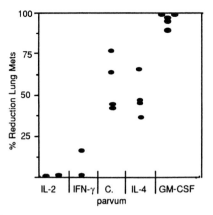

FIGURE 2   Comparison of the adjuvant effect of various cytokines (IL-2, IFN-γ, IL-4, and GM-CSF) elaborated at the site of tumor inoculation for priming pre-effector cells in the tumor-draining lymph nodes. *C. parvum* was also included as a bacterial adjuvant admixed with tumor cells. The antitumor reactivity of the tumor-draining lymph node cells was assessed in adoptive immunotherapy experiments. Each dot represents a group of animals from separate experiments and the percent reduction of pulmonary metastases recorded for each group. GM-CSF was the most potent adjuvant compared to the other agents studied.

Based upon the above preclinical observations, we conducted a clinical study in patients with advanced melanoma to evaluate the immunobiological effects of retrovirally transduced autologous tumor cells given as a vaccine to prime draining lymph nodes (Fig. 3) [32]. Patients were inoculated with both wild-type (WT) and GM-CSF gene-transduced tumor cells in different extremities. Approximately 7 days later, VPLNs were removed. There was an increased infiltration of DCs in the GM-CSF-secreting vaccine sites compared with the WT vaccine sites (Fig. 4 [see also color insert], Table 2).

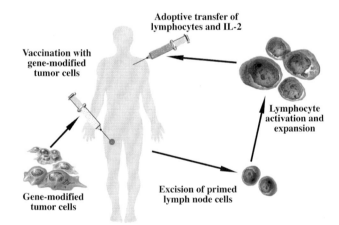

FIGURE 3   Schema of a clinical protocol being performed at the University of Michigan. Melanoma patients with stage IV disease have tumor harvested for transduction with a retrovirus encoding GM-CSF. The gene-modified tumor cells are inoculated intradermally in the thigh, and the vaccine-primed lymph nodes are harvested one week later for *ex vivo* activation and expansion. These cells are subsequently transferred back to the patient intravenously along with the concomitant administration of IL-2.

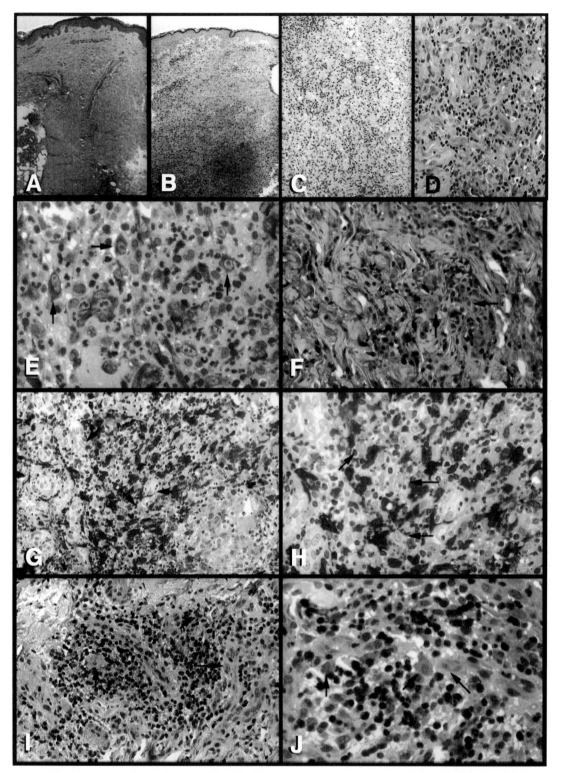

FIGURE 4   Skin vaccine sites sampled from a patient 8 days after injection with either wild-type (A) or granulocyte–macrophage colony-stimulating factor (GM-CSF)-transfected (B to J) autologous melanoma cells. (A) Routine light microscopic appearance of skin injected with mock-transfected melanoma cells with sparse superficial and deep perivascular lymphocyte infiltrate. (B to D) Routine light microscopic appearance demonstrating dense mononuclear cell infiltrate beginning in upper papillary dermis and extending into deep reticular dermis. This inflammatory infiltrate included lymphocyte, neutrophils, and eosinophils. (E) Presence of melanoma cells confirmed by vimentin immunoreactivity. Note that the malignant cells have enlarged nuclei with vesicular chromatic and prominent nucleoli (arrows). (F) Occasional MAC387-positive macrophages are present near melanoma cells (arrows). (G and H) Extensive increase in Factor XIIIa-positive dermal dendrocytes with elongated cytoplasmic processes in mid- and deep dermis closely associated with melanoma tumor cells (arrows). (I and J) Numerous CD45RO-positive "memory" T cells are present throughout the dermis admixed with melanoma tumor cells (arrows). (See color insert.)

TABLE 2   Summary of Immunohistochemical Analysis of Vaccine Sites

Patient no.	Vaccine site[a]	Scoring of infiltrate[b]				Yield of VPLN ($\times 10^8$)
		Overall	DCs	PMNs	Mϕ	
1	GM-CSF	4+	4+	2+	0–1+	7
2	GM-CSF	5+	5+	2+	0	8
	WT	1+	1+	1+	0	2
3	GM-CSF	3+	1+	1–2+	0–1+	8
	WT	1+	1+	1+	0	2
4	GM-CSF	5+	2–3+	1–2+	0	6
	WT	3+	1+	2+	0	1.5
5	GM-CSF	4+	3+	1–2+	0–1+	0.4
	WT	2+	1–2+	1–2+	0–1+	0.2

[a] GM-CSF-transduced (GM-CSF) or nontransduced wild-type (WT) tumor cells were irradiated and inoculated intradermally into opposite thighs. Approximately 7 to 8 days later, the vaccine sites were harvested at the time of VPLN excision.

[b] The infiltrate was scored on a 1- to 5-point scale with 1+ being mild, 3+ moderate, and 5+ severe. The overall infiltrate was scored on the basis of hematoxylin and eosin staining. DCs, dendritic cells; PMNs, polymorphonuclear cells; Mϕ, macrophages).

This resulted in a greater number of cells harvested from the GM-CSF VPLNs compared with the WT VPLNs at a time when serum levels of GM-CSF were not detectable. Four patients proceeded to have the adoptive transfer of GM-CSF VPLN cells secondarily activated and expanded *ex vivo* with anti-CD3 mAb and IL-2. One of these patients has had a durable complete remission of metastatic tumor. In additional animal models, we have utilized the gene gun technology to deliver GM-CSF plasmids to sites of tumor in order to generate tumor-reactive T cells in TDLNs [33]. This procedure provides an alternative nonviral method for introducing cytokine genes into tumor cells to generate effector T cells for adoptive immunotherapy. (For further discussion of gene gun technology, see Chapter 13).

Other investigators have utilized similar models to evaluate the sensitization of TDLN cells. Shiloni *et al.* [34] used the poorly immunogenic MCA 102 sarcoma that was modified with the gene encoding for IFN-γ and reported an increased expression of MHC class I molecules on transduced tumor cells. These cells were inoculated in the flanks of animals to induce an immune response in the TDLNs, which were excised several days later. After secondary *in vitro* activation, these TDLN cells were adoptive transferred into animals with systemic micrometastases and mediated enhanced antitumor efficacy compared with lymph node cells draining wild-type tumor.

In summary, genetic modification of tumor cells to secrete cytokines or express immunomodulatory proteins has shown promise in enhancing the antitumor reactivity of TILs and TDLN cells in animal models. These observations have established the rationale for evaluating these approaches in clinical studies, which our laboratory is currently pursuing.

III. GENETIC MANIPULATION OF T CELLS TO ENHANCE ANTITUMOR REACTIVITY

The adoptive transfer of LAK cells, TILs, or TDLN cells in combination with the systemic administration of IL-2 has resulted in the regression of several types of tumors in both murine models and human clinical trials [27,35–38]. Specific limitations in adoptive immunotherapy were recognized early in its use. First, the antitumor activity of adoptive immunotherapy is directly proportional to the administered cell dose, yet the frequency of antigen-reactive T cells in cancer patients is postulated to be extremely low ($<1/10^5$ to $<1/10^6$). As a result, extended periods of *ex vivo* culture are required to generate therapeutic cell doses, which may be difficult to achieve in a significant proportion of patients [39,40]. Second, a number of functional deficiencies have been identified in *ex vivo* expanded lymphocytes, including defects in cellular trafficking after adoptive transfer [41,42], altered expression of adhesion molecules [43], and defects in signaling pathways or cytokine responses [44–53]. Extended culture or specific culture conditions may explain some of these abnormalities, but there is also evidence that the anatomic source of tumor-specific T cells (i.e., tumor sites or TDLNs) may contribute to functional defects [54]. Third, tumors may develop resistance to T-cell-mediated immune responses through a variety of mechanisms, including downregulation of tumor-antigen processing or expression [55], loss of cell-surface MHC expression [56–62], production of inhibitory cytokines, and Fas-mediated modulation of T-cell survival and function [63,64]. While the overall concept and therapeutic potential of adoptive immunotherapy remain extremely attractive, new strategies are needed to optimize the therapeutic potency and specificity of this treatment approach.

Direct genetic modification of effector cells is one of several methods being tested to improve the therapeutic efficacy of these cells. From a theoretical standpoint, gene transfer could be used to enhance one or more properties of the cells, including survival, trafficking, antigen specificity, immunomodulatory function, and/or cytotoxic or cytolytic potential. The overall success of these approaches will be influenced by the efficiency, specificity, and stability of gene transfer and gene expression.

A. Genetic Transduction of T Cells

While techniques for successful nonviral transduction of T cells have been developed [65], viral vectors remain the most commonly used genetic delivery systems due to their overall efficiency. Early efforts at genetic modification of T cells with oncoretroviral vectors were relatively inefficient, requiring extended periods of positive selection and expansion prior to adoptive transfer [66]. A number of important technical advances now make it possible to routinely transduce up to 50% or more of a T-cell population, reducing or eliminating the need for dominant selection of genetically modified cells for most applications of adoptive immunotherapy.

Oncoretroviral vectors transduce actively dividing cells [67], requiring that T cells be activated and replicating prior to viral infection. Costimulation of T cells through the T-cell receptor (TCR)–CD3 complex and CD28 is necessary to achieve complete activation of T cells [68,69]. CD28 costimulation promotes the long-term survival, proliferation, and functional potential of T cells via several mechanisms, including inhibition of apoptosis [70,71], enhanced production of lymphokines [72,73], and augmentation of the cytolytic potential of cytotoxic T cells. Whereas activation of T cells with anti-CD3 and IL-2 for adoptive immunotherapy is associated with preferential expansion of CD8$^+$ T cells [27], costimulation with anti-CD3/anti-CD28 yields outgrowth of both CD4$^+$ and CD8$^+$ tumor-reactive T cells [74]. Activation of T cells with immobilized anti-CD3 and anti-CD28 mAbs markedly enhances transduction efficiency with recombinant murine leukemia viruses, notably when combined with the use of fibronectin fragment CH296 (Retronectin) in clinically applicable, cell-free transduction protocols (Fig. 5) [75,76]. Retronectin contains binding domains for cellular adhesion molecules and retrovirus and significantly enhances viral transduction [75,77,78], in part through colocalization of target cells and virus. Autologous human T cells, activated with anti-CD3/anti-CD28 and genetically modified with a murine oncoretroviral vector, have been shown to persist *in vivo* at high levels for up to one year following adoptive transfer in a clinical trial [79] (for a further discussion of CD28 and T-cell activation, see Chapter 7).

To circumvent the requirement for activation and cell division prior to gene transfer, alternative viral vector systems

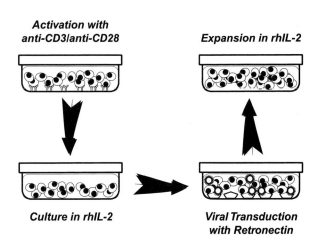

FIGURE 5 Optimized protocol for transduction of T cells with retrovirus. T-cell proliferation is induced by co-activation with immobilized monoclonal anti-CD3 and anti-CD28, followed by culture in recombinant IL-2. Retroviral transduction by cell-free viral supernatant is enhanced by fibronectin fragment CH296 (Retronectin). Retronectin contains binding domains for cellular adhesion molecules and retroviral particles. Following viral transduction, T cells are briefly expanded in media containing IL-2.

with the capacity to transduce resting cell populations have been characterized. Adeno-associated virus (AAV) has been used to successfully transduce primary T cells from rhesus monkeys in a primate model of adoptive T-cell transfer [80]. In this model, AAV vectors appeared to transduce T cells with greater efficiency than murine oncoretroviral vectors when assayed at short intervals following transduction, although retrovirally marked cells persisted for a longer time period following *in vivo* passage. In contrast to the retroviral vectors, the AAV vector did not integrate into the lymphocyte genomic DNA, suggesting that AAV vectors might be considered in clinical applications where long-term vector persistence is unnecessary or undesirable. Lentiviruses have also attracted significant interest as gene delivery vectors based on their ability to transduce nonreplicating cells. A number of lentiviral vectors and packaging cell lines have been developed, including both HIV- and non-HIV-based systems. Recombinant lentiviruses have been used to successfully introduce and express virally encoded transgenes in resting human lymphocytes [81,82]. In contrast to AAV, transduction is associated with stable integration of the virus into genomic DNA [82]. Additional studies are needed to determine if lentiviruses offer an advantage over conventional oncoretroviral vector systems in the context of *in vivo* adoptive transfer models, and safety concerns must be addressed prior to the application of lentiviruses in clinical trials (for a further discussion of lentiviral vectors, see Chapter 6).

B. TIL-Marking Studies

The first human trial involving gene transfer was reported by Rosenberg *et al.* [66] and addressed the issue of the safety of infusion of retrovirally transduced TILs into patients.

The study involved the infusion of TILs transfected with an antibiotic-resistance (neomycin phosphotransferase) gene, which served as a marker to identify adoptively transferred TILs. Ten patients with metastatic melanoma received gene-marked TILs. No toxicities associated with gene transfer were observed. The *in vitro* cytolytic capacity of the TILs was not altered by the genetic transfer. In some of the patients treated with the gene-marked TILs, tumor responses were observed. Infectious retrovirus has not been found in any of the patients. Furthermore, gene-modified TILs were detected using the polymerase chain reaction (PCR) method to detect the neomycin-resistance gene in tumor deposits for up to 64 days and in the circulation for up to 189 days after infusion.

C. T Cells as Delivery Vehicles for Immunoregulatory Molecules

In studies using [111]indium-labeled cells, TIL preferentially localized to tumor sites when compared to peripheral blood lymphocytes in patients with malignant melanoma [83]. This raised the possibility that T cells could be used as vehicles to deliver immunoregulatory molecules directly to sites of metastatic cancer. Early efforts focused on transduction of tumor-specific T cells with cytokine genes that promote T-cell survival or enhance local tumor killing, such as IL-2 [84,85] and TNF-α [86]. While the potential value of cytokine gene-modified T cells was demonstrated in these preclinical models, inefficient gene transfer techniques and suboptimal gene expression inhibited the translation of this approach to clinical trials. With the availability of improved gene-transfer methodology, as well as a more detailed understanding of T-cell biology and regulatory cytokines, this approach still holds considerable promise. The ability of cytokine gene-modified T cells to modulate the *in vivo* course of autoimmune disorders is a striking example of the ability to manipulate local and systemic immune responses through targeted genetic manipulation of effector cell populations [87].

D. Genetic Reprogramming of Effector Cells with Tumor-Antigen-Specific Receptors

A major limitation to the application of adoptive immunotherapy for cancer is the difficulty in isolating and expanding tumor-specific T cells in human subjects. The frequency of these cells is extremely low or undetectable in most patients, and extended periods of culture are generally required to manufacture a therapeutic cell dose. Using genetic approaches, it may be possible to rapidly and efficiently reprogram large numbers of immune effector cells to selectively recognize malignant cells via expression of a tumor-antigen-specific receptor. After engaging the tumor antigen, the receptor activates antitumor effector pathways such as cytokine

release or cellular cytotoxicity. As discussed in greater detail later, both classic $\alpha\beta$ T-cell receptors and chimeric T cell receptors have been used for this purpose.

The choice of antigenic target(s) may be important to the overall effectiveness and safety of this approach. The ideal tumor antigen for immune targeting is specific for the cancer, is not expressed in normal tissues, and is directly involved in the pathogenesis of the disease. Mutated oncogenes, or the oncogenic fusion proteins created by chromosomal translocations, represent attractive potential targets that meet these criteria. However, experimental work in solid tumors, notably malignant melanoma, revealed that tissue differentiation antigens represent important targets of therapeutic cytotoxic T-cell responses to cancer. These antigens are not tumor specific but are overexpressed or selectively expressed in malignant cells. Examples include the melanocyte-associated antigens (MAAs), such as melan-A/MART-1, MAGE-1, MAGE-3, tyrosinase, and gp100, as well as HER2/*neu*, MUC1, CEA, and WT-1. As our understanding of the molecular underpinnings of cancer grows, a range of new potential targets for this approach will emerge.

1. Gene Transfer of Tumor-Antigen-Specific T-Cell Receptors

The TCR is an attractive receptor, as the structure and function of the classic $\alpha\beta$ TCRs are well known and the receptor can recognize processed intracellular antigens (including mutant proteins). With improved ability to culture and subclone antigen-specific T cells, it is now feasible to molecularly clone the cDNAs encoding antigen-specific TCRs. A gene therapy approach based on a classic $\alpha\beta$ TCR requires selection of a TCR that recognizes a true tumor rejection antigen, the MHC restriction of the antigen must be sufficiently common in the population to make the strategy feasible to test, and an efficient method to coordinately express both chains of the TCRs in T cells must be developed (Fig. 6) (for a further discussion of TCR, see Chapter 7).

The MART-1 MAA is an excellent candidate antigen to test the feasibility of this approach. MART-1 is the most common melanoma target antigen recognized by TILs in HLA-A2+ patients, and HLA-A2 is the most common HLA antigen in Caucasian individuals (40%). Nishimura and colleagues at the National Cancer Institute cloned the TCR-α and TCR-β chain cDNAs from a MART-1-reactive, CD8+ TIL cell line (Clone 5), and demonstrated that plasmid-mediated gene transfer could reconstitute the tumor-antigen specificity of the Clone 5 TCRs in a human T-cell line [88]. They subsequently went on to show that retroviral gene transfer could be used to introduce the MART-1-specific TCR in activated primary human T cells and demonstrated MART-1-specific cytokine release and cellular cytotoxicity in selected subclones [89].

αβ *T Cell Receptor*   *Chimeric T Cell Receptor*

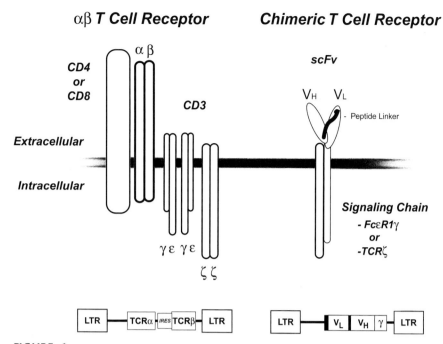

FIGURE 6   Retroviral transduction of a tumor-antigen-specific T-cell receptor or chimeric T-cell receptor. The T-cell receptor complex is formed by the association of multiple proteins encoded by separate genes. The antigenic specificity of the TCR is completely defined by the heterodimeric complex of the TCR-α and TCR-β chains. The mature TCR-α and TCR-β genes are formed by somatic gene rearrangements and are therefore unique to each T-cell clone. To genetically engraft TCR antigenic specificity, coordinate expression of both the TCR-α and TCR-β genes is required. Expression of a functional, antigen-specific TCR is achieved with a bicistronic retroviral vector. The vector produces a single transcript encoding both TCR chains. Translation of the second gene is initiated using an internal ribosome entry site (IRES) element for transcription of the second gene. A chimeric TCR is a recombinant protein composed of an extracellular antigen-recognition domain, coupled to an intracellular signaling chain. The most commonly used antigen-recognition domain is a single-chain Fv (scFv), derived from a mAb by tethering the heavy-chain and light-chain variable (V_H and V_L) regions with a flexible peptide linker. One of several signaling chains, including the FcεR1γ chain and TCR-ζ chain, is used to insert the receptor into the cell membrane and activate effector cell function following engagement of antigen. The signaling chains may heterodimerize with native signaling chains.

We have developed a clinical gene therapy protocol for advanced malignant melanoma at the University of Michigan based on adoptive immunotherapy with autologous T cells genetically modified to express a tumor-antigen-specific TCR. The MART-1-specific TCR cDNAs (obtained from M. I. Nishimura, Department of Surgery, University of Chicago) were inserted into a bicistronic retroviral vector engineered to provide stable gene expression in primary hematolymphoid cells [90]. Using an efficient retroviral transduction protocol [76] incorporating the GALV-based PG13 packaging cell line [91], T-cell activation with anti-CD3/anti-CD28, and viral transduction in the presence of Retronectin, we have been able to generate polyclonal, primary T-cell cultures that express the MART-1 TCRs in a high proportion of cells without the need for dominant selection or isolation of specific subclones. These genetically modified T cells demonstrate MART-1-specific cellular cytotoxicity and T_h1-like cytokine release [92]. This approach to genetic im-

munotherapy of melanoma will be tested in a phase I clinical trial at the University of Michigan.

2. Chimeric T-Cell Receptors

Chimeric T-cell receptors (chTCRs) are recombinant immune receptors composed of two domains: an extracellular antigen-binding domain and an intracellular signaling domain (Fig. 6). While a variety of chTCR designs have been proposed and tested, the most commonly employed antigen targeting domain is a single-chain Fv (scFv), whereby the V_L and V_H domains of a monoclonal antibody are tethered in a single peptide by a flexible linker [93]. Regions of scFv exhibit similar specificities and affinities compared to normal antibody-variable regions [94,95]. The scFv gene is then joined to a signaling chain derived from the TCR or FcR complex. The TCR-ζ chain and Fc-receptor-γ chain are closely related and capable of activating T cells and/or myeloid cells

[96–100], and have been widely adopted in chimeric TCR constructs. Thus, in a single molecule, the scFv–ζ/γ chTCR combines the exquisite antigen specificity of a mAb and signal transduction elements necessary to activate effector cell function.

A significant potential advantage of this design is the lack of MHC restriction and/or need for antigen processing. The antigenic target is an intact cell surface protein defined by an antibody, unlike a "classic" TCR target. This may circumvent potential immune escape mechanisms utilized by tumor cells, including altered processing and presentation of tumor antigens by MHC molecules. Such chTCR molecules have been successfully developed to target a variety of tumor antigens, including the MOv18-defined ovarian tumor antigen [101,102], TAG-72 [103], carcinoembryonic antigen (CEA) [104], and p185HER2 [105–107]. Introduction of the scFv–ζ/γ construct into T-cell lines or hybridomas has resulted in a functioning chTCR with tumor-antigen specificity, as defined by cytokine release and/or cytotoxicity following stimulation of the transduced effector cells with the appropriate antigen-bearing target cell [93,101], and antigen-specific regression of tumor in animal models of T-cell adoptive immunotherapy [102,108,109].

IV. GENETIC MODULATION OF DENDRITIC CELLS

Dendritic cells are highly potent antigen-presenting cells. These cells process antigens and present them to lymphoid cells in association with MHC molecules. In addition, they express a large number of costimulatory molecules on their surface. DCs play a central role in the induction of immune responses and can be isolated from the peripheral blood [110,111]. However, the techniques available to isolate DCs are relatively cumbersome and result in low yields. The combination of GM-CSF and IL-4 was found by Romani *et al.* [112] to be effective in the generation of functional DCs from mobilized stem cells isolated from the peripheral blood of patients undergoing chemotherapy. These studies lend support to the notion that DCs are derived from uncommitted hematopoietic stem cells.

One of the major interests in DCs is their ability to present tumor antigen to resting lymphoid cells, which results in immunocompetent T and B cells. Several animal studies have demonstrated that the *in vitro* "pulsing" of DCs with tumor antigen in the form of whole tumor cells, tumor lysates, or tumor peptides will generate DCs capable of priming naive T cells [113,114]. Moreover, the adoptive transfer of these pulsed DCs by either intravenous or intradermal inoculation can result in regression of established tumors. Hsu *et al.* [115] reported that the administration of human DCs pulsed *ex vivo* with tumor-specific lymphomas idiotype could mediate the regression of recurrent lymphoma. Impressively, some of these patients developed durable complete tumor responses to their therapy.

Methods to enhance the therapeutic efficacy of DCs include genetic modification (Fig. 7). Liposomal transfection, retroviral gene transfer, and the gene gun have all been used to successfully modify DC. Genes encoding for tumor antigens have been successfully transferred into DCs to enhance their ability to sensitize lymphoid cells [116,117]. Reeves *et al.* [117] retrovirally transduced human DCs with the melanoma tumor-associated antigen gene MART-1. These DCs were able to generate specific antitumor CTLs and stimulate greater levels of cytokine release by MART-1-specific

DENDRITIC CELL-BASED VACCINE

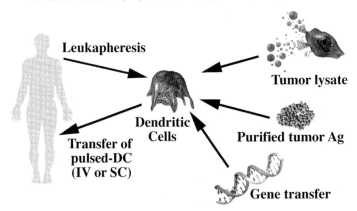

FIGURE 7   Schema of the clinical use of DC-based vaccines. Patients are leukapheresed to obtain DCs from the peripheral circulation. The DCs are "pulsed" with tumor lysate or purified tumor antigen. Alternatively, the DCs can be transfected with genes encoding tumor antigens. The genetic material can be in the form of DNA or RNA.

TILs. In an *in vivo* murine model, Ribas *et al.* [118] demonstrated induction of immunity with MART-1-transduced murine DCs that were capable of successfully treating established tumors expressing the MART-1 antigen. Using an adenoviral vector expressing melanoma-associated antigen, gp100, Wan *et al.* reported that DCs transfected with the vector were capable of eliciting protective immunity and mediated regression of established B16 melanoma upon adoptive transfer in a murine model [119]. This immunity was superior to the administration of adenoviral vector by itself as a vaccine. Of interest, both CD4$^+$ and CD8$^+$ antitumor reactivity could be measured after gene-modified DC administration; however, only CD4$^+$ T cells were necessary to see tumor rejection responses [120]. The latter was initiated in the absence of an IL-12 or CD40 ligand requirement.

Besides modification of DCs with genes encoding tumor-associated antigens, the introduction of genes encoding immunomodulatory cytokines can be considered. In an allograft transplant murine model, the pretransplant infusion of donor DCs transduced to secrete both IL-10 and TGF-β enhanced renal allograft survival compared to mock transfected DCs [121]. There was a correlation of increased graft survival to both inhibition of the induction of CTLs and enhancement of a polarization to produce type 2 cytokines of lymphoid cells upon antigen-specific restimulation *in vitro*. These findings demonstrate that the induction of antigen-specific immunity in an *in vivo* model can be altered by gene-modified DCs. This kind of approach may be useful to enhance the induction of immunity against tumor antigens utilizing alternate cytokine of possibly chemokine genes (for a further discussion of genetically modified DCs, see Chapter 10).

V. SUMMARY

Adoptive cellular therapy remains a powerful method of eradicating established tumor in experimental models. The use of gene transfer techniques has resulted in more effective methods to generate tumor-specific T cells. Another area of tremendous interest is the adoptive transfer of DCs manipulated to present tumor antigen to resting T cells. Gene transfer techniques may offer more optimal ways to generate therapeutic DCs. The application of these methods in clinical studies has been extremely important in identifying new directions to explore in the experimental setting. The ability to genetically alter the host–tumor interaction has dramatically increased our understanding of the immune response to tumor cells. It is anticipated that this increased understanding will lead to more effective therapies of malignancy.

References

1. Gross, L. (1943). Intradermal immunization of C3H mice against a sarcoma that originated in an animal of the same line. *Cancer Res.* **3**, 326–333.

2. Landsteiner, K., and Chase, M. W. (1942). Experiments on transfer of cutaneous sensitivity to simple compounds. *Proc. Soc. Exp. Biol. Med.* **49**, 688–690.

3. Billingham, R. E., Brent, L., and Medawar, P. B. (1954). Quantitative studies on tissue transplantation immunity: the origin, strength and duration of actively and adoptively acquired immunity. *Proc. R. Soc. Biol.* **143**, 58–80.

4. Mitchison, N. A. (1955). Studies on the immunological response to foreign tumor transplants in the mouse: the role of lymph node cells in conferring immunity by adoptive transfer. *J. Exp. Med.* **102**, 157–177.

5. Borberg, H., Oettgen, H. F., Choudry, K., and Beattie, Jr., E. J. (1972). Inhibition of established transplants of chemically induced sarcomas in syngeneic mice by lymphocytes from immunized donors. *Int. J. Cancer* **10**, 539–547.

6. Rosenberg, S. A., and Terry, W. D. (1977). Passive immunotherapy of cancer in animals and man. *Adv. Cancer Res.* **25**, 323–388.

7. Greenberg, P. D. (1990). Adoptive T cell therapy of tumors: mechanisms operative in the recognition and elimination of tumor cells. *Adv. Immunol.* **49**, 281–355.

8. Morgan, D. A., Ruscetti, F. W., and Gallo, R. (1976). Selective in vitro growth of T lymphocytes from normal human bone marrows. *Science* **193**, 1007–1008.

9. Cheever, M. A., Greenberg, P. D., and Fefer, A. (1981). Specific adoptive therapy of established leukemia with syngeneic lymphocytes sequentially immunized in vivo and in vitro non-specifically expanded by culture with interleukin-2. *J. Immunol.* **126**, 1318–1322.

10. Rosenberg, S. A., Lotze, M. T., Muul, L. M., Leitman, S., Chang, A. E., Ettinghausen, S. E., Matory, Y. L., Skibber, J. M., Shiloni, E., and Vetto, V. J. (1985). Observations on the systemic administration of autologous lymphokine-activated killer cells and recombinant interleukin-2 to patients with metastatic cancer. *N. Engl. J. Med.* **313**, 1485–1492.

11. Rosenberg, S. A., Spiess, P., and Lafreniere, P. (1986). A new approach to the adoptive immunotherapy of cancer with tumor-infiltrating lymphocytes. *Science* **233**, 1318–1321.

12. Spiess, P. J., Yang, J. C., and Rosenberg, S. A. (1987). *In vivo* antitumor activity of tumor-infiltrating lymphocytes expanded in recombinant interleukin-2. *J. Natl. Cancer Inst.* **79**, 1067–1075.

13. Rosenberg, S. A., Packard, B. S., Aebersold, P. M., Solomon, D., Topalian, S. L., Toy, S. T., Simon, P., Lotze, M. T., Yang, J. C., and Seipp, C.A. (1988). Use of tumor-infiltrating lymphocytes and interleukin-2 in the immunotherapy of patients with metastatic melanoma: a preliminary report. *N. Engl. J. Med.* **319**, 1676–1680.

14. Aebersold, P., Hyatt, C., Johnson, S., Hines, K., Korcak, L., Sanders, M., Lotze, M., Topalian, S., Yang, J. Y., and Rosenberg, S. A. (1991). Lysis of autologous melanoma cells by tumor-infiltrating lymphocytes: association with clinical response. *J. Natl. Cancer Inst.* **83**, 932–937.

15. Gansbacher, B., Bannerji, R., Daniels, B., Zier, K., Cronin, K., Gilboa, E., Gastle, G., Finstad, C. L., Guarini, A., Bol, G., and Bander, N. H. (1990). Retroviral vector-mediated γ-interferon gene transfer into tumor cells generates potent and long lasting antitumor immunity. *Cancer Res.* **50**, 7820–7825.

16. Golumbek, P. T., Lazenby, A. J., Levitsky, H. I., Jaffee, L. M., Karasuyama, H., Baker, M., and Pardoll, D. M. (1991). Treatment of established renal cancer by tumor cells engineered to secrete interleukin-4. *Science* **254**, 713–716.

17. Tepper, R., Coffman, R., and Leder, P. (1992). An eosinophil-dependent mechanism for the antitumor effect of interleukin-4. *Science* **257**, 548–551.

18. Restifo, N. P., Spiess, P. J., Karp, S. E., Mule, J. J., and Rosenberg, S. A. (1992). A nonimmunogenic sarcoma induced with the cDNA for interferon-γ elicits CD8+ T cells against the wild-type tumor: correlation with antigen presentation capability. *J. Exp. Med.* **175**, 1423–1431.

19. Marincola, F. M., Ettinghausen, S., Cohen, P. A., Cheshire, L. B., Restifo, N. P., Mule, J. J., and Rosenberg, S. A. (1994). Treatment

of established lung metastases with tumor-infiltrating lymphocytes derived from a poorly immunogenic tumor engineered to secrete human TNF-alpha. *J. Immunol.* **152,** 3500–3513.

20. McBride, W. H., Thacker, J. D., Comora, S., Economou, J. S., Kelley, D., Hogge, D., Dubinett, S. M., and Dougherty, G. J. (1992). Genetic modification of a murine fibrosarcoma to produce interleukin-7 stimulates host cell infiltration and tumor immunity. *Cancer Res.* **52,** 3931–3937.

21. Plautz, G. E., Yang, Z. Y., Wu, B. Y., Gao, X., Huang, L., and Nabel, G. J. (1993). Immunotherapy of malignancy by in vivo gene transfer into tumors. *Proc. Natl. Acad. Sci. USA* **90,** 4645–4649.

22. Nabel, G., Gordon, D., Bishop, D. K., Nickoloff, B. J., Yang, Z. Y., Aruga, A., Cameron, M. J., Nabel, E. G., and Chang, A. E. (1996). Immune response in human melanoma after transfer of an allogeneic class I major histocompatibility complex gene with DNA-liposome complexes. *Proc. Natl. Acad. Sci. USA* **93,** 15388–15393.

23. Chou, T., Chang, A. E., and Shu, S. (1988). Generation of therapeutic T lymphocytes from tumor-bearing mice by *in vitro* sensitization: culture requirements and characterization of immunologic specificity. *J. Immunol.* **140,** 2453–2461.

24. Shu, S. Y., Chou, T., and Sakai, K. (1989). Lymphocytes generated by in vivo priming and in vitro sensitization demonstrate therapeutic efficacy against a murine tumor that lacks apparent immunogenicity. *J. Immunol.* **143,** 740–748.

25. Yoshizawa, H., Chang, A. E., and Shu, S. (1991). Specific adoptive immunotherapy mediated by tumor-draining lymph node cells sequentially activated with anti-CD3 and IL-2. *J Immunol* **147,** 729–737.

26. Chang, A. E., Yoshizawa, H., Sakai, K., Cameron, M. J., Sondak, V. K., and Shu, S. Y. (1993). Clinical observations on adoptive immunotherapy with vaccine-primed T-lymphocytes secondarily sensitized to tumor in vitro. *Cancer Res* **53,** 1043–1050.

27. Chang, A. E., Aruga, A., Cameron, M. J., Sondak, V. K., Normolle, D. P., Fox, B. A., and Shu, S. Y. (1997). Adoptive immunotherapy with vaccine-primed lymph node cells secondarily activated with anti-CD3 and interleukin-2. *J. Clin. Oncol.* **15,** 796–807.

28. Geiger, J. D., Wagner, P. D., Cameron, M. J., Shu, S., and Chang, A. E. (1993). Generation of T-cells reactive to the poorly immunogenic B16–BL6 melanoma with efficacy in the treatment of spontaneous metastases. *J. Immunother.* **13,** 153–65.

29. Wahl, W. L., Plautz, G. E., Fox, B. A., Strome, S. E., Nabel, G. J., Cameron, M. J., San, H., Shu, S., and Chang, A. E. (1992). Generation of therapeutic T lymphocytes after in vivo transfection of a tumor with a gene encoding allogeneic class I major histocompatability complex antigen. *Surg. Forum* **63,** 476–478.

30. Arca, M. J., Krauss, J. C., Aruga, A., Cameron, M. J., Shu, S., and Chang, A. E. (1996). Therapeutic efficacy of T cells derived from lymph nodes draining a poorly immunogenic tumor transduced to secrete granulocyte-macrophage colony-stimulating factor. *Cancer Gene Ther.* **3,** 39–47.

31. Arca, M. J., Krauss, J. C., Strome, S. E., Cameron, M. J., and Chang, A. E. (1996). Diverse manifestations of tumorigenicity and immunogenicity displayed by the poorly immunogenic B16–BL6 melanoma transduced with cytokine genes. *Cancer Immunol. Immunother.* **42,** 237–245.

32. Chang, A. E., Li, Q., Bishop, D. K., Normolle, D. P., Redman, B. G., and Nickoloff, B. J. (2000). Immunogenetic therapy of human melanoma utilizing autologous tumor cells transduced to secrete granulocyte-macrophage colony-stimulating factor. *Hum. Gene Ther.* **11,** 839–850.

33. Tanigawa, K., Yu, H., Sun, R., Nickoloff, B. J., and Chang, A. E. (2000). Gene gun application in the generation of effector T cells for adoptive immunotherapy. *Cancer Immunol. Immunother.* **48,** 635–643.

34. Shiloni, E., Karp, W. E., Custer, M. C., Shilyansky, J., Restifo, N. P., Rosenberg, S. A., and Muléy, J. J. (1993). Retroviral transduction of interferon-gamma cDNA into a nonimmunogenic murine fibrosar-

coma: generation of T cells in draining lymph nodes capable of treating established parental metastatic tumor. *Cancer Immunol. Immunother.* **37,** 286–292.

35. Mule, J. J., Shu, S. Y., Schwarz, S. L., and Rosenberg, S. A. (1984). Adoptive immunotherapy of established pulmonary metastases with LAK cells and recombinant interleukin-2. *Science* **225,** 1487–1489.

36. Barth, R. J., Bock, S. N., Mule, J. J., and Rosenberg, S. A. (1990). Unique murine tumor-associated antigens identified by tumor infiltrating lymphocytes. *J. Immunol.* **144,** 1531–1537.

37. Rosenberg, S. A. (1993). *Principles and Applications of Biologic Therapy,* Lippincott, Philadelphia.

38. Rosenberg, S. A. (1994). *Cell Transfer Therapy: Clinical Applications,* Lippincott, Philadelphia.

39. Shornick, Y., and Rosenberg, S. A. (1990). Comparative studies of the long-term growth of lymphocytes from tumor infiltrates, tumor draining lymph nodes, and peripheral blood by repeated in vitro stimulation with autologous tumor. *Biol. Resp. Modif.* **9,** 431.

40. Yannelli, J. R., Hyatt, C., McConnell, S., Hines, K., Jacknin, L., Parker, L., Sanders, M., and Rosenberg, S. A. (1996). Growth of tumor-infiltrating lymphocytes from human solid cancers: summary of a 5-year experience. *Int. J. Cancer* **65,** 413–421.

41. Economou, J. S., Belldegrun, A. S., Glaspy, J., Toloza, E. M., Figlin, R., Hobbs, J., Meldon, N., Kaboo, R., Tso, C. L., Miller, A., Lau, R., McBride, W., and Moen, R. C. (1996). In vivo trafficking of adoptively transferred interleukin-2 expanded tumor-infiltrating lymphocytes and peripheral blood lymphocytes. Results of a double blind gene marking trial. *J. Clin. Invest.* **97,** 515–521.

42. Zhu, H., Melder, R. J., Baxter, L. T., and Jain, R. K. (1996). Physiologically based kinetic model of effector cell biodistribution in mammals: implications for adoptive immunotherapy. *Cancer Res.* **56,** 3771–3781.

43. Stoolman, L. M. (1989). Adhesion molecules controlling lymphocyte migration. *Cell* **56,** 907–910.

44. Nakagomi, H., Peterson, M., Magnusson, I., Juhlin, C., Matsuda, M., Mellstedt, H., Taupin, J. L., Vivier, E., Anderson, P., and Kiessling, R. (1993). Decreased expression of the signal-transducing ζ chains in tumor-infiltrating T-cells and NK cells of patients with colorectal carcinoma. *Cancer Res.* **53,** 5610–5612.

45. Finke, J. H., Zea, A. H., Stanely, J., Longo, D. L., Mizoguchi, H., Tubbs, R. R., Wiltrout, R. H., O'Shea, J. J., Kudoh, S., and Klein, E. (1993). Loss of T-cell receptor ζ chain and p56^{lck} in T-cells infiltrating human renal cell carcinoma. *Cancer Res.* **53,** 5613–5616.

46. Matsuda, M., Peterson, M., Lenkei, R., Taupin, J. L., Magnusson, I., Mellstedt, H., Anderson, P., and Kiessling, R. (1995). Alterations in the signal-transducing molecules of T cells and NK cells in colorectal tumor-infiltrating, gut mucosal and peripheral lymphocytes: correlation with the stage of the disease. *Int. J. Cancer* **61,** 765–772.

47. Wang, Q., Stanley, J., Kudoh, S., Myles, J., Kolenko, V., Yi, T., Tubbs, R., Bukowski, R., and Finke, J. (1995). T cells infiltrating non-Hodgkin's B cell lymphomas show altered tyrosine phosphorylation pattern even though T cell receptor/CD3-associated kinases are present. *J. Immunol.* **155,** 1382–1392.

48. Tartour, E., Latour, S., Mathiot, C., Thiounn, N., Mosseri, V., Joyeux, I., D'Enghien, C. D., Lee, R., Debre, B., and Fridman, W. H. (1995). Variable expression of CD3-zeta chain in tumor-infiltrating lymphocytes (TIL) derived from renal-cell carcinoma: relationship with TIL phenotype and function. *Intl. J. Cancer* **63,** 205–212.

49. Lai, P., Rabinowich, H., Crowley, N. P., Bell, M. C., Mantovani, G., and Whiteside, T. L. (1996). Alterations in expression and function of signal-transducing proteins in tumor-associated and natural killer cells in patients with ovarian carcinoma. *Clin. Cancer Res.* **2,** 161–173.

50. Cardi, G., Heaney, J. A., Schned, A. R., Phillips, D. M., Branda, M. T., and Ernstoff, M. S. (1997). T-cell receptor zeta-chain expression on tumor-infiltrating lymphocytes from renal cell carcinoma. *Cancer Res.* **57,** 3517–3519.

51. Choi, S. H., Chung, E. J., Whang, D. Y., Lee, S. S., Yang, Y. S., and Kim, C. W. (1998). Alteration of signal-transducing molecules in tumor-infiltrating lymphocytes and peripheral blood T lymphocytes from human colorectal carcinoma patients. *Cancer Immunol. Immunother.* **45**, 299–305.

52. Reichert, T. E., Day, R., Wagner, E. M., and Whiteside, T. L. (1998). Absent or low expression of the zeta chain in T cells at the tumor site correlates with poor survival in patients with oral carcinoma. *Cancer Res.* **58**, 5344–5347.

53. Lopez, C. B., Rao, T. D., Feiner, H., Shapiro, R., Marks, J. R., and Frey, A. B. (1998). Repression of interleukin-2 mRNA translation in primary human breast carcinoma tumor-infiltrating lymphocytes. *Cell. Immunol.* **190**, 141–155.

54. Agrawal, S., Marquet, J., Delfau, L. M., Copie, B. C., Jouault, H., Reyes, F., Bensussan, A., and Farcet, J. P. (1998). CD3 hyporesponsiveness and in vitro apoptosis are features of T cells from both malignant and nonmalignant secondary lymphoid organs. *J. Clin. Invest.* **102**, 1715–1723.

55. Restifo, N. P., Esquivel, F., Kawakami, Y., Yewdell, J. W., Mule, J. J., Rosenberg, S. A., and Bennink, J. R. (1993). Identification of human cancers deficient in antigen processing. *J. Exp. Med.* **177**, 265–272.

56. Ruiter, D. J., Mattijssen, V., Broecker, E. B., and Ferrone, S. (1991). MHC antigens in human melanomas. *Semin. Cancer Biol.* **2**, 35.

57. Kourilsky, P., Jaulin, C., and Ley, V. (1991). The structure and function of MHC molecules: possible implications for the control of tumor growth by MHC-restricted T cells. *Semin. Cancer Biol.* **2**, 275.

58. Ruiz-Cabello, F., Perez-Ayala, M., Gomez, O., Redondo, M., Concha, A., and Cabrera, T. (1991). Molecular analysis of MHC-class-I alterations in human tumor cell lines. *Int. J. Cancer* **6**, 123.

59. Bodmer, W. F., Browning, W., Krausa, P., Rowan, A., Bicknell, D. C., and Bodmer, J. G. (1993). Tumor escape from immune response by variation in HLA expression and other mechanisms. *Ann. N.Y. Acad. Sci.* **690**, 42.

60. Garrido, F., Cabrera, T., Concha, A., Glew, S., Ruiz-Cabello, F., and Stern, P. L. (1993). Natural history of HLA expression during tumor development. *Immunol. Today* **14**, 91.

61. Jakobsen, M. K., Restifo, N. P., Cohen, P. A., Marincola, F. M., Cheshire, L. B., and Llinehan, W. M. (1995). Defective major histocompatibility complex class I expression in a sarcomatoid renal cell carcinoma cell line. *J. Immunother. Emphasis Tumor Immunol.* **17**, 222.

62. Restifo, N. P., Marincola, F. M., Kawakami, Y., Taubengerger, J., Yannelli, J. R., and Rosenberg, S. A. (1996). Loss of functional beta 2-microglobulin in metastatic melanomas from five patients receiving immunotherapy. *J. Natl. Cancer Inst.* **88**, 100.

63. Cardi, G., Heaney, J. A., Schned, A. R., and Ernstoff, M. S. (1998). Expression of Fas(APO-1/CD95) in tumor-infiltrating and peripheral blood lymphocytes in patients with renal cell carcinoma. *Cancer Res.* **58**, 2078–2080.

64. Bennett, M. W., O'Connell, J., O'Sullivan, C., Brady, C., Roche, D., Collins, J. K., and Shanahan, F. (1998). The Fast counterattack in vivo: apoptotic depletion of tumor-infiltrating lymphocytes associated with Fas ligand expression by human esophageal carcinoma. *J. Immunol.* **160**, 5669–5675.

65. Jensen, M. C., Clarke, P., Tan, G., Wright, C., Chung-Chang, W., Clar, T. N., Zhang, F., Slovak, M. L., Wu, A. M., Forman, S. J., and Raubitschek, A. (2000). Human T lymphocyte genetic modification with naked DNA. *Mol. Ther.* **1**, 49–55.

66. Rosenberg, S. A., Aebersold, P., Cornetta, K., Kasid, A., Morgan, R. A., Moen, R., Karson, E. M., Lotze, M. T., Yang, J. C., and Topalian, S. L. (1990). Gene transfer into humans: immunotherapy of patients with advanced melanoma, using tumor-infiltrating lymphocytes modified by retroviral gene transduction. *N. Engl. J. Med.* **323**, 570–578.

67. Miller, D. G., Adam, M. A., and Miller, A. D. (1990). Gene transfer by retrovirus vectors occurs only in cells that are actively replicating at the time of infection. *Mole. Cell. Biol.* **10**, 4239–4242.

68. Ho, W. Y., Cooke, M. P., Goodnow, G. C., and Davis, M. M. (1994). Resting and anergic B cells are defective in CD28-dependent costimulation of naive CD4+ T cells. *J. Exp. Med.* **179**, 1539–1549.

69. Dubey, C., Croft, M., and Swain, S. L. (1996). Naive and effector CD4 T cells differ in their requirements for T cell receptor versus costimulatory signals. *J. Immunol.* **157**, 3280–3289.

70. Boise, L. H., Minn, A. J., Noel, P. J., June, C. H., Accavitti, M. A., Lindsten, T., and Thompson, C. B. (1995). CD28 costimulation can promote T cell survival by enhancing the expression of Bcl-xl. *Immunity* **3**, 87–98.

71. Sperling, A. I., Auger, J. A., Ehst, B. D., Rulifson, I. C., Thompson, C. B., and Bluestone, J. A. (1996). CD28/B7 interactions deliver a unique signal to naive T cells that regulates cell survival but not early proliferation. *J. Immunol.* **157**, 3909.

72. Lindsten, T., June, C. H., Ledbetter, J. A., Stella, G., and Thompson, C. B. (1989). Regulation of lymphokine messenger RNA stability by a surface-mediated T cell activation pathway. *Science* **244**, 339–343.

73. Fraser, J. D., Irving, B. A., Crabtree, G. R., and Weiss, A. (1991). Regulation of interleukin-2 gene enhancer activity by the T cell accessory molecule CD28. *Science* **251**, 313–316.

74. Li, Q., Furman, S. A., Bradford, C. R., and Chang, A. E. (1999). Expanded tumor-reactive CD4+ T-cell responses to human cancers induced by secondary anti-CD3/anti-CD28 activation. *Clin. Cancer Res.* **5**, 461–469.

75. Pollok, K. E., Noblitt, T. W., Schroeder, W. L., Kato, I., Emanuel, D., and Williams, D. A. (1998). High-efficiency gene transfer into normal and adenosine deaminase deficient T-lymphocytes is mediated by transduction on recombinant fibronectin fragments. *J. Virol.* **72**, 4882–4892.

76. Friedman, M. S., Yang, J., Fuller, J., Mule, J. J., and McDonagh, K. T. (1999). Highly efficient retroviral transduction of primary human T-cells by centrifugation with GALV supernatant on fibronectin fragment CH296 coated plates. *Blood* **94**, 410b.

77. Kimizuka, F., Taguchi, Y., Ohdate, Y., Kawase, Y., Shimojo, T., Hashino, K., Kato, I., Sekiguchi, K., and Titani, K. (1991). Production and characterization of functional domains of human fibronectin expressed in *Escheria coli*. *J. Biochem.* **110**, 284–291.

78. Hanenberg, H., Xiao, X. L., Dilloo, D., Hashino, K., Kato, I., and Williams, D. A. (1996). Colocalization of retrovirus and target cells on specific fibronectin fragments increases genetic transduction of mammalian cells. *Nature Med.* **2**, 876–882.

79. Mitsuyasu, R. T., Anton, P. A., Deeks, S. G., Scadden, D. T., Connick, E., Downs, M. T., Bakker, A., Roberts, M. R., June, C. H., Jalali, S., Lin, A. A., Pennathur-Das, R., and Hege, K. M. (2000). Prolonged survival and tissue trafficking following adoptive transfer of CD4zeta gene-modified autologous CD4+ and CD8+ T cells in human immunodeficiency virus-infected subjects. *Blood* **96**, 785–793.

80. Hanazono, Y., Brown, K. E., Handa, A., Metzger, M. E., Heim, D., Kurtzman, G. J., Donahue, R. E., and Dunbar, C. E. (1999). In vivo marking of rhesus monkey lymphocytes by adeno-associated viral vectors: direct comparison with retroviral vectors. *Blood* **94**, 2263–2270.

81. Costello, E., Munoz, M., Buetti, E., Meylan, P. R., Diggelmann, H., and Thali, M. (2000). Gene transfer into stimulated and unstimulated T lymphocytes by HIV-1 derived lentiviral vectors. *Gene Ther.* **7**, 596–604.

82. Chinnasamy, D., Chinnasamy, N., Enriquez, M. J., Otsu, M., Morgan, R. A., and Candotti, F. (2000). Lentiviral-mediated gene transfer in human lymphocytes: role of HIV-1 accessory proteins. *Blood* **96**, 1309–1316.

83. Griffith, K. D., Read, E. J., Carrasquillo, J. A., Carter, C. S., Yang, J. C., Fisher, B., Aebersold, P., Packard, B. S., Yu, M. Y., and Rosenberg,

S. A. (1989). *In vivo* distribution of adoptively transferred indium-111-labeled tumor infiltrating lymphocytes and peripheral blood lymphocytes in patients with metastatic melanoma. *J. Natl. Cancer Inst.* **81,** 1709–1717.

84. Nakamura, Y., Wakimoto, H., Abe, J., Kanegae, Y., Saito, I., Aoyagi, M., Hirakawa, K., and Hamada, H. (1994). Adoptive immunotherapy with murine tumor specific T lymphocytes engineered to secrete interleukin-2. *Cancer Res.* **54,** 5757–5760.

85. Treisman, J., Hwu, P., Minamoto, S., and Shafer, G. E. (1995). Interleukin-2 transduced lymphocytes grow in an autocrine fashion and remain responsive to antigen. *Blood* **85,** 139.

86. Hwu, P., Yannelli, J., Kriegler, M., Anderson, W. F., Perez, C., Chiang, Y., Schwarz, S. L., Cowherd, R., and Delgado, C. (1993). Functional and molecular characterization of tumor-infiltrating lymphocytes transduced with tumor necrosis factor-α. *J. Immunol.* **150,** 4104–4115.

87. Costa, G. L., Benson, J. M., and Seroogy, C. M. (2000). Targeting rare populations of murine antigen-specific T lymphocytes by retroviral transduction for potential application in gene therapy for autoimmune disease. *J. Immunol.* **164,** 3581–3590.

88. Cole, D. J., Weil, D. P., Shilyanski, J., Kawakami, Y., Rosenberg, S. A., and Nishimura, M. I. (1995). Characterization of the functional specificity of a cloned T cell receptor heterodimer recognizing the MART-1 melanoma antigen. *Cancer Res.* **55,** 748–752.

89. Clay, T. M., Custer, M. C., Sachs, J., Hwu, P., Rosenberg, S. A., and Nishimura, M. I. (1999). Efficient transfer of a tumor antigen-reactive TCR to human peripheral blood lymphocytes confers anti-tumor reactivity. *J. Immunol.* **163,** 507–513.

90. Friedman, M. S., Fuller, J., Yang, J., Mule, J. J., and McDonagh, K. (1999). pRET: a novel retroviral vector for highly efficient gene transfer and gene expression in human T cells. *Am. Soc. Gene Ther.* 149a.

91. Miller, A. D., Garcia, J. V., von Suhr, N., Lynch, C. M., Wilson, C., and Eiden, M. V. (1991). Construction and properties of retrovirus packaging cells based on gibbon ape leukemia virus. *J. Virol.* **65,** 2220–2224.

92. Friedman, M. S. (1999). Redirecting the immune response to human malignancy via retroviral mediated gene transfer of a tumor antigen specific T-cell receptor. *Blood* **94,** 362a.

93. Eshhar, Z., Waks, T., Gross, G., and Schindler, D. G. (1993). Specific activation and targeting of cytotoxic lymphocytes through chimeric single chains consisting of antibody binding domains and the γ or ζ subunits of the immunoglobulin and T cell receptors. *Proc. Natl. Acad. Sci. USA* **90,** 720–724.

94. Bird, R. E., Hardman, K. D., Jacobson, J. W., Johnson, S., Kaufman, B. M., Lee, S. M., Lee, T., Pope, S. H., Riordan, G. S., and Whitlow, M. (1988). Single-chain antigen-binding proteins. *Science* **242,** 423–426.

95. Raag, R., and Whitlow, M. (1995). Single chain Fvs. *FASEB J.* **9,** 73–80.

96. Orloff, D. G., Ra, C. S., Frank, S. J., Klausner, R. D., and Kinet, J. R. (1990) Family of disulphide-linked dimers containing the zeta and eta chains of the T-cell receptor and the gamma chain of the Fc receptor. *Nature* **347,** 189–191.

97. Letourneur, F., and Klausner, R. D. (1991). T-cell and basophil activation through the cytoplasmic tail of T-cell-receptor zeta family proteins. *Proc. Natl. Acad. Sci. USA* **88,** 8905–8909.

98. Ravetch, J. V., and Kinet, J. P. (1991). Rc receptors. *Annu. Rev. Immunol.* **9,** 457–492.

99. Romeo, C., and Seed, B. (1991). Cellular immunity to HIV activated by CD4 fused to T cell or Fc receptor polypeptides. *Cell* **64,** 1037.

100. Romeo, C., Amiot, M., and Seed, B. (1992). Sequence requirements for induction of cytolysis by the T cell antigen/Fc receptor zeta chain. *Cell* **68,** 889–897.

101. Hwu, P., Shafer, G. E., Treisman, J., Schindler, D. G., Gross, G., Cowherd, R., Rosenberg, S. A., and Eshhar, Z. (1993). Lysis of ovarian cancer cells by human lymphocytes redirected with a chimeric gene composed of an antibody variable region and the Fc receptor γ chain. *J. Exp. Med.* **178,** 361–366.

102. Hwu, P., Yang, T. C., Cowherd, R., Treisman, J., Shafer, G. E., Eshhar, Z., and Rosenberg, S. A. (1995). In vivo antitumor activity of T cells redirected with chimeric antibody/T cell receptor genes. *Cancer Res.* **55,** 3369–3373.

103. Hombach, A., Heuser, C., Sircar, R., Tillman, T., Diehl, V., Kruis, W., Poh, C., and Abken, H. (1997). T cell targeting of TAG72+ tumor cells by a chimeric receptor with antibody-like specificity for a carbohydrate epitopes. *Gastroenterology* **113,** 1163–1170.

104. Darcy, P. K., Kershaw, M. H., Trapani, J. A., and Smyth, M. J. (1998). Expression in cytotoxic T lymphocytes of a single chain anti-carcinoembryonic antigen antibody: redirected Fas ligand mediated lysis of colon carcinoma. *Eur. J. Immunol.* **28,** 1663–1672.

105. Stancovski, I., Schindler, D. G., and Waks, T. (1993). Targeting of T lymphocytes to Neu/HER2 expressing cells using chimeric single chain Fv receptors. *J. Immunol.* **151,** 6577–6582.

106. Moritz, D., Wels, W., Mattern, J., and Groner, B. (1994). Cytotoxic T lymphocytes with a grafted recognition specificity of DRBB2-expressing tumor cells. *Proc. Natl. Acad. Sci. USA* **91,** 4318–4322.

107. Moritz, D., and Groner, B. (1995). A spacer region between the single chain antibody and the CD3 zeta chain domain of chimeric T cell receptor components is required for efficient ligand binding and signalling activity. *Gene Ther.* **2,** 539–546.

108. Hekele, A., Dall, P., Moritz, D., Wels, W., Broner, B., Herrlich, P., and Ponta, H. (1996). Growth retardation of tumors by adoptive transfer of cytotoxic T lymphocytes reprogrammed by CD44V6-specific scFv:zeta-chimera. *Int. J. Cancer* **68,** 232–238.

109. Altenschmidt, U., Klundt, E., and Groner, B. (1997). Adoptive transfer of in vitro targeted, activated T lymphocytes results in total tumor regression. *J. Immunol.* **159,** 5509–5515.

110. Mehta-Damani, A., Markowicz, S., and Engleman, E. G. (1994). Generation of antigen-specific CD8+ CTLs from naive precursors. *J. Immunol.* **153,** 996.

111. Mehta-Damani, A., Markowicz, S., and Engleman, E. G. (1995). Generation of antigen-specific CD4+ T cell lines from naive precursors. *Eur. J. Immunol.* **25,** 1206–1211.

112. Romani, N., Gruner, S., Brang, D., Kampgen, E., Lenz, A., Trockenbacher, B., Konwalinka, G., Fritz, P. O., Steinman, R. M., and Schuler, G. (1994). Proliferating dendritic cell progenitors in human blood. *J. Exp. Med.* **180,** 83–93.

113. Zitvogel, M. T., Mayordomo, J. I., Tjandrawan, T., DeLeo, A. B., Clarke, M. R., Lotze, M. T., and Storkus, W. J. (1996). Therapy of murine tumors with tumor peptide-pulsed dendritic cells: dependence on T cells, B7 costimulation, and T helper cell 1-associated cytokines. *J. Exp. Med.* **183,** 87–97.

114. Coveney, E., Wheatley, G., and Lyerly, H. K. (1997). Active immunization using dendritic cells mixed with tumor cells inhibits the growth of primary breast cancer. *Surgery* **122,** 226–234.

115. Hsu, F. J., Benike, C., Fagnoni, F., Liles, T. M., Czerwinski, D., Taidi, B., Engleman, E. G., and Levy, R. (1996). Vaccination of patients with B-cell lymphoma using autologous antigen-pulsed dendritic cells. *Nature Med.* **2,** 52–58.

116. Henderson, R. A., Nimgaonkar, M. T., Watkins, S. C., Robbins, P. D., Ball, E. D., and Finn, O. J. (1996). Human dendritic cells genetically engineered to express high levels of the human epithelial tumor antigen mucin (MUC-1). *Cancer Res.* **56,** 3763–3770.

117. Reeves, M. E., Royal, R. E., Lam, J. S., Rosenberg, S. A., and Hwu, P. (1996). Retroviral transduction of human dendritic cells with a tumor-associated antigen gene. *Cancer Res.* **56,** 5672–5677.

118. Ribas, A., Butterfield, L. H., and McBride, W. H. (1997). Genetic immunization for the melanoma antigen MART-1/Melan-A using

recombinant adenovirus-transduced murine dendritic cells. *Cancer Res.* **57,** 2865–2869.

119. Wan, Y., Emtage, P., Zhu, Q., Foley, R., Pilon, A., Roberts, B., and Gauldi, J. (1999). Enhanced immune response to the melanoma antigen gp100 using recombinant adenovirus-transduced dendritic cells. *Cell. Immunol.* **198,** 131–138.

120. Wan, Y., Bramson, J., Pilon, A., Zhu, Q., and Gauldie, J. (2000). Genetically modified dendritic cells prime autoreactive T cells through a pathway independent of CD40L and interleukin 12: implications for cancer vaccines. *Cancer Res.* **60,** 3247–3253.

121. Gorczynski, R. M., Bransom, J., Cattral, M., Huang, X., Lei, J., Xiaorong, L., Min, W. P., Wan, Y., and Gauldie, J. (2000). Synergy in induction of increased renal allograft survival after portal vein infusion of dendritic cells transduced to express TGFbeta and IL-10, along with administration of CHO cells expressing the regulatory molecule OX-2. *Clin. Immunol.* **95,** 182–189.

15

Update on the Use of Genetically Modified Hematopoietic Stem Cells for Cancer Therapy

EDSEL U. KIM
Department of Otolaryngology
University of Michigan
Ann Arbor, Michigan 48109

LEE G. WILKE
Department of Surgery
University of Michigan
Ann Arbor, Michigan 48109

JAMES J. MULÉ
Department of Surgery
University of Michigan
Ann Arbor, Michigan 48109

I. INTRODUCTION

Hematopoietic stem cells (HSCs) are rare cells found within the bone marrow (BM), peripheral blood (PB), and cord blood which are self-renewing and capable of differentiating into progeny that multiply to become the erythroid, lymphoid, and myeloid components of the blood. HSCs have been characterized by their expression of the CD34 and Thy antigens and by their lack of mature lymphoid and myeloid lineage markers [1,2]. These primitive progenitor cells are attractive targets for gene therapy (defined as the introduction of new genetic material into somatic cells) because of their potential to self-renew and generate a large population of progeny cells with new genetic material. As evidenced by a number of recent reviews and consensus conferences, gene transfer to HSCs has rapidly become a focus of research into the mechanisms and/or therapeutic options for a variety of inherited, infectious, and neoplastic diseases [3–7].

How can this new technology be applied to the potential treatment of a neoplasm? Originally, the transfer of genetic material into HSCs was conceptualized as a means to correct an inherited disease with a monogenic defect, such as adenosine deaminase deficiency or Gaucher's disease [3]. However, due to the fact that a number of bloodborne and solid cancers are currently being treated with bone marrow transplants (BMTs) or peripheral blood stem cell transplants (PBSCTs), the use of genetically modified HSCs has entered the armamentarium of the clinician treating cancer. The potential use of gene-modified HSCs has branched out into several directions as it has been applied to cancer therapy. First, the introduction of gene-modified HSC into BMTs and/or PBSCTs is being developed as a means to investigate the source of long-term hematopoietic reconstitution in the patient receiving marrow-ablative chemotherapy. It is also conceptualized as a means to determine the source of relapse in the cancer patient undergoing BMT or PBSCT. Second, the introduction of HSCs modified with a drug resistance gene is being investigated as a means to select those cells that express the introduced drug resistance gene and to protect the reinfused marrow from destruction by repeated chemotherapy treatments. An additional strategy might be the use of this drug resistance selection to increase the frequency of gene-modified cells that also contain a therapeutic transgene with antitumor activity. Finally, HSCs have been proposed as targets for the introduction of chimeric receptor gene(s) with reactivity to tumor-associated antigens which would potentially offer the patient receiving a gene-modified HSC transplant a

Gene Therapy of Cancer, Second Edition

long-term antitumor immune response that is biased to recognize autologous tumor.

In this updated chapter, the advantages and disadvantages of HSCs as recipients of new genetic material will be discussed. This discussion will center on the use of viral vectors as vehicles of gene transfer into HSC, including the novel introduction of lentiviruses. Next, the chapter will focus on the preclinical studies that uncovered experimental techniques and the difficulties of genetically manipulating HSCs. The final portion of this chapter will focus on the status of the three applications of genetically manipulated HSCs to the therapy of cancer. While the initial promise of gene therapy of hematopoietic stem cells has waned due to the difficulties inherent in the transfer of the technology from the bench to the bedside, recent developments may lead to a renewed interest and spur further investigation into this theoretically attractive field.

II. HUMAN HEMATOPOIETIC STEM CELLS AS VEHICLES OF GENE TRANSFER

The ideal gene transfer vehicle for HSCs would have the following properties: (1) ability to integrate into chromosomes and be transmitted to progeny, (2) ability to integrate into quiescent cells, (3) lack of replication-competent viral particles, and (4) ability to integrate at a "safe" location in the host [4]. Reports of successful gene integration into HSCs using adeno-associated vectors and a molecular conjugate vector containing steel factor have appeared in the literature. In addition, the recent use of lentiviruses has shown some promise in that these HIV-based vectors are able to replicate in nondividing cells. However, retroviral constructs remain the primary vectors used to genetically modify or transduce HSC [8,9]. Retroviruses are, however, not the optimal gene transfer vehicles, and vector development research continues to search for a more efficient vector for HSC transduction.

Retroviruses are double-stranded RNA viruses of approximately 10,000 base pairs of DNA. The genome undergoes reverse transcription into double-stranded DNA, which then integrates randomly into the cycling host cell. For gene delivery, the structural proteins, or *gag, pol,* and *env* proteins, of the RNA virus are replaced with the gene of interest which can be up to 8 kb [10]. The packaging cell line provides these missing structural proteins *in trans* to enable the retrovirus to be infectious but replication incompetent in the host [6]. Retroviruses satisfy two of the desired requirements for delivery: chromosomal integration and potential for replication incompetence. The production of a replication-competent virus remains a major safety concern when discussing retroviruses. To date, there have been no known deleterious effects from the use of retrovirally modified cells in human clinical trials. Multiple safeguards are in place to prevent replication-competent retrovirus production, including deletions of the packaging signal and 3′-long terminal repeat (LTR) in the packaging

TABLE 1 Retroviral Gene Transfer to Human Hematopoietic Stem Cells: Advantages and Disadvantages

Advantages	Disadvantages
Genetic integration into stem cells enables transmission to multilineage progeny *in vivo*.	Retroviruses require cells in mitoses = a rare event for "true" stem cells.
Large-scale expansion of transduced stem cells is not required *ex vivo*.	Gene integration site unknown in host stem cell chromosome = risk of "insertional oncogenesis".
High transduction efficiency not necessary due to *in vivo* expansion, self-renewal.	Poor *in vitro* assay systems to determine transfer efficacy into stem cell.
Increasing number of hematologic and solid cancers treatable (curable) with hematopoietic stem cell transplants.	Need for myelosuppression with associated risk of infection.

plasmids [11]. In addition, human serum rapidly inactivates those recombinant retroviruses packaged in murine packaging cell lines [12]. While using this system has been safe in both primate and human trials, rhesus monkeys did develop T-cell lymphomas after receiving BMTs with CD34 cells exposed to a replication-competent retrovirus [13,14]. Stringent regulatory practices remain in effect to prevent the potential contamination of vector preparations with replication-competent retroviruses [15–17].

As delineated in Table 1, one of the major disadvantages of retroviruses as the gene delivery mechanism for HSCs is their requirement for actively dividing cells for insertion. Hematopoietic stem cells are primarily nondividing cells. The murine leukemia virus (MLV)-based vectors have been used the most extensively as gene transfer vehicles for HSCs, but their efficiency of transduction remains low due to the requirement for active mitosis in the host cell population [15,17]. In order to improve the efficiency of transduction into HSCs, most investigators use a variety of different cytokines, prior to and/or during exposure to the retrovirus, to induce these target cells to proliferate and become more susceptible to retroviral infection [18–27]. Whether this *ex vivo* manipulation of HSC-containing populations actually results in adequate cycling of HSC or induces their terminal differentiation and loss of repopulating capacity is not currently known and represents an area where continued research into the biology of hematopoiesis is necessary to optimize gene delivery to HSCs [28]. Theoretically, if only a small number of true HSCs are transduced with the transgene of interest, a large number of progeny should be apparent in the host once reconstitution occurs *in vivo*. This expansion property of the HSC also obviates the need for large-scale expansion of genetically modified cells *ex vivo,* a process currently required for genetic manipulation of committed lymphocytes [29].

A second potential disadvantage of the use of retroviruses in HSC is the risk of "insertional mutagenesis" [4,15]. The insertion site for retroviral gene transfer is not defined; therefore, there is the risk that retroviral transduction will cause activation of genes involved in growth control or inactivation of tumor suppressor genes [30]. This potential disadvantage has, again, not been borne out in human trials to date. However, as previously noted, T-cell lymphomas have appeared in rhesus monkeys that received HSCs exposed to a replication-competent murine-leukemia-based retrovirus prior to transplant [13,14].

A disadvantage of the use of HSCs as the host cells for retroviral genetic manipulation is the lack of *in vitro* assay systems to determine the transfer efficacy into true HSCs. The colony-forming unit (CFU) assays in methylcellulose and the long-term marrow culture (LTMC) assays do not definitively study the long-lived pluripotent HSCs [15,17]. Transplantation assays into immunocompromised mice exist and attempt to replicate the reconstitution properties of the true human HSC. They are, however, time consuming, requiring up to 9 months to achieve results, and may not directly predict the success or failure of transfer into a human host [17]. Similar to the need for further research in the areas of vector development and *ex vivo* HSC proliferation, more investigation into the exact microenvironment necessary to study pluripotent HSC maintenance and differentiation must be done to better ease the transition of this strategy into the clinical arena.

The search for alternative viral vectors has led to the use of lentiviruses. These HIV-based vectors are able to mediate efficient and stable transduction in post-mitotic cells as well as in quiescent human HSCs [31]. Recent work by Miyoshi *et al.* showed an 8- to 12-fold higher transduction rate in the number of green fluorescence protein (GFP)+-CFC (colony-forming cell) colonies transduced with the lentivirus versus those transduced by the MLV vector [32]. In addition, when using the NOD/SCID mouse system to evaluated human HSCs *in vivo*, all mice had human cells when engrafted with the lentiviral transplanted GFP+ human CD34+ cells. In contrast, no human cells were detectable in those mice tranplanted with the MLV-vector-transduced CD34+ cells. Another advantage seen with this protocol is that infection of the HSCs is done without the need for cytokine prestimulation as with the retroviral systems. As with any viral vector work, the issue of biosafety is always of importance. The development of third generation lentiviral packaging constructs, which theoretically would not undergo viral recombination by removal of all the accessory proteins not essential for transduction (except for Gag, Pol, and Rev), has moved the use of this system closer to clinical trials [33]. In addition, safety is further enhanced with the self-inactivating viruses in which the HIV-1 LTR is eliminated [33]. While this vector is very encouraging in the murine setting, further investigation is necessary to examine the safety of using HIV vectors for clinical trials in humans.

III. PRECLINICAL STUDIES OF GENE TRANSFER INTO HEMATOPOIETIC STEM CELLS

Successful retroviral transduction of murine HSC has been reported using, as examples, marker genes, the human adenosine deaminase gene (huADA), and a human multidrug resistance (MDR) gene [34–38]. The murine studies uncovered several experimental techniques that improved HSC transduction efficiency and permitted long-term gene expression in progeny cell lineages. These methodologies included pretreatment of the mice with 5-flurouracil (5-FU) prior to bone marrow retrieval in order to increase the number of stem cells in cell cycle [36]. These findings were utilized in subsequent human HSC work. It is now common practice to give patients chemotherapy and/or cytokines prior to obtaining their peripheral blood for stem cell transduction in the hope of improving retroviral gene insertion by increasing the absolute numbers of HSCs and by altering their cell cycle status in the mobilized peripheral blood (MPB) [39].

The findings that the murine HSCs were more effectively transduced on a stroma of retroviral producer cells have been expanded to human HSC studies [34–36]. Multiple groups continue to use cocultivation to maximize human HSC gene transduction [18,20,22–24]. The risk of retroviral producer cell contamination in a HSC sample being used for transplantation and cancer therapy has, however, led others to identify alternative stromal transduction protocols. Moritz *et al.,* using a neomycin (*neo*) marker gene, transduced human cord blood using four different protocols [20]. The most efficient transduction protocol, measured by CFU growth in a neomycin analog (G418), utilized coculture, while the least efficient used supernatant exposure (45 versus 11%). Those protocols that resulted in intermediate transduction efficiencies employed autologous bone marrow stroma or a murine-derived stromal cell line that expressed human steel factor (SF) (32 and 19%) (Table 2).

More recent studies have shown improved transduction of human HSCs on a fragment of the fibronectin molecule [40,41]. Fibronectin contains both target cell and retroviral adhesion moleculess, which can increase the apparent titer of retroviral particles to the target cells. These improvements, however, are in comparison to transduction with a retroviral vector on a bovine serum albumin (BSA)-coated plate. In CFU assays, the number of G418-resistant colonies grown from BM transduced with a *neo* marker gene increased from approximately 2% on BSA to 18% on a fibronectin-coated plate in two studies [40,41]. The transduction efficiency on fibronectin does not necessarily differ significantly from results obtained by other groups who utilize supernatant exposure alone, although direct comparative studies have not been performed (Table 2) [19,21]. The use of a fibronectin molecule, however, does avoid the risk of producer cell contamination and the need for BM aspiration if autologous marrow is to be used as the stomal monolayer for transduction protocols.

TABLE 2　Preclinical Studies of Gene Transfer into Human Progenitor Cells

Progenitor cell source	CD34 isolation	Retroviral vector	Infection protocol	Transduction efficiency	Murine experiments	Ref.
BM, normal donors	No	N2	1. 24-hour preincubation of BM in 15% 5637 human bladder carcinoma conditioned media, IL-6, and LIF 2. 24-hour cocultivation of BM with producer cells	CFU assay: 5–20% growth in G418; PCR for *neo*: approx. 20%+	5 bg/nu/xid mice (BMT with transduced human BM): 0.1% BM and spleen cells neo+ by PCR at 4 months	[16]
BM and MPB from normal and multiple myeloma patients	Yes	LNL6 G1Na. 4	72-hour exposure of progenitor cells to vector supernatant in media with IL-3, IL-6, and SCF	CFU assay: 17% growth in G418 (MPB and BM); PCR for *neo*: 20%+ (MPB), 27%+ (BM)	NA	[17]
MPB breast cancer patients	Yes	LNL6	1. 24-hour prestimulation in media with IFG-1, IL-3, IL-6, SCF, GM-CSF, and erythropoietin 2. 12-hour vector supernatant exposure	CFU assay: 3–30% growth in G418; PCR for *neo*: 67–100% +; RT-PCR for *neo*: 8%+	NA	[19]
BM, normal donors; MPB, breast cancer patients	Yes	MSCV-HSA. NEO	48-hour prestimulation of BM or MPB in media with IL-3, IL-6, and SF 48-hour cocultivation of producer cells with BM or MPB	CFU assay: 11–12% growth in G418 (unsorted); 70–100% growth in G418 (HSA sorted)	NA	[22]
BM, normal donors	Yes	LGsFH	4 "x" vector supernatant exposure on autologous BM stroma in media with IL-3, IL-6, and SCF	FACS: 10% HSA+ PCR for GC gene: 38–50%+ (unsorted), 100%+ (HSA sorted)	NA	[23]
BM and MPB, normal donors	Yes	LN	72-hour vector supernatant exposure on allogenic BM stroma	CFU assay: 35% growth in G418	bnx mice transplanted with transduced CD34+ cells in conjunction with BM stroma engineered to produce human IL-3: 3/24 mice found to have identical proviral fragment sites in T and myeloid cells via PCR	[24]
MPB, normal donors	Yes	L(mCD4tr)SN	3- to 6-day vector supernatant exposure of CD34+ cells on fibronectin-coated flasks in media with IL-1B, IL-3, IL-6, and SCF	FACS: 7% + for mCD4; CFU assay: 8–40% growth in G418 after FACS sort for mCD4	NA	[25]
MPB, cancer patients	Yes	HaMDR/A	24- to 48-hour preincubation with IL-3, IL-6, and SCF 24-hour transduction on fibronection-coated plates with two viral supernatant changes	PCR for MDR: 64.7%+ BFU-E, 77.3%+ CFU-GM; CFU assay: 19–26% BFU-E Taxol resistant, 20–48% CFU-GM Taxol resistant	NA	[38]
Cord blood	No	TK NEO	48-hour prestimulation of cord blood cells in media with IL-6 and SCF Four infection protocols Coculture Vector supernatant exposure on Sl[4]-h220[a] Vector supernatant exposure on allogeneic BM Vector supernatant exposure	CFU assay: 45% growth in G418 (coculture), 32% growth in G418 (allo BM), 19% growth in G418 (Sl[4]-h220[a])	NA	[18]

TABLE 2 (*Continued*)

Progenitor cell source	CD34 isolation	Retroviral vector	Infection protocol	Transduction efficiency	Murine experiments	Ref.
Cord blood	Yes	MFG-mCD2	4-day cocultivation of producer cells with cord blood CD34+ cells in media with IL-3 and SCF	FACS: 40% + for murine CD2	SCID-hu thymic constructs 4–9% thymocytes mCD2+ via FACS in 4/9 SCID chimera (5–10 weeks)	[21]
Fetal liver cells	Yes	LNL6	72-hour cocultivation of producer cells with fetal liver cells in media with IL-3, IL-6, and SCF	PCR for *neo:* 10%+	SCID-hu thymic liver constructs: 3% thymocytes+ for *neo* by PCR (6 weeks); 2% CD4+ and CD8+ cells+ for *neo* by PCR (4 weeks)	[20]
Cord blood/BM	Yes	TK NEO	One-day exposure to retrovirus while adherent to CH-296	Progenitor assay for *neo:* 50–75%+; *in situ* gel analysis of ADA protein: 5/6+ in those cells transduced on CH-296	Lethally irradiated C3H/HeJ mice transplanted with syngeneic BM on CH-296 and retrovirus encoding hADA	[42]
Cord blood	Yes	GFP	5-day cocultivation with SCF, IL-3, and IL-6 Transduced with either MLV or HIV vector	CFU assay for GFP: 8- to 12-fold higher for HIV vector versus MLV	Sublethally irradiated NOD/SCID mice: all HIV transduced CD34+ cells were present in engrafted mice up to 22 weeks; none of the MLV-transduced cells was present at 8 weeks	[32]

Note: BFU-E, burst forming unit–erythroid; BM, bone marrow; BMT, bone marrow transplant; CFU, colony-forming unit; CFU-GM, colony-forming unit granulocyte–macrophage; HIV, human immunodeficiency virus; HSA, heat-stable antigen; LIF, leukemia inhibitory factor; MLV, murine leukemia virus; MPB; mobilized peripheral blood; NA, not applicable; *neo,* neophophotransferase; PCR, polymerase chain reaction; RT-PCR, reverse transcription–polymerase chain reaction; SCF, stem cell factor; SCID, severe combined immunodeficient (mice).

[a] Sl4-h220 is a murine-derived, genetically modified stromal cell line that expresses membrane-bound steel factor (SF).

Williams *et al.* have expanded the use of CH-296, a chimeric fibronectin molecule that contains both the stem cell and retroviral adhesion domains [42]. This group of investigators has shown that colocalization of the retrovirus and target cell on the same chimeric molecule provides for the highest transduction efficiencies (50–75% transduction after a 24-hour infection period versus 5–40% with longer infection periods with or without producer cells in other protocols) [43,44]. Of importance, this protocol obviated the need for cocultivation and extended *in vitro* exposure to growth factors and/or stromal cells.

Another technical contribution from the murine studies was the introduction of cytokines into the culture media during exposure of the HSC to the retrovirus and during a prestimulation phase prior to transduction. As previously noted, efficient retroviral transduction requires the recipient cells to be in active cell cycle. The most commonly used cytokines for human HSC transduction include IL-3, IL-6, and stem cell factor (SCF). Though protocols vary considerably, as

seen in Table 2, most studies utilize cytokines prior to and/or during retroviral vector exposure to attain transduction efficiencies in the 10–30% range as determined by CFU assay. One recent report indicated, however, that cytokine stimulation after transduction decreased the efficiency of gene expression, implying that HSC growth and differentiation could "turn off" the introduced gene of interest [45]. Though this work focused on the potential effects of cytokines after transduction, it is possible that the cytokine protocols currently in use before and during transduction may be affecting the HSC transduction efficiencies. As greater insight into the biology of HSC and its microenvironment is obtained, more efficient, timely transduction protocols for HSC could be designed.

An important finding from the murine studies was the fact that progeny development was not affected by genetic manipulation of HSC [34,35]. Several human studies have used genes encoding for a membrane-bound protein to identify those HSC which have been transduced and, in some protocols, have sorted by flow cytometry for those marked cells to obtain a more pure population of gene-modified HSC. Examples of these surface-expressed marker genes include the heat stable-antigen (HSA), the murine CD4 antigen (mCD4), the murine CD2 antigen (mCD2), and the truncated nerve growth factor receptor (NGFR) [23–25,27,46]. These studies utilized multiple sources of human HSCs— bone marrow, mobilized peripheral blood, and cord blood— indicating the broader feasibility of transduction. An important advantage of these protocols is the ability to perform fluorescence-activated cell sorting (FACS) on the transduced population and obtain more rapid assessment of the HSC transduction efficiency than growth in selection media. Champseix *et al.* transduced CD34$^+$ cells from cord blood via a cocultivation protocol with a retroviral vector containing the mCD2 cDNA [23]. The transduced cells were then injected into human fetal thymic pieces which were then implanted into severe combined immunodeficient mice (SCID). Using this modified *in vivo* system, these investigators obtained 4–9% of the resulting progeny thymocytes with the mCD2 marker. These results support the hypothesis that development of progeny cells from the transduced human HSCs is not altered. Another study described in Table 2 expands this approach to illustrate that transduction of a pluripotent HSC does not alter development of distinct progeny lineages. Nolta *et al.* have described infection of CD34$^+$ cells from BM and MPB with a neomycin-resistant gene marker vector followed by injection into beige/nude mice [26]. Three of 24 of the transplanted mice were found to have identical proviral fragment sites in their progeny T and myeloid cells, confirming that a pluripotent HSC had been transduced, and its ability to differentiate into separate lineages was left intact.

In addition to marker genes, it has been shown that a functional gene, human MDR, can be inserted into murine HSCs and enrichment for transduced progeny bone-marrow-derived lineage cells observed following Taxol^R therapy [35]. These results have been duplicated in human *in vitro* studies [47]. Using the CFU assay, 20–48% of progeny granulocyte–macrophage colonies from transduced MPB CD34$^+$ cells were found to be relatively Taxol^R resistant.

Despite the technical advancements described above and the variety of transgenes capable of being introduced into human HSCs, those studies that have examined the *in vivo* development of the infected HSCs have shown a markedly low transduction efficiency that translates into a small number of progeny cells with the transgene. As shown in Table 2, those investigators who introduced the transduced HSCs into immunocompromised mice, with or without a human fetal hematopoietic microenvironment, found only 0.1–9% of the BM, spleen, or thymic cells to contain the transgene of interest [18,22,23]. Similar results have been obtained with the large animal studies. Table 3 describes several of the primate studies performed to date with retrovirally transduced HSC. Similarly to the *in vitro* preclinical studies, the infection

protocols for the primate studies have been varied, including coculture on producer cells, engineered stroma, simple exposure to vector supernatant, or exposure on long-term marrow culture, with or without cytokine support [48–52]. The results from these studies have, however, been similar, in that the long-term appearance of transduced progeny was minimal. Bienzle *et al.* reported the largest long-term detection of a transgene with 5% of marrow cells from two dogs maintaining G418 resistance in CFU assay after transplantation with BM exposed to the N2 retroviral vector 2 years earlier [50].

The large animal models, in spite of the low numbers of resultant progeny containing the transgene of interest, provided several important results that led many investigators to begin human HSC gene therapy. First, primate HSCs could be transduced with a variety of retroviral genes including the standard neo marker transgene and potentially functional genes, the murine ADA and human glucocerebrosidase gene. Second, the transduction protocols, if a nonreplication-competent retrovirus was used, did not lead to any long-term side effects or significant mortality. Finally, HSCs could be transduced and progeny cells of different lineages could develop, each possessing the transgene of interest, thus indicating that a primitive, pluripotent HSC had incorporated the retroviral vector into its DNA [49,51].

IV. APPLICATIONS OF GENETICALLY MANIPULATED HEMATOPOIETIC STEM CELLS TO THE THERAPY OF HUMAN CANCER

A. Introduction

The technology for gene therapy of stem cells has existed for nearly 15 years, but only recently has its great potential been realized. While the clinical phase I trials have shown the safety of the treatment, the long-term results have been disappointing, as gene transfer into mature blood cells is very low and transient (0.01–1% of marked cells for weeks to month). However, Cavazzana-Calvo *et al.* have recently shown the first full correction of a disease phenotype through gene therapy of HSCs [53]. While this successful attempt was achieved for a single gene defect (severe combined immunodeficiency-X1), a couple of points can be gleaned from their study. The first is the use of the CH-296 fibronectin molecule to enhance gene transfer efficiency in HSCs, as previously discussed. This is the first clinical trial reported to use this molecule with positive results. Another reason for the success of this type of therapy can be attributed to the selective advantage conferred to those cells that are transduced with the gene of interest. Cancer cells typically have a selection advantage over normal cells, thus allowing for growth and metastasis. In order to have gene-modified HSCs develop long-term hematopoiesis, the transgene of interest should be able to offer a selection

TABLE 3   Primate Studies Involving Progenitor Cell Gene Transduction

Progenitor cell source	CD34 isolation	Retroviral vector	Infection protocol	*In vitro* transduction efficiency	*In vivo* outcome	Ref.
Canine BM	No	N2	Long-term marrow culture (LTMC) established from BM and exposed to vector supernatant 3x over 21 days	PCR: 70% + for *neo* on day 21 of LTMC	4 dogs underwent marrow ablation prior to infusion of transduced BM (1 death occurred); 3 dogs had no marrow ablation prior to transduced marrow infusion. 0.1–10% BM + for *neo* via PCR at 3 months; 0.1–1% BM + for *neo* via PCR at 10 to 21 months; 0.1–1% T cells + for *neo* via PCR at 12–19 months	[39]
Canine BM	No	N2	LTMC established from BM and exposed to vector supernatant 3x over 21 days	CFU assay: 44% growth in G418	18 dogs were transplanted with transduced marrow without marrow ablation. Maximum G418-resistant marrow cells in CFU assay was 10–30% at 3 months; 2 dogs maintained 5% G418-resistant marrow cells in CFU assay at 2 years	[41]
Canine MPB	No	LN	24-hour coculture of MPB on producer cells followed by a 10-day incubation of the MPB in LTMC with daily vector supernatant changes	NA	3 dogs received transduced MPB and untransduced BM after total body irradiation. Week 6: 1 dog with PCR + *neo* in peripheral granulocytes/lymphocytes; Week 20: 1 dog 2% CFU-GM G418 resistant	[34]
Rhesus monkey BM	Yes	PGK-mu ADA	96-hour cocultivation of CD34$^+$ cells on Sl^4-h220 stroma with IL-6 and SCF and vector supernatant	NA	3 monkeys were transplanted after marrow ablation. 2% peripheral blood cells (T cells and granulocytes) + for muADA via PCR at 1 year	[40]
Rhesus monkey BM	Yes, with Thy1+ cells	LG4	7-day exposure of CD34$^+$/Thy1+ cells to vector supernatant in media with IL-6 and SCF	NA	4 monkeys were transplanted after marrow ablation (1 death occurred). 1–13% of B cells were PCR+ for the GC gene at days 48–117; 1.2% of CD2$^+$ T cells were GC+ via PCR at >300 days; No long-term gene transfer occurred in the granulocyte population	[42]
Rhesus monkey BM	No	N2	3-day coculture of BM on producer cells followed by 3-day coculture on fresh producer cells	NA	3 monkeys were transplanted with autologous BM which had been exposed to high titer producer cells. PCR+ for *neo* in BM on days 20–99 (approximately 1% of total cells)	[43]

Note: ADA, adenosine deaminase; GC, glucocerebrosidase gene; PGKpr, human phosphoglycerate kinase promoter.

advantage over the cancer cells so that it does not become quiescent or silenced over time.

B. Gene Marking of Human Hematopoietic Stem Cells

The first use of retrovirally transduced HSC was not as a therapy for human cancer but rather as a tool for understanding the biology of BMTs and PBSCTs. BMTs and more recently PBSCTs have been used to restore hematopoiesis in patients who have received ablative chemotherapy for cancer, including a variety of bloodborne leukemias as well as solid tumors such as breast cancer and neuroblastoma [39,43,54–57]. Table 4 delineates those completed and ongoing clinical trials that have utilized a retrovirally inserted transgene to mark the HSCs and study the biology of BMTs and PBSCTs in cancer patients. The first question which investigators sought to answer was whether the reinfused autologous bone marrow and/or the PBSCs provided the source of long-term hematopoietic cells or whether the patient retained

TABLE 4   Clinical Trials Involving Gene Marking of Human Hematopoietic Stem Cells

Patient population	Completed vs. ongoing	Protocol	CD34 selection	Results vs. desired outcomes	Institution	Ref.
Pediatric AML and Pediatric Neuroblastoma	Completed	One third of autologous BM was exposed to either LNL6 or G1N retroviral vectors for 6 hours without cytokine supplementation and reinfused after ablative therapy with two thirds unmanipulated BM.	No	15 of 18 evaluable patients had 0–29% G418-resistant colonies in CFU assay at 1 month. 5 of 5 evaluable patients had 0–15% G418-resistant colonies in CFU assay at 1 year. 1 patient had T and B cells + by PCR for *neo* at 18 months. 4 AML patients relapsed; 2 had leukemic blasts which were G418 resistant in CFU assay 4 neuroblastoma patients relapsed; all 4 had an estimated. 05–1% neuroblasts + for *neo* via limiting cycle PCR.	St Jude's Children's Hospital; Memphis, TN	[45,46,49,52]
Adult multiple myeloma and breast cancer	Completed	One third of BM and MPB was exposed to LNL6 or G1Na. 40 vector supernatant over a 72-hour period in media with IL-3, SCF, and IL-6 (breast cancer patients only). Transduced and untransduced BM and MPB were reinfused into each patient. The BM and MPB for each patient were exposed to distinct vectors for differentiation of the progeny posttransplant.	Yes	Both the MPB and BM were shown to contribute to patient engraftment based on differential recognition of the two *neo* vector sequences. 3 (2 multiple myeloma and 1 breast cancer) of 9 evaluable patients had PB cells positive for the *neo* gene via PCR at 18 months posttransplant.	National Institutes of Health; Bethesda, MD	[50]
Adult CML	Completed	MPB was collected immediately following chemotherapy and 30% of cells were exposed to LNL6 retroviral vector supernatant for 6 hours. Transduced and untransduced MPB were reinfused following ablative therapy.	Yes	2 patients enrolled in study: Patient 1 had return of blast crisis by day 159 and the *neo* gene was detected in leukemic cells by PCR. Patient 2 also relapsed and had Ph+ cells with the *neo* gene by PCR. Patient 1 had normal and leukemic cells + for *neo* via RT-PCR >280 days posttransplant.	M.D. Anderson Cancer Center; Houston, TX	[47]

TABLE 4 (*Continued*)

Patient population	Completed vs. ongoing	Protocol	CD34 selection	Results vs. desired outcomes	Institution	Ref.
Adult AML or ALL	Completed	5–19% of BM obtained during second clinical remission (AML) or first clinical remission (ALL) was exposed to the G1N retroviral vector supernatant for 4 hours, and reinfused with untransduced marrow following ablative therapy.	No	Patient had BM and PB + for *neo* by PCR at 1 year. Neither of the 2 relapsed patients (AML) had *neo* gene marked leukemic blasts.	Indiana University School of Medicine; Indianapolis, IN	[48]
Pediatric AML in first clinical remission	Ongoing	One third of BM exposed to LNL6, a second third of BM exposed to G1N for 6 hours without cytokine support. Each set of BM would then be exposed to a separate *ex vivo* purging regimen.	No	With relapse, assess the presence of a *neo*-specific marker in the leukemic blasts to determine the efficacy of the two purging regimens.	St Jude's Children's Hospital; Memphis, TN	[49,55]
Breast cancer or malignant lymphoma	Ongoing	A portion of MPB to be transduced via a 5-day retroviral vector (LN) supernatant exposure in media with IL-1, IL-3, IL-6, and SCF; transduced and untransduced MPB to be reinfused after ablative therapy.	Yes	Study is designed to determine the ability of MPB to contribute to long-term hematopoiesis.	Fred Hutchinson Cancer Center; Seattle, WA	[34]
Metastatic breast cancer or lymphoma	Ongoing	One third of harvested BM and one half of any 2 MPB harvests will be enriched for CD34 cells, preincubated for 42 hours in media with IL-3 and IL-6, and then incubated for 6 hours in LNL6 or G1Na retroviral vector supernatant.	Yes	Differential gene marking of BM and MPB is used to detemine the relative contributions of either to hematopoeitic reconstitution.	USC Comprehensive Cancer Center; Los Angeles, CA	[50]
Multiple myeloma	Ongoing	10–30% of BM will be placed in LTMC and exposed to GTk1. SvNa. 7 (LN derivative with herpes simplex thymidine kinase gene) retroviral vector supernatant 3 times in 21 days.	No	Study is designed to determine the contribution to relapse of the gene-marked autograft.	University of Toronto; Toronto, Canada	[53]
Patients receiving ABMT for leukemias or solid tumors	Ongoing	Two thirds of BM will undergo CD34$^+$ cell selection. One half of the selected population will be exposed to LNL6 and the other half to G1Na for 6 hours. The transduced cells will then be randomized to either cytokine exposure (IL-3, IL-6, and SCF for 5 to 7 days) or no cytokine exposure.	Yes	Study is designed to investigate the *ex vivo* exposure of CD34$^+$ BM cells to growth factors and the time to hematopoietic engraftment in patients receiving ablative therapy.	St Jude's Children's Hospital; Memphis, TN	[51]
Relapsed follicular non-Hodgkin's lymphoma	Ongoing	A portion of the MPB and BM was exposed to either LNL6 or G1Na retroviral vector supernatant without exposure to cytokines or stroma; both transduced and nontransduced MPB and BM were reinfused into patients after ablative therapy.	Yes	2 patients are currently enrolled in the study: At day 45 both patients have *neo*+ BM cells by PCR (1.5 and 4.3%). The goal of the study is to determine the contribution of contaminated BM or MPB to relapse.	M.D. Anderson Cancer Center, Houston, TX	[54]
Adult recurrent germ cell tumors	Ongoing	CD34$^+$ cells are maintained *ex vivo* with SCF and IL-6 then cultured with MDR-1 vector on plates coated with CH-296 for 4 hours on days 3 and 4.	Yes	Gene transfer seen in: 25% at oneyear in BM. 5.6% at 1 month and 0.5% at 9 months in circulating mature cells.	Indiana University School of Medicine; Indianapolis, IN	[71]

Note: ABMT, autologous bone marrow transplant; ALL, acute lymphoblastic leukemia; AML, acute monocytic leukemia; CML, chronic monocytic leukemia; SCF, stem cell factor; MDR, multiple drug resistance.

a small population of functioning HSCs that could reconstitute their hematopoeitic system despite the myeloablative therapy. Of the completed studies, it can be concluded that human HSCs can be "marked" with a retroviral vector containing a neomycin-resistant gene [43,54–57]. Brenner et al. revealed that five of five evaluable pediatric patients with acute monocytic leukemia (AML) or neuroblastoma had from 0 to approximately 15% of cells in their BM that were G418 resistant in CFU assay at 1 year posttransplant. These investigators thus concluded that harvested autologous BM does contribute to hematopoietic recovery after ablative chemotherapy [54]. Cornetta et al. also demonstrated that in an adult leukemic patient, treated with a BMT in which a portion of the HSC had been exposed to a retroviral vector with the neomycin-resistant gene, both BM and PB cells contained the transgene 1 year posttransplant [56]. Deisseroth et al. found that one of their adult chronic myelogenous leukemia (CML) patients, who had received a PBSCT, had cells in the peripheral blood that were positive via reverse transcription–polymerase chain reaction (RT-PCR) for the neomycin gene, 280 days posttransplant. This finding led to the conclusion that reinfused, retrovirally exposed, mobilized peripheral blood, as well as BM, could contribute to the patient's hematopoietic recovery [57]. Dunbar et al. reported the long-term presence of a marker gene in the circulating granulocytes and T and B cells (0.01–1% of cells) of two patients with myeloma and one patient with breast cancer 18 months after a combined transplant with marked bone marrow and mobilized peripheral blood [4,43]. By using two different vectors with the neomycin gene, the investigators were able to discriminate the source of the engrafted cells after the transplant. Using a transduction protocol similar to that of Brenner et al., in which the HSCs were exposed to vector supernatant for 6 hours without cytokine support, Dunbar and collegues were unable to demonstrate gene marking in adult patients with breast cancer or myeloma who received a BMT or PBSCT [58]. Further study into the conditions necessary for optimal HSC gene transduction for individual cancers at different stages will need to be performed to resolve the discrepancies in results from different laboratories. Despite the differences in outcomes, it has been possible to use the technology of gene marking to address a clinically relevant question about the restoration of human hematopoiesis [59]. As shown in Table 4, several ongoing trials seek to further address the question of the source of hematopoeitic recovery after myeloablation [39,60,61].

A second question of interest to those investigators focusing on the biology of BMT and PBSCT in the cancer setting is that of the origin of relapse after transplantation for leukemia. Brenner et al. and Deisseroth et al. both discovered that, in some of their patients who relapsed from pediatric AML, pediatric neuroblastoma, and adult CML, the leukemic blasts contained the neomycin gene. The authors thus concluded that the reinfused autologous BM or MPB was a contributor to the cancer recurrence [55,57,59,61]. Cornetta et al. reported the relapse of two of four patients with AML who were treated with BMT, in which a portion of the graft was exposed to a retroviral vector with the neomycin resistant gene. However, neither patient had evidence of vector-marked leukemic blasts [56]. Clinical studies are ongoing to further clarify this question concerning the source of relapse in patients with varied forms of cancer at different clinical stages, including multiple myeloma and non-Hodgkin's lymphoma [62,63].

A third area of interest is the role that "purging" of the BMTs and/or PBSCTs with cytokines or chemotherapeutic agents prior to reinfusion into the patient will play in preventing disease recurrence. Brenner et al. have initiated studies using two distinct retroviral vectors with the neomycin-resistant gene in pediatric AML patients to address this question [64]. As improvements in vector design and HSC transduction occur, more clinical studies can be undertaken to further elucidate the biology of BMTs and PBSCTs and those ex vivo and in vivo therapies that will benefit the patient with cancer.

C. Drug Resistance Genes in Hematopoietic Stem Cells

In addition to its role in the study of BMTs and PBSCTs, gene therapy with HSCs has been introduced as a means to potentially improve existing transplant protocols. The dose of chemotherapeutic agents used to treat malignancies is limited to some extent by myelosuppression. If a patient's bone marrow were tolerant to the antineoplastic drugs, increasing doses of the agents could be used to improve disease eradication. The multiple drug resistance gene-1 (MDR1) encodes a 170-kDa P-glycoprotein which functions as an adenosine-triphosphate-dependant efflux pump for lipophilic compounds [65]. Those chemotherapeutic agents that are "removed" by cells with the MDR1 pump include the anthracyclines, the vinca alkaloids, the epipodophyllotoxins, actinomycin D, and Taxol^R [65]. As previously noted, murine HSCs have been transduced with the human MDR1 gene via a retroviral vector, and selection with Taxol^R has been successful [35]. The ability to select for MDR1-containing cells in humans would conceivably permit dose escalation of the antineoplastic agents as well as decreased intervals between chemotherapy administration due to improved hematopoietic recovery [65]. Several human clinical trials, with primarily breast and/or ovarian cancer patients, are currently ongoing to answer the question of whether human HSCs from BM and/or MPB can be transduced without toxicity to the HSCs and in adequate quantities to permit growth selection with Taxol^R [66–69]. Preliminary results from one group indicated that the CD34$^+$ cells from the BM and MPB of ovarian and breast cancer patients, respectively, could be transduced with the MDR1 gene by exposing the cells to the retroviral supernatant on a stromal monolayer with cytokine supplementation in the culture media. The transgene was identified in the BM of five of eight evaluable patients using a solution DNA PCR assay [70].

A more recent trial in patients with recurrent germ cell tumors has shown the highest level of transgene expression using the fibronectin fragment CH-296 in the transduction of HSCs with a retroviral vector containing the MDR1 gene [71]. Approximately 25% of the clonogenic progenitors were transduced in the bone marrow at 1 year with approximately 0.9% at 9 months in the peripheral blood. While the level of transduction was high, expression of the full-length MDR1 cDNA ranged from 10%–50% in the first month. Much of this loss could be attributed to cryptic splice sites that were found in the MDR1 gene, resulting in a nonfunctional product. While the levels of transduction from this trial are the highest reported to date, the subtherapeutic levels of functional expression show that more progress is needed before this technology can enter the clinical arena.

As results from the ongoing trials with MDR1-transduced HSCs become available and indicate the feasibility of BM selection, a second use for the drug-resistance genes may arise. Retroviral constructs containing the MDR1 gene and a second therapeutic gene could be introduced into the HSCs and those cells with the transgene selected for using Taxol^R [65]. This strategy would then increase the frequency of gene-modified cells with a potentially therapeutic antitumor gene (see later discussion). Without the selection process, the antitumor gene might only be present in a small number of HSCs due to low transduction efficiency with current retroviral vectors. With selection, however, the number of cells with the transgene DNA would increase, thereby improving the therapeutic potential after the BMT and/or PBSCT.

D. Chimeric Receptor Genes in Hematopoietic Stem Cells

"Purging" autologous marrow or peripheral blood prior to reinfusion after myeloablative therapy represents one means to address the problem of disease recurrence. Another means of approaching this problem in the cancer patient might be to provide the patient receiving a gene-modified HSC transplant with an immune system that is biased to recognize autologous tumor. Eshhar et al. developed a chimeric single-chain receptor consisting of the Fv domain (scFv) of an antibody linked with the γ or ζ chains, the signal-transducing subunits of the immunoglobulin receptor and the T-cell receptor (TCR) [72]. The scFv domains, which are a combination of the heavy and light variable regions (V_H and V_L) of the antibody, appear to possess the specificity and affinity of the intact Fab′ fragment [72]. Originally, a chimeric gene with scFv from an antitrinitophenyl (TNP) antibody was found to be expressed as a functional surface receptor in a murine cytolytic T-cell line and to specifically secrete IL-2 upon exposure to the TNP antigen [72]. T-cell activation via this

receptor is not major histocompatibility (MHC) restricted because of antibody recognition, as opposed to peptide recognition with the TCR complex. Studies utilizing the chimeric receptor were expanded to investigate antitumor antibodies. Stancoviski et al. introduced a chimeric gene with an anti-HER2/neu antibody into a murine Cytolytic T-lymphocyte (CTL) line and demonstrated specific antigen recognition and lysis of cells overexpressing Neu/HER2 [73]. Moritz et al. [74] similarly utilized an anti-HER2/neu scFv linked to the ζ chain of the TCR complex and transduced a murine CTL line. Target cells that overexpress HER2/neu were lysed in vitro by the transduced CTLs, and the growth of HER2/neu-transformed NIH3T3 cells in nude mice was retarded, though not prevented, by CTLs transduced by the chimeric receptor and subsequently injected into the mice [74]. Moving beyond the stable cell lines, Hwu et al. transduced CD8$^+$ human CTLs with a chimeric receptor designed to recognize a defined human ovarian carcinoma antigen [75]. Again, the transduced cells recognized target cells with the ovarian cancer antigen and secreted granulocyte–macrophage-colony stimulating factor (GM-CSF) upon incubation with the specific antigen [75].

The next step toward applying this technology to the therapy of human cancer involves transduction of human HSCs with a chimeric receptor. Figure 1 depicts a model therapeutic strategy for patients receiving a chimeric receptor-gene-modified HSC transplant. The progeny lymphocytes and myeloid cells (monocytes and neutrophils) from the transduced HSCs would potentially have the ability to recognize and kill in a non-MHC-restricted fashion those cells expressing the specific tumor antigen incorporated into the chimeric gene. Because the HSCs are self-renewing and pluripotent, they would offer the cancer patient receiving a BMT and/or PBSCT a long-term immune system with antitumor function. We have recently introduced into human CD34$^+$ cells, isolated from MPB, the gene encoding for a chimeric receptor to HER2/neu [76]. Following selection, under G418, 81% of the colonies derived from the transduced HSCs (11%) contained the transgene. We are now planning a phase I clinical trial of chimeric receptor-gene-modified HSCs into advanced breast cancer patients receiving PBSCT. Because of the concern that recognition of "normal" levels of HER2/neu on normal tissues, as opposed to "overexpressed" levels on tumor, might lead to toxicity, vector constructs will also contain suicide cytosine deaminase (CD) or thymidine kinase (TK) genes to eliminate expressing effector progeny cells if adverse targeting occurs.

Another strategy based on redirecting the immune response to recognize tumor-associated antigens after a BMT or PBSCT is the introduction of genes encoding "classic" TCRs. Unlike the chimeric receptor approach, this strategy would be limited to progeny T cells only, would be MHC restricted, and would be more specific to tumor cells (as opposed to normal tissue). Recently, TCRs for a cytotoxic CTL-defined peptide

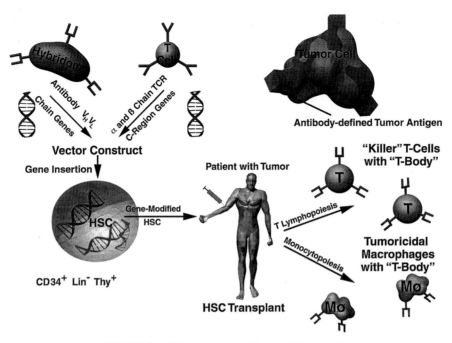

FIGURE 1   HSC gene transfer: chimeric TCR approach.

expressed by HLA-A2 on melanoma has been cloned [77]. The prohibitive limitation of this approach, however, may be the difficulty in successfully expressing two independent genes that encode the separate alpha and beta chains of the TCRs in HSC target cells and their T-lymphoid progeny.

V. CONCLUSIONS

Genetic modification of human HSCs for cancer therapy remains in its infancy but already has shown great potential as a tool to enhance the disease-free survival of patients undergoing a BMT and/or PBSCT for bloodborne or solid tumors. Areas of ongoing and future focus include novel vectors and improvement on existing retroviral vectors for HSC transduction, *ex vivo* HSC proliferation and the cytokines necessary to support survival of the self-renewing pluripotent HSCs, and *in vivo* models to study transduced HSCs. As human clinical trials are completed, more information concerning the biology of BMTs and/or PBSCTs, the feasibility of HSC transduction with multidrug resistance genes, and their ability to provide bone marrow protection will become available. These data will provide information on the safety of this technology and offer a stepping stone toward the use of dual drug-resistance and therapeutic genes, including those encoding receptor molecules that recognize tumor-associated antigens.

References

1. Baum, C. M., Weissman, I. L., Tsukamoto, A. S. *et al.* (1992). Isolation of a candidate human hematopoietic stem-cell population. *Proc. Natl. Acad. Sci. USA* **89**(7), 2804–2808.

2. Morrison, S. J., Uchida, N., and Weissman, I. L. (1995). The biology of hematopoietic stem cells. *Annu. Rev. Cell. Dev. Biol.* **11**, 35–71.

3. Roth, J. A., and Cristiano, R. J. Gene therapy for cancer: what have we done and where are we going? *J. Natl. Cancer Inst.*, (1997). **89**(1), 21–39.

4. Dunbar, C. E. (1996). Gene transfer to hematopoietic stem cells: implications for gene therapy of human disease. *Annu. Rev. Med.* **47**, 11–20.

5. Brenner, M. K. (1996). Gene transfer to hematopoietic cells. *N. Engl. J. Med.* **335**(5), 337–339.

6. Kerr, W. G., and Mule, J. J. (1994). Gene therapy: current status and future prospects. *J. Leukoc. Biol.* **56**(2), 210–214.

7. Bagnis, C., Chabannon, C., and Mannoni, P. (1996). Gene transfer into haemopoietic cells: a challenge for gene therapy. European Concerted Action Workshop, Marseille, 18–19 October, 1995. *Gene Ther.* **3**(4), 362–364.

8. Schwarzenberger, P., Spence, S. E., Gooya, J. M. *et al.* (1996). Targeted gene transfer to human hematopoietic progenitor cell lines through the c-*kit* receptor. *Blood* **87**(2), 472–478.

9. Fisher-Adams, G., Wong, Jr., K. K., Podsakoff, G. *et al.* (1996). Integration of adeno-associated virus vectors in CD34+ human hematopoietic progenitor cells after transduction. *Blood* **88**(2), 492–504.

10. Geraghty, P. J., and Chang, A. E. (1995). Basic principles associated with gene therapy of cancer. *Surg. Oncol.* **4**(3), 125–137.

11. Adam, M. A., and Miller, A. D. (1988). Identification of a signal in a murine retrovirus that is sufficient for packaging of nonretroviral RNA into virions. *J. Virol.* **62**(10), 3802–3806.

12. Rollins, S. A., Birks, C. W., Setter, E. *et al.* (1996). Retroviral vector producer cell killing in human serum is mediated by natural antibody and complement: strategies for evading the humoral immune response. *Hum. Gene Ther.* **7**(5), 619–626.

13. Donahue, R. E., Kessler, S. W., Bodine, D. *et al.* (1992). Helper virus induced T cell lymphoma in nonhuman primates after retroviral mediated gene transfer. *J. Exp. Med.* **176**(4), 1125–1135.

14. Purcell, D. F., Broscius, C. M., Vanin, E. F. *et al.* (1996). An array of murine leukemia virus-related elements is transmitted and expressed in a primate recipient of retroviral gene transfer. *J. Virol.* **70**(2), 887–897.

15. Kohn, D. B. (1997). Gene therapy for haematopoietic and lymphoid disorders. *Clin. Exp. Immunol.* **107**(suppl. 1), 54–57.

16. Mastrangelo, M. J., Berd, D., Nathan, F. E. *et al.* (1996). Gene therapy for human cancer: an essay for clinicians. *Semin. Oncol.* **23**(1), 4–21.

17. Medin, J. A., and Karlsson, S. (1997). Viral vectors for gene therapy of hematopoietic cells. *Immunotechnology* **3**(1), 3–19.

18. Dick, J. E., Kamel-Reid, S., Murdoch, B. *et al.* (1991). Gene transfer into normal human hematopoietic cells using in vitro and in vivo assays. *Blood* **78**(3), 624–634.

19. Cassel, A., Cottler-Fox, M., Doren, S. *et al.* (1993). Retroviral-mediated gene transfer into CD34-enriched human peripheral blood stem cells. *Exp. Hematol.* **21**(4), 585–591.

20. Moritz, T., Keller, D. C., and Williams, D. A. (1993). Human cord blood cells as targets for gene transfer: potential use in genetic therapies of severe combined immunodeficiency disease. *J. Exp. Med.* **178**(2), 529–536.

21. Lu, M., Maruyama, M., Zhang, N. *et al.* (1994). High efficiency retroviral-mediated gene transduction into CD34+ cells purified from peripheral blood of breast cancer patients primed with chemotherapy and granulocyte–macrophage colony-stimulating factor. *Hum. Gene Ther.* **5**(2), 203–208.

22. Akkina, R. K., Rosenblatt, J. D., Campbell, A. G. *et al.* (1994). Modeling human lymphoid precursor cell gene therapy in the SCID-hu mouse. *Blood* **84**(5), 1393–1398.

23. Champseix, C., Marechal, V., Khazaal, I. *et al.* (1996). A cell surface marker gene transferred with a retroviral vector into CD34+ cord blood cells is expressed by their T-cell progeny in the SCID- hu thymus. *Blood* **88**(1), 107–113.

24. Conneally, E., Bardy, P., Eaves, C. J. *et al.* (1996). Rapid and efficient selection of human hematopoietic cells expressing murine heat-stable antigen as an indicator of retroviral-mediated gene transfer. *Blood* **87**(2), 456–464.

25. Medin, J. A., Migita, M., Pawliuk, R. *et al.* (1996). A bicistronic therapeutic retroviral vector enables sorting of transduced CD34+ cells and corrects the enzyme deficiency in cells from Gaucher patients. *Blood* **87**(5), 1754–1762.

26. Nolta, J. A., Dao, M. A., Wells, S. *et al.* (1996). Transduction of pluripotent human hematopoietic stem cells demonstrated by clonal analysis after engraftment in immune-deficient mice. *Proc. Natl. Acad. Sci. USA* **93**(6), 2414–2419.

27. Bauer, Jr., T. R., and Hickstein, D. D. (1997). Transduction of human hematopoietic cells and cell lines using a retroviral vector containing a modified murine CD4 reporter gene. *Hum. Gene Ther.* **8**(3), 243–252.

28. Emerson, S. G., (1996). Ex vivo expansion of hematopoietic precursors, progenitors, and stem cells: the next generation of cellular therapeutics. *Blood* **87**(8), 3082–3088.

29. Arca, M. J., Mule, J. J., and Chang, A. E. (1996). Genetic approaches to adoptive cellular therapy of malignancy. *Semin. Oncol.* **23**(1), 108–117.

30. Herrmann, F. (1995). Cancer gene therapy: principles, problems, and perspectives. *J. Mol. Med.* **73**(4), 157–63.

31. Kafri, T., van Praag, H., Ouyang, L. *et al.* (1999). A packaging cell line for lentivirus vectors. *J. Virol.* **73**(1), 576–584.

32. Miyoshi, H., Smith, K. A., Mosier, D. E. *et al.* (1999). Transduction of human CD34+ cells that mediate long-term engraftment of NOD/SCID mice by HIV vectors *Science* **283**(5402), 682–686.

33. Miyoshi, H., Blomer, U., Takahashi, M. *et al.* (1998). Development of a self-inactivating lentivirus vector. *J. Virol.* **72**(10), 8150–8157.

34. Wilson, J. M., Danos, O., Grossman, M. *et al.* (1990). Expression of human adenosine deaminase in mice reconstituted with retrovirus-transduced hematopoietic stem cells. *Proc. Natl. Acad. Sci. USA* **87**(1), 439–443.

35. Sorrentino, B. P., Brandt, S. J., Bodine, D. *et al.* (1992). Selection of drug-resistant bone marrow cells in vivo after retroviral transfer of human MDR1. *Science* **257**(5066), 99–103.

36. Bodine, D. M., McDonagh, K. T., Seidel, N. E. *et al.* (1991). Survival and retrovirus infection of murine hematopoietic stem cells in vitro: effects of 5-FU and method of infection. *Exp. Hematol.* **19**(3), 206–2012.

37. Fraser, C. C., Eaves, C. J., Szilvassy, S. J. *et al.* (1990). Expansion in vitro of retrovirally marked totipotent hematopoietic stem cells. *Blood* **76**(6), 1071–1076.

38. Sorrentino, B. P., McDonagh, K. T., Woods, D. *et al.* (1995). Expression of retroviral vectors containing the human multidrug resistance 1 cDNA in hematopoietic cells of transplanted mice. *Blood* **86**(2), 491–501.

39. Schuening, F., Miller, A. D., Torok-Storb, B. *et al.* (1994). Study on contribution of genetically marked peripheral blood repopulating cells to hematopoietic reconstitution after transplantation. *Hum. Gene Ther.* **5**(12), 1523–1534.

40. Moritz, T., Patel, V. P., and Williams, D. A. (1994). Bone marrow extracellular matrix molecules improve gene transfer into human hematopoietic cells via retroviral vectors. *J. Clin. Invest.* **93**(4), 1451–1457.

41. Traycoff, C. M., Srour, E. F., Dutt, P. *et al.* (1997). The 30/35 kDa chymotryptic fragment of fibronectin enhances retroviral-mediated gene transfer in purified chronic myelogenous leukemia bone marrow progenitors. *Leukemia* **11**(1), 159–167.

42. Hanenberg, H., Xiao, X. L., Dilloo, D. *et al.* (1996). Colocalization of retrovirus and target cells on specific fibronectin fragments increases genetic transduction of mammalian cells. *Nat. Med.* **2**(8), 876–882.

43. Dunbar, C. E., Cottler-Fox, M., O'Shaughnessy, J. A. *et al.* (1995). Retrovirally marked CD34-enriched peripheral blood and bone marrow cells contribute to long-term engraftment after autologous transplantation. *Blood* **85**(11), 3048–3057.

44. Bordignon, C., Notarangelo, L. D., Nobili, N. *et al.* (1995). Gene therapy in peripheral blood lymphocytes and bone marrow for ADA- immunodeficient patients. *Science* **270**(5235), 470–475.

45. Lu, M. Zhang, N., Maruyama, M. *et al.* (1996). Retrovirus-mediated gene expression in hematopoietic cells correlates inversely with growth factor stimulation. *Hum. Gene Ther.* **7**(18), 2263–2271.

46. Leonard, J., May, C., Gallardo, H. *et al.* (1996). Retroviral transduction of human hematopoietic progenitor cells using a vector encoding a cell surface marker (LNGFR) to optimize transgene expression and characterize transduced cell populations. *Blood* **88**, 433a.

47. Ward, M., Pioli, P., Ayello, J. *et al.* (1996). Retroviral transfer and expression of the human multiple drug resistance (MDR) gene in peripheral blood progenitor cells. *Clin. Cancer Res.* **2**(5), 873–876.

48. Carter, R. F., Abrams-Ogg, A. C., Dick, J. E. *et al.* (1992). Autologous transplantation of canine long-term marrow culture cells genetically marked by retroviral vectors. *Blood* **79**(2), 356–364.

49. Bodine, D. M., Moritz, T., Donahue, R. E. *et al.* (1993). Long-term in vivo expression of a murine adenosine deaminase gene in rhesus monkey hematopoietic cells of multiple lineages after retroviral mediated gene transfer into CD34+ bone marrow cells. *Blood* **82**(7), 1975–1980.

50. Bienzle, D., Abrams-Ogg, A. C., Kruth, S. A. *et al.* (1994). Gene transfer into hematopoietic stem cells: long-term maintenance of in vitro activated progenitors without marrow ablation. *Proc. Natl. Acad. Sci. USA* **91**(1), 350–354.

51. Donahue, R. E., Byrne, E. R., Thomas, T. E. *et al.* (1996). Transplantation and gene transfer of the human glucocerebrosidase gene into immunoselected primate CD34+Thy-1+ cells. *Blood* **88**(11), 4166–4172.

52. Bodine, D. M., McDonagh, K. T., Brandt, S. J. *et al.* (1990). Development of a high-titer retrovirus producer cell line capable of gene transfer into rhesus monkey hematopoietic stem cells. *Proc. Natl. Acad. Sci. USA* **87**(10), 3738–3742.

53. Cavazzana-Calvo, M., Hacein-Bey, S., de Saint Basile, G. *et al.* (2000). Gene therapy of human severe combined immunodeficiency (SCID)-X1 disease [see comments]. *Science* **288**(5466), 669–672.

54. Brenner, M. K., Rill, D. R., Holladay, M. S. *et al.* (1993). Gene marking to determine whether autologous marrow infusion restores long-term haemopoiesis in cancer patients. *Lancet* **342**(8880), 1134–1137.

55. Rill, D. R., Santana, V. M., Roberts, W. M. *et al.* (1994). Direct demonstration that autologous bone marrow transplantation for solid tumors can return a multiplicity of tumorigenic cells. *Blood* **84**(2), 380–383.

56. Cornetta, K., Srour, E. F., Moore, A. *et al.* (1996). Retroviral gene transfer in autologous bone marrow transplantation for adult acute leukemia. *Hum. Gene Ther.* **7**(11), 1323–1329.

57. Deisseroth, A. B., Zu, Z., Claxton, D. *et al.* (1994). Genetic marking shows that Ph+ cells present in autologous transplants of chronic myelogenous leukemia (CML) contribute to relapse after autologous bone marrow in CML. *Blood* **83**(10), 3068–3076.

58. Emmons, R. V., Doren, S., Zujewski, J. *et al.* (1997). Retroviral gene transduction of adult peripheral blood or marrow-derived CD34+ cells for six hours without growth factors or on autologous stroma does not improve marking efficiency assessed in vivo. *Blood* **89**(11), 4040–4046.

59. Heslop, H. E., Rooney, C. M., Rill, D. R. *et al.* (1996). Use of gene marking in bone marrow transplantation. *Cancer Detect. Prev.* **20**(2), 108–113.

60. Douer, D., Levine, A., Anderson, W. F. *et al.* (1996). High-dose chemotherapy and autologous bone marrow plus peripheral blood stem cell transplantation for patients with lymphoma or metastatic breast cancer: use of marker genes to investigate hematopoietic reconstitution in adults. *Hum. Gene Ther.* **7**(5), 669–684.

61. Brenner, M. K., Rill, D. R., Moen, R. C. *et al.* (1993). Gene-marking to trace origin of relapse after autologous bone-marrow transplantation. *Lancet* **341**(8837), 85–86.

62. Stewart, A. K., Dube, I. D., Kamel-Reid, S. *et al.* (1995). A phase I study of autologous bone marrow transplantation with stem cell gene marking in multiple myeloma. *Hum. Gene Ther.* **6**(1), 107–119.

63. Bachier, C., Giles, R., Ellerson, D. *et al.* (1996). Retroviral gene marking in relapsed follicular non-Hodgkin's lumphoma (FNHL) to determine the origin of relapse following autologous bone marrow (ABMT) and peripheral stem cell transplant (PSCT) [meeting abstract]. *Proc. Annu. Meet. Am. Assoc. Cancer Res.* **37,** A1398.

64. Brenner, M., Krance, R., Heslop, H. E. *et al.* (1994). Assessment of the efficacy of purging by using gene marked autologous marrow transplantation for children with AML in first complete remission. *Hum. Gene Ther.* **5**(4), 481–499.

65. Koc, O. N., Allay, J. A., Lee, K. *et al.* (1996). Transfer of drug resistance genes into hematopoietic progenitors to improve chemotherapy tolerance. *Semin. Oncol.* **23**(1), 46–65.

66. Hesdorffer, C., Antman, K., Bank, A. *et al.* (1994). Human MDR gene transfer in patients with advanced cancer. *Hum. Gene Ther.* **5**(9), 1151–1160.

67. Deisseroth, A. B., Kavanagh, J., and Champlin, R. (1994). Use of safety-modified retroviruses to introduce chemotherapy resistance sequences into normal hematopoietic cells for chemoprotection during the therapy of ovarian cancer: a pilot trial. *Hum. Gene Ther.* **5**(12), 1507–1522.

68. Deisseroth, A. B., Holmes, F., Hortobagyi, G. *et al.* (1996). Use of safety-modified retroviruses to introduce chemotherapy resistance sequences into normal hematopoietic cells for chemoprotection during the therapy of breast cancer: a pilot trial. *Hum. Gene Ther.* **7**(3), 401–416.

69. O'Shaughnessy, J. A., Cowan, K. H., Nienhuis, A. W. *et al.* (1994). Retroviral mediated transfer of the human multidrug resistance gene (MDR-1) into hematopoietic stem cells during autologous transplantation after intensive chemotherapy for metastatic breast cancer. *Hum. Gene Ther.* **5**(7), 891–911.

70. Hanania, E. G., Giles, R. E., Kavanagh, J. *et al.* (1996). Results of MDR-1 vector modification trial indicate that granulocyte/macrophage colony-forming unit cells do not contribute to posttransplant hematopoietic recovery following intensive systemic therapy [published *erratum* appears in *Proc. Natl. Acad. Sci. USA* **94**(10), 5495,1997]. *Proc. Natl. Acad. Sci. USA* **93**(26), 15346–15351.

71. Abonour, R., Williams, D. A., Einhorn, L. *et al.* (2000). Efficient retrovirus-mediated transfer of the multidrug resistance 1 gene into autologous human long-term repopulating hematopoietic stem cells. *Nat. Med.* **6**(6), 652–658.

72. Eshhar, Z., Waks, T., Gross, G. *et al.* (1993). Specific activation and targeting of cytotoxic lymphocytes through chimeric single chains consisting of antibody-binding domains and the gamma or zeta subunits of the immunoglobulin and T-cell receptors. *Proc. Natl. Acad. Sci. USA* **90**(2), 720–724.

73. Stancovski, I., Schindler, D. G., Waks, T. *et al.* (1993). Targeting of T lymphocytes to Neu/HER2-expressing cells using chimeric single chain Fv receptors. *J. Immunol.* **151**(11), 6577–6582.

74. Moritz, D., Wels, W., Mattern, J. *et al.* (1994). Cytotoxic T lymphocytes with a grafted recognition specificity for ERBB2-expressing tumor cells. *Proc. Natl. Acad. Sci. USA* **91**(10), 4318–4322.

75. Hwu, P., Shafer, G. E., Treisman, J. *et al.* (1993). Lysis of ovarian cancer cells by human lymphocytes redirected with a chimeric gene composed of an antibody variable region and the Fc receptor gamma chain. *J. Exp. Med.* **178**(1), 361–366.

76. Wilke, L., Reynolds, C., McDonough, K. *et al.* (1997). Engineering immune effector cells to target the HER-2/neu breast cancer antigen (meeting abstract). *SSO Annual Cancer Symposium,* **51,** P4

77. Cole, D. J., Weil, D. P., Shilyansky, J. *et al.* (1995). Characterization of the functional specificity of a cloned T-cell receptor heterodimer recognizing the MART-1 melanoma antigen. *Cancer Res.* **55**(4), 748–752.

ONCOGENE-TARGETED GENE THERAPY

Clinical Applications of Tumor-Suppressor Gene Therapy

RAYMOND D. MENG

Laboratory of Molecular Oncology and Cell Cycle Regulation
Howard Hughes Medical Institute
Departments of Medicine and Genetics
Cancer Center and The Institute for Human Gene Therapy
University of Pennsylvania School of Medicine
Philadelphia, Pennsylvania 19104

WAFIK S. EL-DEIRY

Laboratory of Molecular Oncology and Cell Cycle Regulation
Howard Hughes Medical Institute
Departments of Medicine and Genetics
Cancer Center and The Institute for Human Gene Therapy
University of Pennsylvania School of Medicine
Philadelphia, Pennsylvania 19104

I. INTRODUCTION

Multiple human tumors show mutations or deletions in tumor suppressor genes, which control cellular growth by regulating the cell cycle or by inducing cellular apoptosis. In particular, it has been shown that the p53 tumor suppressor gene is mutated or deleted in over half of all human malignancies (reviewed by Levine [see Ref. 83 in Chapter 18]). Hence, one strategy in cancer gene therapy has focused on the replacement or overexpression of tumor suppressors genes. Over the past decade, extensive research has been conducted *in vitro* with multiple tumor suppressors, including p53, p21$^{WAF1/CIP1}$, p16, *Rb*, p14^{ARF}, p27, *E2F-1, BRCA1, VHL*, and *FHIT*. Overexpression of tumor suppressors in selective cancer cell lines has resulted in either growth suppression or apoptosis. Currently, clinical trials for cancer gene therapy are being conducted with select tumor suppressor genes, most notably the p53 gene. Its dual role in both cell-cycle arrest and in apop-

tosis makes the p53 tumor suppressor an important target for replacement.

II. p53

Currently, of the tumor suppressors being considered for gene replacement in cancer, only the p53 gene is being extensively tested in clinical trials (see Table 1). Much of the pioneering work with p53 gene therapy was conducted by Jack Roth and colleagues at the M.D. Anderson Cancer Center. They reported the results of the first clinical trials using p53 delivered by a retrovirus vector to treat patients with non-small-cell lung cancer who had failed other treatments [28]. The virus was administered intratumorally and caused no toxic side effects up to 5 months later. Wild-type p53 was detected in lung biopsies by *in situ* hybridization and PCR amplification, and apoptosis (as determined by the TUNEL assay) was increased in posttreatment biopsy samples. Of the nine patients in the study, three showed tumor growth stabilization, and three showed slight tumor regression. It was also reported by the same group that tumor growth inhibition was enhanced when p53 gene therapy was combined with systemic cisplatin [26].

In another phase I study of patients with non-small-cell lung cancers, p53-expressing adenovirus (Ad-p53) treatments only stabilized two of 15 patients, with one patient being stable at more than 6 months following his treatments [32]. A comprehensive phase I study using Ad-p53 for recurrent head and neck cancers was also recently reported [3].

TABLE 1   Selected p53 Gene Therapy Clinical Trials in 2000

Cancer	Mode of treatment	Site of trial	Results
Bladder cancer	Adenovirus	Houston, TX	Phase I planned
	Adenovirus	Hamburg, Germany; New Brunswick, New Jersey; San Diego, CA	Phase I completed; intravesical injection more effective than intratumoral injection; both are safe [31]
Colorectal cancer	Adenovirus	Houston, TX	Phase I planned
Glioblastoma	Adenovirus	Houston, TX	Phase I planned
Head–neck, squamous cell carcinoma	Adenovirus; multiple injections	Houston, TX	Phase I completed; 2 of 17 nonresectable had partial regression; 40% of resectable were disease-free for 6 months; Phase II in progress
Head–neck squamous cell cancer, mutant p53	Adenovirus; intratumoral	Pittsburgh, PA; London, England	Phase I planned; 6–18 patients
Liver cancer, primary and metastatic; mutant p53	Adenovirus; hepatic artery infusions	San Francisco, CA; Philadelphia, PA	Phase I planned; 21–42 patients
Liver cancer	Adenovirus	Houston, TX	Phase I planned
Malignant ascites	Adenovirus	Houston, TX	Phase I planned
Non-small-cell lung cancer	Adenovirus and retrovirus; intratumoral	Houston, TX	Phase I completed, Ad-p53 slowed growth at high doses; phase II in progress.
Non-small-cell lung cancer, mutant p53	Adenovirus; intratumoral	Mainz, Germany; Basel, Switzerland	Phase I planned; 6–18 patients
Ovarian cancer	Adenovirus; combined with cisplatin	Houston, TX	Phase I planned
Ovarian, fallopian tube, or peritoneal cancer; mutant p53	Adenovirus; intraperitoneal	Karolinksa, Sweden; Iowa City, IA	Phase I planned; 6–24 patients
Prostate cancer	Adenovirus	Houston, TX	Phase I planned; intraprostatic injections in 30 patients

Source: Data compiled from Roth and Cristiano [39], National Cancer Institute PDQ Clinical Trial Database, and *Genetic Engineering News,* June 15, 1997.

Patients with surgically nonresectable and previously irradiated head and neck cancers were given cycles of treatment consisting of three injections of Ad-p53 per week for 2 weeks followed by 2 weeks of rest. In 18 patients, only two patients experienced a decrease in tumor size of more than 50%, although this remission lasted almost 8 months. Six patients, however, had no change in tumor size. The major side effects encountered in this study were pain at the injection site and fever and headache with the higher doses of Ad-p53 used. Following treatment, the authors were able to detect antibodies to the adenovirus in the blood and could detect p53 in the blood and urine [3].

In addition, a phase I trial using Ad-p53 to treat invasive bladder cancers was recently completed [31]. Patients with locally invasive bladder cancers were administered Ad-p53 either by cystoscopically aided guidance or by intravesical treatment. Major side effects of the treatment consisted of urethral discomfort or abdominal pain. In five of six patients given intravesical administration of Ad-p53, PCR could detect p53 expression 3 days later; however, in the patients with intratumoral injections of Ad-p53, no p53 protein could be detected at that time. Another study examined the utility of using computed tomography (CT) of the chest to document the effectiveness of Ad-p53 gene therapy for non-small-cell lung cancers [24]. From 33 tumors administered Ad-p53, CT scans showed no change in 20 tumors, a decrease in size in six tumors, and an increase in size in seven tumors. A biopsy from the lung tumors revealed that in two of the six tumors that had decreased in size no evidence of malignant cells could be found; in contrast, in the seven tumors that increased in size, the biopsies were all positive for malignant cells. Unfortunately, no significant features from the CT scan, such as the detection of necrosis or of decreased attenuation, were specific enough to document the efficacy of

Ad-p53 treatments for lung tumors. Phase I trials of Ad-p53 have also been planned for metastatic liver tumors, whereby Ad-p53 will be infused through the hepatic artery [10]; for locally advanced prostate cancers, whereby Ad-p53 will be administered directly by intraprostatic injections [33]; and for non-small-cell lung cancers [37].

III. BRCA1

One clinical trial is currently being conducted at Vanderbilt University to examine the feasibility of BRCA1-mediated gene therapy for the treatment of ovarian cancers. Unlike the p53 clinical trials, this study utilized a retrovirus vector to deliver BRCA1. Following injection into patients, the retroviral BRCA1 vector was stable, generated minimal antibody responses against itself, and produced subtle decreases in tumor size [34]. The clinical trials then progressed to a phase II level, but the results were unfortunately not very successful. Retroviral BRCA1 produced no decreases in tumor size and also generated strong antibody responses [34]. The authors felt that the retroviral vector was more immunogenic because it was being packaged in mouse cells; therefore, a new retroviral BRCA1 vector was developed that could be packaged in human 293 cells [35]. Currently, further trials utilizing retrovirus BRCA1 are being conducted.

IV. ONYX-015 ADENOVIRUSES

A. Introduction

Intratumoral delivery of adenoviruses that selectively replicate in p53-deficient tumor cells have been designed that may take full advantage of introducing a small amount of virus in some tumor cells followed by propagation and toxicity only within the tumor mass [1,17] (reviewed by Heise and Kirn [13] and Hermiston [18]). The adenovirus E1B gene region, which binds to and inactivates wild-type p53 in host cells, was deleted from an adenovirus backbone [1]. Consequently, because it can no longer inactivate p53, this vector can only replicate in p53-null cells, effectively targeting this virus to cells with p53 mutations, which are usually cancer cells. It was shown that normal cells are highly resistant to the E1B-deleted adenovirus [17].

Recent studies, however, suggest that the replication of ONYX-015 may also occur in cell lines with wild-type p53 rather than being limited solely to cells with mutant p53 [8,9,11,17,29]. For example, ONYX-015 was able to replicate in and lyse lung cancer cell lines with mutant p53, as expected, but also in cells with wild-type p53 [38]. The apoptosis induced by ONYX-015, however, was ten times less efficient in cell lines with wild-type p53 compared to those with mutant p53 [38]. The virus can replicate in cells

with wild-type p53, although the kinetics of replication are much slower and the total viral load much less than that observed in cells with mutant p53 [27]. This finding was also reported by another group that observed that ONYX replication was threefold slower when a cell line with a temperature-sensitive p53 mutant had wild-type p53 [12]. It has been hypothesized that two modes of cell death may be induced by adenoviruses: a rapid p53-dependent death observed with wild-type adenoviruses that requires E1B 55-KDa protein binding to p53, and a slower p53-independent cell death observed with ONYX [4]. In addition, the presence or absence of p14^{ARF} may be important for efficient apoptosis following infection with ONYX-015. Infection with ONYX–015 induced cell death in mesothelioma cell lines lacking p14^{ARF}, whereas cell lines with wild-type p14^{ARF} were resistant [36]. Transfection of wild-type p14^{ARF} rendered the previously ONYX-015-sensitive cells more resistant to the virus [36]. Despite this controversy over the exact cellular mechanisms underlying the apoptosis induced by ONYX-015, the virus does eliminate tumor cells effectively. It was shown that mixing as few as 5% of tumor cells infected with ONYX-015 along with wild-type tumor cells was enough to inhibit tumor cell growth in vivo for 8 weeks in nude mice [16]. Recent studies have combined ONYX-015 with cisplatin and 5-fluorouracil in head and neck cancer patients [20] and have examined the ability of a mutant E1A adenovirus to lyse tumor cells [14,15].

B. Gene Therapy

Multiple studies have now shown that ONYX-015 can infect and replicate in a wide range of tumor cells with either mutant p53 or wild-type p53, although to different degrees. The therapeutic efficacy of ONYX-015 has been enhanced by combining the virus with either chemotherapy or radiotherapy. For example, in human lung cancer cells, treatment with ONYX-015 and with paclitaxel or cisplatin enhanced the degree of apoptosis [38]. Similarly, the combination of ONYX-015 and radiotherapy caused an additive increase in the degree of cell death, although high doses of γ-irradiation (approximately 20 Gy) began to inhibit ONYX-015 replication [5,27].

Several groups have attempted to improve the efficacy of the ONYX-015 virus by modifying its delivery. First, the schedule of administration appears to be important, as multiple dosings of ONYX-015 decrease tumor burden to a greater extent than a single intratumoral dose [16]. Second, a single intratumoral injection of ONYX-015 was more effective at inhibiting growth when given in a large volume (100 μL) as opposed to a small volume (40 μL) [16]. Finally, the efficiency of ONYX-015 can be improved almost threefold simply by changing the injection carrier. Clinical trials have shown that following an intratumoral injection of ONYX-015, the majority of the virus will remain in the center of the tumor, causing an unequal distribution of necrosis, mostly in the center and not in the periphery. One group hypothesized that the intratumoral spread of the virus could be improved by injecting in a carrier with lidocaine, which would vasodilate the tumor, or with hyaluronidase, which degrades tissue [23]. The most efficient tumor spread of ONYX-015 was observed with lidocaine, which was almost threefold greater than injection in PBS alone.

Finally, the ONYX-015 virus itself is being modified to improve its efficacy. First, ONYX-015/CD contains the cytosine deaminase gene, which is a prodrug that converts 5-fluorocytosine to 5-flurouracil [19]. Infection of human pancreatic tumor cells xenografted on nude mice with ONYX-015/CD significantly inhibited tumor growth [19]. Other adenoviruses are being created that also have deletions within the E1B region. The E1B region contains primarily two adenoviral proteins: 55 kDa, which is deleted in ONYX-015, and 19 kDa, which is hypothesized to delay apoptosis in cells infected with wild-type adenovirus. An adenovirus was created that only had the 19-kDa protein in the E1B region deleted (Ad-337). Infection of human lung cancer cells with Ad-337 induced apoptosis more extensively than wild-type adenovirus [30].

Finally, viruses with deletions in the E1A region rather than the E1B region, like ONYX-015, have also been created. ONYX-838 contains a partial deletion in the E1A region, which usually binds the Rb protein [21]. Infection of human tumor cell lines *in vitro* or *in vivo* with ONYX-838 caused greater apoptosis than that observed with wild-type adenovirus and with fewer toxic side effects [14,15,21]. Intravenous injection of ONYX-838 was especially efficient with decreasing lung and lymph node metastases in an orthotopic mouse model of human breast cancer [14,15]. The ONYX-838 virus could still be detected 2 months after a single intratumoral or intravascular injection. Another E1A-deleted adenovirus (Ad-24) was created in which amino acids 120–127, which bind Rb, were deleted. Infection of glioma cells *in vitro* or *in vivo* with Ad-24 caused greater cell death than wild-type adenovirus but had no effect on normal lung fibroblasts [6]. Resistance to Ad-24 was correlated with the presence of endogenous wild-type Rb protein. In another approach, an E1-deleted adenovirus was modified to allow it to replicate but only on a limited basis [2]. An E1-deleted adenovirus was modified by the addition of an exogenous plasmid containing the E1 region to allow one round of infection. This adenovirus was then used to infect HeLa tumor cells xenografted on nude mice, which resulted in growth suppression.

C. Clinical Trials

The results of a phase I trial using ONYX-015 to treat recurrent head and neck cancers that had failed traditional chemotherapeutic or radiotherapeutic regimens was recently reported [7]. Intratumoral injections of ONYX-015 given to 22 patients produced no significant side effects, with the most common symptom being low-grade fever. Blood samples taken immediately following the injection of virus and up to 29 days later did not reveal ONYX-015, as determined by polymerase chain reaction (PCR) analysis. This suggested that ONYX-015 virus was not being shed into the blood or that it did not persist in the blood. Furthermore, tissue samples taken from the injection site and from the oropharynx of the patients did not reveal the adenoviral hexon protein, as determined by direct fluorescence. In terms of efficacy, three of the 22 patients experienced a decrease in tumor size of more than 50%, and two saw decreases of 25%. Interestingly, of these five patients who experienced a response, four of them had head and neck tumors with mutant p53. Despite these responses, eventually all of the patients in the study developed progression of the original tumor, new tumors, or metastases. In examining a possible reason for failure, the authors showed that 21 of the 22 patients had also developed increasing antibody levels to the adenovirus [7]. Another phase I study examined the utility of ONYX-015 in refractory cancers [25].

Phase II studies examining the efficacy of ONYX-015 for other head and neck cancers are currently being conducted. A phase II trial recently examined the efficacy of ONYX-015 in treating recurrent head and neck cell cancers in combination with the chemotherapeutic agents cisplatin and 5-fluorouracil [20,21]. Out of 30 patients in the study, almost 60% experienced some degree of tumor size decrease, with 30% of the patients experiencing a complete response. Follow-up 5 months later revealed no tumor recurrence in these patients with a complete response. A phase IIB trial with ONYX-015 is also being conducted for hepatic or billiary tumors [22]. ONYX-015 is being administered by CT-guided intratumoral injections in ten patients with locally advanced or metastatic hepatic or billiary tumors.

V. SUMMARY AND FUTURE WORK

The use of tumor suppressors in gene therapy represents an important strategy in the war on cancer. p53 gene therapy remains the most important tumor suppressor strategy being developed, and its combination with chemotherapy or radiotherapy may prove to be even more beneficial. Currently, only p53 gene therapy has progressed to clinical trials. However, p53 may not represent the ideal choice for gene therapy in all cancers. In tumor cells that overexpress *MDM2* or have HPV16 *E6*, other tumor suppressors such as p21 may be more desirable targets of gene therapy because they can bypass the inactivation of p53. In addition to p53 and p21, other tumor suppressors that have been studied for gene replacement, include p16, Rb, p27, p14, *PTEN, BRCA1, VHL,* and *FHIT.* Although significant progress in gene therapy for cancer has been made within the past decade, several problems still need to be resolved. First, an efficient vector needs to be designed that can effect prolonged high expression of the transduced gene while only targeting cancer cells. Second, further criteria need to be established for determining which tumor suppressor to employ for gene therapy. The use of tumor suppressors represents a potentially important anticancer treatment that needs to be further investigated.

References

1. Bischoff, J. R., Kirn, D. H., Williams, A., Heise, C., Horn, S., Muna, M., Ng, L., Nye, J. A., Samspon-Johannes, A., Fattaey, A., and McCormick, F. (1996). An adenovirus mutant that replicates selectively in p53-deficient human tumor cells. *Science* **274,** 373–376.
2. Clarke, M. F., Qian, D., Han, J., Nunez, G., and Wicha, M. (1999). Targeting the programmed cell death pathway for cancer treatment. *Cancer Gene Ther.* **6**(suppl.), S1.
3. Clayman, G. L., El-Naggar, A. K., Lippman, S. M., Henderson, Y. C., Frederick, M., Merritt, J. A., Zumstein, L. A., Timmons, T. M., Liu, T. J., Ginsberg, L., Roth, J. A., Hong, W. K., Bruso, P., and Goepfert, H. (1998). Adenovirus-mediated p53 gene transfer in patients with advanced recurrent head and neck squamous cell carcinoma. *J. Clin. Oncol.* **16,** 2221–2232.
4. Dix, B. R., O'Carroll, S. J., Myers, C. J., Edwards, S. J., and Braithwaite, A. W. (2000). Efficient induction of cell death by adenoviruses requires binding of E1B55k and p53. *Cancer Res.* **60,** 2666–2672.
5. Duque, P. M., Alonso, C., Sanchez-Prieto, R., Lleonart, M., Martinez, C., Conzalez de Buitrago, G., Cano, A., Quintanilla, M., and Ramon y Cajal, S. (1999). Adenovirus lacking the 19-kDa and 55-kDa E1B genes exerts a marked cytotoxic effect in human malignant cells. *Cancer Gene Ther.* **6,** 554–563.
6. Fueyo, J., Gomez-Manzano, C., Alemany, R., Lee, P. S. Y., McDonnell, T. J., Mitlianga, P., Shi, Y.-X., Levin, V. A., Yung, W. K. A., and Kyritsis, A. P. (2000). A mutant oncolytic adenovirus targeting the Rb pathway produces anti-glioma effect in vivo. *Oncogene* **19,** 2–12.
7. Ganley, I., Kirn, D., Eckhardt, S. G., Rodriguez, G. I., Soutar, D. S., Otto, R., Robertson, A. G., Park, O., Gulley, M. L., Heise, C., Von Hoff, D. D., and Kaye, S. B. (2000). A phase I study of ONYX-015, an E1B attenuated adenovirus, administered intratumorally to patients with recurrent head and neck cancer. *Clin. Cancer Res.* **6,** 798–806.
8. Goodrum, F. D., and Ornelles, D. A. (1998). p53 status does not determine outcome of E1B 55-kilodalton mutant adenovirus lytic infection. *J. Virol.* **72,** 9479–9490.
9. Goodrum, F. D., and Ornelles, D. A. (1997). The early region 1B 55-kilodalton oncoprotein of adenovirus relieves growth restrictions imposed on viral replication by the cell cycle. *J. Virol.* **71,** 548–561.
10. Habib, N. A., Hodgson, H. J., Lemoine, N., and Pignatelli, M. (1999). A phase I/II study of hepatic artery infusion with wtp53-CMV-Ad in metastatic malignant liver tumours. *Hum. Gene Ther.* **10,** 2019–2034.
11. Hall, A. R., Dix, B. R., O'Carroll, S. J., and Braithwaite, A. W. (1998). p53-dependent cell death/apoptosis is required for a productive adenoviral infection. *Nat. Med.* **4,** 1068–1072.
12. Harada, J., and Berk, A. (1999). p53-independent, and -dependent requirements for E1B-55K in adenovirus type 5 replication. *J. Virol.* **73,** 5333–5344.
13. Heise, C., and Kirn, D. H. (2000). Replication-selective adenoviruses as oncolytic agents. *J. Clin. Invest.* **105,** 847–851.
14. Heise, C. C., Hermiston, T., Brooks, G., Samspon-Johannes, A., Trown, P., and Kirn, D. H. (2000). An adenovirus E1A mutant, ONYX-838, that demonstrates potent and selective systemic and local anti-tumoral efficacy. *Proc. Annu. Meet. Am. Assoc. Cancer Res.* **41,** 350.
15. Heise, C., Hermiston, T., Johnson, L., Brooks, G., Samspon-Johannes, A., Williams, A., Hawkins, L., and Kirn, D. (2000). An adenovirus E1A mutant that demonstrates potent and selective systemic anti-tumoral efficacy. *Nat. Med.* **6,** 1134–1139.
16. Heise, C. C., Williams, A., Olesch, J., and Kirn, D. H. (1999). Efficacy of a replication-competent adenovirus (ONYX-015) following intratumoral injection: intratumoral spread and distribution effects. *Cancer Gene Ther.* **6,** 499–504.
17. Heise, C., Sampson-Johannes, A., Williams, A., McCormick, F., von Hoff, D. D., and Kirn, D. H. (1997). ONYX-015, an E1B gene attenuated adenovirus, causes tumor-specific cytolysis and antitumoral efficacy that can be augmented by standard chemotherapeutic agents. *Nat. Med.* **3,** 639–645.
18. Hermiston, T. (2000). Gene delivery from replication-selective viruses: arming guided missiles in the war against cancer. *J. Clin. Invest.* **105,** 1169–1172.
19. Hermiston, T., Hawkins, L., Nye, J., Hatfield, J. M., Lemmon, M., Johnson, L., Trown, P., Kirn, D., and Heise, C. (1999). Superior efficacy of a selectively replicating adenovirus, ONYX-015/CD, expressing the cytosine deaminase gene, and development of new replicating adenoviral gene delivery systems. *Cancer Gene Ther.* **6**(suppl.), S13.
20. Khuri, F. R., Nemunaitis, J., Ganly, I., Arseneau, J., Tannock, I. F., Romel, L., Gore, M., Ironside, J., MacDougall, R. H., Heise, C., Randlev, B., Gillenwater, A. M., Bruso, P., Kaye, S. B., Hong, W. K., and Kirn, D. H. (2000). A controlled trial of intratumoral ONYX-015, a selectively-replicating adenovirus, in combination with cisplatin and 5-fluorouracil in patients with recurrent head and neck cancer. *Nat. Med.* **6,** 879–885.
21. Kirn, D., Nemunaitis, J., Heise, C., and Hermiston, T. (1999). Selectively-replicating oncolytic adenoviruses for the treatment of cancer: clinical development of ONYX-015 (p53-targeted) and preclinical studies with ONYX-838 (RB pathway-targeted). *Cancer Gene Ther.* **6**(suppl.), S10.
22. Makower, D., Rozenblit, A., Edelman, M., Haugenlicht, L., Kaufman, H., Haynes, H., Zwiebel, J., and Wadler, S. (1999). Phase IIB trial of ONYX-015 therapy in hepatobiliary tumors. *Cancer Gene Ther.* **6**(suppl.), S16.
23. Morley, S. E., Brown, R., and Kaye, S. (2000). Improvement in distribution of ONYX-015 adenovirus within tumor xenografts by altering the carrier medium. *Proc. Annu. Meet. Am. Assoc. Cancer Res.* **41,** 350–351.
24. Munden, R. F., Truong, M. T., Swisher, S., and Roth, J. A. (1999). CT evaluation of nonsmall cell lung cancer undergoing gene therapy with adenoviral p53. *Radiology* **213P,** 173.
25. Nemunaitis, J., Cunningham, C., Edelman, G., Berman, B., and Kirn, D. (1999). Phase I dose escalation trial of intravenous infusion of ONYX-015 in patients with refractory cancer. *Cancer Gene Ther.* **6**(suppl.), S16–S17.
26. Nguyen, D. M., Spitz, F. R., Yen, N., Cristiano, R. J., and Roth, J. A. (1996). Gene therapy for lung cancer: enhancement of tumor suppression by a combination of sequential systemic cisplatin and adenovirus-mediated p53 gene transfer. *J. Thoracic Cardiovascular Surg.* **112,** 1372–1376.
27. Rogulski, K. R., Freytag, S. O., Zhang, K., Gilbert, J. D., Paielli, D. L., Kim, J. H., Heise, C. C., and Kirn, D. H. (2000). In vivo antitumor activity of ONYX-015 is influenced by p53 status and is augmented by radiotherapy. *Cancer Res.* **60,** 1193–1196.
28. Roth, J. A., Nguyen, D., Lawrence, D. D., Kemp, B. L., Carrasco, C. H., Ferson, D. Z., Hong, W. K., Komaki, R., Lee, J. J., Nesbitt, J. C., Pisters,

K. M., Putnam, J. B., Schea, R., Shin, D. M., Walsh, G. L., Dolormente, M. M., Han, C. I., Martin, F. D., Yen, N., Xu, K., Stephens, L. C., McDonnell, T. J., Mukhopadhyay, T., and Cai, D. (1996). Retrovirus-mediated wild-type p53 gene transfer to tumors of patients with lung cancer. *Nat. Med.* **2,** 974–975.

29. Rothmann, T., Hengstermann, A., Whitaker, N. J., Scheffner, M., and zur Hansen, H. (1998). Replication of ONYX-015, a potential anticancer adenovirus, is independent of p53 status in tumor cells. *J. Virol.* **72,** 9470–9478.

30. Sauthoff, H., Heitner, S., Rom, W. N., and Hay, J. G. (1999). Deletion of the E1B-19-kDa gene enhances the tumoricidal effect of a replicating adenoviral vector. *Cancer Gene Ther.* **6**(suppl.), S10.

31. Schuler, M., Kuball, J., Leibner, J., Atkins, D., Wen, S. F., Engler, H., Meinhardt, P., Uhlenbusch, R., Horowitz, J. A., Hutchins, B., Maneval, D. C., Storkel, S., Thuroff, J. W., and Huber, C. (1999). A phase I study of adenovirus-mediated wild-type p53 gene transfer in patients with invasive bladder cancer. *Cancer Gene Ther.* **6**(suppl.), S2.

32. Schuler, M., Rochlitz, C., Horowitz, J. A., Schlegel, J., Perruchoud, A. P., Kommoss, F., Bolliger, C. T., Kauczor, H. U., Dalquen, P., Fritz, M. A., Swanson, S., Herrmann, R., and Huber, C. (1998). A phase I study of adenovirus-mediated wild-type p53 gene transfer in patients with advanced non-small cell lung cancer. *Hum. Gene Ther.* **9,** 2075–2082.

33. Sweeney, P., and Pisters, L. L. (2000). Ad5CMVp53 gene therapy for locally advanced prostate cancer—where do we stand? *World J. Urol.* **18,** 121–124.

34. Tait, D. L., Obermiller, P. S., Hatmaker, A. R., Redlin-Frazier, S., and Holt, J. T. (1999). Ovarian cancer BRCA1 gene therapy: phase I and II trial differences in immune response and vector stability. *Clin. Cancer Res.* **5,** 1708–1714.

35. Tait, D. L., Obermiller, P. S., Jensen, R. A., and Holt, J. T. (1999). Ovarian cancer gene therapy with a BRCA1 retroviral vector. *Cancer Gene Ther.* **6**(suppl.), S1.

36. Yang, C.-T, You, L., Song, J., Kirn, D. H., McCormick, F., and Jabions, D. M. (1999). Adenoviral therapy of human mesotheliomas. *Cancer Gene Ther.* **6**(suppl.), S17.

37. Yen, N., Ioannides, C. G., Xu, K., Swisher, S. G., Lawrence, D. D., Kemp, B. L., El-Naggar, A. K., Cristiano, R. J., Fang, B., Glisson, B. S., Hong, W. K., Khuri, F. R., Kurie, J. M., Lee, J. J., Lee, J. S., Merritt, J. A., Mukhopadhyay, T., Nesbitt, J. C., Nguyen, D., Perez-Soler, R., Pisters, K. M. W., Putnam, J. B., Jr., Schrump, D. S., Shin, D. M., and Roth, J. A. (2000). Cellular and humoral immune responses to adenovirus and p53 protein antigens in patients following intratumoral injection of an adenovirus vector expressing wild-type p53 (Ad-p53). *Cancer Gene Ther.* **7,** 530–536.

38. You, L., Yang, C.-T., and Jablons, D. M. (2000). ONYX-015 works synergistically with chemotherapy in lung cancer cell lines and primary cultures freshly made from lung cancer patients. *Cancer Res.* **60,** 1009–1013.

39. Roth, J. A., and Cristiano, R. J. (1997). Gene therapy for cancer: what have we done and where are we going? *J. Natl. Cancer Inst.* **89,** 21–39.

17

Cancer Gene Therapy with Tumor Suppressor Genes Involved in Cell-Cycle Control

RAYMOND D. MENG

Laboratory of Molecular Oncology and Cell Cycle Regulation
Howard Hughes Medical Institute
Departments of Medicine and Genetics
Cancer Center and The Institute for Human Gene Therapy
University of Pennsylvania School of Medicine
Philadelphia, Pennsylvania 19104

WAFIK S. EL-DEIRY

Laboratory of Molecular Oncology and Cell Cycle Regulation
Howard Hughes Medical Institute
Departments of Medicine and Genetics
Cancer Center and The Institute for Human Gene Therapy
University of Pennsylvania School of Medicine
Philadelphia, Pennsylvania 19104

I. INTRODUCTION

Tumorigenesis results from genetic alterations, including the mutation or deletion of genes involved in growth inhibition. These tumor suppressor genes, therefore, have become a prime focus in cancer gene replacement. The p53 tumor suppressor has been studied extensively both in the laboratory and in clinical trials because of its dual ability to induce both cell cycle arrest and/or apoptosis. In certain conditions, however, p53 replacement may not be the ideal strategy. Consequently, other tumor suppressors have been studied for possible cancer gene therapy (see Fig. 1), including p21$^{WAF1/CIP1}$, p16, *Rb*, p14^{ARF}, p27, *E2F-1, BRCA1, VHL,* and *FHIT*. The following sections will discuss the background of each gene and its current role in gene therapy for cancer.

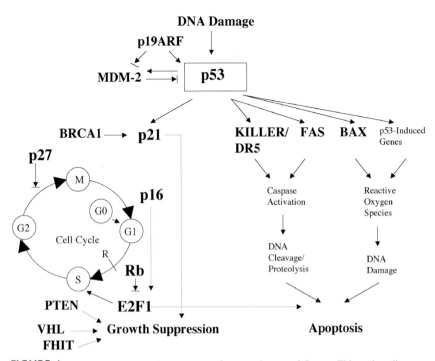

FIGURE 1 Tumor suppressors that are targets for gene therapy of Cancer. This review discusses tumor suppressors involved in cell-cycle regulation that have been studied as potential targets for gene replacement in the treatment of cancer. Loss of these tumor suppressors, most notably p53, results in tumor development and progression. p53 mediates the cellular response to DNA damage, resulting in growth arrest or in apoptosis. p21 is a main effector of p53 that mediates growth arrest and is a CDKI, along with p16 and p27, which help to regulate G1 transition. *Rb* helps to mediate cell-cycle progression from G1 to S phase. In addition, the tumor suppressors *BRCA1* (involved in breast cancer), *VHL* (involved in Von Hippel-Landau familial disease) *FHIT* (involved in chromosomal breakages), and *PTEN* (involved in cell attachment) also suppress growth through novel mechanisms. Likewise, the apoptosis induced by p53 is based on the activation of select targets. The recently cloned novel TRAIL target *KILLER/DR5* and the Fas family of death receptors can be activated by p53 and can induce apoptosis through initiation of a proteolytic caspase cascade. Two other p53-mediated targets involved in apoptosis, *bax* and the p53-induced genes or (PIGs), initiate cell death through reactive oxygen species.

II. p21$^{WAF1/CIP1}$

A. Introduction

p21$^{WAF1/CIP1}$ was originally cloned as a transcriptionally activated target of p53 [40] that was found to be a potent universal inhibitor of cyclin-dependent kinases (CDKs) [57,193]. It was shown that following DNA damage, p21 is required for p53-mediated G1 arrest [33]. p21 also associates with PCNA, which results in inhibition of DNA polymerase δ processivity *in vitro* [44]. The negative growth regulatory effects of p21 are also observed during differentiation [208]. Because p21 is an important downstream target of p53 and because it helps to mediate the growth suppressive effects of p53, its effectiveness as an anticancer treatment in gene therapy replacement has been studied. In addition, p21 has growth regulatory effects independent of p53 (reviewed by El-Deiry [39]), as it has been shown that expression of p21 effectively inhibits cancer cell growth *in vivo* [189].

B. Gene Therapy with *p21*

Most of the studies on p21 gene therapy have used adenovirus vectors. It has been shown that p21-expressing adenovirus (Ad-p21) can infect a variety of cancer cell types and can produce readily detectable p21 protein within 24 hours after infection (see Table 1) [205]. The induction of *p21* was comparable to, if not greater than, that induced by Ad-p53 [38,80,104]. Infection with Ad-p21 was able to inhibit tumor growth both *in vitro* and *in vivo,* causing cell-cycle arrest at G0/G1 and altering tumor morphology. Retroviral infection with p21 also caused growth suppression in mice [17]. Infection of normal tissues *in vivo* with Ad-p21 produced no adverse effects [76,205]. The exogenously transferred p21 was also shown to be functional, as histone H1 kinase assays showed that CDC2 activity and CDK2 activity were both decreased after Ad-p21 infection of glioma cells [20]. The growth inhibition induced by Ad-p21 was found to be greater than that produced by Ad-p53 in endometrial cancer

TABLE 1 Infection of Selected Cell Lines by Ad-p21

Cell type	Name of cell line	p53[a]	MOI[b]	In vitro	In vivo	Ref.
Breast	MCF-7	wt	10	x		[80]
	MDA-MB-231	mut	10	x		[80]
	SKBr3	mut	20	x		[128]
Bladder	RT4	wt	10	x		[54]
	UMUC3	mut	10	x		[54]
Cervical	C33A	wt	N.A.[c]	x		[171]
	HeLA	wt	N.A.	x		[171]
Choriocarcinoma	JEG3	wt	100	x		[104]
Colon	DLD-1	mut	50	x		[79]
	HCT116	wt	20	x		[104]
	LoVo	wt	50	x		[79]
	SW480	mut	20	x		[128]
Endometrial	SPEC-2	mut	50	x		[132]
Esophageal	TE-1	mut	100	x		[76]
	TE-3	wt	100	x		[76]
Glioblastoma	U-87 MG	wt	100	x		[50]
	U-373 MG	mut	10	x	x	[20]
Head and neck	MDA-686-LN	wt	100	x		[27]
	MDA 886	wt	100	x		[27]
	Tu-138	mut	100	x		[27]
	Tu-177	mut	100	x		[27]
Lung	H-358	null	10	x		[80]
	H460	wt	20	x		[128]
	H1299	null	50	x		[79]
Melanoma	7336	wt	100	x		[104]
	A875	wt	100	x		[104]
Prostate	LNCaP	wt	20	x		[51]
Renal cell carcinoma	293	N.A.	200	x		[205]

[a]p53 mutations are being examined because p21 mutations are very rare in human cancers.
[b]MOI indicates >50% transduction.
[c]N.A., not available.

cells [132]. Although cell death was noted in some tumors after Ad-p21 infection, no evidence of massive apoptosis was observed [80,104,128,205]. In contrast, apoptosis was induced by Ad-p21 overexpression in endometrial cancer cells [132], cervical cancer cells [171], and esophageal cancer cells [76]. Infection with Ad-p21 has also been reported to block apoptosis through inhibition of caspases [194].

C. Gene Therapy with a *p21* Mutant Deficient in PCNA Interaction

It has been shown that the N-terminal (cyclin- and CDK-interacting and inhibitory) domain of p21 is sufficient for cancer cell growth inhibition [128]. p21-341 is a p21 mutant in which a premature stop codon has been inserted at nucleotide 341 to delete the C-terminal domain that binds to proliferating cell nuclear antigen (PCNA). Transfection of

p21-341 into human colon cancer cells inhibits their growth more than that of wild-type p21 [128]. In addition, loss of the PCNA-interacting region of p21 contributes to a repair defect [97]. A strategy using p21 lacking the PCNA-interacting domain would be expected to inhibit growth but not stimulate DNA repair due to the absence of interaction with PCNA [97,128]. Therefore, an adenovirus containing *p21*-341 was constructed to evaluate whether it can play a role in growth suppression. It was shown that the p21-341 adenovirus (Ad-p21-341) can infect various human cancer cells *in vitro* like Ad-p21 and produced a significant suppression of DNA synthesis comparable to that of Ad-p53 and independent of p53 status. Furthermore, some DNA fragmentation was observed in lung and colon cancer cells following infection with Ad-p21-341. Interestingly, it has been reported that the N-terminal domain of p21 may suppress growth by a different mechanism than the C-terminal domain, as the N-terminal domain also seemed to inhibit E2F-1 activity [139]. Therefore, Ad-p21-341 may be an effective candidate for gene replacement in cancer.

D. Gene Therapy with p21 as an Alternative to p53

In evaluating the potential for p21 gene therapy, most groups have compared it to Ad-p53. In contrast to Ad-p53, Ad-p21 causes little or no apoptosis following the infection of many cell lines, including head and neck cancer [26], prostate cancer [51], lung cancer [80], gliomas [50], and melanomas [104]. However, p21 is a potent suppressor of cancer cell growth; thus, Ad-p21 may be an important alternative to Ad-p53 in situations where p53 is inactivated (see Fig. 2). For example, in some cell lines, p53 is nonfunctional because it is targeted for degradation by the human papillomavirus (HPV) type 16 or 18 E6 protein or because it is bound to inactivating cellular proteins such as the human homolog of the mouse double minute-2 (MDM2) oncoprotein. In these situations, where overexpression of these proteins may inactivate transduced exogenous p53, p21 may represent a more promising approach.

1. HPV16 *E6* Inactivates p53

Infection by HPV type 16 or 18 has been correlated with a greatly increased risk of cervical cancer worldwide [175]. It was discovered that the E6 protein of HPV targets human p53 for degradation through ubiquitin-mediated proteolysis [145]. In the presence of *E6,* a cellular protein called E6-associated protein (E6AP) binds to p53 and functions as an E3 ubiquitin ligase in mediating the degradation of p53. It has also recently been proposed that the HPV *E6* gene can also downregulate p53 function by targeting CBP/p300, a transcriptional co-activator of p53 [210].

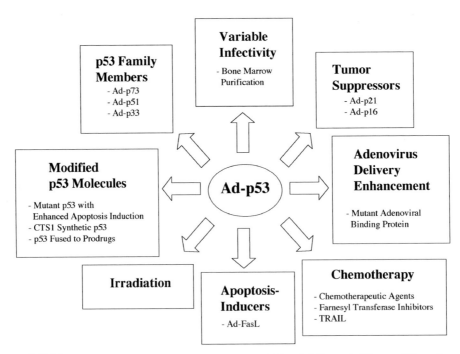

FIGURE 2 Mechanisms of resistance to p53 gene therapy. Although delivery of wild-type p53 by a virus vector suppresses the growth of and induces apoptosis in many human cancer cell lines, some cells are resistant to p53 gene therapy. An important cause of nonresponsiveness to Ad-p53 is target cell resistance to adenovirus infection, although the mechanism for this resistance is currently unknown. In other cell lines, overexpression of MDM2 or SV40 T antigen, two proteins that bind to and inactivate p53, can decrease the effectiveness of exogenously transduced p53. The expression of HPV16 *E6* causes enhanced degradation of p53. In these cases, gene replacement with other tumor suppressors, such as p21, may bypass this resistance.

In studies on the role of p53 in chemosensitivity, ovarian cancer cells that stably express HPV16 E6 protein were engineered, leading to endogenous p53 degradation [190]. Infection of these ovarian cancer cells with Ad-p53 produced only slight inhibition of DNA synthesis [128]. In contrast, infection of *E6*-overexpressing cells with Ad-p21-341 produced a significant suppression of DNA synthesis, which occurred at a much lower multiplicity of infection (MOI) as well. Infection with Ad-p21-341 also caused some DNA fragmentation, indicative of apoptosis. Thus, in HPV-associated cancers in which *E6* may be overexpressed, the inactivation of p53 may be bypassed by Ad-p21.

2. MDM2 Overexpression Inactivates p53

p53 can also be inactivated by binding to MDM2, which targets p53 for degradation by MDM2. The MDM2 oncogene is a target for transcriptional activation by p53 [9], but upon binding MDM2 conceals the transactivation domain of p53 and inhibits p53-dependent transcriptional activation [108]. The importance of MDM2 in development was shown when its targeted disruption in mice led to embryonic lethality [75,113]. Thus, although p53 activates MDM2, it is MDM2 that downregulates p53 in a feedback loop that

inhibits p53 function in both growth arrest and apoptosis. Recently, in addition to inactivating p53, MDM2 was also shown to promote the rapid degradation of p53 [58,87]. Therefore, in tumors where MDM2 is elevated, exogenous p53 gene replacement may not lead to optimal growth suppression.

To test this hypothesis, Ad-p53 was used to infect several human cancer cell lines that have high expression levels of the MDM2 protein [104]. In comparison to cell lines with low levels of MDM2, the tumor cell lines with elevated MDM2 were resistant to the growth inhibitory effects of Ad-p53. Although cancer cells that overexpress MDM2 were still readily infected by Ad-p53 and induced high expression of p53 protein as determined by western immunoblotting, their rate of DNA synthesis was only slightly decreased as compared to mock-infected or Ad-LacZ-infected cells, and they displayed a blunted induction of p21. Because the inhibitory effect of the exogenous p53 was blunted in these cell lines, Ad-p21 was tested to determine if it could bypass this MDM2-mediated inhibition of p53 because p21 is a downstream target of p53. Infection of MDM2-overexpressing cells with Ad-p21, however, resulted in a strong inhibition of cell-cycle progression and of cellular viability. Similar results were obtained with Ad-p21-341, suggesting that the cyclin/CDK-interacting domain is sufficient for bypassing p53 resistance in MDM2-overexpressing cells. Furthermore, persistence of the hyperphosphorylated form of the Rb protein correlated with tumor resistance to Ad-p53 infection, suggesting that the phosphorylation state of *Rb* may be a good indicator of p53-mediated growth inhibition. Therefore, Ad-p21 may be able to effectively bypass MDM2-mediated inactivation of p53 in cancer therapy.

3. SV40 Large T Antigen Inactivates p53

The p53 tumor suppressor was originally discovered in 1979 as a 53-kDa simian virus 40 (SV40) large T antigen (Tag)-associated protein [92]. SV40 Tag is known to bind to and inactivate several tumor suppressor genes, including p53. It was reported that over 60% of human mesotheliomas express SV40-like sequences [16]. In several mesothelioma samples, it was shown that p53 coexpressed with Tag, and Tag coprecipitated with p53, suggesting that these sequences may bind to and inactivate p53 [15]. It was also shown that these SV40-like sequences from mesotheliomas can bind to the *Rb* family members as well [31]. Therefore, in some mesotheliomas that are refractory to standard therapy, experimental treatment with p21 gene therapy may be a useful substitute for p53.

4. Hepatitis B Virus and Cytoplasmic Retention of p53

The hepatitis B virus X protein is negatively regulated by p53, and the X protein, in turn, inhibits p53 function [177]. The binding of X protein to p53 has been reported to exclude p53 from the nucleus [172]. Thus, in hepatitis-virus-dependent liver cancer, p53 may be dysfunctional due to exclusion from the nucleus by X protein [166]. This would be another situation where p21 or other tumor suppressors may be more suitable than p53.

5. Overexpression of TRAIL Decoy Receptor TRUNDD

A recent study suggests that p53-mediated apoptosis may be inhibited by overexpression of an anti-apoptotic TRAIL receptor, TRUNDD (also known as DcR2) [102]. It was previously shown that p53 overexpression can induce the pro-apoptotic TRAIL receptor KILLER/DR5 [188] (reviewed by Wu *et al.* [185]). Recent reports now suggest that Ad-p53 can upregulate the two anti-apoptotic TRAIL receptors, TRUNDD [102] and TRID (or DcR1) [151]. TRUNDD is similar to KILLER/DR5 except that it lacks the cytoplasmic apoptotic-signaling region, the "death domain"; consequently, TRUNDD is hypothesized to attenuate cell death by binding the apoptotic ligand TRAIL and preventing it from binding KILLER/DR5 [115,116,152].

E. Combination of Ad-p21 and Chemotherapy or Radiotherapy

Because p21 helps to induce cell-cycle arrest following DNA damage, loss of p21 in cancer cells might cause a deficiency in DNA repair, leading to chemosensitivity or to radiosensitivity. p21−/−cells have defective repair of damaged DNA *in vitro* and are more sensitive to ultraviolet radiation or to chemotherapeutic agents than cells with wild-type p21 [97,176]. When wild-type p53 in a human tumor is not correlated with radiosensitivity, p21 may be involved, and gene therapy with p21 may further enhance such sensitivity. For example, a retrovirus construct containing p21 was used to infect a rat glioma cell line [64]. In addition to tumor growth suppression, the introduction of p21 but not p53 rendered these cells more radiosensitive. In a colony formation assay following 8-Gy exposure, the number of p21-infected cells was decreased 93% compared to the controls and the p53-infected cells. Also, in some tumor cells p21 may help to mediate chemosensitivity. It has been shown that overexpression of p21 in a human sarcoma cell that lacked both p53 and *Rb* resulted in enhanced chemosensitivity [91]. However, in a human colon cancer cell line with disruption of both alleles of p21, chemosensitivity to DNA crosslinking agents appears to be enhanced [42,97,176]. The schedule of p21 administration may be important, as well, in determining synergy with chemotherapy, as infection with Ad-p21 prior to treatment with etoposide in osteosarcoma cells actually protected the cells from the cytotoxic effects of etoposide [127].

F. p21 Overexpression and Induction of Senescence

Another application for p21 gene therapy may be in the induction of terminal differentiation or of senescence. The expression of p21 was originally found to be elevated in senescent cells, and overexpression of p21 can induce premature senescence [111]. Some groups have hypothesized that the induction of senescence by p21 overexpression may be a mechanism to inhibit the growth of tumor cell lines, which usually have mechanisms to escape senescence. Ad-p21 infection of several human cancer cell lines, including lung, osteosarcoma, and colon, induced growth inhibition, induced morphological features characteristic of senescence, and decreased telomerase activity [79]. These senescent features included an enlarged and flattened cytoplasmic shape, an increased cytoplasmic-to-nuclear ratio, a decreased cell density, and an increase in a β-galactosidase activity characteristically associated with senescence [79]. The induction of senescence was also hypothesized to be a mechanism whereby normal cells may evade the effects of p21 overexpression. Infection of esophageal cancer cells with Ad-p21 eventually resulted in the onset of apoptosis;

however, normal epithelial keratinocytes underwent terminal differentiation following Ad-p21, withdrew from the cell cycle, and escaped cell death [76].

III. p16^{INK4}

A. Introduction

p16$^{INK4/CDKN2}$ is a tumor suppressor gene that encodes a specific inhibitor of cyclin D-CDK4 and CDK6. By controlling the activity of CDK4, p16 helps to control the phosphorylation of *Rb* at late G$_1$ [148]. p16 has been termed a tumor suppressor because it is frequently mutated or homozygously deleted in several types of cancers. p16 is also a major target for hypermethylation, leading to its inactivation in many cancers [61]. Homozygous p16 deletions have been found in over 50% of gliomas [110], but mutations in p16 are also found in other tumors, including esophageal, pancreatic, and non-small-cell lung cancer; lymphomas; and familial melanomas (reviewed by Foulkes *et al.* [45] and Pinyol *et al.* [124]).

B. Gene Therapy with p16

A p16-expressing adenovirus (Ad-p16) was first used to infect non-small-cell lung cancer lines that had homozygous deletions of p16 (see Table 2) [72]. These cell lines were readily infected by Ad-p16 *in vitro* and *in vivo* and expressed p16 at high levels; their growth rates were inhibited up to 90%; and cell-cycle arrest occurred at G$_0$/G$_1$. In contrast, infection of a normal mammary epithelial cell line with Ad-p16 did not cause growth inhibition or cell-cycle arrest. Ad-p16 has also been used to successfully infect and inhibit tumor growth up to 80% in several malignant glioma cells *in vitro*, with either wild-type p16 or homozygous p16 deletions [46]. Although Ad-p16 inhibited the growth of esophageal squamous cell cancers *in vitro*, no effect was observed for esophageal adenocarcinoma because these cells were poorly infected [146]. The utility of Ad-p16 for tumor inhibition has now been extended to other cell lines, including head and neck squamous cell carcinomas [137], prostate cancer [163] (reviewed by Allay *et al.* [4]), ovarian cancer cells [183], and pancreatic carcinoma lines [83].

Overexpression of p16 has also been reported to induce senescence. In prostate cancer cells, Ad-p16 induced growth arrest and cellular senescence only in those lines that expressed wild-type *Rb;* tumor lines containing mutant *Rb* were growth-inhibited by Ad-p16 but did not undergo senescence [162]. Because retroviruses infect hematopoietic cells more efficiently than adenoviruses, a p16 retrovirus was created and used to infect several leukemia cell lines *in vitro*.

TABLE 2   Infection of Selected Cell Lines by Ad-p16

Cell type	Name of cell line	p16	MOI^a	In vitro	In vivo	Ref.
Breast	MCF7	null	300	x		[143]
Cervical	C33A	wt	50	x		[143]
Colon	Lovo	null	30	x		[143]
Glioma	U-87 MG	null	125	x		[46]
	U-251 MG	null	125	x		[46]
Head and neck	JHU012	meth^c	5	x		[90]
	JHU022	wt	5	x		[90]
	Tu-138	wt	100	x		[60]
	Tu-177	wt	100	x		[60]
Hepatocellular	HuH7	null	50	x	x	[143]
Leukemia	HL60	null	N.A.^{b}	x		[131]
	Jurkat	null	N.A.	x		[131]
	K562	null	N.A.	x		[131]
Lung	H322	null	50	x		[72]
	H460	null	50	x	x	[72]
Ovarian	OVCA420	wt	100	x		[183]
	SKOV3	mut	100	x		[183]
Pancreas	MIAPaCa-2	null	30	x		[83]
	NP-9	mut	25	x		[18]
Prostate	DU145	mut	200	x		[4]
	LNCaP	null	200	x		[4]
	PC-3	null	200	x		[4]
	PPC-1	null	200	x	x	[4]
	TSU	null	200	x		[163]

^{a}MOI indicates >50% transduction.
^{b}N.A., not available.
^{c}Cell line has wild-type p16, but it is functionally inactivated by promoter hypermethylation.

Strong growth inhibition was observed in three lines with homozygous deletions of p16, but no inhibition was observed for a leukemia cell line with mutant p16 [131]. In contrast, another study reported that an ovarian cancer cell line with coexpression of both wild-type *Rb* and wild-type p16 was resistant to Ad-p16 [170]. Interestingly, a recent report suggests that intratumoral delivery of Ad-p16 to colon cancers in mice may decrease the incidence of liver metastases [184]. Finally, the feasibility of liposomally delivered p16 has recently been studied. Squamous cell carcinoma lines were transfected with liposomes containing p16, and cell-cycle arrest was subsequently observed with sustained p16 expression [99].

The efficacy of p16 gene therapy has been compared to p21 and p53 in causing growth inhibition in prostate cancer cells [51]. At comparable titers, p53 inhibited prostate cancer cell growth *in vitro* more significantly than either p21 or p16, which were comparable to each other. It was also shown that Ad-p53 induced a higher percentage of apoptosis among infected cells. In an *in vivo* model of prostate cancer in nude mice, all three viruses could inhibit tumor growth when initial tumor size was less than 200 mm^3; however, only Ad-p53 was effective in larger sized tumors.

In terms of chemosensitivity, the results with Ad-p16 are mixed. One group reported that overexpression of p16 in an IPTG-p16-inducible cell line made them more resistant to methotrexate, vinblastine, and cisplatin [164]. In contrast, transfection of p16 into a glioma with a homozygous deletion of p16 did not increase its chemosensitivity to nitrogen mustards [55]. Recent studies now suggest that the schedule of Ad-p16 administration in concert with chemotherapy may explain the efficacy of Ad-p16. It was reported that infection of human osteosarcoma cells with Ad-p16 before or during treatment with etoposide protected the cells from chemotherapy-induced cell death, whereas infection with Ad-p16 after etoposide treatment had no such protective effect [127]. It was hypothesized that pretreatment with Ad-p16 decreases chemosensitivity because the infected cells undergo G$_1$ arrest; hence, there are fewer cells in S phase, which is the cellular phase targeted by etoposide or gemcitabine. Another report suggests that Ad-p16 infection several days after treatment of human prostate cancer cells with cisplatin will actually increase chemosensitivity compared to either treatment alone [2]. Therefore, the timing of Ad-p16 infection and the induction of growth arrest appear to be important in regard to any possible synergy with chemotherapy.

Unlike p53 or even p21, the role or p16 in radiosensitivity remains controversial. One group reported that transfection of p16 into two human malignant melanoma cell lines, one of which had a homozygous deletion of p16, increased the radiosensitivity of both cell lines [100]. In contrast, the levels of p16 in bladder cancer cells were not altered following irradiation [135], and the presence of p16 was not correlated with radiosensitivity nor with the induction of p53 in several tumor cell lines [174].

p16, however, has been combined with Ad-p53, which induces apoptosis, to infect a panel of cancer cells [143]. It was shown that infection of tumor cell lines *in vitro* with both viruses induced apoptosis, whereas neither virus alone, at the MOIs used, caused apoptosis. Such a strategy may prove useful for tumors that have mutations in different tumor suppressors, such as gliomas, which often have deletions of p16 and mutations in p53. Thus, the combination of p16 and p53 may induce apoptotic cell death in cancer cells, although p16 alone has not been shown to induce apoptosis in any system. It is not entirely clear that, *in vivo,* the combination of p53 and p16 offers anything that cannot be achieved by p53 alone, if used at a sufficiently high MOI. It has not been shown, for example, that the combination of p53 and p16 is either more tumor specific or less toxic to normal cells. Ad-p16 has also been combined with Ad-p21 to enhance growth inhibition [107].

To improve upon the efficacy of wild-type p16 for gene therapy, a fusion p16 construct was recently created that tar-

geted glioma cells [1]. To wild-type p16 were added 500 base pairs of antisense uPAR (urokinase-type plasminogen activator receptor), which is a gene involved in malignant glioma invasion. The entire construct was then encoded in an adenovirus. Infection of glioma cells with this adenovirus containing antisense uPAR and p16 inhibited the ability of glioma cells to metastasize in Matrigel and in spheroid assays, presumably by decreasing the levels of uPAR while increasing the expression of p16 [1].

IV. *Rb*

A. Introduction

The retinoblastoma, or *Rb,* gene plays a role in the progression of the cell into the S phase. It was the first tumor suppressor identified and has been shown to be a nuclear phosphoprotein that regulates cell cycle progression by binding to several transcription factors needed for DNA synthesis, most notably the E2F family of transcription factors [180] (reviewed by Harbour and Dean [56]). The function of *Rb* is dependent upon its phosphorylation state. If *Rb* is hypophosphorylated, it can bind to and inactivate E2F, halting progression through the cell cycle. Hyperphosphorylation of Rb, however, releases it from binding to E2F, and the cell enters S phase. It has been hypothesized that *Rb* functions as a "guardian" at the R point, the point at which the cell commits itself in G$_1$ to progress to the M phase [180]. Although *Rb* was originally identified as being deleted in the rare disease retinoblastoma, it can be found mutated in many other tumors, including osteosarcoma, breast cancers, hepatocellular carcinoma, and bladder cancer.

B. Gene Therapy with *Rb*

The initial studies on the role of *Rb* as a tumor suppressor focused on the replacement of *Rb* into various *Rb*-defective human cell lines *in vitro,* which suppressed the tumorigenicity of the cell lines (reviewed in Xu [195]). An adenovirus construct encoding the full-length wild-type *Rb* was created and was used to infect several Rb−/− cell lines, including a non-small-cell lung carcinoma, bladder cancer, breast cancer, and an osteosarcoma [196]. Following infection, the cell lines expressed high levels of exogenous Rb proteins, mostly in the hypophosphorylated or unphosphorylated forms, as determined by immunocytochemistry and by western blot analyses. Infection of established bladder tumors in mice with the *Rb* adenovirus (Ad-Rb) slightly decreased the rate of growth of the tumors. Ad-Rb was also used to treat spontaneous pituitary melanotroph tumors in Rb ± mice *in vivo* [136]. Gene replacement with Ad-Rb decreased the growth of the tumors and increased the survival of the mice, compared to untreated controls. Ad-Rb was recently shown to infect a wide range of human cancer cell lines [32].

C. Gene Therapy for *Rb*-Resistant Tumors

Some tumor cells, however, even after replacement of the *Rb* gene, remain resistant to growth inhibition [65,209], suggesting that the *Rb* pathway may be inactivated in these cells [195]. Two approaches have been used to circumvent *Rb*-resistant tumors with *Rb* gene therapy. First, an N-terminal truncated *Rb* mutant has been reported to cause enhanced tumor suppression *in vitro* compared to the full-length wild-type *Rb* [198]. When this *Rb* mutant was expressed in an adenovirus and used to infect human bladder cancer cells in an *in vivo* mouse model, complete growth inhibition of the tumors was observed, and tumor regression was even noted in 50% of the tumors [196]. Second, another approach to *Rb*-resistant tumors focuses on using the other *Rb* family members, specifically p130. A retrovirus encoding p130 was used to infect a lung tumor cell line and caused growth inhibition both *in vitro* and *in vivo* in a nude mouse model [25].

V. p14^{ARF}

A. Introduction

The INK4A/ARF locus on chromosome 9p21 encodes two alternative transcripts, which have been identified as the tumor suppressors p16^{INK4a} and p14^{ARF} (reviewed by Chin *et al.* [23]). It has been calculated that mutations in p16/ARF are the second most common mutations in human cancers, after p53 perturbations [53]. For example, it has been reported that INK4A/ARF mutations occur in over 70% of mesotheliomas [22]. ARF (alternative reading frame) is the β transcript of the INK4A/ARF locus, utilizing exons 1β, 2, and 3 [37,130]. ARF is hypothesized to stabilize the p53 tumor suppressor protein by degrading MDM2, which targets p53 for ubiquitin-mediated degradation [125,208]. Overexpression of ARF induces growth inhibition through both a reported G_1 and G_2 arrest and eventual apoptosis [130].

B. Gene Therapy with *p14*^{ARF}

Because of the role of ARF as a possible tumor suppressor and its frequency of deletions in human cancers, one group has studied the feasibility of using ARF in cancer gene therapy studies. An adenovirus encoding p14 (Ad-p14) was used to infect a panel of mesothelioma cell lines (with mutant ARF and p16) and caused both G_1 growth arrest and eventual apoptosis [202]. Ad-p14 infection induced both p53 and p21 protein and dephosphorylated Rb. Interestingly, the apoptosis induced by Ad-p14 seemed to be dependent upon p53 as the HCT116 cancer cell line with wild-type p53 was more susceptible to Ad-p14-induced apoptosis than the HCT116 cell line genetically engineered with a homozygous deletion of p53 [202]. However, recent studies suggest that ARF may also have growth suppressive effects that are p53 independent but MDM2 dependent [179].

VI. p27^{Kip1}

A. Introduction

p27^{Kip1} is a universal cyclin-dependent kinase inhibitor that was first identified as a downstream effector of TGF-β and contact inhibition [84]. It belongs to the same family of CDK inhibitors as p21$^{WAF1/CIP1}$ and can also arrest cells in G_1 [126]. p27^{Kip1} is believed to be a tumor suppressor because it maps to a chromosomal site often deleted in leukemias, because it functions as a CDK inhibitor like p21, and because it can be found mutated in some tumors [161]. Recent efforts have uncovered frequent loss of p27 protein expression in colon, breast, and lung cancer through increased ubiquitin-mediated proteolysis of p27 in cancer [41,95,168].

B. Gene Therapy with *p27*^{Kip1}

To study its possible role in growth suppression, two groups constructed adenoviruses encoding p27^{Kip1} [20,28]. The p27 adenovirus (Ad-p27) efficiently infected a glioma and a squamous cell carcinoma cell line *in vitro*, producing profound growth suppression that was caused by cell-cycle arrest at G_0/G_1 [20]. Tumor growth was also inhibited in an *in vivo* glioma model [20]. Growth of a human breast cancer xenograft in nude mice was also inhibited by Ad-p27 [81]. In a human chloangiocarcinoma cell line, Ad-p27 infection induced apoptosis *in vitro*, with upregulation of Fas ligand but not Fas ligand mRNA [199]. Another study compared the efficacy of Ad-p27 to another tumor suppressor, Ad-p21. When compared to Ad-p21 following infection of a breast cancer cell line, Ad-p27 produced greater cytotoxicity and caused G_1/S arrest and decreased CDK2 activity at a lower MOI, suggesting that in some breast cancer cell lines p27 may be a better gene therapy agent than p21 [28].

C. Gene Therapy with a Modified Form of p27

Another group has attempted to improve the growth inhibition induced by p27 by modifying the tumor suppressor itself. p27 is usually degraded by ubiquitination following phosphorylation on residue 187T, which releases p27 from a complex with cyclin E; subsequently, a mutant p27 was created in which 187T was deleted [118]. Human lung cancer cell lines were then infected with Ad-p27 or the adenovirus-encoding mutant p27 (Ad-p27-mut). Ad-p27-mut induced a stronger G_1 arrest than Ad-p27 based on the percentage of cells in S phase as determined by fluorescence-activated cell sorting (FACS) analysis (control adenovirus was 42%, Ad-p27 was 22%, and Ad-p27-mut was 10%). Finally, another group created a chimeric tumor suppressor between p27 and p16 and encoded it in an adenovirus construct (Ad-p27/p16) [120]. In prostate, colon, pancreatic, and lung tumor cells, Ad-p27/p16 caused a stronger growth inhibition than either Ad-p27 or Ad-p16 alone [120]. This growth inhibition was independent of endogenous p53 or *Rb* status of the tumor cells. Interestingly, human cancer cell lines that were resistant to Ad-p16 or Ad-p27 were effectively induced to undergo apoptosis by Ad-p27/p16 [120].

VII. E2F-1

A. Introduction

E2F-1 belongs to the E2F family of transcription factors that regulate the expression of genes involved in cell-cycle progression, especially at the G_1 to S phase boundary (reviewed by Adams and Kaelin [3]). When complexed to the DP family of transcription factors [88], E2F-1 binds to DNA and activates the transcription of numerous genes that are involved in DNA synthesis and cell proliferation. The activity of E2F-1, in turn, is regulated by *Rb,* as binding of hypophosphorylated *Rb* to E2F-1 prevents E2F-1 from activating its target genes [19]. When E2F-1 is overexpressed in the absence of growth factors, it can drive quiescent cells into S phase [74]. Consequently, E2F-1 has been proposed to be an oncogene. When its regulation is disrupted, E2F-1 overexpression can cause uncontrolled growth, leading to tumor development [197]. In cooperation with *ras,* E2F-1 can transform rat embryonic fibroblasts [73,159]. In studies using Rb± and E2F-1 −/− mice, tumor development was delayed compared to growth in Rb± mice [200].

Although it can function as an oncogene, E2F-1 has also been postulated to be a tumor suppressor because it can induce apoptosis. Mice with a homozygous deletion of E2F-1 have a defect in thymocyte apoptosis, show increased cellular proliferation, and eventually develop numerous tumors [43,201]. Overexpression of E2F-1 in quiescent fibroblasts can cause apoptosis [74,129]. The apoptosis induced by E2F-1 overexpression does not seem to depend on the ability of E2F-1 to transactivate target genes or to induce DNA synthesis but rather on its ability to bind to DNA [63,123]. Whether the apoptosis induced by E2F-1 is dependent on the tumor suppressor p53, however, is unclear. In some cell lines, this apoptosis appears to be p53 dependent [191], whereas in other cells it seems to be independent of both p53 and *Rb,* although it can be inhibited by overexpression of *Rb* [63]. A recent study showed that apoptosis mediated by E2F-1 overexpression induces p53 protein, and both the induction of

apoptosis and accumulation of p53 can be blocked by expression of MDM2 [86]. Furthermore, DNA damage was reported to elevate E2F-1 expression independent of p53 status, suggesting a role for E2F-1 upregulation in apoptosis [103]. Recent studies have identified the p53 family member p73 as a transcriptional target of E2F-1 and as a potential mediator of E2F-1-dependent apoptosis [69,93].

B. Gene Therapy with E2F-1

Because of its possible role in the induction of apoptosis, E2F-1 overexpression in tumor cell lines was studied using an adenovirus that encodes full-length wild-type E2F-1 (Ad-E2F-1). Infection of human breast and ovarian cancer cell lines with Ad-E2F-1 produced a strong apoptotic response *in vitro* and *in vivo* in a nude mouse model (see Table 3) [67]. The induction of apoptosis was also observed following Ad-E2F-1 infection in human head and neck cancer cells [94], melanoma cells [34], glioma cells [49], and esophageal cancer cells [203]. The effect was related to a functional overexpression of E2F-1 as E2F-1 infection activated a fusion Rb–CAT reporter with E2F-1 binding sites [94]. The expression of other cell-cycle or apoptotic proteins was then examined following E2F-1 overexpression. In melanoma cells that underwent apoptosis following Ad-E2F-1 infection, the levels of two pro-apoptotic proteins, Bax and Bak, were not altered, although the expression of Bcl-X_l and Mcl-1, two anti-apoptotic proteins, decreased [34]. Similarly, Ad-E2F-1 infection of esophageal cancer cells decreased the protein levels of Bcl-2, Mcl-1, and Bcl-X_L [203]. However, in esophageal cancer cells that were resistant to Ad-E2F-1-mediated apoptosis, the levels of these anti-apoptotic proteins were not altered [203]. In contrast, in glioma cells infected with Ad-E2F-1, the expression of the anti-apoptotic Bcl-2 was not altered, and, in fact, an increase in the protein levels of p21 and p27 was observed [49].

The apoptosis induced by Ad-E2F-1 was independent of p53 in human breast, ovarian, and head and neck cancer cell lines, as these lines readily underwent apoptosis despite having endogenous mutant or null p53 alleles [67,94]. Ad-E2F-1 infection of esophageal cancer cells caused apoptosis but did

TABLE 3 Infection of Selected Cell Lines by Ad-E2F-1

Cell type	Name of cell line	p53	MOI[a]	*In vitro*	*In vivo*	Ref.
Esophageal	Yes-4	mut	100	x		[203]
	Yes-6	mut	100	x		[203]
Head and neck	Tu-138	mut	100	x	x	[94]
	Tu-167	mut	100	x		[94]
Melanoma	SK-MEL-2	mut	100	x		[34]
	SK-MEL-28	wt	100	x		[34]

[a]MOI indicates >50% transduction.

Raymond D. Meng and Wafik S. El-Deiry

not induce the expression of p53 protein [203]. Consequently, Ad-E2F-1 may be a method to eradicate tumor cells that are normally resistant to p53-mediated gene therapy. For example, tumor cell lines that overexpress MDM2 are resistant to infection with Ad-p53 but not with Ad-E2F-1, which induces a strong apoptotic response [204]. These studies, however, do not abrogate a contribution from p53 in the induction of apoptosis by E2F-1. In fact, p53-mediated apoptosis may further contribute to E2F-1-induced cell death. In human esophageal cancer cells infected with Ad-E2F-1, the expression of ARF was increased while the levels of MDM2, which inhibits p53, were decreased [70]. Thus, one group studied the feasibility of combined Ad-p53 and Ad-E2F-1 infection of esophageal cancer cells. Cells were first infected with Ad-p53 and then with Ad-E2F-1, and it was shown that this serial infection strategy was more efficient than simultaneous infection with both adenoviruses in inducing apoptosis. Presumably, infection with Ad-p53 first induces apoptosis but also enhances the levels of MDM2, which helps to downregulate the expression of p53; however, infection with Ad-E2F-1, in turn, downregulates MDM2 [70].

VIII. PTEN

A. Introduction

PTEN (also cloned as MMAC1 or TEP1) was originally identified to have homology to protein tyrosine phosphatases and was localized to human chromosome 10q23.3. The gene was found to be mutated in sporadic breast cancers and glioblastomas and in genetic diseases such as Cowden's syndrome, a hereditary predisposition to breast cancers. Functionally, PTEN dephosphorylated phosphatidylinositols phosphorylated by phosphatidylinositol 3′-kinase (PI3K). It was also reported that PTEN shared homology with a structural protein tensin, creating a role for PTEN in integrin signaling (reviewed by Tamura et al. [167]). Stable transfection of PTEN in glioblastoma or breast cancer cell lines lacking endogenous wild-type PTEN caused growth inhibition and eventual apoptosis [96,169,182]. This mechanism of growth inhibition has been linked to the Rb protein [117]. It has also been reported that overexpression of PTEN in selected glioblastoma cell lines can enhance their radiosensitivity but not chemosensitivity [182].

B. Gene Therapy with PTEN

Several groups have examined the utility of PTEN overexpression for gene therapy by constructing adenoviruses that encode wild-type PTEN (Ad-PTEN). Multiple cancer cell lines have been shown to be growth inhibited in vitro by Ad-PTEN, especially brain tumor cells (see Table 4). Infection of glioblastoma cell lines with Ad-PTEN caused growth

TABLE 4 Infection of Selected Cell Lines by Ad-PTEN

Cell type	Name of cell line	MOI[a]	In vitro	In vivo	Ref.
Endometrial	3H12	N.A.[b]	x	x	[142]
	HEC1-A	N.A.	x		[142]
	KLE	N.A.	x		[142]
	RL95-2	N.A.	x		[142]
Glioblastoma	LN-18	N.A.	x		[182]
	LN229	N.A.	x	x	[11,182]
	U87MG	N.A.	x	x	[11,21,169]
Ovarian	Caov-3	100	x		[106]
	ES-2	100	x		[106]
	MCAS	100	x		[106]
	MDAH 2774	100	x		[106]
	OV-1063	100	x		[106]
	OCAR-3	100	x		[106]
	SKOV3	100	x		[106]
	SW626	100	x		[106]
	TYK-nu	100	x		[106]
Prostate	LNCap	N.A.	x		[29]

[a] MOI indicates >50% transduction.
[b] N.A., not available.

inhibition and anoikis and also disrupted Akt-mediated signaling [30]. Another group showed that the mechanism of growth inhibition was related to the induction of $p27^{Kip1}$ to cyclin E, decreased CDK2 kinase activity, and decreased Rb protein [21]. In U87MG glioblastoma cells, Ad-PTEN induced anoikis, which was enhanced by TGF-β [11]. In ovarian cancer cell lines, Ad-PTEN caused G_1 arrest and apoptosis [106]. Another group showed that the induction of apoptosis in endometrial cancer cell lines with Ad-PTEN was independent of endogenous PTEN status (either wild-type or mutant) [142]. Finally, the efficacy of Ad-PTEN has been compared to that of Ad-p53. In prostate cancer cells that lack endogenous PTEN, Ad-PTEN inhibited growth more effectively than Ad-p53 but did not induce the same degree of apoptosis [29]. The apoptosis induced by Ad-PTEN, however, was blocked by overexpression of the anti-apoptotic protein Bcl-2, which did not have an effect in cells infected with Ad-p53 [29]. In contrast, in PTEN-mutant glioblastoma xenografts on nude mice, tumor growth was inhibited to the same degree by either Ad-PTEN or Ad-p53 [11].

IX. BRCA1

BRCA1, or breast cancer susceptibility gene 1, was originally identified as a tumor suppressor lost in familial breast or ovarian cancers [47,105]. BRCA1, however, does not seem to play an important role in the development of primary breast tumors. Interestingly, it has been reported that overexpression of BRCA1 can induce p53 expression both by transcriptional coactivation of p53 and by upregulation and stabilization of p14^{ARF} [160]. The function of *BRCA1* is currently unknown, although several hypotheses exist. It has been shown that an adenovirus encoding wild-type *BRCA1* (Ad-BRCA1) induces growth arrest following infection of human colon, lung, and breast cancer cell lines by dephosphorylating retinoblastoma protein and decreasing cyclin-dependent kinase activity [98]. In another study, infection of human breast or osteosarcoma cell lines with Ad-BRCA1 induced apoptosis independent of endogenous p53 status [149]. Concomitantly, it was observed that *BRCA1* overexpression elevated the protein levels of GADD45, Fas, FasL, and p21 [149]. These results suggest that *BRCA1* overexpression may help inhibit tumor growth through the activation of proteins involved in either DNA damage or cell-cycle regulation.

X. *VHL*

Von Hippel-Landau (*VHL*) loss is associated with cancers, including renal cell carcinomas [48] and central nervous system hemangioblastomas. Initially, *VHL* was shown to be able to inhibit transcription [35]. Because renal cell tumors and hemangioblastomas are highly vascular in nature, it was hypothesized that *VHL* may be involved in angiogenesis. In fact, it was recently reported that *VHL* may play a role in the proteolysis of hypoxia-inducible factors [101]. Because of its hypothesized tumor suppressor role, the utility of an adenovirus encoding *VHL* (Ad-VHL) for gene therapy of tumors was studied. Infection of renal cell carcinoma and breast cell carcinoma lines with Ad-VHL produced a G$_1$ cell-cycle arrest, resulting in growth inhibition but not apoptosis [82]. It was also shown that *VHL* overexpression increased the protein levels of p27, suggesting that p27 may be involved in *VHL*-mediated cell cycle arrest [82]. The tumor suppressive ability of Ad-p27 was then compared to other tumor suppressors, such as Ad-p53, Ad-p21, Ad-p27, and Ad-p16, in an *in vivo* xenograft model. Although Ad-p53 caused the strongest growth suppression, Ad-VHL was observed to cause the next highest inhibition of growth [173]. Further research may examine how *VHL*-mediated growth suppression can be applied to tumor growth inhibition, perhaps specifically focusing upon renal cell or breast cell carcinomas.

XI. *FHIT*

The fragile histidine triad (*FHIT*) gene, localized on human chromosome 3p14.2, was found to be mutated in early lung cancers and in renal cell carcinomas, suggesting a possible role as a tumor suppressor in select cancers (reviewed by Huebner *et al.* [66]). It has also been reported that mutation of *FHIT* occurs in esophageal squamous cell carcinomas [153], head and neck cancers [52], and endometrial carcinomas [147]. The protein product of *FHIT* was later shown to function as a dinucleoside triphosphate hydrolase [10]. Overexpression of *FHIT* in lung tumor cells was then shown to cause apoptosis *in vitro* [144]. *In vivo*, xenografts of renal cell carcinomas with stable expression of *FHIT* had delayed tumor formation compared to cells without expression of *FHIT* [181]. Because of the tumor-suppressive effects of *FHIT*, several groups have engineered viral constructs that encode *FHIT*. Infection of human lung or head and neck cancer cells with an adenovirus that expresses *FHIT* (Ad-FHIT) suppressed tumor growth both *in vitro* and *in vivo* [71]. Another group inhibited the growth of pancreatic cancer cell lines following infection with an adeno-associated adenovirus encoding *FHIT* [36]. This tumor-suppressive effect of *FHIT*, however, is not always observed in all cell lines. Infection of human cervical cancer cell lines with a retrovirus encoding FHIT did not delay anchorage-independent growth [186]. The ability of *FHIT* to suppress growth of tumors may be limited to specific cell lineages.

XII. APOPTOSIS-INDUCING GENES

A. Introduction

Because many tumor suppressors are currently known to function in the mediation of apoptosis, with p53 being a prime example, another strategy for cancer gene therapy has been to directly introduce pro-apoptotic genes into tumor cell lines. For example, rather than overexpressing p53, which presumably activates downstream targets involved in apoptosis such as *bax* or *Fas*, it may be more advantageous to overexpress these apoptosis-inducing genes directly. Two groups of apoptosis-inducing genes have been extensively studied: the pro-apoptotic members of the *bcl-2* gene family (including *bax*) and the *Fas* death receptor members. One of the difficulties in using apoptosis-inducing genes, however, has been in the construction of viral vectors to encode them (reviewed by Bruder *et al.* [13]). Because the apoptosis-inducing genes cause cell death upon overexpression, production of adenoviruses is difficult because overexpression of the pro-apoptotic genes eliminates the producer cells that package the viral constructs. For example, in creating an adenovirus that encodes *FasL*, it was shown that the cells usually used to propagate adenoviruses, human 293 cells, were eliminated by the shuttle plasmid that encoded murine *FasL*, presumably because the *FasL* was binding the *FasR* on 293 cells [89]. Several strategies have been employed to circumvent overexpression of the pro-apoptotic genes in viral constructs. In some vectors, the pro-apoptotic gene is placed under a promoter that can be regulated, such as the tetracycline-inducible promoter [139], or the myelin basic promoter [154]. Other groups modified the adenoviral vector to contain Cre-LoxP excision sites to keep the pro-apoptotic gene dormant unless it is released by coinfection with an adenovirus that expresses the Cre recombinase [85,140,157,192], or had the pro-apoptotic gene under the control of a GAL4 promoter which was only activated by coinfection with an adenovirus containing the GAL4 activator [119]. For example, bax was cloned into an adenovirus vector containing 5 GAL4-binding sites and a TATA box, and, when used for infections, this virus was coinfected with an adenovirus containing the transactivator GAL4/VP-16 [77,78]. Finally, it has been reported that production of adenoviruses containing pro-apoptotic genes is facilitated using the Ad-Easy system [178], which employs Escherichia coli, rather than mammalian cells, as the viral producer cells [59].

B. Gene Therapy with bax

bax is a pro-apoptotic member of the bcl-2 family, thought to induce apoptosis through enhancing the mitochondrial release of cytochrome c [114]. Loss of bax in genetically engineered mice resulted in increased tumor incidence, suggesting that bax may have a tumor suppressor role in vivo [206]. Frameshift mutations have also been reported in various cancer cell lines, including colon cancer and hematopoietic cancers [12,133]. Furthermore, bax is one of the target genes observed to be upregulated following Ad-p53 infection [8,121,134], suggesting that it may be an important mediator of p53-induced apoptosis. Overexpression of bax in human cancer cell lines has been reported to induce apoptosis independent of endogenous p53 status and to enhance the chemosensitivity or radiosensitivity of select tumor lines [85,141].

Several groups, therefore, have successfully developed adenoviruses that encode bax (Ad-Bax). Ad-Bax infection of human lung cancer cell lines, both in vitro and in vivo, caused apoptosis that was independent of endogenous p53 status [77,78]. Interestingly, lung cancer cell lines resistant to Ad-p53 were susceptible to apoptosis following Ad-Bax. In nude mice, Ad-Bax infection was initially not hepatotoxic, as determined by measuring the levels of the liver-associated enzymes ALT and AST. Using a binary vector system, another group regulated the expression of bax following Ad-Bax infection of prostate cancer cell lines and reported extensive apoptosis, which furthermore, was not inhibited by overexpression of the anti-apoptotic bcl-2 [62]. Infection with Ad-Bax was found to enhance the radiosensitivity of human ovarian cancer cell lines to radiotherapy through an increased induction of apoptosis in vitro or in vivo [6].

The gene therapy potential of Ad-Bax was recently combined with an adenovirus that expresses caspase-8 (Ad-Caspase-8) [156]. The caspases are cysteine proteases activated during apoptosis that are involved in the actual cleavage of intracellular protein targets. The caspases act in a pyramid cascade, whereby activation of caspase-8 induces the activity of other caspases. Ad-p53 has been observed to elevate the expression of various caspases, including caspase-8 [8,150]. Both Ad-Bax and Ad-Caspase-8 were encoded in adenoviruses containing loxP sequences; in order for the genes to be expressed, they had to be coinfected with an adenovirus carrying the Cre recombinase, which excises the genes [156]. In two glioblastoma cell lines, the combination of Ad-Bax and Ad-Caspase-8 induced more cell death (83%) than either Ad-Bax (49%) or Ad-Caspase-8 (55%) alone. The authors suggest that the two adenoviruses may work through separate pathways, as infection with Ad-Bax did not induce the expression of caspase-8 protein, and, likewise, infection with Ad-Caspase-8 did not induce Bax protein [156].

The other pro-apoptotic members of the bcl-2 family, related to bax, have also been tested as potential targets for cancer gene therapy. First, Bak has been shown to be upregulated following Ad-p53 infection in human lung cancer cells [8,121]. Bak was then cloned in an adenoviral vector containing GAL4 binding sites (Ad-Bak) [119]. Ad-Bak was then activated by coinfection with an adenovirus expressing a GAL4/GV16 fusion protein. Infection of several human cancer cell lines in vitro with Ad-Bax caused 40–60% apoptosis, which was induced independent of endogenous p53 or bcl-2 status. Infection with Ad-Bak in vivo also decreased tumor growth [119]. Finally, another group has created an adenovirus that encodes bcl-x_s (Ad-Bcl-x_s) [24], another pro-apoptotic bcl-2 family member. Ad-Bcl-x_s induced apoptosis in various tumor lines, including breast, colon, and stomach cancer cells, but not in normal cells. The virus, however, had only limited penetration into the tumors.

C. Gene Therapy with Fas

The Fas death receptor initiates apoptosis following binding of Fas ligand [165] (reviewed by Nagata and Golstein, [109] and Peter and Krammer [122]). In addition, Fas has been observed to be upregulated following Ad-p53 infection, suggesting it may function as a mediator of p53-induced apoptosis [134]. Currently, several groups have shown that infection of tumor cell lines with an adenovirus encoding Fas death receptor (Ad-Fas) produces a rapid and extensive apoptosis [7,154,158]. Other studies have examined the ability of an adenovirus encoding Fas ligand (Ad-FasL) to induce apoptosis. Prostate cancer cells, previously found to be resistant to apoptosis induced by Fas antibody, were infected with Ad-FasL in vitro and found to undergo extensive cell death [68]. Infection with Ad-FasL also caused extensive apoptosis in breast and brain cancer cells [112]. Similar results were also reported in vivo following injection of Ad-FasL into prostate cancer xenografts on nude mice [68]. It has been suggested that the apoptosis observed in vivo following Ad-FasL may also depend upon a bystander effect, as only 25–50% but not 100% of brain tumor cells were infected with Ad-FasL [112]. To enhance the apoptotic effect of Ad-FasL, it has been combined with an adenovirus that expresses caspase-3 (Ad-Caspase-3) [155]. Coinfection of glioma cells with Ad-FasL and Ad-Caspase-3 synergistically enhances the degree of apoptosis, compared to infection with other adenovirus alone [155].

D. Gene Therapy with the TRAIL Receptors

Currently, the use of other death receptors in cancer gene therapy is being explored. An adenovirus was recently constructed that contained the cytoplasmic domain of DR4 (Ad-DR4-CD) [194]. Infection of human breast, lung, and colon cancer cells with Ad-DR4-CD caused extensive apoptosis independent of cellular p53 status. Normal fibroblasts, however, are resistant to the apoptotic effects of Ad-DR4-CD.

XIII. CONCLUSIONS

Although most tumor suppressor replacement strategies have focused upon the p53 tumor suppressor gene, other genes have been studied. The p53 target, p21$^{WAF1/CIP1}$, is involved in cell-cycle inhibition and has been studied as an alternative to p53, especially in situations where p53 may not be entirely effective. The cell-cycle inhibitors p16, *Rb*, p14^{ARF}, and p27 may prove effective in specific tumors with mutations in those genes. Similarly, the tumor suppressors *BRCA1*, *VHL*, and *FHIT* may also be tumor-specific replacement strategies. Finally, the E2F-1 tumor suppressor may also induce apoptosis, when expressed at high levels, and may also be able to play a role in suppressing growth. Likewise, genes that specifically induce apoptosis, such as *bax, Fas,* or the TRAIL receptors, may prove important in cancer gene therapy. Cancer gene therapy strategies utilizing either tumor suppressor genes or apoptosis-inducing genes may eventually comprise clinical trials.

References

1. Adachi, Y., Liu, T., Mohanam, S., Mohan, P. M., Rajan, M. K., Gokasian, Z. L., Kyritsis, A. P., Sawaya, R., and Rao, J. S. (2000). Adenovirus-mediated delivery of a biscistronic construct containing antisense uPAR and sense p16 gene sequences suppress glioma invasion. *Proc. Annu. Meet. Am. Assoc. Cancer Res.* **41**, 387–388.

2. Adams, C. D., Steiner, M. S., and Allay, J. A. (2000). Adenovirus p16 synergizes with cisplatin to inhibit human prostate xenograft growth. *Proc. Annu. Meet. Am. Assoc. Cancer Res.* **41**, 122.

3. Adams, P. D., and Kaelin, Jr., W. G. (1995). Transcriptional control by E2F. *Semin. Cancer Biol.* **6**, 99–108.

4. Allay, J. A., Steiner, M. S., Zhang, Y., Reed, C. P., Cockroft, J., and Lu, Y. (2000). Adenovirus p16 gene therapy for prostate cancer. *World J. Urol.* **18**, 111–120.

5. Aoki, K., Akyurek, L. M., San, H., Leung, K., Parmacek, M. S., Nabel, E. G., and Nabel, G. J. (2000). Restricted expression of an adenoviral vector encoding Fas ligand (CD95L) enhances safety for cancer gene therapy. *Mol. Ther.* **1**, 555–565.

6. Arafat, W. O., Gomez-Navarro, J., Xiang, J., Barnes, M. N., Mahasrenshti, P., Alvarez, R. D., Siegal, G. P., Badib, A. L., Buschsbaum, D., Curiel, D. T., and Stackhouse, M. A. (2000). An adenovirus encoding proapoptotic bax induces apoptosis and enhances the radiation effect in human ovarian cancer. *Mol. Ther.* **1**, 545–554.

7. Arai, H., Gordon, D., Nabel, E. G., and Nabel, G. J. (1997). Gene transfer of Fas ligand induces tumor regression in vivo. *Proc. Natl. Acad. Sci. USA* **25**, 13862–13867.

8. Atencio, I., and Demers, G. W. (1999). Apoptosis in a hepatocellular carcinoma cell line in response to adenovirus-mediated transfer of wild-type p53 gene occurs by regulation of bcl-2 family members and activation of caspases. *Cancer Gene Ther.* **6**(suppl.), S5.

9. Barak, Y., Juven, T., Haffner, R., and Oren, M. (1993). MDM2 expression is induced by wild-type p53 activity. *EMBO J.* **12**, 461–468.

10. Barnes, L. D., Garrison, P. N., Siprashvili, Z., Guranowski, A., Robinson, A.K., Ingram, S.W., Croce, C.M., Ohta, M., and Huebner, F. (1996). FHIT, a putative tumor suppressor in humans, is a dinucleoside 5', 5'''-P-1,P-3-triphosphate hydrolase. *Biochemistry* **35**, 11529–11535.

11. Bookstein, R, Cheney, W., Neuteboom, S., Levy, A., Anderson, S., Johnson, D., Vaillancourt, M., Ramachandra, M., and Steck, P. (1999). Prospects for MMAC1/PTEN gene therapy of cancer. *Cancer Gene Ther.* **6**(suppl.), S1.

12. Brimmell, M., Mendiola, R., Mangion, J., and Packham, G. (1998). Bax frameshift mutations in cell lines derived from human haemopoietic malignancies are associated with resistance to apoptosis and microsatellite instability. *Oncogene* **16**, 1803–1812.

13. Bruder, J. T., Appiah, A., Kirkman, W. M., Chen, P., Tian, J., Reddy, D., Brough, D. E., Lizonova, A., and Kovesdi, I. (2000). Improved production of adenovirus vectors expressing apoptotic transgenes. *Hum. Gene Ther.* **11**, 139–149.

14. Cairns, P., Polascik, T. J., Eby, Y., and Sidransky, D. (1995). High frequency of homozygous deletion at p16/CDKN2 in primary human tumors. *Nat. Genet.* **11**, 210–212.

15. Carbone, M., Rizzo, P., Grimley, P. M., Procopio, A., Mew, D. J. Y., Shridhar, V., de Bartolomeis, A., Esposito, V., Giuliano, M. T., Steinberg, S. M., Levine, A. S., Giordano, A., and Pass, H.I. (1997). Simian virus-40 large T antigen binds p53 in human mesotheliomas. *Nat. Med.* **3**, 908–912.

16. Carbone, M., Pass, H. I., Rizzo, P., Marinetti, M., Di Muzio, M., Mew, D. J., Levine, A. S., and Procopio, A. (1994). Simian virus 40-like DNA sequences in human pleural mesothelioma. *Oncogene* **9**, 1781–1790.

17. Cardinali, M., Jakus, J., Shah, S., Ensley, J. F., Robbins, K. C., and Yeudall, W. A. (1998). p21(WAF1/CIP1) retards the growth of human squamous cell carcinomas in vivo. *Oral Oncol.* **34**, 211–218.

18. Cascallo, M., Mercade, E., Capella, G., Lluis, F., Fillat, C., Gomez-Foix, A. M., and Mazo, A. (1999). Genetic background determines the response to adenovirus-mediated wild-type p53 expression in pancreatic tumor cells. *Cancer Gene Ther.* **6**, 428–436.

19. Chellappan, S. P., Hiebert, S., Mudryj, M., Horowitz, J. M., and Nevins, J. R. (1991). The E2F transcription factor is a cellular target for the RB protein. *Cell* **65**, 1053–1061.

20. Chen, J., Willingham, T., Shuford, M., Bruce, D., Rushing, E., Smith, Y., and Nisen, P.D. (1996). Effects of ectopic overexpression of p21$^{WAF1/CIP1}$ on aneuploidy and the malignant phenotype of human brain tumor cells. *Oncogene* **13**, 1395–1403.

21. Cheney, I. W., Neuteboom, S. T., Vaillancourt, M. T., Ramachandra, M., and Bookstein, R. (1999). Adenovirus-mediated gene transfer of MMAC1/PTEN to glioblastoma cells inhibits S phase entry by the recruitment of p27^{Kip1} into cyclin E/CDK2 complexes. *Cancer Res.* **59**, 2318–2323.

22. Cheng, J. Q., Jhanwar, S. C., Klein, W. M., Bell, D. , Lee, W. C., Altomare, D. A., Nobori, T., Olopade, O. I., Buckler, A. J., and Testa, J. R. (1994). p16 alterations and deletion mapping of 9p21–p22 in malignant mesothelioma. *Cancer Res.* **54**, 5547–5551.

23. Chin, L., Pomerantz, J., and DePinho, R. A. (1998). The INK4a/ARF tumor suppressor: one gene—two products—two pathways. *Trends Biochem. Sci.* **23**, 291–296.

24. Clarke, M. F., Qian, D., Han, J., Nunez, G., and Wicha, M. (1999). Targeting the programmed cell death pathway for cancer treatment. *Cancer Gene Ther.* **6**(suppl.), S1.

25. Claudio, P. P., Howard, C. M., Pacilio, C., Cinti, C., Romano, G., Minimo, C., Maraldi, N. M., Minna, J. D., Gelbert, L., Leoncini, L., Tosi, G.M., Hicheli, P., Caputi, M., Giordano, G. G., and Giordano, A. (2000). Mutations in the retinoblastoma-related gene RB2/p130 in lung tumors and suppression of tumor growth in vivo by retrovirus-mediated gene transfer. *Cancer Res.* **60**, 372–382.

26. Clayman, G. L., Liu, T.-J., Overholt, S. M., Mobley, S. R., Wang, M., Janot, F., and Goepfert, H. (1996). Gene therapy for head and neck cancer: comparing the tumor suppressor gene p53 and a cell cycle regulator WAF1/CIP1 (p21). *Arch. Otolaryngol. Head Neck Surg.* **122**, 489–493.

27. Clayman G. L., El-Naggar, A. K., Roth, J. A., Zhang, W.-W., Goepfert, H., Taylor, D. L., and Liu, T.-J. (1995). In vivo molecular therapy with p53 adenovirus for microscopic residual head and neck squamous carcinoma. *Cancer Res.* **55**, 1–6.

28. Craig, C., Wersto, R., Kim, M., Ohri, E., Li, Z., Katayose, D., Lee, S. J., Trepel, J., Cowan, K., and Seth, P. (1997). A recombinant adenovirus expressing p27^{Kip1} induces cell cycle arrest and loss of cyclin-Cdk activity in human breast cancer cells. *Oncogene* **14**, 2283–2289.

29. Davies, M. A., Koul, D., Dhesi, H., Berman, R., McDonnell, T. J., McConkey, D., Yung, W. K., and Steck, P. A. (1999). Regulation of Akt/PKB activity, cellular growth, and apoptosis in prostate cancer cells by MMAC/PTEN. *Cancer Res.* **59**, 2551–2556.

30. Davies, M. A., Lu, Y., Sano, T., Fang, X., Tang, P., LaPushin, R., Koul, D. Bookstein, R., Stokoe, D., Yung, W. K., Mills, G. B., and Steck, P. A. (1998). Adenoviral transgene expression of MMAC/PTEN in human glioma cells inhibits Akt activation and induces anoikis. *Cancer Res.* **58**, 5285–5290.

31. DeLuca, A., Baldi, A., Esposito, V., Howard, C. M., Bagella, L., Rizzo, P., Caputi, M., Pass, H. I., Giordano, G. G., Baldi, F., Carbone, M., and Giordano, A. (1997). The retinoblastoma gene family pRb/p105, p107, pRb2/p130 and simian virus-40 large T-antigen in human mesotheliomas. *Nat. Med.* **3**, 839–840.

32. Demers, G. W., Harris, M. P., Wen, S. F., Engler, H., Nielsen, L. L., and Maneval, D.C. (1998). A recombinant adenoviral vector expressing full-length human retinoblastoma susceptibility gene inhibits human tumor cell growth. *Cancer Gene Ther.* **5**, 207–214.

33. Deng, C., Zhang, P., Harper, J. W., Elledge, S. J., and Leder, P. (1995). Mice lacking p21$^{CIP1/WAF1}$ undergo normal development, but are defective in G1 checkpoint control. *Cell* **82**, 675–684.

34. Dong, Y.-B., Yang, H.-L., Elliott, M. J., Liu, T.-J., Stilwell, A., Atienza, Jr., C., and McMasters, K. M. (1999). Adenovirus-mediated E2F-1 gene transfer efficiently induces apoptosis in melanoma cells. *Cancer* **86**, 2021–2033.

35. Duan D. R., Pause, A., Burgess, W. H., Aso, T., Chen, D. Y. T., Garrett, K. P., Conway, R. C., Conaway, J. W., Linehan, W. M., and Klausner, R. D. (1995). Inhibition of transcription elongation by the VHL tumor-suppressor protein. *Science* **269**, 1402–1406.

36. Dumon, K. R., Chakrani, F., During, M. J., Rosato, E. F., Williams, N. N., and Croce, C. C. (2000). Inhibition of tumor growth and induction of apoptosis by adeno-associated virus-mediated fragile histidine triad (FHIT) gene overexpression in pancreatic cancer cell lines. *Gastroenterology* **118** (suppl.), 3271.

37. Duro, D., Bernard, O., Della Valle, V., Berger, R., and Larsen, C. J. (1995). A new type of p16$^{INK4/MTS1}$ gene transcript expressed in B-cell malignancies. *Oncogene* **11**, 21–29.

38. Eastham, J. A., Hall, S. J., Sehgal, I., Wang, J., Timme, T. L., Yang, G., Connell-Crowley, L., Elledge, S.J., Zhang, W.-W., Harper, J. W., and Thompson, T. C. (1995). In vivo gene therapy with p53 or p21 adenovirus for prostate cancer. *Cancer Res.* **55**, 5151–5155.

39. El-Deiry, W. S. (1997). Role of oncogenes in resistance and killing by cancer therapeutic agents. *Curr. Opin. Oncol.* **9**, 79–87.

40. El-Deiry, W. S., Tokino, T., Velculescu, V. E., Levy, D. B., Parsons, R., Trent, J. M., Lin, D., Mercer, W. E., Kinzler, K. W., and Vogelstein, B. (1993). WAF1, a potential mediator of p53 tumor suppression. *Cell* **75**, 817–825.

41. Esposito, V., Baldi, A., De Luca, A., Groger, A. M., Loda, M., Giordano, G. G., Caputi, M., Baldi, F., Pagano, M., and Giordano, A. (1997). Prognostic role of the cyclin-dependent kinase inhibitor p27 in non-small-cell lung cancer. *Cancer Res.* **57**, 3381–3385.

42. Fan, S., Chang, J. K., Smith, M. L., Duba, D., Fornace, Jr., A. J., and O'Connor, P.M. (1997). Cells lacking CIP1/WAF1 genes exhibit preferential sensitivity to cisplatin and nitrogen mustard. *Oncogene* **14**, 2127–2136.

43. Field, S. J., Tsai, F.-Y., Kuo, F., Zubiaga, A. M., Kaelin, W. G., Livingston, D. M., Orkin, S. H., and Greenberg, M. E. (1996). E2F-1 functions in mice to promote apoptosis and suppress proliferation. *Cell* **85**, 549–561.

44. Flores-Rozas, H., Kelman, Z., Dean, F. B., Pan, Z. Q., Harper, J. W., Elledge, S. J., O'Donnell, M., and Hurwitz, J. (1994). CDK-interacting protein 1 directly binds with proliferating cell nuclear antigen and inhibits DNA replication catalyzed by the DNA polymerase delta holoenzyme. *Proc. Natl. Acad. Sci. USA* **91**, 8655–8659.

45. Foulkes, W. D., Flanders, T. Y., Pollock, P. M., and Hayward, N. K. (1997). The CDKN2A (p16) gene and human cancer. *Mol. Med.* **3**, 5–20.

46. Fueyo, J., Gomez-Manzano, C., Yung, W. K. A., Clayman, G. L., Liu, T.-J., Bruner, J., Levin, V. A., and Kyritsis, A. P. (1996). Adenovirus-mediated p16/CDKN2 gene transfer induces growth arrest and modifies the transformed phenotype of glioma cells. *Oncogene* **12**, 103–110.

47. Futreal, P. A., Liu, Q. Y., Shattuckeidens, D., Cochran, C., Harshman, K., Tavtigian, S., Bennett, L. M., Haugenstrano, A., Swensen, J., Miki, Y., Eddington, K., McClure, M., Frye, C., Weaverfeldhaus, J., Ding, W., Gholami, Z., Soderkvist, P., Terry, L., Jhanwar, S., Berchuck, A., Iglehart, J. D., Marks, J., Ballinger, D.G., Barrett, J. C., Skolnick, M. H., Kamb, A., and Wiseman, R. (1994). BRCA1 mutations in primary breast and ovarian carcinomas. *Science* **266**, 120–122.

48. Gnarra, J. R., Tory, K., Weng, Y., Schmidt, L., Wei, M. H., Li, H., Latif, F., Liu, S., Chen, F., Duh, F. M., Lubensky, I., Duan, D.R., Florence, C., Pozzatti, R., Walther, M. M., Bander, N. H., Grossman, H. B., Brauch, H., Pomer, S., Brooks, J. D., Isaacs, W. B., Lerman, M. I., Zbar, B., and Linehan, W.M. (1994). Mutations of the VHL tumor-suppressor gene in renal-carcinoma. *Nat. Genet.* **7**, 85–90.

49. Gomez-Manzano, C., Mitlianga, P., Fueyo, J., Liu, T.-J., Kyritsis, A. P., and Yung, W. K. A. (2000). Transfer of E2F-1 results in upregulation of Bcl-2, p21, and p27: action–reaction? *Proc. Annu. Meet. Am. Assoc. Cancer Res.* **41**, 351.

50. Gomez-Manzano, C., Fueyo, J., Kyritsis, A. P., McDonnell, T. J., Steck, P. A., Levin, V. A., and Yung, W. K. A. (1997). Characterization of p53 and p21 functional interactions in glioma cells en route to apoptosis. *J. Natl. Cancer Inst.* **89**, 1036–1044.

51. Gotoh, A., Kao, C., Ko, S.-C., Hamada, K., Liu, T. J., and Chung, L. W. K. (1997). Cytotoxic effects of recombinant adenovirus p53 and cell cycle regulator genes (p21$^{WAF1/CIP1}$ and p16^{CDKN4}) in human prostate cancers. *J. Urol.* **158,** 636–641.

52. Gotte, K., Hadaczek, P., Coy, J. F., Wirtz, H. W., Riedel, F., Neubauer, J., and Hormann, K. (2000). FHIT expression is absent or reduced in a subset of primary head and neck cancer. *Anticancer Res.* **20,** 1057–1060.

53. Haber, D. A. (1997). Splicing into senescence: the curious case of p16 and p19ARF. *Cell* **91,** 555–558.

54. Hall, M. C., Li, Y., Pong, R.-Y., Ely, B., Sagalowsky, A. I., and Hsieh, J.-T. (2000). The growth inhibitory effect of p21 adenovirus on human bladder cancer cells. *J. Urol.* **163,** 1033–1038.

55. Hama, S., Sadatomo, T., Yoshioka, H., Kurisu, K., Tahara, E., Naruse, I., Heike, Y., and Saijo, N. (1997). Transformation of human glioma cell lines with the p16 gene inhibits cell proliferation. *Anticancer Res.* **17,** 1933–1938.

56. Harbour, J. W., and Dean, D. C. (2000). Rb function in cell-cycle regulation and apoptosis. *Nat. Cell Biol.* **2,** E65–E67.

57. Harper, J. W., Adami, G. R., Wei, N., Keyomarsi, K., and Elledge, S. J. (1993). The p21 CDK-interacting protein Cip1 is a potent inhibitor of G1 cyclin-dependent kinases. *Cell* **75,** 805–816.

58. Haupt, Y., Maya, R., Kazaz, A., and Oren, M. (1997). MDM2 promotes the rapid degradation of p53. *Nature* **387,** 296–299.

59. He, T. C., Zhou, S., Da Costa, L. T., Yu, J., Kinzler, K. W., and Vogelstein, B. (1998). A simplified system for generating recombinant adenoviruses. *Proc. Natl. Acad. Sci. USA* **95,** 2509–2514.

60. Henderson, Y. C., Breau, R. L., Liu, T.-J., and Clayman, G. L. (2000). Telomerase activity in head and neck tumors after introduction of wild-type p53, p21, p16, and E2F-1 genes by means of recombinant adenovirus. *Head Neck* **22,** 347–354.

61. Herman, J. G., Jen, J., Merlo, A., and Baylin, S. B. (1996). Hypermethylation-associated inactivation indicates a tumor suppressor role for p16^{NK4B}. *Cancer Res.* **56,** 722–727.

62. Honda, T., Kagawa, S., Higuchi, M., Spurgers, K. B., Gjertsen, B. T., Bruckheimer, E. M., Brisbay, S. M., Roth, J. A., Fang, B., and McDonnell, T. J. (2000). A recombinant adenovirus expressing wild-type bax induces apoptosis in prostate cancer cells. *Proc. Annu. Meet. Am. Assoc. Cancer Res.* **41,** 50.

63. Hsieh, J.-K., Fredersdorf, S., Kouzarides, T., Martin, K., and Lu, X. (1997). E2F-1-induced apoptosis requires DNA binding but not transactivation and is inhibited by the retinoblastoma protein through direct interaction. *Genes Dev.* **11,** 1840–1852.

64. Hsiao, M., Tse, V., Carmel, J., Costanzi, E., Strauss, B., Haas, M., and Silverberg, G. D. (1997). Functional expression of human p21(WAF1/CIP1) gene in rat glioma cells suppresses tumor growth in vivo and induces radiosensitivity. *Biochem. Biophys. Res. Comm.* **233,** 329–335.

65. Huang, H.-J., Yee, J. K., Shew, J. Y., Chen, P. L., Bookstein, R., Friedmann, T., Lee, E. Y., and Lee, W. H. (1988). Suppression of the neoplastic phenotype by replacement of the retinoblastoma gene product in human cancer cells. *Science* **242,** 1563–1566.

66. Huebner, K., Garrison, P. N., Barnes, L. D., and Croce, C. M. (1998). The role of the FHIT/FRA3B locus in cancer. *Ann. Rev. Genet.* **32,** 7–31.

67. Hunt, K. K., Deng, J., Liu, T.-J., Wilson-Heiner, M., Swisher, S. G., Clayman, G., and Hung, M.-C. (1997). Adenovirus-mediated overexpression of the transcription factor E2F-1 induces apoptosis in human breast and ovarian carcinoma cell lines and does not require p53. *Cancer Res.* **57.**

68. Hyer, M. L., Rubinchik, S., Dong, J.-Y., Voelkel-Johnson, C., Gunhan, M., and Norris, J. S. (2000). Gene therapy of prostate cancer utilizing a FasL adenoviral vector. *Proc. Annu. Meet. Am. Assoc. Cancer Res.* **41,** 50–51.

69. Irwin, M., Marin, M. C., Phillips, A. C., Seelan, R. S., Smith, D. I., Liu, W., Flores, E. R., Tsai, K. Y., Jacks, T., Vousden, K. H., and Kaelin, W. G. (2000). Role for the p53 homologue p73 in E2F-1–induced apoptosis. *Nature* **407,** 645–648.

70. Itoshima, T., Fujiwara, T., Kataoka, M., Kodowaki, Y., Fukazawa, Shao, J., Waku, T., Tanaka, N., and Kodama, M. (2000). Induction of apoptosis in human esophageal cancer cells by sequential p53 and E2F-1 gene transfer: involvement of p53 accumulation via ARF-mediated MDM2 downregulation. *Proc. Annu. Meet. Am. Assoc. Cancer Res.* **41,** 352.

71. Ji, L., Fang, B., Yen, N., Fong, K., Minna, J. D., and Roth, J. A. (1999). Induction of apoptosis and inhibition of tumorigenicity and tumor growth by adenovirus vector-mediated fragile histidine triad (FHIT) gene overexpression. *Cancer Res.* **59,** 3333–3339.

72. Jin, X., Nguyen, D., Zhang, W.-W., Kyritsis, A. P., and Roth, J. A. (1995). Cell cycle arrest and inhibition of tumor cell proliferation by the p16INK4 gene mediated by an adenovirus vector. *Cancer Res.* **55,** 3250–3253.

73. Johnson, D. G., Cress, W. D., Jakoi, L., and Nevins, J. R. (1994). Oncogenic capacity of the E2F1 gene. *Proc. Natl. Acad. Sci. USA* **91,** 12823–12827.

74. Johnson, D. G., Schwarz, J. K., Cress, W. D., and Nevins, J. R. (1993). Expression of E2F-1 induces quiescent cells to enter S-phase. *Nature* **365,** 349–352.

75. Jones, S. N., Roe, A. E., Donehower, L. A., and Bradley, A. (1995). Rescue of embryonic lethality in MDM2-deficient mice by absence of p53. *Nature* **378,** 206–208.

76. Kadowaki, Y., Fujiwara, T., Fukazawa, T., Shao, J., Yasuda, T., Itoshima, T., Kagawa, S., Hudson, L. G., Roth, J. A., and Tanaka, N. (1999). Induction of differentiation-dependent apoptosis in human esophageal squamous cell carcinoma by adenovirus-mediated p21 sdi1 gene transfer. *Clin. Cancer Res.* **5,** 4233–4241.

77. Kagawa, S., Gu, J., Swisher, S. G., Ji, L., Roth, J. A., Lai, D., Stephens, L. C., and Fang, B. (2000). Antitumor effect of adenovirus-mediated bax gene transfer on p53-sensitive and p53-resistant cancer lines. *Cancer Res.* **60,** 1157–1161.

78. Kagawa, S., Pearson, S. A., Ji, L., Xu, K., McDonnell, T. J., Swisher, S., Roth, J. A., and Fang, B. (2000). A binary adenoviral vector for expressing high levels of the proapoptotic gene bax. *Gene Ther.* **7,** 75–79

79. Kagawa, S., Fujiwara, T., Kadowaki, Y., Fukazawa, T., Sok-Joo, R., Roth, J.A., and Tanaka, N. (1999). Overexpression of the p21 SDI1 gene induces senescence-like state in human cancer cells: implication for senescence-directed molecular therapy for cancer. *Cell Death Differ.* **6,** 765–772.

80. Katayose, D., Wersto, R., Cowan, K. H., and Seth, P. (1995). Effects of a recombinant adenovirus expressing WAF1/CIP1 on cell growth, cell cycle, and apoptosis. *Cell Growth Differ.* **6,** 1207–1212.

81. Kim, J., Hwang, E. S., Kim, J. S., You, E.-H., Lee, S. H., and Lee, J.-H. (1999). Intraperitoneal gene therapy with adenoviral-mediated p53 tumor suppressor gene for ovarian cancer model in nude mouse. *Cancer Gene Ther.* **6,** 172–178.

82. Kim, M., Katayose, Y., Li, Q., Rakkar, A. N., Li, Z., Hwang, S. G., Katayose, D., Trepel, J., Cowan, K.H., and Seth, P. (1999). Cancer gene therapy with a recombinant adenovirus expressing a Von Hippel-Landau tumor suppressor gene. *Cancer Gene Ther.* **6**(suppl.), S1–S2.

83. Kobayashi, S., Shirasawa, H., Sashiyama, H., Kawahira, H., Kaneko, K., Asano, T., and Ochiai, T. (1999). p16^{INK4A} expression adenoviral vector to suppress pancreas cancer cell proliferation. *Clin. Cancer Res.* **5,** 4182–4185.

84. Koff, A., Ohtsuki, E., Roberts, J., and Massague, J. (1993). Negative regulator of G1 in mammal cells: inhibition of cyclin E-dependent kinase by TGF-β. *Science* **257,** 1689–1694.

85. Komatsu, K., Suzuki, S., Shimosegawa, T., Miyazaki, J., and Toyota, T. (2000). Cre-loxP-mediated bax gene activation reduces growth rate

294   Raymond D. Meng and Wafik S. El-Deiry

and increases sensitivity to chemotherapeutic agents in human gastric cancer cells. *Cancer Gene Ther.* **7,** 885–892.

86. Kowalik, T. F., DeGregori, J., Leone, G., Jakoi, L., and Nevins, J. R. (1998). *E2F1*-specific induction of apoptosis and p53 accumulation, which is blocked by MDM2. *Cell Growth Differ.* **9,** 113–118.

87. Kubbutat, M. H. G., Jones, S. N., and Vousden, K. H. (1997). Regulation of p53 stability by MDM2. *Nature* **387,** 299–303.

88. La Thangue, N. B. (1994). DP and E2F proteins: components of a heterodimeric transcription factor implicated in cell cycle control. *Curr. Opin. Cell Biol.* **6,** 443–450.

89. Larregina, A. T., Morelli, A. E., Dewey, R. A., Castro, M. G., Fontana, A., and Lowenstein, P. R. (1998). FasL induces Fas/Apo1-mediated apoptosis in human embryonic kidney 293 cells routinely used to generate E1-deleted adenoviral vectors. *Gene Ther.* **5,** 563–568.

90. Li, D., Ling, D., Freimuth, P., and O'Malley, Jr., B. W. (1999). Variability of adenovirus receptor density influences gene transfer efficiency and therapeutic response in head and neck cancer. *Clin. Cancer Res.* **5,** 4175–4181.

91. Li, W. W., Fan, J., Hochhauser, D., and Bertino, J. R. (1997). Overexpression of p21^waf1 leads to increased inhibition of E2F-1 phosphorylation and sensitivity to anticancer drugs in retinoblastoma-negative human sarcoma cells. *Cancer Res.* **57,** 2193–2199.

92. Linzer, D. I. H., and Levine, A. J. (1979). Characterization of a 54K dalton cellular SV40 tumor antigen present in SV40-transformed cells and uninfected embryonal carcinoma cells. *Cell* **17,** 43–52.

93. Lissy, N. A., Davis, P. K., Irwin, M., Kaelin, W. G., and Dowdy, S. F. (2000). A common E2F-1 and p73 pathway mediates cell death induced by TCR activation. *Nature* **407,** 642–645.

94. Liu, T.-J., Wang, M., Breau, R. L., Henderson, Y., El-Naggar, A. K., Steck, K. D., Sicard, M. W., and Clayman, G. L. (1999). Apoptosis induction by E2F-1 via adenoviral-mediated gene transfer results in growth suppression of head and neck squamous cell carcinoma cell lines. *Cancer Gene Ther.* **6,** 163–171.

95. Loda, M., Cukor, B., Tam, S. W., Lavin, P., Fiorentino, M., Draetta, G. F., Jessup, J. M., and Pagano, M. (1997). Increased proteasome-dependent degradation of the cyclin-dependent kinase inhibitor p27 in aggressive colorectal carcinomas. *Nat. Med.* **3,** 152–154.

96. Lu, Y., Lin, Y. Z., LaPushin, R., Cuevas, B., Fang, X., Yu, S. X., Davies, M. A., Khan, H., Furui, T., Mao, M., Zinner, R., Hung, M. C., Steck, P., Siminovitch, K., and Mills, G. B. (1999). The PTEN/MMAC1/TEP tumor suppressor gene decreases cell growth and induces apoptosis and anoikis in breast cancer cells. *Oncogene* **18,** 7034–7045.

97. McDonald, III, E. R., Wu, G. S., Waldman, T., and El-Deiry, W. S. (1996). Repair defect in p21^WAF1/CIP1 –/– human cancer cells. *Cancer Res.* **56,** 2250–2255.

98. MacLachlan, T. K., Somasundaram, K., Sgagias, M., Shifman, Y., Muschel, R. J., Cowan K. H., and El-Deiry, W. S. (2000). BRCA1 effects on the cell cycle and the DNA damage response are linked to altered gene expression. *J. Biol. Chem.* **275,** 2777–2785.

99. Marron-Terada, P. G., Park, J. W., and Belinsky, S. A. (2000). Liposome-mediated delivery of the p16INK4A gene in non-small cell lung cancer. *Proc. Annu. Meet. Am. Assoc. Cancer Res.* **41,** 465–466.

100. Matsumura, Y., Yamagishi, N., Miyakoshi, J., Imamura, S., and Takebe, H. (1997). Increase in radiation sensitivity of human malignant melanoma cells by expression of wild-type p16 gene. *Cancer Lett.* **115,** 91–96.

101. Maxwell, P. H., Wiesener, M. S., Chang, G. W., Clifford, S. C., Vaux, E. C., Cockman, M. E., Wykoff, C. C., Pugh, C. W., Maher, E. R., and Ratcliffe, P. J. (1999). The tumour suppressor protein VHL targets hypoxia-inducible factors for oxygen-dependent proteolysis. *Nature* **399,** 271–275.

102. Meng, R. D., McDonald, E. R., III, Sheikh, M. S., Fornace, A. J., and El-Deiry, W. S. (2000). The TRAIL decoy receptor TRUNDD (DcR2, TRAIL-R4) is induced by adenovirus-p53 overexpression and

can delay TRAIL-, p53-, and KILLER/DR5-dependent colon cancer apoptosis. *Molec. Ther.* **1,** 130–144.

103. Meng, R. D., Phillips, P., and El-Deiry, W. S. (1999). p53-independent increase in E2F-1 expression enhances the cytotoxic effects. *Intl. J. Oncol.* **14,** 5–14.

104. Meng, R., Shih, H., Prabhu, N. S., George, D. L., and El-Deiry, W. S. (1997). Bypass of abnormal MDM2 inhibition of p53-dependent growth suppression. *Clin. Cancer Res.* **4,** 251–259.

105. Miki, Y., Swensen, J., Shattuckeidens, D., Futreal, P. A., Harshman, K., Tavtigian, S., Liu, Q. Y., Cochran, C., Bennett, L. M., Ding, W., Bell, R., Rosenthal, J., Hussey, C., Tran, T., McClure, M., Frye, C., Hattier, T., Phelps, R., Haugenstrano, A., Katcher, H., Yakumo, K., Gholami, Z., Shaffer, D., Stone, S., Bayer, S., Wray, C., Bogden, R., Dayananth, P., Ward, J., Tonin, P., Narod, S., Bristow, P. K., Norris, F. H., Helvering, L., Morrison, P., Rosteck, P., Lai, M., Barrett, J. C., Lewis, C., Neuhausen, S., Cannonalbright, L., Goldgar, D., Wiseman, R., Kamb, A., and Skolnick, M. H. (1994). A strong candidate for the breast and ovarian-cancer susceptibility gene BRCA1. *Science* **266,** 66–71.

106. Minaguchi, T., Mori, T., Kanamori, Y., Matsushima, M., Yoshikawa, H., Taketani, Y., and Nakamura, Y. (1999). Growth suppression of human ovarian cancer cells by adenovirus-mediated transfer of the PTEN gene. *Cancer Res.* **59,** 6063–6067.

107. Mobley, S. R., Liu, T. J., Hudson, J. M., and Clayman, G. L. (1998). In vitro growth suppression by adenoviral transduction of p21 and p16 in squamous cell carcinoma of the head and neck: a research model for combination gene therapy. *Arch. Otolaryngol. Head Neck Surg.* **124,** 88–92.

108. Momand, J., Zambetti, G. P., Olson, D. C., George, D., and Levine, A. J. (1992). The MDM-2 oncogene product forms a complex with the p53 protein and inhibits p53-mediated transactivation. *Cell* **69,** 1237–1245.

109. Nagata, S., and Golstein, P. (1995). The Fas death factor. *Science* **267,** 1449–1456.

110. Nobori, T., Miura, K., Wu, D. J., Lois, A., Takabayashi, K., and Carson, D. A. (1994). Deletions of the cyclin-dependent kinase-4 inhibitor gene in multiple human cancers. *Nature* **368,** 753–756.

111. Noda, A., Ning, Y., Venable, S. F., Pereira, S. D., and Smith, J. R. (1994). Cloning of senescent cell-derived inhibitors of DNA synthesis using an expression screen. *Exp. Cell Res.* **211,** 90–98.

112. Norris, J. S., Hyer, M. L., Rubinchik, S., Wang, D., Ding, R.-Z., Berger, M. S., and Dong, J. Y. (1999). Adenoviral-delivered Fas ligand fusion gene therapy for breast, prostate, and breast cancer. *Cancer Gene Ther.* **6**(suppl.), S3–S4.

113. de Oca Luna, R. M., Wagner, D. S., and Lozano, G. (1995). Rescue of early embryonic lethality in MDM2-deficient mice by deletion of p53. *Nature* **378,** 203–206.

114. Oltavi, Z. N., Milliman, C. L., and Korsmeyer, S. J. (1993). Bcl-2 heterodimerizes in vivo with a conserved homolog, bax, that accelerates programmed cell death. *Cell* **74,** 609–619.

115. Pan, G., Ni, J., Wei, Y. F., Yu, G. L., Gentz, R., and Dixit, V. M. (1997). An antagonist decoy receptor and a death domain-containing receptor for TRAIL. *Science* **277,** 815–818.

116. Pan, G., O'Rourke, K., Chinnaiyan, A. M., Gentz, R., Ebner, R., Ni, J., and Dixit, V. M. (1997). The receptor for the cytotoxic ligand TRAIL. *Science* **276,** 111–113.

117. Paramio, J. M., Navarro, M., Segrelles, C., Gomez-Cassero, E., and Jorcano, J. L. (1999). PTEN tumor suppressor is linked to the cell cycle through the retinoblastoma protein. *Oncogene* **18,** 7462–7468.

118. Park, K.-H., Lee, J. H., Yoo, C. G., Chung, H. S., Shim, Y. W., Han, S. K., Shim, Y. W., and Lee, C.-T. (2000). An adenovirus expressing mutant p27 showed more potent antitumor response than adenovirus-wild p27. *Proc. Annu. Meet. Am. Assoc. Cancer Res.* **41,** 352.

119. Pataer, A., Fang, B., Yu, R., Kagawa, S., Hunt, K. K., McDonnell, T. J., Roth, J. A., and Swisher, S. G. (2000). Adenoviral bak overexpression mediates caspase-dependent tumor killing. *Cancer Res.* **60**, 788–792.

120. Patel, S. D., Tran, A. C., Lampere, L., Gyuris, J., and McArthur, J. G. (1999). Cancer gene therapy with novel chimeric p27/p16 tumor suppressor genes. *Cancer Gene Ther.* **6**(suppl.), S4.

121. Pearson, A. S., Spitz, F. R., Swisher, S. G., Kataoka, M., Sarkiss, M. G., Meyn, R. E., McDonnell, T. J., Cristiano, R. J., and Roth, J.A. (2000). Up-regulation of the proapoptotic mediators bax and bak after adenovirus-mediated p53 gene transfer in lung cancer cells. *Clin. Cancer Res.* **6**, 887–890.

122. Peter, M. E., and Krammer, P. H. (1998). Mechanisms of CD95 (APO-1/Fas)-mediated apoptosis. *Curr. Opin. Immunol.* **10**, 545–551.

123. Phillips, A. C., Bates, S., Ryan, K. M., Helin, K., and Vousden, K. H. (1997). Induction of DNA synthesis and apoptosis are separable functions of E2F-1. *Genes Dev.* **11**, 1853–1863.

124. Pinyol, M., Hernandez, L., Caz Balbin, M., Jares, P., Fernandez, P. L., Montserrat, E., Cardesa, A., Lopez-Otin, C., and Campo, E. (1997). Deletions and loss of expression of p16^{INK4a} and p21^{WAF1} genes are associated with aggressive variants of mantle cell lymphomas. *Blood* **89**, 272–280.

125. Pomerantz, J., Schreiber-Agus, N., Liegeois, N. J., Silverman, A., Alland, L., Chin, L., Potes, J., Chen, K., Orlow, I., Lee, H. W., Cordon-Cardo, C., and DePinho, R. A. (1998). The INK4a tumor suppressor gene product, p19^{ARF}, interacts with MDM2 and neutralizes MDM2's inhibition of p53. *Cell* **92**, 713–723.

126. Ponce-Casteneda, M., Lee, M., Latres, E., Polyak, K., Lacombe, L., Montgomery, K., Matthew, S., Krauter, K., Sheifeld, J., Massague, J. *et al.* (1995). p27^{Kip1}: chromosomal mapping to 12p12.2–12p13.1 and absence of mutations in human tumors. *Cancer Res.* **55**, 1211–1214.

127. Prabhu, N. S., Somasundaram, K., Tian, H., Enders, G. H., Satyamoorthy, K., Herlyn, M., and El-Deiry, W. S. (1999) The administration schedule of cyclin-dependent kinase inhibitor gene therapy in etoposide chemotherapy is a major determinant of cytotoxicity. *Int. J. Oncol.* **15**, 209–216.

128. Prabhu, N. S., Blagosklonny, M. V., Zeng, Y.-X., Wu, G. S., Waldman, T., and El-Deiry, W. S. (1996). Suppression of cancer cell growth by adenovirus expressing p21$^{WAF1/CIP1}$ deficient in PCNA interaction. *Clin. Cancer Res.* **2**, 1221–1229.

129. Qin, X.-Q., Livingston, D. M., Kaelin, W. G., and Adams, P. D. (1994). Deregulated transcription factor E2F-1 expression leads to S-phase entry and p53-mediated apoptosis. *Proc. Natl. Acad. Sci. USA* **91**, 10918–10922.

130. Quelle, D. E., Zindy, F., Ashmun, R. A., and Sherr, C. J. (1995). Alternative reading frames of the INK4a tumor suppressor gene encode two unrelated proteins capable of inducing cell cycle arrest. *Cell* **83**, 993–1000.

131. Quesnel, B., Preudhomme, C., Lepelley, P., Hetuin, D., Vanrumbeke, M., Bauters, F., Velu, T., and Fenaux, P. (1996). Transfer of p16$^{INK4A/CDKN2}$ gene in leukaemic cell lines inhibits cell proliferation. *Br. J. Hematol.* **95**, 291–298.

132. Ramondetta, L., Mills, G. B., Burke, T. W., and Wolf, J. K. (2000). Adenovirus-mediated expression of p53 or p21 in a papillary serous endometrial carcinoma cell line (SPEC-2) results in both growth inhibition and apoptotic cell death: potential application of gene therapy to endometrial cancer. *Clin. Cancer Res.* **6**, 278–284.

133. Rampino, N., Yamamoto, H., Ionov, Y., Li, Y., Sawai, H., Reed, J. C., and Perucho, M. (1997). Somatic frameshift mutations in the BAX gene in colon cancers of the microsatellite mutator phenotype. *Science* **275**, 967–969.

134. Reiser, M., Neumann, I., Schmiegel, W., Wu, P.-C., and Lau, J. Y. N. (2000). Induction of cell proliferation arrest and apoptosis in hepatoma cells through adenoviral-mediated transfer of p53 gene. *J. Hepatol.* **32**, 771–782.

135. Ribeiro, J. C., Hanley, J. R., and Russell, P. J. (1996). Studies of X-irradiated bladder cancer cell lines showing differences in p53 status: absence of a p53-dependent cell cycle checkpoint pathway. *Oncogene* **13**, 1269–1278.

136. Riley, D. J., Nikitin, A. Y., and Lee, W. H. (1996). Adenovirus-mediated retinoblastoma gene therapy suppresses spontaneous pituitary melanotroph tumors in Rb ±/–mice. *Nat. Med.* **2**, 1316–1321.

137. Rocco, J. W., Li, D., Liggett, W. H., Jr., Duan, L., Saunders, Jr., J. K., Sidransky, D., and O'Malley, Jr., B. W. (1998). p16^{INK4A} adenovirus mediated gene therapy for human head and neck squamous cell cancer. *Clin. Cancer Res.* **4**, 1697–1704.

138. Rousseau, D., Cannella, D., Boulaire, J., Fitzgerald, P., Fotedar, A., and Fotedar, R. (1999). Growth inhibition by CDK-cyclin and PCNA binding domains of p21 occurs by distinct mechanisms and is regulated by ubiquitin-proteasome pathway. *Oncogene* **18**, 4313–4325.

139. Rubinchik, S., Ding, R., Qiu, A. J., Zhang, F., and Dong, J. (2000). Adenoviral vector which delivers FasL-GFP fusion protein regulated by the tet-inducible expression system. *Gene Ther.* **7**, 875–885.

140. Sakai, K., Mitani, K., and Miyazaki, J. (1995). Efficient regulation of gene expression by adenovirus vector-mediated delivery of the Cre recombinase. *Biochem. Biophys. Res. Commun.* **217**, 393–401.

141. Sakakura, C., Sweeney, E. A., Shirahama, T., Igarashi, Y., Hakomori, S., Tsujimoto, H., Imanishi, T., Ogaki, M., Ohyama, T., Yamazaki, J., Hagiwara, A., Yamaguchi, T., Sawai, K., and Takahashi, T. (1996). Overexpression of bax sensitizes human breast cancer MCF-7 cells to radiation-induced apoptosis . *Int. J. Cancer* **67**, 101–105.

142. Sakurada, A., Hamada, H., Fukushige, S., Yokoyama, T., Yoshinaga, K., Furukawa, T., Sato, S., Yajima, A, Sato, M., Fujimura, S., and Horii, A. (1999). Adenovirus-mediated delivery of the PTEN gene inhibits cell growth by induction of apoptosis in endometrial cancer. *Int. J. Oncol.* **15**, 1069–1074.

143. Sandig, V., Brand, K., Herwig, S., Lukas, J., Bartek, J., and Strauss, M. (1997). Adenovirally transferred p16$^{INK4/CDKN2}$ and p53 genes cooperate to induce apoptotic tumor cell death. *Nat. Med.* **3**, 313–319.

144. Sard, L., Accornero, P., Tornielli, S., Delia, D., Bunone, G., Campiglio, M., Colombo, M.P., Gramegna, M., Croce, C. M., Pierotti, M. A., and Sozzi, G. (1999). The tumor-suppressor gene FHIT is involved in the regulation of apoptosis and in cell cycle control. *Proc. Natl. Acad. Sci. USA* **96**, 8489–8492.

145. Scheffner, M., Werness, B. A., Huibregtse, J. M., Levine, A. J., and Howley, P. M. (1990). The E6 oncoprotein encoded by human papillomavirus types 16 and 18 promotes the degradation of p53. *Cell* **63**, 1129–1136.

146. Schrump, D. S., Chen, G. A., Consuli, U., Jin, X., and Roth, J. A. (1996). Inhibition of esophageal cancer proliferation by adenovirally mediated delivery of p16^{INK4}. *Cancer Gene Ther.* **3**, 357–364.

147. Segawa, T., Sasagawa, T., Saijoh, K., and Inoue, M. (2000). Clinicopathological significance of fragile histidine triad transcription protein expression in endometrial carcinomas. *Clin. Cancer Res.* **6**, 2341–2348.

148. Serrano, M., Hannon, G. J., and Beach, D. (1993). A new regulatory motif in cell-cycle control causing specific inhibition of cyclin D/CDK4. *Nature* **366**, 704–707.

149. Sgagias, M. K., Kim, M., Rikiyama, T., Yun, J. J., and Cowan, K. H. (2000). Induction of apoptosis by adenovirus-mediated overexpression of BRCA1 in T47D human breast cancer cells. *Proc. Annu. Meet. Am. Assoc. Cancer Res.* **41**, 537.

150. Sheikh, M. S., Fernandes-Salas, E., Chou, J., Brooks, K., Huang, Y., and Fornace, A. J. (2000). p53 mediated apoptosis in a caspase 8-dependent but FADD-independent manner. *Proc. Annu. Meet. Am. Assoc. Cancer Res.* **41**, 555.

151. Sheikh, M. S., Huang, Y., Fernandez-Salas, E. A., El-Deiry, W. S., Friess, H., Amundson, S., Yin, J., Meltzer, S.J., Holbrook, N. J.,

and Fornace, Jr., A. J. (1999). The antiapoptotic decoy receptor TRID/TRAIL-R3 is a p53-regulated DNA damage-inducible gene that is overexpressed in primary tumors of the gastrointestinal tract. *Oncogene* **18**, 4153–4159.

152. Sheridan, J. P., Marsters, S. A., Pitti, R. M., Gurney, A., Skubatch, M., Baldwin, D., Ramakrishnan, L., Gray, C.L., Baker, K., Wood, W. I., Goddard, A. D., Godowski, P., and Ashkenazi, A. (1997). Control of TRAIL-induced apoptosis by a family of signaling and decoy receptors. *Science* **277**, 818–821.

153. Shimada, Y., Sato, F., Watanabe, G., Yamasaki, S., Kato, M., Maeda, M., and Imamura, M. (2000). Loss of fragile histidine triad gene expression is associated with progression of esophageal squamous cell carcinoma, but not with the patient's prognosis and smoking history. *Cancer* **89**, 5–11.

154. Shinoura, N., Ohashi, M., Yoshida, Y., Kirino, T., Asai, A., Hashimoto, M., and Hamada, H. (2000). Adenovirus-mediated overexpression of fas induces apoptosis of gliomas. *Cancer Gene Ther.* **7**, 224–232.

155. Shinoura, N., Muramatsu, Y., Yoshida, Y., Asai, A., Kirino, T., and Hamada, H. (2000). Adenovirus-mediated transfer of caspase-3 with Fas ligand induces drastic apoptosis in U-373MG glioma cells. *Exp. Cell Res.* **256**, 423–433.

156. Shinoura, N., Saito, K., Yoshida, Y., Hashimoto, M., Asai, A., Kirino, T., and Hamada, H. (2000). Adenovirus-mediated transfer of bax with caspase-8 controlled by myelin basic protein promoter exerts an enhanced cytotoxic effect in gliomas. *Cancer Gene Ther.* **7**, 739–748.

157. Shinoura, N., Yoshida, Y., Asai, A., Kirino, T., and Hamada, H. (2000). Adenovirus-mediated transfer of p53 and Fas ligand drastically enhances apoptosis in gliomas. *Cancer Gene Ther.* **7**, 732–738.

158. Shinoura, N., Yoshida, Y., Sadata, A., Hanada, K.I., Yamamoto, S., Kirino, T., Asai, A., and Hamada, H. (1998). Apoptosis by retrovirus- and adenovirus-mediated gene transfer of Fas ligand to glioma cells: implications for gene therapy. *Hum. Gene Ther.* **9**, 1983–1993.

159. Singh, C. J., Wong, S. H., and Hong, W. (1994). Overexpression of E2F1 in rat embryo fibroblasts leads to neoplastic transformation. *EMBO J.* **13**, 3329–3338.

160. Somasundaram, K., MacLachlan, T. K., Burns, T. F., Sgagias, M., Cowan, K. H., Weber, B. L., and El-Deiry, W. S. (1999). BRCA1 signals ARF-dependent stabilization and coactivation of p53. *Oncogene* **18**, 6605–6614.

161. Spirin, K. S., Simpson, J. F., Takeuchi, S., Kawamata, N., Miller, C. W., and Koeffler, H.P. (1996). p27/Kip1 mutation found in breast cancer. *Cancer Res.* **56**, 2400–2404.

162. Steiner, M. S., Wang, Y., Zhang, Y., Zhang, X., and Lu, Y. (2000). p16/MTS1/INK4A suppresses prostate cancer by both pRb dependent and independent pathways. *Oncogene* **19**, 1297–1306.

163. Steiner, M. S., Zhang, Y., Farooq, F., Lerner, J., Wang, Y., and Lu, Y. (2000). Adenoviral vector containing wild-type p16 suppresses prostate cancer growth and prolongs survival by inducing cell senescence. *Cancer Gene Ther.* **7**, 360–372.

164. Stone, S., Dayananth, P., and Kamb, A. (1996). Reversible, p16-mediated cell cycle arrest as protection from chemotherapy. *Cancer Res.* **56**, 3199–3202.

165. Suda, T., Takahashi, T., Golstein, P., and Nagata, S. (1993). Molecular cloning and expression of the Fas ligand, a novel member of the tumor necrosis factor family. *Cell* **75**, 1169–1178.

166. Takada, S., Kaneniwa, N., Tsuchida, N., and Koike, K. (1997). Cytoplasmic retention of the p53 tumor suppressor gene product is observed in the hepatitis B virus X gene-transfected cells. *Oncogene* **15**, 1895–1901.

167. Tamura, M., Gu, J., Tran, H., and Yamada, K. M. (1999). PTEN gene and integrin signaling in cancer. *J. Natl. Cancer Inst.* **91**, 1820–1828.

168. Tan, P., Cady, B., Wanner, M., Worland, P., Cukor, B., Magi-Galluzzi, C., Lavin, P., Draetta, G., Pagano, M., and Loda, M. (1997). The cell cycle inhibitor p27 is an independent prognostic marker in small (T1a,b) invasive breast carcinomas. *Cancer Res.* **57**, 1259–1263.

169. Tian, X. X., Pang, J. C., To, S. S., and Ng, H. K. (1999). Restoration of wild-type PTEN expression leads to apoptosis, induces differentiation, and reduces telomerase activity in human glioma cells. *J. Neuropathol. Exp. Neurol.* **58**, 472–479.

170. Todd, M. C., Sclafani, R. A., and Langan, T. A. (2000). Ovarian cancer cells that coexpress endogenous Rb and p16 are insensitive to overexpression of functional p16 protein. *Oncogene* **19**, 258–264.

171. Tsao, Y. P., Huang, S. J., Chang, J. L., Hsieh, J. T., Pong, R. C., and Chen, S. L. (1999). Adenovirus-mediated p21 (WAF1/SDII/CIP1) gene transfer induces apoptosis in human cervical cancer cell lines. *J. Virol.* **73**, 4983–4990.

172. Ueda, H., Ullrich, S. J., Gangemi, J. D., Kappel, C. A., Ngo, L., Feitelson, M. A., and Jay, G. (1995). Functional inactivation but not structural mutation of p53 causes liver cancer. *Nat. Genet.* **9**, 41–47.

173. Vahanian, N., Reichardt, K., Higginbotham, J., Link, C., and Seth, P. (2000). In vivo study of VHL tumor suppressor gene, its potential role in angiogenesis and cancer gene therapy, Abstract No. 408, 3rd Annual Meeting of the American Society of Gene Therapy. *Mol. Ther.* **1**, 5157.

174. Valenzuela, M. T., Nunez, M. I., Villalobos, M., Siles, E., McMillan, T. J., Pedraza, V., and Ruiz de Almodovar, J. M. (1997). A comparison of p53 and p16 expression in human tumor cells treated with hyperthermia or ionizing radiation. *Int. J. Cancer* **72**, 307–312.

175. Villa, L. L. (1997). Human papillomaviruses and cervical cancer. *Adv. Cancer Res.* **71**, 321–341.

176. Waldman, T., Lengauer, C., Kinzler, K. W., and Vogelstein, B. (1996). Uncoupling of S phase and mitosis induced by anticancer agents in cells lacking p21. *Nature* **381**, 643–644.

177. Wang, X. W., Forrester, K., Yeh, H., Feitelson, M. A., Gu, J., and Harris, C. C. (1994). Hepatitis B virus X protein inhibits p53 sequence-specific DNA binding, transcriptional activity, and association with transcription factor ERCC3. *Proc. Natl. Acad. Sci. USA* **91**, 2230–2234.

178. Watzlik, A., Dufter, C., Jung, M., Opelz, G., and Terness, P. (2000). Fas ligand gene-carrying adeno-5 AdEasy viruses can be efficiently propagated in apoptosis-sensitive human embryonic retinoblast 911 cells. *Gene Ther.* **7**, 70–74.

179. Weber, J. D., Jeffers, J. R., Rehg, J. E., Randle, D. H., Lozano, G., Roussel, M. F., Sherr, C. J., and Zambetti, G.P. (2000). p53-independent functions of the p19^{ARF} tumor suppressor. *Genes Dev.* **14**, 2358–2365.

180. Weinberg, R. (1995). The retinoblastoma protein and cell cycle control. *Cell* **81**, 323–330.

181. Werner, N. S., Siprashvili, Z., Fong, L. Y., Marquitan, G., Schroder, J. K., Bardenheuer, W., Seeber, S., Huebner, K., and Schutte, J., and Opalka, B. (2000). Differential susceptibility of renal carcinoma cell lines to tumor suppression by exogenous FHIT expression. *Cancer Res.* **60**, 2780–2785.

182. Wick, W., Furnari, F. B., Naumann, U., Cavenee, W. K., and Weller, M. (1999). PTEN gene transfer in human malignant glioma: sensitization to irradiation and CD95L-induced apoptosis. *Oncogene* **18**, 3936–3943.

183. Wolf, J. K., Kim, T. E., Fightmaster, D., Bodurka, D., Gershenson, D. M., Mills, G., and Wharton, J. T. (1999). Growth suppression of human ovarian cancer cell lines by the introduction of a p16 gene via recombinant adenovirus. *Gynecol. Oncol.* **73**, 27–34.

184. Wolff, G., Schumacher, A., Karawajew, L., Ruppert, V., Arnold, W., Daniel, P., and Doerken, B. (2000). Adenovirus-mediated gene transfer of p16 INK4/CDKN2 to colorectal cancer, Abstract No. 418, 3rd Annual Meeting of the American Society of Gene Therapy. *Mol. Ther.* **1**, 5161.

185. Wu, G. S., Kim, K., and El-Deiry, W. S. (2000). KILLER/DR5, a novel DNA-damage inducible death receptor gene, links the p53-tumor suppressor to caspase activation and apoptotic death. *Adv. Exp. Med. Biol.* **465,** 143–151.

186. Wu, R., Connolly, D. C., Dunn, R. L., and Cho, K. R. (2000). Restored expression of fragile histidine triad protein and tumorigenicity of cervical carcinoma cells. *J. Natl. Cancer Inst.* **92,** 338–344.

187. Wu, G. S., Burns, T. F., McDonald, E. R., III, Meng, R. D., Kao, G., Muschel, R., Yen, T., and El-Deiry, W. S. (1999). Induction of the TRAIL receptor KILLER/DR5 in p53-dependent apoptosis but not growth arrest. *Oncogene* **18,** 6411–6418.

188. Wu, G. S., Burns, T. F., McDonald III, E. R., Jiang, W., Meng, R., Krantz, I. D., Kao, G., Gan, D.-D., Zhou, J.-Y., Muschel, R., Hamilton, S. R., Spinner, N. B., Markowitz, S., Wu, G., and El-Deiry, W. S. (1997). KILLER/DR5 is a DNA damage-inducible p53-regulated death receptor gene. *Nat. Genet.* **17,** 141–143.

189. Wu, H., Wade, M., Krall, L., Grisham, J., Xiong, Y., and Van Dyke, T. (1996). Targeted in vivo expression of the cyclin-dependent kinase inhibitor p21 halts hepatocyte cell-cycle progression, postnatal liver development, and regeneration. *Genes Dev.* **10,** 245–260.

190. Wu, G. S., and El-Deiry, W. S. (1996). Apoptotic death of tumor cells correlates with chemosensitivity, independent of p53 or bcl-2. *Clin. Cancer Res.* **2,** 623–633.

191. Wu, X., and Levine, A. J. (1994). p53 and E2F-1 cooperate to mediate apoptosis. *Proc. Natl. Acad. Sci. USA* **91,** 3602–3606.

192. Xiang, J., Piche, A., Rancourt, C., Gomez-Navarro, J., Siegal, G. P., Alvarez, R. D., and Curiel, D. T. (1999). An inducible recombinant adenoviral vector encoding bax selectively induces apoptosis in ovarian cancer cells. *Tumor Targeting* **4,** 84–91.

193. Xiong, Y., Hannon, G. J., Zhang, H., Casso, D., Kobayashi, R., and Beach, D. (1993). p21 is a universal inhibitor of cyclin kinases. *Nature* **366,** 701–704.

194. Xu, S. Q., and El-Deiry, W. S. (2000). p21(WAF1/CIP1) inhibits initiator caspase cleavage by TRAIL death receptor DR4. *Biochem. Biophys. Res. Commun.* **269,** 179–190.

195. Xu, H.-J. (1997). Strategies for approaching retinoblastoma tumor suppressor gene therapy. *Adv. Pharmacol.* **40,** 369–397.

196. Xu, H.-J., Zhou, Y. L., Seigne, J., Perng, G. S., Mixon, M., Zhang, C. Y., Li, J., Benedict, W. F., and Hu, S. X. (1996). Enhanced tumor suppressor gene therapy via replication-deficient adenovirus vectors expressing an N-terminal truncated retinoblastoma protein. *Cancer Res.* **56,** 2245–2249.

197. Xu, G., Livingston, D. M., and Krek, W. (1995). Multiple members of the E2F transcription factor family are the products of oncogenes. *Proc. Natl. Acad. Sci. USA* **92,** 1357–1361.

198. Xu, H.-J., Xu, K., Zhou, Y., Li, J., Benedict, W. F., and Hu, S. X. (1994). Enhanced tumor cell growth suppression by an N-terminal truncated retinblastoma protein. *Proc. Natl. Acad. Sci. USA* **91,** 9837–9841.

199. Yamamoto, K., Katayose, Y., Suzuki, M., Unno, M., Endo, E., Takemura, S., Yoshida, H., Cowan, K. H, Seth, P., and Matsuno, S. (2000). Adenovirus expressing cyclin-dependent kinase inhibitor p27Kip1 induces apoptosis on human cholangiocarcinoma cell lines utilizing fas pathway. *Proc. Annu. Meet. Am. Assoc. Cancer Res.* **41,** 90.

200. Yamasaki, L., Bronson, R., Williams, B. O., Dyson, N. J., Harlow, E., and Jacks, T. (1998). Loss of E2F-1 reduces tumorigenesis and extends the lifespan of *Rb1*(+/−) mice. *Nat. Genet.* **18,** 360–364.

201. Yamasaki, L., Jacks, T., Bronson, R., Goillot, E., Harlow, E., and Dyson, N. (1996). Tumor induction and tissue atrophy in mice lacking E2F-1. *Cell* **85,** 537–548.

202. Yang, C.-T., Liang, Y., Yeh, C.-C., Chang, J. W.-C., Zhang, F., McCormick, F., and Jablons, D.M. (2000). Adenovirus-mediated p14ARF gene transfer in human mesothelioma cells. *J. Natl. Cancer Inst.* **92,** 636–641.

203. Yang, H. L., Dong, Y. B., Elliott, M. J., Liu, T. J., and McMasters, K. M. (2000). Caspase activation and changes in bcl-2 family member protein expression associated with E2F-1-mediated apoptosis in human esophageal cancer cells. *Clin. Cancer Res.* **6,** 1579–1589.

204. Yang, H. L., Dong, Y. B., Elliott, M. J., Liu, T. J., Atienza, Jr., C., Stilwell, A., and McMasters, K. M. (1999). Adenovirus-mediated E2F-1 gene transfer inhibits MDM2 expression and efficiently induces apoptosis in MDM2-overexpressing tumor cells. *Clin. Cancer Res.* **5,** 2242–2250.

205. Yang, Z.-Y., Perkins, N. D., Ohno, T., Nabel, E. G., and Nabel, G. J. (1995). The p21 cyclin-dependent kinase inhibitor suppresses tumorigenicity in vivo. *Nat. Med.* **1,** 1052–1056.

206. Yin, C., Kundson, C. M., Korsmeyer, S. J., and Van Dyke, T. (1997). Bax suppresses tumorigenesis and stimulates apoptosis in vivo. *Nature* **385,** 637–640.

207. Zhang, Y., Xiong, Y., and Yarbrough, W. G. (1998). ARF promotes MDM2 degradation and stabilizes p53: ARF-INK4a locus deletion impairs both the Rb and p53 tumor suppression pathways. *Cell* **92,** 725–734.

208. Zhang, W., Grasso, L., McClain, C. D., Gambel, A. M., Cha, Y., Travali, S., Deisseroth, A. B., and Mercer, W. E. (1995). p53-independent induction of WAF1/CIP1 in human leukemia cells is correlated with growth arrest accompanying monocyte/macrophage differentiation. *Cancer Res.* **55,** 668–674.

209. Zhu, L., van der Heuvel, S., Helin, K., Fattaey, A., Ewen, M., Livingston, D., Dyson, N., and Harlow, E. (1993). Inhibition of cell proliferation by p107, a relative of the retinoblastoma protein. *Genes Dev.* **7,** 1111–1125.

210. Zimmermann, H., Degenkolbe, R., Bernard, H. U., and O'Connor, M. J. (1999). The human papillomavirus type 16 E6 oncoprotein can down-regulate p53 activity by targeting the transcriptional coactivator CBP/p300. *J. Virol.* **73,** 6209–6819.

18

Cancer Gene Therapy with the p53 Tumor Suppressor Gene

RAYMOND D. MENG

Laboratory of Molecular Oncology and Cell Cycle Regulation
Howard Hughes Medical Institute
Departments Medicine and Genetics
Cancer Center and The Institute for Human Gene Therapy
University of Pennsylvania School of Medicine
Philadelphia, Pennsylvania 19104

WAFIK S. EL-DEIRY

Laboratory of Molecular Oncology and Cell Cycle Regulation
Howard Hughes Medical Institute
Departments Medicine and Genetics
Cancer Center and The Institute for Human Gene Therapy
University of Pennsylvania School of Medicine
Philadelphia, Pennsylvania 19104

I. INTRODUCTION

Cancer cells accumulate numerous genetic alterations that contribute to tumorigenesis, tumor progression, and chemotherapeutic drug resistance. Most of these alterations affect the regulation of the cell cycle. In normal cells, a balance is achieved between proliferation and cell death by tightly regulating the progression through the cell cycle with cellular checkpoints (reviewed by Hartwell and Kastan [55]).

Before the cell can enter the next phase of the cell cycle, it must pass through a checkpoint that decides if all the previous processes have been completed. The decision to enter the cell cycle is made during the G_1 phase by cyclins and their regulatory units, the cyclin-dependent kinases (CDKs) (reviewed by McDonald and El-Deiry [98]). Control of the CDKs is achieved by phosphorylation on different sites of the protein and by the activity of CDK inhibitors, which are composed of two families [105]. The INK4 proteins (consisting of p15, p16, p18, and p19) bind to CDK4 or to CDK6, whereas the CIP/KIP proteins (including p21, p27, and p57) bind to cyclin-CDK complexes.

Cell-cycle progression is also affected by environmental stimuli. If a eukaryotic cell is deprived of nutrients or growth factors, the cell responds by activation of a checkpoint leading to cell-cycle arrest until conditions become favorable for cell division (reviewed in Murray and Hunt [106]). In mammalian cells, growth factors and mitogens regulate the expression level of cyclin D1, which is involved in driving forward the transition from the G_1 to S phase (reviewed by Sherr [139]). In mammalian cells, the growth-factor-dependent G_1 to S checkpoint is regulated by the retinoblastoma (Rb) protein (reviewed by Weinberg [165] and Harbour and Dean [53]). Once a cell has traversed this so-called "restriction point" [120], it is committed to going through the S, G_1, and M phases, leading to two daughter cells. One of the hallmarks of cancer cells is loss of this checkpoint control. Both the viral oncoproteins (reviewed by Nevins [108]) and mutations in human cancer (reviewed by Sherr [139]) appear to target two parallel, yet related, cell-cycle controlling pathways: (1) the p16–cyclin D1–CDK4–Rb pathway, and (2) the ATM–p53–p21 pathway. The p16–cyclin D1–CDK4–Rb pathway helps the cell progress through the G$_1$ phase. Cyclin D1 binds to and positively regulates CDK4, which helps to phosphorylate Rb. In a hypophosphorylated state, Rb is a negative regulator of cell-cycle progression because it binds to and inactivates the E2F transcription factor family. However, when it is phosphorylated by cyclin D1–CDK4, Rb becomes inactive, and the cell can continue into the S phase. p16 is a negative regulator of CDK4, which prevents it from phosphorylating Rb. Abnormalities of each component of this pathway have been identified in human cancers and are usually mutually exclusive (reviewed by Sherr [139]).

The detection of DNA damage is governed by the tumor suppressor p53. Multiple genes have been designated tumor suppressors because of their ability to limit cancer cell growth; subsequently, they have a high frequency of mutation or of deletion in human tumors (reviewed by Macleod [92]). Following DNA damage, p53 arrests the cell to allow time for repair, but if the damage is extensive enough p53 initiates programmed cell death, or apoptosis. Loss of these various molecular checkpoints has been found to underlie the development of many tumors because cell-cycle progression becomes deregulated. The accumulation of genetic alterations also contributes to enhanced chemoresistance resulting from the loss of the ability to respond to DNA damage (reviewed by El-Deiry [33]). Of the many alterations that have been identified, changes involving the tumor suppressors, such as p53, are the most common. With loss of growth suppression, progression through the cell cycle remains unchecked, and tumorigenesis results. Therefore, a major strategy in gene therapy for cancer has focused on replacing the tumor suppressors in cancer cells that have been lost through deletion or through mutation (see Fig. 1).

II. VECTORS FOR GENE THERAPY

Successful cancer gene therapy depends on successful delivery of the tumor suppressor to the intended target. Several vectors are currently under development for efficacious gene delivery: nonviral systems, such as liposomes, or the more extensively studied viral systems, including retroviruses and adenoviruses.

A. Liposomes

Liposomes are a nonviral gene delivery system in which the DNA is directly mixed with liposomes to form a complex, linked by charge interactions (reviewed by Li and Huang [84]). The DNA–liposome complex is then endocytosed by the cell membranes. The advantages of liposomes are that they are relatively convenient to use, they can be produced

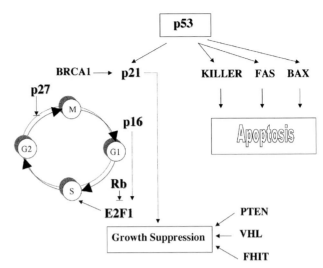

FIGURE 1 Tumor suppressors that are targets for gene therapy of cancer. This review discusses tumor suppressors involved in cell-cycle regulation that have been studied as potential targets for gene replacement in the treatment of cancer. Loss of these tumor suppressors, most notably p53, results in tumor development and progression. p53 mediates the cellular response to DNA damage, resulting in growth arrest or in apoptosis. p21 is a main effector of p53 that mediates growth arrest and is a CDKI, along with p16 and p27, which help to regulate G$_1$ transition. Rb helps to mediate cell-cycle progression from G$_1$ to S phase. In addition, the tumor suppressors *BRCA1* (involved in breast cancer), *VHL* (involved in Von Hippel–Landau familial disease), *FHIT* (involved in chromosomal breakages), and *PTEN* (involved in cell attachment) also suppress growth through novel mechanisms. Likewise, the apoptosis induced by p53 is based on the activation of select targets. The recently cloned novel TRAIL target *KILLER/DR5* and the Fas family of death receptors can be activated by p53 and can induce apoptosis through initiation of a proteolytic caspase cascade. Two other p53-mediated targets involved in apoptosis, *bax* and the p53-induced genes (PIGs), initiate cell death through reactive oxygen species.

on a large scale, and, most importantly, they do not elicit a strong immune response *in vivo*. Liposomes, however, do not have high transfection efficiencies, they lack target specificity, and *in vivo* gene expression is transient because they are rapidly cleared by the reticuloendothelial system. Recent reports also suggest that the unmethylated CpG islands in the DNA plasmid delivered by liposomes may elicit an immune response in mice [84,86,181].

B. Retroviruses

Viruses have often been used as vectors for gene therapy because they can transfer genes efficiently and at high expression. The early gene therapy experiments used retroviruses because they can integrate into the chromosomes of the host cell to provide prolonged gene expression and because they can carry large genes, up to 10 kb (reviewed by Palu *et al.* [119]). Retroviruses, however, have two disadvantages. First, because retroviruses integrate into host chromosomes, they can only efficiently infect cells that are dividing; hence, quiescent cells, such as neurons, are not effectively infected.

Second, it is difficult to prepare sufficiently high titers of retroviruses for *in vivo* gene therapy.

C. Adenoviruses

In contrast to retroviruses, adenoviruses infect a wide range of cells, including dividing and nondividing cells, and they can be prepared at extremely high titers, on the order of 10^{12} plaque-forming units (PFU) per milliliter, which is sufficiently concentrated for *in vivo* use (reviewed by Hitt *et al.* [59]). Although they offer high gene expression, adenoviruses do not undergo chromosomal integration. In cancer gene therapy, however, successful eradication of a tumor would obviate the need for prolonged gene expression. In fact, almost 75% of the patients enrolled in gene therapy clinical trials in 1996 were being treated for malignancies, primarily solid tumors [94]. Adenoviruses also do not infect hematopoietic cells with high efficiency (reviewed by Marini *et al.* [96]). The most significant drawback to adenoviruses, however, is that they elicit a strong host immune response [144,180]. Furthermore, because gene delivery by adenoviruses is transient, multiple treatments may be required, which would further elicit a strong humoral immune response. Consequently, many of the preclinical *in vivo* experiments with adenoviruses have been conducted in immunodeficient hosts, such as nude mice. Currently, efforts are underway to try to lessen the immune response, either by direct suppression of host immunity [133] or by modification of the adenovirus vector to lessen its inherent immunogenicity [64,68]. Although it may seem intuitive that a heightened immune response may be good in cancer gene therapy, it is less desirable on a practical scale because the immune response helps to eliminate the vector and decreases expression of the transduced gene.

Consequently, recent research has focused on improving the gene delivery capability of the adenoviral vectors. First, many groups have now shown that the efficacy of adenoviral gene delivery correlates with the ability of the adenovirus to bind to its receptor on cancer cells which was recently identified as the coxsackie/adenovirus receptor (CAR) [8]. The adenoviral fiber protein binds CAR through a domain in its carboxyl terminus, and another adenoviral protein, the penton base, mediates viral internalization by binding to host-cell integrin proteins $\alpha v\beta3$ and $\alpha v\beta5$ [170]. Subsequently, several groups have shown that the efficiency of adenoviral gene transfer into cancer cell lines depends on the presence of CARs [85] and on the integrins [28,121,151]. It was also shown that the degree of infectivity of head and neck cancer cell lines could be improved by coinfection with an adenovirus expressing CARs, which presumably enhances low CAR levels [85]. Transfection of CARs was also reported to improve the transduction efficiency of an adenovirus encoding the tumor suppressor p21$^{WAF1/CIP1}$ in bladder cancer cells [115].

Finally, several groups have attempted to create adenoviruses that can infect specific targets (reviewed by Miller and Whelan [101] and Wickham [169]). Besides specificity, targeted adenoviruses may also allow systemic rather than intratumoral administration. Although intratumoral injection effectively directs adenovirus to the tumor, it transduces tumor along the needle path rather than the entire tumor, and it does not allow elimination of metastases (reviewed by Wickham [169]). To target selected tumors, the adenovirus backbone has been modified at the transcriptional level by adding tissue-specific promoters. Another strategy adds specific antibody-fragments to the virus vector (reviewed by Bilbao *et al.* [10]). Recently, tetracycline-responsive recombinant adenoviruses have been developed to try to regulate gene expression following infection [61,183].

D. Adeno-Associated Viruses

Adeno-associated viruses (AAVs) are derived from a non-pathogenic and defective human parvovirus that often coexists with adenoviruses (reviewed by Monahan and Samulski [104]). They are relatively safe and nonimmunogenic, provide high titers, and can infect a wide range of cell types, including quiescent cells, hematopoietic cells, and terminally differentiated cells (reviewed by Flotte and Carter [38]). Because the virus can integrate, gene expression induced by AAVs can be long-term, lasting up to 1 year after infection [73,177]. AAVs, however, have several disadvantages. Currently, high-yield AAV production is difficult because of possible contamination with wild-type AAVs or with other helper viruses. Finally, because of the small size of AAVs, only small inserts, up to 5 kb, can be packaged.

E. Lentiviral Vectors

Lentiviruses, such as HIV, can infect nondividing and even growth-arrested cells because virus entry is gained using the host-cell nuclear import system (reviewed by Buchschacher and Wong-Staal [18]). Vectors based on lentiviruses are first deleted of their virulence genes [31,187] or made less virulent by decreasing their transcriptional activity [102]. Lentiviral vectors, however, do not have efficient delivery into some terminally differentiated cells, such as skeletal muscle cells or hepatocytes (reviewed by Trono [157]). Currently, because of the shadow of HIV, no clinical trials have been attempted with attenuated lentiviral vectors.

F. Chimeric Viral Vectors

Finally, new vectors are being designed that incorporate the advantages of different viral systems. For example, a novel chimeric vector was derived from both adenovirus and retrovirus genes [37]. These vectors were then used to infect human cancer cells, causing them to act as transient producer cells which release retroviruses that can then integrate into nearby cells. Recently, a fusion adenovirus and Epstein–Barr virus (EBV) vector was created, which allows high titer production like adenoviruses but stable expression because of the EBV episome [152].

G. Route of Administration

With respect to delivery *in vivo,* infectivity represents the gateway into the target cells when the virus reaches its destination. However, the ability of a particular agent to reach the tumor is influenced by its route of administration. The most direct route is intratumoral, which provides delivery to the target tissue. Furthermore, the existence of a bystander effect obviates the need to deliver the agent into every cell, and local delivery minimizes toxicity to the normal tissues where the agent is not present. Intratumoral delivery, however, is not always entirely efficient, as the spread of virus can be hindered by a pressure gradient within large tumors, by limited distribution from blood vessels, or by the extracellular matrix (reviewed by Jain [67]).

Similarly, delivery of virally expressed genes by intravascular or intracavitary injections also presents barriers. In intravascular administration, instillation into a peripheral vein dilutes the vehicle so only a small portion may ultimately reach the tumor. Intravascular administration also elicits a powerful immune response [122]. Tropism for organs such as the liver (for example, by adenovirus) can be a disadvantage if delivery is intended elsewhere or it can be advantageous if the liver is the target [44]. Even with regional intravascular administration, the virus must traverse the endothelial wall and travel against pressures within an expanding tumor mass. In the case of intracavitary administration (i.e., intrapleural or intraperitoneal), the surface of the tumor mass is coated by virus, but intratumoral delivery within a solid mass represents an important barrier.

III. p53

A. Introduction

Among the tumors suppressors being considered for gene replacement in malignancies, p53 has been the focus of many groups for several reasons. First, p53 plays a pivotal role in the fate of a cell following DNA damage. It determines if the damaged cell will undergo growth arrest in order to repair itself [72], or if the cell will undergo programmed cell death or apoptosis because the damage is too extensive [90,91]. Therefore, loss of p53 or mutations in p53, some of which can act in a dominant-negative manner to inhibit residual wild-type p53, significantly contribute to tumor development, tumor progression, and chemotherapeutic resistance (reviewed by Velculescu and El-Deiry [162] and Levine [83]). A survey of the toxicity of hundreds of anticancer drugs toward over 60 human cancer and leukemia cell lines has indicated that the vast majority of clinically useful drugs are most effective in cells that express wild-type p53 [168]. Second, mutations in p53 are the most common genetic alterations in tumors, being mutated or deleted in over 50% of all human cancers [50]. Germline transmission of a mutant p53 allele predisposes individuals with Li–Fraumeni syndrome to a high risk of cancers [93]. In p53 knockout mice, 75% developed tumors by 6 months of age, and all died by 2 years [30]. Third, loss of p53 results in decreased apoptosis [149] and decreased susceptibility to radiotherapy or chemotherapy [90,91].

With respect to gene replacement therapy, p53 is a potent inducer of cancer cell apoptosis and is effective despite the presence of multiple genetic changes in the cancer cells [5]. Because p53 offers a promising way to regulate the growth of cancers *in vivo,* much work has been directed at developing this form of gene therapy (reviewed by Nielsen and Maneval [112]; Baselga [6]; Roth *et al.* [128]). Cancer gene therapy strategies have focused on replacing or even overexpressing wild-type p53 in the hopes that aberrant cell cycle control can once again be tightly regulated. It is important, however, to realize that, despite its strengths, p53 gene therapy has important limitations that must be considered for its clinical development as an anticancer agent.

B. Gene Therapy with p53

The initial p53 gene therapy experiments used retroviruses to deliver the tumor suppressor gene into various cancer cell lines (see Table 1). Two groups introduced wild-type p53 with a retrovirus vector into a non-small-cell lung cancer line and suppressed the growth of the tumor both *in vitro* [20] and *in vivo* in a nude mouse model [42]. It was also shown in a phase I clinical trial that retrovirus-transferred p53 can be used to infect human non-small-cell lung cancers by intratumoral injection and that it may effectively limit growth in a small minority of these patients with advanced terminal cancer [131].

Although liposomal delivery of p53 is being studied [186], the majority of p53-directed gene replacement strategies have now shifted toward using an adenovirus vector because the adenovirus can infect numerous cell types and because it can be produced in high titers. It has been previously reported that an adenovirus expressing β-galactosidase (Ad-LacZ) is capable of infecting tumor cells from a wide range of tissues [12]. Furthermore, the ability of the adenovirus to infect these tumor cells was independent of the endogenous p53 status. Some cell lines, however, remain inherently resistant to adenovirus infection. For example, it was reported that two leukemia and two lymphoma cell lines showed less than 0.001% infectivity following infection with an Ad-LacZ at a multiplicity of infection (MOI) of 150 [12]. Other adenovirus-resistant cells include a breast cancer cell line MDA-MB-435 with mutant p53 [113] and a choriocarcinoma cell line JEG3 with wild-type p53 [99]. The explanation for cellular resistance to adenovirus

TABLE 1   Infection of Selected Cell Lines by Ad-p53

Cell type	Name of cell line	p53	MOI[a]	In vitro	In vivo	Ref.
Bladder	UMUC3	mut	20	x		[99]
Breast	MCF7	wt	10	x		[113]
	MDA-MB-435	mut	N.I.[b]	x		[113]
	SKBr3	mut	<30	x		[12]
Cervical	HeLa	wt	100	x	x	[146]
	SiHa	wt	100	x		[146]
Choriocarcinoma	JEG3	wt	100	x		[99]
Colon	DLD-1	mut	30	x		[136]
	HCT116	wt	<30	x		[12]
	LoVo	wt	100	x		[136]
	RKO	wt	<30	x		[12]
	SW480	mut	<30	x		[12]
Endometrial	SPEC-2	mut	50	x		[125]
Glioblastoma	A172	mut	100	x		[4]
	ADF	wt	20	x		[11]
	De14A	mut	<30	x		[12]
	G122	mut	30	x		[79]
	T98	mut	N.A.[c]	x		[88]
	U-87 MG	wt	100	x		[4,47]
	U251	mut	100	x		[140]
	U373 MG	mut	100	x		[4,88]
Head and neck	1986LN	wt	100	x		[146]
	A253	mut	100	x	x	[146]
	Det 562	mut	100	x		[146]
	JSQ-3	mut	20	x	x	[23]
	MDA 886	wt	100	x		[26]
	SCC9	mut	100	x		[51,146]
	SCC40	mut	100	x		[146]
	Tu-138	mut	100	x		[56]
	Tu-177	mut	100	x	x	[26]
Hepatoma	Hep3B	null	10	x		[126]
	HepG2	wt	50	x		[126]
	HLF	mut	N.A.	x		[2]
	Huh-7	mut	10	x		[126]
	Mahlavu	mut	10	x		[126]
Leukemia	HL60	null	N.I.	x		[12]
	ML-1	null	N.I.	x		[12]
Lung	A549	wt	<30	x		[12]
	H23	mut	10	x		[60]
	H157	mut	10	x		[117]
	H226	mut	30	x	x	[60,114]
	H322	mut	350	x		[60]
	H358	null	10	x		[117]
	H460	wt	<30	x		[12]
	H1299	null	10	x		[60]
Lymphoma	CA46	mut	N.I.	x		[12]
	RAMOS	mut	N.I.	x		[12]
	SUDHL-1	wt	100	x		[159,160]
Melanoma	7336	wt	100	x		[99]
	A875	wt	100	x		[99]
Nasopharyngeal	RPMI 2650	wt	10	x	x	[184]
	Fadu	mut	250	x		[184]
	Detroit 562	mut	250	x		[184]
Ovarian	2774	mut	1×10^8	x		[165][d]
	HEY	wt	500	x		[172]
	MDAH	mut	300	x		[66]

TABLE 1   (Continued)

Cell type	Name of cell line	p53	MOI[a]	In vitro	In vivo	Ref.
	OCC-1	mut	50	x		[172]
	OVCA420	wt	100	x		[172]
	OVCA423	mut	250	x		[172]
	OVCA433	wt	100	x		[172]
	OVCA429	wt	50	x		[172]
	OVCA2774	mut	20	x	x	[76]
	OVCAR	mut	<30	x		[12]
	PA-1	wt	300	x		[66]
	SKOV3	null	<30	x		[12,76]
Pancreatic	Capan-1	mut	10	x		[32]
	DANG	null	10	x		[32]
	NP-9	mut	25	x		[22]
	NP-18	mut	25	x	x	[22]
	NP-29	wt	25	x		[22]
	NP-31	mut	25	x		[22]
Prostate	DU-145	mut	<30	x		[12,66]
	LNCap	wt	<30	x	x	[12,66]
	PC-3	null	300	x		[66]
Sarcoma	SKLMS-1	mut	100	x	x	[100]
Vulvar	A431	mut	100	x		[146]

[a]MOI indicates >50% transduction.
[b]N.I., not infected by the adenovirus, even at the highest MOI tested.
[c]N.A., not available.
[d]The amount of virus consisted of in vivo injection of virus particles.

infection remains unknown, primarily because the expression of the cellular receptor responsible for adenovirus binding in these cells has not yet been determined. One report suggested that α integrins are required for efficient internalization of adenoviruses [170], but fluorescence-activated cell sorting (FACS) analysis did not show decreased integrin expression in a breast cancer cell line that was infected poorly with a p53 adenovirus (Ad-p53) [113]. Although the mechanism is unknown, inefficient adenovirus infection of cells remains an important cause of nonresponsiveness to p53 treatment.

To help determine successful responsiveness to Ad-p53, markers for productive p53 infection are currently being studied [143]. Unfortunately, examining p53 levels following Ad-p53 infection is difficult because many human cancer cells already overexpress mutant p53. To determine the degree of penetration of Ad-p53 in human xenografts on SCID mice, one group utilized laser scanning cytometry, and the functionality of the p53 was determined by measuring apoptosis [49]. Another common strategy to determine the functionality of the transferred p53 has been to evaluate the levels of known target genes activated by p53 following infection with Ad-p53, such as p21, *MDM2, bax,* or *bcl-2* [111]. For example, the CDK-inhibitor p21 is transcriptionally induced following p53 stabilization after DNA damage [36]. p21 acts a negative growth regulator during the G_1 cell-cycle checkpoint by binding and inhibiting cyclin/CDKs [36,54]. It was hypothesized that p21 would be a good marker for p53 transduction be-cause, in tumor cells expressing mutant p53, p21 is expressed at low levels [35]. Consequently, following Ad-p53 infection in different cancer cell lines, high expression of p21 was in-duced independently of endogenous p53 status both *in vitro* and *in vivo,* as detected by immunocytochemistry, by west-ern blot analysis, or by "real-time" polymerase chain reaction (PCR) [12,111,184].

The responsiveness to p53 gene therapy has also been correlated with the phosphorylation state of the retinoblas-toma (Rb) protein [99]. It was shown that persistence of the hypophosphorylated form of Rb predicted effective growth suppression of cancer cells following Ad-p53 infection. Fi-nally, the levels of *MDM2,* a protooncogene that downregu-lates p53, in cancer cell lines was also evaluated (reviewed by Momand *et al.* [103]). Flow cytometry showed that p53 upregulation increased the levels of *MDM2,* as hypothesized [66]. Therefore, the effectiveness of Ad-p53 infections may be correlated with the expression of p53 and with the levels of its target genes. Rather than checking for each target sep-arately, recent studies have used flow cytometry to gate cells for both p53 and for a p53-target, such as *MDM2* [66].

C. Cell-Cycle Arrest and Apoptosis Induced by Ad-p53

Numerous studies have documented the efficacy of Ad-p53 infection in cancer cell growth inhibition [131] (re-viewed by Roth and Cristiano [130]; Lang *et al.* [81]; Pagliaro [118]; Sweeney and Pisters [148]). The cell-cycle arrest is both dose and time dependent. One study reported that the degree of apoptosis is lessened in cells harboring wild-type p53 when compared to those with mutant p53, suggesting that some cells with wild-type p53 are relatively resistant to Ad-p53 in the absence of DNA damage [12]. In contrast, growth inhibition by Ad-p53 did not seem to depend on p53 status in prostate cancer cells [48].

Although most studies have focused on the effect of Ad-p53 infection on cancer cells, the effect on normal cells has important implications regarding toxicity. Until a vector that specifically targets cancer cells is developed, inadvertent in-fection of normal cells by Ad-p53 could lead to undesired side effects. In comparing the response of a human fibroblast cell line or of carcinoma cell lines to Ad-p53 infection, it was reported that the fibroblasts had lower transfection efficiency and transgene expression following Ad-p53 infection; conse-quently, these fibroblasts remained resistant to p53-mediated cytotoxicity, although some cell death was observed [23,87]. It has also been reported that normal human keratinocytes have poor infectivity to Ad-p53 [12]. It is therefore likely that normal cells may be more resistant to the effects of Ad-p53 due to the poor infectivity and possibly to the fact that they have intact mechanisms to survive despite wild-type p53 expression.

p53 differs from the other tumor suppressors in that it can induce apoptosis. Apoptosis is a highly desired effect in can-cer therapy. What may prove useful in p53 gene therapy is a greater understanding of the apoptotic pathways regulated by p53 (reviewed by Canman and Kastan [21]; Vogelstein *et al.* [164]). Although p53 can regulate the expression of genes involved in apoptosis, such as *bax* or *fas*/APO1, p53-dependent apoptosis can still occur independently of those target genes [41,78]. In a search for other effectors of p53-dependent apoptosis, the *KILLER/DR5* gene was recently identified [176]. *KILLER/DR5* is a novel member of the tumor necrosis factor (TNF) receptor family that is most identical to DR4 (reviewed by Wu *et al.* [174]). *KILLER/DR5* is strongly induced in cell lines with wild-type p53 following DNA dam-age initiated by ionizing radiation, chemotherapy, or Ad-p53 infection [138]. p53 seems to transcriptionally activate *KILLER/DR5* through an intronic-sequence-specific binding site [151]. *KILLER/DR5* is only activated by p53 in the me-diation of apoptosis, not in situations of growth arrest [175]. Thus, *KILLER/DR5* may represent a new class of genes that may be targeted for gene replacement in malignancies, and its induction following Ad-p53 infection may correlate with apoptosis induction. A recent study has proposed combining Ad-p53 (to induce *KILLER/DR5*) with the ligand TRAIL to sensitize cancer cells to apoptosis induction [74].

Another study suggests that Ad-p53 may induce apoptosis by inhibiting anti-apoptotic signals within the cells, such as NF-κB [136]. NF-κB is a transcription factor that has been implicated in cell survival by targeting anti-apoptotic genes following various cellular stresses (reviewed by May and Ghosh [97]). It was recently reported that infection of human colon cancer cells with Ad-p53 decreased the expression of both the p50 and the p65 subunits of NF-κB, decreased the activity of NF-κB, and increased the activity of IκB, which inhibits NF-κB. Conversely, stable transfection of NF-κB in colon cancer cells increased their resistance to apoptosis me-diated by Ad-p53 [136]. These results suggest that Ad-p53 may induce apoptosis not only by activating target genes in-volved in cell death but also by inhibiting pro-survival genes.

D. Applications of p53 Gene Therapy

1. Bystander Effect of Ad-p53

Many groups have observed that the efficacy of Ad-p53-mediated gene therapy is enhanced by a "bystander effect" [40,127]. The bystander effect occurs when infection of only a small portion of a tumor *in vivo* results in a degree of apoptosis that is greater than expected. This effect has also been observed for other tumor suppressors, although the by-stander effect seems to be more prominent for Ad-p53 than for Ad-p21 [80]. When cells infected with Ad-p21 were mixed with uninfected control cells, no bystander effect was observed. When cells infected with Ad-p53, however, were mixed with uninfected cells, many of the control cells underwent cell death [80]. The exact mechanism underlying the bystander effect of p53, however, remains unknown. It has been hypothesized that the bystander effect improves p53-mediated apoptosis either through a local effect or by an immune effect (reviewed by Vile *et al.* [163]). Other groups have hypothesized that the bystander effect of Ad-p53 may be mediated through an anti-angiogenic effect [24,114]. Previously, p53 has been reported to activate a target involved in inhibiting angiogenesis, thrombospondin-1 [29]. Several groups have now observed that infection with Ad-p53 can, in fact, downregulate angiogenic factors in tumor cells [14,179].

Another report suggests that the bystander effect of p53 may be immune mediated, specifically focusing upon natural killer (NK) cells. The efficacy of Ad-p53 infection of human breast cancer xenografts was reduced in SCID-Beige mice (which have defective T, B, and NK cell activity) compared to SCID mice (which have intact NK cell activity) [109]. The immune response against the adenovirus that encodes p53 may be important as well. In a similar hypothesis, one group tested if the effectiveness of p53 gene therapy in mice could be improved by using murine p53 rather than human p53, which is usually employed in the *in vivo* studies [185]. Using the same backbone adenoviral vector, immune-competent mice were infected with either Ad-p53-mouse or Ad-p53-human. The effectiveness of tumor growth inhibition was increased in those mice injected with murine p53, suggesting that the efficacy of foreign p53 molecules may be decreased because of a heightened immune response.

2. *MDR1*-Overexpressing Cancer Cells

One area in which p53 gene therapy may prove useful is the infection of cancer cells that overexpress the multidrug resistance gene, or *MDR1* (reviewed by El-Deiry [34]). MDR1 is a cell-surface membrane glycoprotein that decreases the intracellular concentration of chemotherapeutic agents (reviewed by Ueda *et al.* [161]). It has been reported that colon or breast cancer cell lines that are up to 1000-fold more resistant to adriamycin because of overexpression of *MDR1* are readily infected by Ad-p53 and undergo apoptosis [12]. Furthermore, p53 may be able to regulate *MDR1*, so loss of p53 would result in altered *MDR1* expression and, consequently, an increase in chemoresistance [25,153]. Therefore, in cell lines that are resistant to chemotherapy, Ad-p53 may effectively bypass the effect of *MDR1*.

3. p53 and Chemotherapy

p53 seems to play an important role in mediating DNA damage induced by various drugs because mutations in p53 are associated with decreased susceptibility to chemotherapeutic agents [90,168]. Therefore, replacing p53 in cell lines may enhance their chemosensitivity. It was first reported that human non-small-cell lung cancer cell lines with

mutant p53 became more sensitive to cisplatin following transduction of wild-type p53 [42]. The combination of Ad-p53 with various chemotherapeutic compounds has now been shown to enhance chemosensitivity in multiple cell lines [32,51,60,82,142]. The combination of DNA-damaging agents and p53 infection has been reported to result in synergy in cancer cell killing [12].

These findings may be explained by the observation that the timing of Ad-p53 administration in relation to chemotherapeutic treatment may be an important determinant of therapeutic synergy. It was reported that the use of Ad-p53 enhanced the chemosensitivity of glioblastoma cell lines to BCNU only if Ad-p53 was administered after the chemotherapy [11]. It was hypothesized that pretreatment with 1, 3 bis chloroethyl 2-nitrosourea (BCNU) caused a G_2 arrest, and those cells not growth-inhibited were then induced to undergo apoptosis by p53 overexpression [11]. Similar results were also reported in the treatment of pancreatic cancer cells with Ad-p53 and with gemcitabine or with cisplatin [22]. In another study, human lung cancer cell lines *in vitro* were treated with one of 12 different chemotherapeutic drugs and then administered Ad-p53. It was shown that seven of the drugs in combination with Ad-p53 had additive or supra-additive effects [117].

Recent groups have examined the utility of combining Ad-p53 with farnesyl transferase inhibitors (FTIs), which target the Ras signaling pathway (reviewed by Beaupre and Kurzrock [7]; Hill *et al.* [57]). In many cancer cell lines, activation of *ras* signaling may contribute to cellular transformation. In order for *ras* to be activated, however, it must be translocated to the plasma membrane, which is accomplished by adding a farnesyl group. The FTIs, in turn, prevent the addition of this farnesyl group and have been shown to decrease tumor growth *in vitro* and *in vivo* (reviewed by Hill *et al.* [57]). In human pancreatic, colon, ovarian, and prostate cancer cell lines, the combination of Ad-p53 with a FTI caused synergistic or additive growth inhibition either *in vitro* or *in vivo* [110]. This effect was independent of endogenous p53 status. The addition of a third drug, paclitaxel, to the regimen did not produce further synergy.

Finally, Ad-p53 has been combined with an apoptosis-inducing ligand, TRAIL, to enhance tumor cell killing (see Fig. 2) [74,75]. TRAIL is a TNF-related ligand that binds specific receptors on cell surfaces which then initiate apoptosis (reviewed by Burns and El-Deiry [19]; Wu *et al.* [174]). In selected cell lines, TRAIL treatment can induce extensive apoptosis [75]. Thus, TRAIL was combined with Ad-p53 to enhance apoptosis. The combination of Ad-p53 and TRAIL enhanced apoptosis in cell lines compared to either treatment alone or in cell lines previously resistant to TRAIL. The enhanced apoptosis was correlated with increased induction of the pro-apoptotic TRAIL receptor *KILLER/DR5* [74,75]. Ad-p53 was also reported to enhance the sensitivity of tumor cell lines to another apoptosis ligand, TNF-α [137].

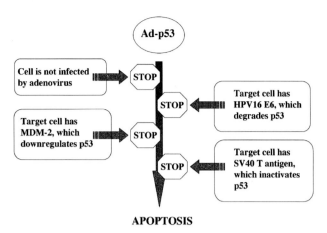

APOPTOSIS

FIGURE 2   p53 cancer gene therapy strategies. Various strategies have been used to enhance the effectiveness of the tumor suppressor p53 in cancer gene therapy. Administration of p53 in an adenovirus vector (Ad-p53) has been combined with both standard radiotherapy and chemotherapy. A novel chemotherapeutic compound currently being tested with Ad-p53 is TRAIL, a recently identified apoptosis-inducing ligand related to TNF-α. Ad-p53 has also been combined with other tumor suppressors delivered by adenoviruses, including p16 and p21. Likewise, Ad-p53 has been delivered with adenoviruses encoding inducers of apoptosis such as Bax or FasL. The p53 molecule itself has been modified to reduce its ability to be downregulated or to enhance its ability to induce apoptosis. Finally, several genes related to p53, such as p73, p51, and p33, are being tested for their potential use in cancer gene therapy.

More work remains to be done on the combination of p53 with apoptosis ligands, but preliminary data suggest that a synergy in apoptosis can be created [74].

4. p53 and Radiotherapy

It was originally shown that thymocytes readily undergo apoptosis following exposure to γ-irradiation but that loss of wild-type p53 dramatically reduced the number of apoptotic cells [91]. In another experiment, transfection of the human papillomavirus (HPV) 16 *E6* gene (which leads to p53 degradation) into human fibroblasts rendered these cells more resistant to irradiation, presumably due to loss of p53 [158]. Therefore, loss of p53 may render cancer cells more resistant to radiotherapy. Efforts to combine p53 gene therapy with radiotherapy have focused on enhancing the radiosensitivity of cancer cells. Thus, p53 has been transduced into human cancer cells that lack functional p53 to render them more radiosensitive [4,17,43]. p53 therapy has also been used to decrease the inherent radioresistance of cancer cells. The administration of Ad-p53 to a radioresistant colon cancer cell line with mutant p53 enhanced susceptibility to subsequent irradiation both *in vitro* and *in vivo* [145]. Irradiation and Ad-p53 infection also decreased the growth of a radioresistant head and neck cancer cell line *in vivo* in a mouse model more effectively than either treatment alone [23].

Unfortunately, it has become apparent that the presence of p53 is neither sufficient nor necessary for radiosensitization [16]. It has been shown that some human colorectal adenoma and carcinoma cell lines lacking wild-type p53 can still undergo γ-irradiation-induced apoptosis [15]. Furthermore, introduction of the HPV16 *E6* gene into several tumor lines did not increase their radioresistance [62]. In conclusion, the role played by p53 in radiosensitivity may be a cell-type-specific phenomenon that needs to be established for each tumor.

5. p53 and Inhibition of Metastases

It has been hypothesized that p53 overexpression may help to inhibit tumor metastasis in melanoma cells. Recent research suggests that several genes may contribute to the development of tumor metastases. For example, matrix metalloproteinase-2 (MMP-2; human type 4 collagenase) and MMP-9 are endopeptidases secreted by melanoma cells to help degrade the extracellular matrix (reviewed by Nelson *et al.* [107]). Upregulation of MMP-2 secretion has been correlated with melanoma invasiveness. Because it was recently reported that p53 might activate MMP-2 [9], one group studied if p53 overexpression would have an effect on metastasis [156]. Infection with Ad-p53 of melanoma cell lines inhibited the invasion capability of a cell line with mutant p53 and also decreased the expression of MMP-2 but not MMP-9. In contrast, another melanoma cell line with wild-type p53 had no change in invasive ability following Ad-p53 infection. It was also reported that p53 represses cathepsin D expression in epithelial cancer cells [175]. These limited results suggest that Ad-p53 infection may help to modify the metastatic potential of tumor cell lines.

6. p53 and Bone Marrow Purification

Some groups have begun to examine the feasibility of using Ad-p53 to purify bone marrow products. The basis for this approach lies in the initial observation that adenoviruses do not efficiently infect hematopoietic cells [166] (reviewed by Marini *et al.* [96]). Therefore, it was hypothesized that Ad-p53 could be used to selectively target cancer cell contaminants in a bone marrow preparation, because Ad-p53 would preferentially infect the tumor cells instead of the hematopoietic cells. Following high-dose chemotherapy, stem cell rescue is often used to replenish the bone marrow of patients; however, it is difficult to completely purify bone marrow to eliminate all tumor cells [155]. *In vitro*, Ad-p53 at an MOI of 10 specifically infected MDA-MB-231 breast cancer cells when they were mixed with CD34$^+$ bone marrow cells [135]. The Ad-p53 did not affect the bone marrow cells as no change occurred in a colony formation assay following infection at an MOI of 10 or even at a tenfold higher concentration (MOI of 100) [135]. Other studies also confirmed that Ad-p53 could purge breast cancer cell contaminants from bone marrow [58,173].

E. Combination of Ad-p53 with Other Adenoviruses

To enhance the therapeutic efficacy of Ad-p53, some groups have advocated the use of Ad-p53 with adenoviruses encoding other tumor suppressors or apoptosis-inducing genes. For example, one group studied the efficacy of coinfecting with Ad-p53 and an adenovirus encoding p16 (Ad-p16) [154]. p16, a tumor suppressor gene, is deleted in multiple cancers, and its overexpression can cause growth inhibition in numerous tumor cell lines (reviewed by Foulkes *et al.* [39]). Coinfection of human hepatocellular carcinoma cell lines either *in vitro* or *in vivo* with the combination of Ad-p53 and Ad-p16 induced apoptosis. Interestingly, when the tumor cells were mixed with normal hepatocytes, apoptosis only occurred in the cancer cells [154]. The combination of Ad-p53 and Ad-p16 has also been used to induce apoptosis in pancreatic cancer cell lines [52]. Further synergy was created by adding chemotherapeutic drugs to the Ad-p53/Ad-p16 regimen [52].

In another example, Ad-p53 was coinfected along with an adenovirus encoding a pro-apoptotic gene *FasL* [140]. FasL is a protein that binds to a specific receptor on cell surfaces, the Fas receptor, which activates a signaling cascade that culminates in the initiation of apoptosis (reviewed by Peter and Krammar [123]). It was observed that Ad-p53 infection of glioma cells induces the expression of *FasL,* so the combination of both adenoviruses was studied. In glioblastoma cell lines, coinfection with Ad-p53 and Ad-FasL induced a significantly greater percentage of apoptosis than either adenovirus alone [140]. These results suggest that Ad-p53 may be combined with adenoviruses encoding other pro-apoptotic genes.

Finally, Ad-p53 has been combined with an adenovirus to enhance its transduction efficiency. Adenoviral binding depends on the fiber knob protein, which helps to bind the virus to the cell surface receptor. It has been reported that a mutant adenoviral fiber knob F/K20 binds with greater efficiency [182]. Therefore, an adenovirus encoding F/K20 was created and used to coinfect tumor cell lines with Ad-p53, enhancing the apoptotic effect of Ad-p53 in glioblastoma cell lines [140].

F. Synthetic p53 Molecules in Gene Therapy

The p53 molecule itself has been altered to expand its efficacy in tumor suppression, as it was previously observed that specific mutants of p53 often developed characteristics more beneficial for therapeutic purposes than wild-type p53 itself. For example, overexpression of a p53 mutant 121F actually induced a greater extent of apoptosis than wild-type p53 and

was not affected by overexpression of *MDM2* [132]. Therefore, one group created a mutant p53 in which substitutions were made in the hydrophobic core of the protein, preventing many of the binding interactions with dominant-negative p53 that would also prevent its downregulation [147].

In another example, previous experiments have demonstrated that p53 activity can be downregulated through structural motifs in the amino- and/or carboxyl termini. Therefore, one group designed a p53 molecule, CTS1 (chimeric tumor suppressor 1), in which the amino- and carboxyl-terminal domains were replaced, reasoning that the deletion of the domains usually responsible for the downregulation of p53 may enhance its activity [27]. Compared to wild-type p53, CTS1 had enhanced transcriptional activity and apoptotic activity. Furthermore, CTS1 had increased resistance to inhibition by dominant-negative mutants of p53, by *MDM2,* or by HPV16 *E6* [27]. An adenovirus encoding the modified p53 (Ad-CTS1) was then created and used to infect several human tumor cell lines. It was reported that the degree of apoptosis induced by Ad-CTS1 was almost sevenfold greater than that induced by Ad-p53 in cell lines that overexpress *MDM2* [13]. This enhanced apoptotic effect of CTS1 was observed both *in vitro* and *in vivo* in a nude mouse model [13]. Ad-CTS1, however, was not more efficacious than Ad-p53 at inducing apoptosis in cell lines with wild-type p53. In fact, Ad-CTS1 was less effective than Ad-p53 in inducing apoptosis in tumor cell lines that have mutated or null p53 status [13]. Recently, a CTS1 vector was designed that is only activated *in vivo* with coexpression of a specific ligand [134]. These results suggest that CTS1 may be more effective than wild-type p53 in cell lines that overexpress *MDM2*. Another synthetic p53 molecule was created in which only the sequences that bind *MDM2* were deleted. An adenovirus encoding this p53 mutant was found to induce higher expression of p53 and increased apoptosis compared to Ad-p53 (wild-type) [171].

In another approach, rather than protecting p53 from degradation, a synthetic p53 molecule was created that had enhanced nuclear localization. Wild-type p53 protein was fused to the herpes simplex VP22 structural protein, which can transport proteins directly to the nucleus from the cytoplasm [3]. It was hypothesized that because p53 is a nuclear protein, enhanced concentration of p53 in the nucleus would increase its therapeutic efficacy. Using an adenovirus, this fusion p53–VP22 protein was delivered to various cell lines. Although the degree of p53-mediated apoptosis was not increased compared to cells infected with Ad-p53 (wild-type), the ability to enhance apoptosis in cell lines previously resistant to wild-type p53 was increased [3]. Finally, the p53 molecule has been fused to chemotherapeutic prodrugs to enhance its chemosensitizing properties. A fusion gene between p53 and either the prodrugs cytosine deaminase or thymidine kinase was created and synthesized into an adenovirus vector, which was used to infect human tumor cell lines [178].

G. Gene Therapy with p53 Family Members

Because of the possible therapeutic efficacy of p53 for cancer gene therapy, other groups have begun to examine the feasibility of using p53-related genes, as several p53 homologs have been recently identified (reviewed by Arrowsmith [1] and Kaelin [70]). This approach may be a treatment for tumors that are resistant to p53-mediated apoptosis. In 1997, the first p53-related gene was described, p73 [71]. p73 shares homology with p53 by having several of the conserved p53 structural regions. Functionally, p73 can activate promoters with p53 binding sites, and overexpression of p73 can induce apoptosis [65,69,11,124]. In 1998, another p53-related gene, p51 (also identified as p40, p63, and p73L), was cloned [116]. Like p73, p51 can also activate promoters with p53 binding sites and can induce apoptosis in cells with mutant p53 [116]. The two p53-related genes differ from p53 in that they have two splicing variants: p51A and p51B and p73α and p73β. Another gene recently cloned, p33^{ING1}, was found not to have structural similarity with p53, but instead seems to help in the transcriptional activity of the target genes of p53 [45,46].

An adenovirus vector expressing p73 (Ad-p73) was recently engineered [89]. It was hypothesized that in cancer cell lines that are resistant to the effects of Ad-p53, p73 may represent an alternative therapeutic approach [95,124,129]. Previously, overexpression of p73β was able to suppress the growth of p53-resistant cancer cell lines that overexpress HPV E6, which targets p53 for degradation [124]. Infection of neural cancer cell lines with Ad-p73 decreased viability independent of the endogenous p53 or p73 status [89].

Similarly, another group constructed an adenovirus encoding both transcripts of p51 (Ad-p51α and Ad-p51β) which was used to infect human lung or breast cancer cell lines [65]. Infection with Ad-p51α caused more extensive apoptosis than that observed with either Ad-p51β or Ad-p53. This apoptosis was also p53 independent. Interestingly, one group has constructed a hybrid p53 molecule that combines the N-terminus of p53 with the C-terminus of p51 [63]. This fusion p51–p53 protein was found to transactivate a BAX luciferase promoter (containing p53 binding sites) 30 times greater than either p53 or p51 alone [63]. An adenovirus encoding this fusion p53–p51 protein was able to inhibit growth of tumor cells both in vitro and in vivo [77]. Finally, an adenovirus was recently created that encodes p33^{ING1} (Ad-p33), a protein that helps to induce p53-mediated growth inhibition [141]. Coinfection of glioblastoma cells with Ad-p33 and Ad-p53 caused more extensive apoptosis than infection with either adenovirus alone [141]. Although preliminary, these results begin to suggest that the p53-related genes may be alternatives to inducing apoptosis in p53-resistant cell lines.

IV. CONCLUSIONS

Because many human malignancies show mutations or deletions in tumor suppressor genes, replacement of these genes has become an important strategy in cancer gene therapy. The p53 tumor suppressor gene has been extensively studied because it can induce either cell-cycle arrest or apoptosis through its differential activation of target genes. Overexpression of p53 in tumor cell lines may then modulate chemosensitivity and/or radiosensitivity. The p53 tumor suppressor may also be combined with other therapeutic modalities, such as other tumor suppressors. Recent research suggests that p53-related genes may also prove beneficial in cancer gene therapy. Finally, some groups are attempting to modify the structure of p53 to improve its capabilities to cause either growth inhibition or apoptosis.

References

1. Arrowsmith, C. H. (1999). Structure and function in the p53 family. Cell Death Diff. 6, 1169–1173.
2. Atencio, I., and Demers, G. W. (1999). Apoptosis in a hepatocellular carcinoma cell line in response to adenovirus-mediated transfer of wild-type p53 gene occurs by regulation of bcl-2 family members and activation of caspases. Cancer Gene Ther. 6(suppl.), S5.
3. Avanzini, J. B., Johnson, D., Atencio, I., Neuteboom, S., Ramachandra, M., Sutjipto, S., Vaillancourt, M. T., Ralston, R., Phelan, A., and Wills, K. N. (2000). Expanded efficacy of tumor growth inhibition by a recombinant adenovirus expressing a p53–VP22 fusion protein, Abstract No. 330, 3rd Annual Meeting of the American Society of Gene Therapy. Mol. Ther. 1, 5129.
4. Badie, B., Goh, C. S., Klaver, J., Herweijer, H., and Boothman, D. A. (1999). Combined radiation and p53 gene therapy of malignant glioma cells. Cancer Gene Ther. 6, 155–162.
5. Baker, S. J., Markowitz, S., Fearon, E. R., Willson, J. K., and Vogelstein, B. (1990). Suppression of human colorectal carcinoma cell growth by wild-type p53. Science 249, 912–915.
6. Baselga, J. (1999). New horizons: gene therapy for cancer. Anticancer Drugs 10(suppl.), S39–S42.
7. Beaupre, D. M., and Kurzrock, R. (1999). RAS and leukemia: from basic mechanisms to gene-directed therapy. J. Clin. Oncol. 17, 1071–1079.
8. Bergelson, J. M., Cunningham, J. A., Drouguett, G., Kurt-Jones, E. A., Krithivas, A., Hong, J. S., Horwitz, M. S., Crowell, R. L., and Finberg, R. W. (1997). Isolation of a common receptor for coxsackie B viruses and adenoviruses 2 and 5. Science 275, 1320–1323.
9. Bian, J. H., and Sun, Y. (1997). Transcriptional activation by p53 of the human type IV collagenase (gelatinase A or matrix metalloproteinase 2) promoter. Mol. Cell. Biol. 17, 6330–6338.
10. Bilbao, G., Conteras, J. L., Gomez-Navarro, J., and Curiel, D. T. (1998). Improving adenoviral vectors for cancer gene therapy. Tumor Targeting 3, 59–79.
11. Biroccio, A., Del Bufalo, D., Ricca, A., D'Angelo, C., D'Orazi, G., Sacchi, A., Soddu, S., and Zupi, G. (1999). Increase of BCNU sensitivity by wt-p53 gene therapy in glioblastoma lines depends on the administration schedule. Gene Ther. 6, 1064–1072.
12. Blagosklonny, M. V., and El-Deiry, W. S. (1996). In vitro evaluation of a p53-expressing adenovirus as an anticancer drug. Int. J. Cancer 67, 386–392.

13. Bougeret, C., Virone-Oddos, A., Adeline, E., Lacroix, F., Lefranc, C., Ferrero, L., and Huet, T. (2000). Cancer gene therapy mediated by CTS1, a p53 derivative: advantage over wild-type p53 in growth inhibition of human tumors overexpressing MDM2. *Cancer Gene Ther.* **7,** 789–798.

14. Bouvet, M., Ellis, L. M., Nishizaki, M., Fujiwara, T., Liu, W., Bucana, C. D., Fang, B., Lee, J. J., and Roth, J. A. (1998). Adenovirus-mediated wild-type p53 gene transfer down-regulates vascular endothelial growth factor expression and inhibits angiogenesis in human colon cancer. *Cancer Res.* **58,** 2288–2292.

15. Bracey, T. S., Miller, J. C., Preece, A., and Paraskeva, C. (1995). γ-radiation-induced apoptosis in human colorectal adenoma and carcinoma cell lines can occur in the absence of wild type p53. *Oncogene* **10,** 2391–2396.

16. Bristow, R. G., Benchimol, S., and Hill, R. P. (1996). The p53 gene as a modifier of intrinsic radiosensitivity: implications for radiotherapy. *Radiother. Oncol.* **40,** 197–223.

17. Broaddus, W. C., Liu, Y., Steele, L. L., Gillies, G. T., Lin, P. S., Loudon, W. G., Valerie, K., Schmidt-Ulrich, R. K., and Fillmore, H. L. (1999). Enhanced radiosensitivity of malignant glioma cells after adenoviral p53 transfer *J. Neurosurg.* **91,** 997–1004.

18. Buchschacher, Jr. G. L., and Wong-Staal, F. (2000). Development of lentiviral vectors for gene therapy for human diseases. *Blood* **95,** 2499–504.

19. Burns, T. F., and El-Deiry, W. S. (1999). The p53 pathway and apoptosis. *J. Cell. Physiol.* **181,** 231–239.

20. Cai, D. W., Mukhopadhyay, T., Liu, Y. J., Fujiwara, T., and Roth, J. A. (1993). Stable expression of the wild-type p53 gene in human lung cancer cells after retrovirus-mediated gene transfer. *Hum. Gene Ther.* **4,** 617–624.

21. Canman, C. E., and Kastan, M. B. (1997). Role of p53 in apoptosis. *Adv. Pharmacol.* **41,** 429–460.

22. Cascallo, M., Mercade, E., Capella, G., Lluis, F., Fillat, C., Gomez-Foix, A. M., and Mazo, A. (1999). Genetic background determines the response to adenovirus-mediated wild-type p53 expression in pancreatic tumor cells. *Cancer Gene Ther.* **6,** 428–436.

23. Chang, E. H., Jang, Y.-J., Hao, Z., Murphy, G., Rait, A., Fee, W. E., Jr., Sussman, H. H., Ryan, P., Chiang, Y., and Pirollo, K. F. (1997). Restoration of the G1 checkpoint and the apoptotic pathway mediated by wild-type p53 sensitizes squamous cell carcinoma of the head and neck to radiotherapy. *Arch. Otolaryngol. Head Neck Surg.* **123,** 507–512.

24. Chen, Q. R., and Mixson, J. A. (1998). Systemic gene therapy with p53 inhibits breast cancer: recent advances and therapeutic implications. *Front. Biosci.* **3,** D997–D1004.

25. Chin, K. V., Ueda, I., Pastan, I., and Gottesman, M. M. (1992). Modulation of activity of the promoter of the human MDR1 gene by Ras and p53. *Science* **255,** 459–462.

26. Clayman G. L., El-Naggar, A. K., Roth, J. A., Zhang, W.-W., Goepfert, H., Taylor, D. L., and Liu, T.-J. (1995). In vivo molecular therapy with p53 adenovirus for microscopic residual head and neck squamous carcinoma. *Cancer Res.* **55,** 1–6.

27. Conseiller, E., Debussche, L., Landais, D., Venot, C., Maratrat, M., Sierra, V., Tocque, B., and Bracco, L. (1998). CTS1: a p53-derived chimeric tumor suppressor gene with enhanced in vitro apoptotic properties. *J. Clin. Invest.* **101,** 120–127.

28. Croyle, M. A., Walter, E., Janich, S., Roessler, B. J., and Amidon, G. L. (1998). Role of integrin expression in adenovirus-mediated gene delivery to the intestinal epithelium. *Hum. Gene Ther.* **9,** 561–573.

29. Dameron, K. M., Voplert, O. V., Tainsky, M. A., and Bouck, N. (1994). Control of angiogenesis in fibroblasts by p53 regulation of thrombospondin-1. *Science* **265,** 1582–1584.

30. Donehower, L. A., Harvey, M., Slagle, B. L., McArthur, M. J., Montgomery, Jr., C. A., Butel, J. S., and Bradley, A. (1992). Mice deficient for p53 are developmentally normal but susceptible to spontaneous tumours. *Nature* **356,** 215–221.

31. Dull, T., Zufferey, R., Kelly, M., Mandel, R. J., Nguyen, M., Trono, D., and Naldini, L. (1998). A third-generation lentivirus vector with a conditional packaging system. *J. Virol.* **72,** 8463–8471.

32. Eisold, S., Ridder, R., Burguete, T., Schmidt, J., Klar, E., Herfarth, C., and Doeberitz, M. (1999). Synergistic effect of p53 adenovirus-mediated gene therapy and 5-FU-chemotherapy against pancreatic cancer cells in vivo. *Cancer Gene Ther.* **6**(suppl.), S5.

33. El-Deiry, W. S. (1998). p21/p53 cellular growth control, and genomic integrity. *Curr. Topics Microbiol. Immunol.* **227,** 121–137.

34. El-Deiry, W. S. (1997). Role of oncogenes in resistance and killing by cancer therapeutic agents. *Curr. Opin. Oncol.* **9,** 79–87.

35. El-Deiry, W. S., Harper, J. W., O'Connor, P. M., Velculescu, V. E., Canman, C. E., Jackman, J., Pietenpol, J. A., Burrell, M., Hill, D. E., Wang, Y., Wiman, K. G., Mercer, W. E., Kastan, M. B., Kohn, K. W., Elledge, S. J., Kinzler, K. W., and Vogelstein, B. (1994). WAF1/CIP1 is induced in p53-mediated G1 arrest and apoptosis. *Cancer Res.* **54,** 1169–1174.

36. El-Deiry, W. S., Tokino, T., Velculescu, V. E., Levy, D. B., Parsons, R., Trent, J. M., Lin, D., Mercer, W. E., Kinzler, K. W., and Vogelstein, B. (1993). WAF1, a potential mediator of p53 tumor suppression. *Cell* **75,** 817–825.

37. Feng, M., Jackson, W. H., Goldman, C. K., Rancourt, C., Wang, M., Dusing, S. K., Siegal, G., and Curiel, D. T. (1997). Stable in vivo gene transduction via a novel adenovirus/retroviral chimeric vector. *Nat. Biotechnol.* **15,** 866–870.

38. Flotte, T. R., and Carter, B. J. (1997). In vivo gene therapy with adeno-associated virus vectors for cystic fibrosis. *Adv. Pharmacol.* **40,** 85–101.

39. Foulkes, W. D., Flanders, T. Y., Pollock, P. M., and Hayward, N. K. (1997). The CDKN2A (p16) gene and human cancer. *Mol. Med.* **3,** 5–20.

40. Frank, D. K., Frederick, M. J., Liu, T. J., and Clayman, G. L. (1998). Bystander effect in the adenovirus-mediated wild-type p53 gene therapy model of human squamous cell carcinoma of the head and neck. *Clin. Cancer Res.* **4,** 2521–2528.

41. Fuchs, E. J., McKenna, K. A., and Bedi, A. (1997). p53-dependent DNA damage-induced apoptosis requires Fas/Apo-1-independent activation of CPP32β. *Cancer Res.* **57,** 2550–2554.

42. Fujiwara, T., Cai, D. W., Georges, R. N., Mukhopadhyay, T., Grimm, E. A., and Roth, J. A. (1994). Therapeutic effect of a retroviral wild-type p53 expression vector in an orthotopic lung cancer model. *J. Natl. Cancer Inst.* **86,** 1458–1462.

43. Gallardo, D., Drazan, K. E., and McBride, W. H. (1996). Adenovirus-based transfer of wild-type p53 gene increases ovarian tumor radiosensitivity. *Cancer Res.* **56,** 4891–4893.

44. Gao, G. P., Yang, Y., and Wilson, J. M. (1996). Biology of adenovirus vectors with E1 and E4 deletions for liver-directed gene therapy. *J. Virol.* **70,** 8934–8943.

45. Garkavtsev, I., Grigorian, I. A., Ossovskaya, V. S., Chernov, M. V., Chumakov, P. M., and Gudkov, A. V. (1998). The candidate tumour suppressor p33^ING1 cooperates with p53 in cell growth control. *Nature* **391,** 295–298.

46. Garkavtsev, I., Kazarov, A., Gudkov, A., and Riabowol, K. (1996). Suppression of the novel growth inhibitor p33(ING1) promotes neoplastic transformation. *Nat. Genet.* **14,** 415–420.

47. Gomez-Manzano, C., Fueyo, J., Kyritsis, A. P., McDonnell, T. J., Steck, P. A., Levin, V. A., and Yung, W. K. A. (1997). Characterization of p53 and p21 functional interactions in glioma cells en route to apoptosis. *J. Natl. Cancer Inst.* **89,** 1036–1044.

48. Gotoh, A., Kao, C., Ko, S.-C., Hamada, K., Liu, T. J., and Chung, L. W. K. (1997). Cytotoxic effects of recombinant adenovirus p53 and

cell cycle regulator genes (p21$^{WAF1/CIP1}$ and p16^{CDKN4}) in human prostate cancers. *J. Urol.* **158**, 636–641.

49. Grace, M. J., Xie, L., Musco, M. L., Cui, S., Gurnani, M., DiGiacomo, R., Chang, A., Indelicato, S., Sved, J., Johnson, R., and Nielsen, L. L. (1999). The use of laser scanning cytometry to assess depth of penetration of adenovirus gene therapy in human xenograft biopsies. *Am. J. Pathol.* **155**, 1869–1878.

50. Greenblatt, M. S., Bennett, W. P., Hollstein, M., and Harris, C. C. (1994). Mutations in the p53 tumor suppressor gene: clues to cancer etiology and molecular pathogenesis. *Cancer Res.* **54**, 4855–4878.

51. Gurnani, M., Lipari, P., Dell, J., Shi, B., and Nielsen, L. L. (1999). Adenovirus-mediated p53 gene therapy has greater efficacy when combined with chemotherapy against human head and neck, ovarian, prostate, and breast cancer cells. *Cancer Chemother. Pharmacol.* **44**, 143–151.

52. Halloran, C., Ghaneh, P., Greenhalf, W., Neoptolemos, J., and Costello, E. (2000). An investigation of the combined effects on pancreatic cancer cells of chemotherapeutic drugs and adenoviral-mediated delivery of p53/p16^{INK4A}, Abstract No. 916, 3rd Annual Meeting of the American Society of Gene Therapy. *Mol. Ther.* **1**, 5325.

53. Harbour, J. W., and Dean, D. C. (2000). Rb functions in cell-cycle regulation and apoptosis. *Nat. Cell Biol.* **2**, E65–E67.

54. Harper, J. W., Adami, G. R., Wei, N., Keyomarsi, K., and Elledge, S. J. (1993). The p21 CDK-interacting protein CIP1 is a potent inhibitor of G1 cyclin-dependent kinases. *Cell* **75**, 805–816.

55. Hartwell, L. H., and Kastan, M. B. (1994). Cell cycle control and cancer. *Science* **266**, 1821–1888.

56. Henderson, Y. C., Breau, R. L., Liu, T.-J., and Clayman, G. L. (2000). Telomerase activity in head and neck tumors after introduction of wild-type p53, p21, p16, and E2F-1 genes by means of recombinant adenovirus. *Head Neck* **22**, 347–354.

57. Hill, B. T., Perrin, D., and Kruczynski, A. (2000). Inhibition of RAS-targeted prenylation: protein farnesyl transferase inhibitors revisited. *Crit. Rev. Oncol. Hematol.* **33**, 7–23.

58. Hirai, M., Kelsey, L. S., Vaillancourt, M., Maneval, D. C., Watanabe, T., and Talmadge, J. E. (2000). Purging of human breast cancer cells from stem cell products with an adenovirus containing p53. *Cancer Gene Ther.* **7**, 197–206.

59. Hitt, M. M., Addison, C. L., and Graham, F. L. (1997). Human adenovirus vectors for gene transfer into mammalian cells. *Adv. Pharmacol.* **40**, 137–206.

60. Horio, Y., Hasegawa, Y., Sekido, Y., Takahashi, M., Roth, J. A., and Shimokata, K. (2000). Synergistic effects of adenovirus expressing wild-type p53 on chemosensitivity of non-small-cell lung cancer cells. *Cancer Gene Ther.* **7**, 537–544.

61. Hu, S. X., Ji, W., Zhou, Y., Logothetis, C., and Xu, H. J. (1997). Development of an adenovirus vector with tetracycline-regulatable human tumor necrosis factor α gene expression. *Cancer Res.* **57**, 3339–3343.

62. Huang, H., Li, C. Y., and Little, J. B. (1996). Abrogation of p53 function by transfection of HPV16 E6 gene does not enhance resistance of human tumour cells to ionizing radiation. *Int. J. Rad. Biol.* **70**, 151–160.

63. Ikawa, S., Obinata, M., and Ikawa, Y. (1999). Human p53–p51 (p53–related) fusion protein: a potent BAX transactivator. *Jpn. J. Cancer Res.* **90**, 596–599.

64. Ilan, Y., Droguett, G., Chowdhury, N.R., Li, Y., Sengupta, K., Thummala, N. R., Davidson, A., Chowdhury, J., and Horwitz, M. S. (1997). Insertion of the adenovirus E3 region into a recombinant viral vector prevents antiviral humoral and cellular immune responses and permits long term gene expression. *Proc. Natl. Acad. Sci. USA* **94**, 2587–2592.

65. Ishida, S., Yamashita, T., Nakaya, U., and Tokino, T. (2000). Adenovirus-mediated transfer of p53–related genes induces apoptosis of human cancer cells. *Jpn. J. Cancer Res.* **91**, 174–180.

66. Jacobberger, J. W., Sramkoski, R. M., Zhang, D., Zumstein, L. A., Doerksen, L. D., Merritt, J. A., Wright, S. A., and Shults, K. E. (1999). Bivariate analysis of the p53 pathway to evaluate Ad-p53 gene therapy efficacy. *Cytometry* **38**, 201–213.

67. Jain, R. K. (1994). Barriers to drug delivery in solid tumors. *Sci. Am.* **271**, 58–65.

68. Ji, L., Fang, B., Yen, N., Fong, K., Minna, J. D., and Roth, J. A. (1999). Induction of apoptosis and inhibition of tumorigenicity and tumor growth by adenovirus vector-mediated fragile histidine triad (FHIT) gene overexpression. *Cancer Res.* **59**, 3333–3339.

69. Jost, C. A., Marin, M. C., and Kaelin, Jr., W. G. (1997). p73 is a human p53-related protein that can induce apoptosis. *Nature* **389**, 191–194.

70. Kaelin, Jr., W. G. (1999). The p53 gene family. *Oncogene.* **18**, 7701–7705.

71. Kaghad, M., Bonnet, H., Yang, A., Creancier, L., Biscan, J.-C., Valent, A., Minty, A., Chalon, P., Lelias, J.-M., Dumont, X., Ferrara, P., McKeon, F., and Caput, D. (1997). Monoallelically expressed gene related to p53 at 1p36, a region frequently deleted in neuroblastoma and other cancers. *Cell* **90**, 809–819.

72. Kastan, M. B., Onyekwere, O., Sidransky, D., Vogelstein, B., and Craig, R. W. (1991). Participation of p53 protein in the cellular response to DNA damage. *Cancer Res.* **51**, 6304–6311.

73. Kessler, P. D., Podsakoff, G. M., Chen, X., McQuiston, S. A., Colosi, P. C., Matelis, L. A., Kurtzman, G. J., and Byrne, B. J. (1996). Gene delivery to skeletal muscle results in sustained expression and systemic delivery of a therapeutic protein. *Proc. Natl. Acad. Sci. USA* **93**, 14082–14087.

74. Kim, K., Takimoto, R., Dicker, D. T., Gazitt, Y., and El-Deiry, W. S. (2001). Enhanced TRAIL sensitivity by p53 overexpression in human cancer but not normal cell lines. *Int. J. Oncol.*, press.

75. Kim, K., Fisher, M. J., Xu, S. Q., and El-Deiry, W. S. (2000). Molecular determinants of response to TRAIL in killing of normal and cancer cells. *Clin. Cancer Res.* **6**, 335–346.

76. Kim, J., Hwang, E. S., Kim, J. S., You, E.-H., Lee, S. H., and Lee, J.-H. (1999). Intraperitoneal gene therapy with adenoviral-mediated p53 tumor suppressor gene for ovarian cancer model in nude mouse. *Cancer Gene Ther.* **6**, 172–178.

77. Kitamura, R., Tani, K., Ikawa, S., Hase, H., Nakazaki, Y., Sugiyama, H., Bai, Y., Tanabe, T., and Asano, S. (2000). p51/p63, a novel p53 homologue, acts as an effective tumor suppressor, and cooperatively inhibits tumor growth with p53 in a murine model, Abstract No. 429, 3rd Annual Meeting of the American Society of Gene Therapy. *Mol. Ther.* **1**, 5164.

78. Knudson, C. M., Tung, K. S., Tourtellotte, W. G., Brown, G. A., and Korsmeyer, S. J. (1995). Bax-deficient mice with lymphoid hyperplasia and male germ cell death. *Science* **270**, 96–99.

79. Koch, H., Harris, M. P., Anderson, S. C., Machemer, T., Hancock, W., Sutjipto, S., Wills, K. N., Gregory, R. J., Shepard, H. M., Westphal, M., and Maneval, D. C. (1996). Adenovirus-mediated p53 gene transfer suppresses growth of human glioblastoma cells in vitro and in vivo. *Int. J. Cancer* **67**, 808–815.

80. Lambright, E. S., Force, S. D., Lanuti, M., El-Deiry, W. S., Kaiser, L. R., Albelda, S. M., and Amin, K. M. (2000). Ad.CMVp21 cancer gene therapy prevents in vivo tumorigenesis but does not elicit any in vivo bystander effect. *Proc. Annu. Meet. Am. Assoc. Cancer Res.* **41**, 465.

81. Lang, F. F., Yung, W. K., Sawaya, R., and Tofilon, P. J. (1999). Adenovirus-mediated p53 gene therapy for human gliomas. *Neurosurgery* **45**, 1093–1040.

82. Lang, D., Miknyoczki, S. J., Huang, L., and Ruggeri, B. A. (1998). Stable reintroduction of wild-type p53 (MTmp53ts) causes the induction of apoptosis and neuroendocrine-like differentiation in human ductal pancreatic carcinoma cells. *Oncogene* **16**, 1593–1602.

83. Levine, A. J. (1997). p53, the cellular gatekeeper for growth and division. *Cell* **88**, 323–331.

84. Li, S., and Huang, L. (2000). Nonviral gene therapy: promises and challenges. *Gene Ther.* **7**, 31–34.

85. Li, D., Ling, D., Freimuth, P., and O'Malley, Jr., B. W. (1999). Variability of adenovirus receptor density influences gene transfer efficiency and therapeutic response in head and neck cancer. *Clin. Cancer Res.* **5**, 4175–4181.

86. Li, S., Wu, S. P., Whitmore, M., Loeffert, E. J., Wang, L., Watkins, S. C., Pitt, B. R., and Huang, L. (1999). Effect of immune response on gene transfer to lung via systemic administration of cationic lipid vectors. *J. Physiol.* **276**, L796–L804.

87. Li, J.-H., Li, P., Klamut, H., and Liu, F.-F. (1997). Cytotoxic effects of Ad5CMV-p53 expression in two human nasopharyngeal carcinoma cell lines. *Clin. Cancer Res.* **3**, 507–514.

88. Liu, Y., Fillmore, H., Lin, P. S., Chen, Z. G., Valerie, C. K., and Broaddus, W. C. (1999). Increased radiosensitivity of human glioma cells after transduction with wild-type p53, independent of cellular p53 mutation status. *Cancer Gene Ther.* **6**(suppl.), S4.

89. Lo, W. D., Zhu, L., Akhmametyeva, E., and Chang, L.-S. (2000). Adenovirus-mediated gene transfer of p53-related p73 and baculovirus-encoded anti-apoptotic p35 into neuronal cells. *Pediatric Res.* **47**(suppl.), 461A.

90. Lowe, S. W., Ruley, H. E., Jacks, T., and Housman, D. E. (1993). p53-dependent apoptosis modulates the cytotoxicity of anticancer agents. *Cell* **74**, 957–967.

91. Lowe, S. W., Schmitt, E. M., Smith, S. W., Osborne, B. A., and Jacks, T. (1993). p53 is required for radiation-induced apoptosis in mouse thymocytes. *Nature* **362**, 786–787.

92. Macleod, K. (2000). Tumor suppressor genes. *Curr. Opin. Genet. Dev.* **10**, 81–93.

93. Malkin, D., Li, F. P., Strong, L. C., Fraumeni, J. J. F., Nelson, C. E., Kim, D. H., Kassel, J., Gryka, M. A., Bischoff, F. Z., Tainsky, M. A., and Friend, S. H. (1990). Germ line p53 mutations in a familial syndrome of breast cancer, sarcomas, and other neoplasms. *Science* **250**, 1233–1238.

94. Marcel, T., and Grausz, J. D. (1997). The TMC worldwide gene therapy enrollment report, end 1996. *Human Gene Ther.* **8**, 775–800.

95. Marin, M. C., Jost, C. A., Irwin, M. S., DeCaprio, J. A., Caput, D., and Kaelin, Jr., W. G. (1998). Viral oncoproteins discriminate between p53 and the p53 homolog p73. *Mol. Cell. Biol.* **11**, 6316–6324.

96. Marini, F. C., Yu, Q., Wickham, T., Kovesdi, I., and Andreeff, M. (2000). Adenovirus as a gene therapy vector for hematopoietic cells. *Cancer Gene Ther.* **7**, 816–825.

97. May, M. J., and Ghosh, S. (1997). Rel/NF-kappa B and I kappa B proteins: an overview. *Semin. Cancer Biol.* **8**, 63–73.

98. McDonald, III, E. R., and El-Deiry, W. S. (2000). Cell cycle control as a basis for cancer drug development. *Int. J. Oncol.* **16**, 871–886.

99. Meng, R., Shih, H., Prabhu, N. S., George, D. L., and El-Deiry, W. S. (1997). Bypass of abnormal MDM2 inhibition of p53-dependent growth suppression. *Clin. Cancer Res.* **4**, 251–259.

100. Milas, M., Yu, D., Lang, A., Ge, T., Feig, B., El-Naggar, A. K., and Pollock, R. E. (2000). Adenovirus-mediated p53 gene therapy inhibits human sarcoma tumorigenicity. *Cancer Gene Ther.* **7**, 422–429.

101. Miller, N., and Whelan, J. (1997). Progress in transcriptionally targeted and regulatable vectors for gene therapy. *Hum. Gene Ther.* **8**, 803–815.

102. Miyoshi, H., Blomer, U., Takahashi, M., Gage, F. H., and Verma, I. M. (1998). Development of a self-inactivating lentivirus vector. *J. Virol.* **72**, 8150–8157.

103. Momand, J., Wu, H. H., and Dasgupta, G. (2000). MDM2—master regulator of the p53 tumor suppressor protein. *Gene* **242**, 15–29.

104. Monahan, P. E., and Samulski, R. J. (2000). AAV vectors: is clinical success on the horizon? *Gene Ther.* **7**, 24–30.

105. Morgan, D. O. (1995). Principles of CDK regulation. *Nature* **374**, 131–134.

106. Murray, A., and Hunt, T. (1993). *The Cell Cycle: An Introduction,* Oxford University Press, London.

107. Nelson, A. R., Fingleton, B., Rothenberg, M. L., and Matrisian, L. M. (2000). Matrix metalloproteinases: biologic activity and clinical implications. *J. Clin. Oncol.* **18**, 1135–1149.

108. Nevins, J. R. (1994). Cell cycle targets of the DNA tumor viruses. *Curr. Opin. Genet. Develop.* **4**, 130–134.

109. Nielsen, L. L. (2000). NK cells mediate the anti-tumor effects of E1-deleted, type 5 adenovirus in a tumor xenograft model. *Oncol. Rep.* **7**, 151–155.

110. Nielsen, L. L., Shi, B., Hajian, G., Yaremko, B., Lipari, P., Ferrari, E., Gurnami, M., Malkowski, M., Chen, J., Bishop, W. R., and Liu, M. (1999). Combination therapy with the farnesyl protein transferase inhibitor SCH66336 and SCH58500 (p53 adenovirus) in preclinical cancer models. *Cancer Res.* **59**, 5896–5901.

111. Nielsen, L. L., Xie, L., McDonald, M., DiGiacomo, R., Chang, A., Gurnani, M., Shi, B., Liu, S., Indelicato, S. R., Hutchins, B., and Wen, S. F. (1999) Quantification of gene expression after p53 gene therapy and paclitaxel in preclinical cancer models. *Cancer Gene Ther.* **6**(suppl.), S5.

112. Nielsen, L. L., and Maneval, D. C. (1998). p53 tumor suppressor gene therapy for cancer. *Cancer Gene Ther.* **5**, 52–63.

113. Nielsen, L. L., Dell, J., Maxwell, E., Armstrong, L., Maneval, D., and Catino, J. J. (1997). Efficacy of p53 adenovirus-mediated gene therapy against human breast cancer xenografts. *Cancer Gene Ther.* **4**, 129–138.

114. Nishizaki, M., Fujiwara, T., Tanida, T., Hizuta, A., Nishimori, H., Tokino, T., Nakamura, Y., Bouvet, M., Roth, J. A., and Tanaka, N. (1999). Recombinant adenovirus expressing wild-type p53 is antiangiogenic: a proposed mechanism for bystander effect. *Clin. Cancer Res.* **5**, 1015–1023.

115. Okegawa, T., Li, Y., Pong, R.-C., and Hsieh, J.-T. (2000). Mechanism of coxsackie and adenovirus receptor (CAR) and its impact on prostatic cancer therapy. *Proc. Annu. Meet. Am. Assoc. Cancer Res.* **41**, 351.

116. Osada, M., Ohba, M., Kawahara, C., Ishioka, C., Kanamaru, R., Katoh, I., Ikawa, Y., Nimura, Y., Nakagawara, A., Obinata, M., and Ikawa, S. (1998). Cloning and functional analysis of human p51, which structurally and functionally resembles p53. *Nat. Med.* **4**, 839–843.

117. Osaki, S., Nakanishi, Y., Takayama, K., Pei, X.-H., Ueno, H., and Hara, N. (2000). Alteration of drug chemosensitivity caused by the adenovirus-mediated transfer of the wild-type p53 gene in human lung cancer cells. *Cancer Gene Ther.* **7**, 300–307.

118. Pagliaro, L. C. (2000). Gene therapy for bladder cancer. *World J. Urol.* **18**, 148–151.

119. Palu, G., Parolin, C., Takeuchi, Y., and Pizzato, M. (2000). Progress with retroviral gene vectors. *Rev. Med. Virol.* **10**, 185–202.

120. Pardee, A. B. (1989). G1 events and recognition of cell proliferation. *Science* **246**, 603–608.

121. Pearson, A. S., Koch, P. E., Atkinson, N., Xiong, M., Finberg, R. W., Roth, J. A., and Fang, B. (1999). Factors limiting adenovirus-mediated gene transfer into human lung and pancreatic cancer cell lines. *Clin. Cancer Res.* **5**, 4208–4213.

122. Peeters, M. J., Patijn, G. A., Lieber, A., Meuse, L., and Kay, M. A. (1996). Adenovirus-mediated hepatic gene transfer in mice: comparison of intravascular and biliary administration. *Hum. Gene Ther.* **7**, 1693–1699.

123. Peter, M. E., and Krammer, P. H. (1998). Mechanisms of CD95 (APO-1/Fas)-mediated apoptosis. *Curr. Opin. Immunol.* **10**, 545–551.

124. Prabhu, N. S., Somasundaram, K., Satyamoorthy, K., Herlyn, M., and El-Deiry, W. S. (1998). p73β, unlike p53, suppresses growth and induces apoptosis of human papillomavirus E6-expressing cancer cells. *Int. J. Oncol.* **13**, 5–9.

125. Ramondetta, L., Mills, G. B., Burke, T. W., and Wolf, J. K. (2000). Adenovirus-mediated expression of p53 or p21 in a papillary serous

endometrial carcinoma cell line (SPEC-2) results in both growth inhibition and apoptotic cell death: potential application of gene therapy to endometrial cancer. *Clin. Cancer Res.* **6,** 278–284.

126. Reiser, M., Neumann, I., Schmiegel, W., Wu, P.-C., and Lau, J. Y. N. (2000). Induction of cell proliferation arrest and apoptosis in hepatoma cells through adenoviral-mediated transfer of p53 gene. *J. Hepatol.* **32,** 771–782.

127. Rizk, N. P., Chang, M. Y., El Kouri, C., Seth, P., Kaiser, L. R., Albelda, S. M., and Amin, K. M. (1999). The evaluation of adenoviral p53-mediated bystander effect in gene therapy of cancer. *Cancer Gene Ther.* **6,** 291–301.

128. Roth, J. A., Swisher, S. G., and Meyn, R. E. (1999). p53 tumor suppressor gene therapy for cancer. *Oncology* **13,** 148–154.

129. Roth, J., Konig, C., Wienzek, S., Weigel, S., Ristea, S., and Dobbelstein, M. (1998). Inactivation of p53 but not p73 by adenovirus type 5 E1B 55-kilodalton and E4 34-kilodalton oncoproteins. *J. Virol.* **11,** 8510–8516.

130. Roth, J. A., and Cristiano, R. J. (1997). Gene therapy for cancer: what have we done and where are we going? *J. Natl. Cancer Inst.* **89,** 21–39.

131. Roth, J. A., Nguyen, D., Lawrence, D. D., Kemp, B. L., Carrasco, C. H., Ferson, D. Z., Hong, W. K., Komaki, R., Lee, J. J., Nesbitt, J. C., Pisters, K. M., Putnam, J. B., Schea, R., Shin, D. M., Walsh, G. L., Dolormente, M. M., Han, C. I., Martin, F. D., Yen, N., Xu, K., Stephens, L. C., McDonnell, T. J., Mukhopadhyay, T., and Cai, D. (1996). Retrovirus-mediated wild-type p53 gene transfer to tumors of patients with lung cancer. *Nat. Med.* **2,** 974–975.

132. Saller, E., Tom, E., Brunori, M., Otter, M., Estreicher, A., Mach, D. H., and Iggo, R. (1999). Increased apoptosis induction by 121F mutant p53. *EMBO J.* **18,** 4424–4437.

133. Scaria, A., St. George, J. A., Gregory, R. J., Noelle, R. J., Wadsworth, S. C., Smith, A. E., and Kaplan, J. M. (1997). Antibody to CD40 ligand inhibits both humoral and cellular immune responses to adenovirus vectors and facilitates repeated administration to mouse airway. *Gene Ther.* **4,** 611–617.

134. Sengupta, S., Ralhan, R., and Wasylyk, B. (2000). Tumour regression in a ligand inducible manner mediated by a chimeric tumor suppressor derived from p53. *Oncogene* **19,** 337–350.

135. Seth, P., Brinkmann, P., Schwartz, G. N., Katayose, D., Gress, R., Pastan, I., and Cowan, K. (1996). Adenovirus-mediated gene transfer to human breast tumor cells: an approach for cancer gene therapy and bone marrow purging. *Cancer Res.* **56,** 1346–1351.

136. Shao, J., Fujiwara, T., Kadowaki, Y., Fukazawa, T., Waku, T., Itoshima, T., Yamatsuji, T., Nishizaki, M., Roth, J. A., and Tanaka, N. (2000). Overexpression of the wild-type p53 gene inhibits NF-κB activity and synergizes with aspirin to induce apoptosis in human colon cancer cells. *Oncogene* **19,** 726–736.

137. Shatrov, V. A., Ameyar, M., Bouquet, C., Cai, Z., Stancou, R., and Haddada, H. Chouaib S. (2000). Adenovirus-mediated wild-type-p53-gene expression sensitizes TNF-resistant tumor cells to TNF-induced cytotoxicity by altering the cellular redox state. *Int. J. Cancer.* **85,** 93–97.

138. Sheikh, M. S., Fernandes-Salas, E., Chou, J., Brooks, K., Huang, Y., and Fornace, A. J. (2000). p53 mediated apoptosis in a caspase 8-dependent but FADD-independent manner. *Proc. Annu. Meet. Am. Assoc. Cancer Res.* **41,** 555.

139. Sherr, C. J. (1996). Cancer cell cycles. *Science* **274,** 1672–1677.

140. Shinoura, N., Yoshida, Y., Asai, A., Kirino, T., and Hamada, H. (2000). Adenovirus-mediated transfer of p53 and Fas ligand drastically enhances apoptosis in gliomas. *Cancer Gene Ther.* **7,** 732–738.

141. Shinoura, N., Muramatsu, Y., Nishimura, M., Yoshida, Y., Saito, A., Yokoyama, T., Furukawa, T., Horii, A., Hashimoto, M., Asai, A., Kirino, T., and Hamada, H. (1999). Adenovirus-mediated transfer of p33^{ING1} with p53 drastically augments apoptosis in gliomas. *Cancer Res.* **59,** 5521–5528.

142. Song, K., Cowan, K. H., and Sinha, B. K. (1999). In vivo studies of adenovirus-mediated p53 gene therapy for *cis*-platinum-resistant human ovarian tumor xenografts. *Oncol. Res.* **11,** 153–159.

143. Song, S., MacLachlan, T. K., Meng, R. D., and El-Deiry, W. S. (1999). Comparative gene expression profiling in response to p53 in a human lung cancer cell line. *Biochem. Biophys. Res. Commun.* **264,** 891–895.

144. Song, W. R., Kong, H. L., Traktman, P., and Crystal, R. G. (1997). Cytotoxic T lymphocyte responses to proteins encoded by heterologous transgenes transferred in vivo by adenoviral vectors. *Hum. Gene Ther.* **8,** 1207–1217.

145. Spitz, F. R., Nguyen, D., Skibber, J. M., Meyn, R. E., Cristiano, R. J., and Roth, J. A. (1996). Adenovirus-mediated wild-type p53 gene expression sensitizes colorectal cancer cells to ionizing radiation. *Clin. Can. Res.* **2,** 1665–1671.

146. St. John, L. S., Sauter, E. R., Herlyn, M., Litwin, S., and Adler-Storthz, K. (2000). Endogenous p53 gene status predicts the response of human squamous cell carcinomas to wild-type p53. *Cancer Gene Ther.* **7,** 749–756.

147. Stavridi, E. S., Chehab, N. H., Caruso, L. C., and Halazonetis, T. D. (1999). Change in oligomerization specificity of the p53 tetramerization domain by hydrophobic amino acid substitutions. *Protein Sci.* **8,** 1773–1779.

148. Sweeney, P., and Pisters, L. L. (2000). Ad5CMVp53 gene therapy for locally advanced prostate cancer—where do we stand? *World J. Urol.* **18,** 121–124.

149. Symonds, H., Krall, L., Remington, L., Saenz-Robles, M., Lowe, S., Jacks, T., and Van Dyke, T. (1994). p53-dependent apoptosis suppresses tumor growth and progression in vivo. *Cell* **78,** 703–711.

150. Takayama, K., Ueno, H., Pei, X. H., Nakanishi, Y., Yatsunami, J., and Hara, N. (1998). The levels of integrin $\alpha v \beta 5$ may predict the susceptibility to adenovirus-mediated gene transfer in human lung cancer cells. *Gene Ther.* **5,** 361–368.

151. Takimoto, R., and El-Deiry, W. S. (2000). Wild-type p53 transactivates the KILLER/DR5 gene through an intronic sequence-specific DNA-binding site. *Oncogene* **19,** 1735–1743.

152. Tan, B. T., Wu, L., and Berk, A. J. (1999). An adenovirus/Epstein–Barr virus hybrid vector that stably transforms cultured cells with high efficiency. *J. Virol.* **73,** 7582–7589.

153. Thottassery, J. V., Zambetti, G. P., Arimori, K., Schuetz, E. G., and Schuetz, J. D. (1997). p53-dependent regulation of MDR1 gene expression causes selective resistance to chemotherapeutic agents. *Proc. Natl. Acad. Sci. USA* **94,** 11037–11042.

154. Tiemann, F., Gruber, C., Schirmacher, P., Arnold, W., Jennings, G., Sandig, V., and Strauss, M. (2000). Efficacy and safety of p53/p16 adenovirus-mediated gene therapy for the treatment of hepatocellular carcinoma. *Proc. Annu. Meet. Am. Assoc. Cancer Res.* **41,** 121.

155. To, L. B., Haylock, D. N., Simmons, P. J., and Juttner, C. A. (1997). The biology and clinical uses of blood stem cells. *Blood* **89,** 2233–2258.

156. Toschi, E., Rota, R., Antonini, A., Melillo, G., and Capogrossi, M. C. (2000). Wild-type p53 gene transfer inhibits invasion and reduces matrix metalloproteinase-2 levels in p53-mutated human melanoma cells. *J. Invest. Dermatol.* **114,** 1188–1194.

157. Trono, D. (2000). Lentiviral vectors: turning a deadly foe into a therapeutic agent. *Gene Ther.* **7,** 20–23.

158. Tsang, N.-M., Nagasawa, H., Li, C., and Little, J. B. (1995). Abrogation of p53 function by transfection of HPV16 E6 gene enhances the resistance of human diploid fibroblasts to ionizing radiation. *Oncogene* **10,** 2403–2408.

159. Turturro, F., Heineke, H. L., Drevyanko, T. F., Link, Jr., C. J. and Seth, P. (2000). Adenovirus-p53-mediated gene therapy of anaplastic large cell lymphoma with t(2;5) in a nude mouse model. *Gene Ther.* **7,** 930–933.

160. Turturro, F., Seth, P., and Link, Jr., C. J. (2000). In vitro adenoviral vector p53-mediated transduction and killing correlates with

expression of coxsackie-adenovirus receptor and $\alpha v\beta 5$ integrin in SUDHL-1 cells derived from anaplastic large-cell lymphoma. *Clin. Cancer Res.* **6,** 185–192.

161. Ueda, K., Yoshida, A., and Amachi, T. (1999). Recent progress in P-glycoprotein research. *Anti-Cancer Drug Design* **14,** 115–121.

162. Velculescu, V. E., and El-Deiry, W. S. (1996). Biological and clinical importance of the p53 tumor suppressor gene. *Clin. Chem.* **42,** 858–868.

163. Vile, R. G., Russell, S. J., and Lemoine, N. R. (2000). Cancer gene therapy: hard lessons and new courses. *Gene Ther.* **7,** 2–8.

164. Vogelstein, B., Lane, D. L., and Levine, A. J. (2000). Surfing the p53 network. *Nature* **408,** 307–310.

165. Von Gruenigen, V. E., Santoso, J. T., Coleman, R. L., Muller, C. Y., Miller, D.S., and Mathis, J.M. (1998). In vivo studies of adenovirus-based p53 gene therapy for ovarian cancer. *Gynecol. Oncol.* **69,** 197–204.

166. Wattel, E., Vanrumbeke, M., Abina, M. A., Cambier, N., Preud-homme, C., Haddada, H., and Fenaux, P. (1996). Differential efficacy of adenoviral-mediated gene transfer into cells from hematological cell lines and fresh hematological malignancies. *Leukemia* **10,** 171–174.

167. Weinberg, R. (1995). The retinoblastoma protein and cell cycle control. *Cell* **81,** 323–330.

168. Weinstein, J. N., Myers, T. G., O'Connor, P. M., Friend, S. H., Fornace, A. J., Jr., Kohn, K. W., Fojo, T., Bates, S. E., Rubinstein, L. V., Anderson, N. L., Buolamwini, J. K., van Osdol, W. W., Monks, A. P., Scudiero, D. A., Sausville, E. A., Zaharevitz, D. W., Bunow, B., Viswanadhan, V. N., Johnson, G. S., Wittes, R. E., and Paull, K. D. (1997). An information-intensive approach to the molecular pharmacology of cancer. *Science* **275,** 343–349.

169. Wickham, T. J. (2000). Targeting adenovirus. *Gene Ther.* **7,** 110–114.

170. Wickham, T. J., Mathias, P., Cheresh, D. A., and Nemerow, G. R. (1993). Integrins alpha v beta 3 and alpha v beta 5 promote adenovirus internalization but not virus attachment. *Cell* **73,** 309–319.

171. Wills, K. N., Avanzini, J., Johnson, D., Neuteboom, S., Ramachandra, M., Sutjipto, S., Vaillancourt, M. T., Ralston, R., and Atnecio, I. (2000). Adenovirus delivery of a p53 variant induces apoptosis in tumor cell resistant to wild-type p53 treatment, Abstract No. 428, 3rd Annual Meeting of the American Society of Gene Therapy. *Mol. Ther.* **1,** 5163.

172. Wolf, J. K., Mills, G. B., Bazzet, L., Bast, Jr., R., Roth, J. A., and Gershenson, D. M. (1999). Adenovirus-mediated p53 growth inhibition of ovarian cancer cells is independent of endogenous p53 status. *Gynecol. Oncol.* **75,** 261–266.

173. Wroblewski, J., Lay, L. T., Van Zant , G., Phillips, G., Seth, P., Curiel, D., and Meeker, T. C. (1996). Selective elimination (purging) of contaminating malignant cells from hematopoietic stem cell auto-grafts using recombinant adenovirus. *Cancer Gene Ther.* **3,** 257–264.

174. Wu, G. S., Kim, K., and El-Deiry, W. S. (2000). KILLER/DR5, a novel DNA-damage inducible death receptor gene, links the p53-tumor suppressor to caspase activation and apoptotic death. *Adv. Exp. Med. Biol.* **465,** 143–151.

175. Wu, G. S., Burns, T. F., McDonald, III, E. R., Meng, R. D., Kao, G., Muschel, R., Yen, T., and El-Deiry, W.S. (1999). Induction of the TRAIL receptor KILLER/DR5 in p53-dependent apoptosis but not growth arrest. *Oncogene* **18,** 6411–6418.

176. Wu, G. S., Burns, T. F., McDonald, III, E. R., Jiang, W., Meng, R., Krantz, I. D., Kao, G., Gan, D.-D., Zhou, J.-Y., Muschel, R., Hamilton, S. R., Spinner, N. B., Markowitz, S., Wu, G., and El-Deiry, W. S. (1997). KILLER/DR5 is a DNA damage-inducible p53-regulated death receptor gene. *Nat. Genet.* **17,** 141–143.

177. Xiao, X., Li, J., and Samulski, R. (1996). Efficient long term gene transfer into muscle tissue of immunocompetent mice by adeno-associated virus vector. *J. Virol.* **70,** 8098–8108.

178. Xie, Y., Gilbert, J. D., Kim, J. H., and Freytag, S. O. (1999). Efficacy of adenovirus-mediated CD/5-FC and HSV-1 thymidine kinase/ganciclovir suicide gene therapies concomitant with p53 gene therapy. *Clin. Cancer Res.* **5,** 4224–4232.

179. Xu, M., Kumar, D., Srinivas, S., Detolla, L. J., Yu, S. F., Stass, S. A., and Mixson, A. J. (1997). Parenteral gene therapy with p53 inhibits human breast tumors in vivo through a bystander mechanism without evidence of toxicity. *Hum. Gene Ther.* **8,** 177–185.

180. Yang, Y., Ertl, H. C., and Wilson, J. M. (1994). MHC class I-restricted cytotoxic T lymphocytes to viral antigens destroy hepatocytes in mice infected with E1-deleted recombinant adenoviruses. *Immunity* **1,** 433–442.

181. Yew, N. S., Wang, K. X., Przybylska, M., Bagley, R. G., Stedman, M., Marshall, J., Scheule, R. K., and Cheng, S. H. (1999). Contribution of plasmid DNA to inflammation of the lung via systemic administration of cationic lipid:pDNA complexes. *Hum. Gene Ther.* **10,** 223–234.

182. Yoshida, Y., Sadata, A., Zhang, W., Saito, K., Shinoura, N., and Hamada, H. (1998). Generation of fiber-mutant recombinant aden-oviruses for gene therapy of malignant glioma. *Hum. Gene Ther.* **9,** 2503–2515.

183. Yoshida, Y., and Hamada, H. (1997). Adenovirus-mediated inducible gene expression through tetracycline-controllable transactivator with nuclear localization signal. *Biochem. Biophys. Res. Comm.* **230,** 426–430.

184. Zeng, Y.-X., Prabhu, N. S., Meng, R., and El-Deiry, W. S. (1997). Adenovirus-mediated p53 gene therapy in nasopharyngeal cancer. *Int. J. Oncol.* **11,** 221–226.

185. Zepeda, M., Kang, D., Tsai, V., Levy, A., Huang, W.-M., Maneval, D., and LaFace, D. (2000). Host immune responses to native or foreign p53 in preclinical models: a comparison of rAd-p53 (mouse) vs. rAd-p53 (human) in immunocompetent mice, Abstract No. 902, 3rd Annual Meeting of the American Society of Gene Therapy. *Mol. Ther.* **1,** 5321.

186. Zou, Y., Zong, G., Ling, Y.-H., and Perez-Soler, R. (2000). Develop-ment of cationic liposome formulations for intratracheal gene therapy of early lung cancer. *Cancer Gene Ther.* **7,** 683–696.

187. Zufferey, R., Nagy, D., Mandel, R. J., Naldini, L., and Trono, D. (1997). Multiply attenuated lentiviral vector achieves efficient gene delivery in vivo. *Nat. Biotechnol.* **15,** 871–875.

19

Antisense Downregulation of the Apoptosis–Related Bcl-2 and Bcl-xl Proteins: A New Approach to Cancer Therapy

IRINA V. LEBEDEVA

Department of Medicine and Pharmacology
Columbia University
College of Physicians and Surgeons
New York, New York 10032

C. A. STEIN

Department of Medicine and Pharmacology
Columbia University
College of Physicians and Surgeons
New York, New York 10032

I. THE Bcl FAMILY OF PROTEINS AND THEIR ROLE IN APOPTOSIS

Apoptosis is a form of cell death critical for the normal development of multicellular organisms [1–3]. It is accompanied by characteristic morphological and biochemical features (e.g., nuclear shrinkage, chromatin condensation, cytoplasmic blebbing, and internucleosomal DNA fragmentation) [1,2]. There are many ways by which cell death via apoptosis can be induced, including growth factor deprivation, cytokine treatment, antigen-receptor engagement, cell–cell interactions, irradiation, glucocorticoids, or treatment with various cytotoxic agents [3]. Disregulation of apoptosis can lead to the aberrant accumulation of cells and can contribute to tumor growth [4]. Apoptosis is an active, irreversible process, and much effort has been made recently to identify the external signals and the genes responsible for its induction and suppression [5].

Many proteins are known to be involved in the control of apoptosis, and members of the evolutionarily conserved *bcl*-2 family are thought to be its central regulators [6,7]. The first identified member of this gene family, *bcl*-2, was discovered by virtue of its involvement in t(14;18) chromosomal translocations commonly found in B-cell lymphomas [8,9]. Normally, the level of *bcl*-2 expression differs for different cell types [10], however, high levels and aberrant patterns of *bcl*-2 expression have been reported in a wide variety of human cancers, including prostate, colorectal, lung, gastric, renal, neuroblastoma, non-Hodgkin's lymphomas, and both acute and chronic leukemias (reviewed in Reed [11]). Overproduction of Bcl-2 protein significantly prolongs cell survival toward classical apoptotic stimuli, including lymphokine deprivation from factor-dependent hematopoetic cells, glucocorticoid treatment of thymocytes and lymphoid leukemia cells, γ-irradiation of thymocytes, and nerve growth factor (NGF) deprivation of fetal sympathetic neurons [5,12,13].

The anti-apoptotic properties of the Bcl-2 protein have been extensively reviewed [5,11,14]. Elevation of Bcl-2 protein expression contributes not only to the development of cancer but also to resistance against a wide variety of anti-cancer agents, including cyclophosphamide, cisplatin, etoposide (VP-16), mitoxantrone, adriamycin, 1-β-D-arabinofuranosil-cytosine (Ara-C), methotrexate, 5-fluorouracil, thapsigargin, staurosporine, dexamethasone, and radiation [15–20].

Since the discovery of the *bcl*-2 gene, at least 16 other homologous proteins have been identified in humans [21,22]. Some of them, like Bcl-2, possess anti-apoptotic properties (Bcl-xL, Mcl-1) [23–26], and others are pro-apoptotic proteins (Bax, Bak, Bcl-xS) [27–29]. Inhibition of apoptosis induced by a variety of anticancer agents by Bcl-2 and Bcl-xL suggests that these two proteins share a common potentially inhibitory pathway [23,26,30,31].

Several mechanisms have been proposed to explain the regulatory functions of Bcl-2. Initial observations suggested its role as an antioxidant [32,33] or as a regulator of intracellular calcium levels [15,34], as well as its having a role in the transport of proteins across cellular membranes (e.g., across the nuclear membrane) [35]. It has also been suggested that heterodimerization between death agonists and death antagonists within the Bcl-2 family regulates their respective functions [36,37]. Recently, it has been proposed that Bcl-2 regulates activation of caspase proteases [26,38–40], which are responsible for the final executionary steps of apoptosis. Bcl-2, Bax, and their related proteins control activation of the downstream group of caspases, probably either by controlling the release of cytochrome c from mitochondria or by modulating the activity of CED-4-type proteins. Other dimerization partners, such as Raf-1 [41], calcineurin [42], and Apaf-1 [43], may also be involved in the ability of Bcl-2 family proteins to regulate cell survival. However, the extensive network of protein–protein interactions between Bcl-2 family members and non-Bcl-2 family members makes it difficult to assess the relative contribution of each of these interactions to the regulation of apoptosis.

Bcl-2 expression can inhibit the release of cytochrome c from the mitochondrial space and can stabilize the mitochondrial potential in response to apoptosis-inducing agents (e.g., staurosporine) [44–46]. Furthermore, due to the proposed pore-forming capability of Bcl-2, Bcl-xL, and Bax in lipid membranes [47–49], they may act as channels for ions, proteins, or both. However, it is probably not possible to suggest a single mechanism to explain the mechanism of action of Bcl-2 family proteins. Most likely, several synergistic mechanisms are involved [50].

The protection afforded by Bcl-2 against chemotherapeutic agents defines a novel type of drug resistance in which Bcl-2 overexpression does not prevent drug entry and accumulation in tumor cells or the interaction of drugs with their primary molecular targets. Rather, Bcl-2 appears to block the transmission of signals originating from the cellular damage to the molecular effectors of apoptosis [11]. This allows cells to survive in the presence of otherwise lethal damage, repairing drug-induced damage after the drugs are withdrawn. Via this mechanism, Bcl-2 essentially converts anti-cancer drugs from cytotoxic to cytostatic. This increases the chances of acquiring genetic alterations and favors the development of a more malignant phenotype.

II. DOWNREGULATION OF Bcl-2 EXPRESSION: ANTISENSE STRATEGIES

The central role that *bcl*-2 family genes play in the control of programmed cell death and responses to chemotherapy suggests that these genes and their encoded proteins define a set of new targets. These targets can potentially be exploited to improve therapeutic outcomes for patients with advanced cancer. The antisense strategy can be efficiently employed to downregulate the expression of oncogenes [51,52]. This strategy is based on the specific interactions of DNA molecules with target mRNAs through Watson–Crick base pairing [53,54]. The formation of a DNA:RNA heteroduplex results in mRNA inactivation and consequent inhibition of synthesis of the protein product. This simple and attractive model has proven to be much more complicated to achieve in practice [55]. The three different types of antisense molecules are (1) relatively short synthetic oligonucleotides [55]; (2) antisense RNA, which is expressed intracellularly following transfection with antisense genes [56]; and (3) ribozymes, which possess an enzyme-like activity [57]. All three models have been used to downregulate *bcl*-2 expression in different cancer models *in vitro* and *in vivo*; however, only antisense oligonucleotide downregulation will be discussed at length in this review. The main consequences of the antisense downregulation of Bcl-2 protein and mRNA are decreased cell survival [58,99], increased drug sensitivity *in vitro* and *in vivo* [59], and, in many cases, the induction of apoptotic cell death [60].

A. Antisense Bcl-2 Vectors

After stable transfection into Jurkat cells, a Bcl-2 antisense expression plasmid ablated tumor formation in irradiated athymic mice [61]. In addition, a Bcl-2 antisense plasmid decreased *in vitro* survival of Jurkat T cells in serum-free medium. A similar plasmid caused reductions in Bcl-2 protein levels in t(14;18)-containing non-Hodgkin's lymphoma (NHL) cell lines [59,62], which, in turn, resulted in markedly enhanced sensitivity to several anticancer drugs.

In the presence of estrogen, MCF-7 breast cancer cells expressing Bcl-2 antisense transcripts were rendered twice as sensitive to adriamycin cytotoxicity as a control clone [63]. Low Bcl-2 levels mediated by antisense transcript expression and/or estrogen withdrawal were associated with regression of MCF-7 tumors in nude mice following estrogen withdrawal [64].

Chicken and human antisense Bcl-2 constructs caused an abrupt decrease in the number of surviving neurons in tissue culture within 24 hours [65]. Similar to neurons that die following neurotropin deprivation, the dying antisense-injected neurons exhibited chromatin condensation and nuclear fragmentation. Multiple chondrocyte cell lines overexpressing antisense Bcl-2 mRNA displayed increased apoptosis not only in response to serum withdrawal and retinoic acid treatment, but even in the presence of 10% serum [66]. An antisense Bcl-2 retroviral vector increased the sensitivity of a human gastric adenocarcinoma cell line to photodynamic therapy and its susceptibility to chemotherapy induced apoptosis [67].

B. Antisense Bcl-2 Ribozymes

Hammerhead ribozymes represent a class of genetically engineered RNA molecules that can specifically cleave other RNA sequences by means of their intrinsic enzyme-like activity [57]. These molecules consist of two important domains: "hammerhead" and antisense. The hammerhead portion cleaves the target RNA sequence by an enzyme-like mechanism in the presence of divalent cations (Mg^{2+}). The flanking antisense sequences provide the specificity of binding to the targeted RNA molecule.

Recently, an anti-*bcl*-2 gene therapeutic reagent based on the ribozyme technology has been developed and tested *in vitro* and *in vivo* [68]. A divalent (two catalytic units in one structure) hammerhead ribozyme was constructed by recombining two catalytic RNA domains into an antisense segment of the coding region for human *bcl*-2 mRNA (codons 162–166). The region chosen has several advantages over other sites on *bcl*-2 mRNA: It does not appear to have a highly ordered RNA structure, there exist two ideal cleavage sites with proper spacing for the ribozymes, and there is no sequence homology with the other *bcl*-2-related genes. This region is also extrinsic to the BH1 and BH2 domains that are responsible for heterodimerization with other members of the *bcl*-2 family. A disabled ribozyme lacking catalytic activity was also constructed as a control for the experiments. The functional, but not the disabled, ribozyme significantly decreased Bcl-2 protein and mRNA levels in LNCaP prostate carcinoma cells within 18 hours after transfection with Lipofectin®. This activity was sufficient to induce apoptosis in low-*bcl*-2-expressing LNCaP cells, but not in a high *bcl*-2 (forced overexpression) LNCaP line. In the latter, however, it did restore the ability to respond to phorbol ester, an apoptotic agent. When transfected with the TransIT polyamine transfection agent, the ribozyme synergized with other apoptosis-inducing agents (serum withdrawal, phorbol ester) and induced apoptosis in the high-*bcl*-2-expressing LNCaP cell line [69]. To increase the efficiency of cellular ribozyme delivery, Dorai *et al.* developed a recombinant defective adenoviral agent capable of expressing the anti-*bcl*-2 ribozyme upon infection [70]. This defective adenovirus-anti-*bcl*-2 ribozyme induced extensive apoptosis in several androgen-sensitive (LNCaP) and androgen-insensitive (LNCaP/*bcl*-2 and PC-3) human prostate cancer cell lines that express variable amounts of Bcl-2. In the PC3 cell line, which has an intermediate level of Bcl-2 expression, the efficacy of the ribozyme was less than in the high-*bcl*-2-expressing LNCaP cells. The androgen-insensitive prostate cancer cell line DU-145 was completely refractory to the effect of the ribozyme. The identical adenovirus-encoded hammerhead ribozyme targeted to Bcl-2 inhibited neointimal hyperplasia in the carotid artery model of balloon injury in rats and induced vascular smooth muscle cell apoptosis [71].

A hammerhead ribozyme was designed to cleave the *bcl*-2 transcript after nucleotide 279 and was confirmed to be effective against a synthetic *bcl*-2 transcript [72]. An adenovirus-delivered ribozyme caused growth reduction in human oral cancer cell lines over 6 days. This inhibition of cell growth could be attributed to apoptosis, as indicated by the detection of histone-associated DNA fragments in an immunoassay. Western blots demonstrated a reduction of Bcl-2 protein 24 hours after infection with the ribozyme-expressing adenovirus vector, although northern blots showed no detectable reduction in the level of *bcl*-2 mRNA.

A ribozyme targeted against positions 497–479 of the *bcl*-2 mRNA was also used in chronic myelogenous leukemia cell lines [73]. An increase in the number of apoptotic cells was shown in BV173 cells after treatment with this ribozyme, but there was no correlation with the level of Bcl-2 protein expressed.

C. Antisense Bcl Oligonucleotides

1. Limitations of the Antisense Biotechnology

To date, antisense oligonucleotides directed against Bcl-2 mRNA have been widely employed as a method to decrease tumor cell viability and chemoresistance and to induce apoptosis *in vitro* and *in vivo*. However, problems remain in the design and interpretation of these experiments. The major drawback of synthetic phosphodiester oligonucleotides is their sensitivity to nuclease digestion [74], which affects their half-life in culture and *in vivo*. Moreover, stepwise release of deoxynucleoside monophosphates by serum exonucleases has been shown to produce an antiproliferative effect [75]. This problem can be at least partially solved by the use of modified nuclease-stable oligonucleotides (e.g., phosphorothioates) [76]. Other problems with the employment of antisense oligonucleotides are their poor uptake and inappropriate intracellular compartmentalization (i.e., sequestration within endosomes/lysosomes) [77]. Many delivery agents have been suggested for improving the nuclear uptake of antisense oligonucleotides *in vitro* and *in vivo* (reviewed in Lebedeva *et al.* [78]). However, several reports of animal studies have suggested that oligonucleotides can penetrate cells, in at least some tissues, in the absence of a delivery agent [79–81].

At this time, the most reliable way to choose an antisense sequence is the "mRNA walking" method, suggested by Wickstrom [82], which has provided encouraging results [83–86]. However, the oligonucleotides used in most works cited herein empirically targeted the translation initiation site of the Bcl-2 or Bcl-x mRNA, assuming that mRNA is single stranded at this site, which may or may not be true. Interestingly, the recent screening of 14 oligonucleotides targeted against the Bcl-2 mRNA demonstrated that those directed against the coding region of the Bcl-2 mRNA are more efficient then those directed against the translation initiation site [83].

Sequence homology of the targeted mRNA with mRNAs of related proteins (e.g., the BH1 or BH3 codons for the bcl-2 family of genes) may lead to unwanted cleavage of nontargeted mRNAs [68]. A related problem is the potential elimination of other nontargeted mRNAs by so-called "irrelevant cleavage" [87], which is due to the fact that RNase H requires only approximately five Watson–Crick base pairs to form before it can recognize and cleave the duplex.

When attempting to validate any proposed specific antisense mechanism of action, multiple studies should be undertaken [88]. These include proof of cellular uptake, direct measurement of target mRNA or protein levels, and the use of multiple control oligonucleotide sequences. *This last step is absolutely critical.* Papers employing only the sense or a single control must be viewed as having diminished credibility. Unfortunately, many experiments reported in the literature lack many of the necessary controls. In addition, the oligonucleotide should not contain four contiguous guanosine residues [89] or CpG motifs [90], as both have led to a bewildering variety of nonspecific effects *in vitro* and *in vivo.* Even disregarding these specific motifs, phosphorothioate oligonucleotides, the currently most popular class of antisense molecules, are in general able to produce a wide spectrum of nonspecific effects, especially at high concentrations (>5 μM) (reviewed in Lebedeva and Stein [91]). Non-antisense effects can be therapeutically useful; however, their unpredictability can confound research applications of these very biologically active molecules.

2. Bcl-2 Antisense Oligonucleotides

a. In Leukemia

Initially, antisense oligonucleotides specific for the *Bcl*-2 mRNA were used to inhibit the growth in culture of the 697 human leukemia cell line [92]. The oligonucleotide targeted the translation-initiation site (Fig. 1). Both phosphodiester and phosphorothioate antisense *bcl*-2 oligonucleotides decreased cell proliferation, although the latter were more potent inhibitors; *bcl*-2 antisense oligonucleotides also led to 697 leukemic cell death through what was claimed to be a sequence-specific mechanism. However, the oligonucleotides were not delivered to the cells by a carrier, and the concentrations were high (25 μM). These problems (i.e., lack of carrier delivery, high concentration, use of only a single control) recur in almost all the papers quoted in this review, unless specifically noted. In addition, this oligonucleotide was empirically chosen to target the translation initiation site of the Bcl-2 mRNA. Unfortunately, the empirical approach is no longer considered to be an appropriate method for identifying active oligonucleotides.

The isosequential phosphorothioate oligonucleotide inhibited *Bcl*-2 protein expression in acute myeloid leukemia (AML) cells, decreased cell survival duration, and decreased the number of clonogenic cells in culture [93]. In this work, two controls were used, but the oligonucleotide concentration was still excessive (50 μM). Exposure to daunorubicin and Ara-C resulted in more effective killing of AML cells in the presence of the antisense oligomers [93,94]. Durrieu *et al.* [94] used the identical oligonucleotide, but the concentration was lower (1 μM), and delivery was accomplished with cationic lipids; however, only a single control was employed.

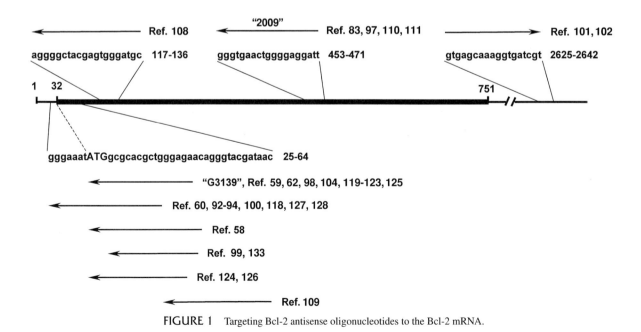

FIGURE 1   Targeting Bcl-2 antisense oligonucleotides to the Bcl-2 mRNA.

Incubation of myeloblasts from AML patients with this antisense oligonucleotide resulted in a significant decrease in the expression of *Bcl*-2 protein in seven of 17 samples [60]. The decreased expression of *Bcl*-2 was accompanied by increased apoptosis in response to Ara-C. The results described in this paper were better controlled than in the Reed *et al.* work [92], and the oligonucleotide concentration was lower (5 μM); however, no carrier was used for delivery, introducing a specificity question.

B-cell chronic lymphocytic leukemia (B-CLL) cells contain high levels of *bcl*-2 and a high *bcl*-2/*bax* ratio, which has been correlated with *in vitro* and *in vivo* drug resistance [95,96]. Bcl-2 antisense oligonucleotides were identified by screening a panel of candidates [83], and one directed against the coding region of the *bcl*-2 mRNA (Fig. 1) decreased Bcl-2 protein and mRNA expression, possibly shifting the balance between pro- and apoptotic proteins in these cells [97]. The reduction in *bcl*-2 gene expression was associated with an increase in *bax* gene expression and increased apoptosis in the B-CLL cells.

b. In Lymphomas and Myelomas

Phosphodiester and phosphorothioate antisense oligonucleotides directed against the first six codons of the Bcl-2 mRNA (denoted G3139; see Fig. 1) were employed in NHL cell lines, where expression of the *bcl*-2 gene had become deregulated as a result of the t(14;18) translocation [59,62,98]. Kitada *et al.* [98] demonstrated a specific reduction in Bcl-2 mRNA level in SU-DHL-4 cells within 1 day of treatment with this oligomer. However, a commensurate reduction in Bcl-2 protein level did not occur until 3 days, presumably because of the long half-life of the Bcl-2 protein. Consistent with this observation, the viability of SU-DHL-4 cells drastically decreased after 4 days of treatment. Antisense treatment of RS11846 lymphoma cells induced a reduction in Bcl-2 expression, resulting in marked enhancement in the sensitivity of these cells to Ara-C and methotrexate [62]. However, here, too, the concentration of the oligonucleotides was excessive (200 μM), no carrier was used for delivery, and the authors employed only a single control sequence.

Smith *et al.* [99] treated the t(14;18)-carrying follicular lymphoma cell line WSU-FSCCL with phosphodiester antisense oligonucleotides directed at codons 2–7 of the Bcl-2 mRNA (Fig. 1) without a carrier. Three controls (sense, scrambled, and two-base mismatched) were employed. The authors found a dose-dependent, sequence-specific downregulation of Bcl-2 protein level along with a corresponding inhibition of cell accumulation. A dose-dependent decrease in Bcl-2 protein expression was observed in the different lymphomas incubated with lipid-incorporated antisense oligonucleotides directed against the translation initiation site of Bcl-2 mRNA [100]. The P-ethoxy oligonucleotides used in this work precluded an RNase H-dependent mechanism of action. Growth inhibition was observed in a transformed follicular lymphoma cell line, which bore the t(14;18) translocation and overexpressed Bcl-2 protein. Antisense oligonucleotides did not induce growth inhibition in lymphoma cells not expressing Bcl-2 protein.

An interesting and unusual example of antisense downregulation of *bcl*-2 expression was purportedly demonstrated in lymphomas expressing a hybrid *bcl*-2-*IgH* transcript in the antisense orientation [101,102]. Oligonucleotides directed to the unique nucleotide sequence in the fusion region [103] or to the *bcl*-2 fragment [101,102] of the hybrid transcript effectively downregulated the overexpression of Bcl-2 protein, activated programmed cell death, and inhibited cell growth in the SU-DHL-4 human follicular B-cell lymphoma line. Curiously, biological activity was exerted exclusively by the *sense* oligonucleotides, which induced early strong inhibition of cell growth and late fulminant cell death [102]. Sense oligonucleotides targeting the common *bcl*-2 segment of the transcript (Fig. 1) were efficient in all t(14;18) cell lines tested, but not in the untranslocated cell lines. Antisense-oriented oligonucleotides complementary to the *bcl*-2-*IgH* mRNA, and control oligonucleotides (scrambled, inverted, or mismatched) were biologically ineffective. However, the antisense transcript was found only by reverse transcription–polymerase chain reaction (RT-PCR) and not by an RNase protection assay, thus raising artifactual questions.

Because overexpression of Bcl-2 seems to play a critical role in multidrug resistance in multiple myeloma, the G3139 antisense oligonucleotide was used to treat the Bcl-2 expressing U266 and RPMI-8226 cell lines (although the conditions were not given). Treatment resulted in a decrease in Bcl-2 mRNA and consequent sensitization of the cells to daunorubicin [104].

c. In Lung Carcinoma Cells

Elevated expression of Bcl-2 has been observed in the majority of small-cell lung cancer specimens and cell lines and has been associated with radiation and drug resistance [105–107]. One of the earlier works in this area employed an antisense oligonucleotide directed against the downstream region of the translation initiation site of the Bcl-2 mRNA (different from G3139; see Fig. 1) to inhibit proliferation after 10 days of treatment of non-small-cell lung cancer (NSCLC) cell lines [108]. However, the authors did not perform measurements of Bcl-2 protein expression and used very high oligonucleotide concentrations and control oligomers of uncertain sequence. Treatment of NSCLC cell lines with oligonucleotides directed against the start codon region of Bcl-2 mRNA resulted in decreased Bcl-2 levels, reduced cell proliferation, decreased cell viability, and increased level of spontaneous apoptosis [109].

In perhaps the most well-controlled effort, Ziegler and colleagues [83] screened 14 antisense oligodeoxynucleotides targeting various regions of the Bcl-2 mRNA in order to identify the most effective sequences for reducing Bcl-2 protein levels. Interestingly, an oligonucleotide directed against the coding region of the Bcl-2 mRNA (denoted 2009; see Fig. 1) and delivered to the cell with Lipofectin® had the greatest effect on Bcl-2 protein levels and cell viability. This is in contrast to previous data, in which only oligonucleotides targeting the translation initiation site or regions immediately downstream were shown to downregulate *bcl*-2 expression [59,98,99]. A dose-dependent reduction in Bcl-2 levels became detectable 24 hours after treatment and persisted up to 96 hours; analysis of cellular morphology demonstrated that viability was reduced through apoptosis. Interactions between the Bcl-2 antisense oligonucleotide 2009 and the chemotherapeutic agents etoposide, doxorubicin, and cisplatin were then investigated [110]. The level of expression of Bcl-2 protein in the cells was inversely correlated with their sensitivities to treatment with oligonucleotide 2009 and the various chemotherapeutic agents. In the NCI-H69 (high basal level of Bcl-2 expression) and SW2 (moderate level of Bcl-2 expression) cell lines, all combinations resulted in synergistic cytotoxicity. However, in the NCI-H82 cell line, which had a low basal level of Bcl-2 expression, most of the combinations were slightly antagonistic. These data suggested the use of oligonucleotide 2009 in combination with chemotherapy for the treatment of small-cell lung cancers that overexpress Bcl-2.

Cotreatment of the A549/CPT multidrug-resistant human lung cancer cell line with oligonucleotide 2009 and a p21$^{WAF1/CIP1}$ antisense oligonucleotides in complex with Lipofectin® restored drug susceptibility in the cells more effectively then either one of them alone [111]. However, the mechanism of the synergism must be questioned, as the p21$^{WAF1/CIP1}$ antisense oligonucleotide was chosen empirically (i.e., not by screening a panel of candidates).

d. In Carcinomas of the Breast and Prostate

Recently, deregulated overexpression of Bcl-2 has also been observed in other malignant tissues, for example, breast [112] and prostate cancer [113]. Although Bcl-2 expression in normal prostatic epithelial cells is low or absent, Bcl-2 is upregulated in 35% of prostate cancer cell specimens from patients after progression to androgen independence [114]. Increased expression of Bcl-2 has been correlated with the emergence of hormone-independent, apoptosis-resistant tumors [31,115,116]. Moreover, stable transfection of LNCaP prostate cancer cells with Bcl-2 increased their *in vivo* tumorigenic potential and resistance to apoptosis [117]. In addition, expression of Bcl-2 protein increased in LNCaP cells that had metastasized in nude mice [31].

Treatment with a Bcl-2 antisense oligonucleotide targeted to the translation initiation codon of the Bcl-2 mRNA (Fig. 1) blocked the protective effect of androgens on etoposide-induced cytotoxicity in LNCaP cells [118]. However, the authors used an empirically identified oligonucleotide at high concentration and with minimal controls. Moreover, measurements of Bcl-2 levels were not performed. Treatment of LNCaP [119] and Shionogi tumor cells [120] *in vitro* with G3139 inhibited Bcl-2 expression in a dose-dependent and sequence-specific manner. The authors used Lipofectin® for delivery and measured Bcl-2 protein and mRNA levels but employed only a single control oligonucleotide. Bcl-2 mRNA levels returned to pretreatment levels by 48 hours after discontinuing treatment [119]. Antisense Bcl-2 oligonucleotide treatment substantially enhanced paclitaxel and docetaxel chemosensitivity in a dose-dependent manner. The characteristic apoptotic changes were demonstrated only after combined treatment [121,122].

DU-145 cell growth in liquid culture was inhibited by the antisense Bcl-2 oligonucleotide G3139 (Fig. 1) compared with a control sense oligonucleotide, although a delivery agent was not employed [123]. The authors claimed a 46% downregulation of the levels of Bcl-2 protein, but the data were not shown. Inhibition of cell growth by this oligonucleotide was significantly enhanced by combination with the synthetic retinoid *N*-(2-hydroxyphenyl)all-*trans*-retinamide. Interestingly, growth inhibition occurred in the absence of apoptosis. The authors suggested that in DU-145 cells, Bcl-2 may not solely function as a regulator of apoptosis but may mediate another pathway of growth inhibition [123].

Treatment of T24 bladder carcinoma cells with 20 μM of a Bcl-2 phosphorothioate antisense oligonucleotide (Fig. 1) reduced the level of Bcl-2 protein in these cells [124]. Combined administration with adriamycin resulted in synergistic cytotoxicity, accompanied by an increase of apoptosis; however, the specificity of the oligonucleotide effect is still somewhat questionable.

G3139 has been delivered to the SK-BR-3 and MCF-7 breast cancer cell lines using polyethylene glycol modified cationic liposomes [125]. The lipid complexes contained conjugated anti-HER2 F(ab′) fragments at the distal termini of poly(ethylene glycol) (PEG) chains to target oligonucleotide delivery to the malignant cells. Such targeted delivery of antisense oligonucleotide reduced *bcl*-2 expression (however, only by 40%) in the cells compared with nontargeted delivery. Unfortunately, no data are yet available regarding the further effects of antisense treatment of breast cancer cells.

e. In Liver Carcinoma Cells

Malignant cholangiocytes express 15-fold more Bcl-2 protein than nonmalignant cholangiocytes [126]. G3139 reduced the expression of Bcl-2 protein by 50% and increased the rate of beauvericin-induced apoptosis more than threefold in the malignant cells [126]. However, a high concentration of the oligonucleotide in addition to long treatment duration, the lack of a delivery agent, and the use of only a single control raise questions of specificity. Bcl-2 antisense oligonucleotides targeted against the translation-initiation site of the Bcl-2 mRNA (similar to G3139; see Fig. 1) easily caused butyrate-induced apoptosis in the HCC-M and HCC-T human hepatoma cells [127] and spontaneous Fas-antibody-mediated apoptosis in HCC-T cells, but not in HepG2 cells [128]. These results suggested that Bcl-2 is essential for survival of the HCC-T and HCC-M cells, whereas other proteins may substitute for it in HepG2 cells. However, no measurements of the Bcl-2 protein or mRNA levels were shown.

3. *In Vivo* Applications of Bcl-2 Antisense Oligonucleotides

The pharmacokinetics of phosphorothioate oligonucleotides after intravenous or intraperitoneal infusion [129,130] has been extensively studied. Phosphorothioate oligonucleotides are stable for 48 hours (15–50%), at which time detectable levels are still present in tissues. Phosphorothioates are widely distributed and slowly eliminated from the body. Plasma clearance is biphasic with half-lives of 15–25 min and 20–40 hours. G3139 demonstrated similar pharmacokinetics after intravenous bolus (approximately 5 mg/kg) with a plasma half-life of elimination of 22 hours [131]. Of note is that subcutaneous infusion of G3139 resulted in less excretion and metabolism of the administered dose compared with a single intravenous bolus. The pharmacokinetics and tissue distribution studies of liposome-delivered P-ethoxy Bcl-2 antisense oligonucleotides demonstrated that daily administration of 20 mg/kg of body weight over 5 consecutive days had no adverse effects on renal or hepatic function or on hematological parameters [132].

The t(14;18) translocation found in the majority of follicular lymphomas (FL) results in deregulation of the *bcl*-2 gene and appears to play a role in oncogenesis [8,9]. Cotter *et al.* xenografted cells from a patient with B-cell lymphoma bearing the t(14;18) translocation into severe combined immunodefficient (SCID) mice model [58]. Pretreatment of the cells with G3139 prior to inoculation resulted in failure to develop lymphoma. Experiments *in vitro* demonstrated downregulation of Bcl-2 protein with subsequent induction of apoptosis. The antisense oligonucleotide two bases longer than G3139 (Fig. 1) had little or no effect on the viability of cell lines not expressing high levels of Bcl-2. Oligonucleotide G3139 almost completely abolished lymphoma growth in 50 out of 60 (83%) treated mice after 2 weeks; however, disease was still present in the remaining mice. Extension of treatment to 3 weeks completely eradicated lymphoma in all animals, even at the PCR level [133]. G3139 contains two CpG sequences and is thus capable of natural killer (NK) cell activation [134]. The authors claimed that this does not account for the main antitumor activity of G3139, as methylation of G3139 eliminated immune system activation but does not eliminate the antitumor activity of G3139. However, the oligonucleotides still may be demethylated *in vivo*. Nevertheless, experiments were repeated in NOD/SCID mice having no NK, B, or T cell activity, and similar efficacy was observed [135].

These studies were extended to a phase I study for lymphoma patients with high Bcl-2 expression [136,137]. No treatment-related toxicity occurred at doses of 5 mg/kg/day [136]. The effects of Bcl-2 antisense oligonucleotide treatment in patients were more prolonged in duration compared to usual chemotherapy responses. Continued reduction in lymphoma bulk was observed in excess of 6 weeks from the end of the infusion. One patient remains in remission 3 years after starting treatment without having further therapy. No other patients achieved a complete response, but in two patients computer tomography scans showed a reduction in tumor size [136]. Nine patients had stabilization of the disease, and at least two of them had symptomatic improvement; however, nine patients had progressive disease [137]. The level of Bcl-2 protein was measured by flow cytometry in peripheral blood samples and was found reduced in seven of 16 assessable patients.

Treatment of melanoma cells *in vitro* with G3139 (200 nM) in the presence of cationic lipids led to a sequence-specific and dose-dependent downregulation of the Bcl-2 mRNA [81]. Two controls were ineffective under the same conditions. G3139 treatment improved the chemosensitivity of human melanoma cells in SCID mice. Treatment enhanced the number of naturally occurring apoptotic cells, and a combination of G3139 and dacarbazine resulted in complete ablation of the tumor in three of six animals. Although Bcl-2 does not play a central role in the oncogenesis of human melanoma [138], the data of Jansen *et al.* [139] suggested that Bcl-2 contributes to the lack of chemosensitivity of human melanoma. In the current phase I/II study, dacarbazine and G3139 are being combined in patients with advanced malignant melanoma, including patients with disease resistant to prior single-agent dacarbazine. To date, one complete response has been observed. Additional initial results from this study indicated that G3139 can reduce Bcl-2 expression in metastatic melanoma and combined therapy with dacarbazine is well tolerated.

Treatment with G3139 resulted in either a dramatic reduction of tumor growth or complete remission in a SCID mouse xenotransplantation model of human Merkel cell carcinoma [140]. Apoptosis was enhanced 2.4-fold in the Bcl-2-antisense-treated tumors compared with the saline-treated group. Reverse sequence and two-base mismatch control oligonucleotides or cisplatin had no significant antitumor effects compared with saline-treated controls.

Miyake *et al.* have tested the efficacy of a G3139, administered adjuvantly after castration, to delay the time to androgen-independent recurrence in the androgen-dependent mouse Shionogi tumor model [119–122]. Systemic administration of an antisense Bcl-2 oligonucleotide beginning one day postcastration in mice bearing the Shionogi [120] or LNCaP [119] tumors resulted in a more rapid regression of tumors and a significant delay of emergence of androgen-independent recurrent tumors. However, despite significant reductions of Bcl-2 expression in tumor tissues, the antisense Bcl-2 oligonucleotide had no effect on Bcl-2 expression in normal mouse organs [120]. Adjuvant *in vivo* administration of an antisense Bcl-2 oligonucleotide and micellar paclitaxel following castration resulted in a statistically significant delay of androgen-independent, recurrent tumors compared with administration of either agent alone [121]. Furthermore, combined treatment of mice bearing androgen-independent recurrent Shionogi tumors with this regimen synergistically induced tumor regression and growth inhibition when compared to treatment with either agent alone [122]. These findings illustrated the potential utility of antisense Bcl-2 therapy for prostate cancer in the adjuvant setting when combined with androgen ablation and taxane treatment.

Oligonucleotide G3139 now is also being evaluated in phase I/IIA trial for patients with androgen-independent prostate cancer and other advanced solid tumor malignancies [141]. No toxicity was observed at a dose of 2.3 mg/kg/day, except in one patient. Two patients with renal cell and one with prostate cancer demonstrated no progression of disease. G3139 in combination with docetaxel is in a phase I trial for patients with advanced breast cancer and other solid tumors and has demonstrated tolerable toxicity [142]. Tumor response was observed in two patients with breast cancer. The safety data from these clinical trials support further clinical development of G3139, both as a single agent and in combination with cytotoxic agents.

4. Antisense Oligonucleotides Targeted to Bcl-xL

Another member of the *bcl*-2 family of genes, *bcl-x*, encodes two proteins in humans, a long form (Bcl-xL) and a short form (Bcl-xS), derived via alternative splicing [25]. The Bcl-xL protein has amino acid and overall structural homology to Bcl-2, and it also inhibits apoptosis as effectively as the Bcl-2 protein [23,33,143–146]. Although details about the biochemical functions of Bcl-xL have not been clarified, recent studies have shown that the mechanisms of Bcl-xL activity may involve dimerization with other members [36,37,147,148] or nonmembers [50] of the Bcl-2 family. In addition, the protein may have pore-forming activity in mitochondria [149], caspase inhibitory properties [40], and the ability to form ion channels in biological membranes [150,151]. The most likely mode of Bcl-xL action is via a combination of several mechanisms [50]. Overexpression of Bcl-xL has been demonstrated in different tissues and cell culture lines, including prostate cancers [116], breast carcinomas [143], glioblastoma cells [147], human gastric adenomas and carcinomas [152], NHL and Reed–Sternberg cells of Hodgkin's disease [153,154], bladder carcinomas [155], and many others [156,157]. Overexpression of Bcl-xL has been shown to delay the onset of apoptosis induced by multiple drugs *in vitro*, similar to Bcl-2 [23,158,159]. This suggests that strategies to downregulate Bcl-xL expression could overcome drug resistance and potentially lead to successful therapeutic intervention.

The first published work employing the Bcl-xL antisense strategy was performed in WEHI-231 B lymphoma cells, where upregulation of *bcl-x* by CD40 plays an important role in CD40-mediated apoptotic rescue [160]. An oligonucleotide directed to the translation initiation site of the *bcl-x* mRNA (Fig. 2) partially blocked this CD40-mediated apoptotic rescue, although the concentration of the oligonucleotide was high, and no delivery agent was employed. In WEHI-231

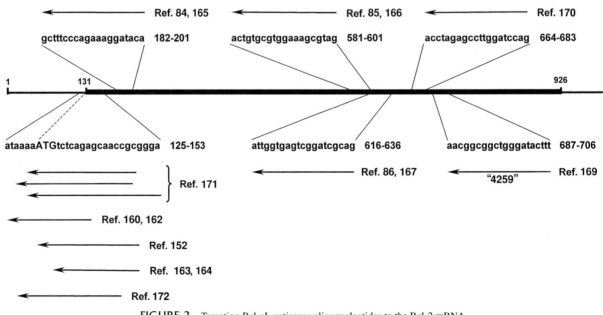

FIGURE 2   Targeting Bcl-xL antisense oligonucleotides to the Bcl-2 mRNA.

B cells, overexpression of *bcl-2* failed to protect against apoptosis, whereas overexpression of *bcl-x* did protect these cells [161]. The authors [161] suggested that *bcl-2* and *bcl-x* may regulate at least partially independent systems that protect B cells from apoptosis.

Fujio *et al.* [162] used the same *bcl-x* antisense oligonucleotide as in [160] in cardiac myocytes. Leukemia inhibitory factor (LIF) promoted survival of cardiac myocytes by increasing the levels of Bcl-xL protein expression. Antisense oligonucleotides targeted to the *bcl-x* mRNA (Fig. 2) inhibited the protective effect of LIF and was accompanied by a reduction in Bcl-xL protein levels.

Treatment of the MKN-45 gastric cancer cell line with a phosphorothioate antisense oligonucleotide directed against the translation initiation site of the *bcl-x* mRNA (Fig. 2) specifically decreased Bcl-xL protein levels and also induced cell death after serum deprivation or Fas-antibody administration [152].

Transformation of the HL-60 pro-myelocytic cell line with Bcr–Abl rendered these cells extremely resistant to apoptosis as induced by a wide variety of agents [163]. Moreover, Bcr–Abl expression induced a dramatic downregulation of Bcl-2 expression and increased the level of Bcl-xL expression. Antisense oligonucleotides targeted to the *bcl-x* mRNA downregulated the expression of Bcl-xL and increased the susceptibility of the HL-60–Bcr chemoresistant cell line to staurosporine [163]. It was suggested that Bcl-xL seemed to participate, at least in part, in Bcr–Abl-mediated resistance to apoptosis in HL-60 cells, and could act independently of Bcl-2. A prevalent role for the Bcl-xL protein in the prevention of apoptosis was also suggested in retinal pigmented epithelial cells [164], where downregulation of *bcl-x* by antisense oligonucleotide treatment inhibited the survival-promoting activity of fibroblast growth factor 2.

Arriola *et al.* suggest that Bcl-2 and Bcl-xL have different abilities to protect against chemotherapy-induced apoptosis in human testicular germ cell tumors normally expressing undetectable levels of Bcl-2 [165]. Bcl-2-overexpressing transformed cells were actually sensitized to chemotherapy-induced apoptosis compared to the parental and vector control cells, perhaps because of reciprocal downregulation of the endogenously expressed Bcl-xL in these clones. In contrast, downregulation of Bcl-xL to the same extent with an antisense oligonucleotide directed to the coding region of *bcl-x* mRNA (Fig. 2) enhanced etoposide-induced apoptosis in human testicular germ cells by twofold.

Several investigators, trying to elaborate an efficient antisense strategy to downregulate *bcl-x* expression, screened a panel of oligonucleotides chosen by random walking along the bcl-x mRNA. In freshly isolated human eosinophils, these oligonucleotides (Fig. 2) downregulated Bcl-xL mRNA and protein levels and partially blocked the cytokine-mediated rescue of apoptotic death [84]. Taylor *et al.* screened a panel of 22 oligonucleotides to find one that would most

efficiently downregulate Bcl-xL mRNA and protein expression in normal human keratinocytes in a concentration-dependent, sequence-specific manner [85]. Treatment of keratinocytes and epithelial cells with the active oligomers in the complex with lipids sensitized these cells to ultraviolet-B radiation and cisplatinum-induced apoptosis. Ackermann *et al.* utilized this molecule to examine the role of Bcl-xL in human umbilical vein endothelial cells [166]. Inhibition of constitutive levels of Bcl-xL caused 10–25% of the cell population to undergo apoptosis and increased cellular susceptibility to treatment with low concentrations of staurosporine or ceramide.

Forty 18- and 20-mer oligonucleotides complementary to human *bcl-x* mRNA were screened to find the most active inhibitor of Bcl-xL expression in prostate and bladder carcinoma cells [86]. The five most active sequences were delivered to the LNCaP, PC3, and T24 cells in culture with Lipofectin® or a cationic porphyrin delivery agent. Downregulation of Bcl-xL protein and mRNA was accompanied by increased chemosensitivity of the tumor cells to variable chemotherapeutic agents. However, oligonucleotide treatment itself did not induce apoptosis in cells. One of the five oligonucleotides was synthesized as nuclease-resistant external guide sequence (EGS) and used in the T24 bladder carcinoma cell line to elicit intracellular mRNA cleavage induced through activation of RNase P [167]. The EGS oligonucleotide has two hybridizing arms, joined by a T-stem and T-loop. This construct mimics structural elements of a precursor tRNA, and the EGS–mRNA duplex can elicit RNase P activity, leading to target cleavage. Use of this EGS construct avoids the "irrelevant cleavage" that occurs when RNase H cleaves the mRNA–DNA duplex in sites of partial complementarity [87]. This strategy may provide specific antisense downregulation of particular members of the *bcl-2* family that possess significant homology with each other.

Most small-cell lung cancer (SCLC) cells overexpress Bcl-xL in addition to Bcl-2, raising the question of which of the anti-apoptotic proteins is the more crucial for the survival of these tumor cells [168]. An antisense phosphorothioate oligonucleotide targeting a sequence unique to the *bcl-xL* coding region and not shared by the pro-apoptotic splice variant *bcl-xS* (denoted 4259; see Fig. 2), was obtained by screening in an effort to override the apoptotic block in lung adenocarcinoma and SCLC cells caused by overexpression of Bcl-xL [169]. 2′-methoxy-ethoxy modifications of selected deoxyribose residues were made to reduce non-antisense-related toxicity antisense effects. Treatment of the adenocarcinoma cell lines A549 and NCI-H125 and the SCLC cell lines SW2 and NCI-H69 with 600 n*M* of oligomer 4529 in the presence of Lipofectin® reduced Bcl-xL protein levels by 70–90%. Oligonucleotide treatment induced apoptosis in the adenocarcinoma cell lines, but not in the SCLC cell lines. These findings imply that Bcl-xL might be a more critical survival factor for NSCLC than for SCLC. It is also possible that

Bcl-xL expression is not crucial for the survival of SCLC cells because of compensation by upregulated Bcl-2.

Differential splicing of the *bcl-x* gene gives rise to two transcripts, one coding for Bcl-xL, an anti-apoptotic protein, and the other for Bcl-xS, a pro-apoptotic protein [25]. Taylor and colleagues used an antisense oligonucleotide to alter the ratio of anti-apoptotic to pro-apoptotic Bcl-x and sensitize cells to undergo apoptosis in response to ultraviolet-B radiation and chemotherapeutic drug treatment [170].

The Bcl-xL antisense strategy was also employed in *in vivo* models. Downregulation of intimal cell Bcl-xL expression in vascular lesions in rabbits with antisense oligonucleotides induced apoptosis and acute regression of the lesions [171]. In contrast to neointimal cells transfected with anti-Bcl-x antisense oligonucleotides, medial vascular smooth muscle cells within the normal vessels do not undergo apoptosis (although they take up oligonucleotides with similar efficiency). Antisense Bcl-xL oligonucleotides prevented neointimal formation in murine cardiac allografts via enhanced apoptosis [172]. However, oligonucleotides in these experiments were not selected from a panel but were empirically directed to the translation initiation site of the *bcl-x* mRNA.

5. Antisense Bcl-2 Oligonucleotides: Some Pros and Cons

To obtain meaningful data from experiments employing the antisense biotechnology, mechanistic questions must be thoroughly considered. Initially, the significance of Bcl-2 expression was questioned when high levels of its expression did not explain the aggressiveness of some types of lymphoblastic leukemia [173], the malignant progression in gliomas [174], or the improved clinical outcome in patients with breast cancers that express high levels of *Bcl-2* [175,176]. Overexpression of the *Bcl-2* gene delays the growth of developing B lymphocytes [177] and leukemia cells [178]. Excessive expression of *Bcl-2* induces apoptosis in glioma [179] and in some other cancer cell lines [180]. In other circumstances, *Bcl-2* can also act like a pro-apoptotic gene (e.g., when it increases the half-life of Bax [181] and promotes the death of normal photoreceptor cells [182]). It has been reported that, via caspase-3 cleavage, Bcl-2 in fact may be transformed to a Bax-like inducer of cell death [183]. Several therapeutic approaches using the anti-apoptotic function of Bcl-2 have already been reported [184–186].

Overexpression of Bcl-xL has now been demonstrated for many different types of cancers [143,154,155,187,188]. Paradoxical associations of high Bcl-2 expression with improved clinical outcome in some types of cancer may also be attributable to compensatory decreases in Bcl-xL expression [175,176,189]. As mentioned previously, Bcl-2 and Bcl-xL have structural homology and localize to similar extracellular sites, suggesting that they inhibit cell death by a similar biochemical mechanism [6,23,26,30]. However, Nunez and

co-workers have recently shown that Bcl-2 and Bcl-xL may differentially block apoptosis induced by chemotherapeutic drugs [190]. Moreover, recent studies demonstrated that in many cases Bcl-xL protected cells against apoptosis more efficiently than Bcl-2 [161,179,188]. While it is still controversial as to whether Bcl-2 prevents CD95/FAS-APO-1-mediated apoptosis [191–194], it is generally accepted that Bcl-xL inhibits this cell death pathway [195]. Moreover, it has been recently shown that Bcl-2 and Bcl-xL interact differently with some intracellular targets *in vivo* [196,197].

Studies on the development of *Bcl-2* and *Bcl-xL* knock-out mice indicate that functional differences exist between these two proteins [198–200]. *Bcl-2* knockout mice fail to maintain a stable immune system after birth, develop polycystic kidneys, and have hypopigmented hair but otherwise have viable and normal [198,199] cells of hematopoetic and neuronal tissues [200]. Together, these studies point out that *Bcl-2* and *Bcl-xL* may be able to regulate apoptosis differentially. In addition, it has been speculated that the presence of an independent Bcl-xL-controlled checkpoint, downstream from the Bcl-2/Bax checkpoint, blocks further propagation of the apoptotic signal [201].

Overexpression of other anti-apoptotic homologs of Bcl-2 (Bcl-xL, Mcl-1, etc.) might partially replace *bcl-2* function when Bcl-2 protein is reduced or eliminated [64,70,169]. Forced overexpression of *Bcl-2* (or *Bcl-xL*) by transfection of the cell line and further selection of overexpressing clones may result in switching anti-apoptotic mechanisms. Recently, it has been shown that an apoptosis-resistant variant of leukemia HL-60 cells switched expression from the anti-apoptotic protein Bcl-2 to Bcl-xL [201]. The role of Bcl-2 and Bcl-xL in regulating lymphoid survival and death and the possibility of switching their mechanisms have been extensively discussed by Nunez and colleagues [202].

In summary, the results obtained from the use of antisense *Bcl-2* and *Bcl-xL* oligonucleotides are encouraging, and clinical trials with these antisense drugs are in progress. However, substantial questions regarding mechanism still remain unresolved. An understanding of the hierarchies of factors promoting survival (*bcl-2, bcl-xL,* and other survival signals) combined with targeting the pivotal molecules may impact dramatically on the fate of these clinical trials and may permit the development of truly selective therapy. Further basic and clinical studies in this field are definitely warranted.

References

1. Kerr, J. F., Wyllie, A. H., and Currie, A. R. (1972). Apoptosis: a basic biological phenomenon with wide-ranging implications in tissue kinetics. *Br. J. Cancer* **26,** 239–257.
2. Wyllie, A. H., Kerr, J. F., and Currie, A. R. (1980). Cell death: the significance of apoptosis. *Int. Rev. Cytol.* **68,** 251–306.
3. Cohen, J. J., Duke, R. C., Fadok, V. A., and Sellins, K. S. (1992). Apoptosis and programmed cell death in immunity. *Annu. Rev. Immunol.* **10,** 267–293.

4. Korsmeyer, S. J. (1992). Bcl-2 initiates a new category of oncogenes: regulators of cell death. *Blood* **80,** 879–886.

5. Reed, J. C. (1994). Bcl-2 and the regulation of programmed cell death. *J. Cell. Biol* **124,** 1–6.

6. Chao, D. T., and Korsmeyer, S. J. (1998). BCL-2 family: regulators of cell death. *Annu. Rev. Immunol.* **16,** 395–419.

7. Kroemer, G. (1997). The proto-oncogene Bcl-2 and its role in regulating apoptosis [published *erratum* appears in *Nat. Med.* **3**(8), 934, 1997]. *Nat. Med.* **3,** 614–620.

8. Tsujimoto, Y., and Croce, C. M. (1986). Molecular genetics of human B-cell neoplasia. *Curr. Top. Microbiol. Immunol.* **132,** 183–192.

9. Tsujimoto, Y., Cossman, J., Jaffe, E., and Croce, C. M. (1985). Involvement of the bcl-2 gene in human follicular lymphoma. *Science* **228,** 1440–1443.

10. Hockenbery, D. M., Zutter, M., Hickey, W., Nahm, M., and Korsmeyer, S. J. (1991). BCL2 protein is topographically restricted in tissues characterized by apoptotic cell death. *Proc. Natl. Acad. Sci. USA* **88,** 6961–6965.

11. Reed, J. C. (1995). Bcl-2: prevention of apoptosis as a mechanism of drug resistance. *Hematol. Oncol. Clin. North Am.* **9,** 451–473.

12. Pegoraro, L., Palumbo, A., Erikson, J., Falda, M., Giovanazzo, B., Emanuel, B. S., Rovera, G., Nowell, P. C., and Croce, C. M. (1984). A 14;18 and an 8;14 chromosome translocation in a cell line derived from an acute B-cell leukemia. *Proc. Natl. Acad. Sci. USA* **81,** 7166–7170.

13. Rodin, C. M., and Thompson, C. B. (1997). Apoptosis and disease. *Annu. Rev. Med.* **48,** 267–281.

14. Reed, J. C., Jurgensmeier, J. M., and Matsuyama, S. (1998). Bcl-2 family proteins and mitochondria. *Biochim. Biophys. Acta.* **1366,** 127–137.

15. Baffy, G., Miyashita, T., Williamson, J. R., and Reed, J. C. (1993). Apoptosis induced by withdrawal of interleukin-3 (IL-3) from an IL-3-dependent hematopoietic cell line is associated with repartitioning of intracellular calcium and is blocked by enforced Bcl-2 oncoprotein production. *J Biol. Chem.* **268,** 6511–6519.

16. Miyashita, T., and Reed, J. C. (1993). Bcl-2 oncoprotein blocks chemotherapy-induced apoptosis in a human leukemia cell line. *Blood* **81,** 151–157.

17. Walton, M. I., Whysong, D., O'Connor, P. M., Hockenbery, D., Korsmeyer, S. J., and Kohn, K. W. (1993). Constitutive expression of human Bcl-2 modulates nitrogen mustard and camptothecin induced apoptosis. *Cancer Res.* **53,** 1853–1861.

18. Kamesaki, S., Kamesaki, H., Jorgensen, T. J., Tanizawa, A., Pommier, Y., and Cossman, J. (1993). Bcl-2 protein inhibits etoposide-induced apoptosis through its effects on events subsequent to topoisomerase II-induced DNA strand breaks and their repair [published *erratum* appears in *Cancer Res.* **54**(11), 3074, 1994]. *Cancer Res.* **53,** 4251–4256.

19. Fisher, T. C., Milner, A. E., Gregory, C. D., Jackman, A. L., Aherne, G. W., Hartley, J. A., Dive, C., and Hickman, J. A. (1993). bcl-2 modulation of apoptosis induced by anticancer drugs: resistance to thymidylate stress is independent of classical resistance pathways. *Cancer Res.* **53,** 3321–3326.

20. Tang, C., Willingham, M. C., Reed, J. C., Miyashita, T., Ray, S., Ponnathpur, V., Huang, Y., Mahoney, M. E., Bullock, G., and Bhalla, K. (1994). High levels of p26BCL-2 oncoprotein retard taxol-induced apoptosis in human pre-B leukemia cells. *Leukemia* **8,** 1960–1969.

21. Reed, J. C. (1998). Bcl-2 family proteins. *Oncogene* **17,** 3225–3236.

22. Boise, L. H., Gottschalk, A. R., Quintans, J., and Thompson, C. B. (1995). Bcl-2 and Bcl-2-related proteins in apoptosis regulation. *Curr. Top. Microbiol. Immunol.* **200,** 107–121.

23. Minn, A. J., Rudin, C. M., Boise, L. H., and Thompson, C. B. (1995). Expression of bcl-xL can confer a multidrug resistance phenotype. *Blood* **86,** 1903–1910.

24. Reynolds, J. E., Yang, T., Qian, L., Jenkinson, J. D., Zhou, P., Eastman, A., and Craig, R. W. (1994). Mcl-1, a member of the Bcl-2 family, delays apoptosis induced by c-Myc overexpression in Chinese hamster ovary cells. *Cancer Res.* **54,** 6348–6352.

25. Boise, L. H., Gonzalez-Garcia, M., Postema, C. E., Ding, L., Lindsten, T., Turka, L. A., Mao, X., Nunez, G., and Thompson, C. B. (1993). bcl-x, a bcl-2-related gene that functions as a dominant regulator of apoptotic cell death. *Cell* **74,** 597–608.

26. Ibrado, A. M., Huang, Y., Fang, G., Liu, L., and Bhalla, K. (1996). Overexpression of Bcl-2 or Bcl-xL inhibits Ara-C-induced CPP32/Yama protease activity and apoptosis of human acute myelogenous leukemia HL-60 cells. *Cancer Res.* **56,** 4743–4748.

27. Yin, C., Knudson, C. M., Korsmeyer, S. J., and Van Dyke, T. (1997). Bax suppresses tumorigenesis and stimulates apoptosis in vivo. *Nature* **385,** 637–640.

28. Krajewski, S., Krajewska, M., and Reed, J. C. (1996). Immunohistochemical analysis of in vivo patterns of Bak expression, a proapoptotic member of the Bcl-2 protein family. *Cancer Res.* **56,** 2849–2855.

29. Sumantran, V. N., Ealovega, M. W., Nunez, G., Clarke, M. F., and Wicha, M. S. (1995). Overexpression of Bcl-XS sensitizes MCF-7 cells to chemotherapy-induced apoptosis. *Cancer Res.* **55,** 2507–2510.

30. Thompson, C. B. (1995). Apoptosis in the pathogenesis and treatment of disease. *Science* **267,** 1456–1462.

31. McConkey, D. J., Greene, G., and Pettaway, C. A. (1996). Apoptosis resistance increases with metastatic potential in cells of the human LNCaP prostate carcinoma line. *Cancer Res.* **56,** 5594–5599.

32. Hockenbery, D. M., Oltvai, Z. N., Yin, X. M., Milliman, C. L., and Korsmeyer, S. J. (1993). Bcl-2 functions in an antioxidant pathway to prevent apoptosis. *Cell* **75,** 241–251.

33. Shimizu, S., Eguchi, Y., Kosaka, H., Kamiike, W., Matsuda, H., and Tsujimoto, Y. (1995). Prevention of hypoxia-induced cell death by Bcl-2 and Bcl-xL. *Nature* **374,** 811–813.

34. Lam, M., Dubyak, G., Chen, L., Nunez, G., Miesfeld, R. L., and Distelhorst, C. W. (1994). Evidence that BCL-2 represses apoptosis by regulating endoplasmic reticulum-associated Ca^{2+} fluxes. *Proc. Natl. Acad. Sci. USA* **91,** 6569–6573.

35. Ryan, J. J., Prochownik, E., Gottlieb, C. A., Apel, I. J., Merino, R., Nunez, G., and Clarke, M. F. (1994). c-myc and bcl-2 modulate p53 function by altering p53 subcellular trafficking during the cell cycle. *Proc. Natl. Acad. Sci. USA* **91,** 5878–5882.

36. Yin, X. M., Oltvai, Z. N., and Korsmeyer, S. J. (1995). Heterodimerization with Bax is required for Bcl-2 to repress cell death. *Curr. Top. Microbiol. Immunol.* **194,** 331–338.

37. Sedlak, T. W., Oltvai, Z. N., Yang, E., Wang, K., Boise, L. H., Thompson, C. B., and Korsmeyer, S. J. (1995). Multiple Bcl-2 family members demonstrate selective dimerizations with Bax. *Proc. Natl. Acad. Sci. USA* **92,** 7834–7838.

38. Monney, L., Otter, I., Olivier, R., Ravn, U., Mirzasaleh, H., Fellay, I., Poirier, G. G., and Borner, C. (1996). Bcl-2 overexpression blocks activation of the death protease CPP32/Yama/apopain. *Biochem. Biophys. Res. Commun.* **221,** 340–345.

39. Srinivasan, A., Foster, L. M., Testa, M. P., Ord, T., Keane, R. W., Bredesen, D. E., and Kayalar, C. (1996). Bcl-2 expression in neural cells blocks activation of ICE/CED-3 family proteases during apoptosis. *J. Neurosci.* **16,** 5654–5660.

40. Clem, R. J., Cheng, E. H., Karp, C. L., Kirsch, D. G., Ueno, K., Takahashi, A., Kastan, M. B., Griffin, D. E., Earnshaw, W. C., Veliuona, M. A., and Hardwick, J. M. (1998). Modulation of cell death by Bcl-XL through caspase interaction. *Proc. Natl. Acad. Sci. USA* **95,** 554–559.

41. Wang, H. G., Rapp, U. R., and Reed, J. C. (1996). Bcl-2 targets the protein kinase Raf-1 to mitochondria [see comments]. *Cell* **87,** 629–638.

42. Shibasaki, F., Kondo, E., Akagi, T., and McKeon, F. (1997). Suppression of signalling through transcription factor NF-AT by interactions between calcineurin and Bcl-2. *Nature* **386,** 728–731.

43. Hu, Y., Benedict, M. A., Wu, D., Inohara, N., and Nunez, G. (1998). Bcl-XL interacts with Apaf-1 and inhibits Apaf-1-dependent caspase-9 activation. *Proc. Natl. Acad. Sci. USA* **95**, 4386–4391.

44. Zamzami, N., Susin, S. A., Marchetti, P., Hirsch, T., Gomez-Monterrey, I., Castedo, M., and Kroemer, G. (1996). Mitochondrial control of nuclear apoptosis [see comments]. *J. Exp. Med.* **183**, 1533–1544.

45. Marzo, I., Brenner, C., Zamzami, N., Susin, S. A., Beutner, G., Brdiczka, D., Remy, R., Xie, Z. H., Reed, J. C., and Kroemer, G. (1998). The permeability transition pore complex: a target for apoptosis regulation by caspases and bcl-2-related proteins. *J. Exp. Med.* **187**, 1261–1271.

46. Yang, J., Liu, X., Bhalla, K., Kim, C. N., Ibrado, A. M., Cai, J., Peng, T. I., Jones, D. P., and Wang, X. (1997). Prevention of apoptosis by Bcl-2: release of cytochrome c from mitochondria blocked [see comments]. *Science* **275**, 1129–1132.

47. Antonsson, B., Conti, F., Ciavatta, A., Montessuit, S., Lewis, S., Martinou, I., Bernasconi, L., Bernard, A., Mermod, J. J., Mazzei, G., Maundrell, K., Gambale, F., Sadoul, R., and Martinou, J. C. (1997). Inhibition of Bax channel-forming activity by Bcl-2. *Science* **277**, 370–372.

48. Brenner, C., Cadiou, H., Vieira, H. L., Zamzami, N., Marzo, I., Xie, Z., Leber, B., Andrews, D., Duclohier, H., Reed, J. C., and Kroemer, G. (2000). Bcl-2 and Bax regulate the channel activity of the mitochondrial adenine nucleotide translocator. *Oncogene* **19**, 329–336.

49. Schlesinger, P. H., Gross, A., Yin, X. M., Yamamoto, K., Saito, M., Waksman, G., and Korsmeyer, S. J. (1997). Comparison of the ion channel characteristics of proapoptotic BAX and antiapoptotic BCL-2. *Proc. Natl. Acad. Sci. USA* **94**, 11357–11362.

50. Minn, A. J., Kettlun, C. S., Liang, H., Kelekar, A., Vander Heiden, M. G., Chang, B. S., Fesik, S. W., Fill, M., and Thompson, C. B. (1999). Bcl-xL regulates apoptosis by heterodimerization-dependent and -independent mechanisms. *EMBO J.* **18**, 632–643.

51. Ho, P. T., and Parkinson, D. R. (1997). Antisense oligonucleotides as therapeutics for malignant diseases. *Semin. Oncol.* **24**, 187–202.

52. Lebedeva, I. V., and Stein, C. A. (2000). Antisense oligonucleotides in cancer: recent advances. *BioDrugs* **13**, 195–216.

53. Stephenson, M. L., and Zamecnik, P. C. (1978). Inhibition of Rous sarcoma viral RNA translation by a specific oligodeoxyribonucleotide. *Proc. Natl. Acad. Sci. USA* **75**, 285–288.

54. Zamecnik, P. C., and Stephenson, M. L. (1978). Inhibition of Rous sarcoma virus replication and cell transformation by a specific oligodeoxynucleotide. *Proc. Natl. Acad. Sci. USA* **75**, 280–284.

55. Stein, C. A., and Cheng, Y. C. (1993). Antisense oligonucleotides as therapeutic agents—is the bullet really magical? *Science* **261**, 1004–1012.

56. Nellen, W., and Lichtenstein, C. (1993). What makes an mRNA antisense-itive? *Trends Biochem. Sci.* **18**, 419–423.

57. Cech, T. R., and Bass, B. L. (1986). Biological catalysis by RNA. *Annu. Rev. Biochem.* **55**, 599–629.

58. Cotter, F. E., Johnson, P., Hall, P., Pocock, C., al-Mahdi, N., Cowell, J. K., and Morgan, G. (1994). Antisense oligonucleotides suppress B-cell lymphoma growth in a SCID-hu mouse model. *Oncogene* **9**, 3049–3055.

59. Reed, J. C., Kitada, S., Takayama, S., and Miyashita, T. (1994). Regulation of chemoresistance by the bcl-2 oncoprotein in non-Hodgkin's lymphoma and lymphocytic leukemia cell lines. *Ann. Oncol.* **5**, 61–65.

60. Keith, F. J., Bradbury, D. A., Zhu, Y. M., and Russell, N. H. (1995). Inhibition of bcl-2 with antisense oligonucleotides induces apoptosis and increases the sensitivity of AML blasts to Ara-C. *Leukemia* **9**, 131–138.

61. Reed, J. C., Cuddy, M., Haldar, S., Croce, C., Nowell, P., Makover, D., and Bradley, K. (1990). BCL2-mediated tumorigenicity of a human T-lymphoid cell line: synergy with MYC and inhibition by BCL2 antisense. *Proc. Natl. Acad. Sci. USA* **87**, 3660–3664.

62. Kitada, S., Takayama, S., De Riel, K., Tanaka, S., and Reed, J. C. (1994). Reversal of chemoresistance of lymphoma cells by antisense-mediated reduction of bcl-2 gene expression. *Antisense Res. Dev.* **4**, 71–79.

63. Teixeira, C., Reed, J. C., and Pratt, M. A. (1995). Estrogen promotes chemotherapeutic drug resistance by a mechanism involving Bcl-2 proto-oncogene expression in human breast cancer cells. *Cancer Res.* **55**, 3902–3907.

64. Pratt, M. A., Krajewski, S., Menard, M., Krajewska, M., Macleod, H., and Reed, J. C. (1998). Estrogen withdrawal-induced human breast cancer tumour regression in nude mice is prevented by Bcl-2. *FEBS Lett.* **440**, 403–408.

65. Allsopp, T. E., Kiselev, S., Wyatt, S., and Davies, A. M. (1995). Role of Bcl-2 in the brain-derived neurotrophic factor survival response. *Eur. J. Neurosci.* **7**, 1266–1272.

66. Feng, L., Precht, P., Balakir, R., and Horton, Jr., W. E. (1998). Evidence of a direct role for Bcl-2 in the regulation of articular chondrocyte apoptosis under the conditions of serum withdrawal and retinoic acid treatment. *J. Cell. Biochem.* **71**, 302–309.

67. Zhang, W. G., Ma, L. P., Wang, S. W., Zhang, Z. Y., and Cao, G. D. (1999). Antisense bcl-2 retrovirus vector increases the sensitivity of a human gastric adenocarcinoma cell line to photodynamic therapy. *Photochem. Photobiol.* **69**, 582–586.

68. Dorai, T., Olsson, C. A., Katz, A. E., and Buttyan, R. (1997). Development of a hammerhead ribozyme against bcl-2. I. Preliminary evaluation of a potential gene therapeutic agent for hormone-refractory human prostate cancer. *Prostate* **32**, 246–258.

69. Dorai, T., Goluboff, E. T., Olsson, C. A., and Buttyan, R. (1997). Development of a hammerhead ribozyme against BCL-2. II. Ribozyme treatment sensitizes hormone-resistant prostate cancer cells to apoptotic agents. *Anticancer Res.* **17**, 3307–3312.

70. Dorai, T., Perlman, H., Walsh, K., Shabsigh, A., Goluboff, E. T., Olsson, C. A., and Buttyan, R. (1999). A recombinant defective adenoviral agent expressing anti-bcl-2 ribozyme promotes apoptosis of bcl-2-expressing human prostate cancer cells. *Int. J. Cancer* **82**, 846–852.

71. Perlman, H., Sata, M., Krasinski, K., Dorai, T., Buttyan, R., and Walsh, K. (2000). Adenovirus-encoded hammerhead ribozyme to Bcl-2 inhibits neointimal hyperplasia and induces vascular smooth muscle cell apoptosis. *Cardiovasc Res.* **45**, 570–578.

72. Gibson, S. A., Pellenz, C., Hutchison, R. E., Davey, F. R., and Shillitoe, E.J. (2000). Induction of apoptosis in oral cancer cells by an anti-bcl-2 ribozyme delivered by an adenovirus vector. *Clin. Cancer Res.* **6**, 213–222.

73. Scheid, S., Heinzinger, M., Waller, C. F., and Lange, W. (1998). Bcl-2 mRNA-targeted ribozymes: effects on programmed cell death in chronic myelogenous leukemia cell lines. *Ann. Hematol.* **76**, 117–125.

74. Wickstrom, E. (1986). Oligodeoxynucleotide stability in subcellular extracts and culture media. *J. Biochem. Biophys. Methods* **13**, 97–102.

75. Vaerman, J. L., Moureau, P., Deldime, F., Lewalle, P., Lammineur, C., Morschhauser, F., and Martiat, P. (1997). Antisense oligodeoxyribonucleotides suppress hematologic cell growth through stepwise release of deoxyribonucleotides. *Blood* **90**, 331–339.

76. Stein, C. A., Subasinghe, C., Shinozuka, K., and Cohen, J. S. (1988). Physicochemical properties of phosphorothioate oligodeoxynucleotides. *Nucleic Acids Res.* **16**, 3209–3221.

77. Beltinger, C., Saragovi, H. U., Smith, R. M., LeSauteur, L., Shah, N., DeDionisio, L., Christensen, L., Raible, A., Jarett, L., and Gewirtz, A. M. (1995). Binding, uptake, and intracellular trafficking of phosphorothioate-modified oligodeoxynucleotides. *J. Clin. Invest.* **95**, 1814–1823.

78. Lebedeva, I. V., Benimetskaya, L., Stein, C. A., and Vilenchik, M. (2000). Cellular delivery of antisense oligonucleotides. *Eur. J. Pharm. Biopharm.* **50**, 101–119.

79. Agrawal, S., and Iyer, R. P. (1997). Perspectives in antisense therapeutics. *Pharmacol. Ther.* **76**, 151–160.
80. Monia, B. P., Johnston, J. F., Geiger, T., Muller, M., and Fabbro, D. (1996). Antitumor activity of a phosphorothioate antisense oligodeoxynucleotide targeted against C-raf kinase [see comments]. *Nat. Med.* **2**, 668–675.
81. Jansen, B., Schlagbauer-Wadl, H., Brown, B. D., Bryan, R. N., van Elsas, A., Muller, M., Wolff, K., Eichler, H. G., and Pehamberger, H. (1998). bcl-2 antisense therapy chemosensitizes human melanoma in SCID mice. *Nat. Med.* **4**, 232–234.
82. Bacon, T. A., and Wickstrom, E. (1991). Walking along human c-myc mRNA with antisense oligodeoxynucleotides: maximum efficacy at the 5′ cap region. *Oncogene Res.* **6**, 13–19.
83. Ziegler, A., Luedke, G. H., Fabbro, D., Altmann, K. H., Stahel, R. A., and Zangemeister-Wittke, U. (1997). Induction of apoptosis in small-cell lung cancer cells by an antisense oligodeoxynucleotide targeting the Bcl-2 coding sequence [see comments]. *J. Natl. Cancer Inst.* **89**, 1027–1036.
84. Dibbert, B., Daigle, I., Braun, D., Schranz, C., Weber, M., Blaser, K., Zangemeister-Wittke, U., Akbar, A. N., and Simon, H.U. (1998). Role for Bcl-xL in delayed eosinophil apoptosis mediated by granulocyte-macrophage colony-stimulating factor and interleukin-5. *Blood* **92**, 778–783.
85. Taylor, J. K., Zhang, Q. Q., Monia, B. P., Marcusson, E. G., and Dean, N. M. (1999). Inhibition of Bcl-xL expression sensitizes normal human keratinocytes and epithelial cells to apoptotic stimuli. *Oncogene* **18**, 4495–4504.
86. Lebedeva, I., and Stein, C. (1999). Bcl-xL-directed oligonucleotides in prostate and bladder carcinoma cell lines. *Antisense Nucleic Acid Drug Dev.* **9**, 381–382.
87. Stein, C. A. (2000). Is irrelevant cleavage the price of antisense efficacy? *Pharmacol Ther.* **85**, 231–236.
88. Stein, C. (1998). How to design an antisense oligodeoxynucleotide experiment: a consensus approach. *Antisense Nucleic Acid Drug Dev.* **8**, 129–132.
89. Benimetskaya, L., Berton, M., Kolbanovsky, A., Benimetsky, S., and Stein, C. A. (1997). Formation of a G-tetrad and higher order structures correlates with biological activity of the RelA (NF-kappaB p65) 'antisense' oligodeoxynucleotide. *Nucleic Acids Res.* **25**, 2648–2656.
90. Krieg, A. M., Yi, A. K., Matson, S., Waldschmidt, T. J., Bishop, G. A., Teasdale, R., Koretzky, G. A., and Klinman, D. M. (1995). CpG motifs in bacterial DNA trigger direct B-cell activation. *Nature* **374**, 546–549.
91. Lebedeva, I. V., and Stein, C. A. (1999). Phosphorothioate oligonucleotides as inhibitors of gene expression: antisense and non-antisense effects, in *Applications of Antisense Therapies to Restenosis* (L. E. Rabbani, ed.), pp. 99–199. Kluwer Academic, Dordrecht.
92. Reed, J. C., Stein, C., Subasinghe, C., Haldar, S., Croce, C. M., Yum, S., and Cohen, J. (1990). Antisense-mediated inhibition of BCL2 protooncogene expression and leukemic cell growth and survival: comparisons of phosphodiester and phosphorothioate oligodeoxynucleotides. *Cancer Res.* **50**, 6565–6570.
93. Campos, L., Sabido, O., Rouault, J. P., and Guyotat, D. (1994). Effects of BCL-2 antisense oligodeoxynucleotides on in vitro proliferation and survival of normal marrow progenitors and leukemic cells. *Blood* **84**, 595–600.
94. Durrieu, F., Belaud-Rotureau, M. A., Lacombe, F., Dumain, P., Reiffers, J., Boisseau, M. R., Bernard, P., and Belloc, F. (1999). Synthesis of Bcl-2 in response to anthracycline treatment may contribute to an apoptosis-resistant phenotype in leukemic cell lines. *Cytometry* **36**, 140–149.
95. Pepper, C., Hoy, T., and Bentley, D. P. (1997). Bcl-2/Bax ratios in chronic lymphocytic leukaemia and their correlation with in vitro apoptosis and clinical resistance [see comments]. *Br. J. Cancer* **76**, 935–938.
96. Thomas, A., El Rouby, S., Reed, J. C., Krajewski, S., Silber, R., Potmesil, M., and Newcomb, E. W. (1996). Drug-induced apoptosis in B-cell chronic lymphocytic leukemia: relationship between p53 gene mutation and bcl-2/bax proteins in drug resistance. *Oncogene* **12**, 1055–1062.
97. Pepper, C., Thomas, A., Hoy, T., Cotter, F., and Bentley, P. (1999). Antisense-mediated suppression of Bcl-2 highlights its pivotal role in failed apoptosis in B-cell chronic lymphocytic leukaemia. *Br. J. Haematol.* **107**, 611–615.
98. Kitada, S., Miyashita, T., Tanaka, S., and Reed, J. C. (1993). Investigations of antisense oligonucleotides targeted against bcl-2 RNAs. *Antisense Res Dev.* **3**, 157–169.
99. Smith, M. R., Abubakr, Y., Mohammad, R., Xie, T., Hamdan, M., and al-Katib, A. (1995). Antisense oligodeoxyribonucleotide down-regulation of bcl-2 gene expression inhibits growth of the low-grade non-Hodgkin's lymphoma cell line WSU-FSCCL. *Cancer Gene Ther.* **2**, 207–212.
100. Tormo, M., Tari, A. M., McDonnell, T. J., Cabanillas, F., Garcia-Conde, J., and Lopez-Berestein, G. (1998). Apoptotic induction in transformed follicular lymphoma cells by Bcl-2 downregulation. *Leuk. Lymphoma* **30**, 367–379.
101. Capaccioli, S., Quattrone, A., Schiavone, N., Calastretti, A., Copreni, E., Bevilacqua, A., Canti, G., Gong, L., Morelli, S., and Nicolin, A. (1996). A bcl-2/IgH antisense transcript deregulates bcl-2 gene expression in human follicular lymphoma t(14;18) cell lines. *Oncogene* **13**, 105–115.
102. Morelli, S., Delia, D., Capaccioli, S., Quattrone, A., Schiavone, N., Bevilacqua, A., Tomasini, S., and Nicolin, A. (1997). The antisense bcl-2–IgH transcript is an optimal target for synthetic oligonucleotides. *Proc. Natl. Acad. Sci. USA* **94**, 8150–8155.
103. Morelli, S., Alama, A., Quattrone, A., Gong, I.., Copreni, E., Canti, G., and Nicolin, A. (1996). Oligonucleotides induce apoptosis restricted to the t(14;18) DHL-4 cell line. *Anticancer Drug Des.* **11**, 1–14.
104. Bloem, A., and Lockhorst, H. (1999). Bcl-2 antisense therapy in multiple myeloma. *Pathol. Biol. (Paris)* **47**, 216–220.
105. Jiang, S. X., Sato, Y., Kuwao, S., and Kameya, T. (1995). Expression of bcl-2 oncogene protein is prevalent in small cell lung carcinomas. *J. Pathol.* **177**, 135–138.
106. Ikegaki, N., Katsumata, M., Minna, J., and Tsujimoto, Y. (1994). Expression of bcl-2 in small cell lung carcinoma cells. *Cancer Res.* **54**, 6–8.
107. Ohmori, T., Podack, E. R., Nishio, K., Takahashi, M., Miyahara, Y., Takeda, Y., Kubota, N., Funayama, Y., Ogasawara, H., Ohira, T. *et al.* (1993). Apoptosis of lung cancer cells caused by some anti-cancer agents (MMC, CPT-11, ADM) is inhibited by bcl-2. *Biochem. Biophys. Res. Commun.* **192**, 30–36.
108. Robinson, L. A., Smith, L. J., Fontaine, M. P., Kay, H. D., Mountjoy, C. P., and Pirruccello, S. J. (1995). c-myc antisense oligodeoxyribonucleotides inhibit proliferation of non-small-cell-lung cancer. *Ann. Thorac. Surg.* **60**, 1583–1591.
109. Koty, P. P., Zhang, H., and Levitt, M. L. (1999). Antisense bcl-2 treatment increases programmed cell death in non-small cell lung cancer cell lines. *Lung Cancer* **23**, 115–127.
110. Zangemeister-Wittke, U., Schenker, T., Luedke, G. H., and Stahel, R. A. (1998). Synergistic cytotoxicity of bcl-2 antisense oligodeoxy nucleotides and etoposide, doxorubicin and cisplatin on small-cell lung cancer cell lines. *Br. J. Cancer* **78**, 1035–1042.
111. Zhang, Y., Fujita, N., and Tsuruo, T. (1999). p21$^{Waf1/Cip1}$ acts in synergy with bcl-2 to confer multidrug resistance in a camptothecin-selected human lung-cancer cell line. *Int. J. Cancer* **83**, 790–797.
112. Barbareschi, M., Caffo, O., Veronese, S., Leek, R. D., Fina, P., Fox, S., Bonzanini, M., Girlando, S., Morelli, L., Eccher, C., Pezzella, F., Doglioni, C., Dalla Palma, P., and Harris, A. (1996). Bcl-2 and p53

expression in node-negative breast carcinoma: a study with long-term follow-up. *Hum. Pathol.* **27,** 1149–1155.

113. McDonnell, T. J., Navone, N. M., Troncoso, P., Pisters, L. L., Conti, C., von Eschenbach, A. C., Brisbay, S., and Logothetis, C. J. (1997). Expression of bcl-2 oncoprotein and p53 protein accumulation in bone marrow metastases of androgen independent prostate cancer [see comments]. *J. Urol.* **157,** 569–574.

114. Bauer, J. J., Sesterhenn, I. A., Mostofi, F. K., McLeod, D. G., Srivastava, S., and Moul, J. W. (1996). Elevated levels of apoptosis regulator proteins p53 and bcl-2 are independent prognostic biomarkers in surgically treated clinically localized prostate cancer. *J. Urol.* **156,** 1511–1516.

115. Apakama, I., Robinson, M. C., Walter, N. M., Charlton, R. G., Royds, J. A., Fuller, C. E., Neal, D. E., and Hamdy, F. C. (1996). bcl-2 overexpression combined with p53 protein accumulation correlates with hormone-refractory prostate cancer. *Br. J. Cancer* **74,** 1258–1262.

116. Krajewska, M., Krajewski, S., Epstein, J. I., Shabaik, A., Sauvageot, J., Song, K., Kitada, S., and Reed, J. C. (1996). Immunohistochemical analysis of bcl-2, bax, bcl-X, and mcl-1 expression in prostate cancers. *Am. J. Pathol.* **148,** 1567–1576.

117. Raffo, A. J., Perlman, H., Chen, M. W., Day, M. L., Streitman, J. S., and Buttyan, R. (1995). Overexpression of bcl-2 protects prostate cancer cells from apoptosis in vitro and confers resistance to androgen depletion in vivo. *Cancer Res.* **55,** 4438–4445.

118. Berchem, G. J., Bosseler, M., Sugars, L. Y., Voeller, H. J., Zeitlin, S., and Gelmann, E. P. (1995). Androgens induce resistance to bcl-2-mediated apoptosis in LNCaP prostate cancer cells. *Cancer Res.* **55,** 735–738.

119. Gleave, M., Tolcher, A., Miyake, H., Nelson, C., Brown, B., Beraldi, E., and Goldie, J. (1999). Progression to androgen independence is delayed by adjuvant treatment with antisense Bcl-2 oligodeoxynucleotides after castration in the LNCaP prostate tumor model. *Clin. Cancer Res.* **5,** 2891–2898.

120. Miyake, H., Tolcher, A., and Gleave, M. E. (1999). Antisense Bcl-2 oligodeoxynucleotides inhibit progression to androgen-independence after castration in the Shionogi tumor model. *Cancer Res.* **59,** 4030–4034.

121. Miayake, H., Tolcher, A., and Gleave, M. E. (2000). Chemosensitization and delayed androgen-independent recurrence of prostate cancer with the use of antisense Bcl-2 oligodeoxynucleotides. *J. Natl. Cancer Inst.* **92,** 34–41.

122. Gleave, M. E., Miayake, H., Goldie, J., Nelson, C., and Tolcher, A. (1999). Targeting bcl-2 gene to delay androgen-independent progression and enhance chemosensitivity in prostate cancer using antisense bcl-2 oligodeoxynucleotides. *Urology* **54,** 36–46.

123. Campbell, M. J., Dawson, M., and Koeffler, H. P. (1998). Growth inhibition of DU-145 prostate cancer cells by a Bcl-2 antisense oligonucleotide is enhanced by N-(2-hydroxyphenyl)all-*trans* retinamide. *Br. J. Cancer* **77,** 739–744.

124. Bilim, V., Kasahara, T., Noboru, H., Takahashi, K., and Tomita, Y. (2000). Caspase involved synergistic cytotoxicity of bcl-2 antisense oligonucleotides and Adriamycin on transitional cell cancer cells. *Cancer Lett.* **155,** 191–198.

125. Meyer, O., Kirpotin, D., Hong, K., Sternberg, B., Park, J. W., Woodle, M. C., and Papahadjopoulos, D. (1998). Cationic liposomes coated with polyethylene glycol as carriers for oligonucleotides. *J. Biol. Chem.* **273,** 15621–15627.

126. Harnois, D. M., Que, F. G., Celli, A., LaRusso, N. F., and Gores, G. J. (1997). Bcl-2 is overexpressed and alters the threshold for apoptosis in a cholangiocarcinoma cell line. *Hepatology* **26,** 884–890.

127. Saito, H., Ebinuma, H., Takahashi, M., Kaneko, F., Wakabayashi, K., Nakamura, M., and Ishii, H. (1998). Loss of butyrate-induced apoptosis in human hepatoma cell lines HCC-M and HCC-T having substantial Bcl-2 expression. *Hepatology* **27,** 1233–1240.

128. Takahashi, M., Saito, H., Okuyama, T., Miyashita, T., Kosuga, M., Sumisa, F., Yamada, M., Ebinuma, H., and Ishii, H. (1999).
Overexpression of Bcl-2 protects human hepatoma cells from Fas-antibody-mediated apoptosis. *J. Hepatol.* **31,** 315–322.

129. Agrawal, S., Temsamani, J., and Tang, J. Y. (1991). Pharmacokinetics, biodistribution, and stability of oligodeoxynucleotide phosphorothioates in mice. *Proc. Natl. Acad. Sci. USA* **88,** 7595–7599.

130. Iversen, P. (1991). In vivo studies with phosphorothioate oligonucleotides: pharmacokinetics prologue. *Anticancer Drug Des.* **6,** 531–538.

131. Raynaud, F. I., Orr, R. M., Goddard, P. M., Lacey, H. A., Lancashire, H., Judson, I. R., Beck, T., Bryan, B., and Cotter, F. E. (1997). Pharmacokinetics of G3139, a phosphorothioate oligodeoxynucleotide antisense to bcl-2, after intravenous administration or continuous subcutaneous infusion to mice. *J. Pharmacol. Exp. Ther.* **281,** 420–427.

132. Gutierrez-Puente, Y., Tari, A. M., Stephens, C., Rosenblum, M., Guerra, R. T., and Lopez-Berestein, G. (1999). Safety, pharmacokinetics, and tissue distribution of liposomal P-ethoxy antisense oligonucleotides targeted to Bcl-2. *J. Pharmacol. Exp. Ther.* **291,** 865–869.

133. Cotter, F. E., Corbo, M., Raynaud, F., Orr, R. M., Pocock, C., Bryan, B., Harper, M., Webb, A., Clarke, P., Judson, I. R., and Cunningham, D. (1996). Bcl-2-antisense therapy in lymphoma: in vitro and in vivo mechanisms, efficacy, pharmacokinetics and toxicity studies. *Ann. Oncol.* **7,** 32.

134. Wooldridge, J. E., Ballas, Z., Krieg, A. M., and Weiner, G. J. (1997). Immunostimulatory oligodeoxynucleotides containing CpG motifs enhance the efficacy of monoclonal antibody therapy of lymphoma. *Blood* **89,** 2994–2998.

135. Cotter, F. E., Waters, J., and Cunningham, D. (1999). Human Bcl-2 antisense therapy for lymphomas. *Biochim. Biophys. Acta.* **1489,** 97–106.

136. Webb, A., Cunningham, D., Cotter, F., Clarke, P. A., di Stefano, F., Ross, P., Corbo, M., and Dziewanowska, Z. (1997). BCL-2 antisense therapy in patients with non-Hodgkins lymphoma. *Lancet* **349,** 1137–1141.

137. Waters, J. S., Webb, A., Cunningham, D., Clarke, P. A., Raynaud, F., di Stefano, F., and Cotter, F. E. (2000). Phase I clinical and pharmacokinetic study of bcl-2 antisense oligonucleotide therapy in patients with non-Hodgkin's lymphoma. *J. Clin. Oncol.* **18,** 1812–1823.

138. Grover, R., and Wilson, G. D. (1996). Bcl-2 expression in malignant melanoma and its prognostic significance. *Eur. J. Surg. Oncol.* **22,** 347–349.

139. Jansen, B., Wacheck, V., Heere-Ress, E., Schlagbauer-Wadl, H., Hollenstein, U., Lucas, T., Eichler, H.-G., Wolff, K., and Pehamberger, H. (1999). A phase I–II study with dacarbazine and BCL-2 antisense oligonucleotide G3139 (GENTA) as a chemosensitizer in patients with advanced malignant melanoma [abstract]. *Proc. Am. Soc. Clin. Oncol.* **18,** 2047.

140. Schlagbauer-Wadl, H., Klosner, G., Heere-Ress, E., Waltering, S., Moll, I., Wolff, K., Pehamberger, H., and Jansen, B. (2000). Bcl-2 antisense oligonucleotides (G3139) inhibit merkel cell carcinoma growth in SCID mice. *J. Invest. Dermatol.* **114,** 725–730.

141. Morris, M. J., Tong, W., Osman, I., Maslak, P., Kelly, W. K., Terry, K., Rosen, N., and Scher, H. (1999). A phase I/IIA dose-escalating trial of bcl-2 antisense (G3139) treatment by 14-day continuous intravenous infusion (CI) for patients with androgen-independent prostate cancer or other advanced solid tumor malignancies [abstract]. *Proc. Am. Soc. Clin. Oncol.* **18,** 1241.

142. Chen, H. X., Marshall, J. L., Trocky, N., Ling, Y., Baidas, S., Rizvi, N., Bhargava, P., Lippman, M. E., Yang, D., and Hayes, D.F. (2000). A phase I study of BCL-2 antisense G3139 (GENTA) and weekly docetaxel in patients with advanced breast cancer and other solid tumors [abstract]. *Proc. Am. Soc. Clin. Oncol.* **19,** 692.

143. Schott, A. F., Apel, I. J., Nunez, G., and Clarke, M. F. (1995). Bcl-XL protects cancer cells from p53-mediated apoptosis. *Oncogene* **11,** 1389–1394.

144. Chao, D. T., Linette, G. P., Boise, L. H., White, L. S., Thompson, C. B., and Korsmeyer, S. J. (1995). Bcl-XL and Bcl-2 repress a common pathway of cell death. *J. Exp. Med.* **182**, 821–828.

145. Robertson, J. D., Datta, K., and Kehrer, J. P. (1997). Bcl-xL overexpression restricts heat-induced apoptosis and influences hsp70, bcl-2, and Bax protein levels in FL5.12 cells. *Biochem. Biophys. Res. Commun.* **241**, 164–168.

146. Zhang, X., Li, L., Choe, J., Krajewski, S., Reed, J. C., Thompson, C., and Choi, Y. S. (1996). Up-regulation of Bcl-xL expression protects CD40-activated human B cells from Fas-mediated apoptosis. *Cell Immunol.* **173**, 149–154.

147. Shinoura, N., Yoshida, Y., Asai, A., Kirino, T., and Hamada, H. (1999). Relative level of expression of Bax and Bcl-XL determines the cellular fate of apoptosis/necrosis induced by the overexpression of Bax. *Oncogene* **18**, 5703–5713.

148. Nouraini, S., Six, E., Matsuyama, S., Krajewski, S., and Reed, J. C. (2000). The putative pore-forming domain of Bax regulates mitochondrial localization and interaction with Bcl-X(L). *Mol. Cell. Biol.* **20**, 1604–1615.

149. Muchmore, S. W., Sattler, M., Liang, H., Meadows, R. P., Harlan, J. E., Yoon, H. S., Nettesheim, D., Chang, B. S., Thompson, C. B., Wong, S. L., Ng, S. L., and Fesik, S. W. (1996). X-ray and NMR structure of human Bcl-xL, an inhibitor of programmed cell death. *Nature* **381**, 335–341.

150. Minn, A. J., Velez, P., Schendel, S. L., Liang, H., Muchmore, S. W., Fesik, S. W., Fill, M., and Thompson, C. B. (1997). Bcl-x(L) forms an ion channel in synthetic lipid membranes. *Nature* **385**, 353–357.

151. Schendel, S. L., Montal, M., and Reed, J. C. (1998). Bcl-2 family proteins as ion-channels. *Cell Death Differ.* **5**, 372–380.

152. Kondo, S., Shinomura, Y., Kanayama, S., Higashimoto, Y., Kiyohara, T., Zushi, S., Kitamura, S., Ueyama, H., and Matsuzawa, Y. (1998). Modulation of apoptosis by endogenous Bcl-xL expression in MKN-45 human gastric cancer cells. *Oncogene* **17**, 2585–2591.

153. Schlaifer, D., March, M., Krajewski, S., Laurent, G., Pris, J., Delsol, G., Reed, J. C., and Brousset, P. (1995). High expression of the bcl-x gene in Reed–Sternberg cells of Hodgkin's disease. *Blood* **85**, 2671–2674.

154. Schlaifer, D., Krajewski, S., Galoin, S., Rigal-Huguet, F., Laurent, G., Massip, P., Pris, J., Delsol, G., Reed, J. C., and Brousset, P. (1996). Immunodetection of apoptosis-regulating proteins in lymphomas from patients with and without human immunodeficiency virus infection. *Am. J. Pathol.* **149**, 177–185.

155. Chresta, C. M., Masters, J. R., and Hickman, J. A. (1996). Hypersensitivity of human testicular tumors to etoposide-induced apoptosis is associated with functional p53 and a high Bax:Bcl-2 ratio. *Cancer Res.* **56**, 1834–1841.

156. Chresta, C. M., Arriola, E. L., and Hickman, J. A. (1996). Apoptosis and cancer chemotherapy. *Behring Inst. Mitt.* 232–240.

157. Dole, M. G., Jasty, R., Cooper, M. J., Thompson, C. B., Nunez, G., and Castle, V. P. (1995). Bcl-xL is expressed in neuroblastoma cells and modulates chemotherapy-induced apoptosis. *Cancer Res.* **55**, 2576–2582.

158. Ibrado, A. M., Liu, L., and Bhalla, K. (1997). Bcl-xL overexpression inhibits progression of molecular events leading to paclitaxel-induced apoptosis of human acute myeloid leukemia HL-60 cells. *Cancer Res.* **57**, 1109–1115.

159. Schmitt, E., Cimoli, G., Steyaert, A., and Bertrand, R. (1998). Bcl-xL modulates apoptosis induced by anticancer drugs and delays DEVDase and DNA fragmentation-promoting activities. *Exp. Cell. Res.* **240**, 107–121.

160. Wang, Z., Karras, J. G., Howard, R. G., and Rothstein, T. L. (1995). Induction of bcl-x by CD40 engagement rescues sIg-induced apoptosis in murine B cells. *J. Immunol.* **155**, 3722–3725.

161. Gottschalk, A. R., Boise, L. H., Thompson, C. B., and Quintans, J. (1994). Identification of immunosuppressant-induced apoptosis in a murine B-cell line and its prevention by bcl-x but not bcl-2. *Proc. Natl. Acad. Sci. USA* **91**, 7350–7354.

162. Fujio, Y., Kunisada, K., Hirota, H., Yamauchi-Takihara, K., and Kishimoto, T. (1997). Signals through gp130 upregulate bcl-x gene expression via STAT1–binding *cis*-element in cardiac myocytes. *J. Clin. Invest.* **99**, 2898–2905.

163. Amarante-Mendes, G. P., McGahon, A. J., Nishioka, W. K., Afar, D. E., Witte, O. N., and Green, D. R. (1998). Bcl-2-independent Bcr-Abl-mediated resistance to apoptosis: protection is correlated with up regulation of Bcl-xL. *Oncogene* **16**, 1383–1390.

164. Bryckaert, M., Guillonneau, X., Hecquet, C., Courtois, Y., and Mascarelli, F. (1999). Both FGF1 and bcl-x synthesis are necessary for the reduction of apoptosis in retinal pigmented epithelial cells by FGF2: role of the extracellular signal-regulated kinase 2. *Oncogene* **18**, 7584–7593.

165. Arriola, E. L., Rodriguez-Lopez, A. M., Hickman, J. A., and Chresta, C. M. (1999). Bcl-2 overexpression results in reciprocal downregulation of Bcl-(XL) and sensitizes human testicular germ cell tumours to chemotherapy- induced apoptosis. *Oncogene* **18**, 1457–1464.

166. Ackermann, E. J., Taylor, J. K., Narayana, R., and Bennett, C. F. (1999). The role of antiapoptotic Bcl-2 family members in endothelial apoptosis elucidated with antisense oligonucleotides. *J. Biol. Chem.* **274**, 11245–11252.

167. Ma, M., Benimetskaya, L., Lebedeva, I., Dignam, J., Takle, G., and Stein, C. A. (2000). Intracellular mRNA cleavage induced through activation of RNase P by nuclease-resistant external guide sequences. *Nat. Biotechnol.* **18**, 58–61.

168. Reeve, J. G., Xiong, J., Morgan, J., and Bleehen, N. M. (1996). Expression of apoptosis-regulatory genes in lung tumour cell lines: relationship to p53 expression and relevance to acquired drug resistance. *Br. J. Cancer* **73**, 1193–1200.

169. Leech, S. H., Olie, R. A., Gautschi, O., Simoes-Wust, A. P., Tschopp, S., Haner, R., Hall, J., Stahel, R. A., and Zangemeister-Wittke, U. (2000). Induction of apoptosis in lung-cancer cells following bcl-xL anti-sense treatment. *Int. J. Cancer* **86**, 570–576.

170. Taylor, J. K., Zhang, Q. Q., Wyatt, J. R., and Dean, N. M. (1999). Induction of endogenous Bcl-xS through the control of Bcl-x pre-mRNA splicing by antisense oligonucleotides [see comments]. *Nat. Biotechnol.* **17**, 1097–1100.

171. Pollman, M. J., Hall, J. L., Mann, M. J., Zhang, L., and Gibbons, G.H. (1998). Inhibition of neointimal cell bcl-x expression induces apoptosis and regression of vascular disease. *Nat. Med.* **4**, 222–227.

172. Suzuki, J., Isobe, M., Morishita, R., Nishikawa, T., Amano, J., and Kaneda, Y. (2000). Antisense Bcl-x oligonucleotide induces apoptosis and prevents arterial neointimal formation in murine cardiac allografts. *Cardiovasc. Res.* **45**, 783–787.

173. Coustan-Smith, E., Kitanaka, A., Pui, C.H., McNinch, L., Evans, W. E., Raimondi, S. C., Behm, F. G., Arico, M., and Campana, D. (1996). Clinical relevance of BCL-2 overexpression in childhood acute lymphoblastic leukemia. *Blood* **87**, 1140–1146.

174. Weller, M., Rieger, J., Grimmel, C., Van Meir, E. G., De Tribolet, N., Krajewski, S., Reed, J. C., von Deimling, A., and Dichgans, J. (1998). Predicting chemoresistance in human malignant glioma cells: the role of molecular genetic analyses. *Int. J. Cancer* **79**, 640–644.

175. Krajewski, S., Krajewska, M., Turner, B. C., Pratt, C., Howard, B., Zapata, J. M., Frenkel, V., Robertson, S., Ionov, Y., Yamamoto, H., Perucho, M., Takayama, S., and Reed, J. C. (1999). Prognostic significance of apoptosis regulators in breast cancer. *Endocr. Relat. Cancer* **6**, 29–40.

176. Schorr, K., Li, M., Krajewski, S., Reed, J. C., and Furth, P. A. (1999). Bcl-2 gene family and related proteins in mammary gland involution and breast cancer. *J. Mammary Gland Biol. Neoplasia* **4**, 153–164.

177. O'Reilly, L. A., Huang, D. C., and Strasser, A. (1996). The cell death inhibitor Bcl-2 and its homologues influence control of cell cycle entry. *EMBO J.* **15**, 6979–6990.

178. Vairo, G., Innes, K. M., and Adams, J. M. (1996). Bcl-2 has a cell cycle inhibitory function separable from its enhancement of cell survival. *Oncogene* **13**, 1511–1519.

179. Shinoura, N., Yoshida, Y., Nishimura, M., Muramatsu, Y., Asai, A., Kirino, T., and Hamada, H. (1999). Expression level of Bcl-2 determines anti- or proapoptotic function. *Cancer Res.* **59**, 4119–4128.

180. Uhlmann, E. J., Subramanian, T., Vater, C. A., Lutz, R., and Chinnadurai, G. (1998). A potent cell death activity associated with transient high level expression of BCL-2. *J. Biol. Chem.* **273**, 17926–17932.

181. Miyashita, T., Kitada, S., Krajewski, S., Horne, W. A., Delia, D., and Reed, J. C. (1995). Overexpression of the Bcl-2 protein increases the half-life of p21Bax. *J. Biol. Chem.* **270**, 26049–26052.

182. Chen, J., Flannery, J. G., LaVail, M. M., Steinberg, R. H., Xu, J., and Simon, M. I. (1996). bcl-2 overexpression reduces apoptotic photoreceptor cell death in three different retinal degenerations. *Proc. Natl. Acad. Sci. USA* **93**, 7042–7047.

183. Cheng, E. H., Kirsch, D. G., Clem, R. J., Ravi, R., Kastan, M. B., Bedi, A., Ueno, K., and Hardwick, J. M. (1997). Conversion of Bcl-2 to a Bax-like death effector by caspases. *Science* **278**, 1966–1968.

184. Offen, D., Beart, P. M., Cheung, N. S., Pascoe, C. J., Hochman, A., Gorodin, S., Melamed, E., Bernard, R., and Bernard, O. (1998). Transgenic mice expressing human Bcl-2 in their neurons are resistant to 6-hydroxydopamine and 1-methyl-4-phenyl-1,2,3,6-tetrahydropyridine neurotoxicity. *Proc. Natl. Acad. Sci. USA* **95**, 5789–5794.

185. Chen, D. F., Schneider, G. E., Martinou, J. C., and Tonegawa, S. (1997). Bcl-2 promotes regeneration of severed axons in mammalian CNS [see comments]. *Nature* **385**, 434–439.

186. Kostic, V., Jackson-Lewis, V., de Bilbao, F., Dubois-Dauphin, M., and Przedborski, S. (1997). Bcl-2: prolonging life in a transgenic mouse model of familial amyotrophic lateral sclerosis. *Science* **277**, 559–562.

187. Krajewska, M., Fenoglio-Preiser, C. M., Krajewski, S., Song, K., Macdonald, J. S., Stemmerman, G., and Reed, J. C. (1996). Immunohistochemical analysis of Bcl-2 family proteins in adenocarcinomas of the stomach. *Am. J. Pathol.* **149**, 1449–1457.

188. Benito, A., Silva, M., Grillot, D., Nunez, G., and Fernandez-Luna, J. L. (1996). Apoptosis induced by erythroid differentiation of human leukemia cell lines is inhibited by Bcl-XL. *Blood* **87**, 3837–3843.

189. Reed, J. C. (1996). Mechanisms of Bcl-2 family protein function and dysfunction in health and disease. *Behring Inst. Mitt.* 72–100.

190. Simonian, P. L., Grillot, D. A., and Nunez, G. (1997). Bcl-2 and Bcl-XL can differentially block chemotherapy-induced cell death. *Blood* **90**, 1208–1216.

191. Itoh, N., Tsujimoto, Y., and Nagata, S. (1993). Effect of bcl-2 on Fas antigen-mediated cell death. *J. Immunol.* **151**, 621–627.

192. Mandal, M., Maggirwar, S. B., Sharma, N., Kaufmann, S. H., Sun, S. C., and Kumar, R. (1996). Bcl-2 prevents CD95 (Fas/APO-1)-induced degradation of lamin B and poly(ADP-ribose) polymerase and restores the NF-kappaB signaling pathway. *J. Biol. Chem.* **271**, 30354–30359.

193. Memon, S. A., Moreno, M. B., Petrak, D., and Zacharchuk, C. M. (1995). Bcl-2 blocks glucocorticoid- but not Fas- or activation-induced apoptosis in a T cell hybridoma. *J. Immunol.* **155**, 4644–4652.

194. Strasser, A., Harris, A. W., Huang, D. C., Krammer, P. H., and Cory, S. (1995). Bcl-2 and Fas/APO-1 regulate distinct pathways to lymphocyte apoptosis. *EMBO J.* **14**, 6136–6147.

195. Boise, L. H. and Thompson, C. B. (1997). Bcl-x(L) can inhibit apoptosis in cells that have undergone Fas-induced protease activation. *Proc. Natl. Acad. Sci. USA* **94**, 3759–3764.

196. Oltvai, Z. N., Milliman, C. L., and Korsmeyer, S. J. (1993). Bcl-2 heterodimerizes in vivo with a conserved homolog, Bax, that accelerates programmed cell death. *Cell* **74**, 609–619.

197. Yang, E., Zha, J., Jockel, J., Boise, L. H., Thompson, C.B., and Korsmeyer, S. J. (1995). Bad, a heterodimeric partner for Bcl-XL and Bcl-2, displaces Bax and promotes cell death. *Cell* **80**, 285–291.

198. Sentman, C. L., Shutter, J. R., Hockenbery, D., Kanagawa, O., and Korsmeyer, S. J. (1991). bcl-2 inhibits multiple forms of apoptosis but not negative selection in thymocytes. *Cell* **67**, 879–888.

199. Veis, D. J., Sorenson, C. M., Shutter, J. R., and Korsmeyer, S. J. (1993). Bcl-2-deficient mice demonstrate fulminant lymphoid apoptosis, polycystic kidneys, and hypopigmented hair. *Cell* **75**, 229–240.

200. Motoyama, N., Wang, F., Roth, K. A., Sawa, H., Nakayama, K., Negishi, I., Senju, S., Zhang, Q., Fujii, S. *et al.* (1995). Massive cell death of immature hematopoietic cells and neurons in Bcl-x-deficient mice. *Science* **267**, 1506–1510.

201. Han, Z., Chatterjee, D., Early, J., Pantazis, P., Hendrickson, E. A., and Wyche, J. H. (1996). Isolation and characterization of an apoptosis-resistant variant of human leukemia HL-60 cells that has switched expression from Bcl-2 to Bcl-xL. *Cancer Res.* **56**, 1621–1628.

202. Nunez, G., Merino, R., Grillot, D., and Gonzalez-Garcia, M. (1994). Bcl-2 and Bcl-x: regulatory switches for lymphoid death and survival. *Immunol Today* **15**, 582–588

20

Gene Therapy for Chronic Myelogenous Leukemia

CATHERINE M. VERFAILLIE

Stem Cell Institute, Cancer Center, and Division of Hematology
Oncology and Transplantation
Department of Medicine
University of Minnesota
Minneapolis, Minnesota 55455

ROBERT CH ZHAO

Stem Cell Institute, Cancer Center, and Division of Hematology
Oncology and Transplantation
Department of Medicine
University of Minnesota
Minneapolis, Minnesota 55455

I. MOLECULAR MECHANISMS UNDERLYING Ph$^+$ LEUKEMIAS

Chronic myelogenous leukemia (CML) is a malignancy of the hematopoietic stem cell [1], characterized by a reciprocal translocation between chromosomes 9 and 22 (Philadelphia chromosome [Ph]) resulting in the formation of a hybrid *bcr–abl* gene on chromosome 22 [2,3]. In >80% of patients with CML, exons b2 or b3 of the *bcr* gene and exon a2 of the *abl* gene are juxtaposed, generating a 210-kD oncoprotein [2–4]. The presence of *bcr–abl* is necessary and sufficient for transformation [5–11]. *bcr–abl* cDNA introduced in hematopoietic cell lines causes growth-factor-independent growth *in vitro* and tumorigenicity *in vivo*; transplantation of murine stem cells transduced with *bcr–abl* cDNA causes

a CML-like syndrome; transgenic expression of *bcr–abl* causes a syndrome with myeloproliferative or acute leukemia characteristics.

Chronic myelogenous leukemia is characterized in its initial chronic phase by abnormal premature circulation of malignant progenitors in the peripheral blood [12]. The pool of malignant progenitors and precursors is also massively expanded [12–14]. However, the proliferative rate of the progenitors is not increased [13]. It is thought that the massive expansion of the malignant cell population is due in part to decreased cell death [15,16] and in part to the fact that, in contrast to normal progenitors [17], CML progenitors are never quiescent and continuously proliferate [18,19]. This can be attributed to the increased tyrosine kinase activity of p210$^{bcr–abl}$ compared with p145^{abl} activating a large number of signal pathways, including Stats, MAPK, and PI3K, and the greater ability of p210$^{bcr–abl}$ compared with p145^{abl} to bind F-actin [20–27].

II. THERAPY

Treatment of CML with intensive chemotherapy alone does not induce persistent cytogenetic remissions [12,28]. Although allografts can cure CML [28], this therapy is available to less than 40% of patients, as CML affects patients over the age of 55. Autografts are being considered as an alternative therapy. There is evidence to suggest that autografts can induce a Ph$^-$ state and may increase survival [29–33]. Unfortunately, most patients will suffer from leukemic relapse after autografting, due in part to persistent disease in the host after the preparative regime [34] and to persistent disease in the graft [31–33]. Novel therapeutic approaches are therefore needed.

III. GENE-DISRUPTION METHODS

As CML is caused by p210$^{bcr-abl}$ [5–11], suppression of the expression of p210$^{bcr-abl}$ or proteins downstream in the *bcr–abl* signal pathway which are specifically activated by p210$^{bcr-abl}$ may restore normal, nonleukemic growth of *bcr–abl*$^+$ cells [35–44]. A large body of work has focused on targeting mRNA to eliminate target gene products [45–47]. These are so-called "antisense" strategies because they form reverse (antisense) Watson–Crick base pairs between the targeting sequence and the mRNA whose function is to be disrupted. Antisense-mediated inhibition of mRNA translation may be accomplished by hybridization between the target mRNA and the antisense sequence which leads to duplex formation that serves as a "block" and prevents the ribosomal complex from translating the message [46]. If the antisense sequence is an RNA molecule, the duplexed RNA may be further modified by RNA-editing enzymes, such as double-stranded RNA adenosine deaminase, which destroy the coding sequence and target the mRNA for clearance; if the antisense sequence is a DNA molecule, the duplexed DNA–RNA may be further modified by RNase H [48,49]. The cell has natural defenses such as ribosomal proofreading enzymes, helicases, and unwindases, which dissociate the RNA–RNA or DNA–RNA complex and cleave phosphodiester bonds in antisense oligonucleotides (AS-ONs) [50]. Several modifications have been made to the phosphodiester backbone to generate AS-ONs that are resistant to the action of phosphodiesterases [51], such as replacement of one oxygen atom in this bond with a phosphorus atom which renders the phosphorothioate oligonucleotide phosphodiesterase resistant [52]. Because of their hydrophilic nature and relatively high nuclease resistance, phosphorothioate oligodeoxynucleotides (ODNs) are the most widely studied AS-ONs [53].

Ribozymes are small RNA molecules capable of catalyzing RNA cleavage reactions in a sequence-specific manner. Synthetic hammerhead ribozymes contain a conserved region of 24 nucleotides which must be flanked by 3′ and 5′ sequences that are at least 8 nucleotides in length and complementary to the target sequence surrounding the cleavage site (GUA, C, or U) sequence [47]. Thus, aside from an antisense effect, the catalytic component of the ribozyme cleaves the mRNA to which it has hybridized, leading to physical destruction of the mRNA. Finally, DNAzymes have been developed, that unlike ribozymes, which cleave their target only at the G–U–C nucleotide sequence, can cleave the target m-RNA at any purine–pyrimidine sequence, which offers them higher target specificity [53].

IV. ANTI-*bcr-abl* TARGETED THERAPIES

Because p210$^{bcr-abl}$ is tumor specific and responsible for the development of leukemia [5–11,54], it is an obvious target to treat CML. The native *cbcr* and *cabl* genes have important functions in the cell [55–57]. Anti-*bcr–abl* therapy is therefore ideally targeted at the *bcr–abl* breakpoint and not the *bcr* or *abl* portions of the oncogene. Alternatively, therapies could be directed at molecules downstream from *bcr–abl* that are involved in the transformation of hematopoietic cells [24,41,43,44,58,59]. This requires that activation of the downstream molecules is *bcr–abl* specific and is not by processes that govern normal progenitor proliferation and differentiation.

Several studies have examined the effect of antisense ODNs or ribozymes directed at the BCR–ABL breakpoint on the behavior of CML cell lines or primary human CML progenitors [15,16,35–38,60]. Exposure of *bcr–abl* cDNA containing cell lines or primary CML blasts, which are cytokine independent and tumorigenic *in vivo,* to ODNs complementary to the b3a2 BCR/ABL junction restores growth-factor dependency and sensitivity to apoptotic stimuli and abrogates leukemogenicity *in vivo.* Similar effects have been seen following exposure of chronic-phase CML progenitors to breakpoint-specific ODNs [15,16,38,54].

Ribozymes of various lengths and compositions have been used to target the *bcr–abl* junctional sequence and have, as for AS-ONs, shown varying degrees of target specificity and efficiency [37,60–64]. Transfection with ribozymes or transgenic expression of ribozymes suppresses *bcr–abl* m-RNA and protein levels by 60–100%, inhibits growth of p210$^{bcr-abl}$-containing cell lines by 43%, enhances cell apoptosis [63,64], and decreases tumorigenicity of *bcr–abl*-containing cells in SCID mice [65]. These results suggest that ribozymes directed against *bcr–abl* m-RNA can effectively suppress *bcr–abl* gene expression and alter the leukemic nature of CML cell lines. Like ODNs, ribozymes are limited by their low resistance to nucleases and less-than-ideal specificity of cleavage of the *bcr–abl* junctional sequence [47].

DNAzymes have similar growth inhibitory effects on *bcr–abl*-containing cell lines [66]. Because DNAzymes can cleave the *bcr–abl* mRNA within one base pair from the breakpoint, they may be more specific and efficient than ribozymes in cleavage of the *bcr–abl* junctional sequence [67,68].

V. ANTI-*bcr-abl* DRUG-RESISTANCE GENE THERAPY FOR CML

A. Anti-*bcr-abl* Component

Introduction of the anti-*bcr–abl* antisense sequence into the cell genome which would then be continuously produced

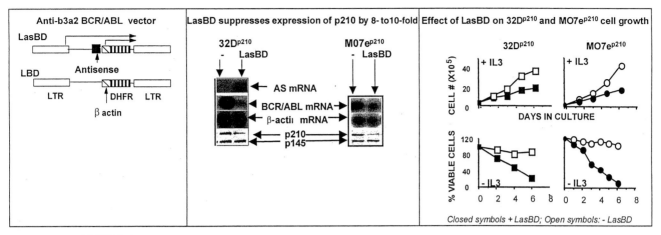

FIGURE 1   The LasBD retroviral vector suppresses BCR/ABL expression by 8- to 10-fold at the RNA and protein level and restores normal growth-factor dependence of BCR/ABL-containing cell lines (32D cells and MO7e cells).

for the life of the cell and its progeny would circumvent the problem that transient exposure to antisense sequences does not eliminate the malignant clone. In addition, this may circumvent the problems caused by ribosomal proofreading enzymes, helicases, and unwindases, which dissociate the DNA–RNA complex and cleave phosphodiester bonds, as the AS sequence would constantly be replenished. To accomplish this, Zhao *et al.* [54] designed a retroviral vector containing two copies of an anti-b3a2 BCR/ABL antisense sequence, termed LasBD (Fig. 1). The selectable marker in the LasBD vector is a tyrosine-22 mutant dihydrofolate reductase (DHFR) gene (*D*) [69,70] transcriptionally regulated by an internal β-actin promoter (*B*). Hematopoietic cell lines containing the b3a2 *bcr–abl* cDNA were transduced with LasBD and selected in methotrexate (MTX) for 14 days. High levels of antisense RNA could be detected which resulted in a 5- to 10-fold suppression of target *bcr–abl* mRNA and protein levels, normalization of cell growth, and a 3- to 4-log reduction in tumorigenicity *in vivo*. LasBD also decreased *bcr–abl* mRNA levels in Ph[+] CD34[+] cells 10-fold. Thus, the AS sequence expressed from the LasBD vector suppresses *bcr–abl* expression sufficiently to restore normal *in vitro* (Fig. 1) and *in vivo* behavior of *bcr–abl*-cDNA-containing cells [54].

One drawback is that current viral or nonviral gene delivery systems are not sufficiently specific or efficient to target all leukemic stem cells [71–76]. Therefore, the practical application of such a genetic approach is in the setting of an autografting protocol. A number of clinical trials in humans have shown that retroviral transfer of genes in stem cells is inefficient and lower than 1% or even 0.1% [71,77,78]. However, recent preclinical and clinical studies suggest that up to 20% of stem cells from baboons or humans may be retrovirally transduced if cytokines [72–74,79–83] such as thrombopoietin, stem cell factor, and fetal liver kinase-2 are used to induce stem cell proliferation and when the likelihood

of stem cell-retrovirus interactions is enhanced, for instance, by coculture with fibronectin or ultra-centrifugation. Even so, it remains unlikely that a 100% transduction efficiency can be achieved. Thus, transfer of antisense genes in the graft will only suppress oncoprotein expression in the small fraction of CML stem cells and their progeny that have been transduced, and restoration of "normal" function will only occur in this small population of CML stem cells. A second problem is that, even if the retroviral vector transduced in the graft eliminates a fraction of malignant stem cells in the graft that contribute to relapse, patients will still relapse due to disease persisting in the host [34]. We hypothesized, therefore, that cotransduction of an anti-BCR/ABL sequence and a drug-resistance gene in the graft may overcome both problems: Following transplantation with drug-resistance-gene-transduced marrow, patients can be treated with chemotherapy to eliminate leukemia persisting in the host and to eliminate the nontransduced leukemic stem cells in the graft while selecting for the drug-resistant, normal stem cells in the graft. Although the chemotherapy will also select for leukemic stem cells that persist in the graft that had been transduced, the presence of the anti-BCR/ABL sequence would render these cells functionally normal.

B. Drug-Resistance Component

Chronic myelogenous leukemia is commonly treated with anthracyclines [29]. The obvious choice for a drug-resistance gene would have been the multidrug resistance (MDR1) gene, which would allow administration of chemotherapy, considered the standard of treatment for CML [84]. We chose a MTX-resistant DHFR gene over the MDR1 gene to avoid the possibility that the Ph[+] clone would be multidrug resistant and no longer sensitive to commonly used chemotherapeutic agents when the antisense strategy fails. Though MTX is not considered a mainstay in the treatment of CML, we showed that MTX suppresses growth of Ph$^+$ colony forming cell (CFC) and long-term culture–initiating cell (LTC-IC) as well as, or better than, that of normal Ph$^-$ progenitors [85]. Several *in vivo* animal studies have shown that introduction of MTX^R DHFR genes in stem cells allows for the safe administration of MTX *in vivo* while allowing selection of the DHFR-containing progenitor population [69,70].

C. *In Vivo* Efficacy of the Antisense/ Drug-Resistance Gene Therapy Approach

Ideally, testing of the efficacy of the this approach would be tested in an *in vivo* model of CML. Although *in vivo* murine models for the chronic phase [10,11,86–89] have been established, the degree of engraftment is not yet completely reproducible and the models favor engraftment of Ph$^-$ stem cells over that of Ph$^+$ stem cells. Use of these models may thus not yet be applicable to testing of certain anti-CML therapies. We believe, therefore, that testing the efficacy of novel anti-*bcr–abl* strategies in the treatment of chronic-phase CML will need to be done in well-controlled human trials.

VI. CONCLUSION

Ph$^-$ progenitors present in the bone marrow and peripheral blood of patients with chronic-phase CML are thought to be polyclonal, indicating that the residual population of stem and progenitor cells is normal [90]. These studies thus suggest that acquisition of the BCR/ABL gene rearrangement likely occurs in polyclonal, normal cells and that elimination of the oncogene may restore normal cell function. Animal studies have demonstrated that transfer of BCR/ABL cDNA in bone marrow leads to the development of a CML-like disorder [5–11], further indicating that the *bcr–abl* gene rearrangement is not only necessary but also sufficient for the malignant transformation of hematopoietic cells. These observations led us to hypothesize that elimination of the *bcr–abl* gene product will restore normal function of Ph$^+$ cells [54]. However, the possibility remains that cells that have acquired the *bcr–abl* gene rearrangement have additional abnormalities that contribute to the malignant behavior of cells. If this is the case, then elimination of the *bcr–abl* mRNA and protein will not completely eliminate the abnormal adhesive and proliferative behavior of Ph$^+$, *bcr–abl*-mRNA-negative progenitors *in vivo* and possibly *in vitro*. Planned clinical trials using the antisense drug-resistant gene therapy approach will shed light on this possibility. Finally, a similar approach could be contemplated in the treatment of other hematological malignancies associated with novel fusion oncoproteins, such as Ph$^+$ acute lymphocytic leukemia or acute promyelocytic leukemia [4,91].

Acknowledgments

Supported in part by RO1-CA-74887 and RO1-CA-79955; CMV is a scholar of the Leukemia Society of America.

References

1. Fialkow, P. J., Jacobson, R. J., and Papayannopoulou, T. H. (1997). Chronic myelocytic leukemia: clonal origin in a stem cell common to the granulocyte, erythrocyte, platelet and monocyte/macrophage. *Am. J. Med.* **63**, 125–131.
2. Rowley, J. (1990). The Philadelphia chromosome translocation: a paradigm for understanding leukemia. *Cancer* **65**, 2178–2184.
3. Ben-Neriah, Y., Daley, G. Q., Mes-Masson, A. M., Witte, O. N., and Baltimore, D. (1986). The chronic myelogenous leukemia-specific p210 protein is the product of the bcr/abl hybrid gene. *Science* **233**, 212–214.
4. Melo, JV. (1997). BCR-ABL gene variants. *Baillieres Clin. Haematol.* **10**, 203–222.
5. Hariharan, I. K., Adams, J. M., and Cory, S. (1988). bcr-abl oncogene renders myeloid cell line factor independent: potential autocrine mechanism in chronic myeloid leukemia. *Oncogene Res.* **3**, 387–399.
6. Gishizky, M. L., Witte, O. N. (1992). Initiation of deregulated growth of multipotent progenitor cells by bcr-abl in vitro. *Science* **256**, 836–839.
7. Carlesso, N., Griffin, J. D., and Druker, B. J. (1994). Use of a temperature-sensitive mutant to define the biological effects of the p210$^{BCR-ABL}$ tyrosine kinase on proliferation of a factor-dependent murine myeloid cell line. *Oncogene* **9**, 149–156.
8. Daley, G. Q., Van Etten, R. A., and Baltimore, D. (1990). Induction of chronic myelogenous leukemia in mice by the p210$^{bcr/abl}$ gene of the Philadelphia chromosome. *Science* **247**, 824–830.
9. Voncken, J. W., Kaartinen, V., Pattengale, P. K., Germeraad, W. T., Groffen, J., and Heisterkamp, N. (1995). BCR/ABL p210 and p190 cause distinct leukemia in transgenic mice. *Blood* **86**, 4603–4611.
10. Honda, H., Oda, H., Suzuki, T., Takahash, T., Witte, O. N., Ozawa, K., Ishikawa, T., Yazaki, Y., and Hirai, H. (1998). Development of acute lymphoblastic leukemia and myeloproliferative disorder in transgenic mice expressing p210$^{bcr/abl}$: a novel transgenic model for human Ph1-positive leukemias. *Blood* **91**, 2067–2075.
11. Pear, W. S., Miller, J. P., Xu, L., Pui, J. C., Soffer, B., Quackenbush, R. C., Pendergast, A. M., Bronson, R., Aster, J. C., Scott, M. L., and Baltimore, D. (1998). Efficient and rapid induction of a chronic myelogenous leukemia-like myeloproliferative disease in mice receiving p210$^{bcr/abl}$-transduced bone marrow. *Blood* **92**, 3780–3792.
12. Kantarjian, H., Giles, F., O'Brien, S., and Talpaz, M. (1998). Clinical course and therapy of chronic myelogenous leukemia with interferon-alpha and chemotherapy. *Hematol. Oncol. Clin. North Am.* **12**, 31.
13. Clarkson, B. D., Strife, A., Wisniewski, D., Lambek, C., and Carpino, N. (1997). New understanding of the pathogenesis of CML: a prototype of early neoplasia. *Leukemia* **11**, 404–428.
14. Marley, S. B., Lewis, J. L., Scott, M. A., Goldman, J. M., and Gordon, M. Y. (1996). Evaluation of "discordant maturation" in chronic myeloid leukaemia using cultures of primitive progenitor cells and their production of clonogenic progeny (CFU-GM). *Br. J. Haematol.* **95**, 299–305.
15. Bedi, A., Zehnbauer, B. A., Barber, J. P., Sharkis, S. J., and Jones, R. J. (1994). Inhibition of apoptosis by BCR-ABL in chronic myeloid leukemia. *Blood* **83**, 2038–2044.
16. McGahon, A., Bissonnette, R., Schmitt, M., Cotter, K. M., Green, D. R., and Cotter, T. G. (1994). BCR-ABL maintains resistance of chronic myelogenous leukemia cells to apoptotic cell death. *Blood* **83**, 1179–1187.
17. Verfaillie, C. (1997). Stem cells in chronic myelogenous leukemia. *Hematol. Oncol. Clin. North Am.* **11**, 1079–1114.

18. Eaves, A. C., Cashman, J. D., Gaboury, L. A., Kalousek, D. K., and Eaves, C. J. (1986). Unregulated proliferation of primitive chronic myeloid leukemia progenitors in the presence of normal marrow adherent cells. *Proc. Natl. Acad. Sci. USA* **83**, 5306–5310.

19. Bhatia, R., McCarthy, J. B., and Verfaillie, C. M. (1996). Interferon-alpha restores normal beta 1 integrin-mediated inhibition of hematopoietic progenitor proliferation by the marrow microenvironment in chronic myelogenous leukemia. *Blood* **87**, 3883–3891.

20. Ilaria, IRJ, and Van Etten, R. A. (1996). p210 and p190(BCR/ABL) induce the tyrosine phosphorylation and DNA binding activity of multiple specific STAT family members. *J. Biol. Chem.* **271**, 31704–31710.

21. Pendergast, A. M., Gishizky, M. L., Havlik, M. H., and Witte, O. N. (1993). SH1 domain autophosphorylation of p210$^{BCR-ABL}$ is required for transformation but not growth factor independence. *Mol. Cell. Biol.* **13**, 1728–1736.

22. Cortez, D., Reuther, G., and Pendergast, A. M. (1997). The BCR-ABL tyrosine kinase activates mitogenic signaling pathways and stimulates G1-to-S phase transition in hematopoietic cells. *Oncogene* **15**, 2333–2342.

23. Shuai, K., Halpern, J., ten Hoeve, J., Rao, X., and Sawyers, C. L. (1996). Constitutive activation of STAT5 by the BCR-ABL oncogene in chronic myelogenous leukemia. *Oncogene* **13**, 247–254.

24. Skorski, T., Kanakaraj, P., Nieborowska-Skorska, M., Ratajczak, M. Z., Wen, S.-C., Zon, G., Gewirtz, A. M., Perussia, B., and Calabretta, B. (1995). Phosphatidylinositol-3 kinase activity is regulated by BCR/ABL and is required for the growth of Philadelphia chromosome-positive cells. *Blood* **86**, 726–736.

25. Jain, S. K., Susa, M., Keeler, M. L., Carlesso, N., Druker, B., and Varticovski, L. (1996). PI 3-kinase activation in BCR/ABL-transformed hematopoietic cells does not require interaction of p85 SH2 domains with p210$^{BCR/ABL}$. *Blood* **88**, 1542–1550.

26. Van Etten, R. A., Jackson, P. K., Baltimore, D., Sanders, M. C., Matsudaira, P. T., and Janmey, P. A. (1994). The COOH-terminus of the c-abl tyrosine kinase contains distinct F-actin and G-actin binding domains with bundling activity. *J. Cell. Biol.* **124**, 325–340.

27. McWhirter, J. R., and Wang, J. Y. (1991). Activation of tyrosine kinase and microfilament-binding functions of c-abl by bcr sequences in bcr/abl fusion proteins. *Mol. Cell. Biol.* **11**, 1553–1564.

28. Enright, H., and McGlave, P. B. (1996). Biology and treatment of chronic myelogenous leukemia. *Oncology* **11**, 1295–1300.

29. Carella, A. M., Celesti, L., Lerma, E., Dejana, A., and Frassoni, F. (1997). Stem-cell mobilization for autografting in chronic myeloid leukemia. *Blood Rev.* **11**, 154–159.

30. McGlave, P. B., De Fabritiis, P., Deisseroth, A., Goldman, J., Barnett, M., Reiffers, J., Simonsson, B., Carella, A., and Aeppli, D. (1994). Autologous transplant therapy for chronic myelogenous leukemia prolongs survival: results from eight transplant centers. *Lancet* **343**, 1486–1490.

31. Talpaz, M., Kantarjian, H., Liang, J., Calvert, L., Hamer, J., Tibbits, P., Durett, A., Claxton, D., Giralt, S., Khouri, I., and Deisseroth, A. B. (1995). Percentage of Philadelphia chromosome (Ph)-negative and Ph-positive cells found after autologous transplantation for chronic myelogenous leukemia depends on percentage of diploid cells induced by conventional-dose chemotherapy before collection of autologous cells. *Blood* **85**, 3257–3263.

32. Deisseroth, A. B., Zu, Z., Claxton, D., Hanania, E. G., Fu, S., Ellerson, D., Goldberg, L., Thomas, M., Janicek, K., and Anderson, W. F. (1994). Genetic marking shows that Ph$^+$ cells present in autologous transplants of chronic myelogenous leukemia (CML) contribute to relapse after autologous bone marrow in CML. *Blood* **83**, 3068–3076.

33. Verfaillie, C., Bhatia, R., Steinbuch, M., DeFor, T., Hirsch, B., Miller, J., Weisdorf, D., and McGlave, P. (1998). Comparative analysis of autografting in chronic myelogenous leukemia: effects of priming regimen and marrow or blood origin of stem cells. *Blood* **92**, 1820–1831.

34. Pichert, G., Alyea, E. P., Soiffer, R. J., Roy, D. C., and Ritz, J. (1994). Persistence of myeloid progenitor cells expressing BCR-ABL mRNA after allogeneic bone marrow transplantation for chronic myelogenous leukemia. *Blood* **84**, 2109–2114.

35. Skorski, T., Nieborowska-Skorska, M., Nicolaides, N. C., Szczylik, C., Iversen, P., Iozzo, R. V., Zon, G., and Calabretta, B. (1994). Suppression of Philadelphia leukemia cell growth in mice by BCR-ABL antisense oligodeoxynucleotide. *Proc. Natl. Acad. Sci. USA* **91**, 4504–4508.

36. Szczylik, C., Skorski, T., Nicolaides, N. C., Manzella, L., Malaguarnera, L., Venturelli, D., Gewirtz, A. M., and Calabretta, B. (1991). Selective inhibition of leukemia cell proliferation by BCR-ABL antisense oligodeoxynucleotides. *Science* **253**, 562–565.

37. Snyder, D. S., Wu, Y., Wang, J. L., Rossi, J. J., Swiderski, P., Kaplan, B. E., and Forman, S. J. (1993). Ribozyme-mediated inhibition of bcr-abl gene expression in a Philadelphia chromosome-positive cell line. *Blood* **82**, 600–605.

38. Bhatia, R., and Verfaillie, C. (1998). Inhibition of BCR-ABL expression with antisense oligodeoxynucleotides restores beta1 integrin-mediated adhesion and proliferation inhibition in chronic myelogenous leukemia hematopoietic progenitors. *Blood* **91**, 3414–3422.

39. Bhatia, R., Munthe, H., and Verfaillie, C. (1998). Tyrphostin AG957, a tyrosine kinase inhibitor with anti-BCR/ABL tyrosine kinase activity restores beta1 integrin-mediated adhesion and inhibitory signaling in chronic myelogenous leukemia hematopoietic progenitors. *Leukemia* **12**, 1708–1717.

40. Druker, B. J., and Lydon, N. B. (2000). Lessons learned from the development of an abl tyrosine kinase inhibitor for chronic myelogenous leukemia. *J. Clin. Invest.* **105**, 5–10.

41. Ratajczak, M. Z., Hijiya, N., Catani, L., DeRiel, K., Luger, S. M., McGlave, P., and Gewirtz, A. M. (1992). Acute- and chronic-phase chronic myelogenous leukemia colony-forming units are highly sensitive to the growth inhibitory effects of c-myb antisense oligodeoxynucleotides. *Blood* **79**, 1956–1964.

42. Gewirtz, A. M. (1994). Treatment of chronic myelogenous leukemia (CML) with c-myb antisense oligodeoxynucleotides. *Bone Marrow Transplant* **14**, S57–S61.

43. Luger, S. M., Ratajczak, J., Ratajczak, M. Z., Kuczynski, W. I., DiPaola, R. S., Ngo, W., Clevenger, C. V., and Gewirtz, A. M. (1996). A functional analysis of protooncogene Vav's role in adult human hematopoiesis. *Blood* **87**, 1326–1373.

44. Ratajczak, M. Z., Luger, S. M., DeRiel, K., Abrahm, J., Calabretta, B., and Gewirtz, A. M. (1992). Role of the KIT protooncogene in normal and malignant human hematopoiesis. *Proc. Natl. Acad. Sci. USA* **89**, 1710–1715.

45. Gewirtz, A. M. (1997). Antisense oligonucleotide therapeutics for human leukemia. *Crit. Rev. Oncogol.* **8**, 93–109.

46. Gewirtz, A. M., Sokol, D. L., and Ratajczak, M. Z. (1998). Nucleic acid therapeutics: state of the art and future prospects. *Blood* **92**, 712–736.

47. James, H. A., and Gibson, I. (1998). The therapeutic potential of ribozymes. *Blood* **91**, 371–375.

48. Crooke, S. T. (1999). Molecular mechanisms of action of antisense drugs. *Biochim. Biophys. Acta* **1489**, 31–44.

49. Wu, H., Lima, W., and Crooke, S. T. (1999). Properties of cloned and expressed human RNase H1. *J. Biol. Chem.* **274**, 28270–28277.

50. Nellen, W., and Lichtenstein, C. (1993). What makes an mRNA antisense-itive? *Trends Biochem. Sci.* **18**, 419–428.

51. Miller, P. S. (1991). Oligonucleoside methylphosphonates as antisense reagents. *Biotechnology (New York)* **9**, 358–363.

52. Zon, G. (1995). Antisense phosphorothioate oligodeoxynucleotides: introductory concepts and possible molecular mechanisms of toxicity. *Toxicol. Lett.* **82–83**, 419–426.

53. Levin, A. A. (1999). A review of the issues in the pharmacokinetics and toxicology of phosphorothioate antisense oligonucleotides. *Biochim. Biophys. Acta* **1489**, 69–74.

54. Zhao, R., McIvor, R., Griffin, J., and Verfaillie, C. (1997). Gene therapy for chronic myelogenous leukemia (CML): a retroviral vector that renders hematopoietic progenitors methotrexate-resistant and CML progenitors functionally normal and nontumorigenic in vivo. *Blood* **90**, 4687–4698.

55. Tybulewicz, V. L., Crawford, C. E., Jackson, P. K., Bronson, R. T., and Mulligan, R. C. (1991). Neonatal lethality and lymphopenia in mice with a homozygous disruption of the c-abl proto-oncogene. *Cell* **65**, 1153–1163.

56. Schwartzberg, P. L., Stall, A. M., Hardin, J. D., Bowdish, K. S., Humaran, T., Boast, S., Harbison, M. L., Robertson, E. J., and Goff, S. P. (1991). Mice homozygous for the ablm1 mutation show poor viability and depletion of selected B and T cell populations. *Cell* **65**, 1165–1175.

57. Voncken, J. W., van Schaick, H., Kaartinen, V., Deemer, K., Coates, T., Landing, B., Pattengale, P., Dorseuil, O., Bokoch, G. M., Groffen, J., and Heisterkamp, N. (1993). Increased neutrophil respiratory burst in bcr-null mutants. *Cell* **80**, 719–728.

58. Skorski, T., Wlodarski, P., Daheron, L., Salomoni, P., Nieborowska-Skorska, M., Majewski, M., Wasik, M., and Calabretta, B. (1998). BCR/ABL-mediated leukemogenesis requires the activity of the small GTP-binding protein Rac. *Proc. Natl. Acad. Sci. USA* **95**, 11858–11862.

59. Skorski, T., Kanakaraj, P., Ku, D. H., Nieborowska-Skorska, M., Canaani, E., Zon, G., Perussia, B., and Calabretta, B. (1994). Negative regulation of p120GAP GTPase promoting activity by p210$^{bcr/abl}$: implication for RAS-dependent Philadelphia chromosome positive cell growth. *J. Exp. Med.* **179**, 1855–1865.

60. Leopold, L. H., Shore, S. K., and Reddy, E. P. (1996). Multi-unit anti-BCR-ABL ribozyme therapy in chronic myelogenous leukemia. *Leuk. Lymphoma* **22**, 365–373.

61. Lange, W., Daskalakis, M., Finke, J., and Dolken, G. (1994). Comparison of different ribozymes for efficient and specific cleavage of BCR/ABL related mRNAs. *FEBS Lett.* **338**, 175–182.

62. Lange, W., Cantin, E. M., Finke, J., and Dolken, G. (1993). In vitro and in vivo effects of synthetic ribozymes targeted against BCR/ABL mRNA. *Leukemia* **7**, 1786–1793.

63. Wright, L., Wilson, S. B., Milliken, S., Biggs, J., and Kearney, P. (1993). Ribozyme-mediated cleavage of the bcr/abl transcript expressed in chronic myeloid leukemia. *Exp. Hematol.* **21**, 1714–1723.

64. Shore, S. K., Nabissa, P. M., and Reddy, E. P. (1993). Ribozyme-mediated cleavage of the BCR/ABL oncogene transcript: in vitro cleavage of RNA and in vivo loss of p210 protein-kinase activity. *Oncogene* **8**, 3183–3191.

65. Mills, K., Walsh, V., and Gilkes, A. (1996). In vitro ribozyme treatment of 32D cells expressing a BCR-ABL construct prolongs the survival of SCID mice [abstract]. *Blood* **88**, 577.

66. Wu, Y., Yu, L., McMahon, R., Rossi, J., Forman, S. J., and Snyder, D. S. (1999). Inhibition of bcr-abl oncogene expression by novel deoxyribozymes (DNAzymes). *Hum. Gene Ther.* **10**, 2847–2856.

67. Warashina, M., Kuwabara, T., Taira, K., Warashina, M., Kuwabara, T. and Taira, K. (1997). Comparison of activities between hammerhead ribozymes and DNA enzymes targeted to L6 BCR-ABL chimeric (b2a2) mRNA. *Nucleic Acids Symp. Ser.* **37**, 213–319.

68. Warashina, M., Kuwabara, T., Nakamatsu, Y., and Taira, K. (1999). Extremely high and specific activity of DNA enzymes in cells with a Philadelphia chromosome. *Chem. Biol.* **6**, 237–245.

69. McIvor, R. S. (1996). Drug-resistant dihydrofolate reductases: generation, expression and therapeutic application. *Bone Marrow Transplant.* **18**(suppl. 3), 50–44.

70. James, R., May, C., Vagt, M., Studebaker, R., and McIvor R.S. (1997). Transgenic mice expressing the tyr22 variant of murine DHFR: protection of transgenic marrow transplant recipients from lethal doses of methotrexate. *Exp. Hematol.* **25**, 1286–1295.

71. Emmons, R. V., Doren, S., Zujewski, J., Cottler-Fox, M., Carter, C. S., Hines, K., O'Shaughnessy, J. A., Leitman, S. F., Greenblatt, J. J., Cowan, K., and Dunba, C. E. (1997). Retroviral gene transduction of adult peripheral blood or marrow-derived CD34+ cells for six hours without growth factors or on autologous stroma does not improve marking efficiency assessed in vivo. *Blood* **89**, 4040–4046.

72. Kiem, H. P., Andrews, R. G., Morris, J., Peterson, L., Heyward, S., Allen, J. M., Rasko, J. E., Potter, J., and Miller, A. D. (1998). Improved gene transfer into baboon marrow repopulating cells using recombinant human fibronectin fragment CH-296 in combination with interleukin-6 stem cell factor FLT-3 ligand and megakaryocyte growth and development factor. *Blood* **92**, 1878–1886.

73. Hennemann, B., Conneally, E., Pawliuk, R., Leboulch, P., Rose-John, S., Reid, D., Chuo, J., Humphries, R., and Eaves, C. (1999). Optimization of retroviral-mediated gene transfer to human NOD/SCID mouse repopulating cord blood cells through a systematic analysis of protocol variables. *Exp. Hematol.* **27**, 817–825.

74. Rebel, V., Tanaka, M., Lee, J. S., Hartnet, T. S., Pulsipher, M., Nathan, D., Mulligan, R., and Sieff, C. (1999). One-day *ex vivo* culture allows effective gene transfer into human nonobese diabetic/severe combined immune-deficient repopulating cells using high-titer vesicular stomatitis virus G protein pseudotyped retrovirus. *Blood* **93**, 2217–2224.

75. Bodine, D. M., Dunbar, C. E., Girard, L. J., Seidel, N. E., Cline, A., Donahue, R. E., and Orlic, D. (1998). Improved amphotropic retrovirus-mediated gene transfer into hematopoietic stem cells. *Ann. N. Y. Acad. Sci.* **850**, 139–150.

76. May, C., Rivella, S., Callegari, J., Heller, G., Gaensler, K. M., Luzzatto, L., and Sadelain, M. (2000). Therapeutic haemoglobin synthesis in beta-thalassaemic mice expressing lentivirus-encoded human beta-globin. *Nature* **406**, 82–86.

77. Hanania, E., Giles, R., Kavanagh, J., Fu, S., Ellerson, D., Zu, Z., Wang, T., Su, Y., Kudelka, A., Rahman, Z., Holmes, F., Hortobagyi, G., Claxton, D., Bachier, C., Thall, P., Cheng, S., Ester, J., Ostrove, J., Bird, R., Chang, A., Korbling, M., Seong, D., Cote, R., Holzmayer, T., and Deisseroth, A. (1996). Results of MDR-1 vector modification trial indicate that granulocyte/macrophage colony-forming unit cells do not contribute to posttransplant hematopoietic recovery following intensive systemic therapy. *Proc. Natl. Acad. Sci. USA* **93**, 15346–15351.

78. Hesdorffer, C., Ayello, J., Ward, M., Kaubisch, A., Vahdat, L., Balmaceda, C., Garrett, T., Fetell, M., Reiss, R., Bank, A., and Antman, K. (1998). Phase I trial of retroviral-mediated transfer of the human MDR1 gene as marrow chemoprotection in patients undergoing high-dose chemotherapy and autologous stem-cell transplantation. *J. Clin. Oncol.* **16**, 165–172.

79. von Kalle, C., Kiem, H. P., Goehle, S., Darovsky, B., Heimfeld, S., Torok-Storb, B., Storb, R., and Schuening, F. G. (1994). Increased gene transfer into human hematopoietic progenitor cells by extended in vitro exposure to a pseudotyped retroviral vector. *Blood* **84**, 2807–2809.

80. Marandin, A., Dubart, A., Pflumio, F., Cosset, F., Cordette, V., Chapel-Fernandes, S., Coulombel, L., Vainchenker, W., and Louache, F. (1998). Retrovirus-mediated gene transfer into human 34+38low primitive cells capable of reconstituting long-term cultures *in vitro* and nonobese diabetic-severe combined immunodeficiency mice *in vivo*. *Hum. Gene Ther.* **9**, 1497–1511.

81. Conneally, E., Eaves, C. J., and Humphries, R. K. (1998). Efficient retroviral-mediated gene transfer to human cord blood stem cells with in vivo repopulating potential. *Blood* **91**, 3487–3493.

82. Cavazzana-Calvo, M., Hacein-Bey, S., de Saint, Basile, G., Gross, F., Yvon, E., Nusbaum, P., Selz, F., Hue, C., Certain, S., Casanova, J. L., Bousso, P., Deist, F. L., and Fischer, A. (2000). Gene therapy of human severe combined immunodeficiency (SCID)-X1 disease. *Science* **288**, 669–672.

83. Abonour, R., William, D. A., Einhorn, L., Hall, K. M., Chen, J., Coffman, J., Traycoff, C. M., Bank, A., Kato, I., Ward, M., Williams, S. D.,

Hromas, R., Robertson, M. J., Smith, F.O., Woo, D., Mills, B., Srour, E. F., Cornett, A. K. (2000). Efficient retrovirus-mediated transfer of the multidrug resistance 1 gene into autologous human long-term repopulating hematopoietic stem cells. *Nat. Med.* **6,** 652–658.

84. Tsuruo, T., and Tomida, A. (1995). Multidrug resistance. *Anticancer Drugs* **6,** 213–218.

85. Zhao, R., McIvor, R., Griffin, J., and Verfaillie, C. (1997). Elimination of tumorigenicity of BCR/ABL positive cells in vivo by a retroviral vector containing an anti-BCR/ABL antisense sequence and confers methotrexate resistance. *Blood* **90,** 4687–4698.

86. Wang, J. C., Lapidot, T., Cashman, J. D., Doedens, M., Addy, L., Sutherland, D. R., Nayar, R., Laraya, P., Minden, M., Keating, A., Eaves, A. C., Eaves, C. J., and Dick, J. E. (1998). High level engraftment of NOD/SCID mice by primitive normal and leukemic hematopoietic cells from patients with chronic myeloid leukemia in chronic phase. *Blood* **91,** 2406–2414.

87. Dazzi, F., Capelli, D., Hasserjian, R., Cotter, F., Corbo, M., Poletti, A., Chinswangwatanakul, W., Goldman, J. M., and Gordon, M. Y. (1998). The kinetics and extent of engraftment of chronic myelogenous leukemia cells in non-obese diabetic/severe combined immunodeficiency mice

reflect the phase of the donor's disease: an in vivo model of chronic myelogenous leukemia biology. *Blood* **92,** 1390–1396.

88. Lewis, I. D., McDiarmid, L. A., Samels, L. M., To, L. B., and Hughes, T. P. (1998). Establishment of a reproducible model of chronic-phase chronic myeloid leukemia in NOD/SCID mice using blood-derived mononuclear or CD34+ cells. *Blood* **91,** 630–640.

89. Van den Berg, D., Wessman, M., Murray, L., Tong, J., Chen, B., Chen, S., Simonetti, D., King, J., Yamasaki, G., DiGiusto, R., Gearing, D., and Reading, C. (1996). Leukemic burden in subpopulations of CD34+ cells isolated from the mobilized peripheral blood of alpha-interferon-resistant or -intolerant patients with chronic myeloid leukemia. *Blood* **87,** 4348–4357.

90. Delforge, M., Boogaerts, M., McGlave, P., and Verfaillie, C. (1999). BCR/ABL-CD34(+)HLA-DR-progenitor cells in early chronic phase, but not in more advanced phases, of chronic myelogenous leukemia are polyclonal. *Blood* **93,** 284–292.

91. Chen, Z., Tong, J. H., Dong, S., Zhu, J., Wang, Z. Y., and Chen, S. J. (1996). Retinoic acid regulatory pathways, chromosomal translocations, and acute promyelocytic leukemia. *Genes Chromosomes Cancer* **15,** 147–159.

PART IV

MANIPULATION OF DRUG RESISTANCE MECHANISMS BY GENE THERAPY

CHAPTER

21

Transfer of Drug-Resistance Genes into Hematopoietic Progenitors

OMER N. KOÇ

*Division of Hematology/Oncology
Department of Medicine and Ireland
Cancer Center at Case Western
Reserve University and University
Hospitals of Cleveland
Cleveland, Ohio 44106*

STEVEN P. ZIELSKE

*Division of Hematology/Oncology
Department of Medicine and Ireland
Cancer Center at Case Western
Reserve University and University
Hospitals of Cleveland
Cleveland, Ohio 44106*

JUSTIN C. ROTH

*Division of Hematology/Oncology
Department of Medicine and Ireland
Cancer Center at Case Western
Reserve University and University
Hospitals of Cleveland
Cleveland, Ohio 44106*

JANE S. REESE

*Division of Hematology/Oncology
Department of Medicine and Ireland
Cancer Center at Case Western
Reserve University and University
Hospitals of Cleveland
Cleveland, Ohio 44106*

STANTON L. GERSON

*Division of Hematology/Oncology
Department of Medicine and Ireland
Cancer Center at Case Western
Reserve University and University
Hospitals of Cleveland
Cleveland, Ohio 44106*

I. INTRODUCTION

Although hematopoietic progenitors were one of the first cell types targeted for therapeutic gene transfer, it has proven to be a difficult task to transduce a sufficient number of hematopoietic stem and progenitor cells to have a clinical impact. Hematopoietic progenitors can be easily obtained in quantities that allow repopulation when transplanted, and these cells can be *ex vivo* manipulated and transduced at levels that should have therapeutic value. Yet, early clinical trials have demonstrated genetic modification of only a low number of hematopoietic cells. Work to improve stem cell transduction efficiency has proceeded along two fronts: improving transduction conditions, vector backbones, and gene insertions and selectively enriching transduced populations of hematopoietic stem and progenitor cells *in vivo*. Although at the onset drug-resistance gene transfer into stem cells had a goal of providing chemotherapy protection from myelosuppression in patients with cancer (considered futile by some) [1], its importance in *in vivo* selection of transduced cells was highlighted by observations of corrective gene therapy studies. These included conditions such as adenosine deaminase defficiency (ADA), deficiency and γC cytokine receptor deficiency, both of which result in severe combined immunodeficiency. Transduction of normal ADA and γC genes into hematopoietic progenitor and stem cells in patients with these deficiencies was associated with modest but gradually increasing clinically significant numbers of functional lymphocytes circulating in recipients due to the survival advantage provided to these lymphocytes by corrective gene therapy [2,3]. This is in contrast to chronic granulomatosis disease (CGD), where correction of a metabolic pathway critical in neutrophil function did not provide a survival advantage for neutrophils or neutrophil progenitors. Therefore, in CGD it has been very difficult to achieve clinically relevant numbers of corrected neutrophils in the circulation of patients [2]. Beacuse the biologic survival advantage has had such an impact, it is reasonable to postulate that *in vivo* selection of transduced hematopoietic progenitors and/or stem cells using a drug-resistance gene would be an effective alternative to transducing a high percentage of stem cells with repopulating capacity. This chapter will review the rationale, basic science, and clinical approach to drug-resistance gene transfer into hematopoietic progenitors.

II. RATIONALE FOR DRUG-RESISTANCE GENE THERAPY

A. Myelosuppression, the dose-Limiting Toxicity of Antineoplastic Agents

Infection and bleeding due to myelosuppression represent the predominant dose-limiting toxicity of a number of chemotherapeutic agents. The mechanisms responsible for bone marrow toxicity may vary among drugs but in general can be explained by the fact that most chemotherapeutic drugs are identified based on their efficacy against highly proliferative tumors. Human marrow is extremely proliferative, producing 4×10^{11} cells every day while maintaining a pool of stem cells sufficient for a lifetime. Thus, it is not surprising to observe sensitivity of marrow progenitors to cell-cycle-specific chemotherapeutic agents. Sensitivity to agents that are not cell cycle specific depends upon the mechanism of action of the drug and the ability of the cell to detoxify the drug. The latter varies with the degree of maturation within the hematopoietic hierarchy. Commonly used cytotoxic drugs such as cyclophosphamide, paclitaxel, anthracyclines, and epipodophyllotoxins are predominantly

cytotoxic to progenitors in the marrow and cause a relatively short pause in hematopoiesis. In contrast, early hematopoietic progenitors are spared from the cytotoxic effects of these agents because of the relatively higher levels of aldehyde dehydrogenase [3] and P-glycoprotein [4]. In addition, early progenitors are quiescent and less susceptible to cell-cycle-specific agents. They also express low topoisomerase I and II and thus are less sensitive to topoisomerase poisons. On the other hand, nitrosoureas are equally cytotoxic to late and early progenitors, in part due to the formation of DNA interstrand crosslinks. These crosslinks are very poorly repaired, regardless of the time given to repair, resulting in prolonged cytopenias or even aplasia with repeated exposures [7,8]. Therefore, chemotherapeutic agents vary in their ability to exert selective pressure on the bone marrow as a whole or on particular cell compartments such as stem cells versus proliferating progenitors versus more mature progenitors. In addition, this dose-limiting toxicity can be overcome by transplantation of either autologous or allogeneic stem cells in sufficient quantities to repopulate the recipient's hematopoietic organs. Drug-resistance genes also bring about the possibility of achieving mixed chimerism between endogenous and transplanted hematopoietic progenitors without the need for myeloablative treatments but with a constant low-level selective pressure favoring transplanted cells.

B. Bone Marrow as a Target for Drug-Resistance Gene Therapy

The major limitation of effective gene therapy *in vivo* is the inefficiency of gene transfer into a sufficient number of cells to create a phenotypic change at the level of a tissue rather than an individual cell. One approach has been to use bone marrow ablation followed by transplantation of transduced bone marrow stem cells. Unless a population of very early hematopoietic progenitors and/or stem cells can be genetically altered, the population of transduced cells will become extinct over time due to terminal differentiation. Therefore, the emphasis of hematopoietic gene therapy has been on targeting stem cells. Although hematopoietic stem cells with long-term repopulating potential are sought after for gene transfer, certain applications may be successful with transduction of short-term repopulating progenitors. In particular, if the interest is to provide transient (1–3 months) of myeloprotection to get patients through chemotherapy without neutropenia, this may be accomplished by transient expression in a pool of myeloid progenitors.

A variety of drug resistance genes are currently being evaluated in preclinical and clinical studies (Fig. 1) [33,87]. The most common method of drug-resistance gene transfer is the retroviral vector, which allows stable integration in hematopoietic progenitors and can be produced relatively easily (Fig. 2). However, this class of retrovirus requires a breakdown of the nuclear membrane for integration, and

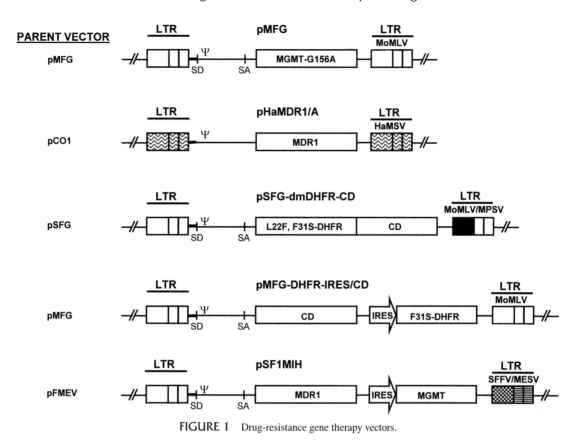

FIGURE 1 Drug-resistance gene therapy vectors.

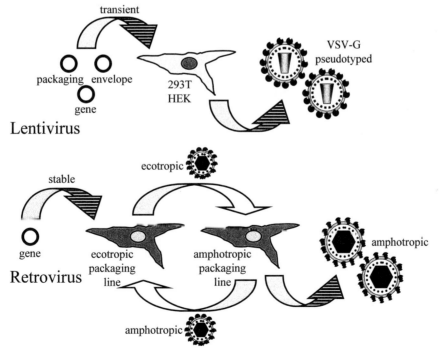

FIGURE 2 Methods of lentiviral and retroviral vector production.

quiescent stem cells are not readily transduced. More recently, HIV-based lentiviral vectors have been shown to stably transduce nondividing cells [88,89], increasing the potential for gene transfer into stem cells.

Despite a number of advances made in the field of drug-resistant gene transfer into hematopoietic progenitors, a number of fundamental issues remain, including:

- Factors that allow efficient gene transfer and expression in hematopoietic stem cells and their differentiated progeny
- Determining the optimal preclinical models for proof of principle
- Clinical trial designs that maintain patient safety while generating important data on gene transfer efficacy

Significant improvements in hematopoietic gene transfer include identification and application of new envelope proteins such as the gibbon ape leukemia virus (GALV) envelope; the use of CH-296 fibronectin fragment or stromal cells or stromal progenitors (mesenchymal stem cells) during transduction; the use of early-acting cytokines such as Flt-3 ligand (FL) and megakaryocyte growth and differentiation factor (MGDF); elimination of inhibitory cytokines from culture conditions; high-titer lentiviral vectors; and an expanded number of drug-resistance genes as well as enhanced drug-resistance activity by identification of mutant drug-resistance genes.

C. MDR1 Leading the Way

Initial studies in the field started with the demonstration that multidrug resistance 1 (MDR1) was an important drug efflux pump that protected early hematopoietic stem cells and progenitors from a variety of drugs that are typically cationic, weakly basic, and hydrophobic but are structurally unrelated. These include the anthracyclines, the vinca alkaloids, the epipodophyllotoxins, actinomycin D, and taxenes. MDR1 cDNA was introduced into murine and human hematopoietic cells by retroviral-mediated gene transfer. A landmark MDR1 gene transfer study demonstrated *in vivo* selection of MDR1-transduced cells in mice treated with paclitaxel [5]. Long-term expression of MDR1 in repopulating stem cells was demonstrated in mice by serially transplanting MDR1-transduced bone marrow cells in six successive cohorts of mice with paclitaxel selection [6]. Each successive transplant produced recipient mice with greater resistance to paclitaxel than the donor mice. MDR1 expression was observed 17 months after the initial transduction, implying that continuous expression can be achieved during *in vivo* selection. Both *in vitro* and *in vivo* selection for transduced MDR1 cells have been demonstrated even though a truncated MDR1 mRNA resulting from cryptic splice sites within the MDR1 cDNA has been detected, possibly resulting in decreased expression of p-Glycoprotein (p-Gp) [7]. MDR1 gene therapy has also led to clinical trials and, as of this writing, four clinical trials have been published.

These trials demonstrated the feasibility of MDR1 gene transfer into repopulating human hematopoietic progenitors and *in vivo* detection of transduced cells over time and form the basis for second-generation trials with more efficient gene transfer techniques with improved vectors. Extensive discusssion of MDR1 gene therapy is provided in Chapter 22. Although not entered in clinical trials yet, dihydrofolate reductase (DHFR) gene-mediated antifolate resistance of hematopoietic progenitors has been extensively studied *in vitro* and in animal models with remarkable success. Full discussion of DHFR-mediated antifolate protection of hematopoietic cells is provided in Chapter 23.

D. Risks

The focus of drug-resistance gene therapy remains on feasibility and efficacy. Development of improved vectors with high transduction efficiency and high expression may also bring about safety issues not recognized thus far. High MDR1 copy numbers and expression were associated with a myeloproliferative disorder in mice that has not been fully explained [8]. In addition, there is a potential risk for extinction of transduced hematopoietic clones selected *in vivo* at the expense of untransduced ones, leading to aplastic anemia and/or myelodysplastic syndrome. Although these may not be relevant concerns for cancer patients with incurable diseases who are treated for palliation, it may limit the use of drug-selection strategies in patients with genetic disorders. There is impetus, therefore, to identify drug-resistance gene and drug combinations that will allow selection without ablating untransduced stem cells to create a stable mixed chimerism state.

III. METHYLTRANSFERASE-MEDIATED DRUG RESISTANCE

Human alkyltransferase (AGT) is a DNA repair protein encoded by the methylguanine-DNA-methyltransferase (MGMT) gene. This 207-amino-acid protein repairs alkyl lesions that form on the O^6 position of guanine and to a lesser extent on the O^4 position of thymine, after exposure to a variety of alkylating agents. This list includes the monofunctional alkylating agents temozolomide, dacarbazine, streptozotocin, and procarbazine and the bifunctional nitrosoureas 1,3-*bis*(2-chloroethyl)-1-nitrosourea (BCNU), 2-chloroethyl-3-sarcosinamide-1-nitrosourea (SarCNU), 1-(4-amino-2-methyl-5-pyrimidinyl)methyl-3-(2-chloro)-3-nitrosourea (ACNU), and 3-cyclohexyl-1-chloroethyl-nitrosourea (CCNU). Alkylation of the O^6 position of guanine is the principal lesion causing cell death [9]. There are other known mechanisms of resistance against agents forming O^6 adducts (specifically, glutathione-*S*-transferase [10] and polyamines [11]), but direct repair of the DNA adducts by AGT is the principal repair mechanism associated with this lesion [15–18]. Expression of AGT and protection from lethal DNA damage induced by alkylating agents form the basis of drug-resistance and gene therapy strategies, which are outlined further in this section.

A. Mechanism of MGMT-Mediated Drug Resistance

Alkyltransferase has a repair mechanism that is unique among DNA repair enzymes. Repair occurs through a transfer of the alkyl group at the O^6 position of guanine to the cysteine residue within the active site of the protein. Covalent attachment of the alkyl group is not reversible, thus inactivating the protein. Accordingly, each AGT protein can repair just a single alkyl lesion, and additional protein must be synthesized if further repair is needed. For this reason AGT is not a "true" enzyme but is considered a "suicide" enzyme.

O^6-chloroethylguanine lesions undergo rapid intramolecular rearrangement to the more stable O^6–N^1–ethanoguanine. AGT can repair this adduct by forming a covalent protein–DNA crosslink using the cysteine residue in the active site [9,12]. If the cell does not repair the O^6–N^1–ethanoguanine adduct, a highly toxic interstrand DNA crosslink will form with the complementary cytosine nucleotide residue [13]. Cytotoxicity by methylating agents is through a process called abortive mismatch repair [14]. The O^6-methylguanine:cytosine or O^6-methylguanine:thymine base mispair formed after one round of replication is recognized by the mismatch repair complex and results in induction of aberrant repair processes. Abortive mismatch repair leads to multiple DNA strand breaks and cell death.

Alkyltransferase expression varies considerable in mammalian tissues. In humans, the highest AGT activity is found in the liver, the lowest in hematopoietic CD34$^+$ cells [15]. This explains the observation of myelosuppression after nitrosourea treatment. The low level of AGT activity in human hematopoietic progenitors compared with high-level expression of other drug-resistance genes, notably MDR1 and ALDH1, suggests that targeting these cells for MGMT gene transfer may result in significant protection from myelosuppression after chemotherapy compared to that observed with other drug-resistance genes.

B. MGMT Gene Transfer

To limit myelosuppression observed with alkylating agent treatment, gene therapy strategies have been devised that are based on transfer of MGMT to hematopoietic stem cells. Transfer of wild-type MGMT into primary murine hematopoietic progenitors has been shown to confer nitrosourea resistance *in vitro* [16–18] and after transplantation into lethally irradiated mice [16]. In addition, after transplantation of MGMT-transduced bone marrow, repeated treatment with BCNU increased the proportion of bone marrow

cells carrying the MGMT provirus [19]. During these initial studies, it was found that MGMT transduction increases BCNU resistance of hematopoietic progenitors twofold compared to untransduced controls. This is probably due to the fact that MGMT expression was only slightly increased in hematopoietic progenitors after gene transfer [15]. With low levels of relative resistance, protection from BCNU is low, and selection pressure for transduced cells is modest.

C. MGMT Mutants

Alkyltransferase can be inactivated by the pseudosubstrate, O^6-benzylguanine (BG). *In vitro* and xenograft studies have shown that the antitumor effect of BCNU, temozolomide, and other alkylating agents can be improved by pretreatment with BG, which depletes AGT in the target tumor cell [20–24]. BG irreversibly forms an S-benzylcysteine moiety at the cysteine acceptor site [25], resulting in inactivation of AGT. This observation led to clinical investigation of BG in combination with BCNU as a way to sensitize tumors to BCNU treatment by depleting endogenous AGT [26]. However, preclinical and phase I clinical trials with BG and BCNU showed increased toxicity to hematopoietic progenitors [15] and myelosuppression [27]. While these studies were ongoing, several AGT mutants were discovered and found to be resistant to inhibition by this new combination. These discoveries caused investigators to shift toward using BG combinations with mutant MGMT drug-resistance gene transfer for protection of hematopoietic stem cells. The most important aspect of these new mutants was their ability to resist inactivation by BG while still maintaining DNA repair activity. While wtAGT is very sensitive to inactivation by BG, some mutants of AGT can withstand up to the maximal plasma concentration of BG attainable and still retain DNA repair activity. This difference means: (1) even low levels of mutant MGMT expression could protect cells against BG plus BCNU, compared to the high expression of wtMGMT required to protect against BCNU; and (2) cancer drug resistance would be overcome at the same time that hematopoietic tissue is being selectively protected.

Several mutants of AGT resistant to inactivation by BG have been described. Most recently studied have been P140A, P140K, and G156A, each with a single amino acid substitution at amino acid position 140 or 156, and P140A/G156A and PVP(138–140)MLK, with two and three amino acid substitutions, respectively. These mutants are thought to resist inactivation by BG due to steric hindrance or disruption of electrostatic interactions which blocks BG entry into the active site. The amino acid changes occur around the active site cysteine at position 145. The G156A mutant, first described by Pegg and co-workers in 1994, is 240-fold more resistant to BG inactivation than the wtAGT [28]. However, the P140K and PVP(138–140)MLK mutants, which were identified through a randomization method [29], exhibit very high-level resistance to BG [30]. Direct comparison of all three of these mutants *in vitro* shows the G156A mutant to be less stable and to have 44% of the activity of wtAGT, while the P140K and PVP(138–140)MLK mutants are more stable and exhibit at least 87% of the activity of wtAGT [30]. In spite of these observations, all three mutants caused similar survival rates of transduced K562 cells treated with BG/BCNU or BG/temozolomide, indicating that all three may have similar efficacy *in vivo*.

New approaches of "directed evolution" have yielded additional mutants that are resistant to BG while maintaining high levels of DNA repair activity [31,32]. These mutants contain multiple amino acid mutations (average 4.7), which could not have been rationally designed. Resistance to BG inactivation was substantial while significant repair activity was maintained, and one mutant displayed greater resistance than any mutant described thus far.

D. Myeloprotection and Tumor Sensitization

Gene transfer of mutant forms of MGMT into hematopoietic progenitors leads to protection from alkylating-agent-induced damage in the presence of BG. This was first demonstrated by Reese *et al.* [33], who showed that retroviral transduction of G156A MGMT into CHO, K562, and primary human CD34+ cells conferred greater *in vitro* resistance to BG/BCNU than transduction of wtMGMT. This work clearly showed the potential advantages of incorporating mutant MGMT into gene transfer studies utilizing BG/nitrosoureas. Subsequently, work on MGMT gene transfer has focused on characterization of various mutants, associated protection from drug-induced damage, and enrichment of hematopoietic progenitors *in vivo*.

The mutants P140A and P140A/G156A have been studied by Hickson *et al.* [34], who showed that transfer of either mutant can protect K562 cells from the toxic effects of temozolomide, mitozolomide, and chlorozotocin. However, when BG was included in the treatments, only the P140A/G156A mutant was able to protect. When the double mutant was transferred into human CD34+ cells, the result was enhanced protection from BG/temozolomide toxicity, with 35% clonogenic survival at a dose that gave only 1% survival of untransduced cells. Somewhat different *in vitro* results were obtained by Maze *et al.* [35], in that they determined the P140A mutant to be protective, while the P140A/G156A mutant was not found to have significant activity at all. One notable difference between this study and the Hickson study is that BCNU was used instead of temozolomide. It is interesting that different MGMT mutants may possess a preference for repair of one type of DNA adduct over another.

Work in a murine model shows that animals are protected from death after BG/BCNU treatment when infused with G156A MGMT-transduced bone marrow [36]. Bone-marrow colony forming units (CFUs) obtained after animal treatment demonstrated marked resistance to BG/BCNU. Protection from death has been found using P140K MGMT, but only modest difference was seen with P140A [37]. Myelosuppression was also prevented after BG/temozolomide treatment after infusion of P140K MGMT-transduced bone marrow [38]. Data with P140A/G156A MGMT and temozolomide showed protection of multiple hematopoietic lineages (colony forming unit–spleen [CFU-S] and granulocyte–macrophage colony forming cell [GM-CFC]) after transplantation of transduced bone marrow [39]. Furthermore, when mice were reconstituted with P140A/G156A-expressing bone marrow, micronucleus formation indicative of clastogenicity was greatly reduced relative to controls when treated with BG and fotemustine or streptozotocin [39]. Thus, although use of BG in combination with alkylating agents may increase the risk of therapy-related malignancy, mutant AGT may provide protection in this regard as well.

Protection of hematopoietic cells while simultaneously sensitizing tumor cells has been demonstrated *in vitro* and in a xenograft model [40,41]. The colon cancer cell line, SW480, expresses high levels of AGT and is resistant to both BCNU and temozolomide (TMZ) [42]. G156A MGMT-transduced human CD34+ cells and SW480 cells were compared in a clonogenic survival assay after treatment with BG/temozolomide or BG/BCNU. Strikingly, transduced CD34+ cells exhibited dramatic resistance to killing compared to that of SW480 cells, while untransduced CD34+ cells remained more sensitive than SW480 cells [40]. A murine xenograft study to determine whether G156A MGMT-transduced hematopoietic progenitors could provide marrow tolerance and allow effective treatment of an AGT-expressing tumor has been done using SW480 cells transplanted into *nu/nu* athymic mice [41]. Murine bone marrow was transduced with G1565A MGMT and transplanted into mice. When tumors reached 100 mm^3, treatment with BG/BCNU was initiated. Mice transplanted with transduced cells were able to survive (70%) the five-dose treatment regimen, while none of the lacZ control mice survived. Peripheral blood counts, bone marrow cellularity, and CFU content of mice indicated tolerance of transduced bone marrow to treatment. There was also a significant tumor growth delay observed in mice treated with a dose-intensive regimen (Fig. 3). These results suggest that G156A MGMT-transduced bone marrow may increase the therapeutic index of BG/BCNU to allow aggressive cancer treatment.

To adequately protect against myelotoxicity associated with alkylating agents, early hematopoietic progenitors must be transduced with mutant MGMT. When resistance of human LTC-IC to BG/temozolomide was investigated, it was found that G156A MGMT-transduced long term culture–initiating cell (LTC-IC) showed similar resistance to that of untransduced LTC-IC, with 21% survival at up to 400 μM temozolomide [43]. In contrast, G156A MGMT-transduced

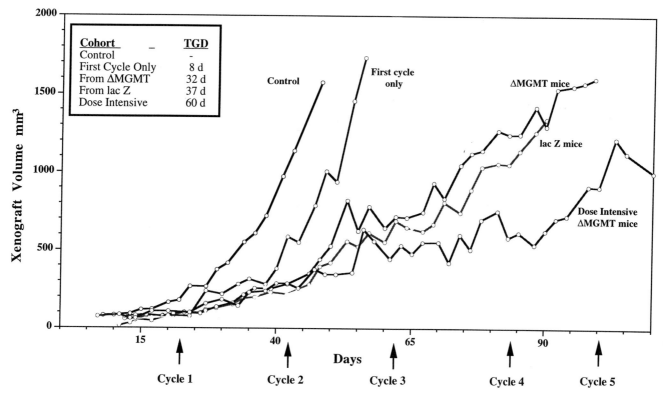

FIGURE 3 Xenograft study demonstrating simultaneous protection of human hematopoietic progenitors and sensitization of tumors cells. SW480 human colon cancer xenograft growth is shown in control (no treatment, $n = 17$ tumors); lacZ or G156A MGMT mice treated with only one cycle of therapy (first cycle only, $n = 7$ tumors); lacZ or G156A MGMT mice treated with multiple cycles of therapy (from lacZ mice $n = 19$ tumors or G156A MGMT mice $n = 18$ tumors); and G156A MGMT mice treated with dose-intensive BG and BCNU regimen (dose intensive, $n = 5$ tumors). Arrows indicate the day of chemotherapy administration. Insert shows days of tumor growth delay (TGD) for each group relative to control. (From Koç, O. N. et al., Hum. Gene Ther. **10**, 1021–1030, 1999. With permission.)

LTC-IC show significant resistance to BG/BCNU [44]. The relative resistance of quiescent cells to temozolomide may be explained by the time required for formation of the cytotoxic lesion [40]. In quiescent cells, regeneration of AGT after BG-induced depletion can result in repair of DNA damage before entry into the cell cycle. But in BCNU-treated quiescent cells, rapid rearrangement of the DNA adduct results in an unrepairable interstrand crosslink, which leads to cell death during eventual replication. These differences may not be significant for protection from myelotoxicity, but they may be important for methods designed to enrich for mutant MGMT-transduced human hematopoietic stem cells (HSCs).

E. *In Vivo* Selection

It has been shown that in a myeloablated setting, that G156A MGMT-transduced murine hematopoietic progenitors could be enriched *in vivo* using BG/BCNU [36]. After a single treatment, the proportion of G156A AGT-expressing bone marrow cells increased from 30–60%. The completion of two treatments caused the proportion of G156A MGMT+ bone-marrow-derived CFUs to increase to 100%. Additional

data by Ragg *et al.* [37] show *in vivo* selection for P140K MGMT-transduced HSCs after BG/BCNU or BCNU treatment. This study also evaluated the P140A MGMT mutant, but found those animals could not survive the dose-intensive treatment. A bicistronic vector carrying green fluorescent protein (GFP) as a convenient marker showed that, after treatment, a large proportion of peripheral blood neutrophils were GFP expressing. Subsequent secondary and tertiary transplants of selected marrow showed that expression was persistent and that selection occurred at the stem cell level. Selection *in vivo* for P140K MGMT-transduced HSCs has also been shown using BG/temozolomide [38]. This study showed high-level enrichment of transduced cells in peripheral blood after two treatment courses. In light of the LTC-IC data concerning temozolomide resistance [40], it will be interesting to find out if human HSCs can be selected *in vivo* using BG/temozolomide. These studies with myeloablation were extended to show enrichment in a nonmyeloablated setting by Davis *et al.* [45]. They showed that G156A MGMT-transduced hematopoietic progenitors could be enriched to nearly 100% of the bone marrow in nonmyeloablated mice after BG/BCNU treatment. Mice receiving transduced bone marrow were protected from myelosuppression, and the proportion of bone-marrow CFUs that were G156A MGMT+ after three cycles of treatment was 97%, which represents 350-fold enrichment. These studies are promising and suggest the opportunity to treat genetic disorders of hematopoietic origin by linking expression of mutant MGMT to a second, therapeutic gene.

IV. CYTIDINE DEAMINASE

Cytidine deaminase (CD) belongs to a group of enzymes involved in pyrimidine metabolism and the salvage of pyrimidine compounds. CD is a homotetramer of 16.2-kDa subunits, each of which catalyzes the hydrolytic deamination of cytidine or deoxycytidine to uridine or deoxyuridine, respectively. CD is also able to inactivate cytosine nucleotide analogs, such as cytosine arabinoside (Ara-C), 5-aza-2'-deoxycytidine (5-AZA-CdR), and 2',2'-difluorodeoxycytidine (dFdC), used as antineoplastic agents. The pronounced hematotoxicity of these drugs limits their use in nonhematologic malignancies.

The low basal levels of CD in hematopoietic cell lines [46] and the elevated CD expression in the leukemic blasts from patients relapsing after Ara-C treatment [47] prompted Momparler *et al.* [48] to investigate its utility as a myeloprotective agent. In these studies, CFU survival of murine hematopoeitic progenitors transduced with human CD was 100%, compared to 2.5% for lacZ-transduced progenitors at an Ara-C concentration typical of the level achieved in the plasma of patients receiving this drug. Eliopoulos *et al.* demonstrated that gene transfer of CD into murine fibroblast protected the cells from Ara-C, 5-AZA-CdR, and dFdC by a colony survival assay. However, addition of 3,4,5,6-tetrahydrouridine, a CD inhibitor, restored sensitivity to these drugs [49]. In subsequent experiments, they investigated CD-mediated protection of murine hematopoietic progenitors from ara-C treatment *in vivo* [50]. Lethally irradiated recipients of $1-2 \times 10^6$ transduced bone marrow cells demonstrated CD activity levels between 5- and 30-fold higher than mice receiving lacZ-transduced marrow, whereas CFU survival of hematopoietic progenitors in the marrow and spleen after 0.5 μM Ara-C treatment was as high as 29 and 23%, respectively, versus <1% in lacZ-transduced progenitors 11–13 months after transplant.

V. GLUTATHIONE-S-TRANSFERASE

The glutathione-S-transferases (GSTs) represent a group of multifunctional enzymes responsible for protecting cells from a broad spectrum of exogenous cytotoxins and oxidative byproducts. The correlation between specific GST expression patterns in human tumors and their role in acquired drug resistance has sparked an interest in understanding the role of these proteins in human malignancy.

The GST superfamily includes at least two membrane-bound and over 20 soluble isoforms. The membrane-bound forms of GST (microsomal GST and leukotrine C4 synthase) are genetically distinct, whereas the soluble isozymes have modest sequence similarities that have been used to subgroup them into GST-alpha, mu, pi, sigma, and theta classes [51]. All of the cytosolic GST proteins catalyze the conjugation of glutathione (GSH, γ-Glu–Cys–Gly) to the electrophilic center of target substrates, forming a thioether linkage that diminishes the reactivity of the foreign compound or improves its excretion from the body. To become active, isozymes within each class must associate as homodimers or heterodimers. Nevertheless, both subunits contain active sites that catalyze independent reactions [52]. The three-dimensional crystal structures for a number of different GSTs have been resolved, improving our understanding of the mechanisms by which key compounds are inactivated [53–60]. The active site of each monomer contains a binding site specific for GSH (G-site) and an adjacent region (H-site) that binds a broad range of hydrophobic substrates, including industrial toxins, pesticides, and carcinogens [61]. Some GSTs also have glutathione peroxidase activity, catalyzing the reduction of organic peroxides and epoxides formed from oxidative stress. Although GSTs have a broad range of substrate specificity, the level and type of GSTs expressed ultimately determines the extent of protection.

Many of the cytosolic proteins exhibit tissue-restricted expression patterns, but the influence of diet, age, and environmental factors on expression levels further complicates our understanding of how these proteins are regulated. Hayes *et al.* [52] have extensively reviewed the types of compounds inactivated by murine, rat, and human GSTs, as well as the mechanisms by which their expression is regulated. In addition to the heterogeneity of GST expression patterns, pronounced sequence polymorphisms in the alpha, mu, and theta GST alleles also exist in the human population [62]. GST-null and increased GST expression patterns have been correlated with the onset of human malignancy and disease progression (62). While the GST-null phenotype may allow an accumulation of genetic lesions and sensitize tissue to chemotherapy drugs, increased GST expression is linked to the onset of drug-resistance cancers. The GST class alpha members inactivate many alkylating agents and are not expressed in hematopoietic cells [63], which might further explain hematopoietic cell sensitivity to the alkylating agents used in chemotherapy. Letourneau *et al.* [64] showed that transduction of human erythroleukemia cells with the rat class alpha GST-Yc gene resulted in approximately two-, three-, and fivefold increases to melphalan, chlorambucil, and mechlorethamine, respectively. The role of GST-pi in acquired drug resistance has also been established. Moscow *et al.* [65] generated a MCF-7 human breast cancer cell line stably expressing a 15-fold increase in human GST-pi; however, these cells were only moderately resistance to ethacrynic acid (3.1- to 4.4-fold) and benzo(*a*)pyrene (1.3- to 4.1-fold) but were not protected from doxorubicin, melphalan, or (*cis*)-platinum treatment. In a separate study, Kuga *et al.* [66] transduced human CD34$^+$ bone marrow progenitors with a human GST-pi retroviral vector. Expression of GST-pi conferred approximately 2.5- and 3-fold resistance to Adriamycin and 4-hydroperoxycyclophosphamide (4HC), respectively. Recently, Matsunaga *et al.* [67] demonstrated GST-pi-mediated protection of murine bone marrow progenitor cells from treatment with cyclophosphamide. Transduction of progenitors used to reconstitute lethally irradiated recipients was 30%, based on transgene detection in colony forming unit–granulocyte macrophage (CFU-GM), and increased to 50% after three high-dose cyclophosphamide (100 mg/kg) treatments given at 6-week intervals. In addition, they were able to detect the transgene in 20% of CFU-GM obtained from secondary recipients 6 months after transplant.

VI. <u>DUAL-DRUG-RESISTANCE APPROACH</u>

Because multi-agent chemotherapy is proven to be more effective in variety of malignancies it may be advantageous to provide resistance against multiple different chemotherapeutics. Consequently, dual-drug-resistance gene vectors are being developed to protect hematopoietic progenitors from multiple agents. Several retroviral vector designs have been employed to attain dual-gene expression and drug resistance at levels that allow dose escalation and hematopoietic protection.

The first dual-drug-resistance gene transfer experiments utilized the addition of strong exogenous promoters, such as those obtained from the simian virus 40 (SV40) and cytomegalovirus (CMV) viruses. In this design, transcription of the first gene is driven by promoter and enhancer sequences within the long terminal repeat (LTR), while a second promoter and enhancer sequence is added to drive expression of the second gene. The orientation of the internal promoter with respect to the LTR is flexible, but the strong transcriptional activity of the LTR often inhibits transcription of the second gene, known as promoter interference [68]. Doroshow *et al.* [69] used a Harvey murine sarcoma virus vector containing hMDR1 and an internal SV40 promoter to drive expression of hGST-pi. Transduction of murine NIH3T3 cells with this vector followed by incremental increases of colchicine selection led to proportional increases in hGST-pi transcripts, protein, and activity. The transduced cells selected in 1.28μg/mL colchicine had a 3.6-fold increase in GST-pi expression compared to untransduced cells. As expected for MDR1 expression, the IC$_{50}$ of doxorubicin and colchicine was increased in the transduced cells by 100- and 10-fold, respectively. Whereas GST-pi had no effect on the level of doxorubicin or

cisplatinum toxicity, it provided a four- to fivefold increase in resistance to ethacrynic acid and 1, chloro-2,4-dinitrobenzene (CDNB).

Another retroviral vector design involves the use of gene fusions, in which two separate drug-resistance genes are fused and translated as a single peptide. There are several limitations to this technique, the most obvious being the ability to translate the protein such that both of the drug-resistance activities are maintained. Strategies for maintaining activity include the addition of flexible and/or protease-cleavable linkers between the two proteins. Differential localization requirements can also negate the function of fusion proteins; a cytosolic protein may not mediate resistance if it is recruited to the nucleus by its fusion partner. Even if one is able to get functional protection from both of the fused proteins, there is still a chance of immunoreactivity towards the fusion product. Despite these challenges, this technique has been used with some success. Hansen *et al.* [70] were able to produce an MGMT–apurinic endonuclease fusion peptide for expanded resistance to alkylating agents. This fusion protein provided a 10-fold increase in HELA cell resistance to 75 μM BCNU alone or in combination with 0.5 mM methylmethane sulfonate (MMS). Another fusion pair was created between an antifolate-resistant dihydrofolate reductase [71] (Phe22–Ser31 DHFR) and cytidine deaminase (Fig. 1; pSFG-dmDHFR-CD) [72]. Strikingly, the fusion protein maintained wild-type activity for both enzymes. Transduction of murine bone marrow cells with the fusion vector followed by treatment with either 500 nM Ara-C or 20 nM methotrexate (MTX) resulted in a significant increase (58 and 17%, respectively) in the number CFU-GM compared to mock or single gene-transduced controls.

The third and most flexible approach for designing dual-gene vectors is to transcriptionally link two genes to one promoter, using an internal ribosome entry site (IRES) sequence. This sequence forms a secondary structure in the transcript that allows cap-independent initiation of translation. Although it is generally accepted that IRES-initiated translation is 20–50% less efficient than cap-dependent translation, this technique has yielded the most reliable results. The encephalomyocarditis and polio virus IRES elements are the most commonly used. However, endogenous mRNA sequences have recently been identified that, when linked as multimers, have enhanced expression and smaller sequence requirements than the viral IRES elements [73]. The transcriptional link between two genes indicates that selection for one gene will indirectly select for the other, unlike the alternative-spliced and internal-promoter vectors.

Galipeau *et al.* [74] constructed a bicistronic Harvey murine sarcoma virus (HaMSV) vector containing a modified hMDR1 and an IRES element driving expression of the L22Y mutant DHFR. Human lymphoblastic cell lines transduced with this vector were resistant to individual drug treatments consisting of Taxol (13-fold), trimetrexate (8.9-fold), vinblastine (5.6-fold), methotrexate (2.5-fold), and etoposide (1.5-fold). In addition, murine myeloid progenitor cells transduced with this vector were resistant to Taxol and trimetrexate (2.9- and 140-fold, respectively). Bicistronic HaMSV vectors containing hMDR1 and MGMT in both orientations with respect to the IRES element (HaMDR–IRES–MGMT and HaMGMT–IRES–MDR1) have also been constructed [75]. Murine bone marrow progenitors transduced with the two vectors demonstrated increased resistance to both vincristine and ACNU after a 2-day preselection in 25 ng/mL vincristine. Notably, enhanced resistance to vincristine and ACNU correlated with MDR and MGMT, respectively. In a similar study, Jelinek *et al.* [76] designed a myeloproliferative sarcoma virus containing a hMDR1–IRES–MGMT insert (Fig. 1; pSF1MIH). K562 cells transduced with this vector and preselected in 20 ng/mL colchicine were significantly resistant to colchicine (26-fold), doxorubicin (28-fold), temozolomide (13-fold), and N-methyl-nitrosourea (MNU) (60-fold) compared to untransduced cells. Moreover, the drug sensitivity of transduced cells was reestablished by the addition of MDR1 and MGMT inhibitors (verapamil and BG, respectively). Beausejour *et al.* [77] have designed a bicistronic vector for the combined resistance to antifolates and cytosine nucleotide analogs. This construct involves a hDHFR(F31S)–IRES–CD cassette inserted into the MFG retrovirus vector (Fig. 1; pMFG-DHFR-IRES/CD). Colony forming unit–cell (CFU-C) survival assays of murine bone marrow progenitor cells transduced with the bicistronic vector demonstrated a 10- to 20-fold increase to the LD_{50} of Ara-C and a 50-fold increase in the LD_{50} to methotrezate. Interestingly, the level of protection mediated by both genes was equivalent to the level obtained when transducing with either gene alone. Additional bicistronic vectors have been designed that contain express rat GST-A3-IRES-CD transcripts [78]. Murine NIH3T3 cells transduced with this vector showed evidence of resistance to Ara-C, chlorambucil, and melphalan, indicating combined resistance to cytosine analogs and nitrogen mustards.

VII. CLINICAL TRIALS

The clinical success of retroviral gene transfer has been limited by low gene transfer rates into hematopoietic stem cells, which results in low frequency and transient detection of transduced cells following reinfusion into the patient. In the last several years, however, efforts have led to improved results using techniques such as colocalization of stem cells and viral particles using the fibronectin fragment CH-296 or autologous stroma, the use of newly discovered early-acting cytokines, and modifications to protocols to enhance gene transfer into long-term repopulating cells.

A number of clinical drug-resistance gene transfer protocols have examined whether the MDR1 gene could be used to protect hematopoietic cells from the toxic effects

of chemotherapy. These trials were based on murine studies that demonstrated that MDR1-transduced hematopoietic cells had preferential survival *in vivo* after treatment with MDR drugs [5,79–81]. This led to the concept that the MDR1 gene could act as an *in vivo* selectable marker and could be used to selectively expand transduced cells. In the first clinical trials, patients received a conditioning regimen of myeloablative therapy and subsequent transplantation with autologous cytokine-stimulated MDR1-transduced CD34+ cells supplemented with at least two thirds of the initial collection of unmanipulated cells in order to prevent engraftment failure. After engraftment, patients were treated with one of the P-glycoprotein substrate drugs such as paclitaxel or doxorubicin. These trials resulted in low numbers of MDR1-marked cells (0.01–1%) in the peripheral blood [82–84] and bone marrow [83] and no enrichment for marked cells with repeated cycles of therapy. The limitation of these studies could be attributed to the following: First, there was a low level of gene transduction into long-term repopulating cells (LTRCs), possibly a result of the cytokine exposure, which may have stimulated primitive stem cells toward differentiation. Second, animal studies have shown that culturing cells under *ex vivo* conditions results in impaired engraftment [85], thus the coinfusion of unmanipulated cells may have had a selective advantage in competition for engraftment with transduced cells.

In an attempt to overcome these issues, a trial was designed in which patients received only cells incubated with MDR1 retrovirus and a less-intensive conditioning regimen to obviate the need for unmanipulated CD34+ cells [86]. This transduction protocol used autologous stroma but no cytokines to optimize gene transfer into LTRCs. After engraftment, MDR1 was not detected initially in some patients but was found in peripheral blood granulocytes and monocytes ($\geq$1%) after repeated cycles of paclitaxel, doxorubicin, or vinblastine, suggesting marking of stem cells with long-term engrafting potential [86].

Most recently, Abonour and colleagues [87] demonstrated retroviral gene marking of LTRCs facilitated by the use of a purified fragment of fibronectin, CH-296, to increase the interaction between viral envelope protein and target cell receptors. Here, transplantation of cytokine-stimulated, transduced, peripheral-blood-mobilized CD34+ cells resulted in enrichment of MDR1 cells in the peripheral blood after etoposide and persistence of up to 15% of marked cells in the bone marrow and up to 2.6% in the peripheral blood 1 year after treatment. Of note, only 1 of 12 patients had evidence of weakly positive ELISA for antibody to CH-296. Overall, these trials have proven the safety and feasibility of this approch with no untoward toxicities of gene transduction, and no trials to date have detected replication-competent virus in patient samples at after more than 1 year. Interestingly, one murine study reported a myeloproliferative syndrome in mice recipients of bone marrow transduced with the MDR gene [8]. This has raised safety concerns over the clinical use of

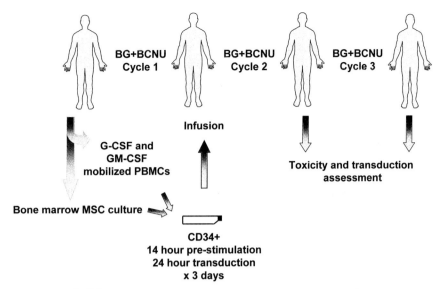

FIGURE 4   Schema of clinical trial using G156A-transduced CD34$^+$ cells.

MDR because it suggests hematopoietic cells expressing high levels may have a proliferative advantage *in vivo*. The murine study differed from human studies in that cells were *ex vivo* expanded for 12 days after transduction; therefore, the phenomenon observed may have been a result of the culture conditions. In addition, transduction efficiency into murine bone marrow was high, resulting in multiple integrations per cell and high MDR1 expression. In general, such transduction efficiency and expression levels have not been achieved for humans, and it will be important to determine if human progenitors also acquire proliferative advantage with high-level MDR1 expression. Alternatively, the myeloproliferative disorder in mice may simply be due to the specific retroviral vector or producer cells, although there is no evidence to support this. Ongoing studies in humans or primates have not duplicated this effect; however, the mechanism leading to MDR1-mediated stem cell expansion must be elucidated.

Clinical trials using hematopoietic cells transduced with MGMT are also underway. In contrast to MDR substrate chemotherapy, nitrosoureas are more toxic to early hematopoietic progenitors and therefore may facilitate strong selection for MGMT-transduced cells. Two trials have been proposed using this approach. The first is a study in pediatric patients with central nervous system (CNS) tumors in which wild-type MGMT-transduced CD34$^+$ cells will be reinfused after each cycle of nitrosourea therapy. Nitrosourea therapy will be then administered in a dose-intensive regimen every 4 weeks (D. Williams, Indiana University), and marrow and peripheral blood will be monitored for the presence of gene transfer. We have proposed a trial in adult patients with solid tumors using G156A (mutant) MGMT-transduced CD34$^+$ cells (Fig. 4). The hypothesis of our study is based on murine studies in which transplantation of as few as 5×10^4 transduced marrow cells into nonmyeloablated mice resulted in

50% MGMT$^+$ bone marrow progenitors after BCNU [45]. Therefore, it is expected that infusion of even a relatively small number of MGMT-transduced progenitors into patients will allow emergence of drug-resistant marrow after repeated cycles of therapy, resulting in amelioration of cumulative myelosuppression previously observed in phase I trials with BG and BCNU [26]. The endpoints of this study include detection of transduced cells after each cycle of chemotherapy, detection of drug-resistant marrow CFUs and therapeutic response to the drug combination.

VIII. CONCLUSION

Significant advances have been made over the last 5 years to improve gene transfer into human hematopoietic progenitor cells. Furthermore, preclinical data using a variety of drug-resistance genes and their mutants have shown remarkable biological effects. It is expected that sustained engraftment with drug-resistance-gene-transduced hematopoietic cells in a substantial mixed chimerism (>10% transduced) will be achieved in the next generation of clinical trials.

Acknowledgments

Supported by Public Health Service Grants RO1CA73062 and P30CA43703.

References

1. Brenner, M. (1999). Resistance is futile. *Gene Ther.* **6,** 1646–1647.
2. Malech, H. L., Maples, P. B., Whiting-Theobald, N., Linton, G. F., Sekhsaria, S., Vowells, S. J., *et al.* (1997). Prolonged production of NADPH oxidase-corrected granulocytes after gene therapy of chronic granulomatous disease. *Proc. Natl. Acad. Sci. USA* **94**(22), 12133–12138.

3. Kastan, M. B., Schlaffer, E., Russo, J., Colvin, O., Civin, C., and Hilton, J. (1990). Direct demonstration of elevated aldehyde dehydrogenase in human hematopoietic progenitor cells. *Blood* **75**, 1947–1950.

4. Chaudhary, P., and Roninson, I. (1991). Expression and activity of P-glycoprotein, a multidrug efflux pump, in human hematopoietic stem cells. *Cell* **66**, 85–94.

5. Sorrentino, B., Brandt, S., Bodine, D., Gottesman, M., Pastan, I., Cline, A. *et al.* (1992). Selection of drug-resistant bone marrow cells in vivo after retroviral transfer of human MDR1. *Science* **257**, 99–103.

6. Hanania, E. G., and Deisseroth, A. B. (1994). Serial transplantation shows that early hematopoietic precursor cells are transduced by MDR-1 retroviral vector in a mouse gene therapy model. *Cancer Gene Ther.* **1**(1), 21–25.

7. Sorrentino, B. P., McDonagh, K. T., Woods, D., and Orlic, D. (1995). Expression of retroviral vectors containing the human multidrug resistance 1 cDNA in hematopoietic cells of transplanted mice. *Blood* **86**(2), 491–501.

8. Bunting, K. D., Galipeau, J., Topham, D., Benaim, E., and Sorrentino, B. P. (1998). Transduction of murine bone marrow cells with an MDR1 vector enables ex vivo stem cell expansion, but these expanded grafts cause a myeloproliferative syndrome in transplanted mice. *Blood* **92**(7), 2269–2279.

9. Gonzaga, P. E., and Brent, T. P. (1989). Affinity purification and characterization of human O^6-alkylguanine-DNA alkyltransferase complexed with BCNU-treated, synthetic oligonucleotide. *Nucleic Acids Res.* **17**(16), 6581–6590.

10. Hansson, J., Edgren, M., Ehrsson, H., Ringborg, U., and Nilsson, B. (1988). Effect of D,L-buthionine-S,R-sulfoximine on cytotoxicity and DNA cross-linking induced by bifunctional DNA-reactive cytostatic drugs in human melanoma cells. *Cancer Res.* **48**(1), 19–26.

11. Seidenfeld, J., and Komar, K. A. Chemosensitization of cultured human carcinoma cells to 1,3-*bis*(2-chloroethyl)-1-nitrosourea by difluoromethylornithine-induced polyamine depletion. *Cancer Res.* **45**(5), 2132–2138.

12. Brent, T. P., Remack, J. S., and Smith, D. G. (1987). Characterization of a novel reaction by human O^6-alkylguanine-DNA alkyltransferase with 1,3-*bis*(2-chloroethyl)-1-nitrosourea-treated DNA. *Cancer Res.* **47**(23), 6185–6188.

13. Tong, W. P., Kirk, M. C., and Ludlum, D. B. (1982). Formation of the cross-link 1-[N3-deoxycytidyl), 2-[N1-deoxyguanosinyl]ethane in DNA treated with N,N′-*bis*(2-chloroethyl)-N-nitrosourea. *Cancer Res.* **42**(8), 3102–3105.

14. Karran, P., Macpherson, P., Ceccotti, S., Dogliotti, E., Griffin, S., and Bignami, M., O^6-methylguanine residues elicit DNA repair synthesis by human cell extracts. *J. Biol. Chem.* **268**(21), 15878–15886.

15. Gerson, S. L., Phillips, W., Kastan, M., Dumenco, L. L., and Donovan, C. (1996). Human CD34+ hematopoietic progenitors have low, cytokine-unresponsive O^6-alkylguanine-DNA alkyltransferase and are sensitive to O^6- benzylguanine plus BCNU. *Blood* **88**(5), 1649–1655.

16. Allay, J. A., Dumenco, L. L., Koc, O. N., Liu, L., and Gerson, S. L. (1995). Retroviral transduction and expression of the human alkyltransferase cDNA provides nitrosourea resistance to hematopoietic cells. *Blood* **85**(11), 3342–3351.

17. Moritz, T., Mackay, W., Glassner, B. J., Williams, D. A., and Samson, L. (1995). Retrovirus-mediated expression of a DNA repair protein in bone marrow protects hematopoietic cells from nitrosourea-induced toxicity in vitro and in vivo. *Cancer Res.* **55**(12), 2608–2614.

18. Maze, R., Carney, J. P., Kelley, M. R., Glassner, B. J., Williams, D. A., and Samson, L. (1996). Increasing DNA repair methyltransferase levels via bone marrow stem cell transduction rescues mice from the toxic effects of 1,3-*bis*(2-chloroethyl)-1-nitrosourea, a chemotherapeutic alkylating agent. *Proc. Natl. Acad. Sci. USA* **93**(1), 206–210.

19. Allay, J. A., Davis, B. M., and Gerson, S. L. (1997). Human alkyltransferase-transduced murine myeloid progenitors are enriched in vivo by BCNU treatment of transplanted mice. *Exp. Hematol.* **25**(10), 1069–1076.

20. Gerson, S. L., and Willson, J. K., O^6-alkylguanine-DNA alkyltransferase. A target for the modulation of drug resistance. *Hematol. Oncol. Clin. North Am.* **9**(2), 431–450.

21. Gerson, S. L., Berger, N. A., Arce, C., Petzold, S. J., and Willson, J. K. (1992). Modulation of nitrosourea resistance in human colon cancer by O^6-methylguanine. *Biochem. Pharmacol.* **43**(5), 1101–1107.

22. Dolan, M. E., Mitchell, R. B., Mummert, C., Moschel, R. C., and Pegg, A. E. (1991). Effect of O^6-benzylguanine analogues on sensitivity of human tumor cells to the cytotoxic effects of alkylating agents. *Cancer Res.* **51**(13), 3367–3372.

23. Gerson, S. L., Zborowska, E., Norton, K., Gordon, N. H., and Willson, J. K. (1993). Synergistic efficacy of O^6-benzylguanine and 1,3-*bis* (2-chloroethyl)-1-nitrosourea (BCNU) in a human colon cancer xenograft completely resistant to BCNU alone. *Biochem. Pharmacol.* **45**(2), 483–491.

24. Dolan, M. E., Pegg, A. E., Moschel, R. C., and Grindey, G. B. (1993). Effect of O^6-benzylguanine on the sensitivity of human colon tumor xenografts to 1,3-*bis*(2-chloroethyl)-1-nitrosourea (BCNU). *Biochem. Pharmacol.* **46**(2), 285–290.

25. Pegg, A. E., Boosalis, M., Samson, L., Moschel, R. C., Byers, T. L., Swenn, K. *et al.* (1993). Mechanism of inactivation of human O^6-alkylguanine-DNA alkyltransferase by O^6-benzylguanine. *Biochemistry* **32**(45), 1998–2006.

26. Spiro, T. P., Gerson, S. L., Liu, L., Majka, S., Haaga, J., Hoppel, C. L. *et al.* (1999). O^6-benzylguanine: a clinical trial establishing the biochemical modulatory dose in tumor tissue for alkyltransferase-directed DNA repair. *Cancer Res.* **59**(10), 2402–2410.

27. Page, J., Giles, H. D., Phillips, W., Gerson, S. L., Smith, A. C., and Tomaszewski, J. E., Preclinical toxicology study of O^6-benzylguanine (NSC-637037) and BCNU (carmustine, NSC-409962) in male and female Beagle dogs. *Proc. Am. Assoc. Cancer Res.* **35**, 328.

28. Crone, T. M., Goodtzova, K., Edara, S., and Pegg, A. E. (1994). Mutations in human O^6-alkylguanine-DNA alkyltransferase imparting resistance to O^6-benzylguanine. *Cancer Res.* **54**(23), 6221–6227.

29. Xu-Welliver, M., Kanugula, S., and Pegg, A. E. (1998). Isolation of human O^6-alkylguanine-DNA alkyltransferase mutants highly resistant to inactivation by O^6-benzylguanine. *Cancer Res.* **58**(9), 1936–1945.

30. Davis, B. M., Roth, J. C., Liu, L., Xu-Welliver, M., Pegg, A. E., and Gerson, S. L. (1999). Characterization of the P140K, PVP(138–140)MLK, and G156A O^6-methylguanine-DNA methyltransferase mutants: implications for drug resistance gene therapy. *Hum Gene Ther.* **10**(17), 2769–2778.

31. Encell, L. P., Coates, M. M., and Loeb, L. A. (1998). Engineering human DNA alkyltransferases for gene therapy using random sequence mutagenesis. *Cancer Res.* **58**(5), 1013–1020.

32. Davis, B. M., Encell, L. P., Zielske, S. P., Christians, F. C., Liu, L., Friebert, S. E. *et al.* (2001). Applied molecular evolution of O^6-benzylguanine-resistant DNA alkyltransferases in human hematopoietic cells. *Proc. Nat. Acad. Sci. USA* **98**(9), 4950–4954.

33. Reese, J. S., Koc, O. N., Lee, K. M., Liu, L., Allay, J. A., Phillips, W. P. *et al.* (1996). Retroviral transduction of a mutant methylguanine DNA methyltransferase gene into human CD34 cells confers resistance to O^6-benzylguanine plus 1,3-*bis*(2-chloroethyl)-1-nitrosourea. *Proc. Natl. Acad. Sci. USA* **93**(24), 14088–14093.

34. Hickson, I., Fairbairn, L. J., Chinnasamy, N., Lashford, L. S., Thatcher, N., and Margison, G. P. *et al.* (1998). Chemoprotective gene transfer. I. Transduction of human haemopoietic progenitors with O^6-benzylguanine-resistant O^6-alkylguanine-DNA alkyltransferase attenuates the toxic effects of O^6-alkylating agents in vitro. *Gene Ther.* **5**(6), 835–841.

35. Maze, R., Kurpad, C., Pegg, A. E., Erickson, L. C., and Williams, D. A. (1999). Retroviral-mediated expression of the P140A, but not P140A/G156A, mutant form of O^6-methylguanine DNA methyltransferase protects hematopoietic cells against O^6-benzylguanine sensitization to chloroethylnitrosourea treatment. *J. Pharmacol. Exp. Ther.* **290**(3), 1467–1474.

36. Davis, B. M., Reese, J. S., Koc, O. N., Lee, K., Schupp, J. E., and Gerson, S. L. (1997). Selection for G156A O^6-methylguanine DNA methyltransferase gene-transduced hematopoietic progenitors and protection from lethality in mice treated with O^6-benzylguanine and 1,3-*bis* (2-chloroethyl)-1-nitrosourea. *Cancer Res.* **57**(22), 5093–5099.

37. Ragg, S., Xu-Welliver, M., Bailey, J., D'Souza, M., Cooper, R., Chandra, S. *et al.* (2001). Direct reversal of DNA damage by mutant methyltransferase protein protects mice against dose-intensified chemotherapy and leads to in vivo selection of hematopoietic stem cells. *Cancer Res.* **60**(18), 5187–5195.

38. Sawai, N., Zhou, S., Vanin, E., Houghton, P., Brent, T. P., and Sorrentino, B. P. (2001). Protection and in vivo selection of hematopoietic stem cells using temozolomide, O^6-benzylguanine, and an alkyltransferase-expressing retroviral vector. *Mol. Ther.* **3**(1), 78–87.

39. Chinnasamy, N., Fairbairn, L. J., Laher, J., Willington, M. A., and Rafferty, J. A., Modulation of O^6-alkylating agent induced clastogenicity by enhanced DNA repair capacity of bone marrow cells. *Mutat. Res.* **416**(1–2), 1–10.

40. Reese, J. S., Davis, B. M., Liu, L., and Gerson, S. L. (1999). Simultaneous protection of G156A methylguanine DNA methyltransferase gene-transduced hematopoietic progenitors and sensitization of tumor cells using O^6-benzylguanine and temozolomide. *Clin. Cancer Res.* **5**(1), 163–169.

41. Koc, O. N., Reese, J. S., Davis, B. M., Liu, L., Majczenko, K. J., and Gerson, S. L. (1999). DeltaMGMT-transduced bone marrow infusion increases tolerance to O^6- benzylguanine and 1,3-*bis*(2-chloroethyl)-1-nitrosourea and allows intensive therapy of 1,3-*bis*(2-chloroethyl)-1-nitrosourea-resistant human colon cancer xenografts. *Hum. Gene Ther.* **10**(6), 1021–1030.

42. Liu, L., Markowitz, S., and Gerson, S. L. (1996). Mismatch repair mutations override alkyltransferase in conferring resistance to temozolomide but not to 1,3-*bis*(2-chloroethyl)nitrosourea. *Cancer Res.* **56**(23), 5375–5379.

43. Platt, G. M., and Price, C. (1997). Isolation of a *Schizosaccharomyces pombe* gene which in high copy confers resistance to the nucleoside analogue 5-azacytidine. *Yeast* **13**(5), 463–474.

44. Koc, O. N., Reese, J. S., Szekely, E. M., and Gerson, S. L. (1999). Human long-term culture initiating cells are sensitive to benzylguanine and 1,3-*bis*(2-chloroethyl)-1-nitrosourea and protected after mutant (G156A) methylguanine methyltransferase gene transfer. *Cancer Gene Ther.* **6**(4), 340–348.

45. Davis, B. M., Koc, O. N., and Gerson, S. L. (2000). Limiting numbers of G156A O(6)-methylguanine-DNA methyltransferase-transduced marrow progenitors repopulate nonmyeloablated mice after drug selection. *Blood* **95**(10), 3078–3084.

46. Nygaard, P., and Sundstrom, C. (1987). Low cytidine deaminase levels in human hematopoietic cell lines. *Leuk. Res.* **11**(8), 681–685.

47. Onetto, N., Momparler, R. L., Momparler, L. F., and Gyger, M. (1987). In vitro biochemical tests to evaluate the response to therapy of acute leukemia with cytosine arabinoside or 5-AZA-2′-deoxycytidine. *Semin. Oncol.* **14**(2, suppl. 1), 231–237.

48. Momparler, R. L., Eliopoulos, N., Bovenzi, V., Letourneau, S., Greenbaum, M., and Cournoyer, D. (1996). Resistance to cytosine arabinoside by retrovirally mediated gene transfer of human cytidine deaminase into murine fibroblast and hematopoietic cells. *Cancer Gene Ther.* **3**(5), 331–338.

49. Eliopoulos, N., Cournoyer, D., and Momparler, R. L. (1998). Drug resistance to 5-AZA-2′-deoxycytidine, 2′,2′-difluorodeoxycytidine, and cy-

tosine arabinoside conferred by retroviral-mediated transfer of human cytidine deaminase cDNA into murine cells. *Cancer Chemother. Pharmacol.* **42**(5), 373–378.

50. Eliopoulos, N., Bovenzi, V., Le, N. L., Momparler, L. F., Greenbaum, M., Letourneau, S. *et al.* (1998). Retroviral transfer and long-term expression of human cytidine deaminase cDNA in hematopoietic cells following transplantation in mice. *Gene Ther.* **5**(11), 1545–1551.

51. Mannervik, B., Alin, P., Guthenberg, C., Jensson, H., Tahir, M. K., Warholm, M. *et al.* (1985). Identification of three classes of cytosolic glutathione transferase common to several mammalian species: correlation between structural data and enzymatic properties. *Proc. Natl. Acad. Sci. USA* **82**(21), 7202–7206.

52. Hayes, J. D., and Pulford, D. J. (1995). The glutathione S-transferase supergene family: regulation of GST and the contribution of the isoenzymes to cancer chemoprotection and drug resistance. *Crit. Rev. Biochem. Mol. Biol.* **30**(6), 445–600.

53. Fu, J. H., Rose, J., Chung, Y. J., Tam, M. F., and Wang, B. C. (1991). Crystals of isoenzyme 3-3 of rat liver glutathione S-transferase with and without inhibitor. *Acta Crystallogr. B* **47**(pt. 5)(2), 813–814.

54. Reinemer, P., Prade, L., Hof, P., Neuefeind, T., Huber, R., Zettl, R. *et al.* (1996). Three-dimensional structure of glutathione S-transferase from *Arabidopsis thaliana* at 2.2 A resolution: structural characterization of herbicide-conjugating plant glutathione S-transferases and a novel active site architecture. *J. Mol. Biol.* **255**(2), 289–309.

55. Neuefeind, T., Huber, R., Dasenbrock, H., Prade, L., and Bieseler, B. (1997). Crystal structure of herbicide-detoxifying maize glutathione S-transferase-I in complex with lactoylglutathione: evidence for an induced-fit mechanism. *J. Mol. Biol.* **274**(4), 446–453.

56. Neuefeind, T., Huber, R., Reinemer, P., Knablein, J., Prade, L., Mann, K. *et al.* (1997). Cloning, sequencing, crystallization and X-ray structure of glutathione S-transferase-III from *Zea mays* var. *mutin*: a leading enzyme in detoxification of maize herbicides. *J. Mol. Biol.* **274**(4), 577–587.

57. Oakley, A. J., Bello, M. L., Battistoni, A., Ricci, G., Rossjohn, J., Villar, H. O. *et al.* (1997). The structures of human glutathione transferase P1-1 in complex with glutathione and various inhibitors at high resolution. *J. Mol. Biol.* **274**(1), 84–100.

58. Krengel, U., Schroter, K. H., Hoier, H., Arkema, A., Kalk, K. H., Zimniak, P. *et al.* (1998). Crystal structure of a murine alpha-class glutathione S-transferase involved in cellular defense against oxidative stress. *FEBS Lett.* **422**(3), 285–290.

59. Board, P. G., Coggan, M., Chelvanayagam, G., Easteal, S., Jermiin, L. S., Schulte, G. K. *et al.* (2000). Identification, characterization, and crystal structure of the omega class glutathione transferases. *J. Biol. Chem.* **275**(32), 24798–24806.

60. Xiao, B., Singh, S. P., Nanduri, B., Awasthi, Y. C., Zimniak, P., and Ji, X. (1999). Crystal structure of a murine glutathione S-transferase in complex with a glutathione conjugate of 4-hydroxynon-2-enal in one subunit and glutathione in the other: evidence of signaling across the dimer interface. *Biochemistry* **38**(37), 11887–11894.

61. Awasthi, Y. C., Sharma, R., and Singhal, S. S. (1994). Human glutathione S-transferases. *Int. J. Biochem.* **26**(3), 295–308.

62. Strange, R. C., Jones, P. W., and Fryer, A. A. (2000). Glutathione S-transferase: genetics and role in toxicology. *Toxicol. Lett.* **112–113**, 357–363.

63. Czerwinski, M., Kiem, H. P., and Slattery, J. T. (1997). Human CD34+ cells do not express glutathione S-transferases alpha. *Gene Ther.* **4**(3), 268–270.

64. Letourneau, S., Greenbaum, M., and Cournoyer, D. (1996). Retrovirus-mediated gene transfer of rat glutathione S-transferase Yc confers in vitro resistance to alkylating agents in human leukemia cells and in clonogenic mouse hematopoietic progenitor cells. *Hum. Gene Ther.* **7**(7), 831–840.

65. Moscow, J. A., Townsend, A. J., and Cowan, K. H. (1989). Elevation of pi class glutathione S-transferase activity in human breast cancer cells by transfection of the GST pi gene and its effect on sensitivity to toxins. *Mol. Pharmacol.* **36**(1), 22–28.

66. Kuga, T., Sakamaki, S., Matsunaga, T., Hirayama, Y., Kuroda, H., Takahashi, Y. *et al.* (1997). Fibronectin fragment-facilitated retroviral transfer of the glutathione-S-transferase pi gene into CD34+ cells to protect them against alkylating agents. *Hum. Gene Ther.* **8**(16), 1901–1910.

67. Matsunaga, T., Sakamaki, S., Kuga, T., Kuroda, H., Kusakabe, T., Akiyama, T. *et al.* (2000). GST-pi gene-transduced hematopoietic progenitor cell transplantation overcomes the bone marrow toxicity of cyclophosphamide in mice. *Hum. Gene Ther.* **11**(12), 1671–1681.

68. Emerman, M., and Temin, H. M. (1984). Genes with promoters in retrovirus vectors can be independently suppressed by an epigenetic mechanism. *Cell* **39**(3, pt. 2), 449–467.

69. Doroshow, J. H., Metz, M. Z., Matsumoto, L., Winters, K. A., Sakai, M., Muramatsu, M. *et al.* (1995). Transduction of NIH3T3 cells with a retrovirus carrying both human MDR1 and glutathione S-transferase pi produces broad-range multidrug resistance. *Cancer Res.* **55**(18), 4073–4078.

70. Hansen, W. K., Deutsch, W. A., Yacoub, A., Xu, Y., Williams, D. A., and Kelley, M. R. (1998). Creation of a fully functional human chimeric DNA repair protein. Combining O^6-methylguanine DNA methyltransferase (MGMT) and AP endonuclease (APE/redox effector factor 1 (Ref 1)) DNA repair proteins. *J. Biol. Chem.* **273**(2), 756–762.

71. Ercikan-Abali, E. A., Mineishi, S., Tong, Y., Nakahara, S., Waltham, M. C., Banerjee, D. *et al.* (1996). Active site-directed double mutants of dihydrofolate reductase. *Cancer Res.* **56**(18), 4142–4145.

72. Sauerbrey, A., McPherson, J. P., Zhao, S. C., Banerjee, D., and Bertino, J. R. (1999). Expression of a novel double-mutant dihydrofolate reductase-cytidine deaminase fusion gene confers resistance to both methotrexate and cytosine arabinoside. *Hum. Gene Ther.* **10**(15), 2495–2504.

73. Chappell, S. A., Edelman, G. M., and Mauro, V. P. (2000). A 9-nt segment of a cellular mRNA can function as an internal ribosome entry site (IRES) and when present in linked multiple copies greatly enhances IRES activity. *Proc. Natl. Acad. Sci. USA* **97**(4), 1536–1541.

74. Galipeau, J., Benaim, E., Spencer, H. T., Blakley, R. L., and Sorrentino, B. P. (1997). A bicistronic retroviral vector for protecting hematopoietic cells against antifolates and P-glycoprotein effluxed drugs. *Hum. Gene Ther.* **8**(15), 1773–1783.

75. Suzuki, M., Sugimoto, Y., and Tsuruo, T. (1998). Efficient protection of cells from the genotoxicity of nitrosoureas by the retrovirus-mediated transfer of human O^6-methylguanine-DNA methyltransferase using bicistronic vectors with human multidrug resistance gene 1. *Mutat. Res.* **401**(1–2), 133–141.

76. Jelinek, J., Rafferty, J. A., Cmejla, R., Hildinger, M., Chinnasamy, D., Lashford, L. S. *et al.* (1999). A novel dual function retrovirus expressing multidrug resistance 1 and O^6-alkylguanine-DNA-alkyltransferase for engineering resistance of haemopoietic progenitor cells to multiple chemotherapeutic agents. *Gene Ther.* **6**(8), 1489–1493.

77. Beausejour, C. M., Le, N. L., Letourneau, S., Cournoyer, D., and Momparler, R. L. (1998). Coexpression of cytidine deaminase and mutant dihydrofolate reductase by a bicistronic retroviral vector confers

78. resistance to cytosine arabinoside and methotrexate. *Hum. Gene Ther.* **9**(17), 2537–2544.

78. Letourneau, S., Palerme, J. S., Delisle, J. S., Beausejour, C. M., Momparler, R. L., and Cournoyer, D. (2000). Coexpression of rat glutathione S-transferase A3 and human cytidine deaminase by a bicistronic retroviral vector confers in vitro resistance to nitrogen mustards and cytosine arabinoside in murine fibroblasts. *Cancer Gene Ther.* **7**(5), 757–765.

79. Podda, S., Ward, M., Himelstein, A., Richardson, C., de la Flor-Weiss, E., Smith, L. *et al.* (1992). Transfer and expression of the human multiple drug resistance gene into live mice. *Proc. Natl. Acad. Sci. USA* **89**(20), 9676–9680.

80. Richardson, C., and Bank, A. (1995). Preselection of transduced murine hematopoietic stem cell populations leads to increased long-term stability and expression of the human multiple drug resistance gene. *Blood* **86**(7), 2579–2589.

81. Hanania, E. G., Fu, S., Roninson, I., Zu, Z., and Deisseroth, A. B. (1995). Resistance to taxol chemotherapy produced in mouse marrow cells by safety-modified retroviruses containing a human MDR-1 transcription unit. *Gene Ther.* **2**(4), 279–284.

82. Devereux, S., Corney, C., Macdonald, C., Watts, M., Sullivan, A., Goldstone, A. H., *et al.* (1998). Feasibility of multidrug resistance (MDR-1) gene transfer in patients undergoing high-dose therapy and peripheral blood stem cell transplantation for lymphoma. *Gene Ther.* **5**(3), 403–408.

83. Hesdorffer, C., Ayello, J., Ward, M., Kaubisch, A., Vahdat, L., Balmaceda, C. *et al.* (1998). Phase I trial of retroviral-mediated transfer of the human MDR1 gene as marrow chemoprotection in patients undergoing high-dose chemotherapy and autologous stem-cell transplantation. *J. Clin. Oncol.* **16**(1), 165–172.

84. Cowan, K. H., Moscow, J. A., Huang, H., Zujewski, J. A., O'Shaughnessy, J., Sorrentino, B. *et al.* (1999). Paclitaxel chemotherapy after autologous stem-cell transplantation and engraftment of hematopoietic cells transduced with a retrovirus containing the multidrug resistance complementary DNA (MDR1) in metastatic breast cancer patients [see comments]. *Clin. Cancer Res.* **5**(7), 1619–1628.

85. Peters, S. O., Kittler, E. L., Ramshaw, H. S., and Quesenberry, P. J. (1996). Ex vivo expansion of murine marrow cells with interleukin-3 (IL-3), IL-6, IL-11, and stem cell factor leads to impaired engraftment in irradiated hosts. *Blood* **87**(1), 30–37.

86. Moscow, J. A., Huang, H., Carter, C., Hines, K., Zujewski, J., Cusack, G. *et al.* (1999). Engraftment of MDR1 and NeoR gene-transduced hematopoietic cells after breast cancer chemotherapy. *Blood* **94**(1), 52–61.

87. Abonour, R., Williams, D. A., Einhorn, L., Hall, K. M., Chen, J., Coffman, J. *et al.* (2000). Efficient retrovirus-mediated transfer of the multidrug resistance 1 gene into autologous human long-term repopulating hematopoietic stem cells [see comments]. *Nat. Med.* **6**(6), 652–658.

88. Sutton, R. E., Reitsma, M. J., Uchida, N., and Brown, P. O. (1999). Transduction of human progenitor hematopoietic stem cells by human immunodeficiency virus type 1-based vectors is cell cycle dependent. *J. Virol.* **73**(5), 3649–3660.

89. Uchida, N., Sutton, R. E., Friera, A. M., He, D., Reitsma, M. J., Chang, W. C. *et al.* (1998). HIV, but not murine leukemia virus, vectors mediate high efficiency gene transfer into freshly isolated G_0/G_1 human hematopoietic stem cells. *Proc. Natl. Acad. Sci. USA* **95**(20), 11939-11944.

22

Multidrug-Resistance Gene Therapy in Hematopoietic Cell Transplantation

RAFAT ABONOUR

Department of Medicine
Indiana University School of Medicine
Indianapolis, Indiana 46202

JAMES M. CROOP

Department of Pediatrics
Indiana University School of Medicine
Indianapolis, Indiana 46202

KENNETH CORNETTA

Department of Medicine
Indiana University School of Medicine
Indianapolis, Indiana 46202

I. INTRODUCTION

The mainstay of cancer treatment continues to be chemotherapy, the use of toxic agents that preferentially deter the growth of cancer cells. Many patients show an initial complete clinical response after chemotherapy in which the tumor appears to be eradicated. In a significant number of these patients, however, cancer recurs. Because many cancer cells exhibit a dose response to chemotherapy agents, dose escalation has been proposed as a method of increasing the durable response after chemotherapy. Unfortunately, dose escalation is associated with increased toxicity, resulting in

damage to normal tissues. One of the most common toxicities is the reduction in the number of normal cells in the bloodstream resulting from chemotherapy-associated bone marrow suppression. This reduction leads to anemia, infections, and bleeding. An attractive strategy to overcome chemotherapy-associated bone marrow suppression is to introduce drug-resistance genes into hematopoietic cells.

A variety of drug-resistance genes have been identified that decrease the hematopoietic toxicity of specific chemotherapeutic agents. Many of these genes were first identified in tumor cells that had developed resistance to chemotherapy. For example, certain cells that had become resistant to the chemotherapy drug methotrexate were found to contain a mutant form of the enzyme dihydrofolate reductase (DHFR). Methotrexate inhibits the growth of tumor cells by interfering with the DHFR gene, and cells that develop the mutant DHFR are less sensitive to the effects of chemotherapy. By placing the mutant form of DHFR into normal bone marrow cells, we can protect these cells from such effects. Other drug-resistant genes include DNA-repair proteins, which correct damages resulting from exposure to ionizing radiation and alkylating agents [1–3]. The list of enzymes and proteins that prevent the toxic effects of various chemotherapeutic agents is expanding daily. Both the cloning of drug-resistance genes and our ability to express them in hematopoietic cells have set the stage for several clinical trials [4].

This chapter will focus on the multidrug resistance gene 1 (MDR1), which encodes a drug efflux pump that removes toxic compounds, including many chemotherapy agents, from the cell. We will review preclinical experience with the MDR1 in conferring chemoprotection to hematopoietic tissues and examine its current potential for clinical applications.

agents is to increase the expression of MDR1 in progenitor cells [9,10].

II. P-GLYCOPROTEIN

The MDR1 gene encodes a 170-kDa membrane glycoprotein [5,6]. Termed the P-glycoprotein, this is a member of a large superfamily of adenosine triphosphate (ATP)-dependent transport proteins that shuttle a wide variety of substrates across cell membranes. The P-glycoprotein prevents cellular accumulation of specific compounds, most of which are lipophilic cations, including anthracyclines, vinca alkaloids, epipodophyllotoxins, dactinomycin, and paclitaxel [7]. P-glycoprotein is differentially expressed in normal mammalian tissues, usually at secretory surfaces such as the biliary canaliculi, proximal tubules of the kidney, and the colonic epithelium. While moderate expression is seen in bone marrow stem cells, P-glycoprotein is expressed at low levels in bone marrow progenitor and peripheral blood cells [8]. This explains, in part, why treatment with certain toxic agents leads to significant bone marrow suppression and a decrease in circulating blood cells. One approach that has been proposed to minimize the hematologic toxicity of chemotherapeutic

III. TARGETING HEMATOPOIETIC PROGENITOR CELLS FOR GENETIC MODIFICATION

Under normal physiological conditions, a strictly regulated number of mature blood cells are continuously produced from a small population of hematopoietic stem cells. Stem cells possess the unique ability to differentiate into developmentally restricted progenitor cells (i.e., are pluripotent) while retaining the capacity for self-renewal. When cells within the stem cell pool commit to differentiation, various progenitor cells are produced and have been identified in the pathway to mature blood cell elements (Fig. 1). These progenitor cells include granulocyte–macrophage colony-forming units (CFU-GMs), erythrocyte burst-forming units (BFU-Es), and granulocyte–erythrocyte–macrophage–megakaryocyte colony-forming units (CFU-GEMMs). Extensive cell proliferation of the progenitor cell pool occurs, allowing a small number of stem cells to generate the large number of mature cells necessary for maintaining normal blood profile.

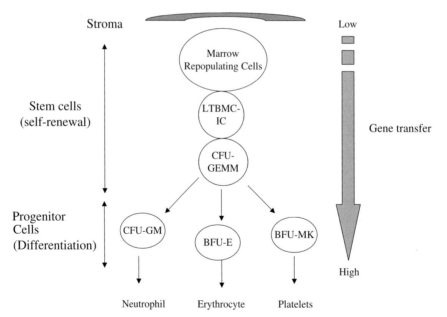

FIGURE 1   Blood cells are continuously produced from a small population of hematopoietic stem cells. Stem cells can differentiate into developmentally restricted progenitor cells and retain the capacity for self-renewal. Progenitor cells lack the capacity to self-renew, as they are committed to differentiate into various blood components. Gene transfer rate is low in the stem cell pool, while it is high in more differentiated cells. To achieve permanent expression of a therapeutic gene, efforts should focus on enhancing gene transfer into stem cells. LTBMC-IC = long-term, bone-marrow-cell-initiated culture; CFU-GM = granulocyte–macrophage colony-forming units; BFU-E = erythrocyte burst-forming units; BFU-MK = burst-forming unit megakaryocyte; and CFU-GEMM = granulocyte–erythrocyte–macrophage–megakaryocyte colony-forming units.

Stem cells are attractive targets for treating genetic disease of the bone marrow because relatively few genetically modified stem cells can repopulate the entire hematopoietic system, conferring the genetic modification to all subsequent progeny [11]. Because committed progenitor cells cannot self-renew, they are less attractive targets for lifelong replacement of a missing or defective gene product. Nevertheless, the use of gene transfer technology to generate drug resistance in hematopoietic cells could be exploited using these committed cells because the duration of chemotherapy administration is not intended to be lifelong. In addition, a major advantage of targeting progenitor cells is the efficient gene transfer achieved using retroviral vectors. Cell cycling is necessary for retroviral gene transfer, a requirement that has limited the transduction rate of stem cells, which are generally quiescent [12]. In contrast, a relatively large portion of the progenitor cell pool actively cycles. Cycling and gene transfer rates can be further increased through the use of hematopoietic growth factors [13].

A major technical advantage of gene transfer into hematopoietic tissue is the ease of collection, *ex vivo* manipulation, and transplantation of blood cell products. A practical consideration when attempting gene transfer in hematopoietic tissue is the relative rarity of the target cells. For example, progenitor cells represent less than 1% of the total nucleated bone marrow cells, and stem cells are estimated at approximately one cell per million. The frequency of these cells in mobilized peripheral blood is even less. To facilitate gene therapy, investigators have sought to obtain an enriched source of progenitor and stem cells. A variety of investigational and commercially available systems have been devised that target cells expressing the CD34 antigen.

CD34 is a cell-surface antigen expressed on primitive cells and has been used to enrich stem and progenitor cells from various sources, including bone marrow, mobilized peripheral blood cells, and umbilical cord blood cells. Establishment of normal hematopoiesis is feasible using CD34+-enriched cells in both autologous and allogeneic transplantation [14–17]. The documentation of donor engraftment more than a year after transplantation consistently attests to the presence of primitive hematopoietic progenitors in CD34-enriched marrow. CD34 selection also provides a potential safety advantage. Bone marrow and peripheral blood have been shown to contain malignant cells from a number of hematologic malignancies and solid tumors, raising concerns that during transduction of progenitor cells the drug resistance gene may be inadvertently introduced into tumor cells. Because most tumor cells do not express the CD34 antigen, enrichment for CD34-expressing cells will decrease tumor cell contamination and the risk of inadvertent transduction of tumor cells with the drug-resistant gene vector. In addition, by using CD34-enriched products as targets for drug-resistance genes, the amount of cytokines and vector required per subjects could be decreased 50- to 100-fold due to the small number of CD34+ cells required per subject.

IV. EXPRESSION OF P-GLYCOPROTEIN IN MURINE HEMATOPOIETIC PROGENITORS

The human MDR1 was first isolated in 1986 by Roninson *et al.* from drug-resistant human carcinoma cell lines [18]. When the DNA sequences of 4.5 kb were introduced to drug-sensitive cell lines using retroviral vectors, the resultant cell lines were drug resistant [19,20].

To begin to assess the clinical potential of MDR1 gene transfer, the protective effects of MDR1 were studied in murine hematopoietic cells *in vivo*. First, the introduction of the MDR1 cDNA into transgenic mice led to chemoprotection *in vitro* and *in vivo* [21,22]. When the human MDR1 cDNA is placed under the control of a truncated chicken β-actin promoter, the MDR1 gene is expressed in cells of many organs it is not normally expressed in, including hematopoietic cells isolated from the bone marrow and spleen. Mice appeared phenotypically normal and no abnormalities could be identified as a result of MDR1 expression. An encouraging finding in regard to clinical application of MDR1, peripheral blood counts are protected from the effects of a number of cytotoxic agents transported by the P-glycoprotein. This effect was specific; no protection was seen against the toxic effects of non-P-glycoprotein-dependent drugs. Protection was particularly obvious when escalated doses of chemotherapeutic drugs such as daunorubicin, vinblastine, etoposide and paclitaxel were used. In these transgenic mice, resistance to P-glycoprotein-dependent drugs was increased tenfold compared to nontransgenic animals. In addition, bone marrow cells harvested from these transgenic animals protected lethally irradiated recipient mice from chemotherapy-induced bone marrow suppression.

Retroviral vectors have been used to confer the MDR1 cDNA to hematopoietic cells in bone marrow transplantation models. In these experiments, syngeneic mice were subjected to irradiation and injected with bone marrow cells, which normally rescue the animals from the lethal effects of irradiation. When vector-transduced marrow was utilized, human MDR1 DNA was detected in up to 78% of peripheral blood cells following transplantation [23]; however, vector–derived cells were short-lived and lost MDR1 expression over time in the absence of *in vivo* selection.

To demonstrate the selective advantage of MDR1 transduced murine bone marrow cells *in vivo*, Sorrentino *et al.* [22] administered paclitaxel after syngeneic transplantation. This treatment increased the percentage of transduced cells and their progenitors, as judged by the increased number of leukocytes expressing P-glycoprotein in the blood. Podda *et al.* [24] confirmed these findings using a retroviral construct containing MDR1 cDNA in a Harvey-virus-based vector [24]. The presence of the MDR gene was assayed by polymerase chain reaction (PCR) analysis using MDR1-specific primers. Eight of nine transduced mice were positive for MDR1 by PCR of peripheral blood 14 and 50 days following transplantation. However, 8 months later, only three of the nine transduced mice were positive. When PCR-negative mice were treated with paclitaxel, transduced cells were subsequently detected by PCR, indicating drug selection of MDR1-transduced bone marrow stem cells.

Unlike committed progenitor and blood cells in which MDR1 expression is very low, primitive hematopoietic cells express a moderate level of MDR1. Therefore, MDR1 vectors must provide a significant increase in P-glycoprotein expression over endogenous levels for drug selection to be feasible in this cell population. Studies by Sorrentino *et al.* [25] demonstrate that expression of MDR1 by retroviral vectors is approximately four times that found in primitive murine bone marrow cells. Such overexpression was associated with increased drug resistance in the animal models.

Further studies showed that long-term engraftment with MDR1 transduced cells could be obtained while maintaining a high level of P-glycoprotein expression [26]. First, using a highly enriched population of hematopoietic stem cells transduced with MDR1 resulted in the highest number of long-term circulating mature white cells expressing P-glycoprotein. This finding indicated that it is possible to efficiently transfer genes into the earliest hematopoietic progenitors, which are capable of providing long-term reconstitution. The number of the transduced cells in the transplanted population of the cells, however, must be high enough to outgrow untransduced cells. Second, the long-term expression of P-glycoprotein was compared between mice transplanted with transduced stem cells that had or had not been selected with cytotoxic agents to enrich for cells expressing P-glycoprotein. Selection was associated with a higher number of circulating white cells expressing the MDR1 gene compared to transplantation with unselected cells. At one year, the MDR1 gene was expressed in 0–5% (median, 1.3%) of granulocytes recovered from mice receiving unselected cells. In mice transplanted with selected cells, 4–50% (median, 20.5%) of circulating granulocytes expressed MDR1. It is possible that the combination of *ex vivo* and *in vivo* selection may lead to better engraftment with drug-resistant cells [27].

To determine if stem cells transduced with MDR1 maintain their capacity for self-renewal and differentiation, mice were transplanted with MDR1-transduced cells. After engraftment, the bone marrow cells were transplanted into a second cohort of animals. The ability to reconstitute animals after successive transplantation is a characteristic of stem cells. The bone marrow and peripheral blood cells of serially transplanted mice displayed an increased level of MDR1 expression and resistance to the toxic effects of paclitaxel. Modified bone marrow cells were serially transplanted into six successive cohorts of BALB/c mice. Paclitaxel-resistant hematopoiesis with little or no bone marrow suppression was observed in all six of the cohorts [28]. These data present strong evidence for the transduction of murine bone marrow stem cells.

V. EXPRESSION OF P-GLYCOPROTEIN IN HUMAN HEMATOPOIETIC PROGENITORS

The success of MDR1 in murine models led to intensified investigations into the possibility of generating drug-resistant human hematopoietic cells. The target cells for gene manipulation initially were bone marrow cells. However, with the wide use of mobilized peripheral blood progenitor cells in autologous transplantation, research interest shifted to these populations of cells (Fig. 2). MDR1 cDNA packaged into retroviral vectors has been used widely to study the feasibility of generating chemotherapy-resistant human progenitor

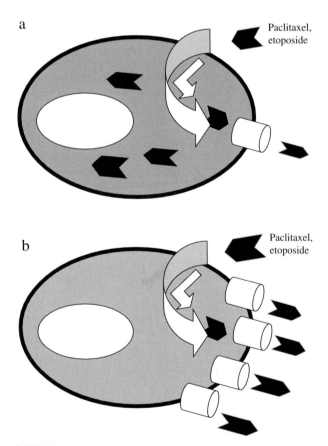

FIGURE 2 P-glycoprotein is expressed at low levels in bone marrow progenitor and peripheral blood cells (a). This leads to significant accumulation of certain toxic agents in these cells, resulting in their death and a decrease in circulating blood cells. One approach that has been proposed to minimize the hematologic toxicity of chemotherapeutic agents is to increase the expression of P-glycoprotein in progenitor cells (b), leading to removal of such toxic agents from progenitor cells.

cells. Several approaches were used to test the efficiency of MDR1 gene transfer into human cells.

Bertolini *et al.* [29] used either cord blood or bone-marrow-derived, low-density or purified CD34+ cells as a target for MDR1 gene transfer. Cells were cultured with an irradiated MDR1 retroviral producer line. Drug selection and culture were performed in the presence of daunorubicin, colchicine, and paclitaxel. In nontransduced controls, 1–2% of CFU-GM and CFU-GEMM were found to be drug resistant, while 14–31% of transduced cells were drug resistant. For efficient transfer of MDR1, prestimulation with cytokines of human cells was required. MDR1 gene expression as evidenced by MDR1 mRNA was very low in cultures of nontransduced cells, whereas after drug selection MDR1 mRNA levels in transduced cells was as high as in the MDR1 retroviral producer line [29].

Hanania *et al.* [30] used a more clinically applicable approach to transfer MDR1-containing retroviral vectors into CD34+ human marrow cells. This group avoided the co culture of human cells and producer cell lines by culturing vector supernatant and bone marrow cells (collected from subjects recovering from chemotherapy-induced myelosuppression) on the stromal layer in the presence of the hematopoietic growth factors interleukin-3 (IL-3) and IL-6. Transduced CD34+ cells were maintained in cultures for longer than 35 days to test the long-term clonogenic potential of long-term, culture-initiating cells (LTCICs). Analysis of the colonies from the CD34+ MDR1-transduced cells showed that the viral MDR1 mRNA levels were much higher than those of the endogenous MDR1 mRNA. The transduction frequency of CD34+ cells was 7–20%. The rhodamine efflux assay and the ability of colony-forming units to grow in the presence of a toxic drug (paclitaxel) confirmed the presence of functional P-glycoprotein in transduced cells [30].

Several preclinical models have studied the AM12M1 vector, which contains the MDR1 cDNA within the Harvey murine sarcoma virus (HaMSV) long terminal repeats, (LTRs) and packaged by the AM12 amphotrophic packaging cell line. This vector has been used to transduce human hematopoietic stem cells from a variety of sources [31]. Vector-expressing cells have been successfully selected with many agents, including daunorubicin, colchicine, and paclitaxel. Enhanced gene transfer of the MDR1 cDNA into human CD34+ marrow cells was demonstrated in the presence of IL-3, IL-6, and stem cell factor [32]. These findings were incorporated into a preclinical trial utilizing modified bone marrow cells obtained from women with breast cancer [33]. *In vitro* analyses of MDR1 cDNA transfer into mobilized peripheral blood progenitor cells obtained from these subjects showed results comparable to those obtained using bone-marrow-derived progenitors with 65% of BFU-Es and 57% of CFU-GMs containing the transgene. Moreover, 19–48% of these progenitors were paclitaxel resistant, indicating that at least a quarter of these cells were resistant to chemotherapy.

VI. RESULTS OF EARLY PHASE I STUDIES USING MDR1-TRANSDUCED HEMATOPOIETIC CELLS

The encouraging results in transferring MDR1 to bone marrow progenitor cells in animal models and the success of *in vitro* MDR1 cDNA transfer to human cells led to optimism that such results could be achieved clinically. Several groups have conducted exploratory clinical trials to test the safety of this approach.

This cancer therapy technique has been used in tumors that are known to be sensitive to chemotherapeutic agents and have demonstrated a steep dose–response curve *in vitro*. The major side effect of high-dose therapy, however, is severe hematopoietic suppression. To circumvent this toxicity, autologous stem cells are harvested from the bone marrow or peripheral blood by stem cell mobilization and cryopreserved. These cells are then given following the administration of high-dose chemotherapy regimens. The progenitor cells in these products shorten the period of hematopoietic suppression after high-dose therapy and decrease toxicity. This approach has increased overall survival when compared to conventional chemotherapy. Unfortunately, a significant number of subjects ultimately relapse. Administering additional chemotherapeutic agents after transplantation could provide further tumor cell reduction but is also associated with significant hematopoietic suppression [34,35]. It is in the setting of autologous transplantation that all clinical trials involving MDR1 have been conducted. All published studies are phase I and are designed to test safety, not efficacy. The ultimate goal of these trials is the prevention of hematopoietic suppression associated with chemotherapy following autologous transplantation (Fig. 3).

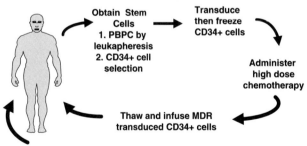

FIGURE 3 Schema for using genetically manipulated stem cells in autologous transplantation. Stem cells are mobilized in the bloodstream using cytokines such as granulocyte colony-stimulating factor (G-CSF). These cells are collected from the bloodstream by a process called leukapheresis. The cells are then purified to enrich for the target cells (CD34+ cells), which are then transduced with MDR1 vector. Following the administration of high-dose chemotherapy, patients receive the manipulated cells. Transduced cells are then selected for *in vivo* using drugs such as paclitaxel and etoposide.

In one of the early trials, peripheral blood progenitor cells from breast cancer subjects and bone marrow cells from ovarian cancer subjects were enriched for CD34$^+$ cells [36]. Two different transduction protocols were used. In the first, CD34$^+$ cells were transduced in suspension in the presence of retroviral vector and protamine sulfate. The second protocol used autologous bone marrow stroma and both (IL-3) and IL-6. It was postulated that these conditions would maintain the growth and proliferation of primitive hematopoietic cells. After infusion of transduced cells, only transient and low engraftment of MDR1-transduced CD34$^+$ cells was observed. None of the subjects transduced without stroma had detectable transgene-positive cells following transplantation. Five out of eight subjects transplanted with the use of stroma showed a low level of engraftment of transduced cells. The transduction conditions led to marking of committed progenitors (GM-CFU), which did not appear to contribute to lasting hematopoietic recovery following high-dose chemotherapy.

Heserdorfer *et al.* [37] used the AM12M1 vector in five subjects transplanted with CD34$^+$-enriched marrow or peripheral blood cells transduced on fibronectin-coated plates. Fibronectin avoids the technical difficulties associated with the use of stroma while providing an extracellular matrix. In addition, these cells were cultured in the presence of IL-3, IL-6, and stem cell factor (SCF) to induce cell cycling of early progenitor cells. Transduction of colony-forming units was high (20–70%), but engraftment of vector-transduced cells was poor. Although all subjects showed evidence of gene marking in the immediate posttransplant period, the latest time that vector-transduced cells were detected was 10 weeks after transplantation. Because subjects received a combination of unmanipulated and transduced cells, it is possible that the unmanipulated cells outcompeted the vector-transduced population. The other possibility is that the culture system only transduced committed progenitors.

Devereux *et al.* [38] used the AM12M1 vector in treating three subjects with lymphoma. CD34$^+$ cells were transduced with retroviral supernatant and protamine sulfate, without additional cytokines. Transduction efficiency was 0–2.5%, and no transgene-containing cells were detected in the blood or the bone marrow of these subjects within 3–6 months of followup. No delay in engraftment was observed, although one subject received additional autologous cells for falling peripheral blood counts. The effect of MDR1 substrate drugs was not tested, as all subjects remained in clinical and radiological remission of their lymphoma. The explanation of these inferior results may be related to two factors: low cell cycling without cytokine stimulation and low multiplicity of infection used in this study [38].

More recently, a group from the National Cancer Institute (NCI) reported the results of two different trials that included MDR1 gene transfer into hematopoietic cells from subjects with breast cancer [39]. In the first study, mobi-

lized peripheral blood CD34$^+$cells from six subjects were transduced with either an MDR1 vector or a marking vector (Neo^R) in the presence of autologous stroma (without the addition of cytokines). In general, neutrophils and platelet recovery after transplantation of manipulated CD34$^+$ cells was prompt, although one subject required the infusion of additional autologous cells to aid in hematopoietic recovery. Gene marking of granulocytes was low; only one of the six subjects had Neo^R-marked granulocytes 6 months following transplantation. Of interest, despite the low rate of gene transfer of MDR1 (mean of 0.26%), MDR1-marked granulocytes were seen in two subjects for about 6 months following chemotherapy, suggesting *in vivo* selection. No long-term data on hematopoietic function and gene marking were presented. In a second study by the NCI group, the effects of cytokine and prestimulation on gene transfer were assessed [40]. Four subjects received a combination of unmanipulated cells and CD34$^+$ cells transduced with the MDR1 vector in the presence of IL-6/IL-3/SCF. Again, gene transfer efficiency was only 0.1–0.5% as determined by semi quantitative analysis. Only one subject had low but detectable MDR1 transgene at 4 months. No long-term data were provided.

VII. OVERCOMING TRANSDUCTION INEFFICIENCY

Despite the impressive results generated *in vitro*, early human studies with MDR1 demonstrated low transduction efficiency. While preclinical work documented increased gene transfer in the presence of stromal elements, the establishment of stroma required for transducing the large number of cells used in human gene therapy protocols poses significant technical problems. In analyzing the role of stroma in enhancing gene transfer, fibronectin was found to assist in the colocalization of marrow cells and vector particles.

Fibronectin is a ubiquitous extracellular matrix molecule that contains adhesion sites where cells can bind via the cell-surface proteoglycans or via the integrins very late antigen 4 (VLA4) or very late antigen 5 (VLA5) [41]. Increased gene transfer efficiency in murine bone marrow and committed human progenitors *in vitro* was observed when gene transfer with retroviral elements was performed in the presence of either chymotryptic or recombinant fragments of human fibronectin [42]. Gene transfer in the presence of the chymotryptic fibronectin fragment into very early nonleukemic progenitors (CD34$^+$ HLA-DR negative) resulted in a 14-fold increase in gene transfer rates when compared to control [43]. Several recombinant chimeric fibronectin fragments that contain such domains have been developed to be used in clinical-grade gene transfer systems. Of these fragments, CH-296 was associated with highest gene transduction efficiency and is superior to intact fibronectin molecules in terms of gene transfer efficiency [44].

VIII. MDR1 GENE TRANSFER INTO HUMANS: RECENT PROGRESS

At Indiana University, we have recently completed a clinical gene therapy trial that evaluated the use of recombinant fibronectin fragments and early-acting hematopoietic cytokines [45]. Eligible patients were those with recurrent germ cell tumors undergoing autologous transplantation. The study was based on the observation that administering maintenance etoposide chemotherapy after autologous transplantation to subjects with recurrent and refractory germ cell tumors improves overall survival. However, maintenance therapy with etoposide is associated with bone marrow suppression, resulting in treatment delay and dose modifications, with decreased dose intensity. We proposed that introduction of the MDR1 gene into peripheral blood progenitor cells (PBPCs) during autologous transplantation for germ cell tumors would protect the transduced cells from etoposide, a P-glycoprotein substrate.

Twelve subjects with advanced germ cell tumors were enrolled on this study. PBPCs were mobilized using granulocyte colony-stimulating factor (G-CSF). After mobilization, CD34+ cells were isolated by immunomagnetic separation. CD34+ cells were cultured for 2 days in the presence of cytokines. Following prestimulation, cells were plated on Petri dishes treated with the recombinant fibronectin fragment CH-296. These cells were then exposed to clinical-grade MDR1 vector (AM12M1) for 4 hours, collected, and resuspended in fresh media with cytokines. The following day the transduction was repeated, and the cells were cultured for an additional 8–16 hours prior to cryopreservation.

Subjects received two cycles of high-dose chemotherapy. The autologous hematopoietic stem cells infused with the first cycle were unmanipulated. During the second cycle, subjects received only cells that had undergone the transduction protocol. Thus, each subject served as his/her own control to determine the rate of hematopoietic recovery following transplantation of manipulated or unmanipulated cells. Hematopoietic recovery is determined by measuring the period of neutropenia (decrease in the number of white blood cells, which can lead to infection) and thrombocytopenia (decrease in the number of platelets, which can lead to bleeding). The duration of both neutropenia and thrombocytopenia was similar for the two cycles of high-dose chemotherapy (unmanipulated vs. gene-transduced). The time between stem cell infusion and the rise in absolute neutrophil counts (ANCs) above 0.5×10^9/L was similar for both transplants (median, 10 and 9 days, respectively). In addition, time to unsupported platelet count above 20×10^9/L was similar for the first and second cycle of high-dose chemotherapy (11 and 12 days, respectively). We concluded that using cytokine-stimulated CD34+ cells exposed to CH-296 and MDR1 retroviral vectors did not adversely affect the capacity of these cells to support hematopoietic recovery. After hema-

tologic recovery from cycle two, etoposide (a chemotherapy agent transported by the P-glycoprotein) was administered orally for 20 out of 28 days. Three cycles of etoposide were planned.

Immediately following transduction and prior to infusion into subjects, the median gene transduction efficiency in hematopoietic progenitor cells was 14% (range, 4–52%). Of BFU-E colonies, 24% contained the AM12M1 vector (4–47%) when assessed for vector DNA by PCR. Of the CFU-GM, 13% were MDR1 positive by PCR (0–67%). MDR1-vector-containing progenitors were detected throughout the first year of followup, with up to 15% of the progenitors containing the vector 1 year after transplantation. Peripheral blood leukocytes were also positive for the transgene in seven subjects, with up to 5.6% of the cells containing MDR1 transgene. In regard to chemotherapy after transplant, 85% of maintenance oral etoposide was administered successfully. With more than 2 years of followup, eight subjects are alive and free of disease with no clinical side effects related to this trial.

IX. IMPLICATION AND FUTURE OF MDR1 GENE THERAPY IN HUMANS

Despite the improved gene transfer obtained with cytokine stimulation and colocalization with extracellular matrix protein (such as fibronectin), two issues continue to hamper progress in MDR1 gene therapy: the integrity of the MDR1 transgene and the safety of these trials.

Despite the high rate of transgene transfer, drug-resistance colony assay of transduced CD34+ cells and postinfusion bone marrow cells has failed to show consistent expression and function of P-glycoprotein. One possible explanation is that the transgene may have been altered at some point during packaging into retroviral vector or during transduction. MDR1 cDNA contains cryptic splice donor and splice acceptor sites. Aberrant splicing of vector-derived transgene within the producer cells can result in the passage of a truncated transgene into CD34+ cells, which may result in a loss of expression and function of P-glycoprotein [46]. To address this concern, we performed PCR analyses on transduced products and posttransplantation blood and marrow samples obtained from subjects in our clinical trial. For PCR analyses we used primers (previously described by Sorrentino [25]) that allow discrimination between full and truncated versions of the transgene. Both truncated and full versions were detected in the transduced products. Following transplantation, we detected more of the truncated version of the transgene than full ones. It is of note that both retroviral vector production and CD34+ cell transduction were carried out without drug selection, which may have allowed for the propagation of cells containing the truncated form of MDR1.

Several strategies can be used to minimize the generation of CD34$^+$ cells containing the truncated MDR1. One is to perform *in vitro* selection of transduced cells in the presence of toxic drugs such as paclitaxel. Another strategy is to select the cells that were transduced with the full version of MDR1 cDNA *in vivo* by initiating maintenance chemotherapy earlier than the 1-month time point used in our trial. Other investigators have suggested that alternative vector systems may lead to more efficient transfer and expression of the full version of MDR1 cDNA than traditional vectors by providing alternative splice donor [47]. Clinical work with such vectors is ongoing in Germany (Baum, pers. comm.).

Regarding the safety of MDR1 gene therapy, we were pleased to see that our data demonstrate that exposure of CD34$^+$ cells to CH-296 did not affect engraftment kinetics. Hematopoietic function has remained normal in the eight surviving subjects with over 2 years of followup. The issue of safety of MDR1 gene expression in hematopoietic stem cells has been raised by Bunting *et al.* [48], who reported the development of a myeloproliferative syndrome in mice transplanted with MDR1-transduced marrow cells. At present, we do not know if the expression of MDR1 was responsible for the hematologic disorder they observed. Our study differs in several ways from that murine study. Bunting *et al.* [48] utilized a mutated version of the MDR1 cDNA that contains a substitution of valine with glycine at position 185, and the transduced cells were expanded *in vitro* for a significantly longer time (up to 12 days). It is unclear whether the function of P-glycoprotein is altered if transcribed by the mutated MDR1. In addition, transduction conditions may have led to multiple integrations, increasing the risk of insertional mutagenesis [49]. In contrast to the murine study by Bunting *et al.*, abnormal hematopoiesis has not been reported in four published human studies utilizing MDR1 or in multiple other murine transplant experiments using the A12M1 vector [36–40,45]. Transgene mice expressing MDR1 in committed hematopoietic cells did not show any evidence of myeloproliferative disorder. Because prior human gene therapy studies of MDR1 attained only lower levels of transduction and our study noted the majority of transduced cells contained a truncated MDR1 gene, the propensity of MDR1 vectors to elicit human myeloproliferative disorders requires further investigation.

The cloning of the human MDR1 gene more than 14 years ago set the stage for the extensive animal and preclinical studies summarized in this chapter. It is clear from these experiments that hematopoietic stem cells containing MDR1 can protect against the toxic effects of many drugs used to combat cancer. To date, no adverse events have been associated with the transplantation of MDR1-transduced cells. With the improved gene transfer obtained with cytokine stimulation and fibronectin fragments, future efforts should focus on *in vivo* selection of MDR1-transduced cells.

Acknowledgments

This work was supported by the National Centers for Research Resources (NIH M01 RR00750), American Cancer Society grant (CRTG-97-042-EDT), and Center for Excellence in Hematopoiesis (P30 DK49218-06).

References

1. McIvor, R. S. (1996). Drug-resistant dihydrofolate reductases: generation, expression and therapeutic application. *Bone Marrow Transplant.* **18**, S50–S54.
2. McIvor, R. S., and Simonsen, C. C. (1990). Isolation and characterization of a variant dihydrofolate reductase cDNA from methotrexate-resistant murine L5178Y cells. *Nucleic Acids Res.* **18**, 7025–7032.
3. Spencer, H. T., Sleep, S. E., Rehg, J. E., Blakley, R. L., and Sorrentino, B. P. (1996). A gene transfer strategy for making bone marrow cells resistant to trimetrexate. *Blood* **87**, 2579–2587.
4. Koc, O. N., Allay, J. A., Lee, K., Davis, B. M., Reese, J. S., and Gerson, S. L. (1996). Transfer of drug resistance genes into hematopoietic progenitors to improve chemotherapy tolerance. *Semin. Oncol.* **23**, 46–65.
5. Croop, J. M. (1998). Evolutionary relationships among ABC transporters. *Meth. Enzymol.* **292**, 101–116.
6. Bosch, I., and Croop, J. M. (1998). P-glycoprotein structure and evolutionary homologies. *Cytotechnology* **27**, 1–30.
7. Pastan, I., and Gottesman, M. M. (1991). Multidrug resistance. *Annu. Rev. Med.* **42**, 277–286.
8. Chaudhary, P. M., and Roninson, I. B. (1991). Expression and activity of P-glycoprotein, a multidrug efflux pump, in human hematopoietic stem cells. *Cell* **66**, 85–94.
9. Galski, H., Sullivan, M., Willingham, M. C., Chin, K. V., Gottesman, M. M., Pastan, I., and Merlino, G. T. (1989). Expression of a human multidrug resistance cDNA (MDR-1) in the bone marrow of transgenic mice: resistance to daunomycin-induced leukopenia. *Mol. Cell. Biol.* **9**, 4357–4363.
10. McLachlin, J. R., Eglitis, M. A., Ueda, K., Kantoff, P. W., Anderson, W. F., and Gottesman, M. M. (1990). Expression of a human complementary DNA for the multidrug resistance gene in murine hematopoietic precursor cells with the use of retroviral gene transfer. *J. Natl. Cancer Inst.* **82**, 1260–1263.
11. Williams, D. A. (1990). Expression of introduced genetic sequences in hematopoietic cells following retroviral-mediated gene transfer, *Human Gene Ther.* **1**, 229–239.
12. Van Beusechem V. W., and Valerio, D. (1996). Gene transfer into hematopoietic stem cells of nonhuman primates. *Hum. Gene Ther.* **7**, 1649–1668.
13. Emerson, S. G. (1996). *Ex vivo* expansion of hematopoietic precursors, progenitors, and stem cells: the next generation of cellular therapeutics. *Blood* **87**, 3082–2088.
14. Abonour, R., Scott, K. M., Kunkel, L. A., Robertson, M. J., Hromas, R., Graves, V., Lazaridis, E. N., Cripe, L., Gharpure, V., Traycoff, C. M., Mills, B., Srour, E. F., and Cornetta, K. (1998). Autologous transplantation of mobilized peripheral blood CD34+ cells selected by immunomagnetic procedures in patients with multiple myeloma. *Bone Marrow Transplant.* **22**, 957–963.
15. Chabannon, C., Cornetta, K., Lotz, J. P., Rosenfeld, C., Shlomchik, M., Yanovitch, S., Marolleau, J. P., Sledge, G., Novakovitch, G., Srour, E. F., Burtness, B., Camerlo, J., Gravis, G., Lee-Fischer, J., Faucher, C., Chabbert, I., Krause, D., Maraninchi, D., Mills, B., Kunkel, L., Oldham, F., Blaise, D., and Viens, P. (1998). High-dose chemotherapy followed by reinfusion of selected CD34+ peripheral blood cells in patients with poor-prognosis breast cancer: a randomized multicentre study. *Br. J. Cancer* **78**, 913–921.

16. Voso, M. T., Hohaus, S., Moos, M., Pforsich, M., Cremer, F. W., Schlenk, R. F., Martin, S., Hegenbart, U., Goldschmidt, H., and Haas R. (1999). Autografting with CD34+ peripheral blood stem cells: retained engraftment capability and reduced tumour cell content. *Br. J. Haematol.* **104,** 382–391.

17. Cornetta, K., Gharpure, V., Mills, B., Hromas, R., Abonour, R., Broun, E. R., Traycoff, C. M., Hanna, M., Wyman, N., Danielson, C., Gonin, R., Kunkel, L. K., Oldham, F., and Srour, E. F. (1998). Rapid engraftment after allogeneic transplantation using CD34 enrich marrow cells. *Bone Marrow Transplant.* **21,** 65–71.

18. Roninson, I. B., Chin, J. E., Choi, K. G., Gros, P., Housman, D. E., Fojo, A., Shen, D. W., Gottesman, M. M., and Pastan, I. (1986). Isolation of human mdr DNA sequences amplified in multidrug-resistant KB carcinoma cells. *Proc. Natl. Acad. Sci. USA* **83,** 4538–4542.

19. Ueda, K., Cardarelli, C., Gottesman, M. M., and Pastan, I. (1987). Expression of a full-length cDNA for the human "MDR1" gene confers resistance to colchicine, doxorubicin, and vinblastine *Proc. Natl. Acad. Sci. USA* **84,** 3004–3008.

20. Pastan, I., Gottesman, M. M., Ueda, K., Lovelace, E., Rutherford, A. V., and Willingham, M. C. (1988). A retrovirus carrying an MDR1 cDNA confers multidrug resistance and polarized expression of P-glycoprotein in MDCK cells. *Proc. Natl. Acad. Sci. USA* **85,** 4486–4490.

21. Choi, K, Frommel, T. O., Kaplan Stern, R., Perez, C. F., Kriegler, M., Tsuruo, T., and Roninson, I. B. (1991). Multidrug resistance after retroviral transfer of the human MDR-1 gene correlates with P-glycoprotein density in the plasma membrane and is not affected by cytotoxic selection. *Proc. Natl. Acad. Sci. USA* **88,** 7386–7390.

22. Sorrentino, B. P., Brandt, S. J., Bodine, D., Gottesman, M., Pastan, I., Cline, A., and Nienhuis, A. W. (1992) Selection of drug-resistant bone marrow cells in vivo after retroviral transfer of human MDR-1. *Science* **257,** 99–103.

23. Licht, T., Aksentijevich, I., Gottersman, M. M., and Pastan, I. (1995). Efficient expression of functional human MDR-1 genc in murine bone marrow after retroviral transduction of purified hematopoietic stem cells. *Blood* **86,** 111–121.

24. Podda, S., Ward, M., Himelstein, A., Richardson, C., de la Flor-Weiss, E., Smith, L., Gottesman, M., Pastan, I., and Bank, A. (1992). Transfer and expression of the human multiple drug resistance gene into live mice. *Proc. Natl. Acad. Sci. USA* **89,** 9676–9680.

25. Sorrentino, B. P., McDonagh, K. T., Woods, D., and Orlic, D. (1995). Expression of retroviral vectors containing the human multidrug resistance 1 cDNA in hematopoietic cells of transplanted mice *Blood* **86,** 491–501.

26. Richardson, C., and Bank, A. (1995). Preselection of transduced murine hematopoietic stem cell populations leads to increased long-term stability and expression of the human multiple drug resistance gene. *Blood* **86,** 2579–2589.

27. Qin, S., Ward, M., Raftopoulos, H., Tang, H., Bradley, B., Hesdorffer, C., and Bank, A. (1999). Competitive repopulation of retrovirally transduced haemopoietic stem cells. *Br. J. Haematol.* **107,** 162.

28. Hanania, E. G., and Deisseroth, A. B. (1994). Serial transplantation shows that early hematopoietic precursor cells are transduced by MDR-1 retroviral vector in a mouse gene therapy model. *Cancer Gene Ther.* **1,** 21–25.

29. Bertolini, F., Battaglia, M., Corsini, C., Lazzari, L., Soligo, D., Zibera, C., and Thalmeier K. (1996). Engineered stromal layers and continuous flow culture enhance multidrug resistance gene transfer in hematopoietic progenitors. *Cancer Res.* **56,** 2566–2572.

30. Hanania, E. G., Fu, S., Roninson, I., Zu, Z., and Deisseroth, A. B. (1995). Resistance to Taxol chemotherapy produced in mouse marrow cells by safety-modified retroviruses containing a human MDR-1 transcription unit. *Gene Ther.* **2,** 279–284.

31. Ward, M., Richardson, C., Pioli, P., Smith, L., Podda, S., Goff, S., Hesdorffer, C., and Bank, A. (1994). Transfer and expression of the human multiple drug resistance gene in human CD34+ cells. *Blood* **84,** 1408–1414.

32. Ward, M., Pioli, P., Ayello, J., Reiss, R., Urzi, G., Richardson, C., Hesdorffer, C., and Bank, A. (1996). Retroviral transfer and expression of the human multiple drug resistance (MDR) gene in peripheral blood progenitor cells. *Clin. Cancer Res.* **2,** 873–876.

33. Hesdorffer, C., Antman, K., Bank, A., Fetell, M., Mears, G., and Begg, M. (1994). Human MDR gene transfer in patients with advanced cancer. *Hum. Gene Ther.* **5,** 1151–1160.

34. Cooper, M. A., and Einhorn, L. H. (1995). Maintenance chemotherapy with daily oral VP-16 following salvage therapy in patients with germ cell tumors *J. Clin. Oncol.* **13,** 1167–1169.

35. Paukovits, W. R., Moser, M. H., and Paukovits, J. B. (1993). Pre-CFU-S quiescence and stem cell exhaustion after cytostatic drug treatment: protective effects of the inhibitory peptide. *Blood* **81,** 1755–1761.

36. Hanania, E. G., Giles, R. E., Kavanagh, J., Fu, S. Q., Ellerson, D., Zu, Z., Wang, T., Su, Y., Kudelka, A., Rahman, Z., Holmes, F., Hortobagyi, G., Claxton, D., Bachier, C., Thall, P., Cheng, S., Hester, J., Ostrove, J. M., Bird, R. E., Chang, A., Korbling, M., Seong, D., Cote, R., Holzmayer, T., and Deisseroth, A. B. (1996). Results of MDR-1 vector modification trial indicate that granulocyte/macrophage colony-forming unit cells do not contribute to posttransplant hematopoietic recovery following intensive systemic therapy. *Proc. Natl. Acad. Sci. USA* **93,** 15346–15351.

37. Hesdorffer, C., Ayello, J., Ward, M., Kaubisch, A., Vahdat, L., Balmaceda, C., Garrett, T., Fetell, M., Reiss, R., Bank, A., and Antman, K. (1998). Phase I trial of retroviral-mediated transfer of the human MDR-1 gene as marrow chemoprotection in patients undergoing high-dose chemotherapy and autologous stem-cell transplantation. *J. Clin. Oncol.* **16,** 165–172.

38. Devereux, S., Corney, C., Macdonald, C., Watts, M., Sullivan, A., Goldstone, A. H., Ward, M., Bank, A., and Linch, D. C. (1998). Feasibility of multidrug resistance (MDR-1) gene transfer in patients undergoing high-dose therapy and peripheral blood stem cell transplantation for lymphoma. *Gene Ther.* **5,** 403–408.

39. Moscow, J. A., Huang, H., Carter, C., Hines, K., Zujewski, J., Cusack, G., Chow, C., Venzon, D., Sorrentino, B., Chiang, Y., Goldspiel, B., Leitman, S., Read, E. J., Abati, A., Gottesman, M. M., Pastan, I., Sellers, S., Dunbar, C., and Cowan, K. H. (1999). Engraftment of MDR-1 and NeoR gene-transduced hematopoietic cells after breast cancer chemotherapy. *Blood* **94,** 52–61.

40. Cowan, K. H., Moscow, J. A., Huang, H., Zujewski, J. A., O'Shaughnessy, J., Sorrentino, B., Hines, K., Carter, C., Schneider, E., Cusack, G., Noone, M., Dunbar, C., Steinberg, S., Wilson, W., Goldspiel, B., Read, E. J., Leitman, S. F., McDonagh, K., Chow, C., Abati, A., Chiang, Y., Chang, Y. N., Gottesman M M., Pastan, I., and Nienhuis, A. (1999). Paclitaxel chemotherapy after autologous stem-cell transplantation and engraftment of hematopoietic cells transduced with a retrovirus containing the multidrug resistance complementary DNA (MDR-1) in metastatic breast cancer patients *Clin. Cancer Res.* **5,** 1619–1628.

41. Williams, D. A., Rios, M., Stephens, C., and Patel, V. (1991). Fibronectin and VLA-4 in hematopoietic stem cell microenvironment interactions. *Nature* **352,** 438–441.

42. Moritz, T., Patel, V. P., and Williams, D. A. (1994). Bone marrow extracellular matrix molecules improved gene transfer into human hematopoietic cells via retroviral vectors. *J. Clin. Invest.* **93,** 1451–1457.

43. Traycoff, C., Srour, E. F., Dutt, P., Fan, Y., and Cornetta, K. (1997). The 30/35 kDa chymotryptic fragment of fibronectin enhances retroviral-mediated gene transfer in purified chronic myelogenous leukemia bone marrow progenitors. *Leukemia* **11,** 159–167.

44. Hanenberg, H., Hashino, K., Konishi, H., Hock, R. A., Kato, I., and Williams D. A. (1997). Optimization of fibronectin-assisted retroviral gene transfer into human CD34+ hematopoietic cells. *Hum. Gene Ther.* **8,** 2193–2206.

45. Abonour, R., Williams, D. A., Einhorn, L., Hall, K. M., Chen, J., Coffman, J., Traycoff, C. M., Bank, A., Kato, I., Ward, M., Williams, S. D., Hromas, R., Robertson, M. J., Smith, F. O., Woo, D., Mills, B., Srour, E. F., and Cornetta, K. (2000). Efficient retrovirus-mediated transfer of the multidrug resistance 1 gene into autologous human long-term repopulating hematopoietic stem cells. *Nat. Med.* **6,** 652–658.

46. Ma, J. F., Grant, G., Staelens, B., Howard, D. L., and Melera, P. W. (1999). In vitro translation of a 2.3-kb splicing variant of the hamster pgp1 gene whose presence in transfectants is associated with decreased drug resistance. *Cancer Chemother. Pharmacol.* **43,** 19–28.

47. Baum, C., Hegewisch-Becker, S., Eckert, H. G., Stocking, C., and Ostertag, W. (1995). Novel retroviral vectors for efficient expression of the multidrug resistance (MDR-1) gene in early hematopoietic cells. *J. Virol.* **69,** 7541–7547.

48. Bunting, K. D., Galipeau, J., Topham, D., Benaim, E., and Sorrentino, B. P. (1998). Transduction of murine bone marrow cells with an MDR1 vector enables ex vivo stem cell expansion, but these expanded grafts cause a myeloproliferative syndrome in transplanted mice. *Blood* **92,** 2269–2279.

49. Cornetta, K. (1992). Safety aspects of gene therapy. *Br. J. Haematol.* **80,** 421–426.

23

Development and Application of an Engineered Dihydrofolate Reductase and Cytidine-Deaminase-Based Fusion Genes in Myeloprotection-Based Gene Therapy Strategies

OWEN A. O'CONNOR
Department of Medicine
Division of Hematologic Oncology,
Lymphoma, and
Developmental Chemotherapy Services
Memorial-Sloan Kettering Cancer Center
New York, New York 10021

TULIN BUDAK-ALPDOGAN
Department of Medicine
Programs of Molecular Pharmacology and
Therapeutics
Memorial-Sloan Kettering Cancer Center
New York, New York 10021

JOSEPH R. BERTINO
Department of Medicine
Division of Hematologic Oncology and Lymphoma,
and Programs of Molecular Pharmacology and
Therapeutics
Memorial-Sloan Kettering Cancer Center
New York, New York 10021

I. INTRODUCTION

A. Myeloprotection: Evolution and Application

The concept of bone marrow protection, or myeloprotection, has emerged directly from our understanding of acquired drug resistance mechanisms in cancer cells. The recognition that malignant cells could acquire resistance to a variety of antineoplastic agents following treatment, while the normal host cells could not, has been identified by many as one of the greatest challenges to effective cancer treatment [1]. As the genetic basis for drug resistance has emerged, the possible exploitation of these molecular mechanisms in a therapeutically beneficial manner was advanced [1]. In addition, over the last two decades, advances in structural biology have facilitated a detailed understanding of many enzyme–substrate interactions at the molecular level. This information has been used in conjunction with techniques in site-directed and random mutagenesis to rationally generate mutant forms of drug resistance genes [2–5]. These mutants typically fail to bind the target drug with any significant affinity and demonstrate relatively uncompromised catalytic activity despite toxic levels of drug exposure. The applicability of this approach in the clinic was initially limited because of the intrinsic difficulties associated with the efficiency and reproducibility of somatic gene transfer. As these methods improved, so to has the potential for novel innovations in gene therapy.

Over the last two decades, three major areas of cancer gene therapy research have dominated the field: (1) the introduction of genes into tumor cells that lead to apoptosis (herpes simplex virus thymide kinase [HSV-TK], cytosine deaminase), cessation of growth, or increased chemosensitivity (p53); (2) the introduction of genes into tumor cells or dendritic cells to stimulate the host's immune response against the cancer cells specifically (tumor antigen, DNA vaccines, cytokine genes); and (3) the introduction of drug-resistance genes into hematopoietic progenitor cells to decrease host toxicity (mutant dihydrofolate reductase [DHFR], multidrug resistance gene encoding for P-glycoprotein [MDR1], and wild-type or mutant forms of the methylguanine–DNA–methyltransferase gene [(Δ) MGMT]). Nearly a decade after the initial application of a novel gene therapy strategy for a nonmalignant disease (adenosine deaminase deficiency), a new repertoire of related tools and strategies has emerged that may complement our current approaches for managing human cancer.

The introduction of genes into cancer cells to promote apoptosis or chemosensitivity generally requires selective transduction of all the cancer cells to be successful, save the poorly misunderstood phenomenon of the "bystander effect" observed in suicide gene therapy (e.g., HSV-TK). Conversely, myeloprotection-based strategies may require transduction of only a minority of host stem cells with only transient expression of the biologically relevant protein in order to confer the desired host protection. The approach offers the advantage of conferring a selectable phenotype to the target cell that can theoretically allow for *in vivo* enrichment of the target cell population. This approach alone attempts to reconcile the well-known barriers to successful implementation of human gene therapy strategies: poor efficiency of gene transfer and transient duration of gene expression.

B. Limitations and Obstacles

Like any new pharmacologic agent or technology, there are theoretical and practical limitations and obstacles to consider. Myeloprotection-based strategies are no different. First and perhaps foremost is establishing the fact that transfer of a drug-resistance gene into a somatic tissue will indeed confer an increase in the drug-resistance phenotype of that tissue. Central to this consideration is the fact that many anti-neoplastic agents can produce several dose-limiting toxicites (DLTs). Protecting one tissue (for example, the hematopoietic cells) may do little to protect the organism as a whole from the other dose-limiting toxicities (for example, gastrointestinal mucositis, cardiomyopathy, hepatotoxicity, nephrotoxicity, pulmonary toxicity). Hence, in developing strategies that may have clinical utility, it is essential to understand the dose-limiting toxicities of the target drugs, as well as the schedules and doses that produce these toxicities.

The second major consideration that revolves around drug-resistance gene transfer strategies pertains to the inadvertent transduction of contaminating tumor cells and the possible emergence in the relapsed state of even more drug-resistant disease. This potential limitation can be overcome, to some extent, by using drug-resistance genes that confer protection against a more narrow spectrum of antineoplastic agents. For example, dihydrofolate reductase, cytidine deaminase, and MGMT all confer resistance to a relatively small number of agents compared to MDR1 and MRP, which can confer resistance against a theoretically larger panoply of agents. Practically, the relapse of methotrexate-resistant lymphoma following peripheral blood stem cell transplant, for example, may not limit therapeutic options as these drugs are poorly tolerated in the posttransplant setting. The emergence of MDR-resistant lymphoma cells, however, may render this disease resistant to epidophyltoxins, anthracyclines, and vinca alkaloids, potentially limiting future therapeutic options. Tailoring the *ex vivo* conditions may also aid in reducing the probability of inadvertent tumor cell transductions. Selection of CD34$^+$ cells and purging of grafts with disease-specific antibodies may help reduce the degree of tumor cell contamination in the graft. Likewise, the selection of oncoretroviruses (such as the Moloney murine leukemia virus [MoMLV]) that selectively infect only replicating cells may further reduce the transduction of tumor cells that most likely will not be stimulated to proliferate in the cytokine cocktails employed to facilitate CD34$^+$ cell replication.

The third, and by no means the last, concern revolves around the theoretical possibility that the transduction of a drug-resistance gene into otherwise normal healthy stem cells could result in a biological reprogramming of that cell. Such events could lead to transformation of these cells, resulting in myelodysplasia or even frank leukemia. Alternatively, such manipulations could trigger premature apoptosis, eventually leading to an aplastic crisis. Regardless, it is prudent to consider the spectrum of adverse outcomes and to integrate contingency safety measures where appropriate when developing clinical applications of the technology.

C. Selection of Drug-Resistance Genes

To date, the majority of data on the transfer of drug resistance genes have revolved around three specific drug-resistance models: (1) the multidrug resistance phenotype (conferred by the MDR1 gene); (2) dihydrofolate reductase mutants conferring resistance to antifolates, in particular methotrexate and trimetrexate; and (3) the methylguanine methyltransferase genes conferring resistance to alkylating agents and O^6-benzylguanine. The MDR1 gene encodes a 170-kDa membrane protein (from a 4.1-kbp cDNA); P-glycoprotein (PGP), which functions to efflux noxious byproducts of metabolism; xenobiotics; and natural products from the intracellular space [6,7]. This model is attractive as it theoretically confers resistance to a large diversity of drugs, including the taxanes, vinca alkaloids, and anthracyclines. However, potential limitations of this gene include the large size of its cDNA, which poses a constraint when cloning into double-copy retroviral vectors, and equivocal evidence regarding the clinical utility of P-glycoprotein in facilitating practical drug resistance to MDR drugs. In addition, a recent study has reported that mice receiving *ex vivo* expanded stem cells transduced with the MDR1 gene developed a myelodysplastic syndrome, manifest as leukocytosis and splenomegaly. While replication of this observation and establishing the mechanism for this effect remain to be achieved, these results have raised questions about the safety of MDR1 vectors for clinical myeloprotection-based strategies [8].

In contrast, resistance to methotrexate (MTX) is known to be multifactorial and can be mediated by any of the following mechanisms: (1) reduced influx of the drug (i.e., decreased expression of the reduced folate carrier [RFC]); (2) decreased expression of folypolyglutamyl synthase (FPGS), leading to decreased intracellular half-life and reduced affinity for the target enzymes; (3) increased γ-glutamyl hydrolase (GGH), which leads to increased cellular escape of the drug; (4) amplification of the target enzyme DHFR; and (5) specific point mutations in the DHFR gene that encode for an enzyme with markedly reduced affinity for MTX [9]. It is this latter mechanism that makes mutant forms of DHFR an attractive tool for myeloprotection-based strategies.

To date, several different mutations of both human and murine DHFR have been described [3,4,10–15]. In general, these proteins carry single-amino-acid substitutions in the active sites of the DHFR protein. The best characterized of these mutants have replaced: (1) the wild-type leucine at position 22 for arginine (Arg-22) or phenylalanine (Phe-22), and/or (2) the wild-type phenylalanine at position 31 for serine (Ser-31) or tryptophan (Trp-31). The Arg-22 mutant has been shown to impart an exceptional level of resistance to MTX (75,000-fold decrease in MTX affinity) but is significantly compromised from a catalytic perspective (250-fold decrease in k_{cat}). The Ser-31 DHFR mutant, however, confers high-level resistance to MTX (100-fold decrease in MTX affinity) without a significant reduction in the catalytic activity. Intentions to develop a DHFR mutant with the favorable features of each of the above enzymes (low MTX affinity with little to no change in k_{cat}) eventually led to the generation of "double" mutants

(dm) containing amino acid substitutions at both position 22 and 31. One of these double-mutant DHFR proteins contains phenylalanine at position 22, and serine at position 31 (Phe-22/Ser-31, or F/S DHFR). This enzyme has been shown to confer marked resistance to MTX with minimal compromise of catalytic activity [3,4]. Experimentally, these mutant enzymes have been shown to impart considerable resistance to methotrexate in tissue culture, granulocyte–macrophage colony-forming unit (CFU-GM) assays, and animal experiments [16–19].

In addition to the drug-resistance mechanisms discussed above, significant resistance can be conferred by genes involved in the normal detoxification and catabolism of the administered drug. Cytarabine (or Ara-C), for example, undergoes deamination by the enzyme cytidine deaminase (CD), leading to the formation of the inactive metabolite arabinosyl uracil (Ara-U). Recently, the cDNA for human cytidine deaminase was cloned [20], and subsequent retroviral transfection and expression of the enzyme in murine fibroblasts have resulted in increased resistance to both Ara-C and gemcitibine (2,2′-difluorodeoxy-cytidine), another substrate for CD [21,22].

One of the practical limitations presented by the design of vectors with very specific drug-resistance qualities lies in the inability to integrate multiply different drug-resistance genes into the same vector without compromising gene expression. The ability to tailor high-expression drug-resistance vectors would be of significance because it would allow the ability to: (1) apply specific combination chemotherapy regimens to the treatment of a particular disease; (2) exploit known synergistic drug combinations; and (3) augment gene expression by facilitating the generation of a single transcript without the need for intervening sequences. Currently, the most common mechanism for creating vectors with different drug-resistance genes employs the use of an internal ribosomal entry site (IRES element) that has been cloned from the encephalomyocarditis virus (ECMV). The IRES sequence permits Cap-independent translation of the downstream gene. These vectors have been widely used to facilitate the expression of two genes simultaneously within the same cell [23,24]. However, recent studies of these bicistronic vectors has revealed that the cDNA located on the 3′ end of the IRES element may not always be expressed to the same degree as the upstream gene [25]. An alternative strategy to facilitate the simultaneous expression of two enzymatic activities in the same cell is to facilitate the expression of both activities encoded as a single fusion protein that retains the enzymatic properties of its component proteins. Such drug-resistance fusion proteins are known to exist naturally and have even been shown to confer resistance to somatic cells *in vitro* [26,27]. This paradigm offers the possibility of cloning genes serially into a vector with or without an IRES element, allowing for the expression of several drug-resistance genes in the same cell following a single retroviral transduction.

II. FUSION GENES

A. Drug-Resistance Fusion Genes in Nature

There is little doubt that discrete proteins, in response to various evolutionary and environmental stresses, have broadly expanded their range of biological functions through their fusion and rearrangement with other ancestral genes. While the fraction of such bifunctional proteins is relatively small, a variety of naturally occurring multifunctional enzymes have evolved through their *in vivo* fusion with other cellular genes. To date, a number of these enzymes have been described and characterized, including aspartokinase-homoserine dehydrogenase [28], anthranilate synthases [29,30], DNA polymerase-1 [31], and tryptophan synthase [32] found in mammals. However, one of the best-characterized naturally occurring drug-resistance fusion genes is the bifunctional dihydrofolate reductase–thymidylate synthase fusion gene, which has evolved in parasitic protozoa. Thymidylate synthase (TS) and DHFR catalyze sequential enzymatic reactions in the *de novo* biosynthesis of deoxythymidylate (dTMP), as seen in the schematized thymidylate cycle in Fig. 1. TS catalyzes the methylation of deoxyuridylate using the cofactor 5,10-methylene-tetrahydrofolate reductase (CH_2H_4–folate). Aside from generating dTMP, the reaction also produces dihydrofolate (H_2–folate), which itself cannot be used as a cofactor by TS until it has undergone reduction back to tetrahydrofolate (H_4–folate). The conversion of tetrahydrofolate back to 5,10-methylene tetrahydrofolate occurs through DHFR, which catalyzes the reduction of H_2–folate by NADPH to regenerate the tetrahydrofolate (H_4–folate). Serine transhydroxymethylase (STH) then regenerates the

CH_2H_4–folate required for further thymidylate biosynthesis. Thymidylate synthase catalyzes the rate-limiting step in the *de novo* synthesis of thymidylate, which is subsequently phosphorylated to the nucleotide dTTP. In most living organisms, both TS and DHFR exist as distinct monfunctional enzymes. This is true in bacteriophage, viruses, fungi, and mammals [33]. In most of these organisms, TS exists as a dimer of two identical 35-kDa subunits, with one monomer being encoded by each of the two allelic forms of the thymidylate synthase gene. DNA sequencing of TS from many species has now revealed that this enzyme is one of the most highly conserved proteins in all of biology. In contrast, DHFR is a monofunctional enzyme with a molecular weight of about 20 kDa and, based upon DNA sequence data from many organisms, is very heterologous in nature.

Early investigations into the enzymology of TS and DHFR in protozoa demonstrated that these two proteins possessed unusually high molecular weights compared with the enzymes isolated from other organisms. It was not until Ferone and Roland [34] observed that TS and DHFR from *Crithidia fasciculata* copurified as a 110-kDa dimer of two subunits of identical size that these two enzymes were then believed to exist as a single fusion protein. Subsequent studies from the laboratory of Santi [35–37] clearly established that TS and DHFR from diverse protozoal species in fact exist as a bifunctional protein. Both enzymes exist on a single polypeptide chain, with the DHFR domain on the amino terminus and TS on the carboxyl terminus. These domains are separated by a junction peptide of varying size depending on the protozoal species from which it is derived. The native DHFR–TS protein is actually a dimer of two subunits which has a molecular weight of approximately 110–140 kDa. To date, a significant literature has evolved regarding the enzymology of the bifunctional protein in numerous parasitic protozoal species, including *Leishmania, Plasmodium, Toxoplasma, Trypanosoma, Cryptosporidium,* and *Crithidia,* and even the higher plant *Arabisdopsis thaliana.*

Many insights into the biological relevance of a bifunctional DHFR–TS have been advanced [38]. However, at present it seems that the major biological advantage of this fusion relates to the metabolic channeling of H_2–folate. Thymidylate synthase is relatively unique among enzymes that utilize folic acid cofactors in that for each mole of dTMP formed, 1 mole of H_4–folate is consumed. Dihydrofolate reductase must therefore rapidly reduce the H_2–folate formed in order to avoid rapid depletion of the H_4–folate required for thymidylate biosynthesis. Because H_2–folate generated by the TS subunits of the fusion protein is channeled to and reduced by DHFR faster than if it was released into the medium [33,36], the net rate of the sequential reactions becomes exclusively dependent on the catalytic rate of thymidylate synthase. As a result, there is minimal accumulation of inhibitory dihydrofolate and virtually no depletion of reduced folates. These evolutionary modifications to the normally

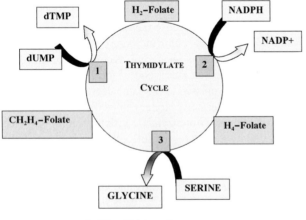

1 = Thymidylate synthase
2 = Dihydrofolate reductase
3 = Serine transhydroxymethylase

FIGURE 1   The thymidylate cycle. See text for details.

monofunctional enzymes have likely optimized the synthesis of dTMP in parasitic protozoa.

Perhaps one of the more intriguing aspects of the bifunctional DHFR–TS protein relates to studies regarding its role in acquired drug-resistance in various protozoal species. To date, two notably different mechanisms of acquired drug resistance have been found for the DHFR–TS fusion enzymes in the *Leishmania* sp. and *Plasmodium* sp. Much of the insight into acquired drug resistance of the bifunctional DHFR–TS began with the finding that exposure of *L. major* to MTX resulted in overproduction of the fusion protein which was eventually attributed to gene amplification [39]. In fact, protracted exposure of *Leishmania* to inhibitors of DHFR (e.g., MTX) or TS (e.g., 10-propargyl-5,8-dideazafolate) resulted in amplification of a 30- to 40-kb segment on chromosome 3 that corresponded to the locus for the DHFR–TS gene. This amplified DNA is replicated and circularized to give an extrachromosomal element referred to as R-DNA, which is believed to undergo autonomous replication and acquisition of mitotic stability upon continued exposure to methotrexate. This R-DNA can be present at a copy number up to 100 per cell. In fact, it was early studies in *L. major* containing DHFR–TS R-DNA and excess fusion DHFR–TS that led to the initial purification and characterization of the fusion protein. Interestingly, this model of gene amplification was established almost 5 years earlier in AT-300 cells (murine sarcoma) when Shimke and colleagues discovered that resistant cell variants could amplify up to a 300-fold increase in dihydrofolate reductase activity following exposure to progressively increasing concentrations of methotrexate [40–43]. This report was among the first to describe gene amplification as an important mechanism of drug resistance.

Resistance of *Plasmodium falciparum* to pyrimethamine, one of the few clinically effective antimalarial drugs, has now become a major global infectious disease problem. The DHFR–TS enzyme is the target for the antimalarial drug pyrimethamine. To date, several DHFR–TS genes have been sequenced from several pyrimethamine-resistant *P. falciparum* isolates [44–46]. In contrast to the gene amplification mechanisms discussed above for *Leishmania* sp., *P. falciparum* resistance appears to be primarily attributed to point mutations in the DHFR–TS fusion gene. These mutations, as mentioned earlier, result in a protein that has a markedly reduced affinity for pyrimethamine. In all of the resistant isolates studied to date, the threonine or serine amino acids at position 108 (depending upon the species studied) are converted to asparagine (Asn). Like the human dihydrofolate reductase double mutants, more resistant *P. falciparum* strains have been shown to possess a second Asn-51-to-isoleucine (Ile) mutation and/or a cysteine-59-to-arginine (Arg) mutation [47]. Specific resistance to the antifolate cycloguanil results from an Ile-164-to-leucine mutation.

While the primary structures of DHFR have been reported to be less than 25% homologous between species, comparison

of the three-dimensional X-ray structure has revealed a remarkable level of conservation of amino acids in the active site of the enzyme [48–50]. Binding sites analogous to the Phe-31 residue in human and murine DHFR have been mapped to the methionine-53 (Met-53) site in *Leishmania* sp. and the leucine-28 (Leu-28) residue in *Escherichia coli*. Replacement of the Phe-31 in murine DHFR with tryptophan or arginine or the Phe-31 site in human DHFR with serine can confer marked resistance to methotrexate [5,10,12,14]. Because this area of the DHFR binding site is known to form a hydrophobic binding site for the *P*-aminobenzoic acid moiety of MTX, it has been hypothesized that the change from methionine to arginine at position 53 of the DHFR in *L. major* may reduce the affinity for methotrexate, a scenario reminiscent of how the modifications described above alter the affinity of MTX in murine and human DHFR [51].

Recombinant DNA technology can now allow one to construct an almost endless variety of fusion genes. Such strategies can theoretically allow for the design and construction of novel fusion proteins with very focused clinical applications. These fusion genes, which have not been selected for over the course of protein evolution, might nonetheless possess unique and tailored properties that could allow their integration into novel biotherapeutic applications. Minimally, these products of natural selection may offer a useful paradigm for thinking about the development of new gene constructs.

B. Novel Synthetic Drug-Resistance Fusion Genes

In an effort to develop a drug-resistance fusion gene that might have an application in the treatment of non-Hodgkin's lymphoma, the possibility of fusing a previously generated double mutant of DHFR with a second gene that confers resistance to cytidine analogs (e.g., cytidine deaminase) was advanced. Building on the observations of Momparlear and Blau, as well as earlier work published by Bertino and colleagues, a novel fusion gene was developed that contained the F/S DHFR linked to CD through a thrombin cleavage site [52]. The F/S DHFR–CD fusion protein as well as the F/S DHFR and wild-type CD proteins were all overexpressed and purified from *E. coli* BL21 cells after IPTG induction. Essentially homogenous enzyme preparations of each enzyme were obtained based upon classic SDS-PAGE and Coomassie blue staining [52]. The purified recombinant F/S DHFR–CD fusion protein migrated at the expected size of 37 kDa (the molecular mass of DHFR is 21 kDa and CD is 16.1 kDa). Steady-state enzyme kinetic properties for the recombinant fusion protein as well as the F/S DHFR and CD enzymes were evaluated and compared (see Tables 1 and 2). The K_m values for 7,8-dihydrofolate and for NADPH were similar for F/S DHFR and F/S DHFR–CD (0.77 vs. 1.3 μM and 1.7 vs. 1.5 μM, respectively). Kinetic data for CD and F/S DHFR–CD were also similar in that the K_m values for the substrate Ara-C

TABLE 1   Enzyme Kinetic Parameters for Purified CD, F/S DHFR, and the Fusion Enzyme Protein F/S DHFR–CD

	K_m (μM) (FH2)	K_m (μM) (NADPH)	K_m (μM) (Ara-C)	K_m (μM) (cytidine)	K_m (S^{-1}) (FH2)	k_{cat} (S^{-1}) (Ara-C)	k_{cat} (S^{-1}) (cytidine)
CD	—	—	97.9 ± 8.03	22.3 ± 0.22	—	3.05 ± 0.26	4.77 ± 0.1
F/S DHFR	0.77 ± 0.01	1.7 ± 0.02	—	—	5.01 ± 0.47	—	—
F/S DHFR–CD	1.3 ± 0.25	1.5 ± 0.28	96 ± 1.7	24.4 ± 0.20	3.95 ± 39	2.2 ± 0.10	3.13 ± 0.09

Source: Adapted from Sauerbrey *et al.* [52].

were 97.9 vs. 96.0 μM, respectively. Likewise, K_m values for the natural substrate cytidine were 22.2 and 24.4 μM for CD and F/S DHFR–CD, respectively (see Tables 1 and 2).

The F/S DHFR–CD fusion gene is unique in its ability to confer resistance to a variety of antimetabolites, including methotrexate, trimetrexate, cytarabine, and gemcitabine. Host resistance conferred by this fusion gene offers the possibility of using several antineoplastic agents in combination. In particular, it could allow the simultaneous administration of two drugs (methotrexate and cytarabine) that have independent activity in the treatment of lymphoma, both of which have myelotoxicty as the principle dose-limiting toxicity. Additionally, these agents are potently synergistic against lymphoma. Additional data on the preclinical evaluation of this construct are presented subsequently.

C. Tailored Drug-Resistance Fusion Genes

One of the strengths of using recombinant techniques to generate novel drug-resistance genes lies in the potential of tailoring strategies for particular applications. Figure 2 presents a partial list of some theoretically possible constructs in various stages of development. As our knowledge of bone marrow transplantation biology expands, the feasibility of integrating myeloprotection-based strategies into not only myeloablative but also nonmyeloablative settings may become increasingly more attractive. Table 3 presents a list of some theoretical fusion gene constructs and their application to several types of malignancy.

TABLE 2   DHFR and CD Enzyme Activities in 3T3 Cells Following Infection with F/S DHFR, CD, CD-IRES F/S DHFR, or F/S DHFR–CD

Gene Transfected	DHFR (FH2) (U/μg)	CD (cytidine) (U/μg)
SFG-mock	9.2 ± 2.0	3.6 ± 0.9
SFG F/S DHFR	35.0 ± 5.2	3.9 ± 0.7
SFG CD	9.2 ± 1.2	35.0 ± 9.0
SFG CD IRES F/S DHFR	21.2 ± 4.8	23.0 ± 3.6
SFG F/S DHFR–CD	28.1 ± 5.0	22.0 ± 4.0

Source: Adapted from Sauerbrey *et al.* [52].

For example, fusion genes that unite mutants of DHFR and thymidylate synthase could be used to mitigate the myelotoxicity associated with fluoropyrimidines and antifolates in the management of breast and colorectal cancer, where combinations of MTX and 5-fluorouracil are commonly employed. Likewise, fusion constructs that employ CD can confer resistance to not only cytarabine but gemcitabine as well. These vectors could be used for any number of diseases, as presented in Table 3.

III. DEVELOPMENT OF CLINICALLY APPLICABLE GENE TRANSFER APPROACHES

One of the greatest obstacles in the development and successful application of effective gene-therapy-based technologies has revolved around the inability to efficiently and reproducibly deliver prescribed genes into target cell

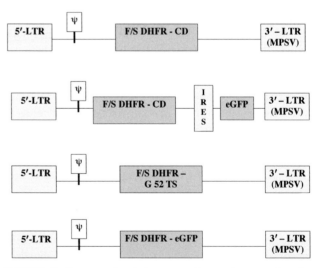

FIGURE 2   Representative fusion genes. Drug resistance and marker genes are represented as follows: F/S DHFR is the double mutant of dihydrofolate reductase; CD is the cytidine deaminase gene; G 52 TS is a mutant of thymidylate synthase; eGFP is the enhanced green fluorescent protein. The elements of the viral SFG backbone are represented as follows: LTR is the long terminal repeat; Ψ is the packaging signal; IRES is the internal ribosomal entry site; and the MPSVs represent the enhancing sequence from the myeloproliferative sarcoma virus. See text for additional details.

TABLE 3 Theoretical Drug-Resistance Fusion Genes and Their Clinical Application

Disease	Gene construct	Drugs
Hodgkin's disease non-Hodgkin's lymphoma	F/S DHFR–CD	Methotrexate, trimetrexate, cytosine arabinoside, gemcitabine
Osteosarcoma, choriocarcinoma	F/S DHFR	Methotrexate, Trimetrexate
Colorectal cancer, breast cancer	F/S DHFR–G52S TS F/S DHFR–CD	Methotrexate, trimetrexate, fluoropyrimidines, gemcitabine
Bladder cancer, lung cancer	F/S DHFR–MRP1	Methotrexate, taxanes, doxorubicin
Pancreatic cancer	G/52S TS–CD	Gemcitabine, fluoropyrimidines

populations. At present, many methods of gene delivery are routinely employed in both the laboratory and clinical settings, including both nonviral (cationic liposomes, cationic polyamines, ballistic delivery, calcium phosphate coprecipitation, electroporation) and viral (adenoviruses, adeno-associated viruses, lentiviruses, herpes viruses, and retroviruses) technologies. Each has demonstrated advantages and disadvantages depending upon the cells targeted and conditions studied. Unfortunately, many of the non-viral-based approaches suffer from several drawbacks that limit their widespread clinical application. These limitations generally include poor efficiency of gene transfer, variable expression of the transduced gene, variability in the number of gene copies transduced, excessive toxicity from the nonviral vector itself, and technical difficulties when batch scale-up is essential.

While viral-based vectors have become the principle tools for delivering genes of interest in both clinical and laboratory pursuits, each demonstrates particular advantages or disadvantages depending upon the experimental question being posed. For example, while adenovirus (AV) and adeno-associated viruses (AAVs) have been shown to reliably transduce many different types of resting cells, they have been of minimal value in the transduction of hematopoietic stem cells. Retroviruses, in contrast, have proven to be powerful tools for facilitating hematopoieitc stem cell gene transfer. Current retrovirus-mediated gene transfer techniques have been shown to produce stable proviral integration in the target cell genome, assuring safe and faithful transmission of the transduced gene into subsequent generations of cells. However, oncoretroviruses (in contrast to lentiviruses) obligately require their target cells to be in cycle for transduction, a cumbersome limitation when working with hematopoietic stem cells. While more technical discussions regarding the

subtleties of viral vector design are presented elsewhere, it is clear that the development of any hematopoietic progenitor cell gene transfer strategy will be inexorably linked to a sophisticated understanding of virus structure and function.

The construct proposed for the clinical studies discussed later is depicted in Fig. 2. The vector contains all of the elements described above, including the Myeloproliferative sarcoma virus (MPSV)-enhancing sequence in the U3 region of the 3' long terminal repeat (LTR) and the F/S DHFR–CD fusion gene. The packaging cell line used for these studies will be the AM12 line developed by Bank and colleagues [53]. These cells generate amphotrophic retroviral particles, which have been shown to efficiently transduce peripheral blood mononuclear cells (PBMCs) and long-term, culture-initiating cells (LTCICs) derived from CD34$^+$ cell selection.

IV. PRECLINICAL EVIDENCE FOR MYELOPROTECTION STRATEGIES

A. Evidence for Drug-Resistance Gene Transfer Using Mutants of DHFR

Mutations in the active binding site of dihydrofolate reductase (DHFR) have been shown to markedly reduce the affinity of methotrexate (MTX) and trimetrexate (TMTX) for DHFR. Two such mutations in the active site of the enzyme, Phe-34 $\rightarrow$ Ser-34 and Phe-31 $\rightarrow$ Ser-31, have been shown to confer significant resistance to MTX [54]. Myeloprotection with these mutant forms of DHFR was originally demonstrated using the N2A-based double-copy vector containing the simian virus 40 (SV40) promoter. Transduction of the mutant DHFR cDNA into murine hematopoietic progenitor cells (mHPCs) conferred significant resistance to 100 nM MTX (mock-transfected progenitor cells formed no colonies in the presence of MTX, while Ser-31 transfectants formed approximately 27% of the colonies observed in the control plates containing no MTX). These observations were extended to human hematopoietic progenitor cells (hHPC) in which CD34$^+$/CD38$^-$ selected cells from peripheral blood and umbilical cord blood were transduced with a retroviral vector derived from the spleen-focus virus genome (SFG)-based moloney murine leukemia virus (MoMLV). Infection-competent/replication-incompetent retrovirus carrying the Ser-31 mutant DHFR was generated using the amphotrophic cell line AM12 [55]. These results demonstrated both the feasibility and efficiency of ex vivo CD34$^+$ stem cell transduction with this retroviral vector and in vivo selection in the presence of MTX. The cytokines required for successful ex vivo expansion and transduction were subsequently refined in further studies using the same gene transfer system in HPCs from patients undergoing autologous stem cell transplantation for solid tumors [56].

Experiments using CFU-GM with these single mutants of DHFR led to whole animal studies using the double-copy (DC) retroviral vector (SFG) derived from the MoMLV. This vector carried the mutant DHFR with the Ser-31 mutation under the control of the SV40 promoter. In these experiments, transduction of mHPC with the Ser-31 DHFR conferred resistance to lethally irradiated mice transplanted with transduced marrow and challenged with lethal or low-dose MTX. These data revealed that the degree of myelosuppression was significantly reduced in the animals receiving low-dose MTX and transduced bone marrow compared to nontransduced marrow or mock-transduced controls. In addition, 75% of the animals receiving transduced bone marrow survived a high-dose MTX challenge, while none of the control animals (nontransduced or mock transduced) could survive the MTX administration [18]. These observations were extended when the long-term protection of recipient mice was demonstrated using secondary and tertiary murine transplant animal models [19]. These results demonstrated the ability to successfully transduce a "true" pluripotent stem cell capable of reconstituting murine hematopoiesis.

For example, retroviral transduction of bone marrow with the mutant S31F DHFR has been shown to confer long-term protection to mice receiving MTX [19]. Using the MoMLV-based vector carrying both the neomycin phosphotransferase gene (neo^R) and murine mutant DHFR S31F gene, lethally irradiated CBA/J mice (900 rad) were transplanted with 2×10^6 mononuclear cells transduced with the mutant DHFR. The animals were then given MTX in escalating doses twice weekly. Those animals that received marrow transduced with the mock producer cell supernatant (negative control) all died within 22 days of irradiation, while those animals not receiving any transplant died within 14 days. The deaths among these animals were attributed to severe anemia, gastrointestinal bleeding, and anorexia. Approximately 80% of those animals treated with the mutant DHFR retrovirus survived longer than 30 days (the termination of the experiment). Subsequent secondary and tertiary transplants were also performed. Again, approximately 80% of these animals survived longer that 30 days, while the control animals died by day 22. Both PCR and DNA sequence data confirmed the presence of the transfected DHFR mutant in plucked colonies from CFU assays [19].

Clearly, one of more important objectives in the clinical development of a myeloprotection transplant strategies in cancer is to be able to demonstrate therapeutic merit. Using a tumor model in which the flanks of mice were implanted with a murine mammary adenocarcinoma (E077), animals were treated in one of several fashions. All animals were initially treated with a sublethal dose of cyclophosphamide (cytoreductive therapy). Then, cohorts of animals were treated with weekly posttransplant MTX after receiving a bone marrow transplant (BMT) with or without HPC transduction with mutant DHFR. Approximately 50% of the an-

imals receiving transduced marrow survived the high-dose CTX and weekly MTX. At necrosy, all of these animals had no evidence of residual tumor. All mice receiving mock-transduced marrow died of MTX toxicity (anemia, gastrointestinal bleeding, anorexia), while 10 out of 10 and 7 out of 8 mice treated with untransduced marrow died of tumor progression. Of those animals that did not receive a BMT, approximately 90% died as a result of CTX and MTX toxicity [57]. Interestingly, this and other studies [58] confirm the findings that bone marrow transduction with mDHFR confers not only myeloprotection but also gastrointestinal protection to MTX. While the mechanisms for this remain unknown at present, it is hypothesized that preventing the depth and duration of neutropenia may contribute to the reduced mucositis.

Most of the research presented above describes the use of the Ser-31 mutant of human DHFR. Over the last several years, our understanding of DHFR biochemistry has facilitated the development of other DHFR mutants that can confer an even greater level of resistance to MTX, with even more favorable catalytic properties. One such double mutant contains phenylalanine at position 22, and serine at position 31 (Phe-22/Ser-31, or F/S DHFR). This enzyme has been shown to confer marked resistance to MTX with minimal compromise of the natural catalytic activity of the enzyme [3] and has been targeted for use in future clinical studies.

B. The Double-Mutant DHFR–CD Fusion Gene: Tissue Culture and CFU-GM

As mentioned earlier, confirmation of retroviral expression of F/S DHFR and the F/S DHFR–CD fusion protein were confirmed by western blot analysis in NIH3T3 cells. Detection of human protein resulting from vector expression was performed with an antibody specific to human DHFR that does not cross-react with endogenous murine DHFR. DHFR was not detectable in mock-transfected (AM12 producer cell line not previously transfected with any transgene) and CD-transfected cells. DHFR, and CD activities were measured in cell extracts prepared from 3T3 cells transfected with mock, SFG CD, SFG F/S DHFR, SFG CD IRES F/S DHFR, and SFG F/S DHFR-CD retrovirus. In lysates from mock-transfected cells, the endogenous (murine) DHFR activity was found to be about 9.2 U/μg of protein (where 1 unit = 1 nmol of substrate per minute), whereas the endogenous CD activity was found to be about 3.6 U/μg of protein. Extracts from the cells transduced with the SFG F/S DHFR displayed a 3.8-fold increase in DHFR activity (35 U/μg of protein) above background levels with no measurable change in CD activity. Lysates prepared from cells transduced with vectors carrying the SFG CD transgene displayed a 10-fold increase in CD activity (35 U/μg protein) above background, with no measurable increase in DHFR activity (Table 2). A 2.3-fold increase in DHFR activity (21.2 U/μg protein) and a 6.4-fold increase in CD activity (23 U/μg protein) were observed in lysates obtained from cells transduced with the SFG CD IRES F/S DHFR retrovirus. Extracts from cells infected with SFG F/S DHFR–CD displayed a 3-fold increase in DHFR activity (28.2 U/μg protein) and a 6.1-fold increase in CD activity (22 U/μg protein). These data helped to establish that DHFR and CD enzyme activities were essentially equivalent in lysates from cells transfected with either SFG CD IRES F/S DHFR or SFG F/S DHFR-CD (Table 2). However, the levels of functional resistance conferred by the gene on the 3' side of the IRES element tend to be lower than those found in vectors without the IRES sequence.

Evidence for the integration of transfected cDNA into the genome of NIH3T3 cells was based upon the detection of the transgene by PCR using genomic DNA. Using primers specific for human CD, a 465-bp CD fragment was amplified from genomic DNA isolated from cells transduced with SFG CD, SFG CD IRES F/S DHFR, and SFG F/S DHFR–CD. Using primers specific for human DHFR cDNA, a 400-bp fragment was amplified from cells infected with SFG F/S DHFR, SFG CD IRES F/S DHFR, and SFG F/S DHFR–CD [52]. The F/S DHFR–CD gene was detected as a 498-bp product amplified from genomic DNA isolated from NIH3T3 cells transduced with the SFG F/S DHFR–CD vector only, using a forward primer for human DHFR and a reverse primer for human CD [52].

Following the procurement of molecular and protein evidence for successful transgene expression, experiments were directed toward establishing hematopoietic cell protection. Initially, CFU-GM assays in the presence and absence of MTX were performed following retroviral transduction of mouse bone marrow cells (Table 4). After exposure to 500-nM Ara-C, no colony growth was noted in mock-transduced hematopoietic cells and 20% colony formation was observed in SFG-F/S-DHFR-transduced cells. In contrast, bone marrow cells transduced with SFG CD (53% colony formation) and SFG F/S DHFR–CD (58% colony formation) showed significantly higher cell survival after 500-nM Ara-C exposure. Hematopoietic cells transduced with SFG CD IRES F/S DHFR also demonstrated resistance against Ara-C (38%

colony formation). Genomic DNA isolated from mouse bone marrow colonies was used to demonstrate detection of the DHFR–CD transgene [52]. Using primers specific for human CD, a 465-bp fragment was detected in cells transfected with SFG CD and SFG CD IRES F/S DHFR. A 400-bp fragment of human DHFR cDNA was amplified from cells transfected with SFG F/S DHFR and SFG CD IRES F/S DHFR. The F/S DHFR–CD fusion gene was detected in SFG-F/S-DHFR–CD-transfected cells by amplification of a 498-bp fragment using a forward primer within the DHFR sequence and a reverse primer binding in the CD sequence. Using these primers, no bands could be detected when genomic DNA from mock-transfected murine bone marrow cells served as the template [52].

Employing a standard *ex vivo* coculture transduction method, the efficiency of gene transfer in canine mononuclear cells following retroviral transduction with the SFG[MPSV] F/S DHFR–CD transgene was observed to increase over twofold with increasing frequency of transduction. Following three separate transductions every 7 days, the percent colonies growing in the presence of 100-nM MTX ranged from 52–92%, with greater than 70% of all colonies demonstrating PCR evidence for the DHFR transgene [59]. These studies and others have begun to explore the identification of other large animal models that may serve as a more meaningful surrogate for assessing hematopoietic progenitor cell gene transfer [59].

V. CLINICAL APPLICATIONS OF MYELOPROTECTION STRATEGIES

A. Clinical Applications of Drug-Resistance Fusion Genes: Background

There are generally two major applications of hematopoietic stem cell gene transfer approaches. The first pertains to its integration as a therapeutic modality in existing cancer treatment regimens. Such approaches can be used to

TABLE 4 CFU-GM Assay After Retroviral Infection of Mouse Bone Marrow with F/S DHFR, CD IRES F/S DHFR, or F/S DHFR-CD.

	Number of colonies			
	Control	Ara-C (500 nM)	MTX (20 nM)	Ara-C (500 nM) + MTX (20 nM)
SFG-mock	280	56 (20%)	0	0
SFG F/S DHFR	302	0	42 (14%)	0
SFG CD	310	164 (53%)	0	0
SFG CD IRES F/S DHFR	308	117 (38%)	46 (15%)	22 (7%)
SFG F/S DHFR–CD	316	183 (58%)	54 (17%)	46 (15%)

allow for the administration of chemotherapy in the immediate posttransplant period in an effort to help eradicate minimal residual disease. This approach offers the possibility of protecting a newly engrafting stem cell product from the myelotoxic effects of chemotherapy, in particular antimetabolites. Other applications in cancer treatment include the possibility of markedly reducing the major dose-limiting toxicity (e.g., myelosuppression), potentially allowing for an increase in the dose intensity of the chemotherapeutic agent. The second major application of a drug-resistance-based gene therapy strategy involves the selection of nonselectable phenotypes. For example, correction of inherited hemoglobinopathies or coagulopathies in pluripotent stem cells may soon be a possibility. However, because the "corrected" phenotype may not offer a selective advantage of the "disease" phenotype, it will be difficult to enrich for these subpopulations of cells *in vivo*. Integration of a drug-resistance gene transfer component can potentially allow for the enrichment of these less-selectable phenotypes. For the purpose of this discussion, however, the remainder of the text will focus on the theoretical merits of using myeloprotection strategies to allow for the positioning of chemotherapy at novel points in a patient's treatment, as well as dose intensification.

B. Clinical Applications of Drug-Resistance Fusion Genes: Dose-Intensification Strategies

Perhaps our best understanding of cytokinetic modeling and the application of dose-intensification strategies (that is, sequential therapy) has emerged directly from an understanding of breast cancer biology and tumor growth. A reevaluation of several cytokinetic models of breast cancer cell proliferation has suggested that human solid tumors probably do not exhibit exponential growth (cells increase by a constant percentage per constant unit time regardless of the number of cells present), as was appreciated in the early studies of the transplantable L1210 leukemia in DBA mice [60–62]. Rather, solid tumors such as breast cancer more closely fit Gompertizian models (S-shaped pattern of growth) of cell proliferation [63]. Fundamental to this pattern of growth are the following features: (1) tumor proliferation slows with time, reaching a plateau at approximately 10^{12} cells; (2) doubling time constantly increases, eventually reaching infinity; and (3) tumor regression in response to effective antineoplastic therapy results in a log-kill that is larger when the tumor volume is smaller, though the rebound following the cessation of therapy is more rapid. This latter feature may be particularly germane to the integration of drug-resistance gene therapy strategies. In order to improve the overall impact of any cancer treatment, the tumor must be completely eradicated, or all residual neoplastic cells must enter a permanent G_0 state. Given the tools at our disposal today, it would appear that the most practical approach to preventing tumor regrowth is

to facilitate complete eradication of all cancer cells, a most formidable challenge.

Recently, some of these mathematical concepts of tumor cell growth have been utilized in order to design treatment schedules that might favorably exploit the fundamental features of Gompertizian growth [64,65]. One emerging aspect of tumor biology that is particularly relevant to the design of new therapeutic strategies concerns the observation that most cancers are remarkably heterogeneous [66]. While this heterogeneity can be envisioned in a variety of different ways, one practical interpretation involves visualizing any cancer as consisting of several different sublines of cells with different rates of proliferation. These sublines, based upon their intrinsic rate of cell proliferation and/or other acquired mutations in growth-regulating genes, can exhibit different sensitivities (or resistance) to different antineoplastic agents. Eradicating some sublines with chemotherapy would leave others to regrow rapidly following treatment, as would the incomplete eradication of any given subline [66].

One commonly employed way to facilitate "complete eradication" of all sublines has been to increase the dose intensification of the combination regimen (mg drug per m^2 per unit time, usually expressed as weeks), the logical extension of which can also include high-dose chemotherapy (HDC) with peripheral blood progenitor cell (PBPC) support. This approach, which has both laboratory and clinical support (primarily in the management of hematologic malignancies), is based upon the tenet that a linear relationship between cell kill and dose exists for most drugs (a dose response), with each drug in the combination adding its own cell-kill fraction. An often complementary strategy to enhance the eradication of residual sublines of cancer cells employs the use of non-cross-resistant sequential chemotherapy. This approach would increase not only dose intensity but also the "dose density" of a treatment regimen by allowing maximally tolerated doses of any one drug (as opposed to the reduced doses employed when drugs are given in combination). In addition, this approach, it is hypothesized, should overcome some of the inherent limitations of traditionally delivered chemotherapy by overcoming intrinsic drug resistance. Clinical evidence for the value of sequential adjuvant therapy in high-risk breast cancer has been demonstrated by a randomized trial performed by Bonadonna in Milan [67,68].

One of the major impediments to dose-intensity- or dose-density-based treatment paradigms is the dose-limiting toxicity (DLT). The ability to attenuate the major DLT, without the confounding prospect of rescuing tumor cells as well, would theoretically afford new opportunities to eradicate intrinsically resistant sublines. In addition, it would allow the integration of additional agents with different mechanisms of action without adding further toxicity. Antimetabolites have been the cornerstone of many chemotherapeutic regimens for over 25 years. Drugs such as MTX and the fluoropyrimidines disrupt DNA synthesis by reducing thymidylate pools, while drugs such as cytarabine and gemcitabine result in misincorporation, leading to increased DNA strand breakage and apoptosis. Because they are S-phase-specific agents, they are most effective against cells that are rapidly synthesizing DNA. Myeloprotection-based strategies could favorably exploit many of the cytokinetic principles discussed above by integrating the approach into already established and promising sequential-based treatment paradigms under study.

If a Gompertizian model of tumor cell growth is accurate, then incompletely eradicated sublines of tumor cells will regrow rapidly immediately following cessation of the prior therapy. This discrete period of accelerated regrowth would suggest that these cells, which are in a phase of rapid DNA synthesis, should be very sensitive to S-phase-specific agents. In addition, if the assumptions of dose density and tumor heterogeneity are correct, then the addition of new agents to the treatment regimen would eradicate sublines of cells insensitive to the previously employed agents. Given the steep nature of the dose–response curves for many of these agents, a mere doubling of the maximally tolerated dose could substantially improve both the log-kill fraction and dose density of the treatment. It is well recognized that dose-limiting toxicities other than myelosuppression may limit the dose escalation of antimetabolites such as MTX and the fluoropyrimidines, especially mucositis. However, it is interesting to note, as already mentioned above, that animal studies with transgenic animals expressing mutant transgenes of DHFR conferred simultaneous resistance to both hematopoietic and mucosal tissues of the gastrointestinal tract [58]. While the mechanisms for this have not been entirely elucidated, the answer may revolve around the depth and duration of neutropenia or the preservation of submucosal lymphoid aggregates from the cytotoxic therapy.

C. Clinical Applications of Drug Resistance Fusion Genes: Minimal Residual Disease Therapy

The second theoretical advantage of drug-resistance gene therapy strategies includes the possibility of delivering treatment during remission phase, or minimal residual disease treatment without incurring unacceptable toxicity. Such an approach might enhance the probability of catching cells as they exit G_0 and enter the cell cycle. Collectively, the theoretical advantages of a myeloprotection-based approach suggest that such a strategy, which confers meaningful resistance to the most important target tissue to the most biologically relevant drugs, could lead to an augmented eradication of all additional sublines of cells.

The concept of administering posttransplant chemotherapy has been regarded with skepticism secondary to concerns about post-hematopoietic recovery and engraftment. Damage to sensitive engrafting hematopoietic cells, leading to

myelosuppression, has raised concerns regarding the possibility of irreversible bone marrow suppression and aplasia. While the practice of posttransplant therapy has been tested only sparingly in the clinic, to date, several studies have explored the issue using both immunomodulatory therapies as well as cytotoxic agents. While none of these studies has been conducted in a randomized and controlled fashion, they are collectively suggestive of a therapeutic merit.

Recently, Powles et al. [69] evaluated the use of maintenance chemotherapy after autotransplantation in adult acute lymphoblastic leukemia (ALL). They enrolled 50 consecutive patients with ALL in first remission to receive myeloablation followed by peripheral blood progenitor cell transplant (PBPCT). After hematologic recovery, 6-mercaptopurine (6-MP) and MTX were administered for 2 years. Approximately 91.7% of PBPCT recipients received a median dose of 44 mg/m^2 MTX beginning 32 days posttransplant, while 75% of PBSCT recipients received weekly MTX. No graft failures were observed in any patients receiving posttransplant chemotherapy. The actuarial 5-year probabilities of overall survival and relapse were 56 and 30%, respectively. Furthermore, the relapse rate for patients who received 6-MP for less than 4 months was approximately 50%, while those patients who continued therapy for longer than 4 months had a relapse rate of only 20%. The authors concluded, in the absence of any randomized control, that posttransplant therapy was not only feasible but was also likely to have improved the relapse-free survival. In a slightly different study design, Tallman et al. [70] evaluated posttransplant 5-fluorouracil (5-FU) and cisplatin in 48 patients undergoing autologous transplant for metastatic breast cancer. Comparing their results to actuarial survival probabilities, the authors showed a statistically significant difference in the event-free survival (EFS) for patients receiving four cycles of 5-FU. The actuarial EFS for those patients who did not receive posttransplant 5-FU was 17% at 4 years, while it was 27% for those patients who received the 5-FU. Interestingly, those patients who achieved a complete response (CR) at the time of transplant had an EFS of 44% at 4 years. The overall survival doubled from 20%–40% in favor of those patients receiving posttransplant 5-FU. In lymphoma and myeloma, there is also limited experience with posttransplant treatment. Gryn et al. [71] treated patients who developed relapsed or refractory non-Hodgkin's lymphoma with α-interferon (1.5 million units daily for 2 years). They reported a disease-free survival of 77% at 2 years, with only 13% of patients developing a relapse posttransplant.

deMagalhaes-Silverman et al. [72] similarly showed that posttransplant doxorubicin and paclitaxel following high-dose chemotherapy with PBPCT did not adversely affect engraftment and was associated with only mild toxicity. Again, while these studies are clearly small and uncontrolled, there is a consistent suggestion that treatment in the minimal residual disease is feasible and safe and may have therapeutic merit.

Recently, a MDR1-based myeloprotection protocol for advanced breast cancer demonstrated that patients could tolerate posttransplant paclitaxel up to 225 mg/m^2 administered every 21 days for up to 12 cycles [73], despite the lack of detectable MDR1 in hematopoietic cells by PCR. Three patients who experienced only a partial remission from the transplant were converted to complete clinical responses following paclitaxel. In this study however, one half of the transplanted CD34$^+$ cell graft was transduced secondary to safety concerns. Molecular studies revealed that three of eight patients had PCR evidence for the MDR1 gene following the first paclitaxel, while no patients demonstrated molecular evidence of the gene by the end of the paclitaxel therapy. Moscow et al. [74] have published their experience with a similar kind of study using retroviral gene transfer of the MDR1 gene into the hematopoietic progenitor cells of patients undergoing high-dose chemotherapy for metastatic breast cancer. Patients received sequential paclitaxel (175 mg/m^2 every 3 weeks × 4 doses) and doxorubicin (75 mg/m^2 every 3 weeks × 4 total doses) following engraftment of HPCs transduced with the MDR1 and Neo^R genes. An average of 0.75 × 10^6 CD34$^+$ cells were transduced with viral supernatants containing the MDR1 or Neo^R transgene, with each patient receiving approximately 1.5 × 10^6 CD34$^+$-cell/kg-containing cells derived from each of the viral supernatants. Interestingly, the transduction efficiencies for the MDR1 gene ranged from 0.19–0.33% (average ~0.26%), while the efficiency for Neo^R ranged from 8–52% (average 30%) [74]. While the detection of the transgenes was considered low or nonexistent in the bone marrow and peripheral blood, the difference in transduction efficiencies was attributed to significant differences in the titer of the producer cell lines, which was at least one log greater for the Neo^R producer cell line. More recently, Abonour et al. [75] completed a clinical trial in which the MDR1 gene was transduced into the HPCs of patients with relapsed germ cell tumors who were treated in a high-dose chemotherapy program with PBPCT followed by oral posttransplant etoposide. These investigators showed that by using a fibronectin-based transduction method with stem cell factor, megakaryocyte-derived growth factor, and G-CSF they achieved the highest levels of gene transfer reported thus far in any clinical study based upon a CFU-GM assay on marrow progenitor cells (about 25% at 1 year). They also noted that a significant number of circulating differentiated hematopoietic cells were resistant to etoposide at 1 month (approximately 5.6%) and 9 months (0.5%) posttransplant. One of the most encouraging features of this study was that the authors provided strong evidence for in vivo selection, noting a 10-fold increase in drug resistant progenitor cells following oral etoposide. These study designs, and others, have begun to demonstrate, albeit in painfully incremental steps, the feasibility of genetically modifying pluripotent stem cells in the clinical setting [76].

The construction of a fusion gene such as the F/S DHFR–CD, described earlier, potentially allows one to exploit the favorable pharmacological features of several drugs simultaneously. Because methotrexate and cytarabine are cell-cycle-specific agents, the duration of drug exposure is critical in determining the fraction of cells killed. In addition, several in vitro studies have shown that Ara-C has significant synergistic antitumor activity when administered with MTX [77,78]. Simultaneous administration of Ara-C and MTX is associated with greater retention of Ara-cytosine triphosphate (CTP) in murine lymphoma cell lines which was associated with an improved therapeutic outcome when compared to either drug alone or when used in sequence separated by >24 hours [79]. Because antimetabolites are S-phase specific, their cytotoxic influence is greatest when the rate of DNA synthesis is highest. Several experimental systems have shown that cell kill from Ara-C in culture is maximized when DNA synthesis is at a peak, as is often the case in the recovery period following exposure to other classes of cytotoxic agents [80]. For example, some studies have exploited the kinetic patterns of leukemic cell recovery after Ara-C exposure to optimize the sequential dosing of drug [81,82]. Retreatment 8–10 days after initial treatment with Ara-C has been shown to improve the duration of remission in adult leukemic patients [82]. Both experimental and clinical studies seem to suggest that optimal treatment with some antimetabolites may depend on the timing and duration of subsequent treatments in order to take advantage of recruitment of residual cells into S-phase of the cell cycle [80].

In one ongoing application of this strategy, patients with poor risk relapsed/refractory non-Hodgkin's lymphoma (international prognostic index [IPI] >3) will be integrated directly into an established transplant paradigm. This approach will involve transduction of stem cells with the F/S DHFR–CD fusion gene, followed by posttransplant Ara-C and MTX. The proposed posttransplant treatment schedule is designed, first, to test the limits of myeloprotection conferred by the fusion gene and, second, to attempt to exploit some of the favorable pharmacodynamic and pharmacokinetic properties of both cytarabine and methotrexate in the treatment of residual lymphoma following HDC with PBSCT, as discussed previously.

D. Integration of a Myeloprotection Strategy Using the F/S DHFR–CD Fusion Gene into the Treatment of Relapsed or Refractory High-Risk Non-Hodgkin's Lymphoma

The best conventional regimen for the initial management of intermediate-grade lymphomas is CHOP chemotherapy: cyclophosphamide (Cytoxan), hydroxydaunorubicin (doxorubicin [Adriamycin]), vincristine (Oncovin), and prednisone [83]. However, despite a high remission rate, the 3- to 5-year progression-free survival remains only 40–50%. For those patients who develop primary refractory or relapsed disease, conventional salvage chemotherapy produces remissions in approximately 50–70% of patients, although long-term remissions are seen in less than 10% of patients [84,85]. For select patients, however, curative treatment can still be obtained with high-dose chemotherapy followed by bone marrow or peripheral blood stem cell transplantation [86–90]. HDC with peripheral blood stem cell (PBSC) transplant is now considered the best option for patients with chemosensitive intermediate- or high-grade non-Hodgkin's lymphoma (NHL), and all eligible patients should be directed toward transplant at the first sign of relapse.

The PARMA study was a prospectively randomized trial that clearly established the benefit of HDC with autologous bone marrow transplantation (ABMT) over conventional salvage treatment for patients with relapsed NHL [91]. The study included 215 patients with intermediate-grade lymphoma (IGL) or high-grade lymphoma (HGL) in first relapse. Patients with primary refractory disease, or those with evidence of central nervous system or bone marrow involvement were ineligible to participate in the study. All patients received two cycles of DHAP (dexamethasone, high-dose cytarabine, and cisplatin), after which they were randomized to four additional cycles of DHAP or HDC with BEAM (BCNU, etoposide, Ara-C, and melphalan) and ABMT, plus involved field radiotherapy for bulky disease. Patients randomized to ABMT had a significantly higher response rate (84 vs. 44%) and event-free (46 vs. 12%) and overall survival (53 vs. 32%) at 5 years. A recent update and analysis of outcome using the age-adjusted international prognostic index (IPI) demonstrated that, for patients treated with chemotherapy alone, there is a progressive decline in disease-free survival with increasing IPI [92]. Patients in risk group 0 had an 8-year survival of 36%, compared to 20% for those in risk groups 1–5. In contrast, no difference in survival based on IPI was seen in the transplant group, where the 8-year survival was reported to be approximately 48% for those in both risk group 0 and risk groups 1–5.

The efficacy of ABMT in the treatment of chemoresistant relapsed disease has been an area of considerable controversy. Philip et al. [93] documented a 14% progression-free survival (PFS) for 22 patients with resistant disease. More recently, Stiff et al. [94] found that 22% of patients with chemoresistant disease achieved PFS. In this retrospective analysis, responding patients transplanted with less than 1 cm of residual disease had a 2-year survival of 86%, compared to 7% for those patients with greater than 1 cm of residual disease (in CR). While these are both small studies, they raise the possibility that some fraction of patients with documented chemoresistant relapsed disease may benefit from HDC with ABMT.

However, increasingly, many investigators feel these patients should be enrolled in research studies exploring alternative approaches.

A number of different antineoplastic agents are known to be active in the treatment of relapsed NHL. These drugs typically include ifosphamide (I), etoposide or VP-16 (E), cytarabine (A) or high-dose cytarabine (HA), cisplatin (P), carboplatin (C), methotrexate (M), and dexamethasone (D). Initially, one combination of drugs was used at Memorial-Sloan Kettering Cancer Center in a regimen to cytoreduce patients prior to myeloablation. The regimen, ICEMAN (N = neupogen), utilized two non-cross-resistant treatment programs given sequentially: ICE every 2 weeks for three cycles and MA weekly for five cycles. Because MA was found to significantly increase the toxicity of the regimen in general and did not change the number of patients eventually reaching ABMT, present induction regimens were modified to employ ICE alone [95]. In the clinical study alluded to earlier, poor-risk patients with relapsed or refractory NHL (IPI $\geq$ 3) entered into this trial will receive a standard transplant treatment [95]. However, following peripheral blood progenitor cell mobilization, their CD34$^+$ cells will be transduced with the F/S DHFR–CD fusion gene. Following transplantation, patients will receive escalating doses of posttransplant MTX and Ara-C in an effort to eradicate minimal residual disease and facilitate in vivo selection of the transduced stem cells.

VI. CHALLENGES

The recent results from several clinical trials have prompted investigators to highlight some of the short- and long-term failings of hematopoietic stem cell gene transfer [96]. In general, four concerns have been underscored: (1) the (in)ability to select for transduced cells in vivo; (2) the (in)ability to demonstrate successful transduction of the long-term repopulating stem cell; (3) concerns regarding the inadvertent transduction of contaminating malignant cells [97]; and (4) the practical merits of widening the therapeutic window. At the heart of these issues lie the fundamental difficulties associated with genetic modification of only a quiescent, long-term, repopulating hematopoietic stem cell using cell-cycle-dependent oncoretroviruses and ways in which investigators can measure, in surrogate systems, the efficiency of transduction and stem cell homing. The use of long-term, culture-initiating cell (LTCIC) assays with secondary CFU cultures and the NOD/SCID repopulating models are currently the "gold standards." Increasingly, however, many investigators are beginning to sense that there is no optimal substitute and that the best assay may in fact involve testing in competitive repopulating large animal models, including the rhesus monkey, baboon, dog, and possibly even pilot human clinical trials. The process required to implement successful HSC gene transfer is inherently complex, with any given protocol involving a seemingly infinite array of methodological permutations. Variables such as the best cytokine cocktail, duration of *ex vivo* incubation, frequency of transduction, type of vector (with and without different enhancing elements, envelopes, etc.), multiplicity of infection, and media composition, for example, require careful study and examination using the most meaningful surrogate assay conditions available. Coupling these variables to the ever-increasing pace of new developments (vector design issues, insulating genetic elements, cytokine fusion proteins, etc.) makes optimization intrinsically complicated, though no less interesting. As many in the field begin to confront these issues directly, it now appears that some investigators are beginning to report evidence of long-term repopulating cell transduction in large animal models [98,99] and even in a preliminary human clinical trial [76].

Many of the recent advancements in hematopoietic stem cell gene transfer can be attributed to a better understanding of the underlying stem cell biology and its interactions with the stromal compartment in the intramedullary compartment. For example, several independent investigators have shown that either fibronectin-coated plates or fibronectin fragments can enhance the retroviral transduction of human hematopoietic progenitor cells [100–102]. Fibronectin is a bone marrow extracellular matrix protein that mediates adhesion of human hematopoietic progenitor cells through cell surface proteoglycans and the β_1-integrins. A recombinant human fibronectin fragment (CH-296 or Retronectin) contains the essential domains required for both cell adhesion and viral binding [103], enhancing retroviral transduction. A second major area of improvement revolves around an improved understanding of cytokine priming strategies for stem cell cycle activation. While this topic is covered extensively elsewhere [104], it is clear that stimulating stem cells into cell cycle and mitosis to facilitate oncoretroviral transduction without differentiation may be one of the greatest challenges. Recent data suggest that cytokine combinations containing thrombopoietin (TPO), stem cell factor (c-kit ligand), and Flk-2 ligand (Flk-2L) can increase the number of cobblestone-area-forming cells (CAFC) in LTCIC, also resulting in hematopoietic reconstitution in NOD/SCID models [105–107]. These data are complicated by recent evidence that prolonged culture of CD34$^+$ cells has a negative effect on competitive repopulation in rhesus monkeys [108]. Hence, issues related not only to the cytokine cocktail but also the duration of *ex vivo* incubation and exposure can have profound effects on the efficiency of transduction and homing to the intrameduallary compartment.

It has now become apparent that the future direction and success of this field will depend upon our ability to test important questions under the conditions that are most likely to provide information that can impact translational develop-

ment. In addition, it will be imperative to begin correlating our observations from these *in vivo* experiments in large animal models and pilot clinical trials with the results from surrogate assays, so that we can better determine the merits and limitations of the latter.

References

1. Bertino, J. R. (1990). Turning the tables—making normal marrow resistant to chemotherapy. *J. Natl. Cancer Inst.* **82**(15), 1234–1235.

2. Tong, Y., Liu-Chen, X., Ercikan-Abali, E. A., Capiaux, G. M., Zhao, S. C., Banerjee, D., and Bertino, J. R. (1998). Isolation and characterization of thymitaq (AG337) and 5-fluoro-2-deoxyuridylate-resistant mutants of human thymidylate synthase from ethyl methanesulfonate-exposed human sarcoma HT1080 cells. *J. Biol. Chem.* **273**(19), 11611–11618.

3. Ercikan-Abali, E. A., Mineishi, S., Tong, Y., Nakahara, S., Waltham, M. C., Banerjee, D., Chen, W., Sadelain, M., and Bertino, J. R. (1996). Active site-directed double mutants of dihydrofolate reductase. *Cancer Res.* **56**(18), 4142–4145.

4. Ercikan-Abali, E. A., Waltham, M. C., Dicker, A. P., Schweitzer, B. I., Gritsman, H., Banerjee, D., and Bertino, J. R. (1996). Variants of human dihydrofolate reductase with substitutions at leucine-22: effect on catalytic and inhibitor binding properties. *Mol. Pharmacol.* **49**(3), 430–437.

5. Schweitzer, B. I., Srimatkandada, S., Gritsman, H., Sheridan, R., Venkataraghavan, R., and Bertino, J. R. (1989). Probing the role of two hydrophobic active site residues in the human dihydrofolate reductase by site-directed mutagenesis. *J. Biol. Chem.* **264**(34), 20786–20795.

6. Pastan, I., and Gottesman, M. M. (1991). Multidrug resistance. *Annu. Rev. Med.* **42**, 277–286.

7. Ward, M., Richardson, C., Pioli, P., Smith, L., Podda, S., Goff, S., Hesdorffer, C., and Bank, A. (1994). Transfer and expression of the human multiple drug resistance gene in human CD34+ cells. *Blood* **84**(5), 1408–1414.

8. Bunting, K. D., Galipeau, J., Topham, D., Benaim, E., and Sorrentino, B. P. (1998). Transduction of murine bone marrow cells with an MDR1 vector enables ex vivo stem cell expansion, but these expanded grafts cause a myeloproliferative syndrome in transplanted mice. *Blood* **92**(7), 2269–2279.

9. Gorlick, R., Goker, E., Trippett, T., Waltham, M., Banerjee, D., and Bertino, J. R. (1996). Intrinsic and acquired resistance to methotrexate in acute leukemia. *N. Engl. J. Med.* **335**(14), 1041–1048.

10. Thillet, J., Absil, J., Stone, S. R., and Pictet, R. (1988). Site-directed mutagenesis of mouse dihydrofolate reductase. Mutants with increased resistance to methotrexate and trimethoprim. *J. Biol. Chem.* **263**(25), 12500–12508.

11. Haber, D. A., Beverley, S. M., Kiely, M. L., and Schimke, R. T. (1981). Properties of an altered dihydrofolate reductase encoded by amplified genes in cultured mouse fibroblasts. *J. Biol. Chem.* **256**(18), 9501–9510.

12. Srimatkandada, S., Schweitzer, B. I., Moroson, B. A., Dube, S., and Bertino, J. R. (1989). Amplification of a polymorphic dihydrofolate reductase gene expressing an enzyme with decreased binding to methotrexate in a human colon carcinoma cell line, HCT-8R4, resistant to this drug. *J. Biol. Chem.* **264**(6), 3524–3528.

13. Dicker, A. P., Volkenandt, M., Schweitzer, B. I., Banerjee, D., and Bertino, J. R. (1990). Identification and characterization of a mutation in the dihydrofolate reductase gene from the methotrexate-resistant Chinese hamster ovary cell line Pro-3 MtxRIII. *J. Biol. Chem.* **265**(14), 8317–8321.

14. McIvor, R. S., and Simonsen, C. C. (1990). Isolation and characterization of a variant dihydrofolate reductase cDNA from methotrexate-resistant murine L5178Y cells. *Nucleic Acids Res.* **18**(23), 7025–7032.

15. Simonsen, C. C., and Levinson, A. D. (1983). Isolation and expression of an altered mouse dihydrofolate reductase cDNA. *Proc. Natl. Acad. Sci. USA* **80**(9), 2495–2499.

16. Williams, D. A., Hsieh, K., DeSilva, A., and Mulligan, R. C. (1987). Protection of bone marrow transplant recipients from lethal doses of methotrexate by the generation of methotrexate-resistant bone marrow. *J. Exp. Med.* **166**(1), 210–218.

17. Corey, C. A., DeSilva, A. D., Holland, C. A., and Williams, D. A. (1990). Serial transplantation of methotrexate-resistant bone marrow: protection of murine recipients from drug toxicity by progeny of transduced stem cells. *Blood* **75**(2), 337–343.

18. Li, M. X., Banerjee, D., Zhao, S. C., Schweitzer, B. I., Mineishi, S., Gilboa, E., and Bertino, J. R. (1994). Development of a retroviral construct containing a human mutated dihydrofolate reductase cDNA for hematopoietic stem cell transduction. *Blood* **83**(11), 3403–3408.

19. Zhao, S. C., Li, M. X., Banerjee, D., Schweitzer, B. I., Mineishi, S., Gilboa, E., and Bertino, J. R. (1994). Long-term protection of recipient mice from lethal doses of methotrexate by marrow infected with a double-copy vector retrovirus containing a mutant dihydrofolate reductase. *Cancer Gene Ther.* **1**(1), 27–33.

20. Laliberte, J., and Momparler, R. L. (1994). Human cytidine deaminase: purification of enzyme, cloning, and expression of its complementary DNA. *Cancer Res.* **54**(20), 5401–5407.

21. Neff, T., and Blau, C. A. (1996). Forced expression of cytidine deaminase confers resistance to cytosine arabinoside and gemcitabine. *Exp. Hematol.* **24**(11), 1340–1346.

22. Momparler, R. L., Eliopoulos, N., Bovenzi, V., Letourneau, S., Greenbaum, M., and Cournoyer, D. (1996). Resistance to cytosine arabinoside by retrovirally mediated gene transfer of human cytidine deaminase into murine fibroblast and hematopoietic cells. *Cancer Gene Ther.* **3**(5), 331–338.

23. Sorrentino, B. P., Brandt, S. J., Bodine, D., Gottesman, M., Pastan, I., Cline, A., and Nienhuis, A. W. (1992). Selection of drug-resistant bone marrow cells in vivo after retroviral transfer of human MDR1. *Science* **257**(5066), 99–103.

24. Aran, J. M., Gottesman, M. M., and Pastan, I. (1998). Construction and characterization of bicistronic retroviral vectors encoding the multidrug transporter and beta-galactosidase or green fluorescent protein. *Cancer Gene Ther.* **5**(4), 195–206.

25. Sugimoto, Y., Aksentijevich, I., Murray, G. J., Brady, R. O., Pastan, I., and Gottesman, M. M. (1995). Retroviral coexpression of a multidrug resistance gene (MDR1) and human alpha-galactosidase A for gene therapy of Fabry disease. *Hum. Gene Ther.* **6**(7), 905–915.

26. Inselburg, J., and Zhang, R. D. (1988). Study of dihydrofolate reductase-thymidylate synthase in *Plasmodium falciparum. Am. J. Trop. Med. Hyg.* **39**(4), 328–336.

27. Fantz, C. R., Shaw, D., Moore, J. G., and Spencer, H. T. (1998). Retroviral coexpression of thymidylate synthase and dihydrofolate reductase confers fluoropyrimidine and antifolate resistance. *Biochem. Biophys. Res. Commun.* **243**(1), 6–12.

28. Cohen, G. N., and Dautry-Varsat, A. (1980). The aspartokinase-homoserine dehydrogenase of *Escherichia coli*, In *Multidomain Proteins—Structure and Evolution* (D. G. Hardies and J. R. Coggins, eds.), pp. 49–121. Elsevier, Amsterdam.

29. Arroyo-Begovich, A., and DeMoss, J. A. (1973). The isolation of the components of the anthranilate synthetase complex from *Neurospora crassa. J. Biol. Chem.* **248**(4), 1262–1267.

30. Hankins, C. N., and Mills, S. E. (1976). Anthranilate synthase-amidotransferase (combined). A novel form of anthranilate synthase from *Euglena gracilis. J. Biol. Chem.* **251**(24), 7774–7778.

31. Huang, W. M., and Lehman, I. R. (1972). On the exonuclease activity of phage T4 deoxyribonucleic acid polymerase. *J. Biol. Chem.* **247**(10), 3139–3146.

32. Matchett, W. H., and DeMoss, J. A. (1975). The subunit structure of tryptophan synthase from *Neurospora crassa. J. Biol. Chem.* **250**(8), 2941–2946.

33. Trujillo, M., Donald, R. G., Roos, D. S., Greene, P. J., and Santi, D. V. (1996). Heterologous expression and characterization of the bifunctional dihydrofolate reductase-thymidylate synthase enzyme of *Toxoplasma gondii. Biochemistry.* **35**(20), 6366–6374.

34. Ferone, R., and Roland, S. (1980). Dihydrofolate reductase: thymidylate synthase, a bifunctional polypeptide from *Crithidia fasciculata. Proc. Natl. Acad. Sci. USA* **77**(10), 5802–5806.

35. Garrett, C. E., Coderre, J. A., Meek, T. D., Garvey, E. P., Claman, D. M., Beverley, S. M., and Santi, D. V. (1984). A bifunctional thymidylate synthetase–dihydrofolate reductase in protozoa. *Mol. Biochem. Parasitol.* **11**, 257–265.

36. Meek, T. D., Garvey, E. P., and Santi, D. V. (1985). Purification and characterization of the bifunctional thymidylate synthetase–dihydrofolate reductase from methotrexate-resistant *Leishmania tropica. Biochemistry.* **24**(3), 678–686.

37. Grumont, R., Washtien, W. L., Caput, D., and Santi, D. V. (1986). Bifunctional thymidylate synthase–dihydrofolate reductase from *Leishmania tropica*: sequence homology with the corresponding monofunctional proteins. *Proc. Natl. Acad. Sci. USA* **83**(15), 5387–5391.

38. Ivanetich, K. M., and Santi, D. V. (1990). Thymidylate synthase–dihydrofolate reductase in protozoa. *Exp. Parasitol.* **70**(3), 367–371.

39. Coderre, J. A., Beverley, S. M., Schimke, R. T., and Santi, D. V. (1983). Overproduction of a bifunctional thymidylate synthetase–dihydrofolate reductase and DNA amplification in methotrexate-resistant *Leishmania tropica. Proc. Natl. Acad. Sci. USA* **80**(8), 2132–2136.

40. Schimke, R. T., Alt, F. W., Kellems, R. E., Kaufman, R. J., and Bertino, J. R. (1978). Amplification of dihydrofolate reductase genes in methotrexate-resistant cultured mouse cells. *Cold Spring Harbr Symp. Quant. Biol.* **42**(pt 2), 649–657.

41. Dolnick, B. J., Berenson, R. J., Bertino, J. R., Kaufman, R. J., Nunberg, J. H., and Schimke, R. T. (1979). Correlation of dihydrofolate reductase elevation with gene amplification in a homogeneously staining chromosomal region in L5178Y cells. *J. Cell. Biol.* **83**(2, pt 1), 394–402.

42. Alt, F. W., Kellems, R. E., Bertino, J. R., and Schimke, R. T. (1978). Selective multiplication of dihydrofolate reductase genes in methotrexate-resistant variants of cultured murine cells. *J. Biol. Chem.* **253**(5), 1357–1370.

43. Kaufman, R. J., Bertino, J. R., and Schimke, R. T. (1978). Quantitation of dihydrofolate reductase in individual parental and methotrexate-resistant murine cells. Use of a fluorescence activated cell sorter. *J. Biol. Chem.* **253**(16), 5852–5860.

44. Cowman, A. F., Morry, M. J., Biggs, B. A., Cross, G. A., and Foote, S. J. (1988). Amino acid changes linked to pyrimethamine resistance in the dihydrofolate reductase-thymidylate synthase gene of *Plasmodium falciparum. Proc. Natl. Acad. Sci. USA* **85**(23), 9109–9113.

45. Peterson, D. S., Walliker, D., and Wellems, T. E. (1988). Evidence that a point mutation in dihydrofolate reductase–thymidylate synthase confers resistance to pyrimethamine in *Falciparum malaria. Proc. Natl. Acad. Sci. USA* **85**(23), 9114–9118.

46. Hyde, J. E. (1989). Point mutations and pyrimethamine resistance in *Plasmodium falciparum. Parasitol. Today.* **5**, 252–255.

47. Ivanetich, K. M., and Santi, D. V. (1990). Bifunctional thymidylate synthase–dihydrofolate reductase in protozoa. *FASEB J.* **4**(6), 1591–1597.

48. Matthews, D. A., Alden, R. A., Bolin, J. T., Freer, S. T., Hamlin, R., Xuong, N., Kraut, J., Poe, M., Williams, M., and Hoogsteen, K. (1977).

Dihydrofolate reductase: X-ray structure of the binary complex with methotrexate. *Science* **197**(4302), 452–455.

49. Oefner, C., D'Arcy, A., and Winkler, F. K. (1988). Crystal structure of human dihydrofolate reductase complexed with folate. *Eur. J. Biochem.* **174**(2), 377–385.

50. Bolin, J. T., Filman, D. J., Matthews, D. A., Hamlin, R. C., and Kraut, J. (1982). Crystal structures of *Escherichia coli* and *Lactobacillus casei* dihydrofolate reductase refined at 1.7 A resolution. I. General features and binding of methotrexate. *J. Biol. Chem.* **257**(22), 13650–13662.

51. Arrebola, R., Olmo, A., Reche, P., Garvey, E. P., Santi, D. V., Ruiz-Perez, L. M., and Gonzalez-Pacanowska, D. (1994). Isolation and characterization of a mutant dihydrofolate reductase–thymidylate synthase from methotrexate-resistant *Leishmania* cells. *J. Biol. Chem.* **269**(14), 10590–10596.

52. Sauerbrey, A., McPherson, J. P., Zhao, S. C., Banerjee, D., and Bertino, J. R. (1999). Expression of a novel double-mutant dihydro-folate reductase–cytidine deaminase fusion gene confers resistance to both methotrexate and cytosine arabinoside. *Hum Gene Ther.* **10**(15), 2495–2504.

53. Markowitz, D., Goff, S., and Bank, A. (1988). A safe packaging line for gene transfer: separating viral genes on two different plasmids. *J. Virol.* **62**(4), 1120–1124.

54. Banerjee, D., Schweitzer, B. I., Volkenandt, M., Li, M. X., Waltham, M., Mineishi, S., Zhao, S. C., and Bertino, J. R. (1994). Transfection with a cDNA encoding a Ser31 or Ser34 mutant human dihydrofolate reductase into Chinese hamster ovary and mouse marrow progenitor cells confers methotrexate resistance. *Gene* **139**(2), 269–274.

55. Flasshove, M., Banerjee, D., Bertino, J. R., and Moore, M. A. (1995). Increased resistance to methotrexate in human hematopoietic cells after gene transfer of the Ser31 DHFR mutant. *Leukemia* **9**(Suppl. 1), S34–S7.

56. Flasshove, M., Banerjee, D., Mineishi, S., Li, M. X., Bertino, J. R., and Moore, M. A. (1995). Ex vivo expansion and selection of human CD34+ peripheral blood progenitor cells after introduction of a mutated dihydrofolate reductase cDNA via retroviral gene transfer. *Blood* **85**(2), 566–574.

57. Zhao, S. C., Banerjee, D., Mineishi, S., and Bertino, J. R. (1997). Post-transplant methotrexate administration leads to improved curability of mice bearing a mammary tumor transplanted with marrow transduced with a mutant human dihydrofolate reductase cDNA. *Hum. Gene Ther.* **8**(8), 903–909.

58. May, C., Gunther, R., and McIvor, R. S. (1995). Protection of mice from lethal doses of methotrexate by transplantation with transgenic marrow expressing drug-resistant dihydrofolate reductase activity. *Blood* **86**(6), 2439–2448.

59. Budak-Alpdogan, T., Ng, S., Farrelly, J., McKnight, J., Charney, S., Hohenhaus, A., Bergman, P., Bertino, J. R., and O'Connor, O. A., Retroviral gene transfer of novel drug resistance fusion genes into canine hematopoeitic progenitor cells (HPC) under long-term culture conditions (LTC), in *13th Int. Symp. on Treatment of Leukemia and Cancer,* New York, NY.

60. Skipper, H. E., Shabel, F. M., and Wilcox, W. (1967). Experimental evaluation of potential anticancer agents. XIII: On the criteria and kinetics associated with "curability" of experimental leukemia. *Cancer Chemother. Rep.* **51**, 125–165.

61. Norton, L., and Simon, R. (1977). Growth curve of an experimental solid tumor following radiotherapy. *J. Natl. Cancer Inst.* **58**(6), 1735–1741.

62. Norton, L., and Simon, R. (1977). Tumor size, sensitivity to ther-apy, and design of treatment schedules. *Cancer Treat. Rep.* **61**(7), 1307–1317.

63. Norton, L. (1988). A Gompertzian model of human breast cancer growth [see comments]. *Cancer Res.* **48**(24, pt 1), 7067–7071.

64. Hudis, C., Fornier, M., Riccio, L., Lebwohl, D., Crown, J., Gilewski, T., Surbone, A., Currie, V., Seidman, A., Reichman, B., Moynahan, M., Raptis, G., Sklarin, N., Theodoulou, M., Weiselberg, L., Salvaggio, R., Panageas, K. S., Yao, T. J., and Norton, L. (1999). 5-year results of dose-intensive sequential adjuvant chemotherapy for women with high-risk node-positive breast cancer: a phase II study. *J. Clin. Oncol.* **17**(4), 1118.

65. Hudis, C., Seidman, A., Baselga, J., Raptis, G., Lebwohl, D., Gilewski, T., Moynahan, M., Sklarin, N., Fennelly, D., Crown, J. P., Surbone, A., Uhlenhopp, M., Riedel, E., Yao, T. J., and Norton, L. (1999). Sequential dose-dense doxorubicin, paclitaxel, and cyclophosphamide for resectable high-risk breast cancer: feasibility and efficacy. *J. Clin. Oncol.* **17**(1), 93–100.

66. Gilewski, T., and Norton, L. (1996). Cytokinetics and breast cancer chemotherapy. in *Diseases of the Breast* (J. R. Harris, M. E. Lippman, and M. Morrow, eds.), pp. 751–768. Lippincott-Raven, Philadelphia.

67. Bonadonna, G., Valagussa, P., Moliterni, A., Zambetti, M., and Brambilla, C. (1995). Adjuvant cyclophosphamide, methotrexate, and fluorouracil in node-positive breast cancer: the results of 20 years of follow-up [see comments]. *N. Engl. J. Med.* **332**(14), 901–906.

68. Bonadonna, G., Zambetti, M., and Valagussa, P. (1995). Sequential or alternating doxorubicin and CMF regimens in breast cancer with more than three positive nodes. Ten-year results [see comments]. *JAMA* **273**(7), 542–547.

69. Powles, R., Mehta, J., Singhal, S., Horton, C., Tait, D., Milan, S., Pollard, C., Lumley, H., Matthey, F., Shirley, J. *et al.* (1995). Autologous bone marrow or peripheral blood stem cell transplantation followed by maintenance chemotherapy for adult acute lymphoblastic leukemia in first remission: 50 cases from a single center. *Bone Marrow Transplant.* **16**(2), 241–247.

70. Tallman, M. S., Rademaker, A. W., Jahnke, L., Brown, S. G., Bauman, A., Mangan, C., Kelly, C., Rubin, H., Kies, M. S., Shaw, J., Kiel, K., Gordon, L. I., Gradishar, W. J., and Winter, J. N. (1997). High-dose chemotherapy, autologous bone marrow or stem cell transplantation and post-transplant consolidation chemotherapy in patients with advanced breast cancer [published *erratum* appears in *Bone Marrow Transplant.* **21**(8), 861, 1998]. *Bone Marrow Transplant.* **20**(9), 721–729.

71. Gryn, J., Johnson, E., Goldman, N., Devereux, L., Grana, G., Hageboutros, A., Fernandez, E., Constantinou, C., Harrer, W., Viner, E., and Goldberg, J. (1997). The treatment of relapsed or refractory intermediate grade non-Hodgkin's lymphoma with autologous bone marrow transplantation followed by cyclosporine and interferon. *Bone Marrow Transplant.* **19**(3), 221–226.

72. deMagalhaes-Silverman, M., Hammert, L., Lembersky, B., Lister, J., Rybka, W., and Ball, E. (1998). High-dose chemotherapy and autologous stem cell support followed by post-transplant doxorubicin and Taxol as initial therapy for metastatic breast cancer: hematopoietic tolerance and efficacy. *Bone Marrow Transplant.* **21**(12), 1207–1211.

73. Rahman, Z., Kavanagh, J., Champlin, R., Giles, R., Hanania, E., Fu, S., Zu, Z., Mehra, R., Holmes, F., Kudelka, A., Claxton, D., Verschraegen, C., Gajewski, J., Andreeff, M., Heimfeld, S., Berenson, R., Ellerson, D., Calvert, L., Mechetner, E., Holzmayer, T., Dayne, A., Hamer, J., Bachier, C., Ostrove, J., Deisseroth, A. *et al.* (1998). Chemotherapy immediately following autologous stem-cell transplantation in patients with advanced breast cancer. *Clin. Cancer Res.* **4**(11), 2717–2721.

74. Moscow, J. A., Huang, H., Carter, C., Hines, K., Zujewski, J., Cusack, G., Chow, C., Venzon, D., Sorrentino, B., Chiang, Y., Goldspiel, B., Leitman, S., Read, E. J., Abati, A., Gottesman, M. M., Pastan, I., Sellers, S., Dunbar, C., and Cowan, K. H. (1999). Engraftment of MDR1 and NeoR gene-transduced hematopoietic cells after breast cancer chemotherapy. *Blood* **94**(1), 52–61.

75. Abonour, R., Williams, D. A., Einhorn, L., Hall, K. M., Chen, J., Coffman, J., Traycoff, C. M., Bank, A., Kato, I., Ward, M., Williams,

S. D., Hromas, R., Robertson, M. J., Smith, F. O., Woo, D., Mills, B., Srour, E. F., and Cornetta, K. (2000). Efficient retrovirus-mediated transfer of the multidrug resistance 1 gene into autologous human long-term repopulating hematopoietic stem cells. *Nat. Med.* **6**(6), 652–658.

76. Cavazzana-Calvo, M., Bagnis, C., Mannoni, P., and Fischer, A. (1999). Peripheral blood stem cell and gene therapy. *Baillieres Best Pract. Res. Clin. Haematol.* **12**(1-2), 129–138.

77. Cadman, E., and Eiferman, F. (1979). Mechanism of synergistic cell killing when methotrexate precedes cytosine arabinoside: study of L1210 and human leukemic cells. *J. Clin. Invest.* **64**(3), 788–797.

78. Hoovis, M. L., and Chu, M. Y. (1973). Enhancement of the antiproliferative action of 1-β-D-arabinofuranosylcytosine by methotrexate in murine leukemic cells (L5178Y). *Cancer Res.* **33**(3), 521–525.

79. Roberts, D., Peck, C., Hilliard, S., and Wingo, W. (1979). Methotrexate-induced changes in the levels of 1-beta-D-arabinofuranosylcytosine triphosphate in L1210 cells. *Cancer Res.* **39**(10), 4048–4054.

80. Chabner, B. A. (1996). Cytidine analogues. in *Cancer Chemotherapy and Biotherapy* (B. A. Chabner and D. L. Longo, eds.), Lippincott-Raven, New York.

81. Burke, P. J., Karp, J. E., Vaughan, W. P., and Sanford, P. L. (1982). Recruitment of quiescent tumor by humoral stimulatory activity: requirements for successful chemotherapy. *Blood Cells* **8**(3), 519–533.

82. Vaughan, W. P., Karp, J. E., and Burke, P. J. (1984). Two-cycle timed-sequential chemotherapy for adult acute nonlymphocytic leukemia. *Blood* **64**(5), 975–980.

83. Fisher, R. I., Gaynor, E. R., Dahlberg, S., Oken, M. M., Grogan, T. M., Mize, E. M., Glick, J. H., Coltman, Jr, C. A., and Miller, T. P. (1993). Comparison of a standard regimen (CHOP) with three intensive chemotherapy regimens for advanced non-Hodgkin's lymphoma. *N. Engl. J. Med.* **328**(14), 1002–1006.

84. Cabanillas, F., Velasquez, W. S., McLaughlin, P., Jagannath, S., Hagemeister, F. B., Redman, J. R., Swan, F., and Rodriguez, M. A. (1988). Results of recent salvage chemotherapy regimens for lymphoma and Hodgkin's disease. *Semin. Hematol.* **25**(2, suppl. 2), 47–50.

85. Velasquez, W. S., Cabanillas, F., Salvador, P., McLaughlin, P., Fridrik, M., Tucker, S., Jagannath, S., Hagemeister, F. B., Redman, J. R., Swan, F. *et al.* (1988). Effective salvage therapy for lymphoma with cisplatin in combination with high-dose Ara-C and dexamethasone (DHAP). *Blood* **71**(1), 117–122.

86. Cohen, S. C., and Krigel. R. L. (1995). High-dose therapy with stem cell infusion in lymphoma. *Semin. Oncol.* **22**(3), 218–229.

87. Bierman, P. J., and Armitage, J. O. (1994). Autologous bone marrow transplantation for non-Hodgkin's lymphoma, in *Bone Marrow Transplantation* (S. J. Forman, K. G. Blume, and E. D. Thomas, eds.), pp. 683–695. Blackwell Scientific, Boston.

88. Phillips, G. L. (1994). Transplantation in Hodgkins' disease, in *Bone Marrow Transplantation* (S. J. Forman, K. G. Blume, and E. D. Thomas, eds.), pp. 952–962. Blackwell Scientific, Boston.

89. Bolwell, B. J. (1994). Autologous bone marrow transplantation for Hodgkin's disease and non-Hodgkin's lymphoma. *Semin. Oncol.* **21** (4, suppl. 7), 86–95.

90. Vose, J. M., and Armitage, J. O. (1993). Role of autologous bone marrow transplantation in non-Hodgkin's lymphoma. *Hematol. Oncol. Clin. North. Am.* **7**(3), 577–590.

91. Philip, T., Guglielmi, C., Hagenbeek, A., Somers, R., Van der Lelie, H., Bron, D., Sonneveld, P., Gisselbrecht, C., Cahn, J. Y., Harousseau, J. L. *et al.* (1995). Autologous bone marrow transplantation as compared with salvage chemotherapy in relapses of chemotherapy-sensitive non-Hodgkin's lymphoma. *N. Engl. J. Med.* **333**(23), 1540–1545.

92. Philip, T., Gomez, F., and Guglelmi, C. (1998). Long-term outcome of relapsed non-Hodgkin's lymphoma (NHL) patients included in

the PARMA trial: incidence of late relapses, long-term toxicity, and impact of the international prognostic index (IPI) at relapse. *Proc. Am. Soc. Clin. Oncol.* **17**(16a).

93. Philip, T., Armitage, J. O., Spitzer, G., Chauvin, F., Jagannath, S., Cahn, J. Y., Colombat, P., Goldstone, A. H., Gorin, N. C., Flesh, M. *et al.* (1987). High-dose therapy and autologous bone marrow transplantation after failure of conventional chemotherapy in adults with intermediate-grade or high-grade non-Hodgkin's lymphoma. *N. Engl. J. Med.* **316,** 1493–1498.

94. Stiff, P. J., Dahlberg, S., Forman, S. J., McCall, A. R., Horning, S. J., Nademanee, A. P., Blume, K. G., LeBlanc, M., and Fisher, R. I. (1998). Autologous bone marrow transplantation for patients with relapsed or refractory diffuse aggressive non-Hodgkin's lymphoma: value of augmented preparative regimens—a Southwest Oncology Group trial. *J. Clin. Oncol.* **16**(1), 48–55.

95. Moskowitz, C. H., Bertino, J. R., Glassman, J. R., Hedrick, E. E., Hunte, S., Coady-Lyons, N., Agus, D. B., Goy, A., Jurcic, J., Noy, A., O'Brien, J., Portlock, C. S., Straus, D. S., Childs, B., Frank, R., Yahalom, J., Filippa, D., Louie, D., Nimer, S. D., and Zelenetz, A. D. (1999). Ifosfamide, carboplatin, and etoposide: a highly effective cytoreduction and peripheral-blood progenitor-cell mobilization regimen for transplant-eligible patients with non-Hodgkin's lymphoma. *J. Clin. Oncol.* **17**(12), 3776–3785.

96. Brenner, M. (1999). "Resistance is futile" [comment]. *Gene Ther.* **6**(10), 1646–1647.

97. Brenner, M. K., Rill, D. R., Moen, R. C., Krance, R. A., Mirro, Jr. J., Anderson, W. F., and Ihle, J. N. (1993). Gene-marking to trace origin of relapse after autologous bone-marrow transplantation. *Lancet* **341**(8837), 85–86.

98. Kiem, H. P., Heyward, S., Winkler, A., Potter, J., Allen, J. M., Miller, A. D., and Andrews, R. G. (1997). Gene transfer into marrow repopulating cells: comparison between amphotropic and gibbon ape leukemia virus pseudotyped retroviral vectors in a competitive repopulation assay in baboons. *Blood* **90**(11), 4638–4645.

99. Dunbar, C. E., Seidel, N. E., Doren, S., Sellers, S., Cline, A. P., Metzger, M. E., Agricola, B. A., Donahue, R. E., and Bodine, D. M. (1996). Improved retroviral gene transfer into murine and Rhesus peripheral blood or bone marrow repopulating cells primed in vivo with stem cell factor and granulocyte colony-stimulating factor. *Proc. Natl. Acad. Sci. USA* **93**(21), 11871–11876.

100. Bahner, I., Kearns, K., Hao, Q. L., Smogorzewska, E. M., and Kohn, D. B. (1996). Transduction of human CD34+ hematopoietic progenitor cells by a retroviral vector expressing an RRE decoy inhibits human immunodeficiency virus type 1 replication in myelomonocytic cells produced in long-term culture. *J. Virol.* **70**(7), 4352–4360.

101. Moritz, T., Patel, V. P., and Williams, D. A. (1994). Bone marrow extracellular matrix molecules improve gene transfer into human hematopoietic cells via retroviral vectors. *J. Clin. Invest.* **93**(4), 1451–1457.

102. Moritz, T., Dutt, P., Xiao, X., Carstanjen, D., Vik, T., Hanenberg, H., and Williams, D. A. (1996). Fibronectin improves transduction of reconstituting hematopoietic stem cells by retroviral vectors: evidence of direct viral binding to chymotryptic carboxy-terminal fragments. *Blood* **88**(3), 855–862.

103. Hanenberg, H., Xiao, X. L., Dilloo, D., Hashino, K., Kato, I., and Williams, D. A. (1996). Colocalization of retrovirus and target cells on specific fibronectin fragments increases genetic transduction of mammalian cells. *Nat. Med.* **2**(8), 876–882.

104. Moore, M. A., and MacKenzie, K. L. (1999). Optimizing conditions for gene transfer into human hematopoietic cells. *Prog. Exp. Tumor Res.* **36**, 20–49.

105. Conneally, E., Cashman, J., Petzer, A., and Eaves, C. (1997). Expansion in vitro of transplantable human cord blood stem cells demonstrated using a quantitative assay of their lympho-myeloid

repopulating activity in nonobese diabetic-scid/scid mice. *Proc. Natl. Acad. Sci. USA* **94**(18), 9836–9841.

106. Spence, S. E., Keller, J. R., Ruscetti, F. W., McCauslin, C. S., Gooya, J. M., Funakoshi, S., Longo, D. L., and Murphy, W. J. (1998). Engraftment of ex vivo expanded and cycling human cord blood hematopoietic progenitor cells in SCID mice. *Exp. Hematol.* **26**(6), 507–514.

107. Luens, K. M., Travis, M. A., Chen, B. P., Hill, B. L., Scollay, R., and Murray, L. J. (1998). Thrombopoietin, kit ligand, and flk2/flt3

ligand together induce increased numbers of primitive hematopoietic progenitors from human CD34+Thy-1+Lin– cells with preserved ability to engraft SCID-hu bone. *Blood* **91**(4), 1206–1215.

108. Tisdale, J. F., Hanazono, Y., Sellers, S. E., Agricola, B. A., Metzger, M. E., Donahue, R. E., and Dunbar, C. E. (1998). Ex vivo expansion of genetically marked rhesus peripheral blood progenitor cells results in diminished long-term repopulating ability. *Blood* **92**(4), 1131–1141.

CHAPTER

24

Protection from Antifolate Toxicity by Expression of Drug-Resistant Dihydrofolate Reductase

R. SCOTT McIVOR

Gene Therapy Program, Institute of Human Genetics
Department of Genetics, Cell Biology and Development
University of Minnesota
Minneapolis, Minnesota 55455

I. INTRODUCTION

The introduction of new gene sequences into normal cells or tumor cells has become an important approach in the development of new treatments of neoplastic disease [15,51]. One strategy for the application of gene transfer in the treatment of cancer is the introduction of new genes into normal tissues that are sensitive to the toxic side effects of cancer chemotherapy [41]. Protection of these normal cells and tissues by the introduction and expression of genes that confer drug resistance may thus allow for more effective use of chemotherapeutic agents in the treatment of tumors, either through administration of increased doses of the agent or by reduced morbidity at currently administered doses. Several

different combinations of agents and genes conferring drug resistance have been explored for the potential of chemoprotection. The multidrug resistance (MDR) gene has been tested extensively in experimental animals [31,64,76] and in clinical trials for its ability to confer resistance to Taxol as well as other chemotherapeutic agents [1,17,32,34,35]. O^6-methylguanine–DNA–methyltransferase activity (MGMT) has also been used as a detoxifying enzyme in conjunction with protection from alkylating agents such as 1,3-*bis*(2-chloroethyl)-1-nitrosourea (BCNU) [15,49,56,67]. A third system, and the subject of this chapter, is the use of drug-resistant forms of dihydrofolate reductase (DHFR) which exhibit reduced affinity and inhibition kinetics mediated by antifolates such as methotrexate (MTX) [50].

Dihydrofolate reductase (EC 1.5.1.3) catalyzes the NADPH-dependent reduction of dihydrofolate to tetrahydrofolate and in mammalian systems also catalyzes the formation of dihydrofolate from folic acid [8,9]. Tetrahydrofolate is the precursor to several different one-carbon donor molecules essential for several metabolic functions, including the biosynthesis of purines, thymidylate, methionine, and glycine (Fig. 1). DHFR is effectively inhibited by any one of several different compounds (antifolates) such as aminopterin, methotrexate (MTX; amethopterin), and trimetrexate, resulting in the intracellular depletion of tetrahydrofolate and its metabolic derivatives and a substantial antiproliferative effect against rapidly dividing cells and tissues [10,46]. Antifolates such as MTX have been of great utility in the treatment of several different highly proliferative tumors, especially acute lymphocytic leukemia, choriocarcinoma, Ewing's sarcoma, non-Hodgkins's lymphoma, and osteosarcoma [8,40,71]. MTX has also been used for graft-versus-host

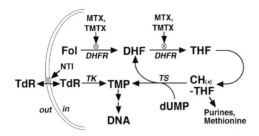

FIGURE 1 Antifolate inhibition of cellular metabolism. DHFR activity provides reduced folate derivatives important for biosynthesis of purines, methionine, and thymine nucleotides. Antifolates such as MTX and TMTX inhibit DHFR, depleting cells of reduced folates, the basis of antifolate toxicity for rapidly dividing cells. Exogenous nucleosides such as thymidine or purine nucleosides can circumvent antifolate toxicity by providing an alternate source of these materials that does not require reduced folates. However, salvage of nucleosides can be blocked by the use of a nucleoside transport inhibitor (NTI), thus restoring antifolate toxicity even in the presence of nucleosides. Other abbreviations: Fol, folic acid; DHF, dihydrofolic acid; THF, tetrahydrofolic acid; TdR, thymidine; TMP, thymidine monophosphate; $CH_{(x)}$-THF, methyl-, methylene-, and methenyl-tetrahydrofolic acids; dUMP, deoxyuridine monophosphate; TS, thymidylate synthase; TK, thymidine kinase.

disease prophylaxis following allogeneic bone marrow transplant [81,82], and in the treatment of arthritis [2]. However, the effectiveness of antifolate therapy is limited by toxicity for normal tissues, primarily cells of the bone marrow and of the gastrointestinal tract [46,70].

Methotrexate and other antifolates might be more effectively used as antitumor agents if normal cells and tissues could be rendered resistant to the toxic side effects of antifolate chemotherapy [50]. Resistance of cells to MTX has been associated with several different mechanisms, including: (1) reduced transport of MTX across the plasma membrane [59,73,75], (2) reduced polyglutamylation of MTX [13], (3) increased breakdown of polyglutamylated MTX by gamma-glutamyl hydrolase [68], (4) increased expression of DHFR associated with amplification of the DHFR gene [6], and (5) expression of drug-resistant DHFR enzyme activity resulting from a missense mutation in the DHFR coding sequence [74]. Introduction and expression of a mutant DHFR gene encoding drug-resistant DHFR enzyme thus constitutes one potential way of counteracting antifolate toxicity in normal cells and tissues associated with antifolate chemotherapy [50].

Numerous drug-resistant, mutant forms of mammalian DHFRs have been isolated and characterized. Such DHFRs have been demonstrated to confer drug resistance when expressed in cultured mammalian cells, thus allowing DHFR to be used as a dominant selectable marker. Expression of drug-resistant DHFR activity has been demonstrated to result in increased antifolate dose tolerance in mice. The increased doses of antifolate afforded by drug-resistant DHFR expression in normal tissues may allow for more effective use of antifolates as antitumor chemotherapeutic agents. This chapter describes

the generation and characteristics of drug-resistant DHFRs, their use to render experimental animals resistant to antifolate toxicity, and the potential of this system for improved anti-tumor chemotherapy using antifolates. Also discussed is the possibility of using drug-resistant DHFR expression to allow for *in vivo* selection and expansion of rare, genetically modified cells in a therapeutic setting. In this way, DHFR gene transfer may be clinically applicable not only as an approach to improved anti-tumor chemotherapy, but also to the treatment of other diseases by co-introduction and expansion of therapeutic genes.

II. DRUG-RESISTANT DIHYDROFOLATE REDUCTASES

The existence of drug-resistant forms of DHFR first became apparent through numerous reports describing drug-resistant enzyme activity present in extracts of cell lines adapted to grow in a medium containing high levels of MTX [3,27,28,29,30,37]. The molecular genetic basis for variant, drug-resistant DHFR activity was eventually provided when Simonsen and Levinson reported the cloning of a murine DHFR cDNA [74] generated from a MTX-resistant 3T6 cell line established in Schimke's laboratory [30]. The identified mutation, a leucine-to-arginine substitution at codon position 22 in the active site of the enzyme, plus the ability of Arg-22 DHFR expression constructs to render transfectant cells resistant to MTX, confirmed its role in mediating drug resistance [74]. Several other variant, drug-resistant forms of DHFR were subsequently cloned from different mammalian cell lines and characterized (Table 1). These initially identified mutations from mouse, hamster, and human sources were found at active-site codon positions 22 or 31, although non-active-site mutation has also been observed [19]. In general, the enzymatic character of these different mutant DHFR forms is such that the greater the perturbation of the active site, the greater the degree of drug resistance. However, these resistances come at a kinetic price in the form of reduced overall catalytic activity (Table 1), particularly for the Arg-22 mutation.

The isolation of naturally occurring mutant forms of DHFR led inevitably to the engineering of mutant DHFRs based on the prediction that modification of certain active-site residues known to be involved in substrate and/or inhibitor binding would result in an enzyme exhibiting reduced drug sensitivity [66,72,84,85]. Residues 22 and 31 in particular served as focal points for modifications of the human DHFR gene [11,22,43], resulting in an array of rationally designed mutant DHFR forms potentially useful for drug-resistance studies (Table 1). In addition to these rational approaches, our laboratory carried out randomization studies in which oligonucleotide mixtures were used in conjunction with the polymerase chain reaction (PCR) to introduce all possible

TABLE 1   Drug-Resistant Dihydrofolate Reductases

Mutation (wt)	Source	Resistance[a]	Activity (% wt)	Ref.
DHFR mutants cloned from drug-resistant cell lines				
Arg-22 (Leu)	Mouse	270-fold[b]	5	[30,74]
Phe-22 (Leu)	Hamster	100-fold[c]		[18,55]
Ser-31 (Phe)	Human	8-fold[b]		[79]
Trp-31 (Phe)	Mouse	20-fold[d]	15	[53,58]
Trp-15 (Gln)	Mouse	165-fold[c]	Unstable	[19]
Some genetically engineered DHFR mutants				
Arg-22 (Leu)	Mouse	7.5×10^5-fold[c]	1.5	[84]
	Mouse	2000-fold[d]	1.3–2	[58]
	Human	1300-fold[c]	0.4[e]	[43]
Phe-22 (Leu)	Mouse	44-fold[d]	26	[58]
	Human	134-fold[c]	216[e]	[43]
	Human	88-fold[c]	58[e]	[22]
Trp-22 (Leu)	Mouse	300-fold[d]	18	[58]
	Human	1270-fold[c]	38[e]	[43]
Tyr-22 (Leu)	Mouse	420-fold[d]	17	[58]
	Human	3200-fold[c]	56[e]	[43]
	Human	1650-fold[c]	12[e]	[22]
Arg-24 (Trp)	Mouse	7.5×10^4-fold[c]	2.5	[84]
Ser-31 (Phe)	Mouse	1100-fold[c]	83	[84]
	Mouse	5.6-fold[d]	390	[58]
	Mouse	4.8-fold[c]	300[e]	[23]
	Human	92-fold[b]	65[e]	[72]
	Human	70-fold[c]	64[e]	[43]
His-31 (Phe)	Mouse	50-fold[d]	64	[58]
Asn-31 (Phe)	Mouse	83-fold[d]	45	[58]
Leu-34 (Phe)	Mouse	200-fold[d]	95	[84]
Ser-34 (Phe)	Human	8×10^4-fold[b]	34	[72]

[a] Compared to wt enzyme.
[b] Determined by equilibrium dialysis.
[c] Increase in K_i.
[d] Increase in I_{50} (concentration of MTX required to inhibit reaction by 50%).
[e] Decrease in k_{cat}.

codons at positions 22 and 31 [58], subsequently screening for functional, drug-resistant DHFRs by transfection into DHFR-deficient Chinese hamster ovary cells and selection in MTX.

One of the goals of DHFR enzyme engineering has been to generate new mutant forms that are highly drug resistant and yet retain substantial residual enzyme activity. In this sense, both random and rational approaches have been successful in identifying DHFR variants with a more effective combination of drug resistance and retained catalytic activity. Several different substitutions at codon position 22 have yielded enzymes that are characterized by a high degree of drug resistance while retaining a significant level of catalytic activity, such as Tyr-22 and Trp-22. Some substitu-

tions at position 31 are also highly drug resistant, such as the murine His-31 and Asn-31 variants [58] and the human Ser-31, Asp-31, and Gly-31 variants [11], but not nearly so as position 22 substitutions, while position 31 substitutions retain a greater degree of catalytic activity [58]. One surprise in the murine system was the apparent fourfold increase in catalytic activity for the Ser-31 variant enzyme when compared to wild-type. This increased catalytic activity for the Ser-31 variant was subsequently verified using highly purified enzyme expressed in *Escherichia coli* [23]. Such an increase was not reported for the human Ser-31 variant [11,72], however, and is thus apparently unique to the murine enzyme. Several DHFRs bearing substitutions at both positions 22 and 31 have also been generated which reportedly exhibit improved combinations of drug resistance and catalytic activity [21,64].

A variety of mutant DHFR forms have thus been either isolated or generated (Table 1), providing an extensive array of antifolate-resistant enzymes with which the physiological and pharmacological effectiveness of drug-resistance gene transfer and expression can be explored and evaluated. Of these mutants, the murine Arg-22 enzyme, the human Ser-31 enzyme, and the human and murine Tyr-22 enzymes have been most noteworthy in their characterization and application to chemoprotection and antifolate-mediated *in vivo* selection, as described below.

III. PROTECTION FROM ANTIFOLATE TOXICITY *IN VITRO*

The ability of newly introduced variant DHFR genes to confer resistance to antifolate-mediated toxicity was initially demonstrated by Simonsen and Levinson [74], who showed that transfection of murine Arg-22 DHFR expression constructs resulted in the formation of drug-resistant transfectant colonies for several different cell lines. Thus, the murine Arg-22 DHFR, along with the neomycin phosphotransferase gene [77] and the *EcoGPT* gene [60], constituted one of the first dominant selectable markers established for cultured mammalian cells and the only one of the three that was not of prokaryotic origin. As a dominant selectable marker, DHFR is versatile in the sense that it can also serve as an amplifiable marker as well, by adapting transfected cells to grow in medium containing higher and higher concentrations of MTX [69]. Cells selected at higher concentrations of MTX usually contain a higher level of DHFR due to an increase in DHFR gene copy number [6]. This gene amplification approach has been used extensively to overproduce recombinant proteins in cultured mammalian cells. Because the murine Arg-22 variant DHFR was first characterized as a dominant-selectable marker [74], numerous other DHFRs have been demonstrated to function in this capacity.

IV. PROTECTION FROM ANTIFOLATE TOXICITY *IN VIVO:* RETROVIRAL TRANSDUCTION STUDIES

While variant DHFR gene transfer and expression have been used and reported extensively for *in vitro* selection, extension of this approach to the *in vivo* setting for protection of cells and tissues from the toxicity of antifolates has presented a formidable challenge. Hock and Miller [36] initially demonstrated retroviral transduction of the murine Arg-22 DHFR gene and MTX resistance in primary human hematopoietic cultures as proof of principle for the possibility of applying this approach for *in vivo* chemoprotection of hematopoietic cells. Williams and co-workers [12,88] then demonstrated significant protection from MTX toxicity in mice transplanted with normal marrow retrovirally transduced with vectors encoding the murine Arg-22 cDNA transcriptionally regulated by the retroviral long-terminal repeat (either Moloney leukemia virus or spleen focus-forming virus). This seminal work included the demonstration that DHFR-transduced cells were transplantable from primary to secondary recipients [12], with the implication that the initial exposure successfully introduced retroviral vector sequences into stem cells capable of engraftment in serially transplanted animals. Zhao *et al.* [90] subsequently demonstrated protection from MTX toxicity in animals serially transplanted with Arg-22-DHFR-transduced marrow. Protection from MTX toxicity was also demonstrated after transduction of the human Ser-31 DHFR cDNA into murine hematopoietic stem cells [44]. More recently, Sorrentino and co-workers have demonstrated retroviral transduction of the human Tyr-22 DHFR variant into murine hematopoietic cells to protect against toxicity associated with administration of trimetrexate [78]. The effectiveness of retroviral DHFR gene transfer in large animal studies has been hampered by a low level of expression [80], or low level of gene transfer into primitive hematopoietic stem cells [20]. Nonetheless, the progress of DHFR retroviral transduction studies in the mouse, plus recent advances in DHFR gene transfer into human hematopoietic cells [24–26,89], have demonstrated the potential for *ex vivo* somatic modification of hematopoietic cells to protect recipient animals from antifolate toxicity.

V. DIHYDROFOLATE REDUCTASE TRANSGENIC MOUSE SYSTEM FOR *IN VIVO* DRUG-RESISTANCE STUDIES

To provide a model system in which to study drug resistance associated with mutant DHFR expression, our laboratory has established several lines of transgenic mice that express Arg-22, Trp-31, or Tyr-22 drug-resistant DHFRs [38,57]. These animals were established on an inbred (FVB/N) background to allow transplantation of marrow from animals found to express the drug-resistant DHFR gene into normal syngeneic recipients. Marrow from Trp-31 DHFR transgenics protected recipient animals from low doses of MTX (1 mg/kg/day) under conditions that were lethal for animals transplanted with normal FVB/N marrow (i.e., lethal total-body irradiation followed by transplantation with 10^6 donor marrow cells) [47,48]. However, transplantation with Arg-22 or Tyr-22 DHFR transgenic marrow protected animals from higher doses of MTX (4 mg/kg/day and in some cases up to 6 mg/kg/day) that were lethal for animals transplanted with 10^7 normal donor marrow cells [38,47]. In parallel control groups, the maximum tolerated dose of MTX in normal animals or in animals transplanted with 10^7 normal donor marrow cells was 2 mg/kg/day [48]. Transplantation with DHFR (Arg-22 or Tyr-22) transgenic marrow thus afforded a two- to three-fold increase in the maximum tolerated dose of MTX.

The ability of transplantation with DHFR transgenic marrow to confer drug resistance at a dose of 4 mg/kg/day MTX was surprising because toxicity at this level of drug administration is observed not only in hematopoietic cells and tissues but in the gastrointestinal tract as well [46,47,57]. These results imply that transplantation with drug-resistant DHFR transgenic marrow not only protects recipients from hematopoietic toxicity, but also protects gastrointestinal tissues from the toxicity associated with daily, long-term (up to 60 days) MTX administration. This possibility was supported by further histologic studies in which a statistically significant increase in villus length was observed in animals transplanted with DHFR transgenic marrow in comparison with normal animals or animals transplanted with normal marrow and subsequently administered 4 mg/kg/day MTX [47].

There are several potential mechanisms by which transplanted hematopoietic cells may contribute to the protection of gastrointestinal (GI) tissues from MTX-mediated toxicity: (1) animals engrafted with drug-resistant hematopoietic cells may clear MTX more readily than normal animals or animals transplanted with normal marrow; (2) drug-resistant phagocytic cells may contribute to recovery from MTX-mediated GI tissue damage in spite of continued drug administration; (3) drug-resistant hematopoietic cells may contribute to the maintenance of normal GI tissue structure and function (for example, through the secretion of growth factors); and (4) drug-resistant lymphoid cells may contribute to the maintenance of a more well-regulated immune response and an intact epithelial barrier. While the mechanism(s) contributing to protection of GI tissues by transplantation with DHFR transgenic marrow have yet to be determined, some results do address these possibilities. First, we conducted pharmacokinetic studies to determine the levels of MTX in plasma and in GI tissues at several times posttransplant and found that MTX levels in animals transplanted with drug-resistant marrow were comparable to levels found in normal animals or animals transplanted with normal marrow [7]. These results indicate that animals transplanted with drug-resistant marrow do not eliminate MTX faster than animals transplanted with normal marrow and that the drug-resistance conferred by transplanted drug-resistant marrow must be based on some other cellular or molecular mechanism. Also, quantitative Southern analysis of DNA extracted from the small intestine of animals transplanted with Tyr-22 DHFR transgenic marrow indicated the presence of 20–30% donor DHFR transgenic cells in the total GI cellular material from this organ [38]. These results demonstrate a significant level of hematopoietic cell engraftment in the GI tract several months posttransplant. Characterization of these newly engrafted, drug-resistant transgenic cells may provide insight not only into the mechanism by which transplantation with drug-resistant marrow protects animals from MTX-mediated GI toxicity, but also the extent to which the hematopoietic toxicity of MTX (or other antiproliferative agents, for that matter) exacerbates toxicity for the GI tract.

The use of donor marrow from animals that have been genetically modified in the germline does not address the clinically important issue of gene transfer efficiency in somatic, hematopoietic cell targets. Furthermore, this transgenic system is highly artificial with respect to the level of DHFR gene expression achievable, as individual transgenic animals were screened for DHFR transgene expression before their use in transplant studies. Donor marrow from these animals is thus homogeneous with respect to the level of gene expression from one cell to the next. This contrasts with genetically manipulated (i.e., retrovirally transduced) normal marrow cells, in which different insertion events may be expected to result in different levels of gene expression in a general or in a temporal manner. However, while side-stepping the issues of somatic gene transfer efficiency and gene expression after somatic cell gene transfer, the use of transgenic animals as a source of donor drug-resistant marrow offers several experimental opportunities. Engraftment levels can be readily engineered for groups of recipient animals either in mixing experiments with normal marrow or by sublethal irradiation [39]. With respect to engraftment levels and the level of gene expression in engrafted cells, there is thus a high degree of reproducibility among similarly treated animals from group to group and from experiment to experiment. This provides the opportunity for experimental focus on the hematologic and pharmacologic characteristics and potential of mutant DHFR expression, allowing conclusions to be drawn among groups of animals similarly engrafted and expressing drug-resistant DHFR.

One way in which the reproducibility of DHFR transgenic cell engraftment in recipient animals has been particularly useful has been in evaluating the effectiveness of the engraftment level achieved in animals preconditioned with nonmyeloablative doses of irradiation in protecting animals from MTX toxicity [39]. In this series of experiments we found that transgenic marrow engraftment levels correlated directly with the dose of irradiation administered as well as the number of marrow cells subsequently transplanted into recipient animals. Additionally, animals administered the lowest dose of total body irradiation (TBI) (100 rads, approx. 1/8 lethal dose for FVB/N mice) and transplanted with a modest number of donor cells (1×10^6) were protected from MTX toxicity (4 mg/kg/day, a lethal dose for normal animals or animals transplanted with nontransgenic marrow) at the low donor cell engraftment level of about 1%. The clinical implication of these results is that a low level of DHFR gene transduction into hematopoietic cells may be sufficient to protect patients from antifolate toxicity, thus potentiating the use of antifolates as antitumor agents to prevent tumor relapse after bone marrow transplant.

VI. ANTITUMOR STUDIES IN ANIMALS EXPRESSING DRUG-RESISTANT DIHYDROFOLATE REDUCTASE

The primary goal of rendering recipient animals more tolerant to toxic doses of antifolates is to allow more effective administration of these drugs as antitumor agents. The key question here is whether the increased dose tolerance brought about by DHFR gene transfer and expression allows for administration of higher levels of MTX and an improvement in chemotherapeutic outcome. Zhao *et al.* [91] first reported such preclinical studies, using a combination of bone marrow transplant and transduction with a retroviral vector encoding the Ser-31 variant of human DHFR to allow posttransplant administration of MTX in the treatment of an experimental breast tumor in mice. Animals transplanted with control marrow succumbed to the toxic effects of MTX administration, while animals transplanted with Ser-31-DHFR-transduced marrow were resistant to MTX administration, which protected these animals from tumor formation. Using the DHFR transgenic model system, our laboratory has also shown that expression of drug-resistant (Tyr-22) DHFR activity allows for tolerance to increased antifolate doses and reduced growth of FMC, a syngeneic, transplantable mammary adenocarcinoma established in our laboratory from an FVB/N mouse [54]. Similar studies are currently being carried out in a murine model of chronic myeloid leukemia [83,89]. These studies represent initial forays into the preclinical application of drug-resistance gene transfer for improved antitumor chemotherapy. Future studies will doubtless lead to the testing of other drug resistance genes in combination with various antifolates, thus expanding the group of conditions toward which this approach may prove to be effective.

VII. ANTIFOLATE-MEDIATED *IN VIVO* SELECTION OF HEMATOPOIETIC CELLS EXPRESSING DRUG-RESISTANT DIHYDROFOLATE REDUCTASE

One of the most formidable challenges currently facing the gene therapy field is the low frequency of gene transfer into target cells, including hematopoietic cells and in particular hematopoietic stem cells (HSCs). Novel vectors such as lentivirus vectors [61] or modifications in stem cell culture and retroviral transduction conditions [42,45] may ultimately provide an increased gene transfer frequency into HSCs. Another approach to dealing with the low frequency of gene transfer is to apply selective pressure for *in vivo* expansion of HSCs that have been successfully transduced with a drug-resistance gene. Selective expansion of MDR- and DHFR-transduced hematopoietic cells in peripheral blood was demonstrated in Taxol- and MTX-administered animals [12,76], but these studies did not extend to assessment of selective HSC expansion. Sorrentino's group subsequently found that the sensitivity of murine hematopoietic stem cells to the antifolate trimetrexate (TMTX) is apparently compromised by salvage of nucleosides (Fig. 1), as co-administration of a nucleoside transport inhibitor (nitrobenzylmercaptopurine riboside-phosphate, NBMPR-P) resulted in the inability of stem cells from a treated animal to engraft in recipient congenic recipients [5]. Allay *et al.* then went on to elegantly demonstrate that co-administration of TMTX + NBMPR-P resulted in the selective expansion of HSC transduced with a retroviral vector expressing the human Tyr-22 DHFR along with a green fluorescene protein (GFP) gene, thus allowing quantification of transduction and antifolate-mediated expansion in all tested hematopoietic lineages for both primary and secondary transplant recipients [4]. These results provided the first convincing evidence for *in vivo* selection of hematopoietic stem cells using a dominant-acting genetic function. Similar evidence for expansion of DHFR expressing HSCs using TMTX + NBMPR-P has been generated using the DHFR transgenic system [52,86]. *In vivo* expansion of HSCs expressing MGMT has also been demonstrated in animals administered benzylguanine and BCNU [14]. The ability to selectively expand stem cells expressing a drug-resistance gene such as DHFR potentiates the co-expansion of other therapeutic genes co-introduced along with the DHFR gene as part of a bifunctional vector, such as those encoding glucocerebrosidase [33], iduronidase [63], or an antisense sequence designed for downregulation of the BCR/ABL oncogene [89].

These advances in the field of drug-resistance gene transfer and expression provide support for their potential application to increase the representation of transduced hematopoeitic cells in cases where low-level gene transfer limits therapeutic effectiveness. With respect to the use of antifolates to expand DHFR-expressing cells, it remains to be seen whether inhibition of nucleoside rescue (using drugs such as NBMPR-P) will be necessary in large animals and in humans, as the levels of thymidine and inosine (nucleosides that can rescue cells from antifolate toxicity [87]) are reportedly much higher in mice than in humans [62], making the mouse perhaps a challenging model to work with in this system.

VIII. SUMMARY AND FUTURE CONSIDERATIONS

Several conclusions can be drawn from the foregoing discussion: (1) A range of DHFR variants are available as tools that can be used to explore resistance to antifolates as a selection system or as a means of conferring drug resistance *in vitro* or *in vivo*. (2) Expression of several different drug-resistant DHFRs in hematopoietic cells, either through germline transgenesis or by retrovirus-mediated gene transfer, can render experimental animals significantly resistant to antifolate toxicity. (3) This increased resistance can be used for the purpose of dose tolerization and chemoprotection of sensitive normal tissues and thus improved antitumor therapy using antifolates. (4) Evidence from experiments in mice suggests that antifolate administration can be used for *in vivo* selection of hematopoietic stem cells expressing drug-resistant DHFR activity when used in conjunction with a nucleoside transport inhibitor to prevent nucleoside salvage and circumvention of antifolate mediated starvation for nucleotide precursors. Overall, these advances lay the groundwork for further characterization of DHFR-mediated drug resistance and its clinical application.

What are the future challenges facing the application of DHFR gene transfer for improved antitumor therapy or its application as a selectable function for other genetic therapies? With respect to applying DHFR gene transfer and expression toward improved antitumor therapy, the success of this approach will depend on the extent to which the process provides a chemoprotective effect in the patient. More directly stated, what is the precise increase in dose tolerance to antifolate administration (i.e., how does the gene transfer and expression process actually affect quantitative toxicity parameters such as LD_{50} or the maximum tolerated dose)? Our laboratory was able to address this question using the DHFR transgenic mouse system (i.e., a two- to three-fold increase in maximum tolerated dose resulting from transgenic DHFR marrow transplant), but as described above the transgenic system is highly optimized and does not closely model the somatic modification of hematopoietic cells to confer drug resistance that will be required in the human clinical setting. Thus, a quantitative assessment of the degree to which vector-mediated DHFR gene transfer into hematopoietic cells renders recipient animals resistant to antifolate administration, measured as an increase in the LD_{50}, maximum tolerated dose or some other quantitative parameter, rather than simply resistance vs. sensitivity at a given dose, would contribute greatly to the clinical application of DHFR gene transfer. It will also be necessary to determine preclinically that such chemoprotection can, in fact, be effectively utilized for improved antitumor chemotherapy using an animal model relevant for the human clinical application proposed [54,83,90].

Several issues also face the clinical application of *in vivo* selection using DHFR. Antifolate-mediated expansion of hematopoietic stem cells in the murine system required co-administration of a nucleoside transport inhibitor [4,52,86], presumably because circulating levels of nucleosides rescue stem cells from antifolate toxicity. However, circulating levels of nucleosides are much lower in humans [62] than in mice, perhaps explaining the toxicity of antifolates for humans [70]. Thus, antifolate-mediated *in vivo* selection in humans may not require inhibition of nucleoside transport, a possibility that will require human clinical trials to assess. Another important question that remains to be addressed is the relatively low frequency of gene transfer into hematopoietic cells that can subsequently be established as a graft in recipients. It has not yet been determined what engraftment level or DHFR expression level will be necessary in order to protect animals from antifolate toxicity. This issue is then compounded, as antifolate administration that exceeds the maximum tolerated dose in normal animals may be necessary in order to provide *in vivo* selection pressure sufficient in its demand for an increase in DHFR-transduced stem cells. DHFR-mediated *in vivo* selection would be most useful in circumstances where the initial level of gene transfer is very low (i.e., less than 0.1–1.0%), thus modeling what has so far been achieved in humans and nonhuman primates using retroviral vectors. At these levels of transgenic cell engraftment, it is not known whether doses of antifolate sufficient to mediate expansion of HSCs can be safely administered. It has yet to be determined whether antifolate-mediated expansion of stem cells from such low initial levels of DHFR-transduced cell engraftment can be achieved. Thus, while there clearly are reasons to be optimistic about the potential clinical application of DHFR gene transfer and expression, several key questions remain to be addressed in preclinical studies or await human clinical trials. Finally, for both antitumor as well other therapeutic strategies, a greater understanding of the physiologic and pharmacologic mechanisms that underlie the systemic effectiveness or limitations of DHFR gene transfer and expression will optimize the potential of this system for future clinical application.

Acknowledgments

I thank Dr. Lalitha Belur for her critical review of the manuscript, and Drs. Roland Gunther, Canston R. Wagner, Cheryl Zimmerman, and all past and current members of the McIvor laboratory for their contributions to this work. Many of the described studies were supported by USPHS grant CA 60803 from the National Institutes of Health.

References

1. Abonour, R., Williams, D. A., Einhorn, L., Hall, K. M., Chen, J., Coffman, J., Traycoff, C. M., Bank, A., Kato, I., Ward, M., Williams, S. D., Hromas, R., Robertson, M. J., Smith, F. O., Woo, D., Mills, B., Srour, E. F., and Cornetta, K. (2000). Efficient retrovirus-mediated transfer of the multidrug resistance 1 gene into autologous human long-term repopulating hematopoietic stem cells. *Nat. Med.* **6**, 652–658.
2. Alarcon, G. S. (2000). Methotrexate use in rheumatoid arthritis. A clinician's perspective. *Immunopharmacology* **47**, 259–271.
3. Albrecht, A. M., Biedler, J. L., and Hutchison, D. J. (1972). Two different species of dihydrofolate reductase in mammalian cells differentially resistant to amethopterin and methasquin. *Cancer Res.* **32**, 1539–1546.
4. Allay, J. A., Persons, D. A., Galipeau, J., Riberdy, J. M., Ashmun, R. A., Blakley, R. L., and Sorrentino, B. P. (1998). In vivo selection of retrovirally transduced hematopoietic stem cells. *Nat. Med.* **4**, 1136–1143.
5. Allay, J. A., Spencer, H. T., Wilkinson, S. L., Belt, J. A., Blakley, R. L., and Sorrentino, B. P. (1997). Sensitization of hematopoietic stem and progenitor cells to trimetrexate using nucleoside transport inhibitors. *Blood* **90**, 3546–3554.
6. Alt, F. W., Kellems, R. E., Bertino, J. R., and Schimke, R. T. (1978). Selective multiplication of dihydrofolate reductase genes in methotrexate-resistant variants of cultured murine cells. *J. Biol. Chem.* **253**, 1357–1370.
7. Belur, L., Boelk-Galvan, D., Diers, M. D., McIvor, R. S., and Zimmerman, C. L. (2001). Methotrexate accumulates to similar levels in animals transplanted with normal vs. drug-resistant transgenic marrow. *Cancer Res.,* **61**, 1522–1526.
8. Bertino, J. R. (1993). Ode to methotrexate. *J. Clin. Oncol.* **11**, 5–14.
9. Blakley, R. L. (1995). Eukaryotic dihydrofolate reductase. *Adv. Enzymol. and Related Areas Mol. Biol.* **70**, 23–102.
10. Blakley, R. L., and Benkovic, S. J. (1984). *Chemistry and Biochemistry of Folates and Pterins.* Vol. 1., John Wiley & Sons, New York.
11. Chunduru, S. K., Cody, V., Luft, J. R., Pangborn, W., Appleman, J. R., and Blakley, R. L. (1994). Methotrexate-resistant variants of human dihydrofolate reductase. *J. Biol. Chem.* **269**, 9547–9555.
12. Corey, C. A., DeSilva, A. D., Holland, C. A., and Williams, D. A. (1990). Serial transplantation of methotrexate-resistant bone marrow: protection of murine recipients from drug toxicity by progeny of transduced stem cells. *Blood* **75**, 337–343.
13. Cowan, K. H., and Jolivet, J. (1984). A methotrexate-resistant human breast cancer cell line with multiple defects, including diminished formation of methotrexate polyglutamates. *J. Biol. Chem.* **259**, 10793–10800.
14. Davis, B. M., Koc, O. N., and Gerson, S. L. (2000). Limiting numbers of G156A $O(6)$-methylguanine-DNA methyltransferase-transduced marrow progenitors repopulate nonmyeloablated mice after drug selection. *Blood* **95**, 3078–3084.
15. Davis, B. M., Koc, O. N., Lee, K., and Gerson, S. L. (1996). Current progress in the gene therapy of cancer. *Curr. Opin. Oncol.* **8**, 499–508.
16. Davis, B. M., Reese, J. S., Koc, O. N., Lee, K., Schupp, J. E., and Gerson, S. L. (1997). Selection for G156A O^6-methylguanine DNA methyltransferase gene-transduced hematopoietic progenitors and protection from lethality in mice treated with O^6-benzylguanine and 1,3-*bis*(2-chloroethyl)-1-nitrosourea. *Cancer Res.* **57**, 5093-5099.
17. Devereux, S., Corney, C., Macdonald, C., Watts, M., Sullivan, A., Goldstone, A. H., Ward, M., Bank, A, and Linch, D. C. (1998). Feasibility of multidrug resistance (MDR-1) gene transfer in patients undergoing

high-dose therapy and peripheral blood stem cell transplantation for lymphoma. *Gene Ther.* **5,** 403–408.

18. Dicker, A. P., Volkenandt, M., Schweitzer, B. I., Banerjee, D., and Bertino, J. R. (1990). Identification and characterization of a mutation in the dihydrofolate reductase gene from the methotrexate-resistant Chinese hamster ovary cell line Pro-3 MTX. *J. Biol. Chem.* **265,** 8317–8321.

19. Dicker, A. P., Waltham, M. C., Volkenandt, M., Schweitzer, B. I., Otter, G. M., Schmid, F. A., Sirotnak, F. M., and Bertino, J. R. (1993). Methotrexate resistance in an in vivo mouse tumor due to a non-active-site dihydrofolate reductase mutation. *Proc. Nat. Acad. Sci. USA* **90,** 11797–11801.

20. Donahue, R. E., Wersto, R. P., Allay, J. A., Agricola, B. A., Metzger, M. E., Nienhuis, A. W., Persons, D. A., and Sorrentino, B. P. (2000). High levels of lymphoid expression of enhanced green fluorescent protein in nonhuman primates transplanted with cytokine-mobilized peripheral blood CD34(+) cells. *Blood* **95,** 445–452.

21. Ercikan-Abali, E. A., Mineishi, S., Tong, Y., Nakahari, S., Waltham, M. C., Banerjee, D., Chen, W., Sadelain, M., and Bertino, J. R. (1996). Active site-directed double mutants of dihydrofolate reductase. *Cancer Res.* **56,** 4142–4145.

22. Ercikan-Abali, E. A., Waltham, M. C., Dicker, A. P., Schweitzer, B. I., Gritsman, H., Banerjee, D., Bertino, J. R. (1996). Variants of human dihydrofolate reductase with substitutions at leucine-22: effect on catalytic and inhibitor binding properties. *Mol. Pharmacol.* **49,** 430–437.

23. Evenson, D. A., Adams, J., McIvor, R. S., and Wagner, C. R. (1996). Methotrexate resistance of mouse dihydrofolate reductase: effect of substitution of phenylalanine-31 by serine or tryptophan. *J. Med. Chem.* **39,** 1763–1766.

24. Flasshove, M., Banerjee, D., Bertino, J. R., and Moore, M. A. (1995). Increased resistance to methotrexate in human hematopoietic cells after gene transfer of the Ser31 DHFR mutant. *Leukemia* **9,** S34–S37.

25. Flasshove, M., Banerjee, D., Leonard, J. P., Mineishi, S., Li, M. X., Bertino, J. R., and Moore, M. A. (1998). Retroviral transduction of human CD34+ umbilical cord blood progenitor cells with a mutated dihydrofolate reductase cDNA. *Hum. Gene Ther.* **9,** 63–71.

26. Flasshove, M., Banerjee, D., Mineishi, S., Li, M-X, Bertino, J. R., and Moore, M. A. (1995). Ex vivo expansion and selection of human CD34⁺ peripheral blood progenitor cells after introduction of a mutated dihydrofolate reductase cDNA via retroviral gene transfer. *Blood* **85,** 566–574.

27. Flintoff, W. F., and Essani, K. (1980). Methotrexate-resistant Chinese hamster ovary cells contain a dihydrofolate reductase with an altered affinity for methotrexate. *Biochemistry* **19,** 4321–4327.

28. Goldie, J. H., Krystal, G., Hartley, D., Gudauskas, G., and Dedhar, S. (1980). A methotrexate insensitive variant of folate reductase present in two lines of methotrexate-resistant L5178Y cells. *Eur. J. Cancer (Oxford)* **16,** 1539–1546.

29. Gupta, R. S., Flintoff, W. F., and Siminovitch, L. (1977). Purification and properties of dihydrofolate reductase from methotrexate-sensitive and methotrexate-resistant Chinese hamster ovary cells. *Can. J. Biochem.* **55,** 445–452.

30. Haber, D. A., Beverly, S. M., Kiely, M. L., and Schimke, R. T. (1981). Properties of an altered dihydrofolate reductase encoded by amplified genes in cultured mouse fibroblasts. *J. Biol. Chem.* **256,** 9501–9510.

31. Hanania, E. G., Fu, S., Roninson, I., Zu, Z., and Deisseroth, A. B. (1995). Resistance to Taxol chemotherapy produced in mouse marrow cells by safety-modified retroviruses containing a human MDR-1 transcription unit. *Gene Ther.* **2,** 279–284.

32. Hanania, E. G., Giles, R. E., Kavanagh, J., Fu, S. Q., Ellerson, D., Zu, Z., Wang, T., Su, Y., Kudelka, A., Rahman, Z., Holmes, F., Hortobagyi, G., Claxton, D., Bachier, C., Thall, P., Cheng, S., Hester, J., Ostrove, J. M., Bird, R. E., Chang, A., Korbling, M., Seong, D., Cote, R., Holzmayer, T., and Deisseroth, A. B. (1996). Results of MDR-1 vector modification trial indicate that granulocyte/macrophage colony-forming unit cells

do not contribute to posttransplant hematopoietic recovery following intensive systemic therapy. *Proc. Natl. Acad. Sci. USA* **93,** 15346–15351.

33. Havenga, M., Valerio, D., Hoogerbrugge, P., and Es, H. (1999). In vivo methotrexate selection of murine hemopoietic cells transduced with a retroviral vector for Gaucher disease. *Gene Ther.* **6,** 1661–1669.

34. Hesdorffer, C., Antman, K., Bank, A., Fetell, M., Mears, G., and Begg, M. (1994). Human MDR gene transfer in patients with advanced cancer. *Hum. Gene Ther.* **5,** 1151–1160.

35. Hesdorffer, C., Ayello, J., Ward, M., Kaubisch, A., Vahdat, L., Balmaceda, C., Garrett, T, Fetell, M., Reiss, R., Bank, A., and Antman, K. (1998). Phase I trial of retroviral-mediated transfer of the human MDR1 gene as marrow chemoprotection in patients undergoing high-dose chemotherapy and autologous stem-cell transplantation. *J. Clin. Oncol.* **16,** 165–172.

36. Hock, R., and Miller, A. (1986). Retrovirus-mediated transfer and expression of drug resistance genes in human hematopoietic progenitor cells. *Nature* **320,** 275–277.

37. Jackson, R. C., and Niethammer, D. (1977). Acquired methotrexate resistance in lymphoblasts resulting from altered kinetic properties of dihydrofoltate reductase. *Euro. J. Cancer (Oxford)* **13,** 567–575.

38. James, R. I., May, C., Vagt, M. D., Studebaker, R., and McIvor, R. S. (1997). Transgenic mice expressing the tyr22 variant of murine DHFR: protection of transgenic marrow transplant recipients from lethal doses of methotrexate. *Exp. Hematol.* **25,** 1286–1295.

39. James, R. I., Warlick, C. A., Diers, M. D., Gunther, R., and McIvor, R. S. (2000). Mild preconditioning and low-level engraftment confer methotrexate resistance in mice transplanted with marrow expressing drug-resistant dihydrofolate reductase activity. *Blood* **96,** 1334–1341.

40. Jolivet, J., Cowan, K. H., Curt, G. A., Clendeninn, N. J., and Chabner, B. A. (1983). The pharmacology and clinical use of methotrexate. *N. Engl. J. Med.* **309,** 1094–1104.

41. Koc, O. N., Allay, J. A., Lee, K., Davis, B. M., Reese, J. S., and Gerson, S. L. (1996). Transfer of drug resistance genes into hematopoietic progenitors to improve chemotherapy tolerance. *Sem. Oncol.* **23,** 46–65.

42. Kohn, D. B. (1995). The current status of gene therapy using hematopoietic stem cells. *Curr. Opin. Pediat.* **7,** 56–63.

43. Lewis, W. S., Cody, V., Galitsky, N., Luft, J. R., Pangborn, W., Chunduru, S. K., Spencer, H. T., Appleman, J. R., and Blakley, R. L. (1995). Methotrexate-resistant variants of human dihydrofolate reductase with substitutions of leucine 22. *J. Biol. Chem.* **270,** 5057–5064.

44. Li, M-X, Banerjee, D., Zhao, S-C, Schweitzer, B. I., Mineishi, S., Gilboa, E., and Bertino, J. R. (1994). Development of a retroviral construct containing a human mutated dihydrofolate reductase cDNA for hematopoietic stem cell transduction. *Blood* **83,** 3401–3408.

45. Liu, H., Hung, Y., Wissink, S. D., and Verfaillie, C. M. (2000). Improved retroviral transduction of hematopoietic progenitors by combining methods to enhance virus-cell interaction. *Leukemia* **14,** 307–311.

46. Margolis, S., Philips, F. S., and Sternberg, S. S. (1971). The cytotoxicity of methotrexate in mouse small intestine in relation to inhibition of folic acid reductase and of DNA synthesis. *Cancer Res.* **31,** 2037–2046.

47. May, C., Gunther, R., and McIvor, R. S. (1995). Protection of mice from lethal doses of methotrexate by transplantation with transgenic marrow expressing drug-resistant dihydrofolate reductase activity. *Blood* **86,** 2439–2448.

48. May, C., James, R. I., Gunther, R., and McIvor, R. S. (1996). Methotrexate dose-escalation studies in transgenic mice and marrow transplant recipients expressing drug-resistant dihydrofolate reductase activity. *J. Pharmacol. & Exptl. Therapeut.* **278,** 1444–1451.

49. Maze, R., Carney, J. P., Kelley, M. R., Glassner, B. J., Williams, D. A., and Samson, L. (1996). Increasing DNA repair methyltransferase levels via bone marrow stem cell transduction rescues mice from the toxic effects of 1,3-*bis*(2-chloroethyl)-1-nitrosourea, a chemotherapeutic alkylating agent. *Proc. Natl. Acad. Sci. USA* **93,** 206–210.

50. McIvor, R. S. (1996). Drug-resistant dihydrofolate reductases: generation, expression and therapeutic application. *Bone Marrow Transplant.* **18,** S50–S54.

51. McIvor, R. S. (1999). Gene therapy of genetic diseases and cancer. *Pediatric Transplant.* **3**(suppl. 1), 116–121.

52. McIvor, R. S., Belur, L., Vagt, M., and Warlick, C. A. (1999). An inter-strain transgenic mouse system for tracking hematopoietic cells during selective, antifolate-mediated expansion in vivo. *Keystone Symp. Mol. Cell. Biol. Gene Ther.* **A4,** 320.

53. McIvor, R. S., and Simonsen, C. C. (1990). Isolation and characterization of a variant dihydrofolate reductase cDNA from methotrexate-resistant murine L5178Y cells. *Nucleic Acids Res.* **18,** 7025–7032.

54. McIvor, R. S., Weigel, B., Gunther, R., Diers, M. D., and Frandsen, J. (2000). Methotrexate chemotherapy of a murine mammary adeno-carcinoma in animals expressing drug-resistant dihydrofolate reductase activity. *Mol. Ther.* **1,** S166.

55. Melera, P. W., Davide, J. P., Hession, C. A., and Scotto, K. W. (1984). Phenotypic expression in *escherichia coli* and nucleotide sequence of two chinese hamster lung cell cDNAs encoding different dihydrofolate reductases. *Mol. Cell. Biol.* **4,** 38–48.

56. Moritz, T., Mackay, W., Glassner, B. J., Williams, D. A., and Samson, L. (1995). Retrovirus-mediated expression of a DNA repair protein in bone marrow protects hematopoietic cells from nitrosourea-induced toxicity in vitro and in vivo. *Cancer Res.* **55,** 2608–2614.

57. Morris, J. A., May, C., Kim, H. S., Ismail, R., Wagner, J. E., Gunther, R., and McIvor, R. S. (1996). Comparative methotrexate resistance of transgenic mice expressing two distinct dihydrofolate reductase variants. *Transgenics* **2,** 53–67.

58. Morris, J. A., and McIvor, R. S. (1994). Saturation mutagenesis at dihydrofolate reductase codons 22 and 31. A variety of amino acid substitutions conferring methotrexate resistance. *Biochem. Pharmacol.* **47,** 1207–1220.

59. Moscow, J. A. (1998). Methotrexate transport and resistance. *Leukemia Lymphoma* **30,** 215–224.

60. Mulligan, R. C., and Berg, P. (1981). Selection for animal cells that express the *Escherichia coli* gene coding for xanthine-guanine phos-phoribosyltransferase. *Proc. Natl. Acad. Sci. USA* **78,** 2072–2076.

61. Naldini, L., Blomer, U., Gallay, P., Ory, D., Mulligan, R., Gage, F. H., Verma, I. M., and Trono, D. (1996). In vivo gene delivery and stable transduction of nondividing cells by a lentiviral vector. *Science* **272,** 263–267.

62. Nottebrock, H., and Then, R. (1977). Thymidine concentrations in serum and urine of different species and man. *Biochem. Pharmacol.* **26,** 2175–2179.

63. Pan, D., Aronovich, E., McIvor, R. S., and Whitley, C. B. (2000). Retro-viral vector design studies toward hematopoietic stem cell gene therapy for mucopolysaccharidosis type I. *Gene Ther.* **7,** 1875–1883.

64. Patel, M., Sleep, S. E., Lewis, W. S., Spencer, H. T., Mareya, S. M., Sorrentino, B. P., and Blakley, R. L. (1997). Comparison of the protec-tion of cells from antifolates by transduced human dihydrofolate reduc-tase mutants. *Hum. Gene Ther.* **8,** 2069–2077.

65. Podda, S., Ward, M., Himelstein, A., Richardson, C., de la Flor-Weiss, E., Smith, L., Gottesman, M., Pastan, I., and Bank, A. (1992). Transfer and expression of the human multiple drug resistance gene into live mice. *Proc. Natl. Acad Sci. USA* **89,** 9676–9680.

66. Prendergast, N. J., Appleman, J. R., Delcamp, T. J., Blakley, R. L., and Freisheim, J. H. (1989). Effects of conversion of phenylalanine-31 to leucine on the function of human dihydrofolate reductase. *Biochemistry* **28,** 4645–4650.

67. Reese, J. S., Koc, O. N., Lee, K. M., Liu, L., Allay, J. A., Phillips, Jr., W. P., and Gerson, S. L. (1996). Retroviral transduction of a mutant methylguanine DNA methyltransferase gene into human CD34 cells confers resistance to O^6-benzylguanine plus 1,3-*bis*(2-chloroethyl)-1-nitrosourea. *Proc. Natl. Acad. Sci. USA* **93,** 14088–14093.

68. Rhee, M. S., Wang, Y., Nair, M. G., and Galivan, J. (1993). Acquisition of resistance to antifolates caused by enhanced gamma-glutamyl hydrolase activity. *Cancer Res.* **53,** 2227–2230.

69. Ringold, G., Dieckmann, B., and Lee, F. (1981). Co-expression and amplification of dihydrofolate reductase cDNA and the *Escherichia coli* XGPRT gene in Chinese hamster ovary cells. *J. Mol. Appl. Genet.* **1,** 165–175.

70. Rivera, B. K., Evans, W. E., Kalwinski, D. R., Mirro, J., Ochs, J., Dow, L. W., Abromowitch, M., Pui, C-H., Dahl, G. V., Look, A. T., Crone, M., and Murphy, S. B. (1985). Unexpectedly severe toxicity from intensive early treatment of childhood lymphoblastic leukemia. *J. Clin. Oncol.* **3,** 201–206.

71. Schornagel, J. H., and McVie, J. G. (1983). The clinical pharmacology of methotrexate. *Cancer Treat. Rev.* **10,** 53–75.

72. Schweitzer, B. I., Srimatkandada, S., Gritsman, H., Sheridan, R., Venkataraghavan, R., and Bertino, J. R. (1989). Probing the role of two hydrophobic active site residues in the human dihydrofolate reductase by site-directed mutagenesis. *J. Biol. Chem.* **264,** 20786–20795.

73. Sierra, E. E., Goldman, I. D. (1999). Recent advances in the understand-ing of the mechanism of membrane transport of folates and antifolates. *Semin. Oncol.* **26,** 11–23.

74. Simonsen, C. C., and Levinson, A. D. (1983). Isolation and expression of an altered mouse dihydrofolate reductase cDNA. *Proc. Natl. Acad. Sci. USA* **90,** 2495–2499.

75. Sirotnak, F. M., Moccio, D. M., Kelleher, L. E. and Goutas, L. J. (1981). Relative frequency and kinetic properties of transport-defective pheno-types among methotrexate-resistant L1210 clonal cell lines derived in vivo. *Cancer Research* **41,** 4447–4452.

76. Sorrentino, B. P., Brandt, S. J., Bodine, D., Gottesman, M., Pastan, I., Cline, A., and Nienhuis, A. W. (1992). Selection of drug-resistant bone marrow cells in vivo after retroviral transfer of human MDR1. *Science* **257,** 99–103.

77. Southern, P. J., and Berg, P. (1982). Transformation of mammalian cells to antibiotic resistance with a bacterial gene under control of the SV40 early region promoter. *J. Mol. Appl. Genet.* **1,** 327–341.

78. Spencer, H. T., Sleep, S. E., Rehg, J. E., Blakley, R. L., and Sorrentino, B. P. (1996). A gene transfer strategy for making bone marrow cells resistant to trimetrexate. *Blood* **87,** 2579–2587.

79. Srimatkandada, S., Schweitzer, B. I., Morosan, B. A., Dube, S., and Bertino, J. R. (1989). Amplification of a polymorphic dihydrofo-late reductase gene expressing an enzyme with decreased binding to methotrexate in a human colon carcinoma cell line, HCT-8R4, resistant to this drug. *J. Biol. Chem.* **264,** 3524–3528.

80. Stead, R. B., Kwok, W. W., Storb, R., and Miller, A. D. (1988). Canine model for gene therapy: inefficient gene expression in dogs reconstituted with autologous marrow infected with retroviral vectors. *Blood* **71,** 742–747.

81. Storb, R., Deeg, H. J., Whitehead, J., Appelbaum, F., Beatty, P., Bensinger, W., Buckner, C. D., Clift. R., Doney, K., and Farewell, V. (1986). Methotrexate and cyclosporine compared with cyclosporine alone for prophylaxis of acute graft versus host disease after marrow transplantation for leukemia. *N. Engl. J. of Med.* **314,** 729–735.

82. Storb, R., Leisenring, W., Anasetti, C., Appelbaum, F. R., Deeg, H. J., Doney, K., Martin, P., Sullivan, K. M., Witherspoon, R., Pettinger, M., Bensinger, W., Buckner, C. D., Clift, R., Flowers, M. E., Hansen, J. A., Pepe, M., Chauncey, T., Sanders, J., and Thomas, E. D. (1997). Methotrexate and cyclosporine for graft-vs.-host disease prevention: what length of therapy with cyclosporine? *Biol. Blood Marrow Trans-plant.* **3,** 194–201.

83. Sweeney, C., Frandsen, J., Zhao, R., Verfaillie, C., and McIvor, R. S. (2000). Effect of methotrexate on tumor progression in a mouse model of chronic myeloid leukemia. *Mol. Ther.* **1,** S167.

84. Thillet, J., Absil, J., Stone, S. R., and Pictet, R. (1988). Site-directed mutagenesis of mouse dihydrofolate reductase. *J. Biol. Chem.* **263,** 12500–12508.

85. Thompson, P. D., and Freisheim, J. H. (1991). Conversion of arginine to lysine at position 70 of human dihydrofolate reductase: generation of a methotrexate-insensitive mutant enzyme. *Biochemistry* **30,** 8124–8130.

86. Warlick, C. A., Diers, M., and McIvor, R. S. (2002). *In vivo* selection of antifolate-resistant transgenic hematopoietic stem cells in a murine bone marrow transplant model. *J. Pharmacol. Exptl. Therapeut.,* in press.

87. Warlick, C. A., Sweeney, C. L., and McIvor, R. S. (2000). Maintenance of differential methotrexate toxicity between cells expressing drug-resistant and wild-type dihydrofolate reductase activities in the presence of nucleosides through nucleoside transport inhibition. *Biochem. Pharmacol.* **59,** 141–151.

88. Williams, D. A., Hsieh, K., DeSilva, A., and Mulligan, R. C. (1987). Protection of bone marrow transplant recipients from lethal doses of methotrexate by the generation of methotrexate-resistant bone marrow. *J. Exp. Med.* **166,** 210–218.

89. Zhao, R. C., McIvor, R. S., Griffin, J. D., and Verfaillie, C. M. (1997). Gene therapy for chronic myelogenous leukemia (CML): a retroviral vector that renders hematopoietic progenitors methotrexate-resistant and CML progenitors functionally normal and nontumorigenic in vivo. *Blood* **90,** 4687–4698.

90. Zhao, S.-C., Li, M.-X., Banerjee, D., Schweitzer, B. I., Mineishi, S., Gilboa, E., and Bertino, J. R. (1994). Long-term protection of recipient mice from lethal doses of methotrexate by marrow infected with a double-copy vector retrovirus containing a mutant dihydrofolate reductase. *Cancer Gene Ther.* **1,** 27–33.

91. Zhao, S. C., Banerjee, D., Mineishi, S., and Bertino, J. R. (1997). Post-transplant methotrexate administration leads to improved curability of mice bearing a mammary tumor transplanted with marrow transduced with a mutant human dihydrofolate reductase cDNA. *Human Gene Ther.* **8,** 903–909.

A Genomic Approach to the Treatment of Breast Cancer

K.V. CHIN

Departments of Medicine and Pharmacology
The Cancer Institute of New Jersey
UMDNJ—Robert Wood
Johnson Medical School
Piscataway, New Jersey 08901

DEBORAH TOPPMEYER

Department of Medicine
The Cancer Institute of New Jersey
UMDNJ—Robert Wood
Johnson Medical School
Piscataway, New Jersey 08901

THOMAS KEARNEY

Department of Surgery
The Cancer Institute of New Jersey
UMDNJ—Robert Wood
Johnson Medical School
Piscataway, New Jersey 08901

MICHAEL REISS

Department of Medicine
The Cancer Institute of New Jersey
UMDNJ—Robert Wood
Johnson Medical School
Piscataway, New Jersey 08901

EDMUND LATTIME

Departments of Medicine and
Molecular & Microbiology
The Cancer Institute of New Jersey
UMDNJ—Robert Wood
Johnson Medical School
Piscataway, New Jersey 08901

WILLIAM N. HAIT

Departments of Medicine
and Pharmacology
The Cancer Institute of New Jersey
UMDNJ—Robert Wood
Johnson Medical School
Piscataway, New Jersey 08901

I. INTRODUCTION

The age of empiric treatment of cancer is ending and is being supplanted by a more rationale, targeted approach. Perhaps no better symbol of this evolution is the developing use of signal transduction inhibitors in the treatment of chronic myelogenous leukemia and antibodies that target oncogene products such as members of the epidermal growth factor receptor family (e.g., Her1 and Her2) for the treatment of breast and other carcinomas. These new targets are either unique or uniquely overexpressed in the disease; the drugs are likely to be effective in the subset of patients in which this abnormality represents the mechanism required by the cancer cell to maintain viability in the face of genetic changes that favor cell death. For example, a subset of breast cancers that express estrogen receptors maintain dependence on this pathway for cell viability and therefore are effectively treated with an estrogen receptor modulator such as tamoxifen. Although estrogen receptors are expressed in 40–60% of human breast cancers, only 40–60% (16–36%) of these individuals will respond to tamoxifen, a drug that blocks estrogen-receptor mediated transcription [1]. Similarly, whereas ~30% of breast cancers overexpress HER2/*neu,* only 25% of these patients respond to trastuzumab [2]. Why so many patients who express the estrogen receptor do not respond to hormonal therapy and why the majority of patients whose breast cancer overexpress HER2/*neu* do not respond to trastuzumab remain two of the most important questions facing the cancer pharmacologist.

II. TOWARD A GENOMIC APPROACH TO THERAPY

The above examples highlight an evolution in the treatment of cancer that is based on a detailed understanding of the underlying pathophysiology of the disease. Much of our

---

newer thinking resulted from the world's investment in cancer research. For example, a more detailed understanding of programmed cell death, first elucidated in the flatworm, *caenorhabditis elegans* [3], is leading to new approaches to overcoming resistance to currently available drugs and to new medications that will shorten the life expectancies of malignant cells. For example, overexpression of the antiapoptotic protein, Bcl-2, produces resistance to a variety of anticancer treatments [4]. Numerous laboratories are attempting to overcome this downstream mechanism of multidrug resistance. For example, DiPaola *et al.* [5,6] found that the combination of alpha-interferon with retinoic acid inhibited the function of Bcl-2 and restored the apoptotic response as measured by caspase activation. This approach has now been applied in a variety of clinical settings including breast, prostate, and kidney cancer.

We now appreciate that cells with wild-type p53 are more sensitive to apoptotic signals than cells in which this tumor suppressor gene is mutated [7]. Thus, DNA-damaging agents, such as topoisomerase poisons and radiation, produce greater cell kill in cell lines and animal models in which the tumor contains wtp53 [7,8]. Lutzker and Levine [9] demonstrated that the unique sensitivity of germ cell cancers to topoisomerase II poisons such as etoposide was due to the presence of presynthesized p53, poised to respond to a damaged genome.

Unexpectedly, mutant p53 appears to predict sensitivity to taxanes both in retrospective analyses of clinical trials [10,11] and in a variety of experimental models [12,13]. The mechanism by which p53 increases the sensitivity to taxanes was revealed when Murphy *et al.* [14] found that the wild-type protein suppressed the expression of a microtubule associated protein (MAP4) that regulates the polymerization of microtubules [14]. Zhang *et al.* [15] then demonstrated that overexpression of MAP4, which increased the polymerization state of microtubules, reproduced the sensitivity to taxanes seen in cell lines with mutant p53. Furthermore, it was found that cells with wtp53 could be sensitized to microtubule depolymerizing drugs (e.g., vinca alkaloids) by induction of the wild-type protein through DNA damage [16]. This provided a rationale for the sequential administration of doxorubicin, a DNA-damaging agent, followed be vinorelbine, a vinca alkaloid currently under clinical investigation in the treatment of breast cancer. p53 has also been shown to transcriptionally repress the multidrug resistance gene MDR1 [17] and MRP1 [18] so that when this gene is nonfunctional these drug resistance proteins may be overexpressed.

Alterations in the function of the retinoblastoma gene product can have unexpected effects on drug sensitivity. For example, Banerjee *et al.* [19] demonstrated that mutant Rb led to the transcriptional activation of several targets of chemotherapies, including dihydrofolate reductase, thymidylate synthase, and ribonucleotide reductase.

ras plays a central role in transmitting growth factor signals from the cytoplasm to the nucleus. For example, activation of the epidermal growth factor receptor family leads to the activation of mitogen-activated protein kinases through both the activation of phospholipase C and the activation of *ras*. Mutations in *ras* trap the molecule in the perpetually "on" position, thereby overwhelming growth inhibitory signals with proliferative cues. For *ras* to function, it must insert itself into the plasma membrane, a step requiring a posttranslational modification with isoprenoids of the farnesyl or geranyl type. Farnesyl transferase inhibitors currently under investigation may provide future opportunities to selectively interrupt signaling in several forms of cancer. Several lines of evidence suggest that the farnesyl transferase inhibitors target additional proteins which may give these types of drugs unexpectedly broad anticancer activity. In addition, overexpression of mutant *ras* can alter drug sensitivity. For example, Fan and colleagues demonstrated that overexpression of mutant H-Ras produced marked resistance to cisplatin and mitomycin-C and moderate resistance to trimetrexate and methotrexate, without changing sensitivity to drugs such as paclitaxel, doxorubicin, and etoposide [20].

These and similar data led investigators at The Cancer Institute of New Jersey to develop an algorithm for the treatment of breast cancer, shown in Fig. 1. In this model, patients with metastatic breast cancer are assigned to clinical trials based on an analysis of the expression of several known drug sensitivity and resistance factors. For example, tumors expressing hormone receptors receive estrogen receptor (ER) modulators or other hormonal therapies. Tumors that do not express hormone receptors are further analyzed for the expression of HER2/*neu,* which occurs in ~15% of ER-positive patients. Patients with HER2/*neu*-positive tumors are assigned to herceptin-based protocols.

Those cancers not expressing HER2/*neu* are further analyzed for the status of p53. Those harboring mutations are assigned to taxane-based protocols, unless these tumors also express multidrug resistance gene products such as P-glycoprotein.[1] For p53 wild type tumors, we have designed a protocol to test whether induction of the wild-type protein will sensitize tumors to vinca alkaloids [21,22]. In this study, patients receive doxorubicin on day one and vinorelbine either 24 or 48 hours later. Biopsies are performed immediately before treatment with doxorubicin and before the first dose of vinorelbine. To date, we have obtained preliminary evidence that p53 retains its ability to suppress the expression of MAP4 in some but not all patients [21,22]. This sequential therapy designed to manipulate the transcriptional profiles of tumors by inducing the transcriptionally active p53 is proving to be both safe and effective. For tumors that harbor mutant p53, we are testing taxane-based protocols combined with reagents designed to inactivate Bcl-2. For example, in

---

[1] Because MDR modulators have failed to live up to expectations does not negate the likely impact of P-glycoprotein or MRP on the effectiveness of natural product chemotherapeutics.

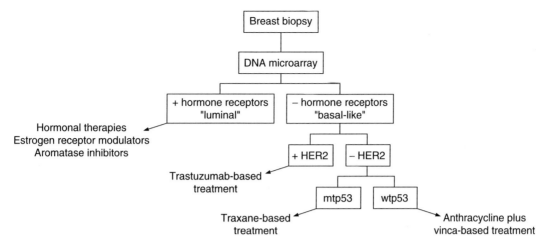

FIGURE 1   Genomics-based paradigm for selection of therapies based on gene expression profiles.

preclinical and phase I studies, DiPaola *et al.* [5,6] found that the combination of interferon with retinoic acid inhibited the function of Bcl-2 and restored the apoptotic pathway as measured by caspase activation. Current protocols are in progress to test the efficacy of *cis*-retinoic acid, interferon, and taxanes in refractory breast cancer.

For MDR-positive tumors (either P-glycoprotein or MRP), our algorithm directs choices to drugs that are not affected by these efflux pumps such as fluoropyrimidines (5-fluoruracil, capecitabine), methotrexate, or gemcitabine. A choice among these drugs could be made by understanding the status of Rb-based signaling pathways, which has been shown to regulate the targets of antimetabolites including dihydrofolate reductase, ribonucleotide reductase, and thymidylate synthase [19]. Thus, in Rb-mutated cells that overexpress dihydrofolate reductase (DHFR), RR, and or TS, antimetabolites may not be a good choice, whereas in Rb wt cells, where DHFR, ribonucleotide reductase (RR), and thymidylate synthase (TS) may not be increased, any or all of these drugs could be tested.

These concepts must be validated through well-controlled clinical trials. In the meantime, we believe it would be unwise for the savvy oncologist to ignore these emerging data, which provide compelling evidence for rational rather than empiric selection of anticancer drugs.

What we have described here is a rudimentary glimpse at what could be the future of cancer treatment (i.e., the rational selection of currently available chemotherapeutic agents based on limited phenotyping of individual tumors). This approach is already being applied with early indications of success in the treatment of colorectal cancer with 5-fluorouacil [23,24]. In this case, samples are analyzed for the target of drug action, thymidylate synthase, and the enzymes of activation and deactivation of the drug including dihydropyrimidine dehydrogenase. A similar analysis could be applied to the use of capecitabine, where the activity of thymidine phosphorylase in the tumor might be an indicator of respon-

siveness to this drug, which requires thymidine phosphorylase for activation.

However, this approach may quickly be supplanted by a more robust analysis of the transcription profile of an individual's tumor. Using DNA microarray or proteomic analysis, it should be possible to gain an understanding of the unique susceptibilities of an individual's cancers (if they exist). Already, the feasibility of this approach to analyzing human breast cancer has been reported [25], and a new concept has been suggested for the origins and classification of breast cancer. In this analysis, breast cancers fell into at least three groups: those of luminal cell origin (ER-positive), basal epithelial cell origin (ER-negative), and HER2/*neu* (Erb-2)-positive cells. Each one of these was characterized by high expression of a specific subset of genes. Furthermore, it has recently been shown that mutations in *BRCA1* or *BRCA2* can independently influence the transcription profile of breast cancers [26]. Therefore, it should be possible to sort these data by viewing all of the pathways involved in the ultimate outcome of a drug–target interaction. In this way, it may be possible to select a drug that is likely to be effective over one that is not.

Following relapse from first-line treatments, patients with breast and most other solid malignancies fail to respond meaningfully to second and third line forms of therapy. Several reasons for this have been discovered over the years, including the presence of gene products that confer broad-spectrum resistance to cancer chemotherapeutics. Examples include P-glycoprotein [27,28], MRP [29,30], glutathione [31], *bcl*-2 [32], and downregulation of topoisomerases [33,34] and other proteins that either prevent the drug from reaching meaningful concentrations at its target or block the death signal that occurs following an effective drug–target interaction. Attempts at overcoming these individual mechanisms of multidrug resistance have been relatively unsuccessful, perhaps because of the lack of effective resistance modulators. An alternative hypothesis is that overlapping pathways of drug resistance acquired by cancer cells preclude effectively targeting a single protein. Below, we will summarize work with the MCF-7 human breast cancer cell line in which DNA microarrays were done before and after treatment with doxorubicin, one our most effective drugs for the treatment of breast cancer.

III. THE USE OF DNA MICROARRAYS TO UNDERSTAND DRUG RESISTANCE

The development of cDNA microarrays, whereby tens of thousands of cDNAs are arrayed on a solid support for simultaneous probing for the expression of a large set of genes, has revolutionized the approach of gene expression analysis [35,36]. Ultimately, the estimated 30–40,000 genes in the human genome [37–39] will be arrayed on high-density solid surfaces, allowing expression monitoring of the entire genome in a single hybridization. This approach, coupled with detailed biochemical analysis of the individual gene products, will greatly facilitate functional analysis of the human genome. These discoveries have advanced gene analysis from the cumbersome, serial, single-gene analysis to the high-throughput, simultaneous analysis of tens of thousands of genes in a single hybridization.

Genomic technologies can now be applied to the problem of drug resistance to cancer chemotherapy. Drug resistance in cancer may arise spontaneously or may be acquired as a result of exposure to chemotherapy [40]. The development of spontaneous resistance may be associated with tumor progression, whereby genetic alterations of oncogenes, tumor suppressor genes, and other growth-controlling genes alter cellular transcription profiles [41]. In acquired resistance, exposure to chemotherapy may alter gene expression, enabling a small fraction of cancer cells to escape the killing by chemotherapy. The surviving cells may undergo clonal expansion and produce cancers that are refractory to further treatment. Global alterations in the transcription program during tumorigenesis or following chemotherapy underlie the complexity and heterogeneity of drug resistance encountered in the clinic.

Regulatory genes that are altered during tumorigenesis may also influence sensitivity to chemotherapy [42]. As shown in Table 1, these genetic alterations involve a diverse group of growth regulators that include tumor suppressors, oncogenes, cell-cycle regulators, transcription factors, growth factor receptors, DNA repair factors, and cell death regulators (Bcl-2, Bcl-xL, and BAX).

In view of the complex changes underlying drug resistance, use of the cDNA microarray to examine the entire transcription profile or "transcriptomes" of drug-resistant tumors might reveal the several pathways that contribute to resistance. We have recently studied whether or not there are differences in the genes that confer resistance derived by either drug exposure or drug selection during chemother-

TABLE 1 Genetic Alterations That Influence Drug Sensitivity

Genetic alteration	Ref.
Apoptosis	
Bcl-2, Bcl-xL, BAX	[55,56]
Oncogenes/growth factor receptors	
HER2/*neu*	[57,58]
EGFR	[59]
ras	[60,61]
Tumor suppressor genes/cell cycle checkpoints	
p53	[39,62]
p21	[63,64]
p16	[16]
pRb	[65]
DNA repair	
hMLH1, hMSH	[66,67]
Transcription factors	
YB-1	[68]
NF-κB	
GPI-anchored gene	[70]

apy. To address this, we used the DNA microarray to follow the expression profiles of MCF-7 breast cancer cells in response to doxorubicin [43]. We compared the profiles obtained in the sensitive, parental MCF-7 line to those of MCF-7 cells that were selected for resistance to doxorubicin (MCF-7/D40). Doxorubicin is an inhibitor of topoisomerase II that produces DNA damage, which in certain contexts leads to programmed cell death or apoptosis [44]. Expression profiles obtained following treatment of MCF-7 cells with doxorubicin showed alterations over time in the expression of approximately 500 genes (Fig. 2 [see also color insert]). (The entire dataset of genes analyzed in this experiment can be found at a website maintained by our laboratory at *http://cinj.umdnj.edu/drug_resistance*.) The biochemical functions of the altered genes are diverse and include transcription factors, protein kinases and phosphatases, cell-cycle regulators, proteases, and apoptotic and antiapoptotic factors as well as a large number of metabolic genes. Altered expression of the transcription factors, including downregulation of the general transcription factor RNA polymerase II, the transcription corepressor Dr1-associated protein [45], and the enhancer-binding proteins AP-3 and AP-4, is indicative of a general repression of transcription in response to the cytotoxic effects of doxorubicin. In addition, increased expression of cytochrome c, which triggers apoptosis by activating the caspases [46], and downregulation of Bcl-2, an antiapoptotic factor [47], were consistent with the induction of apoptosis by doxorubicin.

Genes involved in proteolysis also showed striking changes after doxorubicin treatment. Protein degradation is recognized to be critical in the regulation of the cell cycle, transcription, and signal transduction [48]. We found that

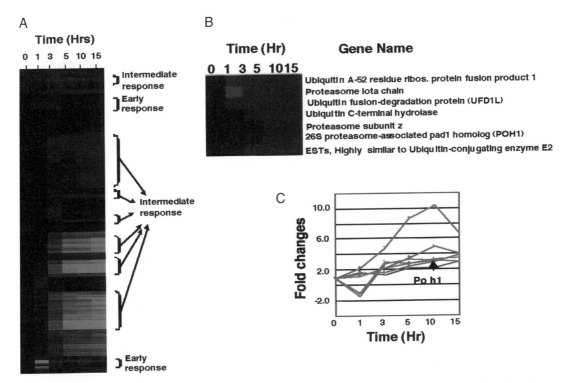

FIGURE 2   Cluster image showing distinct patterns of gene expression. (A) Approximately 500 genes that showed alterations in expression in response to doxorubicin were selected for hierarchical clustering analysis using a software suite developed by Eisen *et al*. The altered genes were clustered into groups on the basis of their similarity in expression patterns and the results are displayed using TreeView® software. Transition of color for each gene from brown to green indicates a gradual decrease in expression with time, and changes in color from brown to red indicate increased expression. (B) The insert shows a TreeView® display of the ubiquitin-proteasome gene cluster with altered expression following exposure to doxorubicin. (C) Plot of the ubiquitin-proteasome genes showing their gradual changes with time. The entire data set analyzed in this experiment can be found at: *http://cinj.umdnj.edu/drug* resistance. (See color insert.)

some ubiquitin-associated factors (subunits of the 26S proteasome including Poh1 and the regulatory subunit 4), were upregulated after doxorubicin treatment. It has been shown previously that overexpression of Poh1 confers multidrug resistance [49,50].

Comparison of the expression profiles between parental and resistant MCF-7 cells treated with doxorubicin revealed a subset of overlapping genes (Table 2). Based on their functions, overexpression of some of these genes suggests that they may have roles in drug resistance. For example, increased expression of epoxide hydrolase, a drug-metabolizing enzyme, may enhance the metabolism of doxorubicin [51,52]. The single-strand DNA repair protein, XRCC1, is also overexpressed after doxorubicin in the drug-resistant MCF-7/D40 cells. XRCC1 forms a repair complex with DNA-β-polymerase, ligase III, and poly(ADP-ribose) polymerase [53] and also binds to single-stranded DNA breaks (gaps and nicks) in the DNA-β-polymerase complex. XRCC1 may also bind to and repair double-stranded DNA gap lesions produced by topoisomerase II. Thus, increased expression of XRCC1 may increase the efficiency of the repair of lesions generated by topoisomerase II.

The regulatory subunit 4 of the 26S proteasome is also overexpressed in the doxorubicin-resistant cell line. Because it has already been shown that overexpression of the Poh1 gene in the ubiquitin/proteasome pathway confers multidrug resistance [49,50], it will be interesting to determine if increased expression of the 26S proteasome regulatory subunit 4 gene in MCF-7/D40 cells may also contribute to an increase in resistance to doxorubicin.

These data indicate that the mechanisms of drug resistance in a breast cancer cell line selected for doxorubicin resistance are exceedingly complex and go well beyond the expression of P-glycoprotein [54]. The coordinated expression of a subset of genes including epoxide hydrolase, XRCC1, regulatory subunit 4 of the 26S proteasome, and others (Table 2) may represent a distinct signature for doxorubicin resistance in breast cancer. The role of these genes in drug resistance needs to be verified by functional expression studies. Alternatively, the genetic background of an individual's tumor may alter the response to chemotherapeutic drugs, thereby creating unique signatures for each patient. Whether by induction or selection, our results suggest that there is a convergence in the genes and pathways that are critical for drug resistance in

TABLE 2   Subset of Genes transiently Induced by Doxorubicin in MCF-7 Cells
that Overlap with Genes Overexpressed in Doxorubicin-Resistant MCF-7/D40 Cells

Name	Accession number	Change in gene expression 15 hr after exposure of MCF-7 to doxorubicin (-fold)	Change in gene expression in MCF/D40 cells (-fold)
Cell cycle genes			
CDC28 kinase 1	L29222	−7.3	−2.3
XRCC1	M36089	9.8	7.6
Protein CDC27	AA489098	5.9	5.5
Neuronal genes			
Ataxin 2	U70323	−7.7	−2.8
Human neuronal apoptosis inhibitory protein	U19251	4.6	3.7
Synaptotagmin	M55047/J05710	14.0	10.4
Signal transduction genes			
Fms-related protein tyrosine kinase	D00133	23.0	9.5
Granulin	M75161	17.1	11.6
Recoverin	AB001838	5.1	7.6
Transferrin receptor protein	M11507	7.4	4.6
Collapsin response mediator protein 1	U17278	20.7	13.5
Transcription factors			
RNA polymerase II	L37127	−7.8	−2.0
Sigma 3B	X99459	27.5	16.4
Metabolic genes			
Cellular retinoic acid binding protein	M68867	29.4	17.0
Epoxide hydrolase	J03518	15.8	11.3
Ubiquitin-proteasome			
26S proteasome regulatory subunit 4	AA622905	18.6	11.6
Protein secretion			
Pescadillo	U78310	11.9	8.8

breast cancer. Therefore, analysis of drug resistance in cancer by DNA microarray should produce important insights into the mechanisms of resistance and ultimately provide information that will lead to the circumvention of resistance to chemotherapy.

IV. EFFECTS OF GENOMIC-BASED APPROACHES ON THE MANAGEMENT OF BREAST CANCER PATIENTS

At the turn of the 20th century, the treatment of patients with breast cancer was surgical. The development of the radical mastectomy afforded many patients the chance for cure, even for those patients with positive nodes [71]. We now recognize that the cured patients were those with true local or regional disease. Patients who died from breast cancer after radical mastectomy were patients who died from systemic metastases. With the realization that many patients have oc-

cult systemic disease at the time of clinical presentation, therapy became multidisciplinary. Patients still required adequate local and regional therapy but now systemic adjuvant therapy was added [72]. This newer systemic approach was combined with lesser surgical procedures (lumpectomy), plus regional radiation therapy. The role of the surgeon had changed from being the sole treating physician to that of a multimodality team member.

The 21st century surgeon is responsible for the evaluation of patients with new breast complaints, some diagnostic procedures, and local/regional treatment of early-stage breast cancer. Local/regional treatment includes axillary node sampling in order to define the stage of disease. Newer techniques such as sentinel node mapping [73,74] may allow accurate axillary staging with lower morbidity. The surgeon also initially evaluates patients with locally advanced breast cancer and must initiate appropriate diagnostic maneuvers along with prompt referral for preoperative systemic therapy. The surgeon may also provide palliative procedures for patients presenting with systemic disease combined with local control problems.

The current staging system for breast cancer is anatomic [75]. The surgeon ultimately provides the tumor to the pathologist for inspection following a lumpectomy or mastectomy. Axillary lymph nodes are removed by axillary dissection or sentinel node mapping and biopsy. The current anatomic staging is quite crude. A size measurement is made of the tumor and the number of lymph nodes containing cancer is recorded. In the new paradigm described in this chapter, the genetic "fingerprint" of the tumor will be examined. The overexpression and suppression of gene activity is measured and a treatment plan is devised that is individualized for the genetic characteristics of the patient's tumor. This technology demands the acquisition of fresh tissue in such a manner that sophisticated tissue testing is not compromised. Thus, the modern surgeon now assumes a new responsibility: coordinated tissue acquisition. Patient tissue must be removed in such a way that ischemia is minimized. Standard pathologic examination is still done but must be performed expeditiously to allow prompt tissue freezing in liquid nitrogen. A dedicated tissue retrieval service with trained technicians greatly facilitates this goal. The tissue retrieval service comes to the operating room to receive the tissue as soon as it is available. They transport the tissue on ice to the pathologist and stand by while initial studies are done. Within a few minutes, portions of viable tumor must be frozen in liquid nitrogen. This effort requires a coordinated effort between the surgeon and the service.

In addition to intraoperative tissue acquisition, the surgeon must develop skills in other forms of tissue acquisition. Fine-needle cytology skills and minimally invasive core biopsy skills are essential for tissue acquisition in patients with locally advanced breast cancer. Repeat tissue acquisition during treatment, often facilitated by imaging capabilities, allows the multidisciplinary team to measure tumor response on a genetic level. Thus, a patient's individual treatment plan can be continued or altered based on genetically based tumor-response criteria.

Acknowledgments

This work was supported by grants from the Public Health Service, including CA 78695, CA 72720, and CA 82607, and from the Department of Defense DAMD17–98–1–8043.

References

1. Furr, B. J. A., and Jordan, V. C. (1984). The pharmacology and clinical uses of tamoxifen. *Pharmacol. Ther.* **25**, 127–205.

2. Vogel, C., Cobleigh, M., Tipathy, D. *et al.* (1998). Efficacy and safety of herceptin (trastuzumab, humanized anti-Her2 antibody) as a single agent in first line treatment of Her2 overexpressing metastatic breast cancer. *Breast Cancer Res. Treat.* **50**, 232.

3. Liu, Q. A., and Hegartner, M. O. (2000). The molecular mechanism of programmed cell death in *C. elegans. Ann. N.Y. Acad. Sci.* **887,** 92–104.

4. Reed, J. C. (1997). Bcl-2 family proteins: regulators of apoptosis and chemoresistance in hematologic malignancies. *Semin. Hematol.* **54** (4, suppl. 5), 9–19.

5. DiPaola, R. S., Rafi, M. M., Vyas, V., Toppmeyer, D., Rubin, E., Patel, J., Goodin, S., Medina, M., Medina, P., Zamek, R., Zhang, C., White, E., Gupta, E., and Hait, W. N. (1999). Phase I clinical study of 13–xiaretioic acid, interferon alpha, and paclitaxel in patients with prostate cancer and other advanced malignancies. *J. Clin. Oncol.* **17**, 2213–2218.

6. DiPaola, R. W. S., and Aisner, J. (1999). Overcoming *bcl-2* and *p53* mediated resistance in prostate cancer. *Semin. Oncol.* **26,** 112–116.

7. Lowe, S. W., Ruley, H. E., Jacks, T., and Housman, D. E. (1993). p53-dependent apoptosis modulated the cytotoxicity of anticancer agents. *Cell* **74**, 957–967.

8. Bunz, F., Hwang, P. M., Torrance, C., Waldman, T., Zhang, Y., Dillehay, L., Williams, J., Lengauer, C., Kinzler, K. W., and Vogelstein, B. (1999). Disruption of p53 in human cancer cells alters the responses to therapeutic agents. *J. Clin. Invest.* **104**, 263–269.

9. Lutzker, S. G., and Levine, A. J. (1996). A functionally inactive p53 protein in teratocarcinoma cells is activated by either DNA damage or cellular differentiation. *Nat. Med.* **2**, 804–810.

10. Kandioler-Eckersberger, D., Ludwig, C., Rudas, M., Kapperl, S., Janschek, E., and Wenzel, C. (2000). Tp53 mutation and p53 overexpression for prediction of response to neoadjuvant treatment in breast cancer patients. *Clin. Cancer Res.* **6**, 50–56.

11. Kiang, T. C., Arkerley, W., Fan, A. C., Moore, T., Mangray, S., Hsiu, C. M., and Safran, H. (2000). p53 mutations do not predict response to paclitaxel in metastatic nonsmall cell lung carcinoma. *Cancer* **89,** 769–773.

12. Rakovitch, E., Mellado, W., Hall, E. J., Pandita ,T. K., Sawant, S., and Geard, C. R. (1999). Paclitaxel senstivity correlates with p53 status and DNA fragmentation, but not G2/M accumulation. *Int. J. Radiat. Oncol. Biol. Phys.* **44**, 1119–1124.

13. Wahl, A. F., Donaldson, K. L., Fairchild, C., Lee, F. Y., Foser, S. A., Demers, G. W., and Galloway, D. A. (1996). Loss of normal p53 function confers sensitization to Taxol by increasing G2/M arrest and apoptosis. *Nat. Med.* **2**, 72–79.

14. Murphy, M., Hinman, A. and Levin, A. J. (1996). Wild-type p53 negatively regulates the expression of a microtubule-associated protein. *Genes Dev.* **10**, 2971–2980.

15. Zhang, C. C., Yang, J. M., White, E., Murphy, M., Levine, A. J., and Hait, W. N. (1997). The role of MAP4 expression in the sensitivity to paclitaxel and resistance to vinca alkaloids in p53 mutant cells. *Oncogene.* **16**, 1617–1624.

16. Zhang, C. C., Yang, J. M., Bash-Babula, J., White, E., Murphy, M., Levine, A. J., and Hait, W. N. (1999). DNA damage increases sensitivity to vinca alkaloids and decreases sensitivity to taxanes through p53-dependent repression of microtubule associated protein 4. *Cancer Res.* **59**, 3663–3370.

17. Chin, K. V., Ueda, K., Pastan, I., and Gottesman, M. M. Modulation of activity of the promoter of the human MDR1 gene by Ras and p53. *Science* **255**, 459–462.

18. Sullivan, G. F., Yang, J. M., Vassil, A., Yang, J., Bash-Babula, J., and Hait, W. N. (2000). Regulation of expression of the multidrug resistance protein MRP1 by p53 in human prostate cancer cells. *J. Clin. Invest.* **105**, 1261–1267.

19. Banerjee, D., Schneider, B., Fu, J. Z., Adhikari, D., Zhao, S. C., and Bertino, J. R. (1998). Role of E2F in chemosensitivity. *Cancer Res.* **58**, 4292–4296.

20. Fan, J., Banerjee, D., Stambrook, P. J., and Bertino, J. R. (1997). Modulation of cytotoxicity of chemotherapeutic drugs by activate H-ras. *Biochem. Pharmacol.* **53**, 1203–1209.

21. Bash-Babula, J., Alli, E., Hait, W. N., and Toppmeyer, E. (2001). A phase I/II clinical trial of doxorubicin and vinorelbine: effects on

p53 and microtubule associated protein 4 expression in patients with advanced breast cancer. *Proc. Am. Assoc. Cancer Res.* **42,** 119.

22. Toppmeyer, D., Bash-Babula, J., Alli, E., and Hait, W. N. (2001). Sensitization of breast cancer patients to vinorelbine (V) following p53 induction and MAP4 repression by doxorubicin (DOX): A phase I/II study. *Proc. Am. Soc. Clin. Oncol.* **20,** 46a.

23. Salonga, D., Danenberg, K. D., Johnson, M., Metzger, R., Groshen, S., Tsao-Wei, D. D. *et al.* (2000). Colorectal tumors responding to 5-fluorouracil have low gene expression levels of dihydropyrimidine dehydrogenase, thymidylate synthase, and thymidine phosphorylase. *Clin. Cancer Res.* **6,** 1322–1327.

24. Copur, S., Aiba, K., Drake, J. C., Allegra, C. J., and Chu, E. (1995). Thymidylate synthase gene amplification in human colon cancer cell lines resistant to 5-fluorouracil. *Biochem. Pharmacol.* **49,** 1419–1426.

25. Perou, C. M., Sorlle, T., Eisen, M., van de Rijn, M., Jeffrey, S. S., Rees, C. A., Pollack, J. R. *et al.* (2000). Molecular portraits of human breast tumours. *Nature* **406,** 747–752.

26. Hedenfalk, I., Duggan, D., Chen, Y., Radmacher, M., Bittner, M., Simon, R., Meltzer, P., *et al.* (2001). Gene-expression profiles in hereditary breast cancer. *N. Engl. J. Med.* **344,** 539–548.

27. Biedler, J. L., and Rehm, H. (1970). Cellular resistance to actinomycin D in Chinese hamster ovary cells in vitro: cross resistance, radiographic, and cytogenetic studies. *Cancer Res.* **30,** 1174–1184.

28. Julian, R. L., and Ling, V. (1976). A surface glycoprotein modulating drug permeability in Chinese hamster ovary cell mutants. *Biochem. Biophys. Acta.* **455,** 1252–1262.

29. Cole, S. P. C., Bhardwaj, G., Gerlach, J. H., Mackie, J. E., Grant, C. E., Almquist, K. C. *et al.* (1992). Overexpression of a transporter gene in a multidrug-resistant human lung cancer cell line. *Science* **258,** 1650.

30. Marsh, W., and Center, M. (1987). Adriamycin resistance in HL60 cells and accompanying modification of a surface membrane protein continued in drug-sensitive cells. *Cancer Res.* **47,** 5080–5086.

31. Raha, A., and Tew, K. D. (1996). Glutathione S-transferases, *in Drug Resistance* (W. N. Hait, ed.), pp. 83–122. Kluwer Academic, Dordrecht.

32. Walton, M. I., Whysong, D., O'Connor, P. M., Hockenbery, D., Korsmeyer, S. J., and Kohn, K. W. (1993). Constitutive expression of human Bcl-2 modulates nitrogen mustard and camptothecin induced apoptosis. *Cancer Res.* **53,** 1853–1861.

33. Beck, W. T., Cirtain, M. C., Danks, M. K., Felsted, R. L., Safa, A. R., Wolverton, J. S., Suttle, D. P., and Trent, J. M. (1987). Pharmacological, molecular, and cytogenetic analysis of "atypical" multidrug-resistant human leukemic cells. *Cancer Res.* **47,** 5455–5460.

34. Rubin, E. H., Li, T. K., Duann, P., and Liu, L. F. (1996). Cellular resistance to topoisomerase poisons. in *Drug Resistance* (W. N. Hait, ed.), pp. 243–262. Kluwer Academic, Dordrecht.

35. Schena, M., Shalon, D., Davis, R. W., and Brown, P. O. (1995). Quantitative monitoring of gene expression patterns with a complementary DNA microarray. *Science,* **270,** 467–470.

36. Shalon, D., Smith, S. J., and Brown, P. O. (1996). A DNA microarray system for analyzing complex DNA samples using two-color fluorescent probe hybridization. *Genome Res.* **6,** 639–645.

37. Antequera, F., and Bird, A. (1994). Predicting the total number of human genes. *Nat. Genet.* **8,** 114.

38. Fields, C., Adams, M. D., White, O., and Venter, J. C. (1994). How many genes in the human genome? *Nat. Genet.* **7,** 345–346.

39. Nowak, R. (1994). Mining treasures from 'junk DNA.' *Science* **263,** 608–610.

40. Chin, K.-V., Pastan, I., and Gottesman, M. M. (1993). Function and regulation of the human multidrug resistance gene. *Adv. Cancer Res.* **60,** 157–180.

41. Vogelstein, B., and Kinzler, K. W. (1993). The multistep nature of cancer. *Trends Genet.* **9,** 138–141.

42. el-Deiry, W. S. (1997). Role of oncogenes in resistance and killing by cancer therapeutic agents. *Curr. Opin. Oncol.* **9,** 79–87.

43. Kudoh, K., Ramanna, M., Elkahloun, A. G., Bittner, M. L., Meltzer, P. S., Trent, J. M., Dalton, W. S., and Chin, K.-V. (2000). Analysis of the mechanisms of drug resistance in cancer by cDNA microarray. *Cancer Res.* **60,** 4161–4166.

44. Burden, D. A., and Osheroff, N. (1998). Mechanism of action of eukaryotic topoisomerase II and drugs targeted to the enzyme. *Biochim. Biophys. Acta* **1400,** 139–154.

45. Kim, S., Na, J. G., Hampsey, M., and Reinberg, D. (1997). The Dr1/DRAP1 heterodimer is a global repressor of transcription in vivo. *Proc. Natl. Acad. Sci. USA,* **94,** 820–825.

46. Yang, J., Liu, X., Bhalla, K., Kim, C. N., Ibrado, A. M., Cai, J., Peng, T. I., Jones, D. P., and Wang, X. (1997). Prevention of apoptosis by Bcl-2: release of cytochrome c from mitochondria blocked. *Science* **275,** 1129–1132.

47. Reed, J. C. (1998). Bcl-2 family proteins. *Oncogene* **17,** 3225–3236.

48. Baumeister, W., Walz, J., Zuhl, F., and Seemuller, E. (1998). The proteasome: paradigm of a self-compartmentalizing protease. *Cell* **92,** 367–380.

49. Spataro, V., Toda, T., Craig, R., Seeger, M., Dubiel, W., Harris, A. L., and Norbury, C. (1997). Resistance to diverse drugs and ultraviolet light conferred by overexpression of a novel human 26S proteasome subunit. *J. Biol. Chem.* **272,** 30470–30475.

50. Spataro, V., Norbury, C., and Harris, A. L. (1998). The ubiquitin-proteasome pathway in cancer. *Br. J. Cancer* **77,** 448–455.

51. Murray, G. I., Weaver, R. J., Paterson, P. J., Ewen, S. W., Melvin, W. T., and Burke, M. D. (1993). Expression of xenobiotic metabolizing enzymes in breast cancer. *J. Pathol.* **169,** 347–353.

52. Murray, G. I., Paterson, P. J., Weaver, R. J., Ewen, S. W., Melvin, W. T., and Burke, M. D. (1993). The expression of cytochrome P-450, epoxide hydrolase, and glutathione S-transferase in hepatocellular carcinoma. *Cancer* **71,** 36–43.

53. Marintchev, A., Mullen, M. A., Maciejewski, M. W., Pan, B., Gryk, M. R., and Mullen, G. P. (1999). Solution structure of the single-strand break repair protein XRCC1 N-terminal domain. *Nat. Struct. Biol.* **6,** 884–893.

54. Taylor, C. W., Dalton, W. S., Parrish, P. R., Gleason, M. C., Bellamy, W. T., Thompson, F. H., Roe, D. J., and Trent, J. M. (1991). Different mechanisms of decreased drug accumulation in doxorubicin and mitoxantrone resistant variants of the MCF-7 human breast cancer cell line. *Br. J. Cancer* **63,** 923–929.

55. Dole, M. G., Jasty, R., Cooper, M. J., Thompson, C. B., Nunez, G., and Castle, V. P. (1995). Bcl xL is expressed in neuroblastoma cells and modulates chemotherapy-induced apoptosis. *Cancer Res.* **55,** 2576–2582.

56. Lowe, S. W., Bodis, S., McClatchey, A., Remington, L., Ruley, H. E., Fisher, D. E., Housman, D. E., and Jacks, T. (1994). p53 status and the efficacy of cancer therapy in vivo. *Science* **266,** 807–810.

57. Yu, D., Liu, B., Tan, M., Li, J., Wang, S., and Hung, M.-C. (1996). Overexpression of c-erbB-2/neu in breast cancer cells confers increased resistance to Taxol via MDR1-independent mechanisms. *Oncogene* **13,** 1359–1365.

58. Pegram, M. D., Finn, R. S., Arzoo, K., Beryt, M., Pietras, R. J., and Slamon, D. J. (1997). The effect of HER-2/*neu* overexpression on chemotherapeutic drug sensitivity in human breast and ovarian cancer cells. *Oncogene* **15,** 537–547.

59. Strobel, T., Swanson, L., Korsmeyer, S., and Cannistra, S. A. (1996). BAX enhances paclitaxel induced apoptosis through a p53-independent pathway. *Proc. Natl. Acad. Sci. USA* **93,** 14094–14099.

60. Jansen, B., Schlagbauer-Wadl, H., Eichler, H., Wolff, K., van Elsas, A., Schrier, P. I., and Pehamberger, H. (1997). Activated N-ras contributes to the chemoresistance of human melanoma in severe combined immunodeficiency (SCID) mice by blocking apoptosis. *Cancer Res.* **57,** 362–365.

61. Koo, H., Monks, A., Mikheev, A., Rubinstein, L. V., Gray-Goodrich, A., McWilliams, M. J., Alvord, W. G., Oie, H. K., Gazdar, A. F., Paull, K. D., Zarbl, H., and Vande Woude, G. F. (1996). Enhanced sensitivity

to 1-B-D-arabinofuranosylcytosine and topoisomerase II inhibitors in tumor cell lines harboring activated *ras* oncogenes. *Cancer Res.* **56,** 5211–5216.

62. Hawkins, D. S., Demers, D. W., and Galloway, D. A. (1996). Inactivation of p53 enhances sensitivity to multiple chemotherapeutic agents. *Cancer Res.* **56,** 892–898.

63. McDonald, III, E. R., Wu, G. S., Waldman, T., and El-Deiry, W. S. (1996). Repair defect in p21WAF1/CIP1–/– human cancer cells. *Cancer Res.* **56,** 2250–2255.

64. Fan, S., Chang, J. K., Smith, M. L., Duba, D., Fornace, Jr., A. J., and O'Connor, P. M. (1997). Cells lacking CIP/WAF1 genes exhibit preferential sensitivity to cisplatin and nitrogen mustard. *Oncogene* **14,** 2127–2136.

65. Li, W., Fan, J., Hochhauser, D., Banerjee, D., Zielinski, Z., Almasan, A., Yin, Y., Kelly, R., Wahl, G. M., and Bertino, J. R. (1995). Lack of functional retinoblastoma protein mediates increased resistance to antimetabolites in human sarcoma cell lines. *Proc. Natl. Acad. Sci. USA* **92,** 10436–10440.

66. Brown, R., Hirst, G. L., Gallagher, W. M., McIlwrath, A. J., Margison, G. P., van der Zee, A. G. J., and Anthoney, D. A. (1997). HMLH1 expression and cellular responses of ovarian tumor cells to treatment with cytotoxic anticancer agents. *Oncogene* **15,** 45–52.

67. Fink, D., Nebel, S., Aebi, S., Zheng, H., Cenni, B., Nehme, A., Christen, R. D., and Howell, S. B. (1996). The role of DNA mismatch repair in platinum drug resistance. *Cancer Res.* **56,** 4881–4886.

68. McCurrach, M. E., Connor, T. M., Knudson, C. M., Korsmeyer S. J., and Lowe, S. W. (1997). Bax deficiency promotes drug resistance and oncogenic transformation by attenuating p53 dependent apoptosis. *Proc. Natl. Acad. Sci. USA* **94,** 2345–2349.

69. Baldwin, A. S. (2001). Control of oncogenesis and cancer therapy resistance by the transcription factor NF-kappaB. *J. Clin. Invest.* **107,** 241–246.

70. Furuhata, T., Tokino, T., Urano, T., and Nakamura, Y. (1996). Isolation of a novel GPI anchored gene specifically regulated by p53: correlation between its expression and anticancer drug sensitivity. *Oncogene* **13,** 1965–1970.

71. Quiet, C. A., Ferguson, D. J., Weichselbaum, R. R., and Hellman, S. (1996). Natural history of node-positive breast cancer: the curability of small cancers with a limited number of positive nodes. *J. Clin. Oncol.* **14,** 3105–3111.

72. Early Breast Cancer Trialist's Collaborative Group. (1992). Systemic treatment of early breast cancer by hormonal, cytotoxic, or immune therapy. *Lancet* **339,** 71.

73. Krag, D. *et al.* (1998). The sentinel node in breast cancer—a multicenter validation trial. *N. Eng. J. Med.* **339,** 941–946.

74. Giuliano, A. E., Kirgan, D. M., Guenthler, J. M., and Morton, D. L. (1994). Lymphatic mapping and sentinel lymphadenectomy for breast cancer. *Ann. Surgery* **220,** 391–401.

75. American Joint Commission on Cancer. (1997). *AJCC Cancer Staging Manual,* Lippincott–Raven, Philadelphia.

PART V

ANTI-ANIOGENESIS
AND PRO-APOPTOTIC
GENE THERAPY

26

Antiangiogenic Gene Therapy

STEVEN K. LIBUTTI

Surgery Branch
National Cancer Institute
Bethesda, Maryland 20892

ANDREW L. FELDMAN

Surgery Branch
National Cancer Institute
Bethesda, Maryland 20892

I. INTRODUCTION

The discovery that tumors require angiogenesis for growth has spurred substantial investigation into the possibility of inhibiting angiogenesis as a novel treatment strategy for patients with cancer. More than 40 endogenous inhibitors of angiogenesis have been discovered [34]. Difficulties in the stability, mass manufacture, and chronic administration of recombinant forms of these inhibitors, however, have hampered efforts to test these agents in a clinical setting; therefore, the strategy of using gene therapy to induce *in vivo* production of these inhibitors by the host is an attractive alternative.

This chapter will summarize the results of preclinical antiangiogenic gene therapy investigations and discuss potential clinical applications of this treatment strategy.

II. ANGIOGENESIS AND ITS ROLE IN TUMOR BIOLOGY

Angiogenesis is the development of new capillary growth from a previously established vasculature. Several studies in the first half of the 1900s noted the abnormalities of tumor-associated blood vessels and suggested that tumors might produce substances that promote blood vessel growth and permeability (reviewed in Beckner [8]). The isolation of such a substance by Folkman [43] was followed by the discovery of numerous tumor-derived proangiogenic cytokines, including the vascular endothelial growth factor (VEGF) and fibroblast growth factor (FGF) families [8,26]. A key advance in the practicability of antiangiogenic therapy as a treatment for cancer was the demonstration that tumors depend upon angiogenesis for sustained growth [42]. It also appeared, however, that endogenous inhibitors of angiogenesis might be present in the body. Tumor cells introduced into the cornea were noted to retain their viability but did not produce a growing, vascularized tumor [46]. It was further hypothesized that some of these inhibitors might be generated by primary tumors themselves, as some patients with cancer have been noted to demonstrate rapid growth of metastatic disease after their primary tumor is removed [99]. This phenomenon was later demonstrated in experimental models [58,96].

III. ANTIANGIOGENIC THERAPY OF CANCER AND THE ROLE OF GENE THERAPY

The existence of endogenous inhibitors of angiogenesis has led to intensive research efforts to identify these agents and to generate them recombinantly for testing as potential anticancer therapeutics. These agents include antiangiogenic proteolytic fragments, immunomodulatory cytokines with antiangiogenic properties, tissue inhibitors of matrix metalloproteinases (TIMPs), and other molecules. A number of potential advantages to the use of antiangiogenic agents in the treatment of cancer have been described. Side effects associated with the systemic administration of these endogenous antiangiogenic agents have not been reported [9]. Furthermore, a wide variety of solid tumors may be sensitive to these agents, as the target of therapy is the tumor neovasculature rather than the tumor cells themselves [54]. Finally, tumors do not appear to develop traditional resistance to the effects of antiangiogenic agents [14], probably because of the low mutagenic potential of endothelial cells.

There are, however, several potential disadvantages of antiangiogenic cancer therapy. Angiogenesis inhibitors do not appear to be cytotoxic to tumor cells themselves [94,96]. Because tumor micrometastases may remain dormant but viable for long periods of time [23,28,58], antiangiogenic therapy will likely need to be administered on a chronic basis [41]. In addition, some recombinant antiangiogenic agents are unstable *in vitro* and require high therapeutic doses, thus posing manufacturing and economic constraints on their widespread clinical use in recombinant form [24,125]. Finally, many endogenous cytokines have limited half-lives, and recent evidence has suggested that the peak/trough kinetics resulting from bolus administration may not yield the optimal antiangiogenic effect on tumor neovasculature, compared with delivery systems that maintain continuously elevated levels of antiangiogenic agents [24,31,54]. These potential drawbacks to the chronic delivery of recombinant antiangiogenic proteins have led investigators to address the feasibility of delivering antiangiogenic agents by means of gene therapy (reviewed in Feldman [34], Folkman [40], Kong [67], and Lau [69]). Antiangiogenic gene therapy might obviate or lessen the need to give recombinant biologics, with their attendant manufacturing difficulties, *in vitro* instability, and significant cost at the high doses required to achieve therapeutic benefit. Furthermore, steady-state local and/or circulating concentrations resulting from constant transgene expression might improve antitumor efficacy or achieve similar efficacy with lower protein concentrations. Finally, posttranslational processing events such as glycosylation might yield proteins that are more effective when they occur in the host rather than in *in vitro* protein production systems such as yeast.

Two basic strategies have been proposed for the antiangiogenic gene therapy of cancer: tumor-directed gene therapy and systemic gene therapy. Tumor-directed antiangiogenic gene therapy represents those strategies that utilize a vector that has some degree of tumor specificity in order to facilitate selective transgene expression at the tumor site(s) [67]. There are two principal advantages to this strategy: First, the risk of systemic toxicity from the transgene product is decreased, and, second, the amount of transgene product necessary to have a therapeutic effect when it acts locally in a paracrine or autocrine fashion is likely to be less than when the transgene product is produced at a site distant from the tumor. Much progress has been made in developing progressively better gene delivery vectors with increasing tumor specificity; however, at this time the optimal tumor-directed vector has not yet been developed. Another potential difficulty with tumor-directed gene therapy as it relates to angiogenesis is that, for the most part, gene delivery vectors require an established vasculature in order to be delivered to tumors. It is unclear, therefore, whether a tumor-specific gene delivery vector would be successful in maintaining dormancy of micrometastases in a prevascular state [35].

Because toxicity of recombinant forms of endogenous antiangiogenic agents has not been demonstrated, a systemic approach to antiangiogenic gene therapy appears feasible [34–36,40]. Using this strategy, the patient's normal tissues act as a "factory" for the production of increased circulating levels of an antiangiogenic agent. Examples of successful preclinical studies using this approach include utilizing the host liver as the target of systemic delivery of recombinant adenovirus [36], and utilizing skeletal muscle to produce therapeutic protein after local injection of DNA [13].

The angiogenic phenotype of a tumor has been shown to reflect a combination of the influences of the pro- and antiangiogenic cytokines present. Therefore, antiangiogenic gene therapy strategies may be classified into those that increase local or systemic concentrations of antiangiogenic proteins and those that decrease local or systemic concentrations of proangiogenic proteins. The majority of preclinical models designed to increase concentrations of antiangiogenic proteins involve delivering the genes encoding these proteins to the tumor and/or to normal host tissues. However, because a significant proportion of endogenous antiangiogenic agents are proteolytic cleavage products of larger proteins, several recent models have been reported in which the genes delivered encode proteases that cleave these antiangiogenic fragments from their parent proteins [52,82]. Although this strategy has been successful in inhibiting tumor growth, some of these proteases (e.g., elastase) also have been demonstrated to promote tumor invasion. This approach thus will necessitate caution in its clinical application. It should be noted, however, that the proteases that cleave many of the known endogenous antiangiogenic protein fragments have not been identified, rendering this strategy a potentially fertile area for future investigation.

Gene therapy designed to decrease concentrations of proangiogenic cytokines such as VEGF may target the production of these cytokines or their downstream effects on the tumor. Production of proangiogenic cytokines is regulated in part by tumor suppressor genes, thus the restoration of wild-type copies in tumors that exhibit mutations in these genes represents one antiangiogenic gene therapy strategy. In addition, several oncogenes have been demonstrated to lead to increased transcription of genes encoding proangiogenic cytokines [100], so another potential gene therapy strategy would be to inhibit translation of oncogene mRNA (e.g., by delivering antisense constructs). Finally, a number of soluble growth factor receptors have been identified which inhibit tumor growth by competing with the tumor for binding of proangiogenic growth factors. Delivery of the genes encoding these soluble receptors represents a third gene therapy approach to diminishing the influence of proangiogenic cytokines on tumor tissue [49].

A detailed description of the vectors currently used in preclinical antiangiogenic gene therapy approaches is beyond the scope of this chapter, and the majority of these vectors are described elsewhere in this book. It is important to note, however, that vector development is an active and advancing field of research and that the vectors available for clinical trials can be expected to improve in the future. We therefore view the preclinical models described in this chapter as prototypes documenting the feasibility of antiangiogenic gene therapy for the treatment of malignant diseases, although the vectors used in some studies currently may be considered too toxic or immunogenic to warrant clinical trials.

IV. PRECLINICAL MODELS OF ANTIANGIOGENIC GENE THERAPY

A. Antiangiogenic Proteolytic Fragments

A number of endogenous inhibitors of angiogenesis have been identified as naturally occurring fragments of larger proteins (see Table 1). Many of these parent molecules play a role in tumor invasion or the coagulation system, raising interesting questions about the physiologic homeostasis between these processes and their perturbation in patients with cancer [56]. As previously mentioned, the antiangiogenic properties of some proteolytic fragments of larger molecules also has led to the strategy of delivering genes encoding the proteinases that generate these fragments.

1. Angiostatin

Angiostatin is an antiangiogenic proteolytic fragment of plasminogen. Plasminogen contains five triple loop structures known as kringles, and angiostatin represents a 38-kDa

internal fragment containing kringles 1–4. Angiostatin has been shown to have antiangiogenic and antitumor properties in mice [95,96]. Other angiostatin-like plasminogen fragments also have antiangiogenic and antitumor activity, including kringles 1–3 [61], kringles 1–5 [16], and kringle 5 alone [19,78]. Supernatants from tumor cells transduced with an adeno-associated viral vector containing the angiostatin gene have been demonstrated to inhibit endothelial cell proliferation in vitro [92]. Chen et al. [20] showed that supernatant from cells transfected in vitro with liposome–plasmid DNA complexes inhibited angiogenesis in an in vivo Matrigel model. Cao et al. [18] demonstrated that in vitro transfection of murine fibrosarcoma cells with angiostatin cDNA inhibited the subsequent growth of primary and metastatic tumors in vivo. This phenomenon has also been demonstrated using transfected B16F10 murine melanoma cells [3]. Tanaka et al. [121] showed that retroviral transduction of glioma cells with the angiostatin gene in vitro demonstrated similar growth inhibition in mice. Furthermore, stereotactic injection of adenovirus containing angiostatin cDNA into intracerebral gliomas inhibited tumor growth, demonstrating that antiangiogenic gene therapy could be effective in treating preestablished tumors. Adenovirus carrying the angiostatin gene was shown to be an effective systemic gene therapy vector by Griscelli et al. [53], who demonstrated dose-dependent inhibition of establishment and growth of C6 rat gliomas in nude mice after pretreatment with intravenously administered adenovirus. In a study examining the effect of various antiangiogenic gene plasmids complexed to cationic liposomes, Liu et al. [77] demonstrated that intravenous delivery of liposome–DNA complexes containing the angiostatin gene reduced B16F10 murine melanoma metastasis when injected up to 7 days after tumor cell inoculation. Recently, Sacco et al. [102] have demonstrated the efficacy of angiostatin gene therapy in the mouse mammary tumor virus (MMTV)-neu transgenic mouse model, in which female mice predictably develop mammary tumors. Intramammary injection of liposome–DNA complexes inhibited the subsequent growth of primary mammary tumors, as well as the development of spontaneous lung metastases.

A number of proteases have been reported to cleave angiostatin or angiostatin-like fragments from the parent plasminogen molecule, including urokinase- and tissue-type plasminogen activators [45,130], plasmin [16], elastase [82,96], and matrix metalloproteinases (MMPs) 2 (gelatinase A) [93], 3 (stromelysin-1) [72], 7 (matrilysin) [97], 9 (gelatinase B/type IV collagenase) [97], and 12 (macrophage metalloelastase) [21,30]. In a gene therapy approach, Matusda et al. [82] used a retrovirus to transduce murine NIH3T3 fibroblasts and Lewis lung carcinoma (LLC) cells with the gene for porcine pancreatic elastase 1. NIH3T3 cells were used to verify the ability of elastase 1 to generate a functionally active plasminogen fragment (kringles 1–3) in vitro. Elastase 1-transduced LLC cells

TABLE 1   Gene Therapy Strategies Using Antiangiogenic Proteolytic Fragments

Gene	Vector	Findings	Ref.
Angiostatin	Adeno-associated virus	Supernatant from transduced tumor cells inhibited endothelial cell proliferation *in vitro*	[92]
	In vitro transfection	Supernatant from transfected tumor cells inhibited angiogenesis in the *in vivo* Matrigel model	[20]
	In vitro transfection	Transfection of murine fibrosarcoma cells inhibited their growth as primary or metastatic tumors in mice	[18]
	In vitro transfection	Transfection of murine B16F10 melanoma cells inhibited their growth as primary or metastatic tumors in mice	[3]
	Retrovirus	*In vitro* transduction of glioma cells inhibited their growth in mice	[121]
	Adenovirus	Stereotactic injection into intracerebral gliomas inhibited tumor growth in mice	[121]
	Adenovirus	Pretreatment with i.v. adenovirus inhibited establishment and growth of C6 rat gliomas in mice	[53]
	Liposome	Intravenous injection of liposome–DNA complexes reduced B16F10 murine melanoma metastasis	[77]
	Liposome	Intramammary injection inhibited growth of primary breast tumors and spontaneous lung metastases in MMTV-*neu* transgenic mice	[102]
Elastase 1	Retrovirus	*In vitro* transduction of murine Lewis lung carcinoma cells inhibited s.c. growth, prolonged survival, and inhibited formation of lung metastases	[82]
MMP-12	*In vitro* transfection	Transfection of murine B16 melanoma cells inhibited angiogenesis and s.c. tumor growth in syngeneic mice	[52]
Endostatin	Adeno-associated virus	Supernatant from transduced tumor cells inhibited endothelial cell proliferation *in vitro*	[92]
	In vitro transfection	Stable transfection of mouse and human tumor cell lines inhibited formation of lung and liver metastases in mice	[131]
	Polymerized plasmid DNA	Intramuscular injection of endostatin plasmid inhibited syngeneic tumor growth and tumor metastases in mice	[13]
	Cationic liposome–DNA complex	Intravenous injection of liposome–DNA complexes inhibited growth of human breast cancer in nude mice	[20]
	Adenovirus	Intravenous injection of recombinant adenovirus inhibited growth of subcutaneous tumors in nude mice	[36]
	Adenovirus	Intravenous adenovirus inhibited subcutaneous tumor growth and prevented lung metastases in nude mice	[107]

Abbreviations: i.v., intravenous; MMP, matrix metalloproteinase; MMTV, mouse mammary tumor virus; s.c., subcutaneous.

yielded an inhibition of tumor growth and prolongation of survival when inoculated subcutaneously into C57BL/6 mice compared to parental LLC cells. In addition, the transduced cells produced less aggressive experimental lung metastases than their nontransduced counterparts. Gorrin-Rivas *et al.* [52] transfected murine B16 melanoma cells with the gene for mouse macrophage elastase (MME, or MMP-12). These clones cleaved plasminogen into an angiostatin-like fragment *in vitro*. MME-transfected cells demonstrated inhibited growth when injected subcutaneously in syngeneic mice. These effects were associated with angiostatin production, a decrease in microvessel density, and disruption of vascular morphology within the tumors.

2. Endostatin

Endostatin is a 20-kDa fragment derived from the C-terminal noncollagenous domain of the basement membrane constituent collagen XVIII [94]. Nguyen *et al.* [92]

utilized an adeno-associated viral vector to transduce various tumor cell lines with the endostatin gene *in vitro,* demonstrating inhibition of endothelial cell proliferation by transduced tumor cell supernatant. Yoon *et al.* [131] have reported that stable transfection of murine and human tumor cell lines with the murine endostatin gene inhibited the ability of these cells to form lung and liver metastases in mice. Blezinger *et al.* [13] reported inhibition of syngeneic tumor growth and tumor metastases using polymerized plasmid DNA containing the endostatin gene injected into the skeletal muscle of mice, although circulating levels of endostatin were low (8 ng/mL). Chen *et al.* [20] have demonstrated growth inhibition of the human breast cancer cell line MDA-MB-435 in the mammary fat pads of nude mice after intravenous injection of endostatin plasmid DNA complexed to a cationic liposome. This effect may have been due partly to delivery of liposome–DNA complexes to tumor tissue [35]. In a systemic gene therapy model, we have shown that intravenous administration of a recombinant adenovirus carrying the endostatin

TABLE 2 Gene Delivery of Immunomodulatory Cytokines with Antiangiogenic Properties

Gene	Vector	Findings	Ref.
Interferons			
IFN-α	Retrovirus, packaging cells	*In vitro* transduction of Kaposi's sarcoma cells or coinjection of packaging cells inhibited s.c. tumor growth in nude mice	[2]
IFN-β	Retroviral packaging cells	Coinjection of packaging cells and Kaposi's sarcoma cells inhibited s.c. tumor growth in nude mice	[2]
	Retrovirus	*In vitro* transduction of human prostate cancer cells inhibited their ability to form tumors and lymph node metastases in nude mice	[29]
Interleukins			
IL-4	Retroviral packaging cells	Stereotactic injection inhibited angiogenesis and growth of intracerebral rat gliomas and was immunostimulatory	[103]
IL-10	*In vitro* transfection	Transfection of human melanoma cells inhibited angiogenesis, tumor growth, and establishment of metastases	[59]
	In vitro transfection	Transfection of human prostate cancer cells inhibited their growth in nude mice	[116]
IL-12	*In vitro* transfection	Transfection of SCK murine mammary carcinoma cells inhibited tumor formation in A/J mice; synergism with IL-18	[22]
	In vitro transfection	Transfection of human pancreatic cancer cells inhibited angiogenesis and s.c. tumor growth in SCID mice	[118]
	Coinjected fibroblasts	Coinjection of *in vitro* transduced fibroblasts with human pancreatic cancer cells inhibited angiogenesis and s.c. tumor growth in SCID mice	[32]
	Coinjected fibroblasts	Coinjection of *in vitro* transduced fibroblasts with murine ovarian cancer cells prolonged survival in a murine peritoneal carcinomatosis model	[106]
	Semliki Forest virus	Intratumoral injection inhibited B16 melanoma angiogenesis and growth in mice independent of immune response	[5]
	Liposome–DNA complexes	Intravenous injection resulted in transfection of tumor endothelium and inhibition of s.c. squamous cell carcinoma growth in C3H mice	[4]
IL-18	Retrovirus	*In vitro* transduction of SCK murine mammary carcinoma cells inhibited their ability to form tumors in A/J mice; synergism with IL-12	[22]
CXC chemokines			
IP-10	*In vitro* transfection	Transfection of A549 human non-small-cell lung cancer cells inhibited s.c. tumor growth in SCID mice	[1]
	Retrovirus	*In vitro* transduction of A375 human melanoma cells inhibited their ability to form s.c. tumors in nude mice (unpublis. observ.)	
	Adenovirus	Supernatant from adenovirally infected cells inhibited angiogenesis in the corneal pocket angiogenesis assay	[1]
MIG	*In vitro* transfection	Transfection of A549 human non-small-cell lung cancer cells inhibited angiogenesis and s.c. tumor growth in SCID mice	[1]
	Adenovirus	Peritumoral injection of adenovirus inhibited angiogenesis and s.c. growth of A549 human non-small-cell lung cancer cells in SCID mice	[1]
PF-4	Adenovirus	Coinjection of adenovirus with glioma cells under renal capsule of nude mice inhibited tumor growth and vascularity	[122]

Abbreviations: IFN, interferon; IL, interleukin; IP-10, interferon-inducible protein-10; MIG, monokine induced by interferon-γ; PF-4, platelet factor-4; s.c., subcutaneous.

gene leads to high circulating levels of endostatin (2038 ng/mL) in nude mice [36]. We noted a 40% inhibition of the growth of subcutaneous MC38 murine colon adenocarcinomas, despite the resistance of these cells to direct adenoviral infection. Sauter *et al.* [107] recently have reported a 78% reduction in subcutaneous murine LIC growth using a similar strategy, as well as prevention of the formation of lung metastases.

B. Immunomodulatory Genes with Antiangiogenic Properties (Table 2)

1. Interferons

The interferons (IFN)-α, -β, and -γ, represent a family of endogenous glycoproteins initially identified by their antiviral properties [7]. Further research revealed multiple antitumor effects of the interferons, including the ability to inhibit angiogenesis [44,101,113].

Albini *et al.* [2] developed retroviral packaging cells containing the genes for IFN-α and IFN-β. Both packaging cell lines inhibited angiogenesis when incorporated in subcutaneous Matrigel plugs in mice. The Kaposi's cell line KS-IMM was transduced using the IFN-α-containing retroviral supernatant, and these cells demonstrated subcutaneous tumor growth inhibition in nude mice. Coinjection of parental KS-IMM cells with packaging cells containing either the IFN-α or IFN-β gene also inhibited growth.

Dong *et al.* [29] transduced the human prostate cancer cell line PC-3M with a retrovirus carrying the gene for IFN-β. Intraprostatic or subcutaneous injection of these cells into nude mice demonstrated inhibited growth and regional lymph node metastasis compared to control-transduced PC-3M cells. Tumor growth inhibition by IFN-β-transduced cells was associated with decreased blood vessel density and also inhibited the tumorigenicity of coinjected nontransduced PC-3M cells. The authors conclude that these effects were due in part to activation of host effector cells, as well as inhibition of angiogenesis. The effective use of interferon gene therapy will depend on an understanding of the multiple cellular and immune effects of these proteins, as well as their antiangiogenic properties.

2. Interleukins

The interleukins are a family of leukocyte-derived proteins with broad-ranging effects on multiple physiologic processes, including angiogenesis. Interleukins bearing an N-terminal glu–leu–arg (ELR) motif, such as interleukin (IL)-8 tend to display proangiogenic properties, whereas those lacking this motif have been found to inhibit angiogenesis [117]. Because of the varied functions of the interleukins, their selection as antiangiogenic gene therapy agents must be done carefully to simultaneously derive benefit from their chemotactic and immunologic properties. Further investigation of the properties of these multifunctional cytokines may reveal ties between the mechanisms regulating angiogenesis and antitumor immunity.

While the role of IL-4 as a critical modulator of the humoral immune system and potential antitumor agent has long been recognized, Volpert *et al.* [128] demonstrated the ability of IL-4 to inhibit corneal neovascularization when delivered locally in rats or systemically in mice. Saleh *et al.* [103] recently have shown that stereotactic delivery of retroviral packaging cells producing retrovirus encoding IL-4 to cerebral gliomas in rats inhibited tumor angiogenesis and growth. Furthermore, the treatment elicited a CD8$^+$ T cell and macrophage response, and subsequent delivery of glioma cells to the contralateral hemisphere resulted in tumor rejection.

Interleukin-10 was first identified as a product of murine Th2 cells that inhibited cytokine production by Th1 lymphocytes [39]. Huang *et al.* [59] demonstrated that human melanoma cells transfected with the IL-10 gene grew more slowly and developed fewer lung metastases than control transfected tumor cells, and this inhibition was associated with a decrease in neovascularity. Stearns *et al.* [116] recently have shown a similar phenomenon using IL-10-transfected human prostate tumor cells. Both studies suggest that the antiangiogenic properties of IL-10 may relate to upregulation of TIMP expression and downregulation of the production of MMPs.

Interleukin-12 has known stimulatory effects on natural killer (NK) cells and cytotoxic T lymphocytes (CTLs) associated with antitumor activity [15]. Voest *et al.* [127] demonstrated that IL-12 has potent antiangiogenic properties *in vivo* mediated through IFN-γ. This effect appears to be due to increased expression of the potent antiangiogenic molecule interferon-inducible protein (IP)-10 [111]. Coughlin [22] demonstrated that transfection of the IL-12 gene (consisting of p35 and p40 subunits) into SCK murine mammary carcinoma cells inhibited their ability to form tumors in mice. Treating the mice with antibodies against either IL-12 or IFN-γ blocked this effect. Inhibition of tumor formation also was noted when SCK cells were retrovirally transduced with the gene for IL-18, another immunomodulatory cytokine with antiangiogenic effects [17]. Both cell types demonstrated antiangiogenic effects in a subcutaneous Matrigel model and were synergistic with each other. *In vitro* transfection of the human pancreatic adenocarcinoma cell line PK-1 with the IL-12 gene inhibited angiogenesis and tumor growth after subcutaneous injection in SCID mice [118]. Coinjecting parental PK-1 cells with NIH3T3 murine fibroblasts retrovirally transduced with the IL-12 gene also inhibited tumor angiogenesis and growth *in vivo* [32]. In a similar approach [106], murine MC57 fibroblasts were retrovirally transduced with the IL-12 gene. Intraperitoneal injection of these fibroblasts prolonged survival in a syngeneic model of peritoneal carcinomatosis utilizing murine ID8 ovarian cancer cells. Asselin-Paturel *et al.* [5] engineered a Semliki Forest virus vector carrying the gene for IL-12. Intratumoral injection of this vector into B16 murine melanomas inhibited angiogenesis and tumor growth. These effects appeared independent of an immune response and were accompanied by increased expression of IFN-γ, IP-10, and monokine induced by IFN-γ (MIG). Recently, Anwer *et al.* [4] reported intravenous injection of cationic lipids complexed to plasmid DNA carrying the IL-12 gene in C3H mice. This approach inhibited the growth of subcutaneous squamous cell carcinomas. Interestingly, the authors demonstrated using anti-CD31 immunostaining and fluorescein-labeled plasmid that at least some of the *in vivo* transfection involved the tumor endothelium.

TABLE 3 Delivery of Genes Encoding TIMPs and Other Antiangiogenic Proteins

Gene	Vector	Findings	Ref.
TIMP-1	Adenovirus	Adenoviral infection of endothelial cells inhibited their migration *in vitro*	[37]
TIMP-2	*In vitro* transfection	Transfection of B16F10 murine melanoma cells inhibited angiogenesis, tumor growth, and metastasis in mice	[126]
TSP-1	*In vitro* transfection	Transfection of FRO human thyroid carcinoma cells inhibited angiogenesis and tumor growth and promoted tumor cell apoptosis in nude mice	[91]
	Liposome–DNA complexes	Intravenous injection of DNA complexes encoding a TSP-1 fragment inhibited human breast cancer growth in mice	[132]
p53	*In vitro* transfection	Transfection of p53-null FRO human thyroid carcinoma cells inhibited angiogenesis and tumor growth in nude mice	[91]
	Liposome–DNA complexes	i.v. injection inhibited human breast cancer growth associated with decreased blood vessel density in nude mice	[132,133]
	Liposome–DNA complexes	i.v. injection reduced angiogenesis and inhibited establishment of B16F10 lung metastases in C57BL/6 mice	[77]
p16	Adenovirus	Pretreatment with virus inhibited the ability of glioma cells to induce angiogenesis in a dorsal air sac model in nude mice	[57]
EMAP-II	Vaccinia	i.v. injection of vaccinia enhanced tumor sensitivity to TNF associated with endothelial cell TNF receptor upregulation	[47]

Abbreviations: i.v., intravenous; TIMP, tissue inhibitor of matrix metalloproteinases; TSP, thrombospondin; TNF, tumor necrosis factor.

3. ELR(−) CXC Chemokines

CXC chemokines are chemotactic cytokines with an amino-terminal CXC motif. The presence of an additional ELR motif, ELR(+), is associated with proangiogenic activity (e.g., IL-8), while the absence of this motif, ELR(−), is associated with antiangiogenic activity [117]. ELR(−) CXC chemokines include IP-10 and MIG, as well as platelet factor-4 (PF-4).

Our lab recently has demonstrated pronounced subcutaneous growth inhibition of A375 human melanomas in nude mice after retroviral transduction with the IP-10 gene [unpubl. observ.]. Addison *et al.* [1] reported decreased growth of subcutaneous A549 human lung adenocarcinoma tumors in SCID mice when stably transfected with the gene for either IP-10 or MIG. When A549 cells were infected *in vitro* with recombinant adenoviruses expressing either of these chemokines, the lyophilized cell supernatants inhibited angiogenesis in the corneal pocket angiogenesis assay. In addition, peritumoral injection of the adenovirus carrying the MIG gene in mice with subcutaneous A549 tumors caused growth inhibition associated with reduced tumor vessel densities.

Platelet factor-4 is a platelet-derived chemokine released during platelet aggregation which has antiangiogenic properties *in vitro* and *in vivo* [80]. Tanaka [122] developed a recombinant adenovirus containing the PF-4 gene. This vector inhibited tumor growth and vascularity when coinjected with U87MG glioma cells into the subcapsular space of the kidney in nude mice. In addition, survival was prolonged when the

adenovirus was injected stereotactically into preestablished intracerebral gliomas.

C. TIMPs and Other Antiangiogenic Proteins (Table 3)

1. TIMPs

Tissue inhibitors of matrix metalloproteinases were first described by Moses *et al.* [90], who demonstrated the antiangiogenic properties of a collagenase inhibitor isolated from cartilage. Four such inhibitors have been identified in humans (TIMP-1 to TIMP-4). Valente *et al.* [126] transfected B16F10 murine melanoma cells with the human TIMP-2 gene and demonstrated inhibition of tumor growth in mice, reduced angiogenesis when injected subcutaneously in Matrigel, and decreased formation of lung metastases. Fernandez *et al.* [37] demonstrated inhibition of migration in endothelial cells infected with a recombinant adenovirus containing the TIMP-1 gene.

2. Thrombospondins

Thrombospondins (TSPs) are 450-kDa trimeric glycoproteins associated with stabilization of platelets during blood clotting. Good *et al.* [50] demonstrated antiangiogenic properties of TSP-1. TSP-2 [129], as well as fragments of TSP-1 [124], also have been shown to inhibit angiogenesis. Nagayama *et al.* [91] transfected FRO human thyroid carcinoma cells with the TSP-1 gene and showed inhibited angiogenesis, decreased tumor growth, and increased tumor cell apoptosis in nude mice. Xu *et al.* [132] injected liposomes complexed to plasmid DNA encoding an antiangiogenic TSP-1 fragment into nude mice and demonstrated growth inhibition of MDA-MB-435 human breast tumors. Coadministration of liposome–DNA complexes containing the wild-type p53 gene, which increases TSP-1 expression [25], enhanced this effect.

3. p53

p53 is a tumor suppressor gene commonly mutant or absent in solid tumors. Wild-type p53 is associated with decreased angiogenesis; it enhances TSP-1 expression, as mentioned above, and decreases expression of VEGF [65] (see following section). In addition to TSP-1, Nagayama *et al.* [91] transfected the wild-type p53 gene into FRO human thyroid carcinoma cells, which lack p53 expression. These cells demonstrated decreased subcutaneous tumor growth and angiogenesis in nude mice. Interestingly, this effect was partially reversed by contransfecting the FRO cells with wild-type p53 and the VEGF gene. As mentioned above, Xu *et al.* [132,133] administered liposome-conjugated p53 DNA to nude mice with MDA-MB-435 human breast tumors. This approach inhibited tumor growth and decreased tumor blood vessel density. Liu *et al.* [77] also used liposome:p53 DNA complexes, and demonstrated inhibition of the formation of B16F10 murine melanoma lung metastases.

4. p16

(See Table 3.) p16 is a tumor suppressor gene often deleted in patients with gliomas [108]. In a recent gene delivery model, Harada *et al.* [57] used a dorsal air sac model of angiogenesis, in which an implantable chamber containing human glioma cells induced vessel formation in the surrounding fascia. Reestablishing wild-type p16 expression by *in vitro* adenoviral infection inhibited the angiogenesis observed in this model.

5. Endothelial Monocyte-Activating Polypeptide II (EMAP-II)

Endothelial monocyte-activating polypeptide II (EMAP-II) is a tumor-derived cytokine with effects on the inflammatory cascade and endothelial cells [10]. In addition to its antiangiogenic properties [11,109], EMAP-II upregulates tumor necrosis factor (TNF) receptor expression on endothelial cells [12]. In a gene therapy approach, Gnant *et al.* [47] used a recombinant vaccinia virus to deliver the EMAP-II gene specifically to tumors. This treatment rendered the tumors TNF-sensitive, and systemic administration of TNF led to tumor regression due to coagulative necrosis.

V. INHIBITING PROANGIOGENIC CYTOKINES (TABLE 4)

Numerous proangiogenic cytokines have been described (reviewed in Beckner [8] and Desai [26]); this chapter focuses on those that have served as targets for antiangiogenic gene therapy approaches.

A. Vascular Endothelial Growth Factor

Vascular endothelial growth factor is the best characterized proangiogenic cytokine and probably the most clinically relevant in patients with cancer [8,26]. VEGF initially was described as a tumor-derived cytokine that increased the permeability of tumor vasculature and was called vascular permeability factor (VPF) [110]. Its mitogenic effect on endothelial cells was subsequently demonstrated [64]. Ferrara *et al.* [38] later cloned the VEGF gene and further characterized its proangiogenic properties. Two VEGF receptors (VEGFR-1 and -2) appear to be expressed specifically on vascular endothelial cells. VEGFR-1 is also known as Flt1 (fms-like tyrosine kinase) [27]; VEGFR-2 is known as KDR (kinase-insert-domain-containing receptor) in humans and Flk1 (fetal liver kinase) in mice [86,123].

A number of antiangiogenic gene therapy approaches have targeted the VEGF receptors. One strategy has aimed to deliver an alternatively spliced soluble form of Flt1 (sFlt1), which is present endogenously and has been shown to inhibit endothelial cell proliferation by binding VEGF *in vitro* [66]. Goldman *et al.* [49] demonstrated that transfecting human sarcoma and glioblastoma cells with the gene for sFlt1 inhibited subcutaneous tumor growth and the formation of lung metastases; this transfection also conferred a survival advantage when the transfected glioblastoma cells were injected intracranially. Using intraperitoneal delivery of cationic liposomes encapsulating sFlt1 DNA, Mori *et al.* [89] showed prolonged survival in a peritoneal metastasis model of human gastric cancer in nude mice. An antitumor effect was also demonstrated using HT1080 human fibrosarcoma cells stably transfected to overexpress VEGF. Kong *et al.* [68] developed a recombinant adenovirus carrying the sFlt1 gene. Intravenous administration of this vector decreased tumor burden in mice with lung or liver metastases, and intratumoral injection inhibited subcutaneous tumor growth. Using intramuscular administration of a similar adenoviral vector, Takayama *et al.* [120] showed inhibition of subcutaneous growth in five of six human lung cancer cell lines tested in nude mice. Finally, in a recent report, intratumoral injection of an adenovirus carrying the sFlt1 gene inhibited B16 murine melanoma growth in an eyelid model in nude mice [112].

Millauer *et al.* [85] showed that transfer of a dominant-negative VEGFR-2 gene could inhibit endogenous wild-type

TABLE 4 Gene Therapy Strategies to Inhibit Production or Action of Proangiogenic Cytokines

Target	Gene transferred	Vector	Findings	Ref.
VEGF	Soluble Flt1	*In vitro* transfection	Transfection of glioblastoma cells prolonged survival in an intracranial xenograft model	[49]
		Liposome–DNA complexes	i.p. injection inhibited growth of peritoneal metastases from human gastric cancer and fibrosarcoma cells in nude mice	[89]
		Adenovirus	i.v. injection inhibited growth of liver metastases and prolonged survival in mice	[67,68]
		Adenovirus	i.m. injection inhibited subcutaneous growth in 5 of 6 human lung cancer cell lines tested in nude mice	[120]
		Adenovirus	Intratumoral injection inhibited growth of B16 murine melanoma in an eyelid model in nude mice	[112]
	VEGF receptor-2	Retrovirus	Subcutaneous coinjection of retroviral packaging cells inhibited growth of glioblastoma in mice	[85]
		Retrovirus	Intracerebral coinjection of retroviral packaging cells with glioma cells prolonged survival in rats	[79]
	Antisense VEGF	Adeno-associated virus	Conditioned media from transduced tumor cells inhibited endothelial cell proliferation *in vitro*	[92]
		Liposome–DNA complexes	Intratumoral injection inhibited growth of SK-HEP1 human hepatomas in nude mice	[62]
		Adenovirus	Intratumoral injection inhibited growth of human gliomas in nude mice	[60]
	p73	*In vitro* transfection	Transfection of osteosarcoma cells decreased VEGF production	[105]
	VHL gene	Retrovirus	*In vitro* transduction of renal cancer cells decreased VEGF production	[48]
uPA	uPA/uPAR antagonist	Adenovirus	i.v. injection inhibited metastases of Lewis lung carcinoma and human colon cancer	[70,71]
Angiopoietin-1	Soluble Tie2 receptor	Adenovirus	i.v. injection inhibited growth of primary murine tumors and lung metastases	[73]

Abbreviations: i.m., intramuscular; i.p., intraperioneal; i.v., intravenous; VEGF, vascular endothelial growth factor; VHL, von Hippel–Lindau; uPA, urokinase-type plasminogen activator; uPAR, uPA receptor.

VEGFR-2 function. Retroviral packaging cells carrying the mutant VEGFR-2 gene inhibited subcutaneous tumor growth in nude mice when coinjected with C6 glioblastoma cells. This approach conferred a survival advantage to rats receiving intracerebral coinjection of packaging cells with GS-9L gliosarcoma cells [79]; intratumoral injection of packaging cells also caused an inhibition of subcutaneous tumor growth associated with a decrease in vessel density.

Several studies have aimed to inhibit VEGF production by delivering an antisense VEGF construct. Nguyen *et al.* [92] developed an adeno-associated virus containing antisense VEGF-165 cDNA; tumor cell lines infected with this vector *in vitro* demonstrated decreased VEGF expression, and their supernatants caused a relative inhibition of endothelial cell proliferation. Kang *et al.* [62] used peritumoral injection of liposomes carrying antisense VEGF cDNA to inhibit the subcutaneous growth of SK-HEP1 human hepatoma cells in nude mice. Finally, Im *et al.* [60] developed a recombinant adenovirus encoding antisense VEGF-165 cDNA. *In vitro* infection of human glioma cells decreased VEGF production,

and intratumoral injection inhibited the subcutaneous growth of these cells in nude mice.

The decrease in VEGF expression associated with restoration of the wild-type p53 tumor suppressor gene was noted in the previous section. Other tumor suppressor genes also contribute to the regulation of VEGF expression. p73 is a tumor suppressor gene with homology to p53 whose expression may be silenced by hypermethylation in lymphoid neoplasms [63]. Salimath *et al.* [105] showed that transfection of Saos-2 osteosarcoma cells, which lack p73 expression, with the p73 gene inhibits VEGF production by these cells. The Von Hippel–Lindau (VHL) gene also is involved in regulating VEGF production. Mutations in this gene can lead to highly angiogenic neoplasms, including renal cancer [75]. Gnarra *et al.* [48] demonstrated that retroviral transduction of renal cancer cells with the wild-type VHL gene decreased VEGF expression. *In vivo* gene delivery of the wild-type forms of these genes may be a practical approach to treating cancer in patients with decreased expression of tumor suppressor genes.

TABLE 5   Gene Delivery Approaches that Target Areas of Active Angiogenesis

Targeting element(s)	Gene(s) transferred	vector	Findings	Ref.
NGR motif (9th type III repeat of fibronectin)	None	Retrovirus	Incorporation of NGR motif in MoMLV envelope escort proteins enhanced EC binding and transduction *in vitro*	[76]
vWF-derived collagen-binding motif	Mutant cyclin G1	Retrovirus	Portal vein infusion led to EC transduction and decreased growth of human pancreatic cancer liver metastases in nude mice	[51]
Pepro-endothelin-1 promoter	β-galactosidase	Retroviral packaging cells	Coinjection of irradiated packaging cells and human Kaposi's sarcoma or colorectal cancer cells yielded EC-specific expression in nude mice	[84]
Hypoxia response element, NF-κB binding site, KDR promoter	TNF-α, luciferase	Retrovirus	*In vitro* transduction led to EC-specific gene expression in various EC lines under hypoxic conditions	[88]

Abbreviations: MoMLV, Moloney murine leukemia virus; EC, endothelial cell; vWF, von Willebrand factor.

B. Other Proangiogenic Cytokines

1. Plasminogen Activators

Bacharach *et al.* [6] have shown that endothelial cells upregulate urokinase-type plasminogen activator (uPA) expression in the *in vitro* aortic ring assay. Interestingly, neighboring cells expressed PA inhibitor type 1 (PAI-1) in response to uPA. This phenomenon may contribute to the regulation of physiologic angiogenesis. uPA receptor antagonists were later shown to inhibit tumor angiogenesis in mice [87]. Li *et al.* [71] used a recombinant adenovirus encoding a secreted N-terminal fragment of uPA to compete with uPA for binding to the cell surface uPA receptor (uPAR). Intratumoral injection of the adenovirus into preestablished MDA-MB-231 human breast cancers in nude mice or Lewis lung carcinomas in syngeneic mice yielded inhibition of tumor growth. Systemic delivery of the virus inhibited the formation of liver metastases in a human colon carcinoma model. This inhibition was associated with improved survival [70].

2. Angiopoietin-1

Knockout mice lacking the gene for angiopoietin-1 fail to form normal blood vessels, demonstrating the critical role of this cytokine in angiogenesis [119]. Angiopoietin-1 is an endogenous stimulatory ligand for Tie2 (Tek), an endothelium-specific receptor tyrosine kinase. An alternate ligand, angiopoietin-2, inhibits the Tie2-mediated effects of angiopoietin-1 [81]. Local administration of recombinant, soluble Tie2 receptor has been shown to inhibit tumor angiogenesis and growth in mice [74]. Lin *et al.* [73] developed an adenoviral vector containing the gene for the soluble Tie2 receptor. Intravenous administration of this vector inhibited growth of subcutaneous primary tumors, as well as experimental and spontaneously occurring lung metastases.

VI. ENDOTHELIAL CELL-SPECIFIC GENE DELIVERY (TABLE 5)

The concept of eliciting a biologic effect only in a specific cell type is common to most forms of cancer treatment and has been used extensively in approaches to cancer gene therapy [55]. In addition to the various strategies reviewed above designed to alter peritumoral concentrations of angiomodulatory cytokines, another gene therapy strategy is to target tumor endothelial cells with gene delivery vectors.

Altering the host range of gene delivery vectors by inducing changes in viral envelope proteins has been one way of achieving specific targeting. Liu *et al.* [76] incorporated the NGR motif, derived from the 9th type III repeat of fibronectin, into Moloney murine leukemia virus envelope escort proteins. Retroviruses generated using this construct demonstrated enhanced binding to and transduction of endothelial cells *in vitro*. Gordon *et al.* [51] altered an amphotropic retroviral envelope protein to contain motifs from von Willebrand factor known to bind exposed collagen in the extracellular matrix. Retroviruses bearing this envelope protein and a mutant cyclin G1 transgene then were infused into the portal vein of nude mice. This strategy led to transduction of endothelial cells, stromal cells, and tumor cells and inhibited tumor growth of experimental liver metastases derived from human pancreatic cancer cells. Mavria *et al.* [84] have reported replacing a retroviral murine leukemia virus enhancer with elements from the human pepro-endothelin-1 promoter. When irradiated retroviral packaging cells incorporating these modifications were coinjected into nude mice with human Kaposi's sarcoma or colorectal cancer cells, endothelial cell-specific expression of the reporter gene β-galactosidase was observed. Finally, Modlich *et al.* [88] have evaluated the endothelial cell specificity of various retroviral constructs containing the hypoxia response element of the murine phosphoglycerate kinase-1 promoter, the murine vascular cell adhesion molecule-1 NF-κB binding site, and the KDR promoter. *In vitro* transduction led to endothelial cell-specific gene expression in various endothelial cell lines under hypoxic conditions. Ongoing efforts to identify molecular targets specific to tumor endothelium [104] will advance the therapeutic possibilities in this relatively new area of investigation.

VII. FUTURE DIRECTIONS IN ANTIANGIOGENIC GENE THERAPY

The toxicity and immunogenicity associated with most currently available viral vectors have raised concerns regarding their clinical use. However, advances in viral gene delivery systems continue to be made. Meanwhile, the preclinical data discussed above suggest that a number of nonviral gene delivery approaches may be suitable for early clinical trials designed to evaluate the safety and efficacy of antiangiogenic gene therapy. DNA complexed to polymers or liposomes have shown efficacy in both systemic and tumor-directed gene therapy models. The delivery of electric pulses to the DNA injection site has been reported to improve DNA uptake and gene expression [83], and may enhance some of these nonviral gene delivery methods. The ability to transduce autologous cells *ex vivo* with antiangiogenic genes [106] demonstrates the possibility of adoptively transferring these cells back to the host to achieve systemic or locoregional antiangiogenic effects. Finally, novel vectors continue to be developed; for example, *Salmonella* can be altered to express therapeutic transgenes and appears to demonstrate tumor specificity *in vivo* [98].

The optimal antiangiogenic gene for use in early clinical trials is not clear. The above review indicates that most preclinical data are available for angiostatin, endostatin, and interleukin-12. Ongoing clinical trials using the recombinant forms of these proteins will provide data regarding the toxicity and therapeutic potential of these agents. Also, gene therapy approaches have not been described for many endogenous antiangiogenic proteins [34]. Agents with extremely high potency may be good candidates for gene delivery, as the amount of local or systemic gene expression may limit the efficacy of some gene therapy strategies. Finally, further studies on the effects of these agents on physiologic processes that require angiogenesis, such as wound healing [9], are required.

It should be noted that some data regarding the efficacy of angiogenesis inhibitors already are available from clinical trials. Thalidomide has generated the most promising efficacy data to date, with evidence of activity against advanced myeloma [114] and possibly renal cancer [33]. Other synthetic agents, such as the fumagillin derivative TNP-470 [115], have shown less therapeutic promise. Ongoing clinical trials evaluating recombinant angiogenesis inhibitors and antiangiogenic monoclonal antibodies will help focus the potential clinical role of antiangiogenic gene therapy. Gene therapy might be unnecessary in certain clinical situations, while it might offer an alternative to using recombinant biologics in others. Finally, gene therapy may represent a useful adjuvant to prolong or improve responses achieved by antiangiogenic or other agents.

References

1. Addison, C. L., Arenberg. D. A., Morris, S. B., Xue, Y.-Y., Burdick, M. D., Mulligan, M. S., Iannettoni, M. D., and Strieter, R. M. (2000). The CXC chemokine, monokine induced by interferon-γ, inhibits non-small cell lung carcinoma tumor growth and metastasis. *Hum. Gene Ther.* **11,** 247–261.
2. Albini, A., Marchisone, C., Del, Grosso, F., Bennelli, R., Masiello, L., Tacchetti, C., Bono, M., Ferrantini, M., Rozera, C., Truini, M., Belardelli, F., Santi, L., and Noonan, D. M. (2000). Inhibition of angiogenesis and vascular tumor growth by interferon-producing cells: a gene therapy approach. *Am. J. Pathol.* **156,** 1381–1393.
3. Ambs, S., Dennis, S., Fairman, J., Wright, M., and Papkoff, J. (1999). Inhibition of tumor growth correlates with the expression level of a human angiostatin transgene in transfected B16F10 melanoma cells. *Cancer Res.* **59,** 5773–5777.
4. Anwer, K., Meaney, C., Kao, G., Hussain, N., Shelvin, R., Earls, R. M., Leonard, P., Quezada, A., Rolland, A. P., and Sullivan, S. M. (2000). Cationic lipid-based delivery system for systemic cancer gene therapy. *Cancer Gene Ther.* **7,** 1156–1164.
5. Asselin-Paturel, C., Lassau, N., Guinebretiere, J.-M., Zhang, J., Gay, F., Bex, F. *et al.* (1998). Transfer of the murine interleukin-12 gene in vivo by a Semliki Forest virus vector induces B16 tumor regression through inhibition of tumor blood vessel formation monitored by Doppler ultrasonography. *Gene Ther.* **5,** 606–615.
6. Bacharach, E., Itin, A., and Keshet, E. (1992). *In vivo* patters of expression of urokinase and its inhibitor PAI-1 suggest a concerted role in regulating physiological angiogenesis. *Proc. Natl. Acad. Sci. USA* **89,** 10686–10690.
7. Baron, S., and Dianzani, F. (1994). The interferons: a biological system with therapeutic potential in viral infections. *Antiviral Res.* **24,** 97–110.
8. Beckner, M. E. (1999). Factors promoting tumor angiogenesis. *Cancer Invest.* **17,** 594–623.
9. Berger, A. C., Feldman, A. L., Gnant, M. F., Kruger, E. A., Sim, B. K., Hewitt, S., Figg, W. K., Alexander, H. R., and Libutti, S. K. (2000). The angiogenesis inhibitor, Endostatin, does not affect murine cutaneous wound healing. *J. Surg. Res.* **91,** 26–31.
10. Berger, A. C., Tang, G., Alexander, H. R., and Libutti, S. K. (2000). Endothelial monocyte activating polypeptide II (EMAP-II), a tumor-derived cytokine that plays an important role in inflammation, apoptosis, and angiogenesis *J. Immunother.* **23,** 519–527.
11. Berger, A. C., Alexander, H. R., Tang, G., Wu, P. C., Hewitt, S. M., Turner, E. *et al.* (2000). Endothelial monocyte activating polypeptide II (EMAP-II) induces endothelial cell specific apoptosis and may be an inhibitor of tumor angiogenesis. *Microvasc. Res.* **60,** 70–80.
12. Berger, A. C., Alexander, H. R., Wu, P. C., Tang, G., Gnant, M. F., Mixon, A., Turner, E. S., and Libutti, S. K. (2000). Tumour necrosis factor receptor I (p55) is upregulated on endothelial cells by exposure to the tumour-derived cytokine endothelial monocyte-activating polypeptide II (EMAP-II). *Cytokine* **12,** 992–1000.
13. Blezinger, P., Wang, J., Gondo, M., Quezeda, A., Mehrens, D., French, M. *et al.* (1999). Systemic inhibition of tumor growth and tumor metastases by intramuscular administration of the endostatin gene. *Nature Biotechnol.* **17,** 343–348.

14. Boehm, T., Folkman, J., Browder, T., and O'Reilly, M. S. (1997). Antiangiogenic therapy of experimental cancer does not induce acquired drug resistance. *Nature (London)* **390,** 404–407.

15. Brunda, M. J., Luistro, L., Warrier, R. R., Wright, R. B., Hubbard, B. R., Murphy, M. *et al.* (1993). Antitumor and antimetastatic activity of interleukin 12 against murine tumors. *J. Exp. Med.* **178,** 1223–1230.

16. Cao, R., Wu, H.-L., Veitonmäki, N., Linden, P., Farnebo, J., Shi, G.-Y., and Cao, Y. (1999). Suppression of angiogenesis and tumor growth by the inhibitor K1-5 generated by plasmin-mediated proteolysis. *Proc. Natl. Acad. Sci. USA* **96,** 5728–5733.

17. Cao, R., Farnebo, J., Kurimoto, M., and Cao, Y. (1999). Interleukin-18 acts as an angiogenesis and tumor suppressor. *FASEB J.* **13,** 2195–2202.

18. Cao, Y., O'Reilly, M. S., Marshall, B., Flynn, E., Ji, R.-W., and Folkman, J. (1998). Expression of angiostatin cDNA in a murine fibrosarcoma suppresses primary tumor growth and produces long-term dormancy of metastases. *Cell* **5,** 1055–1063.

19. Cao, Y., Chen, A., An, S. S. A., Ji, R. W., Davidson, D., and Llinas, M. (1997). Kringle 5 of plasminogen is a novel inhibitor of endothelial cell growth. *J. Biol. Chem.* **272,** 22924–22928.

20. Chen, Q.-R., Kumar, D., Stass, S. A., and Mixson, A. J. (1999). Liposomes complexed to plasmids encoding angiostatin and endostatin inhibit breast cancer in nude mice. *Cancer Res.* **59,** 3308–3312.

21. Cornelius, L. A., Nehring, L. C., Harding, E., Bolanowski, M., Welgus, H. G., Kobayashi, D. K., Pierce, R. A., and Shapiro, S. D. (1998). Matrix metalloproteinases generate angiostatin: effects on neovascularization. *J. Immunol.* **161,** 6845–6852.

22. Coughlin, C. M., Salhany, K. E., Wysocka, M., Aruga, E., Kurzawa, H., Chang, A. E., Hunter, C. A., Fox, J. C., Trinchieri, G., and Lee, W. M. (1998). Interleukin-12 and interleukin-18 synergistically induce murine tumor regression which involves inhibition of angiogenesis. *J. Clin. Invest.* **101,** 1441–1452.

23. Crowley, N. J., and Siegler, H. F. (1992). Relationship between disease-free interval and survival in patients with recurrent melanoma. *Arch. Surg.* **127,** 1303–1308.

24. Crystal, R. G. (1999). The body as a manufacturer of endostatin. *Nature Biotechnol.* **17,** 336–337.

25. Dameron, K. M., Volpert, O. V., Tainsky, M. A., and Bouck, N. (1994). Control of angiogenesis in fibroblasts by p53 regulation of thrombospondin-1. *Science* **265,** 1582–1584.

26. Desai, S. B., and Libutti, S. K. Tumor angiogenesis and endothelial cell modulatory factors. *J. Immunother.* **22,** 186–211.

27. de Vries, C., Escobedo, J. A., Ueno, H., Houck, K., Ferrara, N., and Williams, L. T. (1992). The fms-like tyrosine kinase, a receptor for vascular endothelial growth factor. *Science* **255,** 989–991.

28. Demicheli, R., Terenziani, T., Valagussa, P., Moliterni, A., Zambetti, M., and Bonadonna, G. (1994). Local recurrences following mastectomy: support for the concept of tumor dormancy. *J. Natl. Cancer. Inst.* **86,** 45–48.

29. Dong, Z., Greene, G., Pettaway, C., Dinney, C. P. N., Eue, I., Lu, W. *et al.* (1999). Suppression of angiogenesis, tumorigenicity, and metastasis by human prostate cancer cells engineered to produce interferon-β. *Cancer Res.* **59,** 872–879.

30. Dong, Z., Kumar, R., Yang, X., and Fidler, I. J. (1997). Macrophage-derived metallo-elastase is responsible for the generation of angiostatin in Lewis lung carcinoma. *Cell* **88,** 801–810.

31. Drixler, T. A., Rinkes, I. H., Ritchie, E. D., van Vroonhoven, T. J., Gebbink, M. F., and Voest, E. E. (2000). Continuous administration of angiostatin inhibits accelerated growth of colorectal liver metastases after partial hepatectomy. *Cancer Res.* **60,** 1761–1765.

32. Duda, D. G., Sunamura, M., Lozonschi, L., Kodama, T., Egawa, S., Matsumoto, G., Shimamura, H., Shibuya, K., Takeda, K., and Matsuno, S. (2000). Direct *in vitro* evidence and *in vivo* analysis of the antiangiogenesis effects of interleukin 12. *Cancer Res.* **60,** 1111–1116.

33. Eisen, T., Boshoff, C., Mak, I., Sapunar, F., Vaughan, M. M., Pyle, L. *et al.* (2000). Continuous low dose Thalidomide: a phase II study in advanced melanoma, renal cell, ovarian and breast cancer. *Br. J. Cancer* **82,** 812–817.

34. Feldman, A. L., and Libutti, S. K. (2000). Progress in antiangiogenic gene therapy of cancer. *Cancer* **89,** 1181–1194.

35. Feldman, A. L., and Libutti, S. K. (2000). Correspondence re: Q.-R. Chen *et al.* Liposomes complexed to plasmids encoding angiostatin and endostatin inhibit breast cancer in nude mice [letter]. *Cancer Res.* **60,** 1463.

36. Feldman, A. L., Restifo, N. P., Alexander, H. R., Bartlett, D. L., Hwu, P., Seth, P. *et al.* (2000). Antiangiogenic gene therapy of cancer utilizing a recombinant adenovirus to elevate systemic endostatin levels in mice. *Cancer Res.* **60,** 1503–1506.

37. Fernandez, H. A., Kallenbach, K., Seghezzi, G., Grossi, E., Colvin, S., Schneider, R. *et al.* (1999). Inhibition of endothelial cell migration by gene transfer of tissue inhibitor of metalloproteinases-1. *J. Surg. Res.* **82,** 156–162.

38. Ferrara, N., Houck, K., Jakeman, L., and Leung, D. W. (1992). Molecular and biological properties of the vascular endothelial growth factor family of proteins. *Endocr. Rev.* **13,** 18–32.

39. Fiorentino, D. F., Bond, M. W., and Mosmann, T. R. (1989). Two types of mouse helper T cell. IV. Th2 clones secrete a factor that inhibits cytokine production by Th1 clones. *J. Exp. Med.* **170,** 2081–2095.

40. Folkman, J. (1998). Antiangiogenic gene therapy. *Proc. Natl. Acad. Sci. USA* **95,** 9064–9066.

41. Folkman, J. (1995). The influence of angiogenesis research on management of patients with breast cancer. *Breast Cancer Res. Treat.* **36,** 109–118.

42. Folkman, J. (1990). What is the evidence that tumors are angiogenesis dependent? *J. Natl. Cancer Inst.* **82,** 4–6.

43. Folkman, J. (1971). Tumor angiogenesis: therapeutic implications. *N. Engl. J. Med.* **285,** 1182–1186.

44. Friesel, R., Komoriy, A., and Maciag, T. (1987). Inhibition of endothelial cell proliferation by gamma-interferon. *J. Cell. Biol.* **104,** 689–696.

45. Gately, S., Twardowski, P., Stack, M. S., Cundiff, D. L., Grella, D., Castellino, F. J., Enghild, J., Kwaan, H. C., Lee, F., Kramer, R. A., Volpert, O., Bouck, N., and Soff, G. A. (1997). The mechanism of cancer-mediated conversion of plasminogen to the angiogenesis inhibitor angiostatin. *Proc. Natl. Acad. Sci. USA* **94,** 10868–10872.

46. Gimbrone, Jr., M. A., Leapman, S. B., Cotran, R. S., and Folkman, J. (1972). Tumor dormancy in vivo by prevention of neovascularization. *J. Exp. Med.* **136,** 261–276.

47. Gnant, M. F., Berger, A. C., Huang, J., Puhlmann, M., Wu, P. C., Merino, M. J. *et al.* (1999). Sensitization of tumor necrosis factor alpha-resistant human melanoma by tumor-specific in vivo transfer of the gene encoding endothelial monocyte-activating polypeptide II using recombinant vaccinia virus. *Cancer Res.* **59,** 4668–4674.

48. Gnarra, J. R., Zhou, S., Merrill, M. J., Wagner, J. R., Krumm, A., Papvassiliou, E. *et al.* (1996). Post-transcriptional regulation of vascular endothelial growth factor mRNA by the product of the *VHL* tumor suppressor gene. *Proc. Natl. Acad. Sci. USA* **93,** 10589–10594.

49. Goldman, C. K., Kendall, R. L., Cabrera, G., Soroceanu, L., Heike, Y., Gillespie, G. Y. *et al.* (1998). Paracrine expression of a native soluble vascular endothelial growth factor receptor inhibits tumor growth, metastasis, and mortality rate. *Proc. Natl. Acad. Sci. USA* **95,** 8795–8800.

50. Good, D. J., Polverini, P. J., Rastinejad, F., Le Beau, M. M., Lemons, R. S., Frazier, W. A. *et al.* (1990). A tumor suppressor-dependent inhibitor of angiogenesis is immunologically and functionally indistinguishable from a fragment of thrombospondin. *Proc. Natl. Acad. Sci. USA* **87,** 6624–6628.

51. Gordon, E. M., Liu, P. X., Chen, Z. H., Liu, L., Whitley, M. D., Gee, C., Groshen, S., Hinton, D. R., Beart, R. W., and Hall, F. L. (2000).

Inhibition of metastatic tumor growth in nude mice by portal vein infusions of matrix-targeted retroviral vectors bearing a cytocidal cyclin G1 construct. *Cancer Res.* **60,** 3343–3347.

52. Gorrin-Rivas, M. J., Arii, S., Furutani, M., Mizumoto, M., Mori, A., Hanaki, K., Maeda, M., Furuyama, H., Kondo, Y., and Imamura, M. (2000). Mouse macrophage metalloelastase gene transfer into a murine melanoma suppresses primary tumor growth by halting angiogenesis. *Clin. Cancer Res.* **6,** 1647–1654.

53. Griscelli, F., Li, H., Bennaceur-Griscelli, A., Soria, J., Opolon, P., Soria, C. *et al.* (1998). Angiostatin gene transfer: inhibition of tumor growth *in vivo* by blockage of endothelial cell proliferation associated with a mitosis arrest. *Proc. Natl. Acad. Sci. USA* **95,** 6367-6372.

54. Hahnfeldt, P., Panigrahy, D., Folkman, J., and Hlatky, L. (1999). Tumor development under angiogenic signaling: a dynamical theory of tumor growth, treatment response, and postvascular dormancy. *Cancer Res.* **59,** 4770–4775.

55. Hallenbeck, P. L., and Stevenson, S. C. (2000). Targetable gene delivery vectors, in *Cancer Gene Therapy: Past Achievements and Future Challenges* (N. A. Habib, ed.), pp. 37–46. Kluwer Academic, New York.

56. Hanahan, D., and Folkman, J. (1996). Patters and emerging mechanisms of the angiogenic switch during tumorigenesis. *Cell* **86,** 353–364.

57. Harada, H., Nakagawa, K., Iwata, S., Saito, M., Kumon, Y., Sakaki, S. *et al.* (1999). Restoration of wild-type p16 down-regulates vascular endothelial growth factor expression and inhibits angiogenesis in human gliomas. *Cancer Res.* **59,** 3783–3789.

58. Holmgren, L., O'Reilly, M. S., and Folkman, J. (1995). Dormancy of micrometastases: balanced proliferation and apoptosis in the presence of angiogenesis suppression. *Nature Med.* **1,** 149–153.

59. Huang, S., Xie, K., Bucana, C. D., Ullrich, S. E., Bar-Eli, M. (1996). Interleukin 10 suppresses tumor growth and metastasis of human melanoma cells: potential inhibition of angiogenesis. *Clin. Cancer Res.* **2,** 1969–1979.

60. Im, S.-A., Gomez-Manzano, C., Fueyo, J., Liu, T.-J., Ke, L. D., Kim, J.-S. *et al.* (1999). Antiangiogenesis treatment for gliomas: transfer of antisense-vascular endothelial growth factor inhibits tumor growth *in vivo. Cancer Res.* **59,** 895–900.

61. Joe, Y.-A., Hong, Y.-K., Chung, D.-S., Yang, Y.-J., Kang, J.-K., Lee, Y.-S. *et al.* (1999). Inhibition of human malignant glioma growth *in vivo* by human recombinant plasminogen kringles 1-3. *Int. J. Cancer* **82,** 694–699.

62. Kang, M. A., Kim, K. Y., Seol, J. Y., Kim, K. C., and Nam, M. J. (2000). The growth inhibtion of hepatoma by gene transfer of antisense vascular endothelial growth factor. *J. Gene Med.* **2,** 289–296.

63. Kawano, S., Miller, C. W., Gombart, A. F., Bartram, C. R., Matsuo, Y., Asou, H., Sakashita, A., Said, J., Tatsumi, E., and Koeffler, H. P. (1999). Loss of p73 gene expression in leukemias/lymphomas due to hypermethylation. *Blood* **94,** 1113–1120.

64. Keck, P. J., Hauser, S. D., Krivi, G., Sanzo, K., Warren, T., Feder, J. *et al.* (1989). Vascular permeability factor, an endothelial cell mitogen related to PDGF. *Science* **246,** 1309–1312.

65. Keiser, A., Weich, H. A., Brandner, G., Marme, D., and Kolch, W. (1994). Mutant p53 potentiates protein kinase C induction of vascular endothelial cell growth factor expression. *Oncogene* **9,** 363–369.

66. Kendall, R. L., and Thomas, K. A. (1993). Inhibition of vascular endothelial cell growth factor activity by an endogenously encoded soluble receptor. *Proc. Natl. Acad. Sci. USA* **90,** 10705–10709.

67. Kong, H.-L., and Crystal, R. G. (1998). Gene therapy strategies for tumor antiangiogenesis. *J. Natl. Cancer Inst.* **90,** 273–286.

68. Kong, H.-L., Hecht, D., Song, W., Kovesdi, I., Hackett, N. R., Yayon, A. *et al.* (1998). Regional suppression of tumor growth by *in vivo* transfer of a cDNA encoding a secreted form of the extracellular domain of the *flt*-1 vascular endothelial growth factor receptor. *Hum. Gene Ther.* **9,** 823–833.

69. Lau, K., and Bicknell, R. (1999). Antiangiogenic gene therapy. *Gene Ther.* **6,** 1793–1795.

70. Li, H., Griscelli, F., Lindenmeyer, F., Opolon, P., Sun, L.-Q., Connault, E. *et al.* (1999). Systemic delivery of antiangiogenic adenovirus AdmATF induces liver resistance to metastasis and prolongs survival of mice. *Hum. Gene Ther.* **10,** 3045–3053.

71. Li, H., Lu, H., Griscelli, F., Opolon, P., Sun, L.-Q., Ragot, T. *et al.* (1998). Adenovirus-mediated delivery of a uPA/uPAR antagonist suppresses angiogenesis-dependent tumor growth and dissemination in mice. *Gene Ther.* **5,** 1105–1113.

72. Lijnen, H. R., Ugwu, F., Bini, A., and Collen, D. (1998). Generation of an angiostatin-like fragment from plasminogen by stromelysin-1 (MMP-3). *Biochemistry* **37,** 4699–4702.

73. Lin, P., Buxton, J. A., Acheson, A., Radziejewski, C., Maisonpierre, P. C., Yancopoulos, G. D. *et al.* (1998). Antiangiogenic gene therapy targeting the endothelium-specific receptor tyrosine kinase Tie2. *Proc. Natl. Acad. Sci. USA* **95,** 8829–8834.

74. Lin, P., Polverini, P., Dewhirst, M., Shan, S., Rao, P. S., and Peters, K. G. (1997). Inhibition of tumor angiogenesis using a soluble receptor establishes a role for Tie2 in pathologic vascular growth. *J. Clin. Invest.* **100,** 2072–2078.

75. Linehan, W. M., Lerman, M. I., and Zbar, B. (1995). Identification of the von Hippel–Lindau (VHL) gene: its role in renal cancer. *J. Am. Med. Assoc.* **273,** 564–570.

76. Liu, L., Liu, L., Anderson, W. F., Beart, R. W., Gordon, E. M., and Hall, F. L. (2000). Incorporation of tumor vasculature targeting motifs into Moloney murine leukemia virus Env escort proteins enhances retrovirus binding and transduction of human endothelial cells. *J. Virol.* **74,** 5320–5328.

77. Liu, Y., Thor, A., Shtivelman, E., Cao, Y., Tu, G., and Heath, T. D. *et al.* (1999). Systemic gene delivery expands the repertoire of effective antiangiogenic agents. *J. Biol. Chem.* **274,** 13338–13344.

78. Lu, H., Dhanabal, M., Volk, R., Waterman, M. J., Ramchandran, R., Knebelmann, B. *et al.* (1999). Kringle 5 causes cell cycle arrest and apoptosis of endothelial cells. *Biochem. Biophys. Res. Commun.* **258,** 668–673.

79. Machein, M. R., Risau, W., and Plate, K. H. (1999). Antiangiogenic gene therapy in a rat glioma model using a dominant-negative vascular endothelial growth factor receptor 2. *Hum. Gene Ther.* **10,** 1117–1128.

80. Maione, T. E., Gray, G. S., Petro, J., Hunt, A. J., Donner, A. L., Bauer, S. I. *et al.* (1990). Inhibition of angiogenesis by recombinant human platelet factor-4 and related peptides. *Science* **247,** 77–79.

81. Maisonpierre, P. C., Suri, C., Jones, P. F., Bartunkova, S., Wiegand, S. J., Radziejewski, C. *et al.* (1997). Angiopoietin-2, a natural antagonist for Tie2 that disrupts in vivo angiogenesis. *Science* **277,** 55–60.

82. Matsuda, K. M., Madoiwa, S., Hasumi, Y., Kanazawa, Y., Saga, Y., Kume, A., Mano, H., Ozawa, K., and Matsuda, M. (2000). A novel strategy for tumor angiogenesis-targeted gene therapy: generation of angiostatin from endogenous plasminogen by protease gene transfer. *Cancer Gene Ther.* **7,** 589–596.

83. Mir, L. M., Bureau, M. F., Gehl, J., Rangara, R., Rouy, D., Caillaud, J.-M. *et al.* (1999). High-efficiency gene transfer into skeletal muscle mediated by electric pulses. *Proc. Natl. Acad. Sci. USA* **96,** 4262–4267.

84. Mavria, G., Jäger, U., and Porter, C. D. (2000). Generation of a high titre retroviral vector for endothelial cell-specific gene expression in vivo. *Gene Ther.* **7,** 368–376.

85. Millauer, B., Shawver, L. K., Plate, K. H., Risau, W., and Ullrich, A. (1994). Glioblastoma growth inhibited in vivo by a dominant-negative Flk-1 mutant. *Nature (London)* **367,** 576–579.

86. Millauer, B., Wizigmann-Voss, S., Schnurch, H., Martinez, R., Moller, N. P., Risau, W. et al. (1993). High affinity VEGF binding and development expression suggest flk-1 as a major regulator of vasculogenesis and angiogenesis. Cell **72,** 835–846.

87. Min, H. Y., Doyle, L. V., Vitt, C. R., Zandonella, C. L., Stratton-Thomas, J. R., Shuman, M. A., et al. Urokinase receptor antagonists inhibit angiogenesis and primary tumor growth in syngeneic mice. Cancer Res. **56,** 2428–2433.

88. Modlich, U., Pugh, C. W., and Bicknell, R. (2000). Increasing endothelial cell specific expression by the use of heterologous hypoxic and cytokine-inducible enhancers. Gene Ther. **7,** 896–902.

89. Mori, A., Arii, S., Furutani, M., Mizumoto, M., Uchida, S., Furuyama, H., Kondo, Y., Gorrin-Rivas, M. J., Furumoto, K., Kaneda, Y., and Imamura, M. (2000). Soluble Flt-1 gene therapy for peritoneal metastases using HVJ-cationic liposomes. Gene Ther. **7,** 1027–1033.

90. Moses, M. A., Sudhalter, J., and Langer, R. (1990). Identification of an inhibitor of neovascularization from cartilage. Science **248,** 1408–1410.

91. Nagayama, Y., Shigematsu, K., Namba, H., Zeki, K., Yamashita, S., and Niwa, M. (2000). Inhibition of angiogenesis and tumorigenesis, and induction of dormancy by p53 in a p53-null thyroid carcinoma cell line in vivo. Anticancer Res. **20,** 2723–2728.

92. Nguyen, J. T., Wu, P., Clouse, M. E., Hlatky, L., and Terwilliger, E. F. (1998). Adeno-associated virus-mediated delivery of antiangiogenic factors as an antitumor strategy. Cancer Res. **8,** 5673–5677.

93. O'Reilly, M. S., Wiedershain, D., Stetler-Stevenson, W. G., Folkman, J., and Moses, M. A. (1999). Regulation of angiostatin production by matrix metalloproteinase-2 in a model of concomitant resistance. J. Biol. Chem. **274,** 29568–29571.

94. O'Reilly, M. S., Boehm, T., Shing, Y., Naomi, F., Vasios, G., Lane, W. S. et al. (1997). Endostatin: an endogenous inhibitor of angiogenesis and tumor growth. Cell **88,** 277–275.

95. O'Reilly, M. S., Holmgren, L., Chen, C., and Folkman, J. (1996). Angiostatin induces and sustains dormancy of human primary tumors in mice. Nature Med. **2,** 689–692.

96. O'Reilly, M. S., Holmgren, L., Shing, Y., Chen, C., Rosenthal, R. A., Moses, M. et al. (1994). Angiostatin: a novel angiogenesis inhibitor that mediates the suppression of metastases by a Lewis lung carcinoma. Cell **79,** 315–328.

97. Patterson, B. C., and Sang, Q. A. (1997). Angiostatin-converting enzyme activities of human matrilysin (MMP-7) and gelatinase B/type-IV collagenase (MMP-9). J. Biol. Chem. **272,** 28823–28825.

98. Pawelek, J. M., Low, K. B., and Bermudes, D. (1997). Tumor-targeted Salmonella as a novel anticancer vector. Cancer Res. **57,** 4537–4544.

99. Prehn, R. T. (1993). Two competing influences that may explain concomitant tumor resistance. Cancer Res. **53,** 3266–3269.

100. Rak, J., Filmus, J., Finkenzeller, G., Grugel, S., Marmé, D., and Kerbel, R. S. (1995). Oncogenes as inducers of tumor angiogenesis. Cancer Metastasis Rev. **14,** 263–277.

101. Ribatti, D., Vacca, A., Iurlaro, M., Ria, R., Roncali, L., and Dammacco, F. Human recombinant interferon alpha-2a inhibits angiogenesis of chick area vasculosa in shell-less culture. Int. J. Mircorcirc. **16,** 165–169.

102. Sacco, M. G., Caniatti, M., Catò, E. M., Frattini, A., Chiesa, G., Ceruti, R., Adorni, F., Zecca, L., Scanziani, E., and Vezzoni, P. (2000). Liposome-delivered angiostatin strongly inhibits tumor growth and metastatization in a transgenic model of spontaneous breast cancer. Cancer Res. **60,** 2660–2665.

103. Saleh, M., Wiegmans, A., Malone, Q., Stylli, S. S., and Kaye, A. H. (1999). Effect of in situ retroviral interleukin-4 transfer on established intracranial tumors. J. Natl. Cancer Inst. **91,** 438–445.

104. St. Croix, B., Rago, C., Velculescu, V., Traverso, G., Romans, K. E., Montgomery, E., Lal, A., Riggins, G. J., Lengauer, C., Vogelstein, B., and Kinzler, K. W. (2000). Genes expressed in human tumor endothelium. Science **289,** 1197–1202.

105. Salimath, B., Marmé, D., and Finkenzeller, G. (2000). Expression of the vascular endothelial growth factor gene is inhibited by p73. Oncogene **19,** 3470–3476.

106. Sanches, R., Kuiper, M., Penault-Llorca, F., Aunoble, B., D'Incan, C., and Bignon, Y.-J. (2000). Antitumoral effect of interleukin-12-secreting fibroblasts in a mouse model of ovarian cancer: implications for the use of ovarian cancer biopsy-derived fibroblasts as a vehicle for regional gene therapy. Cancer Gene Ther. **7,** 707–720.

107. Sauter, B. V., Martinet, O., Zhang, W.-J., Mandeli, J., and Woo, S. L. (2000). Adenovirus-mediated gene transfer of endostatin in vivo results in high level of transgene expression and inhibition of tumor growth and metastases. Proc. Natl. Acad. Sci. USA **97,** 4802–4807.

108. Schmidt, E. E., Ichimura, K., Reifenberger, G., and Collins, V. P. (1994). CDKN2 (p16/MTS1) gene deletion or CDK4 amplification occurs in the majority of glioblastomas. Cancer Res. **54,** 6321–6324.

109. Schwarz, M. A., Kandel, J., Brett, J., Li, J., Hayward, J., Schwarz, R. E. et al. (1999). Endothelial-monocyte activating polypeptide II, a novel antitumor cytokine that suppresses primary and metastatic tumor growth and induces apoptosis in growing endothelial cells. J. Exp. Med. **190,** 341–354.

110. Senger, D., Galli, S. J., Dvorak, A. M., Perruzzi, C. A., Harvey, V. S., and Dvorak, H. F. (1983). Tumor cells secrete a vascular permeability factor which promotes ascites fluid accumulation. Science **291,** 983–985.

111. Sgadari, C., Angiolillo, A. L., and Tosato, G. (1996). Inhibition of angiogenesis by interleukin-12 is mediated by the interferon-inducible protein 10. Blood **87,** 3877–3882.

112. Shiose, S., Sakamoto, T., Yoshikawa, H., Hata, Y., Kawano, Y., Ishibashi, T., Inomata, H., Takayama, K., and Ueno, H. (2000). Gene transfer of a soluble receptor of VEGF inhibits the growth of experimental cyelid malignant melanoma. Invest. Ophthalmol. Vis. Sci. **41,** 2395–2403.

113. Singh, R. K., Gutman, M., Bucana, C. D., Sanchez, R., Llansa, N., and Fidler, I. J. (1995). Interferons alpha and beta downregulate the expression of basic fibroblast growth factor in human carcinomas. Proc. Natl. Acad. Sci. USA **92,** 4562–4566.

114. Singhal, S., Mehta, J., Desikan, R., Ayers, D., Roberson, P., Eddlemon, P. et al. (1999). Antitumor activity of thalidomide in refractory multiple myeloma. N. Engl. J. Med. **341,** 1565–1571.

115. Stadsler, W. M., Kuzel, T., Shapiro, C., Sosman, J., Clark, J., and Vogelzang, N. J. (1999). Multi-institutional study of the angiogenesis inhibitor TNP-470 in metastatic renal carcinoma. J. Clin. Oncol. **17,** 2541–2545.

116. Stearns, M. E., Garcia, F. U., Fudge, K., Rhim, J., and Wang, M. (1999). Role of interleukin 10 and transforming growth factor $\beta 1$ in the angiogenesis and metastasis of human prostate primary tumor lines from orthotopic implants in severe combined immunodeficiency mice. Clin. Cancer Res. **5,** 711–720.

117. Strieter, R. M., Polverini, P. J., Kunkel, S. L., Arenberg, D. A., Burkick, M. D., Kasper, J., et al. (1995). The functional role of the ELR motif in CXC chemokine-mediated angiogenesis. J. Biol. Chem. **270,** 27348–27357.

118. Sunamura, M., Sun, L., Lozonschi, L., Duda, D. G., Kodama, T., Matsumoto, G., Shimamura, H., Takeda, K., Kobari, M., Hamada, H., and Matsuno, S. (2000). The antiangiogenesis effect of interleukin 12 during early growth of human pancreatic cancer in SCID mice. Pancreas **20,** 227–233.

119. Suri, C., Jones, P. F., Patan, S., Bartunkova, S., Maisonpierre, P. C., Davis, S. et al. (1996). Requisite role of angiopoietin-1, a ligand for the TIE2 receptor, during embryonic angiogenesis. Cell **87,** 1171–1180.

120. Takayama, K., Ueno, H., Nakanishi, Y., Sakamoto, T., Inoue, K., Shimizu, K. et al. (2000). Suppression of tumor angiogenesis and

growth by gene transfer of a soluble form of vascular endothelial growth factor receptor into a remote organ. *Cancer Res.* **60,** 2169–2177.

121. Tanaka, T., Cao, Y., Folkman, J., and Fine, H. A. (1998). Viral vector-targeted antiangiogenic gene therapy utilizing an angiostatin complementary DNA. *Cancer Res.* **58,** 3362–3369.

122. Tanaka, T., Manome, Y., Wen, P., Kufe, D. W., and Fine, H. A. (1997). Viral vector-mediated transduction of a modified platelet factor 4 cDNA inhibits angiogenesis and tumor growth. *Nature Med.* **3,** 437–442.

123. Terman, B. I., Dougher-Vermaaen, M., Carrion, M. E., Kimitrov, D., Armellino, D. C., Gospodarowicz, D. *et al.* (1992). Identification of the KDR tyrosine kinase as a receptor for vascular endothelial cell growth factor. *Biochem. Biophsy. Res. Com.* **187,** 1579–1586.

124. Tolsma, V. S., Volpert, O. V., Good, D. J., Frazier, W. A., Polverini, P. J., and Bouck, N. (1993). Peptides derived from two separate domains of the matrix protein thrombospondin-1 have anti-angiogenic activity. *J. Cell. Biol.* **122,** 497–511.

125. Tomlinson, E. (1992). Impact of the new biologies on the medical and pharmaceutical sciences. *J. Pharm. Pharmacol.* **44**(suppl. 1) 147–159.

126. Valente, P., Fassina, G., Melchiori, A., Masiello, L., Cilli, M., Vacca, A. *et al.* (1998). TIMP-2 over-expression reduces invasion and angiogenesis and protects B16F10 melanoma cells from apoptosis. *Int. J. Cancer* **75,** 246–253.

127. Voest, E. E., Kenyon, B. M., O'Reilly, M. S., Truitt, G., D'Amato, R. J., and Folkman, J. (1995). Inhibition of angiogenesis in vivo by interleukin 12. *J. Natl. Cancer Inst.* **87,** 581–586.

128. Volpert, O. V., Fong, T., Koch, A. E., Peterson, J. D., Waltenbaugh, C., Tepper, R. I. *et al.* (1998). Inhibition of angiogenesis by interleukin 4. *J. Exp. Med.* **188,** 1039–1046.

129. Volpert, O. V., Tolsma, S. S., Pellerin, S., Feige, J. J., Chen, H., and Mosher, D. F. (1995). Inhibition of angiogenesis by thrombospondin-2. *Biochem. Biophys. Res. Commun.* **217,** 326–332.

130. Westphal, J. R., Van't, Hullenaar, R., Geurts-Moespot, A., Seep, F. C., Verheijen, J. H., Bussemakers, M. M., Askaa, J., Clemmensen, I., Eggermont, A. A., Ruiter, D. J., and De Waal, R. M. (2000). Angiostatin generation by human tumor cell lines: involvement of plasminogen activators. *Int. J. Cancer* **86,** 760–767.

131. Yoon, S. S., Eto, H., Lin, C., Nakamura, H., Pawlik, T. M., Song, S. U. *et al.* (1999). Mouse endostatin inhibits the formation of lung and liver metastases. *Cancer Res.* **59,** 6251–6256.

132. Xu, M., Kumar, D., Stass, S. A., and Mixson, A. J. (1998). Gene therapy with p53 and a fragment of thrombospondin I inhibits human breast cancer *in vivo. Molec. Genet. Metab.* **63,** 103–109.

133. Xu, M., Kumar, D., Srinivas, S., Detolla, L. J., Yu, S. F., Stass, S. A. *et al.* (1997). Parenteral gene therapy with p53 inhibits human breast tumors *in vivo* through a bystander mechanism without evidence of toxicity. *Hum. Gene Ther.* **8,** 177–185.

CHAPTER

27

VEGF-Targeted Antiangiogenic Gene Therapy

CALVIN J. KUO

Departments of Surgery and Genetics
Children's Hospital
Harvard Medical School
Boston, Massachusetts 02115

FILIP A. FARNEBO

Departments of Surgery and Genetics
Children's Hospital
Harvard Medical School
Boston, Massachusetts 02115

CHRISTIAN M. BECKER

Department of Surgery
Children's Hospital
Harvard Medical School
Boston, Massachusetts 02115

JUDAH FOLKMAN

Department of Surgery
Children's Hospital
Harvard Medical School
Boston, Massachusetts 02115

I. INTRODUCTION

Antiangiogenic therapy constitutes a new approach for cancer therapeutics that targets the tumor vasculature. The vascular endothelial growth factor (VEGF) is a polypeptide angiogenic factor whose interaction with high-affinity receptors plays an essential role in both physiologic and pathologic angiogenesis. Gene therapy approaches to inhibit

VEGF activity and tumor angiogenesis have assumed diverse forms, from intratumoral administration of retroviruses to the local and systemic administration of adenoviruses. In this review, the advantages and disadvantages of different VEGF-targeted antiangiogenic gene therapy approaches are summarized and surrogate biologic endpoints and clinical translation discussed.

II. ANGIOGENESIS AND TUMOR GROWTH

Angiogenesis, the formation of new blood vessels, is a rate-limiting step for numerous pathologic processes, including cancer and many ophthalmologic disorders [19]. With particular regard to cancer, the concept of an "angiogenic switch" has been proposed by Folkman and Hanahan [30], whereby neovascularization both precedes and is necessary for tumor progression and metastasis. The development of tumor blood vasculature is mechanistically quite complex, involving angiogenesis into initially avascular tumor masses [20], the early co-option of vasculature from neighboring tissue [31], and the contribution of circulating endothelial stem cells [4]. Given the anticancer therapeutic potential of angiogenesis blockade, significant efforts have been devoted towards understanding the molecular mechanisms underlying blood vessel formation.

The process of angiogenesis is appropriately under tight positive and negative regulation. At least three families of receptor tyrosine kinases have been implicated in positive angiogenic regulation, the VEGF receptors (Flk1, Flt1), the TIE receptors (Tie1, Tie2), and the ephB4/ephrin B2 system [15,25]. On the other hand, putative negative angiogenic regulators have been recently identified (angiostatin, endostatin) whose mechanisms of action are less well understood [47,48]. The existence of these two classes of molecules has led to strategies to inhibit angiogenesis and tumor growth by either inhibiting positively acting agents or supplying exogenous negatively acting agents.

The hypothesis by Folkman [19] that cancer is angiogenesis dependent has been repeatedly affirmed by experimental treatment of tumors with angiogenesis inhibitors. For example, striking inhibition of tumor growth in animals can be achieved not by direct treatment of the tumor but rather by selective inhibition of the endothelial growth factor VEGF. Indeed, anti-VEGF monoclonal antibodies, VEGF receptor small molecule kinase inhibitors or antibodies, or soluble VEGF receptors have all been utilized to inhibit tumor growth via angiogenesis inhibition, as will be described later [15,22,38,53]. Similarly, negatively acting factors such as endostatin or angiostatin have also demonstrated efficacy in animal models [47,48]. An observation common to the use of either strategy has been broad-spectrum inhibition of very diverse tumor types, consistent with the commonality of angiogenesis-dependent tumor growth regardless of tissue of origin.

III. GENE THERAPY FOR DELIVERY OF ANTIANGIOGENIC FACTORS

Polypeptide antiangiogenic factors can be delivered by "gene therapy"—a broad term that encompasses either nonviral strategies (such as naked DNA or liposome-mediated gene transfer) or viral strategies. Nonviral strategies offer minimal toxicity at the expense of transient and/or low level expression and will not be covered in this review. Viral vectors occur in many flavors, each with their own intrinsic profile of advantages and disadvantages (Table 1). Retroviruses, while relatively easy to manufacture, can only infect dividing cells which often restricts their utility for *ex vivo* approaches and can produce potentially mutagenic integration sites. Adenoviruses (Ads), on the other hand, infect quiescent cells quite readily and provide extremely high level expression but trigger neutralizing immune responses that limit transgene expression to several weeks and manifest potentially dangerous cytotoxicities during the initial infection phase. The adeno-associated viruses (AAVs) are comparatively nonimmunogenic and can provide impressive transgene expression exceeding 1 year's duration, however, they are relatively difficult to produce and typically produce protein levels that are several logs less than adenoviral infections [46]. Newer technologies include lentiviruses and "gutless" adenoviruses [2], which are beyond the scope of the current discussion.

Ideally, gene therapy approaches for delivery of antiangiogenic agents would convert the body into its own "factory" for producing these proteins [18,39]. In this manner, direct infection of the tumor cell would not be required, as opposed to gene therapy to restore defective copies of a tumor suppressor such as *p53*. Rather, the infection could occur at a site distant from the tumor (say, in liver or muscle), and these remote sites would produce systemic levels of angiogenesis

TABLE 1   Characteristics of Viral Vectors

Virus type	Infection of quiescent cells	Genome integration	Duration of expression	Antiviral immunity	Expression levels	Difficulty of production
Retrovirus	No	Yes	Years	No	+	+
Adenovirus	Yes	No	Weeks	Yes	++++	+
Adeno-associated virus (AAV)	Yes	±	Years	No	+	+++
Lentivirus	Yes	Yes	Years	No	+	+++
Gutless adenovirus	Yes	No	Years	No	++++	++++

inhibitors capable of suppressing growth even of widespread metastases. Through such an approach, even inefficient infection of the target would be tolerated, as long as sufficient circulating levels were achieved.

IV. ANTIANGIOGENIC GENE THERAPY IN THE EXPERIMENTAL AND CLINICAL SETTINGS

Why deliver antiangiogenic gene products by gene therapy? Two major settings exist—experimental and clinical. In the experimental setting, a need clearly exists for technology enabling the rapid and large-scale production of antiangiogenic agents in quantities sufficient for animal studies. The proliferation of vascular biology research has uncovered an ever-expanding catalog of proteins with potential utility for regulating tumor vasculature. Unfortunately, the evaluation of an antiangiogenic agent currently requires laborious production of recombinant protein merely to obtain quantities sufficient for animal experimentation, a difficulty compounded and amplified by experiments comparing or combining different agents. As described below, gene therapy approaches in the experimental setting could be used to create an array of easily administered antiangiogenic therapeutics to investigate questions of efficacy, comparison, and synergy. In the clinical setting, these approaches could create antiangiogenic treatments characterized by favorable economics of production, single-injection pharmacokinetics, and continuous secretion.

The use of gene therapy for rapid experimental assessment of antiangiogenic agents and in the clinic provides several theoretical advantages versus the use of recombinant proteins. First, virus production could occur more rapidly and more economically than recombinant protein production. For example, an experiment using 1 mg of recombinant protein per day in 10 mice for 10 days' duration would require the fairly large quantity of 100 mg of protein per experiment. On the other hand, a single research-scale adenoviral maxiprep in our facility can yield $1–5 \times 10^{11}$ plaque forming units (PFU), or enough for 100–500 animal injections. The difficulties in producing sufficient quantities of protein to treat animals are compounded exponentially for the treatment of humans with over 1000 times the body mass and would be more severe for antiangiogenic agents with poor pharmacokinetic properties.

Second, gene therapy can produce sustained circulating transgene levels that are not subject to the peak-and-trough pharmacokinetics of bolus administration of recombinant proteins. Such sustained levels would be theoretically quite appropriate for antiangiogenic agents, which have been proposed to require chronic administration [5,18]. These proteins are continuously delivered without daily oscillations to compound the repeated administration of recombinant proteins, which would be optimal for chronic antiangiogenic therapy.

A third advantage of gene therapy is the potential convenience of single-injection dosing. As opposed to repeated administration of recombinant proteins, the creation of a depot for chronic protein production at the vector injection site would circumvent the requirement for laborious daily administration. Single intravenous (i.v.) injections of adenoviral vectors are capable of producing substantial levels of circulating protein for 2–4 weeks, while intramuscular injections of adeno-associated virus can produce therapeutic levels for over a year.

V. VASCULAR ENDOTHELIAL GROWTH FACTOR AND RECEPTORS

The polypeptide growth factor VEGF is intricately associated with both physiologic and pathologic angiogenesis. These characteristics render VEGF an attractive target for antiangiogenic gene therapy approaches, as will be described in this review. Although produced by a number of different cells, VEGF appears to acts selectively on endothelial cells, stimulating angiogensis both *in vitro* and *in vivo*. The direct actions of VEGF are numerous, including promotion of endothelial cell permeability, growth, and migration and serving as a survival factor for newly formed blood vessels. In comparison to histamine, VEGF is 50,000 times more potent in inducing vascular leakage [13]. Topical administration of VEGF results in fenestration of the endothelium of small venules and capillaries through a c-*src*-dependent mechanism [14]. VEGF also stimulates the expression of tissue plasminogen activator, urokinase plasminogen activator, collagenases, and matrix metalloproteinases, which are involved in the degradation of the extra cellular matrix needed for endothelial cell migration [52,54,64,73].

Expression of VEGF is highly regulated by hypoxia, providing a physiologic feedback mechanism to accommodate insufficient tissue oxygenation by promoting blood vessel formation. Hypoxic regulation of VEGF gene expression is mediated by a family of hypoxia-inducible transcription factors (HIFs) [7]. In addition, hypoxia upregulates VEGF levels by message stabilization [42]. In both breast and prostate cancer, VEGF levels are augmented by the presence of sex hormones [33,56]. Cytokines, such as epidermal growth factor (EGF) and transforming growth factor beta (TGF-β), may also stimulate the expression of VEGF [62]. Both VEGF mRNA and protein are markedly upregulated in the vast majority of human tumors, and VEGF overexpression in cancer patients is associated with poor prognosis and low survival [51]. In tumors, VEGF is not produced by endothelial cells but instead by tumor cells or tumor stroma, consistent with a paracrine mode of action [15,24].

Vascular endothelial growth factor occurs as different isoforms encoded by splice variants of a single gene containing eight exons. In the human, at least five different isoforms

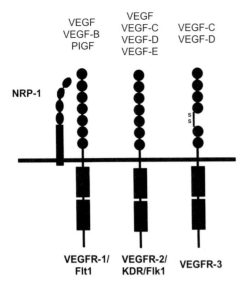

VEGF
VEGF-B
PIGF

VEGF
VEGF-C
VEGF-D
VEGF-E

VEGF-C
VEGF-D

NRP-1

VEGFR-1/
Flt1

VEGFR-2/
KDR/Flk1

VEGFR-3

FIGURE 1 Schematic of the VEGF and VEGF receptor families. VEGF receptors associate with a distinct and overlapping subset of VEGF family members. VEGFR-1 to -3 are presumed to associate with exons 1–3 of VEGF-A, while the nonclassical VEGF receptor neuropilin associates with exon 7.

exist: $VEGF_{121}$, $VEGF_{143}$, $VEGF_{165}$, $VEGF_{189}$, and $VEGF_{206}$ [67]. Although the different isoforms exhibit identical biological activity, they differ in their binding to heparin and to the extracellular matrix. The smaller isoforms are secreted in a soluble form, whereas the larger ones remain cell-associated and their bioavailability is regulated by proteolysis [32]. VEGF (VEGF-A) is the founding member of a growing family of growth factors, which today includes six different proteins, PIGF, VEGF-B, VEGF-C, VEGF-D, and orf virus VEGF (VEGF-E) (Fig. 1). The VEGFs mediate angiogenic signals to the vascular endothelium via high-affinity receptor tyrosine kinases, designated VEGFR-1 (Flt1), VEGFR-2 (Flk1/KDR), and VEGFR-3 (Flt4), characterized by seven immunoglobulin-like domains in the extracellular region, a single transmembrane domain, and an intracellular split tyrosine-kinase domain [67]. These VEGFRs bind distinct but nonoverlapping subsets of VEGF family members; for example, Flt1 binds PIGF, VEGF-A, and VEGF-B, while Flk1 binds VEGF-A, -C, and -D (Fig. 1). VEGF-A may bind to VEGFR-1/Flt1 on cell membranes with higher affinity than VEGFR-2/Flk1/KDR [68], although this differential affinity may be less pronounced with truncated soluble versions of Flk1 and Flt1 [36].

More recently, the cell surface glycoprotein neuropilin-1 has been identified as a nonclassical VEGF receptor that binds exon 7 of $VEGF_{165}$ as opposed to the more N-terminal motifs recognized by Flk1 and Flt1. These data suggest a very versatile function for neuropilin, which was originally implicated as a semaphorin receptor controlling developmental axon guidance. By virtue of this distinct binding epitope,

neuropilin-1 has been proposed to function as a VEGFR-2/Flk1/KDR coreceptor capable of enhancing the biological effect of $VEGF_{165}$ on endothelial cells [60].

VI. VASCULAR ENDOTHELIAL GROWTH FACTOR AND ANGIOGENESIS

Gene targeting experiments have provided important insights into the biological functions of VEGF-A and the VEGF receptors during development [25]. VEGF-A is critically involved in early stages of vascular development. Targeted inactivation of a single VEGF allele results in haplo-insufficiency with embryonic lethality due to abnormal blood vessel development around 9 days after gestation. Various developmental vascular phenotypes are observed, including smaller lumen of the dorsal aorta, reduced sprouting of vessels, abnormal connection of large blood vessels with the heart, and an irregular plexus of enlarged capillaries in yolk sac and placenta. VEGF thus seems not to be essential for the initial differentiation of angioblasts to early endothelial cells but is required for further blood vessel development [16]. Blood vessel development is more affected in homozygous VEGF-deficient embryos than in heterozygous VEGF-deficient embryos, suggesting a tight dose dependency. In neonatal mice, as opposed to embryos, conditional inactivation of the VEGF gene or administration of a soluble VEGF receptor (Flt-Fc) results in impaired vascular development and endothelial apoptosis, leading to increased mortality, growth, and impaired organ development, most severe in liver, heart, and kidney. This dependency on VEGF-A decreases with age, however, with older animals being less affected by the soluble VEGF receptor [26,27].

The inactivation of the two receptors for VEGF-A both produce embryonic lethality around mid-gestation, although the phenotypes indicate distinct functions. Mice lacking Flk1 show the most dramatic phenotype: a complete failure of vasculogenesis, with blockade of differentiation of both endothelial and haematopoetic cells [57]. In contrast, Flt1-deficient embryos develop an excess of endothelial cells, but these fail to organize into normal vascular channels [21]. Flt1 and Flk1 may therefore exert opposing functions, with Flk1 signaling required for the emergence of mature endothelial cells, while Flt1 negatively regulates the commitment of angioblasts during vasculogenesis [23]. Disruption of the third VEGF receptor, VEGFR-3/Flt4, is not required for primitive vasculogenesis but leads to a defective remodeling of the primary vascular plexus and cardiovascular failure after embryonic day 9.5 [12].

The role of VEGF in pathologic, as opposed to physiologic, angiogenesis has been emphasized by potent inhibition of both tumor angiogenesis and tumor growth by systemic administration of VEGF antagonists. A direct link between VEGF activity and tumor angiogenesis was first demonstrated by the suppression of tumor growth in mice after injection of an anti-VEGF neutralizing monoclonal antibody [38]. Similarly, monoclonal antibodies that inhibit the association of VEGF with Flk1/KDR potently antagonize angiogenesis and tumor growth in parallel in a wide variety of murine tumor models [53]. Recently, multiple small molecule antagonists of the KDR kinase ATP-binding site have been developed that produce striking inhibition of tumor growth and angiogenesis upon systemic administration [22,69].

VII. VASCULAR ENDOTHELIAL GROWTH FACTOR INHIBITION BY GENE TRANSFER

A. Direct Transduction of Tumor Endothelium with Dominant-Negative Receptors

Multiple theoretical approaches exist for the inhibition of VEGF function by gene transfer, including dominant-negative kinase and soluble receptor approaches. In the first demonstration of feasibility, Ullrich and colleagues employed retroviruses encoding dominant-negative alleles of Flk1 lacking the kinase domain. These kinase-deficient proteins were capable of heterodimerizing with and inactivating full-length Flk1, thereby disrupting VEGF signalling. The coinjection of cell lines producing dominant-negative Flk1 retroviruses along with glioblastoma, lung, breast, or colon tumor lines into mice resulted in strong inhibition of tumor growth, presumably from intratumoral retroviral production, infection of tumor endothelium, and blockade of VEGF function [44,45]. Decreased tumor vascularity was also observed in these studies, consistent with an antiangiogenic effect [45].

B. Direct Transduction of Tumor Cells with Soluble VEGF Receptors

In a different approach, genes encoding "soluble VEGF receptors" have been directly introduced into tumor cells. Most commonly employed has been the ligand-binding ectodomain of the Flt1 VEGF receptor. Intriguingly, an alternative splice form of Flt1 has been described (sFlt1) which encodes the N-terminal of six of seven extracellular Ig domains but truncates the transmembrane and kinase domains. This sFlt1 protein binds VEGF with affinity identical to native Flt1 and is expressed in vivo by vascular endothelial cells [28,35].

The use of soluble VEGF receptors in gene transfer strategies has several potential advantages. These proteins can potentially diffuse freely from their site of synthesis and thus bind VEGF in the tumor microenvironment or in the circulation, sequestering it from native endothelial cell receptors. Additionally, as with their full-length counterparts, these soluble VEGF receptors retain the ability to bind an entire spectrum of VEGF family members versus the presumed restricted target specificity of anti-VEGF-A monoclonals. Indeed, Flk1 binds VEGF-A, -C, and -D, while Flt1 binds PIGF, VEGF-A, and VEGF-B [25]. This broader binding spectrum may be of therapeutic utility as VEGF-B, -C, and -D are potent angiogenic agents in vivo and in vitro [50,55,65,72].

Curiel and co-workers [28] transfected HT-1080 human fibrosarcoma cells with sFlt1 cDNA and observed significant retardation of tumor growth rate either in an intravenous lung metastasis model or in a subcutaneous implantation model. Similarly, transfection of glioblastoma cells with sFlt1 significantly prolonged survival when these were implanted intracranially into mice. These data strongly suggest that paracrine expression of a soluble VEGF receptor could impede tumor growth, presumably by sequestration of VEGF and inhibition of endothelial cell function.

C. Regional Control of Tumor Angiogenesis by Adenoviruses Encoding Soluble VEGF Receptors

In a different approach, Crystal and colleagues [40] utilized adenovirus-mediated gene transfer to express soluble ectodomains of the VEGF receptor Flt1. Mice bearing 3-day-old preestablished colon carcinomas in the spleen which could metastasize to liver received intravenous injections of adenoviruses encoding a soluble Flt ectodomain. Because of the tropism of intravenously administered adenovirus for liver, transgene production presumably occurred in the vicinity of the primary tumor and metastasis, and a striking (>90%) reduction in liver metastases was observed. Similarly, intratracheal administration of the soluble FLT adenovirus to 3-day-old lung metastases produced a statistically significant reduction in tumor burden, again by a regional mechanism [40].

D. Systemic Inhibition of Minimal Tumor Burden by Intramuscular Adenoviral Administration

While establishing an important role for VEGF during tumor angiogenesis, the aforementioned approaches introducing genes encoding dominant-negative alleles or soluble receptors required direct transduction of tumor cells, the implantation of retroviral producer lines in the vicinity of tumor cells, or regional adenoviral administration to minimal tumor burdens. The efficacy of such strategies thus critically depends on the ability to achieve substantial genetic manipulation of tumor cells or the tumor microenvironment, which is difficult to accomplish ex vivo, much less in vivo with distantly metastatic malignancies. Similarly, local administration of viruses cannot always be achieved, again as in the case of diffusely metastatic disease.

To overcome the requirement for transduction of the tumor cell/tumor microenviroment, the production of circulating antiangiogenic proteins at a distance from the tumor has been attempted, thus providing systemic inhibition of tumor endothelium while circumventing the need for direct action on the tumor milieu. Although potentially encumbering unforeseen consequences of systemic inhibition of angiogenesis, this approach theoretically affords control of distant and diffusely metastatic disease.

Systemic inhibition of distant malignant disease was initially attempted with an adenovirus encoding a soluble Flt ectodomain [40]. While regional control of liver or lung metastases was achieved by local administration of the vector (see previous discussion), neither subcutaneous tumors nor lung metastases were inhibited at a distance by intravenous vector delivery to the liver, although circulating levels of the transgene were not documented [40].

More recently, Takayama *et al.* [63] demonstrated the ability of an adenovirus encoding soluble Flt to provide distant suppression of tumor angiogenesis and tumor growth. Intramuscular administration of an adenovirus (5×10^8 PFU) encoding a fusion of the Flt extracellular domain to an antibody Fc fragment produced systemic levels of approximately 4 nM, or roughly 20 ng/mL. The coadministration of this virus by an intramuscular route at the same time as subcutaneous implantation of a wide variety of tumor lines produced impressive suppression of tumor growth ranging from partial to complete inhibition over a 120-day period. Interestingly, the efficacy of distant tumor suppression correlated with the VEGF secretion rate of the tumor lines and was accompanied by decreased tumor microvessel density [63].

E. Systemic Inhibition of Larger Preexisting Tumor Burdens by Intravenous Adenoviral Administration

In our experience, either soluble Flt1 or Flk1 receptor ectodomains can be delivered by adenoviruses to produce distant inhibition of tumor growth. Using adenoviruses encoding the ligand-binding Ig domains 1–3 of Flt or the entire Flk ectodomain fused to an antibody Fc fragment, systemic levels of >3 mg/mL (Flk1-Fc) or 30–50 μg/mL (Flt1(1–3)) can be achieved with single intravenous injections of 1×10^9 infectious particles. These systemic levels display pharmacokinetics typical of adenoviral transgene products, with an early peak at 2–3 days and decremental expression thereafter (Fig. 2), presumably resulting from a combination of immune response and virus-mediated cytotoxicity [46].

We have found that suppression of angiogenesis in the corneal micropocket assay correlates strongly with and appears to be predictive of the ability to suppress tumor growth. In this assay, mice receive an intravenous injection of adenovirus, primarily transducing the liver, followed after 2–3 days by implantation of a slow-release hydron

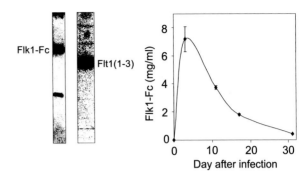

FIGURE 2 Pharmacokinetics of expression of Flk1-Fc after intravenous injection of adenovirus. C57Bl/6 mice received an i.v. injection of 10^9 PFU of Ad Flk1-Fc by tail vein, followed by periodic sampling for transgene expression. Flk1-Fc plasma concentrations were determined by sandwich ELISA. The transient expression is typical of adenoviral infection of immunocompetent mice.

VEGF pellet into the cornea. While the cornea is normally avascular, VEGF pellet implantation elicits a brisk angiogenic response from the limbus within 3–5 days. Under these conditions, both Flk1-Fc and Flt1(1–3) adenoviruses transducing the liver after i.v. injection produced potent and remote inhibition of VEGF-induced vascularization in the cornea (Fig. 3; see color insert). These assays demonstrate the ability of adenoviruses producing soluble VEGF receptors to provide systemic inhibition of angiogenesis.

Single intravenous injections of Ad Flk1-Fc or Ad Flt1 (1–3) also strongly inhibit the distant growth of preestablished tumors. Indeed, when tumor cells implanted subcutaneously on the dorsum are allowed to form palpable tumors of 100–200 mm^3 for 10 days, subsequent i.v. administration of Ad Flk1-Fc or Ad Flt1(1–3) produces >80% suppression of growth over a >14-day period relative to null adenovirus controls (Fig. 4). Intramuscular injections of an Ad Flt1-Fc virus have been shown to inhibit the distant growth of tumor cells when simultaneously implanted subcutaneously [63]. However, in our experience, intramuscular (i.m.) injections of Ad Flk1-Fc or Ad Flt1(1–3) are unable to inhibit growth of preestablished tumors, suggesting that the suppression of preexisting disease may require transduction of larger tissue masses such as liver, with concomitant higher circulating levels of transgene product.

Several characteristics of tumor suppression by Ad Flk1-Fc and Ad Flt1(1–3) are worthy of mention. First, this inhibition likely occurs by inhibition of angiogenesis, given the decreased microvessel density of treated tumors and the ability of the recombinant Flk1-Fc and Flt1(1–3) proteins to inhibit VEGF-induced endothelial proliferation *in vitro*. Second, these adenoviruses produce an extremely broad-spectrum inhibition of a wide range of tumors implanted at remote subcutaneous or orthotopic positions or in transgenic mice (Table 2). This broad potency parallels results obtained with an anti-Flk1 monoclonal antibody and is consistent

TABLE 2   Activity of Adenoviruses Encoding Soluble VEGF Receptors Against Tumor Models in Mice

Tumor type	Mouse strain	Suppression (%)	
		Ad Flk1-Fc	Ad Flt1(1–3)
Murine Lewis lung carcinoma	C57Bl/6J	79	81
Murine T241 fibrosarcoma	C57Bl/6J	83	87
Human BxPc3 pancreatic carcinoma	CB17 SCID	87	90
Human LS174T colon carcinoma	CB17 SCID	80	ND
Human U87 glioblastoma	CB17 SCID	70	ND
Human LNCaP	CB17 SCID	72	ND
TRAMP transgenic murine prostate	C57Bl/6J	50–80	ND

Note: ND, not determined.

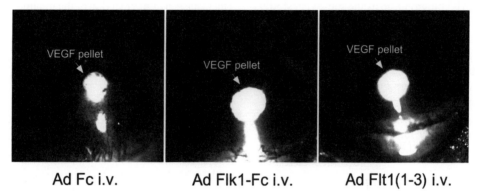

Ad Fc i.v.   Ad Flk1-Fc i.v.   Ad Flt1(1-3) i.v.

FIGURE 3   Systemic inhibition of angiogenesis produced by remote infection with anti-angiogenic adenoviruses. C57Bl/6 mice received injections of 10^9 plaque-forming units of Ad Flk1–Fc, Ad Flt1(1–3), or the control virus Ad Fc by tail vein, a procedure that predominantly produces infection of the liver. Two days later, vascular endothelial growth factor (VEGF)-containing hydron pellets were implanted in the cornea, and corneal neovascularization was assessed 5 days post virus administration. Note the robust angiogenic response toward the pellet in Ad Fc mice but not in Ad Flk1–Fc or Ad Flt1(1–3) mice. (See color insert.)

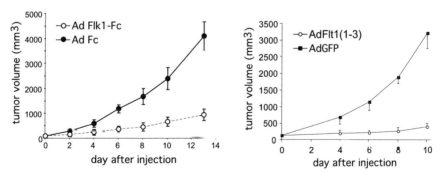

FIGURE 4   Inhibition of T241 fibrosarcoma tumor growth by remote infection with antiangiogenic adenoviruses. T241 fibrosarcoma was implanted on the dorsum of C57Bl/6 mice. After 10–14 days, after tumors had reached a size of 100 mm³, mice received i.v. injections of 10^9 P^fu of the control virus Ad Fc, or the antiangiogenic adenoviruses Ad Flk1-Fc (left panel) or Ad Flt(1–3) (right panel). Both viruses produce strong suppression of tumor growth over a 10- to 14-day period.

with a dependence of most, if not all, tumor vasculature on VEGF function, regardless of tumor type. Third, although tumor suppression can be obtained with either Ad FLK1-Fc or Ad Flt1(1–3) for >14 days in immnocompetent mice (i.e., C57Bl/6J), tumor growth eventually supervenes.

Why might tumor growth escape after adenoviral soluble VEGF receptor therapy? Theoretically, this could result from either a decrease in viral efficacy or an acquired resistance in the tumor and/or vasculature. The expression kinetics of either Ad Flk1-Fc or Ad Flt1(1–3) decrease substantially by 14 days postinjection, and it is thus conceivable that escape results from subtherapeutic transgene expression levels. In support of this possibility, we have observed prolonged expression from Ad Flk1-Fc in immunodeficient SCID mice, which are less able to mount an antiviral immune response. This longer duration of expression correlates with more prolonged suppression of human tumor xenografts, although this potential conclusion is compromised by the somewhat slower xenograft growth rate. Alternatively, acquired resistance to VEGF therapy could arise either in the tumor or the tumor vasculature. It is unproven and perhaps less likely that tumors would lose their dependence on angiogenesis. Our preliminary results indicate that repeat injections of Ad Flk1-Fc into tumor-bearing SCID mice may blunt tumor escape (B. Swearingen and C. J. Kuo, unpubl. observ.), arguing against acquired resistance in either tumor or vasculature and suggesting that vectors capable of longer lasting expression could be more efficacious.

It is notable that the ability to rapidly produce diverse adenoviruses greatly facilitates the comparative analysis of diverse antiangiogenic agents, a formidable task using a recombinant protein approach. Taking advantage of the ease with which adenoviruses can be constructed, we have also produced viruses encoding the angiogenesis inhibitors angiostatin and endostatin and a soluble form of the nonclassical VEGF receptor neuropilin. In our initial results, the Flk1-Fc and Flt1(1–3) viruses were more effective (>80% suppression) than angiostatin, endostatin, and neuropilin viruses (20–30% suppression) in both VEGF corneal micropocket assays and tumor growth suppression assays, suggesting the predictive nature of the former assay. These comparative data suggest that soluble VEGF receptors may be particularly amenable for use in systemic antiangiogenic gene therapy, while viral approaches with endostatin and angiostatin will require further optimization.

VIII. ISSUES REGARDING CLINICAL TRANSLATION OF ANTIANGIOGENIC GENE THERAPY

A. Safety of Vectors

A foremost concern regarding the use of antiangiogenic gene therapy in the oncology clinic is the safety of the viral vectors themselves. Current nonviral vectors, while poten-

tially less dangerous, are unlikely to achieve systemic levels of soluble VEGF receptors sufficient to inhibit significant tumor burdens. On the other hand, while first-generation adenoviruses lacking E1 and E3 can produce extremely robust and therapeutic levels, these viruses have been associated with significant toxicities, likely related to systemic adenoviral infection, as well as cytotoxicity from viral protein production in transduced cells (i.e., liver) [46]. Certainly, these toxicities would be substantially reduced if infection could be accomplished in a remote, although peripheral, site in which the graft could be physically removed or chemically inactivated, such as muscle, skin, or intratumoral. Our data, however, indicate that suppression of bulky preestablished tumors using current adenoviral vectors may require expression levels achievable only by transduction of a substantial tissue mass such as liver. Thus, while current adenoviral vectors may suffice for applications such as the suppression of metastasis by i.m. injection or the treatment of accessible disease by intratumoral injection, the substantial disease often seen in the oncology clinic may require robust but safer vectors delivered by systemic or portal infusion, such as gutless adenoviruses.

B. Safety of Transgene Products

Little published information currently exists regarding the safety of long-term VEGF inhibition in adult animals. VEGF-A knockout mice exhibit an embryonic lethal phenotype manifest even in heterozygotes, hampering the analysis of loss of function in the adult [16]. The administration of a Flt1 (1–3)–Fc fusion protein to neonatal mice has been reported to produce growth retardation with hepatic and renal dysfunction in a manner decreasing with age, as well as elevations of hematocrit and decreases in platelet counts [26]. Morerover, the Flt1(1–3)–Fc fusion protein has been reported to inhibit endochondral bone formation in 24-day-old mice [27] and corpus luteum development in rats [17]. In adult mice, prolonged administration of anti-Flk1 monoclonal antibodies has not been reported to induce histologic evidence of toxicity in any organs surveyed [53]. On the other hand, angiogenesis-dependent processes in the adult include wound healing and menses [19], and the blockade of angiogenesis would seem to be contraindicated in patients with cerebrovascular insufficiency, cardiac ischemia, or peripheral vascular disease. In our experience, although both Ad Flk1-Fc and Ad Flt1 (1–3) potently inhibit tumor growth over a 2-week period, Ad Flt(1–3) mice develope ascites with approximately 30% penetrance and exhibit frequent lethality after 22–28 days, while Ad Flk1-Fc mice are grossly asymptomatic for >1 year (C. J. Kuo, F. Farnebo, and E. Yu, unpublished data). Certainly, more evidence regarding the safety of both transient and prolonged VEGF blockade from phase I trials using small molecule inhibitors or monoclonal antibodies would be desirable prior to proceding with unregulated gene therapy approaches with current vectors.

C. Context of Translation

1. Control of Local Disease

Mutiple contexts can be envisioned in which VEGF-directed gene therapy could be used in the clinic, such as in the treatment of regional disease with local virus administration. Such a strategy could take the form of intratumoral delivery or intratracheal delivery for disease limited to the lung. This might have several safety advantages compared to systemic/intravenous viral administration, including decreased risk of side effects from systemic inhibition of angiogenesis and the ability to remove the graft by surgery or chemical inactivation.

The efficacy of local virus delivery, though, has perhaps not been as rigorously demonstrated as possible. While Crystal and colleages [40] have obtained locoregional activity by intratracheal or portal vein infusions of Ad sFlt, these have only been tested against 3-day-old tumor burdens that are not grossly visible and are much smaller than the typical preexisting disease seen in the oncology clinic. Intratumoral injections of self-replicating adenoviruses which selectively multiply in *p53*-deficient tumors [37] have demonstrated impressive efficacy against head and neck cancer. By analogy to this strategy, it should prove interesting to evaluate the efficacy of intratumoral injections of Ad Flk-Fc or Ad Flt, although it is possible that the nonreplicative nature of first-generation adenoviruses will hamper the ability to infect a sufficiently large percentage of the tumor mass to achieve local therapeutic levels. Because of the soluble nature of these VEGF receptors, finite although smaller risks exist for side effects of systemic angiogenesis inhibition from local vector administration. Nevertheless, because of safety factors, local administration for regional control of tumor angiogenesis appears to be a promising initial avenue for clinical translation.

2. Systemic Control of Distant Disease

Data from Takayama *et al.* [63] and from our group have also established the systemic efficacy of adenoviral delivery of soluble VEGF receptors. The intramuscular administration of Ad Flt1-Fc quite effectively inhibits the growth of tumor cells injected at the same time at a remote site. These data indicate that i.m. delivery, which would be anticipated to be safer than i.v. routes, can be efficacious against very small tumor burdens or perhaps in the prevention of metastasis. On the other hand, we have not found intramuscular administration of Ad Flt1(1–3) or Ad Flk1-Fc to be effective against pre-established day 10–14 tumor burdens of >100 mm^3, in contrast to i.v. injection, which produces >80% suppression. Consequently, therapeutic inhibition of distant, bulky, and preestablished disease may require the >2- to 3-log higher circulating levels achieved by transduction of large tissue masses during i.v. injection and accompanying transduction of >50% of hepatocytes.

The impressive systemic antiangiogenic and antitumor activity of i.v. administration are certainly accompanied by increased toxicity risks from both virus and transgene, as discussed above. In contrast, although potentially less toxic, intramuscular injections may be effective only in the treatment of very small tumor burdens or in prevention of metastasis. Possibly, reasonable strategies for clinical translation of systemic VEGF receptor treatment by gene therapy would be to use i.m. routes as adjuvant treatment for prevention of metastasis or i.v. routes for administering regulated, less toxic "gutless" adenoviruses against bulky tumor burdens.

3. Combination with Conventional Chemotherapy and Radiotherapy

A distinct use of VEGF-directed antiangiogenic gene therapy would be in the combination with conventional modalities such as chemotherapy or radiotherapy. Experimental evidence suggests that anti-KDR antibodies can produce additive-to-synergistic effects with either chemotherapy or radiotherapy against experimental tumors in mice. Our preliminary data in murine tumor models suggest that intravenous treatment with Ad Flk1-Fc in combination with radiotherapy results in impressive gains in tumor suppression and survival relative to either modality alone (F. Farnebo, K. Camphausen, and C. J. Kuo, unpubl. observ.). The limited duration of expression from conventional adenoviruses (i.e., 2–4 weeks) may well be suited to the duration of administration of conventional radiotherapy and chemotherapy. Clinical trials involving conventional radiotherapy or chemotherapy with or without administration of VEGF-directed antiangiogenic gene therapy could therefore be a reasonable option in patient populations in which it would not be ethical to withhold standard treatments.

D. Assessment of Response to Antiangiogenic Gene Therapy

1. Microvessel Density

The cytostatic nature of most antiangiogenic agents has several implications for the design of clinical trials, including the need for longer follow-up periods than traditional therapies [19] and the need to measure potential tumor response to treatment before a change in anatomical tumor volume can be detected. Indeed, given these characteristics, the ability to document functional rather than dimensional changes in tumor progression may be quite important during phase I trials in which overt changes in tumor size may only occur after prolonged treatment periods.

One biologic endpoint commonly employed for determination of efficacy of antiangiogenic agents is the microvessel density of tumor biopsy samples. Weidner and associates [71] established methodology by which tumor sections are stained with specific antibodies against endothelial antigens (vWF, CD31, CD34), and microvessel "hot spots" are counted by a microscope under 200-fold magnification. This microvessel density of histologic samples of breast cancer has been shown to correlate with patient survival [71], a finding that has been extended to various solid neoplasms (reviewed in Weidner, [70]). Other investigators have been less successful at establishing such a correlation [8,29,41], perhaps reflecting differences in quantitation techniques or from antibody cross-reactivity with other antigens such as on plasma cells (CD31) or perivascular stromal cells (CD34).

Microvessel density has been successfully applied to the assessment of biologic response to anti-KDR monoclonal antibodies and dominant-negative KDR, among others [45,53]. In our experience, antiangiogenic gene therapy with Ad Flk1-Fc (Fig. 5; see color insert) or Ad Flt1(1–3) (data not shown) results in decreased microvessel density as measured by vWF staining of tumors from treated versus mock-virus-infected controls. Overall, the use of histological microvessel count as a biologic endpoint to assess efficacy of VEGF-targeted therapy seems reasonable if performed by experienced investigators and well-defined protocols.

2. MRI Imaging of Tumor Blood Flow and Vascularity

While technically feasible, the repeated biopsy of a tumor for microvessel density determination may often be impractical because of the invasive nature of the sampling procedure, the requirement for easy tissue accessibility, and sampling errors due to tumor heterogeneity. To circumvent these shortcomings, noninvasive imaging techniques have been developed with potential application for detecting changes in tumor vasculature and blood flow in response to antiangiogenic agents .

Magnetic resonance imaging (MRI) is a cross-sectional imaging technique utilizing strong magnetic fields and multiple radiofrequency pulses to generate an image with outstanding spatial resolution and tissue contrast. Other than for the detection of tumor masses, MRI has recently been used to detect fluid motion such as blood flow. Furthermore, functional changes in various tissues can be evaluated with this method.

a. Functional Magnetic Resonance Imaging

By measuring the transverse relaxation time (T2) of nuclei activated by a radiofrequency pulse, functional changes in tissues can be measured. Functional MRI (FMRI) was first used to map regions of cortical brain activity [43,49], with activated and nonactivated regions having different imaging properties depending the oxygenation status of local hemoglobin due to differential metabolism. The same principles can be applied in fMRI of tumor tissue. Inhalation of a gas mixture of 95% oxygen and 5% carbon dioxide (carbogen) leads to changes in T2 signal as a result of local vasodilatation and a stimulatory effect on central respiratory regions of the brainstem. Vasodilatation and increased oxygenation of hemoglobin and tumor tissue can be monitored by MRI [1].

b. Dynamic Contrast-Enhanced Magnetic Resonance Imaging

In the 1980s, low-molecular-weight gadolinium-based contrast media for MRI were introduced which greatly facilitated the detection of neoplastic lesions due to their hypervascularity. This method, dynamic contrast-enhanced magnetic resonance imaging (DCE-MRI), is now a well-established adjunct to mammography and ultrasound in breast cancer

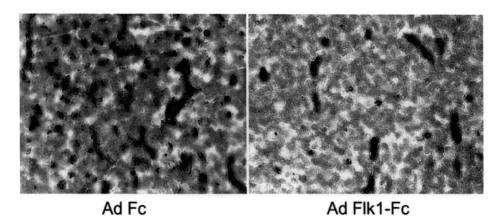

Ad Fc   Ad Flk1-Fc

FIGURE 5   Reduction in microvessel density in tumors treated with Ad Flk1–Fc. Mice bearing Lewis lung carcinoma tumors of approximately 100 mm³ received intravenous injections of 10⁹ plaque-forming units of the control virus Ad Fc or the anti-angiogenic adenoviruses Ad Flk1–Fc. Tumors were harvested for immunohistochemistry with anti-CD31 antibody after 4 days. The density of CD31 immunostaining is reduced in the tumor treated with Ad Flk1–Fc (right panel) relative to the Ad Fc control virus (left panel). (See color insert.)

detection [34] and has also been used to detect numerous neoplasms including hepatic [66] and brain [61].

Low-molecular-contrast agents are particularly useful in enhancing MRI detection of microvessels during the early postinjection phase. Unfortunately, gadolinium-based contrast agents exhibit a tendency to leak into the extravascular space, making it almost impossible to assess blood volume in the tumor microvasculature with low molecular contrast agents. This leak is presumably mediated by the vascular permeability activity of intratumoral VEGF, which prior to vessel sprouting induces a strong increase in transendothelial permeability and leakage of large molecules (e.g., fibrin, albumin) into the extravascular space. More importantly, gadolinium-based contrast agents permeate into the extravascular space of normal tissues, making differentiation of benign and malignant lesions problematic [11]. New macromolecular contrast agents with a higher probability of staying intravascular are under investigation in preclinical trials [6].

Nevertheless, by virtue of the permeability-modulating activity of VEGF, the therapeutic efficacy of VEGF antagonism could be indirectly measured by alterations in permeability for MRI contrast agents. For example, a single dose of anti-VEGF antibody decreases transendothelial leakage of a macromolecular contrast agent in a human breast cancer by 98% after 24 hours [6]. Additionally, new macromolecular contrast agents may greatly facilitate measure changes in perfusion and microvessel density in tumor tissues over the course of various antiangiogenic treatments.

The sensitivity of DCE-MRI may be increased by specific targeting of the contrast agents to the tumor vasculature. Sipkins *et al.* [59] coupled paramagnetic liposomes to antibodies directed against the $\alpha_V\beta_3$ integrin expressed in tumor vasculature. Using these paramagnetic liposome–antibody complexes in DCE-MRI, enhancement and distinct localization of tumor angiogenesis were observed in a rabbit carcinoma model with sensitivity exceeding conventional MRI. Certainly, this enhanced *in vivo* imaging method could be used in the evaluation of VEGF-targeted antiangiogenic therapy.

3. PET Imaging of Tumor Blood Flow and Vascularity

Few diagnostic fields in oncology have undergone as comparable an expansion as positron emission tomography (PET). In this imaging modality, radionucleotides with a short half-life are administered either intravenously or via inhalation. The unstable nucleus emits a proton that collides with an electron, creating energy in the form of two gamma rays traveling in opposite directions. Subsequently, sites of accumulated radioactivity can be detected with a camera rotating around the patient.

Position emission tomography is well adapted for the measurement of blood flow and volume and tumor metabolism. To assess increased blood flow in areas with disproportionate amounts of microvasculature such as tumors,

radioactive oxygen (^{15}O) in the form of $H_2^{15}O$ is injected intravenously. $H_2^{15}O$ has a half-life of approximately 2 minutes and is able to diffuse freely in various tissues. By measuring the rate of delivery to the tissue or organ, the extent of diffusion, and the speed of washout from the tissue, it is possible to calculate blood flow in milliliters per minute per gram of tissue and any alterations after antiangiogenic therapy.

Radioactive carbon monoxide (^{11}CO) can be used for blood volume measurements. Only minute amounts of the radioactive gas are needed for this procedure so that toxic effects are negligible. Radiolabeled red blood cells are detected depending on the blood volume in the specific tissue or organ, any background signals are subtracted by computer to enhance contrast, and the effects of antiangiogenic therapy on blood volume in the tumor tissue are calculated.

Another very elegant method used with PET is based on the fact that tumors metabolize glucose but often lack sufficient amounts of enzymes to metabolize the intermediate product glucose-6-PO_4. By radiolabeling glucose with radioactive fluorine (^{18}F) and injecting this compound intravenously (fluorodeoxyglucose, FDG) an accumulation of FDG-6-PO_4 in tumor tissue gives a strong signal in the PET scan. A correlation between high glucose uptake and microvessel density in human gliomas has recently been described [3]. Although FDG is not highly selective for tumor tissue uptake, the pyrimidine analogue (3'-deoxy-3'-fluorothymidine, Flt) appears to have superior discrimination [58].

4. Ultrasound Imaging of Tumor Blood Flow and Vascularity

Ultrasound is an inexpensive, nonisotopic imaging modality often used in oncology to verify pelvic or testicular masses or to clarify irregularities found by other imaging techniques. The development and improvement of color and power Doppler ultrasound permits real-time assessment of blood flow in various organs and tissues. Tumor vessels often lack multiple smooth muscle cell layers and are therefore easily distensible [9], resulting in increased diastolic flow and a low resistance index.

Doppler ultrasound is based on the technique that the Doppler signal can be color encoded, thereby allowing visualizion of not only blood flow but also velocity. A specific combination of speed and direction of blood flow is assigned a color designation. With conventional Doppler ultrasound, blood flow can be detected in vessels with a diameter of approximately 100 μm or more. The resulting ratio of colored pixels with the tumor section to the total number of pixels in that section is defined as the color Doppler vascularity index (CDVI). Chen *et al.* [10] used transabdominal Doppler ultrasound in patients with colon carcinoma to find a positive correlation between the CDVI and neovascularization to predict distant metastasis and survival. The authors hypothesized that by conventional Doppler ultrasound the CDVI indirectly accounts for intratumoral neovascularization as the number of capillaries is directly proportional to the detectable larger vessels. Additionally, intratumoral Doppler ultrasound signals can be amplified by injecting a perfluorocarbon-based contrast agent intravenously into tumor-bearing mice. These microbubbles increase signal intensity up to 10,000-fold and are small enough to pass freely through capillaries, allowing determination of CDVI based upon the actual microvascular network rather than only the larger supplying and draining vessels (C. Becker and G. Taylor, unpubl.). Further study will be required to evaluate the feasibility of ultrasound determination of vascularity given the need to standardize variables such as the pressure of application of the transducer or the timing of measurement after application of contrast.

IX. CONCLUSION

The use of VEGF-directed antiangiogenic gene therapy derives strong mechanistic rationale from abundant experimentation supporting the physiologic role of VEGF in angiogeneisis and the broad-spectrum antiangiogenic and antitumor activities of VEGF-or KDR/Flk-1-targeting monoclonal antibodies and small molecule kinase inhibitors. In its current state, VEGF-directed antiangiogenic gene therapy represents a powerful experimental tool to affirm the therapeutic potential of VEGF inhibition, in which vectors encoding soluble Flk1 or Flt1 ectodomains are easily propagated and conveniently administered. These characteristics have allowed our group and others to rapidly evaluate local and systemic inhibition of VEGF function, to perform comparative analyses, and to begin to assess combinations with conventional chemotherapy and radiotherapy.

At the same time, the experimental use of VEGF-directed antiangiogenic gene therapy should provide substantial preclinical information guiding eventual translation into the oncology clinic. Indeed, these viruses should greatly facilitate the study of toxicity of both viral vectors and their transgene products such as soluble VEGF receptors. Additionally, such studies will also allow validation of surrogate endpoints such as microvessel density or imaging correlates of tumor blood flow and vascularity. The inherent safety profile of first-generation adenoviral vectors may restrict their current utility to the adjuvant setting or in the treatment of minimal disease. However, in the future, different vector systems may allow robust and systemic suppression of diffuse bulky disease, fulfilling the promise of a long-term, single-injection, and economically advantageous antiangiogenic agent.

Acknowledgments

We thank George Taylor and Bruce Zetter for allowing us to cite unpublished observations. We are indebted to Cecile Chartier for helpful comments. This work was supported by CaPCURE, the Radley Family Foundation, Deutsche Forschungsgemeinschaft, HHMI, and the National Institutes of Health.

References

1. Abramovitch, R., Frenkiel, D., and Neeman, M. (1998). Analysis of subcutaneous angiogenesis by gradient echo magnetic resonance imaging. *Magn. Reson. Med.* **39**, 813–824.
2. Anderson, W. F. (1998). Human gene therapy. *Nature* **392**, 25–30.
3. Aronen, H. J., Pardo, F. S., Kennedy, D. N., Belliveau, J. W., Packard, S. D., Hsu, D. W., Hochberg, F. H., Fischman, A. J., and Rosen, B. R. (2000). High microvascular blood volume is associated with high glucose uptake and tumor angiogenesis in human gliomas. *Clin. Cancer Res.* **6**, 2189–2200.
4. Asahara, T., Masuda, H., Takahashi, T., Kalka, C., Pastore, C., Silver, M., Kearne, M., Magner, M., and Isner, J. M. (1999). Bone marrow origin of endothelial progenitor cells responsible for postnatal vasculogenesis in physiological and pathological neovascularization. *Circ. Res.* **85**, 221–228.
5. Boehm, T., Folkman, J., Browder, T., and O'Reilly, M. S. (1997). Antiangiogenic therapy of experimental cancer does not induce acquired drug resistance. *Nature* **390**, 404–407.
6. Brasch, R., and Turetschek, K. (2000). MRI characterization of tumors and grading angiogenesis using macromolecular contrast media: status report. *Eur. J. Radiol.* **34**, 148–155.
7. Carmeliet, P., and Collen, D. (2000). Molecular basis of angiogenesis. Role of VEGF and VE-cadherin. *Ann. N.Y. Acad. Sci.* **902**, 249–262. discussion 262–264.
8. Carnochan, P., Briggs, J. C., Westbury, G., and Davies, A. J. (1991). The vascularity of cutaneous melanoma: a quantitative histological study of lesions 0.85–1.25 mm in thickness. *Br. J. Cancer* **64**, 102–107.
9. Catellino, R. A. (1997). Imaging techniques in cancer management. In *Cancer: Principles and Practice of Oncology* (V. DeVita, ed.), pp. 633–689. Lippincott-Raven, New York.
10. Chen, C. N., Cheng, Y. M., Liang, J. T., Lee, P. H., Hsieh, F. J., Yuan, R. H., Wang, S. M., Chang, M. F., and Chang, K. J. (2000). Color Doppler vascularity index can predict distant metastasis and survival in colon cancer patients. *Cancer Res.* **60**, 2892–2897.
11. Daldrup, H., Shames, D. M., Wendland, M., Okuhata, Y., Link, T. M., Rosenau, W., Lu, Y., and Brasch, R. C. (1998). Correlation of dynamic contrast-enhanced MR imaging with histologic tumor grade: comparison of macromolecular and small-molecular contrast media. *Am. J. Roentgenol.* **171**, 941–949.
12. Dumont, D. J., Jussila, L., Taipale, J., Lymboussaki, A., Mustonen, T., Pajusola, K., Breitman, M., and Alitalo, K. (1998). Cardiovascular failure in mouse embryos deficient in VEGF receptor-3. *Science* **282**, 946–949.
13. Dvorak, H. F., Brown, L. F., Detmar, M., and Dvorak, A. M. (1995). Vascular permeability factor/vascular endothelial growth factor, microvascular hyperpermeability, and angiogenesis. *Am. J. Pathol.* **146**, 1029–1039.
14. Eliceiri, B. P., Paul, R., Schwartzberg, P. L., Hood, J. D., Leng, J., and Cheresh, D. A. (1999). Selective requirement for Src kinases during VEGF-induced angiogenesis and vascular permeability. *Mol. Cell.* **4**, 915–924.

15. Ferrara, N., and Alitalo, K. (1999). Clinical applications of angiogenic growth factors and their inhibitors. *Nat. Med.* **5,** 1359–1364.

16. Ferrara, N., Carver-Moore, K., Chen, H., Dowd, M., Lu, L., O'Shea, K. S., Powell-Braxton, L., Hillan, K. J., and Moore, M. W. (1996). Heterozygous embryonic lethality induced by targeted inactivation of the VEGF gene. *Nature* **380,** 439–442.

17. Ferrara, N., Chen, H., Davis-Smyth, T., Gerber, H. P., Nguyen, T. N., Peers, D., Chisholm, V., Hillan, K. J., and Schwall, R. H. (1998). Vascular endothelial growth factor is essential for corpus luteum angiogenesis. *Nat. Med.* **4,** 336–340.

18. Folkman, J. (1998). Antiangiogenic gene therapy. *Proc. Natl. Acad. Sci. USA* **95,** 9064–9066.

19. Folkman, J. (1995). Seminars in Medicine of the Beth Israel Hospital, Boston. Clinical applications of research on angiogenesis [see comments]. *N. Engl. J. Med.* **333,** 1757–1763.

20. Folkman, J., and D'Amore, P. A. (1996). Blood vessel formation: what is its molecular basis? [comment]. *Cell* **87,** 1153–1155.

21. Fong, G. H., Rossant, J., Gertsenstein, M., and Breitman, M. L. (1995). Role of the Flt-1 receptor tyrosine kinase in regulating the assembly of vascular endothelium. *Nature* **376,** 66–70.

22. Fong, G. H., Zhang, L., Bryce, D. M., and Peng, J. (1999). Increased hemangioblast commitment, not vascular disorganization, is the primary defect in Flt-1 knock-out mice. *Development* **126,** 3015–3025.

23. Fong, T. A., Shawver, L. K., Sun, L., Tang, C., App, H., Powell, T. J., Kim, Y. H., Schreck, R., Wang, X., Risau, W., Ullrich, A., Hirth, K. P., and McMahon, G. (1999). SU5416 is a potent and selective inhibitor of the vascular endothelial growth factor receptor (Flk-1/KDR) that inhibits tyrosine kinase catalysis, tumor vascularization, and growth of multiple tumor types. *Cancer Res.* **59,** 99–106.

24. Fukumura, D., Xavier, R., Sugiura, T., Chen, Y., Park, E. C., Lu, N., Selig, M., Nielsen, G., Taksir, T., Jain, R. K., and Seed, B. (1998). Tumor induction of VEGF promoter activity in stromal cells. *Cell* **94,** 715–725.

25. Gale, N. W., and Yancopoulos, G. D. (1999). Growth factors acting via endothelial cell-specific receptor tyrosine kinases: VEGFs, angiopoietins, and ephrins in vascular development. *Genes Dev.* **13,** 1055–1066.

26. Gerber, H. P., Hillan, K. J., Ryan, A. M., Kowalski, J., Keller, G. A., Rangell, L., Wright, B. D., Radtke, F., Aguet, M., and Ferrara, N. (1999). VEGF is required for growth and survival in neonatal mice. *Development* **126,** 1149–1159.

27. Gerber, H. P., Vu, T. H., Ryan, A. M., Kowalski, J., Werb, Z., and Ferrara, N. (1999). VEGF couples hypertrophic cartilage remodeling, ossification and angiogenesis during endochondral bone formation [see comments]. *Nat. Med.* **5,** 623–628.

28. Goldman, C. K., Kendall, R. L., Cabrera, G., Soroceanu, L., Heike, Y., Gillespie, G. Y., Siegal, G. P., Mao, X., Bett, A. J., Huckle, W. R., Thomas, K. A., and Curiel, D. T. (1998). Paracrine expression of a native soluble vascular endothelial growth factor receptor inhibits tumor growth, metastasis, and mortality rate. *Proc. Natl. Acad. Sci. USA* **95,** 8795–8800.

29. Hall, N. R., Fish, D. E., Hunt, N., Goldin, R. D., Guillou, P. J., and Monson, J. R. (1992). Is the relationship between angiogenesis and metastasis in breast cancer real? *Surg. Oncol.* **1,** 223–229.

30. Hanahan, D., and Folkman, J. (1996). Patterns and emerging mechanisms of the angiogenic switch during tumorigenesis. *Cell* **86,** 353–364.

31. Holash, J., Maisonpierre, P. C., Compton, D., Boland, P., Alexander, C. R., Zagzag, D., Yancopoulos, G. D., and Wiegand, S. J. (1999). Vessel cooption, regression, and growth in tumors mediated by angiopoietins and VEGF. *Science* **284,** 1994–1998.

32. Houck, K. A., Leung, D. W., Rowland, A. M., Winer, J., and Ferrara, N. (1992). Dual regulation of vascular endothelial growth factor bioavailability by genetic and proteolytic mechanisms. *J. Biol. Chem.* **267,** 26031–26037.

33. Joseph, I. B., and Isaacs, J. T. (1997). Potentiation of the antiangiogenic ability of linomide by androgen ablation involves down-regulation of vascular endothelial growth factor in human androgen-responsive prostatic cancers. *Cancer Res.* **57,** 1054–1057.

34. Kaiser, W. A., and Zeitler, E. (1989). MR imaging of the breast: fast imaging sequences with and without Gd-DTPA. Preliminary observations. *Radiology* **170,** 681–686.

35. Kendall, R. L., Wang, G., and Thomas, K. A. (1996). Identification of a natural soluble form of the vascular endothelial growth factor receptor, Flt-1, and its heterodimerization with KDR. *Biochem. Biophys. Res. Commun.* **226,** 324–328.

36. Keyt, B. A., Nguyen, H. V., Berleau, L. T., Duarte, C. M., Park, J., Chen, H., and Ferrara, N. (1996). Identification of vascular endothelial growth factor determinants for binding KDR and Flt-1 receptors. Generation of receptor-selective VEGF variants by site-directed mutagenesis. *J. Biol. Chem.* **271,** 5638–5646.

37. Khuri, F. R., Nemunaitis, J., Ganly, I., Arseneau, J., Tannock, I. F., Romel, L., Gore, M., Ironside, J., MacDougall, R. H., Heise, C., Randlev, B., Gillenwater, A. M., Bruso, P., Kaye, S. B., Hong, W. K., and Kirn, D. H. (2000). A controlled trial of intratumoral ONYX-015, a selectively-replicating adenovirus, in combination with cisplatin and 5-fluorouracil in patients with recurrent head and neck cancer. *Nat. Med.* **6,** 879–885.

38. Kim, K. J., Li, B., Winer, J., Armanini, M., Gillett, N., Phillips, H. S., and Ferrara, N. (1993). Inhibition of vascular endothelial growth factor-induced angiogenesis suppresses tumour growth in vivo. *Nature* **362,** 841–844.

39. Kong, H. L., and Crystal, R. G. (1998). Gene therapy strategies for tumor antiangiogenesis [see comments]. *J. Natl. Cancer. Inst.* **90,** 273–286.

40. Kong, H. L., Hecht, D., Song, W., Kovesdi, I., Hackett, N. R., Yayon, A., and Crystal, R. G. (1998). Regional suppression of tumor growth by in vivo transfer of a cDNA encoding a secreted form of the extracellular domain of the Flt-1 vascular endothelial growth factor receptor. *Hum. Gene Ther.* **9,** 823–833.

41. Leedy, D. A., Trune, D. R., Kronz, J. D., Weidner, N., and Cohen, J. I. (1994). Tumor angiogenesis, the p53 antigen, and cervical metastasis in squamous carcinoma of the tongue. *Otolaryngol. Head Neck Surg.* **111,** 417–422.

42. Levy, N. S., Chung, S., Furneaux, H., and Levy, A. P. (1998). Hypoxic stabilization of vascular endothelial growth factor mRNA by the RNA-binding protein HuR. *J. Biol. Chem.* **273,** 6417–6423.

43. Libutti, S. K., Choyke, P., Carrasquillo, J. A., Bacharach, S., and Neumann, R. D. (1999). Monitoring responses to antiangiogenic agents using noninvasive imaging tests. *Cancer J. Sci. Am.* **5,** 252–256.

44. Millauer, B., Longhi, M. P., Plate, K. H., Shawver, L. K., Risau, W., Ullrich, A., and Strawn, L. M. (1996). Dominant-negative inhibition of Flk-1 suppresses the growth of many tumor types in vivo. *Cancer Res.* **56,** 1615–1620.

45. Millauer, B., Shawver, L. K., Plate, K. H., Risau, W., and Ullrich, A. (1994). Glioblastoma growth inhibited in vivo by a dominant-negative Flk-1 mutant. *Nature* **367,** 576–579.

46. Mountain, A. (2000). Gene therapy: the first decade. *Trends Biotechnol.* **18,** 119–128.

47. O'Reilly, M. S., Boehm, T., Shing, Y., Fukai, N., Vasios, G., Lane, W. S., Flynn, E., Birkhead, J. R., Olsen, B. R., and Folkman, J. (1997). Endostatin: an endogenous inhibitor of angiogenesis and tumor growth. *Cell* **88,** 277–285.

48. O'Reilly, M. S., Holmgren, L., Shing, Y., Chen, C., Rosenthal, R. A., Moses, M., Lane, W. S., Cao, Y., Sage, E. H., and Folkman, J. (1994). Angiostatin: a novel angiogenesis inhibitor that mediates the suppression of metastases by a Lewis lung carcinoma. *Cell* **79,** 315–328.

49. Ogawa, S., Lee, T. M., Nayak, A. S., and Glynn, P. (1990). Oxygenation-sensitive contrast in magnetic resonance image of rodent brain at high magnetic fields. *Magn. Reson. Med.* **14,** 68–78.

50. Olofsson, B., Pajusola, K., Kaipainen, A., von Euler, G., Joukov, V., Saksela, O., Orpana, A., Pettersson, R. F., Alitalo, K., and Eriksson, U. (1996). Vascular endothelial growth factor B, a novel growth factor for endothelial cells. *Proc. Natl. Acad. Sci. USA* **93,** 2576–2581.

51. Paley, P. J., Staskus, K. A., Gebhard, K., Mohanraj, D., Twiggs, L. B., Carson, L. F., and Ramakrishnan, S. (1997). Vascular endothelial growth factor expression in early stage ovarian carcinoma. *Cancer* **80,** 98–106.

52. Pepper, M. S., and Montesano, R. (1990). Proteolytic balance and capillary morphogenesis. *Cell. Differ. Dev.* **32,** 319–327.

53. Prewett, M., Huber, J., Li, Y., Santiago, A., O'Connor, W., King, K., Overholser, J., Hooper, A., Pytowski, B., Witte, L., Bohlen, P., and Hicklin, D. J. (1999). Antivascular endothelial growth factor receptor (fetal liver kinase 1) monoclonal antibody inhibits tumor angiogenesis and growth of several mouse and human tumors. *Cancer Res.* **59,** 5209–5218.

54. Rifkin, D. B., Moscatelli, D., Bizik, J., Quarto, N., Blei, F., Dennis, P., Flaumenhaft, R., and Mignatti, P. (1990). Growth factor control of extracellular proteolysis. *Cell. Differ. Dev.* **32,** 313–318.

55. Salven, P., Lymboussaki, A., Heikkila, P., Jaaskela-Saari, H., Enholm, B., Aase, K., von Euler, G., Eriksson, U., Alitalo, K., and Joensuu, H. (1998). Vascular endothelial growth factors VEGF-B and VEGF-C are expressed in human tumors. *Am. J. Pathol.* **153,** 103–108.

56. Scott, P. A., Gleadle, J. M., Bicknell, R., and Harris, A. L. (1998). Role of the hypoxia sensing system, acidity and reproductive hormones in the variability of vascular endothelial growth factor induction in human breast carcinoma cell lines. *Int. J. Cancer* **75,** 706–712.

57. Shalaby, F., Rossant, J., Yamaguchi, T. P., Gertsenstein, M., Wu, X. F., Breitman, M. L., and Schuh, A. C. (1995). Failure of blood-island formation and vasculogenesis in Flk-1-deficient mice. *Nature* **376,** 62–66.

58. Shields, A. F., Grierson, J. R., Dohmen, B. M., Machulla, H. J., Stayanoff, J. C., Lawhorn-Crews, J. M., Obradovich, J. E., Muzik, O., and Mangner, T. J. (1998). Imaging proliferation in vivo with [F-18]FLT and positron emission tomography. *Nat. Med.* **4,** 1334–1336.

59. Sipkins, D. A., Cheresh, D. A., Kazemi, M. R., Nevin, L. M., Bednarski, M. D., and Li, K. C. (1998). Detection of tumor angiogenesis in vivo by alpha$_v$beta$_3$–targeted magnetic resonance imaging. *Nat. Med.* **4,** 623–626.

60. Soker, S., Takashima, S., Miao, H. Q., Neufeld, G., and Klagsbrun, M. (1998). Neuropilin-1 is expressed by endothelial and tumor cells as an isoform-specific receptor for vascular endothelial growth factor. *Cell* **92,** 735–745.

61. Sze, G., Shin, J., Krol, G., Johnson, C., Liu, D., and Deck, M. D. (1988). Intraparenchymal brain metastases: MR imaging versus contrast-enhanced CT. *Radiology* **168,** 187–194.

62. Takahashi, Y., Bucana, C. D., Cleary, K. R., and Ellis, L. M. (1998). p53, vessel count, and vascular endothelial growth factor expression in human colon cancer. *Int. J. Cancer* **79,** 34–38.

63. Takayama, K., Ueno, H., Nakanishi, Y., Sakamoto, T., Inoue, K., Shimizu, K., Oohashi, H., and Hara, N. (2000). Suppression of tumor angiogenesis and growth by gene transfer of a soluble form of vascular endothelial growth factor receptor into a remote organ. *Cancer Res.* **60,** 2169–2177.

64. Tolnay, E., Kuhnen, C., Wiethege, T., Konig, J. E., Voss, B., and Muller, K. M. (1998). Hepatocyte growth factor/scatter factor and its receptor c-Met are overexpressed and associated with an increased microvessel density in malignant pleural mesothelioma. *J. Cancer Res. Clin. Oncol.* **124,** 291–296.

65. Tsurusaki, T., Kanda, S., Sakai, H., Kanetake, H., Saito, Y., Alitalo, K., and Koji, T. (1999). Vascular endothelial growth factor-C expression in human prostatic carcinoma and its relationship to lymph node metastasis. *Br. J. Cancer* **80,** 309–313.

66. Unger, E. C., Winokur, T., MacDougall, P., Rosenblum, J., Clair, M., Gatenby, R., and Tilcock, C. (1989). Hepatic metastases: liposomal Gd-DTPA-enhanced MR imaging. *Radiology* **171,** 81–85.

67. Veikkola, T., and Alitalo, K. (1999). VEGFs, receptors and angiogenesis. *Semin. Cancer Biol.* **9,** 211–220.

68. Waltenberger, J., Claesson-Welsh, L., Siegbahn, A., Shibuya, M., and Heldin, C. H. (1994). Different signal transduction properties of KDR and Flt1, two receptors for vascular endothelial growth factor. *J. Biol. Chem.* **269,** 26988–26995.

69. Wedge, S. R., Ogilvie, D. J., Dukes, M., Kendrew, J., Curwen, J. O., Hennequin, L. F., Thomas, A. P., Stokes, E. S., Curry, B., Richmond, G. H., and Wadsworth, P. F. (2000). ZD4190: an orally active inhibitor of vascular endothelial growth factor signaling with broad-spectrum antitumor efficacy. *Cancer Res.* **60,** 970–975.

70. Weidner, N. (1995). Intratumor microvessel density as a prognostic factor in cancer. *Am. J. Pathol.* **147,** 9–19.

71. Weidner, N., Semple, J. P., Welch, W. R., and Folkman, J. (1991). Tumor angiogenesis and metastasis—correlation in invasive breast carcinoma. *N. Engl. J. Med.* **324,** 1–8.

72. Yamada, Y., Nezu, J., Shimane, M., and Hirata, Y. (1997). Molecular cloning of a novel vascular endothelial growth factor, VEGF-D. *Genomics* **42,** 483–488.

73. Zucker, S., Mirza, H., Conner, C. E., Lorenz, A. F., Drews, M. H., Bahou, W. F., and Jesty, J. (1998). Vascular endothelial growth factor induces tissue factor and matrix metalloproteinase production in endothelial cells: conversion of prothrombin to thrombin results in progelatinase A activation and cell proliferation. *Int. J. Cancer* **75,** 780–786.

28

Strategies for Combining Gene Therapy with Ionizing Radiation to Improve Antitumor Efficacy

DAVID H. GORSKI

The Cancer Institute of New Jersey
UMDNJ—Robert Wood Johnson
Medical School
New Brunswick, New Jersey 08901

HELENA J. MAUCERI

Department of Radiation and Cellular Oncology
University of Chicago Hospitals
Chicago, Illinois 60637

RALPH R. WEICHSELBAUM

Department of Radiation and Cellular Oncology
University of Chicago Hospitals
Chicago, Illinois 60637

I. INTRODUCTION

The utility of gene therapy in the treatment of cancer results from its ability to deliver therapeutic genes to tumor cells in order to alter the malignant phenotype or to induce tumor cell cytotoxicity. Ionizing radiation (IR) is a conventional and effective local treatment for many different tumors. Unfortunately, many human tumors remain refractory to treatment with IR. Several gene products (p53 and p21, for example) that have been proposed for gene therapy approaches to treat cancer are also involved in determining tumor cell sensitivity or resistance to IR, making the concept of combining anticancer gene therapy with IR an attractive one. Strategies that involve employing gene therapy to improve the antitumor effect of IR fall into two general categories: (1) the use of gene therapy vectors to deliver genes whose protein products improve the antitumor effect of IR, and (2) the use of radiation to enhance the antitumor effect of replication-competent viruses such as herpes virus. The first strategy generally involves the use of gene products that, when expressed within the radiation field, result in either radiosensitization or improved antitumor effects compared with IR or gene therapy alone. The second strategy relies on the observation that some viruses are capable of replicating preferentially in tumor cells, thus killing them, and that radiation enhances this viral proliferation. Finally, the variant of the first strategy, which we term "genetic radiotherapy," relies on the existence of promoters whose activity is inducible by IR. When introduced into a tumor to be irradiated, administration of IR results in the enhanced production of a toxic gene product. This approach thus provides both temporal and spatial targeting of the toxic gene product, as well as the possibility of additive or even synergistic effects between the gene product and IR. In this chapter, we discuss these approaches to using gene therapy to improve the efficacy of IR.

II. STRATEGIES USING GENE THERAPY TO INCREASE THE EFFICACY OF RADIATION THERAPY

A. Introduction

Generally, viruses are the method of choice in most gene therapy applications because they are capable of delivering genes of interest to the largest number of cells most efficiently and drive production of the desired gene product. In most of these strategies, gene transfer to a target tissue or tumor is accomplished using replication-incompetent viral shuttle vectors such as retroviruses or replication-deficient adenoviruses. Genes inserted into these vectors can include immunomodulatory cytokines (e.g., interleukins such as IL-6, IL-12, IL-13) that recruit immune cells to the site of the tumor [1–5]; tumor suppressor genes that slow tumor growth, reverse the activity of mutant tumor suppressor genes, or induce tumor cell apoptosis [6–8]; or prodrug converting enzymes (e.g., herpes simplex virus thymidine kinase, cytosine deaminase) that result in the metabolism of intravenously administered nontoxic prodrugs to cytotoxic compounds within the target tissue [9–11]. More recently, genes encoding antiangiogenic peptides have been inserted into viral vectors and used to block tumor-associated angiogenesis in experimental models [12–16]. Gene therapy strategies in which such genes are expressed in tumor cells and that show promise of increasing the effectiveness of radiation therapy through either radiosensitization or synergistic antitumor effects are discussed below.

B. p53 Gene-Transfer-Mediated Radiosensitization

Among the tumor suppressors being considered for gene replacement therapy in cancer, the p53 protein is especially attractive, because it plays a critical role in regulating the cellular response to DNA damage [17]. When DNA is damaged by an agent such as IR or cytotoxic chemotherapy, p53, a transcription factor capable of activating multiple downstream genes [17,18], mediates several processes critical to preventing the propagation of the DNA damage when the cell divides: (1) G_1 cell-cycle arrest through the induction of the cyclin-dependent kinase p21, which allows the cell time to repair its DNA before entering the cell cycle [17]; (2) DNA repair and synthesis through the activation of the growth arrest and DNA damage-dependent (GADD) genes and proliferating cell nuclear antigen [19,20]; and (3) inducing apoptosis in cells whose DNA is too damaged to be successfully repaired [21] (Fig. 1). Mutations in p53 are the most common genetic alterations in tumors, with p53 mutations or deletions being present in over 50% of all human cancers. Loss of or mutations in p53 significantly contribute to tumor development, progression, and chemotherapy resistance [22,23]. Indeed,

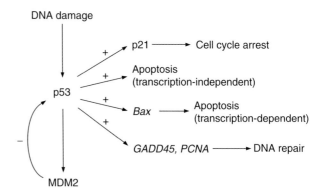

FIGURE 1   p53-mediated responses to DNA damage. p53 activates multiple processes in response to various stimuli, especially DNA damage. It is also downregulated by mouse double minute-2 (MDM2), which accelerates its degradation.

a survey of the toxicity of hundreds of anticancer drugs indicates that the vast majority of clinically useful drugs are most effective in cells that express wild-type p53 [24]. Moreover, inhibition of p53 action is associated with radioresistance [25]. Thus, the rationale for transducing tumor cells with wild-type p53 depends on the observation that apoptosis induced by chemotherapy and IR is at least partially dependent on the expression of wild-type p53 and that mutated p53 is associated with radioresistance [26,27]. In theory, at least, expressing wild-type p53 in tumor cells that either lack it or express mutant p53 represents a rational strategy for overcoming tumor cell resistance to IR.

Ionizing radiation activates p53 through posttranslational modifications, including the phosphorylation of N-terminal serine residues in the transactivating domain of p53 and C-terminal modifications such as lysine acetylation that increase transcriptional activation of downstream mediators by p53. It is not clear whether p53 status correlates with radiosensitivity. For instance, it has been shown that thymocytes from mice transgenic for a p53 null mutation were radioresistant, whereas thymocytes from wild-type mice were radiosensitive [28]. The clear implication was that radiation-induced apoptosis depends on the presence of functional p53. Other studies link radiosensitivity to p53 status in normal and transformed fibroblasts [29] and cells of hematopoietic origin [30]. For instance, transfection of the human papillomavirus 16 (HPV-16) E6 gene, which binds and inactivates p53, into human diploid fibroblasts rendered these cells more resistant to irradiation, presumably due to loss of p53 [29]. In contrast, other studies failed to find a conclusive correlation between p53 function and radiosensitivity in tumor cell lines [31–34].

Several studies have addressed the question of whether p53 gene therapy can improve the efficacy of IR in experimental tumor models. Most reports have focused on enhancing the radiosensitivity of tumor cells, although the results appear to depend upon the specific cell type. p53 has been transduced into a homozygous mutant p53 ovarian tumor cell line. Tumor xenograft data showed that 45% of mice treated with IR and a replication-deficient adenovirus encoding p53 (Ad-p53) had long-term cures, compared with the mice treated with IR or Ad-p53 alone [35]. In p53$^{-/-}$ human colorectal tumor cell lines, delivery of Ad-p53 did not result in a significant increase in apoptosis in response to IR, but Ad-p53-infected cells underwent significantly more apoptosis than control cells. Tumor cell apoptosis was further enhanced by IR [36,37]. In the same study, tumor xenografts injected with Ad-p53 and then treated with a single 5-Gy dose of radiation demonstrated a significant increase in apoptotic cells and tumor growth delay [36]. Similar evidence supporting the potential benefit of combining IR and p53 gene therapy comes from studies utilizing head and neck squamous cell xenografts [38] and glioma cell lines and xenografts [39].

The presence of p53 is not always sufficient or necessary for radiosensitization [32]. Some human colorectal adenoma and carcinoma cell lines lacking wild-type p53 still undergo apoptosis in response to IR [31]. In addition, introduction of the HPV-16 E6 gene, which binds and inactivates p53, into several tumor cell lines did not increase their radioresistance [33], in contrast to the results seen with human diploid fibroblasts, where E6 was observed to increase radioresistance [29]. Moreover, replacement of wild-type p53 is not always sufficient to reverse the cellular defects caused by the presence of a mutant p53 because certain mutated forms of p53 appear to act in a dominant fashion [40]. The role of p53 in modulating radiosensitivity and radioresistance is likely a cell-type-specific phenomenon, making it necessary to establish whether exogenous p53 alters radiosensitivity for each tumor type before using p53 gene therapy as a radiosensitizer.

C. p21 Gene Therapy and Ionizing Radiation

The cyclin-dependent kinase inhibitor p21 is an immediate downstream target of p53 and is responsible for the p53-dependent checkpoint that results in G_1 arrest after DNA damage [41–43]. As such, it has been examined as a potential gene therapy target in conjunction with IR. In contrast to p53, p21 causes much less apoptosis following introduction into many cell lines, including head and neck cancer [6], lung cancer [44], prostate cancer [45], gliomas [46], and melanomas [47]. Although p21 is not as strongly proapoptotic as p53, its overexpression has been shown to promote radiosensitivity in a glioma tumor model [48]. Similarly, p21 overexpression has also been shown to promote chemosensitivity in tumor cells [49]. One hypothesis to explain these observations is that, because p21 is important in causing cell-cycle arrest in response to DNA damage, loss of p21 may cause a deficiency in repair leading to chemosensitivity or radiosensitivity [50]. However, this has not been a universal finding, and more

recent studies have contradicted these results and postulated a protective role against radiation damage. For instance, in a colon cancer cell line (HCT116), it was found that a lack of p21 expression produced an increase in apoptosis in vitro but no decrease in clonogenic survival and no radiosensitization. However, in HCT116 xenografts, loss of p21 led to an increased sensitivity to killing by IR that was independent of induction of cell-cycle arrest and apoptosis. More interestingly, this effect is specific to cells growing as a tumor and is not observed in vitro, implying that the tumor microenvironment likely influences whether p21 affects the radiosensitivity of a tumor cell [51]. Consistent with this observation is a recent study reporting that p21 antisense therapy sensitizes a colon carcinoma cell line to IR [52]; therefore, it is possible that the role of p21 (radiation sensitizer or protector) may depend upon cell type or the specific genetic derangements present in each tumor type that effect downstream or upstream effectors of p21 leading to cell cycle arrest and/or apoptosis.

D. Prodrug Converting Enzyme Suicide Gene Therapy Radiosensitizes Tumor Cells

1. Introduction

The treatment of cancer is different from treatment of genetic diseases because effective antitumor therapy requires the complete eradication of all tumor cells. One obstacle that gene therapy must overcome, therefore, is the requirement that the therapeutic gene be introduced into every tumor cell. One strategy to overcome this problem is to use prodrug converting enzymes, which rely on the transfer of non-mammalian genes encoding enzymes to convert nontoxic, systemically administered prodrugs to toxic antimetabolites [53,54]. These strategies aim to increase intratumoral concentration of the toxic metabolite in order to kill tumor cells. When used in conjunction with IR, the enzymes used are selected in order to generate drugs that are radiosensitizers. The two most commonly used prodrug converting enzyme/prodrug strategies that have been used in conjunction with IR include herpes simplex virus thymidine kinase (HSV-TK)/ganciclovir (GCV) and cytosine deaminase (CD)/5-fluorocytosine (5-FC).

2. HSV-TK Radiosensitization

HSV-TK phosphorylates the nucleoside analogs (E)-5-(2-bromovinyl)-2′-deoxyuridine (BvdUrd), acyclovir, and GCV to toxic antimetabolites. This reaction is the basis for the effectiveness of acyclovir or GCV in the treatment of HSV infections. Monophosphorylated forms of acyclovir and GCV are then phosphorylated to nucleotide triphosphates by cellular kinases. These aberrant nucleotide triphosphates disrupt DNA replication at the level of DNA chain elongation by interfering with DNA polymerase α. The effects of phosphorylated BvdUrd are caused by its inhibition of thymidylate synthase, which results in a depletion of thymidine pools within the cell [55]. Based on this knowledge, studies have been performed to demonstrate the efficacy of transferring HSV-TK to tumor cells with subsequent systemic administration of GCV. Only cells transfected with the HSV-TK gene convert GCV to its toxic phosphorylated form, resulting in tumor cell death. Several studies have demonstrated the efficacy of such an antitumor approach [10,56–58].

Radiosensitization combining antiviral nucleoside analogs with IR interferes with potential lethal damage repair or the modification of DNA to a more radiosensitive form [59]. Studies with 5-bromodeoxyuridine (BrdU) and acyclovir combined with IR have demonstrated a radiosensitizing effect in experimental systems, but with doses of IR or acyclovir too high to be clinically applicable [60,61]. Several studies have shown that HSV-TK/prodrug treatment of tumor cells results in radiosensitization and enhanced tumor regression in experimental tumor xenografts. These effects have been demonstrated in glioma cells transduced with HSV-TK followed by acyclovir administration and IR. Cells transfected retrovirally with HSV-TK followed by administration of either BvdUrd or acyclovir and IR resulted in a radiosensitizing effect, with a sensitizing enhancement ration of 1.3–1.6 [62,63].

BvdUrd and acyclovir are hypothesized to radiosensitize cells by different mechanisms. BvdUrd radiosensitizes HSV-TK glial cells only when it is administered prior to IR administration. This is necessary because the phosphorylated form of BvdUrd inactivates thymidylate synthase, resulting in depleted intracellular thymidine pools within the cell. In contrast, acyclovir radiosensitizes cells if administered before or after IR. A potential mechanism to account for this observation is that acyclovir may radiosensitize because its metabolite is incorporated in to DNA before IR, thus making the DNA more susceptible to IR. Alternatively, its metabolite may inhibit repair of DNA damage, thus increasing the toxicity of IR to the cell. The efficacy of this strategy has been demonstrated in animal tumor models. In one study, glial cells were infected with HSV-TK in cell culture and then implanted in rats. Systemic administration of prodrug and a single 20-Gy dose of IR resulted in a threefold increase in rat survival compared with IR alone [64].

3. Cytosine Deaminase Radiosensitization

A second strategy for using a prodrug converting enzyme for radiosensitization involves cytosine deaminase (CD) [65], an enzyme found in many bacteria and fungi that catalyzes the deamination of cytosine to uracil, providing uracil for the organism in time of nutritional stress [65]. It is employed to convert the nontoxic drug 5-FC to the antitumor drug 5-fluorouracil (5-FU) [9, 66–69]. 5-FU and its metabolites kill tumor cells by interfering with both DNA and RNA metabolism through incorporation in nucleic acids and through their inhibition of thymidylate synthase and depletion of the cellular TTP pool [70,71]. 5-FU has activity against some solid tumors and is a mainstay of adjuvant therapy of colorectal cancer. It is also used as a radiosensitizer to treat a variety of human tumors [72,73]. The mechanism of radiosensitization by CD/5-FU appears to be inhibition of DNA repair due to inhibition of thymidylate synthase by the monophosphorylated form of 5-FU, as its radiosensitizing effects can be abrogated by exogenously administered thymidine [70,71,74]. Also, triphosphorylated 5-FU is incorporated into RNA, disrupting protein translation [70,71].

Because systemically administered 5-FU has dose-limiting toxicities of mucositis, diarrhea, and myelosuppression, attempts have been made to use gene therapy with CD/5-FC to produce high intratumoral concentrations of 5-FU, thus providing the benefit of its radiosensitization effect in the tumor bed but sparing patients the systemic toxicities associated with 5-FU administration. Also, 5-FU is diffusible, which would allow CD-transduced tumor cells to convert 5-FC to 5-FU, which could then diffuse into surrounding untransduced tumor cells. Several studies have combined CD/5-FC therapy with IR to enhance tumor cell killing. For example, in cell culture, the transduction of human colorectal tumor cells with a retrovirus encoding CD followed by 5-FC and irradiation produced markedly increased tumor cell killing, although the 5-FC had to be administered at least 24 hours prior to IR [67]. Pederson et al. [75,76] treated cholangiocarcinoma cells with CD delivered by a replication-deficient adenovirus and IR and demonstrated specific radiosensitization with CD/5-FC. Next, in xenograft models of colon cancer and cholangiocarcinoma, adenovirus-delivered CD plus 5-FC resulted in improved tumor growth delay when these xenografts were treated with IR [77]. Hanna et al. [78] also demonstrated similar results treating human squamous cell carcinoma xenografts grown in athymic nude mice and treated with intratumoral injections of a replication-defective adenovirus expressing CD. Xenografts treated with Ad.CD/5-FC and IR showed a significant tumor growth delay compared with IR or Ad.CD/5-FC alone.

Because tumors consist of a heterogeneous population of cells, it is likely that certain subclones of cells will be more resistant to either HSV-TK/GCV or CD/5-FC. Rogulski et al. [79] have constructed a bifunctional fusion gene expressing both HSV-TK and CD. The CD gene was fused to HSV-TK through a polyglycine linker to allow for proper folding of both prodrug converting enzymes, producing the CDglyHSV–TK construct. Transducing gliosarcoma cells with a retrovirus expressing this fusion protein, then treating with 5-FC and BvdU followed by IR resulted in two- to threefold greater cell killing than would be expected if the two prodrugs interacted with IR in an additive fashion. These results were confirmed in vivo in human tumor xenograft models [80,81]. Because it will be necessary to transfer antitumor genes to a large percentage of tumor cells to achieve clinically relevant antitumor effects in humans, Freytag *et al.* [82] placed the CDglyHSV-TK construct into a replication-conditional adenovirus, ONYX-015. Because preliminary data show that only 20–30% of tumor cells within a xenograft show infection by ONYX-015, further work must validate that this viral construct combined with both 5-FC/GCV and IR results in synergistic control of tumor xenografts due to virus replication and transduction of a large percentage of the tumor [82].

E. Enhancement of the Cytotoxic Effects of Ionizing Radiation by Antiangiogenic Gene Therapy

1. Angiogenesis and Antiangiogenesis

Tumors are critically dependent upon inducing the ingrowth of blood vessels from the host to supply their oxygen and nutrient needs [83]. To this end they secrete proangiogenic factors such as basic fibroblast growth factor (bFGF) [84] and vascular endothelial growth factor (VEGF) [85]. Consequently, inhibition of angiogenesis, either by blocking these proangiogenic factors or treatment with antiangiogenic factors has emerged as a promising strategy to treat primary and metastatic tumors [86–89]. Strategies to block the activity of proangiogenic factors include the administration of neutralizing antibodies to proangiogenic cytokines, such as VEGF [86,90–93] or bFGF [94–96]; antisense against VEGF [97,98] or bFGF [99]; and the engineering and expression of soluble receptors that bind to VEGF and inactivate it [92,100]. It has also become apparent that tumors can induce the production of antiangiogenic peptides that directly inhibit vascular endothelial cell proliferation and angiogenesis. The most potent and specific of these are angiostatin, a proteolytic fragment of plasminogen containing its first four kringle domains [89,101,102], and endostatin, a proteolytic fragment of collagen XVIII [88,103]. Finally, a number of smaller molecules are under study that act with varying degrees of specificity on endothelial cells to block angiogenesis. Among these are drugs such as VEGF receptor tyrosine kinase inhibitors [104,105] and TNP-1470 [106,107]. All of these strategies target tumor endothelium and disrupt angiogenesis.

2. Antiangiogenic Therapy Potentiates the Antitumor Activity of Ionizing Radiation

Antiangiogenic proteins, although effective at shrinking tumors, are not tumoricidal. Tumor regrowth frequently occurs once treatment with the angiogenesis inhibitor is terminated [87,108], although there is evidence that antiangiogenic therapy can be used to induce tumor dormancy

[87,108]. IR is a major cytotoxic therapeutic modality that is primarily effective in the treatment of relatively small tumors while large tumors respond only with considerable toxicity to normal tissues. One strategy to overcome these therapeutic limitations is to combine angiogenesis inhibitors with cytotoxic therapies. Such an approach has been tried and has shown promise thus far. In our laboratory, we have demonstrated that combining IR with angiostatin derived from the proteolytic digestion of human plasminogen produces a greater than additive antitumor effect (Fig. 2A) [109,110]. Moreover, this effect requires that angiostatin be present in the circulation at the time IR is administered [109] and involves sensitization of the tumor endothelial cells to the cytotoxic effects of IR [110]. Similarly, in several mouse tumor models, we have also observed that some tumors secrete increased levels of VEGF in response to IR and that blocking that response by pretreatment of the mouse with a neutralizing antibody to VEGF results in greatly increased antitumor efficacy of IR treatment (Fig. 2B) [111]. These observations suggest that combining antiangiogenic peptides with IR or other cytotoxic therapies may well represent the most promising potential use of these potent new compounds.

3. Ionizing Radiation: A Means of Targeting Antiangiogenic Gene Therapy

Unfortunately, large antiangiogenic peptides, especially angiostatin and endostatin, present several practical problems to overcome for clinical use. Of these, aberrant folding of the recombinant peptides when they are synthesized *in vitro* represents the main difficulty encountered in making active angiostatin and endostatin. This problem has hindered the ability of pharmaceutical companies to manufacture sufficient quantities of pharmaceutical-grade material for use in humans, and it is only recently that clinical trials involving endostatin have gotten under way. In addition, because these peptides are not tumoricidal, continuous administration for long periods will be necessary if they are to be used as single agents. Consequently, there has been great interest in developing gene therapy approaches for the *in situ* production of antiangiogenic peptides such as angiostatin and endostatin, as well other proteins such as the soluble VEGF receptor. In several tumor models, it has been shown that delivery of the angiostatin or endostatin cDNA by various means, including viral vectors [12,13,15,16,112,113], liposome-mediated methods [13,14,112,114,115], and even injection of naked DNA into skeletal muscle [116] can result in antitumor effects and marked systemic inhibition of angiogenesis. Inhibition of tumor growth and angiogenesis has also been achieved using a variation of this strategy, in which a vector expressing one of the proteases responsible for generating angiostatin *in vivo* is used to inhibit tumor growth [117]. Similarly, constructs expressing the extracellular domain of the VEGF receptor [92,100] or antisense to VEGF [118] can also inhibit tumor

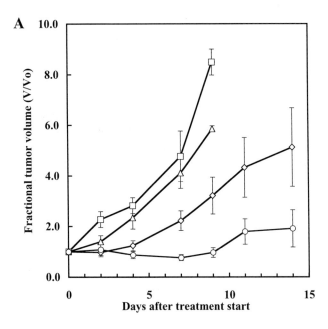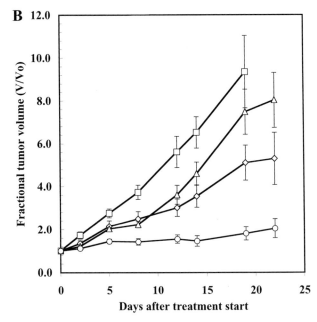

FIGURE 2 Improved antitumor effect by combining antiangiogenic therapy with ionizing radiation. The effect of combining antiangiogenic therapy with ionizing radiation was examined in different tumor models. (A.) Lewis lung carcinoma (LLC) and angiostatin: C57BL6 mice were inoculated in the hindleg with LLC cells, and tumors were allowed to grow as subcutaneous tumors to a starting volume of 1012–111 mm^3 (approximately 5% of the mouse body weight) prior to the commencement of treatment. Mice were then treated with either IR (20 Gy × 2 doses on days 0 and 1) or angiostatin (25 mg/kg/d) throughout the time course of the experiment, or both. (B.) SQ20B and anti-VEGF antibody: Athymic nude mice were inoculated in the hindleg with SQ20B squamous cell carcinoma cells (derived from a radioresistant human head and neck tumor), and the cells were allowed to grow as subcutaneous xenografts to a starting volume of 372–16 mm^3 prior to the commencement of treatment. Mice were then treated with either IR (10 Gy on days 0, 1, 2, and 3) or a neutralizing monoclonal anti-VEGF antibody (10 g on days 0, 1, 2, and 3), or a combination of both, with the anti-VEGF antibody administered 3 hours prior to IR. The combination of blocking VEGF activity and treating with IR produced superior tumor growth delay. Squares = untreated controls; diamonds = IR alone; triangle = angiostatin (A) or anti-VEGF antibody (B) alone; circle = combination therapy. (Graphs adapted from data in Mauceri et al.[110] and Gorski et al.[111])

growth and angiogenesis. Strategies such as these form the basis for combining antiangiogenic gene therapy with other cytotoxic modalities.

Given the success in demonstrating the efficacy of combining at least two different antiangiogenic strategies with IR, a logical next step would be to combine antiangiogenic gene therapy with IR. Such an approach would theoretically produce much higher intratumoral levels of antiangiogenic peptide than is possible by exogenous administration and, therefore, presumably a greater antitumor effect when combined with IR. In addition, given that some antiangiogenic gene therapy strategies can generate systemic levels of angiostatin or endostatin and inhibit angiogenesis at distant sites [12,14,114–116], such strategies may have the potential advantage of also inhibiting the growth of metastatic disease. In one study, Griscelli et al. [13] reported the use of a replication-defective adenovirus expressing the secretable angiostatin-like molecule K3 (AdK3). K3 contains the first three kringle domains of plasminogen and has antiangiogenic activity comparable to angiostatin [13,119]. C6 glioma xenografts implanted in athymic nude mice were treated with either AdK3 alone or IR alone, or a combination of the two. The combination produced a significantly higher antitumor effect that tightly correlated with a marked decrease in intra-

tumoral vascularization. Seetharam et al. [120] have reported that the addition of an adenovirus expressing IL-12, which, in addition to its effects on the immune system, is also antiangiogenic, to IR not only enhances the local antitumor effect of IR but also suppresses microscopic growth of tumors at distant sites, probably through an immune enhancement [120]. These experiments suggest that the combination of antiangiogenic gene therapy with IR shows promise as a means of increasing the efficacy of IR. Further experiments will be necessary to verify the general utility of this approach and determine the best strategies for applying it to human tumors.

III. ENHANCING THE REPLICATIVE POTENTIAL OF ANTITUMOR VIRUSES WITH IONIZING RADIATION

Antitumor replication-competent viruses derive their antitumor effect from direct tumor cell lysis after completion of the viral replicative cycle. Ideally, such viruses replicate preferentially in tumor cells relative to normal tissue. One strategy to abrogate reproduction in normal but not tumor cells is to delete or mutate genes necessary for replication in normal cells but not tumor cells. The last two decades have witnessed explosive growth in our knowledge of molecular biology. Many viruses have been characterized and completely sequenced, and the specific gene products for necessary viral functions have been identified. These include genes encoding proteins involved in cell cycle, pathogenesis, and avoidance of cellular immunity. This knowledge, coupled with the ability to genetically construct viruses to reduce their pathogenicity or target tumor cells, has led to strategies for herpes, adenoviruses, and reoviruses. Such strategies show the most promise for success for treating tumors of the central nervous system, where, in contrast to growing tumor cells, the neurons are quiescent and genes responsible for the neurovirulence of various viruses have been identified [121,122]. Herpes simplex virus-1 is a 152-kb DNA virus that encodes the $\gamma_1 134.5$ gene, which has reportedly been involved in herpes neurovirulence [121,122]. $\gamma_1 134.5$-deleted herpesviruses are severely attenuated in their ability to replicate in neurons and thereby cause encephalitis in murine models. Wild-type-HSV-1(F) has an LD_{50} of 10^2 PFU upon intracranial injection into mice. However, $\gamma_1 134.5$-deleted virus has an LD_{50} of 10^7 PFU, thus accounting for its observed decreased neurovirulence compared with wild-type HSF-1(F) [121,123,124].

R3616 is an HSV with both copies of $\gamma_1 134.5$ inactivated. The Roizman laboratory has focused on the use of R3616 with IR in the treatment of gliomas [125]. One of the problems demonstrated by attenuated herpesviruses is their inability to cause tumor xenograft regression. Instead, these herpesviruses have resulted solely in tumor growth delay. The relative lack of antitumor efficacy of attenuated HSV is largely based on the failure of attenuated viruses to replicate within the tumor. However, the combination of R3616 with IR results in significant tumor regression with greater than 60% of combined treated subcutaneous glioma xenografts regressing completely [125]. Additional studies have shown that IR results in two- to fivefold greater viral recovery from irradiated tumors xenografts than unirradiated infected xenografts. An orthotopic intracranial glioma model has confirmed the enhancement of mouse survival in gliomas treated with both R3616 and IR, as well as prolonged detection of R3616 within irradiated tumors compared with unirradiated tumor [126]. These results implicate a role for radiotherapy in enhancing attenuated viral replication within tumors.

Our current hypothesis for the mechanism by which IR results in enhanced herpesvirus proliferation is that IR results in a cellular environment more conducive to HSV replication through the induction of cellular proteins that promote HSV replication. Evidence from another system supporting this general hypothesis includes the observation that transfection of cells with a construct in which the luciferase reporter gene is linked to the cytomegalovirus (CMV) promoter results in increased luciferase activity after transfected cells are irradiated [127], implying that IR is inducing proteins

that activate the CMV promoter. This effect is not observed in all cell types, however. For instance, Cheng *et al.* [128] observed no induction of CMV promoter activity after COS-7 cells transfected with plasmids in which the chloramphenicol acetyl transferase gene was linked to the CMV promoter were irradiated.

IV. TRANSCRIPTIONAL TARGETING OF GENE THERAPY WITH IONIZING RADIATION (GENETIC RADIOTHERAPY)

A. Introduction

The utility of gene therapy as a cancer therapy is often limited by inherent tumor resistance to the gene product, difficulty introducing the gene into a sufficient number of tumor cells to cause a therapeutic effect, or poor diffusion of gene product. One of the more daunting challenges in delivering gene therapy to tumors is spatial and temporal control of the expression and effect of exogenously delivered gene. This is important in strategies using genes whose products are toxic, where it is necessary to deliver these toxic gene products to tumor cells selectively and spare normal cells as much as possible. Such strategies require precise spatial targeting of the gene to the appropriate cells, because widespread expression or diffusion could be detrimental to the patient, such as in the case of cytotoxic proteins such as ricin and *Pseudomonas* endotoxin. Antitumor cytokines, such as tumor necrosis factor-α, (TFN-α), can also be toxic if they diffuse away from the tumor site and reach high systemic levels. Strategies for keeping gene expression localized to the tumor have included techniques as simple as intratumoral injection [129] and techniques as sophisticated as engineering constructs in which the gene of interest is under the control of a tissue- or tumor-specific promoter [53,130]. Temporal control is also difficult to achieve. When using plasmid or viral constructs in which expression of the therapeutic gene is driven by a strong constitutive promoter, this is usually accomplished simply by injecting the gene when expression is desired. The drawback of this approach is that the length of time the gene is expressed is highly variable, depending upon the method of gene delivery (naked plasmid, liposome-based methods, adenovirus, vaccinia virus, etc.).

B. Induction of the Immediate Early Gene *Egr*-1 Following Exposure to Ionizing Radiation

The study of gene induction by IR is important to the understanding of how cells and organisms respond to radiation exposure. IR activates the transcription of a number of genes, implying the existence of radiation-responsive elements residing upstream of IR-induced genes. These elements could be exploited in genetic constructs to activate gene expression after exposure to IR. Among the genes induced soon after cells are exposed to IR are the tissue plasminogen activator (t-PA) gene [131] and the immediate early genes such as c-*jun* and the early growth response-1 (*Egr*-1) gene [132–134]. *Egr*-1, also known as *zif*/268, *NGFI-A*, *Krox*-24, and *TIS*-8, encodes a nuclear phosphoprotein with a cysteine/histidine zinc finger motif that is partially homologous to the corresponding domain of the Wilm's tumor susceptibility gene, and its expression is rapidly induced after cells are stimulated or reenter the cell cycle [135,136].

Despite the relatively large number of genes that are induced after exposure to IR, relatively few radiation-inducible promoters or enhancers have been characterized. DNA sequences that activate transcription after irradiation include AP-1 [137], the NF-κB binding sequence [132], and the CArG element [138]. The *Egr*-1 promoter has been examined as an inducible promoter for gene therapy because it is inducible by radiation in several types of human tumor cells [133,138]. Datta *et al.* [138] have studied deletion mutants of the 5′ promoter region of the *Egr*-1 promoter to identify elements responsible for radiation inducibility. The radiation response element was identified as the CArG box [CC(A + T rich)GG], a DNA sequence motif originally identified in the serum response element and found in the promoters of several immediate early genes, as well as muscle-specific promoters [139–141]. By linking CArG boxes together and placing them upstream of a CAT reporter gene, Datta *et al.* demonstrated a three- to fourfold increase in the expression of the CAT reporter gene after irradiation, and promoter deletion analysis revealed that the first three 5′ CArG boxes were the most important for the induction of *Egr*-1 promoter activation [142]. Not surprisingly, the mechanism of CArG box activation depends upon the generation of free radical intermediates by IR [142]. Overall, this work defined a radiation response element (RRE) that could be placed upstream of a cDNA encoding a therapeutic protein and used to turn on expression of this gene in irradiated cells.

C. Tumor Necrosis Factor-α: A Toxin for Radiation-Inducible Gene Therapy

Tumor necrosis factor-α is a polypeptide cytokine that activates a wide variety of biological responses, predominately in the immune system [143]. TNF was first identified based on its ability to induce hemorrhagic necrosis in murine tumors and damage to tumor vasculature [144]. The 55-kDa TNF receptor initiates a signaling cascade that results in the apoptosis of some tumor cells [145,146]. Clinical trials have demonstrated that the levels of TNF-α protein achieved in animal studies could not be achieved in human subjects due to systemic toxicity, including hypotension and respiratory insufficiency. A phase I trial using TNF and concomitant

radiotherapy was done to determine the maximal tolerated dose of TNF that could be used to enhance radiation effects on tumors and to establish patterns of both in-field and systemic toxicity [147]. Radiotherapy combined with human recombinant TNF-α were administered for 5 days for each consecutive week until completion of the planned course of radiotherapy. When locally advanced primary tumors were treated to doses $\geq$60 Gy (given as 1.80–2.25 Gy/d), minimal in-field toxicity was observed; however, acute systemic toxicity, including rigors, fever, and nausea, was observed in nearly all patients. Response to treatment was evaluated in 20 of the 31 patients. Complete regression was observed in four patients. It was proposed that tumor localization of TNF-α using gene therapy combined with radiotherapy might eliminate the observed systemic toxicity and enhance the antitumor effects of IR through the production of high local levels of intratumoral TNF-α [148].

D. Ad.Egr-TNF-α: Gene Therapy Spatially and Temporally Controlled by Ionizing Radiation

Viral-mediated transfer of cytotoxic genes whose expression is controlled by RRE allows for spatial and temporal control of gene expression using IR as the means of throwing the molecular "switch" (Fig. 3). Hallahan *et al.* [149] synthesized a genetic construct in which the *Egr*-1 promoter including the CArG elements responsible for radiation inducibility was placed upstream from a cDNA encoding TNF-α

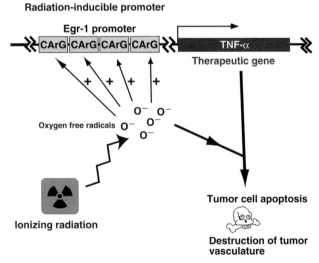

FIGURE 3   Genetic radiotherapy. The *Egr*-1 promoter is placed upstream of a cDNA encoding TNF-α. Ionizing radiation activates the CArG elements in the *Egr*-1 promoter, driving enhanced expression of the TNF-α gene in the tumor bed. The combination of the enhanced TNF-α expression and ionizing radiation results in increased tumor cell apoptosis and vascular destruction, as well as improved antitumor activity.

(Egr–TNF) [149]. Replication-deficient adenovirus type 5 (Ad-5) virus was then employed to deliver the Egr–TNF construct to radioresistant head and neck squamous cell carcinoma cells growing as xenografts in athymic nude mice in order to study the interaction of radiation-targeted TNF gene therapy in tumor cell lines and mouse xenograft models. For the human tumor xenograft SQ20B derived from an oropharyngeal squamous cell carcinoma, the treatment protocol consisted of Ad.Egr–TNF (four injections) and IR given as 5-Gy fractions to a total dose of 50 Gy. Xenografts treated with both Ad.Egr–TNF and IR demonstrated significantly greater tumor shrinkage and growth delay than xenografts treated with either modality alone. TNF expression in irradiated tumor xenografts was elevated threefold at 7 days after initiation of treatment and by eightfold by day 21. Staba *et al.* [150] have demonstrated similar efficacy of combined therapy in human glioma xenografts. D54 glioma cells are resistant to the cytotoxic effects of TNF-α, and no enhancement of radiation killing was observed following treatment with TNF-α and IR *in vitro*. However, when nude mice bearing D54 xenografts received intratumoral injections of Ad.Egr–TNF or null adenovirus (Ad-null) with and without fractionated IR (5 Gy/d, total 30 Gy), combined treatment produced complete tumor regression in 71% of xenografts, as opposed to the 7% observed in those treated with IR alone and 0% in those treated with Ad.Egr–TNF alone. Combined treatment also resulted in a significantly longer growth delay and produced marked tumor vessel thrombosis, an effect not seen with either therapy alone, suggesting that Ad.Egr–TNF and IR target the tumor vasculature [149–152]. Results similar to those observed in SQ20B and D54 xenografts have also been noted in xenografts of the prostate tumor cell line PC3 [153].

This strategy has also been applied to another therapeutic gene, HSV-TK. Joki *et al.* [154] combined transcriptional regulation with converting enzyme/prodrug strategies to further regulate the interaction with IR. When the *Egr*-1 promoter was linked to the HSV-TK gene, not only has irradiation of transfected cells resulted in enhanced HSV-TK expression driven by the *Egr*-1 promoter, but elevated HSV-TK also has allowed for more complete activation of GCV. Phosphorylated GCV then acts as a radiosensitizer upon subsequent IR administration. The results of these experiments show that GCV is phosphorylated to radiosensitizing levels in transfected irradiated cells, but that the basal transcription rate of the *Egr*-1–HSV-TK construct without IR was insufficient to phosphorylate GCV to its toxic antimetabolite. Therefore, HSV-TK expression can be regulated both temporally and spatially by IR. Taken together, all these experiments demonstrate the feasibility of using radiation-inducible promoters linked to therapeutic genes or prodrug converting enzymes to control gene expression both temporally and spatially and to enhance tumor response to IR. They also show some of the many potential strategies for exploiting such promoters for therapy.

Efforts are presently under way to improve upon this system by developing synthetic promoters whose activity is more tightly regulated by radiation exposure than the *Egr*-1 promoter. Marples *et al.* [155], for instance, have reported that a synthetic promoter made of multiple CArG elements is at least as effectively induced by low doses of IR as the *Egr*-1 promoter. More recently, in another variation on this approach, Scott *et al.* [156] have developed a promoter that combines the CArG elements from the *Egr*-1 promoter and the *cre*-Lox-P site-specific recombination system of the P1 bacteriophage. In this system, a single, minimally toxic dose of radiation induces *cre*-mediated excision of a Lox-P flanked stop cassette in a silenced expression vector, resulting in amplified levels of CMV promoter-driven expression of HSV-TK [156]. Experiments such as these demonstrate the feasibility of making promoters whose activity is very tightly regulated by exposure of the cell to IR. Once such promoters are developed, genetic radiotherapy with more tumoricidal or toxic genes will become possible.

V. SUMMARY AND FUTURE DIRECTIONS

Combining radiation and gene therapy has multiple advantages. Both gene therapy and radiation therapy are used in the treatment of local disease and kill tumor cells by independent mechanisms, thus minimizing the likelihood of the tumor developing treatment-resistant clones during treatment. Moreover, in some cases, the gene therapy can impact systemic disease as well as local disease, as is the case when antiangiogenic gene therapy is combined with radiation. In theory, the locally administered antiangiogenic peptide will have a greater than additive local antitumor effect [109,110] and suppress distant metastases while the peptide is being expressed [12,14,114–116]. Moreover, spatial and temporal control can be achieved through conforming radiotherapy to the virally inoculated tumor bed expressing the therapeutic gene. Viruses delivering radiosensitizing agents or antiangiogenic peptides may allow for higher intratumoral concentrations of these drugs than is possible by systemic administration, thus theoretically enhancing the interaction between these drugs in the tumor itself and minimizing systemic toxicity due to drug. With radiosensitization, enhanced local tumor control may be achieved in radioresistant tumors, and radiosensitive tumors may be controlled with lower doses of radiation, thereby minimizing radiation-induced damage to surrounding normal tissue as much as possible. As an adjunct to this, it is possible to imagine the use of additional gene therapy using antioxidant proteins such as manganese superoxide dismutase to protect surrounding normal tissue further [157,158], thus increasing the therapeutic ratio of radiation even further. Finally, viral replication enhancement by IR can be confined to the tumor by conformal radiotherapy, allowing for high titers of virus in the tumor. In light of the rapid advances in the development of these approaches to combining gene therapy with radiation in experimental models, it will be of great interest to begin to move these approaches into clinical use by developing clinical trials to test their efficacy.

References

1. Bramson, J. L., Hitt, M., Addison, C. L., Muller, W. J., Gauldie, J., and Graham, F. L. (1996). Direct intratumoral injection of an adenovirus expressing interleukin-12 induces regression and long-lasting immunity that is associated with highly localized expression of interleukin-12. *Hum. Gene Ther.* **7**, 1995–2002.

2. Chen, L., Chen, D., Block, E., O'Donnell, M., Kufe, D. W., and Clinton, S. K. (1997). Eradication of murine bladder carcinoma by intratumor injection of a bicistronic adenoviral vector carrying cDNAs for the IL-12 heterodimer and its inhibition by the IL-12 p40 subunit homodimer. *J. Immunol.* **159**, 351–359.

3. Drozdzik, M., Qian, C., Xie, X., Peng, D., Bilbao, R., Mazzolini, G., and Prieto, J. (2000). Combined gene therapy with suicide gene and interleukin-12 is more efficient than therapy with one gene alone in a murine model of hepatocellular carcinoma. *J. Hepatol.* **32**, 279–286.

4. Kasaoka, Y., Nakamoto, T., Wang, J., Usui, T., and Hamada, H. (2000). Gene therapy for murine renal cell carcinoma using genetically engineered tumor cells to secrete interleukin-12. *Hiroshima J. Med. Sci.* **49**, 29–35.

5. Lotze, M. T., Shurin, M., Esche, C., Tahara, H., Storkus, W., Kirkwood, J. M., Whiteside, T. L., Elder, E. M., Okada, H., and Robbins, P. (2000). Interleukin-2: developing additional cytokine gene therapies using fibroblasts or dendritic cells to enhance tumor immunity. *Cancer J. Sci. Am.* **6**, S61–66.

6. Clayman, G. L., Liu, T. J., Overholt, S. M., Mobley, S. R., Wang, M., Janot, F., and Goepfert, H. (1996). Gene therapy for head and neck cancer. Comparing the tumor suppressor gene p53 and a cell cycle regulator WAF1/CIP1 (p21). *Arch. Otolaryngol. Head Neck Surg.* **122**, 489–493.

7. Roth, J. A., Swisher, S. G., and Meyn, R. E. (1999). p53 tumor suppressor gene therapy for cancer. *Oncology (Huntington)* **13**, 148–154.

8. Takeda, S., Nakao, A., Miyoshi, K., and Takagi, H. (1998). Gene therapy for pancreatic cancer. *Semin. Surg. Oncol.* **15**, 57–61.

9. Mullen, C. A., Kilstrup, M., and Blaese, R. M. (1992). Transfer of the bacterial gene for cytosine deaminase to mammalian cells confers lethal sensitivity to 5-fluorocytosine: a negative selection system. *Proc. Natl. Acad. Sci. USA* **89**, 33–37.

10. Oldfield, E. H., Ram, Z., Culver, K. W., Blaese, R. M., DeVroom, H. L., and Anderson, W. F. (1993). Gene therapy for the treatment of brain tumors using intra-tumoral transduction with the thymidine kinase gene and intravenous ganciclovir. *Hum. Gene Ther.* **4**, 39–69.

11. Singhal, S., and Kaiser, L. R. (1998). Cancer chemotherapy using suicide genes. *Surg. Oncol. Clin. N. Am.* **7**, 505–536.

12. Feldman, A. L., Restifo, N. P., Alexander, H. R., Bartlett, D. L., Hwu, P., Seth, P., and Libutti, S. K. (2000). Antiangiogenic gene therapy of cancer utilizing a recombinant adenovirus to elevate systemic endostatin levels in mice. *Cancer Res.* **60**, 1503–1506.

13. Griscelli, F., Li, H., Cheong, C., Opolon, P., Bennaceur-Griscelli, A., Vassal, G., Soria, J., Soria, C., Lu, H., Perricaudet, M., and Yeh, P. (2000). Combined effects of radiotherapy and angiostatin gene therapy in glioma tumor model. *Proc. Natl. Acad. Sci. USA* **97**, 6698–6703.

14. Liu, Y., Thor, A., Shtivelman, E., Cao, Y., Tu, G., Heath, T. D., and Debs, R. J. (1999). Systemic gene delivery expands the repertoire of effective antiangiogenic agents. *J. Biol. Chem.* **274**, 13338–13344.

15. Nguyen, J. T., Wu, P., Clouse, M. E., Hlatky, L., and Terwilliger, E. F. (1998). Adeno-associated virus-mediated delivery of antiangiogenic factors as an antitumor strategy. *Cancer Res.* **58**, 5673–5677.

16. Nguyen, J. T. (2000). Adeno-associated virus and other potential vectors for angiostatin and endostatin gene therapy. *Adv. Exp. Med. Biol.* **465**, 457–466.

17. Agarwal, M. L., Taylor, W. R., Chernov, M. V., Chernova, O. B., and Stark, G. R. (1998). The p53 network. *J. Biol. Chem.* **273**, 1–4.

18. Wang, Y., Schwedes, J. F., Parks, D., Mann, K., and Tegtmeyer, P. (1995). Interaction of p53 with its consensus DNA-binding site. *Mol. Cell. Biol.* **15**, 2157–2165.

19. Fornace, A. J., Jackman, J., Hollander, M. C., Hoffman-Liebermann, B., and Liebermann, D. A. (1992). Genotoxic-stress-response genes and growth arrest genes: *gadd, MyD,* and other genes induced by treatments eliciting growth arrest. *Ann. N.Y. Acad. Sci.* **663**, 139–153.

20. Kastan, M. B., Zhan, Q., El-Deiry, W. S., Carrier, F., Jacks, T., Walsh, W. V., Plunkett, B. S., Vogelstein, B., and Fornace, A. J. (1992). A mammalian cell cycle checkpoint pathway utilizing p53 and *GADD45* is defective in ataxia-telangiectasia. *Cell* **71**, 587–597.

21. Neubauer, A., Thiede, C., Huhn, D., and Wittig, B. (1996). p53 and induction of apoptosis as a target for anticancer therapy. *Leukemia* **10**, S2–S4.

22. Greenblatt, M. S., Bennett, W. P., Hollstein, M., and Harris, C. C. (1994). Mutations in the p53 tumor suppressor gene: clues to cancer etiology and molecular pathogenesis. *Cancer Res.* **54**, 4855–4878.

23. Velculescu, V. E., and El-Deiry, W. S. (1996). Biological and clinical importance of the p53 tumor suppressor gene. *Clin. Chem.* **42**, 858–868.

24. Weinstein, J. N., Myers, T. G., O'Connor, P. M., Friend, S. H., Fornace, Jr., A. J., Kohn, K. W., Fojo, T., Bates, S. E., Rubinstein, L. V., Anderson, N. L., Buolamwini, J. K., van Osdol, W. W., Monks, A. P., Scudiero, D. A., Sausville, E. A., Zaharevitz, D. W., Bunow, B., Viswanadhan, V. N., Johnson, G. S., Wittes, R. E., and Paull, K. D. (1997). An information-intensive approach to the molecular pharmacology of cancer. *Science* **275**, 343–349.

25. Komarov, P. G., Komarova, E. A., Kondratov, R. V., Christov-Tselkov, K., Coon, J. S., Chernov, M. V., and Gudkov, A. V. (1999). A chemical inhibitor of p53 that protects mice from the side effects of cancer therapy. *Science* **285**, 1733–1737.

26. Fan, S., el-Deiry, W. S., Bae, I., Freeman, J., Jondle, D., Bhatia, K., Fornace, Jr., A. J., Magrath, I., Kohn, K. W., and O'Connor, P. M. (1994). p53 gene mutations are associated with decreased sensitivity of human lymphoma cells to DNA damaging agents. *Cancer Res.* **54**, 5824–5830.

27. Lowe, S. W., Ruley, H. E., Jacks, T., and Housman, D. E. (1993). p53-dependent apoptosis modulates the cytotoxicity of anticancer agents. *Cell* **74**, 957–967.

28. Lowe, S. W., Schmitt, E. M., Smith, S. W., Osborne, B. A., and Jacks, T. (1993). p53 is required for radiation-induced apoptosis in mouse thymocytes. *Nature* **362**, 847–849.

29. Tsang, N. M., Nagasawa, H., Li, C., and Little, J. B. (1995). Abrogation of p53 function by transfection of HPV16 E6 gene enhances the resistance of human diploid fibroblasts to ionizing radiation. *Oncogene* **10**, 2403–2408.

30. Su, L. N., and Little, J. B. (1992). Transformation and radiosensitivity of human diploid skin fibroblasts transfected with SV40 T-antigen mutants defective in RB and p53 binding domains. *Int. J. Radiat. Biol.* **62**, 461–468.

31. Bracey, T. S., Miller, J. C., Preece, A., and Paraskeva, C. (1995). Gamma-radiation-induced apoptosis in human colorectal adenoma and carcinoma cell lines can occur in the absence of wild type p53. *Oncogene* **10**, 2391–2396.

32. Bristow, R. G., Benchimol, S., and Hill, R. P. (1996). The p53 gene as a modifier of intrinsic radiosensitivity: implications for radiotherapy. *Radiother. Oncol.* **40**, 197–223.

33. Huang, H., Li, C. Y., and Little, J. B. (1996). Abrogation of p53 function by transfection of HPV16 E6 gene does not enhance resistance of human tumour cells to ionizing radiation. *Int. J. Radiat. Biol.* **70**, 151–160.

34. Kohli, M., and Jorgensen, T. J. (1999). The influence of SV40 immortalization of human fibroblasts on p53-dependent radiation responses. *Biochem. Biophys. Res. Commun.* **257**, 168–176.

35. Gallardo, D., Drazan, K. E., and McBride, W. H. (1996). Adenovirus-based transfer of wild-type p53 gene increases ovarian tumor radiosensitivity. *Cancer Res.* **56**, 4891–4893.

36. Spitz, F. R., Nguyen, D., Skibber, J. M., Meyn, R. E., Cristiano, R. J., and Roth, J. A. (1996). Adenoviral-mediated wild-type p53 gene expression sensitizes colorectal cancer cells to ionizing radiation. *Clin. Cancer Res.* **2**, 1665–1671.

37. Spitz, F. R., Nguyen, D., Skibber, J. M., Cusack, J., Roth, J. A., and Cristiano, R. J. (1996). In vivo adenovirus-mediated p53 tumor suppressor gene therapy for colorectal cancer. *Anticancer Res.* **16**, 3415–3422.

38. Chang, E. H., Jang, Y. J., Hao, Z., Murphy, G., Rait, A., Fee, Jr., W. E., Sussman, H. H., Ryan, P., Chiang, Y., and Pirollo, K. F. (1997). Restoration of the G1 checkpoint and the apoptotic pathway mediated by wild-type p53 sensitizes squamous cell carcinoma of the head and neck to radiotherapy. *Arch. Otolaryngol. Head Neck Surg.* **123**, 507–512.

39. Badie, B., Kramar, M. H., Lau, R., Boothman, D. A., Economou, J. S., and Black, K. L. (1998). Adenovirus-mediated p53 gene delivery potentiates the radiation-induced growth inhibition of experimental brain tumors. *J. Neurooncol.* **37**, 217–222.

40. Vinyals, A., Peinado, M. A., Gonzalez-Garrigues, M., Monzo, M., Bonfil, R. D., and Fabra, A. (1999). Failure of wild-type p53 gene therapy in human cancer cells expressing a mutant p53 protein. *Gene Ther.* **6**, 22–33.

41. el-Deiry, W. S., Harper, J. W., O'Connor, P. M., Velculescu, V. E., Canman, C. E., Jackman, J., Pietenpol, J. A., Burrell, M., Hill, D. E., Wang, Y., and *et al.* (1994). WAF1/CIP1 is induced in p53-mediated G1 arrest and apoptosis. *Cancer Res.* **54**, 1169–1174.

42. Waldman, T., Kinzler, K. W., and Vogelstein, B. (1995). p21 is necessary for the p53-mediated G1 arrest in human cancer cells. *Cancer Res.* **55**, 5187–5190.

43. Xiong, Y., Hannon, G. J., Zhang, H., Casso, D., Kobayashi, R., and Beach, D. (1993). p21 is a universal inhibitor of cyclin kinases. *Nature* **366**, 701–704.

44. Katayose, D., Wersto, R., Cowan, K. H., and Seth, P. (1995). Effects of a recombinant adenovirus expressing WAF1/Cip1 on cell growth, cell cycle, and apoptosis. *Cell Growth Differ.* **6**, 1207–1212.

45. Gotoh, A., Kao, C., Ko, S. C., Hamada, K., Liu, T. J., and Chung, L. W. (1997). Cytotoxic effects of recombinant adenovirus p53 and cell cycle regulator genes (p21 WAF1/CIP1 and p16CDKN4) in human prostate cancers. *J. Urol.* **158**, 636–641.

46. Gomez-Manzano, C., Fueyo, J., Kyritsis, A. P., McDonnell, T. J., Steck, P. A., Levin, V. A., and Yung, W. K. (1997). Characterization of p53 and p21 functional interactions in glioma cells en route to apoptosis. *J. Natl. Cancer Inst.* **89**, 1036–1044.

47. Meng, R. D., Shih, H., Prabhu, N. S., George, D. L., and el-Deiry, W. S. (1998). Bypass of abnormal MDM2 inhibition of p53-dependent growth suppression. *Clin. Cancer Res.* **4**, 251–259.

48. Hsiao, M., Tse, V., Carmel, J., Costanzi, E., Strauss, B., Haas, M., and Silverberg, G. D. (1997). Functional expression of human p21(WAF1/CIP1) gene in rat glioma cells suppresses tumor growth in vivo and induces radiosensitivity. *Biochem. Biophys. Res. Commun.* **233**, 329–335.

49. Li, W. W., Fan, J., Hochhauser, D., and Bertino, J. R. (1997). Overexpression of p21waf1 leads to increased inhibition of E2F-1 phosphorylation and sensitivity to anticancer drugs in retinoblastoma-negative human sarcoma cells. *Cancer Res.* **57**, 2193–219.

50. Sheikh, M. S., Chen, Y. Q., Smith, M. L., and Fornace, Jr., A. J. (1997). Role of p21Waf1/Cip1/Sdi1 in cell death and DNA repair as studied using a tetracycline-inducible system in p53-deficient cells. *Oncogene* **14**, 1875–1882.

51. Wouters, B. G., Giaccia, A. J., Denko, N. C., and Brown, J. M. (1997). Loss of p21Waf1/Cip1 sensitizes tumors to radiation by an apoptosis-independent mechanism. *Cancer Res.* **57**, 4703–4706.

52. Tian, H., Wittmack, E. K., and Jorgensen, T. J. (2000). p21WAF1/CIP1 antisense therapy radiosensitizes human colon cancer by converting growth arrest to apoptosis. *Cancer Res.* **60**, 679–684.

53. Dachs, G. U., Dougherty, G. J., Stratford, I. J., and Chaplin, D. J. (1997). Targeting gene therapy to cancer: a review. *Oncol. Res.* **9**, 313–325.

54. Martin, L. A., and Lemoine, N. R. (1996). Direct cell killing by suicide genes. *Cancer Metastasis Rev.* **15**, 301–316.

55. Balzarini, J., Bohman, C., and De Clercq, E. (1993). Differential mechanism of cytostatic effect of (E)-5-(2-bromovinyl)-2′- deoxyuridine, 9-(1,3-dihydroxy-2-propoxymethyl)guanine, and other antiherpetic drugs on tumor cells transfected by the thymidine kinase gene of herpes simplex virus type 1 or type 2. *J. Biol. Chem.* **268**, 6332–6337.

56. Culver, K. W., Ram, Z., Wallbridge, S., Ishii, H., Oldfield, E. H., and Blaese, R. M. (1992). In vivo gene transfer with retroviral vector-producer cells for treatment of experimental brain tumors. *Science* **256**, 1550–1552.

57. Ezzeddine, Z. D., Martuza, R. L., Platika, D., Short, M. P., Malick, A., Choi, B., and Breakefield, X. O. (1991). Selective killing of glioma cells in culture and in vivo by retrovirus transfer of the herpes simplex virus thymidine kinase gene. *New Biol.* **3**, 608–614.

58. Moolten, F. L., and Wells, J. M. (1990). Curability of tumors bearing herpes thymidine kinase genes transferred by retroviral vectors. *J. Natl. Cancer Inst.* **82**, 297–300.

59. Kinsella, T. J., Mitchell, J. B., Russo, A., Morstyn, G., and Glatstein, E. (1984). The use of halogenated thymidine analogs as clinical radiosensitizers: rationale, current status, and future prospects: non-hypoxic cell sensitizers. *Int. J. Radiat. Oncol. Biol. Phys.* **10**, 1399–1406.

60. Bagshaw, M. A., Doggett, R. L., Smith, K. C., Kaplan, H. S., and Nelsen, T. S. (1967). Intra-arterial 5-bromodeoxyuridine and X-ray therapy. *Am. J. Roentgenol. Radium Ther. Nucl. Med.* **99**, 886–894.

61. Sougawa, M., Akagi, K., Murata, T., Kawasaki, S., Sawada, S., Yoshii, G., and Tanaka, Y. (1986). Enhancement of radiation effects by acyclovir. *Int. J. Radiat. Oncol. Biol. Phys.* **12**, 1537–1540.

62. Kim, J. H., Kim, S. H., Brown, S. L., and Freytag, S. O. (1994). Selective enhancement by an antiviral agent of the radiation-induced cell killing of human glioma cells transduced with HSV-tk gene. *Cancer Res.* **54**, 6053–6056.

63. Kim, J. H., Kim, S. H., Kolozsvary, A., Brown, S. L., Kim, O. B., and Freytag, S. O. (1995). Selective enhancement of radiation response of herpes simplex virus thymidine kinase transduced 9L gliosarcoma cells in vitro and in vivo by antiviral agents. *Int. J. Radiat. Oncol. Biol. Phys.* **33**, 861–868.

64. Kim, S. H., Kim, J. H., Kolozsvary, A., Brown, S. L., and Freytag, S. O. (1997). Preferential radiosensitization of 9L glioma cells transduced with HSV- tk gene by acyclovir. *J. Neurooncol.* **33**, 189–194.

65. Andersen, L., Kilstrup, M., and Neuhard, J. (1989). Pyrimidine, purine and nitrogen control of cytosine deaminase synthesis in *Escherichia coli* K 12. Involvement of the glnLG and purR genes in the regulation of codA expression. *Arch. Microbiol.* **152**, 115–118.

66. Blaese, R. M., Ishii-Morita, H., Mullen, C., Ramsey, J., Ram, Z., Oldfield, E., and Culver, K. (1994). In situ delivery of suicide genes for cancer treatment. *Eur. J. Cancer* **8**, 1190–1193.

67. Khil, M. S., Kim, J. H., Mullen, C. A., Kim, S. H., and Freytag, S. O. (1996). Radiosensitization by 5-fluorocytosine of human colorectal carcinoma cells in culture transduced with cytosine deaminase gene. *Clin. Cancer Res.* **2**, 53–57.

68. Mullen, C. A., and Blaese, R. M. (1994). Gene therapy of cancer. *Cancer Chemother. Biol. Response Modif.* **15**, 176–189.

69. Mullen, C. A., Coale, M. M., Lowe, R., and Blaese, R. M. (1994). Tumors expressing the cytosine deaminase suicide gene can be eliminated in vivo with 5-fluorocytosine and induce protective immunity to wild type tumor. *Cancer Res.* **54**, 1503–1506.

70. Parker, W. B., and Cheng, Y. C. (1990). Metabolism and mechanism of action of 5-fluorouracil. *Pharmacol. Ther.* **48**, 381–395.

71. Weckbecker, G. (1991). Biochemical pharmacology and analysis of fluoropyrimidines alone and in combination with modulators. *Pharmacol. Ther.* **50**, 367–424.

72. Moertel, C. G., Gunderson, L. L., Mailliard, J. A., McKenna, P. J., Martenson, Jr., J. A., Burch, P. A., and Cha, S. S. (1994). Early evaluation of combined fluorouracil and leucovorin as a radiation enhancer for locally unresectable, residual, or recurrent gastrointestinal carcinoma. The North Central Cancer Treatment Group. *J. Clin. Oncol.* **12**, 21–27.

73. O'Connell, M. J., Martenson, J. A., Wieand, H. S., Krook, J. E., Macdonald, J. S., Haller, D. G., Mayer, R. J., Gunderson, L. L., and Rich, T. A. (1994). Improving adjuvant therapy for rectal cancer by combining protracted- infusion fluorouracil with radiation therapy after curative surgery. *N. Engl. J. Med.* **331**, 502–507.

74. Bruso, C. E., Shewach, D. S., and Lawrence, T. S. (1990). Fluorodeoxyuridine-induced radiosensitization and inhibition of DNA double strand break repair in human colon cancer cells. *Int. J. Radiat. Oncol. Biol. Phys.* **19**, 1411–1417.

75. Pederson, L. C., Buchsbaum, D. J., Vickers, S. M., Kancharla, S. R., Mayo, M. S., Curiel, D. T., and Stackhouse, M. A. (1997). Molecular chemotherapy combined with radiation therapy enhances killing of cholangiocarcinoma cells in vitro and in vivo. *Cancer Res.* **57**, 4325–4332.

76. Pederson, L. C., Vickers, S. M., Buchsbaum, D. J., Kancharla, S. R., Mayo, M. S., Curiel, D. T., and Stackhouse, M. A. (1998). Combined cytosine deaminase expression, 5-fluorocytosine exposure, and radiotherapy increases cytotoxicity to cholangiocarcinoma cells. *J. Gastrointest. Surg.* **2**, 283–291.

77. Stackhouse, M. A., Pederson, L. C., Grizzle, W. E., Curiel, D. T., Gebert, J., Haack, K., Vickers, S. M., Mayo, M. S., and Buchsbaum, D. J. (2000). Fractionated radiation therapy in combination with adenoviral delivery of the cytosine deaminase gene and 5–fluorocytosine enhances cytotoxic and antitumor effects in human colorectal and cholangiocarcinoma models. *Gene Ther.* **7**, 1019–1026.

78. Hanna, N. N., Mauceri, H. J., Wayne, J. D., Hallahan, D. E., Kufe, D. W., and Weichselbaum, R. R. (1997). Virally directed cytosine deaminase/5–fluorocytosine gene therapy enhances radiation response in human cancer xenografts. *Cancer Res.* **57**, 4205–4209.

79. Rogulski, K. R., Kim, J. H., Kim, S. H., and Freytag, S. O. (1997). Glioma cells transduced with an *Escherichia coli* CD/HSV-1 TK fusion gene exhibit enhanced metabolic suicide and radiosensitivity. *Hum. Gene Ther.* **8**, 73–85.

80. Rogulski, K. R., Wing, M. S., Paielli, D. L., Gilbert, J. D., Kim, J. H., and Freytag, S. O. (2000). Double suicide gene therapy augments the antitumor activity of a replication-competent lytic adenovirus through enhanced cytotoxicity and radiosensitization. *Hum. Gene Ther.* **11**, 67–76.

81. Rogulski, K. R., Zhang, K., Kolozsvary, A., Kim, J. H., and Freytag, S. O. (1997). Pronounced antitumor effects and tumor radiosensitization of double suicide gene therapy. *Clin. Cancer Res.* **3**, 2081–2088.

82. Freytag, S. O., Rogulski, K. R., Paielli, D. L., Gilbert, J. D., and Kim, J. H. (1998). A novel three-pronged approach to kill cancer cells selectively: concomitant viral, double suicide gene, and radiotherapy. *Hum. Gene Ther.* **9**, 1323–1333.

83. Folkman, J. (1995). Angiogenesis in cancer, vascular, rheumatoid and other disease. *Nat. Med.* **1**, 27–31.

84. Klein, S., Roghani, M., and Rifkin, D. B. (1997). Fibroblast growth factors as angiogenesis factors: new insights into their mechanism of action. *Exs* **79**, 159–192.

85. Thomas, K. A. (1996). Vascular endothelial growth factor, a potent and selective angiogenic agent. *J. Biol. Chem.* **271**, 603–606.

86. Asano, M., Yukita, A., Matsumoto, T., Kondo, S., and Suzuki, H. (1995). Inhibition of tumor growth and metastasis by an immunoneutralizing monoclonal antibody to human vascular endothelial growth factor/vascular permeability factor-121. *Cancer Res.* **55**, 5296–5301.

87. Boehm, T., Folkman, J., Browder, T., and O'Reilly, M. S. (1997). Antiangiogenic therapy of experimental cancer does not induce acquired drug resistance. *Nature* **390**, 404–407.

88. O'Reilly, M. S., Boehm, T., Shing, Y., Fukai, N., Vasios, G., Lane, W. S., Flynn, E., Birkhead, J. R., Olsen, B. R., and Folkman, J. (1997). Endostatin: an endogenous inhibitor of angiogenesis and tumor growth. *Cell* **88**, 277–285.

89. O'Reilly, M. S., Holmgren, L., Shing, Y., Chen, C., Rosenthal, R. A., Moses, M., Lane, W. S., Cao, Y., Sage, E. H., and Folkman, J. (1994). Angiostatin: a novel angiogenesis inhibitor that mediates the suppression of metastases by a Lewis lung carcinoma. *Cell* **79**, 315–328.

90. Asano, M., Yukita, A., Matsumoto, T., Matsuo, K., Kondo, S., and Suzuki, H. (1995). Isolation and characterization of neutralizing monoclonal antibodies to human vascular endothelial growth factor/vascular permeability factor121 (VEGF/VPF121). *Hybridoma* **14**, 475–480.

91. Borgstrom, P., Hillan, K. J., Sriramarao, P., and Ferrara, N. (1996). Complete inhibition of angiogenesis and growth of microtumors by anti-vascular endothelial growth factor neutralizing antibody: novel concepts of angiostatic therapy from intravital videomicroscopy. *Cancer Res.* **56**, 4032–4039.

92. Lin, P., Sankar, S., Shan, S., Dewhirst, M. W., Polverini, P. J., Quinn, T. Q., and Peters, K. G. (1998). Inhibition of tumor growth by targeting tumor endothelium using a soluble vascular endothelial growth factor receptor. *Cell Growth Differ.* **9**, 49–58.

93. Yuan, F., Chen, Y., Dellian, M., Safabakhsh, N., Ferrara, N., and Jain, R. K. (1996). Time-dependent vascular regression and permeability changes in established human tumor xenografts induced by an anti-vascular endothelial growth factor/vascular permeability factor antibody. *Proc. Natl. Acad. Sci. USA* **93**, 14765–14770.

94. Coppola, G., Atlas-White, M., Katsahambas, S., Bertolini, J., Hearn, M. T., and Underwood, J. R. (1997). Effect of intraperitoneally, intravenously and intralesionally administered monoclonal anti-beta-FGF antibodies on rat chondrosarcoma tumor vascularization and growth. *Anticancer Res.* **17**, 2033–2039.

95. Hori, A., Sasada, R., Matsutani, E., Naito, K., Sakura, Y., Fujita, T., and Kozai, Y. (1991). Suppression of solid tumor growth by immunoneutralizing monoclonal antibody against human basic fibroblast growth factor. *Cancer Res.* **51**, 6180–6184.

96. Takahashi, J. A., Fukumoto, M., Kozai, Y., Ito, N., Oda, Y., Kikuchi, H., and Hatanaka, M. (1991). Inhibition of cell growth and tumorigenesis of human glioblastoma cells by a neutralizing antibody against human basic fibroblast growth factor. *FEBS Lett.* **288**, 65–71.

97. Belletti, B., Ferraro, P., Arra, C., Baldassarre, G., Bruni, P., Staibano, S., De Rosa, G., Salvatore, G., Fusco, A., Persico, M. G., and Viglietto, G. (1999). Modulation of in vivo growth of thyroid tumor-derived cell lines by sense and antisense vascular endothelial growth factor gene. *Oncogene* **18**, 4860–4869.

98. Im, S. A., Gomez-Manzano, C., Fueyo, J., Liu, T. J., Ke, L. D., Kim, J. S., Lee, H. Y., Steck, P. A., Kyritsis, A. P., and Yung, W. K. (1999). Antiangiogenesis treatment for gliomas: transfer of antisense-vascular endothelial growth factor inhibits tumor growth in vivo. *Cancer Res.* **59**, 895–900.

99. Wang, Y., and Becker, D. (1997). Antisense targeting of basic fibroblast growth factor and fibroblast growth factor receptor-1 in human melanomas blocks intratumoral angiogenesis and tumor growth. *Nat. Med.* **3**, 887–893.

100. Goldman, C. K., Kendall, R. L., Cabrera, G., Soroceanu, L., Heike, Y., Gillespie, G. Y., Siegal, G. P., Mao, X., Bett, A. J., Huckle, W. R., Thomas, K. A., and Curiel, D. T. (1998). Paracrine expression of a

native soluble vascular endothelial growth factor receptor inhibits tumor growth, metastasis, and mortality rate. *Proc. Natl. Acad. Sci. USA* **95**, 8795–8800.

101. Gately, S., Twardowski, P., Stack, M. S., Cundiff, D. L., Grella, D., Castellino, F. J., Enghild, J., Kwaan, H. C., Lee, F., Kramer, R. A., Volpert, O., Bouck, N., and Soff, G. A. (1997). The mechanism of cancer-mediated conversion of plasminogen to the angiogenesis inhibitor angiostatin. *Proc. Natl. Acad. Sci. USA* **94**, 10868–10872.

102. Stathakis, P., Fitzgerald, M., Matthias, L. J., Chesterman, C. N., and Hogg, P. J. (1997). Generation of angiostatin by reduction and proteolysis of plasmin: catalysis by a plasmin reductase secreted by cultured cells. *J. Biol. Chem.* **272**, 20641–20645.

103. Hohenester, E., Sasaki, T., Olsen, B. R., and Timpl, R. (1998). Crystal structure of the angiogenesis inhibitor endostatin at 1.5 A resolution. *EMBO J.* **17**, 1656–1664.

104. Laird, A. D., Vajkoczy, P., Shawver, L. K., Thurnher, A., Liang, C., Mohammadi, M., Schlessinger, J., Ullrich, A., Hubbard, S. R., Blake, R. A., Fong, T. A., Strawn, L. M., Sun, L., Tang, C., Hawtin, R., Tang, F., Shenoy, N., Hirth, K. P., McMahon, G., and Cherrington, J. M. (2000). SU6668 is a potent antiangiogenic and antitumor agent that induces regression of established tumors. *Cancer Res.* **60**, 4152–4160.

105. Mendel, D. B., Laird, A. D., Smolich, B. D., Blake, R. A., Liang, C., Hannah, A. L., Shaheen, R. M., Ellis, L. M., Weitman, S., Shawver, L. K., and Cherrington, J. M. (2000). Development of SU5416, a selective small molecule inhibitor of VEGF receptor tyrosine kinase activity, as an anti-angiogenesis agent. *Anticancer Drug Des.* **15**, 29–41.

106. Castronovo, V., and Belotti, D. (1996). TNP-470 (AGM-1470): mechanisms of action and early clinical development. *Eur. J. Cancer* **32A**, 2520–2527.

107. Gervaz, P., and Fontolliet, C. (1998). Therapeutic potential of the anti-angiogenesis drug TNP-470. *Int. J. Exp. Pathol.* **79**, 359–362.

108. O'Reilly, M. S., Holmgren, L., Chen, C., and Folkman, J. (1996). Angiostatin induces and sustains dormancy of human primary tumors in mice. *Nat. Med.* **2**, 689–692.

109. Gorski, D. H., Mauceri, H. J., Salloum, R. M., Gately, S., Hellman, S., Beckett, M. A., Sukhatme, V. P., Soff, G. A., Kufe, D. W., and Weichselbaum, R. R. (1998). Potentiation of the antitumor effect of ionizing radiation by brief concomitant exposures to angiostatin. *Cancer Res.* **58**, 5686–5689.

110. Mauceri, H. J., Hanna, N. N., Beckett, M. A., Gorski, D. H., Staba, M. J., Stellato, K. A., Bigelow, K., Heimann, R., Gately, S., Dhanabal, M., Soff, G. A., Sukhatme, V. P., Kufe, D. W., and Weichselbaum, R. R. (1998). Combined effects of angiostatin and ionizing radiation in antitumour therapy. *Nature* **394**, 287–291.

111. Gorski, D. H., Beckett, M. A., Jaskowiak, N. T., Calvin, D. P., Mauceri, H. J., Salloum, R. M., Seetharam, S., Koons, A., Hari, D. M., Kufe, D. W., and Weichselbaum, R. R. (1999). Blockade of the vascular endothelial growth factor stress response increases the antitumor effects of ionizing radiation. *Cancer Res.* **59**, 3374–3378.

112. Griscelli, F., Li, H., Bennaceur-Griscelli, A., Soria, J., Opolon, P., Soria, C., Perricaudet, M., Yeh, P., and Lu, H. (1998). Angiostatin gene transfer: inhibition of tumor growth in vivo by blockage of endothelial cell proliferation associated with a mitosis arrest. *Proc. Natl. Acad. Sci. USA* **95**, 6367–6372.

113. Tanaka, T., Cao, Y., Folkman, J., and Fine, H. A. (1998). Viral vector-targeted antiangiogenic gene therapy utilizing an angiostatin complementary DNA. *Cancer Res.* **58**, 3362–3369.

114. Chen, Q. R., Kumar, D., Stass, S. A., and Mixson, A. J. (1999). Liposomes complexed to plasmids encoding angiostatin and endostatin inhibit breast cancer in nude mice. *Cancer Res.* **59**, 3308–3312.

115. Sacco, M. G., Caniatti, M., Cato, E. M., Frattini, A., Chiesa, G., Ceruti, R., Adorni, F., Zecca, L., Scanziani, E., and Vezzoni, P. (2000). Liposome-delivered angiostatin strongly inhibits tumor growth and metastatization in a transgenic model of spontaneous breast cancer. *Cancer Res.* **60**, 2660–2665.

116. Blezinger, P., Wang, J., Gondo, M., Quezada, A., Mehrens, D., French, M., Singhal, A., Sullivan, S., Rolland, A., Ralston, R., and Min, W. (1999). Systemic inhibition of tumor growth and tumor metastases by intramuscular administration of the endostatin gene. *Nat. Biotechnol.* **17**, 343–348.

117. Matsuda, K. M., Madoiwa, S., Hasumi, Y., Kanazawa, T., Saga, Y., Kume, A., Mano, H., Ozawa, K., and Matsuda, M. (2000). A novel strategy for the tumor angiogenesis-targeted gene therapy: generation of angiostatin from endogenous plasminogen by protease gene transfer. *Cancer Gene Ther.* **7**, 589–596.

118. Saleh, M., Stacker, S. A., and Wilks, A. F. (1996). Inhibition of growth of C6 glioma cells in vivo by expression of antisense vascular endothelial growth factor sequence. *Cancer Res.* **56**, 393–401.

119. Cao, Y., Ji, R. W., Davidson, D., Schaller, J., Marti, D., Sohndel, S., McCance, S. G., O'Reilly, M. S., Llinas, M., and Folkman, J. (1996). Kringle domains of human angiostatin. Characterization of the antiproliferative activity on endothelial cells. *J. Biol. Chem.* **271**, 29461–29467.

120. Seetharam, S., Staba, M. J., Schumm, L. P., Schreiber, K., Schreiber, H., Kufe, D. W., and Weichselbaum, R. R. (1999). Enhanced eradication of local and distant tumors by genetically produced interleukin-12 and radiation. *Int. J. Oncol.* **15**, 769–773.

121. Chou, J., Kern, E. R., Whitley, R. J., and Roizman, B. (1990). Mapping of herpes simplex virus-1 neurovirulence to gamma 134.5, a gene nonessential for growth in culture. *Science* **250**, 1262–1266.

122. Martuza, R. L., Malick, A., Markert, J. M., Ruffner, K. L., and Coen, D. M. (1991). Experimental therapy of human glioma by means of a genetically engineered virus mutant. *Science* **252**, 854–856.

123. Advani, S. J., Chung, S. M., Yan, S. Y., Gillespie, G. Y., Markert, J. M., Whitley, R. J., Roizman, B., and Weichselbaum, R. R. (1999). Replication-competent, nonneuroinvasive genetically engineered herpes virus is highly effective in the treatment of therapy-resistant experimental human tumors. *Cancer Res.* **59**, 2055–2058.

124. Markert, J. M., Gillespie, G. Y., Weichselbaum, R. R., Roizman, B., and Whitley, R. J. (2000). Genetically engineered HSV in the treatment of glioma: a review. *Rev. Med. Virol.* **10**, 17–30.

125. Advani, S. J., Sibley, G. S., Song, P. Y., Hallahan, D. E., Kataoka, Y., Roizman, B., and Weichselbaum, R. R. (1998). Enhancement of replication of genetically engineered herpes simplex viruses by ionizing radiation: a new paradigm for destruction of therapeutically intractable tumors. *Gene Ther.* **5**, 160–165.

126. Bradley, J. D., Kataoka, Y., Advani, S., Chung, S. M., Arani, R. B., Gillespie, G. Y., Whitley, R. J., Markert, J. M., Roizman, B., and Weichselbaum, R. R. (1999). Ionizing radiation improves survival in mice bearing intracranial high-grade gliomas injected with genetically modified herpes simplex virus. *Clin. Cancer Res.* **5**, 1517–1522.

127. Tang, D. C., Jennelle, R. S., Shi, Z., Garver, Jr., R. I., Carbone, D. P., Loya, F., Chang, C. H., and Curiel, D. T. (1997). Overexpression of adenovirus-encoded transgenes from the cytomegalovirus immediate early promoter in irradiated tumor cells. *Hum. Gene Ther.* **8**, 2117–2124.

128. Cheng, X., and Iliakis, G. (1995). Effect of ionizing radiation on the expression of chloramphenicol acetyltransferase gene under the control of commonly used constitutive or inducible promoters. *Int. J. Radiat. Biol.* **67**, 261–267.

129. Mauceri, H. J., Seung, L. P., Grdina, W. L., Swedberg, K. A., and Weichselbaum, R. R. (1997). Increased injection number enhances adenoviral genetic radiotherapy. *Radiat. Oncol. Invest.* **5**, 220–226.

130. Robertson, 3rd, M. W., Wang, M., Siegal, G. P., Rosenfeld, M., Ashford, 2nd, R. S., Alvarez, R. D., Garver, R. I., and Curiel, D. T. (1998). Use of a tissue-specific promoter for targeted expression of the herpes simplex virus thymidine kinase gene in cervical carcinoma cells. *Cancer Gene Ther.* **5**, 331–336.

131. Boothman, D. A., Lee, I. W., and Sahijdak, W. M. (1994). Isolation of an X-ray-responsive element in the promoter region of tissue-type

plasminogen activator: potential uses of X-ray-responsive elements for gene therapy. *Radiat. Res.* **138,** S68–S71.

132. Brach, M. A., Hass, R., Sherman, M. L., Gunji, H., Weichselbaum, R., and Kufe, D. (1991). Ionizing radiation induces expression and binding activity of the nuclear factor kappa B. *J. Clin. Invest.* **88,** 691–695.

133. Hallahan, D. E., Sukhatme, V. P., Sherman, M. L., Virudachalam, S., Kufe, D., and Weichselbaum, R. R. (1991). Protein kinase C mediates x-ray inducibility of nuclear signal transducers EGR1 and JUN. *Proc. Natl. Acad. Sci. USA* **88,** 2156–2160.

134. Sherman, M. L., Datta, R., Hallahan, D. E., Weichselbaum, R. R., and Kufe, D. W. (1990). Ionizing radiation regulates expression of the c-jun protooncogene. *Proc. Natl. Acad. Sci. USA* **87,** 5663–5666.

135. Chavrier, P., Zerial, M., Lemaire, P., Almendral, J., Bravo, R., and Charnay, P. (1988). A gene encoding a protein with zinc fingers is activated during G0/G1 transition in cultured cells. *EMBO J.* **7,** 29–35.

136. Sukhatme, V. P. (1990). Early transcriptional events in cell growth: the Egr family. *J. Am. Soc. Nephrol.* **1,** 859–866.

137. Hallahan, D. E., Gius, D., Kuchibhotla, J., Sukhatme, V., Kufe, D. W., and Weichselbaum, R. R. (1993). Radiation signaling mediated by Jun activation following dissociation from a cell type-specific repressor. *J. Biol. Chem.* **268,** 4903–4907.

138. Datta, R., Rubin, E., Sukhatme, V., Qureshi, S., Hallahan, D., Weichselbaum, R. R., and Kufe, D. W. (1992). Ionizing radiation activates transcription of the EGR1 gene via CArG elements. *Proc. Natl. Acad. Sci. USA* **89,** 10149–10153.

139. Muscat, G. E., Gustafson, T. A., and Kedes, L. (1988). A common factor regulates skeletal and cardiac alpha-actin gene transcription in muscle. *Mol. Cell. Biol.* **8,** 4120–4133.

140. Liu, Z. J., Moav, B., Faras, A. J., Guise, K. S., Kapuscinski, A. R., and Hackett, P. (1991). Importance of the CArG box in regulation of beta-actin-encoding genes. *Gene* **108,** 211–217.

141. Treisman, R. (1992). The serum response element. *TIBS* **17,** 423–427.

142. Datta, R., Taneja, N., Sukhatme, V. P., Qureshi, S. A., Weichselbaum, R., and Kufe, D. W. (1993). Reactive oxygen intermediates target CC(A/T)6GG sequences to mediate activation of the early growth response 1 transcription factor gene by ionizing radiation. *Proc. Natl. Acad. Sci. USA* **90,** 2419–2422.

143. Weichselbaum, R. R. (1995). Growth factors alter the therapeutic ratio in radiotherapy. *Cancer J. Sci. Am.* **1,** 28.

144. Carswell, E. A., Old, L. J., Kassel, R. L., Green, S., Fiore, N., and Williamson, B. (1975). An endotoxin-induced serum factor that causes necrosis of tumors. *Proc. Natl. Acad. Sci. USA* **72,** 3666–3670.

145. Tartaglia, L. A., Ayres, T. M., Wong, G. H., and Goeddel, D. V. (1993). A novel domain within the 55 kD TNF receptor signals cell death. *Cell* **74,** 845–853.

146. Tartaglia, L. A., Rothe, M., Hu, Y. F., and Goeddel, D. V. (1993). Tumor necrosis factor's cytotoxic activity is signaled by the p55 TNF receptor. *Cell* **73,** 213–216.

147. Hallahan, D. E., Vokes, E. E., Rubin, S. J., O'Brien, S., Samuels, B., Vijaykumar, S., Kufe, D. W., Phillips, R., and Weichselbaum, R. R. (1995). Phase I dose-escalation study of tumor necrosis factor-alpha and concomitant radiation therapy. *Cancer J. Sci. Am.* **1,** 204.

148. Weichselbaum, R. R., Hallahan, D. E., Sukhatme, V. P., and Kufe, D. W. (1992). Gene therapy targeted by ionizing radiation. *Int. J. Radiat. Oncol. Biol. Phys.* **24,** 565–567.

149. Hallahan, D. E., Mauceri, H. J., Seung, L. P., Dunphy, E. J., Wayne, J. D., Hanna, N. N., Toledano, A., Hellman, S., Kufe, D. W., and Weichselbaum, R. R. (1995). Spatial and temporal control of gene therapy using ionizing radiation. *Nat. Med.* **1,** 786–791.

150. Staba, M. J., Mauceri, H. J., Kufe, D. W., Hallahan, D. E., and Weichselbaum, R. R. (1998). Adenoviral TNF-alpha gene therapy and radiation damage tumor vasculature in a human malignant glioma xenograft. *Gene Ther.* **5,** 293–300.

151. Mauceri, H. J., Hanna, N. N., Staba, M. J., Beckett, M. A., Kufe, D. W., and Weichselbaum, R. R. (1999). Radiation-inducible gene therapy. *C. R. Acad. Sci. III* **322,** 225–228.

152. Mauceri, H. J., Hanna, N. N., Wayne, J. D., Hallahan, D. E., Hellman, S., and Weichselbaum, R. R. (1996). Tumor necrosis factor alpha (TNF-alpha) gene therapy targeted by ionizing radiation selectively damages tumor vasculature. *Cancer Res.* **56,** 4311–4314.

153. Chung, T. D., Mauceri, H. J., Hallahan, D. E., Yu, J. J., Chung, S., Grdina, W. L., Yajnik, S., Kufe, D. W., and Weichselbaum, R. R. (1998). Tumor necrosis factor-alpha-based gene therapy enhances radiation cytotoxicity in human prostate cancer. *Cancer Gene Ther.* **5,** 344–349.

154. Joki, T., Nakamura, M., and Ohno, T. (1995). Activation of the radiosensitive EGR-1 promoter induces expression of the herpes simplex virus thymidine kinase gene and sensitivity of human glioma cells to ganciclovir. *Hum. Gene Ther.* **6,** 1507–1513.

155. Marples, B., Scott, S. D., Hendry, J. H., Embleton, M. J., Lashford, L. S., and Margison, G. P. (2000). Development of synthetic promoters for radiation-mediated gene therapy. *Gene Ther.* **7,** 511–517.

156. Scott, S. D., Marples, B., Hendry, J. H., Lashford, L. S., Embleton, M. J., Hunter, R. D., Howell, A., and Margison, G. P. (2000). A radiation-controlled molecular switch for use in gene therapy of cancer. *Gene Ther.* **7,** 1121–1125.

157. Gorecki, M., Beck, Y., Hartman, J. R., Fischer, M., Weiss, L., Tochner, Z., Slavin, S., and Nimrod, A. (1991). Recombinant human superoxide dismutases: production and potential therapeutical uses. *Free Radic. Res. Commun.* **12–13,** 401–410.

158. Zwacka, R. M., Dudus, L., Epperly, M. W., Greenberger, J. S., and Engelhardt, J. F. (1998). Redox gene therapy protects human IB-3 lung epithelial cells against ionizing radiation-induced apoptosis. *Hum. Gene Ther.* **9,** 1381–1386.

29

Virotherapy with Replication-Selective Oncolytic Adenoviruses: A Novel Therapeutic Platform for Cancer

DAVID KIRN

Imperial Cancer Research Fund
Program for Viral and Genetic Therapy of Cancer
Imperial College School of Medicine
Hammersmith Hospital
London, W11 OHS, United Kingdom

I. INTRODUCTION

Most currently available therapies for metastatic solid tumors fail as a result of inadequate antitumoral potency and/or an overly narrow therapeutic index between cancerous and normal cells. Countless changes in dose, frequency and/or combinations of standard cytotoxic chemotherapies or radiotherapy have had at best a modest impact on patient outcome in the metastatic setting. Although standard chemotherapies and radiotherapy target a variety of different structures within cancer cells, almost all of them kill cancer cells through the induction of apoptosis. Apoptosis-resistant clones almost universally develop following standard therapy for metastatic solid epithelial cancers (e.g., nonsmall-cell lung, colon, breast, prostate, pancreatic), even if numerous high-dose chemotherapeutic agents are used in combination. The overall survival rates for most metastatic solid tumors have changed relatively little despite decades of work with this approach (adjuvant therapy, in contrast, applied by definition when tumor burden is low, has resulted in clinically significant improvements in mortality). Novel therapeutic approaches must therefore have greater potency and greater selectivity than currently available treatments, and they should have novel mechanisms of action that will not lead to cross-resistance with existing approaches (i.e., do not rely exclusively on apoptosis induction in cancer cells).

Tumor-targeted oncolytic viruses (virotherapy with replication-selective viruses) appear to have these characteristics. Viruses have evolved over millions of years to infect target cells, multiply, cause cell death and release of viral particles, and finally spread in human tissues. Their ability to replicate in tumor tissue allows for amplification of the input dose (e.g., 1000- to 10,000-fold increases) at the tumor site, while their lack of replication in normal tissues results in efficient clearance and reduced toxicity (Fig. 1 [see also color insert]). This selective replication within tumor tissue can theoretically increase the therapeutic index of these agents dramatically over standard replication-incompetent approaches. Also, viruses lead to infected cell death through a number of unique and distinct mechanisms. In addition, to direct lysis at the conclusion of the replicative cycle, viruses can kill cells through expression of toxic proteins, induction of both inflammatory cytokines and T-cell-mediated immunity, and enhancement of cellular sensitivity to their effects. Therefore, because activation of classical apoptosis pathways in the cancer cell is not the exclusive mode of killing, cross-resistance with standard chemotherapeutics or radiotherapy is much less likely to occur.

Revolutionary advances in molecular biology and genetics have led to a fundamental understanding of both (1) the replication and pathogenicity of viruses and (2) carcinogenesis. These advances have allowed novel agents to be engineered to enhance their safety and/or their antitumoral potency. Over the past decade, genetically engineered viruses in development have included adenoviruses, herpesviruses, and vaccinia. Inherently tumor-selective viruses such as reovirus, autonomous parvoviruses, Newcastle disease virus, measles virus strains, and vesicular stomatitis virus have each been characterized. Each of these agents has shown tumor selectivity *in vitro* and/or *in vivo,* and efficacy has been demonstrated in murine tumor models, with many of these agents following intratumoral, intraperitoneal, and/or intravenous routes of administration.

Although preclinical data reported with these agents has been encouraging, many critical questions have awaited results from clinical trials. Viral agents such as adenovirus have complex biologies, potentially including species-specific interactions with host-cell machinery and/or immune response effectors [1,2]. Antitumoral efficacy and safety studies with these viruses have been performed in rodent or primate

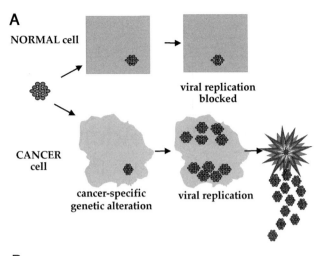

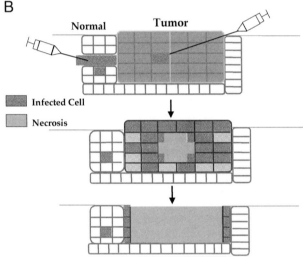

FIGURE 1 Schematic representation of tumor-selective viral replication and cell killing (panel A) and tumor-selective tissue necrosis (panel B). (See color insert.)

models, and all published animal tumor model data with replication-selective adenoviruses have come from immunodeficient mouse–human tumor xenograft models [3–5]; therefore, data from cancer patients have been eagerly awaited. Now, after over 5 years of clinical development with *dl*1520, roughly 15 clinical trials have been completed and recently analyzed involving approximately 250 patients. This article will review the discovery and development of replication-selective oncolytic adenoviruses, with an emphasis on recently acquired data from phase I and II clinical trials. The goal will be to summarize: (1) the genetic targets and mechanisms of selectivity for these agents; (2) clinical trial data and what they have taught us to date about the promise but also the potential hurdles to be overcome with this approach; and (3) future approaches to overcome these hurdles.

II. ATTRIBUTES OF REPLICATION-SELECTIVE ADENOVIRUSES FOR CANCER TREATMENT

A number of efficacy, safety, and manufacturing issues must be assessed when considering a virus species for development as an oncolytic therapy. First, by definition the virus must replicate in and destroy human tumor cells. An understanding of the genes modulating infection, replication, or pathogenesis is necessary if rational engineering of the virus is to be possible. Because most solid human tumors have relatively low growth fractions, the virus should ideally infect noncycling cells. In addition, receptors for viral entry must be expressed on the target tumor(s) in patients [6]. From a safety standpoint, the parental wild-type virus should ideally cause only mild, well-characterized human disease(s). Nonintegrating viruses have potential safety advantages, as well. A genetically stable virus is desirable from both safety and manufacturing standpoints. Finally, the virus must be amenable to high-titer production and purification under good manufacturing practices (GMP) guidelines for clinical studies. Human adenoviruses have these characteristics and are therefore excellent candidates for therapeutic development.

III. BIOLOGY OF HUMAN ADENOVIRUS

Adenovirus biology is reviewed in detail elsewhere [7]. Roughly 50 different serotypes of human adenovirus have been discovered; the two most commonly studied are types 2 and 5 (group C). All adenoviruses have linear, double-stranded DNA genomes of approximately 38 kb. The capsid is nonenveloped and is comprised of the structural proteins hexon, penton (binds $\alpha_V\beta_{3,5}$ integrins for virus internalization), and fiber (binds coxsackie and adenovirus receptor, CAR) (Fig. 2 [see also color insert]). The adenovirus life-

cycle includes the following steps: (1) virus entry into the cell following CAR and integrin binding, (2) release from the endosome and entry into the nucleus, (3) expression of early region gene products, (4) cell entry into S-phase, (5) prevention of p53-dependent and –independent apoptosis, (6) shut-off of host cellular protein synthesis, (7) viral DNA replication, (8) viral structural protein synthesis, (9) virion assembly in the nucleus, (10) cell death, and (11) virus release. The E3 region encodes a number of gene products responsible for immune response evasion [8,9]. The gp-19-kDa protein inhibits major histocompatibility complex (MHC) class I expression on the cell surface (i.e., avoidance of cytotoxic T-lymphocyte-mediated killing) [10], and the E3 10.4/14.5-kDa (RID complex) and 14.7-kDa proteins inhibit apoptosis mediated by FasL or tumor necrosis factor (TNF) [9,11].

IV. MECHANISMS OF ADENOVIRUS-MEDIATED CELL KILLING

Adenovirus replication within a target tumor cell can lead to cell destruction by several mechanisms (Table 1). Viral proteins expressed late in the course of infection are directly cytotoxic, including the E3-11.6 adenovirus death protein [12] and E4orf4. Deletion of these gene products results in a significant delay in cell death. In addition, E1A expression early during the adenovirus lifecycle induces cell sensitivity to TNF-mediated killing [13]. This effect is inhibited by the E3 proteins 10.4/14.5 and 14.7; deletion of these E3 proteins leads to an increase in TNF expression *in vivo* and enhanced cell sensitivity to TNF [2]. Finally, viral replication in and lysis of tumor cells has been shown to promote the induction

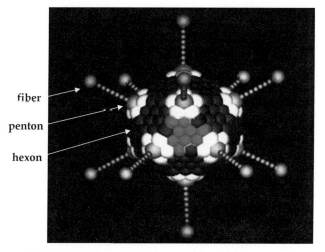

FIGURE 2   Human adenovirus coat structure. (See color insert.)

TABLE 1   Potential Mechanisms of Antitumoral Efficacy with Replication-Selective Adenoviruses

Mechanism	Examples of adenoviral genes modulating effect
Direct cytotoxicity due to viral proteins	E3-11.6-kDa, E4ORF4
Augmentation of antitumoral immunity	
CTL infiltration, killing	E3-gp-19-kDa[a]
Tumor cell death, antigen release	E3-11.6-kDa
Immunostimulatory cytokine induction	E3-10.4/14.5, 14.7 kDa[a]
Antitumoral cytokine induction (e.g., TNF)	E3-10.4/14.5, 14.7 kDa[a]
Enhanced sensitivity to cytokines (e.g., TNF)	E1A
Sensitization to chemotherapy	Unknown (? E1A, others)
Expression of exogenous therapeutic genes	NA

[a]Viral protein may inhibit antitumoral mechanism.
Note: CTL, cytotoxic T-lymphocyte; TNF, tumor necrosis factor; NA, not applicable.

of cell-mediated immunity to uninfected tumor cells in model systems with other viruses [14,15]; whether this will occur in patients and with adenovirus remains to be determined.

V. APPROACHES TO OPTIMIZING TUMOR-SELECTIVE ADENOVIRUS REPLICATION

Two broad approaches are currently being used to engineer tumor-selective adenovirus replication. One is to limit the expression of the E1A gene product to tumor tissues through the use of tumor- and/or tissue-specific promoters. E1A functions to stimulate S-phase entry and to transactivate both viral and cellular genes that are critical for a productive viral infection [16]. A second broad approach to optimizing tumor selectivity is to delete gene functions that are critical for efficient viral replication in normal cells but not in tumor cells (described later).

Tissue- or tumor-specific promoters can replace endogenous viral sequences in order to restrict viral replication to a particular target tissue. For example, the prostate-specific antigen (PSA) promoter/enhancer element has been inserted upstream of the E1A gene; the result is that viral replication correlates with the level of PSA expression in a given cell [3]. This virus, CN706 (Calydon Pharmaceuticals, CA), is currently in a phase I clinical trial of intratumoral injection for patients with locally recurrent prostate carcinoma. A second prostate-specific enhancer sequence has been inserted upstream of the E1B region in the CN706 virus; the use of these two prostate-specific enhancer elements to drive separate early gene regions has led to improved selectivity over the first generation virus [17]. A similar approach has been pursued by other groups using tissue-specific promoters to drive E1A expression selectively in specific carcinomas (e.g., alpha-fetoprotein, carcinoembryonic antigen, MUC-1) [18] (D. Kufe, in press).

A second general approach is to complement loss-of-function mutations in cancers with loss-of-function mutations within the adenovirus genome. Many of the same critical regulatory proteins that are inactivated by viral gene products during adenovirus replication are also inactivated during carcinogenesis [19,20–22]. Because of this convergence, the deletion of viral genes that inactivate these cellular regulatory proteins can be complemented by genetic inactivation of these proteins within cancer cells [23,24]. The deletion approach was first described with herpesvirus. Martuza et al. [25] deleted the thymidine kinase gene (dlsptk) and subsequently the ribonucleotide reductase gene (G207) [26] to engineer replication selectivity. Two adenovirus deletion mutants have been described. The first, dl1520 (ONYX-015) was hypothesized to replicate selectively in p53-deficient tumor cells (see later discussion). A second class of deletion mutants has now been described in E1A. Mutants in the E1A con-

served region 2 are defective in pRB binding. These viruses are being evaluated for use against tumors with pRB pathway abnormalities [24,27]. With dl922/947, for example, S-phase induction and viral replication are reduced in quiescent normal cells, whereas replication and cytopathic effects are not reduced in tumor cells; interestingly, dl922/947 demonstrates significantly greater potency than dl1520 both in vitro and in vivo [24]. In a nude mouse–human tumor xenograft model, intravenously administered dl922/947 had significantly superior efficacy to even wild-type adenovirus [28]. Unlike the complete deletion of E1B-55-kDa in dl1520, these mutations in E1A are targeted to a single conserved region and may therefore leave intact other important functions of the gene product.

VI. BACKGROUND: dl1520 (ONYX-015)

One approach to engineering replication selectivity is to delete viral genes that are necessary for efficient replication in normal cells but are expendable in tumor cells. This pioneering approach was first described with herpesvirus. Martuza et al. [25] deleted the thymidine kinase gene (dlsptk) and subsequently the ribonucleotide reductase gene (G207) [26] to engineer replication-selectivity. dl1520 (ONYX-015) was the first adenovirus described to mirror this approach. McCormick hypothesized that an adenovirus with deletion of a gene encoding a p53-inhibitory protein, E1B-55-kDa, would be selective for tumors that already had inhibited or lost p53 function [34]. p53 function is lost in the majority of human cancers through mechanisms including gene mutation, overexpression of p53-binding inhibitors (e.g., MDM2, human papillomavirus E6) and loss of the p53-inhibitory pathway modulated by p14^{ARF} [29–31]. However, the precise role of p53 in the inhibition of adenoviral replication has not been defined to date. In addition, other adenoviral proteins also have direct or indirect effects on p53 function (e.g., E4orf6, E1B-19-kDa, E1A) [32]. Finally, E1B-55-kDa itself has important viral functions that are unrelated to p53 inhibition (e.g., viral mRNA transport, host cell protein synthesis shut-off) (Fig. 3) [33].

Not surprisingly, therefore, the exact role of p53 in the replication-selectivity of dl1520 has been difficult to confirm despite extensive in vitro experimentation by many groups. E1B-55-kDa gene deletion was associated with decreased replication and cytopathogenicity in p53(+) tumor cells versus matched p53(−) tumor cells, relative to wild-type adenovirus, in RKO and H1299 cells [34,35]. However, conflicting data on the role of p53 in modulating dl1520 replication and/or cytopathic effects (cpe) have come from different cell systems; no p53 effect was demonstrated in matched U2OS cells, for example [36]. It is clear that many other cellular factors independent of p53 play critical roles in determining the sensitivity of cells to dl1520 [35,37–39] and that the role of

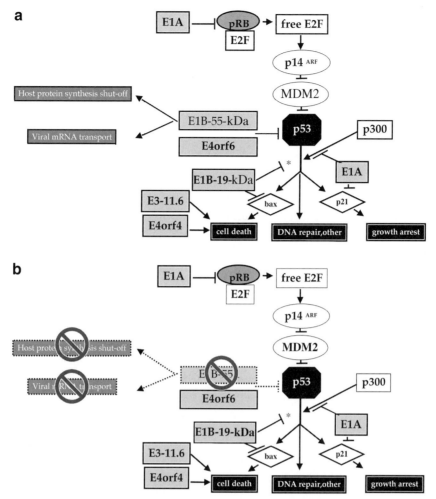

FIGURE 3   Diagram of both p53 pathway interactions with adenoviral gene products and functions of E1B-55-kDa: complexity of cancer cell and adenoviral biology. Note that adenoviral proteins target multiple components of this pathway at sites upstream of p53, downstream of p53, and at the level of p53 itself. Examples of p53-regulated cell functions are shown, as are the known functions of E1B-55-kDa. In addition to the loss of p53 binding when E1B-55-kDa was deleted in *dl*1520 (ONYX-015), these functions are also lost.

p53 in regulating the replication of *dl*1520 can vary depending on the cell line. In addition, the significant attenuation of this virus relative to wild-type adenovirus in most tumor cells studied to date, presumably due to the loss of other critical viral functions, is a potential drawback to this therapeutic candidate. Clinical trials were ultimately necessary to determine the selectivity and clinical utility of *dl*1520.

VII. CLINICAL TRIAL RESULTS WITH WILD-TYPE ADENOVIRUS: FLAWED STUDY DESIGN

Over the last century a diverse array of viruses were injected into cancer patients by various routes, including adenovirus, *Bunyamwara,* coxsackie, dengue, feline panleukemia,

Ilheus, mumps, Newcastle disease, vaccinia, and West Nile [40–43]. These studies illustrated both the promise and the hurdles to overcome with oncolytic viral therapy. Unfortunately, these previous clinical studies were not performed to current clinical research standards; therefore, none gives interpretable and definitive results. At best, these studies are useful in generating hypotheses that can be tested in future trials.

Although suffering from many of the trial design flaws listed below, a trial with wild-type adenovirus is one of the most useful for hypothesis generation, as well as for illustrating how clinical trial design flaws severely curtail the utility of the study results. The knowledge that adenoviruses could eradicate a variety of tumor cells *in vitro* led to a clinical trial in the 1950s with wild-type adenovirus. Ten different serotypes were used to treat 30 cervical cancer patients [43]. Forty total treatments were administered by either direct intratumoral injection ($n = 23$), injection into the artery perfusing the tumor ($n = 10$), treatment by both routes ($n = 6$), or intravenous administration ($n = 1$). Characterization of the material injected into patients was minimal. The volume of viral supernatant injected is reported, but actual viral titers/doses are not; injection volumes (and by extension doses) varied greatly. When possible, the patients were treated with a serotype to which they had no neutralizing antibodies present. Corticosteroids were administered as nonspecific immunosuppressive agents in roughly half of the cases. Therefore, no two patients were treated in identical fashion.

Nevertheless, the results are intriguing. No significant local or systemic toxicity was reported. This relative safety is notable, given the lack of preexisting immunity to the serotype used and concomitant corticosteroid use in many patients. Some patients reported a relatively mild viral syndrome lasting 2–7 days (severity not defined); this viral syndrome resolved spontaneously. Infectious adenovirus was recovered from the tumor in two thirds of the patients for up to 17 days postinoculation.

Two thirds of the patients had a "marked to moderate local tumor response" with necrosis and ulceration of the tumor (definition of "response" not reported). None of the seven control patients treated with either virus-free tissue culture fluid or heat-inactivated virus had a local tumor response (statistical significance not reported). Therefore, clinically evident tumor necrosis was only reported with viable virus. Neutralizing antibodies increased within 7 days after administration. Although the clinical benefit to these patients is unclear, and all patients eventually had tumor progression and died, this study did demonstrate that wild-type aden-

oviruses can be safely administered to patients and that these viruses can replicate and cause necrosis in solid tumors despite a humoral immune response. The maximally tolerated dose, dose-limiting toxicity, objective response rate, and time to tumor progression, however, remain unknown for any of these serotypes by any route of administration.

VIII. A NOVEL STAGED APPROACH TO CLINICAL RESEARCH WITH REPLICATION-SELECTIVE VIRUSES: *dl*1520 (ONYX-015)

For the first time since viruses were first conceived as agents to treat cancer over a century ago, we now have definitive data from numerous phase I and II clinical trials with a well-characterized and –quantitated virus. *dl*1520 (ONYX-015, Onyx Pharmaceuticals, Richmond, CA) is a novel agent with a novel mechanism of action. This virus was to become the first virus to be used in humans that had been genetically engineered for replication selectivity. We predicted that both toxicity and efficacy would be dependent on multiple factors, including: (1) the inherent ability of a given tumor to replicate and shed the virus; (2) the location of the tumor to be treated (e.g., intracranial vs. peripheral), and (3) the route of administration of the virus. In addition, we felt it would be critical to obtain biological data on viral replication, antiviral immune responses, and their relationship to antitumoral efficacy in the earliest phases of clinical development.

We therefore designed and implemented a staged clinical research and development approach (Fig. 4). The goal of

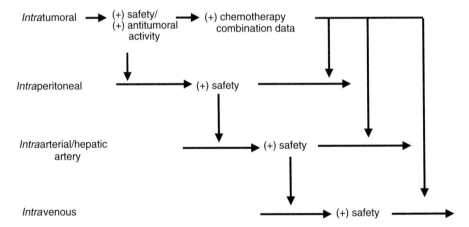

FIGURE 4   Staged clinical research and development approach for a replication-selective agent in cancer patients. Following demonstration of safety and biological activity by the intratumoral route, trials were sequentially initiated to study intracavitary instillation (initially intraperitoneal), intraarterial infusion (initially hepatic artery), and eventually intravenous administration. In addition, only patients with advanced and incurable cancers were initially enrolled on trials. Only after safety had been demonstrated in terminal cancer patients were trials initiated for patients with premalignant conditions. Finally, clinical trials of combinations with chemotherapy were initiated only after the safety of *dl*1520 as a single agent had been documented by the relevant route of administration.

this approach was to sequentially increase systemic exposure to the virus only after safety with more localized delivery had been demonstrated. Following demonstration of safety and biological activity by the intratumoral route, trials were sequentially initiated to study intracavitary instillation (initially intraperitoneal), intraarterial infusion (initially hepatic artery), and eventually intravenous administration. In addition, only patients with advanced and incurable cancers were initially enrolled on trials. Only after safety had been demonstrated in terminal cancer patients were trials initiated for patients with premalignant conditions. Finally, clinical trials of combinations with chemotherapy were initiated only after the safety of dl1520 as a single agent had been documented by the relevant route of administration.

A. Intratumoral Indications

Cancer patients can benefit from the effective local therapy of an established tumor mass if the target tumor mass causes morbidity or death before other masses do. For example, patients with recurrent glioblastoma multiforme or head and neck cancer frequently die from local tumor progression without evidence of distant metastases. In contrast, eradication of a localized skin lesion in a patient with widespread pulmonary or CNS metastases is unlikely to be of benefit. Patients with recurrent head and neck carcinomas were enrolled in the initial clinical trials because most suffer severe morbidity, and even mortality, from the local/regional progression of treatment-refractory tumors; therefore, intratumoral administration had the potential to cause substantial palliation and even survival prolongation. This population was also chosen because of the accessibility of superficial tumors for direct injection and biopsy in the outpatient clinic setting. Finally, patients with tumors in superficial neck and oral locations would presumably better tolerate peritumoral inflammation and swelling than patients with intraparenchymal tumors (e.g., intracranial, intrapulmonary, or intrahepatic). Once the safety of intratumoral injection was demonstrated in the superficial neck and oral regions, trials of intratumoral injection in solid organs (pancreas, liver) were carried out.

B. Intracavitary Indications

Tumor types that spread and/or cause complications primarily within specific body cavities are potentially amenable to intracavitary administration of therapeutic agents. Examples include mesothelioma (pleural cavity), ovarian carcinoma (peritoneal cavity), and recurrent superficial bladder carcinoma (bladder). In addition, several premalignant conditions are also amenable to superficial intracavitary administration, including Barrett's esophagus and oral dysplasias (e.g., oral leukoplakia). Intraperitoneal administration to patients with advanced, refractory ovarian carcinoma was followed by intraesophageal instillation in patients with Barrett's esophagus. The virus was sequestered within the affected region of the esophagus following instillation through a Wilson–Cook catheter by occlusive proximal and distal balloons. Finally, oral dysplasias were targeted through administration as a mouthwash.

C. Vascular Delivery: Intraarterial and Intravenous Administration

Although patients with the indications listed above can potentially benefit from local–regional therapy, systemic antitumoral efficacy can have a much greater impact on overall cancer-related mortality. Preclinical studies proved that intravenous adenovirus could infect, replicate within, and inhibit the growth of established metastatic tumor [4,23]. However, nude mouse–human tumor xenograft models were of unknown relevance to human cancer patients from both safety and efficacy standpoints. Targeted intraarterial infusions were studied in the first intravascular trial. Colorectal carcinoma metastases to the liver cause morbidity and death in a large proportion of these patients, and these metastases receive $\geq$90% of their bloodflow from the hepatic artery. Hepatic artery infusions had therefore been used previously to target colorectal liver metastases with a variety of agents [44]. Once safety by this route of administration had been demonstrated, intravenous trials were initiated in patients with lung metastases.

IX. RESULTS FROM CLINICAL TRIALS WITH dl1520 (ONYX-015)

A. Toxicity

No maximally tolerated dose or dose-limiting toxicities were identified at doses up to 2×10^{12} particles administered by intratumoral injeciton. This safety is true not only for tumors injected in superficial neck and oral sites, but also for intrahepatic and intrapancreatic tumor masses, as well. No clinically significant hepatitis or pancreatitis was demonstrated. Flu-like symptoms were the most common associated toxicities. No clear association between flu-like symptoms and viral dose was demonstrable. Phase I/II and phase II trials reported a similar lack of clinically significant toxicities. This safety is remarkable given the daily or even twice-daily dosing that was repeated every 1–3 weeks in the head and neck region or pancreas. Local complications of intratumoral injections in the pancreas appeared to be related to the endoscopic ultrasound procedure rather than to the agent itself; these included bacteremia, cyst formation, and a tear of the doudenal wall (45). Each of these complications was avoided once procedural changes were made and prophylactic antibiotic treatment was mandated.

Intraperitoneal, intraarterial, and intravenous administration were also remarkably well tolerated, in general. Intraperitoneal administration was feasible at doses up to 10^{13}

particles divided over 5 days [46]. The most common toxicites included fever, abdominal pain, nausea/vomiting and bowel motility changes (diarrhea, constipation). The severity of the symptoms appeared to correlate with tumor burden. Patients with heavy tumor burdens reached a maximally tolerated dose at 10^{12} particles (dose-limiting toxicities were abdominal pain and diarrhea), whereas patients with a low tumor burden tolerated 10^{13} without significant toxicity.

No dose-limiting toxicities were reported following repeated intravascular injection at doses up to 2×10^{12} particles (hepatic artery) [47] or 2×10^{13} particles (intravenous) [48,61]. Fever, chills, and asthenia following intravascular injection were more common and more severe than after intratumoral injections (grade 2–3 fever and chills vs. grade 1). Dose-related transaminitis was reported infrequently. The transaminitis was typically transient (<10 days) and low grade (grade 1–2) and was not clinically relevant. At the highest intravenous doses administered, aspartate transaminase/alanine transaminase (AST/ALT) levels reached approximately 3–5 times the upper limit of normal for the assay; although the maximally tolerated intravenous dose has not yet been defined, it may be close to this dose. Further dose escalation was limited by supply of the virus. A single patient on the hepatic artery trial had a systemic proinflammatory cytokine response on cycle 4 which led to transient vascular leak at tumor sites and metabolic acidosis; unusually high levels of TNF and interferon-γ were associated with an idiosyncratic inhibition of IL-10 [47].

B. Viral Replication

Viral replication was assessed by two methods during clinical trials with dl1520 (Table 2). The first was *in situ* hybridization for adenoviral DNA in tumor biopsy samples (direct), while the second tested blood samples for adenoviral genomes by quantitative PCR. Histologic analysis of tumor biopsies is attractive because the nature and distribution of the infected cells can be evaluated; however, the evaluation of biopsy samples to assess replication has severe limitations. First, false negatives are possible given the small amount of tissue obtained. Posttreatment biopsies can also be inevaluable if necrotic tissue is obtained. Ethical and practical considerations also limit the number of samples that can be obtained over time, particularly in tissues that are not superficial and therefore require invasive procedures to access them. Quantitative PCR is far more practical. Blood samples can be conveniently obtained and analyzed at multiple time points. Replication is suggested if the initial clearance of the input genomes is followed by a delayed rise in genome concentration (e.g., >10-fold increase over the lower-limit-of-detection after 48–72 hours). False-negative results are possible, however, as the lower limit of detection is 10^4 genomes/mL, and viral shedding into the bloodstream

is required for detection. It was encouraging that the frequency of replication detected in head and neck cancer trials by biopsy staining was nearly identical to that determined by indirectly plasma sample polymerase chain reaction (PCR) testing. At this time, it appears that these two approaches each have merits and that they are complementary.

Viral replication has been documented at early time points after intratumoral injection in head and neck cancer patients by both tests (Fig. 5). Roughly 70% of patients with either biopsy analysis or plasma testing by PCR had evidence of replication on days 1–3 after the last treatment. In contrast, day 14–17 samples were uniformly negative. This time-course for replication mirrors closely the clinical evidence for biological activity (e.g., local inflammation and necrosis). Intratumoral injection of liver metastases (primarily colorectal) led to similar PCR results at the highest doses of a phase I trial; high-quality biopsy samples could not be collected given the location of these tumors. Patients with injected pancreatic tumors, in contrast, showed no evidence of viral replication by plasma PCR or fine-needle aspiration. Similarly, intraperitoneal dl1520 could not be shown to reproducibly infect ovarian carcinoma cells within the peritoneum. None of the plasma PCR samples was positive, and none of the 12 peritoneal fluid samples was positive. Therefore, different tumor types can vary dramatically in their permissiveness for viral infection and replication.

Proof of concept for tumor infection following intra-arterial or intravenous administration with human adenovirus has been achieved. Following initial clearance of input genomes, approximately half of the roughly 25 patients receiving hepatic artery infusions of 2×10^{12} particles were positive by PCR 3–5 days following treatment [60]. Three of four patients with metastatic carcinoma to the lung treated intravenously with $\geq 2 \times 10^{12}$ particles were positive for genomes by PCR on day 3 (± 1). A single lung metastasis biopsy was positive for viral replication [61]. Therefore, it is feasible to infect distant tumor nodules following intravenous or intraarterial administration.

C. Immune Response

Neutralizing antibody titers to the coat (Ad-5) of dl1520 were positive but relatively low in roughly 50–60% of all clinical trial patients at baseline. Antibody titers increased uniformly following administration of dl1520 by any of the routes tested, in some cases to levels >1:80,000. Antibody increases occurred regardless of evidence for replication or shedding into the bloodstream. Flu-like symptoms (fevers, rigors) were significantly more frequent and severe with intravascular administration than with intratumoral injections. The acute inflammatory cytokine response to hepatic arterial infusion was evaluated using reverse-transcription PCR (RT-PCR) for specific cytokine mRNAs from buffy coat

TABLE 2 Viral Replication Data from Phase I and II Trials of *dl1520* (ONYX-015): Intratumoral, Intraperitoneal, Intraarterial or Intravenous Injection

Route of administration	Tumor type	Phase	Dose/cycle (particles)	Regimen (cycle frequency)	Tumor biopsy				Blood quantitative (PCR)			
					Days posttreatment	No. positive	No. evaluable	Percent positive	Days posttreatment	No. positive	No. evaluable	Percent positive
Intratumoral	Head and neck	I	2×10^8–2×10^{12}	Single dose (q 4 wk)	5	4	16	25	Not done			
					1–3	5	7	71				
Intratumoral	Head and neck	II	10^{12}	Daily × 5 (q 3 wk)	7–10	2	4	50	10	2	19	11
					14–17	0	10	0	17	0	19	0
Intratumoral	Gastrointestinal, primarily colorectal	I	2×10^8–2×10^{12}	Single dose (q 4 wk)					5–10: Overall	7	19	37
									High dose	5	6	83
Intratumoral	Pancreatic	I	2×10^8–2×10^{12}	Single dose (q 4 wk)					15	0	19	0
Intratumoral	Pancreatic	I/II	2×10^{11}	Single dose (day 1, 5, 8, 15)	8				5	0	22	0
									15	0	22	0
Intraperitoneal	Ovarian	I	10^{11}–10^{13}	Daily × 5 (q 3 wk)	1–4	0	12	0	Not done			
Intraarterial (hepatic artery)	Gastrointestinal, primarily colorectal	I	2×10^9–2×10^{11}	Single dose (q 1 wk; 2 wk on, 2 off)	4–7				5	0	16	0
									15	0	16	0
		II	6×10^{11}–2×10^{12}						3	6	15	40
Intravenous	Metastatic carcinoma in lung	I	2×10^8–2×10^{12}	Single dose (q 1 wk; 3 wk on, 1 off)	4: Overall	1	9	11	8: Low dose	0	4	0
					High dose	1	3	33	High dose	3	5	60

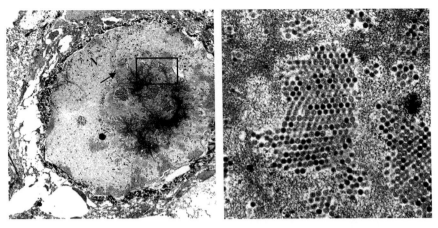

FIGURE 5   Replication of adenoviral agent (*dl*1520) in the nucleus of a squamous carcinoma cell from the head and neck region of a patient 3 days after intratumoral virus injection. The dark-staining individual particles (arrow) and clusters within the nucleus (box) are adenoviral particles.

leukocyte samples. The levels of the following were determined prior to treatment, 3 hours post- and 18 hours post-treatment: IL-1, IL-6, IL-10, interferon-gamma (IFN-γ), and TNF. Significant increases were demonstrated within 3 hours for IL-1, IL-6, and TNF and to a lesser extent IFN-γ; all cytokines were back down to pretreatment levels by 18 hours. In contrast, IL-10 did not increase until 18 hours. Future analyses will attempt to correlate clinical outcomes with cytokine levels.

D. Efficacy with *dl*1520 (ONYX-015) as a Single Agent

The single-agent efficacy of ONYX-015 is outlined in Table 3. Two phase II trials enrolled a total of 40 patients with recurrent head and neck cancer. Tumors were treated very aggressively with 6–8 daily needle passes for 5 consecutive days (30–40 needle passes per 5- day cycle; $n = 30$) and 10 to 15 per day on a second trial (50–75 needle passes per cycle; $n = 10$). The median tumor volume on these studies was approximately 25 cm³; an average cubic centimeter, of tumor therefore, received an estimated 4–5 needle passes per cycle. Despite the intensity of this treatment, the unconfirmed response rate was only 14%. Therefore, even in a tumor that can be extensively and repeatedly injected, the majority of injected tumors did not respond. Interestingly, there was no correlation between evidence of antitumoral activity and neutralizing antibody levels at baseline or post-treatment.

Phase I and I/II data are available for other tumor types. No objective responses were demonstrated in patients with tumor types that could not be so aggressively injected (due to their deep locations). Although some evidence of minor shrinkage or necrosis was obtained, no objective responses were documented with intratumoral injection of either

pancreatic cancer (phase I and II trials; $n = 43$ patients) or gastrointestinal carcinomas (phase I trial, primarily colorectal; $n = 19$ patients). Similarly, no responses were seen following intraperitoreal (i.p.) administration in 16 ovarian cancer patients (phase I) or intravenous (i.v.) administration to 10 patients with metastatic carcinomas (phase I). Although some of these patients were treated during phase I portions of these trials, during which segment tumor response is not a primary endpoint, the lack of responses is notable. In summary, single-agent responses across all studies were rare; therefore, combinations with chemotherapy were explored.

E. Efficacy in Combination with Chemotherapy: Potential Synergy Discovered

Evidence for a potentially synergistic interaction between adenoviral therapy and chemotherapy have been obtained on multiple trials (Table 4). Encouraging clinical data have been obtained in patients with recurrent head and neck cancer treated with intratumoral *dl*1520 in combination with intravenous cisplatin and 5-fluorouracil [49]. Of the 37 patients treated, 19 responded (54%, intent to treat; 63%, evaluable); this compares favorably with response rates to chemotherapy alone in previous trials (30–40%, generally). The time to tumor progression was also superior to previously reported studies; however, comparisons to historical controls are unreliable. We therefore used patients as their own controls whenever possible ($n = 11$ patients). Patients with more than one tumor mass had a single tumor injected with *dl*1520, while the other mass was left uninjected. Because both masses were exposed to chemotherapy, the effect of the addition of viral therapy to chemotherapy could be assessed. The *dl*1520-injected tumors were significantly more likely to respond

TABLE 3   Antitumoral Efficacy Data of *dl*1520 (ONYX-015) as a Single Agent: Intratumoral, Intraperitoneal, Intra-Arterial or Intravenous Injection

Route of administration	Tumor type	Phase	Dose/cycle (particles)	Regimen/ cycle frequency	≥50% tumor regression[a] number of responders/total (%)
Intratumoral	Head and neck	I	2×10^8–2×10^{12}	Single dose/ q 4 week	3/22 (14)
Intratumoral	Head and neck	II	10^{12}	Daily × 5/ q 3 week	Intent-to-treat: 4/30 (13) Confirmed,intent-to-treat[b]: 2/30 (7) Unconfirmed, evaluable: 4/19 (20)
Intratumoral	Gastrointestinal Liver metastases- colorectal, gastric, pancreatic	I	2×10^9–2×10^{12}	Single dose/ q 4 week	0/19 (0)
Intratumoral	Pancreatic (CT-guided)	I	2×10^8–2×10^{12}	Single dose/ q 4 week	0/22 (0)
Intratumoral	Pancreatic (endoscopic US)	I	2×10^{10} (n = 3)	Single dose/ days 1, 5, 8, 15	0/3 (0)
		II	2×10^{11} (n = 18)		0/18 (0)[c]
Intraperitoneal	Ovarian	I	10^{11}–10^{13}	Daily × 5/ q 3 wk.	0/16 (0)
Intravenous	Carcinoma metastatic to lung	I	2×10^8–2×10^{12}	Single dose/ q 1 wk. (3 wk. on, 1 off)	0/ 9 (0)

[a] Non-necrotic cross-sectional area used for response assessment (i.e., necrotic area subtracted from total cross-sectional area). All responses refer to shrinkage of the injected tumor mass only (i.e., distant, noninjected tumors not included. All responses were in tumors with a p53 gene mutation.

[b] Evaluable patients defined as those receiving >1 cycle of therapy and measurable tumor at baseline and at least one occasion > 6 weeks after treatment initiation (i.e., patients without follow-up tumor measurements after 1+cycles of treatment were excluded). Intent-to-treat analysis includes all patients receiving at least one dose of ONYX-015. The confirmed responses reflect those that were confirmed to be durable for ≥4 weeks on an intent-to-treat basis.

[c] Responses of single agent ONYX-015 determined after 4 cycles (on day 35) on the pancreatic EUS phase I/II trial. Subsequent cycles given with chemotherapy.

($p = 0.017$) and less likely to progress ($p = 0.06$) than were noninjected tumors. Noninjected control tumors that progressed on chemotherapy alone were subsequently treated with ONYX-015 in some cases; two of the four injected tumors underwent complete regressions. These data illustrate the potential of viral and chemotherapy combinations. The clinical utility of *dl*1520 in this indication will be definitively determined in an on-going phase III randomized trial.

A phase I/II trial of *dl*1520 administered by hepatic artery infusion in combination with intravenous 5-fluorouracil and leukovorin was carried out ($n = 33$ total) [47,60]. Following phase I dose escalation, 15 patients with colorectal carcinoma who had previously failed the same chemotherapy were treated with combination therapy after failing to respond to *dl*1520 alone; one patient underwent a partial response and roughly 10 had stable disease (2–7+ months). Chemotherapy-refractory tumors can therefore respond following the same chemotherapy in combination with hepatic artery infusions of adenovirus; the magnitude and frequency of this effect re-

main to be determined. In contrast, data from a phase I/II trial studying the combination of *dl*1520 and gemcitabine chemotherapy were disappointing ($n = 21$); the combination resulted in only two responses, and these patients had not received prior gemcitabine [45]. Therefore, potential synergy was demonstrated with *dl*1520 and chemotherapy in two tumor types that supported viral replication (head and neck, colorectal) but not in a tumor type that was resistant to viral replication (pancreatic).

X. RESULTS FROM CLINICAL TRIALS WITH *dl*1520 (ONYX-015): SUMMARY

*dl*1520 has been extremely well-tolerated at the highest practical doses that could be administered (2×10^{12} to 2×10^{13}) by intratumoral, intraperitoneal, intraarterial, and intravenous routes. The lack of clinically significant toxicity in the liver or other organs was remarkable. Flu-like symptoms (fever, rigors, asthenia) were the most common toxicities

TABLE 4   Evidence for Potential Synergy[a] Between *dl*1520 (ONYX-015) and Chemotherapy from Clinical Trials

Route of administration	Tumor type	Phase	Dose/cycle (particles)	Regimen (cycle frequency)	Evidence for potential synergy[a]
Intratumoral	Head and neck	II	10^{12}	ONYX-015 daily × 5 + cisplatin day 1 i.v.b; 5-FU days 1–5 c.i. (q3 wk)	ONYX-015-injected tumors significantly more likely to respond than matched, noninjected control tumors ($p = 0.017$; McNemar's test) ONYX-015-injected tumors less likely to progress than matched, noninjected control tumors ($p = 0.06$; log rank test) 2 of 4 tumors progressing on chemotherapy responded to same chemotherapy plus ONYX-015 Uncontrolled: response rate 63% vs. historical 30–40% with chemotherapy and 14% with ONYX-015 alone
Intratumoral	Pancreatic (endoscopic US)	I II	2×10^{10} ($n = 3$) 2×10^{11} ($n = 18$)	ONYX-015 single dose + gemcitabine i.v.b. (q 1 wk)	None; 2 of 21 patients responded to combination
Intraperitoneal	Ovarian	I	10^{11}–10^{13}	ONYX-015 daily × 5 (q 3 wk)	One patient had tumor responses (>50% reduction in CA-125) on platinum-based chemotherapy following ONYX-015, despite previous tumor progression on platinum-based chemotherapy alone and on ONYX-015 alone
Intraarterial (hepatic artery)	Gastrointestinal, liver metastases, primarily colorectal	I II	2×10^{9}–6×10^{11} 2×10^{12}	ONYX-015 single dose + 5-FU/leucovorin i.v.b. (q 4 wk)	One partial regression, approx. 10 stable disease (2–7+ months) to combination ONYX-015 plus 5-FU/leucovorin in patients with tumor progression on both single-agent ONYX-015 and on 5-FU/leucovorin alone
Intravenous	Metastatic carcinoma	I	2×10^{10}–2×10^{13}	Single dose (q 1 wk; 3 wk on, 1 off) then with weekly carboplatin/paclitaxel	N.A.

[a] Although synergy cannot be definitively proven in phase II clinical trials, these clinical trial results are consistent with synergy and/or a positive interaction between ONYX-015 and chemotherapy with cisplatin and 5-FU.

Note: i.v.b., intravenous bolus; c.i., continuous infusion; 5-FU, 5-fluorouracil; N.A., not available.

and were increased in patients receiving intravascular treatment. Acute inflammatory cytokines (including IL-1, IL-6, TNF, and interferon-γ) increased within hours following intraarterial infusion. Neutralizing antibodies increased in all patients, regardless of dose, route, or tumor type. Viral replication was documented in head and neck and colorectal tumors following intratumoral or intraarterial administration. Neutralizing antibodies did not block antitumoral activity in head and neck cancer trials of intratumoral injection; however, viral replication/shedding into the blood was inhibited by neutralizing antibodies. Single-agent antitumoral activity was minimal ($\cong$15%) in head and neck cancers that could be repeatedly and aggressively injected. No objective responses were documented with single-agent therapy in phase I or I/II trials in patients with pancreatic, colorectal, or ovarian carcinomas; these were not definitive efficacy studies. A favorable

and potentially synergistic interaction with chemotherapy was discovered in some tumor types and by different routes of administration.

XI. FUTURE DIRECTIONS

A. Why Has *dl*1520 ONYX-015 Failed as a Single Agent for Refractory Solid Tumors?

Future improvements with this approach will be possible if the reasons for *dl*1520 failure as a single agent and success in combination with chemotherapy are uncovered. Factors specific to this adenoviral mutant, as well as factors that may be generalizable to other viruses, should be considered. Regarding this particular adenoviral mutant, it is important to remember that this virus is significantly attenuated relative to wild-type adenovirus in most tumor cell lines *in vitro* and *in vivo,* including even p53 mutant tumors [35,36,38,50]. This is not an unexpected phenotype, as this virus has lost critical E1B-55-kDa functions that arc unrelated to p53, including viral mRNA transport. This attenuated potency is not apparent with other adenovirus mutants such as *dl*922/947 [28].

In addition, a second deletion in the E3 gene region (10.4/14.5 complex) may make this virus more sensitive to the antiviral effects of TNF; an immunocompetent animal model will need to be identified in order to resolve this issue. Factors likely to be an issue with any virus include barriers to intratumoral spread, antiviral immune responses, and inadequate viral receptor expression (e.g., CAR, integrins). Viral coat modifications may be beneficial if inadequate CAR expression plays a role in the resistance of particular tumor types [51].

B. Improving the Efficacy of Replication-Selective Agents

Given the high degree of safety, but to date disappointing single-agent efficacy, of *dl*1520 (ONYX-015) against advanced solid tumors, second-generation viruses will clearly be engineered for greater potency [71]. Mutations in the adenoviral genome can enhance selectivity and/or potency. For example, a promising adenoviral E1A CR-2 mutant (*dl*922/947) has been described that demonstrates not only tumor selectivity (based on the G1-S checkpoint status of the cell) but also significantly greater antitumoral efficacy *in vivo* compared to *dl*1520 (all models tested) and even wild-type adenovirus (in a breast cancer metastasis model) [24]. Another E1A mutant adenovirus has demonstrated replication and cytopathic effects based on the pRB status of the target cell [27]. Deletion of the E1B-19-kDa gene (antiapoptotic bcl-2 homolog) is known to result in a "large plaque" phenotype due to enhanced speed of cell killing [52]. This observation has now been extended to multiple tumor cell lines and primary tumor cell cultures [53,54]. A similar phenotype resulted from overexpression of the E3-11.6 adenovirus death protein [55]. It remains to be seen whether these *in vitro* observations are followed by evidence for improved efficacy *in vivo* over wild-type adenovirus.

Potency can also be improved by arming viruses with therapeutic genes (e.g., prodrug-activating enzymes and cytokines) [56–59]. Viral coat modifications may be beneficial

TABLE 5   Replication-Selective Microbiological Agents in Clinical Trials for Cancer Patients

Parental strain	Agent	Cell phenotype allowing selective replication	Genetic alterations
Engineered			
Adenovirus (2/5 chimera)	ONYX-015	Cells lacking p53 function (e.g., deletion, mutation, HPV infection)	E1B55kDa gene deletion
Herpes simplex virus -1	G207	Proliferating cells	• Ribonucleotide reductase disruption (lac-Z insertion into ICP6 gene) • Neuropathogenesis gene mutation (γ34.5 gene)
Adenovirus (serotype 5)	CN706	Prostate cells (malignant, normal)	E1A expression driven by PSE element
Adenovirus (2/5 chimera)	Ad5-CD/tk-rep	Cells lacking p53 function (e.g., deletion, mutation, HPV infection)	• E1B-55kD gene deletion • Insertion of HSV-TK/*CD* fusion
Vaccinia virus	Wildtype ± GM-CSF	Unknown	• None for selectivity • Immunostimulatory gene (GM-CSF) insertion
Salmonella typhimurium	Vion /VNP20009	Extracellular proliferation in tumor milleu (mechanism unknown: ? nutrient, hypoxia, immune clearance differences)	• Deletion of *msbB* (lipidA metabolism) • Deletion of *purI* (purine synthesis)
Nonengineered			
Newcastle Disease virus	73-T	Unknown	Unknown (serial passage on tumor cells)
Autonomous parvoviruses	H-1	Transformed cells • ↑ proliferation • ↓ differentiation • ras, p53 mutation	None
Reovirus	Reolysin[a]	Ras-pathway activation (e.g., ras mutation, EGFR signaling) and loss of interferon responsiveness	None

[a]not yet in clinical trials; to enter clinical trials in 2000.

Note: HPV, human papillomavirus; PSE, prostate-specific enhancer; LPS, lipopolysaccharide; EGFR, epidermal growth factor receptor.

if inadequate CAR expression plays a role in the resistance of particular tumor types [51]. Improved systemic delivery may require novel formulations or coat modifications, as well as suppression of the humoral immune response. Determination of the viral genes (e.g., E3 region) and immune response parameters mediating efficacy and toxicity will lead to immunomodulatory strategies. Finally, identification of the mechanisms leading to the potential synergy between replicating adenoviral therapy and chemotherapy may allow augmentation of this interaction. This understanding may then allow us to bolster this interaction.

XII. SUMMARY

Adenovirus has a number of attractive features as a replication-selective agent for cancer treatment. Clinical studies have demonstrated that replication-competent adenovirus treatment can be well tolerated and that tumor necrosis can result. The feasibility of adenovirus delivery to tumors through the bloodstream has also been demonstrated [4,60,61]. The inherent ability of replication-competent adenoviruses to sensitize tumor cells to chemotherapy was a novel discovery that has led to chemosensitization strategies. These data will support the further development of adenoviral agents, including second-generation constructs containing exogenous therapeutic genes to enhance both local and systemic antitumoral activity [56,62,63]. In addition to adenovirus, other viral species are being developed including herpesvirus, vaccinia, reovirus, and measles virus (Table 5). [15,25,40,64–68]. Because intratumoral spread also appears to be a substantial hurdle for viral agents, inherently motile agents such as bacteria may hold great promise for this field (Table 5) [69,70].

Given the limited ability of *in vitro* cell-based assays and murine tumor model systems to accurately predict the efficacy and therapeutic index of replication-selective adenoviruses in patients, we believe that the timely translation of encouraging adenoviral agents into well-designed clinical trials with relevant biological endpoints is critical [71]. Only then can the true therapeutic potential of these agents be realized. The clinical development of the first-generation adenovirus ONYX-015 (*dl*1520) has taught us a great deal about the hurdles to be overcome with the replication-selective adenovirus approach. It has also demonstrated, however, the potential of this novel therapeutic platform to improve and prolong the lives of cancer patients.

Acknowledgments

The following individuals have been instrumental in making this manuscript possible: Frank McCormick, John Nemunaitis, Stan Kaye, Tony Reid, Fadlo Khuri, James Abruzzesse, Eva Galanis, Joseph Rubin, Antonio Grillo-Lopez, Carla Heise, Larry Romel, Chris Maack, Sherry Toney, Nick LeMoine, Britta Randlev, Patrick Trown, Fran Kahane, and Margaret Uprichard.

References

1. Wold, W. S., Hermiston, T. W., and Tollefson, A. E. (1994). Adenovirus proteins that subvert host defenses. *Trends Microbiol.* **2**, 437–443.
2. Sparer, T. E., Tripp, R. A., Dillehay, D. L., Hermiston, T. W., Wold, W. S., and Gooding, L. R. (1996). The role of human adenovirus early region 3 proteins (gp19K, 10.4K, 14.5K, and 14.7K) in a murine pneumonia model. *J. Virol.* **70**, 2431–2439.
3. Rodriguez, R., Schuur, E. R., Lim, H. Y., Henderson, G. A., Simons, J. W., and Henderson, D. R. (1997). Prostate attenuated replication competent adenovirus (ARCA) CN706: a selective cytotoxic for prostate-specific antigen-positive prostate cancer cells. *Cancer Res.* **57**, 2559–2563.
4. Heise, C., Williams, A., Xue, S., Propst, M., and Kirn, D. (1999). Intravenous administration of ONYX-015, a selectively-replicating adenovirus, induces antitumoral efficacy. *Cancer Res.* **59**, 2623–2628.
5. Heise, C., Williams, A., Olesch, J., and Kirn, D. (1999). Efficacy of a replication-competent adenovirus (ONYX-015) following intratumoral injection: intratumoral spread and distribution effects. *Cancer Gene Ther.* **6**, 499–504.
6. Wickham, T. J., Segal, D. M., Roelvink, P. W., Carrion, M. E., Lizonova, A., Lee, G. M., and Kovesdi, I. (1996). Targeted adenovirus gene transfer to endothelial and smooth muscle cells by using bispecific antibodies. *J. Virol.* **70**, 6831–6838.
7. Shenk, T. (1996). Adenoviridae: the viruses and their replication, in *Fields Virology.* (K. Fields, ed.), pp. 2135–2137. Lippincott-Raven, Philadelphia.
8. Wold, W. S., Tollefson, A. E., and Hermiston, T. W. (1995). E3 transcription unit of adenovirus. *Curr. Top. Microbiol. Immunol.* **199**, 237–274.
9. Dimitrov, T., Krajcsi, P., Hermiston, T. W., Tollefson, A. E., Hannink, M., and Wold, W. S. (1997). Adenovirus E3-10.4K/14.5K protein complex inhibits tumor necrosis factor-induced translocation of cytosolic phospholipase A2 to membranes. *J. Virol.* **71**, 2830–2837.
10. Hermiston, T. W., Tripp, R. A., Sparer, T., Gooding, L. R., and Wold, W. S. (1993). Deletion mutation analysis of the adenovirus type 2 E3-gp19K protein: identification of sequences within the endoplasmic reticulum lumenal domain that are required for class I antigen binding and protection from adenovirus-specific cytotoxic T lymphocytes. *J. Virol.* **67**, 5289–5298.
11. Shisler, J., Duerksen, H. P., Hermiston, T. M., Wold, W. S., and Gooding, L. R. (1996). Induction of susceptibility to tumor necrosis factor by E1A is dependent on binding to either p300 or p105-Rb and induction of DNA synthesis. *J. Virol.* **70**, 68–77.
12. Tollefson, A. E., Ryerse, J. S., Scaria, A., Hermiston, T. W., and Wold, W. S. (1996). The E3-11.6-kDa adenovirus death protein (ADP) is required for efficient cell death: characterization of cells infected with ADP mutants. *Virology* **220**, 152–162.
13. Gooding, L. R. (1994). Regulation of TNF-mediated cell death and inflammation by human adenoviruses. *Infect. Agents Dis.* **3**, 106–115.
14. Toda, M., Rabkin, S., Kojima, H., and Martuza, R. (1999). Herpes simplex virus as an in situ cancer vaccine for the induction of specific antitumo immunity. *Hum. Gene Ther.* **10**, 385–393.
15. Martuza, R. (2000). Conditionally replicating herpes viruses for cancer therapy. *J. Clin. Invest.* **105**, 841–846.
16. Whyte, P., Ruley, H., and Harlow, E. (1988). Two regions of the adenovirus early region 1A proteins are required for transformation. *J. Virol.* **62**, 257–265.

17. Yu, D., Sakamoto, G., and Henderson, D. R. (1999). Identification of the transcriptional regulatory sequences of human kallikrein 2 and their use in the construction of calydon virus 764, an attenuated replication competent adenovirus for prostate cancer therapy. *Cancer Res.* **59**, 1498–1504.

18. Hallenback, P. L., Chang, Y. N., Hay, C., Golightly, D., *et al.* (1999). A novel tumor-specific replication-restricted adenoviral vector for gene therapy of hepatocellular carcinoma. *Hum. Gene Ther.* **10**, 1721–1733.

19. Barker, D. D., and Berk, A. J. (1987). Adenovirus proteins from both E1B reading frames are required for transformation of rodent cells by viral infection and DNA transfection. *Virology* **156**, 107–121.

20. Nielsch, U., Fognani, C., and Babiss, L. E. (1991). Adenovirus E1A–p105(Rb) protein interactions play a direct role in the initiation but not the maintenance of the rodent cell transformed phenotype. *Oncogene* **6**, 1031–1036.

21. Sherr, C. J. (1996). Cancer cell cycles. *Science* **274**, 1672–1677.

22. Olson, D. C., and Levine, A. J. (1994). The properties of p53 proteins selected for the loss of suppression of transformation. *Cell Growth Differ.* **5**, 61–71.

23. Heise, C., Sampson-Johannes, A., Williams, A., McCormick, F., Von Hoff, D. D., and Kirn, D. H. (1997). ONYX-015, an E1B gene-attenuated adenovirus, causes tumor-specific cytolysis and antitumoral efficacy that can be augmented by standard chemotherapeutic agents [see comments]. *Nat. Med.* **3**, 639–645.

24. Kirn, D., Heise, C., Williams, M., Propst, M., and Hermiston, T. (1998). Adenovirus E1A CR2 mutants as selectively-replicating agents for cancer, in *Cancer Gene Therapy* (E. C. Lattime and S. L. Gerson, eds.). San Diego.

25. Martuza, R. L., Malick, A., Markert, J. M., Ruffner, K. L., and Coen, D. M. (1991). Experimental therapy of human glioma by means of a genetically engineered virus mutant. *Science* **252**, 854–856.

26. Mineta, T., Rabkin, S. D., Yazaki, T., Hunter, W. D., and Martuza, R. L. (1995). Attenuated multi-mutated herpes simplex virus-1 for the treatment of malignant gliomas. *Nat. Med.* **1**, 938–943.

27. Fueyo, J., Gomez-Manzano, C., Alemany, R., Lee, P., McDonnell, T., Mitlianga, P., Shi, Y., Levin, V., Yung, W., and Kyritsis, A. (2000). A mutant oncolytic adenovirus targeting the Rb pathway produces anti-glioma effect in vivo. *Oncogene* **19**, 2–12.

28. Heise, C., Hermiston, T., and Kirn, D. (2000). An adenovirus E1A mutant that has potent and selective antitumoral efficacy following intravenous administration in murine tumor models. *Proc. Am. Assoc. Cancer Res.* **41**, 350.

29. Scheffner, M., Munger, K., Byrne, J. C., and Howley, P. M. (1991). The state of the p53 and retinoblastoma genes in human cervical carcinoma cell lines. *Proc. Natl. Acad. Sci. USA* **88**, 5523–5527.

30. Zhang, Y., Xiong, Y., and Yarbrough, W. G. (1998). ARF promotes MDM2 degradation and stabilizes p53: ARF-INK4a locus deletion impairs both the Rb and p53 tumor suppression pathways. *Cell* **92**, 725–734.

31. Hollstein, M., Sidransky, D., Vogelstein, B., and Harris, C. C. (1991). p53 mutations in human cancers. *Science* **253**, 49–53.

32. Dobner, T., Horikoshi, N., Rubenwolf, S., and Shenk, T. (1996). Block-age by adenovirus E4orf6 of transcriptional activation by the p53 tumor suppressor. *Science* **272**, 1470–1473.

33. Yew, P. R., Liu, X., and Berk, A. J. (1994). Adenovirus E1B oncoprotein tethers a transcriptional repression domain to p53. *Genes Dev.* **8**, 190–202.

34. Bischoff, J. R., Kirn, D. H., Williams, A., Heise, C., Horn, S., Muna, M., Ng, L., Nye, J. A., Sampson-Johannes, A., Fattaey, A., and McCormick, F. (1996). An adenovirus mutant that replicates selectively in p53-deficient human tumor cells [see comments]. *Science* **274**, 373–376.

35. Harada, J., and Berk, A. (1999). p53-independent and -dependent requirements for E1B-55kD in adenovirus type 5 replication. *J. Virol.* **73**, 5333–5344.

36. Rothmann, T., Hengstermann, A., Whitaker, N. J., Scheffner, M., and zur Hausen, H. (1998). Replication of ONYX-015, a potential anti-cancer adenovirus, is independent of p53 status in tumor cells. *J. Virol.* **72**, 9470–9478.

37. Heise, C., Sampson, J. A., Williams, A., McCormick, F., Von, H. D., and Kirn, D. H. (1997). ONYX-015, an E1B gene-attenuated adenovirus, causes tumor-specific cytolysis and antitumoral efficacy that can be augmented by standard chemotherapeutic agents [see comments]. *Nat. Med.* **3**, 639–645.

38. Goodrum, F. D., and Ornelles, D. A. (1997). The early region 1B 55-kilodalton oncoprotein of adenovirus relieves growth restrictions imposed on viral replication by the cell cycle. *J. Virol.* **71**, 548–561.

39. Goodrum, F. D., and Ornelles, D. A. (1998). p53 status does not determine outcome of E1B 55-kilodalton mutant adenovirus lytic infection. *J. Virol.* **72**, 9479–9490.

40. Kirn, D. (2000). Replication-selective micro-organisms: fighting cancer with targeted germ warfare. *J. Clin. Invest.* **105**, 836–838.

41. Southam, C. M., and Moore, A. E. (1952). Clinical studies of viruses as antineoplastic agents, with particular reference to Egypt 101 virus. *Cancer* **5**, 1025–1034.

42. Asada, T. (1974). Treatment of human cancer with mumps virus. *Cancer* **34**, 1907–1928.

43. Smith, R., Huebner, R. J., Rowe, W. P., Schatten, W. E., and Thomas, L. B. (1956). Studies on the use of viruses in the treatment of carcinoma of the cervix. *Cancer* **9**, 1211–1218.

44. Kemeny, N., Huang, Y., Cohen, A., Shi, W., Conti, J., Brennan, M., Bertino, J., Turnbull, A., Sullivan, D., Stockman, J., Blumgart, L., and Fong, Y. (1999). Hepatic arterial infusion of chemotherapy following resection of hepatic metastases from colorectal cancer. *N. Engl. J. Med.* **341**, 2039–2048.

45. Hecht, R., Abbruzzese, J., Bedford, R., Randlev, B., Romel, L., Lahodi, S., and Kirn, D. (2000). Endoscopic ultrasound-guided intratumoral injection of pancreatic carcinomas with a replication-selective adenovirus: a phase I/II clinical trial. *Proc. Am. Soc. Clin. Oncol.* **19**, 1039 (abstract).

46. Vasey, P., Shulman, L., Gore, M., Kirn, D., and Kaye, S. (2000). A phase I trial of an E1B-55kD gene-deleted adenovirus administered by intraperitoneal injection into patients with advanced, refractory ovarian carcinoma. *Proc. Am. Soc. Clin. Oncol.*

47. Reid, T., Galanis, E., Abbruzzese, J., Randlev, B., Romel, L., Rubin, J., and Kirn, D. (2000). Hepatic arterial infusion of a replication-selective adenovirus, ONYX-015: a phase I/II clinical trial. *Proc. Am. Soc. Clin. Oncol.* **19**, 953 (abstract).

48. Kirn, D. (2001). Clinical trial results with the replication-selective adenovirus dl1520 (Onyx-015): What have we learned? *Gene Therapy* **8**, 89–98.

49. Khuri, F., Nemunaitis, J., Ganly, I., Gore, M., MacDougal, M., Tannock, I., Kaye, S., Hong, W., and Kirn, D. (2000). A controlled trial of ONYX-015, an E1B gene-deleted adenovirus, in combination with chemotherapy in patients with recurrent head and neck cancer. *Nat. Med.* **6**, 879–885.

50. Kirn, D., Hermiston, T., and McCormick, F. (1998). ONYX-015: clinical data are encouraging [letter; comment]. *Nat. Med.* **4**, 1341–1342.

51. Roelvink, P., Mi, G., Einfeld, D., Kovesdi, I., and Wickham, T. (1999). Identification of a conserved reseptor-binding site on the fiber proteins of CAR-recognizing adenoviridae. *Science* **286**, 1568–1571.

52. Chinnadurai, G. (1983). Adenovirus 2 Ip+ locus codes for a 19 kd tumor antigen that plays an essential role in cell transformation. *Cell* **33**, 759–766.

53. Sauthoff, H., Heitner, S., Rom, W., and Hay, J. (2000). Deletion of the adenoviral E1B-19kD gene enhances tumor cell killing of a replicating adenoviral vector. *Hum. Gene Ther.* **11**, 379–388.

54. Medina, D. J., Sheay, W., Goodell, L., Kidd, P., White, E., Rabson, A. B., and Strair, R. K. (1999). Adenovirus-mediated cytotoxicity of chronic lymphocytic leukemia cells. *Blood* **94,** 3499–3508.

55. Doronin, K., Toth, K., Kuppuswamy, M., Ward, P., Tollefson, A., and Wold, W. (2000). Tumor-specific, replication-competent adenovirus vectors overexpressing the adenovirus death protein. *J. Virol.* **74,** 6147–6155.

56. Hermiston, T. (2000). Gene delivery from replication-selective viruses: arming guided missiles in the war against cancer. *J. Clin. Invest.* **105,** 1169–1172.

57. Hawkins, L., Nye, J., Castro, D., Johnson, L., Kirn, D., and Hermiston, T. (1999). Replicating adenoviral gene therapy. *Proc. Am. Assoc. Cancer Res.* **40,** 476.

58. Freytag, S. O., Rogulski, K. R., Paielli, D. L., Gilbert, J. D., and Kim, J. H. (1998). A novel three-pronged approach to kill cancer cells selectively: concomitant viral, double suicide gene, and radiotherapy [see comments]. *Hum. Gene Ther.* **9,** 1323–1333.

59. Wildner, O., Blaese, R. M., and Morris, J. M. (1999). Therapy of colon cancer with oncolytic adenovirus is enhanced by the addition of herpes simplex virus-thymidine kinase. *Cancer Res.* **59,** 410–413.

60. Reid, A., Galanis, E., Abbruzzese, J., Romel, L., Rubin, J., and Kirn, D. (1999). A phase I/II trial of ONYX-015 administered by hepatic artery infusion to patients with colorectal carcinoma, EORTC-NCI-AACR Meeting on Molecular Therapeutics of Cancer. **19,** 953 (abstract).

61. Nemunaitis, J., Cunnungham, C., Buchanan, A., Blackburn, A., Edelman, G., Maples, P., Netto, G., Tong, A., Olson, S., and Kirn, D. (2001). Intravenous infusion of a replication-selective adenovirus (Onyx-015) in cancer patients: safety, feasibility, and biological activity. *Gene Therapy* **8**(10), 746–759.

62. Heise, C., and Kirn, D. (2000). Replication-selective adenviruses as oncolytic agents. *J. Clin. Invest.* **105,** 847–851.

63. Agha-Mohammadi, S., and Lotze, M. (2000). Immunomodulation of cancer: potential use of replication-selective agents. *J. Clin. Invest.* **105,** 1173–1176.

64. Norman, K., and Lee, P. (2000). Reovirus as a novel oncolytic agent. *J. Clin. Invest.* **105,** 1035–1038.

65. Mastrangelo, M., Eisenlohr, L., Gomella, L., and Lattime, E. (2000). Poxvirus vectors: orphaned and underappreciated. *J. Clin. Invest.* **105,** 1031–1034.

66. Coffey, M., Strong, J., Forsyth, P., and Lee, P. (1998). Reovirus therapy of tumors with activated ras pathway. *Science* **282,** 1332–1334.

67. Kirn, D. (2000). A tale of two trials: selectively replicating herpesviruses for brain tumors. *Gene Ther.* **7,** 815–816.

68. Lattime, E. C., Lee, S. S., Eisenlohr, L. C., and Mastrangelo, M. J. (1996). In situ cytokine gene transfection using vaccinia virus vectors. *Semin. Oncol.* **23,** 88–100.

69. Low, K., Ittensohn, M., Le, T., Platt, J., Sodi, S., Amoss, M., Ash, O., Carmichael, E., Chakraborty, A., Fischer, J., Lin, S., Luo, X., Miller, S., Zheng, L., King, I., Pawelek, J., and Bermudes, D. (1999). Lipid A mutant *Salmonella* with suppressed virulence and TNF-alpha induction retain tumor-targeting in vivo. *Nat. Biotechnol.* **17,** 37–41.

70. Sznol, M., Lin, S., Bermudes, D., Zheng, L., and King, I. (2000). Use of preferentially replicating bacteria for the treatment of cancer. *J. Clin. Invest.* **105,** 1027–1030.

71. Kirn, D., Martuza, R., Zwiebel, J. (2001). Replication-selective virotherapy for cancer: biological principles, risk management and future directions. *Nature Med.* **7**(7), 781–787.

E1A Cancer Gene Therapy

DUEN-HWA YAN

*Departments of Molecular and Cellular
Oncology and Surgical Oncology
M. D. Anderson Cancer Center
The University of Texas
Houston, Texas 77030*

RUPING SHAO

*Department of Molecular and
Cellular Oncology
M. D. Anderson Cancer Center
The University of Texas
Houston, Texas 77030*

MIEN-CHIE HUNG

*Departments of Molecular and Cellular
Oncology and Surgical Oncology
M. D. Anderson Cancer Center
The University of Texas
Houston, Texas 77030*

I. INTRODUCTION

The E1A gene products of human adenovirus type 5—12S (243 amino acids) and 13S (289 amino acids)—are known to activate viral gene transcription and regulate the host gene expression as viruses propagate inside the cell [1–3]. In contrast to adenovirus type-12 E1A, a potent oncogene that can trans-

form established cell lines [4], adenovirus type-5 or type-2 E1A cannot transform established cell lines [5] but could cooperate with other viral and cellular oncogenes to transform primary culture cells [6]. Therefore, adenovirus type-5 and type-2 E1A were considered as immortalization oncogenes. In this review, "E1A" is refers to the nontransforming E1A, and most of the experimental results described here were based on the use of type-5 E1A. In the last decade, E1A was found to be associated with multiple antitumor activities [7–1] (Fig. 1). Multicenter E1A clinical trials on ovarian, breast, and head and neck cancers are currently underway. Cancer model studies have confirmed the E1A-mediated antitumor activity. Study on the molecular mechanisms underlying the E1A-mediated antitumor activity has been an enlightening endeavor that has yielded many insightful observations (Fig. 2). In this review, we attempt to summarize these results with an emphasis on the observations obtained during the past decade.

II. HER2 OVEREXPRESSION AND E1A-MEDIATED ANTITUMOR ACTIVITY

HER2 (also known as *neu* or c-*erb*B-2) overexpression in breast, ovarian, and head and neck tumors is known to be the indicator for poor prognosis and poor survival of these cancer patients [12–17]. Although HER2 overexpression alone is not sufficient to confer a chemoresistance in normal mammary epithelial cells [18], our results and others indicate that HER2 overexpression found in human tumors usually correlates

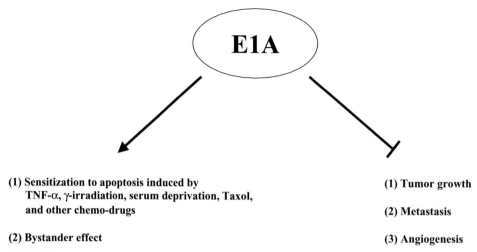

FIGURE 1   E1A-mediated antitumor activities. E1A enhances sensitization to apoptosis and produces bystander effect. E1A also inhibits tumor growth and suppresses metastasis and angiogenesis.

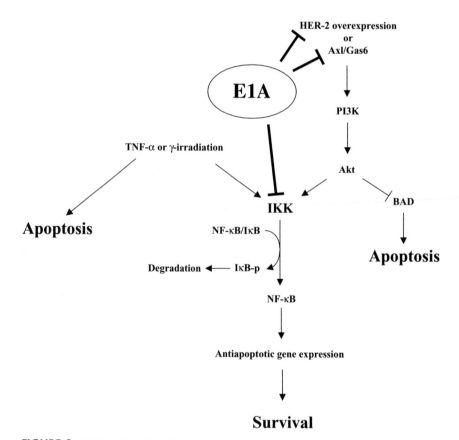

FIGURE 2   E1A-mediated signaling pathways involved in antitumor activities. E1A targets HER2 overexpression, Ax1/Gas6, and IKK in Akt and NF-κB survival pathways to sensitize cells to apoptosis-induced by stress signals.

with chemoresistance [19–23] and insensitivity to radiation treatment [24,25]. These results argue that, in addition to HER2 overexpression, other genetic changes during tumorigenesis are required to render a tumor cell chemoresistant. One should also keep in mind that the level of HER2 expression in cancer cells may also play a role in this process; that is, a certain level of HER2 overexpression is required to exhibit a chemoresistance phenotype [21]. It is therefore conceivable that E1A may convert a chemoresistant phenotype of HER2-overexpressing cells to a chemosensitive phenotype by downregulating HER2 expression.

Although HER2 mRNA stability may contribute in part to HER2 overexpression in certain cancer cells [26], gene amplification [12–14] and transcriptional upregulation [27–31] are the primary causes for HER2 overexpression in cancer cells. One of the best ways to achieve an antitumor activity in HER2-overexpressing tumor is to turn off HER2 gene transcription that leads to HER2 downregulation. Our initial effort to identify nuclear factors that could do just that led us to the adenovirus E1A [32,33]. We showed that E1A could readily downregulate the transcription of the HER2 gene *in vitro*. Importantly, we demonstrated *in vivo* that E1A expression could reduce tumorigenecity and suppress metastatic potential of murine fibroblast cells transformed by murine HER2 gene [34–36]. We subsequently demonstrated the tumor suppressor activity of E1A in HER2-overexpressing human cancer cells derived from breast [37], ovary [38,39], and lung [40]. Thus, an antitumor activity appears to be one of the multifunctional features of E1A. In particular, E1A seems to exert its antitumor activity in HER2-overexpressing cancer cells by reducing their metastatic potentials [35,36,41] and by increasing their sensitivity to the chemotherapy [42–44] and radiation [45] treatments. The above observations form a scientific basis for the development of an E1A-based gene therapy strategy against HER2-overexpressing tumors, and, as we shall see, the E1A gene therapy could be applied to the low HER2-expressing cancer cells, as well.

III. MECHANISMS OF E1A-MEDIATED ANTI-TUMOR ACTIVITY

A. HER2 Downregulation

1. HER2 Gene Transcriptional Repression

E1A expression represses the steady-state HER2 mRNA and protein expression [33] by downregulating HER2 promoter activity [32]. To investigate the mechanism of the E1A-mediated HER2 transcriptional repression, we showed that a forced expression of an E1A-binding protein, p300 [46], could override the E1A-mediated HER2 transcriptional repression. Conversely, E1A expression inhibits p300-mediated transcriptional activation via a p300 consensus binding sequence on the HER2 promoter [47]. These results

strongly suggest that p300 is a coactivator for HER2 promoter activity, and that E1A targets p300, resulting in HER2 transcriptional repression. This idea was further supported by the finding that the p300-binding domain of E1A (4–25 or 40–80 amino acids) is required for HER2 transcriptional repression [48]. However, it may not be the only mechanism that accounts for the HER2 transcriptional repression by E1A. Our earlier data indicated that a HER2 promoter element containing a consensus sequence responsive to E1A-mediated transcriptional repression (i.e., TGGAATG [49–52]) and an E-box sequence (i.e., CAGTTG [53]) could alleviate the E1A-mediated HER2 transcriptional repression when provided exogenously [32]. This result suggests that E1A may repress HER2 promoter activity by targeting the *trans*-acting factors interacting with these promoter elements. Further experiments are required to pinpoint the protein/DNA interaction on the HER2 promoter that is the target for E1A-mediated transcriptional repression. However, it appears that multiple *cis*-acting elements and *trans*-acting factors on the HER2 promoter are involved in the E1A-mediated transcriptional repression.

2. Sensitization to TNF-α-Induced Apoptosis

Functionally speaking, downregulation of HER2 overexpression has another expected consequence: sensitization to tumor necrosis factor-alpha (TNF-α)-induced apoptosis [54]. Although the molecular mechanism responsible for this sensitization was not known at that time, recent study on the pathway of HER2 overexpression showed that Akt (a serine/threonine kinase) activation is one of the main events leading to TNF-α resistance (Fig. 2) [55]. In short, HER2 overexpression activates phosphatidylinositol 3-kinase (PI3K) [56], which, in turn, activates Akt by phosphorylation [57]. Under certain condition, Akt activation turns on multiple downstream signal pathways [58], including the nuclear factor kappa B (NF-κB) pathway leading to cell survival in response to stress signals [59–63]. In other cases, however, NF-κB is not activated in response to Akt activation [64,65]. The difference between these two observations could be due to different cell types used in their respective systems. In our experimental system, however, it makes sense that HER2 overexpression could protect the cells from TNF-α-induced apoptosis by activating the NF-κB pathway. Intriguingly, introducing a dominant-negative Akt (dnAkt) that abolishes Akt function in HER2-overexpressing cells could restore their sensitivity to TNF-α killing. This result clearly establishes a positive relationship between HER2 overexpression and Akt activation. Indeed, this observation was confirmed by the immunohistochemical analysis on tumor tissues that showed a positive correlation between the level of active Akt and the level of HER2 expression [55]. Given that, one would predict a NF-κB activation in HER2-overexpressing and Akt-activating tumor cells even in the absence of TNF-α. It is indeed the case, as the activity of a NF-κB-activated promoter is higher in HER2-overexpressing cells than in low-HER2-expressing cells. Furthermore, study on the Akt/NF-κB pathway showed an increased level in HER2-overexpressing cells of the phosporylated form of a NF-κB inhibitor, IκB-α (the phosporylation of which leads to subsequent ubiquitination and degradation of IκB-α protein [66]). Thus, IκB-α inactivation correlates well with a higher NF-κB activity in HER2-overexpressing cells than that in low-HER2-expressing cells. The fact that the activity of IκB kinases (IKKs), which phosphorylate IκB-α and result in IκB-α degradation [67,68], is higher in HER2-overexpressing cells than in dnAkt-transfected cells further confirms the notion that activation of the HER2/Akt/NF-κB pathway is responsible for cell survival in response to TNF-α.

Taken together, HER2 downregulation by E1A would predict a sensitization of HER2-overexpressing cells to apoptosis induced by stress signals such as TNF-α that triggers the NF-κB pathway. As will be discussed later, it is indeed the case. Moreover, we shall see that, in addition to HER2, the molecules involved in the NF-κB pathway turn out to be targets for E1A action as well. In that light, E1A-mediated sensitization to certain stress-signals-induced apoptosis may be a general effect independent of HER2 expression level (Fig. 2).

3. Sensitization to Taxol-Induced Apoptosis

Is HER2 overexpression a hallmark for chemoresistance in cancer cells? Little agreement regarding the answer to this question can be found among investigators in the HER2 cancer biology field. The controversy surrounding this issue in both clinical and laboratory settings has been discussed recently [21,69]. The possibility of a threshold HER2 expression required for chemoresistance in cancer cells (e.g., Taxol resistance) appears to be a plausible explanation for the discrepancies between the seemingly conflicting results [70,71]. To further complicate the matter, in the case of "normal" human mammary epithelial cell lines (e.g., MCF-10A [72,73] and human mammary epithelial cells (HMEC) [18]), enforced HER2 expression causes Taxol resistance in MCF-10A cells [22], but not in the HMECs [18]. Obviously, the different genetic background associated with these cell lines will no doubt contribute to the apparent opposite results. The caveat, however, is that it is possible for a different HER2 expression level between these cell lines to account for the differences in their sensitivity to Taxol.

Given the precaution mentioned above, one would expect that E1A-mediated HER2 downregulation may sensitize HER2-overexpressing cancer cells to Taxol-induced apoptosis, and that is indeed the case [42–44]. Either in E1A-stable transfectants [42,44] or with adenovirus infection [43], E1A expression in HER2-overexpressing cancer cells makes them more susceptible to Taxol killing. E1A-expressing SKOV3.ip1 ovarian cancer cells (ip1-E1A) are sensitive to Taxol killing as well as cisplatin, doxorubicin, TNF-α, or serum-withdrawl-induced killing [44]. In contrast, the E1A-expressing, low-HER2-expressing ovarian [42] or breast [43] cancer cells do not exhibit such chemosensitization. Although these data may argue for the importance of HER2 overexpression in chemoresistance, it cannot exclude the possibility that E1A could sensitize cells to apoptosis independent of the HER2 expression level. In fact, previous reports support that possibility [74,75]. At any rate, these results are in agreement with the E1A action in repressing HER2 transcription and inactivating the NF-κB survival pathway in response to stress signals.

B. Inhibition of Metastasis

The ability of E1A to inhibit metastasis *in vitro* and *in vivo* has been demonstrated in different laboratories [35,37,39, 76–80]. The E1A-regulated molecules the involved in reducing the metastatic potential of cancer cells include upregulated E-cadherins [81,82], a nucleoside diphosphate kinase (NM23) [78], and tissue inhibitors of metalloproteinase (TIMPs) [83], as well as downregulated matrix metalloproteinases (MMPs) (e.g., MMP-1 [80], MMP-3 [84,85], MMP-9 [80,86–89]) urokinase-type plasminogen activator (uPA) [80], adhesion molecules (e.g., CD44s) [90], and HER2 [35,41].

Because HER2 overexpression enhances metastatic potential of the cancer cell [35,91], it is conceivable that E1A may inhibit metastasis by downregulating HER2 expression. Indeed, E1A expression in HER2-overexpressing cancer cells rendered the cells less metastatic [35,41]. In addition, enforced HER2 expression, which results in MMP upregulation, partially restored the metastatic potential of the otherwise less metastatic cancer cells *in vitro* and *in vivo* [92], suggesting that HER2 overexpresion is involved in the metastatic phenotype of cancer cells. As in E1A-mediated sensitization, it is possible that E1A could suppress metastasis of cancer cells regardless of HER2 expression level. In support of this idea, reexpressing HER2 in E1A-expressing cancer cells could restore their tumorigenecity but failed to restore the suppressed metastatic potential or the repressed MMP expression [41]. Again, this result demonstrates the multifunctional features of E1A in suppressing metastasis that include the downregulation of HER2 and MMP, and the fact that E1A may suppress metastasis through other mechanisms independent of HER2 downregulation.

C. Axl Downregulation

Using a differential display technique based on polymerase chain reaction (PCR) to analyze the receptor tyrosine kinase (RTK) expression profiles in the E1A-expressing cancer cells, Axl, a member of the UFO membrane receptor family [93–95], was found consistently down-regulated in

E1A-expressing cells [96]. Similar to HER2 downregulation, Axl transcription was repressed by E1A. Using an E1A-expressing cell line in which Axl expression was enforced (E1A–Axl), the functional significance of Axl downregulation in the E1A-mediated antitumor activity could be probed. E1A–Axl cells have a growth rate similar to that of the E1A-expressing cells (albeit much reduced as compared with that of the parental cells) in the absence of the Axl-specific ligand, Gas6 [97–99], whereas in the presence of Gas6 E1A–Axl cells resumed their growth rate comparable to that of the parental cells. These results suggest that Axl downregulation may be one of the critical events in the process of the E1A-mediated growth retardation. Moreover, in contrast to the E1A-expressing cells, for which Gas6 has no effect on the E1A-mediated sensitization to serum-deprivation-induced apoptosis, E1A–Axl cells kept that sensitization in the absence of Gas6 but lost it when Gas6 is present. This observation confirms the notion that the Axl/Gas6 signal pathway is one of the cellular survival pathways in response to stress such as serum starvation, and the inhibition of this pathway by E1A sensitizes cells to apoptosis induced by these stress signals (Fig. 2).

Further elucidation of the molecular mechanisms underlying the E1A-mediated inhibition of Axl/Gas6 pathway led to the finding that a critical Axl/Gas6 downstream molecule, Akt [100], was inactivated as a result of E1A expression (Lee and Hung, unpubl. results). In E1A–Axl cells, however, Akt is reactivated in the presence of Gas6, indicating a direct relationship between Axl/Gas6 signal transduction and Akt activation leading to cell survival. Functional confirmation of this relationship came from the observation that either blocking Akt by a dominant-negative Akt or blocking PI3K, an upstream molecule of Akt, by a specific inhibitor could render E1A–Axl cells sensitive to serum-deprivation-induced apoptosis in the presence of Gas6. Upon examining the Akt downstream molecules, including BAD (a proapoptotic molecule) [101,102], IKK [63,103], and FKHRL [104], the data showed an increased level of phosphorylated BAD (which leads to BAD inactivation) in E1A–Axl cells in the presence of Gas6. However, no detectable biochemical changes in NF-κB activation (resulting from IKK activation by Akt) or Fas ligand downregulation (resulting from FKHRL activation by Akt) (Lee and Hung, unpubl. results) were observed in these cells. These results strongly suggest a scenario that E1A represses Axl transcription leading to BAD activation by inhibiting Akt activity through the Gas6/Axl/PI3K/Akt/BAD pathway. As a corollary, cells become sensitive to apoptosis induced by serum deprivation (Fig. 2).

D. NF-κB Inactivation

It has been known that E1A could sensitize cells to apoptosis induced by TNF-α [105,106] or ionizing radiation [75,107]. After investigating the E1A action in the TNF-α

and γ-irradiation pathways, the finding suggests that NF-κB, a common molecule shared by these pathways [60,108], may be a critical target of E1A to mediate apoptosis induced by TNF-α or γ-irradiation [45,109] (Fig. 2).

Nuclear factor-κB can be activated by inflammatory cytokines such as TNF-α and a host of other stimuli including ionizing radiation [66,110], however, the role of the active NF-κB is to prevent apoptosis induced by these stimuli [59–62]. Indeed, in most cells, NF-κB activation prevents apoptosis through activating antiapoptotic genes, though under certain conditions and in certain cell types, active NF-κB could also induce apoptosis [111,112]. Importantly, aberrant NF-κB activation is involved in a variety of human diseases [113], including cancer [114,115]. These findings provide a possible explanation for the reason why a majority of cancer cells are resistant to TNF-α, radiation, or chemotherapy treatment. Thus, NF-κB has become an attractive target for various cancer gene therapy designs.

ip1-E1A cells are highly susceptible to γ-irradiation-induced apoptosis [45] as compared with the parental cells that do not express E1A. When a NF-κB/DNA binding assay was performed, it became apparent that ip1-E1A cells lose the ability to generate active NF-κB in response to γ-irradiation, as indicated by the lack of p50/p65 heterodimer (the active form of NF-κB) binding to DNA. This result suggests that E1A expression may block the activation of NF-κB upon γ-irradiation in ip1-E1A cells. That enforced expression of NF-κB in ip1-E1A partially rescued the γ-irradiation-induced apoptosis confirms this observation [45]. Thus, E1A-mediated NF-κB inactivation may be responsible for the sensitization to γ-irradiation-induced apoptosis. But, how does E1A inactivate NF-κB?

The answer to this question came from the study of the effect of TNF-α on E1A-expressing cells [109]. Similar to γ-irradiation, TNF-α could preferentially induce ip1-E1A cells to apoptosis as compared with cells that do not express E1A. Again, the active form of NF-κB is missing in ip1-E1A cells in response to TNF-α treatment, suggesting that NF-κB inactivation may also play a role in the E1A-mediated sensitization to TNF-α-induced apoptosis. One possible mechanism for NF-κB inactivation by E1A is that E1A may downregulate NF-κB protein expression. Although the NF-κB protein level was not downregulated by E1A in response to TNF-α, the phosporylated form of IκB-α was concurrently reduced in ip1-E1A cells. This phenomenon takes place without any changes on IκB-α protein level either before or after TNF-α treatment in ip1-E1A cells, suggesting that E1A may inactivate NF-κB by keeping IκB-α underphosporylated. Because IKK phosphorylates IκB-α, it is possible that E1A may inhibit IKK activity so that IκB-α could not be properly phosporylated. This possibility was supported by the observation that the endogenous IKK (both α and β forms) activity was inhibited in TNF-α-treated ip1-E1A cells as determined by its ability to phosphorylate IκB-α, whereas IKK was active in the TNF-α-treated, non-E1A-expressing cells. Interestingly, E1A could also inhibit the exogenous IKK activity when IKK was transiently transfected into ip1-E1A cells. Taken together, these results suggest a scenario in which E1A may inactivate NF-κB by inhibiting IKK activity, leading to the stablization of IκB-α protein and retention of the IκB-α/NF-κB complex in the cytoplasm. Thus, in ip1-E1A cells, NF-κB is prevented from entering the nucleus and from activating the antiapoptotic genes. In this fashion, E1A mediates the TNF-α (and very likely γ-irradiation)-induced apoptosis (Fig. 2). Despite vigorous effort to elucidate how E1A inactivates IKK activity, no direct interaction between E1A and IKK could be found, suggesting that E1A may inactivate IKK in an indirect manner. Determining the molecular mechanism underlying IKK inactivation by E1A presents an exciting area of study into the function of E1A as a cancer therapeutic gene.

E. Bystander Effect

Bystander effect is one of the important features of a useful therapeutic gene for cancer gene therapy. The classical example is the bystander effect generated by the herpes simplex virus thymidine kinase (TK) gene. In the presence of the prodrug ganciclovir, TK expression kills not only the TK-transfected cells but also the nearby untransfected cells [116]. The gap junction that mediates the intercellular communication appears to be responsible for the TK-induced bystander killing [117].

The evidence that E1A may possess a bystander effect came from a tumorigenicity assay that coimplanted ip1-E1A cells with the control cell lines that do not express E1A [118]. The result was surprising in that ip1-E1A cells (which possess a reduced tumorigenecity) could somehow suppress the tumor growth of two highly tumorigenic cancer cell lines: HER2-overexpressing SKOV3 and low-HER2-expressing MDA-MB-435. The histological sections of tumors generated by coimplantation showed a significant reduction of microvessels determined by the staining of a blood vessel marker (i.e., Factor VIII) as compared with that in SKOV3 tumors. This result indicates a reduced angiogenesis taken place in the coimplanted tumors. Using TdT (terminal deoxynucleotidy transferase)-mediated dUTP nick end labeling (TUNEL) assay to detect apoptotic cells in tumor sections, the coimplanted tumors have a level of apoptosis comparable to that found in the ip1-E1A tumors. In contrast, SKOV3 tumors have a minimum level of apoptosis. These data suggest that reduced angiogenesis and enhanced apoptosis may cause the reduced tumorigenecity observed in the coimplanted tumors.

One possible mechanism for the E1A-mediated bystander effect is that ip1-E1A cells may secret certain factor(s) that suppresses tumor growth by its ability to generate antiproliferative and proapoptotic effects on the neighboring cells. Indeed, when SKOV3 cells were cultured in the medium that has been used to culture ip1-E1A cells, the growth of SKOV3 cells was significantly inhibited. A similar result was also observed when MDA-MB-435 cells were cultured in ip1-E1A medium. Furthermore, as determined by TUNEL assay, more apoptotic SKOV3 cells were observed when cultured in ip1-E1A medium than in SKOV3 medium. These results strongly suggest the existence of a secreted factor generated from ip1-E1A cells, and this secreted factor could suppress proliferation and induce apoptosis in the neighboring, non-E1A-expressing cells. The identification and characterization of this E1A-induced factor would certainly facilitate our understanding about the E1A-mediated bystander effect.

IV. E1A GENE THERAPY: PRECLINICAL MODELS

We have so far described the encouraging *in vitro* results regarding the phenomenon of E1A-mediated HER2 downregulation and chemosensitization. We also proposed several mechanisms by which E1A may act to achieve these functions. To test the efficacy of an E1A-based gene therapy in mice bearing human HER2-overexpressing tumors, three orthotopic cancer xenograft models—ovarian, breast, and lung—were established and two E1A delivery systems were used: a cationic liposome, DC-Chol:DOPE {3β[N-(N'-dimethylaminoethane)-carbamoyl] cholesterol:dioleoylphatidylethanolamine (3:2)} (DC-Chol) [119] and adenovirus E1A (Ad.E1A) [33]. A toxicity study was subsequently conducted in immunocompetent mice to ensure the safety of the procedure and to determine the minimum side effects associated with the E1A gene therapy treatment.

A. Ovarian Cancer Model

The orthotopic ovarian cancer model was established by injecting human HER2-overexpressing ovarian cancer cells (SKOV3) intraperitonealy (i.p.) into female nu/nu mice. The implanted ovarian tumors obtained from mesentery and the inside of the peritoneal cavity showed HER2-positive staining [39]. The tumor-bearing mice received i.p. injection of E1A expression vector complexed with either DC-Chol (E1A/DC-Chol) [39] or Ad.E1A [38].

1. E1A/DC-Chol Treatment

Necropsy analysis showed that some of E1A/DC-Chol-treated mice, though dying of tumor-related symptoms, had no detectable tumor invasion and metastasis as commonly seen in mice in the control groups (i.e., no treatment, mutant E1A/DC-Chol, E1A alone, or DC-Chol alone) [39]. Upon examination of tumor tissues excised from the E1A/DC-Chol-treated mice, it became clear that E1A expression correlated well with downregulation of HER2 protein but there was no decrease in HER2 protein level in the tumors obtained from the control groups. Remarkably, 70% of the E1A/DC-Chol-treated mice survived more than a year, while the controls all died within 200 days [39]. The surviving mice appeared normal and healthy, as there were no detectable tumors inside the mice or any obvious side effects associated with the treatment. These results showed that (1) by i.p. injection, E1A/DC-Chol complex is a useful vehicle to transduce E1A into ovarian cancer cells *in vivo;* and (2) E1A/DC-Chol treatment could repress HER2 expression, suppress tumor growth, reduce metastasis, increase survival, and have no obvious side effects. This observation became one of the first indications of the efficacy and feasibility of using E1A/DC-Chol-based gene therapy to effectively treat ovarian cancer in a xenograft model.

2. Ad.E1A Treatment

The efficacy of using Ad.E1A in the above orthotopic ovarian cancer model appeared similar to that obtained from the E1A/DC-Chol treatment [38,39]. In addition to SKOV3, the Ad.E1A study also included a low-HER2-expressing human ovarian cancer cell line, 2774. Intriguingly, while Ad.E1A could effectively increase survival in the SKOV3 tumor model, it failed to do so in the 2774 tumor model [38,120]. This result was not due to a difference of viral infection efficiency between SKOV3 and 2774 cell lines, as both could be infected equally as determined by adenovirus carrying β-galactosidase gene (Ad.LacZ) [38]. This observation raises a possibility that E1A may mediate a preferential antitumor effect on HER2-overexpressing ovarian cancer cells but not on ovarian cancer cells with low HER2 expression. Alternatively, a more rigorous treatment may be needed for the low-HER2-expressing cancer cells such as 2774 to achieve a efficacy similar to that seen in treating the HER2-overexpressing cancer cells. SKOV3 tumors excised from Ad.E1A-infected mice showed a positive staining for E1A proteins and a concurrent reduction of HER2 protein expression on the same tumor samples. This result, therefore, confirms *in vivo* a causal relationship between E1A expression and HER2 downregulation. Using Ad.LacZ to monitor the E1A expression spectrum in the SKOV3 tumor model, it is encouraging to know that a high LacZ expression was found in malignant ascites and tumors as compared with that in other tissues and organs, suggesting that Ad.E1A may preferentially target these tumor sites [38].

3. E1A/DC-Chol and Taxol Combined Treatment

As mentioned before, E1A could sensitize HER2-overexpressing ovarian cancer cells to Taxol-induced apoptosis [42,44], and this phenomenon seems to be HER2 overexpression specific, as no such sensitization was observed in Taxol-treated, E1A-expressing 2774 cells [42]. To test a possible enhancement of E1A-mediated antitumor activity in conjunction with Taxol treatment, E1A/DC-Chol + Taxol was i.p. injected to treat mice bearing SKOV3 tumors. The E1A/DC-Chol + Taxol treatment yielded the best survival result among all treatment groups including the E1A/DC-Chol alone treatment [42]. This observation is congruous to the *in vitro* data [42,44] and suggests that the E1A/DC-Chol + Taxol treatment may enhance the E1A-mediated antitumor activity in mice bearing HER2-overexpressing ovarian tumors.

B. Breast Cancer Model

Ad.E1A infection preferentially inhibited the growth of HER2-overexpressing breast cancer cells (e.g., MDA-MB-361 and SKBR3), whereas there was little or no E1A-mediated growth inhibitory effect on the low-HER2-expressing cancer cells (e.g., MDA-MB-435 and MDA-MB-231) [37]. Based on this observation, both Ad.E1A and E1A/DC-Chol were used to assess the potential efficacy in an orthotopic, HER2-overexpressing breast cancer model. MDA-MB-361 cells were transplanted into the mammary fat pads of female nu/nu mice. The mammary tumors become palpable usually about 45 days after implantation. Ad.E1A or E1A/DC-Chol was intratumor injected. Six months of E1A treatment by either Ad.E1A or E1A/DC-Chol prolonged survival (the mean survival was greater than 2 years as opposed to less than 15 months in the control groups) and inhibited tumor growth. The Ad.E1A treatment appeared slightly better than E1A/DC-Chol treatment. Remarkably, no metastasis was found in intraperitoneal organs such as liver, intestine, spleen, and kidney [37]. These results are consistent with the ability of E1A to inhibit metastasis and are reminiscent of the E1A-mediated antitumor effect on HER2-overexpressing ovarian tumors, for which no detectable metastasis was found in E1A-treated mice [39]. The mammary tumor suppression correlated well with the expression of E1A and the downregulation of HER2 protein as determined by western blot and immunohistochemical analysis on the tumor samples [37]. The above data suggest the feasibility of an E1A-based gene therapy (either by Ad.E1A or by E1A/DC-Chol) against HER2-overexpressing breast cancer *in vivo.*

C. Safety Studies

To ensure the safety of E1A/DC-Chol administration by intraperitoneal injection in clinical trials, it is imperative that a safe and tolerable dosage of E1A/DC-Chol is well defined. A series of studies that evaluate any adverse effects associated with a range of E1A and DC-Chol combinations were conducted in immunocompetent ICR female mice [121]. In an acute toxicity study, a range of E1A/DC-Chol doses that were 0.5–10 times the starting dose (10 nmol of DC-Chol complexed with 1 μg E1A DNA, 10:1) proposed in the phase I clinical trial did not cause apparent acute or residual toxic effects on mice. Hepatic and renal functions appeared normal and other major organs showed normal pathology. These results suggest that the E1A/DC-Chol dosage proposed in the clinical trial may be safe, as it is much lower than that used in the animal toxicity study. A repeated E1A/DC-Chol treatment by i.p. injection did not show significant lesions on major organs in treated mice even 6 weeks after the injections. Again, this observation supports the idea that the proposed E1A/DC-Chol dosage in the human clinical trial may be safe.

Although the DC-Chol/E1A DNA ratio of 10:1 was proposed for the clinical trial, a ratio of 13:1 has been used *in vitro* and *in vivo* studies and successfully demonstrated the gene delivery efficiency and treatment efficacy [39,119]. Can a comparable treatment efficacy be achieved by a minimum E1A/DC-Chol dosage so that a potential toxicity associated with a high concentration of DC-Chol could be avoided? To find that minimum effective dosage, a dose study using i.p.-injected DC-Chol/LacZ DNA to monitor the β-galactosidase (β-gal) expression in SKOV3 tumor model was conducted [120]. There was no significant difference in β-gal expression between the DC-Chol/LacZ ratios of 0.5:1 and 26:1. Thus, it is possible that a much lower concentration of DC-Chol than was proposed in the clinical trial could be used without compromising the efficiency of gene delivery.

Another concern of using E1A-based gene therapy in cancer treatment is the ability of E1A to immortalize and transform the otherwise normal cells under certain circumstances [4,122]. Different from retroviral vector that usually integrates into the chromosomes of the recipient cells, the E1A/DC-Chol or Ad.E1A delivery system ordinarily expresses E1A transiently. Therefore, the issue of E1A-mediated immortalization may not be a serious problem using the above-mentioned delivery systems. However, if it is necessary to circumvent this potential complication, we generated a modified E1A, named mini-E1A, that lacks the CR2 region responsible for immortalization and Rb protein binding. Mini-E1A remains competent in tumor suppression, as shown by its ability to repress HER2 transcription and suppress HER2-mediated transformation phenotype and tumorigenecity [48]. To test the efficacy of mini-E1A/DC-Chol treatment in SKOV3 model, mini-E1A/DC-Chol was i.p. injected into mice that bore SKOV3 tumors. Similar to wild-type E1A, mini-E1A/DC-Chol treatment prolonged survival at the two DC-Chol/DNA ratios tested: 1:1 and 13:1 [120]. Thus, it is may be possible to substitute wild-type E1A with mini-E1A in gene therapy treatment. By doing so, we may avoid the potentially undesirable complications associated with wild-type E1A.

Although E1A/DC-Chol treatment generally allows the E1A gene to be expressed transiently, it is possible that a small percentage of the transfected E1A gene may be integrated into the host chromosome. This potential problem could be especially serious if E1A remains in the cells of the reproductive organs. If that was the case, it is likely that E1A may be transmitted to the next generation. For this reason, the organs of E1A/DC-Chol-treated mice (tumor free and surviving for 1.5 years) were analyzed by PCR technique to detect the presence of E1A DNA [120]. Only two organs consistently contained E1A DNA: lungs and kidneys. Other organs such as liver, heart, spleen, brain, uterus, or ovaries had no detectable E1A DNA. This result suggests that, under the E1A/DC-Chol treatment conditions, lungs and kidneys are the most susceptible organs for chromosomal integration of E1A DNA, and, more importantly, E1A DNA is undetectable in the uterus and ovaries.

V. E1A GENE THERAPY: CLINICAL TRIALS

A. Phase I Breast and Ovarian Cancer with Intracavity Administration

To evaluate the feasibility of using E1A gene therapy for patients with HER2-overexpressing cancer, a phase I clinical trial was conducted in a group of patients with advanced breast and ovarian cancers. An E1A/DC-Chol cationic liposome complex was injected weekly into the thoracic or peritoneal cavity of 18 patients. The most common toxicity were fever, nausea, vomiting, and/or discomfort at the injection sites. E1A gene expression was readily detectable in tumor cells that showed a concurrent HER2 downregulation. Analyzing the intracavitary fluid from patients over the course of E1A/DC-Chol treatment, the total number of tumor clumps was significantly decreased after the treatment. In addition, the expression of Ki-67, a nuclear antigen expressed on all human proliferating cells [123], was also decreased, suggesting a decreased proliferation of tumor cells as a result of the E1A/DC-Chol treatment. When apoptosis was analyzed by TUNEL assay, a significant increase in the percentage of apoptosis in tumor cells was seen in all patients analyzed. Thus, the E1A/DC-Chol gene therapy suppresses tumor cell growth by both reducing cell replication and enhancing apoptosis. It is interesting to note that E1A/DC-Chol treatment caused a more drastic effect on the tumor cell clump reduction than on the HER2 downregulation. It is likely, therefore, that other molecular mechanisms also contribute to the E1A-mediated antitumor activity. In the treated patient fluids, TNF-α expression was significantly enhanced, which may account for the increased apoptosis seen in patient tumors. Thus, the phase I data clearly indicate the feasibility of using E1A/DC-Chol gene therapy to treat patients with HER2-overexpressing tumors, and the study successfully proves the working concept developed from our preclinical studies [11].

B. Phase II Head and Neck Cancer with Intratumor Administration

A multicenter phase II study of E1A gene therapy on head and neck cancers has recently been completed [124]. E1A/DC-Chol complex was used as a single agent and administered by intratumor injection. Among 20 treated patients,

5% (1 out of 20) showed a complete response and 45% (9 out of 20) showed an objective response or reaching a state of stable disease. The most common side effect was pain at the injection site, but there were no serious adverse events relating to E1A/DC-Chol administration. Based on the encouraging results of the phase II trials, a possible combined E1A/DC-Chol therapy with ionizing radiation and/or chemotherapy should be feasible in the near future.

VI. CONCLUSION

The E1A-mediated antitumor activity has manifested itself in the ability of E1A to inhibit tumor growth, suppress metastasis and angiogenesis, and sensitize tumor cells to apoptosis induced by therapeutic treatments such as TNF-α, γ-irradiation, and Taxol. The utility of an E1A-based gene therapy was validated by the apparent success of such application in ovarian, breast, and head and neck cancer models without obvious toxic side effect. The study on the molecular mechanisms underlying the E1A-mediated antitumor activity revealed that E1A represses HER2 and Axl transcription and targets the NF-κB pathway that connects HER2 overexpression and the Axl/Gas6-mediated signal pathway and the TNF-α (or γ-irradiation) signal pathway. In addition, E1A inactivates IKK, leading to NF-κB inactivation and subsequent shutting down of the survival program, which sensitizes cells to apoptosis.

Based on the success of the phase I breast and ovarian cancer clinical trials using E1A/DC-Chol via intracavity administration, a phase II trial in ovarian cancer patients is currently underway. Also, because advanced breast cancer patients usually have distant metastasis in other organs such as bone and brain, to treat metastatic breast cancer it is imperative that E1A be delivered systemically. Thus, a phase I trial of E1A/DC-Chol by intravenous administration for metastatic breast cancer, patients has been proposed. Moreover, a tumor-specific E1A gene therapy strategy will be valuable to enhance targeting specificity and reduce potential side effects. To that end, several approaches have been developed. For examples, taking advantage of high HER2 promoter activity in many HER2-overexpressing tumors, a HER2 promoter-driven E1A could specifically express E1A in HER2-overexpressing tumor cells [125]. Or, using HER2 antisense iron-responsible element, one could preferentially direct E1A expression in cancer cells that overexpress HER2 mRNA [126]. Another approach is to use HER2-specific binding filamentous bacteriophage to deliver E1A gene into HER2-overexpressing cancer cells [127].

It has been a long journey from the first demonstration of the ability of E1A to repress HER2 transcription up to today when E1A gene therapy is in phase II clinical trials. We now know more about the antitumor activity of E1A than when we first witnessed such activity in HER2-overexpressing cancer cells about a decade ago, when the

predominant view about E1A was that it was an "oncogene." In light of the E1A-mediated sensitization effects, the challenge in future E1A gene therapy is the development of effective combined E1A/chemo- or radiation-therapy strategies supported by solid *in vitro* and *in vivo* studies. With a better understanding of the mechanism by which E1A suppresses tumors and a better design of the delivery vehicle, it is hoped that an E1A-based gene therapy could become an effective treatment for cancer patients.

References

1. Nevins, J. R. (1995). Adenovirus E1A: transcription regulation and alteration of cell growth control. *Curr. Top. Microbiol. Immunol.* **199,** 25–32.
2. Jones, N. (1995). Transcriptional modulation by the adenovirus E1A gene. *Curr. Top. Microbiol. Immunol.* **199,** 59–80.
3. Brockmann, D., and Esche, H. (1995). Regulation of viral and cellular gene expression by E1A proteins encoded by the oncogenic adenovirus type 12. *Curr. Top. Microbiol. Immunol.* **199,** 81–112.
4. Moran, E., and Mathews, M. B. (1987). Multiple functional domains in the adenovirus E1A gene. *Cell* **48,** 177–178.
5. Houweling, A., van den Elsen, P. J., and van der Eb, A. J. (1980). Partial transformation of primary rat cells by the leftmost 4.5% fragment of adenovirus 5 DNA. *Virology* **105,** 537–550.
6. Ruley, H. E. (1990). Transforming collaborations between ras and nuclear oncogenes. *Cancer Cells* **2,** 258–268.
7. Chinnadurai, G. (1992). Adenovirus E1A as a tumor-suppressor gene. *Oncogene* **7,** 1255–1258.
8. Frisch, S. M. (1996). Reversal of malignancy by the adenovirus E1A gene. *Mutat. Res.* **350,** 261–266.
9. Mymryk, J. S. (1996). Tumour suppressive properties of the adenovirus 5 E1A oncogene. *Oncogene* **13,** 1581–1589.
10. Yu, D., and Hung, M.-C. (1998). The erbB2 gene as a cancer therapeutic target and the tumor- and metastasis-suppressing function of E1A. *Cancer Metastasis Rev.* **17,** 195–202.
11. Hung. M.-C., Wang, S.-C., and Hortobagyi, G. (1999). Targeting HER-2/neu-overexpressing cancer cells with transcriptional repressor genes delivered by cationic liposome, in *Nonviral Vectors for Gene Therapy,* Huang, L., Hung, M.-C., and Wagner, E., eds., pp. 357–375. Academic Press, San Diego.
12. Slamon, D. J., Clark, G. M., Wong, S. G., Levin, W. J., Ullrich, A., and McGuire, W. L. (1987). Human breast cancer: correlation of relapse and survival with amplification of the HER-2/neu oncogene. *Science* **235,** 177–182.
13. Slamon, D. J., and Clark, G. M. (1988). Amplification of c-erbB-2 and aggressive human breast tumors? *Science* **240,** 1795-1798.
14. Slamon, D. J., Godolphin, W., Jones, L. A., Holt, J. A., Wong, S. G., Keith, D. E., Levin, W. J., Stuart, S. G., Udove, J., Ullrich, A., and Press, M. F. (1989). Studies of the HER-2/neu proto-oncogene in human breast and ovarian cancer. *Science* **244,** 707–712.
15. Xia, W., Lau, Y. K., Zhang, H. Z., Liu, A. R., Li, L., Kiyokawa, N., Clayman, G. L., Katz, R. L., and Hung, M.-C. (1997). Strong correlation between c-ErbB-2 overexpression and overall survival of patients with oral squamous cell carcinoma. *Clin. Cancer Res.* **3,** 3–9.
16. Xia, W., Lau, Y. K., Zhang, H. Z., Xiao, F. Y., Johnston, D. A., Liu, A. R., Li, L., Katz, R. L., and Hung, M.-C. (1999). Combination of EGFR, HER-2/neu, and HER-3 is a stronger predictor for the outcome of oral squamous cell carcinoma than any individual family members. *Clin. Cancer Res.* **5,** 4164–4174.
17. Berchuck, A., Kamel, A., Whitaker, R., Kerns, B., Olt, G., Kinney, R., Soper, J. T., Dodge, R., Clarke-Pearson, D. L., and Marks, P. (1990). Overexpression of HER-2/neu is associated with poor

survival in advanced epithelial ovarian cancer. *Cancer Res.* **50,** 4087–4091.

18. Orr, M. S., O'Connor, P. M., and Kohn, K. W. (2000). Effects of c-erbB2 overexpression on the drug sensitivities of normal human mammary epithelial cells. *J. Natl. Cancer Inst.* **92,** 987–994.

19. Tsai, C. M., Chang, K. T., Perng, R. P., Mitsudomi, T., Chen, M. H., Kadoyama, C., and Gazdar, A. F. (1993). Correlation of intrinsic chemoresistance of non-small-cell lung cancer cell lines with HER-2/neu gene expression but not with ras gene mutations. *J. Natl. Cancer Inst.* **85,** 897–901.

20. Gusterson, B. A., Gelber, R. D., Goldhirsch, A., Price, K. N., Säve-Söderbörgh, S. J., Anbazhagan, R., Styles, J., Rudenstam, C. M., Golouh, R., Reed, R., Marinez-Tello, F., Tiltman, F., Torhorst, J., Grigolato, P., Bettelheim, R., Neville, A. M., Bürki, K., Castiglione, M., Collins, J., Lindtner, J., and Senn, H. J., (1992). Prognostic importance of c-erbB-2 expression in breast cancer. International (Ludwig) Breast Cancer Study Group. *J. Clin. Oncol.* **10,** 1049–1056.

21. Yu, D., and Hung, M.-C., (2000). Role of erbB2 in breast cancer chemosensitivity. *Bioessays* **22,** 673–680.

22. Ciardiello, F., Caputo, R., Pomatico, G., De Laurentiis, M., De Placido, S., Bianco, A. R., and Tortora, G. (2000). Resistance to taxanes is induced by c-erbB-2 overexpression in human MCF-10A mammary epithelial cells and is blocked by combined treatment with an antisense oligonucleotide targeting type I protein kinase. A. *Int. J. Cancer* **85,** 710–715.

23. Baselga, J., Tripathy, D., Mendelsohn, J., Baughman, S., Benz, C. C., Dantis, L., Sklarin, N. T., Seidman, A. D., Hudis, C. A., Moore, J., Rosen, P. P., Twaddell, T., Henderson, I. C., and Norton, L. (1999). Phase II study of weekly intravenous trastuzumab (Herceptin) in patients with HER2/neu-overexpressing metastatic breast cancer. *Semin. Oncol.* **26,** 78–83.

24. Pirollo, K. F., Hao, Z., Rait, A., Ho, C. W., and Chang, E. H., (1997). Evidence supporting a signal transduction pathway leading to the radiation-resistant phenotype in human tumor cells. *Biochem. Biophys. Res. Commun.* **230,** 196–201.

25. Burke, H. B., Hoang, A., Iglehart, J. D., and Marks, J. R. (1998). Predicting response to adjuvant and radiation therapy in patients with early stage breast carcinoma. *Cancer* **82,** 874–877.

26. Doherty, J. K., Bond, C. T., Hua, W., Adelman, J. P., and Clinton, G. M., (1999). An alternative HER-2/neu transcript of 8 kb has an extended 3′UTR and displays increased stability in SKOV-3 ovarian carcinoma cells. *Gynecol. Oncol.* **74,** 408–415.

27. Miller, S. J., and Hung, M.-C. (1995). Regulation of HER-2/neu gene expression. *Oncol. Rep.* **2,** 497–503.

28. Kraus, M. H., Popescu, N. C., Amsbaugh, S. C., and King, C. R. (1987). Overexpression of the EGF receptor-related proto-oncogene erbB-2 in human mammary tumor cell lines by different molecular mechanisms. *EMBO J.* **6,** 605–610.

29. Bosher, J. M., Williams, T., and Hurst, H. C. (1995). The developmentally regulated transcription factor AP-2 is involved in c-erbB-2 overexpression in human mammary carcinoma. *Proc. Natl. Acad. Sci. USA* **92,** 744–747.

30. Bosher, J. M., Totty, N. F., Hsuan, J. J., Williams, T., and Hurst, H. C. (1996). A family of Ap-2 proteins regulates c-erbB-2 experssion in mammary carcinoma. *Oncogene* **13,** 1701–1707.

31. Hollywood, D. P., and Hurst, H. C. (1993). A novel transcription factor, OB2-1, is required for overexpression of the proto-oncogene c-erbB-2 in mammary tumour lines. *EMBO J.* **12,** 2369–2375.

32. Yu, D., Suen, T. C., Yan, D.-H., Chang, L. S., and Hung, M.-C. (1990). Transcriptional repression of the neu protooncogene by the adenovirus 5 E1A gene products. *Proc. Natl. Acad. Sci. USA* **87,** 4499–4503.

33. Yan, D.-H., Chang, L. S., and Hung, M.-C. (1991). Repressed expression of the HER-2/c-erbB-2 proto-oncogene by the adenovirus E1A gene products. *Oncogene* **6,** 343–345.

34. Yu, D., Scorsone, K., and Hung, M.-C. (1991). Adenovirus type 5 E1A gene products act as transformation suppressors of the neu oncogene. *Mol. Cell. Biol.* **11,** 1745–1750.

35. Yu, D., Hamada, J., Zhang, H., Nicolson, G. L., and Hung, M.-C. (1992). Mechanisms of c-erbB2/neu oncogene-induced metastasis and repression of metastatic properties by adenovirus 5 E1A gene products. *Oncogene* **7,** 2263—2270.

36. Yu, D., Wolf, J. K., Scanlon, M., Price, J. E., and Hung, M.-C. (1993). Enhanced c-erbB-2/neu expression in human ovarian cancer cells correlates with more severe malignancy that can be suppressed by E1A. *Cancer Res.* **53,** 891–898.

37. Chang, J. Y., Xia, W., Shao, R., Sorgi, F., Hortobagyi, G. N., Huang, L., and Hung, M.-C. (1997). The tumor suppression activity of E1A in HER-2/neu-overexpressing breast cancer. *Oncogene* **14,** 561–568.

38. Zhang, Y., Yu, D., Xia, W., and Hung, M.-C. (1995). HER-2/neu-targeting cancer therapy via adenovirus-mediated E1A delivery in an animal model. *Oncogene* **10,** 1947–1954.

39. Yu, D., Matin, A., Xia, W., Sorgi, F., Huang, L., and Hung, M.-C. (1995). Liposome-mediated in vivo E1A gene transfer suppressed dissemination of ovarian cancer cells that overexpress HER-2/neu. *Oncogene* **11,** 1383–1388.

40. Chang, J. Y., Xia, W. Y., Shao, R. P., and Hung, M.-C. (1996). Inhibition of intratracheal lung cancer development by systemic delivery of E1A. *Oncogene* **13,** 1405–1412.

41. Yu, D., Shi, D., Scanlon, M., and Hung, M.-C. (1993). Reexpression of neu-encoded oncoprotein counteracts the tumor-suppressing but not the metastasis-suppressing function of E1A. *Cancer Res.* **53,** 5784–5790.

42. Ueno, N. T., Bartholomeusz, C. L., Hermann, J. L., Estrov, Z., Shao, R., Andreeff, M., Price, J., Paul, R. W., Anklesaria, P., Yu, D., and Hung, M.-C. (2000). *E1A*-mediated paclitaxel sensitization in HER-2/*neu*-overexpressing ovarian cancer through apoptosis involving caspase-3 pathway. *Clin. Cancer Res.* **6,** 250–259.

43. Ueno, N. T., Yu, D., and Hung, M.-C. (1997). Chemosensitization of HER-2/*neu*-overexpressing human breast cancer cells to paclitaxel (Taxol) by adenovirus type 5 *E1A*. *Oncogene* **15,** 953–960.

44. Brader, K. R., Wolf, J. K., Hung, M. C., Yu, D., Crispens, M. A., van Golen, K. L., and Price, J. E. (1997). Adenovirus E1A expression enhances the sensitivity of an ovarian cancer cell line to multiple cytotoxic agents through an apoptotic mechanism. *Clin. Cancer Res.* **3,** 2017–2024.

45. Shao, R., Karunagaran, D., Zhou, B. P., Li, K., Lo, S. S., Deng, J., Chiao, P., and Hung, M.-C. (1997). Inhibition of nuclear factor-kappaB activity is involved in E1A-mediated sensitization of radiation-induced apoptosis. *J. Biol. Chem.* **272,** 32739–32742.

46. Rikitake, Y., and Moran, E. (1992). DNA-binding properties of the E1A-associated 300-kilodalton protein. *Mol. Cell. Biol.* **12,** 2826–2836.

47. Chen, H., and Hung, M.-C. (1997). Involvement of co-activator p300 in the transcriptional regulation of the HER-2/neu gene. *J. Biol. Chem.* **272,** 6101–6104.

48. Chen, H., Yu, D., Chinnadurai, G., Karunagaran, D., and Hung, M.-C. (1997). Mapping of adenovirus 5 E1A domains responsible for suppression of neu-mediated transformation via transcriptional repression of neu. *Oncogene* **14,** 1965–1971.

49. Borrelli, E., Hen, R., and Chambon, P. (1984). Adenovirus-2 E1A products repress enhancer-induced stimulation of transcription. *Nature* **312,** 608–612.

50. Velcich, A., Kern, F. G., Basilico, C., and Ziff, E. B., (1986). Adenovirus E1A proteins repress expression from polyomavirus early and late promoters. *Mol. Cell. Biol.* **6,** 4019–4025.

51. Hen, R., Borrelli, E., and Chambon, P. (1985). Repression of the immunoglobulin heavy chain enhancer by the adenovirus-2 E1A products. *Science* **230,** 1391–1394.

52. Stein, R. W., and Ziff, E. B. (1987). Repression of insulin gene expression by adenovirus type 5 E1A proteins. *Mol. Cell. Biol.* **7,** 1164–1170.

53. Funk, W. D., Ouellette, M., and Wright, W. E. (1991). Molecular biology of myogenic regulatory factors. *Mol. Biol. Med.* **8,** 185–195.

54. Hudziak, R. M., Lewis, G. D., Winget, M., Fendly, B. M., Shepard, H. M., and Ullrich, A. (1989). p185HER2 monoclonal antibody has antiproliferative effects in vitro and sensitizes human breast tumor cells to tumor necrosis factor. *Mol. Cell. Biol.* **9,** 1165–1172.

55. Zhou, B. P., Hu, M. C., Miller, S. A., Yu, Z., Xia, W., Lin, S. Y., and Hung, M.-C. (2000). HER-2/neu blocks tumor necrosis factor-induced apoptosis via the Akt/NF-kappaB pathway. *J. Biol. Chem.* **275,** 8027–8031.

56. Hu, P., Margolis, B., Skolnik, E. Y., Lammers, R., Ullrich, A., and Schlessinger, J. (1992). Interaction of phosphatidylinositol 3-kinase-associated p85 with epidermal growth factor and platelet-derived growth factor receptors. *Mol. Cell. Biol.* **12,** 981–990.

57. Franke, T. F., Kaplan, D. R., and Cantley, L. C. (1997). PI3K: downstream AKTion blocks apoptosis. *Cell* **88,** 435–437.

58. Downward, J. (1998). Mechanisms and consequences of activation of protein kinase B/Akt. *Curr. Opin. Cell. Biol.* **10,** 262–267.

59. Beg, A. A., and Baltimore, D. (1996). An essential role for NF-kappaB in preventing TNF-alpha-induced cell death. *Science* **274,** 782–784.

60. Wang, C. Y., Mayo, M. W., and Baldwin, Jr., A. S. (1996). TNF- and cancer therapy-induced apoptosis: protentiation by inhibition of NF-kappaB. *Science* **274,** 784–787.

61. Van Antwerp, D. J., Martin, S. J., Kafri, T., Green, D. R., and Verma, I. M. (1996). Suppression of TNF-alpha-induced apoptosis by NF-kappaB. *Science* **274,** 787–789.

62. Liu, Z. G., Hsu, H., Goeddel, D. V., and Karin, M. (1996). Dissection of TNF receptor 1 effector functions: JNK activation is not linked to apoptosis while NF-kappaB activation prevents cell death. *Cell* **87,** 565–576.

63. Ozes, O. N., Mayo, L. D., Gustin, J. A., Pfeffer, S. R., Pfeffer, L. M., and Donner, D. B. (1999). NF-kappaB activation by tumour necrosis factor requires the Akt serine-threonine kinase. *Nature* **401,** 82–85.

64. Delhase, M., Li, N., and Karin, M. (2000). Kinase regulation in inflammatory response. *Nature* **406,** 367–368.

65. Madge, J. A., and Pober, J. S. (2000). A phosphatidylinositol 3-kinase/Akt pathway, activated by tumor necrosis factor or interleukin-1, inhibits apoptosis but does not activate NFkappaB in human endothelial cells. *J. Biol. Chem.* **275,** 15458–15465.

66. Baldwin, A. S. J. (1996). The NF-κB and IκB proteins: New discoveries and insights. *Annu. Rev. Immunol.* **14,** 649–681.

67. Mercurio, F., Zhu, H., Murray, B. W., Shevchenko, A., Bennett, B. L., Li, J., Young, D. B., Barbosa, M., Mann, M., Manning, A., and Rao, A. (1997). IKK-1 and IKK-2: cytokine-activated IkappaB kinases essential for NF-kappaB activation. *Science* **278,** 860–866.

68. Traenckner, E. B., Pahl, H. L., Henkel, T., Schmidt, K. N., Wilk, S., and Baeuerle, P. A. (1995). Phosphorylation of human I kappa B-alpha on serines 32 and 36 controls I kappa B-alpha proteolysis and NF-kappa B activation in response to diverse stimuli. *EMBO J.* **14,** 2876–2883.

69. Hung, M.-C., and Yu, D. (2000). Therapeutic resistance of erb B-2-overexpressing cancers and strategies to overcome this resistance, in *DNA Alterations in Cancer* (M. Ehrlich, ed.), pp. 457–470. Eaton publishing, Natick, MA.

70. Pegram, M. D., Finn, R. S., Arzoo, K., Beryt, M., Pietras, R. J., and Slamon, D. J. (1997). The effect of HER-2/neu overexpression on chemotherapeutic drug sensitivity in human breast and ovarian cancer cells. *Oncogene* **15,** 537–547.

71. Yu, D., Liu, B. L., Tan, M., Li, J. Z., Wang, S. S., and Hung, M.-C. (1996). Overexpression of c-erbB-2/neu in breast cancer cells confers increased resistance to taxol via MDR-1-independent mechanisms. *Oncogene* **13,** 1359–1365.

72. Ciardiello, F., McGeady, M. L., Kim, N., Basolo, F., Hynes, N., Langton, B. C., Yokozaki, H., Saeki, T., Elliott, J. W., Masui, H. *et al.* (1990). Transforming growth factor-alpha expression is enhanced in human mammary epithelial cells transformed by an activated c-Ha-ras protooncogene but not by the c-neu protooncogene, and overexpression of the transforming growth factor-alpha complementary DNA leads to transformation. *Cell Growth Differ.* **1,** 407–420.

73. Ciardiello, F., Gottardis, M., Basolo, F., Pepe, S., Normanno, N., Dickson, R. B., Bianco, A. R., and Salomon, D. S. (1992). Additive effects of c-erbB-2, c-Ha-ras, and transforming growth factor-alpha genes on in vitro transformation of human mammary epithelial cells. *Mol. Carcinog.* **6,** 43–52.

74. de Stanchina, E., McCurrach, M. E., Zindy, F., Shieh, S. Y., Ferbeyre, G., Samuelson, A. V., Prives, C., Roussel, M. F., Sherr, C. J., and Lowe, S. W. (1998). E1A signaling to p53 involves the p19 (ARF) tumor suppressor. *Genes Dev.* **12,** 2434–2442.

75. Frisch, S. M., and Dolter. K. E. (1995). Adenovirus E1A-mediated tumor suppression by a c-erbB-2/neu-independent mechanism. *Cancer Res.* **55,** 5551–5555.

76. Pozzatti, R., McCormick, M., Thompson, M. A., and Khoury, G. (1988). The E1A gene of adenovirus type 2 reduces the metastatic potential of ras-transformed rat embryo cells. *Mol. Cell. Biol.* **8,** 2984–2988.

77. Pozzatti, R., Muschel, R., Williams, J., Padmanabhan, R., Howard, B., Liotta, L., and Khoury, G. (1986). Primary rat embryo cells transformed by one or two oncogenes show different metastatic potentials. *Science* **232,** 223–227.

78. Steeg, P. S., Bevilacqua, G., Pozzatti, R., Liotta, L. A., and Sobel, M. E. (1988). Altered expression of NM23, a gene associated with low tumor metastatic potential, during adenovirus 2 E1A inhibition of experimental metastasis. *Cancer Res.* **48,** 6550–6554.

79. Steeg, P. S., Bevilacqua, G., Kopper, L., Thorgeirsson, U. P., Talmadge, J. E., Liotta, L. A., and Sobel, M. E. (1988). Evidence for a novel gene associated with low tumor metastatic potential. *J. Natl. Cancer Inst.* **80,** 200–204.

80. Frisch, S. M., Reich, R., Collier, I. E., Genrich, L. T., Martin, G., and Goldberg, G. I. (1990). Adenovirus E1A represses protease gene expression and inhibits metastasis of human tumor cells. *Oncogene* **5,** 75–83.

81. Frisch, S. M. (1994). E1A induces the expression of epithelial characteristics. *J. Cell. Biol.* **127,** 1085–1096.

82. Hennig, G., Behrens, J., Truss, M., Frisch, S., Reichmann, E., and Birchmeier, W. (1995). Progression of carcinoma cells is associated with alterations in chromatin structure and factor binding at the E-cadherin promoter in vivo. *Oncogene* **11,** 475–484.

83. Santoro, M., Battaglia, C., Zhang, L., Carlomagno, F., Martelli, M. L., Salvatore, D., and Fusco, A. (1994). Cloning of the rat tissue inhibitor of metalloproteinases type 2 (TIMP-2) gene: analysis of its expression in normal and transformed thyroid cells. *Exp. Cell. Res.* **213,** 398–403.

84. Offringa, R., Smits, A. M., Houweling, A., Bos, J. L., and van der Eb, A. J. (1988). Similar effects of adenovirus E1A and glucocorticoid hormones on the expression of the metalloprotease stromelysin. *Nucleic Acids Res.* **16,** 10973–10984.

85. Linder, S., Popowicz, P., Svensson, C., Marshall, H., Bondesson, M., and Akusjarvi, G. (1992). Enhanced invasive properties of rat embryo fibroblasts transformed by adenovirus E1A mutants with deletions in the carboxy-terminal exon. *Oncogene* **7,** 439–443.

86. Garbisa, S., Pozzatti, R., Muschel, R. J., Saffiotti, U., Ballin, M., Goldfarb, R. H., Khoury, G., and Liotta, L. A. (1987). Secretion of type IV collagenolytic protease and metastatic phenotype: induction by transfection with c-Ha-ras but not c-Ha-ras plus Ad2-E1A. *Cancer Res.* **47,** 1523–1528.

87. Offringa, R., Gebel, S., van Dam, H., Timmers, M., Smits, A., Zwart, R., Stein, B., Bos, J. L., van der Eb, A., and Herrlich, P. (1990). A novel function of the transforming domain of E1A: repression of AP-1 activity. *Cell* **62**, 527–538.

88. Bernhard, E. J., Muschel, R. J., and Hughes, E. N. (1990). Mr 92,000 gelatinase release correlates with the metastatic phenotype in transformed rat embryo cells. *Cancer Res.* **50**, 3872–3877.

89. Bernhard, E. J., Hagner, B., Wong, C., Lubenski, I., and Muschel, R. J. (1995). The effect of E1A transfection on MMP-9 expression and metastatic potential. *int. J. Cancer* **60**, 718–724.

90. Hofmann, M., Rudy, W., Gunthert, U., Zimmer, S. G., Zawadzki, V., Zoller, M., Lichtner, R. B., Herrlich, P., and Ponta, H. (1993). A link between ras and metastatic behavior of tumor cells: ras induces CD44 promoter activity and leads to low-level expression of metastasis-specific variants of CD44 in CREF cells. *Cancer Res.* **53**, 1516–1521.

91. Yu, D., Wang, S. S., Dulski, K. M., Tsai, C. M., Nicolson, G. L., and Hung, M.-C. (1994). C-erbB-2/neu overexpression enhances metastatic potential of human lung cancer cells by induction of metastasis-associated properties. *Cancer Res.* **54**, 3260–3266.

92. Tan, M., Yao, J., and Yu, D. (1997). Overexpression of the c-erbB-2 gene enhanced intrinsic metastasis potential in human breast cancer cells without increasing their transformation abilities. *Cancer Res.* **57**, 1199–1205.

93. O'Bryan, J. P., Frye, R. A., Cogswell, P. C., Neubauer, A., Kitch B., Prokop, C., Espinosa, R. D., Le Beau, M. M., Earp, H. S., and Liu, E. T. (1991). Axl, a transforming gene isolated from primary human myeloid leukemia cells, encodes a novel receptor tyrosine kinase. *Mol. Cell. Biol.* **11**, 5016–5031.

94. Fantl, W. J., Johnson, D. E., and Williams, L. T. (1993). Signalling by receptor tyrosine kinases. *Annu. Rev. Biochem.* **62**, 453–481.

95. van der Geer, P., Hunter, T., and Lindberg, R. A. (1994). Receptor protein-tyrosine kinases and their signal transduction pathways. *Annu. Rev. Cell. Biol.* **10**, 251–337.

96. Lee, W.-P., Liao, Y., Robinson, D., Kung, H. J., Liu, E. T., and Hung, M.-C. (1999). Axl–gas6 interaction counteracts E1A-mediated cell growth suppression and proapoptotic activity. *Mol. Cell. Biol.* **19**, 8075–8082.

97. Manfioletti, G., Brancolini, C., Avanzi, G., and Schneider, C. (1993). The protein encoded by a growth arrest-specific gene (gas6) is a new member of the vitamin K-dependent proteins related to protein S, a negative coregulator in the blood coagulation cascade. *Mol. Cell. Biol.* **13**, 4976–4985.

98. Stitt, T. N., Conn, G., Gore, M., Lai, C., Bruno, J., Radziejewski, C., Mattsson, K., Fisher, J., Gies, D. R., Jones, P. F. *et al.* (1995). The anticoagulation factor protein S and its relative, Gas6, are ligands for the Tyro 3/Axl family of receptor tyrosine kinases. *Cell* **80**, 661–670.

99. Varnum, B. C., Young, C., Elliott, G., Garcia, A., Bartley, T. D., Fridell, Y. W., Hunt, R. W., Trail, G., Clogston, C., Toso, R. J. *et al.* (1995). Axl receptor tyrosine kinase stimulated by the vitamin K-dependent protein encoded by growth-arrest-specific gene 6. *Nature* **373**, 623–626.

100. Goruppi, S., Ruaro, E., Varnum, B., and Schneider, C. (1997). Requirement of phosphatidylinositol 3-kinase-dependent pathway and Src for Gas6-Axl mitogenic and survival activities in NIH 3T3 fibroblasts. *Mol. Cell. Biol.* **17**, 4442–4453.

101. Datta, S. R., Dudek, H., Tao, X., Masters, S., Fu, H., Gotoh, Y., and Greenberg, M. E. (1997). Akt phosphorylation of BAD couples survival signals to the cell-intrinsic death machinery. *Cell* **91**, 231–241.

102. del Peso, L., Gonzalez-Garcia, M., Page, C., Herrera, R., and Nunez, G. (1997). Interleukin-3-induced phosphorylation of BAD through the protein kinase Akt. *Science* **278**, 687–689.

103. Romashkova, J. A., and Makarov, S. S. (1999). NF-kappaB is a target of AKT in anti-apoptotic PDGF signalling. *Nature* **401**, 86–90.

104. Brunet, A., Bonni, A., Zigmond, M. J., Lin, M. Z., Juo, P., Hu, L. S., Anderson, M. J., Arden, K. C., Blenis, J., and Greenberg, M. E. (1999). Akt promotes cell survival by phosphorylating and inhibiting a Forkhead transcription factor. *Cell* **96**, 857–868.

105. Chen, M. J., Holskin, B., Strickler, J., Gorniak, J., Clark, M. A., Johnson, P. J., Mitcho, M., and Shalloway, D. (1987). Induction by E1A oncogene expression of cellular susceptibility to lysis by TNF. *Nature* **330**, 581–583.

106. Duerksen-Hughes, P., Wold, W. S., and Gooding, L. R. (1989). Adenovirus E1A renders infected cells sensitive to cytolysis by tumor necrosis factor. *J. Immunol.* **143**, 4193–4200.

107. Lowe, S. W., and Ruley, H. E. (1993). Stabilization of the p53 tumor suppressor is induced by adenovirus 5 E1A and accompanies apoptosis. *Genes Dev.* **7**, 535–545.

108. Foo, S. Y., Nolan, G. P. (1999). NF-κB to the rescue: REL$_s$, apoptosis and cellular transformation. *TIG* **15**, 229–235.

109. Shao, R., Hu, M. C., Zhou, B. P., Lin, S. Y., Chiao, P. J., von Lindern, R. H., Spohn, B., and Hung M.-C. (1999). E1A sensitizes cells to tumor necrosis factor-induced apoptosis through inhibition of IkappaB kinases and nuclear factor kappaB activities. *J. Biol. Chem.* **274**, 21495–21498.

110. Siebenlist, U., Franzoso, G., and Brown, K. (1994). Structure, regulation and function of NF-kappa B. *Annu. Rev. Cell. Biol.* **10**, 405–455.

111. Baichwal, V. R., and Baeuerle, P. A. (1997). Activate NF-kappa B or die? *Curr. Biol.* **7**, R94–R96.

112. Sonenshein, G. E. (1997). Rel/NF-kappa B transcription factors and the control of apoptosis. *Semin. Cancer Biol.* **8**, 113–119.

113. Karin, M., and Ben-Neriah, Y. (2000). Phosphorylation meets ubiquitination: the control of NF-[kappa]B activity. *Annu. Rev. Immunol.* **18**, 621–663.

114. Gilmore, T. D., Koedood, M., Piffat, K. A., and White, D. W. (1996). Rel/NF-kappaB/IkappaB proteins and cancer. *Oncogene* **13**, 1367–1378.

115. Luque, I., and Gelinas, C. (1997). Rel/NF-kappa B and I kappa B factors in oncogenesis. *Semin. Cancer Biol.* **8**, 103–111.

116. Culver, K. W., Ram, Z., Wallbridge, S., Ishii, H., Oldfield, E. H., and Blaese, R. M. (1992). In vivo gene tansfer with retroviral vector-producer cells for treatment of experimental brain tumors. *Science* **256**, 1550–1552.

117. Mesnil, M., Piccoli, C., Tiraby, G., Willecke, K., and Yamasaki, H. (1996). Bystander killing of cancer cells by herpes simplex virus thymidine kinase gene is mediated by connexins. *Proc. Natl. Acad. Sci. USA* **93**, 1831–1835.

118. Shao, R., Xia, W., and Hung, M. C. (2000). Inhibition of angiogenesis and induction of apoptosis are involved in E1A-mediated bystander effect and tumor suppression. *Cancer Res.* **60**, 3123–3126.

119. Gao, X., and Huang, L. (1991). A novel cationic liposome reagent for efficient transfection of mammalian cells. *Biochem. Biophys. Res. Commun.* **179**, 280–285.

120. Xing, X., Zhang, S., Chang, J. Y., Tucker, S. D., Chen, H., Huang, L., and Hung, M.-C. (1998). Safety study and characterization of E1A-liposome complex gene delivery in an ovarian cancer model. *Gene Ther.* **5**, 1538–1544.

121. Xing, X., Liu, V., Xia, W., Stephens, L. C., Huang, L., Lopez-Berestein, G., and Hung, M.-C. (1997). Safety studies of the intraperitoneal injection of E1A–liposome complex in mice. *Gene Ther.* **4**, 238–243.

122. Ruley, H. E. (1983). Adenovirus early region 1A enables viral and cellular transforming genes to transform primary cells in culture. *Nature* **304**, 602–606.

123. Gerdes, J., Schwab, U., Lemke, H., and Stein, H. (1983). Production of a mouse monoclonal antibody reactive with a human nuclear antigen associated with cell proliferation. *Int. J. Cancer* **31**, 13–20.

124. Reynolds, T. C., Alberts, D., Gershenson, D., Gleich, L., Glisson, B., Hanna, E., Huang, L., Hung, M.-C., Kenady, D., Ueno, N., Villaret, D.,

and Yoo, G. (2000). Activity of E1A in human clinical trials, in *ASCO* **19,** 461a.

125. Pandha, H. S., Martin, L. A., Rigg, A., Hurst, H. C., Stamp, G. W., Sikora, K., and Lemoine, N. R. (1999). Genetic prodrug activation therapy for breast cancer: a phase I clinical trial of erbB-2-directed suicide gene expression. *J. Clin. Oncol.* **17,** 2180–2189.

126. Yan, D.-H. (1998). Targeting human breast cancer cells that overexpress HER-2/neu mRNA by an antisense iron responsive element. *Biochem. Biophys. Res. Commun.* **246,** 353–358.

127. Poul, M. A., and Marks, J. D. (1999). Targeted gene delivery to mammalian cells by filamentous bacteriophage. *J. Mol. Biol.* **288,** 203–211.

PRODRUG ACTIVATION STRATEGIES FOR GENE THERAPY OF CANCER

31

Preemptive and Therapeutic Uses of Suicide Genes for Cancer and Leukemia

FREDERICK L. MOOLTEN

Edith Nourse Rogers Memorial Veterans Hospital
Bedford, Massachusetts
Boston University School of Medicine
Boston, Massachusetts 02118

PAULA J. MROZ

Edith Nourse Rogers Memorial Veterans Hospital
Bedford, Massachusetts 02188

I. INTRODUCTION

The emergence of cancer gene therapy as a new discipline bears testimony to a need unmet by conventional therapies: selectivity. Cytokine gene therapy, suppressor genes, and antisense/ribozymes each aim at targeting cancer cells selectively. Implicit in these approaches is the presumption that there will be something about neoplastic cells that distinguishes them sufficiently from vital normal cells to permit therapeutic modalities to suppress or kill them without subjecting their normal counterparts to intolerable host toxicity. The presumption is probably true for some cancers but false for others, perhaps for a majority.

Suicide genes constitute an alternative approach. Rather than manipulating, positively or negatively, existing cellular functions, they introduce new functions that sensitize cells to drugs at concentrations that would otherwise be innocuous. Most suicide genes encode enzymes that catalyze the conversion of prodrugs into cytotoxic antimetabolites. The best known among these genes, the herpes thymidine kinase (HSV-TK) gene, sensitizes cells to the guanosine analog ganciclovir (GCV) as a consequence of HSV-TK-catalyzed phosphorylation of GCV to intermediates that lethally inhibit DNA synthesis. Since the initial reports introducing the suicide gene concept [1,2], many animal studies have demonstrated that systemically administered GCV can eradicate transplanted tumors bearing transduced HSV-TK genes (reviewed in Moolten [3] and Tiberghien [4])(Fig. 1). A serendipitous property of the HSV-TK/GCV combination is the bystander effect, a phenomenon that manifests itself as an ability of GCV to kill not only HSV-TK transduced cells but also untransduced cells in their proximity. The mechanism probably involves the transfer of activated GCV metabolites [1,5–8], at least *in vitro,* although stimulation of host immune/inflammatory reactions and damage to tumor blood vessels may also play a role *in vivo.* To the extent that immune phenomena are involved, systemic antitumor effects may sometimes be observable.

Numerous other suicide gene/prodrug systems have since been described [9–23]. Some of the better characterized combinations are listed in Table 1. Of interest, p450-2B1 and nitroreductase genes generate products that are not antimetabolites but alkylating agents and therefore potentially more effective than antimetabolites in quiescent cells. The products of the Fas/FKBP and caspase/FKBP fusion genes

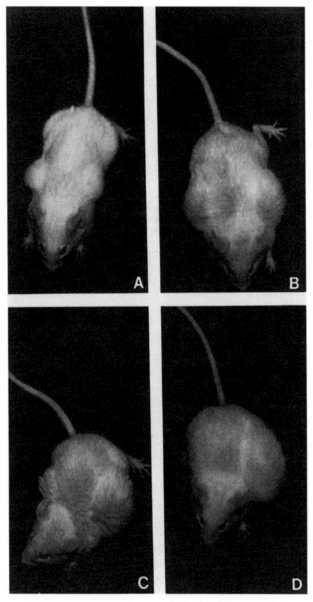

FIGURE 1 Differential effects of GCV on HSV-TK-positive and -negative tumors in the same mouse. HSV-TK-positive sarcoma cells were injected subcutaneously into the right flank and HSV-TK-negative cells into the left flank. (A) At 13 days, small tumors were observed at each site. (B) At day 16, the tumors were growing progressively; an 8-day course of GCV administered intraperitoneally was begun. (C) By day 23, the gene-positive tumor had shrunk while the negative tumor had enlarged. (D) At day 37, the positive tumor had regressed completely, while the negative tumor continued to grow. (From Moolten, F. L., *Cancer Res.*, **46,** 5276–5281, 1986. With permission.)

induce apoptosis. Both components of each pair are human proteins; similarly, p450-2B1, although of rat origin, has human p450 counterparts. Fas/FKBP, caspase/FKBP, and p450 genes are therefore less likely to provoke host immune reactions against transduced cells than other suicide genes that generate proteins of bacterial or viral origin.

II. THERAPEUTIC USES OF SUICIDE GENES

In theory, suicide genes can be used both *therapeutically* in cancer patients and *preemptively* in individuals not yet afflicted with a cancer, as described below. Clinical trials to date, however, have been limited to patients with established malignancies. These trials have principally involved tumors at limited sites, including brain tumors [24,25], ovarian cancer that has extended to the peritoneal cavity [26–28], and mesotheliomas [29]. Most of the trials utilize the HSV-TK/GCV combination, with the majority of these employing a modified virus as a vehicle ("vector") for introducing the HSV-TK gene into tumor cells after intratumoral injection or other instillation techniques that restrict the gene to the known location of the tumor. HSV-TK transduction is then followed by systemic GCV administration. Suicide gene trials are included in a comprehensive listing of gene therapy trials compiled as of late 1999 [30].

Most ongoing clinical trials have utilized one of two different vector systems to transduce the tumor cells [31]. In each case, the vector consists of a virus capable of infecting human cells that has been genetically engineered to eliminate genes responsible for viral replication and cellular pathology, substituting in their place the suicide gene to be used for therapy. The first system entails the use of vectors derived from murine retroviruses. Retroviral vectors mediate transduction that is relatively stable, at least in the short term, as a consequence of the integration of vector sequences into the DNA of the host genome, but to date it has been difficult to produce cell-free suspensions containing these vectors at titers sufficient to yield more than minimal transduction levels *in vivo*. Because of this limitation, most protocols have not attempted to introduce the vectors themselves into tumors, but rather *producer cells,* which are murine fibroblasts that generate and release the HSV-TK vectors at their *in vivo* injection site to yield a continuous supply until the cells are rejected by the host or killed by the administration of GCV.

The second system entails the use of vectors derived from human adenoviruses. Because adenoviral vectors do not integrate into genomic DNA, they mediate only transient transduction but possess the advantage of high titers that obviate the need for producer cells. Neither the adenoviral nor retroviral vector system, however, is currently capable of transducing suicide genes into more than a minority of tumor cells *in vivo*. A major component of the rationale underlying current trials is the expectation that bystander effects might permit GCV to eradicate untransduced cells by virtue of their proximity to transduced cells.

Most of the clinical trials are in early stages. To date, reported results include signs of tumor regression in some individuals [32], but few patients have experienced significant clinical benefit. A clear limitation is the difficulty of delivering suicide genes to all areas of a large tumor, even if the tumor has not metastasized [33]. This problem is not

TABLE 1   Suicide Gene/Prodrug Combinations

Gene	Prodrug	Active product	Ref.
HSV-TK	GCV	GCV mono- and diphosphates	[3]
Cytosine deaminase	5-Fluorocytosine	5-Fluorouracil	[9]
Gpt	6-Thioxanthine	6-Thioxanthine ribonucleotide	[10]
p450-2B1	Cyclophosphamide	Phosphoramide mustard	[11,12]
Purine nucleoside phosphorylase	6-Methylpurine deoxyribonucleoside	6-Methylpurine	[13]
Deoxycytidine kinase	Ara-C	Ara-C monophosphate	[14]
Nitroreductase	CB1954	5-Azaridin-1-yl-4-hydroxylamino-2-nitrobenzamide	[15]
Fas-FKBP	AP1903[a]	Multimerized fas	[16,17]
Caspase-FKBP	AP1903[a]	Multimerized caspase	[18,19]
Sodium/iodide symporter	Radioiodide	Concentrated intracellular radioiodide	[20,21]
Carboxypeptidase	Peptide-linked alkylating agent or methotrexate	Free alkylating agent or methotrexate	[22,23]

[a] Strictly speaking, AP1903 is not a prodrug, as it is not activated by the product of the suicide gene but rather activates that product by cross-linking it to form the multimers needed for the Fas or caspase proteins to trigger apoptotic pathways. FKBP is an abbreviation for FK506 binding protein.

fully solved by bystander effects, as these effects tend to be powerful only at short ranges.

Another obvious limitation of a localized injection approach stems from the fact that the lethality of most cancer results from metastatic rather than localized disease. Metastatic disease will require systemic approaches that expose normal as well as neoplastic cells to the therapeutic modality. Attempts to address this problem include the linkage of suicide genes to promoters that might be highly active in tumor cells, with little or no activity in vital normal tissues. These include a tyrosinase promoter for melanomas [34,35], an alpha-fetoprotein promoter for hepatomas [36,37], Ebstein–Barr virus (EBV)-encoded transcriptional regulatory elements for EBV-related lymphomas and other EBV-associated malignancies [38,39], an osteocalcin promoter for osteosarcomas [40], and an ErbB2 promoter for breast carcinomas (based on evidence that a subset of breast cancers may exhibit ErbB2 promoter hyperactivity [41]). Promising initial evidence for therapeutic specificity has been reported in murine systems involving transplanted tumors [34,35,37,42], including a reduction in lung metastases of a melanoma after intravenous administration of a retroviral HSV-TK vector followed by GCV therapy [35]. It remains to be determined how much specificity might be achievable with these promoters in human cancers that arise endogenously and whether or not these genes can be delivered in bulk to metastatic deposits in sufficient quantity and uniformity to ensure tumor eradication by prodrug therapy. In addition, these cancers are exceptional; most cancers have yet to exhibit evidence of promoter activities unshared by vital normal stem cells.

III. PREEMPTIVE USES OF SUICIDE GENES IN CANCER

Our recent work in murine systems has focused on exploring the feasibility of a different application of suicide genes: their *preemptive* use before a cancer develops, with particular emphasis on individuals at excessive risk for cancer. The goal of preemption is to achieve selectivity without requiring neoplastic cells to possess the one property whose frequent absence has confounded other approaches to cancer therapy, genetic or otherwise — a targetable difference from normal cells. To obviate the need for targetability, preemption is designed to exploit the *clonal* (i.e., single cell) origin of human cancers [43–45] by introducing suicide genes not into an established cancer but into a tissue from which cancers may arise. Because it is clonal, any cancer that subsequently arises within that tissue from a transduced cell should uniformly carry the suicide gene in all its cells as a clonal property, including metastases. Within a transduced clone of cancer cells, suicide gene expression might be lost in an occasional cell through mutations that delete or inactivate the gene, but in theory such cells might be susceptible to bystander killing by their proximity to gene-positive cells. The several studies that report the curability of tumors that arise from transplanted clonal populations of HSV-TK-positive tumor cells [1,2,6,46], even when the tumors are known to harbor gene-negative mutants [1,46,47], are consistent with this expectation.

In a nonvital tissue such as breast or prostate epithelium, preemption can aim at transducing a chosen suicide gene into as many cells as possible to maximize the probability that a subsequent cancer will arise from a transduced cell. Cells outside the transduced tissue would remain unsensitized, and measures to promote selectivity within the transduced tissue itself would be unnecessary, as loss of nonneoplastic breast or prostate epithelial cells during cancer therapy would not be life threatening. For preemptive sensitization of a vital tissue such as bone marrow or gastrointestinal epithelium, selectivity must be achieved differently by introducing one or more suicide genes in *mosaic* rather than homogeneous fashion [1,48]. Mosaicism creates selectivity by ensuring that whatever cell later spawns a cancer will share its clonal sensitivity pattern with only a fraction of the normal cells (Fig. 2).

As a first step in testing the preemption paradigm, we have asked whether suicide gene transduction into cells that are not yet malignant might permit effective therapy of cancers that later arose from them [49]. TM4 is a line of preneo-

plastic murine mammary epithelial cells that can be propagated in tissue culture for subsequent *in vivo* transplantation [50,51]. A retroviral vector, STK [2], was used to transduce the HSV-TK gene into these cells *in vitro*. The cells were then injected subcutaneously into syngeneic BALB/c mice, where they formed small, nongrowing nodules from which cancers later arose in 40% of the mice. When the mice were treated with GCV, 7 out of 20 responded with complete and durable tumor regressions, and the remainder exhibited a significant retardation of tumor growth (Table 2). Control tumors (transduced and untreated, or untransduced and GCV-treated) invariably exhibited progressive growth. In comparison with controls, the HSV-TK gene by itself exerted no adverse effects on cancer incidence or growth rates, indicating that its presence was not a liability for the preemptively transduced preneoplastic cells and that its observable therapeutic effects operated through GCV.

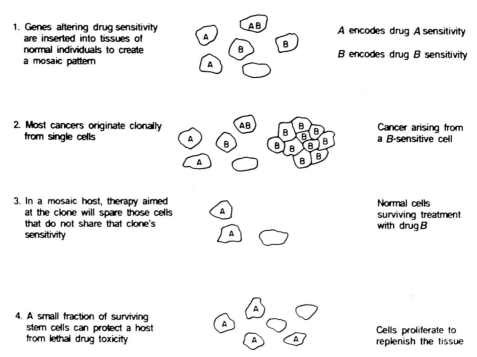

FIGURE 2 Preemptive introduction of suicide genes in *mosaic* fashion to ensure that later cancers are sensitized to eradication by a prodrug while their tissue of origin is protected by the presence of cells that do not share their sensitivity [48]. In the diagram, genes A and B are suicide genes that sensitize cells to prodrugs A and B, respectively. The clonal origin of a cancer arising from one of the B-Sensitized cells renders it uniformly sensitive to the prodrug B. Normal cells with the same sensitivity are also killed, but the remaining normal cells in the mosaic survive to repopulate the tissue. It should be noted that the principle of mosaicism illustrated in the diagram with suicide genes can also be implemented with resistance genes [48]. Thus, for example, if genes A and B encode resistance to drugs A and B, administration of drug A will selectively kill the cells that lack gene A, and drug B will kill cells that lack gene B. Unlike sensitivity mosaicism, however, mosaicism created by resistance genes cannot utilize drug doses that fully exploit the difference between sensitive and resistant cells without risking excessive drug toxicity to other tissues that have not acquired the genes. A potential advantage of resistance mosaicism is stability. Thus, a cell that carries a suicide gene can rid itself of its selecive sensitivity through loss or inactivation of the gene via a one-step mutation, whereas a cell that is selectively sensitive because it lacks a gene for high-level drug resistance will in many cases require multiple independent mutations to achieve that level of resistance.

TABLE 2   GCV Therapy of Tumors Arising from
Preneoplastic Mammary Epithelial Cells

Preneoplastic cells injected	GCV therapy of subsequent tumors[a]	Mice with durable tumor regressions/total[b]	Median survival (days)
HSV-TK transduced[c]	0	0/25	46
	+	7/20	152
Untransduced	0	0/8	70
	+	0/11	72
Transduced with control vector[c]	0	0/8	38
	+	0/7	53

[a]Tumor-bearing mice received 150 mg/kg GCV twice daily for 5 days by intraperitoneal injection.

[b]Seven GCV-treated mice in the HSV-TK group exhibited durable regressions, defined as complete tumor regressions without recurrence over a 300-day observation interval; tumor regressions were not observed in any of the other mice.

[c]The STK vector used to transduce the HSV-TK gene was constructed by inserting this gene into vector LNL6 [2]; the latter was used as an HSV-TK negative control.

Source: Adapted from Moolten *et al.* [49].

The results represent the first experimental validation of the principle of preemption, demonstrating that a process applied to premalignant cells could alter the response to therapy of a future cancer. They also illustrate a number of obstacles that stand between this principle and its effective human implementation. The majority of mice were not cured. In tumors that were not eradicated, HSV-TK enzyme activity was low, consistent with an *in vivo* downregulation of gene expression that occurred during the brief (weeks to months) interval of experimental observation. Durable regressions were limited to small tumors; larger ones responded only with growth delays. Finally, the study was feasible only because the epithelial cells at risk for cancer could be cultivated and transduced *in vitro* and later reintroduced into host mice, thus obviating the need to reach mammary epithelial cells *in situ*. These limitations define issues that must ultimately be addressed to convert the principle of preemption into a modality that can be applied to individuals at risk for breast cancer or other malignancies. Paramount among them is the need to achieve high-efficiency integration of suicide genes into the genomic DNA of tissues *in vivo* and the need to improve the long-term stability of suicide functions in cells harboring the integrated genes beyond what is currently achievable with retroviral transduction. Stable chemosensitivity is threatened not only by changes in gene regulation that cells experience consequent to exposure to an *in vivo* environment but also by mutations that permanently delete or inactivate transduced genes. An additional concern, the possibility of long-term ill effects of transduction by retroviral vectors (including oncogenesis), has been ameliorated by theoretical calculations [52] and by the absence of vector-related cancers over the course of multiple gene therapy studies in animals and humans [53,54].

The efficiency problem is one that has long vexed much of the gene therapy field and may yield only to the eventual development of redesigned, and perhaps synthetic, vectors. Given the constraints of available technology, we have recently focused on the second issue: long-term stability, limiting our current efforts to cells that can be manipulated *ex vivo*. Cells in this category include lymphocytes that might be transfused into recipients after *ex vivo* manipulation, embryonic or tissue-specific stem or progenitor cells of various lineages (hematopoietic, mesenchymal, neural, etc.) that might be cultured as a source of tissue replacement, and cells transplanted as a source of therapeutic genes (for example, genes for Factor IX in some hemophiliacs, growth hormone in deficient individuals, insulin in diabetics, angiogenic factors for cardiovascular disease, or antiangiogenic factors for cancer). Introduction of a suicide gene into such cells is an attractive prospect as a "fail-safe" maneuver to permit their subsequent ablation if they later exhibit malignant or other aberrant behavior [52,55]. Critical to the prospective use of genes in this fashion is the requirement that all, or almost all sensitized cells and their progeny maintain normal viability and retain their chemosensitivity over intervals that may range from months to years and encompass many cell generations. Steps to preserve the gene-bearing cells include the use of genes that encode nonimmunogenic proteins or the use of conditions that promote tolerance to potentially immunogenic proteins. Steps to preserve chemosensitivity include strategies to maximize the persistence of suicide genes by transducing multiple gene copies or to preserve their expression through the use of vectors constructed to render promoter regions insusceptible to methylation or other "silencing" mechanisms. As described later, an additional approach we have explored is designed to maintain stable chemosensitivity by using suicide genes as replacements for, rather than additions to, selected cellular functions.

IV. CREATION OF STABLE SUICIDE FUNCTIONS BY COMBINING SUICIDE GENE TRANSDUCTION WITH ENDOGENOUS GENE LOSS

A. Loss of Purine or Thymidine Salvage Pathways Creates Chemosensitivity

Our effort to maximize the stability of suicide functions exploited the observation that when endogenous cellular functions are lost through mutation, the frequency with which they are regained is typically much lower than the frequency with which the functions of exogenously transduced genes are lost; that is, most *loss of function* mutations are highly stable. Two well-characterized loss of function mutations are those involving the genes for hypoxanthine/guanine phosphoribosyltransferase (HPRT) and cellular thymidine kinase (not to be confused with the HSV-TK gene, which encodes a different enzyme). The salient feature of HPRT and cellular TK is that mutational loss of either creates chemosensitivity, with HPRT deficiency sensitizing cells to inhibitors of purine synthesis and thymidine kinase deficiency creating sensitivity to inhibitors of thymidylate synthesis. A regimen that inhibits both of these biosynthetic pathways is hypoxanthine/aminopterin/thymidine (HAT) [56]. HAT is well tolerated by normal cells because, unlike HPRT-deficient cells, they can utilize the hypoxanthine and, unlike TK-deficient cells, can utilize the thymidine to circumvent the respective blocks in purine and thymidylate synthesis imposed by the antifolate drug aminopterin. HPRT-negative cells can be selected by virtue of their resistance to 6-thioguanine (6TG) [56], and TK-negative cells can be selected for their resistance to iododeoxyuridine or bromodeoxyuridine [57].

The advantage of exploiting the stability that might characterize HAT-sensitizing mutations is offset by a significant potential limitation. The normal role of the HPRT and TK enzymes is to incorporate hypoxanthine and thymidine, respectively, into salvage pathways that reclaim these compounds for nucleic acid synthesis. Although loss of these pathways is not lethal, it appears to create subtle growth disadvantages in certain cell populations, eventually resulting, for example, in the loss of detectable HPRT-negative cells in cell populations of hematopoietic origin in women who begin life with both HPRT-positive and -negative cells [58]. The growth disadvantage appears not to be universal, as some other tissues (fibroblasts, hair follicles) retain their mosaic character in these women. Nevertheless, in the disadvantaged tissues, HAT sensitivity will ultimately prove unstable, not because cells lose sensitivity but because the cells themselves will fail to persist in the absence of a substitute means of accomplishing salvage pathway functions.

To determine whether stable suicide functions could be created, a two-pronged approach was utilized that combined HPRT or TK deficiency with the addition of a new gene that replaced the lost salvage pathway functions and also mediated a suicide function of its own. HPRT-negative cells were obtained by 6TG selection and then exposed to a retroviral vector [10] that transduced the *Escherichia coli gpt* gene, which sensitizes cells to 6-thioxanthine (6TX). Like HPRT, the enzyme encoded by the *gpt* suicide gene, xanthine/guanine phosphoribosyltransferase (XGPRT), is capable of catalyzing the incorporation of hypoxanthine or guanine into nucleotide synthesis salvage pathways; its suicide function derives from its additional ability to use 6TX as a substrate. The *gpt* gene thus serves a dual purpose. It adds to the already stable chemosensitivity of HPRT-deficient cells by introducing an additional suicide function that must be lost by mutation for cells to lose all chemosensitivity. At the same time, it preserves the salvage pathway competency that HPRT-deficient cells would otherwise lack.

TABLE 3   Growth Rates of Wild-Type Cell Lines and Subclones

Cells[a]	Salvage pathway competency	Doubling time (hours)[b]
K3T3 H$^+$G$^-$	+	15.6
K3T3 H$^-$G$^-$	−	16.2, 18.6
K3T3 H$^-$G$^+$	+	15.3, 15.8, 17.1, 19.2, 20.6, 22.6, 24.0
CLS1 H$^+$G$^-$	+	19.5
CLS1 H$^-$G$^-$	−	21.7
CLS1 H$^-$G$^+$	+	13.7, 15.7, 16.0, 24.4, 29.1, 29.4, 41.0, 43.1
LY18 H$^+$G$^-$	+	20.1
LY18 H$^-$G$^-$	−	21.5, 27.3
LY18 H$^-$G$^+$	+	18.5, 20.0, 21.6, 22.1, 22.6, 23.8, 25.1, 27.5

[a] The cell lines tested were of fibroblastic (K3T3), epithelial (CLS1), and pre-B-lymphocytic (LY18) origin.
[b] Doubling times represent the mean calculated from one to four replicates of duplicate cultures for each line and subclone tested.
[c] H, hprt; G, *gpt*.

Based on the same rationale, the stability of the suicide function was also examined in cells that were deficient in cellular TK and had acquired the HSV-TK gene. Before transduction, each of five HPRT-deficient clones tested exhibited a growth rate that was slightly to moderately slower than that of their wild-type parents (Table 3). Transduction of the *gpt* gene into HPRT-deficient cells yielded a new population that substituted 6TX sensitivity for their previously acquired HAT sensitivity (Table 4). Unlike their predecessors, these *gpt*-transduced clones exhibited a broad range of doubling times, with some growing slowly and others growing as fast or faster than parental cells. The observed variations may reflect position effects or other attributes of the integrated *gpt* vector.

B. Stability of Suicide Functions in HPRT-Negative/*gpt*-Positive Cells

When four HPRT-deficient, *gpt*-transduced subclones of murine K3T3 fibrosarcoma cells were exposed to 6TX, surviving colonies ranged in number from 1 to 14 per 1.2×10^3 cells, representing mutant frequencies (corrected for plating efficiency) of 1.3×10^{-3} to 1.9×10^{-2}. The loss of 6TX sensitivity was accompanied in each case by reacquisition of HAT sensitivity, consistent with loss of *gpt*-mediated salvage functions (Table 4).

To determine the frequency with which both suicide functions were lost, expanded populations of three 6TX-resistant

TABLE 4 Frequency of Acquisition and Subsequent Loss of Phenotypes Associated with Chemosensitivity

Transition process	Phenotype	Isolation frequency	Sensitivity profile		
			HAT	6TX	6TG
None (wild type)	H^+G^-	—	R	R	S
6TG selection	H^-G^-	5×10^{-6}–1.3×10^{-5}	S	R	R
gpt transduction	H^-G^+	~0.5	R	S	S
Natural mutation	H^-G^-	1.3×10^{-3}–1.9×10^{-2}	S	R	R
Natural mutation	HAT^R	3.0×10^{-8}	R	R	NT

Note. The parental K3T3, CLS1, and LY18 cells used in the study were subjected to 6TG selection to yield clones that lacked HPRT enzyme activity and were sensitive to HAT, thus manifesting a suicide function consistent with their loss of hypoxanthine salvage capacity. Their 6TX ID_{50} exceeded 200 μM for K3T3 and LY18, and 250 μM for CLS1. After *gpt* transduction, the ID_{50} of the tested clones ranged from 0.5 to 10 μM for K3T3, from 1 to 3 μM for CLS1 cells, and from 2 to 10 μM for LY18. Data on transitions from H^+G^- to H^-G^- to H^-G^+ were obtained for all three cell lines; further transitions to H^+G^- and then HAT^R were measured only for K3T3. In addition, an H^+G^- subclone of CLS1 cells that had never been subjected to *gpt* transduction was tested and yielded no HAT-resistant colonies from 2×10^7 cells.

H, hprt; G, *gpt,* S, sensitive (all cells destroyed except for resistant mutants); R, resistant (cells grow at normal or near normal rates in the presence of the drug); NT, not tested; ID_{50}, 6TX concentration reducing cell numbers to 50% of the numbers in untreated control cultures during a 6-day assay interval.

subclones, each grown in 17 T-75 flasks containing 10^6 cells/flask, were tested for the presence of HAT-resistant mutants. Of the 51 flasks, 50 yielded no HAT-resistant colonies, and the remaining flask yielded a single colony. The result corresponds to a corrected mutant frequency of 3.0×10^{-8}. The subclones originated from an HPRT-negative, *gpt*-transduced clone that yielded 6TX-resistant mutants at a frequency of 5.4×10^{-3}. When this figure is multiplied by the frequency of HAT-resistant mutants, the resulting value,

$$5.4 \times 10^{-3} \times 3.0 \times 10^{-8} = 1.6 \times 10^{-10}$$

constitutes an estimate of the predicted frequency of the combined loss of both suicide functions. The same calculations applied to the clones with the greatest and poorest *gpt* stability yielded a frequency range of 3.9×10^{-11} to 5.7×10^{-10}. The rarity of HAT-resistant colonies among HPRT-negative K3T3 cells appeared to be matched by an HPRT-negative subclone of CLS1 cells, which yielded no surviving colonies among 20 flasks totaling 2×10^7 cells exposed to HAT.

C. Stability of Suicide Functions in TK-Negative/HSV-TK-Positive Cells

NIH3T3 fibroblasts that lack cellular TK but had been transduced with the HSV-TK gene were exposed to 8.8 μM GCV to select for mutants that had lost the HSV-TK suicide function. GCV-resistant clones were obtained from replicate cultures in numbers that corresponded to mutant frequencies of 1.5×10^{-4} to 3.4×10^{-3}. Subsequent reexposure

to GCV confirmed their resistant status. Analysis of two GCV-resistant clones confirmed that they were now HAT sensitive, as expected, and revealed TK enzyme levels that were only 1.5 and 1.8% of the levels of the GCV-sensitive cells from which they were derived, a decline consistent with their loss of GCV sensitivity and reacquisition of HAT sensitivity. When the two clones were subsequently exposed to HAT, HAT-resistant mutants were obtained at a frequency of 2×10^{-7} and 1.2×10^{-6}. The combined frequencies, representing the frequency with which a GCV-chemosensitive population would be expected to revert to a GCV-insensitive, HAT-insensitive, wild-type phenotype, thus ranged from 3.0×10^{-11} to 4.1×10^{-9}. This implies a stability similar to the stability of suicide functions observed in HPRT-negative, *gpt*-transduced K3T3 cells.

V. PREEMPTIVE USES OF SUICIDE GENES TO CONTROL GRAFT-VERSUS-HOST DISEASE IN LEUKEMIA

The relevance of suicide genes to neoplastic disease extends beyond their direct presence in neoplastic cells. A promising area currently under active investigation involves the use of the HSV-TK gene to impart GCV sensitivity not to malignant cells but rather to cells used to treat the malignancy: donor T lymphocytes administered in conjunction with allogeneic bone marrow in patients with leukemia and related diseases. Allogeneic bone marrow transplantation (allo-BMT) is currently associated with long-term remissions in a substantial number of patients with acute leukemia, chronic myelogenous leukemia, multiple myeloma, and myelodysplasia at a frequency that may exceed 50% in favorable circumstances; many of these remissions are thought to represent cures [59,60]. Most of the reduction in leukemic cell numbers is accomplished by the intensive chemoradiotherapy that precedes the allo-BMT, but donor T cells play a critical role in eradicating residual cells. This achievement comes at a cost—the frequent occurrence of graft-versus-host disease (GVHD) severe enough to result, directly or indirectly, in substantial treatment-related mortality. In addition, the severity of GVHD reflects in part the degree of antigenic disparity between donor and recipient and thus limits the availability of suitable donors; HLA mismatching poses the greatest threat of lethal GVHD, and, despite HLA matching, unrelated donors represent a greater hazard than HLA-matched sibling donors. Because immunosuppressive drugs have often failed to control GVHD adequately and impose hazards of their own, T-depleted marrow has been employed in an effort to avoid this complication. Unfortunately, the absence of T cells has been associated with poor leukemia control, reduced marrow engraftment, and a serious immunodeficiency that renders patients vulnerable to a variety of infections. Among the infectious sequelae are severe cytomegalovirus infections and potentially lethal EBV-induced lymphoproliferative disease [60].

One approach to preserving T-cell function involves allo-BMT with T-depleted marrow followed later by infusions of donor peripheral blood leukocytes, a rich source of T-cells. In some cases, the infusions have been delayed until specifically necessitated by leukemia relapse or viral sequelae [60]. Delayed infusion of T cells appears to reduce the threat of GVHD but does not eliminate it. An additional advantage of utilizing separate marrow and T cell infusions, however, is the opportunity to manipulate the T cells. In particular, this opportunity has been exploited to transduce the HSV-TK gene into donor T cells to sensitize them to GCV and thereby permit their subsequent ablation for severe GVHD [61–63]. The administration of HSV-TK-transduced T cells thus extends the benefits of T cells to all patients while later eliminating the cells only in those patients in whom they induce life-threatening pathology. This rationale is the basis for ongoing clinical trials in Milan [61,63] and more recently in this country and elsewhere [30], as well as additional protocols that have been approved or are under review, all involving patients receiving allo-BMT for leukemia or myeloma [30]. In each case, donor peripheral blood leukocytes are isolated, stimulated to proliferate, transduced with retroviral vectors bearing the HSV-TK gene plus a selectable marker, subjected to selection, and infused into patients after further growth to achieve adequate cell numbers.

Results reported to date from the first clinical trial (from Milan) are preliminary but encouraging [63]. Transduced cells retained their ability to exert antileukemic effects in most cases, including complete remissions in three out of eight patients. Two patients developed acute GVHD; in each case, administration of GCV quickly eliminated the transduced cells from the circulation and induced nearly complete resolution of the clinical and biochemical signs of GVHD. The T cells thus appear to have responded as expected. In an additional patient who developed chronic GVHD, GCV resulted in only partial amelioration. The lesser efficacy may reflect the existence in chronic GVHD of a substantial fraction of cells that are not proliferating at the time of GCV administration and are therefore insusceptible to the inhibitory effects of GCV phosphates on DNA synthesis [64].

VI. FUTURE PROSPECTS FOR PREEMPTIVE USE OF SUICIDE GENES

Until *in vivo* transduction efficiency improves, the failsafe use of suicide genes is likely to remain a phenomenon that can only be applied to cells that are manipulated *in vitro* and later reintroduced into human hosts. Potential applications include their use as a precaution against either malignant behavior of the reintroduced cells or immune pathology that they might induce [1,2,4,55]. Additionally, suicide genes added to cells transplanted to supply a missing function constitute a potential mechanism to control hyperactivity of the transplanted cells, such as hyperinsulinemia resulting from excessive growth or function of cells expressing native or transduced insulin genes [65].

If suicide genes of nonhuman origin are to be used preemptively, their success will require that their presence not provoke host immune reactions that result in the elimination of the transduced cells. Such reactions have been observed in some [55,66] but not other [49,67] studies involving cells transduced with the HSV-TK gene. The development of improved methods for inducing immune tolerance, the use of genes transcribed from inducible promoters that remain inactive until an appropriate stimulus is applied, or the creation of suicide genes that are expressed at the level of nucleic acid rather than protein (e.g., as catalytic RNA) [68,69] are possible approaches to this problem. If reliable methods for controlling immune rejection are developed to the point that they permit the use of xenografted tissues in humans, the introduction of suicide genes as transgenes into animals used as a source of the xenografts constitutes a further fail-safe use of suicide genes, one designed to protect against undesired effects of the grafted cells.

An intriguing application of the HSV-TK gene that is likely to be tested soon in clinical trials is its use as a marker for *in vivo* gene transduction. In addition to its phosphorylation of GCV, HSV-TK phosphorylates a number of other nucleoside analogs that are poor substrates for cellular kinases, including halogenated pyrimidine analogs such as 5-iodo-2'-fluoro-2'-deoxy-1-β-D-arabinofuranosyluracil (FIAU). Tjuvajev *et al.* [70] have shown that when ^{131}I-labeled FIAU is administered to mice bearing tumors carrying transduced HSV-TK genes, the location and extent of HSV-TK expression can be precisely delineated by *in vivo* imaging with a gamma camera and single-photon emission tomography (SPECT). Extending this concept, they have also demonstrated that when the HSV-TK vector also transduces a separate gene (lacZ), the imaging analysis not only correlates with HSV-TK expression but also locates and quantifies expression of the linked gene [71]. This use of HSV-TK as a marker in conjunction with FIAU or other substrates that are currently under investigation [72] harbors the potential for it to serve a dual purpose: measuring the function of whatever therapeutic gene it might be linked to in a gene therapy subject and additionally serving to protect that subject against unwanted behavior by the transduced cells.

A final prospect relates to the possibility, discussed above, that efficient incorporation of one or more suicide genes into one or more tissues might eventually permit cancers that arise later to be treated effectively, based on their clonal origin from a sensitized cell. The previous discussion emphasized the prospect that clonality might ensure the presence of a suicide gene even in metastatic or disseminated cancers (i.e., the late stages of a cancer/host relationship). It is also possible, however, that early, preclinical stages might be targetable as well. Recent evidence indicates that DNA derived from cancer cells is sometimes detectable in blood or secretions by PCR analysis. Thus, mutant K-*ras* [73–76] genes have been detected in both plasma [73,74] and feces [75,76] of patients with colorectal [74,75] and pancreatic [73,76] carcinomas, mutant p53 genes have been demonstrated in the urine of patients with bladder cancer [77], and specific microsatellite DNA alterations have been detected in the plasma [78] and sputum [79] of lung cancer patients and in serum from patients with head and neck cancer [80]. Some of the detected alterations represented changes that were also present in premalignant lesions that accompanied the cancer or in one case were found in the absence of a cancer [76]. In theory, suicide genes harbored by the cells of cancers that arose in preemptively transduced tissues would also be detectable, and analysis of flanking genomic sequences could be used to determine whether they represented the monoclonal pattern of a neoplasm or the polyclonal pattern of nonneoplastic tissue. If the detection sensitivity of this type of DNA analysis increases to the point where incipient clonal proliferations are detectable in individuals who harbor suicide genes in various vulnerable tissues (breast, lung, bone marrow, etc.), then detection would permit early action, such as a search for the neoplasm, biopsy, and surgery or radiotherapy as indicated. If the neoplasm is found, prodrug administration could be added to surgery or radiotherapy in an adjuvant role. If the

neoplasm is small enough to elude attempts to locate it, administration of a prodrug could be used to ablate it before it surfaces clinically, in essence exploiting preemption as a form of cancer prevention.

If *in vivo* transduction efficiency in nonvital tissues such as breast or prostate eventually improves to the point where a suicide gene can be transduced into almost all the epithelial cells of these tissues, prevention should also be feasible at an even earlier stage, if desired. Thus, individuals at high risk for breast or prostate cancer might, at some stage in their life, choose to receive a prodrug as a form of molecular "epitheliectomy" in preference to surgical bilateral mastectomy or prostatectomy.

References

1. Moolten, F. L. (1986). Tumor chemosensitivity conferred by inserted herpes thymidine kinase genes: paradigm for a prospective cancer control strategy. *Cancer Res.* **46**, 5276–5281.
2. Moolten, F. L., and Wells, J. M. (1990). Curability of tumors bearing herpes thymidine kinase genes transferred by retroviral vectors. *J. Natl. Cancer Inst.* **82**, 297–300.
3. Moolten, F. L. (1994). Drug sensitivity ("suicide") genes for selective cancer chemotherapy. *Cancer Gene Therap.* **1**, 279–287.
4. Tiberghien, P. (1994). Use of suicide genes in gene therapy. *J. Leukocyte Biol.* **56**, 203–209.
5. Bi, W. L., Parysek, L. M., Warnick, R., and Stambrook, P. J. (1993). In vitro evidence that metabolic cooperation is responsible for the bystander effect observed with HSV tk retroviral gene therapy. *Human Gene Therap.* **4**, 725–731.
6. Freeman, S. M., Abboud, C. N., Whartenby, K. A., Packman, C. H., Koeplin, D. S., Moolten, F. L., and Abraham, G. N. (1993). The "bystander effect": tumor regression when a fraction of the tumor mass is genetically modified. *Cancer Res.* **53**, 5274–5283.
7. Hooper, M. L., and Subak-Sharpe, J. H. (1981). Metabolic cooperation between cells. *Int. Rev. Cytol.* **69**, 45–104.
8. Culver, K. W., Ram, Z., Wallbridge, S., Ishii, H., Oldfield, E. H., and Blaese, R. M. (1992). In vivo gene transfer with retroviral vector producer cells for treatment of experimental brain tumors. *Science* **256**, 1550–1552.
9. Mullen, C. A., Kilstrup, M., and Blaese, R. M. (1992). Transfer of the bacterial gene for cytosine deaminase to mammalian cells confers lethal sensitivity to 5-fluorocytosine: a negative selection system. *Proc. Natl. Acad. Sci. USA* **89**, 33–37.
10. Mroz, P. J., and Moolten, F. L. (1993). Retrovirally transduced *Escherichia coli gpt* genes combine selectability with chemosensitivity capable of mediating tumor eradication. *Hum. Gene Ther.* **4**, 589–595.
11. Wei, M. X., Tamiya, T., Chase, M., Boviatsis, E. J., Chang, T. K. H., Kowall, N. W., Hochberg, F. H., Waxman, D. J., Breakefield, X. O., and Chiocca, E. A. (1994). Experimental tumor therapy in mice using the cyclophosphamide-activating cytochrome P450 2B1 gene. *Human Gene Ther.* **5**, 969–978.
12. Chen, L., Waxman, D. J., Chen, D., and Kufe, D. W. (1996). Sensitization of human breast cancer cells to cyclophosphamide and ifosfamide by transfer of a liver cytochrome P450 gene. *Cancer Res.* **56**, 1331–1340.
13. Sorscher, E. J., Peng, S., Bebok, Z., Allan, P. W., Bennett, L. L., and Parker, W. B. (1994). Tumor cell bystander killing in colonic carcinoma utilizing the *Escherichia coli* DeoD gene to generate toxic purines. *Gene Ther.* **1**, 233–238.

14. Manome, Y., Wen, P. Y., Dong, Y., Tanaka, T., Mitchell, B. S., Kufe, D. W., and Fine, H. A. (1996). Viral vector transduction of the human deoxycytidine kinase cDNA sensitizes glioma cells to the cytotoxic effects of cytosine arabinoside *in vitro* and *in vivo*. *Nat. Med.* **2,** 567–573.

15. Bridgewater, J. A., Knox, R. J., Pitts, J. D., Collins, M. K., and Springer, C. J. (1997). The bystander effect of the nitroreductase/CB1954 enzyme/prodrug system is due to a cell-permeable metabolite. *Human Gene Ther.* **8,** 709–717.

16. Spencer, D. M., Belshaw, P., Chen, L., Ho, S. N., Randazzo, F., Crabtree, G. R., and Schreiber, S. L. (1996). Functional analysis of Fas signaling *in vivo* using synthetic inducers of dimerization. *Current Biol.* **6,** 839–847.

17. Freiberg, R. A., Spencer, D. M., Choate, K. A., Peng, P. D., Schreiber, S. L., Crabtree, G. R., and Khavari, P. A. (1996). Specific triggering of the Fas signal transduction pathway in normal human keratinocytes. *J. Biol. Chem.* **271,** 31666–31669.

18. Fan, L., Freeman, K. W., Khan, T., Pham, E., and Spencer, D. M. (1999). Improved artificial death switches based on caspases and FADD. *Hum. Gene Ther.* **10,** 2273–2285.

19. Amara, J. F., Courage, N. L., and Gilman, M. (1999). Cell surface tagging and a suicide mechanism in a single chimeric human protein. *Hum. Gene Ther.* **10,** 2651–2655.

20. Mandell, R. B., Mandell, L. Z., and Link, Jr., C. J. (1999). Radioisotope concentrator gene therapy using the sodium/iodide symporter gene. *Cancer Res.* **59,** 661–668.

21. Spitzweg, C., Zhang, S., Bergert, E. R., Castro, M. R., McIver, B., Heufelder, A. E., Tindall, D. J., Young, C. Y. F., and Morris, J. C. (1999). Prostate-specific antigen (PSA) promoter-driven androgen-inducible expression of sodium iodide symporter in prostate cancer cell lines. *Cancer Res.* **59,** 2136–2141.

22. Stribbling, S. M., Friedlos, F., Martin, J., Davies, L., Spooner, R. A., Marais, R., and Springer, C. (2000). Regressions of established breast carcinoma xenografts by carboxypeptidase G₂ suicide gene therapy and the prodrug CMDA are due to a bystander effect. *Hum. Gene Ther.* **11,** 285–292.

23. Hamstra, D. A., Pagé, M., Maybaum, J., and Rehemtulla, A. (2000). Expression of endogenously activated secreted or cell surface carboxypeptidase A sensitizes tumor cells to methotrexate-α-peptide prodrugs. *Cancer Res.* **60,** 657–665.

24. Oldfield, E. H. (1993). Gene therapy for the treatment of brain tumors using intra-tumoral transduction with the thymidine kinase gene and intravenous ganciclovir. *Hum. Gene Ther.* **4,** 39–69.

25. Culver, K. W., and van Gilder, J. (1994). Gene therapy for the treatment of malignant brain tumors with in vivo tumor transduction with the herpes simplex thymidine kinase gene/ganciclovir system. *Hum. Gene Ther.* **5,** 343–379.

26. Freeman, S. M., McCune, C., Angel, C., Abraham, G. N., and Abboud, C. N. (1992). Treatment of ovarian cancer using HSV-TK gene modified vaccine — regulatory issues. *Hum. Gene Ther.* **3,** 342–349.

27. Link, C. J., and Moorman, D. (1996). A phase I trial of *in vivo* gene therapy with the herpes simplex thymidine kinase/ganciclovir system for the treatment of refractory or recurrent ovarian cancer. *Hum. Gene Ther.* **7,** 1161–1179.

28. Alvarez, R. D., and Curiel, D. T. (1997). A phase I study of recombinant adenovirus vector-mediated intraperitoneal delivery of herpes simplex virus thymidine kinase (HSV-TK) gene and intravenous ganciclovir for previously treated ovarian and extraovarian cancer patients. *Hum. Gene Ther.* **8,** 597–613.

29. Treat, J., Kaiser, L. R., Sterman, D. H., Litzky, L., Davis, A., Wilson, J. M., and Albelda, S. M. (1996). Treatment of advanced mesothelioma with the recombinant adenovirus H5.010RSVTK: a phase I trial (BB-IND 6274). *Hum. Gene Ther.* **7,** 2047–2057.

30. Anon. (2000). Human gene marker/therapy clinical protocols (complete updated listings). *Human Gene Ther.* **11,** 919–979.

31. Jolly, D. (1994). Viral vector systems for gene therapy. *Cancer Gene Ther.* **1,** 51–64.

32. Freeman, S. M., Whartenby, K. A., Freeman, J. L., Abboud, C. N., and Marrogi, A. J. (1996). *In situ* use of suicide genes for cancer therapy. *Semin. Oncol.* **23,** 31–45.

33. Harsh, G. R., Deisboeck, T. S., Louis, D. N., Hilton, J., Colvin, M. Silver, J. S., Qureshi, N. H., Kracher, J. Finkelstein, D., Chiocca, E. A., and Hochberg, F. H. (2000). Thymidine kinase activation of ganciclovir in recurrent malignant gliomas: a gene-marking and neuropathological study. *J. Neurosurg.* **92,** 804–811.

34. Vile, R. G., and Hart, I. R. (1993). Use of tissue-specific expression of the herpes simplex virus thymidine kinase gene to inhibit growth of established murine melanomas following direct intratumoral injection of DNA. *Cancer Res.* **53,** 3860–3864.

35. Vile, R. G., Nelson, J. A., Castleden, S., Chong, H., and Hart, I. R. (1994). Systemic gene therapy of murine melanoma using tissue specific expression of the HSVtk gene involves an immune component. *Cancer Res.* **54,** 6228–6234.

36. Huber, B. E., Richards, C. A., and Krenitsky, T. A. (1991). Retroviral-mediated gene therapy for the treatment of hepatocellular carcinoma: an innovative approach for cancer therapy. *Proc. Natl. Acad. Sci. USA* **88,** 8039–8043.

37. Macri, P., and Gordon, J. W. (1994). Delayed morbidity and mortality of albumin/SV40 T-antigen transgenic mice after insertion of an alphafetoprotein/herpes virus thymidine kinase transgene and treatment with ganciclovir. *Hum. Gene Ther.* **5,** 175–182.

38. Judde, J.-G., Spangler, G., MacGrath, I., and Bhatia, K. (1996). Use of Epstein–Barr virus nuclear antigen-1 in targeted therapy of EBV-associated neoplasia. *Hum. Gene Ther.* **7,** 647–653.

39. Franken, M., Estabrooks, A., Cavacini, L., Sherburne, B., Wang, F., and Scadden, D. T. (1996). Epstein–Barr virus-driven gene therapy for EBV-related lymphomas. *Nat. Med.* **2,** 1379–1382.

40. Ko, S.-C., Cheon, J., Kao, C., Gotoh, A., Shirakawa, T., Sikes, R. A., Karsenty, G., and Chung, L. W. K. (1996). Osteocalcin promoter-based toxic gene therapy for the treatment osteosarcoma in experimental models. *Cancer Res.* **56,** 4614–4619.

41. Harris, J. D., Gutierrez, A. A., Hurst, H. C., Sikora, K., and Lemoine, N. R. (1994). Gene therapy for cancer using tumour-specific prodrug activation. *Gene Ther.* **1,** 170–175.

42. Kaneko, S., Hallenbeck, P., Kotani, T., Nakabayashi, H., McGarrity, G., Tamaoki, T., Anderson, W. F., and Chiang, Y. L. (1995). Adenovirus-mediated gene therapy of hepatocellular carcinoma using cancer-specific gene expression. *Cancer Res.* **55,** 5283–5287.

43. Fialkow, P. J. (1976). Clonal origin of human tumors. *Biochim. Biophys. Acta* **458,** 283–321.

44. Fearon, E. R., Hamilton, S. R., and Vogelstein, B. (1987). Clonal analysis of human colorectal tumors. *Science* **238,** 193–197.

45. Fujii, H., Marsh, C., Cairns, P., Sidransky, D., and Gabrielson, E. (1996). Genetic divergence in the clonal evolution of breast cancer. *Cancer Res.* **56,** 1493–1497.

46. Moolten, F. L., Wells, J. M., Heyman, R. A., and Evans, R. M. (1990). Lymphoma regression induced by ganciclovir in mice bearing a herpes thymidine kinase transgene. *Hum. Gene Ther.* **1,** 125–134.

47. Moolten, F., Wells, J. M., and Mroz, P. J. (1992). Multiple transduction as a means of preserving ganciclovir chemosensitivity in sarcoma cells carrying retrovirally transduced herpes thymidine kinase genes. *Cancer Lett.* **64,** 257–263.

48. Moolten, F. L. (1990). Mosaicism induced by gene insertion as a means of improving chemotherapeutic selectivity. *Crit. Rev. Immunol.* **10,** 203–233.

49. Moolten, F. L., Vonderhaar, B. K., and Mroz, P. J. (1996). Transduction of the herpes thymidine kinase gene into premalignant murine

mammary epithelial cells renders subsequent breast cancers responsive to ganciclovir therapy. *Hum. Gene Ther.* **7,** 1197–1204.

50. Jerry, D. J., Ozbun, M. A., Kittrell, F. S., Lane, D. P., Medina, D., and Butel, J. S. (1993). Mutations in p53 are frequent in the preneoplastic stage of mouse mammary tumor development. *Cancer Res.* **53,** 3374–3381.

51. Kittrell, F. S., Oborn, C. J., and Medina, D. (1992). Development of mammary preneoplasias *in vivo* from mouse mammary epithelial cell lines *in vitro. Cancer Res.* **52,** 1924–1932.

52. Moolten, F. L., and Cupples, L. A. (1992). A model for predicting the risk of cancer consequent to retroviral gene therapy. *Hum. Gene Ther.* **3,** 479–486.

53. Cornetta, K., Morgan, R. A., and Anderson, W. F. (1991). Safety issues related to retroviral-mediated gene transfer in humans. *Hum. Gene Ther.* **2,** 5–14.

54. Cornetta, K., Morgan, R. A., Gillio, A., Sturm, S., Baltrucki, L., O'Reilly, R., and Anderson, W. F. (1991). No retroviremia or pathology in long–term follow-up of monkeys exposed to a murine amphotropic retrovirus. *Hum. Gene Ther.* **2,** 215–220.

55. Riddell, S. R., Elliott, M., Lewinsohn, D. A., Gilbert, M. J., Wilson, L., Manley, S. A., Lupton, S. D., Overell, R. W., Reynolds, T. C., Corey, L., and Greenberg, P. D. (1996). T-cell mediated rejection of gene-modified HIV-specific cytotoxic T lymphocytes in HIV-infected patients. *Nat. Med.* **2,** 216–223.

56. Sybalski, W., and Sybalska, E. H. (1961). A new chemotherapeutic principle for the treatment of drug resistant neoplasms. *Cancer Chemotherapy Rep.* **11,** 87–89.

57. Littlefield, J. W. (1966). The use of drug-resistant markers to study the hybridization of mouse fibroblasts. *Exp. Cell Res.* **41,** 190–196.

58. Rossiter, B. J. F., and Caskey, C. T. (1995). Hypoxanthine-guanine phosphoribosyltransferase deficiency: Lesch–Nyhan syndrome and gout, in *The Metabolic and Molecular Bases of Inherited Disease,* 7th ed., (C. R. Beaudet *et al.,* eds.), pp. 1679–1706. McGraw-Hill, New York.

59. Beutler, E., Lichtman, M. K., Coller, B. S., and Kipps, T. J., eds. (1995). *Williams Hematology,* 5th ed., McGraw-Hill, New York.

60. Shlomchik, W. D., and Emerson, S. G. (1996). The immunobiology of T cell therapies for leukemias. *Acta Haematol.* **96,** 189–213.

61. Bordignon, C., and Bonini, C. (1995). Transfer of the HSV-TK gene into donor peripheral blood lymphocytes for in vivo modulation of donor anti-tumor immunity after allogeneic bone marrow transplantation. *Hum. Gene Ther.* **6,** 813–819.

62. Tiberghien, P. (1997). Use of donor T-lymphocytes expressing herpes simplex-thymidine kinase in allogeneic bone marrow transplantation: a phase I–II study. *Hum. Gene Ther.* **8,** 615–624.

63. Bonini, C., Ferrari, G., Verzeletti, S., Servida, P., Zappone, E., Ruggieri, L., Ponzoni, M., Rossini, S., Mavilio, F., Traversari, C., and Bordignon, C. (1997). HSV-TK gene transfer into donor lymphocytes for control of allogeneic graft-versus-leukemia. *Science* **276,** 1719–1724.

64. Bonini, C., and Bordignon, C., (1997). Potential and limitations of HSV-TK-transduced donor peripheral blood lymphocytes after allo-BMT. *Hematol. Cell. Ther.* **39,** 273–274,

65. Yoshimoto, K., Murakami, R., Moritani, M., Ohta, M., Iwahana, H., Nakauchi, H., and Itakura, M. (1996). Loss of ganciclovir sensitivity by exclusion of thymidine kinase gene from transplanted proinsulin-producing fibroblasts as a gene therapy model for diabetes. *Gene Ther.* **3,** 230–234.

66. Tapscott, S. J., Miller, A. D., Olson, J. M., Berger, M. S., Groudine, M., and Spence, A. M. (1994). Gene therapy of rat 9L gliosarcoma tumors by transduction with selectable genes does not require drug selection. *Proc. Natl. Acad. Sci. USA* **91,** 8185–8189.

67. Pavlovic, J., Nawrath, M., Tu, R., Heinicke, T., and Moelling, K. (1996). Anti-tumor immunity is involved in the thymidine kinase-mediated killing of tumors induced by activated Ki-ras(G12V). *Gene Ther.* **3,** 635–643.

68. Prudent, J. R., Uno, T., and Schultz, P. G. (1994). Expanding the scope of RNA catalysis. *Science* **264,** 1924–1927.

69. Wilson, C., and Szostak, J. W. (1995). In vitro evolution of a self-alkylating ribozyme. *Nature* **374,** 777–782.

70. Tjuvajev, J. G., Finn, R., Watanabe, K., Joshi, R., Oku, T., Kennedy, J., Beattie, B., Koutcher, J., Larson, S., and Blasberg, R. G. (1996). Noninvasive imaging of herpes virus thymidine kinase gene transfer and expression: a potential method for monitoring clinical gene therapy. *Cancer Res.* **56,** 4087–4095.

71. Tjuvajev, J., Safer, M., Sadelain, M., Avril, N., Oku, T., Joshi, R., Finn, R., Larson, S., and Blasberg, R. (1997). Noninvasive imaging of the HSV1-tk marker gene for monitoring the expression of other target genes *in vivo. J. Neuro-Oncol.* **35** (suppl. 1), S45.

72. Gambhir, S. S., Bauer, E., Black, M. E., Liang, Q., Kokoris, M. S., Barrio, J. R., Iyer, M., Namavari, M., Phelps, M. E., and Herschman, H. R. (2000). A mutant herpes simplex virus type 1 thymidine kinase reporter gene shows improved sensitivity for imaging reporter gene expression with positron emission tomography. *Proc. Natl. Acad. Sci. USA* **97,** 2785–2790.

73. Sorenson, G. D., Pribish, D. M., Valone, F. H., Memoli, V. A., and Yao, S. L. (1993). Mutated K-ras sequences in plasma from patients with pancreatic carcinoma. *Proc. Am. Assoc. Cancer Res.* **34,** A174.

74. Lefort, L., Anker, P., Vasioukhin, V., Lyautey, J., Lederrey, C., and Stroun, M. (1995). Point mutations of the K-ras gene present in the DNA of colorectal tumors are found in the blood plasma DNA of the patients. *Proc. Am. Assoc. Cancer Res.* **36,** A3319.

75. Sidransky, D., Tokino, T., Hamilton, S. R., Kinzler, K. W., Levin, B., Frost, P., and Vogelstein, B. (1992). Identification of ras oncogene mutations in the stool of patients with curable colorectal tumors. *Science* **256,** 102–105.

76. Caldas, C., Hahn, S., Hruban, R. H., Yeo, C., and Kern, S. (1994). Detection of K-ras mutations (mut) in the stool of patients (pts) with pancreatic adenocarcinoma (PCa). *Proc. Am. Soc. Clin. Oncol.* **13,** A294.

77. Sidransky, D., Von Eschenbach, A., Tsai, Y. C., Jones, P., Summerhayes, I., Marshall, F., Meera, P., Green, P., Hamilton, S. R., Frost, P., and Vogelstein, B. (1991). Identification of p53 gene mutations in bladder cancers and urine samples. *Science* **252,** 706–709.

78. Chen, X. Q., Stroun, M., Magnenat, J. -L., Nicod, L. P., Kurt, A-M, Lyautey, J., Lederrey, C., and Anker, P. (1996). Microsatellite alterations in plasma DNA of small cell lung cancer patients. *Nature Med.* **2,** 1033–1035.

79. Miozzo, M., Sozzi, G., Musso, K., Pilotti, S., Incarbone, M., and Pastorino, U. (1996). Microsatellite alterations in bronchial and sputum specimens of lung cancer patients. *Cancer Res.* **56,** 2285–2288.

80. Nawroz, H., Koch, W., Anker, P., Stroun, M., and Sidransky D. (1996) Microsatellite alterations in serum DNA of head and neck cancer patients. *Nat. Med.* **2,** 1035–1037.

CHAPTER

32

Treatment of Mesothelioma Using Adenoviral-Mediated Delivery of Herpes Simplex Virus Thymidine Kinase Gene in Combination with Ganciclovir

DANIEL H. STERMAN

Thoracic Oncology Research Laboratory
Pulmonary, Allergy, and Critical Care Division
University of Pennsylvania Medical Center
Philadelphia, Pennsylvania 19104

STEVEN M. ALBELDA

Thoracic Oncology Research Laboratory
Pulmonary, Allergy, and Critical Care Division
University of Pennsylvania Medical Center
Philadelphia, Pennsylvania 19104

I. INTRODUCTION

One prominent approach in cancer gene therapy is the introduction of toxic or suicide genes into tumor cells to facilitate their destruction. One such suicide gene approach involves the transduction of a neoplasm with a cDNA encoding for an enzyme that would render its cells sensitive to a benign drug by converting the prodrug to a toxic metabolite [1]. As described in previous chapters, these enzymes encoded

by the suicide gene are often of non-human origin, such as the *Escherichia coli* cytosine deaminase (CDA) gene [2] or the herpes simplex virus-1–thymidine kinase (HSV-TK) gene [3]. The latter was shown by Moolten *et al.* [4,5] to kill tumor cells when combined with administration of the antiviral agent, ganciclovir (GCV).

The drug ganciclovir (9-[1,3-dihydroxy-2-propoxy)-methyl]-guanine) is an acyclic nucleoside that is poorly metabolized by mammalian cells and is therefore generally nontoxic. However, after being converted to GCV-*mono*-phosphate by herpes virus family (herpes simplex virus-1, cytomegalovirus, vaccinia virus) thymidine kinases, it is rapidly converted to GCV-*tri*phosphate by mammalian kinases [6]. Ganciclovir triphosphate is a potent inhibitor of viral DNA polymerase and is also a toxic analog that competes with normal mammalian nucleosides for DNA replication [6]. In addition, incorporation of GCV-monophosphate into the DNA template has also been demonstrated to induce significant cytotoxicity [7].

The antitumor effect of HSV–thymidine kinase/ganciclovir (HSV-TK/GCV) gene therapy was assayed originally in animal models where producer cells containing a retroviral construct encoding for HSV-TK were stereotactically injected into brain tumors. In these models, tumor regression was observed after GCV administration [8,9]. Subsequently, similar antineoplastic properties were described in *in vivo* studies involving direct intratumoral delivery of HSV-TK by an adenoviral vector [10,11].

A. Bystander Effects: Intercellular Passage of GCV Metabolites and/or Immunologic Effects

Given the limited gene transfer efficiency of current vector systems, the primary reason for the success of *in vivo* HSV-TK experiments appeared to be the finding that HSV-TK expression in every cell was not required for complete tumor regression. This so-called bystander effect was demonstrated in *in vitro* mixing experiments using retrovirally infected tumor cells. Subsequently, *in vivo* experiments involving tumors where only 10–20% of the cells expressed the HSV-TK gene demonstrated that complete tumor regression was noted in animals after ganciclovir treatment [5,9,12–14]. The nature of this bystander effect is complex and appears to involve passage of toxic GCV metabolites from transduced to nontransduced cells via gap junctions or apoptotic vesicles [15,16]; and induction of antitumor immune responses capable of killing distant, non-HSV-TK-transduced cells [17].

B. Adenoviral Delivery Systems

The transfer of HSV-TK DNA to target tumor cells can be accomplished in a variety of ways including the use of viral vectors, liposomes, cellular delivery systems, and naked DNA electrocorporation [18]. Early *in vitro* and *in vivo* studies utilized retroviral vectors to facilitate HSV-TK DNA transfer into tumors, including the injection of producer cell lines that secrete retrovirus containing the suicide gene [8,19]. These retroviral-based approaches have been used successfully in animal models of brain tumor, ovarian cancer, and hepatocellular carcinoma [20].

Retroviruses have several limitations as delivery vehicles of therapeutic genes for cancer gene therapy insofar as they infect only actively dividing cells, carry risks of insertional mutagenesis, and are difficult to produce in large scale for human clinical trials. Contrastingly, adenoviruses are able to infect both dividing and nondividing cells, do not carry the theoretical risk of insertional mutagenesis (they deliver their DNA episomally), and are much easier to produce in lots large enough for use in clinical studies [21]. For these reasons, adenoviruses have become the vector of choice for delivery of the HSV-TK gene, as well as other therapeutic genes, in many cancer gene therapy experimental models. Based on these factors, our group and others have produced recombinant, replication-deficient adenoviral vectors encoding the HSV-TK gene (Fig. 1A) and have shown that this vector, in combination with GCV, could eradicate tumor cells *in vitro* and in *in vivo* models of various tumors such as malignant mesothelioma, lung cancer, brain tumors, colon carcinoma, hepatocellular carcinoma, glioma, and melanoma [10,22–25].

II. CLINICAL USE OF HSV-TK IN THE TREATMENT OF LOCALIZED MALIGNANCIES

Based upon the success of *in vivo* studies from multiple laboratories, several centers have conducted, or are in the process of conducting, human trials of adenovirally delivered HSV-TK in combination with GCV in advanced malignancies [26–28]. Because current vector technology does not yet allow for systemic administration, the initial clinical trials have primarily focused on localized malignancies, where directed instillation of vector (in conjunction with a bystander effect) could have some potential for therapeutic efficacy. The primary targets have included brain tumors, ovarian carcinoma, melanoma, prostate carcinoma, and malignant mesothelioma.

A. Malignant Mesothelioma: Paradigm for HSV-TK/GCV Gene Therapy

Our group has focused on malignant pleural mesothelioma as a primary target, as we feel it has many features that can serve as a paradigm for other localized malignancies. Several characteristics make mesothelioma an attractive target for gene therapy: (1) There is no standard, effective therapy for the disease; (2) mesothelioma is readily accessible in the pleural space for vector delivery, biopsy, and subsequent analysis of treatment effects; (3) local extension of disease, rather than distant metastases, is responsible for much of the morbidity and mortality of mesothelioma; and (4) current treatment options are very limited. Thus, unlike other neoplasms that metastasize earlier in their course, in patients with mesothelioma small increments of improvement in local control could engender significant improvements in palliation or survival. Accordingly, a number of gene therapy trials aimed at treating mesothelioma have begun or are in the planning stages. At least two of the active programs University of Pennsylvania and Louisiana State University are investigating delivery of the HSV-TK gene to mesothelioma cells in combination with systemic GCV, although the delivery systems differ: adenovirus and PA-1 ovarian carcinoma cell line, respectively [29,30].

B. Preclinical Data: Animal and Toxicity Studies

Initial experiments demonstrated that replication-deficient adenoviral HSV-TK vectors efficiently transduced mesothelioma cells both in tissue culture and in animal models and facilitated HSV-TK-mediated killing of human mesothelioma cells in the presence of low concentrations of GCV [31,32]. Subsequently, the Ad.HSV-TK vector was used to treat established, intraperitoneal human mesothelioma tumors and lung cancers in SCID mice [10,23]. Following GCV therapy, macroscopic tumor was eradicated in 90% of animals, and

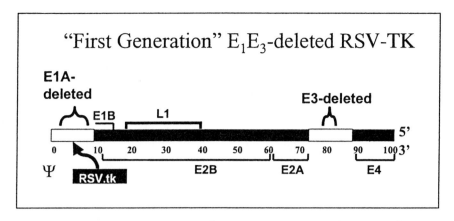

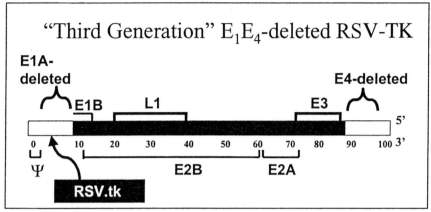

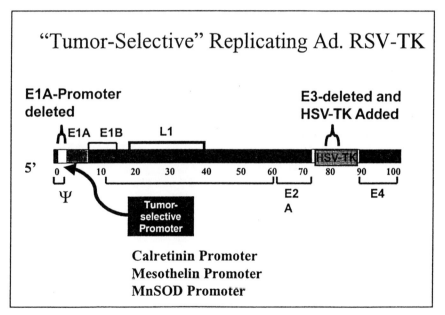

FIGURE 1   (Top) Illustration of the adenoviral (Ad) vector used in the initial phase I trial (H5.010RSV-TK). This so-called first-generation replication-incompetent Ad is deleted in the early genes E1 and E3 with the HSV-TK gene inserted in the E1 region. (Middle) A third-generation Ad vector containing deletions in the E1 and E4 regions with preservation of E3. E1/E4-deleted Ad vectors offer theoretical advantages-over first-generation vectors due to their diminished cytopathic effects and hepatoxicity and reduced cellular immune responses. (Bottom) A tumor-selective replicating Ad. TK vector with a tumor-selective promoter (Calretinin, MnSOD, Mesothelin) substituted for the Ad.E1 promoter. This would potentially allow for greater HSV-TK delivery to solid tumors with decreased collateral injury to normal tissues.

microscopic tumor was undetectable in 80% of animals. Tumor reduction was accompanied by a significant increase in survival. Marked decreases in tumor size have also been seen in an intrapleural rat model of syngeneic mesothelioma with more modest increases in survival [24]. The *in vitro* and *in vivo* sensitivites of human mesothelioma cells to HSV-TK/GCV gene therapy have been confirmed by other independent investigators [33].

Based on the efficacy data in animals, we conducted preclinical toxicity testing for submission to the Recombinant DNA Advisory Committee (RAC) and the Food and Drug Administration (FDA). The trials were designed to mimic the proposed clinical trials. Rats were given high doses of virus intrapleurally followed by intraperitoneal administration of GCV at the same dose proposed for initial use in the clinical trial (10 mg/kg/day). Toxicity was limited to localized inflammation of the pleural and pericardial surfaces. Formal toxicology studies were also done in three non-human primates given high-dose virus (10^{12} PFU) and GCV [34]. No adverse clinical effects were seen, nor any hematological or biochemical abnormalities. Necropsy findings were limited to inflammatory changes in the chest wall and intrathoracic serosa.

C. Initial Phase I Clinical Trial

A phase I clinical trial for patients with mesothelioma began in November 1995 at the University of Pennsylvania Medical Center in conjunction with Penn's Institute for Human Gene Therapy. In this dose-escalation protocol, patients with mesothelioma who met strict inclusion criteria (including patent pleural cavities) underwent intrapleural administration of a single dose of Ad.HSV-TK vector followed by 2 weeks of intravenous GCV [26,29]. The initial adenoviral vector used was a so-called first-generation replication-incompetent virus, deleted in the early genes E1 and E3 with the HSV-TK gene inserted in the E1 region (H5.010RSV-TK; Fig. 1A). The protocol was designed as a dose-escalation study, starting with a vector dose of 1×10^9 plaque forming units (PFU) and increasing in half-log increments to the maximal dose level of 1×10^{12} PFU. At the completion of the 14-day GCV course, patients were discharged to home to continue outpatient follow-up that included serial radiographic, biochemical, and hematological testing. Throughout the study, the patients were carefully evaluated for evidence of toxicity, viral shedding, immune responses to the virus, and radiographic evidence of tumor response.

As summarized in Table 1, 26 patients (21 male, 5 female), ranging in age from 37 to 81, were enrolled in the study between November 1995 and November 1997 [26]. The overall goal of this trial was to determine the toxicity, efficacy of gene transfer, and immune responses generated in response to the intrapleural instillation of Ad.HSV-TK. Clinical toxicities of the Ad.TK/GCV gene therapy were minimal and a maximal tolerated dose (MTD) was not achieved. Intratumoral HSV-TK gene transfer was documented in 17 of 25 evaluable

TABLE 1   Mesothelioma Gene Therapy Trial #1 Results (As of 8/15/2000)

No. of patients	Stage/cell type	Vector dose (PFU)	Status	Survival (months)	Gene transfer score
62/M	IA/E[a]	1×10^9	*Progressed*	57	0
56/M	III/E	1×10^9	Deceased	8	0
69/M	III/B	1×10^9	Deceased	20	5
66/M	II/E	3.2×10^9	Deceased	11	0
71/M	IA/E	3.2×10^9	*Progressed*	53	0
71/M	II/B	1×10^{10}	Deceased	4	3
70/M	II/E	1×10^{10}	Deceased	6	0
60/M	II/E	1×10^{10}	Deceased	27	1
74/M	II/B	3.2×10^{10}	Deceased	2	—[a]
60/M	III/E	3.2×10^{10}	Deceased	9	0
37/F	IV/E	1×10^{11}	Deceased	16	0
37/M	III[b]	1×10^{11}	Deceased	2	0
65/F	III/E	1×10^{11}	Deceased	10	4
66/F	IA/E	3.2×10^{11}	*Progressed*	46	3
60/M	IV/B	3.2×10^{11}	Deceased	5	3
69/M	IB/E	3.2×10^{11}	Deceased	8	6
70/F	IB/E	3.2×10^{11}	Deceased	16	4
69/F	IB/E	3.2×10^{11}	Deceased	26	5
74/M	III/E	(S) 3.2×10^{11}	Deceased	7	4
68/M	III/S	(S) 3.2×10^{11}	Deceased	0.5	6
71/M	IB/E	(S) 3.2×10^{11}	*Progressed*	39	4
75/M	IB/E	(S) 3.2×10^{11}	Deceased	31	5
81/M	II/E	(S) 3.2×10^{11}	Deceased	15	6
72/M	II/E	1×10^{12}	Deceased	21	4
65/M	II/E	1×10^{12}	Deceased	6.5	6
67/M	IA/S	1×10^{12}	Deceased	23	6

[a]Patient 009 was unable to have the follow-up thoracoscopic biopsy.

[b]Patient 012 had a pseudomesotheliomatous adenocarcinoma. Patients 19–23 (S) received high-dose corticosteroids at time of vector instillation.

Note: Gene transfer scoring system: 0 = No gene transfer by any method, 1 = + DNA PCR in single sample, 2 = + DNA PCR in >1 sample, 3 = + RT-PCR or + *in situ* hybridization, 4 = + immunohistochemical detection of HSV-TK in few cells or positive immunoblot within single biopsy, 5 = + immunohistochemistry in few cells on multiple biopsies, 6 = + immunohistochemistry in many cells on multiple biopsies.

patients in a dose-related fashion by DNA-polymerase chain reaction (PCR), reverse transcription–PCR (RT-PCR), *in situ* hybridization, and immunohistochemistry (IHC) utilizing a murine monoclonal antibody directed against HSV-TK. All patients treated at dose levels of 3.2×10^{11} PFU or greater demonstrated evidence of intratumoral HSV-TK expression via IHC [26]. In general, the treatment protocol was well tolerated at all dosage levels. Toxicities were non-dose-limiting and included mild liver function test abnormalities, anemia, fever, and bullous exanthem at the instillation site. No MTD level was attained. At the highest dose level of

1×10^{12} PFU, two of three patients developed transitory hypotension and hypoxemia within hours after vector instillation that resolved with supplemental oxygen and intravenous fluids. Miscellaneous toxicities included atrial tachyarrhythmias, lymphopenia, and migratory polyarthralgias, each in a single patient [26].

Strong antiadenoviral humoral and cellular immune responses were noted, including neutrophil-predominant intratumoral inflammation in the posttreatment biopsy, generation of high titers of antiadenoviral neutralizing antibodies in serum and pleural fluid, significant increases in inflammatory cytokine production (TNF-α, IL-6) in pleural fluid, generation of serum antibodies against adenoviral structural proteins, and increased peripheral blood mononuclear cell proliferative responses to adenoviral proteins [35].

In a small substudy, five patients (patients 19–23) underwent administration of intravenous corticosteroids prior to and immediately following vector delivery [36]. This pilot trial was designed to preliminarily assess the effects of immunosuppression upon the degree of intratumoral gene transfer and antiadenoviral immune responses and was based on animal experiments showing that immunosuppression with dexamethasone augmented antitumor efficacy [37]. Results indicated a decreased incidence of fever and hypoxemia in the corticosteroid-treated cohort but an increased incidence of reversible mental status changes ("steroid psychosis"), particularly with a higher dose of methylprednisolone [36]. No diminution in humoral or cellular immune responses to the adenoviral vector was demonstrated in the group receiving corticosteroids, nor were there any detectable differences in the degree of intratumoral gene transfer.

As a phase I trial, the focus of this initial study was on safety issues and establishment of a MTD. Because of the heterogeneity of the patient population in terms of age, stage, histology, and vector dose, the clinical efficacy of Ad.RSV-TK/GCV gene therapy in malignant pleural mesothelioma was difficult to assess. Of the 26 patients enrolled in the initial phase I trial, 22 have died, with a median survival posttreatment of approximately 11 months and no fatal complications attributable to the gene therapy protocol (see Table 1). One patient (20) in the corticosteroid group who had stage IV mesothelioma at the time of enrollment died in the intensive care unit 2 weeks after completion of the protocol from rapid progression of his mesothelioma with malignant involvement of the contralateral hemithorax.

Four of the 26 patients enrolled in this initial protocol were alive and available for evaluation as of August 2000. All four had stage IA or IB disease at the time of enrollment, and all have had clinical and/or radiographic evidence of progression of disease. The median survival of the four surviving patients posttreatment is 50 months, significantly longer than the median survival of 8–14 months for mesothelioma patients in general. Of the trial participants who are deceased, all had progressive mesothelioma as their primary cause of death, typically with invasion of mediastinum, contralateral

hemithorax, and transdiaphragmatic extension, as well as widespread metastatic disease, a fairly common finding in advanced-stage mesothelioma. Only one of the 26 patients (patient 26) had radiographic evidence of intrathoracic tumor regression posttreatment on follow-up chest computed tomography (CT) scan. This patient eventually died from intraperitoneal disease progression. At autopsy, extensive intraabdominal tumor was observed but relatively minimal disease in the treated thoracic cavity.

D. Adjunctive Phase I Clinical Trials

1. Ad.HSV-TK Gene Therapy for Mesothelioma with Third-Generation Vector

We demonstrated in our first phase I trial that intrapleural Ad.HSV-TK gene therapy was safe, could effectively deliver transgene to superficial areas of mesothelioma tumor nodule, and induced significant humoral and cellular responses to the Ad vector [26,35]. Nevertheless, we felt that in order to achieve significant clinical responses warranted for phase II studies, improved intratumoral gene transfer was necessary. We decided to achieve this goal initially by increasing the vector dose, but doing so with the first-generation vector became problematic because of high levels of homologous recombination during large-scale production for clinical-grade lots, producing unacceptable levels of replication-competent adenovirus. In addition, there were some concerns regarding the hepatotoxicity and systemic inflammatory responses of first-generation adenoviral vectors as doses were increased, consistent with our findings in the highest dose cohort from the first trial.

For these reasons, in June 1998 we started a new phase I clinical trial employing an advanced-generation adenoviral vector, with the goal of maximizing vector dose with minimal toxicity [38]. This new vector contained deletions in the E1 and E4 regions with preservation of the E3 region (Fig. 1B). The presence of an intact E4 region, unlike E3, is critical to the late phase of the viral life cycle. E4 deletions engender decreased viral DNA synthesis and late gene expression as well as instability of late mRNAs [39]. Therefore, adenoviral vectors with lethal deletions in E1 and E4 purportedly offer a significant advantage over first-generation vectors with only a single lethal deletion in the E1 region and thus have diminished cytopathic effects and reduced cellular immune responses [40]. In addition, because two replication-necessary genes are deleted, simple recombination could not produce a replication-competent virus, allowing for production of large amounts of clinical-grade vector at lower cost.

The primary goals of the second phase I clinical trial were to determine the toxicity, gene transfer efficiency, and immune responses associated with the intrapleural injection of high titers of the E1/E4-deleted Ad.RSV-TK combined with systemic ganciclovir. To date, five patients have been treated, starting at a dose 1 log lower than the highest dose used with the E1/E3-deleted Ad vector. The first two patients were treated at a dose of 1.5×10^{13} viral particles. At this dose, we saw minimal toxicity, primarily transitory fever (grade 1) developing approximately 24 hours after vector instillation. Patients treated at this dose level did not exhibit other adverse systemic reactions to vector instillation nor did they develop elevated liver enzymes or bullous skin lesions. The next three patients were treated with a dose of 5.0×10^{13} viral particles with evidence of increased but non-dose-limiting toxicity. All three patients experienced acute febrile responses (grade 1) after vector instillation, with rapid defervescence. One patient (29) developed hypotension and hypoxemia (grade 2) within hours after vector administration which resolved with supplemental oxygen and intravenous fluids. Patient 29 also developed elevated serum transaminases to levels approximately two to three times normal (grade 2) after vector delivery, peaking during the first week of ganciclovir therapy, but returning to normal levels by completion of the protocol. The patient had no associated elevations in serum bilirubin or prothrombin time and no clinical evidence of hepatic dysfunction. The third patient treated at the higher dose level (patient 31) developed low-grade fever (grade 1) after intrapleural vector instillation, as well as a contralateral inflammatory pleural effusion associated with moderate pleuritic chest pain (grade 2). The latter was suggestive of an induced immune response directed against mesothelial antigens. Patient 31 had no signs of hepatotoxicity. Overall, there appeared to be equal or lower hepatoxicity in the patients treated with the E1/E4-deleted vector compared to patients treated with equivalent doses of the E1/E3-deleted adenovirus but a similar pattern of increased but non-dose-limiting systemic side effects at higher dose levels [38].

Gene transfer was detected in all patients at both dose levels via immunohistochemistry using a murine monoclonal antibody directed against the HSV-TK protein. As in the initial phase I trial, gene transfer appeared to be dose related, with the patients at the higher dose level having more extensive staining on their posttreatment biopsies. As in the initial phase I trial, significant humoral responses to the recombinant adenoviral vector were seen, with the development of high titers of total and neutralizing antiadenoviral antibodies within 15–20 days of vector instillation in all five patients. Deletion of the E3 region, therefore, did not seem to impact on the immunogenicity of the vector, at least in this small group of patients [38].

Of the five patients treated, two are surviving (patients 29, 30), both of them treated at the higher dose level of 5.0×10^{13} particles of Ad.HSV-TK. Each of the patients had evidence of stable disease for at least 12 months after treatment. Patient 29, a 34-year-old female with stage I epithelioid mesothelioma, demonstrated evidence of decreased tumor metabolic activity on follow-up 18-fluorodeoxyglucose (^{18}FDG) PET scan performed at day 80. She had an additional ^{18}FDG PET scan at the University of Adelaide, Australia, 10 months after completion of the protocol which demonstrated minimal

pleural FDG uptake. Concomitantly, the patient's clinical status remained stable without other antitumor therapy, and serial chest CT scans have shown no evidence of progression of pleural thickening or nodularity. This delayed decrease in tumor metabolic activity several months after completion of the Ad.RSV-TK/GCV gene therapy protocol suggests the development of an induced antitumor immune response. She has had no antineoplastic therapy other than this gene therapy protocol. Patient 30, a 57-year-old with stage I pleural mesothelioma, had stable disease clinically and radiographically 12 months post completion of the protocol despite refraining from other antineoplastic treatment. At approximately 18 months post treatment, the patient developed increasing chest wall discomfort associated with slowly progressing ipsilateral pleural thickening consistent with progressive disease [38].

2. Ad.HSV-TK Gene Therapy for Mesothelioma with Dose Escalation of Ganciclovir

One other approach to augment Ad.HSV-TK gene therapy is to increase the dose of administered ganciclovir (Fig. 2, top section). *In vitro* and animal experiments clearly show that after tumor transduction with HSV-TK, the cytotoxic

Strategies to Augment HSV–TK Efficacy

1. Increase [Substrate]

2. Alter Enzyme Affinity or Kinetics

Mutant HSV-TK Enzymes (M. Black)		Relative Activities		
		Thymidine	GCV	ACV
HSV-TK-wt	LIFDRHPIAALLCYP	100%	100%	100%
HSV-TK-30	ILADRHPIAYFLCYP	2%	61%	64%
HSV-TK-75	LLLFDRHPIVMLCYP	63%	71%	434%

FIGURE 2   Strategies to augment HSV-TK efficacy.

response is directly related to the GCV dose [23,25–28,40]. The dose of GCV used in all of the human trials (5 mg/kg i.v. b.i.d.) was chosen based upon the *in vitro* sensitivity testing of viral isolates and *in vivo* pharmacological measurements [42], as well as clinical experience with AIDS-related cytomegalovirus (CMV) retinitis. Based upon this hypothesis, we initiated a phase I clinical trial in July 1999 involving intrapleural delivery of the E1/E4-deleted Ad vector followed by intravenous GCV with gradual dose-escalation of the nucleoside analog. We have so far completed the first of four prospective cohorts in this study, with the first group of three patients being treated with 3.0×10^{13} particles of Ad.RSV-TK and 7.5 mg/kg ganciclovir i.v. b.i.d. (15 mg/kg/day). All three patients tolerated the treatment well. The most common toxicity was transitory fever within 24 hours post vector instillation. Other adverse events included grade 3 lymphopenia in patient 102 (discussed later), elevated gamma-glutamyl transferase (GGT) and lactic dehydrogenase (LDH) in one patient (grade 2), hyponatremia and hypokalemia in one patient (grade 2), and thrombocytosis in two patients. These toxicities were all non-dose-limiting and should not preclude advancement to the next GCV dose level of 10 mg/kg/dose (20 mg/kg/day × 14 days).

The initial patient in cohort 1, patient 101, was a 73-year-old male diagnosed with stage I epithelioid mesothelioma who underwent Ad.HSV-TK instillation followed by 28 doses of GCV at 7.5 mg/kg/dose. Successful HSV-TK gene transfer to tumor was confirmed by immunohistochemical evaluation of biopsy samples. He had minimal side effects from the GCV infusion. Thoracostomy tube drainage and subsequent vector infusion engendered full lung expansion and sclerosis of the pleural space with significant improvement in the patient's exercise capacity and performance status. Review of the patient's day 80 postprotocol chest CT scan demonstrated increased diffuse pleural thickening consistent with a postvector infusion inflammatory reaction. No new pleural or parenchymal nodules or masses were noted, nor was there involvement of the chest wall, mediastinum, abdominal cavity, or contralateral lung to indicate disease progression. In addition, as per protocol, the patient underwent pretreatment, day 80, and Day 170 [18]FDG PET scans. His pretreatment scan demonstrated intense [18]FDG uptake in the mediastinal pleural and right hilar regions as well as in the posterior parietal pleura. This uptake was reduced on the day 80 scan consistent with a decrease in tumor metabolic activity (Figure 3). Subsequent [18]FDG PET at day 170 showed a dramatic increase in tracer uptake consistent with increased tumor metabolic activity. This correlated with the patient's increasing shortness of breath and right anterior chest wall fullness and discomfort, as well as with the findings of his repeat chest CT scan, which revealed progressive pleural thickening and nodularity with encasement of the right lung. Patients 102 and 103 both demonstrated increased [18]FDG uptake on their follow-up day 80 PET studies and also had clear evidence of progression on standard chest CT studies. Clear assessment of antitumor

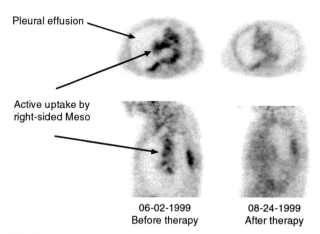

FIGURE 3 Pre- and posttreatment 18-fluorodeoxyglucose (FDG) positron emission tomography (PET) scans from patient 101 in Phase I clinical trial of intrapleural Ad.TK followed by escalating doses of systemic GCV. The initial scan on the left demonstrates a rightward shift of the mediastinum caused by the large pleural effusion with intense FDG uptake in the mediastinal pleural and right hilar regions as well as the posterior parietal pleura. This uptake was dramatically reduced on the day 80 scan consistent with a marked decrease in tumor metabolic activity.

activity awaits determination of the MTDs of both GCV and Ad.HSV-TK, as well as conduct of phase II studies.

III. CHALLENGES AND FUTURE DIRECTIONS

Evidence of showing limited toxicity and detectable gene transfer, as well as our anecdotal experience with tumor responses, suggest that the Ad.HSV-TK approach has exciting potential for the treatment of malignant mesothelioma, as well as other localized malignancies. In addition, one of the most valuable aspects of our trial has been the identification of specific challenges that must be addressed to make this system more useful. These include limitations in gene transfer efficiency and difficulties in noninvasively assessing gene transfer.

A. Strategies To Optimize Patient Selection

Although we obtained gene transfer in areas below the surface of a tumor, penetration was limited. Thus, using the current strategy, therapeutic efficacy could only be expected in patients with relatively small tumor loads (small nodules or diffuse, thin tumors). There are at least two ways in which we could create this clinical situation. First, once a MTD is reached, patients with only small amounts of pleural disease (nodules less than 5 mm) can be treated. Second, and probably more practical, patients with more than minimal disease could undergo a surgical "debulking" to minimize tumor burden. Gene therapy could be administered in the operating room as an adjuvant therapy after most of the tumor has been removed.

B. Strategies To Augment Gene Transfer

There is a subgroup of patients who are not good candidates for debulking surgery because of visceral pleural involvement. Because our data from the clinical trial suggest that gene transfer is possible even in patients with titers of anti-Ad neutralizing antibodies of up to 1:500, we postulate that repeated administration of vector and GCV (e.g., three doses over a 3-week period) will lead to augmented gene transfer. Animal data support this hypothesis. Recently completed studies in immunocompetent mice with established peritoneal tumors by our group [43] and others [44] showed marked increases in efficacy after multiple courses of intraperitoneal injections of Ad.HSV-TK, each followed by a course of GCV therapy. In our study, efficacy was increased equally well in those mice who had previously been immunized with adenovirus and had developed neutralizing antibodies [43].

Another approach to the gene transfer problem is to maximize the efficacy of any of the HSV-TK enzyme that is expressed (Fig. 2, lower section). The underlying principal of our suicide gene approach is that the herpes simplex virus–thymidine kinase-1 enzyme has a relaxed specificity (in comparison to mammalian thymidine kinase) that allows it to phosphorylate not only thymidine but also other nucleoside analogs such as ganciclovir (GCV) and acyclovir (ACV). Unfortunately, HSV-TK has a high affinity for thymidine ($K_m = 0.5 \mu M$), whereas the affinity for GCV ($K_m = 45 \mu M$) and ACV ($K_m = >400 \mu M$) are much lower [45]. This relationship suggests at least two ways in which the efficacy of HSV-TK could be augmented. First, higher levels of GCV could be provided to drive the equilibrium away from thymidine (Fig. 2, top section). This approach is already being used in our ongoing clinical trial (see earlier discussion). Second, "molecular remodeling" of the HSV-TK enzyme has been performed with the goal of increasing the substrate specificity towards GCV and ACV and concomitantly to decreasing thymidine utilization (Fig. 2, lower section). As described in detail [45], a segment of the HSV-TK gene at the putative nucleoside-binding site was substituted with random nucleotide sequences. Mutant enzymes that demonstrated preferential phosphorylation of GCV or ACV were selected from more than one million *Escherichia coli* transformants (Fig. 2, lower section). These mutants show enhanced acyclovir and ganciclovir killing and bystander effects [46]. We are currently producing and testing adenoviral vectors containing the mutated HSV-TK and anticipate they will enhance cell killing and augment the bystander effect.

A growing body of evidence supports the hypothesis that, in most models tested, treatment with HSV-TK/GCV results in immunologic reactions against tumor cells that enhance killing efficacy [14,17,47–49]. One reason for this antitumor immune reaction may be that in many cases, HSV-TK/GCV-mediated cell killing occurs through a nonapoptotic (i.e., necrotic) pathway, a type of cell death that effectively generates appropriate "danger signals" which then trigger significant immune responses [49,50]. With this rationale in mind, a number of investigators have conducted experiments showing that when the HSV-TK gene *plus* a cytokine gene are transduced into malignant cells, augmented tumor killing efficacy is achieved. To provide a few examples, augmented tumor killing effects have been reported with HSV-TK plus IL-2 in mouse liver metastasis from colon carcinoma [51], a mouse squamous cell carcinoma model [52], a murine melanoma model [53], and a rat intraperitoneal colon cancer model [54]. Synergistic effects have also been reported with HSV-TK and interferon-alpha in Friend erythroleukemia cells [55] and with HSV-TK and granulocyte–macrophage colony-stimulating factor (GM-CSF) in mouse liver metastasis from colon carcinoma [56]. Animal studies are underway in mouse models of mesothelioma to determine the best combination of cytokines with HSV-TK, as well as the best way to combine these therapies (i.e., direct injection of cytokine versus delivery of cytokine using gene therapy).

Finally, we hypothesize that a vector capable of replication in tumor cells (even only one to two rounds of replication) would allow much greater gene transfer. In this system, tumor killing could occur via two mechanisms: direct tumor lysis due to viral replication and by HSV-TK-mediated killing after administration of GCV. We anticipate a host immune response will limit viral replication and prevent widespread dissemination. However, it is likely that the generation of a tumor-selective replicating virus would be an important safety feature.

We therefore plan to develop and evaluate tumor-selective replicating adenovirus-HSV-TK vectors [57]. To do this, we will substitute the adenoviral E1 promoter with tumor-selective promoters (Fig. 1C). This is an approach that has been successfully used with the prostate-specific antigen (PSA) promoter to create a virus that selectively replicates in prostate cancer cells [58]. A number of promising choices for mesothelioma include the manganese-superoxide dismutase (MnSOD) promoter. Recent work by the Kinnula group in Finland has shown that MnSOD is very highly expressed in human malignant mesothelioma tissues and cell lines in contrast to normal lung or pleural tissues [59]. Two alternative mesothelioma "selective" promoters are those for the genes calretinin or mesothelin. Calretinin is a 29-kDa calcium-binding protein that is expressed primarily in the central and peripheral nervous system. Interestingly, high levels of calretinin expression have also been noted in mesothelial and mesothelioma cells, with very low expression levels in almost every other peripheral tissue studied [60,61]. Mesothelin is a 40-Kda surface protein of unknown function that is expressed only on the tissues forming the pleural, pericardial, and peritoneal membranes [62]. Other more general tumor-selective promoters, such as promoters responsive to the transcription factor E2F [63] or the survivin gene [64], would also be candidates.

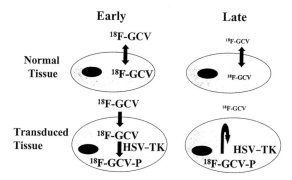

FIGURE 4 Schema of noninvasive imaging of HSV-TK gene transfer utilizing GCV analogs labeled with radioactive tags that can be visualized by PET scanning. Trapping of the radiolabeled substrate will occur in tissues expressing HSV-TK and allow visualization of functional gene transfer.

C. Strategies To Assess Gene Transfer Noninvasively

To date, the only method available to assess gene transfer is to biopsy tumor tissue with subsequent analysis for transgene DNA, RNA, or protein. The ability to measure gene transfer in a quantitative, noninvasive manner would have significant benefits for our clinical trials, as well as others. We would be able to rationally compare different treatment regimens, dosing schedules, and new vectors. Accordingly, we and others [65,66] are developing an approach using GCV analogs labeled with radioactive tags that can be visualized by positron emission tomography (PET) scanning. Trapping of the radiolabeled substrate will occur in tissues expressing HSV-TK and allow visualization of functional gene transfer (Fig. 4.). This methodology has been validated in animal models and will be tested in our clinical trials as a method of noninvasively assessing transgene expression in later trials.

References

1. Tiberghien, P. (1994). Use of suicide genes in gene therapy. *J. Leukoc. Biol.* **56**, 203–209.
2. Huber, B. E., Austin, E. A., Richards, C. A., Davis, S. T., and Good, S. S. (1994). Metabolism of 5-fluorocytosine to 5-fluorouracil in human colorectal tumor cells transduced with the cytosine deaminase gene: significant antitumor effects when only a small percentage of tumor cells express cytosine deaminase. *Proc. Natl. Acad. Sci. USA* **91**, 8302–8306.
3. Hoganson, D. K., Batra, R. K., Olsen, J. C., and Boucher, R. C. (1996). Comparison of the effects of three different toxin genes and their levels of expression on cell growth and bystander effect in lung adenocarcinoma. *Cancer Res.* **56**, 1315–1323.
4. Moolten, F. L., and Wells, J. M. (1990). Curability of tumors bearing herpes thymidine kinase genes transferred by retroviral vectors. *J. Natl. Canc. Inst.* **82**, 297–300.
5. Moolten, F. L., Wells, J. M., and Mroz, P. J. (1992). Multiple transduction as a means of preserving ganciclovir chemosensitivity in sarcoma cells carrying retrovirally transduced herpes thymidine kinase genes. *Cancer Lett.* **64**, 257–263.
6. Matthews, T., and Boehme, R. (1988). Antiviral activity and mechanism of action of ganciclovir. *Rev Infect Dis.* **10**, S490–S494.

7. Rubsam, L. Z., Davidson, B. L., and Shewach, D. S. (1998). Superior cytotoxicity with ganciclovir compared with acyclovir and 1-β-D-arabinofuranosylthymine in herpes simplex virus–thymidine kinase expressing cells: a novel paradigm for cell killing. *Cancer Res.* **58**, 3873–3882.
8. Culver, K. W., Ram, Z., Wallbridge, S., Ishii, I., Oldfield, E. H., and Blaese, R. M. (1992). In vivo gene transfer with retroviral vector-producer cells for treatment of experimental brain tumors. *Science* **256**, 1550–1552.
9. Ram, Z., Culver, K. W., Walbridge, B., Blaese, R. M., and Oldfield, E. H. (1993). *In situ* retroviral-mediated gene transfer for the treatment of brain tumors in rats. *Cancer Res.* **53**, 83–88.
10. Smythe, W. R., Hwang, H. C., Elshami, A. A., Amin, K., Eck, S., Davidson, B., Wilson, J., Kaiser, L. R., and Albelda, S. M. (1995). Successful treatment of experimental human mesothelioma using adenovirus transfer of the herpes simplex–thymidine kinase gene. *Ann. Surg.* **222**, 78–86.
11. Bonnekoh, B., Greenhalgh, D., Bundman, S., Eckhardt, J. N., Longley, M. A., Chen, S. H., Woo, S. L., and Roop, D. R. (1995). Inhibition of melanoma growth by adenoviral-mediated HSV thymidine kinase gene transfer in vivo. *J. Invest. Dermatol.* **104**, 313–319.
12. Freeman, S. M., Abboud, C. N., Whartenby, K. A., Packman, C. H., Koeplin, D. S., Moolten, F. L., and Abraham, G. N. (1993). The "bystander effect": tumor regression when a fraction of the tumor mass is genetically modified. *Cancer Res.* **53**, 5274–5283.
13. Hasegawa, Y., Emi, N., Shimokata, K., Abe, A., Kawabe, T., Hasegawa, T., Kirioka, T., and Saito, H. (1993). Gene transfer of herpes simplex virus type I thymidine kinase gene as a drug sensitivity gene into human lung cancer lines using retroviral vectors. *Am. J. Resp. Cell. Mol. Bio.* **8**, 655–661.
14. Caruso, M., Panis, Y., Gagandeep, S., Houssin, D., Salzmann, J. L., and Klatzmann, D. (1993). Regression of established macroscopic liver metastases after in situ transduction of a suicide gene. *Proc. Natl. Acad. Sci. USA* **90**, 7024–7028.
15. Elshami, A. A., Saavedra, A., Zhang, H. B., Kucharczuk, J. C., Spray, D. C., Fishman, G. I., Kaiser, L. R., and Albelda, S. M. (1996). Gap junctions play a role in the bystander effect of the herpes simplex virus thymidine kinase/ganciclovir system in vitro. *Gene Ther.* **3**, 85–92.
16. Mesnil, M., and Yamasaki, H. (2000). Bystander effect in herpes simples virus-thymidine kinase/ganciclovir cancer gene therapy: role of gap-junctional intercellular communication. *Cancer Res.* **60**, 3989–3999.
17. Pope, I. M., Poston, G. J., and Kinsella, A. R. (1997). The role of the bystander effect in suicide gene therapy. *Eur. J. Cancer* **33**, 1005–1016.
18. Vile, R. G., and Hart, I. R. (1993). Use of tissue-specific expression of the herpes simplex virus thymidine kinase gene to inhibit growth of established murine melanomas following direct intratumoral injection of DNA. *Cancer Res.* **53**, 3860–3864.
19. Vile, R. G., Nelson, J. A., Castleden, S., Chong, H., and Hart, I. R. (1994). Systemic gene therapy of murine melanoma using tissue specific expression of the HSV*tk* gene involves an immune component. *Cancer Res.* **54**, 6228–6234.
20. Moolten, F. L. (1994). Drug sensitivity ("suicide") genes for selective cancer therapy. *Cancer Gene Ther.* **1**, 279–287.
21. Kozarsky, K., and Wilson, J. M. (1993). Gene therapy: adenovirus vectors. *Curr. Opin. Genet. Dev.* **3**, 499–503.
22. Chen, S. H., Shine, H. D., Goodman, J. C.,. Grossman, R., and Woo, S. L. C. (1994). Gene therapy for brain tumors: regression of experimental gliomas by adenovirus-mediated gene transfer in vivo. *Proc. Natl. Acad. Sci. USA* **91**, 3054–3057.
23. Hwang, H. C., Smythe, W. R., Elshami, A. A., Kucharczuk, J. C., Amin, K., Williams, J. P., Litzky, L. A., Kaiser, L. R., and Albelda, S. M. (1995). Gene therapy using adenovirus carrying the herpes simplex thymidine kinase gene to treat in vitro models of human malignant mesothelioma and lung cancer. *Am. J. Resp. Cell Molec. Biol.* **13**, 7–16.

24. Elshami, A., Kucharczuk, J., Zhang, H., Smythe, W., Huang, H., Amin, K., Litzky, L., Kaiser, L. R., and Albelda, S. M. (1996). Treatment of pleural mesothelioma in an immunocompetent rat model utilizing adenoviral transfer of the HSV–thymidine kinase gene. *Hum. Gene Ther.* **7,** 141–148.

25. Perez-Cruet, M. J., Trask, T. W., Chen, S. H., Goodman, J. C., Woo, S. L. C., Grossman, R. G., and Shine, H. D. (1994). Adenovirus-mediated gene therapy of experimental gliomas. *J. Neurosci. Res.* **39,** 506–511.

26. Sterman, D. H., Treat, J., Litzky, L. A., Amin, K., Molnar-Kimber, K., Wilson, J., Albelda, S. M., and Kaiser, L. R. (1998). Adenovirus-mediated herpes simplex virus thymidine kinase gene delivery in patients with localized malignancy: results of a phase I clinical trial in malignant mesothelioma. *Hum. Gene Ther.* **9,** 1083–1092.

27. Alavi, J. B., and Eck, S. L. (1998). Gene therapy for malignant gliomas. *Hematol. Oncol. Clin. N. Am.* **12,** 617–629.

28. Morris, J. C., Ramsey, W. J., Wildner, O., Muslow, H. A., Aguilar-Cordova, E., and Blaese R. M. (2000). A phase I study of intralesional administration of an adenovirus vector expressing the HSV-1 thymidine kinase gene (AdV.RSV-*TK*) in combination with escalating doses of ganciclovir in patients with cutaneous metastatic melanoma. *Hum. Gene Ther.* **11,** 487–503.

29. Treat, J., Kaiser, L. R., Sterman, D. H., Litzky, L. A., Davis, A., Wilson, J. M., and Albelda, S. M. (1996). Treatment of advanced mesothelioma with the recombinant adenovirus H5.010RSV*TK*: a phase I trial (BB-IND 6274). *Hum. Gene Ther.* **7,** 2047–2057.

30. Schwarzenberger, P., Harrison, L., Weinacker, A., Marrogi, A., Byrne, P., Ramesh, R., Theodossiou, C., Gaumer, R., Summer, W., Freeman, S. M., and Kolls, J. K. (1998). The treatment of malignant mesothelioma with a gene modified cancer cell line: a phase I study. *Hum. Gene Ther.* **9,** 2641–2649.

31. Smythe, W. R., Hwang, H. C., Amin, K. M., Eck, S., Wilson, J., Kaiser, L. R., and Albelda, S. M. (1994). Use of recombinant adenovirus to transfer the herpes simplex virus thymidine kinase (HSVtk) gene to thoracic neoplasms: an effective in vitro drug sensitization system. *Cancer Res.* **54,** 2055–2059.

32. Smythe, W. R., Kaiser, L. R., Amin, K. M., Pilewski, J., Eck, S., Wilson, J., and Albelda, S. M. (1994). Successful adenovirus-mediated gene transfer in an in vivo model of human malignant mesothelioma. *Ann. Thor. Surg.* **57,** 1395–1401.

33. Esandi, M. C., van Someren, G. D., Vincent, A. J., van Bekkum, D. W., Valerio, D., Bout, A., and Noteboom, J. L. (1997). Gene therapy of experimental malignant mesothelioma using adenovirus vectors encoding the HSV*tk* gene. *Gene Ther.* **4,** 280–287.

34. Kucharczuk, J. C., Raper, S., Elshami, A. A., Amin, K., Sterman, D. H., Litzky, L. A., Kaiser, L. R., and Albelda, S. M. (1996). Safety of adenoviral-mediated transfer of the herpes simplex thymidine kinase cDNA to the pleural cavity of rats and non-human primates. *Hum. Gene Ther.* **7,** 2225–2233.

35. Molnar-Kimber, K. L., Sterman, D. H., Chang, M., Elbash, M., Elshami, A., Roberts, J. R., Treat, J., Wilson, J. M., Kaiser, L. R., and Albelda, S. M. (1998). Humoral and cellular immune responses induced by adenoviral-based gene therapy for localized malignancy: results of a phase I clinical trial for malignant mesothelioma. *Hum. Gene Ther.* **9,** 2121–2133.

36. Sterman, D. H., Molnar-Kimber, K., Iyengar, T., Chan, M., Lanuti, M., Amin, K. M., Pierce, B. K., Kang, E., Treat, J., Recio, A., Litzky, L. A., Wilson, J. M., Kaiser, L. R., and Albelda, S. M. (2000). A pilot study of systemic corticosteroid administration in conjunction with intrapleural adenoviral vector administration in patients with malignant pleural mesothelioma. *Cancer Gene Ther.* **7,** 1511–1518.

37. Elshami, A., Kucharczuk, J., Sterman, D., Smythe, W., Hwang, H., Amin, K., Litzky, L., Albelda, S., and Kaiser, L. (1995). The role of immune suppression in the efficacy of cancer gene therapy using aden-

38. Sterman, D. H., Recio, A., Molnar-Kimber, K., Knox, L., Hughes, J., Alavi, A., Lanuti, M., Litzky, L. A., Albelda, S. M., and Kaiser, L. R. (1999). Herpes simplex virus thymidine kinase (HSV*tk*) gene therapy utilizing an E1/E4-deleted adenoviral vector: preliminary results of a phase I clinical trial for pleural mesothelioma. *Am. J. Resp. Crit. Care Med.* **159,** A237.

39. Wang, Q., and Finer, M. H. (1996). Second-generation adenoviral vectors. *Nat. Med.* **2,** 714–716.

40. Gao, G. P., Yang, Y., and Wilson, J. M. (1996). Biology of adenovirus vectors with E1 and E4 deletions for liver-directed gene therapy. *J. Virol.* **70,** 8934–8943.

41. Kato, K., Yoshida, J., Mizuno, M., Sugita, K., and Emi, N. (1994). Retroviral transfer of herpes simplex thymidine kinase gene into glioma cells causes targeting of ganciclovir cytotoxic effect. *Neurol. Med. Chir. (Tokyo)* **34,** 339–344.

42. Jacobson, M. A. (1997). Treatment of cytomegalovirus retinitis in patients with the acquired immunodeficiency syndrome. *N. Engl. J. Med.* **337,** 105–114.

43. Lambright, E. S., Force, S. D., Lanuti, M., Wasfi, D. S., Amin, K., Albelda, S. M., and Kaiser, L. R. (2000). Efficacy of repeated adenoviral suicide gene therapy in a localized murine tumor model. *Ann. Thor. Surg.* **70,** 1865–1871.

44. Al-Hendy, A., Magliocco, A. M., Al-Tweigeri, T., Braileanu, G., Crellin, N., Li, H., Strong, T., Curiel, D., and Chedrese, P. J. (2000). Ovarian cancer gene therapy: repeated treatment with thymidine kinase in an adenovirus vector and ganciclovir improves survival in a novel immunocompetent murine model. *Am. J. Obstet. Gynecol.* **182,** 553–559.

45. Black, M. E., Newcomb, T. G., Wilson, H. M. P., and Loeb, L. A. (1996). Creation of drug-specific herpes simplex virus type 1 thymidine kinase mutants for gene therapy. *Proc. Natl. Acad. Sci. USA* **93,** 3525–3529.

46. Qiao, H. J., Black, M. E., and Caruso, M. (2000). Enhanced ganciclovir killing and bystander effect of human tumor cells transduced with retroviral vector carrying a herpes simplex thymidine kinase gene mutant. *Hum. Gene Ther.* **11,** 1569–1576.

47. Hall, S. J., Sanford, M. A., Atkinson, G., and Chen, S. H. (1998). Induction of potent antitumor natural killer cell activity by herpes simplex virus–thymidine kinase and ganciclovir therapy in an orthotopic mouse model of prostate cancer. *Cancer Res.* **58,** 3221–3225.

48. Freeman, S. M., Ramesh, R., and Marogi, A. J. (1997). Immune system in suicide gene therapy. *Lancet* **349,** 2–3.

49. Vile, R. G., Castleden, S., Marshall, J., Camplejohn, R., Upton, C., and Chong, H. (1997). Generation of an anti-tumor immune response in a non-immunogenic tumour: HSV*tk* killing *in vivo* stimulates a mononuclear cell infiltrate and a Th1-like profile of intratumoural cytokine expression. *Int. J. Cancer* **71,** 267–274.

50. Melcher, A., Todryk, S., Hardwick, N., Ford, M., Jacobson, M., and Vile, R. (1998). Tumor immunogenicity is determined by the mechanism of cell death via induction of heat shock protein expression. *Nat. Med.* **4,** 581–587.

51. Chen, S. H., Li Chen, X. H., Wang, Y., Kosai, K. I., Finegold, M. J., Rich, S. S., and Woo, S. C. (1995). Combination gene therapy for liver metastasis of colon carcinoma in vivo. *Proc. Natl. Acad. Sci. USA* **92,** 2577–2581.

52. O'Malley, Jr., B., Cope, K. A., Chen, S. H., Li, D., Schwartz, M., and Woo, S. L. C. (1996). Combination gene therapy for oral cancer in a murine model. *Cancer Res.* **56,** 1737–1741.

53. Castleden, S. A., Chong, H., Garcia-Ribas, I., Melcher, A. A., Hutchinson, G., Roberts, B., Hart, I. R., and Vile, R. G. (1997). A family of bicistronic vectors to enhance both local and systemic antitumor

effects of HSV*tk* or cytokine expression in a murine melanoma model. *Hum. Gene Ther.* **8,** 2087–2102.

54. Coll, J., Mesnil, M., Lefebvre, M., Lancon, A., and Favrot, M. (1997). Long-term survival of immunocompetent rats with intraperitoneal colon carcinoma tumors using herpes simplex thymidine kinase/ganciclovir and IL-2 treatments. *Gene Ther.* **4,** 1160–1166.

55. Santodonato, L., D'Agostino, G., Santini, S., Carlei, D., Musiani, P., Modesti, A., Signorelli, P., Belardelli, F., and Ferrantini, M. (1997). Local and systemic antitumor response after combined therapy of mouse metastatic tumors with tumor cells expressing IFN-alpha and HSVtk: perspectives for the generation of cancer vaccines. *Gene Ther.* **4,** 1246–1255.

56. Hayashi, S., Nobuhiko, E., Yokoyama, I., Namii, Y., Uchida, K., and Takagi, H. (1997). Inhibition of establishment of hepatic metastasis in mice by combination gene therapy using both herpes simplex virus–thymidine kinase and granulocyte macrophage-colony stimulating factor genes in murine colon cancer. *Cancer Gene Ther.* **4,** 339–344.

57. Alemany, R., Balague, C., and Curiel, D. T. (2000). Replicative adenoviruses for cancer therapy. *Nat. Biotechnol.* **18,** 723–727.

58. Rodriguez, R., Schuur, E. R., Lim, H. Y., Henderson, G. A., Simons, J. W., and Henderson, D. R. (1997). Prostate attenuated replication competent adenovirus (ARCA) CN706: a selective cytotoxic for prostate-specific antigen-positive prostate cancer cells. *Cancer Res.* **57,** 2559–2563.

59. Kahlos, K., Anttila, S., Asikainen, T., Kinnula, K., Raivio, K. O., Mattson, K., Linnainmaa, K., and Kinnula, V. L. (1998). Manganese superoxide dismutase in healthy human pleural mesothelium and in malignant pleural mesothelioma. *Am. J. Respir. Cell. Mol. Biol.* **18,** 579–580.

60. Doglioni, C., Dei Tos, A. P., Laurino, L., Iuzzolino, P., Chiarelli, C., Celio, M. R., and Viale, G. (1996). Calretinin: a novel immunocytochemical marker for mesothelioma. *Am. J. Surg. Pathol.* **20,** 1037–1046.

61. Gotzos, V., Vogt, P., and Celio, M. (1996). The calcium binding protein calretinin is a selective marker for malignant pleural mesotheliomas of the epithelial type. *Pathol. Res. Pract.* **192,** 137–147.

62. Chang, K. and Pastan, I. (1996). Molecular cloning of mesothelin, a differentiation antigen present on mesothelium, mesotheliomas, and ovarian cancers. *Proc. Natl. Acad. Sci. USA* **93,** 136–140.

63. Amin, K. M., Tsukuda, K., Odaka, M., Molnar-Kimber, K., Kaiser, L. R., and Albelda, S. M. (2001). The development and characterization of a mutant oncolytic adenovirus that replicates selectivity in ovarian and lung cancer cells over-expressing E2F-1 protein (abstr). *Am. Assoc. Cancer Res. Annu. Meet.,* p. 3716.

64. Ambrosini G., Adid, C., and Altieri, D. C. (1997). A novel anti-apoptosis gene, surviving, expressed in cancer and lymphoma. *Nat. Med.* **3,** 917–921.

65. Gambhir, S. S., Barrio, J. R., Phelps, M. E., Iyer, M., Namavari, M., Satyamurthy, N., Wu, L., Green, L. A., Bauer, E., MacLaren, D. C., Nguyen, K., Berk, A. J., Cherry. S. R., and Herschman, H. R. (1999). Imaging adenoviral-directed repoerter gene expression in living animals with positron emission tomography. *Proc. Natl. Acad. Sci. USA* **96,** 2333–2338.

66. Gambhir, S. S., Bauer, E., Black, M. E., Liang, Q., Kokoris, M. S., Barrio, J. R., Iyer, M., Namavari, M., Phelps, M. E., and Herschman H. R. (2000). A mutant herpes simplex virus type 1 thymidine kinase reporter gene shows improved sensitivity for imaging reporter gene expression with positron emission tomography. *Proc. Natl. Acad. Sci. USA* **97,** 2785–2790.

The Use of Suicide Gene Therapy for the Treatment of Malignancies of the Brain

KEVIN D. JUDY

HUP–Department of Neurosurgery
The University of Pennsylvania Medical Center
Philadelphia, Pennsylvania 19004

STEPHEN L. ECK

HUP–Department of Neurosurgery
The University of Pennsylvania Medical Center
Philadelphia, Pennsylvania 19004

I. INTRODUCTION

High-grade gliomas (anaplastic astrocytoma and glioblastoma multiforme) are the most common and unfortunately the most lethal tumors of the brain that occur in adults. Survival for patients with anaplastic astrocytoma (WHO grade III) is usually less than 3 years and for patients with glioblastoma multiforme (WHO grade IV) is less than 1 year. Surgical resection of these tumors will reduce the mass effect of the tumor, thus improving quality of life and time to clinical progression of the tumor. It is impossible to completely resect high-grade gliomas due to the diffuse nature of the glioma cells dispersed throughout the surrounding "normal"-appearing brain [1]. These neoplastic cells contribute to tumor regrowth in the same location as the original tumor [2] as well as to migration of malignant cells to distant parts of the central nervous system (CNS). Chemotherapy has shown only modest success improving the survival of patients with these tumors [3,4]. The limited benefits of systemic chemotherapy have been attributed in part to inherent resistance of the tumor cells to the chemotherapy due to expression of alkylguanine-DNA alkyltransferase [5,6] and inability of the drugs to cross the blood–brain barrier [7].

Radiation therapy directed to the tumor bed, either alone or following maximal surgical resection, has been the most effective treatment to delay local regrowth [4,8].

Following aggressive treatment with surgery, radiation therapy, and chemotherapy, these high-grade gliomas invariably relapse. Growth of gliomas is restricted to the CNS, as these tumors do not metastasize to visceral organs or bone. The absence of widespread metastases or significant organ failure makes these patients excellent subjects for local experimental therapies. Patients with a good performance status can be treated at the time of recurrence with a second craniotomy combined with additional local therapy. Because the delivery of the genetic vector is the major limitation of current gene therapy technology, brain tumors are more attractive targets for this type of treatment compared to more common cancers that metastasize widely.

Local therapies to the tumor bed have shown success in tumor control and reduce the systemic adverse effects from the agents. Implantable polymer wafers containing carmustine chemotherapy will deliver extremely high doses of chemotherapy directly to the tumor with minimal systemic exposure to the drug [9]. Disruption of the blood–brain barrier using osmotic diuretics or bradykinin analogs can enhance the penetration of chemotherapy agents into the tumor [7,10]. Local delivery of interferon and interleukin-2 (IL-2) through a tumor-embedded catheter (e.g., Ommaya reservoir) has been utilized to overcome the limitation of systemic administration of these short-lived and systemically toxic cytokines. Optimal local therapies must have the ability to kill both dividing and nondividing tumor cells, as the majority of tumor cells are not actively dividing at the time of treatment. The agent must be able to penetrate deeply into the tumor and surrounding tissues to destroy the tumor cells, which

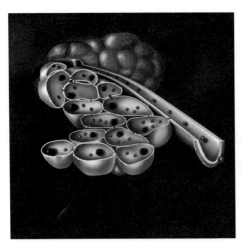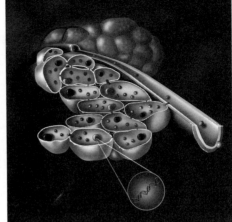

FIGURE 1   Conventional chemotherapy for brain tumors (left panel) requires that a systemically distributed drug reach the tumor in concentrations sufficient to exert a tumoricidal effect. This approach is typically limited by systemic toxicity that prohibits dose escalation to levels sufficient for tumor eradication. Gene-directed enzyme prodrug therapy (right panel) permits the systemic administration of relatively nontoxic drugs (e.g., ganciclovir) which are only converted to their active form in cells that have been transduced to express the enzyme (e.g., HSV-TK) needed to activate them. Moreover, the activated drug can be locally redistributed within the tumor to nontransduced cells, achieving a "bystander effect." This limits systemic exposure to the active form of the drug, which accumulates selectively within the tumor. Prodrugs can be selected for their ability to cross the blood–brain barrier, even though their activated forms may lack this ability. (See color insert.)

infiltrate the surrounding "normal" brain along white matter tracts.

Gene-directed enzyme prodrug therapy (GDEPT) for brain tumors uses gene transfer as a drug delivery system. The genetic vector (adenovirus, retrovirus, and liposome vectors have been used clinically) delivers the suicide gene (typically herpes simplex virus–thymidine kinase, HSV-TK) into the tumor cell, where it phosphorylates ganciclovir (or other suitable substrate), creating a toxic metabolite that leads to abortive DNA synthesis and cell death [11]. Success of this therapy is dependent upon efficient penetration of the tumor by the vector and having rates of cell division that render the cells sensitive to DNA synthesis inhibitors. The requirement for efficient tumor cell transduction is to some degree lessened by the "bystander effect" (Fig. 1 [See also color insert]), which permits activated drug to pass from transduced to nontransduced cells [12]. Recombinant retroviruses and adenoviruses have been the most extensively studied vectors in clinical trials of GDEPT for brain tumors [13]. Recombinant herpes viruses carrying the endogenous HSV-TK gene have been more recently introduced into clinical studies [14].

II. RETROVIRUS VECTOR FOR HSV-TK

Replication-defective recombinant retroviruses were the first vectors to be used in human clinical trials of GDEPT. They had the perceived advantage of selectively transducing only dividing cells which would restrict therapeutic gene ex-

pression (e.g., HSV-TK) to tumor cells and spare the more slowly dividing normal brain parenchyma. The immediate limitation of retroviral vectors was the inability to highly concentrate the vector and thereby limit the volume of infusate to a reasonable size for instillation into the brain. Moreover, the vectors survive *in vivo* for only short periods of time relative to the rate of cell divisions (a requirement for retroviral entry into the nucleus). To circumvent these limitation, vector-producing cells (VPCs) derived from murine fibroblasts were constructed to release the retroviral vector containing the HSV-TK gene [15–17]. In concept, multiple focal deposits of these VPCs would produce the retroviral vector over a sustained period of time (several days) before being cleared by the host immune systems (rejection of a xenograft). Multiple sites of the tumor (up to 50) have been injected with the VPCs to distribute the vector throughout the tumor [18]. Subsequent systemic administration of ganciclovir (GCV) leads to conversion by HSV-TK to GCV–monophosphate within the tumor cells and VPCs that are then rapidly phosphorylated by human cellular kinases to GCV–triphosphate [11]. This metabolite inhibits DNA replication, leading ultimately to the death of dividing tumor cells. Mammalian thymidine kinase is much less efficient in performing the initial phosphorylation which accounts for the low toxicity of systemically administered GCV (e.g., mild, reversible myelosuppression after prolonged use) [11].

A large number of *in vitro* and *in vivo* studies have demonstrated that a significant proportion of nontransduced cells in the vicinity of cells undergoing gene transfer are also killed by a bystander effect (metabolic cooperation). As noted, this effect arises in part from the intercellular transport of GCV–phosphates by way of gap junctions between adjacent tumor cells [19]. Other pathways have been proposed to contribute to the 'bystander effect,' including the release of apoptotic vesicles containing GCV metabolites and an induced immune response to nontransduced cells by released tumor antigens [16,20]. Whether or not such immune mechanisms contribute to eradication of brain tumors in patients remains to be determined. However, it would seem unlikely that an immune bystander effect plays a significant role given the immunosuppressive effects of the human brain tumors themselves and the concomitant use of high doses of immunosuppressive glucocorticoids that are routinely employed to control cerebral edema. Prior *in vitro* and *in vivo* studies have shown that only 10–50% of tumor cells need to be transduced (in laboratory models) to achieve complete tumor destruction. In addition, there is some evidence from animal and human studies to suggest that the HSV-TK gene delivery results in damage to blood vessels within the tumor microenvironment, potentially contributing to the therapeutic effect [21–23]. Ram *et al.* postulated that transduction of endothelial cells may have contributed to tumor response, because in some patients microhemorrhages were seen in the tumors by magnetic resonance imaging (MRI) scan during the first week of ganciclovir treatment [23]. Vascular injury has also been documented in histologic specimens obtained in non-human primate studies [21] and from post-gene-therapy resection specimens.

The first application of the retroviral HSV-TK–GCV system in human brain tumors was performed by Ram *et al.* [18] (see Table 1). Fifteen patients, 12 with gliomas and three with metastases (two melanoma and one breast carcinoma), harboring 19 lesions were treated with 1×10^8 to 1×10^9 VPCs per treatment by stereotactic injection. Intravenous GCV was begun 7 days after the VPC injection to allow for vector release and transduction of the surrounding tumor cells. The

GCV was given for a total of 14 days. Two patients had their tumors, resected 7 days following VPC injection, after which the tumor bed was reinjected with VPCs, and GCV was continued for 7 days following the tumor resection. Four patients showing antitumor activity received a second treatment. The principal adverse events noted were seizures and hemorrhage from the multiple injections. Not surprisingly, injection of the murine VPCs into patients stimulated production of anti-VPC antibodies in the majority of patients. This immune response did not appear to contribute to toxicity or efficacy. The inadvertant introduction of replication-competent retrovirus (RCR) has been a subject of some concern following the induction of lymphoma in monkeys after RCR administration [24]. However, no RCR or vector DNA was identified in the patients' peripheral blood samples. Two patients underwent biopsy of the tumor 7 days following treatment. *In situ* hybridization revealed expression of HSV-TK in the VPCs and in tumor cells immediately surrounding the VPCs. However, the most striking observation was the limited spread of the vector from the VPC sites [18].

Objective responses (i.e., 50% or more reduction in tumor volume) were seen in five lesions in four patients. Two of these patients with gliomas have remained alive with tumor control for several years. These five responding tumors had volumes of 1.4 ± 0.5 mL, suggesting that smaller tumors in which a high density of VPCs could be administered were most responsive to the treatment. This is consistent with the limited distribution of the HSV-TK gene expression in most patients and indicates that techniques to improve delivery and distribution of the therapeutic gene must be developed if clinical utility is to be achieved with this approach.

A similar retroviral–HSV-TK construct was developed by Izquierdo *et al.* utilizing the previously described vector producing cell approach [25]. Five patients with recurrent glioblastoma multiforme were treated with multiple injections of the VPCs into the tumor followed by GCV. The treatment was tolerated well by all five patients. One patient had a significant reduction in the size of the treated tumor but not an adjacent tumor that was not treated. A second patient had a small reduction in size of the treated tumor. The remaining three patients did not have an observable response to the GDEPT. The same investigators proceeded with a second trial for recurrent glioblastoma multiforme (GBM) in which the recurrent tumor was resected, the tumor bed injected with VPCs, and an Ommaya reservoir placed into the tumor bed [26]. Three patients were entered into the trial, but one suffered an intracranial hemorrhage and never received the GDEPT. A second dose of VPCs was given to the two remaining patients through the Ommaya reservoir 1 week following the first injection, followed by 2 weeks of GCV beginning 1 week following the second injection. This allowed the VPCs deposits greater time to transduce the surrounding tumor cells. The patients were then retreated with VPCs through the Ommaya reservoir for the next 3–6 months

TABLE 1   Outcome of Human Brain Tumor Trials Using Viral Vectors Expressing HSV-TK

Vector	Patients alive >1 year (number/total)	Investigators	Ref.
RV–HSV-TK	3/15	Ram *et al.*	[18]
RV–HSV-TK	1/15	Izquierdo *et al.*	[25,26]
RV–HSV-TK	3/12	Klatzmann *et al.*	[27]
RV–HSV-TK	13/48	Shand *et al.*	[28]
RV–HSV-TK	62/124	Rainov *et al.*	[32,59]
Adeno–HSV-TK	3/13	Trask *et al.*	[35]
Adeno–HSV-TK	5/13	Eck *et al.*	[34,60]
Mutant HSV-1	2/21	Martuza *et al.*	[40]

Summary of patients surviving longer than 1 year from the time of gene therapy in clinical trials using viral vectors expressing HSV-TK.

at times when follow-up MRI showed evidence of enhancement suggestive of recurrent tumor. The two patients survived 11 and 17 months with persistent enhancing tumor, suggesting incomplete eradication of all tumor cells. The apparent increase in survival suggests a potential benefit from this therapy; however, these findings have yet to be reproduced in a larger patient population.

Klatzmann *et al.* [27] completed a phase I/II study of retrovirus expressing HSV-TK using VPCs in 12 patients with recurrent glioblastoma multiforme who underwent resection of the tumor with infiltration of the tumor bed with the VPC [27]. GCV was instituted 7 days postoperatively and continued for 14 days as in previous studies. In general, the treatment was tolerated well, with adverse events consisting of bacterial infections, hemorrhage following the surgery, and one patient experiencing progressive aseptic meningitis and death 2 months following treatment. The progressive deterioration in the latter patient was found at autopsy to be due to massive gliomatous invasion of the ventricular cavities. The adverse events were not directly attributable to the use of VPCs. A weak HSV-TK signal was detected in the blood of one patient 1 hour following injection of the VPCs using a polymerase chain reaction (PCR) technique. The remaining PCR investigations in that patient as well as the other patients did not detect transgene dissemination. Four of the 12 patients showed a lack of tumor progression by MRI after 4 months of follow-up. Three of the patients survived more than 1 year, with one patient being alive with no evidence of disease 2.8 years from treatment. MRI scan evidence of progressive tumor growth did not correlate with histological evidence of minimal tumor growth in two patients.

The European/Canadian study group studied 48 patients with recurrent glioblastoma multiforme who were treated with HSV-TK VPCs following tumor resection [28]. The VPCs were directly injected into the tumor cavity, and the patients were given GCV for 14 days. The median survival was 8.6 months, with four patients having no evidence of disease at 12 months. Retrovirus could not be cultured from blood specimens. Retroviral DNA was detected in peripheral white blood cells in some patients, suggesting that retrovirus may have transduced reactive lymphocytes in the tumor bed in the brain [29]. An autopsy study of 32 patients that had received the VPC in clinical trials has been performed [30]. Twenty-four brain tumor biopsies from this group of patients were examined for RCV and vector DNA sequences by PCR assays. RCV was not detected in any of the samples, including normal tissue samples at autopsy; however, vector DNA was found in the scalp, kidney, liver, and lung. Vector DNA was found in brain tumor specimens (55%), adjacent brain tissue (22%), and contralateral brain (6.7%), with increased detection in those patients receiving multiple, rather than single, injections of vector. There was no evidence that detection of retroviral sequences by the sensitive PCR assay was in any way related to clinical outcome.

A study of serial MRI scans in patients receiving two or more cycles of VPCs with GCV was performed by Deliganis *et al.* [31]. They evaluated seven patients receiving between two and four infusions of VPCs via an Ommaya reservoir that was placed into the resected tumor cavity. These patients were followed with MRI scans every 40 days following the initial treatment. The changes in the areas of enhancing tissues were variable. One patient had a transient increase in tissue enhancement followed by a sustained decrease in enhancement. Two patients developed an early increase in enhancing tissue followed by a transient plateau in one and a transient decrease in the other. Another patient had a stable volume of enhancement for 132 days before developing a progressive increase in enhancing tissue. The remaining three patients showed serial increases in enhancing tissue and edema indicative of progressive disease. Patients having an initial increase in enhancing tissue were thought to be experiencing an acute inflammatory reaction to the VPC/retroviral treatment. These and other observations illustrate the potential shortcomings of conventional MRI which, using areas of enhancement, frequently cannot readily distinguish tumor from effects of the gene transfer itself.

A large multicenter trial sponsored by Novartis Pharma and Genetic Therapy, Inc., has now been completed [32]. In this study, patients with newly diagnosed glioblastoma were randomized to receive either standard tumor resection and radiation therapy, or the same plus the addition of VPC injection at surgery, followed by 2 weeks of ganciclovir. The median times to tumor progression and survival were not different in the two treatment arms. One fatal complication occurred: an infection related to ganciclovir-induced neutropenia. No RCR was found in peripheral blood or autopsy specimens.

Taken together, these studies using the retrovirus producer cells and the HSV-TK ganciclovir system demonstrate that the treatment can be given safely. There has been no evidence of systemic toxicity of the virus. The virus has been detected by PCR in a very small percentage of normal tissues distant from the brain. However, the limited efficacy in these highly selected patients suggests that it is probably not very effective in patients with large glial tumors, due to limited distribution of the retrovirus within the tumor and, at best, only a small contribution of the bystander effect. This treatment is more likely to be effective in patients who have had a tumor debulking prior to the vector injection. Although the procedure itself is well tolerated and uses standard neurosurgical techniques, the use of VPCs remains cumbersome from both a manufacturing and pharmacy point of view and affords little if any benefit in the currently employed applications.

As noted above, a potential way to augment to effects of gene transfer is by eliciting an immune response to the tumor. One approach has been to coexpress immune modulatory agents along with the HSV-TK gene. A retroviral vector producing both HSV-TK and human IL-2 has been developed in an effort to combine the cytotoxic effects of the HSV-TK with a cellular immune response to tumor antigens [33]. Four patients with recurrent glioblastoma multiforme were treated by stereotactic implantation of HSV-TK retroviral VPCs. Cerebrospinal fluid levels of IL-2 were sequentially followed in one of the patients as evidence of IL-2 production. Transduction of circulating peripheral blood mononuclear cells was observed in another patient. Two patients had posttreatment biopsies of the tumors. The endothelial cells were intensely stained by the HSV-TK, probe indicating that the vector is expressed in the neovascular component of the tumor. There was no evidence of antitumor immunity in this small study which serves only to illustrate the potential feasibility of the approach.

Where examined, retrovirus from the VPCs has been detected by PCR in only trace amounts in normal tissues and only in proximity to the VPCs within the tumor in these clinical studies. There is no evidence of systemic toxicity from the retroviral vector or the VPCs. Despite the ample evidence of clinical safety, the efficacy has been limited. The retroviral vector has restricted distribution in gliomas and so has shown antitumor activity predominantly in tumors ≤1.5 mL in volume. This size limitation would restrict clinical applications to tumors that have been reduced in size by surgical debulking, radiation therapy, or chemotherapy or any combination of these.

III. ADENOVIRUS VECTOR FOR HSV-TK

The retrovirus HSV-TK GDEPT system established a proof of principle for the antitumor activity and safety of HSV-TK ganciclovir in patients with high-grade gliomas. Subsequent work with recombinant adenovirus vectors sought to overcome some of the inherent limitations of the retroviral vectors [13]. Adenoviruses have the potential advantage of being prepared in high titer, obviating the need for injection of producer cells. They do not integrate into the host genome and thereby lack the risk of insertional mutagenesis (a concern with retroviruses, especially those containing RCR). Adenoviruses transduce both dividing and nondividing cells and, therefore, can achieve a high level of HSV-TK expression shortly after injection, a perceived advantage in gliomas where the majority of tumor cells are not actively dividing. Preclinical studies have demonstrated low neurotoxicity despite the anticipated immune response to adenoviral vectors in the central nervous system [21]. This might be expected to be worse in humans who have preexisting immunity to the serotypes of adenovirus used as gene delivery vectors; however, this has not been seen in clinical trials [34,35].

The E1/E3-deleted, replication-defective adenoviral vectors [36] expressing HSV-TK (Ad. HSV-TK) have been evaluated in several clinical trials for the treatment of malignant gliomas. In a study by Trask et al. [35], 13 patients with recurrent glioma were treated with a single stereotactic injection of Ad. HSV-TK. This phase I study evaluated escalating doses from 10^8 to 10^{11} PFU followed by 14 days of GCV. Patients were followed by clinical examination and serial MRI scans. One patient receiving 10^{10} PFU of vector deteriorated rapidly, apparently due to aggressive tumor growth and without an apparent effect of vector administration. Two patients developed significant toxicity at the 10^{11} PFU dose. These toxicities included an injection-site hematoma and brain edema in one patient and obtundation, hyponatremia, and hydrocephalus in the other patient that required placement of a ventriculoperitoneal shunt. The authors concluded that the latter patient probably suffered from inadvertent injection of the adenovirus into the lateral ventricle. Three patients from this study remained alive at more than 3 years following the treatment.

We have used a similar Ad. HSV-TK in a trial of 13 patients with high-grade gliomas [34,37]. One patient with an unresectable thalamic tumor received a single stereotactic injection of the vector followed by 14 days of GCV infusion. The remaining 12 patients were treated with a stereotactic injection of vector into the tumor followed by 6 days of daily GCV, craniotomy to resect the tumor, and reinjection with vector followed by 14 days of GCV. Two patients with glioblastoma multiforme were treated at the time of first diagnosis with Ad. HSV-TK followed by standard radiation therapy, whereas the remaining 11 patients had recurrent high-grade gliomas. The dose of vector ranged from 10^8 to 10^{11} PFU for each treatment to give a total dose of 2×10^8 to 2×10^{11} PFU for patients undergoing craniotomy and resection of the tumor. Dose-limiting toxicity occurred in two patients at the 10^{11} PFU dose level. Three patients experienced transient increased intracranial pressure (diagnosed by direct measurement or clinical presentation) manifested as severe headache, and one of them developed an altered mental status. One patient experienced altered mental status, agitation, headache, and hypertension after vector administration during the second surgery. In all cases, the patients recovered within 24 hours with routine medical management. Other toxicities included mild reversible elevation of transaminases and transient fever. The median time to tumor progression in these patients was 3 months, and the median survival was 10 months. Five patients lived 12 months or longer, and one patient remained alive without tumor for 3 years following Ad.HSV-TK treatment before suffering a local recurrence. Despite the preexisting immunity to adenovirus, serious adverse events did not correlate with immune response to virus as assessed by changes in adenovirus antibody titers or T-cell responses [34].

Because gene distribution is critical to the success of GDEPT in brain tumors, Puumalainen et al. [38] examined the transfer of the lacZ gene (which produces β-galactosidase as a marker protein) in patients about to undergo surgery for recurrent glioma [38]. A catheter was implanted into the tumor and vector injected for 3 days using either a retrovirus (no VPCs were used) or adenovirus expressing β-galactosidase. The tumor and catheter were resected several days later. They found that gene transfer efficiency varied from <0.01 to 4% with retrovirus and <0.01 to 11% with adenovirus. Focal areas of maximal gene transfer reached 30% after adenovirus vector administration and could be detected up to 2 cm from the catheter. Consistent with other reports, both glioma and endothelial cells were transduced by both vectors. This study would appear to substantiate the previously stated advantages of adenoviral vectors over retrovirus vectors in terms of superior vector distribution. In our own study, HSV-TK vector distribution was not directly determined; however, the ability to detect the vector genome by PCR following resection increased with dose escalation [34].

These adenovirus vector studies demonstrate that the vector is safe in the doses given thus far and the gene transfer takes place predominately in the tumor tissue with some gene transfer in the surrounding tissues. Toxicity does not appear to be due to increased cerebral edema in the tumor and surrounding tissues based on MRI findings. The vector may induce an acute inflammatory effect, causing an alteration in cerebrospinal fluid (CSF) production or reabsorption, leading to an increased intracranial pressure as evidenced by a transiently elevated opening pressure on lumbar puncture in one patient experiencing headache in our study [34]. Such an effect would be expected to be more severe if significant amounts of virus were injected directly into the CSF consistent with previous observation [35]. Median survivals for these gene therapy trials are similar to those seen in other highly selected patients in other glioma therapy trials [9,39]. However, it is perhaps significant that in each of the trials there have been a few long-term survivors. This is indeed very encouraging and suggests that improvements in the vector delivery system may improve the clinical response rate and overall survival.

IV. HERPES SIMPLEX VIRUS VECTORS EXPRESSING ENDOGENOUS HSV-TK

Because one limitation of the currently available replication-defective viral vectors is tumor penetration, several groups have sought to alleviate this problem by using replication-competent vectors that might achieve greater tumor penetration by selectively replicating within the tumor. This approach has been most thoroughly developed as a GDEPT using the G207 conditionally replicating herpes simplex virus mutant [40]. This mutant form of HSV lacks functional copies of the γ-34.5 gene, as well as lacking expression of the viral ribonucleotide reductase enzyme. A recent study by Martuza and colleagues [40] demonstrated that up to 3×10^9 PFU could be safely innoculated into high-

grade gliomas without serious untoward events. A total of 21 patients were treated with doses of vector ranging from 106 PFU to 3×10^9 PFU without reaching dose-limiting toxicity [40,41]. Necrotizing encephalitis, characteristic of the wild-type HSV-1 virus was not observed. This study was able to demonstrate antitumor activity based on radiographic findings, although viral replication was not directly demonstrated. Two patients (one with an anaplastic astrocytoma and one with a glioblastoma multiforme) survived more than 1 year. Although the technique was not part of this clinical study, prior experimental models have demonstrated that the effects of G207 can be improved by the combined administration of ganciclovir [42]. The HSV-1-derived vectors, whether replication competent or not, can be engineered to carry multiple therapeutic genes, potentially allowing for multimodality GDEPT [43].

V. PROMISING PRECLINICAL STUDIES

These retroviral and adenoviral clinical studies of GDEPT have established that this approach is safe with risks that are not substantially different from those encountered in the current standards of care for patients with high-grade gliomas. There remains the problem of efficient tumor transduction given the limited vector distribution following standard intracerebral injection techniques. Transvascular delivery of these agents with disruption of the blood–brain barrier holds promise in providing for greater distribution of the vector throughout the tumor [44]. Pharmacologic measures to enhance the HSV-TK approach include the administration of lovastatin to enhance gap junction expression within tumors and thereby enhance the bystander effect [45] or the use of newer analogs of HSV-TK enzyme to enhance catalytic activity in converting GCV to the cytotoxic phosphorylated form [46–48]. In addition, newer prodrugs such as penciclovir may have performance superior to that of GCV due to better bioavailability in the CNS or more rapid conversion to active drug.

Other prodrug activating systems are being studied in an effort to improve on the HSV-TK experience. One such example is the cytosine deaminase/5-fluorocytosine (CD/5-FC) combination, based on the ability of *Escherichia coli* or yeast CD to deaminate 5-FC to produce 5-fluorouracil (5-FU). 5-FC is an orally administered antifungal agent with excellent biodistribution and biosafety properties. An adenoviral vector has been used to deliver the CD to 9L glioma tumors *in vivo* with antitumor effect [49]. The bystander effect may be a result of diffusion of the cytotoxic 5-FU through the tumor without the need for gap-junction-mediated transfer (as is the case with phosphorylated GCV) and thus may be more efficient. HSV-TK/GCV and CD/5-FC have been combined to achieve synergistic antitumor effects [50] and may be further augmented by the concomitant use of radiation therapy

[51]. HSV-TK/GCV and CD/5-FC have been combined with other prodrug activating systems to create an enhanced antitumor effect. CD and uracil phosphoribosyl transferase genes have been transduced into rat glioma tumors, which were then treated with 5-FC [52]. These two enzymes, both involved in the metabolism of 5-FU, demonstrated an enhanced antitumor effect.

Direct measurement of gene expression has been specifically developed to study the distribution and duration of HSV-TK gene expression. This is made possible through the development of fluorinated analogs of GCV that can be imaged using positron emission tomography (PET) [53–55]. Fluorine-18-GCV (or other suitable positron-emitting HSV-TK substrate) is selectively retained in HSV-TK-expressing tumors because phosphorylation of the labeled GCV by thymidine kinase prevents transport of the labeled drug out of the cells. PET images of animal models of gene transfer have demonstrated that HSV-TK activity can be imaged with this technique [56–58]. This type of real-time, noninvasive assessment of gene transfer will undoubtedly speed the development of this overall strategy.

References

1. Burger, P. C., and Kleihues, P. (1989). Cytologic composition of the untreated glioblastoma multiforme: a postmortem study of eighteen cases. *Cancer* **63**, 2014–2023.
2. Hochberg, F. H., and Pruitt, A. (1980). Assumptions in the radiotherapy of glioblastoma. *Neurolog* **30**, 907–911.
3. Black, P. M. (1991). Brain tumors (first of two parts). *N. Eng. J. Med.* **324**, 1471–1476.
4. Black, P. M. (1991). Brain tumors (second of two parts). *N. Eng. J. Med.* **324**, 1555–1564.
5. Cai, Y., Wu, M. H., Xu-Welliver, M. *et al.* (2000). Effect of O^6-benzylguanine on alkylating agent-induced toxicity and mutagenicity in Chinese hamster ovary cells expressing wild-type and mutant O^6-alkylguanine-DNA alkyltransferases. *Cancer Res.* **60**, 5464–5469.
6. Esteller, M., Garcia-Foncillas, J., Andion, E. *et al.* (2000). Inactivation of the DNA-repair gene MGMT and the clinical response of gliomas to alkylating agents. *N. Engl. J. Med.* **343(19)**, 1350–1354.
7. Kroll, R. A., and Neuwelt, E. A. (1998). Outwitting the blood–brain barrier for therapeutic purposes: osmotic opening and other means. *Neurosurgery* **42**, 1083–1100.
8. DeAngelis, L. M. (2001). Brain tumors. *N. Engl. J. Med.* **344(2)**, 114–123.
9. Brem, H., Piantadosi, S., Burger, P. C. *et al.* (1995). Placebo-controlled trial of safety and efficacy of intraoperative controlled delivery by biodegradable polymers of chemotherapy for recurrent gliomas. The Polymer-Brain Tumor Treatment Group. *Lancet* **345**, 1008–1012.
10. McAllister, L. D., Doolittle, N. D., Guastadisegni, P. E. *et al.* (2000). Cognitive outcomes and long-term follow-up results after enhanced chemotherapy delivery for primary central nervous system lymphoma. *Neurourgery* **46(1)**, 51–61.
11. Hayden, F. (1995). Animicrobial agents, antiviral agents, in *The Pharmacological Basis of Therapeutics* (H. J. Goodman *et al.*, eds.), pp. 1200–1202. McGraw-Hill, New York.
12. Bi, W. L., Parysek, W. M., Warnick, R., Stambrook, P. J. (1993). In vitro evidence that metabolic cooperation is responsible for the bystander effect observed with HSV tk retroviral gene therapy. *Hum. Gene Ther.* **4(6)**, 725–731.
13. Alavi, J. B., and Eck, S. L. (1998). Gene therapy of malignant gliomas, in *Hematology/Oncology Clinics of Norh America: Gene Therapy* (S. L. Eck, ed.), pp. 617–630. W.B. Saunders, Philadelphia.
14. Miyatake, S., Martuza, R. L., and Rabkin, S. D. (1997). Defective herpes simplex virus vectors expressing thymidine kinase for the treatment of malignant glioma. *Cancer Gene Ther.* **4(4)**, 222–228.
15. Short, M. P., Choi, B. C., Lee, J. K. *et al.* (1990). Gene delivery to glioma cells in rat brain by grafting of a retrovirus packaging cell line. *J. Neurosci. Res.* **27(3)**, 427–439.
16. Berenstein, M., Adris, S., Ledda, F. *et al.* (1999). Different efficacy of in vivo herpes simplex virus thymidine kinase gene transduction and ganciclovir treatment on the inhibition of tumor growth of murine and human melanoma cells and rat glioblastoma cells. *Cancer Gene Ther.* **6(4)**, 358–366.
17. Moolten, F., and Wells, J. M. (1990). Curability of tumors bearing herpes thymidine kinase genes transferred by retroviral vectors. *J. Natl. Cancer Inst.* **82**, 297–300.
18. Ram, Z., Culver, K. W., Oshiro, E. M. *et al.* (1997). Therapy of malignant brain tumors by intratumoral implantation of retroviral vector-producing cells. *Nat. Med.* **3(12)**, 1354–1361.
19. Elshami, A. A., Saavedra, A., Zhang, H. *et al.* (1996). Gap junctions play a role in the 'bystander effect' of the herpes simplex virus thymidine kinase/ganciclovir system in vitro. *Gene Ther.* **3(1)**, 85–92.
20. Barba, D., Hardin, J., Sadelain, M. *et al.* (1994). Development of anti-tumor immunity following thymidine kinase-mediated killing of experimental brain tumors. *Proc. Natl. Acad. Sci. USA* **91**, 4348–4352.
21. Smith, J. G., Raper, S. E., Wheeldon, E. *et al.* (1997). Intracranial administration of adenovirus expressing HSVTK in combination with ganciclovir produces a self-limited, dose-dependent inflammatory response. *Hum. Gene Ther.* **8(8)**, 943–954.
22. Goodman, J. C., Trask, T. W., Chen, S.-H. *et al.* (1996). Adenoviral-mediated thymidine kinase gene transfer into the primate brain followed by systemic ganciclovir: pathologic, radiologic, and molecular studies. *Hum. Gene Ther.* **7**, 1241–1250.
23. Ram, Z., Culver, K. W., Oshiro, E. M. *et al.* (1997). Therapy of malignant brain tumors by intratumoral implantation of retroviral vector-producing cells. *Nat. Med.* **3(12)**, 1354–1361.
24. Donahue, R. E., Kessler, S. W., Bodine, D. *et al.* (1992). Helper virus induced T cell lymphoma in nonhuman primates after retroviral mediated gene transfer. *J. Exp. Med.* **176(4)**, 1125–1135.
25. Izquierdo, M., Martin, V., de Felipe, P. *et al.* (1996). Human malignant brain tumor response to herpes simplex thymidine kinase (HSVtk)/ganciclovir gene therapy. *Gene Ther.* **3(6)**, 491–495.
26. Izquierdo, M., Cortes, M. L., Martin, V. *et al.* (1997). Gene therapy in brain tumours: implications of the size of glioblastoma on its curability. *Acta Neurochirurgica Suppl.* **68**, 111–117.
27. Klatzmann, D., Valery, C. A., Bensimon, G. *et al.* (1998). A phase I/II study of herpes simplex virus type 1 thymidine kinase "suicide" gene therapy for recurrent glioblastoma. Study Group on Gene Therapy for Glioblastoma. *Hum. Gene Ther.* **9(17)**, 2595–2604.
28. Shand, N., Weber, F., Mariani, L. *et al.* (1999). A phase I–II clinical trial of gene therapy for recurrent glioblastoma multiforme by tumor transduction with the herpes simplex thymidine kinase gene followed by ganciclovir. GLI328 European–Canadian Study Group. *Hum Gene Ther.* **10(14)**, 2325–2335.
29. Long, Z., Li, L. P., Grooms, T. *et al.* (1998). Biosafety monitoring of patients receiving intracerebral injections of murine retroviral vector producer cells. *Hum Gene Ther.* **9(8)**, 1165–1172.
30. Long, Z., Lu, P., Grooms, T. *et al.* (1999). Molecular evaluation of biopsy and autopsy specimens from patients receiving in vivo retroviral gene therapy. *Hum. Gene Ther.* **10(5)**, 733–740.

31. Deliganis, A. V., Baxter, A. B., Berger, M. S. *et al.* (1997). Serial MR in gene therapy for recurrent glioblastoma: initial experience and work in progress. *Am. J. Neuroradiol.* **18(8)**, 1401–1406.

32. Rainov, N. G. (2000). A phase III clinical evaluation of herpes simplex type 1 thymidine kinase and ganciclovir gene therapy as an adjuvant to surgical resection and radiation in adults with previously untreated glioblastoma multiforme. *Hum. Gene Ther.* **11**, 2389–2401.

33. Palu, G., Cavaggioni, A., Calvi, P. *et al.* (1999). Gene therapy of glioblastoma multiforme via combined expression of suicide and cytokine genes: a pilot study in humans. *Gene Ther.* **6**, 330–337.

34. Alavi, J. B., Judy, K., Alavi, A. *et al.* (1998). Phase I trial of gene therapy in primary brain tumors. *Proc. Am. Soc. Clini. Oncol.* **17**, 379a.

35. Trask, T. W., Trask, R. P., Aguilar-Cordova, E. *et al.* (2000). Phase I study of adenoviral delivery of the HSV-tk gene and ganciclovir administration in patients with recurrent malignant brain tumors. *Mol. Ther.* **1(2)**, 195–203.

36. Wivel, N. A., and Wilson, J. M. (1998). Methods of gene delivery, in *Gene Therapy* (S.L. Eck, ed.), pp. 483–502. Saunders, Philadelphia.

37. Eck, S. L., Alavi, J. B., Alavi, A. *et al.* (1996). Clinical protocol: treatment of advanced CNS malignancies with the recombinant adenovirus H5.010RSVTK: a phase I trial. *Hum. Gene Ther.* **7**, 1465–1482.

38. Puumalainen, A. M., Vapalahti, M., Agrawal, R. S. *et al.* (1998). Beta-galactosidase gene transfer to human malignant glioma in vivo using replication-deficient retroviruses and adenoviruses. *Hum Gene Ther.* **9(12)**, 1769–1774.

39. Subach, B. R., Witham, T. F., Kondziolka, D. *et al.* (1999). Morbidity and survival after 1, 3-*bis*(2-chloroethyl)-1-nitrosourea wafer implantation for recurrent glioblastoma: a retrospective case-matched cohort series. *Neurosurgery* **45(1)**, 17–22.

40. Markert, J. M., Medlock, M. D., Rabkin, S. D. *et al.* (2000). Conditionally replicating herpes simplex virus mutant, G207, for the treatment of malignant glioma: results of a phase I trial. *Gene Ther.* **7(10)**, 867–874.

41. Todo, T., Feigenbaum, F., Rabkin, S. D. *et al.* (2000). Viral shedding and biodistribution of G207, a multimutated, conditionally replicating herpes simplex virus type 1, after intracerebral inoculation in aotus. *Mol. Ther.* **2(6)**, 588–595.

42. Todo, T., Rabkin, S. D., and Martuza, R. L. (2000). Evaluation of ganciclovir-mediated enhancement of the antitumoral effect in oncolytic, multimutated herpes simplex virus type 1 (G207) therapy of brain tumors. *Cancer Gene Ther.* **7(6)**, 939–946.

43. Glorioso, J. C., DeLuca, N. A., and Fink, D. J. (1995). Development and application of herpes simplex virus vectors for human gene therapy. *Annu. Rev. Microbiol.* **49**, 675–710.

44. Nilaver, G., Muldoon, L. L., Kroll, R. A. *et al.* (1995). Delivery of herpesvirus and adenovirus to nude rat intracerebral tumors after osmotic blood-brain barrier disruption. *Proc. Natl. Acad. Sci. USA* **92(21)**, 9829–9833.

45. Touraine, R. L., Vahanian, N., Ramsey, W. J. *et al.* (1998). Enhancement of the herpes simplex virus thymidine kinase/ganciclovir bystander effect and its antitumor efficacy in vivo by pharmacologic manipulation of gap junctions. *Hum. Gene Ther.* **9(16)**, 2385–2391.

46. Kokoris, M. S., Sabo, P., Adman, E. T. *et al.* (1999). Enhancement of tumor ablation by a selected HSV-1 thymidine kinase mutant. *Gene Ther.* **6(8)**, 1415–1426.

47. Qiao, J., Black, M. E., and Caruso, M. (2000). Enhanced ganciclovir killing and bystander effect of human tumor cells transduced with a retroviral vector carrying a herpes simplex virus thymidine kinase gene mutant. *Hum. Gene Ther.* **11(11)**, 1569–1576.

48. Loubiere, L., Tiraby, M., Cazaux, C. *et al.* (1999). The equine herpes virus 4 thymidine kinase is a better suicide gene than the human herpes virus 1 thymidine kinase. *Gene Ther.* **6(9)**, 1638–1642.

49. Dong, Y., Wen, P., Manome, Y. *et al.* (1996). In vivo replication-deficient adenovirus vector-mediated transduction of the cytosine deaminase gene sensitizes glioma cells to 5-fluorocytosine. *Hum. Gene Ther.* **7(6)**, 713–720.

50. Aghi, M., Kramm, C. M., Chou, T. C. *et al.* (1998). Synergistic anticancer effects of ganciclovir/thymidine kinase and 5-fluorocytosine/cytosine deaminase gene therapies. *J. Natl. Cancer Inst.* **90(5)**, 370–380.

51. Rogulski, K. R., Wing, M. S., Paielli, D. L. *et al.* (2000). Double suicide gene therapy augments the antitumor activity of a replication-competent lytic adenovirus through enhanced cytotoxicity and radiosensitization. *Hum. Gene Ther.* **11(1)**, 67–76.

52. Adachi, Y., Tamiya, T., Ichikawa, T. *et al.* (2000). Experimental gene therapy for brain tumors using adenovirus-mediated transfer of cytosine deaminase gene and uracil phosphoribosyltransferase gene with 5-fluorocytosine. *Hum. Gene Ther.* **11(1)**, 77–89.

53. Tjuvajev, J. G., Chen, S. H., Joshi, A. *et al.* (1999). Imaging adenoviral-mediated herpes virus thymidine kinase gene transfer and expression in vivo. *Cancer Res.* **59(20)**, 5186–5193.

54. Tjuvajev, J. G., Avril, N., Oku, T. *et al.* (1998). Imaging herpes virus thymidine kinase gene transfer and expression by positron emission tomography. *Cancer Res.* **58(19)**, 4333–4341.

55. Hustinx, R., Shiue, C.-Y., Alavi, A. *et al.* (2000). Monitoring herpes simplex virus thymidine kinase gene transfer to tumors with ^{18}F-FHPG and PET. *Eur. J. Nuclear Med.* **28(1)**, 5–12.

56. Alauddin, M. M., Shahinian, A., Kundu, R. K. *et al.* (1999). Evaluation of 9-[(3-^{18}F-fluoro-1-hydroxy-2-propoxy)methyl]guanine ([^{18}F]-FHPG) in vitro and in vivo as a probe for PET imaging of gene incorporation and expression in tumors. *Nucl. Med. Biol.* **26(4)**, 371–376.

57. Gambhir, S. S., Barrio, J. R., Phelps, M. E. *et al.* (1999). Imaging adenoviral-directed reporter gene expression in living animals with positron emission tomography. *Proc. Natl. Acad. Sci. USA* **96(5)**, 2333–2338.

58. Gambhir, S. S., Bauer, E., Black, M. E. *et al.* (2000). A mutant herpes simplex virus type 1 thymidine kinase reporter gene shows improved sensitivity for imaging reporter gene expression with positron emission tomography. *Proc. Natl. Acad. Sci. USA* **97(6)**, 2785–2790.

59. Rainov, N. G., Kramm, C. M., Banning, U. *et al.* (2000). Immune response induced by retrovirus-mediated HSV-*tk*/GCV pharmacogene therapy in patients with glioblastoma multiforme. *Gene Ther.* **7**, 1853–1858.

60. Eck, S. L., Alavi, J. B., Alavi, A. *et al.* (1996). Treatment of advanced CNS malignancies with the recombinant adenovirus H5.010RSVTK: a phase I trial. *Hum. Gene Ther.* **7**, 1465–1482.

34

CASE STUDY OF COMBINED GENE AND RADIATION THERAPY AS AN APPROACH IN THE TREATMENT OF CANCER

BIN S. TEH
Department of Radiology
Baylor College of Medicine
Houston, Texas 77030

MARIA T. VLACHAKI
Department of Radiology and
Veterans Affairs Medical Center
Baylor College of Medicine
Houston, Texas 77030

LAURA K. AGUILAR
Harvard Gene Therapy Initiative
Harvard Medical School
Boston, Massachusetts 02115

BRIAN MILES
Department of Urology
Baylor College of Medicine
Houston, Texas 77030

GUSTAVO AYALA
Department of Pathology
Baylor College of Medicine
Houston, Texas 77030

DOV KADMON
Department of Urology
Baylor College of Medicine
Houston, Texas 77030

THOMAS WHEELER
Department of Pathology
Baylor College of Medicine
Houston, Texas 77030

TIMOTHY C. THOMPSON
Department of Urology
Baylor College of Medicine
Houston, Texas 77030

E. BRIAN BUTLER
Department of Radiology
Baylor College of Medicine
Houston, Texas 77030

ESTUARDO AGUILAR-CORDOVA
Department of Radiology
Baylor College of Medicine
Houston, Texas 77030
and
Harvard Gene Therapy Initiative
Harvard Medical School
Boston, Massachusetts 02115

I. INTRODUCTION

Combined gene therapy (herpes simplex virus thymidine–kinase [HSV–TK] + antiherpetic prodrug) and radiation therapy is a novel approach in the armament against cancer. This radio/gene therapy combination may create a new spatial cooperation whereby two local treatment modalities, radiotherapy and *in situ* gene therapy, can lead to improved local and systemic control. Preclinical data have demonstrated the enhanced antitumor effects of this combined approach in local tumor control, prolongation of survival, and systemic control. This combined radio/gene therapy has now entered the clinical phase in prostate cancer. The current approach is to add a novel therapy (gene therapy) to the standard-of-care therapy (radiotherapy). These treatment modalities have very distinct spectra of toxicity profiles. The goal of this combined approach is to enhance the cancer cure without an increase in treatment-related toxicity. Early clinical results have shown the safety of this approach. Combined radio/gene therapy carries promise as a novel approach in oncology.

II. BACKGROUND OF THE FIELD

Radiation therapy, surgery, and chemotherapy are considered as standard treatments in oncology. Advances in each of these modalities have had a positive impact on the cure of cancer and quality of life in cancer patients. Over the last decade, significant progress has taken place in the field of radiation oncology, from technology to radiobiology and radiophysics. Radiation dose plays an important role in tumor control probability as well as normal tissue damage. Figure 1 shows that as the radiation dose escalates, the tumor control probability improves but the normal tissue damage also increases. The advances in radiotherapeutic technology have enabled the improvement of therapeutic index in the treatment of cancer. More conformal radiation therapy has allowed dose escalation leading to the improvement of a cancer cure. However, normal tissue damage is always a concern with high radiation dose escalation. A good example is the use of radiation therapy to treat prostate cancer. Radiation dose response relationship with tumor control has been

demonstrated [1–3]. The delivery of high-dose radiation is however limited by the proximity of critical normal structures. Higher radiation dose is associated with higher incidence of treatment-related toxicity [4,5]. Even with highly conformal radiation therapy, high-dose escalation to the prostate is difficult without causing rectal damage [6,7]. Various methods have been used to overcome this problem such as a rectal shield after reaching radiation tolerance of the rectum [8]. However, this method defeats the purpose of dose escalation as the majority (74%) of prostate carcinoma foci were in the peripheral zone which is in close proximity with the rectum [9]. By shielding the rectum, the peripheral zones will be shielded, especially in view of daily prostate motion. Without high-dose delivery to the peripheral zones, the tumor control is not going to be improved.

Our approach has been to combine radiotherapy (conventional) with *in situ* gene therapy (novel), two different therapies with different spectra of toxicities profile, to achieve better cure without increased side effects.

There are various potential benefits of combining radiation therapy and gene therapy. Gene therapy may cause radiosensitization or additive cell killing [10–13]. The mechanisms of the enhanced antitumor effects elicited by this combined approach may be:

1. Ionizing radiation may improve transfection/transduction efficiency and transgene integration [14–16].
2. Gene therapy and radiation therapy normally target different parts of the cell cycle, i.e., HSV-TK gene

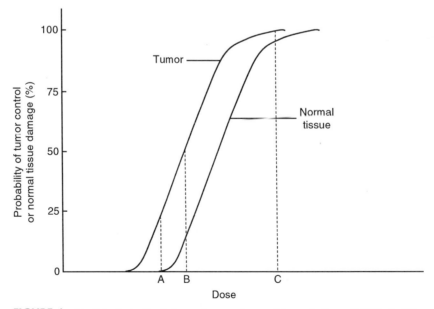

FIGURE 1   Parallel effects of radiation on tumor and normal tissue. As the probability for tumor control increases, the probability of collateral damage also increases. A is the hypothetical point at which there is 25% tumor control; B, 50%; and C, 100%. Note the corresponding increase in normal cell damage with increasing tumor control. The desired consequence of a radiation adjuvant is to shift the tumor curve to the left.

therapy requires "S" phase of the cell cycle while the "M" and "G$_2$" phases are most radiosensitive [17].

3. Phosphorylated prodrugs are incorporated into the newly synthesized DNA, causing termination of DNA synthesis and thus cell death. This may increase the DNA susceptibility to radiation damage.

4. By incorporation into the DNA, phosphorylated prodrugs may interfere with repair of radiation induced DNA damage.

5. Radiation may enhance the bystander effect of gene therapy. This may be due to the release of debris from the radiation-damaged cells and efficient uptake and presentation of tumor antigens by immune effector cells attracting immunocytes and mediating an antitumor response.

III. RECENT ADVANCES IN HERPES SIMPLEX VIRUS–THYMIDINE KINASE SUICIDE GENE THERAPY

A. Preclinical Trials

Gene therapy is one of the novel approaches in the armament against cancer. Three different prototypes include: (1) transfer or insertion of gene(s) that results in activation of a prodrug to produce selective cytotoxicity [18,19]; (2) delivery of gene(s) to stimulate the immune system [20]; and (3) replacement of defective or inactivated tumor suppression gene(s) [21]. The toxicity and efficacy of adenovirus-mediated HSV-TK + ganciclovir (prodrug) therapy have been assessed using many *in vitro* and *in vivo* cancer models. We have demonstrated that adenoviral vectors could transduce prostate cancer cells and that HSV-TK + ganciclovir was highly effective against mouse and human prostate cancer cell lines *in vitro* [22]. In a mouse *in vivo* model, RM-1 prostate cancer cells were injected subcutaneously. Intratumoral injection of Ad-HSV-TK was performed after 6 days. This was then followed by ganciclovir injection intraperitoneally twice daily for 6 days. Using this *in vivo* model, we set our goals to evaluate the effect of Ad-HSV-TK + ganciclovir gene therapy on:

1. Local control and survival
2. Spontaneous or preexisting metastatic disease
3. Induced metastatic disease

Ad-HSV-TK + ganciclovir gene therapy demonstrated significant growth suppression of the local tumor and survival advantage when compared to controls [22]. Increased areas of necrosis, increased numbers of apoptotic bodies, and extensive immunocytic infiltration relative to controls were noted. In contrast to this type of *in situ* delivery therapy, systemic gene therapy for prostate cancer may not be as effective [23,24].

When the mice were sacrificed, the lungs were examined carefully for any metastatic disease. Ad-HSV-TK + ganciclovir gene therapy significantly decreased the number of pulmonary metastatic lesions when compared to controls. To further confirm the antimetastatic effects of gene therapy, an induced metastatic prostate cancer model injecting RM-1 cells into the tail vein in addition to the subcutaneous injection was used. Intratumoral injection of Ad-HSV-TK and intraperitoneal injection of ganciclovir were the same as above. Marked suppression of the induced lung metastatic lesions was achieved by the *in situ* gene therapy [25]. The most likely explanation of the antimetastatic activity was the induction of an antitumor immunity by Ad-HSV-TK + ganciclovir gene therapy. Other possible explanations include the release of an antiangiogenic substance, targeting of angiogenesis, and killing of premetastatic cells.

Prostate cancer cells have been well known to be sensitive to androgen ablation. Studies were performed to investigate the effects of HSV-TK + ganciclovir gene therapy and androgen ablation. Mice were castrated to achieve androgen ablation. The local tumor-control effects of gene therapy were enhanced by concomitant androgen ablation. However, the antimetastatic effects of the combined gene and androgen ablation therapies were not better to the effects of gene therapy alone [26].

These early efficacy studies were followed by extensive rodent toxicity and distribution studies. Little systemic spread and minimal local spread of the adenoviral vector within the genitourinary system were found in our mouse model [27].

B. Clinical Trials

The promising results from our preclinical studies led to phase 1 clinical protocols to evaluate potential toxicity from this type of gene therapy in patients with locally recurrent prostate cancer after initial radiotherapy. The trial consisted of escalating doses of direct intraprostatic injection of an adenoviral vector containing HSV-TK gene under transrectal ultrasound (TRUS) guidance followed by intravenous administration of ganciclovir. A total of 18 patients were enrolled. They were treated with four escalating doses from 1×10^8 to 1×10^{11} infectious units (IU) (2×10^{10} to 2×10^{12} vector particles). Treatment-related toxicity was graded according to the *Cancer Therapy Evaluation Program (CTEP) Common Toxicity* published by the National Cancer Institute. Four patients experienced mild toxicity of grades 1–2. One patient at the highest dose level developed spontaneously reversible grade 4 thrombocytopenia and grade 3 hepatotoxicity. Cultures of nasal mucus, blood, and urine from all the patients were consistently negative for adenoviral growth. Three patients achieved an objective response, one each at the three highest dose levels, documented by a fall in serum prostate-specific antigen (PSA) levels by 50% or more sustained for

6 weeks to 1 year. The trial demonstrated the safety of HSV-TK + ganciclovir gene therapy in human prostate cancer with encouraging efficacy indications [28].

IV. COMBINED HERPES SIMPLEX VIRUS–THYMIDINE KINASE SUICIDE GENE THERAPY AND RADIOTHERAPY

A. Preclinical Study

The prostate cancer cell line RM-1 was used for subcutaneous injection into syngeneic C57BL/6 mice [29,30]. Tumors measuring 50–60 mm^3 were established in the hind flank rather than orthotopically to avoid complications of scatter radiation damage to viscera. Ad-HSV-TK was then injected intratumorally, followed by ganciclovir intraperitoneally twice daily for 6 days. Seventy-two hours after gene vector injection, the tumors were irradiated at 5 Gy using an orthovoltage X-ray generator. To establish lung nodules, RM-1 cells were also injected into the tail vein of the mice. The endpoints of these studies were:

1. Local tumor control
2. Survival
3. Local immunological response
4. Systemic effect on lung metastases

Figures 2 and 3 show that both monotherapy (either gene therapy of radiotherapy alone) modalities decreased the tumor growth compared to controls. The combined gene therapy and radiotherapy group showed the most significant inhibition of tumor growth compared to controls and single treatment groups.

Similarly, as shown in Figure 4, the combined treatment modality group elicited significant prolongation of survival compared to the control groups and the monotherapy groups. Iimprovement of survival may be translated from better local tumor control in the combined gene therapy and radiotherapy groups.

In evaluating local immunological response, lymphocytic infiltrates were noted to be most pronounced in the combined gene therapy and radiotherapy group. Infiltrating lymphocytes were accentuated at the tumor/native tissue interface and percolated throughout the tumor, focally forming lymphocytic clusters. The combined therapy group had a significantly increased frequency of infiltrating CD4$^+$ T cells as compared to controls and monotherapy groups.

In the lung metastases model, gene therapy alone decreased the number of lung nodules by 37% but the addition of radiotherapy to gene therapy further reduced the metastatic lesions by 50% (Fig. 5). Radiation therapy alone had no effect on lung metastases. Figure 6 illustrates the autopsy specimens showing that the lungs from animals treated with combined gene therapy and radiotherapy had the least number of lung metastatic nodules, followed by the gene therapy alone group. The radiotherapy alone group had the same number of lung metastases as the controls. It is not surprising to observe no systemic effect in the radiotherapy-alone group as radiotherapy is a local therapy. We also know from the experiments, detailed above, that in-situ gene therapy had some systemic effects. However, it was surprising to find that when a local therapy, i.e., radiotherapy was added to in-situ Ad-HSV-TK + ganciclovir gene therapy, not only local control but also systemic effects were significantly improved. This may create a new form of spatial cooperation—two local therapies interacting to enhance both local and systemic control.

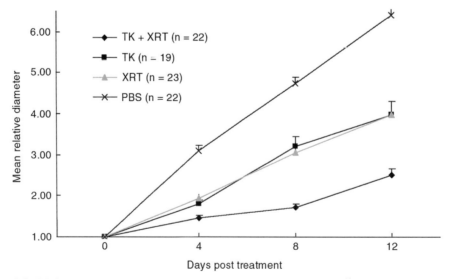

FIGURE 2   Tumor growth analysis after single and combination therapy. 3×10^{10} AdV-*tk* or AdV-β-gal v.p. were injected intratumoraly. Irradiated tumors received a single dose of 5 Gy.

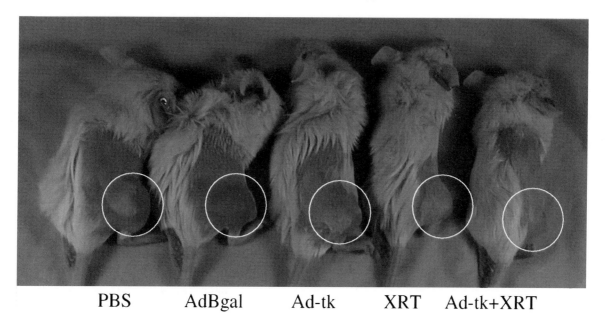

PBS   AdBgal   Ad-tk   XRT   Ad-tk+XRT

FIGURE 3 Typical gross tumor growth observed at 10 days posttreatment. Animals received direct intratumor injection of saline (PBS), a marker gene adenoviral vector (AdV-β-gal), an HSV–thymidine-kinase-expressing vector (AdV-*tk*), and external beam irradiation (XRT).

B. Current Clinical Trial

From these preclinical and clinical this phase I/II trial combining a standard-of-care therapy (radiation therapy with or without hormonal therapy) with a novel therapy (Ad-HSV-TK) was initiated. The hypothesis is that combined radio/gene therapy will lead to better efficacy without additional toxicity to the patients. This hypothesis is based on: (1) preclinical trials showing the added benefits of the com-

bined approach compared to monotherapy, (2) a phase I clinical trial demonstrating minimal toxicity of this type of gene therapy, and (3) each therapy has a different toxicity profile.

1. Patient Selection

The patient eligibility criteria for the study included histologically proven adenocarcinoma of prostate; no prior surgical, hormonal, or radiotherapy prostate treatment; no

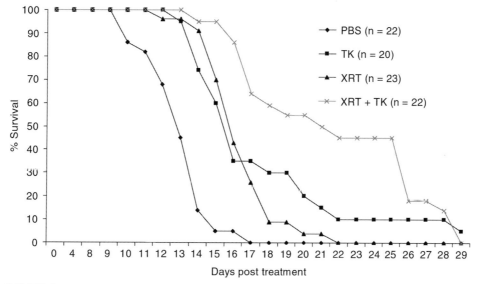

FIGURE 4 Survival curve analysis after single and combination therapy. 3×10^{10} AdV-*tk* or AdV-β-gal v.p. were injected intratumoraly followed 24 hr later by external beam irradiation and 6 days of ganciclovir at a dose of 20 mg/kg intraperitoneally. Irradiated tumors received a single dose of 5 Gy.

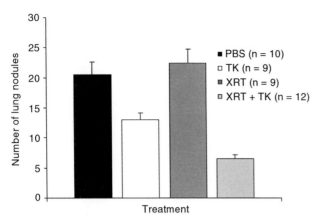

FIGURE 5 Effect of single or combined local treatment of subcutaneous tumors on lung metastases. Lung nodules developed from tail-vein-injected tumor cells. The primary subcutaneous tumors were treated as described in Fig. 4.

evidence of metastatic disease (as observed by chest X-ray [CXR], abdominal and pelvic computerized tomography, and bone scan) or other malignancy (except squamous or basal cell skin cancers). Any area causing suspicion of metastatic disease on bone scan was valuated by appropriate plain radio-

graphs and/or magnetic resonance imaging (MRI) studies of the area. The patient must have had a PSA within 3 months of entry; must have adequate baseline organ function, including serum creatinine <1.5 mg/dL, total bilirubin <2.5 mg/dL, and normal liver enzymes including ALT, AST, GGT, and AP < 2X normal; platelets >100,000/mL, ANC > 1500/mL, hemoglobin > 10 g/dL, as well as normal coagulation profile including prothrombin time (PT) and partial thromboplastin time (PTT). Exclusion criteria also included the prolonged use of corticosteroids or any immunosuppressive drugs, HIV positivity, liver cirrhosis, and acute infections, including viral, bacterial and fungal infections requiring therapy.

2. Treatment Arms

There are three arms to this study: Arm A includes patients with PSA < 10, Gleason's score < 7 and clinical stage T1–T2a. Arm B patients should have one of the following characteristics: PSA ≥ 10, Gleason's score ≥ 7 or Clinical stage T2b–T3. Arm C patients should have pathologically proven regional (pelvic) lymph node involvement of prostate cancer. Table 1 shows the treatment schema in each arm.

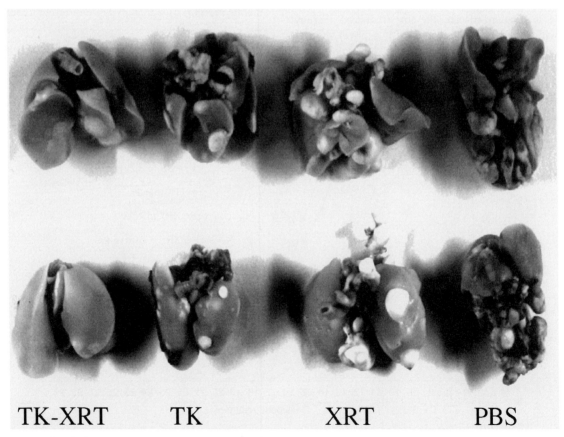

FIGURE 6 Gross morphological examples of lungs from control and treated animals. Lung nodules developed from tail-vein-injected tumor cells. The primary subcutaneous tumors were treated as described in Fig. 4.

TABLE 1   Patient Characteristics (n = 30)

Age (years)	
Median	68
Range	39–85
Race	
Caucasian	26 (86.7%)
African American	1 (3.3%)
Latin American	3 (10.0%)
Arm	
A	13 (43.3%)
B	14 (46.7%)
C	3 (10.0%)
PSA (ng/mL)	
Median	6.5
Range	2.5–335
Gleason score	
<7	15 (50.0%)
7	11 (36.7%)
>7	4 (13.3%)
T-stage	
1c	10 (33.3%)
2a	16 (53.3%)
2b	2 (6.7%)
3a	2 (6.7%)

3. Gene Vector

The vector used for the *in situ* gene therapy is an adenovirus of serotype 5 that contains the herpes simplex virus–thymidine kinase gene and a RSV long terminal repeat (LTR) promoter in the region of the excised E1 wild-type adenoviral genes. The adenoviral vector was constructed as described previously [31]. It was produced at the Baylor College of Medicine gene vector laboratory, in accordance with good manufacturing practice (21 CFR210 and 211). The vector was characterized for purity and potency for clinical use. It was found to be free of adventitious contaminants, including replication-competent adenoviruses at up to 1×10^{10} viral particles. Once produced, it was stored at $-80°C$.

4. Delivery of Ad. HSV-TK

Prior to injection, all patients received a broad-spectrum oral antibiotic such as ciprofloxacin 500 mg b.i.d. beginning the day prior to injection. This continued for 3 days. Four intraprostatic injections, two at each lobe, were performed under transrectal ultrasound guidance. On day 0, a total of 2 mL of Adv-HSV-TK was injected using a 20-gauge needle. Based on the toxicity results from the phase 1 dose escalation trial [28], a total of 5×10^{11} viral particles were injected. To prevent possible outflow obstruction associated with the prostatic injection, a Foley catheter may be placed following the injection. For Arm A patients, the intraprostatic injections were repeated on day 14. For Arm B and C patients, the injections were repeated on days 56 and 70.

5. Radiotherapy

A mean dose of 76 Gy (prescribed dose of 70 Gy in 2 Gy/fraction) is delivered to the prostate utilizing the NOMOS Peacock intensity modulated radiation therapy (IMRT) system (Sewickley, PA). In addition, Arm C patients also received 45 Gy in 1.8 Gy/fraction to the draining pelvic lymphatics. 15 MV photons were used. The techniques have been described in details previously [33,34]. Briefly, the patients were treated prone in a customized vacuum bag fitted into a treatment box for the purpose of immobilization. A rectal balloon was placed each day during treatment to minimize prostate motion. Radiotherapy was initiated 48 hours following the first gene vector injection in Arm A and 48 hours after the second gene vector injection in Arms B and C.

6. Androgen Deprivation

For Arm B (high risk) and Arm C (stage D1) patients, androgen deprivation began concomitantly with the first gene vector injection on day 0. Hormonal therapy consisited of one intramuscular injection of a 4-month (30-mg) leuprolide acetate (Lupron) depot (TAPP Pharmaceuticals), a lutenizing-hormone-releasing hormone (LHRH) agonist, and flutamide (Eulexin) (provided by Schering Oncology), an antiandrogen given p.o. in 125 mg × 2 t.i.d. for 14 days.

7. Patient Evaluation

A complete history, physical examination including digital rectal examination, and complete blood cell count with differential, serum biochemistry, coagulation profile, PSA, and urine analysis were obtained at baseline for all patients. In addition, bone scan, chest X-ray, and CT scans of the abdomen and pelvis were done for Arm B and C patients to rule out any metastatic disease. Patients were monitored weekly throughout treatment by physical examination, recording of treatment-related toxicity, and complete blood cell count with differential and the serum chemistry including renal and liver function tests. Patients were seen 6 weeks following completion of radiotherapy, then every 3–4 months for the first year and 6 months thereafter.

The Cancer Therapy Evaluation Program (CTEP) common toxicity criteria by the National Cancer Institute were used to assess the toxicity related to the gene therapy and hormonal therapy. Radiation Therapy Oncology Group (RTOG) morbidity score [35] were used to evaluate the toxicity related to radiotherapy especially lower gastrointestinal (GI) and genitourinary (GU) systems.

To assess treatment response, PSA testing was performed on weeks 2 and 6 during radiotherapy, 6 weeks following completion of radiotherapy, every 3–4 months for the first year, and 6 months thereafter. Digital rectal examinations were performed at the same time during follow-up. Serial prostate biopsies were performed on day 14 for Arm A patients and on days 56 and 70 for Arm B and C patients during treatment. Further biopsies were scheduled to be performed at the same intervals as the PSA during follow-up visits. Local failure, distant failure, biochemical control, disease-free survival, and overall survival are being assessed.

8. Treatment Cessation Criteria

Any patient with permanent grade 3 toxicity or recurrent grade 4 toxicity as specified in CTEP or RTOG was asked to discontinue the treatment protocol.

9. Patient Characteristics

From September 1999 to October 2000, 30 patients were enrolled in the protocol. Their median age was 68 (range 39–85). The majority of patients were Caucasian (86.7%). The patients were distributed evenly between Arms A and B, 13 and 14 patients respectively. There were three patients in Arm C. Median PSA was 6.5 (range 2.5–335). Twenty-six patients (86%) had PSA of 7 or lower. Twenty-six patients (86.7%) had T1c–T2a disease. Median follow-up was 5.5 months. The patient characteristics are shown in Table 2.

10. Treatment Cessation or Delays

No patient developed any urinary outflow obstruction needing a Foley catheter. Gene vector injection was delayed 1 week in a patient in Arm B because he developed cellulitis requiring antibiotics. No reductions in gene vector viral particles or valacyclovir dosage were performed. There were no other delays or interruptions of treatment due to severe toxicity. No patient has been withdrawn from the study because of toxicity or other reason.

11. Hematologic Toxicity

In the evaluation of hematologic toxicity, there has been no patient with grade 3 or higher anemia, leukopenia, or thrombocytopenia. As shown in Table 3, only two patients devel-

TABLE 2 Hematologic Toxicity (n = 30)

	CTEP grade		
	0	1	2
Anemia	4	26	0
Neutropenia	13	15	2
Thrombocytopenia	14	16	0

TABLE 3 Constitutional Symptoms (n = 30)

	CTEP grade		
	0	1	2
Fever	24	4	2
Rigors/Chills	19	11	0
Fatigue	19	10	1
Weight Loss	30	0	0

oped a grade 2 leukopenia. There has been no associated leukopenic fever and the grade 2 leukopenia resolved spontaneously without any overwhelming infection. Grade 1 anemia, leukopenia, and thrombocytopenia were experienced commonly by the patients but they were transient and mild, not causing any clinical consequences and resolving spontaneously.

12. Constitutional Symptoms

Two patients in Arm A developed grade 2 fever (temperature of 103°F) during the evening after the gene vector injection. One patient (the third patient enrolled) was admitted to the hospital. His fever subsided with acetaminophen and ibuprofen even prior to hospitalization. In the hospital, intravenous antibiotics were given and all the infectious disease cultures were negative. The other patient's temperature responded well to acetaminophen and was not admitted to the hospital. The other four patients had mild transient elevation in temperature, which again responded well to acetaminophen or resolved spontaneously.

Eleven patients (58%) developed flu-like symptoms such as rigors/chills and fatigue (Table 4). However, with the exception of one patient with grade 2 fatigue, the flu-like symptoms were very mild (grade 1). Patients usually experienced the flu-like symptoms in the evening after the gene vector injection. The symptoms did not last long and generally resolved the next morning. There was no documented weight loss in the patient population.

TABLE 4 Hepatic and Renal Toxicity (n = 30)

	CTEP grade			
	0	1	2	3
AST	19	10	1	0
ALT	16	11	2	1
ALP	24	6	0	0
GGT	23	7	0	0
Cr	27	3	0	0

TABLE 5 Genitourinary (GU) and Lower Gastrointestinal (GI) Toxicity (n = 30)

| | RTOG grade | | | |
	0	1	2	3
GU	9	8	12	1
Lower GI	20	6	4	0

TABLE 6 Treatment Schema

Arm A (low risk)	**Gene therapy** (HSV-tk (5 × 10^{11} v.p.) injection on Days 0 and 14, followed by 14 days of valacyclovir). **Radiotherapy** (Starting on Day 2, mean dose 76 Gy in 35 fractions).
Arm B (high risk)	**Gene therapy** (HSV-tk (5 × 10^{11} v.p.) injection on Days 0, 56, and 70, followed by 14 days of valacyclovir). **Radiotherapy** (Starting on day 58, mean dose 76 Gy in 35 fractions). **Hormonal therapy** (Leuprolide 4-month injection on Day 0 and flutamide for 14 days starting on Day 0).
Arm C (stage D$_1$)	Similar to Arm B with additional 45 Gy to pelvic lymphatice.

13. Hepatic and Renal Toxicity

Elevation in liver enzymes was commonly observed as shown in Table 5. Liver enzymes assayed included AST, ALT, ALP, and GGT. A majority of the patients had grade 1 elevation in these enzymes. One patient in Arm B developed a grade 3 elevation in ALT. He also drank moderate amounts of alcohol. His ALT level declined appropriately within 2 days and completely normalized in 2 weeks. Two other patients had grade 2 elevation in ALT while one patient had grade 2 elevation in AST. Despite the abnormality in the liver function tests, there were no documented clinical symptoms/signs or liver dysfunction such as jaundice, pruritus, dark urine, persistent anorexia, right upper quadrant tenderness, asterixis, or hepatic encephalopathy. All the elevations in liver enzymes resolved spontaneously on conservative management without any adjustment in the delivery of gene vector or valacyclovir dosages.

Three patients developed grade 1 elevation in the creatinine level as shown in Table 5. These patients were found to have some urinary irritative symptoms such as urinary frequency, urgency, or nocturia as well as slight dehydratiom. Once these symptoms were managed their creatinine normalized appropriately.

14. Genitourinary and Lower Gastrointestinal Toxicity

Table 6 shows the acute toxicity systems using RTOG acute morbidity scoring criteria. One patient developed hourly urinary frequency (grade 3 acute GU toxicity). This patient was also noted to have gene vector extravasated into the bladder neck region during the injection. His symptoms improved with the help of alpha-adrenoceptor blocker and phenazopyridine. Twelve patients (40.0%) and eight patients (26.7%) developed grade 2 and grade 1 acute GU toxicity respectively.

Similar to the patients treated with IMRT alone [33,34], the acute GI toxicity was of very low incidence and of mild severity. There was no grade 3 or above toxicity. Twenty patients (66.7%) had no GI toxicity at all, with six patients having grade 1 and four patients having grade 2 GI toxicity.

This aims to expand the therapeutic index of radiotherapy by combining in situ gene therapy. Initial experience has demonstrated the safety of this approach. There is no added toxicity to each therapy used alone. Long term follow-up and larger cohort studies are warranted to evaluate long-term toxicity and efficacy.

V. ISSUES REGARDING CLINICAL TRIALS, TRANSLATION INTO CLINICAL USE, PRECLINICAL DEVELOPMENT, EFFICACY, ENDPOINTS, AND GENE EXPRESSION

A replication-deficient adenovirus has been used for the transfer of HSV-TK into prostate cells. Despite the apparent efficacy of this therapeutic regimen in vitro and in mice, injecting a genetically altered virus into humans carries numerous therapeutic concerns, including toxic viremia, viral effects on liver and other organs, viral transfer to other individuals, severe local effects, etc. Another concern was that these side effects may have been worsened by the addition of radiation therapy. The initial experience of this trial demonstrated the safety of this combined radio/gene therapy approach. With the exception of one patient with grade 3 elevation in ALT and another patient with grade 3 acute GU toxicity, no other grade 3 or higher toxicity was observed. Alcohol intake and flutamide use may also have contributed to the rise in ALT. On the other hand, extravasation of the virus-containing fluid outside prostate into the bladder neck and surrounding tissues may have contributed to the grade 3 acute GU toxicity observed after radiation therapy. Overall, there seemed to be no added toxicity in patients receiving combined radio/gene therapy when compared to patients receiving either therapy alone [28,33,34]. However, this was still very early in the follow-up (median of 5.5 months) and a small cohort (30 patients). Longer term follow-up and larger cohorts are warranted to evaluate lone term toxicity.

The incidence of fever was higher than that routinely encountered after TRUS-guided biopsy of the prostate.

Although fever is estimated to occur in 1.7% of patients treated prophylactically with antibiotics before biopsy [36], it developed in six patients (25%) in this trial. The high incidence of fever may be attributed to the introduction of virus or bacteria from the rectum through the prostate and into the bloodstream caused by the pressure exerted during injection. This is in contrast to the mechanism of biopsy, which is based on aspiration or extraction. The hypothesis of viremia- or bacteriemea-induced fever was further evidenced by the coexisting flu-like symptoms of rigors, chills, and fatigue. However, the etiology remains uncertain because all the viral or bacterial cultures have been negative. The other etiology could be due to the host immune system. Currently, all patients receive antibiotics for 3 days starting the day prior to injection (similar to patients undergoing prostate biopsy), as well as administering prophylactic acetaminophen and ibuprofen for 48 hours postviral vector injection. Febrile episode and flu-like symptoms were transient and generally resolved with 24 hours postinjection. No prolonged use of antibiotics was necessary.

Adenovirus mediated HSV-TK prodrug gene therapy is associated with abnormality in liver function and cytopenia [28,32]. The prodrug used may have contributed to abnormal liver function tests, anemia, leukopenia, and thrombocytopenia. In this trial, a majority of the hematologic and hepatic abnormalities were mild (grade 1 and 2) and resolved spontaneously. Three patients in this trial were noted to have grade 1 elevation in creatinine. These patients had irritative GU symptoms such as frequency and nocturia and were found to have some degree of dehydration. Valacyclovir has also been reported to cause elevation in creatinine in 1% of patients. Once their urinary symptoms improved and their hydration status corrected, their creatinine normalized.

Radiation-related side effects, including GU and GI toxicity, were low, similar to those observed in patients receiving only IMRT [33,34]. We previously reported our experience on acute toxicity in prostate cancer patients treated with IMRT utilizing a rectal balloon for prostate immobilization. The favorable toxicity profile could be attributed to a number of factors: (1) IMRT allowing higher dose to the prostate and minimizing the radiation dose to the rectum and bladder, (2) the use of a rectal balloon allowing the rectal mucosa dose reduction because of the air cavity and dose build-up, and (3) the rectal distension by the balloon improving the dosimetry by reducing the rectal volume receiving high-dose radiation. It is very encouraging that the addition of *in situ* gene therapy did not worsen the radiation-induced lower GI and GU toxicity. This paves the way for future use of advanced forms of radiation with this advanced form of biologic therapy. However, it is very important to evaluate the long-term side effects of this combined approach with longer term follow-up.

It is too early to look at the efficacy of this trial, especially in view of prostate cancer treatment, which requires long-term follow-up. We are currently gathering clinical (DRE),

biochemical (PSA), and pathological (biopsy) data. The endpoints of efficacy for cancer gene therapy are currently receiving a lot of attention [37,38]. In our previous phase 1 trial involving salvage gene therapy for patients who have failed initial radiotherapy, we noted that gene therapy caused stabilization of PSA rather than a true PSA nadir achieved in patients treated with radiotherapy alone. Newer serum markers such as caveolin-1 may be more representative of a response to gene therapy. These markers are currently being evaluated. It is also known that the efficacy of this type of gene therapy depends on the "bystander" effects and the host immune system stimulation. Currently, various histological, immunological, and molecular assessments such as apoptosis, necrosis, p53, p21, inflammatory response, local immunological response, cytokine gene expression, and others will be performed on the biopsy specimens. Evaluation of the host immune response will also be carried out. This includes cytokine profiles, such as IL-6, TGF-β, TNF-α, characterization of lymphoid population, proportion of activated T-cells, and functional activities of monocytes and natural killer cells.

VI. POTENTIAL NOVEL USES AND FUTURE DIRECTIONS

The reported phase I/II trial represents one of the first clinical trials using combined radio/gene therapy in the treatment of previously untreated cancer. Early results showed the safety of this approach. Longer term follow-up in larger cohorts are awaited to evaluate the efficacy and potential late toxicity. If this approach shows efficacy, it may be ideal for cancers which have a high propensity for local recurrence and distant metastases.

Future directions include the refinement of various issues regarding gene therapy and the best timing of radio/gene therapy. This type of gene therapy elicits its antitumor effects via direct cytotoxicity, bystander effects, as well as stimulated immulogical responses. More detailed assessment of the mechanisms, especially in humans, is required. Also, the gene vector distribution *in vivo* needs to be further elucidated. As prostate cancer is a multifocal disease and the location of cancer within the prostate cannot be precisely determined, optimal therapy will likely require a uniform distribution of the vector throughout the gland. There is also the physical consideration of the size of the prostate gland which varies a great deal among the patients. We are also planning to utilize the 3-dimensional planning based on reconstructed prostate volume to aid in gene vector delivery and distribution. Initial pathologic volumetric studies from a neoadjuvant gene therapy followed by prostatectomy trial showed that only portions of the tumor show morphologic effects as well as an inverse relationship between the percentage of the affected tumor and the prostate and tumor size [39]. In addition to delivery, the most effective sequencing of radiation therapy and gene therapy requires further investigation in order to achieve the best radiosensitization, maximal cytotoxicity, and optimal new spatial cooperations—a combination of local therapies leading to enhanced local control and systemic effects. We can also explore the combination of radiation therapy with other types of gene therapy using this trial as a baseline for safety.

Acknowledgement

This work was supported by a Specialized Program of Research Excellence (SPORE) grant (CA58204) from the National Cancer Institute, the Methodist Hospital Foundation, the General Clinical Research Center (GCRC), Advantagene, Glaxo-Welcome, and Schering Oncology.

References

1. Hanks, G. E., Martz, K. L., and Diamond, J. J. (1988). The effect of dose on local control of prostate cancer. *Int. J. Radiat. Oncol. Biol. Phys.* **15,** 1299–1305.

2. Pollack, A., and Zagars, G. K. (1997). External beam radiotherapy dose response of prostate cancer. *Int. J. Radiat. Oncol. Biol. Phys.* **39(5),** 1011–1018.

3. Hanks, G. E., Lee, W. R., Hanlon, A. L. *et al.* (1996). Conformal technique dose escalation for prostate cancer: biochemical evidence of improved cancer control with higher doses in patients with pre-treatment prostate-specific antigen >10 ng/ml. *Int. J. Radiat. Oncol. Biol. Phys.* **35,** 862–868.

4. Smit, W. G. J. M., Helle, P. A., Van Putten, W. L. J. *et al.* (1990). Late radiation damage in prostate cancer patients treated by high dose external radiotherapy in relation to rectal dose. *Int. J. Radiat. Oncol. Biol. Phys.* **18,** 23–29.

5. Lawton, C. A., Wonb, M., Pilepich, M. V. *et al.* (1991). Long-term treatment sequelae following external beam irradiation for adenocarcinoma of the prostate: analysis of RTOG studies 7506 and 7706. *Int. J. Radiat. Oncol. Biol. Phys.* **21,** 935–939.

6. Hanlon, A. L., Schultheiss, T. E., Hunt, M. A., Movsas, B., Peter, R. S, Hands, G. E. (1997). Chronic rectal bleeding after high-dose conformal treatment of prostate cancer warrants modification of existing morbidity scales. *Int. J. Radiat. Oncol. Biol. Phys.* **38(1),** 59–63.

7. Zelefsky, M. J., Cowen, D., Fuks, Z. *et al.* (1999). Long term tolerance of high dose three dimensional conformal radiotherapy for patients with localized prostate carcinoma. *Cancer* **85,** 2460–2468.

8. Lee, W. R., Hanks, G. E., Hanlon, A. L. *et al.* (1996). Lateral rectal shielding reduces late rectal morbidity following high dose three-dimensional conformal radiation therapy for clinically localized prostate cancer: further evidence for a significant dose effect. *Int. J. Radiat. Oncol. Biol. Phys.* **35,** 251–257.

9. Chen, M. E., Johnston, D. A., Tang, K., Babaian, R. J., and Troncoso, P. (2000). Detailed mapping of prostate carcinoma foci: biopsy strategy implication. *Cancer* **89,** 1800–1809.

10. Kawashita, Y., Ohtsuru, A., Kaneda, Y. *et al.* (1999). Regression of hepatocellular carcinoma in vitro and in vivo by radiosensitizing suicide gene therapy under the inducible and spatial control of radiation. *Hum. Gene Ther.* **10,** 1509–1519.

11. Kim, J. H., Kim, S. H., Brown, S. L. *et al.* (1994). Selective enhancement by an antiviral agent of the radiation-induced cell killing of human glioma cells transduced with HSV-tk gene. *Cancer Res.* **54,** 6053–6056.

12. Nishihara, E., Nagayama, Y., Mawatari, F. *et al.* (1997). Retrovirus-mediated herpes simplex virus thymidine kinase gene transduction renders human thyroid carcinoma cell lines sensitive to ganciclovir and radiation in vitro and in vivo. *Endocrinology* **138(11),** 4577–4583.

13. Atkinson, G., and Hall, S. (1999). Prodrug activation gene therapy and external beam irradiation in the treatment of prostate cancer *Urology* **54,** 1098–1104.

14. Stevens, C. W., Zeng, M., and Cerniglia, G. J. (1996). Ionizing radiation greatly improves gene transfer efficiency in mammalian cells. *Hum. Gene Ther.* **7,** 1727–1734.

15. Zeng, M., Cerniglia, G. J., Eck, S. L., and Stevens, C. W. (1997). High-efficiency stable gene transfer of adenovirus into mammalian cells using ionizing radiation. *Hum. Gene Ther.* **8,** 1025–1032.

16. Jain, P. T., and Gerwirtz, D. A. (1999). Sustained enhancement of liposome-mediated gene delivery and gene expression in human breast tumor cells by ionizing radiation. *Int. J. Radiat. Oncol. Biol. Phys.* **75(2),** 217–223.

17. Simon, W. J., and Marshall, F. F. (1998). The future of gene therapy in the treatment of urologic malignancies. *Urol. Clin. North Am.* **25,** 23–38.

18. Moolten, F. L. (1986). Tumor chemosensitivity conferred by inserted herpes thymidine kinase genes: paradigm for a prospective cancer control strategy. *Cancer Res.* **46,** 5276–5281.

19. Moolten, F. L., and Wells, J. M. (1990). Curability of tumors bearing herpes thymidine kinase transferred by retroviral vectors. *J. Natl. Cancer. Inst.* **82,** 297–300.

20. Vieweg, J., Rosenthal, F. M., Bannerji, R., Heston, W. D. W., Fair, W. R., Gansbacher, B., and Gilboa, E. (1994). Immunotherapy of prostate cancer in the Dunning rat model: use of cytokine gene modified tumor vaccines. *Cancer Res.* **54,** 1760–1765.

21. Zhang, W. W., Fang, X., Mazur, W., French, B. A., George, R. N., and Roth, J. A. (1994). High-efficiency gene transfer and high level expression of wild-type p53 in human lung cancer cells mediated by recombinant adenovirus. *Cancer Gene Ther.* **1,** 5–13.

22. Eastham, J. A., Chen, S.-H., Sehgal, I., Yang, G., Timme, T. L., Hall, S. H., Woo, S. L. C., and Thompson, T. C. (1996). Prostate cancer gene therapy: herpes simplex virus thymidine kinase gene transduction followed by ganciclovir in mouse and human prostate cancer models. *Hum. Gene Ther.* **7,** 515–523.

23. Prince, M. H. (1998). Gene transfer: a review of methods and applications. *Pathology* **30,** 335–347.

24. Van der Eb, M. M., Cramer, S. J., Vergouwe, Y., Schagen, F. H., Van Krieken, J. H., Van de Eb, A. J., Rinkes, I. H., Van de Velde, C. J., and Hoeben, R. C. (1998). Severe hepatic dysfunction after adenovirus-mediated transfer of the herpes simplex virus thymidine kinase gene and ganciclovir administration. *Gene Ther.* **5,** 451–458.

25. Hall, S. J., Mutchnik, S. E., Chen, S.-H., Woo, S. L. C., and Thompson, T. C. (1997). Adenovirus-mediated herpes simplex virus thymidine kinase gene and ganciclovir therapy leads to systemic activity against spontaneous and induced metastasis in an orthotopic mouse model of prostate cancer. *Int. J. Cancer* **70(b),** 183–187.

26. Hall, S. J., Mutchnik, S. E., Yang, G., Timme, T. L., Nasu, Y., Bangma, C. H., Woo, S. L. C., Shaker, M., and Thompson, T. C. (1999). Cooperative therapeutic effects of androgen ablation and adenovirus-mediated herpes simplex virus–thymidine kinase gene and ganciclovir therapy in experimental prostate cancer. *Cancer Gene Ther.* **6,** 54–63.

27. Timme, T. L., Hall, S. J., Barrios, R., Woo, S. L. C., Aguilar-Cordova, E., and Thompson, T. C. (1998). Local inflammatory response and vector spread after direct intraprostatic injection of a recombinant adenovirus containing the herpes simplex virus thymidine kinase gene and ganciclovir therapy in mice. *Cancer Gene Ther.* **5,** 74–82.

28. Herman, J. R., Adler, H. L., Aguilar-Cordova, E., Rojas-Martinez, A., Woo, S., Timme, T. L., Wheeler, T. M., Thompson, T. C., and Scardino, P. T. (1999). *Hum. Gene Ther.* **10,** 1239–1249.

29. Chhikara, M., Zhu, X., Teh, B. S., Vlachaki, M. T., Chiu, J. K., Woo, S. Y., Berner, B. M., Thompson, T. C., Butler, E. B., and Aguilar-Cordova, E. (2000). Radio-gene therapy enhanced reduction of induced metastases of prostate cancer. First International Conference on Translational Research and Pre-clinical Strategies in Radio-Oncology 2000. *Int. J. Radiat. Oncol. Biol. Phys.* **46**, 786–787.

30. Chhikara, M., Huang, H., Vlachaki, M. T., Zhu, X., Teh, B., Chiu, K. J., Woo, S., Berner, B., Smith, E. O., Oberg, K. C., Aguilar, L. K., Thompson, T. C. Butler, E. B., and Aguilar-Cordova, E. (2001). Enhanced therapeutic effect of HSV-tk+GCV gene therapy and ionizing radiation for prostate cancer. *Mol. Ther.* **3(4)**, 536–542.

31. Chen, S. H., Shine, H. D., Goodman, J. C., Grossman, R. G., and Woo, S. L. C. (1994). Gene therapy for brain tumors: regression of experimental gliomas by adeno-virus-mediated gene transfer in vivo. *Proc. Natl. Acad. Sci. USA* **91**, 3054–3057.

32. Hasenburg, A., Tong, X. W., Rojas-Martinez, A., Myberg-Hoffman, C., Kieback, C. C., Kaplan, A., Kaufman, R. H., Ramzy, I., Aguilar-Cordova, E., and Kieback, D. G. (2000). Thymidine kinase gene therapy with concomitant topotecan chemotherapy for recurrent ovarian cancer. *Cancer Gene Ther.* **7(6)**, 839–844.

33. Teh, B. S., Mai, W. Y., Uhl, B. M., Augspurger, M. E., Grant, W. H., Lu, H. H., Woo, S. Y., Carpenter, L. S., Chiu, J. K., and Butler, E. B. (2001). Intensity-modulated radiation therapy (IMRT) for prostate cancer with the use of a rectal balloon for prostate immobilization: acute

34. Teh, B. S., Woo, S. Y., and Butler, E. B. (1999). Intensity modulated radiation therapy (IMRT): a new frontier in radiation oncology. *Oncologist* **4**, 433–442.

35. Cox, J. D., Stetz, J., and Pajak, T. F. (1995). Toxicity criteria of the Radiation Therapy Oncology Group (RTOG) and the European Organization for Research and Treatment of Cancer (EORTC). *Int. J. Radiat. Oncol. Biol. Phys.* **31**, 1341–1346.

36. Rodriguez, L. V., and Terris, M. K. (1998). Risks and complications of transrectal ultrasound guided prostate needle biopsy: a prospective study and review of the literature. *J. Urol.* **160**, 2115–2120.

37. Zagars, G. K., Pollack, A., Kavadi, V. S., and Eschenbach, A. (1995). Prostate specific antigen and radiation therapy for clinically localized prostate cancer: an update and review of the M. D. Anderson experience. *Int. J. Radiat. Oncol. Biol. Phys.* **32**, 293–306.

38. Zietman, A., Coen, J. J., Shipley, W. U., Willett, C. G., and Efird, J. T. (1994). Radical radiation therapy in the management of prostatic adenocarcinoma: the initial prostate specific antigen value as a predictor of treatment outcome. *J. Urol.* **141**, 640–645.

39. Ayala, G., Wheeler, T. M., Shalev, M., Thompson, T. C., Miles, B., Aguilar-Cordova, E., Chakraborty, S., and Kadmon, D. (2000). Cytopathic effect of insitu gene therapy in prostate cancer. *Hum. Pathol.* **31**, 866–870.

toxicity and dose-volume analysis. *Int. J. Radiat. Oncol. Biol. Phys.* **49(3)**, 705–712.

Index

A

AAV, *see* Adeno-associated viruses
Accessory proteins, lentiviruses, 111
Acid-eluted peptides, genetic immunization, 187
Ad.E1A, ovarian cancer model, 471
Ad.Egr-TNF-α, genetic radiotherapy, 442–443
Adeno-associated viruses
  advantages, 59
  antitumor immunity–tumor vaccines, 67–68
  biology, 54–56
  characterization, 58–59
  DNA, 60
  gene removal, 59–60
  gene therapy, 301
  gene transfer to hematopoietic cells, 62–64
  immune responses, 60
  packaging, 56–57
  purification, 58
  safety issues, 70–71
  titration, 57–59
Adenoviruses
  Ad-p53 combinations in gene therapy, 307
  gene therapy, 301
  HSV-TK, 494, 509–510
  human, biology, 451
  mediated cell killing, mechanisms, 451–452
  p21, *see* p21-expressing adenovirus
  p53, *see* p53-expressing adenovirus
  replication, 451–452
  soluble VEGF, tumor angiogenesis control, 425
  tumor burden inhibition, 425–428
  wild-type, clinical trials, 453–454
Adoptive immunotherapy
  antitumor-reactive T cells
    sensitized lymph nodes, 243–246
    tumor-infiltrating lymphocytes, 242–243
  DC genetic modulation, 250–251
  T cell manipulation
    chimeric TCRs, 249–250
    genetic transduction, 247
    immunoregulatory molecule delivery, 248
    overview, 246–247
    TCR gene transfer, 248–249
    TIL-marking studies, 247–248

Ad-p21, *see* p21-expressing adenovirus
Ad-p53, *see* p53-expressing adenovirus
AdVMART1, genetic immunization, 188
AFP, *see* alpha-Fetoprotein
AGT, *see* Alkyltransferase
Alkyltransferase, methyltransferase-mediated drug resistance, 344
alpha-Fetoprotein, expression, 182
Alternative splicing, retroviral *cis* elements, 21
Androgens, HSV-TK suicide gene therapy, 519
Angiogenesis
  antiangiogenic gene therapy, 439
  ribozymes in cancer models, 100–101
  tumor biology role, 405–406
  tumor growth, 422
  VEGF, 424–425
  VEGF receptor adenovirus control, 425
Angiopoietin-1, antiangiogenic gene therapy, 414
Angiostatin, antiangiogenic gene therapy, 407–408
Animals
  drug-resistant DHFR antitumor studies, 387
  tumor immunotherapy, 172
Antiangiogenic gene therapy
  angiogenesis, 439
  angiostatin, 407–408
  delivery of factors, 422–423
  ELR(−) CXC chemokines, 411
  EMAP-II, 412
  endostatin, 408–409
  endothelial cell-specific gene delivery, 414–415
  experimental *vs.* clinical settings, 423
  future directions, 415
  interferons, 409–410
  interleukins, 410
  ionizing radiation
    antitumor activity, 439
    targeting, 439–440
  p16, 412
  p53, 412
  proangiogenic cytokines
    angiopoietin-1, 414
    plasminogen activators, 414

    VEGF, 412–413
  response assessment
    microvessel density, 429–430
    MRI imaging, 430–431
    PET imaging, 431
    ultrasound imaging, 431–432
  thrombospondins, 411–412
  TIMPs, 411
  transgene product safety, 428
  translation context, 429
  vector safety, 428
  VEGF, 423–424
  VEGF and angiogenesis, 424–425
  VEGF receptors, 423–424
Antiangiogenic proteolytic fragments
  angiostatin, 407–408
  endostatin, 408–409
Antibodies
  antitumor immunity, 137
  conjugated, 83–84
  fragment characteristics, 81–82
  monoclonal, *see* Monoclonal antibodies
  unconjugated, 82–83
Antibody-mediated immune responses, cellular response comparison, 128–129
Antifolates
  drug-resistant DHFR hematopoietic cells, 388
  toxicity protection *in vitro,* drug-resistant DHFR, 385–387
Antigen-presenting cells
  major histocompatibility complex, 146
  peptide-pulsed professional, immunization, 137
  skin penetration in PMGT, 234
Antigens
  cancer rejection, 179–180
  carcinoembryonic
    cancer gene therapy, 34
    expression, 181–182
  HLA, dendritic cell loading, 199–200
  HLA-A, MHC restriction, 180–181
  HLA-A2, mutant *ras* peptides binding, 150–151
  pathways, 146–147
  prostate-specific, expression, 182